A GLENCOE PROGRAM

SCIENCE INTERACTIONS

Bill Aldridge
Jack Ballinger
Linda Crow
Albert Kaskel
Edward Ortleb

Russell Aiuto
Anne Barefoot
Ralph M. Feather, Jr.
Craig Kramer
Susan Snyder

Paul W. Zitzewitz

With Features by:

NATIONAL
GEOGRAPHIC
SOCIETY

GLENCOE

McGraw-Hill

New York, New York   Columbus, Ohio   Mission Hills, California   Peoria, Illinois

A GLENCOE PROGRAM
SCIENCE INTERACTIONS

Student Edition
Teacher Wraparound Edition
Science Discovery Activities
Teacher Classroom Resources
Laboratory Manual
Study Guide
Section Focus Transparencies
Teaching Transparencies
Computer Test Bank

Spanish Resources
Performance Assessment
Performance Assessment in the
Science Classroom
Science and Technology
Videodisc Series
Integrated Science Videodisc
Program
MindJogger Videoquizzes

Glencoe/McGraw-Hill
A Division of The McGraw-Hill Companies

Send all inquiries to:

Glencoe/McGraw-Hill
936 Eastwind Drive
Westerville, OH 43081
ISBN 0-02-828055-5

Printed in the United States of America.
3 4 5 6 7 8 9 10 071/043 03 02 01 00 99

With Features by:

NATIONAL GEOGRAPHIC SOCIETY

Table of Contents

TEACHER GUIDE

Table of Contents

Table of Contents

SCIENCE INTERACTIONS
Contents and Primary Science Emphasis

Course 1		Course 2		Course 3	
Unit 1	Observing the World Around You	Unit 1	Forces in Action	Unit 1	Electricity and Magnetism
Chapter 1	Viewing Earth and Sky	Chapter 1	Forces and Pressure	Chapter 1	Electricity
Chapter 2	Light and Vision	Chapter 2	Forces In Earth	Chapter 2	Magnetism
Chapter 3	Sound and Hearing	Chapter 3	Circulation	Chapter 3	Electromagnetic Waves
Unit 2	Interactions in the Physical World	Unit 2	Energy at Work	Unit 2	Atoms and Molecules
Chapter 4	Describing the Physical World	Chapter 4	Work and Energy	Chapter 4	Structure of the Atom
Chapter 5	Matter in Solution	Chapter 5	Machines	Chapter 5	The Periodic Table
Chapter 6	Acids, Bases, and Salts	Chapter 6	Thermal Energy	Chapter 6	Combining Atoms
Unit 3	Interactions in the Living World	Chapter 7	Moving the Body	Chapter 7	Molecules in Motion
Chapter 7	Describing the Living World	Chapter 8	Controlling the Body Machine	Unit 3	Our Fluid Environment
Chapter 8	Viruses and Simple Organisms	Unit 3	Earth Materials and Resources	Chapter 8	Weather
Chapter 9	Animal Life	Chapter 9	Discovering Elements	Chapter 9	Ocean Water and Life
Chapter 10	Plant Life	Chapter 10	Minerals and Their Uses	Chapter 10	Organic Chemistry
Chapter 11	Ecology	Chapter 11	The Rock Cycle	Chapter 11	Fueling the Body
Unit 4	Changing Systems	Chapter 12	The Ocean Floor and Shore Zones	Chapter 12	Blood: Transport and Protection
Chapter 12	Motion	Chapter 13	Energy Resources	Unit 4	Changes in Life and Earth Over Time
Chapter 13	Motion Near Earth	Unit 4	Air: Molecules in Motion	Chapter 13	Reproduction
Chapter 14	Moving Water	Chapter 14	Gases, Atoms, and Molecules	Chapter 14	Heredity
Chapter 15	Shaping the Land	Chapter 15	The Air Around You	Chapter 15	Moving Continents
Chapter 16	Changing Ecosystems	Chapter 16	Breathing	Chapter 16	Geologic Time
Unit 5	Wave Motion	Unit 5	Life at the Cellular Level	Chapter 17	Evolution of Life
Chapter 17	Waves	Chapter 17	Basic Units of Life	Unit 5	Observing the World Around You
Chapter 18	Earthquakes and Volcanoes	Chapter 18	Chemical Reactions	Chapter 18	Fission and Fusion
Chapter 19	The Earth-Moon System	Chapter 19	How Cells Do Their Jobs	Chapter 19	The Solar System
▓ = Physics ▓ = Life Science ▓ = Earth Science ▓ = Chemistry				Chapter 20	Stars and Galaxies

INTRODUCING THE AUTHOR TEAM

Bill Aldridge was Executive Director of the National Science Teachers Association from 1980-1995. After retiring from NSTA, Mr. Aldridge assumed the position of President, Division of Science Education Solutions, of ARS, Inc., Fredericksburg, VA. Prior to becoming NSTA's Executive Director, he served 3 years as a program officer in science education at the National Science Foundation. He received his B.S. and M.S. degrees in physics from the University of Kansas, where he also received an M.S. degree in Educational Evaluation. Mr. Aldridge also received an M.Ed. degree from Harvard University. He taught physics and mathematics at the high school level for 6 years and at the college level for 17 years. He has authored numerous publications, including five textbooks, 13 monographs, and articles in science and science education magazines and journals. As Executive Director of NSTA, Mr. Aldridge worked with the U.S. Congress and with government agencies in designing and producing support programs for science education. He is the recipient of awards and recognition from the National Science Foundation and the American Association of Physics Teachers, and he is a fellow of the American Association for the Advancement of Science. In 1995, he received the Distinguished Public Service Medal from NASA, the highest recognition given to a non-NASA employee.

Russell Aiuto recently retired from the position of Senior Project Officer for the Council of Independent Colleges. Dr. Aiuto is past Director of Research and Development for the National Science Teachers Association in Washington, DC. Throughout his career, Dr. Aiuto has held several prominent positions, including Director of the Division of Teacher Preparation and Enhancement of the National Science Foundation, President of Hiram College in Hiram, Ohio, and Provost of Albion College in Albion, Michigan. He also has 30 years experience teaching biology and genetics at the high school and college levels. Dr. Aiuto received B.A. degrees from Eastern Michigan University and the University of Michigan, and his M.A. and Ph.D. degrees from the University of North Carolina. He has received numerous awards, including the Phi Beta Kappa Faculty Scholar Award, Campus Teaching Award, and Honors Program Faculty Award from Albion College.

Jack Ballinger is a professor in the chemistry department at St. Louis Community College in St. Louis, Missouri, where he has taught for 27 years. He received his B.S. degree in chemistry from Eastern Illinois University, his M.S. degree in organic chemistry from Southern Illinois University, and his Ed.D. in science education from Southern Illinois University. Professor Ballinger has authored numerous articles and several books, including a reference handbook for chemical technicians. Dr. Ballinger has received many awards for his outstanding teaching: the Manufacturing Chemists Association Regional Award for Excellence in Chemistry Teaching, the Emerson Electric Company's "Excellence in Teaching Award," the Missouri Governor's Award for Excellence in Teaching, and the Outstanding Teacher of the Year Award in 1994 at St. Louis Community College at Florissant Valley.

Anne Barefoot is a veteran physics and chemistry teacher with 35 years of teaching experience. Her career also includes teaching biology and physical science, and working with middle school teachers in the summer at Purdue University in the APAST program. A past recipient of the Presidential Award for Outstanding Science Teaching, Ms. Barefoot holds B.S. and M.S. degrees from East Carolina University, and a Specialist Certificate from the University of South Carolina. Other awards include Whiteville City Schools Teacher of the Year, Sigma Xi Award, and the North Carolina Business Award for Science Teaching. Ms. Barefoot is the former District IV Director of the National Science Teachers Association and the former president of the Association of Presidential Awardees in Science Teaching.

Linda Crow is an associate professor in the Department of Natural Sciences at the University of Houston-Downtown. She is the project leader of the Texas Scope, Sequence and Coordination (SS&C) Project. In addition to 24 years as an award-winning science teacher at college and high school levels, Dr. Crow is a recognized speaker at education workshops both in the United States and abroad. Dr. Crow received her BS, M.Ed., and Ed.D. degrees from the University of Houston. She has been twice named the OHAUS Winner for Innovations in College Science Teaching.

Ralph M. Feather, Jr. teaches geology, astronomy, Earth Science, and integrated science, and serves as Science Department Chair in the Derry Area School District in Derry, PA. Mr. Feather has 26 years of teaching experience in secondary education. He holds a B.S. in geology and an M.Ed. in geoscience from Indiana University of Pennsylvania and is currently completing work on his Ph.D. in "Writing Across the Curriculum" at the University of Pittsburgh. Mr. Feather has received the Presidential Award for Excellence in Earth Science Teaching and the Award for Excellence in Earth Science Teaching from the Geological Society of America. Mr. Feather has also received the Outstanding Earth Science Teacher Award from NAGT and the Kevin Burns Citation from the Spectroscopy Society of Pittsburgh.

Albert Kaskel has 32 years experience teaching science, the last 25 at Evanston Township High School, Evanston, IL. His teaching experience includes biology, A.P. biology, physical science, and chemistry. He holds a B.S. in biology from Roosevelt University in Chicago and an M.Ed. degree from DePaul University. Mr. Kaskel has received the Outstanding Biology Teacher Award for the State of Illinois and the Teacher Excellence Award from Evanston Township High School. He is the major author of a leading high school biology textbook and has contributed to more than 30 different science textbook related publications. He was recently a staff member at the Center for Talent Development at Northwestern University.

Craig Kramer has been a physics teacher for 20 years. He is past chairperson for the Science Department at Bexley High School in Bexley, Ohio. Mr. Kramer received a B.A. in physics and a B.S. in science and math education, and an M.A. in outdoor and science education from The Ohio State University. He has received numerous awards, including the Award for Outstanding Teaching in Science from Sigma Xi. In 1987, the National Science Teachers Association awarded Mr. Kramer a certificate for secondary physics, making him the first nationally certified teacher in physics.

Edward Ortleb serves as a science consultant for the St. Louis, Missouri, Public Schools and has 37 years teaching experience. He holds an A.B. in biology education from Harris Teachers College, an M.A. in science education and an Advanced Graduate Certificate from Washington University, St. Louis. Mr. Ortleb is a lifetime member of the National Science Teachers Association, having served as its president in 1978-79. He has also served as Regional Director for the National Science Supervisors Association. Mr. Ortleb is the recipient of several awards, including the Distinguished Service to Science Education Award (NSTA), the Outstanding Service to Science Education Award, and the Outstanding Achievement in Conservation Education Award.

Susan Snyder is a teacher of Earth science at Jones Middle School, Upper Arlington School District, Columbus, OH. Ms. Snyder received a B.S. in comprehensive science from Miami University, Oxford, OH. and an M.S. in entomology from the University of Hawaii. She has 24 years teaching experience and is author of numerous educational materials. Ms. Snyder has been a state recipient of the Presidential Award for Excellence in Science and Math Teaching, a finalist for National Teacher of the Year, and Ohio Teacher of the Year. She also won the Award for Excellence in Earth Science Teaching from the Geological Society of America.

Paul W. Zitzewitz is Professor of Physics at the University of Michigan-Dearborn. He received his B.A. from Carleton College and his M.A. and Ph.D. from Harvard University, all in physics. Dr. Zitzewitz has taught physics to undergraduates for 25 years and is an active experimenter in the field of atomic physics with more than 50 research papers. He has memberships in several professional organizations, including the American Physical Society, American Association of Physics Teachers, and the National Science Teachers Association. Among his awards are the University of Michigan-Dearborn Distinguished Faculty Research Award.

National Geographic Society The National Geographic Society, founded in 1888 for the increase and diffusion of geographic knowledge, is the world's largest nonprofit scientific and educational organization. Since its earliest days, the Society has used sophisticated communication technology, from color photography to holography, to convey geographic knowledge to a worldwide membership. The Education Products Division supports the Society's mission by developing innovative educational programs—ranging from traditional print materials to multimedia programs including CD-ROMs, videodiscs, and software.

Responding to Changes IN SCIENCE EDUCATION

THE NEED FOR NEW DIRECTIONS IN SCIENCE EDUCATION

By today's projections, seven out of every ten American jobs will be related to science, mathematics, or electronics by the year 2000. And according to the experts, if junior high and middle school students haven't grasped the fundamentals, they probably won't go further in science and may not have a future in a global job market. Studies also reveal that high school students are avoiding taking "advanced" science classes.

SCIENCE INTERACTIONS ANSWERS THE CHALLENGE!

At Glencoe Publishing, we believe that *SCIENCE INTERACTIONS* will help you bring science reform to the front lines—the classrooms of America. But more important, we believe it will help students succeed in middle school and junior high science so that they will continue learning science through high school and into adulthood.

When you compare *SCIENCE INTERACTIONS* to a traditional science program, you'll see fewer terms. But you'll also see more questions and more activities to draw your students in. And you'll find broad themes repeated over and over, rather than hundreds of unrelated topics.

SCIENCE INTERACTIONS FITS YOUR CLASSROOM

SCIENCE INTERACTIONS has the right ingredients to help you ensure your students' future. But it also has to work in today's classroom. That's why, on first glance, *SCIENCE INTERACTIONS* may look like a traditional science textbook.

Glencoe knows you have local curriculum requirements. You teach a variety of students with varying ability levels. And you have limited time, space, and support for doing hands-on activities.

No matter. Unlike a purely hands-on program, *SCIENCE INTERACTIONS* lets you offer the perfect balance of content and activities. Your students will be eager to get their hands on science. But *SCIENCE INTERACTIONS* also gives you the flexibility to use only the activities you choose ... without sacrificing anything.

SCIENCE INTERACTIONS IS LOADED WITH ACTIVITIES

You'll choose from hundreds of activities in all three courses. These easy to set-up and manage activities will allow you to teach using a hands-on, inquiry-based approach to learning.

The **Find Out, Explore, Investigate,** and **Design Your Own Investigation** activities are integrated with the text narrative, complete with transitions into and out of the activity. This aids comprehension for students by building continuity between text and activities.

Your teaching methods will include asking questions such as: *How do we know? Why do we believe? What does it mean?* Throughout both the narrative and activities, you'll invite your students to relate what they learn to their own everyday experiences.

SCIENCE INTERACTIONS TEACHES CONCEPTS IN A LOGICAL SEQUENCE
From the Concrete to the Abstract

Research shows that students learn better when they deal with descriptive matters in science for a reasonable portion of their school years before proceeding to the more quantitative, and eventually the more theoretical, parts of science. *SCIENCE INTERACTIONS* helps your students learn in this manner.

Let's look at the way you'll teach the topic of movement within and on the surface of Earth. In Course 1, students learn about the nature of Earth movements by observing that those movements bring about earthquakes and volcanoes. Learning to observe the results of Earth movement gets students to think about it in a concrete way.

In Course 2, students learn more about Earth movements by first observing how simple, everyday objects react to forces. This study of force is extended to the forces inside Earth that result in various observable phenomena—faults, seismic waves, volcanic eruptions.

In Courses 1 and 2, students have learned a lot about Earth movement by observing its results. In Course 3, they gain exposure to theoretical applications of Earth movement when they study plate tectonics.

SCIENCE INTERACTIONS INTEGRATES SCIENCE AND MATH

Mathematics is a tool that all students, regardless of their career goals, will use throughout their lives. *SCIENCE INTERACTIONS* provides opportunities to hone mathematics skills while learning about the natural world. **Skill Builders**, as well as **Find Out, Explore, Investigate,** and **Design Your Own Investigation** activities, offer numerous options for practicing math, including making

and using tables and graphs, measuring in SI, and calculating. **Across the Curriculum** strategies in the *Teacher Wraparound Edition* provide additional connections between science and mathematics.

SCIENCE INTERACTIONS INTEGRATES TOPICS FOR UNDERSTANDING

According to the experts, by using an integrated approach like *SCIENCE INTERACTIONS*, students will experience dramatic gains in comprehension and retention. For instance, the series helps you teach some of the basic concepts from physical science early on. This, in turn, makes it easier for your students to understand other concepts in life and Earth science.

But you'll be doing more than simply showing how the sciences interconnect. You'll also take numerous "side trips" with your students. Connect one area of science to another. Relate science to technology, society, issues, hobbies, and careers. Show your students again and again how history, the arts, and literature can be part of science. And help your students discover the science behind things they see every day.

IT'S TIME FOR NEW DIRECTIONS IN SCIENCE EDUCATION

The need for new directions in science education has been established by the experts. America's students must prepare themselves for the high-tech jobs of the future.

We at Glencoe believe that *SCIENCE INTERACTIONS* answers the challenge of the 21st century with its new, innovative approach of "connecting" the sciences. We believe *SCIENCE INTERACTIONS* will assist you better in preparing your students for a lifetime of science learning.

Your Questions Answered!

How is *SCIENCE INTERACTIONS* an integrated science program?

Although each chapter has a primary science emphasis, integration of other disciplines occurs throughout the program. Students are more likely to learn and remember a concept because they see it applied to other disciplines. This science integration is evident not only in the narrative of the core part of each chapter, but also in the **Expand Your View** features at the end of each chapter, in the *Teacher Wraparound Edition*, and in the supplements.

How is *SCIENCE INTERACTIONS* different from general science?

There's really no comparing the *SCIENCE INTERACTIONS* program to a traditional general science text. In a general science program, the sequence of the topics and their relationship to one another is of little importance. For example, all the physics chapters in a general science text would be grouped together in a unit at the back of the book and probably have no relationship to the life, Earth, or chemistry units.

SCIENCE INTERACTIONS is different from general science in that chapters from different disciplines are intermixed and sequenced so that what is learned in one discipline can be applied to another.

How is *SCIENCE INTERACTIONS* different from the "layering" or "block" approach?

In a traditional three-year course, students study life science in sixth grade, Earth science in seventh, and physical science in eighth grade. *SCIENCE INTERACTIONS* is different because it contains all of these disciplines in each course. It is true integrated science, where life, Earth, and physical science are integrated throughout the year.

Will *SCIENCE INTERACTIONS* prepare my students for high school science?

The national reform projects agree that the best preparation is a deep understanding of important science concepts. This, rather than requiring students to memorize facts and terms, will keep your students interested in science.

Science will come alive for your students each time they pick up their textbooks. *SCIENCE INTERACTIONS'* visual format lends excitement to the study of integrated science. Students can see fundamental science concepts in living color. Learning concepts by visualizing them also increases cognitive awareness, thus giving students a more solid foundation for future science courses.

SCIENCE INTERACTIONS will help your students frame questions, derive concepts, and obtain evidence. When your students have mastered this language of science, they will be ready for further study.

In addition, *SCIENCE INTERACTIONS* offers your students plenty of reasons to stick with science—including unexpected career choices and examples of women and minorities achieving in science.

Science Interactions Supports
The National Science Education Standards

The *National Science Education Standards*, published by the National Research Council and representing the contribution of thousands of educators and scientists, offer a comprehensive vision of a scientifically literate society. The standards describe not only what students should know but also offer guidelines for science teaching and assessment. Bill Aldridge, a *Science Interactions* author, helped originate this standards initiative and served on its Chair's Advisory Committee. If you are using, or plan to use, the standards to guide changes in your science curriculum, you can be assured that *Science Interactions* aligns with the *National Science Education Standards.*

Science Interactions is an example of how Glencoe's commitment to effective science education is changing the materials used in science classrooms today. More than just a collection of facts in a textbook, *Science Interactions* is a program that provides numerous opportunities for students, teachers, and school districts to meet the *National Science Education Standards.*

Content Standards
The accompanying table shows the close alignment between *Science Interactions* and the content standards. The integrated approach of *Science Interactions* allows students to discover concepts within each of the content standards, giving them opportunities to make connections between the science disciplines. Our hands-on activities and inquiry-based lessons reinforce the science processes emphasized in the standards.

Teaching Standards
Alignment with the *National Science Education Standards* requires much more than alignment with the outcomes in the content standards. The way in which concepts are presented is critical to effective learning. The teaching standards within the *National Science Education Standards* recommend an inquiry-based program facilitated and guided by teachers. *Science Interactions* provides such opportunities through activities and discussions that allow students to discover by inquiry critical concepts and apply the knowledge they've constructed to their own lives. Throughout the program, students are building critical skills that will be available to them for lifelong learning. *The Teacher Wraparound Edition* helps you make the most of every instructional moment. It offers an abundance of effective strategies and suggestions for guiding students as they explore science.

Assessment Standards
The assessment standards are supported by many of the components that make up the *Science Interactions* program. *The Teacher Wraparound Edition* and *Teacher Classroom Resources* provide multiple chances to assess students' understanding of important concepts as well as their abilities to perform a wide range of skills. Ideas for portfolios, performance activities, written reports, and other assessment activities accompany every lesson. For more ideas for assessment and resources, see pages 30T-31T. Rubrics and performance task assessment lists can be found in Glencoe's Professional Series booklet *Performance Assessment in the Science Classroom.*

Program Coordination
The scope of the content standards requires students to meet the outcomes over the course of their education. *Science Interactions Courses 1, 2,* and *3* provide an integrated middle school science curriculum that is aligned not only across a grade level, but also vertically, through several years of instruction. The correlation on the following pages demonstrates the close alignment of this course of *Science Interactions* with the content standards.

Correlation of Science Interactions, Course 1, to the National Science Standards, Content Standards, Grades 5-8

Content Standard	Page Numbers
(UCP) UNIFYING CONCEPTS AND PROCESSESS	
1. Systems, order, and organization	34-37, 43-45, 101-103, 118-123, 126-130, 154-157, 186-206, 218-221, 225-238, 248-251, 255-265, 284-288, 313-321, 331-333, 346-351, 356-360, 455-463, 608-617
2. Evidence, models, and explanation	2-15, 30-33, 40-41, 54-61, 66-71, 86-90, 113, 118-123, 126, 213, 225-234, 252-253, 286-287, 349, 356-360, 382-387, 390-397, 400-404, 414-434, 455-463, 474-481, 484-487, 490-496, 540-542, 546-551, 556-558, 574-577, 580-589, 601-605, 618-620
3. Change, constancy, and measurement	22-28, 32-33, 38-39, 42-45, 86, 88-90, 94-97, 118-123, 126-131, 134-144, 160-162, 166-167, 173-176, 194-195, 198-202, 289-295, 322-327, 336, 364-370, 382-387, 390-397, 400-404, 414-434, 444-451, 455-457, 460-463, 474-481, 484-487, 490-496, 506-507, 510-513, 516-525, 535, 540-543, 546-551, 554-564, 574-592, 608-617, 621-624
4. Evolution and equilibrium	296-299, 302-303, 364-370, 506-507, 510-513, 516-525
5. Form and function	75-77, 92-93, 96-97, 100, 248-251, 255-262, 281, 284-299, 302-303, 313-327, 331-333, 336, 346-351, 364-370, 377
(A) SCIENCE AS INQUIRY	
1. Abilities necessary to do scientific inquiry	2-15, 29-31, 34, 40-41, 56, 58, 60, 66, 85, 98-99, 104-105, 113, 118, 122, 124-125, 129, 132-133, 139-141, 146-147, 155, 158-159, 163, 165, 168-169, 173, 186, 192, 194-195, 200-201, 215, 218, 222-223, 248, 255, 263, 266-267, 282-283, 289, 300-301, 313, 318, 320, 322, 325, 328-329, 331, 334-335, 354-355, 362-363, 379, 388-389, 398-399, 415, 423, 434, 449, 455, 458-459, 482-483, 488-489, 508-509, 544-545, 552-553, 578-579, 618-619
2. Understandings about scientific inquiry	2-15, 42, 46, 53, 55-56, 62-63, 72-73, 87, 91, 94-95, 117, 120-121, 135, 155, 160-161, 168-169, 194-195, 200-201, 225, 239, 242, 247, 266-267, 282-283, 328-329, 377, 416-417, 432-434, 514-515, 535, 590-591, 606-607
(B) PHYSICAL SCIENCE	
1. Properties and changes of properties in matter	27-138
2. Motions and forces	382-387, 390-397, 400-404, 414-415, 418-431, 434-435, 446-451, 454, 456-457, 474-481, 484-487, 490-495, 540-543, 546-551, 574-577, 604-605, 610-613, 621-624
3. Transfer of energy	92-93, 103-105, 137, 144, 284, 336-337, 356-357

Content Standard	Page Numbers
(C) LIFE SCIENCE	
1. Structure and function in living systems	219, 234-235, 248-249, 254, 259-261, 263, 278-281, 284-288, 290-304, 312-327, 330-335, 369
2. Reproduction and heredity	289-295, 322-330
3. Regulation and behavior	277, 298-303
4. Populations and ecosystems	346-351, 356-370
5. Diversity and adaptations of organisms	26-27, 221, 242, 296-303, 368-370, 512, 527, 535
(D) EARTH AND SPACE SCIENCE	
1. Structure of the Earth system	22-28
2. Earth's history	24-25, 500, 524-525
3. Earth in the solar system	38-45, 600-624
(E) SCIENCE AND TECHNOLOGY	
1. Abilities of technological design	80, 107, 109, 137, 145, 269-270, 407, 494-495, 531, 542-543, 560-561, 565, 593, 625-626
2. Understandings about science and technology	12, 47, 80, 107, 494-495, 497, 531, 568, 594-595, 626
(F) SCIENCE IN PERSONAL AND SOCIAL PERSPECTIVES	
1. Personal health	78, 154-155, 170-171, 198, 204-206, 249, 254, 256-257, 262, 269-270, 530, 566-567
2. Populations, resources, and environments	26-27, 46, 48, 101, 107, 178-179, 186-187, 207-208, 240-242, 258, 272, 337-338, 358-360, 364-365, 367, 371-373, 464-467, 476-477, 494-495, 498-499, 518-519, 524-525, 528-529, 566-567, 624
3. Natural hazards	185-187, 207-208, 272, 466-467, 476-477, 498-499, 518-519, 574-577, 582-589, 592, 594-595
4. Risks and benefits	46, 107, 178-179, 240-241, 337-338, 372-373, 494-495, 528-529, 566-567
5. Science and technology in society	2-15, 46, 80, 107, 146-147, 187-191, 204-205, 208, 240-242, 254, 258, 269-270, 272, 304, 337-338, 371-373, 464-467, 476-477, 497-499, 565, 593-595

Content Standard	Page Numbers
(G) HISTORY AND NATURE OF SCIENCE	
1. Science as a human endeavor	2-15, 22-23, 26-27, 30-31, 34-39, 44-45, 47-48, 55, 58, 66, 68-69, 72-73, 75-79, 95-97, 108-109, 124-125, 127, 134, 145-148, 152-154, 162-163, 166, 170-171, 177-179, 185, 189-191, 193, 196, 198-199, 204-205, 208-209, 213, 215, 224-237, 240-242, 254-255, 269-271, 276-277, 304-306, 310-311, 326-327, 332-333, 338, 340, 344-345, 371, 380-381, 383, 408, 412-413, 418-419, 426-427, 436-437, 442-443, 468, 472-473, 504-505, 526-527, 537-539, 572-573, 596, 600-601, 627
2. Nature of science	2-15, 55, 57-59, 64-65, 70-71, 74, 88-90, 95-97, 101-103, 118-123, 126-131, 134, 136, 139-144, 156-157, 166-167, 170-175, 198-199, 218-221, 248-251, 304, 312, 344, 368-369, 393-395, 414, 444, 474, 506, 540, 568, 574, 602
3. History of science	12, 44-45, 57, 67, 74-75, 77, 79, 95, 108, 146-148, 174, 180, 204-205, 207, 227, 230-232, 251, 298-299, 317, 339-340, 405-406, 435, 438, 500, 524-525, 560-561, 576-577, 586-587, 589, 592, 615

Flex Your Brain

A key element in the coverage of problem solving and critical thinking skills in *SCIENCE INTERACTIONS* is a critical thinking matrix called **Flex Your Brain**.

Flex Your Brain provides students with an opportunity to explore a topic in an organized, self-checking way, and then identify how they arrived at their responses during each step of their investigation. The activity incorporates many of the skills of critical thinking. It helps students to consider their own thinking and learn about thinking from their peers.

WHERE IS FLEX YOUR BRAIN FOUND?

In the introductory chapter, "Science: A Tool for Solving Problems," is an introduction to the topics of critical thinking and problem solving. **Flex Your Brain** accompanies the text section as an activity in the introductory chapter. Brief student instructions are given, along with the matrix itself. A worksheet for **Flex Your Brain** appears in the *Critical Thinking/Problem Solving* book of the *Teacher Resources*. This version provides spaces for students to write in their responses.

In the *Teacher Wraparound Edition*, suggested topics are given in each chapter for the use of **Flex Your Brain**. You can either refer students to the introductory chapter for the procedure, or photocopy the worksheet master from the *Teacher Resources*.

USING FLEX YOUR BRAIN

Flex Your Brain can be used as a whole-class activity or in cooperative groups, but is primarily designed to be used by individual students within the class. There are three basic steps.

1. Teachers assign a class topic to be investigated using **Flex Your Brain**.
2. Students use **Flex Your Brain** to guide them in their individual explorations of the topic.
3. After students have completed their explorations, teachers guide them in a discussion of their experiences with **Flex Your Brain**, bridging content and thinking processes.

Flex Your Brain can be used at many different points in the lesson plan.

Introduction: Ideal for introducing a topic, **Flex Your Brain** elicits students' prior knowledge and identifies misconceptions, enabling the teacher to formulate plans specific to student needs.

Development: Flex Your Brain leads students to find out more about a topic on their own, and develops their research skills while increasing their knowledge. Students actually pose their own questions to explore, making their investigations relevant to their personal interests and concerns.

Review and Extension: Flex Your Brain allows teachers to check student understanding while allowing students to explore aspects of the topic that go beyond the material presented in class.

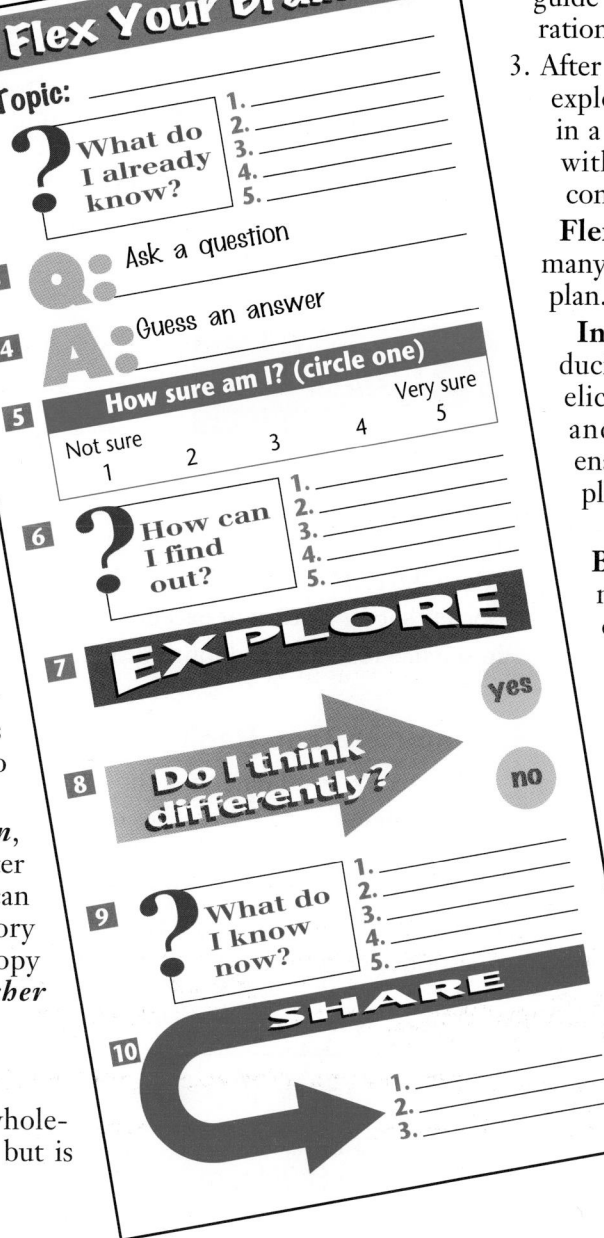

Using Technology in the Classroom

Technology helps you adapt your teaching methods to the needs of your students. Glencoe classroom technology products provide many pathways to help you match students' different learning styles. To make your lesson planning easier, all of the technology products listed below are correlated to the student text.

Videodiscs

Glencoe's *Integrated Science Videodiscs* are designed to be used interactively in the classroom. Barcodes in this book allow you to step through the programs and pause to discuss and answer on-screen questions. Barcodes for the following videodiscs also appear throughout this teacher edition: *Mr. Wizard's Science and Technology Videodisc, Infinite Voyage, Newton's Apple, and National Geographic Society.*

MindJogger Videoquiz

A videoquiz for each chapter can be used to assess prior knowledge or review content before an exam. Student teams work cooperatively to answer three rounds of questions posed by the video host. Each round requires higher level thinking skills than the previous round.

Glencoe Interactive CD-ROMs

The Glencoe *Life, Earth,* and *Physical Science* CD-ROMs are correlated to *Science Interactions*.

You can:
- use a CD-ROM interaction as a whole class presentation;
- allow 2-3 small groups to use the CD-ROMs at once;
- rotate student groups through a single computer station;
- place the materials in a computer lab; or
- use CD-ROMs as a library resource.

Glencoe also offers three National Geographic Society CD-ROMs. These image-rich, reference CD-ROMs are faithful to the Society's long history of excellence in science teaching and journalism.

Computer Test Bank

Glencoe's Test Generator for Macintosh and for DOS makes creating, editing, and printing tests quick and easy. You also can edit questions or add your own favorite questions and graphics.

Computer Competency Activities

This software is unique to the *Science Interactions* series. Closely related to the content of each chapter, these activities are designed to help students master core computer competencies: word processing, graphing, using spreadsheets, and manipulating databases.

English/Spanish Audiocassettes

Audio chapter summaries in English and in Spanish are a way for auditory learners, lower-level readers, and LEP students to review key chapter concepts. Students can listen individually during class or check out tapes and use them at home. You may find them useful for reviewing the chapter as you plan lessons.

Glencoe Software Support Hotline   1-800-437-3715

Should you encounter any difficulty when setting up or running Glencoe software, contact the Software Support Center at Glencoe Publishing between 8:30 a.m. and 4:30 p.m. Eastern Time.

Using the Internet

If you're already familiar with the Internet, skip to the sites listed at the bottom of this page. If you need some tips on how to get started, keep reading.

The Internet is an enormous reference library and a communication tool. You can use it to quickly retrieve information from computers around the globe. Like any good reference, it has an index so you can locate the right piece of information. An Internet index entry is called a *Universal Resource Locator*, or URL. Here's an example:

http://www.glencoe.com/intro/index.html

The first part of the URL tells the computer how to display the information. The second part, after the double slash, names the organization and the computer where the information is stored. The part after the first single slash tells the computer which directory to search and which file to retrieve. File locations change frequently. If you can't find what you're looking for, use the first part of the address only, for ex-

ample, http://www.glencoe.com/ from the URL shown, and follow links to what you need.

The World Wide Web

The World Wide Web (WWW), a subset of the Internet, began in 1992. Unlike regular text files, web files can have links to other text files, images, and sound files. By clicking on a link, you can see or hear the linked information.

How do I get access?

To use the Internet, you need a computer, a modem, a telephone line, and a connection to the Internet. If your school doesn't have a connection, contact your local public library or a university; they often give free access to students and educators.

CAUTION: Contents may shift!

The sites referenced in Glencoe's Internet Connections are not under Glencoe's control. Therefore,

Glencoe can make no representation concerning the content of these sites. Extreme care has been taken to list only reputable links by using educational and government sites whenever possible. Internet searches have been used that return only sites that contain no content apparently intended for mature audiences.

Where to Start

The brief list of science Internet sites below may prove useful. You can also find Internet addresses throughout this book correlated to selected features.

Useful tools for searching the Internet include:
http://www.yahoo.com/search.html
http://www.altavista.digital.com and
http://www.msn.com/access/
allinone/hv1

Science Internet Site	Description
http://ericir.syr.edu/	**Ask ERIC**, an ask-the-expert service for K-12 teachers
http://www.enc.org/	**The Eisenhower National Clearinghouse for Math and Science**, instructional materials
http://medinfo.wustl.edu/~ysp/MSN/	**The Mad Scientist Information Network**, an ask-the-expert service for science students

For more information about Glencoe Technology, call Customer Service at **1-800-334-7344.**

Themes & Scope & Sequence

SCIENCE INTERACTIONS, three science textbooks for middle school, is unique in that it integrates all the natural sciences, presenting them as a single area of study. Our society is becoming more aware of the interrelationship of the disciplines of science. For most people, the ideas that unify the sciences and make connections between them are the most valid.

Themes are the constructs that unify the sciences. Woven throughout *SCIENCE INTERACTIONS*, themes integrate facts and concepts. They are the "big ideas" that link the structures on which the science disciplines are built. While there are many possible themes around which to unify science, we have chosen four: Energy, Systems and Interactions, Scale and Structure, and Stability and Change.

ENERGY

Energy is a central concept of the physical sciences that pervades the biological and geological sciences. In physical terms, energy is the ability of an object to change itself or its surroundings, the capacity to do work. In chemical terms, it forms the basis of reactions between compounds. In biological terms, it gives living systems the ability to maintain themselves, to grow, and to reproduce. Energy sources are crucial in the interactions among science, technology, and society.

SYSTEMS AND INTERACTIONS

A system can be incredibly small, such as an atom's nucleus and electrons, or unbelievably large, as the stars in a galaxy. By defining the boundaries of the system, one can study the interactions among its parts. The interactions may be a force of attraction between the positively charged nucleus and negatively charged electron. In an ecosystem, however, the interactions may be between the predator and its prey, or among the plants and animals. Animals in such a system have many subsystems (circulation, respiration, digestion, etc.) with interactions among them.

SCALE AND STRUCTURE

Used as a theme, "structure" emphasizes the relationship among different structures. "Scale" defines the focus of the relationship. As the focus is shifted from a system to its components, the properties of the structure may remain constant. In other systems, an ecosystem for

Course 1

Themes	Chapter																		
	1	2	3	4	5	6	7	8	9	10	11	12	13	14	15	16	17	18	19
Scale and Structure	P			P			S	P	P										S
Energy		S	S												S		P	P	
Stability and Change			P	S	P	S				S	S	P	S	S	P	S		S	
Systems and Interactions	S	P			S	P	P	S		P	P	S	P	P		P	S		P

P = PRIMARY THEME   **S = SECONDARY THEME**

example, which includes a change in scale from interactions between prey and predator to the interactions among systems inside an animal, the structure changes drastically. In *SCIENCE INTERACTIONS*, explanations remain on the macroscopic level until students have the background needed to understand how the microscopic structure was determined.

STABILITY AND CHANGE

A system that is stable is constant. Often the stability is the result of a system being in equilibrium. If a system is not stable, it undergoes change. Changes in an unstable system may be characterized as trends (position of falling objects), cycles (the motion of planets around the sun), or irregular changes (radioactive decay).

THEME DEVELOPMENT

These four major themes, as well as several others, are developed within the student material and discussed throughout the *Teacher Wraparound Edition*. Each chapter of *SCIENCE INTERACTIONS* incorporates a primary and secondary theme. These themes are interwoven throughout each level and are developed as appropriate to the topic presented.

The *Teacher Wraparound Edition* includes a **Theme Development** section for each unit opener. This section discusses the upcoming unit's key themes and explains how they are supported Throughout the chapters, **Theme Connections** show specifically how a topic in the student edition relates to the themes.

Course 2

Themes	Chapter																		
	1	2	3	4	5	6	7	8	9	10	11	12	13	14	15	16	17	18	19
Scale and Structure									P	S		P		P			P		
Energy		P	S	P	S	P	S						P		P	S	S	S	S
Stability and Change	S	S						S			P	S							
Systems and Interactions	P		P	S	P	S	P	P	S	P	S		S	S	S	P		P	P

Course 3

Themes	Chapter																			
	1	2	3	4	5	6	7	8	9	10	11	12	13	14	15	16	17	18	19	20
Scale and Structure	S	S		S	S	P				P	S	P	P	S					P	P
Energy	P	S	P	P			P		P		P				P			P		
Stability and Change		P			S	S	S	S		S		S	S	P		P	P	S	S	S
Systems and Interactions						P		P	S						S	S	S			

Constructivism

Strategies suggested in *SCIENCE INTERACTIONS* support a constructivist approach to science education. The role of the teacher is to provide an atmosphere where students design and direct activities. To develop the idea that science investigation is not made up of closed-end questions, the teacher should ask guiding questions and be prepared to help his or her students draw meaningful conclusions when their results do not match predictions. Through the numerous activities, cooperative learning opportunities, and a variety of critical thinking exercises in *SCIENCE INTERACTIONS*, you can feel comfortable taking a constructivist approach to science in your classroom.

Activities

A constructivist approach to science is rooted in an activities-based plan. Students must be provided with sensorimotor experiences as a base for developing abstract ideas. *SCIENCE INTERACTIONS* utilizes a variety of learning-by-doing opportunities. **Find Out** and **Explore** activities allow students to consider questions about the concepts to come, make observations, and share prior knowledge. **Find Out** and **Explore** activities require a minimum of equipment, and students may take responsibility for organization and execution.

Investigates and **Design Your Own Investigations** develop and reinforce or restructure concepts as well as develop the ability to use process skills. **Design Your Own Investigation** formats are structured to guide students to make their own discoveries. Students collect real evidence and are encouraged through open-ended questions to reflect and reformulate their ideas based on this evidence.

Cooperative Learning

Cooperative learning adds the element of social interaction to science learning. Group learning allows students to verbalize ideas, and encourages the reflection that leads to active construction of concepts. It allows students to recognize the inconsistencies in their own perspectives and the strengths of others'. By presenting the idea that there is no one, "ready-made" answer, all students may gain the courage to try to find a viable solution. **Cooperative Learning** strategies appear in the *Teacher Wraparound Edition* margins whenever appropriate.

And More ...

Flex Your Brain, a self-directed critical thinking matrix, is introduced in the introductory chapter, "Science: A Tool for Solving Problems." This activity, referenced wherever appropriate in the *Teacher Wraparound Edition* margins, assists students in identifying what they already know about a subject, then in developing independent strategies to investigate further. **Uncovering Misconceptions** in the chapter opener suggests strategies the teacher may use to evaluate students' current perspectives.

Students are encouraged to discover the pleasure of solving a problem through a variety of features. **Apply** questions that require higher-level, divergent thinking appear in **Check Your Understanding**. The **Expand Your View** features in each chapter invite students to confront real-life problems. **You Try It** and **What Do You Think?** questions encourage students to reflect on issues related to technology and society. The **Skill Handbook** gives specific examples to guide students through the steps of acquiring thinking and process skills. **Skill Builder** activities give students a chance to assess and reinforce the concepts just learned through practice. **Developing Skills**, **Critical Thinking**, and **Problem Solving** sections of the **Chapter Review** allow the teacher to assess and reward successful thinking skills.

Thinking Processes

Science is not just a collection of facts for students to memorize. Rather it is a process of applying those observations and intuitions to situations and problems, formulating hypotheses and drawing conclusions. This interaction of the thinking process with the content of science is the core of science and should be the focus of study.

THINKING PROCESSES

Observing

The most basic thinking process is observing. Through observation—seeing, hearing, touching, smelling, tasting—the student begins to acquire information about an object or event.

Organizing Information

Students can begin to organize the information acquired through observation. This process of organizing information encompasses *ordering*, *organizing*, and *comparing*.

Communicating

Once all the information is gathered, it is necessary to communicate the findings so that they can be considered and shared by others. Information can be presented in tables, charts, graphs, or models.

Inferring

This leads to another process—*inferring*. Inferences are logical conclusions based on observations and are made after careful evaluation of all the available facts or data. They can be tested and evaluated.

Relating

Relating cause and effect focuses on how events or objects interact with one another. It also involves examining dependencies and relationships between objects and events.

CRITICAL THINKING SKILLS

Making Generalizations

Identifying similarities among events or processes and then applying that knowledge to new events involves *making generalizations*.

Evaluating Information

Developing ability in several categories of information evaluation is important to critical thinking: differentiating

fact from opinion, identifying weaknesses in logic or in the interpretation of observations, differentiating between relevant and irrelevant data or ideas.

Applying

Applying is a process that puts scientific information to use. Sometimes the findings can be applied in a practical sense, or they can be used to tie together complex data.

Problem Solving

Using available information to develop an appropriate solution to a complex, integrated question is the essence of *problem solving*.

Decision Making

Decision making involves choosing among alternative properties, issues, or solutions. Making informed decisions is not a random process, but requires knowledge, experience, and good judgment.

Inquiry

The process of *inquiry* involves asking questions or predicting outcomes of future situations. Skills used include the ability to make generalizations, problem solve, and distinguish between relevant and irrelevant information.

INTERACTION OF CONTENT AND PROCESS

SCIENCE INTERACTIONS encourages the interaction between science content and thinking processes by offering hundreds of hands-on activities that are easy to set up and do. In the student text, the **Explore** and **Find Out** activities require students to make observations, and collect and record a variety of data. **Investigates** and **Design Your Own Investigations** connect the activity with the content information.

At the end of each chapter, students use the thinking processes as they complete **Developing Skills**, **Critical Thinking**, **Problem Solving**, and **Connecting Ideas** questions. **Expand Your View** connects the science content to other disciplines.

SKILL HANDBOOK

The **Skill Builder/Skill Handbook** provides the student with another opportunity to practice the thinking processes relevant to the material they are studying. The **Skill Handbook** provides examples of the processes which students may refer to as they do the **Skill Builder** exercises.

Developing Thinking Processes

THINKING PROCESSES	Intro	1	2	3	4	5	6	7	8	9	10	11	12	13	14	15	16	17	18	19
ORGANIZING INFORMATION																				
Classifying	✓	✓			✓	✓	✓	✓	✓	✓	✓	✓		✓	✓		✓	✓		
Sequencing		✓		✓						✓		✓	✓	✓	✓		✓			✓
Concept Mapping		✓	✓	✓	✓	✓	✓	✓	✓	✓	✓	✓	✓	✓	✓	✓	✓	✓	✓	✓
Making and Using Tables	✓		✓	✓	✓	✓	✓	✓		✓	✓	✓	✓	✓	✓	✓	✓		✓	✓
Making and Using Graphs	✓			✓		✓	✓			✓	✓		✓	✓						✓
THINKING CRITICALLY																				
Observing and Inferring	✓	✓	✓	✓	✓	✓	✓	✓	✓	✓	✓	✓	✓	✓	✓	✓	✓	✓	✓	✓
Comparing and Contrasting	✓		✓	✓	✓	✓	✓	✓	✓	✓	✓	✓	✓	✓	✓	✓	✓	✓	✓	✓
Recognizing Cause and Effect	✓	✓	✓	✓	✓	✓	✓	✓	✓	✓	✓	✓	✓	✓	✓	✓	✓	✓	✓	✓
Forming Operational Definitions		✓		✓	✓	✓	✓	✓	✓	✓	✓	✓	✓	✓	✓	✓	✓			✓
Measuring in SI	✓	✓		✓	✓	✓	✓			✓			✓	✓	✓	✓			✓	✓
PRACTICING SCIENTIFIC PROCESSES																				
Observing		✓		✓	✓	✓	✓	✓		✓	✓	✓	✓	✓	✓		✓			
Forming a Hypothesis	✓	✓	✓	✓	✓	✓	✓	✓	✓	✓	✓	✓	✓	✓	✓	✓	✓	✓	✓	✓
Designing an Experiment to Test a Hypothesis	✓	✓	✓	✓	✓	✓	✓	✓	✓	✓	✓	✓	✓	✓	✓	✓	✓	✓	✓	✓
Separating and Controlling Variables	✓	✓						✓	✓	✓	✓	✓	✓	✓	✓			✓	✓	
Interpreting Data	✓	✓	✓	✓	✓	✓	✓	✓	✓	✓	✓	✓	✓	✓	✓		✓	✓	✓	✓
REPRESENTING AND APPLYING DATA																				
Interpreting Scientific Illustrations	✓	✓	✓	✓	✓	✓	✓	✓	✓	✓	✓	✓	✓	✓	✓	✓	✓	✓	✓	✓
Making Models	✓	✓	✓	✓	✓	✓	✓		✓			✓	✓	✓	✓	✓	✓	✓	✓	✓
Predicting			✓	✓		✓	✓	✓		✓	✓	✓	✓	✓	✓	✓		✓	✓	✓
Sampling and Estimating					✓	✓				✓	✓	✓		✓				✓		✓

Thinking Processes

Multicultural Perspectives

American classrooms reflect the rich and diverse cultural heritages of the American people. Students come from different ethnic backgrounds and different cultural experiences into a common classroom that must assist all of them in learning. The diversity itself is an important focus of the learning experience.

Diversity can be repressed, creating a hostile environment; ignored, creating an indifferent environment; or appreciated, creating a receptive and productive environment. Responding to diversity and approaching it as a part of every curriculum is challenging to a teacher, experienced or not. The goal of science is understanding. The goal of multicultural education is to promote the understanding of how people from different cultures approach and solve the basic problems all humans have in living and learning. *SCIENCE INTERACTIONS* addresses this issue. In the **Multicultural Perspectives** sections of the *Teacher Wraparound Edition*, information is provided about people and groups who have traditionally been misrepresented or omitted. The intent is to build awareness and appreciation for the global community in which we all live.

The *SCIENCE INTERACTIONS Teacher Classroom Resources* also includes a *Multicultural Connections* booklet that offers additional opportunities to integrate multicultural materials into the curriculum. By providing these opportunities, *SCIENCE INTERACTIONS* is helping to meet the four major goals of multicultural education:

1. promoting the strength and value of cultural diversity
2. promoting human rights and respect for those who are different from oneself
3. promoting social justice and equal opportunity for all people
4. promoting equity in the distribution of power among groups

Two books that provide additional information on multicultural education are:

Atwater, Mary, et al. *Multicultural Education: Inclusion of All.* Athens, Georgia: University of Georgia, Press, 1994.

Banks, James A. (with Cherry A. McGee Banks) *Multicultural Education: Issues and Perspectives.* Boston: Allyn and Bacon, 1989.

School to Work
Tech-Prep

WHAT IS TECH-PREP?

Tech-prep is a rigorous and focused program of study that aims to create a work force in the United States that is technically literate. It is designed to prepare students enrolled in a general curriculum for the demands of further education or for employment by providing them with essential academic and technical foundations, along with problem-solving, group-process, and lifelong-learning skills.

Characteristics of the Tech-Prep Curriculum

The Secretary's Commission on Achieving Necessary Skills (SCANS), published a report in June 1991 that outlined several competencies that characterize successful workers. The Tech-Prep curriculum seeks to address these competencies, which include the

- ability to use resources productively
- ability to use interpersonal skills effectively, including fostering teamwork, teaching others, serving customers, leading, negotiating, and working well with individuals from culturally diverse backgrounds.

- ability to acquire, evaluate, interpret, and communicate data and information.
- ability to understand social, organizational, and technological systems.
- ability to apply technology to specific tasks.

The middle school years provide an opportunity to identify those students who might benefit from a tech-prep curriculum once they reach high school. They also provide an opportunity to introduce unfamiliar students to technological applications leading to career opportunities, or to provide practice for students who already have some knowledge of how technology is used.

Glencoe *Science Interactions* and Tech-Prep Issues

SCIENCE INTERACTIONS helps you develop scientific and technological literacy in your students through a variety of performance-based activities that emphasize problem solving, critical thinking skills, and teamwork. (Each chapter contains an activity with a technological application. Look for the Tech-Prep logo to identify these opportunities.)

CONCEPT MAPS

In science, concept maps make abstract information concrete and useful, improve retention of information, and show students that thought has shape.

Concept maps are visual representations or graphic organizers of relationships among particular concepts. Concept maps can be generated by individual students, small groups, or an entire class. *SCIENCE INTERACTIONS* develops and reinforces four types of concept maps—the **network tree**, **events chain**, **cycle concept map**, and **spider concept map**—that are most applicable to studying science. Examples of the four types and their applications are shown on this page.

Students can learn how to construct each of these types of concept maps by referring to the **Skill Handbook**. Throughout the course, students will have many opportunities to practice their concept mapping skills through **Skill Builder** activities, and **Developing Skills** questions in the **Chapter Review**.

BUILDING CONCEPT MAPPING SKILLS

The **Skill Builders** in each chapter and the **Developing Skills** section of the **Chapter Review** provide opportunities for practicing concept mapping. A variety of concept mapping approaches is used. Students may be directed to make a specific type of concept map and be provided the terms to use. At other times, students may be given only general guidelines. For example, concept terms to be used may be provided and students will be required to select the appropriate model to apply, or vice versa. Finally, students may be asked to provide both the terms and type of concept map to explain relationships among concepts. When students are given this flexibility, it is important for you to recognize that, while sample answers are provided, student responses may vary. Look for the conceptual strength of student responses, not absolute accuracy. You'll notice that most network tree maps provide connecting words that explain the relationships between concepts. We recommend that you not require all students to supply these words, but many students may be challenged by this aspect.

NETWORK TREE

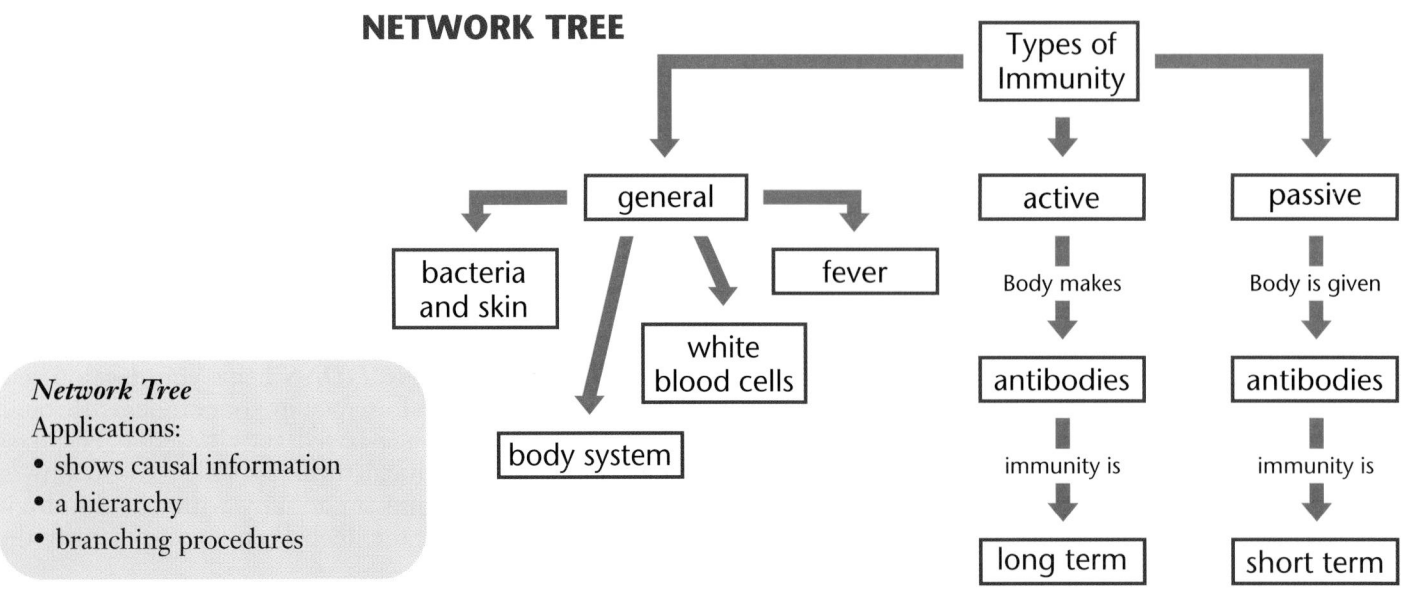

Network Tree
Applications:
• shows causal information
• a hierarchy
• branching procedures

CONCEPT MAPPING BOOKLET

The *Concept Mapping* book of the *Teacher Classroom Resources*, too, provides a developmental approach for students to practice concept mapping.

As a teaching strategy, generating concept maps can be used to preview a chapter's content by visually relating the concepts to be learned and allowing the students to read with purpose. Using concept maps for previewing is especially useful when there are many new key science terms for students to learn. As a review strategy, constructing concept maps reinforces main ideas and clarifies their relationships. Construction of concept maps using cooperative learning strategies as described in this Teacher Guide will allow students to practice both interpersonal and process skills.

CYCLE CONCEPT MAP

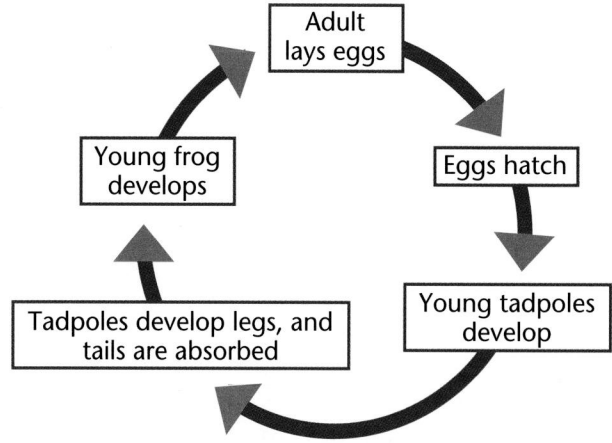

Cycle Concept Map
Application:
• shows how a series of events interact to produce a set of results again and again

EVENTS CHAIN

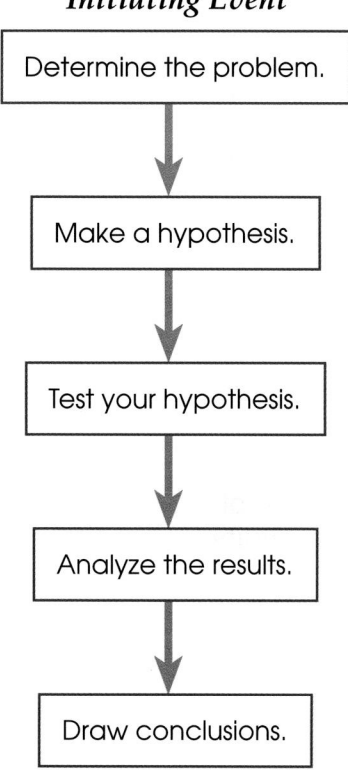

Events Chain
Applications:
• describes the stages of a process
• the steps in a linear procedure
• a sequence of events

SPIDER CONCEPT MAP

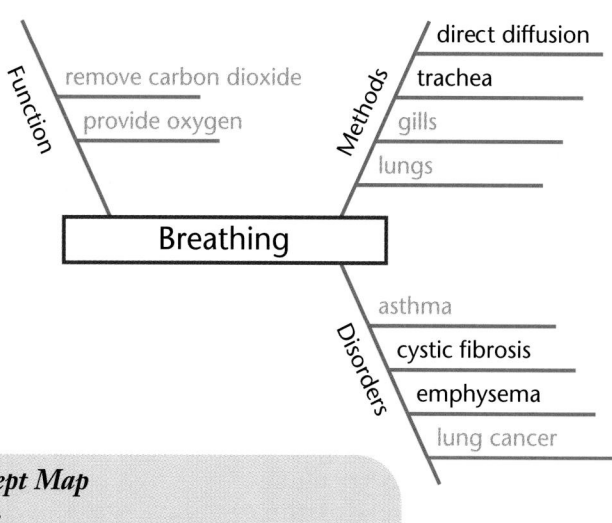

Spider Concept Map
Applications:
• nonhierarchical, except within a category
• unparallel categories

Planning Your Course

SCIENCE INTERACTIONS provides flexibility in the selection of topics and content that allows teachers to adapt the text to the needs of individual students and classes. In this regard, the teacher is in the best position to decide what topics to present, the pace at which to cover the content, and what material to give the most emphasis. To assist the teacher in planning the course, a planning guide has been provided.

SCIENCE INTERACTIONS may be used in a full-year course that is comprised of 180 periods of approximately 45 minutes each. This type of schedule is represented in the table under the heading of Single-class scheduling.

To build flexibility into the curriculum, many schools are introducing a block scheduling, approach. In the table shown here, it is assumed that for block scheduling, the course will be taught for 90 periods of approximately 90 minutes each. If you follow a block schedule, you may want to consider either combining lessons or eliminating certain topics and spending more time on the topics you do cover.

Please remember that the planning guide is provided as an aid in planning the best course for your students. You should use the planning guide in relation to your curriculum and the ability levels of the classes you teach, the materials available for activities, and the time allotted for teaching.

Unit	Chapter/Section	Single-class (180 days)	Block (90 days)
Introduction	Science: A Tool for Solving Problems	3	1
UNIT 1	**Observing the World Around You**	**27**	**13**
1	Viewing Earth and Sky	9	4
	1-1 Viewing Earth	2	1
	1-2 Using Maps	3	1.5
	1-3 Viewing the Sky	3	1
	Chapter Review and Test	1	0.5
2	Light and Vision	9	5
	2-1 The Nature of Light	2	1
	2-2 Reflection and Refraction	3	1.5
	2-3 Color	3	2
	Chapter Review and Test	1	0.5
3	Sound and Hearing	9	4
	3-1 Sources of Sound	2	0.5
	3-2 Frequency and Pitch	3	1.5
	3-3 Music and Resonance	3	1.5
	Chapter Review and Test	1	0.5
UNIT 2	**Interactions in the Physical World**	**28**	**15**
4	Describing the Physical World	10	5
	4-1 Composition of Matter	2	1
	4-2 Describing Matter	2	1
	4-3 Physical and Chemical Changes	3	1.5
	4-4 States of Matter	2	1
	Chapter Review and Test	1	0.5
5	Matter in Solution	9	5
	5-1 Types of Solutions	3	2
	5-2 Solubility and Concentration	3	1.5
	5-3 Colloids and Suspensions	2	1
	Chapter Review and Test	1	0.5
6	Acids, Bases, and Salts	9	5
	6-1 Properties and Uses of Acids	3	1.5
	6-2 Properties and Uses of Bases	2	1.5
	6-3 An Acid or a Base?	2	1
	6-4 Salts	1	0.5
	Chapter Review and Test	1	0.5
UNIT 3	**Interactions in the Living World**	**48**	**24**
7	Describing the Living World	9	5
	7-1 What Is the Living World?	2	1.5
	7-2 Classification	2	1
	7-3 Modern Classification	4	2
	Chapter Review and Test	1	0.5
8	Viruses and Simple Organisms	9	5
	8-1 The Microscopic World	3	1.5
	8-2 Monerans and Protists	2	1
	8-3 Fungus Kingdom	3	2
	Chapter Review and Test	1	0.5
9	Animal Life	11	5
	9-1 What Is an Animal?	3	1.5
	9-2 Reproduction and Development	4	2
	9-3 Adaptations for Survival	3	1
	Chapter Review and Test	1	0.5

Unit	Chapter/Section	Single-class (180 days)	Block (90 days)
10	Plant Life	10	5
	10-1 What Is a Plant?	2	1
	10-2 Classifying Plants	2	1
	10-3 Plant Reproduction	3	1.5
	10-4 Plant Processes	2	1
	Chapter Review and Test	1	0.5
11	Ecology	9	4
	11-1 What Is an Ecosystem?	2	1
	11-2 Organisms in Their Environments	3	1.5
	11-3 How Limiting Factors Affect Organisms	3	1
	Chapter Review and Test	1	0.5
UNIT 4	**Changing Systems**	**47**	**24**
12	Motion	10	5
	12-1 Position, Distance, and Speed	3	1.5
	12-2 Velocity	3	1.5
	12-3 Acceleration	2	1
	12-4 Motion Along Curves	1	0.5
	Chapter Review and Test	1	0.5
13	Motion Near Earth	9	5
	13-1 Falling Bodies	3	1.5
	13-2 Projectile Motion	1	0.5
	13-3 Circular Orbits of Satellites	2	1
	13-4 The Motion of a Pendulum	2	1.5
	Chapter Review and Test	1	0.5
14	Moving Water	9	4
	14-1 Water Recycling	2	1
	14-2 Streams and Rivers	3	1.5
	14-3 Groundwater in Action	3	1
	Chapter Review and Test	1	0.5
15	Shaping the Land	9	5
	15-1 Gravity	1	0.5
	15-2 Running Water	3	1.5
	15-3 Glaciers	2	1.5
	15-4 Wind Erosion	2	1
	Chapter Review and Test	1	0.5
16	Changing Ecosystems	10	5
	16-1 Succession—Building New Communities	3	1.5
	16-2 Interactions in an Ecosystem	4	2
	16-3 Extinction—A Natural Process	2	1
	Chapter Review and Test	1	0.5
UNIT 5	**Wave Motion**	**27**	**13**
17	Waves	9	5
	17-1 Waves and Vibrations	3	1.5
	17-2 Wave Characteristics	2	1.5
	17-3 Adding Waves	2	1
	17-4 Sound as Waves	1	0.5
	Chapter Review and Test	1	0.5
18	Earthquakes and Volcanoes	9	4
	18-1 Earthquakes, Volcanoes, and You	3	1.5
	18-2 Earthquake and Volcano Destruction	2	1
	18-3 Measuring Earthquakes	3	1
	Chapter Review and Test	1	0.5
19	The Earth-Moon System	9	4
	19-1 Earth's Shape and Movements	3	1.5
	19-2 Motions of the Moon	3	1
	19-3 Tides	2	1
	Chapter Review and Test	1	0.5

ASSESSMENT

What criteria do you use to assess your students as they progress through a course? Do you rely on formal tests and quizzes? To assess students' achievement in science, you need to measure not only their knowledge of the subject matter, but also their ability to handle apparatus, to organize, predict, record, and interpret data, to design experiments, and to communicate orally and in writing. *SCIENCE INTERACTIONS* has been designed to provide you with a variety of assessment tools to help you develop a clearer picture of your students' progress.

PERFORMANCE ASSESSMENT

Performance assessment is based on judging the quality of a student's response to a performance task. A performance task is constructed to require the use of important concepts with supporting information, work habits important to science, and one or more of the elements of scientific literacy. The performance task attempts to put the student in a "real world" context so that the class learning can be put to authentic uses.

Performance Assessment in *SCIENCE INTERACTIONS*

Many performance task assessment lists can be found in Glencoe's *Performance Assessment in the Science Classroom*. These lists were developed for the summative and skill performance tasks in the *Performance Assessment* book that accompanies the *SCIENCE INTERACTIONS* program. The Performance Assessment book contains a mix of skill assessments and summative assessments that tie together major concepts of each chapter. Software programs for the assessment lists are also available. The lists can be used to support **Find Out** activities; **Explore** activities; **Investigates; Design Your Own Investigations;** and the **You Try It** and **Going Further** sections of **A Closer Look, Science Connections**, and **Expand Your View**. The assessment lists and rubrics can also be used for the **Project** ideas

described in **Reviewing Main Ideas**. Glencoe's Alternate Assessment in the Science Classroom provides additional background and examples of performance assessment. Activity sheets in the Activity Masters book provide yet another vehicle for formal assessment of student products. The **MindJogger Videoquiz** series offers interactive videos that provide a fun way for your students to review chapter concepts. You can extend the use of the videoquizzes by implementing them in a testing situation. Questions are at three difficulty levels: basic, intermediate, and advanced.

ASSESSING STUDENT WORK WITH RUBRICS

A rubric is a set of descriptions of the quality of a process and/or a product. The set of descriptions includes a continuum of quality from excellent to poor. Rubrics for various types of assessment products are in *Performance Assessment in the Science Classroom.*

GROUP PERFORMANCE ASSESSMENT

Recent research has shown that cooperative learning structures produce improved student learning outcomes for students of all ability levels. *SCIENCE INTERACTIONS* provides many opportunities for cooperative learning and, as a result, many opportunities to observe group work processes and products. *SCIENCE INTERACTIONS: Cooperative Learning Resource Guide* provides strategies and resources for implementing and evaluating group activities. In cooperative group assessment, all members of the group contribute to the work process and the products it produces. An example, along with information about evaluating cooperative work, is provided in the booklet *Alternate Assessment in the Science Classroom.*

SCIENCE JOURNALS AND PORTFOLIOS

A science journal is intended to help the student organize his or her thinking. It is not a lecture or laboratory notebook. It is a place for students to make their thinking explicit in drawings and writing. It is the place to explore what makes science fun and what makes it hard.

The portfolio should help the student see the "big picture" of how he or she is performing in gaining knowledge and skills and how effective his or her work habits are. The portfolio is a way for students to see how individual performance tasks fit into a pattern that reveals the overall quality of their learning. The process of assembling the portfolio should be both integrative (of process and content) and reflective.

OPPORTUNITIES FOR USING SCIENCE JOURNALS AND PORTFOLIOS

SCIENCE INTERACTIONS presents a wealth of opportunities for performance portfolio development. Each chapter in the student text contains projects, enrichment activities, investigations, skill builders, and connections with life, society, and literature. Each of the student activities results in a product. A mixture of these products can be used to document student growth during the grading period. Descriptions, examples, and assessment criteria for portfolios are discussed in *Alternate Assessment in the Science Classroom*. Glencoe's *Performance Assessment in the Science Classroom* contains even more information on using science journals and making portfolios. Performance task assessment lists and rubrics for both journals and portfolios are found there.

CONTENT ASSESSMENT

While new and exciting performance skill assessments are emerging, paper-and-pencil tests are still a mainstay of student evaluation. Students must learn to conceptualize, process, and prepare for traditional content assessments. Presently and in the foreseeable future, students will be required to pass pencil-and-paper tests to exit high school, and to enter college, trade schools, and other training programs.

SCIENCE INTERACTIONS contains numerous strategies and formative checkpoints for evaluating student progress toward mastery of science concepts. Throughout the chapters in the student text, **Check Your Understanding** questions and application tasks are presented. This spaced review process helps build learning bridges that allow all students to confidently progress from one lesson to the next.

For formal review that precedes the written content assessment, *SCIENCE INTERACTIONS* presents a **Chapter Review** at the end of each chapter. By evaluating student responses to this extensive review, you can determine whether any substantial reteaching is needed.

For the formal content assessment, a one-page review and a three-page **Chapter Test** are provided for each chapter. Using the review in a whole class session, you can correct any misperceptions and provide closure for the text. If your individual assessment plan requires a test that differs from the **Chapter Test** in the resource package, customized tests can be easily produced using the **Computer Test Bank**.

MANAGING ACTIVITIES
AND PREPARING SOLUTIONS SAFELY

SCIENCE INTERACTIONS engages students in a variety of hands-on experiences to provide all students with an opportunity to learn by doing. The many hands-on activities throughout *SCIENCE INTERACTIONS* require simple, common materials, making them easy to set up and manage in the classroom.

Find Out and **Explore** activities are intended to be short and occur many times throughout the text. The integration of these activities with the core material provides for thorough development and reinforcement of concepts.

SCIENCE INTERACTIONS provides more than the same "cookbook" activities you've seen hundreds of times before. Students work cooperatively to develop their own experimental designs in the **Investigates** and **Design Your Own Investigations.** They discover firsthand that developing procedures for studying a problem is not as hard as they thought it might be. If you want to give your students additional opportunities to experience self-directed activities, the *Science Discovery Activities* in the *Teacher Classroom Resources* will allow you to do just that. The three activities per chapter challenge students to use their critical thinking skills in developing experimental procedures to test hypotheses.

Preparing Students for Open-ended Labs

To prepare students for the investigations, you should follow the guidelines in the *Teacher Wraparound Edition*, especially in the sections titled Possible Procedures and Teaching Strategies. Your introduction to an **Investigate** or **Design Your Own Investigation** will be very different from traditional activity introductions in that it will be designed to focus students on the problem without giving them directions for how to set up their experiment. Different groups of students will develop alternative hypotheses and alternative procedures. Check their proce-

dures before they begin. In contrast to some "cookbook" activities, there may not always be just one right answer.

Preparation of Solutions

It is most important to use safe laboratory techniques when handling all chemicals. Many substances may appear harmless but are, in fact, toxic, corrosive, or very reactive. Always check with the manufacturer. Chemicals should never be ingested. Be sure to use proper techniques to smell solutions or other agents. Always wear safety goggles and an apron. The following general cautions should be used.

1. Poisonous/corrosive liquid and/or vapor. Use in the fume hood. Examples: *acetic acid, hydrochloric acid, ammonia hydroxide, nitric acid.*

2. Poisonous and corrosive to eyes, lungs, and skin. Examples: *acids, limewater, iron(III) chloride, bases, silver nitrate, iodine, potassium permanganate.*

3. Poisonous if swallowed, inhaled, or absorbed through the skin. Examples: *glacial acetic acid, copper compounds, barium chloride, lead compounds, chromium compounds, lithium compounds, cobalt(II) chloride, silver compounds.*

4. Always add acids to water, never the reverse.

5. When sulfuric acid or sodium hydroxide is added to water, a large amount of thermal energy is released. Sodium metal reacts violently with water. Use extra care if handling any of these substances.

Unless otherwise specified, solutions are prepared by adding the solid to a small amount of distilled water and then diluting with water to the volume listed. If you use a hydrate that is different from the one specified in a particular preparation, you will need to adjust the amount of the hydrate to obtain the required concentration.

Cooperative Learning

WHAT IS COOPERATIVE LEARNING?

In cooperative learning, students work together in small groups to learn academic material and interpersonal skills. Group members learn that they are responsible for accomplishing an assigned group task as well as for each learning the material. Cooperative learning fosters academic, personal, and social success for all students.

ESTABLISHING A COOPERATIVE CLASSROOM

Cooperative groups in the middle school usually contain from two to five students. Heterogeneous groups that represent a mixture of abilities, genders, and ethnicity expose students to ideas different from their own and help them to learn to work with different people.

Initially, cooperative learning groups should only work together for a day or two. After the students are more experienced, they can work with a group for longer periods of time. Students must understand that they are responsible for group members learning the material.

Before beginning, discuss the basic rules for effective cooperative learning—(1) listen while others are speaking, (2) respect other people and their ideas, (3) stay on tasks, and (4) be responsible for your own actions.

The **Teacher Wraparound Edition** uses the code **COOP LEARN** at the end of activities and teaching ideas where cooperative learning strategies are useful.

USING COOPERATIVE LEARNING STRATEGIES

The **Cooperative Learning Resource Guide** of the **Teacher Classroom Resources** provides help for selecting cooperative learning strategies, as well as methods for troubleshooting and evaluation.

EVALUATING COOPERATIVE LEARNING

At the close of the lesson, have groups share their products or summarize the assignment. You can evaluate group performance during a lesson by frequently asking questions to group members picked at random or having each group take a quiz together. Assess individual learning by your traditional methods.

Meeting Individual Needs

Each student brings his or her own unique set of abilities, perceptions, and needs into the classroom. It is important that the teacher try to make the classroom environment as receptive to these differences as possible.

Recognize that individual learning styles are different and that learning style does not reflect a student's ability level. The chart on pages 34T-35T gives additional tips you may find useful in structuring the learning environment in your classroom to meet students' special needs.

In an effort to provide all students with a positive science experience, this text offers a variety of ways for students to interact with materials so that they can utilize their preferred method of learning. This approach allows students to become familiar with other learning styles as well.

ABILITY LEVELS

The activities are broken down into three levels to accommodate all student ability levels. **SCIENCE INTERACTIONS Teacher Wraparound Edition** designates the activities as follows:

L1 basic activities are designed to be within the ability range of all students.

L2 application activities are designed for students who have mastered the concepts presented.

L3 challenging activities are designed for the students who are able to go beyond the basic concepts presented.

LIMITED ENGLISH PROFICIENCY

In providing for the student with limited English proficiency, the focus needs to be on overcoming a language barrier. Once again it is important not to confuse ability in speaking/reading English with academic ability or "intelligence." Look for this symbol **LEP** in the teacher margin for specific strategies for students with limited English proficiency.

In the options margins of the **Teacher Wraparound Edition** there are two or more **Meeting Individual Needs** strategies for each chapter.

Meeting Individual Needs

	DESCRIPTION	SOURCES OF HELP/INFORMATION
Learning Disabled	All learning disabled students have an academic problem in one or more areas, such as academic learning, language, perception, social-emotional adjustment, memory, or attention.	*Journal of Learning Disabilities* *Learning Disability Quarterly*
Behaviorally Disordered	Children with behavior disorders deviate from standards or expectations of behavior and impair the functioning of others and themselves. These children may also be gifted or learning disabled.	*Exceptional Children* *Journal of Special Education*
Physically Challenged	Children who are physically disabled fall into two categories—those with orthopedic impairments and those with other health impairments. Orthopedically impaired children have the use of one or more limbs severely restricted, so the use of wheelchairs, crutches, or braces may be necessary. Children with other health impairments may require the use of respirators or other medical equipment.	Batshaw, M.L. and M.Y. Perset. *Children with Handicaps: A Medical Primer.* Baltimore: Paul H. Brooks, 1981. Hale, G. (Ed.). *The Source Book for the Disabled.* New York: Holt, Rinehart & Winston, 1982. *Teaching Exceptional Children*
Visually Impaired	Children who are visually disabled have partial or total loss of sight. Individuals with visual impairments are not significantly different from their sighted peers in ability range or personality. However, blindness may affect cognitive, motor, and social development, especially if early intervention is lacking.	*Journal of Visual Impairment and Blindness* *Education of Visually Handicapped* American Foundation for the Blind
Hearing Impaired	Children who are hearing impaired have partial or total loss of hearing. Individuals with hearing impairments are not significantly different from their hearing peers in ability range or personality. However, the chronic condition of deafness may affect cognitive, motor, and social development if early intervention is lacking. Speech development also is often affected.	*American Annals of the Deaf* *Journal of Speech and Hearing Research* *Sign Language Studies*
Limited English Proficiency	Multicultural and/or bilingual children often speak English as a second language or not at all. The customs and behavior of people in the majority culture may be confusing for some of these students. Cultural values may inhibit some of these students from full participation.	*Teaching English as a Second Language Reporter* R.L. Jones (Ed.). *Mainstreaming and the Minority Child.* Reston, VA: Council for Exceptional Children, 1976.
Gifted	Although no formal definition exists, these students can be described as having above-average ability, task commitment, and creativity. Gifted students rank in the top 5% of their class. They usually finish work more quickly than other students, and are capable of divergent thinking.	*Journal for the Education of the Gifted* *Gifted Child Quarterly* *Gifted Creative/Talented*

TIPS FOR INSTRUCTION
With careful planning, the needs of all students can be met in the science classroom.

1. Provide support and structure; clearly specify rules, assignments, and duties.
2. Establish situations that lead to success.
3. Practice skills frequently. Use games and drills to help maintain student interest.
4. Allow students to record answers on tape and allow extra time to complete tests and assignments.
5. Provide outlines or tape lecture material.
6. Pair students with peer helpers, and provide class time for pair interaction.

1. Provide a clearly structured environment with regard to scheduling, rules, room arrangement, and safety.
2. Clearly outline objectives and how you will help students obtain objectives. Seek input from them about their strengths, weaknesses, and goals.
3. Reinforce appropriate behavior and model it for students.
4. Do not expect immediate success. Instead, work for long-term improvement.
5. Balance individual needs with group requirements.

1. Openly discuss with the student any uncertainties you have about when to offer aid.
2. Ask parents or therapists and students what special devices or procedures are needed, and if any special safety precautions need to be taken.
3. Allow physically disabled students to do everything their peers do, including participating in field trips, special events, and projects.
4. Help non-disabled students and adults understand physically disabled students.

1. As with all students, help the student become independent. Some assignments may need to be modified.
2. Teach classmates how to serve as guides.
3. Limit unnecessary noise in the classroom.
4. Encourage students to use their sense of touch. Provide tactile models whenever possible.
5. Describe people and events as they occur in the classroom.
6. Provide taped lectures and reading assignments.
7. Team the student with a sighted peer for laboratory work.

1. Seat students where they can see your lip movements easily, and avoid visual distractions.
2. Avoid standing with your back to the window or light source.
3. Use an overhead projector so you can maintain eye contact while writing.
4. Seat students where they can see speakers.
5. Write all assignments on the board, or hand out written instructions.
6. If the student has a manual interpreter, allow both student and interpreter to select the most favorable seating arrangements.

1. Remember, students' ability to speak English does not reflect their academic ability.
2. Try to incorporate the student's cultural experience into your instruction. The help of a bilingual aide may be effective.
3. Include information about different cultures in your curriculum to help build students' self-image. Avoid cultural stereotypes.
4. Encourage students to share their cultures in the classroom.

1. Make arrangements for students to take selected subjects early and to work on independent projects.
2. Let students express themselves in art forms such as drawing, creative writing, or acting.
3. Make public services available through a catalog of resources, such as agencies providing free and inexpensive materials, community services and programs, and people in the community with specific expertise.
4. Ask "what if" questions to develop high-level thinking skills. Establish an environment safe for risk taking.
5. Emphasize concepts, theories, ideas, relationships, and generalizations.

LS Learning Styles

Kinesthetic
Student learns from touch, movement, and manipulating objects

Visual-Spatial
Student learns by responding to images and illustrations

Logical-Mathematical
Student learns by using numbers and reasoning

Interpersonal
Student learns by interacting with others

Intrapersonal
Student learns by working alone

Linguistic
Student learns by using and understanding words

Auditory-Musical
Student learns by listening to the spoken word and to tones and rhythms

NATIONAL GEOGRAPHIC SOCIETY

and Glencoe Science We're a Team

Glencoe Science and National Geographic Society

have teamed up to bring exciting new features and technologies to *Science Interactions*. By incorporating National Geographic's world-renowned photographs, illustrations, and content features, *Science Interactions* will engage students as never before.

Engaging new Unit Openers

feature National Geographic photographs that attract students and help them focus on the unit to come.

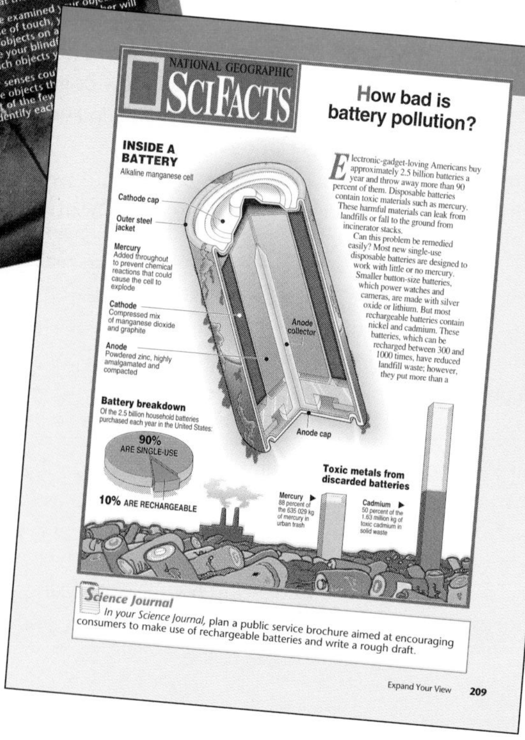

UNIT 1

Observing the World Around You

This golden-eyed snake seems to be always on the lookout. It never closes its eyes, even when it sleeps. Open your eyes to the world around you. In Unit I, begin a journey of discovery and observation. You'll see Earth from a distance and up close as you study the sounds, the light, and the colors of your world.

NATIONAL GEOGRAPHIC try it!

In the world that we can see and hear, we rely on observations to gain information. The process of observing involves using one or more of our senses to learn about the world around us. What if you were limited to only one sense—the sense of touch? How do you think this would change your perception of the world around you?

What To Do

1. Find a partner to work with. Your teacher will provide each of you with a blindfold to wear and several objects. Can you determine what the objects are by touch?

2. Once you have examined your objects with the sense of touch, place all the objects on a table, then remove your blindfold and identify which objects you...

3. What other senses cou... identify the objects th... Make a list of the few... need to identify each...

Fascinating SciFacts

features enrich and extend chapter content with National Geographic illustrations and graphics.

NATIONAL GEOGRAPHIC SciFacts

How bad is battery pollution?

Electronic-gadget-loving Americans buy approximately 2.5 billion batteries a year and throw away more than 90 percent of them. Disposable batteries contain toxic materials such as mercury. These harmful materials can leak from landfills or fall to the ground from incinerator stacks.

Can this problem be remedied easily? Most new single-use disposable batteries are designed to work with little or no mercury. Smaller button-size batteries, which power watches and cameras, are made with silver oxide or lithium. But most rechargeable batteries contain nickel and cadmium. These batteries, which can be recharged between 300 and 1000 times, have reduced landfill waste; however, they put more than a

INSIDE A BATTERY
Alkaline manganese cell

Cathode cap

Outer steel jacket

Mercury
Added throughout to prevent chemical reactions that could cause the cell to explode

Cathode
Compressed mix of manganese dioxide and graphite

Anode
Powdered zinc, highly amalgamated and compacted

Anode collector

Anode cap

Battery breakdown
Of the 2.5 billion household batteries purchased each year in the United States

90% ARE SINGLE-USE

10% ARE RECHARGEABLE

Toxic metals from discarded batteries

Mercury ►
88 percent of the 635,029 kg of mercury in urban trash

Cadmium ►
50 percent of the 1.63 million kg of toxic cadmium in solid waste

Science Journal
In your *Science Journal*, plan a public service brochure aimed at encouraging consumers to make use of rechargeable batteries and write a rough draft.

Expand Your View **209**

Helpful Teacher's Corner

feature in the Teacher's Wraparound Edition correlates *National Geographic Magazine* articles, technology, and other helpful teaching resources available from National Geographic Society.

Correlated Technology

In addition, **Glencoe** offers a wide variety of National Geographic videodiscs and **CD-ROMs** fully correlated to the textbooks. In fact, **National Geographic Videodisc** bar codes are placed in this Teacher Edition for your convenience at the point of use in each chapter.

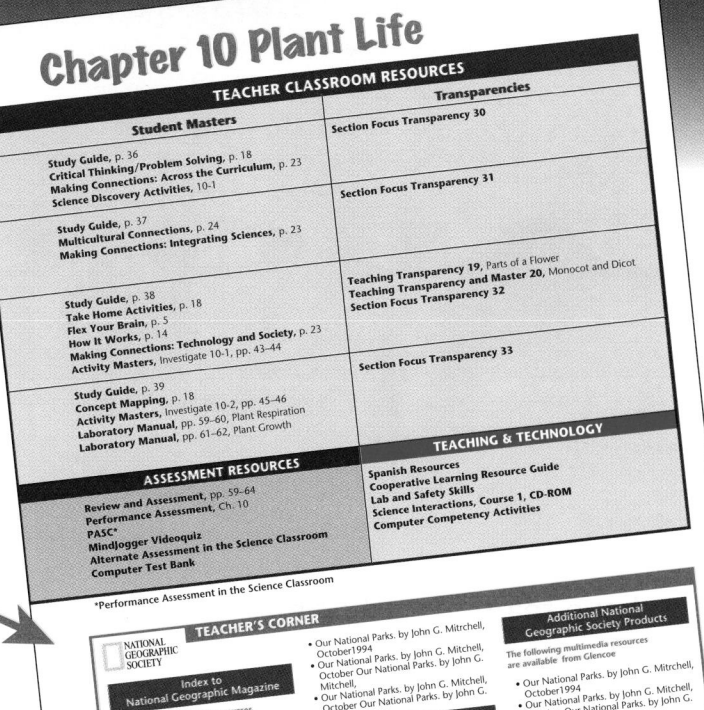

Chapter 10 Plant Life

TEACHER CLASSROOM RESOURCES

Student Masters	Transparencies
Study Guide, p. 36 Critical Thinking/Problem Solving, p. 18 Making Connections: Across the Curriculum, p. 23 Science Discovery Activities, 10-1	Section Focus Transparency 30
	Section Focus Transparency 31
Study Guide, p. 37 Multicultural Connections, p. 24 Making Connections: Integrating Sciences, p. 23	
Study Guide, p. 38 Take Home Activities, p. 18 Flex Your Brain, p. 5 How It Works, p. 14 Making Connections: Technology and Society, p. 23 Activity Masters, Investigate 10-1, pp. 43–44	Teaching Transparency 19, Parts of a Flower Teaching Transparency and Master 20, Monocot and Dicot Section Focus Transparency 32
	Section Focus Transparency 33
Study Guide, p. 39 Concept Mapping, p. 18 Activity Masters, Investigate 10-2, pp. 45–46 Laboratory Manual, pp. 59–60, Plant Respiration Laboratory Manual, pp. 61–62, Plant Growth	

ASSESSMENT RESOURCES

Review and Assessment, pp. 59–64
Performance Assessment, Ch. 10
PASC*
Mindjogger Videoquiz
Alternate Assessment in the Science Classroom
Computer Test Bank

TEACHING & TECHNOLOGY

Spanish Resources
Cooperative Learning Resource Guide
Lab and Safety Skills
Science Interactions, Course 1, CD-ROM
Computer Competency Activities

*Performance Assessment in the Science Classroom

TEACHER'S CORNER

NATIONAL GEOGRAPHIC SOCIETY

Index to National Geographic Magazine

The following multimedia resources are available from Glencoe

• **Poster** Parks. by John G. Mitchell, October1994
• Our National Parks. by John G. Mitchell, October Our National Parks. by John G. Mitchell,
• **Transparency** by John G. Mitchell, October Our National Parks. by John G. Mitchell,
• Our National Parks. by John G. Mitchell,

• Our National Parks. by John G. Mitchell, October1994
• Our National Parks. by John G. Mitchell, October Our National Parks. by John G. Mitchell,
• Our National Parks. by John G. Mitchell, October Our National Parks. by John G.

National Geographic Society Products Available From Glencoe

The following multimedia resources are available from Glencoe

• Our National Parks. by John G. Mitchell, October1994
• Our National Parks. by John G. Mitchell,

Additional National Geographic Society Products

The following multimedia resources are available from Glencoe

• Our National Parks. by John G. Mitchell, October1994
• Our National Parks. by John G. Mitchell, October Our National Parks. by John G. Mitchell,
• Our National Parks. by John G. Mitchell, October Our National Parks. by John G. Mitchell,
• Our National Parks. by John G. Mitchell, October Our National Parks. by John G. Mitchell,

Chapter 10   Plant Life   **310B**

Safety is of prime importance in every classroom. However, the need for safety is even greater when science is taught. The activities in **SCIENCE INTERACTIONS** are designed to minimize dangers in the laboratory. Even so, there are no guarantees against accidents. Careful planning and preparation as well as being aware of hazards can keep accidents to a minimum. Numerous books and pamphlets are available on laboratory safety with detailed instructions on preventing accidents. In addition, the **SCIENCE INTERACTIONS** program provides safety guide-lines in several forms. The ***Lab and Safety Skills*** booklet contains detailed guidelines, in addition to masters you can use to test students' lab and safety skills. The ***Student Edition*** and ***Teacher Wraparound Edition*** provide safety precautions and symbols designed to alert students to possible dangers. Know the rules of safety and what common violations occur. Know the **Safety Symbols** used in this book. Know where emergency equipment is stored and how to use it. Practice good laboratory housekeeping and management to ensure the safety of your students.

DISPOSAL ALERT This symbol appears when care must be taken to dispose of materials properly.	**FUME SAFETY** This symbol appears when chemicals or chemical reactions could cause dangerous fumes.	**CLOTHING PROTECTION SAFETY** This symbol appears when substances used could stain or burn clothing.
BIOLOGICAL HAZARD This symbol appears when there is danger involving bacteria, fungi, or protists.	**ELECTRICAL SAFETY** This symbol appears when care should be taken when using electrical equipment.	**FIRE SAFETY** This symbol appears when care should be taken around open flames.
OPEN FLAME ALERT This symbol appears when use of an open flame could cause a fire or an explosion.	**SKIN PROTECTION SAFETY** This symbol appears when use of caustic chemicals might irritate the skin or when contact with microorganisms might transmit infection.	**EXPLOSION SAFETY** This symbol appears when the misuse of chemicals could cause an explosion.
THERMAL SAFETY This symbol appears as a reminder to use caution when handling hot objects.	**ANIMAL SAFETY** This symbol appears whenever live animals are studied and the safety of the animals and the students must be ensured.	**EYE SAFETY** This symbol appears when a danger to the eyes exists. Safety goggles should be worn when this symbol appears.
SHARP OBJECT SAFETY This symbol appears when a danger of cuts or punctures caused by the use of sharp objects exists.	**RADIOACTIVE SAFETY** This symbol appears when radioactive materials are used.	**POISON SAFETY** This symbol appears when poisonous substances are used.
		CHEMICAL SAFETY This symbol appears when chemicals used can cause burns or are poisonous if absorbed through the skin.

CHEMICAL Storage & DISPOSAL

GENERAL GUIDELINES

Be sure to store all chemicals properly. The following are guidelines commonly used. Your school, city, county, or state may have additional requirements for handling chemicals. It is the responsibility of each teacher to become informed as to what rules or guidelines are in effect in his or her area.

1. Separate chemicals by reaction type. Strong acids should be stored together. Likewise, strong bases should be stored together and should be separated from acids. Oxidants should be stored away from easily oxidized materials, and so on.

2. Be sure all chemicals are stored in labeled containers indicating contents, concentration, source, date purchased (or prepared), any precautions for handling and storage, and expiration date.

3. Dispose of any outdated or waste chemicals properly according to accepted disposal procedures.

4. Do not store chemicals above eye level.

5. Wood shelving is preferable to metal. All shelving should be firmly attached to the wall and should have anti-roll edges.

6. Store only those chemicals that you plan to use.

7. Hazardous chemicals require special storage containers and conditions. Be sure to know what those chemicals are and the accepted practices for your area. Some substances must even be stored outside the building.

8. When working with chemicals or preparing solutions, observe the same general safety precautions that you would expect from students. These include wearing an apron and goggles. Wear gloves and use the fume hood when necessary. Students will want to do as you do whether they admit it or not.

9. If you are a new teacher in a particular laboratory, it is your responsibility to survey the chemicals stored there and to be sure they are stored properly or disposed of. Consult the rules and laws in your area concerning what chemicals can be kept in your classroom. For disposal, consult up-to-date disposal information from the state and federal governments.

DISPOSAL OF CHEMICALS

Local, state, and federal laws regulate the proper disposal of chemicals. These laws should be consulted before chemical disposal is attempted. Although most substances encountered in high school biology can be flushed down the drain with plenty of water, it is not safe to assume that is always true. It is recommended that teachers who use chemicals consult the following book from the National Research Council:

Prudent Practices in the Laboratory: Handling and Disposal of Chemicals. Washington, DC: National Academy Press, 1995.

DISCLAIMER

Glencoe Publishing Company makes no claims to the completeness of this discussion of laboratory safety and chemical storage. The material presented is not all-inclusive, nor does it address all of the hazards associated with handling, storage, and disposal of chemicals, or with laboratory management.

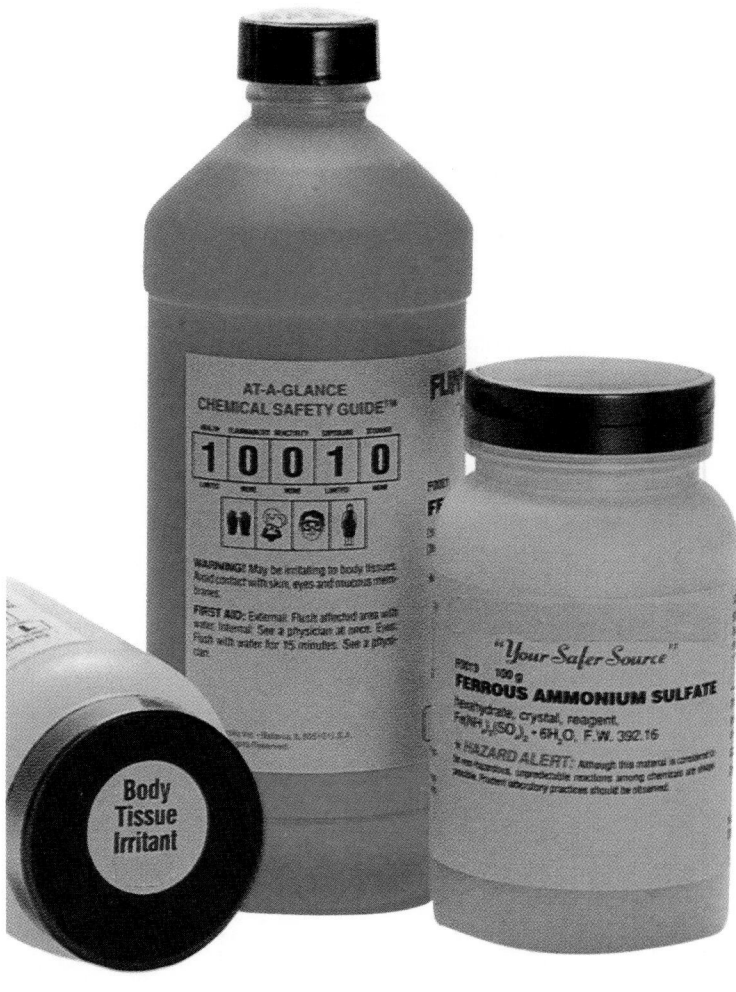

Equipment List

Non-Consumables

Item	INVESTIGATE!	DESIGN YOUR OWN INVESTIGATION	Explore!	Find Out!
Animals, or pictures of animals (15 sets)		40, 98, 194, 222, 266		
Apron (30)	200			
Atlas (15)			22, 601	
Audiotapes (5 to 10)			225	
Balance and set of masses (15)	132	40, 168, 282		154, 155, 186
Ball, nylon, 3/8" (6)		40		
Ball, wood, 3/4" (6)		40		
Ball, polystyrene (15)	252, 618			
Ball, plain, no pattern (15)			34	
Basketball (15)				602, 610
Beaker (15)	30	168	279	135
Beaker, 250-mL (45)				186
Beaker, 300-mL (15)				192
Beaker, 400-mL (60)		98	165	155
Beaker, 500-mL (60)	10, 300	458	129	
Beaker, large (15)	158			444
Beaker, small (30)	158			444
Binoculars (as needed)	514			
Bleachers or wooden floor			573	
Blindfold (15)			573	
Bolts and nuts (30 each)	252		415	
Books (30)	62		58, 582	584
Bowl (15)	354		140, 311	
Bowl, transparent (15)			173	
Box, clear plastic, and lid (15)	30			
Box, plastic storage (15)		282		
Burner (15)				135
Button (15)	398		127	
Can (45)				135, 543, 559
Can opener (1)				543
Clamps, screw (30 or as needed)	452	482		
Coat hanger, metal (30)	590			89
Coins (30)				423
Containers, clear, different shapes (45)				141
Coverslip (several boxes)	334	222, 508	260, 268	255
Cup (15)	354			507
Cup, opaque (15)			64	
Dish (15)	334		142, 279, 289	135
Dish, clear glass, square (15)	552			555
Dishcloth (15)				154
Dishpan (as needed)	10			
Dropper (30)		98, 222, 508		118, 154, 203, 318
Dropping bottles (105)	200			
Erasers, chalkboard (30)			56	

Non-Consumables

Item	INVESTIGATE!	DESIGN YOUR OWN INVESTIGATION	Explore!	Find Out!
Feather (15)			140	
Flashlight (45)	62, 300	72	53, 56, 58, 66, 117, 156, 173	55, 68, 559
Forceps (15)	334, 354	194, 282	136	
Fork (15)			140, 311	
Freezer (1)				135, 491
Funnel (30)				155
Glass (75) 322			153,175	118, 160, 161,
Glass, frosted, piece (15)			58	
Glass, piece (15)			58	
Globe (5 to 15)	618			605
Goggles, safety (30)	132, 158, 200	168, 194, 222, 362, 544	473	136
Graduated cylinder, 10-mL (15)		98, 168		141, 203
Graduated cylinder, 25-mL (15)		458		
Graduated cylinder, 100-mL (15)	132, 200			
Graduated cylinder, 1000-mL (15)	104, 158			141
Hand lens (15)	300, 514	282, 362	263, 268, 289, 320	325, 507
Hole punch (1)				584
Hot plate (15)		168		
Ice cube tray (15)				491
Ingredient list (15)			189, 193	
Jar with lid (30)	300	508	142	136, 196, 218, 318
Jar, heat proof, with lid (30)			117	
Knife, kitchen (15)		266	311	318, 495
Lamp (15)	488, 618			444, 605
Lamp, gooseneck (15)	552	606	555	
Landform, model (15)	30			
Loudspeaker (as needed)			559	
Magnet (15)				122
Magnifying glass (15)	354			192
Marbles, different sizes (about 200)		40		444,523
Measuring cup (15)				141,331
Measuring spoons (6)		222		
Meterstick (15)		388, 432	614	423
Microscope (15)	334, 354	222, 508	247, 259, 260, 268	118, 122, 248, 255, 261
Microscope slide (120 or as needed)	334, 354	222, 508	247, 259, 260, 268	118, 122, 248, 255, 261
Mirror, small (60)	62		142	559
Mortar and pestle (15)				122, 154, 160
Oven mitt (15)		168, 606		
Overhead projector (1)			383	
Pails (up to 30, as needed)	452, 488	482		
Pan, rectangular (30 or 40)	300	40	574, 582	584

Equipment List

Non-Consumables

Item	INVESTIGATE!	DESIGN YOUR OWN INVESTIGATION	Explore!	Find Out!
Paper clip (several boxes)		72	140	55
Penlight (15)				248
Penny (15)			140	136
Pictures, animal (15 sets)			234	
Pie pan, aluminum (75)	10		523	
Pitcher, clear (15)			156	141
Plastic foam (60 strips)	552			
Plastic hose (up to 30 segments as needed)	452	482		
Posterboard		72		
Prism (15)			66	
Protractor (15)	398	606		401, 580
Radio with round speaker (several)				543
Record turntable (1)			403	
Ring stand and ring (30)	590			155, 523
Rocks (45)			129, 140	
Rope (15-4m lengths)			539	
Rubber mallet (15)	104		102	
Ruler, metric (15)	10, 30, 104, 488, 552, 590	40, 98, 432, 458	22, 85, 449	55, 349, 401, 580, 584, 602
Ruler, metal			94	
Scalpel (15)				318, 325
Scissors (15)		72		332, 495, 523, 559, 584, 602
Scoop (15)		168		154
Shoes, boots, sandals		388		
Spoon (15)				495
Spray bottle (15)		362	479	161
Spring, coiled metal (15)		544		87
Sprinkling can (15)			449	
Stirrer (90)	158	194	153, 156, 165, 175	160, 161, 186, 192, 203, 318
Stirrer, copper wire (15)		168		
Stopper, test-tube (15 solid, 15 1-hole, 15 2-hole)		168	403	
Stopwatch (15)		388, 606		160
Stream table (as needed)	452, 488	482	449, 455	
Stringed instrument, non-electrical (several)				95
Sunglasses (15)			58	
Tablespoon (15)				192
Table tennis ball (15)			574	
Teaspoon (15)			153	161, 196
Test-tube holder (15)		168		
Test tubes (120)	200	98, 168, 194, 222	403	122, 203
Test-tube rack (15)	200	98, 194, 222		122, 203
Thermometer (15)		168		
Thermometer, Celsius (15)		606		
Timer, with second hand (1 large or 15 small)		432, 458		
Tongs (15)				135

Non-Consumables

Item	INVESTIGATE!	DESIGN YOUR OWN INVESTIGATION	Explore!	Find Out!
Toothbrush (15)				154
Toy car (battery-powered) (15)			393	
Transparency of House of Terror map (1)		383		
Tray, flat (15)		140		
Tube, plastic or glass, 2.5 cm diameter, 45 cm long (15)	104			
Tuning fork, 2 frequencies (15 each)	104	102		
Vial, small (15)	124			293
Washers (metal) (120)		432		
Watering can (as needed)	10		449	
Wire (15 m)		124, 252		89
Wood (45 blocks)	452	482	449	491
Wood, light piece (15)				415
World map (15)		578		

Living Organisms

Item	INVESTIGATE!	DESIGN YOUR OWN INVESTIGATION	Explore!	Find Out!
Amoeba (15 samples)				261
Bread mold (15 samples)			268	
Diatomite (15 samples)			260	
Earthworms (15)	300			
Euglena (15 samples)				248
Fish tank scrapings (15 samples)			260	
Flowering plant (15)			217	
Flowers (15)				325
Fruit fly cultures (15)				293
Goldfish or guppy (15)			279	
Leaves, variety of (15 sets)			226	
Mealworms (300)		282		
Mold source (10 samples)		362		
Paramecium (15 samples)				261
Philodendron (as needed)				322
Planarian (15 samples)				248
Plants (variety)			313	
Pond vegetation, dried (15 samples)		508		
Seedlings in pots (30)				331
Snail (15)			279	

Equipment List

Consumables

Item	INVESTIGATE!	DESIGN YOUR OWN INVESTIGATION	Explore!	Find Out!
Aluminum foil (several rolls)		124	493	60
Antacid tablet		194		
Balloon (90)			142	163, 559
Box, pizza (15)			601	
Bran flakes (1 box)		282		
Can (15)				523
Candle (15)			217	
Cardboard (60 pieces), some white	354		58, 142, 443	602
Celery stalk (15)				318
Cellophane (15 clear, 15 red, 15 blue, 15 green)			58	68
Cereal, dry, sugarless (1 box)		362		
Chalk, piece (15)		124	56	
Chicken egg, raw (15)			289	
Cheesecloth (several yards)		282		
Clay (several boxes)			58, 413	55
Coffee, hot (15 cups)				507
Coffee, instant (1 jar)			479	
Cornmeal (1 box)			153	
Cornstarch (1 box)				192
Corn syrup (1 bottle)				154
Cotton (1 bag balls)			140	
Cotton swab (1 box)	300	362		
Crackers, dry, unsalted (1 box)		362		
Crackers, wheat (1 box)		282		
Crayons or markers (15 sets)			66	349, 602
Cup, paper (as needed)		328, 362		
Cup, plastic, clear (75)				192
Deodorant		194		
Filter paper (1 box)				155, 186
Flour (several bags)		40	153	495
Food, vial of (15)				293
Fruit (15)		266		
Gelatin dessert mix (several boxes)			173	
Glue (15 bottles)	354		601	559
Gravel (as needed)		458	473	218, 491
House of Terror map (15 copies)			383	
Ice block containing sand, clay, and gravel (15 or as needed)		488		
Index cards (45)				55
Labels (several boxes)		222, 362		192, 261, 331
Leaves or grass clippings	10			
Lemonade		124		
Lemon juice (1 bottle)		194		
Lettuce (1 head)	334			
Litmus paper (several vials)		194		203
Macaroni, dry (1 box)		362		

Consumables

Item	INVESTIGATE!	DESIGN YOUR OWN INVESTIGATION	Explore!	Find Out!
Marker (15), black; transparency (15)	30, 590	98, 458	34	218, 331
Matches (60)			117, 217	
Milk, spoiled and unspoiled (15 samples each)			255	
Mushroom, store-bought (15)			263	
Mustard grains (1 box)				218
Netting (15)				523
Newspaper (as needed)	10			
Oatmeal (dry) (1 box)		282		
Oil, vegetable		124		
Orange slices		194		
Owl pellet (15)	354			
Paper, black (30)		606		325
Paper, construction (as needed)		72	53	
Paper, graph (45)			381	160
Paper, large sheets (as needed)				14
Paper, lined (15)			381	
Paper, plain white (as needed)	236, 552, 590		53, 66, 225, 381, 393, 601, 349	68, 192, 555
Paper, tissue (15)			58	
Paper, tracing (15)		578		349
Paper, waxed (several rolls)			58, 443, 473	
Peanut, in shell (30)			320	
Pebbles (2 bags or as needed)	10	482		
Pencil, grease (15)	200	194		
Peppermint oil (1 bottle)				154
Pipe cleaner (15)	252			
Plaster of paris (5 kg)				580
Plastic bag (90)		266		136, 331
Plastic lids, circular (45)				584
Plastic wrap (several rolls)		362		444, 543
Plate, paper (120)		266	311, 479	580
Potato flakes, dry (1 box)		362		
Rice, uncooked (1 box)			153, 479	543
Rock, granite		124		
Rock Salt		124		
Rubber band (200)	590	72		68, 444, 523, 543, 559
Salad items and dressing			311	
Sand (as needed)	452, 488	458, 482	140, 449, 455 473, 582	155, 491, 580, 584
Seeds (several packets of 3 different types)		328		
Shoe box and lid (15)				495
Soda water (1 L)				186
Soft drink, carbonated, bottled (30)	200	194		163
Soft drink mix, clear (15 packages)			156	
Soil (several bags)/clay soil as required	10, 488	458, 482		

Consumables

Item	INVESTIGATE!	DESIGN YOUR OWN INVESTIGATION	Explore!	Find Out!
Straw, drinking (90)	98		53, 156	
String (1 large ball)/colored as required	398, 590	40, 432	117	87, 415, 584
String, light (15 m)				89
String, thick, water-absorbent (1 ball)	158			
Sugar (10 lbs or as needed)		124, 168, 222	153, 165	161, 192, 580
Sugar, cubes (1 box)				160
Sugar, powdered (1 bag)				154
Tape, cellophane or masking, as needed (several rolls)	30, 552, 590	388, 432, 606	53, 403	163, 523, 602, 610
Toothpicks, flat-ended (2 boxes)	300	222		
Towel, paper (several rolls)	300, 334	266, 328	443	136, 495
Transparency (15)	30			
Wooden splint (15)			117	
Yeast (dry) (30 packages)		222	247	

Chemical Supplies

Item	INVESTIGATE!	DESIGN YOUR OWN INVESTIGATION	Explore!	Find Out!
Alcohol, rubbing (2L)	132		142	
Ammonia (2L)	200	194		136, 196, 203
Baking soda (1 box)	200	124, 194, 222		192
Calcium carbonate (500 g)				154
Castile soap (100 g)				154
Epsom salt (1 box)	158			196
Food coloring (1 bottle)				141, 318
Graphite, piece (15)		124		
Hydrochloric acid (100 mL)	200			
Iron powder (1 bottle)				122
Laundry detergent (1 box)				192
Marble, chips (100 g)				186
Methylene blue stain (100 mL)				255
Petroleum jelly (1 jar)				331
Red cabbage juice indicator (120 mL)	200			
Salt, solution (1 L)	132, 334			
Salt, table (several boxes)		40, 124, 194, 222	153, 165, 479	155, 192
Sodium hydroxide (50 g)	200			
Solder, piece (15)	124			
Sulfur powder (1 bottle)				122
Sulfuric acid, dilute (1 L)				186
Vinegar (1 gallon)	200, 300	124, 194		186, 203
Water (distilled)	as required	as required	as required	as required
Water, pond (100 mL)			259	

References

BIBLIOGRAPHY

GENERAL SCIENCE CONTENT

Cash, Terry. 175 *More Science Experiments To Amuse and Amaze Your Friends: Experiments! Tricks! Things to Make!* New York: Random House, 1991.

Churchill, E. Richard. *Amazing Science Experiments with Everyday Materials.* New York: Sterling Publishing Co., Inc., 1991.

Lewis, James. *Hocus Pocus Stir and Cook, The Kitchen Science-Magic Book.* New York: Meadowbrook Press, Division of Simon and Shuster, Inc., 1991.

Mandell, Muriel. *Simple Science Experiments with Everyday Materials.* New York: Sterling Publishing Co., Inc., 1989.

Roberts, Royston. *Serendipity: Accidental Discoveries in Science.* New York: John Wiley and Sons, Inc., 1989.

Schultz, Robert F. *Selected Experiments and Projects.* Washington, DC: Thomas Alva Edison Foundation, 1988.

Strongin, Herb. *Science on a Shoestring.* Menlo Park, CA: Addison-Wesley Publishing Co., 1985.

PHYSICS

Arons, A.B. *A Guide to Introductory Physics Teaching.* New York: John Wiley and Sons, 1990.

Aronson, Billy. "Water Ride Designers Are Making Waves." *3-2-1 Contact.* August, 1991, pp. 14-16.

Cash, Terry. *Sound.* New York: Warwick Press, 1989.

Hajda, Joey and Lisa B. Hajda. "Sparking Interest in Electricity." *Science Scope,* Nov./Dec., 1994, pp. 36-39.

Hardy, John W. "Adaptive Optics." *Scientific American,* June 1994, pp. 60-65.

Heiligman, Deborah. "There's a Lot More to Color Than Meets the Eye." *3-2-1 Contact.* November, 1991, pp. 16-20.

McGrath, Susan. *Fun with Physics.* Washington, DC: National Geographic Society, 1986.

Taylor, Barbara. *Sound and Music.* New York: Warwick Press, 1990.

Terres, John K. *How Birds Fly.* Mechanicsburg, PA: Stackpole Books, 1994.

Ward, Allen. *Experimenting with Batteries, Bulbs, and Wires.* New York: Chelsea House, 1991.

CHEMISTRY

Barber, Jacqueline. *Of Cabbage and Chemistry.* Washington, DC: Lawrence Hall of Science, NSTA, 1989.

Barber, Jacqueline, *Chemical Reactions.* Washington, DC: Lawrence Hall of Science, NSTA, 1986.

Cornell, John. *Experiments with Mixtures.* New York: Wiley, John and Sons, Inc., 1990.

Joesten, Melvin. *World of Chemistry.* Philadelphia, PA: Saunders College Publishing, 1991.

Laidler, Keith J. *The World of Physical Chemistry.* New York: Oxford University Press, 1993.

Mitchell, Sharon and Juergens, Frederick. *Laboratory Solutions for the Science Classroom.* Batavia, IL: Flinn Scientific, Inc., 1991.

Snyder, Carl H. *The Extraordinary Chemistry of Ordinary Things,* 2nd ed. New York: Wiley, 1995.

LIFE SCIENCE

Children's Atlas of the Environment. Chicago: Rand McNally, 1991.

Dewey, Jennifer Owings. *A Day and Night In the Desert.* Boston, MA: Little Brown, 1991.

Hancock, Judith M. *Variety of Life: A Biology Teacher's Sourcebook.* Portland, OR: J. Weston Walch, 1987.

Johnson, Cathy. *Local Wilderness.* New York: Prentice Hall, 1987.

McGrath, Susan. *The Amazing Things Animals Do.* Washington, DC: National Geographic Society, 1989.

Markmann, Erika. *Grow It! An Indoor/Outdoor Gardening Guide for Kids.* New York: Random House, 1991.

VanCleave, Janice Pratt. *Biology for Every Kid: 101 Easy Experiments that Really Work.* New York: Wiley, 1990.

Wilson, Edward O. *The Diversity of Life.* New York: Norton, 1993.

EARTH SCIENCE

Ardley, Neil. *The Science Book of Air.* New York: Gulliver Books, Harcourt, Brace, Jovanovich, Publishers, 1991.

Barrow, Lloyd H. *Adventures with Rocks and Minerals: Geology Experiments for Young People.* Hillsdale, NJ: Enslow, 1991.

Booth, Basil. *Volcanoes and Earthquakes.* Englewood Cliffs, NJ: Silver Burdett Press, 1991.

Javna, John. *50 Simple Things Kid Can Do to Save the Earth.* Kansas City: The Earth Works Group, Andrews and McMeel, a Universal Press Syndicate Co., 1990.

Norman, David. *Dinosaur!* London: Boxtree Limited, 1991.

Robinson, Andrew. *Earth Shock: Hurricanes, Volcanoes, Earthquakes, Tornadoes and Other Forces of Nature.* Thames and Hudson, 1993.

Seeds, Michael A. *Horizons, Exploring the Universe.* Belmont, CA: Wadsworth Publishing Company, 1995.

VanCleave, Janice. *Earth Science for Every Kid.* New York: John Wiley and Sons, Inc., 1991.

Wood, Robert W. *Science for Kids: 39 Easy Geology Activities.* Blue Ridge Summit, PA: Tab Books, 1992.

SUPPLIER ADDRESSES

SCIENTIFIC SUPPLIERS

Science Kit & Boreal Laboratories
777 East Park Drive
Tonawanda, NY 14150-6748

Carolina Biological Supply Co.
2700 York Road
Burlington, NC 27215

Fisher Scientific Co.
1600 W. Glenlake
Itasca, IL 60143

Flinn Scientific Co.
P.O. Box 219
Batavia, IL 60510

Frey Scientific
100 Paragon Parkway
Mansfield, OH 44903

Kemtec Educational Corp.
9889 Cresent Drive
West Chester, OH 45069

Sargent-Welch Scientific Co.
P.O. Box 5229
Buffalo Grove, IL 60089

Ward's Natural Science
Establishment, Inc.
P.O. Box 92912
Rochester, NY 14692

SOFTWARE DISTRIBUTORS

(AIT) Agency for Instructional
Technology
Box A
Bloomington, IN 47402-0120

Cambridge Development Lab (CDL)
1696 Massachusetts Avenue
Cambridge, MA 02138

COMPress
P.O. Box 102
Wentworth, NH 03282

Earthware Computer Services
P.O. Box 30039
Eugene, OR 97403

Educational Activities, Inc.
1937 Grand Avenue
Baldwin, NY 11510

Educational Materials and Equipment
Company (EME)
P.O. Box 2805
Danbury, CT 06813-2805

Gemstar (Classroom Consortia
Media, Inc.)
P.O. Box 050228
Staten Island, NY 10305

IBM Educational Systems
Department PC
4111 Northside Parkway
Atlanta, GA 30327

McGraw-Hill Webster Division
1221 Avenue of the Americas
New York, NY 10020

Microphys
1737 W. Second Street
Brooklyn, NY 11223

Minnesota Educational Computing
Corporation (MECC)
3490 Lexington Avenue N.
Saint Paul, MN 55126

Queue, Inc.
562 Boston Avenue
Bridgeport, CT 06610

Texas Instruments, Data Systems Group
P.O. Box 1444
Houston, TX 77251

Ventura Educational System
3440 Brokenhill Street
Newbury Park, CA 91320

AUDIOVISUAL DISTRIBUTORS

Aims Media
9710 Desoto Avenue
Chatsworth, CA 91311-4409

BFA Educational Media
468 Park Avenue S.
New York, NY 10016

Churchill Films
662 N. Robertson Blvd.
Los Angeles, CA 90069

Coronet/MTI Film and Video
Distributors of LCA
108 Wilmot Road
Deerfield, IL 60015

CRM Films
2233 Faraday Avenue
Suite F
Carlsbad, CA 92008

Diversified Education Enterprise
725 Main Street
Lafayette, IN 47901

Encyclopaedia Britannica Educational
Corp. (EBEC)
310 S. Michigan Avenue
Chicago, IL 60604

Focus Media, Inc.
839 Stewart Avenue
P.O. Box 865
Garden City, NY 11530

Hawkill Associates, Inc.
125 E. Gilman Street
Madison, WI 53703

Journal Films, Inc.
930 Pitner Avenue
Evanston, IL 60202

Lumivision
1490 Lafayette
Suite 305
Denver, CO 80218

National Earth Science Teachers
c/o Art Weinle
733 Loraine
Grosse Point, MI 48230

National Geographic Society Educational
Services
17th and "M" Streets, NW
Washington, DC 20036

Science Software Systems
11890 W. Pico Blvd.
Los Angeles, CA 90064

Time-Life Videos
Time and Life Building
1271 Avenue of the Americas
New York, NY 10020

Universal Education & Visual Arts
(UEVA)
100 Universal City Plaza
Universal City, CA 91608

Video Discovery
1515 Dexter Avenue N.
Suite 400
Seattle, WA 98109

Glencoe

SCIENCE INTERACTIONS

Course 1

GLENCOE
McGraw-Hill

New York, New York
Columbus, Ohio
Mission Hills, California
Peoria, Illinois

With Features by:

NATIONAL
GEOGRAPHIC
SOCIETY

Science Interactions

Science Interactions

Student Edition

Teacher Wraparound Edition

Science Discovery Activities

Teacher Classroom Resources

Laboratory Manual

Study Guide

Section Focus Transparencies

Teaching Transparencies

Performance Assessment

Performance Assessment in the Science Classroom

Computer Test Bank: IBM and Macintosh Versions

Spanish Resources

English/Spanish Audiocassettes

Science and Technology Videodisc Series

Integrated Science Videodisc Program

MindJogger Videoquizzes

Glencoe/McGraw-Hill

A Division of The McGraw-Hill Companies

Series Design: DECODE, Inc.

Send all inquiries to:

Glencoe/McGraw-Hill
936 Eastwind Drive
Westerville, OH 43081

ISBN 0-02-828054-7

Printed in the United States of America.

4 5 6 7 8 9 10 11 12   071/043   06 05 04 03 02 01 00 99

With Features by:

NATIONAL
GEOGRAPHIC
SOCIETY

Authors

Bill Aldridge, M.S.
Director—Division of Science Education Solutions
Airborne Research and Services, Inc.
Fredericksburg, Virginia

Russell Aiuto, Ph.D.
Education Consultant
Frederick, Maryland

Albert Kaskel, M.Ed.
Biology Teacher, Emeritus
Evanston Township High School
Evanston, Illinois

Jack Ballinger, Ed.D.
Professor of Chemistry
St. Louis Community College at Florissant Valley
St. Louis, Missouri

Craig Kramer, M.A.
Physics Teacher
Bexley High School
Bexley, Ohio

Anne Barefoot, A.G.C.
Physics and Chemistry Teacher, Emeritus
Whiteville High School
Whiteville, North Carolina

Edward Ortleb, A.G.C.
Science Consultant
St. Louis Board of Education
St. Louis, Missouri

Linda Crow, Ed.D.
Associate Professor
University of Houston—Downtown
Houston, Texas

Susan Snyder, M.S.
Earth Science Teacher
Jones Middle School
Upper Arlington, Ohio

Ralph M. Feather, Jr., M.Ed.
Science Department Chair
Derry Area School District
Derry, Pennsylvania

Paul W. Zitzewitz, Ph.D.
Professor of Physics
University of Michigan-Dearborn
Dearborn, Michigan

With Features by:

NATIONAL GEOGRAPHIC SOCIETY

The National Geographic Society, founded in 1888 for the increase and diffusion of geographic knowledge, is the world's largest nonprofit scientific and educational organization. Since its earliest days, the Society has used sophisticated communication technologies, from color photography to holography, to convey geographic knowledge to a worldwide membership. The Education Products Division supports the Society's mission by developing innovative educational programs—ranging from traditional print materials to multimedia programs including CD-ROMs, videodiscs, and software.

Consultants

Chemistry

Richard J. Merrill
Director,
Project Physical Science
Associate Director,
Institute for Chemical
Education
University of California
Berkeley, California

Robert W. Parry, Ph.D.
Dist. Professor of Chemistry
University of Utah
Salt Lake City, Utah

Earth Science

Allan A. Ekdale, Ph.D.
Professor of Geology
University of Utah
Salt Lake City, Utah

Janifer Mayden
Aerospace Education Specialist
NASA
Washington, DC

James B. Phipps, Ph.D.
Professor of Geology
and Oceanography
Gray's Harbor College
Aberdeen, Washington

Life Science

David M. Armstrong, Ph.D.
Professor of Environmental,
Population, and
Organismic Biology
University of Colorado-Boulder
Boulder, Colorado

David Futch, Ph.D.
Professor of Biology
San Diego State University
San Diego, California

Richard D. Storey, Ph.D.
Associate Professor of Biology
Colorado College
Colorado Springs, Colorado

Physics

David Haase, Ph.D.
Professor of Physics
North Carolina State University
North Carolina

Patrick Hamill, Ph.D.
Professor of Physics
San Jose State University
San Jose, California

Middle School Science

Garland E. Johnson
Science and Education Consultant
Fresno, California

Barbara Sitzman
Chatsworth High School
Tarzana, California

Multicultural

Thomas Custer
Coordinator of Science
Anne Arundel County Schools
Annapolis, Maryland

Francisco Hernandez
Science Department Chair
John B. Hood Middle School
Dallas, Texas

Carol T. Mitchell
Instructor
Elementary Science Methods
College of Teacher Education
University of Omaha at Omaha
Omaha, Nebraska

Karen Muir, Ph.D.
Lead Instructor
Department of Social and
Behavioral Sciences
Columbus State
Community College
Columbus, Ohio

Reading

Elizabeth Gray, Ph.D.
Reading Specialist
Heath City Schools
Heath, Ohio
Adjunct Professor
Otterbein College
Westerville, Ohio

Timothy Heron, Ph.D.
Professor, Department
of Educational
Services & Research
The Ohio State University
Columbus, Ohio

Barbara Pettegrew, Ph.D.
Director of Reading
Study Center
Assistant Professor of Education
Otterbein College
Westerville, Ohio

LEP

Ross M. Arnold
Magnet School Coordinator
Van Nuys Junior High
Van Nuys, California

Linda E. Heckenberg
Director
Eisenhower Program
Van Nuys, California

Harold Frederick Robertson, Jr.
Science Resource Teacher
LAUSD Science Materials Center
Van Nuys, California

Safety

Robert Tatz, Ph.D.
Instructional Lab Supervisor
Department of Chemistry
The Ohio State University
Columbus, Ohio

Reviewers

basketball firm? Why does cake batter pour but baked cake crumble? Learn how substances in your surroundings interact with one another.

UNIT 3

UNIT 1 — Observing the World Around You   **18**

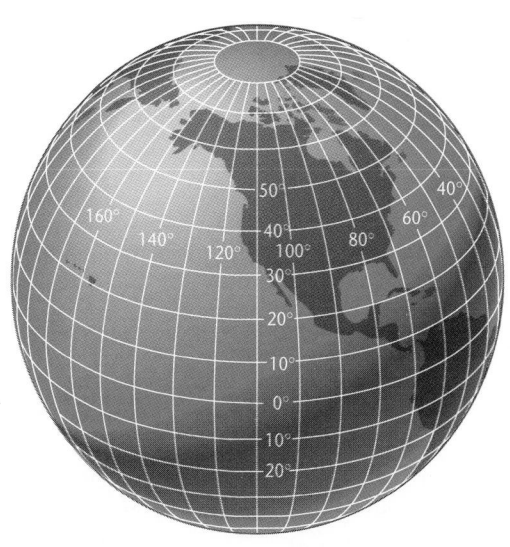

Chapter 2   Light and Vision   52

Contents   **xi**

UNIT
2

Interactions in the Physical World   114

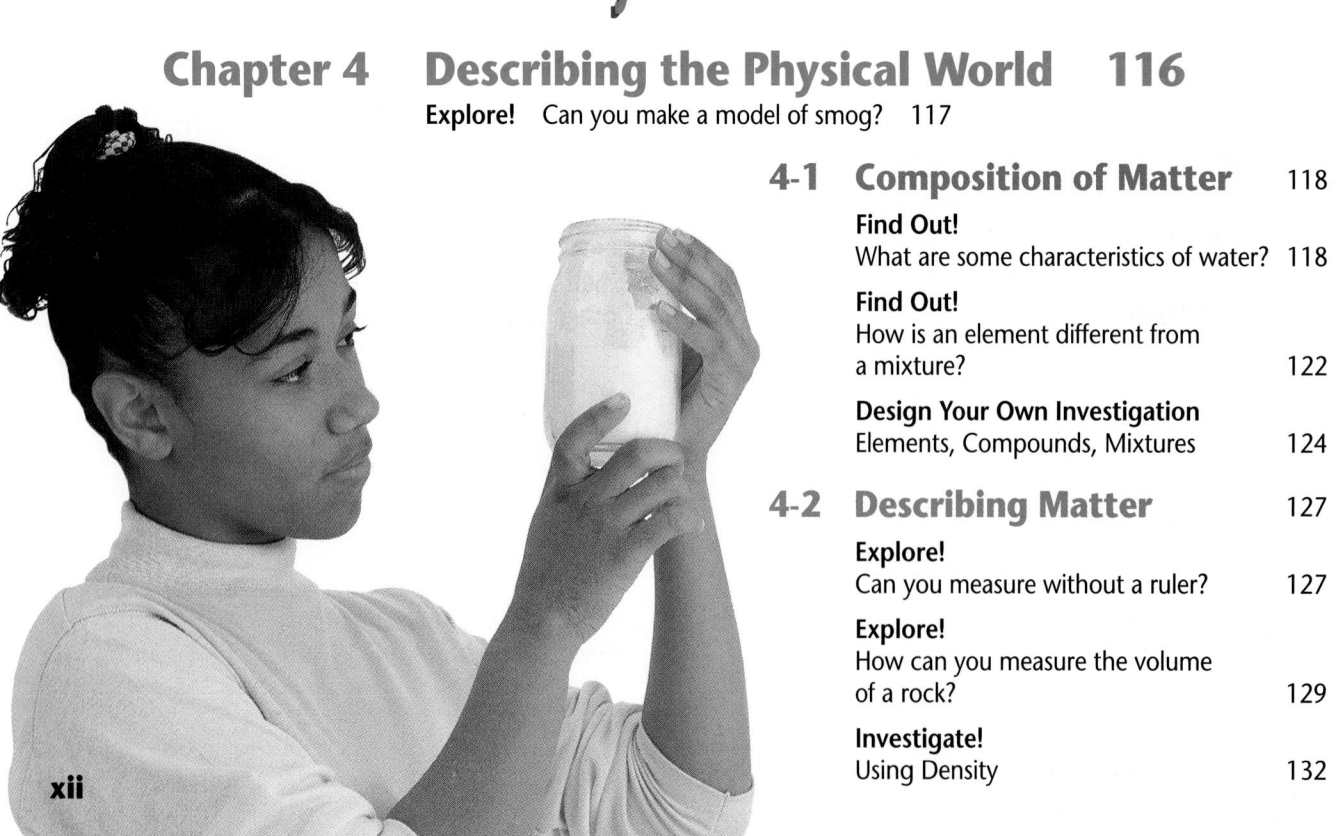

Chapter 5 Matter in Solution 152

UNIT 3   Interactions In the Living World   214

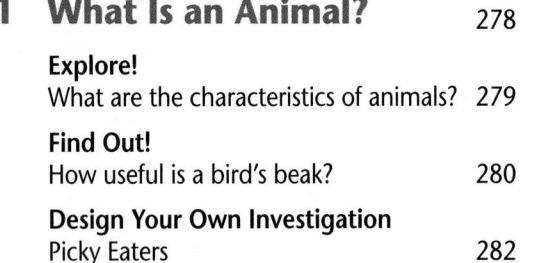

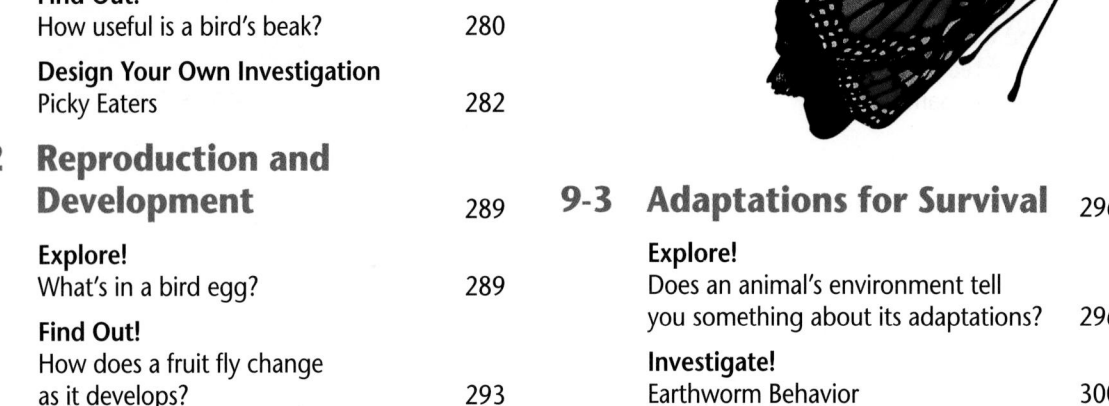

UNIT 5 wave Motion   536

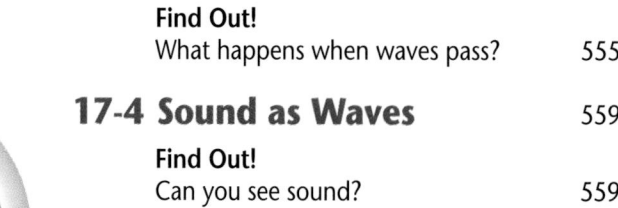

SCIENCE CONNECTIONS

Can you talk about the speed at which an animal runs without mentioning velocity? Is one science related to another? Expand your view of science through A CLOSER LOOK and Science Connections features in each chapter.

Earth Science

Life Science

Physics and Chemistry

A CLOSER LOOK

SCIENCE CONNECTIONS

Science is something that refuses to stay locked away in a laboratory. In both the Science and Society and the Technology features, you'll learn how science impacts the world you live in today. You will also be asked to think about science-related questions that may affect your life fifty years from now.

Science and Society

Technology Connection

CROSS-CURRICULUM CONNECTIONS

With the EXPAND YOUR VIEW features at the end of each chapter, you'll quickly become aware that science is an important part of every subject you'll ever encounter in school. Read these features to learn how science has affected history, health, and the ground you stand on.

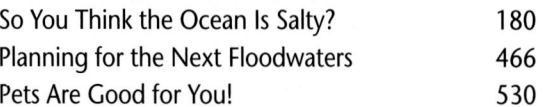

NATIONAL GEOGRAPHIC
CONNECTIONS

As you begin each unit of Science Interactions, start by envisioning the big picture with the help of an exciting National Geographic photograph. Then look for the National Geographic SciFacts article in each unit to enrich and extend your understanding of science in the real world.

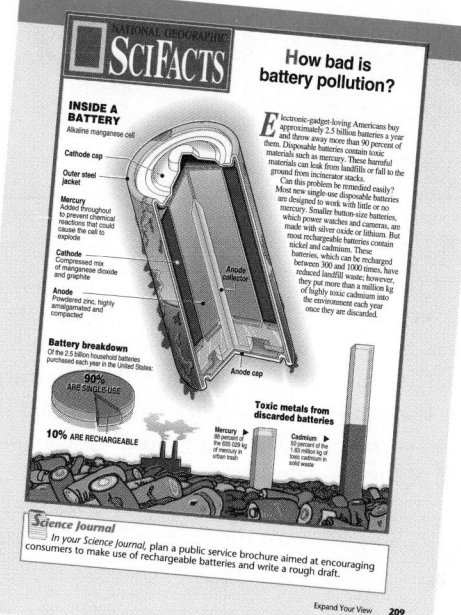

NATIONAL GEOGRAPHIC SOCIETY

Introducing Science: A Tool for Solving Problems

National Content Standards: (5–8) UCP2, A1–2, E2, F5, G1–G3

THEME DEVELOPMENT

The primary theme of this chapter is stability and change. Scientific methods help us understand both stability and change in the natural world. When we understand a process, we are often able to improve upon it, leading to technological advances.

The chapter's secondary theme of systems and interactions is illustrated as scientific methods help us define and predict the interactions of elements within a system and of systems with other systems. Learning how systems interact enables us to modify our environment. One example of this modification is changing a barren lot into a productive garden.

CHAPTER OVERVIEW

In this chapter, a science class learns the purpose of scientific methods by using them to plan how to reclaim an abandoned lot. First, students explore questions science can and cannot answer. Then they carry out a survey as a scientific way to gather information about uses for the lot. The Flex Your Brain approach helps the class choose models as a way to study methods of reducing erosion on the lot. Then students develop charts to analyze growing conditions on the lot and help decide what to plant there.

Consider having your students follow the procedures outlined in this chapter to plan how to reclaim a neglected area near your school. They might spend a week or longer planning how to bring new life to an empty lot or an abandoned flower bed. Through hands-on application, your students, like the class in this chapter, will begin to appreciate the practical uses of science.

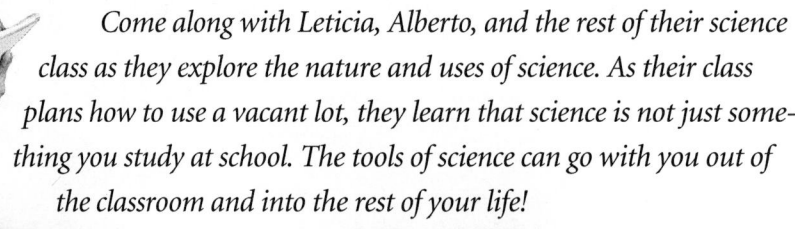

SCIENCE: A Tool for Solving Problems

What is science? Can ordinary people use it? Or do all real scientists look a little like Albert Einstein and spend all their time in laboratories? Do scientists ever guess at something—or make discoveries by accident?

How do scientists do their work? What kinds of tools do they use? You probably know about microscopes and telescopes, but there are other tools scientists use also—tools that help them put their investigations into order. These thought-tools help scientists solve problems and find answers to new questions. How could you use these tools?

Come along with Leticia, Alberto, and the rest of their science class as they explore the nature and uses of science. As their class plans how to use a vacant lot, they learn that science is not just something you study at school. The tools of science can go with you out of the classroom and into the rest of your life!

2

INTRODUCING THE CHAPTER

Have the students read the chapter title and name some tools they use in science. *Possible answers include test tubes, beakers, scales, microscopes, and other equipment.* Explain that in this chapter, they will learn about more tools that can help them put science to work in their lives every day. Point out that they already use some of these tools without realizing it.

Uncovering Preconceptions

Help students recognize ways they solve problems scientifically by asking them to pretend they are on a softball team. What changes could they try to improve their batting averages? *Changing the way they stand or hold the bat.* Point out that each suggestion is a hypothesis, a prediction about what will happen under certain conditions. Students can test each hypothesis. In this way, they are using a scientific tool (approach) to solve a problem.

What Questions Can Science Answer?

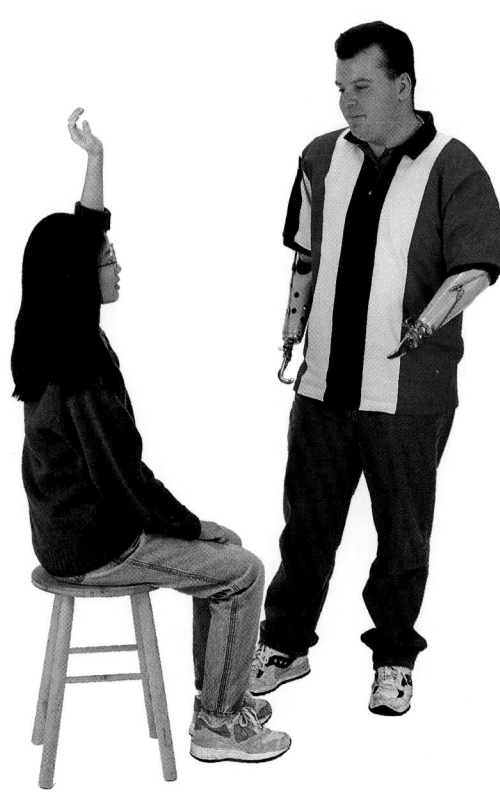

Leticia was surprised. "Mr. Steinmetz, did you just say our science class is going to clean up the empty lot next to the school?" she said. What does science have to do with cleaning up that mess, she wondered.

"Are we going to put all that trash in test tubes and study it?" Alberto asked. "And then fill out a million tables and graphs?"

Mr. Steinmetz smiled. "Maybe only two or three hundred tables and graphs. Actually, we're going to let science help us find a use for the lot. Scientific tools will help us figure out what to do and how to do it."

"What to do is easy," Jennifer said. "We'll just pick up all the garbage and make the lot into a softball diamond. This school needs another one, since the boys always hog the one we have!"

"What about using the lot for a garden?" Kevin suggested.

"Or we could make it into a park or a playground for the kids in the neighborhood," Hiroko said. "Or..."

Leticia raised her hand. "Mr. Steinmetz, you said we'd use science to figure out what to do. How can it help us decide what to do with the lot?"

"Well, science can help us answer a lot of questions, but not all of them," he explained. "Science can help us learn the facts about a situation, but the answers to some questions depend on what we value or think is important.

"For example, we can use science to learn what kind of soil is in that lot, but science won't make the decision as to whether we should use the lot for a garden or a softball diamond. What are some other questions that you think science can help us answer? What are some questions you think science can't answer?"

Introduction   Science: A Tool for Solving Problems   **3**

PREPARATION

Student Journal

Ask each student to bring in a small notebook or booklet that will serve as a journal. Students will use their journals to record their observations, charts and results of activities, and impressions. You may wish to review these journals for evaluation and assessment.

Concepts Developed

The goal throughout this chapter is not for students to reclaim an abandoned lot, but to gain experience in using scientific methods. The answers they obtain in the activities are not nearly so important as the methods they use to obtain them.

1 MOTIVATE

Activity

Have students observe a nearby vacant lot or a close up photo of one and make lists of things they see. Then have students devise classification schemes for these things. Expect many different classification schemes.

2 TEACH

Tying to Previous Knowledge

Divide the class into small groups. Ask each group to list three things they think they will learn from this chapter. Have groups share their lists with the class. Make a master list and save it to refer to at the end of the chapter. Student responses may reveal preconceptions that you can address during this chapter.

Explore!

Sorting Out Questions

Time needed 15–20 minutes

Materials No special materials are required for this activity.

Thinking Processes classifying, comparing and contrasting, hypothesizing

Purpose To compare and contrast questions science can answer and those it cannot.

Teaching the Activity

Science Journal Students should record their group's answers in their own journals.
L1

Expected Outcome

Students should list questions that their groups have classified as answerable by science or not answerable by science. Upon comparing the two lists, students should recognize that science can answer questions based on objective interpretation of data, but that it cannot answer questions based on values.

✔ Assessment

Process Have students work individually to write the answers to the following questions. What kinds of questions can science answer? When is a question not answerable by science? L1

Explore! ACTIVITY

Sorting Out Questions

Have you ever stopped to think about what kinds of questions science can answer? This activity will give you an opportunity to do so.

What To Do

1. Work in small groups to make two lists of questions: those you think science can answer and those you think science cannot answer.

2. Aim for at least five questions in each list. Your questions can relate to cleaning up a vacant lot or other topics. Record your lists of questions *in your Journal.*

3. Each group will take turns sharing its questions with the class. Be ready to explain the reasons for your choices.

■ How Does Science Help Answer Questions?

Hiroko said. "Michael and I thought of a good example of the two types of questions. Suppose a company dumped toxic chemicals into a river. Science could help us find out how much of the river was polluted and exactly what was killing the fish. And science could probably tell us how to clean up the pollution. But what if the company would go out of business if it had to spend a lot of money cleaning up the river? And what if most of the people in the town worked for that company? They would lose their jobs if the company closed."

"So we can't decide whether the company should clean up the river based just on scientific information," Michael added. "We have to decide what's more important to us: clean water or jobs."

After the class had discussed the two kinds of questions, Alberto raised his hand. "I think I can see the difference now. Science can help us find out facts about our world and understand why something happens, but it can't answer questions that depend on what we think is important."

"In that case, how can we figure out what to do with the empty lot?" Leticia asked. "Choosing between a garden or a softball diamond may depend on what's important to us."

Meeting Individual Needs

Learning Disabled To help students distinguish between questions science can and cannot answer, show them two books with illustrated covers. Ask, "What is shown on each cover?" Point out that they can answer this question through observation, a scientific method. Then ask, "Which cover is more interesting?" After some discussion, explain that the answer to this question depends on personal preference. Science cannot answer questions that rely on personal preference because this is not a measurable physical trait. Encourage them to suggest other questions about the books that science can and cannot answer.

"Why don't we ask our friends and see how many want a garden and how many want a softball field?" Jennifer suggested.

"Oh, sure! You'd just ask the girls on your softball team," Kevin teased. "We know what they'd say!"

Leticia had been thinking. "Maybe we could take a survey of everyone in the school and neighborhood. We could list the choices we've thought of and have them pick the one they want."

Mr. Steinmetz agreed. "A survey is a good way to gather information. Then we can analyze our findings. But we don't want our opinions to influence the results of the survey. So, we need to survey as many people as possible, not just the softball team or the garden club. Scientists do the same thing when they collect and study

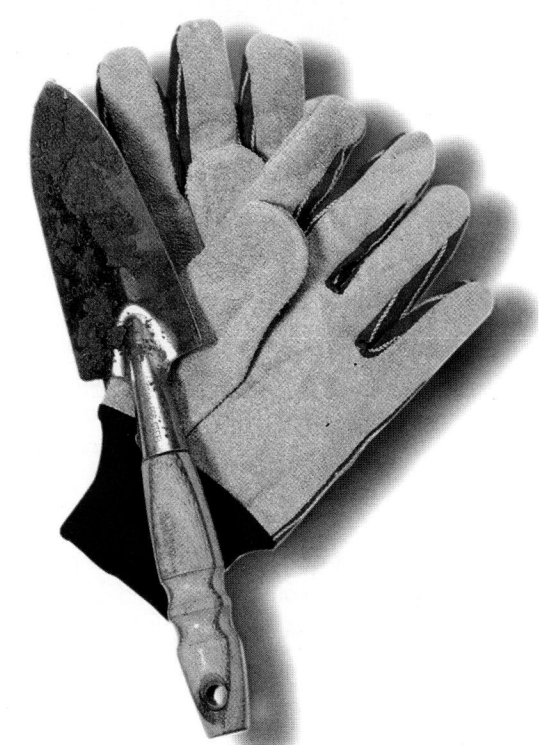

many samples before they analyze their findings and draw conclusions."

"We have to be careful how we list the choices," Hiroko pointed out, "because that could influence the results, too. For example, if we ask 'Wouldn't a garden be better than a softball diamond?' people would know which one we like better. We wouldn't find out what they really wanted."

"We could leave a space on the survey so people can write in their own ideas for ways to use the lot," Michael suggested.

"Then we could count up the votes for each choice," Alberto said, "and put them in a table." He smiled. "I knew there would be a table in here somewhere!"

Science: A Tool for Solving Problems **5**

Survey Service

Time needed 45 minutes and out of classtime to conduct the survey

Materials No special materials are needed for this activity.

Thinking Processes observing and inferring, collecting and organizing data, interpreting data, making and using tables, making and using graphs, separating and controlling variables, designing an experiment, comparing and contrasting

Purpose To collect data in a survey in order to recognize the uses of surveys, tables, and graphs in scientific study

Preparation If you do not plan to work as a class, divide the students into groups.

Teaching the Activity

Science Journal Students should make tables and graphs reflecting their survey data in their journals. [L1]

Expected Outcome

After identifying appropriate target groups and formulating appropriate questions, students should carry out a survey. Their resulting tables and graphs should accurately represent the data collected. Students should prepare a report detailing their procedures and conclusions.

Conclude and Apply

1. Anwers will vary depending on the survey composed, but students may note that the survey questions were similar but percentage returns were different.

2. Students should note that larger numbers from the whole class will average out and yield different results than for individual groups.

3. You might rephrase questions, eliminate questions that caused confusion or add questions that result from any comments on the previous survey.

Survey Service

How is a survey an example of scientific study? How can a survey help you find answers? Find out how by doing the activity.

What To Do

1. Work as a class or in small groups to create and carry out a survey.

2. For example, you might survey parents to find out which fund-raising activities they would be willing to support. You might ask students which afterschool activities should be added or discontinued. These are only two ideas. Base your survey on questions that are important to you and your community.

3. Describe each choice so that your opinion does not affect the outcome of the survey.

4. Decide which groups of people should be included in your survey in order to get a wide sampling of opinions.

5. Show your teacher your survey and get your teacher's approval. Then, carry out the survey.

6. Afterward, make a table and bar graph of the results *in your Journal*.

7. As the last step, let others know what you've found out by writing a report describing your survey procedure and your conclusions.

Conclude and Apply

1. How did the results of your survey differ from that of other groups? How were they similar?

2. If the whole class did a survey on the same question, try combining your group's results with the results from all the other groups. Compare and contrast the class results and your individual results. Which gave you a better idea of what people think? Why?

3. How would you redesign your survey to get more accurate results?

6 Science: A Tool for Solving Problems

✔ Assessment

Performance Have students make the changes they suggested for their surveys in *Conclude and Apply*, #3. Then have them carry out their revised surveys, comparing their new results to their original results. [L1]

■ Analyzing Data

"Here's what the survey says people are interested in," Kevin announced. "There were two main ideas. Over 44 percent of the people thought we should use the lot for a garden. But 21 percent thought it should be used as a softball field."

"Well," Jennifer said, "21 percent is a lot."

"Yes," Hiroko said, "But many of those votes came from outside the neighborhood. More than half of them."

"What about those who wanted a garden?" Jennifer asked.

"About 88 percent of those people lived in the neighborhood," Hiroko said. "Some said they'd even help clean up and take care of the garden."

"Then maybe we should build something everyone can contribute to," Jennifer said, smiling.

■ Observations: Defining the Problem

The school building next door shaded one side of the lot. A cement wall ran along the back of the lot. The back half of the lot was higher than the front half, so the ground sloped down toward the street. Erosion was a big problem. The students could see deep cracks in the soil, from the middle of the lot all the way to the street. It hadn't rained for a while, so some of the cracks looked like tiny, dry creek beds.

The students decided to hold a workday that weekend, when they and their families would get rid of the

trash. They wanted to reuse or recycle as much as possible. For example, they decided to keep the old tires and plant flowers in them.

When the class visited the lot the following Monday, everyone agreed that it looked much better. But now the erosion was even more noticeable.

We sure are going to need a lot of science to stop this erosion," Leticia said. "I wonder how we can do it."

"Let's just fill in the cracks with dirt and plant seeds," suggested Hiroko. "The roots will hold the dirt in place and stop the erosion."

"I think the rain would wash away the dirt and the seeds before the plants could grow," Kevin said. "Let's use these rocks to build a wall across the lot, like a dam, to keep the water from running off."

Flex Your Brain The Flex Your Brain strategy helps students think about what they already know and clearly define the question they want to answer. Here is one way students might use the Flex Your Brain strategy as they plan how to test methods to stop erosion:

1. Topic: erosion

2. What do I know?

Students already know:

a. Erosion is a problem.

b. Plant roots can help stop erosion, but they take a while to grow.

3. Ask a Question:

Possible question: Besides plants, what is the most effective way to stop erosion?

4. Guess an Answer:

Encourage students to brainstorm answers, which will be their hypotheses.

5. How sure am I?

Ask students to mark how sure they are that the answer they suggest will stop erosion.

6. How can I find out?

Brainstorm with the class ways they could test their answers (hypotheses). *Examples: consulting reference books, calling an agricultural extension agent, observing the rate of erosion on the lot and experimenting with different methods of reducing it, and/or making models of the methods and seeing which works best in a controlled experiment*

Mr. Steinmetz looked at the dry soil. "We could try that or several other methods. Let's think about how we can approach this problem."

■ **Thinking About the Problem**

When they had returned to the classroom, Mr. Steinmetz posted a large form.

"This *Flex Your Brain* form is sort of an outline about how to think about a problem. It's a step by step procedure to help you develop a procedure for finding an answer. So let's give it a try."

Leticia raised her hand and said, "Well, the problem seems to be erosion."

Mr. Steinmetz smiled and wrote the word erosion into the blank near the word topic. "We're off to a great start; we've identified the problem."

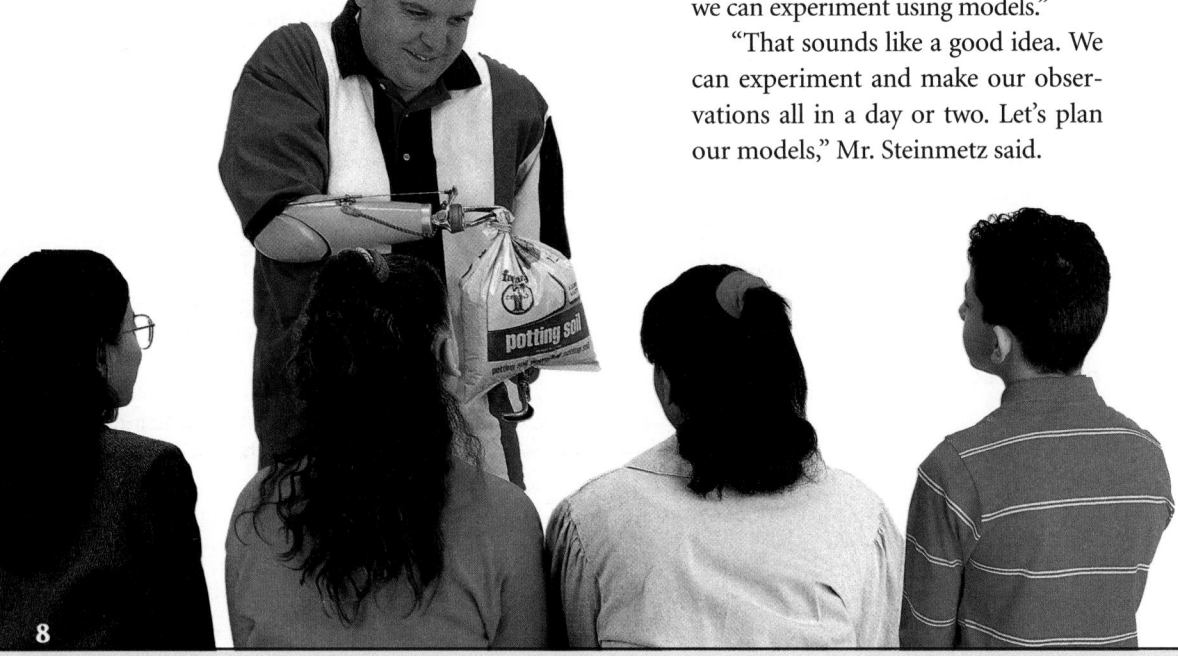

Alberto said, "Well, we already know that plant growth can stop erosion, but that takes a while. Maybe our question should be something like, 'Besides plants, what's the most effective way to stop erosion?'"

The class had several possible answers for how to control erosion. "Okay," Mr. Steinmetz said, "when you have several possible answers to a question, what do you do with each possibility?"

Jennifer suggested, "You test them?" Mr. Steinmetz said, "Great, but how do you test them?"

Hiroko said, "You look at it. You test it by observing."

"Good idea. But in this case we need results that are a little faster."

Kevin said, "Experiment."

Mr. Steinmetz said, "Yes, you can experiment, but it seems like that might be difficult here. How are we going to erode a land area like our garden?"

Alberto said, "Well, maybe we can experiment using models."

"That sounds like a good idea. We can experiment and make our observations all in a day or two. Let's plan our models," Mr. Steinmetz said.

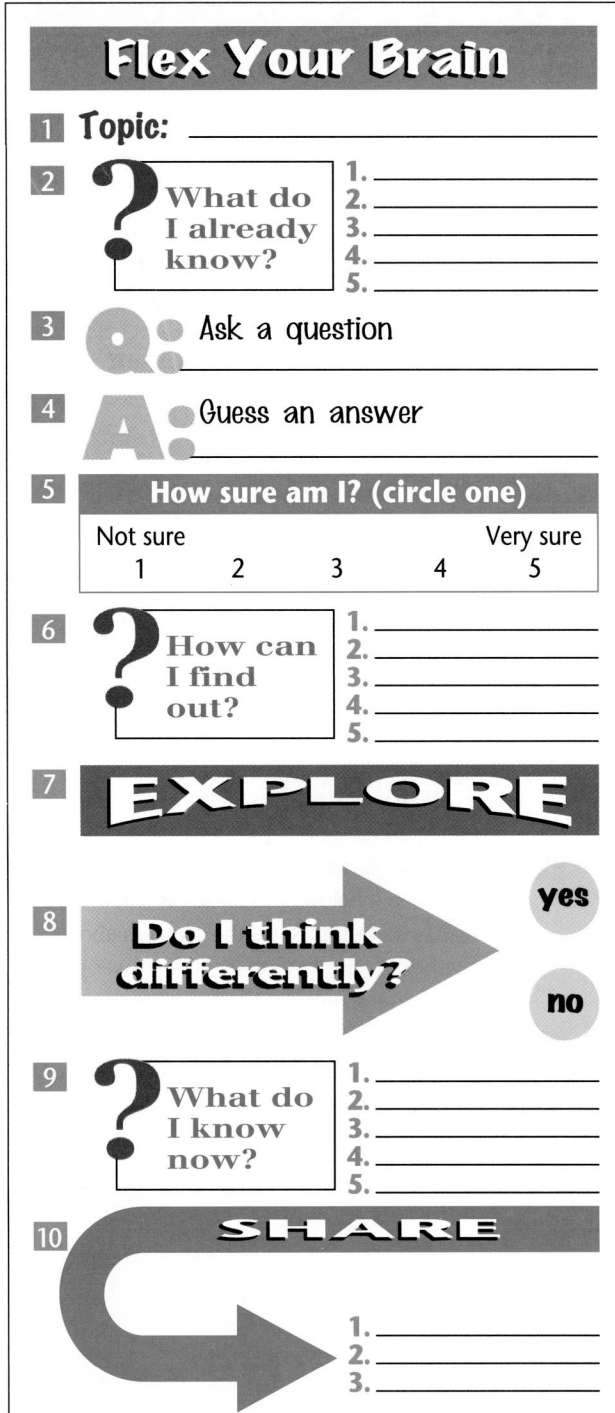

Flex Your Brain

1 Topic: _____

2 ❓ What do I already know?
1. _____
2. _____
3. _____
4. _____
5. _____

3 Q: Ask a question
_____

4 A: Guess an answer
_____

5 How sure am I? (circle one)

Not sure				Very sure
1	2	3	4	5

6 ❓ How can I find out?
1. _____
2. _____
3. _____
4. _____
5. _____

7 EXPLORE

8 Do I think differently? yes no

9 ❓ What do I know now?
1. _____
2. _____
3. _____
4. _____
5. _____

10 SHARE
1. _____
2. _____
3. _____

---

1 Fill in the topic.

2 Jot down what you already know about the topic.

3 Using what you already know (Step 2), form a question about the topic. Are you unsure about one of the items you listed? Do you want to know more? Do you want to know what, how, or why? Write down your question.

4 Guess an answer to your question. In the next few steps, you will be exploring the reasonableness of your answer. Write down your guesses.

5 Circle the number in the box that matches how sure you are of your answer in Step 4. This is your chance to rate your confidence in what you've done so far and, later, to see how your level of sureness affects your thinking.

6 How can you figure out more about your topic? You might want to read a book, ask an expert, or do an experiment. Write down ways you can find out more.

7 Make a plan to explore your answer. Use the resources you listed in Step 6. Then, carry out your plan.

8 Now that you've explored, go back to your answer in Step 4. Would you answer differently? Mark one of the boxes.

9 Considering what you learned in your exploration, answer your question again, adding new things you've learned. You may completely change your answer.

10 It's important to be able to talk about thinking. Choose three people to tell about how you arrived at your response in every step. For example, don't just read what you wrote down in Step 2. Try to share how you thought of those things.

---

7. EXPLORE

In this case, guide students to choose making models as the quickest, most direct way to test their hypotheses. During the Investigate! activity, they will make models that test four hypotheses (possible ways to reduce erosion).

8. Do I think differently?

Ask students to review the answers they proposed in Step 4. Has gathering more information changed their mind?

9. What do I know now?

Students' responses should indicate that they understand how each model (hypothesis) affects the rate of erosion.

10. Share

Ask the class to explain how the Flex Your Brain approach helped them think through and solve this problem.

Encourage students to keep a record of the Flex Your Brain exercises in their journals and later add them to their portfolios.

INVESTIGATE!

Saving the Soil

Time needed 45 minutes

Thinking Processes observing and inferring, comparing and contrasting, recognizing cause and effect, hypothesizing, separating variables, constants, and controls, collecting and organizing data, interpreting data, experimenting, analyzing data, making and using graphs, measuring in SI

Purpose To formulate a hypothesis regarding soil erosion and to test that hypothesis experimentally

Materials See reduced student text.

Teaching the Activity

Process Reinforcement In their journals, have students compare and contrast the four types of erosion-control methods they will test in the experiment. They should focus on aspects they think will affect their respective abilities to control soil loss. This exercise will also help them formulate a well-reasoned hypothesis as to the experiment's results. **L1**

Possible Hypotheses Students may hypothesize that one method of erosion control will work better than the others; hypotheses will vary.

Troubleshooting During step 5, students should be sure that they hold pans C and D so that the water runs *across* the pebble walls and contour grooves, rather than along them.

INVESTIGATE!

Saving the Soil

You've seen eroded soil and you're probably aware of the problems erosion can cause. But how do you begin to explore how to solve these problems?

Problem
Which of five methods is the best way to control soil erosion?

Safety Precautions

Materials

5 aluminum 8 or 9-inch pie pans	metric ruler
leaves or grass clippings	newspaper
	pebbles
500-mL beaker	potting soil
water	watering can
	dishpan

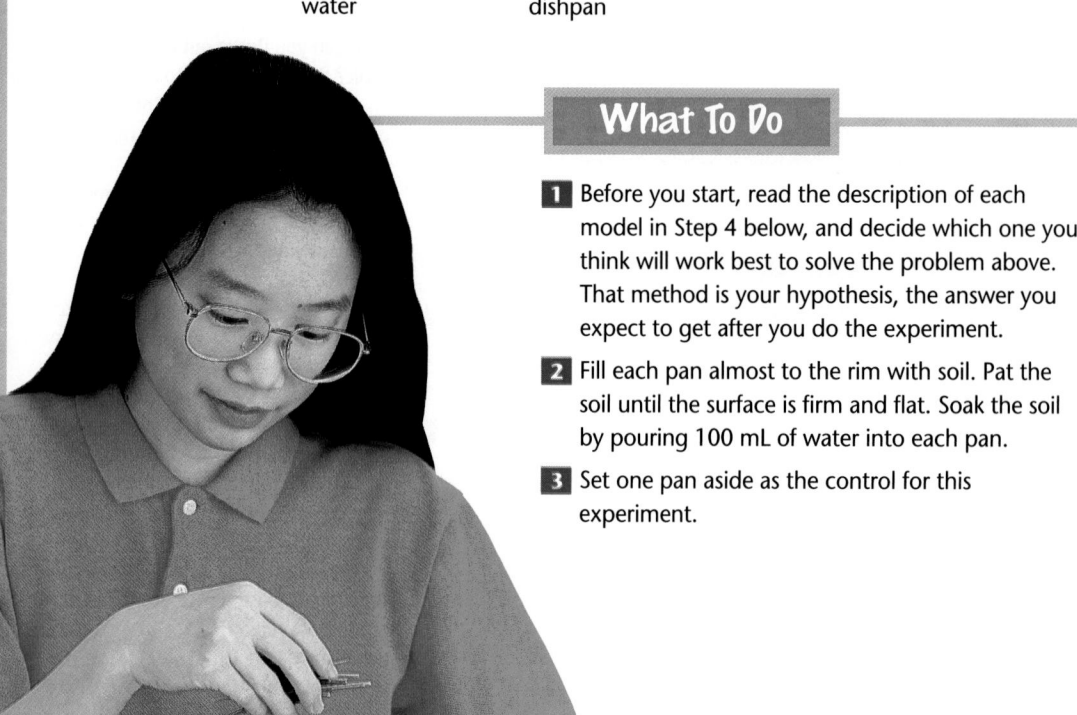

What To Do

1. Before you start, read the description of each model in Step 4 below, and decide which one you think will work best to solve the problem above. That method is your hypothesis, the answer you expect to get after you do the experiment.

2. Fill each pan almost to the rim with soil. Pat the soil until the surface is firm and flat. Soak the soil by pouring 100 mL of water into each pan.

3. Set one pan aside as the control for this experiment.

A   **B**   **C**

4 Use the other four pans to set up these conditions:
  a. Mulching: Cover the surface with a layer of grass clippings or leaves.
  b. Other soil cover: Tear the newspaper into small strips. Lay the strips across the surface, leaving about 5 mm between each strip.
  c. Terracing: Build two small walls of pebbles.
  d. Contour plowing: Use your finger to make a curved groove across the surface.

5 Follow this procedure with all five pans, starting with the control:
  a. Measure 200 mL of water into the watering can.
  b. Hold the pan so that one side touches the edge of the dishpan and the opposite edge is about 10 cm higher.
  c. Slowly pour the water onto the soil at the top of the pan. Wait until the excess water runs across the soil and into the dishpan. Note: Be sure to hold the terracing and contour plowing pans so the water runs across the pebble walls and the grooves.
  d. Pour the water and soil from each pie pan into the beaker. Measure it and record the results. When the soil settles to the bottom, measure and record its height in the beaker.

6 As a class, decide which method was most effective at preventing erosion.

Analyzing

1. Construct a graph to show results for all five pans.

2. Compare and contrast the results for the pans.

3. Was your hypothesis correct?

Concluding and Applying

4. ~~Going Further~~ One test or experiment isn't enough to prove a hypothesis. Most scientists perform numerous tests. Write an outline for another test of this hypothesis that does not use a model.

Expected Outcomes
  Students will find their hypotheses confirmed or refuted by the data which they gather in their experimentation. Their graphs should accurately represent their data.

Answers to Analyzing/ Concluding and Applying

1. Answers will vary, but contour plowing should show some of the best results.

2. Students should observe that some methods are more effective than others. Results will vary among groups. All "treated" pans should show less erosion than the control.

3. Answers will vary depending on individual trials. However, students may say that some results were equally good and further experiments are necessary.

4. Accept all reasonable proposals. A possible idea might be to observe areas of erosion and areas not eroded around the school or neighborhood. Students may propose an experiment in the school yard that could involve gardening or seeding an area in grass.

✔ Assessment

Process Have students design and carry out experiments to find out if soil composition affects which soil cover best prevents erosion. Provide students with potting soil, sand, and half-sand, half-potting-soil mixtures to test, and two or three of the most successful erosion-control methods from the first experiment. (Clay soil could also be modeled by adding flour.) You may wish to have students work in larger groups for this experiment. **L1 COOP LEARN**

Content Background

Most U.S. farmers used to plant crops in straight rows regardless of the contours of the field, resulting in widespread erosion. In the 1930s, the U.S. Soil Conservation Service convinced many farmers to use contour plowing on sloping fields. In contour plowing, furrows are plowed around hills instead of up and down them. The horizontal furrows hold rainwater and allow it to sink in instead of washing the soil away.

Strip cropping also reduces erosion. Crops planted in widely spaced rows, such as corn and cotton, are alternated with crops that grow close to the soil, such as grass and soybeans. Rainwater running off the widely spaced rows is caught by the close-growing plants and sinks into the ground. Each year, the crops in the strips are rotated.

Concept Development

Ask students what scientific methods might have helped farmers discover on their own that contour plowing would reduce erosion. *They could use observation to understand what was happening to their topsoil. Then the farmers might have experimented with different erosion-control methods to see which worked best.*

Across the Curriculum

Daily Life

Ask pairs of students to locate and describe examples of erosion in your community. Have them propose ways this erosion could be reduced. If the erosion is on school or other public land, students might write to the school board, city council, or other responsible group and offer suggestions. The class might help reduce erosion in one area as a service project.

■ Making Discoveries— Accidents in Science

During the experiment, Alberto accidentally let a whole pie pan of soil slide into the dishpan. "Some scientist you are!" Jennifer teased.

"You never know," Mr. Steinmetz said. "Some of science's most important discoveries were made by accident, including the discovery of penicillin. When Sir Alexander Fleming was growing dishes of bacteria for an experiment, some mold accidentally got into one dish. Lucky for us, Fleming didn't just throw that dish away because it was contaminated. He noticed that the mold had killed the bacteria around it. The penicillin he accidentally discovered has helped cure millions of people worldwide."

Alberto nodded. "So maybe we should save this muddy mess!" he told Jennifer.

"And maybe not," she said with a smile.

■ Expanding What You Know

After analyzing the results of the experiment, the class planned how they could use the method that worked the best. They hoped it would slow down the erosion in the lot until the garden was planted.

Hiroko was getting impatient. "Let's start planting!"

"Okay," Mr. Steinmetz said. "Let's plant a garden of bananas." He looked around the classroom. "I see some puzzled faces. What's the problem?"

"Bananas don't grow here," Michael said. "It's not warm enough."

"I see," Mr. Steinmetz said slowly. "So what do we need to do before we start our garden?"

"Figure out what will grow there," Leticia said. Then she smiled at Mr. Steinmetz. "I bet science can help us do that, right?"

"Right! We need to do some scientific observing and analyzing before we decide what to plant in our

Science on the Run

Can you imagine playing basketball or running track in a pair of loafers? We can wear sneakers instead, thanks to two accidental discoveries, one by a scientist and one by a track coach. Charles Goodyear had been trying for nine years to make rubber that would stay flexible at both high and low temperatures. One day in 1839, he accidentally allowed some rubber and sulfur to touch on a hot stove. He had discovered a flexible rubber that had all kinds of uses!

In the 1960s, a college track coach wanted his runners to get more traction from their rubber-soled shoes. He put a piece of rubber on a waffle iron and got a pattern that may have changed the history of track!

Thanks to two accidental discoveries, athletes today, such as Florence Griffith-Joyner, pictured here, have running shoes that help improve traction.

ENRICHMENT

Ask students to review the questions they listed for the Explore activity on page 4. Have small groups each select a question and plan a way they could use scientific methods to answer it. Ask groups to share their questions and plans with the class and explain why their chosen method will help answer the question.

garden," he said. "Science has helped us understand how different plants grow. We can gather and analyze this information and choose plants that will grow best in that lot. What if no one before us had done any observing or experimenting, and we didn't know anything about how plants grow?"

Kevin shook his head. "I guess we'd have to do a lot of experimenting ourselves before we discovered which plants will grow best on that lot."

"True! Now, what kind of growing conditions will our plants have?" Mr. Steinmetz asked.

"Well, the soil is kind of hard and rocky and dry," Jennifer offered. "But after we stop the erosion, maybe the soil will be able to soak up more water. And I noticed a definite wet area back by the wall."

"Part of the lot is in the sun and part is in the shade," Alberto added. "You know, these growing conditions are getting complicated. Maybe we could make a big table." He started smiling again. "We could fill it in with plants that grow in different kinds of soil and light. Then we could decide what to grow."

"You're starting to think like a scientist!" Mr. Steinmetz said. "A lot of science begins with careful observation, and your table will give us a good start. We can use it to analyze information, draw conclusions, and make our decisions based on facts. Now let's divide into groups and see how many ways we can think of to plan this garden!"

Concept Development

Remind students that this chapter's activities are helping them learn scientific ways to solve problems. Choosing the plants is not nearly so important as learning how to select the most appropriate items for certain conditions.

Across the Curriculum

Health

DDT was once a common insecticide, used even in home gardens. Divide the class into three groups. Ask one group to research how DDT moves through the food chain.

Ask the second group to research the current laws governing the use of DDT. Have the third group identify natural ways to control unwanted insects in a garden. (If your class is planning a garden, students might apply some of these techniques.)

ENRICHMENT

Activity Guide the class to create a butterfly garden, perhaps in a school flower bed. Plant a packet of wildflower seeds or garden flowers such as daisies, verbena, phlox, alyssum, lobelia, and mustard. Ask students to record the species of butterflies that visit and which flowers attract each species.

Have the class select and use a scientific approach to determine the flower characteristic that draws the butterflies. Is it the flowers' size, shape, smell, color, or another attribute?

Have interested students work in small groups to develop a strategy. Use the Performance Task Assessment List for Carrying Out a Strategy and Collecting Data in **PASC**, p. 25.

L2 **COOP LEARN**

Find Out!

Find Out! ACTIVITY

Planning Before Planting

Use the information in the table below to plan a garden 60 feet long and 40 feet wide.

What To Do

Begin by drawing a diagram that shows the different conditions in the garden. (Make 1 inch equal 1 foot.)

1. Keep in mind that a 10-foot-wide strip along the left side of the garden is shaded during most of the day by the school building next door. Most of the soil will have average moisture after the erosion is stopped. However, the wall along the back of the lot traps rain water, so a 15-foot-wide section there is a lot wetter.

2. Be sure to take into account the way your class has decided to cut down on erosion.

3. Then, plan your garden. Some of the possible choices for your garden are available in the chart.

4. Copy the chart *into your Journal.* Leave enough space for everyone in your group to add one kind of plant. Use reference books to find the needed information.

5. Be prepared to share with the class your group's plan and the reasons for your choices. You'll need to explain how each of your selections is based on scientific reasoning.

Conclude and Apply

1. How important was each factor (sunlight/soil and plant height) in helping you make your decision?

2. How did you use other people's observations in planning your garden?

3. What did you already know about the garden that helped you plan? What other information would have been helpful to know in making your plan?

Plant	Plant height			Light/soil requirements			
	6-12"	1-3'	over 3'	sun/ average	sun/ moist	shade/ average	shade/ moist
wax begonia	✓			✓		✓	
sunflower			✓	✓			
petunia	✓	✓		✓		✓	
impatiens	✓	✓				✓	✓
butterfly flower	✓	✓		✓	✓	✓	✓
black-eyed Susan vine			✓	✓	✓	✓	✓
peas			✓	✓			
peppers		✓		✓			
lettuce	✓			✓		✓	
cucumbers	✓			✓			
corn			✓	✓			

■ A Solution to the Problem

After each group presented its plan for the garden, the class combined the best ideas.

Michael and Hiroko showed the class a diagram of paths the class could build. They wanted to line the paths with rocks cleared out of the areas where the plants would go.

Leticia and Kevin thought of planting ivy and ferns in the area at the back of the lot to take advantage of the wet conditions and to cover the ugly cement wall. Alberto and Jennifer suggested setting aside part of the garden in full sunlight for tomatoes, corn, and green peppers, which could be donated to the food bank. Hiroko suggested planting herbs to take advantage of the full sunlight and to attract butterflies.

The garden took a lot of work—on the part of the class and the neighborhood. Finally there was little to do but wait for the plants to grow. A photographer from the local newspaper came to take pictures as the first tiny leaves broke through the ground. The class posed around a sign they had posted: "A Garden Built with Science!"

Alberto, Leticia, and their classmates learned some of the processes of science to help solve a problem. As you study this book, you can use similar processes to solve problems. But remember, scientific thinking is not just for the classroom, it can be used in your everyday life as well!

A Garden Built with Science

15

Have students work in small groups to answer the review questions. As the groups finish, have them do the below activity.

Teaching Strategy

Have the different groups share their responses to question number 1. Then make a class list of the ways students have used science in their everyday lives. Leave the list up and encourage students to pursue these and other approaches. Throughout the year, students may add to the list giving their classmates a brief report on any new entry.

Answers to Questions

1. figuring out the best way to throw a basketball or season a favorite dish

2. Answers may vary, but look for specifically *observed* things rather than *inferred* things. For instance, some places have high and low tides twice a day whereas others have them only once.

3. Because the control allows you to see if your "treatment" is actually *doing* anything.

REVIEWING MAIN IDEAS

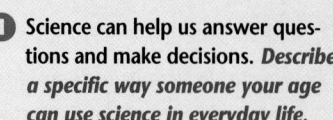

Science Journal

Review the statements below about the big ideas presented in this chapter, and try to answer the questions. Then, re-read your answers to the Did You Ever Wonder questions at the beginning of the chapter. *In your Science Journal*, write a paragraph about how your understanding of the big ideas in the chapter has changed.

1 Science can help us answer questions and make decisions. *Describe a specific way someone your age can use science in everyday life.*

2 Scientists use many methods to increase knowledge. They observe, experiment, make models, and apply old observations to new situations to help solve problems. *What are two specific things scientists have learned about the natural world as a result of observing it?*

3 Sometimes scientists perform experiments under different conditions to see which condition works best. *Why is it important, whenever possible, to have a control in your experiments?*

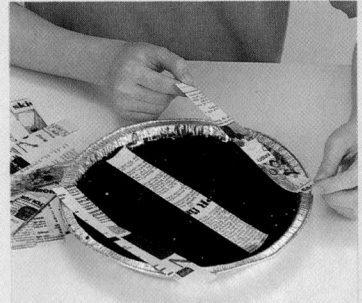

16   Science: A Tool for Solving Problems

Critical Thinking

In your Journal, *answer each of the following questions.*

1. Why is an open mind essential for a scientist?
2. When do scientists make guesses? How do they test their guesses?
3. Why are tables and graphs used so often in science?
4. How could scientists apply observations and information learned during space shuttle flights to building a space station on the moon? What other observations might help?

Problem Solving

Read the following problem and explain your answers.

The deer population at a certain park has greatly increased. The deer have eaten most of the vegetation and are beginning to starve. One group of people wants to shoot some of the deer. That way, the remaining deer will have enough to eat. The deer meat will be given to a homeless shelter. Another group wants to pay $1000 each to move some of the deer to another park. Some deer, maybe half of them, will die from stress during the move.

1. What are at least two questions science can answer relating to this problem?
2. What is at least one question science cannot answer?
3. What would you do in this situation? Explain the reasons for your decision. Discuss which of your reasons is based on science and which is based on what you value.

Understanding Ideas

1. When would you choose to make a model to find the solution to a problem?
2. How can observation be a scientific method?
3. Why is it important to plan your investigation before doing it?
4. Compare and contrast observation and experimentation as scientific methods.

Science: A Tool for Solving Problems **17**

Critical Thinking

1. To allow them to find new things to ask about and new ways of looking at old things

2. When all the best evidence provides no clue to the answer. They test guesses using controlled experiments.

3. Because tables and graphs help summarize information in a much more succinct (shorter) form.

4. They could use the information obtained to design lunar architecture, spaces, and gardens in accordance with lunar conditions. They could use all observations, particularly those related to mental and physical health.

Problem Solving

1. What would be the effects on the present ecosystem of removing the deer? What ways of transporting the deer would cause the least physical injury? Accept all reasonable answers.

2. Should we move the deer? Is it better to move them or kill them? Accept all reasonable answers.

3. Look for a discussion of the scientific approach and the "values" approach and some blending of the two.

Understanding Ideas

1. When the object under investigation is too large, too small, or in some other way too difficult to investigate in other ways.

2. Observation is recording in detail what occurs in natural settings. It is a scientific method because it requires that all information be recorded and analyzed before a conclusion can be made.

3. It's important to eliminate unnecessarily dangerous steps and to set up a clear plan of approach.

4. Both are scientific methods. Both involve observations. However observation usually involves examining a natural occurrence where experimentation is not possible (i.e. geologic strata) and experimentation involves setting up an artificial situation to observe.

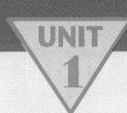

UNIT 1

Observing the World Around You

Chapter 1 Viewing Earth and Sky
Chapter 2 Light and Vision
Chapter 3 Sound and Hearing

UNIT FOCUS

In Unit 1, students will learn about how they use their senses of sight and hearing to observe patterns and features on Earth and in the sky.

THEME DEVELOPMENT

The focus of this unit is discovery and observation using the themes of scale and structure, systems and interactions, change and stability, and energy as the framework.

Scale and structure is developed in the treatment of landforms and their scale as viewed from a distance in an airplane. Earth is a part of the Earth-moon-sun system and interacts with other objects in the solar system. The theme of energy is developed in the treatment of light and sound, which are forms of energy. Senses are used to observe change and stability in the world around us.

Connections to Other Units

The concepts in this unit provide a foundation for developing observational experiences. These experiences will be used in describing materials, animals, and plants in the physical world in future units.

UNIT 1

Observing the World Around You

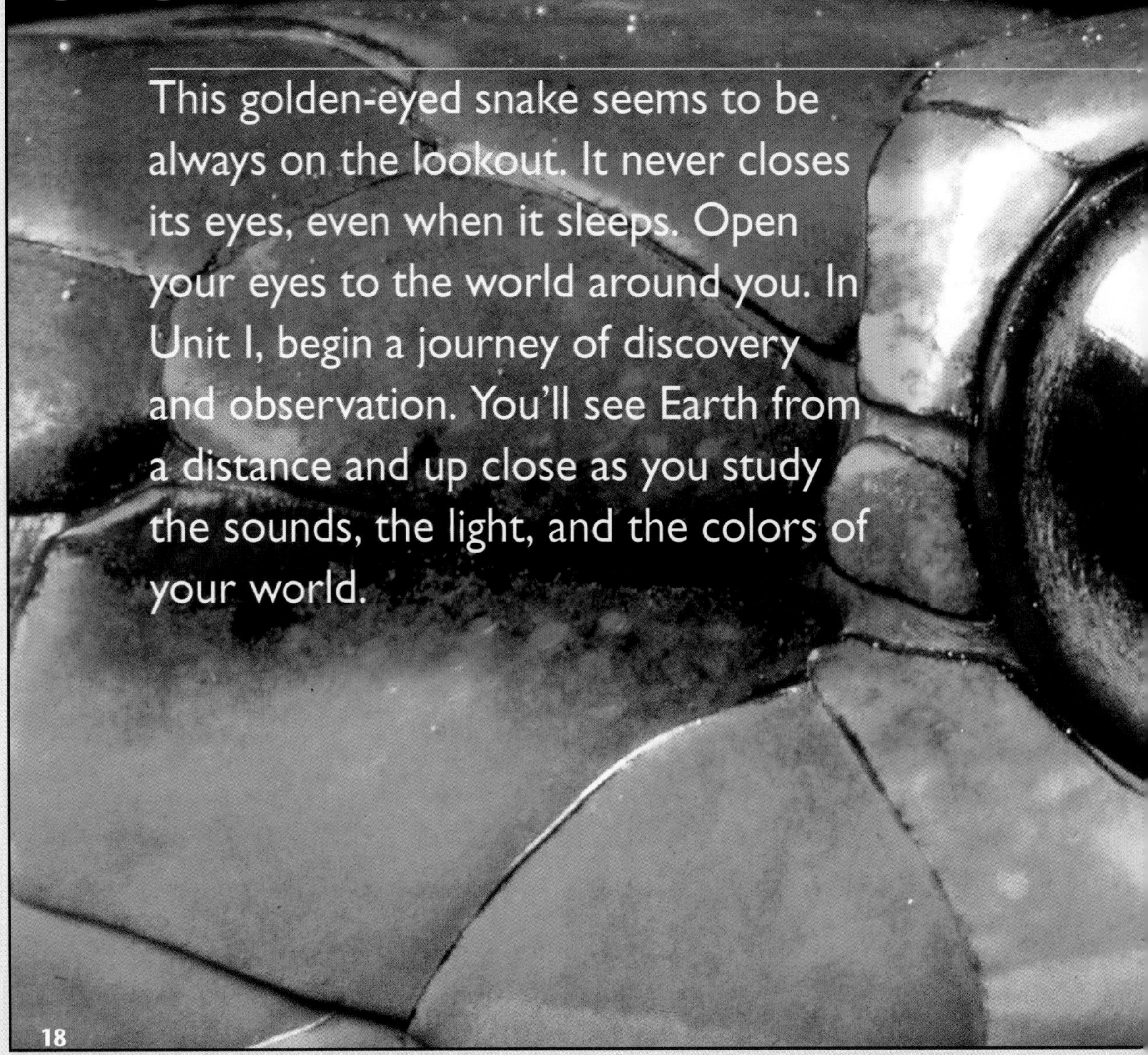

This golden-eyed snake seems to be always on the lookout. It never closes its eyes, even when it sleeps. Open your eyes to the world around you. In Unit I, begin a journey of discovery and observation. You'll see Earth from a distance and up close as you study the sounds, the light, and the colors of your world.

18

GLENCOE TECHNOLOGY

Videodisc

Use the *Science Interactions, Course 1* **Integrated Science Videodisc** lesson, *Observation*, with Chapter 1 of this unit. This videodisc explores the role of observation in science.

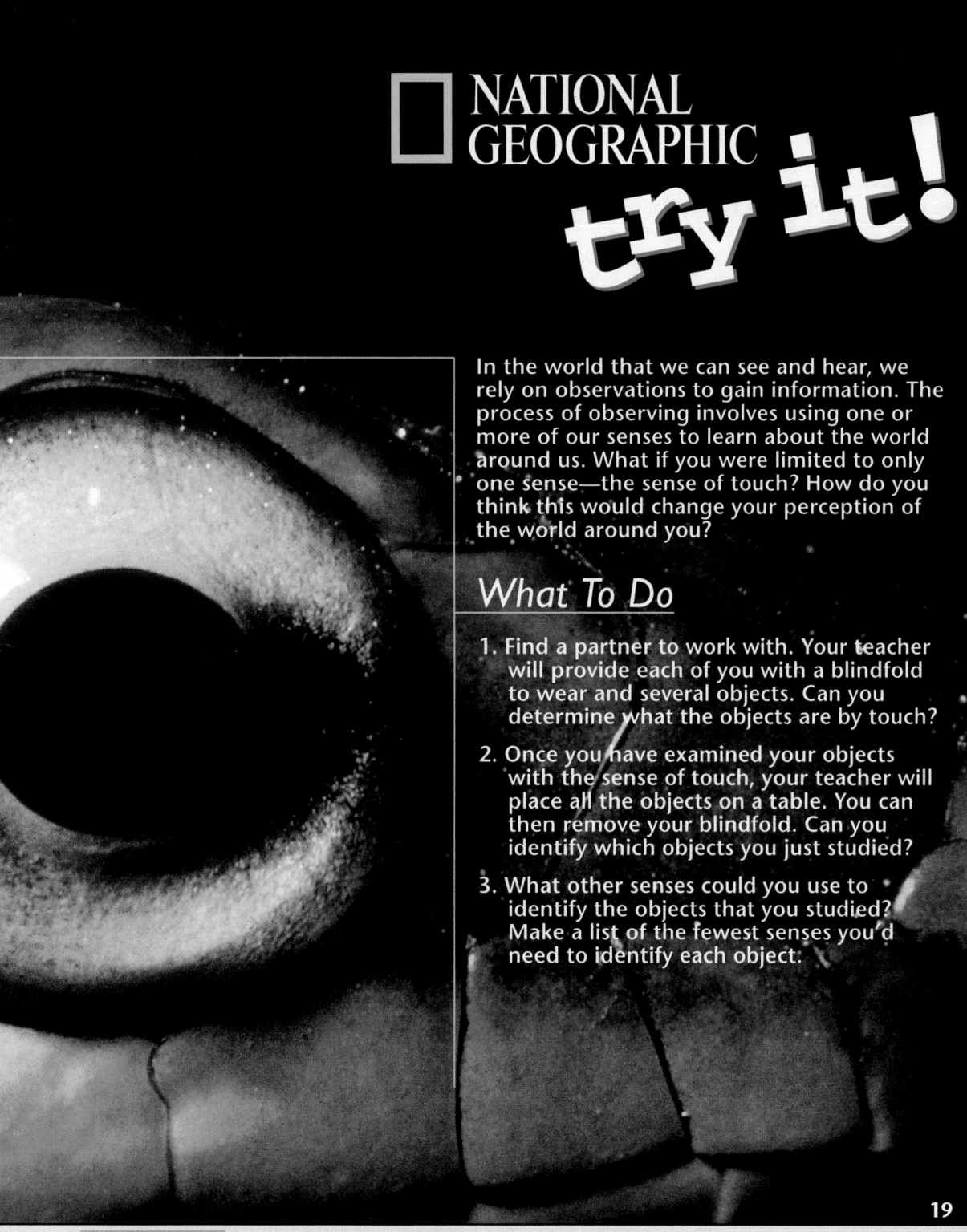

NATIONAL GEOGRAPHIC

try it!

In the world that we can see and hear, we rely on observations to gain information. The process of observing involves using one or more of our senses to learn about the world around us. What if you were limited to only one sense—the sense of touch? How do you think this would change your perception of the world around you?

What To Do

1. Find a partner to work with. Your teacher will provide each of you with a blindfold to wear and several objects. Can you determine what the objects are by touch?

2. Once you have examined your objects with the sense of touch, your teacher will place all the objects on a table. You can then remove your blindfold. Can you identify which objects you just studied?

3. What other senses could you use to identify the objects that you studied? Make a list of the fewest senses you'd need to identify each object.

19

GETTING STARTED

Discussion Some questions that you may want to ask your students are:

1. What evidence is there that Earth and the moon move? Students may have difficulty suggesting evidence from their observations. They may rely instead on what they have been told.

2. How can light "light up" objects? Students may believe that light from a source such as the sun shines on an object and stays there.

3. How can the color of light affect the color of an object? Students may believe that it is the color of the object that changes rather than the light illuminating the object.

Student responses to these questions will help you identify misconceptions they may have.

Try It!

Purpose Students will develop an understanding and appreciation of their senses—particularly sight.

Background Information

Students will rely on their sense of touch to identify objects through their texture and shape. Because they are used to making observations using their sense of sight, most students will have difficulty identifying some of the objects.

Materials blindfolds; a variety of different objects for students to identify, such as sandpaper, paper clips, coins, various fabrics

✔ Assessment

Oral Give students new objects to identify by touch. Then have students discuss which objects were most difficult to identify, which were easiest, and which characteristics were most useful in identifying objects. Use the Performance Task

Assessment List for Oral Presentation in **PASC,** p. 71.

COOP LEARN **L1**

Chapter Organizer

SECTION	OBJECTIVES	ACTIVITIES & FEATURES
Chapter Opener		**Explore!,** p. 21
1-1 Viewing Earth (2 sessions, 1 block)	1. **Describe** basic landforms such as mountains, plains, and plateaus. 2. **Recognize** the kind of landform on which you live. **National Content Standards: (5–8) UCP3, C5, D1–2, F2, G1**	**Explore!,** p. 22 **Life Science Connection,** pp. 26–27
1-2 Using Maps (3 sessions, 1.5 blocks)	1. **Identify** landforms using a topographic map. 2. **Demonstrate** how elevation is shown on a topographic map. 3. **Compare** and **contrast** latitude and longitude. **National Content Standards: (5–8) UPC1–3, A1, G1**	**Find Out!,** p. 29 **Investigate 1-1:** pp. 30–31 **Explore!,** p. 33 **Explore!,** p. 34 **Skillbuilder:** p. 36 **Science and Society,** p. 46
1-3 Viewing the Sky (3 sessions, 1 block)	1. **Describe** the position and appearance of the sun and the moon as viewed from Earth. 2. **Explain** the use of a star map in locating stars in the sky. **National Content Standards: (5–8) UCP1–3, A1–2, D3, E2, F2, F4–5, G1, G3**	**Explore!,** p. 39 **Design Your Own Investigation 1-2:** pp. 40–41 **Explore!,** p. 44 **A Closer Look,** pp. 44–45 **Technology Connection,** p. 47 **SciFacts,** p. 48

ACTIVITY MATERIALS

EXPLORE!

p. 21 pencils with erasers

p. 22* atlas, pencil, ruler

p. 33* No special materials are required.

p. 34* ball with no markings on it, black marker

p. 39 No special materials are required.

p. 44 No special materials are required.

INVESTIGATE!

pp. 30–31 metric ruler, transparency marker, clear plastic box and lid, transparency, tape, plastic model landform, beaker, water

DESIGN YOUR OWN INVESTIGATION

pp. 40–41 flour, salt, pan (about 10×12 in.), marbles of different sizes, metric ruler

FIND OUT!

p. 29* No special materials are required.

KEY TO TEACHING STRATEGIES

The following designations will help you decide which activities are appropriate for your students.

- **L1** Basic activities for all students
- **L2** Activities for average to above-average students
- **L3** Challenging activities for above-average students
- **LEP** Limited English Proficiency activities
- **COOP LEARN** Cooperative Learning activities for small group work
- **P** Student products that can be placed into a best-work portfolio
- Activities and resources recommended for block schedules

Need Materials? Call Science Kit (1-800-828-7777).

[00:00] OUT OF TIME? We recommend that students do the activities with an asterisk.

Chapter 1 Viewing Earth and Sky

TEACHER CLASSROOM RESOURCES

Student Masters	Transparencies
Study Guide, p. 7 **Critical Thinking/Problem Solving**, p. 5 **How It Works**, p. 5 **Making Connections: Across the Curriculum**, p. 5 **Making Connections: Technology and Society**, p. 5 **Multicultural Connections**, p. 5 **Science Discovery Activities**, 1-1	**Section Focus Transparency 1**
Study Guide, p. 8 **Activity Masters**, Investigate 1-1, pp. 7–8 **Critical Thinking/Problem Solving**, p. 9 **Take Home Activities**, p. 6 **Concept Mapping**, p. 9 **Multicultural Connections**, p. 6 **Science Discovery Activities**, 1-2 **Laboratory Manual**, pp. 1–4, Determining Latitude	**Teaching Transparency 1**, Latitude/Longitude **Teaching Transparency 2**, U.S. Physiographic Regions **Section Focus Transparency 2**
Study Guide, p. 9 **Making Connections: Integrating Sciences**, p. 5 **Activity Masters**, Design Your Own Investigation 1-2, pp. 9–10 **Science Discovery Activities**, 1-3 **Laboratory Manual**, pp. 5–6, Time Zones	**Section Focus Transparency 3**

ASSESSMENT RESOURCES	TEACHING & TECHNOLOGY
Review and Assessment, pp. 5–10 **Performance Assessment**, Ch. 1 **PASC*** **MindJogger Videoquiz** **Alternate Assessment in the Science Classroom** **Computer Test Bank**	**Spanish Resources** **Cooperative Learning Resource Guide** **Lab and Safety Skills** **Science Interactions, Course 1, CD-ROM** **Computer Competency Activities**

*Performance Assessment in the Science Classroom

NATIONAL GEOGRAPHIC TEACHER'S CORNER

Index to National Geographic Magazine

The following articles may be used for research relating to this chapter:

- "Orbit," by Jay Apt, November 1996.
- "Seventy-Five Years of Cartography: A Love Affair With Maps," by John Garver, November 1990.

National Geographic Society Products Available From Glencoe

To order the following products for use with this chapter, contact your local Glencoe sales representative or call Glencoe at 1-800-334-7344:

- *Geography of North America* (Transparencies)
- *STV: North America* (Videodisc)

Additional National Geographic Society Products

To order the following products for use with this chapter, call the National Geographic Society at 1-800-368-2728:

- *Our Fifty States* (Book)
- "United States Physical" (Map)
- *Reflecting on the Moon* (Video)
- *Geography: Five Themes for Planet Earth* (Videodisc, Video)
- *Latitude and Longitude* (Videodisc, Video)

Teacher Classroom Resources

These are key components of the classroom resources package.

TEACHING AIDS

Section Focus Transparencies

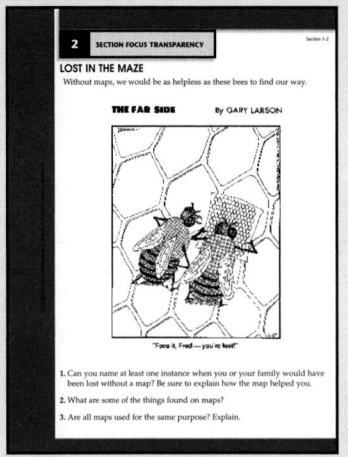

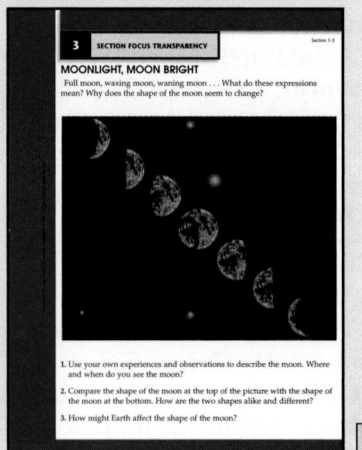

Teaching Transparencies

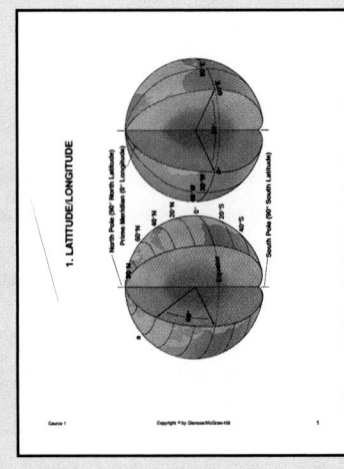

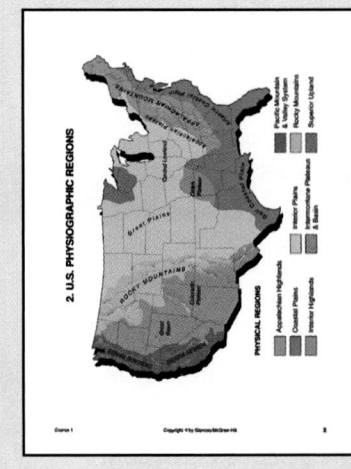

HANDS-ON LEARNING

Science Discovery Activity*

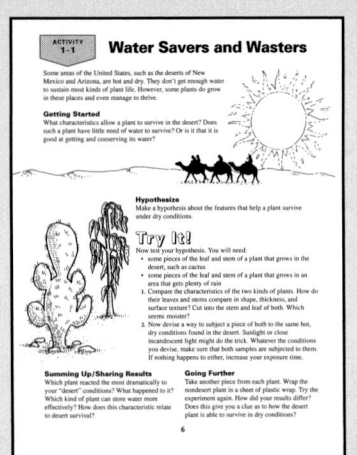

Laboratory Manual*

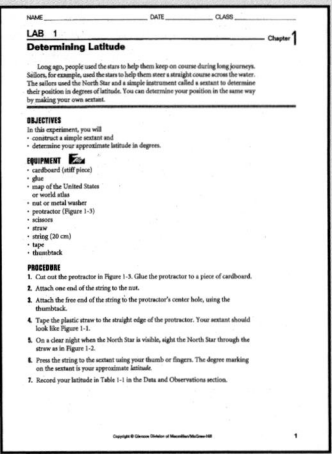

Take Home Activity

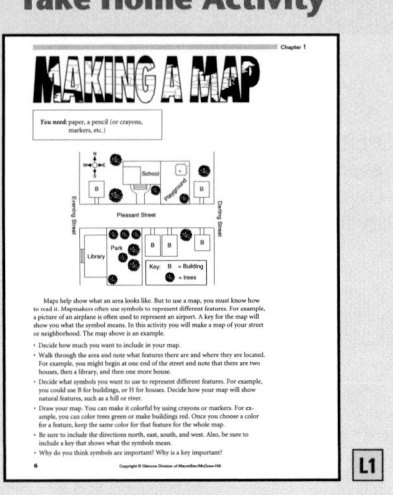

Chapter 1 Viewing Earth and Sky

Study Guide*

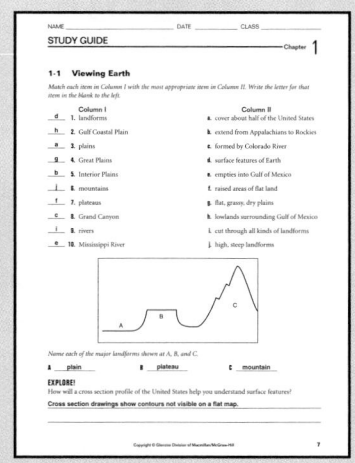

Concept Mapping

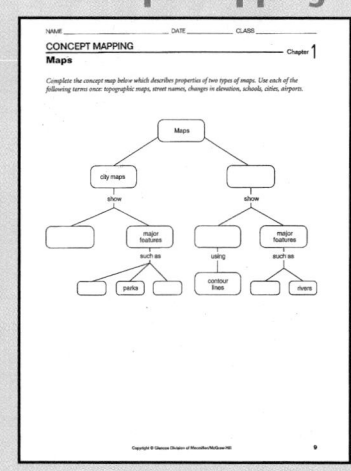

Critical Thinking/ Problem Solving

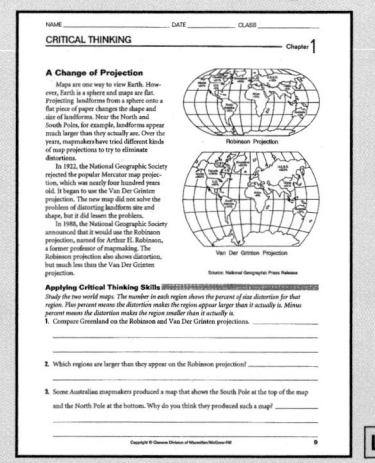

Integrating Sciences

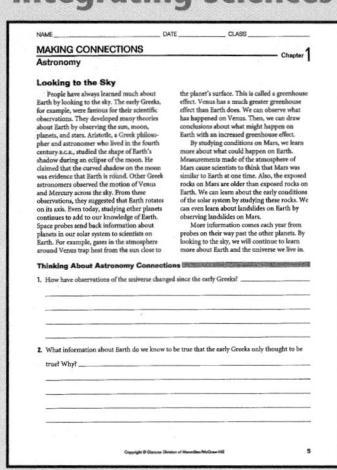

Across the Curriculum

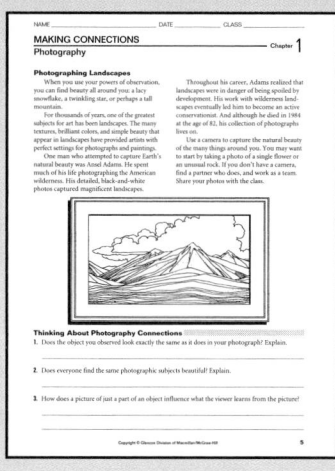

Technology and Society

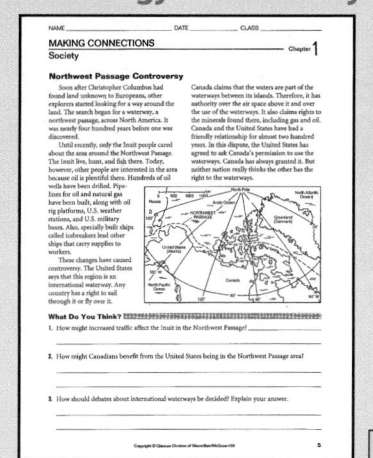

Multicultural Connection**

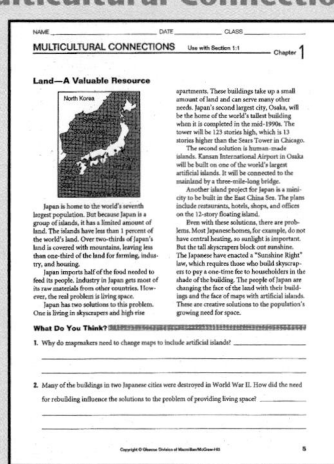

Performance Assessment

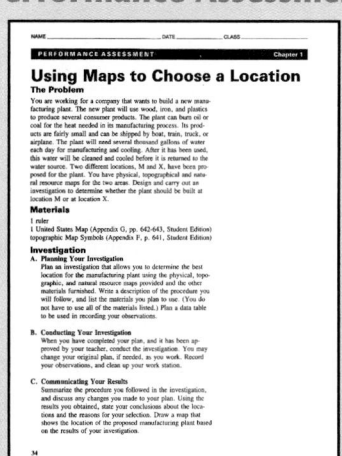

Review and Assessment

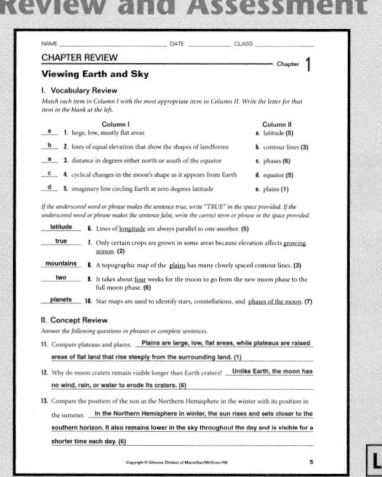

CHAPTER 1

Viewing Earth and Sky

THEME DEVELOPMENT

The major theme of this chapter is scale and structure. Students learn how landforms on Earth are related and how we can represent them on a smaller scale using maps. This chapter also explores the interaction of landscapes and systems.

CHAPTER OVERVIEW

In this chapter, students will study the major landforms. Students will also explore structures in space, the sun, and the moon.

Tying to Previous Knowledge

Have students describe ways they have viewed Earth and the sky, such as from an airplane or through a telescope.

INTRODUCING THE CHAPTER

Invite students to share memories of views or landscapes they found impressive. The landforms they will probably name are explored in this chapter.

Uncovering Preconceptions

Students might think that to view Earth and the sky, people must use scientific equipment. Have students examine a scene from close up and far away. Have them compare and contrast what they observed.

Viewing Earth & Sky

Did you ever wonder...

✓ **How high the mountains are?**
✓ **What the surface of the moon looks like?**
✓ **What the Big Dipper is?**

Science Journal

Before you begin to study what's under your feet and over your head, think about these questions and answer them *in your Science Journal.* When you finish the chapter, compare your journal write-up with what you have learned.

Answers are on page 49.

Have you ever taken a long trip by car, bus, train, or plane? If so, then you know there's not much to do while traveling except look out the window. But when you do, what a view! You can see miles and miles of sweeping fields or desert areas; high, snow-capped mountains; green, rolling hills; gentle valleys; and deep canyons. You might ask yourself how so many different landscapes can exist!

Making observations and asking questions about them is an excellent way to learn about your world. When you are traveling, it's easy to sit back and observe the world going by you. But you don't have to travel to observe. You can start right where you are—at home, in your classroom, around your neighborhood.

▶ *In the next activity, you will be looking up, down, and all around as you observe your physical surroundings.*

20

00:00 OUT OF TIME?

If time does not permit teaching the entire chapter, use the Chapter Overview on this page, Reviewing Main Ideas at the end of the chapter, and the Chapter 1 audiocassette to point out the main ideas of the chapter.

Learning Styles		
Kinesthetic	Activity, pp. 25, 26; Explore, p. 34; Visual Learning, p. 36; Design Your Own Investigation, pp. 40–41; Demonstration, p. 43
Visual-Spatial	Explore, pp. 21, 22, 33, 39, 44; Visual Learning, pp. 23, 24, 27, 32, 42; Activity, pp. 24, 29, 33; Research, p. 25; Find Out, p. 29; A Closer Look, pp. 44–45
Interpersonal	Activity, pp. 22, 28, 35; Multicultural Perspectives, p. 41; Project, p. 49
Logical-Mathematical	Activity, pp. 23, 38, 42; Across the Curriculum, pp. 23, 26, 43; Life Science Connection, pp. 26–27; Investigate, pp. 30–31; Visual Learning, p. 35
LS **Linguistic**	Multicultural Perspectives, p. 24; Across the Curriculum, p. 32; Research, p. 35; Discussion, p. 37

Explore! ACTIVITY

What do the landscape and sky look like where you live?

What To Do

1. Find a spot outdoors, and sit and make yourself comfortable. Facing west, draw *in your Journal* what you see on the land and in the sky. Pay attention to any shadows cast by the objects in your view.

2. Then, face north and draw what you see.

3. Continue on until you have drawn the east and south views as well. What features are in your drawings?

Explore!

What do the landscape and sky look like where you live?

Time needed 50–60 minutes

Materials pencils with erasers

Thinking Processes observing, classifying

Purpose To observe and identify landforms and features of the sky.

Preparation This activity is designed for a clear, mild day. Students should be situated so they can see all the landforms in the surrounding area. It would be helpful if students could draw from the top of a hill.

Teaching the Activity

Sketch one of the visible landforms to demonstrate to students the amount of detail they will need in their drawings. **L1**

Troubleshooting Be sure students know how to find west from the position of the sun.

Science Journal Have students label each view after they have sketched it in their journals. **L1**

Expected Outcome

Students' sketches should demonstrate a general sense of scale, topography, and relative location.

Answer to Question

Depending on your location, features in the sketches might include buildings, roads, motor vehicles, mountains, trees, vegetation, rivers, and valleys.

✔ **Assessment**

Performance Have students work in small groups to write original songs that describe their observations of the landscape and the sky. Use the Performance Task Assessment List for Song with Lyrics in **PASC**, p. 79. **COOP LEARN** **L1**

ASSESSMENT PLANNER

PORTFOLIO
Refer to p. 51 for suggested items that students might select for their portfolios.

PERFORMANCE ASSESSMENT
Process, pp. 31, 39
Skillbuilder, p. 36
Explore! Activities, pp. 21, 22, 33, 34, 39, 44
Find Out! Activity, p. 29
Investigate, pp. 30–31, 40–41

CONTENT ASSESSMENT
Oral, p. 29
Check Your Understanding pp. 28, 37, 45
Reviewing Main Ideas, p. 49
Chapter Review, pp. 50–51

GROUP ASSESSMENT
Opportunities for group assessment occur with Cooperative Learning Strategies.

PREPARATION

Planning the Lesson

Refer to the Chapter Organizer on pages 20A–D.

Concepts Developed

Students will identify various landforms and compare and contrast them. This chapter relates indirectly to Chapters 15 and 18 in Unit 4, which deal with landforms and their causes.

1 MOTIVATE

Bellringer

 Before presenting the lesson, display **Section Focus Transparency 1** on an overhead projector. Assign the accompanying **Focus Activity** worksheet. **L1**

LEP

Activity Display a relief map of the United States. Have students list all the landforms they can find. Then lead a discussion comparing and contrasting various landforms. **COOP LEARN** **L1**

Explore!

What does a profile of the United States look like?

Time needed 20 to 25 minutes

Materials atlases, pencils, rulers

Thinking Processes observing and inferring, making models, forming operational definitions

Purpose To practice drawing cross-sectional profiles.

Preparation See Appendices G and H, pp. 642–645 for maps.

Teaching the Activity

Science Journal Have students make their cross-sections approximately along the

---

Viewing Earth

Section Objectives

■ Describe basic landforms such as mountains, plains, and plateaus.

■ Recognize the kind of landform on which you live.

Key Terms

landforms

Over the Plains, Mountains, and Plateaus

Imagine that you have tickets for an early morning flight from Washington, D.C. to California. You will be flying for about six hours. Thank goodness you were able to get a window seat! You want to see the land unfold beneath you. What will you observe as you fly over the United States? The Explore activity will give you some clues.

Explore! ACTIVITY

What does a profile of the United States look like?

What To Do

1. Find a physical map of the United States that shows its land features.

2. Use the map to guide you as you draw *in your Journal* a cross-section profile of the United States. A profile shows the surface features as they would look if you cut the United States along a line from Washington, D.C. to San Francisco, California and looked at the surface from the cut edge.

3. Mark and label the different features. What do you think those features would look like from the air?

Now that you have drawn a profile, you will see how your profile compares with the surface as seen from the air. The surface features you will see are called **landforms**. Mountains, plains, and plateaus are three common landforms in the United States.

The big day for your airplane trip arrives. You board the plane, settle yourself into your window seat, and fasten your seat belt. You're eager to compare your profile with the landforms you see on the trip. The plane taxis down the runway, and soon you're in the air. Good-bye, Washington! California, here you come!

22 Chapter 1 Viewing Earth and Sky

40th parallel. Guide them to make a vertical scale to plot the relative elevations of the features they observe. These profiles should indicate topographic variation. **L1**

Expected Outcome

Students will create a profile of landforms of the United States.

Answer to Question

Plains would look flat and wide. Mountains would look bumpy, round, and sharp. Plateaus would look flat on top with steep sides.

✔ Assessment

Content Have students use their profiles to operationally define *plain, plateau,* and *mountain.* Use the Performance Task Assessment List for Analyzing the Data in **PASC,** p. 27. **L1**

Across the Low Plains

What do you think of when you hear the word plains? You might think of endless flat fields of wheat or grass. That's often what plains look like because many of them are used to grow crops. Plains are large, low, mostly flat areas. In fact, about half of all the land in the United States is plains.

You will see plains along many of the coastlines and in the interior of the country. The lowland areas along the coastlines are called coastal plains. These areas are characterized by low, rolling hills, swamps, and marshes. One coastal plain is the Gulf Coastal Plain. It includes the lowlands surrounding the Gulf of Mexico. Look at **Figure 1-1**. Where do you think this area is?

As you fly, you see that a large part of the middle of the United States is also made up of plains. They are called the interior plains. You may remember from the Explore activity you just did that the interior plains extend from the Appalachian Mountains in the east, to the Rocky Mountains in the west, and to the Gulf Coastal Plain in the south. The first interior plains you see are the very low, rolling hills of the Great Lakes area and the lowlands around the Mississippi and Missouri rivers.

To the west of the Mississippi lowlands, you see the Great Plains. These are flat, grassy, dry plains with few trees. The Great Plains are covered with nearly horizontal layers of dirt and small rocks. Where did these sediments come from? They washed down from the landform that you will fly over next on your trip.

Figure 1-1

This map shows the location of the coastal plains and the interior plains.

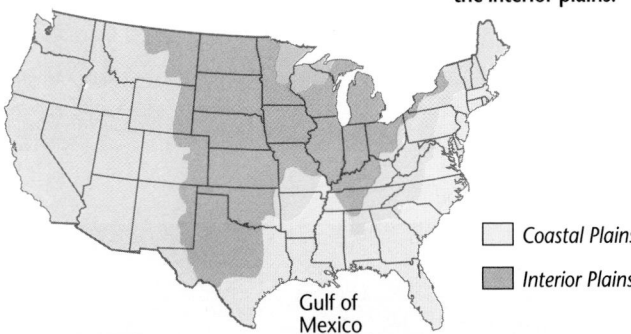

☐ *Coastal Plains*
▨ *Interior Plains*

Gulf of Mexico

Coastal plains on the east coast

Interior plains in southeast Colorado

1-1 Viewing Earth **23**

2 TEACH

Tying to Previous Knowledge

Have students describe the most amazing landform feature they have seen or would like to see. Then have students find a picture in the text of a similar landform.

Visual Learning

Figure 1-1 Have students compare and contrast the plains shown in the two photographs. *Both are vast flat regions on which vegetation is capable of growing. The coastal plain is near water; the interior plain is inland.* L1

Student Text Question
Look at Figure 1-1. Where do you think this area is? *The Gulf Coastal Plain includes parts of Texas, Louisiana, Alabama, Florida, and Mississippi.*

Activity Have students use an atlas to count all the states that border on the Atlantic Ocean, Gulf of Mexico, and Pacific Ocean. *Atlantic—14, Gulf—5, Pacific—5.* Then ask questions such as: **Which state borders an ocean and a gulf?** *Florida* **Which state borders on two oceans?** *Alaska* L1

Across the Curriculum

Math

Have students use the answers from the activity above to write ratios such as: Atlantic states to all states is $\frac{14}{50}$

Connect to

Life Science

Answer: Answers will vary depending upon the animals that are chosen.

---

24 Chapter 1 Viewing Earth and Sky

Mountains

Much of the dirt and rock that helped form the Great Plains washed down from the Rocky Mountains, pictured in **Figure 1-2** below. Mountains tower above the surrounding land, providing a spectacular view of Earth. The world's highest mountain peak, however, is not in the Rockies, or even in the United States. It is Mount Everest in the Himalaya in Asia, which rises more than 8800 meters above sea level. Mountain peaks in the part of the United States that you are traveling over reach just a little more than 4000 meters high.

As you might expect, mountains vary greatly in size and shape. Look at all the pictures of mountains in **Figure 1-2**, and you will see that they appear different. This is because they were formed in different ways. Also, some mountains are older than

Figure 1-2

This map shows the location of the Appalachian, Rocky, and Pacific Mountains.

Appalachian Mountains

Cascades

Rocky Mountains

Sierra Nevadas

Sierra Nevadas in California

Cascades in Washington

24 Chapter 1 Viewing Earth and Sky

Multicultural Perspectives

Living on the Land
The Anasazi built their homes directly into the sides of cliffs to utilize the natural shelter, protection, and cooling that canyon walls offer. Have students research these Southwestern Native Americans who were cliff dwellers and present reasons they chose to live in canyon walls. **L1**

Program Resources

Multicultural Connections, p. 5, Land—A Valuable Resource **L1**

Making Connections: Technology and Society, p. 5, Northwest Passage Controversy **L2**

Critical Thinking/Problem Solving, p. 5, Flex Your Brain **L2**

others. The older ones are often more like large hills, rounder and lower in comparison with younger mountains. They have been worn down to some degree by wind, rain, and running water. Which of the mountains shown in **Figure 1-2** are older mountains?

Appalachian Mountains in West Virginia and Tennessee

Rocky Mountains in Colorado

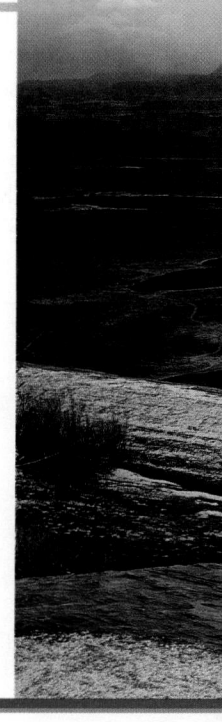

To compare and contrast plains and plateaus, have pairs of students use modeling clay to make three-dimensional models. *Students' models will vary somewhat but they should demonstrate that plains are low-lying flat areas with little topography; plateaus are also flat areas, but unlike plains, they have been uplifted and incised by Earth forces and processes.* Use the Performance Task Assessment List for Model in **PASC**, p. 51.

COOP LEARN L1

Across the Curriculum

Math

Direct students to find the volume of material that was removed from the Bingham mine discussed in Did You Know on this page. Use the formula: $l \times w \times h = V$. *4 km × 4 km × 0.8 km = 12.8 km³*

To help students visualize 12.8 km³, you will need 12 meter sticks. Form a square of four meter sticks on the floor. Choose two students to hold four more meter sticks (two sticks each) perpendicular to the corners of the square. Choose two more students each to hold two sticks to form a square at the top of the perpendiculars. You now have one cubic meter. Ask students how many of those cubic meters would fit into 1 km³. *1000 m × 1000 m × 1000 m = 1 000 000 000 m³ or 1 billion cubic meters.* Point out that the material removed from the Bingham mine was almost 13 times this amount. L2

DID YOU KNOW?

Some landforms are made by humans. The Bingham mine near Salt Lake City, Utah, is the largest hole ever made by humans. The Bingham mine is an open-pit copper mine that is almost 800 m deep and 4 km wide at the surface.

On the Flat Plateaus

As you continue your trip west, you see other highlands. These raised areas of fairly flat land are plateaus. Plateaus are made up of nearly horizontal rock layers. Unlike the plains, they rise steeply from the land around them. Long ago, plateaus were plains. But over time, they were pushed upward by forces inside Earth.

Figure 1-3 shows the Colorado Plateau, which lies just west of the Rocky Mountains. The Colorado River has cut deep into the rock layers of this plateau, forming the Grand Canyon. Other rivers, such as the Green River and the San Juan River, have also created canyonlands on the vast Colorado Plateau.

Plateaus are often found near mountains and can have very high altitudes. The Plateau of Tibet borders the Himalaya in Asia. Its elevation is about 4300 meters above sea level! Plateaus with lower elevations are often used as pasture lands for live-stock.

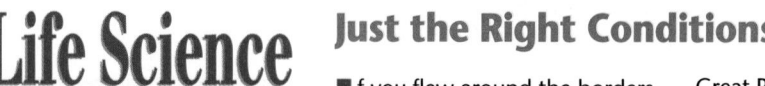

Just the Right Conditions

If you flew around the borders of our country, you would see many distinctly different farms growing a variety of crops in each area.

For instance, passing over North Carolina, you would notice a lot of tobacco farms. Farther up the coast, in New Jersey, farmers are growing vegetables such as tomatoes and green beans. Crossing the northern parts of the states, you see apple orchards on the interior plains around the Great Lakes, wheat on the

Great Plains, potatoes in Idaho, and more (and different kinds of) apples in Washington State. In California, citrus fruit trees—grapefruit, oranges, lemons, and limes—would be plentiful, while in Arizona and Texas, there would be cotton farms.

Why is there such diversity over the continental United States? Why are many crops grown in one place but not another? Crops need a growing season, a time free from frost. It takes a certain amount of time for an apple, an orange, or a cotton plant to grow. If the weather does not act as expected, something will

Purpose

Life Science Connection attributes the variety of crops grown in the United States to the presence of different landforms, soils, and climates.

Content Background

American scientists have been cross-breeding plants for years in an effort to fight world hunger. For example, in 1967, high-yield dwarf varieties of rice and wheat were introduced into Mexico, India, and other developing nations. Because of their shorter growing season, if they have enough water, such crops can be grown and harvested two or more times a year. More plant nutrients also went into the growth of the grain rather than into the growth of leaves and stalks. Within ten years, however, crop yields began to decrease due to floods, droughts, and the need for expensive fertilizers.

Figure 1-3

This map shows the location of the Colorado Plateau and the Ozark Plateau.

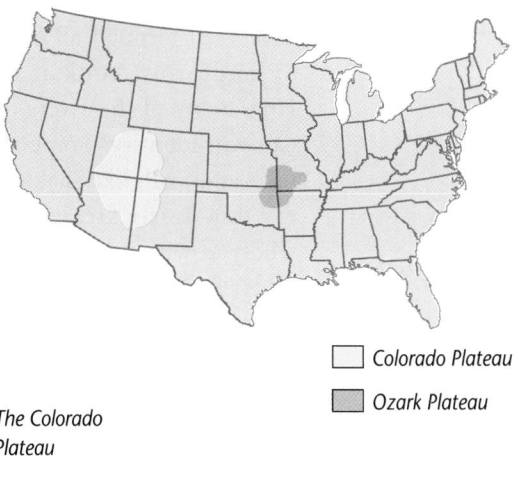

☐ Colorado Plateau

■ Ozark Plateau

The Colorado Plateau

Visual Learning

Figure 1-3 Have students determine in which states the Colorado Plateau and the Ozark Plateau are located. *The Colorado Plateau occupies parts of Colorado, Utah, Arizona, and New Mexico. The Ozark Plateau is located in parts of Arkansas, Oklahoma, Missouri, and Kansas.* **L1**

Demonstration To demonstrate how plateaus are formed, use a large piece of cloth, a bicycle pump, masking tape, and a large resealable freezer bag. Put the hose of the pump into the bag and seal it securely with masking tape. Place the cloth over the bag and inflate the bag slowly. Students will observe that the cloth remains horizontal as its elevation increases.

Content Background

Due to farming and ranching, little natural prairie still exists. The variety of species of grass and plant life are being replaced by uniform single-crop plantings. To find out more about prairie conservation, write to: The Morton Arboretum, Route 53, Lisle, IL 60532.

Teacher F.Y.I.

The Grand Canyon, a mile-deep gorge, is not the deepest land canyon in the world. Challenge students to find out about a gorge in the Andes mountains that is deeper than the Grand Canyon. (Answer: *Colca Gorge*; 4360 m deep)

probably happen to the crops. If it is too hot or too cold for too long, or if there is not enough rain or too much rain, crops will die.

Soil and land formation are as important as climate in growing crops. Mountains are rocky and steep, with a thin layer of soil. They are not good for growing food or grazing cattle. Desert plains or plateaus have little vegetation because of little rain, but they can often produce crops if water is provided through irrigation.

You Try It!

Imagine that you are going to start a farm. Choose a plant or crop and find out about its

ideal growing conditions. For instance, corn needs a lot of water, and tomatoes need sun. Would you start your farm in central Arizona, which has desert conditions, or in northern Ohio, bordering the Great Lakes? Back up your choice with details about the plants and the conditions in the area you choose.

Teaching Strategy

Have students decide which crops are essential for human needs. Make a chart on the chalkboard of the crops they choose, the regions of the United States where the crops are grown, and reasons the crops were chosen. Have students rank these crops in order of most to least important for human needs. **L1**

Answers to
You Try It!

Students should choose appropriate soil, water, and sun conditions for plants of their choice. Be sure they consider possibilities such as irrigation in making their choices.

Going Further ▪▪▪▪▶

Assign three to four neighboring states to each pair of students. Have each pair research the agricultural resources for their states, including the production tonnage for each state. Make a master table, listing the states on one axis and crops on the other and have students enter the crop yield(s) for their states. Have students identify the states that are primarily producers and the states that are primarily consumers. Use the Performance Task Assessment List for Data Table in **PASC**, p. 46. **COOP LEARN** **L2**

3 ASSESS

Check For Understanding

1. Have students name and list the three major landform types from lowest to highest relief and give an example of each. *plains: Great Plains, Coastal Plains; plateaus: Colorado; mountains: Appalachian, Rocky, Cascade, Coastal mountains*

2. Have students answer Check Your Understanding questions 1–3 individually and question 4 in groups.

Reteach

Demonstration Use wet sand, pebbles, and a clear plastic storage box to make a model landscape that includes mountains, plains, and plateaus. Have students study the model from the top (map view) and from the side (cross-sectional view) to compare and contrast the viewpoints. L1 LEP

Extension

Activity Have pairs of students use a U.S. map to draw a cross-section profile of the United States from the southern tip of Florida to the northwestern tip of Oregon. L3

4 CLOSE

Activity

Divide the class into groups of four. Give each student 4 slips of paper. Have students write a question about landforms on one side of each slip and the answer on the other side. Have groups trade slips and play a game. One group member draws a question for the person seated to the right. If a student answers incorrectly, then the student to the left tries to answer. COOP LEARN L1

Rivers

One feature that mountains and plains share with plateaus is rivers. Rivers cut through all different kinds of landforms. Look at **Figure 1-4**. How many of the rivers on the map can you identify? Some rivers curve back and forth like snakes, while others flow fairly straight. Some run wildly down mountain slopes, while others move slowly along.

It's time to prepare for landing. Your flight is nearly over. It was great seeing landforms from the airplane, but such a view is not usually possible. You can't often observe land directly. Most of the time, you use maps to identify landforms. Can a map show how high a mountain is or how flat a plain is? You will find out that maps can do these things and even more.

Figure 1-4

This map shows the major river systems in the U. S. A.

The Mississippi River in Wisconsin

The Colorado River in Arizona

check your UNDERSTANDING

1. How are plains and plateaus alike? How do they differ?

2. If you live near the Gulf Coast, on what landform would you live?

3. Why are the Rocky Mountains in the west higher with sharper peaks than the more rounded Appalachian Mountains in the east?

4. Apply On which kind of landform do you live?

check your UNDERSTANDING

1. Both plains and plateaus are relatively flat. Plains are large, low, mostly flat areas that cover most of the United States. Plateaus are raised flat areas with steep sides.

2. If you live near the Gulf Coast, you live on a coastal plain.

3. The mountains are higher in the western U.S. than in the east because the eastern mountains are older and have been eroded over time.

4. The answer to the Apply question depends on your location.

1-2 Using Maps

Road Maps

Your plane has landed in California, where a good friend has moved, and you'd like to visit him at his school. You find your way from the airport to a beautiful city park, but you wonder where your friend's school is. You know a map would be useful to you. Maps are useful for locating places, finding your way around, and getting a clearer picture of an area. How have you used maps to help you? Use the map in the Find Out activity to help you find your friend's school.

Find Out! ACTIVITY

Where is Central School located?

You can find the school by studying the map. You're at the city park, Balmoral Park, right now. How can you get to the school?

What To Do

1. Look at the different names of the streets and avenues on the map shown on the right. Now notice the location of the city park. *In your Journal*, describe the location of the park.

2. Next, notice the location of Central School. How would you describe its location?

Conclude and Apply

1. Suppose you wanted to get from the park to the school. *In your Journal*, describe how you would do it.

With the road map, you were able to get from the park to the school. Does the map tell you anything about landforms in the area? Are you able to find the highest and lowest points in the city? No, you need a different kind of map to get this information. One kind of map that shows landforms is a topographic map. As you'll discover in the next Investigate, showing hills and valleys on a flat map isn't as hard as it may seem.

Section Objectives

- Identify landforms using a topographic map.
- Demonstrate how elevation is shown on a topographic map.
- Compare and contrast latitude and longitude.

Key Terms

elevation, contour lines, latitude, longitude

PREPARATION

Planning the Lesson

Refer to the Chapter Organizer on pages 20A–D.

Concepts Developed

In this section, students will identify landforms on a topographic map, find elevations, and locate points on a map using latitude and longitude.

1 MOTIVATE

Bellringer

Before presenting the lesson, display **Section Focus Transparency 2** on an overhead projector. Assign the accompanying **Focus Activity** worksheet. **L1**
LEP

Activity Have pairs of students draw a map of the halls inside the school. Have students sketch their journeys on paper. **L1**

shown, actual distances cannot be used. **L1**

Expected Outcome

Most students will be able to plot a route from the park to the school.

Conclude and Apply

Students' routes will depend on their supposed location in the park as well as their perhaps wanting a more "scenic" rather than direct route.

✔ Assessment

Oral Have partners take turns asking each other to describe the most direct route between two points on the map (for example, between the Essex Street entrance of the park and the north entrance of the school). Use the Performance Task Assessment List for Scientific Drawing in **PASC**, p. 55. **COOP LEARN** **L1**

Find Out!

Where is Central School located?

Time needed   15 to 20 minutes

Materials   No special materials or preparation are required for this activity.

Thinking Process   interpreting data

Purpose   To use a map to plot the route from one location to another.

Teaching the Activity

Discussion   Ask a volunteer to describe the route. Ask students to agree or disagree and add comments. Point out that the route is in part dependent on where you are in the park. **L2**   **COOP LEARN**

Science Journal   Students will describe the locations of the park and the school in terms of other features on the map. Since a scale is not

1-1 Using Contour Lines

Planning the Activity

TECH PREP

Time needed 30 minutes

Purpose To demonstrate how contour lines represent elevation.

Process Skills observing and inferring, measuring in SI, making and using models, interpreting scientific illustrations

Materials metric ruler, transparency marker, clear plastic box and lid, clear plastic transparency, tape, plastic model landform, beaker, water

Teaching the Activity

Demonstrate, if necessary, how to carry out the procedure. Circulate throughout the class to be sure that students are proceeding properly. Students should position themselves directly over the model while drawing the contour lines. Best results are obtained if students close one eye and look straight down. [L1]

Troubleshooting Adding a drop of dark food coloring will make the border between water and landform more easily visible. Use cone-shaped cups or plastic deli containers if landform models are not available.

Process Reinforcement Some students may have difficulties visualizing a three-dimensional feature—the mountain—in two-dimensions on the transparency. Obtain some topographic maps of areas familiar to students and allow students time to study the maps and relate what they see on the maps to the actual features with which they are familiar such as a nearby hill, valley, and so on.

Science Journal Have students attach their transparencies to a page in their journals opposite the answers to the questions posed in this activity. [L1]

INVESTIGATE!

Using Contour Lines

How can a map show the shape of a landform? In the following activity, you will show the elevations of a landform by drawing contour lines, which are lines of equal elevation that show the shape of the landform.

Problem
How can elevation of a landform be indicated on a map?

Materials

metric ruler	transparency marker
clear plastic box and lid	transparency
	tape
model landform	beaker
water	

Safety Precautions

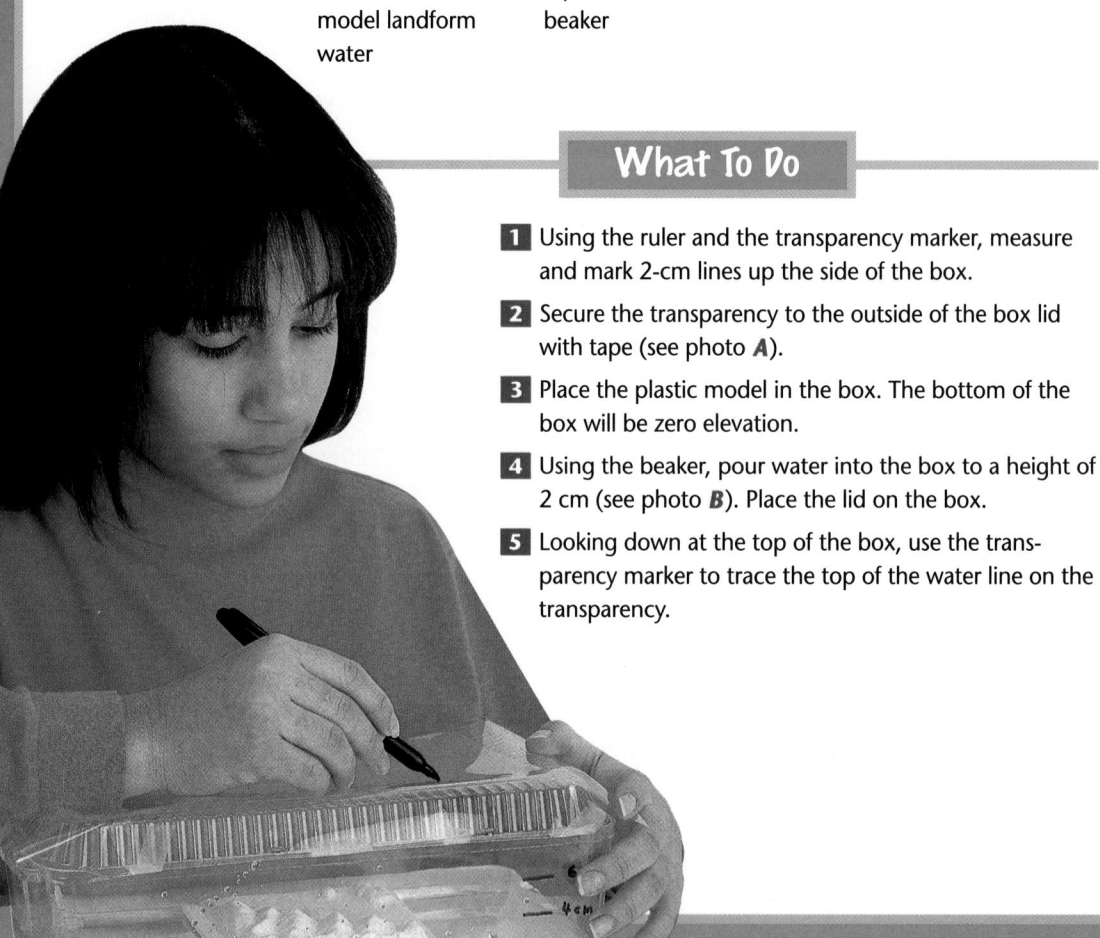

What To Do

1 Using the ruler and the transparency marker, measure and mark 2-cm lines up the side of the box.

2 Secure the transparency to the outside of the box lid with tape (see photo **A**).

3 Place the plastic model in the box. The bottom of the box will be zero elevation.

4 Using the beaker, pour water into the box to a height of 2 cm (see photo **B**). Place the lid on the box.

5 Looking down at the top of the box, use the transparency marker to trace the top of the water line on the transparency.

Program Resources

Activity Masters, pp. 7–8, Investigate 1-1
Science Discovery Activities, 1-2, To Grid or Not to Grid
Section Focus Transparency 2

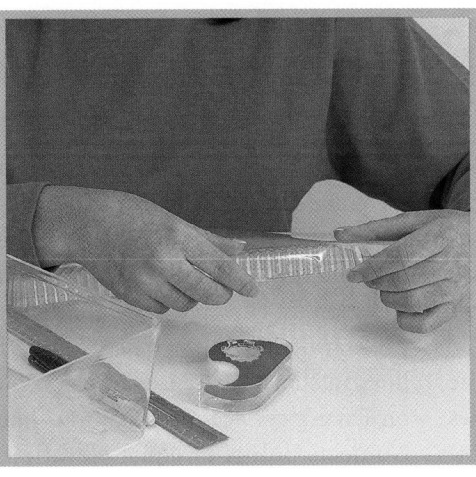

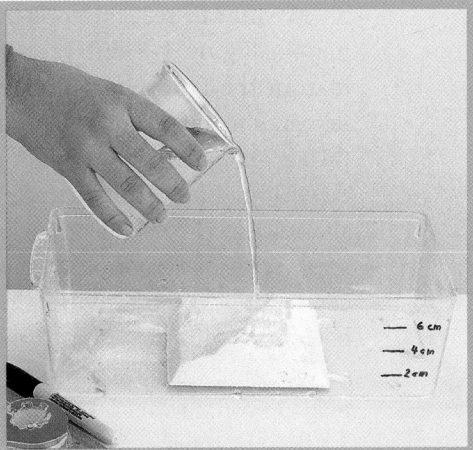

A   B

6 Using the scale 2 cm = 5 ft, mark the elevation on the line.

7 Repeat Steps 4–6, adding water to the next 2-cm level and tracing until you have mapped the landform by means of contour lines.

8 Transfer the tracing of the contours of the landform onto paper.

Analyzing

1. What is the contour interval of this topographic map?

2. *Interpret* the relationship between contour lines on the map and the steepness of the slope on the landform.

3. *Calculate* the total elevation of the landform.

Concluding and Applying

4. How are elevations shown on topographic maps?

5. *Explain* whether all topographic maps must have a 0-ft elevation contour line.

6. ~~Going Further~~ How would the contour interval of an area of steep mountains *compare* with the interval of an area of flat plains?

Expected Outcomes
Students will understand how height is represented by contour lines on a topographic map. Students should be able to draw a simple topographic map.

Answers to Analyzing/ Concluding and Applying

1. 5 ft

2. The closer the contour lines, the steeper the slope. Increased distances between contour lines indicate relatively flat areas.

3. Elevations will vary, but every 2 cm of the model are equal to 5 ft.

4. by contour lines

5. They do not because all parts of the world aren't near sea level or 0 ft.

6. Going Further: Maps with steep mountains will have a larger contour interval because the linear distance between contour lines is very small.

✔ Assessment

Process Obtain permission to take students to an area near your school where they can observe features of the landscape. Have students explain how these features compare to topographic maps of the area. (Topographic maps can be obtained from your state geological survey.) Use the Performance Task Assessment List for Scientific Drawing in **PASC**, p. 55. [L1]

Meeting Individual Needs

Visually Impaired Add a few drops of food coloring to the water to make it easier for visually impaired students to construct the map view of the model landform.

Learning Disabled If students have problems grasping the intent of this activity, show them a collapsible camping cup and make the analogy between the parts of the cup and contour lines used to represent elevation.

Physically Challenged Some students with physical limitations may not be able to manipulate the items used in this activity. Allow those students to observe and record the information gathered during this investigation.

Tying to Previous Knowledge

Ask students to think about times in their lives when they have used maps other than those they used in school, such as road maps, maps of a theme park, ski trail maps, subway maps, park maps, and campground maps. Lead into a discussion of a special kind of map called a topographic map, which they will be learning about in this section.

Visual Learning

Figure 1-5 Ask students to compare and contrast a topographic map and a street map. *Street maps and topographic maps show rivers, bridges, and railroads. Both have map scales to represent distances. Street maps show all the streets; topographic maps usually don't. Topographic maps show elevation and landforms; street maps don't.* Have students study the figure. **Where is the steepest area on the map?** *The west slope of Coon Hill and the east slope of Todd Mountain are the steepest of the long slopes, although there are some shorter, steeper areas.* **Where is the land flattest?** *Areas surrounding the Deerfield River are flattest.* L1 LEP

Across the Curriculum

Language Arts

Have students use dictionaries to find the origins of the word *cartography*. Students will discover that cartography is the art of making maps. The word has two roots (*cart* means "chart" or "leaf of paper" and *ography* means "to write.") Ask students what the technical name for a mapmaker is. *cartographer*

Topographic Maps

As you can see, the topographic map in **Figure 1-5** differs from the map in the Find Out activity. The topographic map shows the shape of the area by giving information about elevation. **Elevation** is the height above sea level or the depth below sea level. With a topographic map, you can tell how steep a mountain is or how deep a canyon is.

The thin lines on a topographic map are contour lines. **Contour lines** are lines of equal elevation that show the shapes, or contours, of landforms. The contour lines represent three-dimensional contours on a two-dimensional map. They show the vertical rise and fall of the land.

Before the use of contour lines, mapmakers used shading to suggest the contours of the land. While shading can give you a good idea of surface contours, contour lines are based on elevation data and are more accurate.

Mapmakers today use contour lines to show differences in elevation on Earth's surface. Each contour line connects points of equal elevation. Between every two contour lines, the land changes elevation.

Figure 1-5

A This topographic map shows a portion of northwestern Massachusetts near the Vermont border. The contour interval on this map is 20 feet. In other words, there is a 20-foot change in elevation between lines.

B If you went from the top of Clark Mountain to the Florida Bridge on the Deerfield River, you would walk down a 1280-foot drop.

C The closer the lines, the steeper the change in elevation. Where is the steepest area on the map? Where is the land flattest?

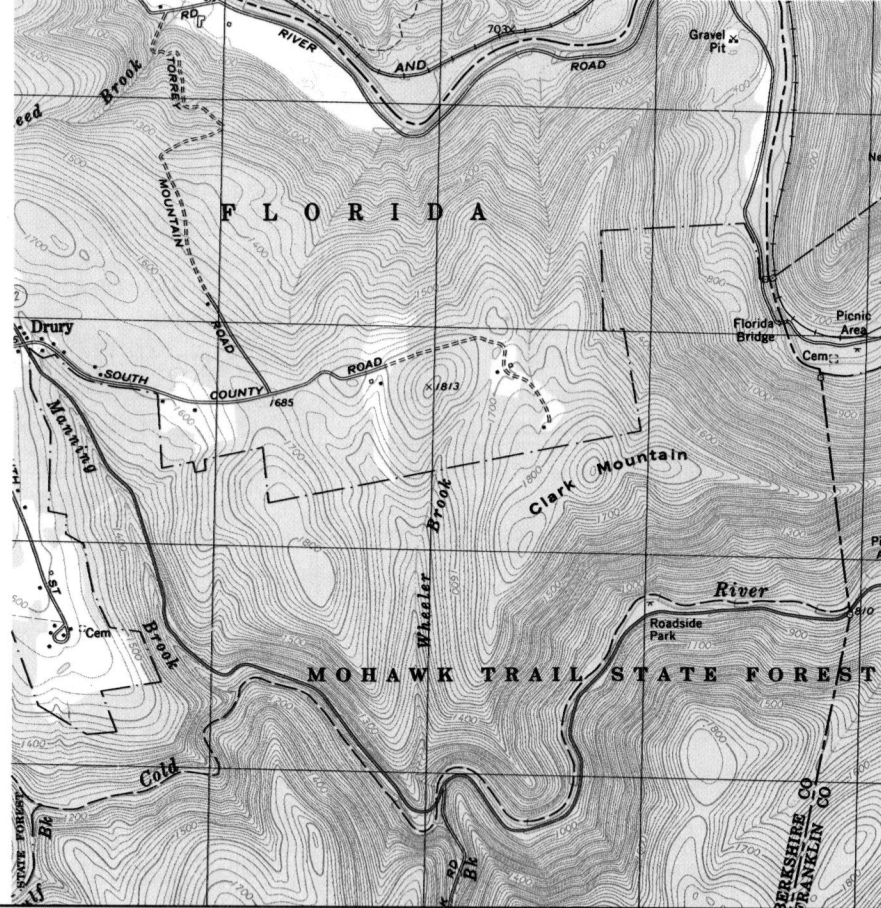

32

Meeting Individual Needs

Learning Disabled Draw a contour map with 5 to 10 concentric contours, each spaced about 1 cm or more apart. Draw a straight reference line across the contours from the center of the map to any edge or corner. Photocopy the map and distribute it, cardboard, scissors, and paste. Have students cut along the outermost contour line and then trace the circular pattern on a piece of cardboard. Tell them to place an X where the reference line meets the cardboard. Direct students to cut out the cardboard along the traced line. Have them repeat the process, each time cutting along the next contour inward. Be sure they point the reference line to an X for every cut. When all the contours have been cut, have students paste the cardboard cutouts together, lining up all the X's. The stair-step result is a model of a hill. L1

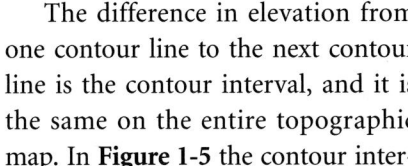

Explore! ACTIVITY

How can you identify a landform without actually seeing it?

What To Do

1. Study the topographic map in **Figure 1-5** and notice its contour interval.

2. *In your Journal*, describe the kind of landform you think occurs where the contour lines are close together. What kind of landform occurs where the lines are far apart?

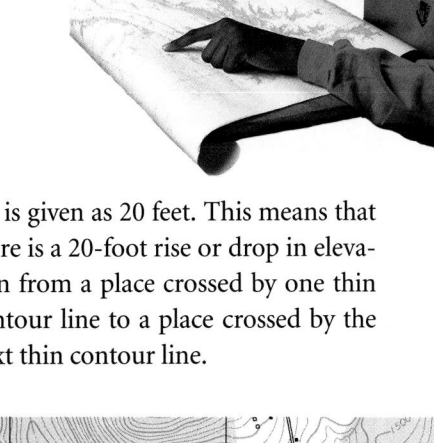

The difference in elevation from one contour line to the next contour line is the contour interval, and it is the same on the entire topographic map. In **Figure 1-5** the contour interval is given as 20 feet. This means that there is a 20-foot rise or drop in elevation from a place crossed by one thin contour line to a place crossed by the next thin contour line.

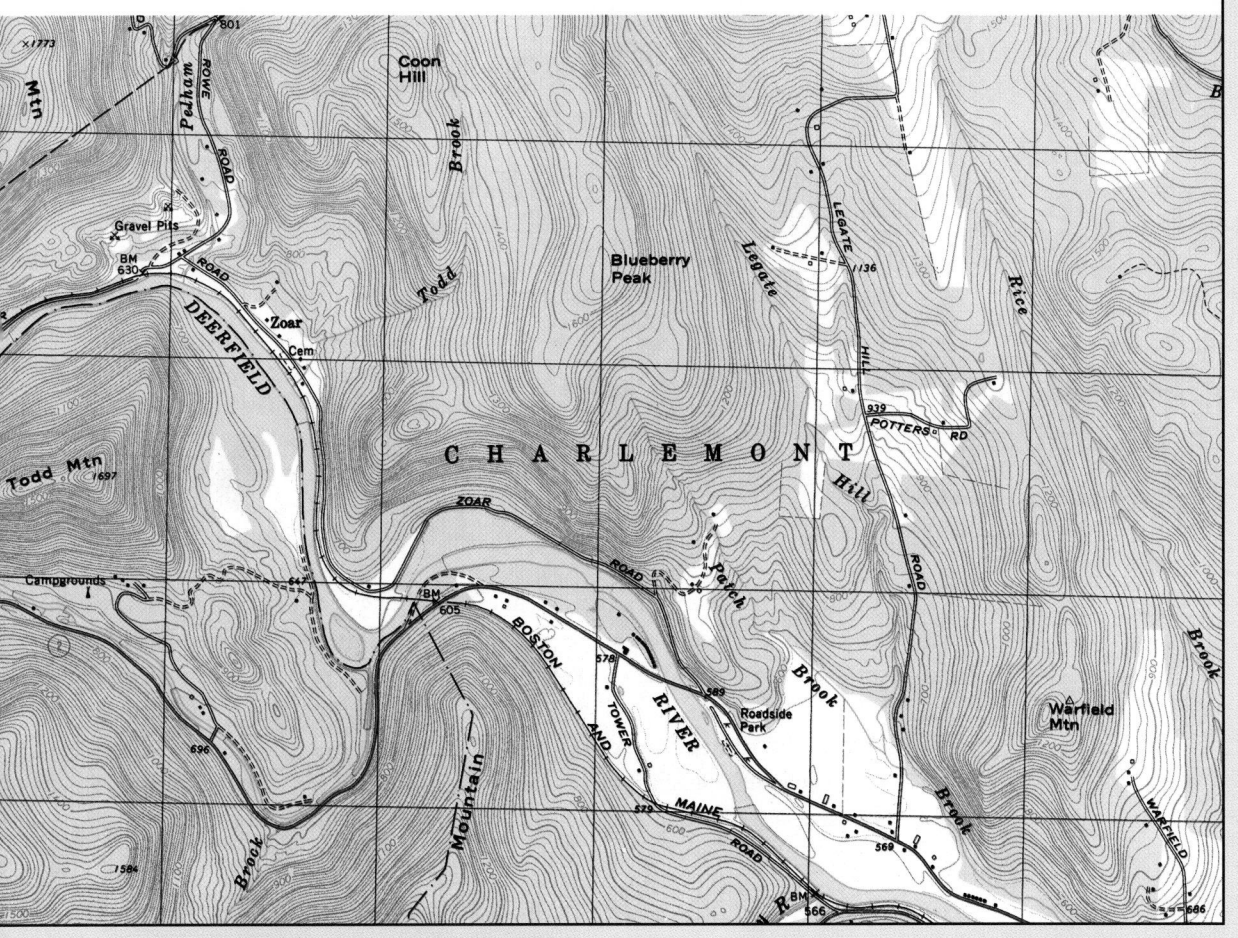

Program Resources

Explore!

How can you identify a landform without actually seeing it?

Time needed 10–15 minutes

Materials No special materials or preparation are required for this activity.

Thinking Processes observing and inferring, interpreting scientific diagrams

Purpose To observe the spacing of contour lines and infer the actual topography.

Teaching the Activity

Troubleshooting Provide magnifying glasses for students to facilitate observations of the map.

Science Journal Students may infer that the closely spaced contour lines indicate a hilly area, a valley, or a mountain. **L1**

Expected Outcome

Students will interpret relative steepness of contour line spacing on a map and infer the actual topography of an area.

Answer to Question

Larger distances between lines indicate a gentle slope or relatively flat area.

✓ Assessment

Process Have groups of students create ficticious topographic maps, exchange them and write three questions about the maps. Use the Performance Task Assessment List for Group Work in **PASC,** p. 97. **L1**

Latitude

Recall how you reached your friend's school from the park. You used a road map. The road map gave you the reference points of streets and avenues to help you get to the school. But suppose you wanted to get from one place on Earth to another. What reference points would you use? Do the following Explore activity to discover just how difficult it would be to describe a place without the use of reference points.

Explore! ACTIVITY

Can you describe the location of a dot on a ball?

Work with a friend to do this activity.

What To Do

1. Obtain a ball that does not have any markings on it. Use a marker to place a black dot on the ball.

2. Now, without showing the ball to your friend, try to describe the location of the dot. Can your friend tell you where the dot is located from your description?

3. With your partner, think of a way to make it easier to describe where the point on the ball is. Describe your system *in your Journal*. Test your system to see how well it works.

As you have discovered, it is almost impossible to describe the location of a dot on a ball. Reference points are needed. Like the ball, Earth is a sphere. To provide reference points, mapmakers have given Earth a grid system that is something like the lines on graph paper.

Look at **Figure 1-6**. The North Pole is the northernmost point on Earth; the South Pole is the southernmost. The equator is an imaginary line that circles Earth exactly halfway between the North and South poles. It

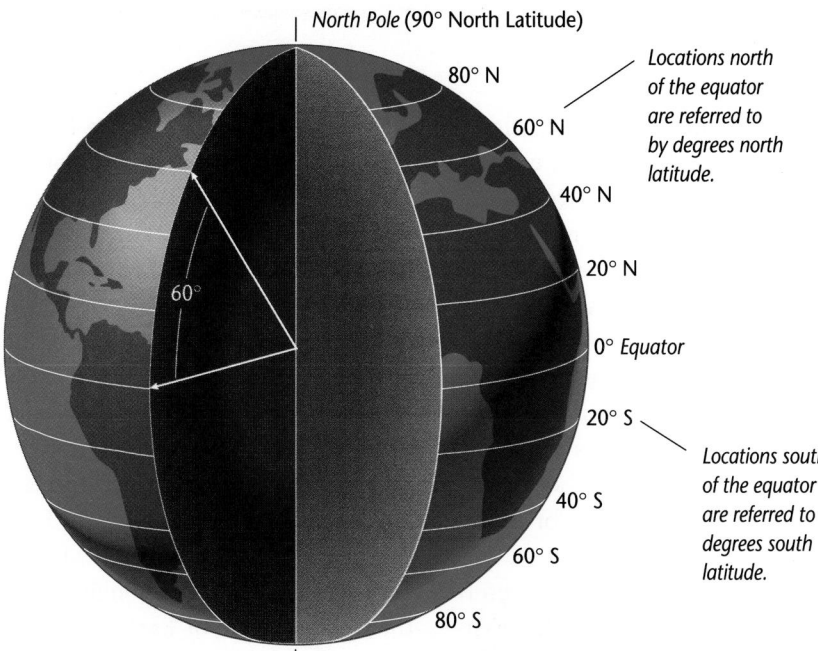

North Pole (90° North Latitude)

80° N

60° N

40° N

20° N

0° Equator

20° S

40° S

60° S

80° S

South Pole (90° South Latitude)

60°

Locations north of the equator are referred to by degrees north latitude.

Locations south of the equator are referred to by degrees south latitude.

Figure 1-6

The equator is 0° latitude. Each of the poles is 90°. Therefore, latitude is measured from 0° at the equator to 90° at the poles.

Cairo, Egypt is located at 30.03° north latitude.

separates Earth into two equal halves called the Northern Hemisphere and the Southern Hemisphere.

The lines circling Earth parallel to the equator are lines of latitude. **Latitude** refers to distance in degrees either north or south of the equator. These degrees are not like the degrees of temperature. Instead, the degree value used for the latitude of a place is the measurement of the imaginary angle created by the equator, the center of Earth, and the location of that place.

Look at **Figure 1-6**. What do you notice about the distance between each of the lines? All of the lines of latitude are parallel to each other. That means each line is always the same distance from the next line. Use a map or globe to find the approxi-

mate latitude of where you live. Follow that latitude around the world to find other places that are at the same latitude as you.

Activity Use this activity to introduce the concepts of latitude and longitude as reference points. Tell students to think of a specific place in their city or town. Have students write the name of the place on a slip of paper, then fold it. Pair the students. One student has five minutes to explain a route from the school to the chosen place to his or her partner without identifying the place. The student must use only directions such as right, left, 5 blocks, or 3 miles. The partner must discover the name of the place from the directions. The first student does not tell if the answer is correct at this point. The first student then gives directions using only reference points, such as next to the bakery on Main Street. Have the student reveal the chosen place, then have students switch roles. Discuss with students the importance of reference points in getting from one place to another. **COOP LEARN** **L2**

Visual Learning

Figure 1-6 Use this activity to help students understand latitude. Pairs of students will need $8\frac{1}{2}'' \times 11''$ paper, a compass, a ruler marked in $\frac{1}{4}''$ graduations, and scissors. Have students draw a heavy horizontal line halfway down the paper and label the line *equator*. Next, have them draw nine parallel lines above and below the "equator" line at intervals of $\frac{1}{4}$ inch. Have students label all of the lines up to 90°N and down to 90°S in 10° increments. Then, have students put the point of the compass in the center of the equator line and open the compass so that the pencil end touches 90°N. Have them draw a circle and then cut it out. Ask: **Which lines are shorter?** *those near the North and South poles* Ask: **Which is the longest line?** *equator*
L2 **COOP LEARN** **P**

ENRICHMENT

Research Have students interview a local geologist, hydrologist, city planner, or surveyor. Students should ask the person to bring in various types of maps of the local area and explain what they show and how they are made.
L1

1. Moscow, Russia
2. Cape Town, South Africa
3. Havana, Cuba
4. Bangkok, Thailand
5. Lisbon, Portugal **L1**

Visual Learning

Figure 1-7 Have pairs of students draw a horizontal line across the center of a sheet of paper, representing the equator. Label this line *E*. Then have students draw a line that bisects the equator at a right angle and label it *Prime Meridian*. Have students sketch six longitude lines on either side of the line. The curved lines should run from pole to pole and the two outside lines should form a circle. Have students cut along outside lines. Have students number each line to the right in 15° increments for the east longitudes and to the left for the west longitudes. When they finish, they should turn the model to the other side and draw a center longitude line and six curved lines again. Label the center line *International Date Line, 180°*. Label each longitude line west to the right and east to the left, subtracting 15° each time. Lead students to see that longitude lines are not parallel, and that they intersect at the poles. **L2**

COOP LEARN **P**

Figure 1-8 Have students locate Hawaii. **How could you describe its location without lines of latitude and longitude?** *by indicating how far and in which direction Hawaii is from some point of reference* **Which figure tells longitude?** *The larger value is longitude as indicated by its value and its direction (west).*

Longitude

Latitude lines are used for locations north and south of the equator, but what reference points do mapmakers use for directions east and west? Vertical lines called meridians indicate east and west directions.

Just as the equator is used as a reference point for north-south grid lines, there's also a reference point for east-west grid lines. This reference point, known as the prime meridian, is shown in **Figure 1-7**. In 1884, scientists agreed that the prime meridian should pass through the Greenwich Observatory near London, England. This imaginary line represents 0° longitude. **Longitude** refers to distance in degrees east or west of the prime meridian.

Look at **Figure 1-7** again. Notice that the prime meridian does not circle Earth as the equator does. Instead, it extends from the North Pole through Greenwich, England, to the South Pole. The line of longitude on the opposite side of Earth from the prime meridian—where east lines of longitude meet west lines of longitude—is the 180° meridian.

Interpreting Scientific Illustrations

Use a world map that shows latitude and longitude to identify the cities that have the following coordinates:
1. 56° N; 38° E
2. 34° S; 18° E
3. 23° N; 82° W
4. 13° N; 101° E
5. 38° N; 9° W
Compare your answers with those of your classmates. If you need help, refer to the **Skill Handbook** on page 667.

Figure 1-7

The degree value used for the longitude of a place is the measurement of the imaginary angle created by the prime meridian, the center of Earth, and the location of that place.

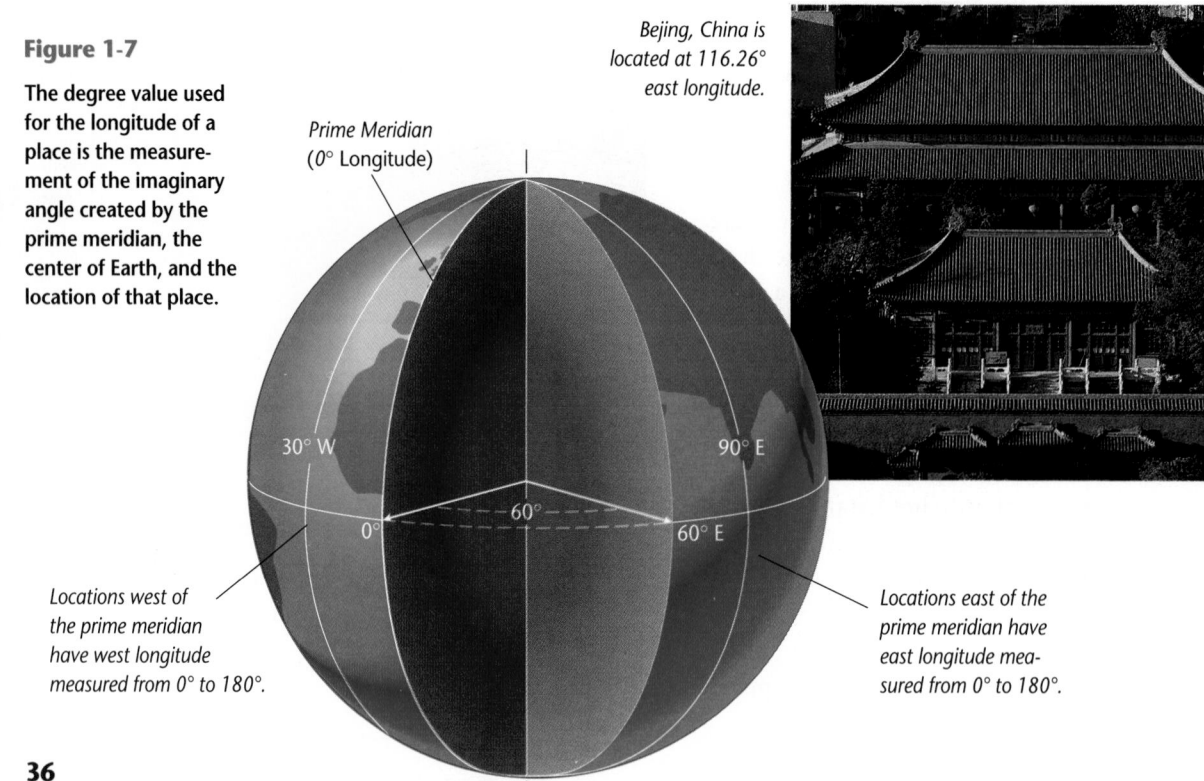

Bejing, China is located at 116.26° east longitude.

Prime Meridian (0° Longitude)

30° W

90° E

0°

60°

60° E

Locations west of the prime meridian have west longitude measured from 0° to 180°.

Locations east of the prime meridian have east longitude measured from 0° to 180°.

36

Meeting Individual Needs

Limited English Proficiency Have globes of Earth available for students to observe while you present the information on longitude and latitude. Have students use their index fingers to trace a single line from pole to pole and then from east to west. Lead them to conclude that lines of longitude converge at the poles whereas lines of latitude are equidistant. **LEP**

L1

Figure 1-8

A Hawaii is located at 20° north latitude and about 155° west longitude. How could you describe its location without lines of latitude and longitude?

Hawaii

B Diamond Head is a famous extinct volcano located in Honolulu, the capital city and chief port of Hawaii. Honolulu's location is 21.19° N and 157.52° W. Which figure depicts longitude?

Think about locating a spot on a ball again. Using the lines of latitude and longitude, you could locate that spot easily.

As you found out, maps can be informative and helpful if you know how to use them. Maps are tools for observing the world at a distance.

They are particularly helpful when you cannot visit or observe the locations directly. In this section, you have learned about road maps, topographic maps, and world maps with latitude and longitude. How do you think this information can be useful to you in the future?

check your UNDERSTANDING

1. Suppose you are looking for an easy hiking route through a particular area. As you look at a topographic map, you notice the area has many closely spaced brown lines. How would you describe the area? Would it be a good place for an easy hike?

2. How would a plateau be represented on a topographic map?
3. Which lines are parallel to the equator?
4. **Apply** How can the approximate elevation of a place be obtained by looking at the contour lines on a topographic map?

1-2 Using Maps **37**

check your UNDERSTANDING

1. The area is steep and not suitable for easy hiking.

2. A plateau would be represented by very few widely spaced contour lines because it is relatively flat. However, near the edge of the plateau, the lines would become close.

3. lines of latitude
4. by estimating the elevation between the labeled contour lines

1-2 Using Maps **37**

Student Text Question
How do you think this information can be useful to you in the future? *Accept all reasonable answers, which might include traveling, wanting to learn more about distant places, planning a hiking adventure, studying the climate of faraway lands, and so on.*

3 ASSESS

Check for Understanding

To help students answer the Apply question, have them draw a topographic map that shows a hill 900 m high, using a contour interval of 100 m.

Reteach

Demonstration It may help students who do not understand latitude and longitude to try to locate points on a grid. Prepare a graph with a hypothetical continent on it. Use letters to label the *x*-axis and numbers to label the *y*-axis. Label points on the graph with symbols and have students give the locations of the symbols. [L1]

Extension

Activity Direct students to look on a world map and write the latitude and longitude of 10 to 15 capital cities. Then have them find the latitude and longitude of the place in which they live. [L3]

4 CLOSE

Discussion

Discuss with students how flight and navigation of boats and planes would be different if there were no longitude and latitude grid system. **What might people use instead?** *stars, landmarks* [L1]

1-3

PREPARATION

Planning the Lesson

Refer to the Chapter Organizer on pages 20A–D.

Concepts Developed

Students will learn about the relative positions of the sun, the moon, and Earth in space and discover the use of star maps. This section relates to Chapter 19, The Earth-Moon System.

1 MOTIVATE

Bellringer

Before presenting the lesson, display **Section Focus Transparency 3** on an overhead projector. Assign the accompanying **Focus Activity** worksheet. ⬜L1

⬜LEP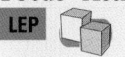

Activity Tell students that the moon is an average distance of 386 000 km from Earth. To put this in perspective, have them figure out the time it would take to drive there at 88 km/h. *182 d, 18 h, 21 min, 49 sec* ⬜L1

2 TEACH

Tying to Previous Knowledge

Many students may be familiar with the notion that the moon looks as if it has a face on it. Most students will know that craters, or depressions, make the moon appear to have eyes, a nose, and a mouth. As they learn about maria, have students describe how these large, dark areas on the moon add to the image of a face. Then have students describe the different shapes of the moon. Listen for terms like *crescent, full moon, half moon,* or *new moon.*

1-3 ## Viewing the Sky

Section Objectives

■ Describe the position and appearance of the sun and the moon as viewed from Earth.

■ Explain the use of a star map in locating stars in the sky.

Key Terms

phase
constellations

Look Up in the Sky! It's a …

So far, you've learned about many of the features of Earth's surface. It's fascinating to observe the beautiful rivers, plains, and mountains that shape the land we live on. Now, let's look a little higher. We're going to explore a couple of things you can find in the sky.

What things come to mind when you think about the sky? When you look up, do you see the same thing during the day as you do at night? Does the sky always look the same from day to day?

Puffy clouds, lightning, the sun, and the moon, as shown in **Figure 1-9**, are just a few of the things you might observe when you gaze upward. They are probably so familiar to you, that you don't always notice them. But how much do you really know about the sun and the moon? What causes their movements across the sky? Why does the moon appear to change shape?

Ancient cultures explained the movements and changes in the moon and sun through various myths and legends involving gods, heroes, and monsters. Today, through the observations of many astronomers, we know there are different reasons behind these phenomena.

Figure 1-9

Program Resources

Study Guide, p. 9
Making Connections: Integrating Sciences, p. 5, Looking to the Sky ⬜L2
Laboratory Manual, pp. 5–6, Time Zones
Section Focus Transparency 3

The Moon

On your cross-country plane trip, you would probably notice the sky as well as the landforms below. What things in the sky do you notice? Perhaps you see white clouds or the bright sun. At nighttime, you might observe the stars twinkling in the sky.

You might notice the position and shape of Earth's moon, too.

You've probably observed these objects in the sky before, but you may not have taken the time to notice how they change. Do the following activity to observe the changes in the moon.

Explore! ACTIVITY

What can you find out by observing the moon?

What To Do

1. Observe the moon at the same time every night for a week.

2. Record *in your Journal* as many observations as you can about the moon. Here are some questions you should try to answer: What does the moon look like? Describe it in detail. In what general direction do you see the moon? How does the moon's position change? Does the shape of the moon appear to change? Describe any changes.

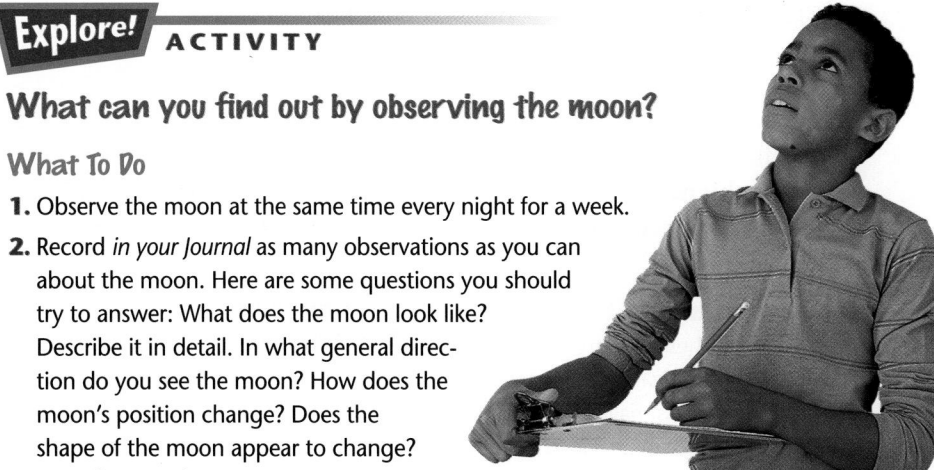

Because the moon is easy to see, it has been observed throughout human history. These observations have led to many interesting stories and superstitions about the moon.

What do you see when you look at the moon? Do you see a "man in the moon," as some people do? Look at the picture of the moon in **Figure 1-9**. You can see dark areas, which may sometimes create a pattern that looks like a face. What are these dark spots, and what are some of the other features that people have discovered on the moon?

1-3 Viewing the Sky **39**

1-2 Formation of Moon Craters

Preparation

Purpose To create craters in the laboratory setting to determine what factors influence their depth.

Process Skills making models, separating and controlling variables, forming a hypothesis, recognizing cause and effect, observing and inferring, measuring in SI

Time Required 30 minutes

Materials See reduced student text. Sand can be substituted for the flour-salt mixture. The string is for students who relate size or circumference to crater depth. This is not a direct factor unless density is taken into consideration.

Possible Hypotheses Students may hypothesize that a heavier ball will make a deeper crater or the farther the ball falls, the deeper the crater it makes. Students may also mention other factors.

The Experiment

Process Reinforcement Students must realize that only one variable can be tested at a time. If students choose to test the mass of the ball as a factor in the depth of a crater, then they need to measure the mass of the ball before they begin the tests. If they choose to test height, they need to drop the ball from premeasured heights. Measuring the craters is hard. Students can eyeball relative depth, use a small ruler, or cut an index card into the flour until it reaches the bottom of the pan. On the card, they can mark the depth of the craters and the depth of the undisturbed flour mixture and extract data from this.

Possible Procedures Experiment by dropping the same marble or ball from various heights. Pre-measure these heights and compare the craters accurately. If balls of different mass are used, measure the mass. Realize that the density or the mass per volume of the ball is a more precise measure than mass alone. To find density, use the equation Density = Mass/Volume. The volume of a sphere is $V = 4\pi r^3$, where $\pi = 3.14$ and r is the radius.

Teaching Strategies

Troubleshooting This lab can be messy. Have the students put newspaper over the lab tables.

Formation of Moon Craters

A crater is made when an object from space strikes the surface of Earth, the moon, or another planet, leaving a dent behind. Before you begin this investigation, make a few craters in a mixture of flour and salt as described under materials. Observe what happens when you drop different "celestial objects" from different heights.

Preparation

Problem
What factors determine how deep a crater will be?

Form a Hypothesis
A hypothesis is a prediction or explanation that can be tested. From your observations, have your group list factors that might determine how deep a crater will be. Each group should make a hypothesis about one factor. A hypothesis might be that a heavier ball will make a deeper crater.

Objectives
• Observe what factors are important in creating a deep crater.
• Analyze your data to find what factors determine the depth of a crater.

Materials
25 × 30-cm pan, one-third full of table salt and two-thirds full of flour
marbles and small balls
metric ruler
pan balance
string

Safety

Program Resources

Activity Masters, pp. 9–10, Design Your Own Investigation 1-2

Science Discovery Activities, 1-3, Star Search

DESIGN YOUR OWN
INVESTIGATION

Plan the Experiment

1 Your group should test only one hypothesis. List specific steps needed to test it. Each experiment should test only one variable. A variable is the one factor that changes in each experiment, such as the size of the marbles.

2 Examine the materials provided. Decide which balls and marbles to use.

3 Design data tables *in your Science Journal* or on a computer word processor or database. Record information about each object before you start the test. What is it? How heavy is it? What height will it be dropped from?

Check the Plan

1 In what order will you test the variables? Does your data table match the test order?

2 Make sure your teacher approves your experiment before you proceed.

3 Carry out your experiment. Record your observations.

The Barringer Meteorite Crater in Arizona is 1300 m in diameter and nearly 200 m deep.

Analyze and Conclude

1. **Observe and Infer** Was your hypothesis supported by the data? Use your data to explain why or why not.

2. **Interpret Data** What factors proved most important in determining how deep a crater could be? What factors did not seem to matter?

3. **Draw a Conclusion** From the information in question 2, make a general statement about how to make the deepest crater.

Going Further

Design a similar investigation to find out what determines the diameter of a crater.

Mythology

A northern Australian Aboriginal myth describes the paths of the sun and the moon through the sky. Wuriupranili, the Sun-woman, and Japara, the Moon-man, travel through the sky at different times carrying a torch made from flaming bark. When each of the creatures reaches the western skies, he or she extinguishes the flames and uses the smoldering ends of the bark to light the journey eastward. Have pairs of students act out the myth. L1 LEP

Discussion Tell students that the craters on the moon were formed as the result of impacts with meteorites. Ask them what they think a crater on Earth might look like.

Science Journal Make certain that the data tables are designed correctly. Each test should be numbered and each height or mass recorded before the ball is dropped. For more advanced classes, you can insist that students record the radius or circumference of every ball tested so that students can realize that density is a more precise indicator of crater depth than mass alone.

Expected Outcome

Students will demonstrate how impact craters can be formed by collisions between objects in space, and they should understand that the greater the mass (or density) of an object, the deeper the crater that forms. The greater the height from which an object falls, the deeper the crater.

Analyze and Conclude

1. Student data should support the hypothesis.

2. The mass (or density) of the ball and the height from which it was dropped were important factors. Color and texture of the ball were unimportant.

3. You can make the deepest crater by dropping the heaviest (most dense) ball from the greatest height.

Going Further

The diameter of a crater is affected by the circumference of the ball. It is also affected by mass because a big, lightweight ball will not sink far into the flour and therefore would not create a very wide crater.

✔ Assessment

Performance Have students graph the results of their tests. Then ask them to use their graphs to answer question 1. Select an appropriate Performance Task Assessment List in **PASC**.

Moon Features

If you could look through a telescope at the moon, you would observe depressions called craters. In fact, there are so many craters that some of them overlap. **Figure 1-10B** shows what these craters look like.

How is a crater formed? As you might guess, the craters on the moon were formed in much the same way as you formed craters in the Investigate. The craters were formed by large objects striking the surface of the moon. In space, rock fragments called meteorites sometimes hit other objects, such as the moon. Meteorites also strike Earth.

Few craters remain visible on Earth because they are worn away by wind, rain, and water. Because these forces do not exist on the moon, craters remain for very long periods of time. Some of the moon's craters are also very large—one moon crater has a diameter of 226 kilometers.

The large dark areas on the moon, called maria, are very flat, low-lying areas. Maria formed when very large meteorites collided with the moon, forming very large depressions that have been filled with lava. If water existed in great amounts on the moon, as it does on Earth, the maria would be covered by water.

Craters and maria are prominent features of the moon. Next, you'll read about one of the more interesting aspects of the moon—its phases.

Figure 1-10

A The moon's surface is covered by a layer of dust and rocks. This is a rock obtained during the *Apollo 16* lunar landing mission.

This lunar rock weighs 443 grams and measures 11 x 7.5 x 5 centimeters.

B Craters in the moon were formed by rock fragments, called meteorites, striking the surface of the moon.

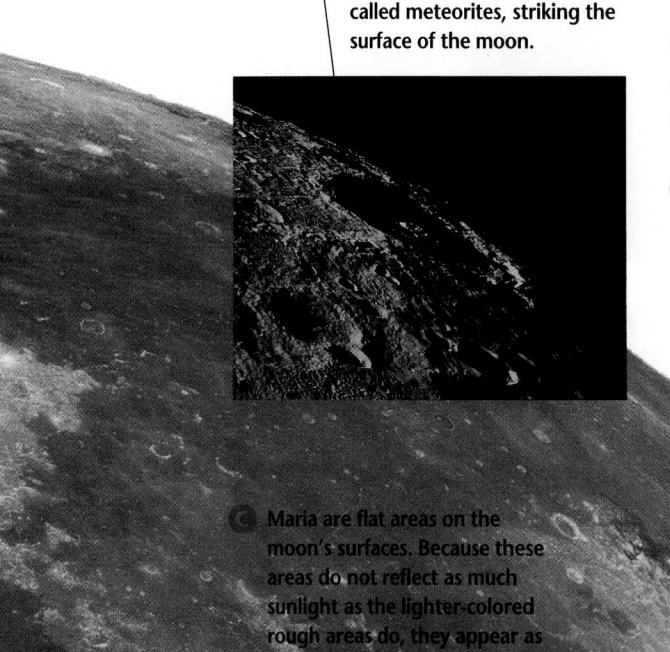

C Maria are flat areas on the moon's surfaces. Because these areas do not reflect as much sunlight as the lighter-colored rough areas do, they appear as dark spots when viewed from Earth.

D In July 1969, Neil Armstrong climbed down from the *Apollo 11* lunar module to become the first person to step onto the moon. In July 1971, astronaut David R. Scott, shown in the photograph to the right, was one of the first U.S. astronauts to walk on the moon.

Moon Phases

Sometimes the moon looks like a thin wedge of melon. At other times it looks like a silver disc. If you have observed the moon over a period of time, you have noticed such changes in its shape. You may have noticed that the changes occur in a cycle. That is, it slowly changes from a silver disc to a black disc and then back again to a silver disc.

Each stage in the cycle is known as a **phase**. Each cycle from full moon to new moon and back to full moon again takes a little over four weeks, or about a month. You'll discover in Chapter 19 why the moon appears to change. Until then, observe the moon periodically to see how it changes. Try to develop a hypothesis to explain your observations.

Figure 1-11

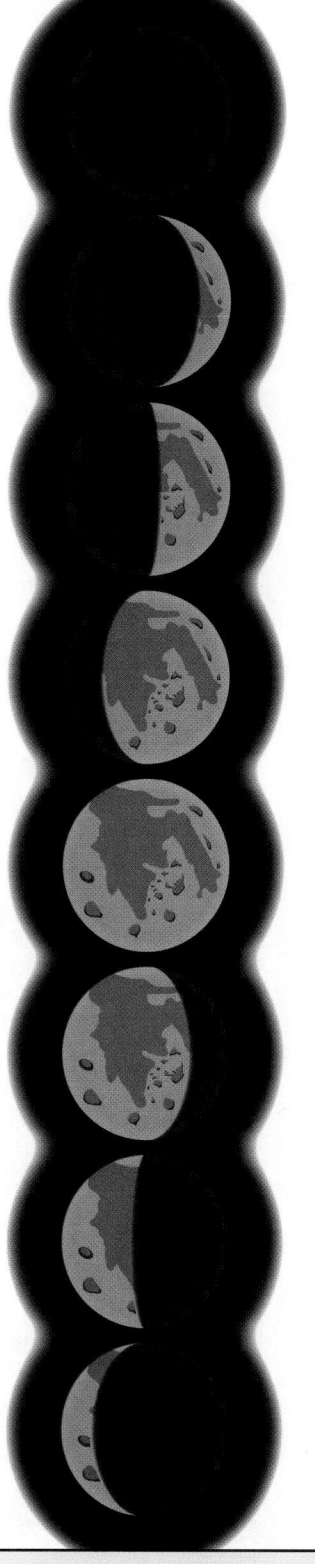

Ⓐ The moon has no light of its own. What we call moonlight is simply sunlight reflected from the moon's surface. As the moon revolves around Earth, various portions of the moon's surface appear to be lighted to an observer on Earth. During a new moon, none of the moon's surface is visible.

Ⓑ The lighted portion visible on Earth starts as a small slice and grows larger over two weeks until a full circle of light or full moon can be seen.

Ⓒ During the next two weeks, the circle of visible light grows smaller until no reflected light can be seen. Then the cycle begins again. The changing pattern of moonlight visible on Earth repeats every 29.5 days.

Demonstration To demonstrate how the moon's position in relation to the sun determines its phases, you will need a large polystyrene ball or all-purpose ball, a flashlight, and some aluminum foil.

Cover the ball with foil. Turn on the flashlight. Have a student hold the ball overhead. Turn off the lights and let students get accustomed to the darkness. Sit on a stool between the student and the class. Shine the flashlight directly on the ball. **What does the flashlight represent?** *sun* Adjust the distance so that the beam spot of the light and the circumference of the ball are the same. **Which phase of the moon does this represent?** *full* Have the student with the ball move around the "sun" while keeping the flashlight beam directly on the ball. Ask students to describe the phases they are seeing. [L2]

Uncovering Preconceptions Some students may think that the far side of the moon is not seen because it is dark. Explain that the side of the moon facing away from Earth receives as much light as the side that faces Earth. We never see the far side because the same side of the moon always faces Earth.

Across the Curriculum

History

In ancient times, people used the moon to mark the beginnings and ends of months. A calendar of months based on the moon's phases is called a lunar calendar. Native Americans had names for the moon at particular times of the year. Have students determine the amount of time of one moon month. *about 1 month; 29 days, 12 h, 44.05 min*

Multicultural Perspectives

Lunar Calendars
Some Middle Eastern and Far Eastern cultures use the lunar calendar. Obtain a calendar from one such culture and show it to students. Using the calendar, have students correlate the beginnings of the months to the new moon. Students will find that the beginnings of the lunar months coincide within two or three days of the new moon. They do not necessarily coincide with the months used in the United States. Students will also find that each month contains either 29 or 30 days and that there are sometimes 13 months in a lunar year.

The stars and moon are visible in the night sky. What do you see in the daytime sky? You see the sun, of course. How does the sun move during the day? Do the following activity to find out.

Explore! ACTIVITY

What can you learn by observing the sun?

What To Do

1. Observe the sun every hour or so from sunrise to sunset. **CAUTION:** *Do not look directly at the sun. The brightness of the sun will damage your eyes. Make general observations of the sun based on shadows or reflections.*

2. Record *in your Science Journal* as many observations as you can about the sun. Here are some questions you should try to answer. What are the shadows like at various times during the day? In what direction do you see the sun? How does the sun's position change?

Starry, Starry Night

Have you ever spent a summer afternoon gazing up at the clouds, imagining them as faces, or sailing ships slowly drifting by? If you have, you know it can be quite an entertaining display. But did you know there's an equally impressive show on a clear night? The night sky is full of wonders such as meteors and planets, and if you use your imagination, the stars can become an endless parade of people and objects.

Throughout human history, people have been enjoying the night sky. As they connected the stars with imaginary lines, the characters of myths and legends earned a place in the starry sky.

The patterns formed by groups of stars are called **constellations**. Cultures in China, the Middle East, Europe, and North and South America each have their own collection of constellations and legends to accompany them. A constellation may represent a hero, maiden, animal, or simply a common object such as a broken pot, or scales.

The patterns in the night sky have also served a more practical purpose. Sailors use them to easily recognize certain stars to help them in navigation. They relate the changing position of their ship to the position

The sun looks as if it rises in the east and sets in the west. It appears to move through the sky because Earth is spinning. As the seasons change in the Northern Hemisphere from summer, to autumn, and into winter, the sun appears to set farther south. Also, the sun doesn't get as high in the winter sky as it does in the summer, and the length of time that the sun is visible is shorter in the winter.

Viewing Earth and the sky is just the beginning of your study of the world around you. In the next chapter, you will find out about light—which helps you make observations.

check your UNDERSTANDING

1. Why does the moon appear to change its shape?
2. What is the difference in the position of the sun in the sky between winter and summer?
3. **Apply** You've just received a telescope and you want to view the full moon to see as many maria and craters as possible. Tonight there will be a new moon. About how long will you have to wait to observe the full moon with your telescope?

of certain constellations.

To learn what some constellations are, you need a star map. Just as maps of Earth are useful in locating cities, star maps are helpful in finding the locations of stars. Star maps show where stars, constellations and planets are at certain times of the year.

You Try It!

At night, go outside and look up at the night sky. Find a pattern of stars and record the pattern *in your Journal.* Now connect the stars in different ways, trying to imagine what the shapes could represent. Decide the one you like best and make a drawing of your constellation, and name it.

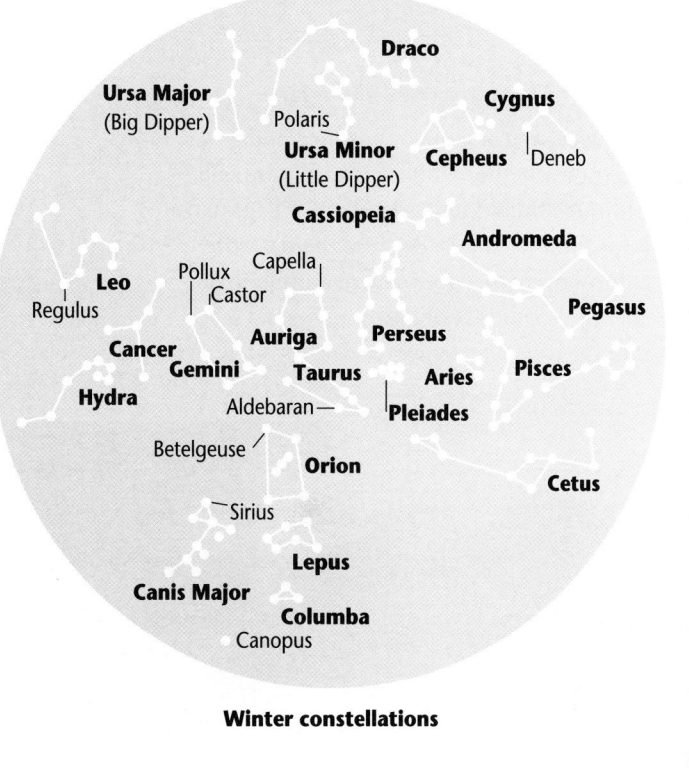

Winter constellations

because our sun is so close to us, it appears very bright and makes the other stars in the sky seem to disappear.

Going Further ▶

Obtain some maps of the stars visible in your area during the current season. Demonstrate how to use them. Make copies of the maps and have students take them home and use them to observe the night skies. Use the Performance Task Assessment List for Making Observations and Inferences in **PASC**, p. 17. **L2**

check your UNDERSTANDING

1. The moon's shape appears to change because we see different portions of its lighted side.

2. In winter, the sun is lower in the sky than in summer.

3. Since the time from new moon to new moon is about 29½ days, the time from new moon to full moon is about 14 or 15 days.

GLENCOE TECHNOLOGY

 Videodisc

The Infinite Voyage: Unseen Worlds

Introduction

Chapter 3
The Very Large Array: Monitoring Radio Waves

3 ASSESS

Check for Understanding

Assign Check Your Understanding questions 1 to 3 to groups of students. Be sure students understand that from new moon to new moon is about 1 month and that from new moon to full moon is about half that time. **COOP LEARN**

Reteach

Demonstration The mechanics of the moon's phases may be difficult for some students. Help students perform the ball and flashlight demonstration (see page 43). **L1**

Extension

Activity Have students look at the moon with binoculars and describe the moon's surface. Tell them to sketch a crater or two and compare it to the craters they made in the Design Your Own Investigation. **L3**

4 CLOSE

Activity Have students draw their zodiac constellation and connect the stars to form a "shape." Ask them to state whether they can see the shape described by ancient people. **L1**

P **LEP**

Science and Society

Purpose As an extension of the study of land contours in Section 1-2, this Science and Society excursion discusses sanitary landfills, an environmental problem that changes the topography of the land.

Content Background

Although landfills did solve some of the problems created by open dumps, landfill dumps may have been a shortsighted cure for solving solid-waste problems. Unfortunately, new environmental problems caused by sanitary landfills are just surfacing. Some materials in landfills may never completely decompose. Intact phone books and disposable diapers from twenty years ago can be dug up at some landfills. Hazardous materials leach into the soil below some landfills and permeate the groundwater.

Many municipalities are beginning to recycle materials like paper, plastics, metals, yard waste, and glass, cutting down on the volume of garbage dumped into landfills while preserving natural resources.

Teaching Strategy Have students, working in small groups, create posters or other displays that encourage people to reduce their solid wastes and to recycle. COOP LEARN P·

Science and Society

What To Do with All That Garbage?

Until the mid-1960s, most communities disposed of waste in the town dump—an open, smelly place. Today, the sanitary landfill has taken the place of the dump. It's more than just a fancy new name for a familiar place. Modern engineering and construction techniques are used to confine wastes to the smallest possible space. Even so, a sanitary landfill can cover acres of land.

Dangerous Dumps

Once seen as a major advance, today we question whether sanitary landfills are safe. Although buried under layers of soil, garbage is far from harmless. Because there is little control over what goes into a landfill, toxic chemicals and other hazardous wastes seep into the soil, threatening underground water supplies. This risk has closed many landfills, while many communities are refusing to allow landfills to be constructed near them. This has led to a shortage of sites for new landfills.

Using Computers

As a society, we produce huge amounts of garbage every day. Imagine you are part of a community that's forming on a small island. With a group, come up with a plan for dealing with the garbage your community will produce. Consider the long-term effects of your actions, such as the impact your solution has on food sources and water supplies. Also consider how much you think it might cost and whether your community would be willing to pay that amount. Prepare a word-processed report describing your plans.

The Fresh Kills Landfill in New York

Going Further ⫸

Have students work in small groups to make a list of items they personally discard as garbage. Have them expand their lists to include items that are regularly disposed of as garbage by other members of their household, by stores, and by restaurants. Then hold a class discussion of the types of items that appear most often on the lists. Most will probably be packaging materials and plastics. Have students suggest alternative methods of garbage disposal for these items which would not harm the environment. Use the Performance Task Assessment List for Group Work in **PASC**, p. 97. COOP LEARN L1

Technology Connection

Franklin Ramon Chang-Diaz

Franklin Ramon Chang-Diaz was born April 5, 1950, in San Jose, Costa Rica. As early as age seven he remembers wanting to go into space. In grade school, he wrote to Dr. Werner von Braun of the United States Space Program asking how to become an astronaut. Dr. von Braun wrote back that Franklin should come to the United States and study science. Franklin Chang-Diaz did just that in 1967. He moved in with distant relatives in Hartford, Connecticut. Chang-Diaz spoke no English when he arrived, and had no money, but he didn't let either of these things stop him. He attended public school to learn English. He studied hard in school to get a college scholarship and worked odd jobs to make extra money.

The "Right Stuff"

In 1973, Franklin Chang-Diaz received a degree in mechanical engineering from the University of Connecticut. Later, in 1977, he obtained a doctorate from the Massachusetts Institute of Technology (MIT) in applied plasma physics. After graduation, he worked on fusion reactor projects at a lab in Cambridge, Massachusetts. In 1978, he applied for astronaut selection but did not even receive a response. But Dr. Chang-Diaz was not about to give up. In 1980, he was selected as one of 19 astronauts admitted to the NASA program. He completed his mission specialist training in 1981 and began working on various research projects. On January 12, 1986, Franklin Chang-Diaz finally journeyed into space on the Space Shuttle *Columbia*.

Help Wanted: Astronauts

Dr. Chang-Diaz believes there is a need for astronauts who can do their own research as well as carrying out someone else's experiments. He sees space as an opportunity for everyone. "In 20 to 30 years, there will be all kinds of people in space and that makes a lot of sense."

*inter*NET CONNECTION

Use the World Wide Web to find out what research will be carried out on upcoming shuttle missions. Select a research problem that you would like to help solve.

Going Further ▸

Dr. Chang-Diaz has also helped develop closer links between the scientific community and the astronaut corps. Have students investigate the Astronaut Science Colloquium Program and the Astronaut Science Support Group, both of which were founded by Dr. Chang-Diaz. Students can make booklets or pamphlets displaying the information gathered. Use the Performance Task Assessment List for Booklet or Pamphlet in **PASC**, p. 57. L2

Technology Connection

Purpose
This Technology Connection examines a United States astronaut's goal to "view the sky," as discussed in Section 1-3.

Content Background
During the space shuttle flight, Dr. Chang-Diaz helped launch the SATCOM KU communications satellite and conducted experiments in astrophysics to learn more about the physical nature and origin of the universe. He spent a total of 146 hours in space on this flight.

He flew a second mission in October 1989, on the shuttle orbiter *Atlantis*. He helped send the *Galileo* spacecraft on its way to Jupiter, took radiation measurements, mapped atmospheric ozone, conducted research on lightning, and performed an experiment on ice crystal growth in space.

Teaching Strategy Emphasize to students that Dr. Chang-Diaz accomplished what he set out to do in spite of many obstacles. Ask students to list what qualifications they consider important for an astronaut and why Dr. Chang-Diaz would meet these qualifications. Have students list two or three of their own attributes that would make them "astronaut material."

*inter*NET CONNECTION

You can access information on upcoming shuttle missions at **http://shuttle.nasa.gov**

What is the largest crater on Earth?

Purpose

The moon's surface is marked by large depressions called craters, which are formed when meteorites collide with the moon. These SciFacts expand on the discussion about moon craters in Section 1-3 by letting students know that meteorites have helped shaped Earth's surface, too.

Content Background

Objects from space, such as comets and asteroids, rarely enter Earth's atmosphere intact. Instead, they tend to break apart into smaller pieces called meteoroids. A meteoroid that burns up in Earth's atmosphere is called a meteor—known to many as a shooting star. Some meteoroids are large enough to withstand Earth's atmosphere. When they strike Earth, they are called meteorites. Meteorites are responsible for the craters that form on the surfaces of planetary bodies.

Discussion

Ask students why they think that the same meteorite that created the Chicxulub crater may have led to the extinction of the dinosaurs. Some students may think that the impact itself killed off the dinosaurs. Encourage them to consider the long-term effects of the impact, such as climatic changes brought about by dust in Earth's upper atmosphere. This dust, in turn, blocked sunlight. Without sunlight, many plants and animals could not survive.

Science Journal

Students should mention pertinent details about the craters, such as name, location, and size. They should also discuss the meteorites that formed these craters. An example might be the meteorite called Hoba West, which struck Grootfontein, Namibia, centuries ago. Hoba West weighed approximately 60 tons.

NATIONAL GEOGRAPHIC SCIFACTS

Where is the largest crater on Earth?

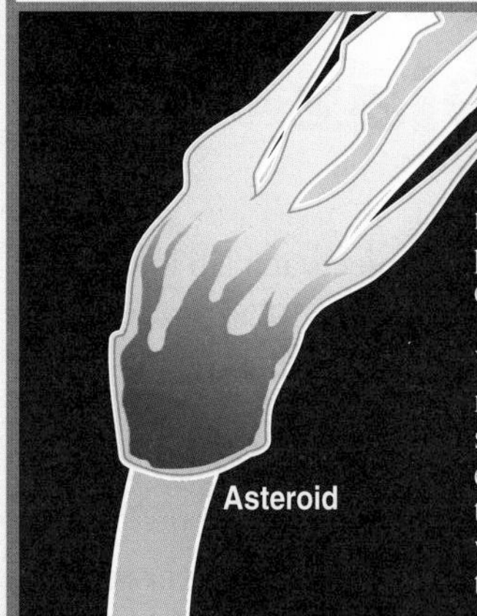

Buried on Mexico's Yucatán Peninsula is the Chicxulub crater, whose diameter measures 300 kilometers at its widest point. It is thought to be the largest crater on Earth.

Geologists hypothesize that the Yucatán crater was formed about 65 million years ago, possibly by the same asteroid or comet that led to the extinction of the dinosaurs. At the time of impact, the Gulf of Mexico was much larger than it is now, and that portion of the Yucatán Peninsula was a shallow sea.

Chicxulub crater was discovered in the early 1980s when geologists detected a circular pattern of gravitational irregularities beneath more than one kilometer of sediment. About 130 other major impact craters have been discovered on Earth, most of them in Canada, the United States, Europe, and Australia. Chicxulub's closest rival in size is the Sudbury crater, located far to the north in the Canadian province of Ontario.

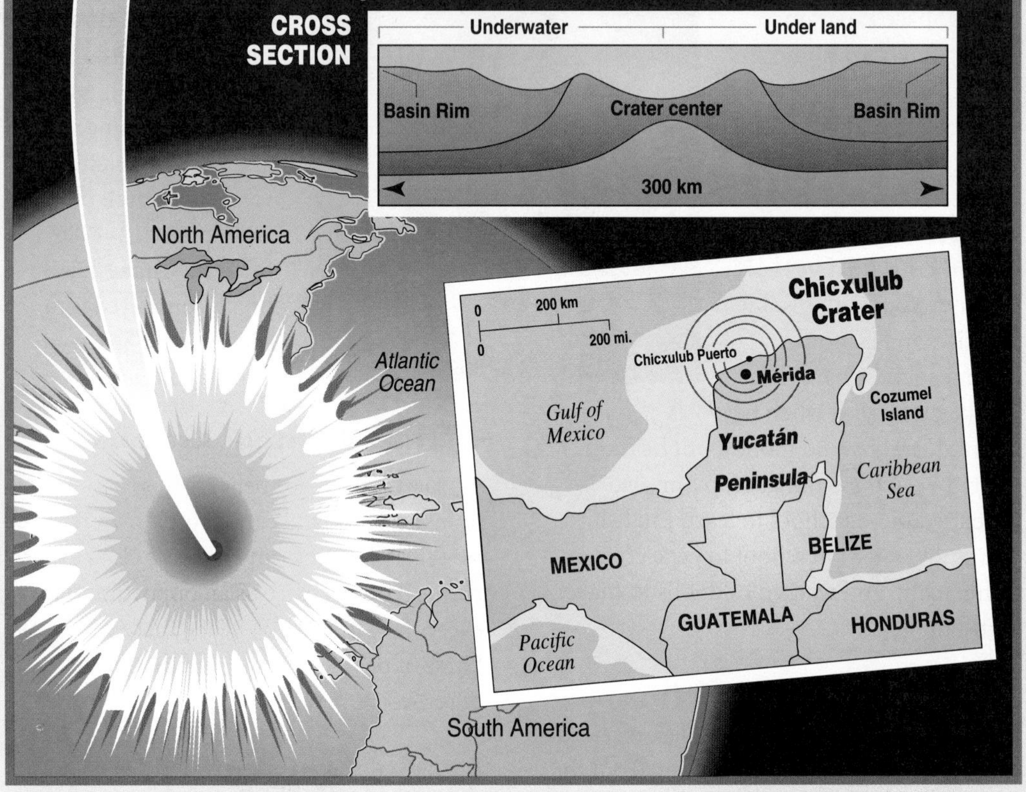

Asteroid

North America

Atlantic Ocean

South America

CROSS SECTION

Underwater		Under land
Basin Rim	Crater center	Basin Rim

300 km

Chicxulub Crater

0   200 km
0   200 mi.

Chicxulub Puerto
Mérida

Gulf of Mexico

Yucatán Peninsula

Cozumel Island

Caribbean Sea

BELIZE

MEXICO

GUATEMALA

HONDURAS

Pacific Ocean

Science Journal

Research information about meteorites that are thought to have caused major craters on Earth, and write your findings *in your Science Journal.*

Science Journal

Review the statements below about the big ideas presented in this chapter, and answer the questions. Then, re-read your answers to the Did You Ever Wonder questions at the beginning of the chapter. *In your Science Journal*, write a paragraph about how your understanding of the big ideas in the chapter has changed.

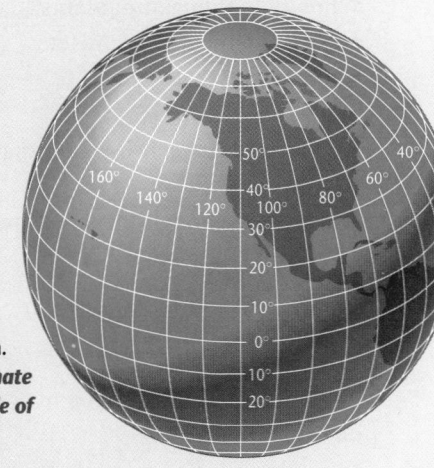

1 When we observe our surroundings, we see landforms such as mountains, plains, and plateaus. We also see the sun appear to move through the sky. *What landforms are found in your area?*

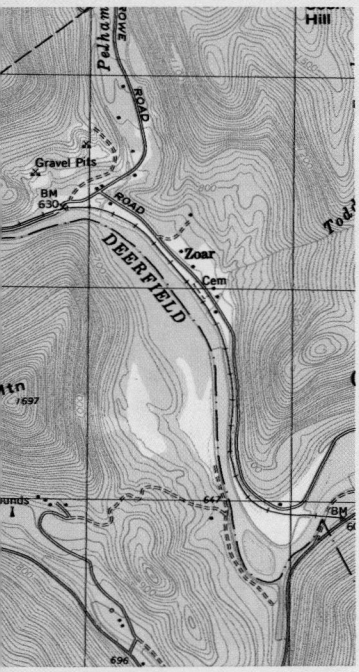

2 Topographic maps record our observations about landforms and their elevations through the use of contour lines. *What does it mean when the contour lines on part of a topographic map are very close together?*

3 Lines of latitude and longitude were created as reference points to help people locate places on Earth. *What is the approximate latitude and longitude of where you live?*

4 From Earth, we can observe large dark areas on the moon called maria. With a telescope, we can observe smaller depressions called craters. The moon's shape appears to change during a cycle that lasts about a month. *Why do craters remain on the moon for a very long time?*

Science Journal

Did you ever wonder...

• The Rocky Mountains are over 4000 meters above sea level; the highest mountain in the world, Mount Everest, towers 8800 meters above sea level. (p. 24)

• The moon's surface is marked by craters, rocks, dust, and large, flat regions called maria. (p. 42)

• The Big Dipper is a group of stars, or a constellation, shaped like a ladle. (pp. 44–45)

Project

Have half the class chart the moon's position for a week at the same time every night. Have the other half record the tides in the nearest ocean for a week, using the newspaper, *Farmer's Almanac*, or direct observation. Have them correlate their findings to explain the relationship between the moon and tides. *Students will discover that interactions among Earth, the sun, and the moon cause tides and most shores have two high tides and two low tides each day.* **L2** **COOP LEARN**

---

chapter 1
REVIEWING MAIN IDEAS

Have students look at the four pictures on this page. Direct them to read the statements to review the main ideas of this chapter.

Teaching Strategy

Divide the class into four equal groups and assign each group one of the illustrations. Have groups write a brief paragraph explaining what their photograph or diagram means. Students can read their paragraphs aloud or write them on the chalkboard. Sample responses include:

1. Earth and sky can be observed from the ground. During the day we see landforms like mountains, plateaus, plains, lakes, and streams on the ground and the sun in the sky.

2. Topographic maps give us a two-dimensional view of three-dimensional surfaces, such as landforms. Contour lines indicate elevations.

3. Latitude and longitude lines form a grid over the surface of Earth. They intersect and are reference points that can be used to locate positions on Earth. Latitude lines are counted from north-south and run east-west. Longitude lines are counted from east-west and run north-south.

4. Maria are large, flat areas of the moon's surface. Craters are formed when meteorites crash into the moon.

Answers to Questions

1. Answers will vary according to your location.

2. When contour lines are close together, the elevation is changing rapidly in that area, indicating there must be a steep hill or cliff.

3. Answers will vary depending on your location.

4. The agents of weathering that exist on Earth, such as wind and rain, are not found on the moon.

Using Key Science Terms

1. constellations
2. latitude
3. landforms
4. elevation
5. phases
6. longitude
7. contour lines

Understanding Ideas

1. mountains, plains, and plateaus; Mountains are landforms that extend above the surrounding land. They vary greatly in size and shape. Plains are large, low, mostly flat areas. Plains are seen along many coastlines and in the interiors of continents. Plateaus are raised flat areas with steep sides.

2. Contour lines are lines of equal elevation that show the topography of the landscape.

3. Latitude and longitude lines form an imaginary grid system that enables points on Earth to be exactly located.

4. Constellations are used to organize some stars as reference points in the sky.

Developing Skills

1. See the reduced student page for the completed concept map.

2. Answers will vary depending on the areas students have chosen.

3. Students should observe and record a sequence of the moon's phases for a two-week period.

GLENCOE TECHNOLOGY

MindJogger Videoquiz

Chapter 1 Have students work in groups as they play the Videoquiz game to review key chapter concepts.

Using Key Science Terms

constellations	latitude
contour lines	longitude
elevation	phase
landform	

An analogy is a relationship between two pairs of words generally written in the following manner: a:b::c:d. The symbol : is read "is to," and the symbol :: is read "as." For example, cat:animal::rose:plant is read "cat is to animal as rose is to plant." In the analogies that follow, a word is missing. Complete each analogy by providing the missing word from the list above.

1. Big Dipper: _____ ::Pacific:oceans
2. east:longitude::north: _____
3. houses:building::mountains: _____
4. degrees:temperature::meters: _____
5. innings:baseball game:: _____ :moon cycle
6. equator:latitude::prime meridian: _____
7. latitude and longitude:direction:: _____ :elevation

Understanding Ideas

Answer the following questions in your Journal using complete sentences.

1. List three common landforms and give a brief description of each.
2. Describe how contour lines are used on a topographic map.
3. How do latitude and longitude lines help describe location?
4. Explain how constellations are used.

Developing Skills

Use your understanding of the concepts developed in this chapter to answer each of the following questions.

1. **Concept Mapping** Using the following terms, complete the concept map of landforms: *mountains, plains, plateaus, Rocky, Great Plains*

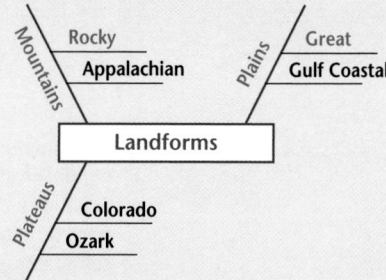

2. **Using Models** After doing the Investigate on pages 30-31, select a specific area in your community and create a topographic map of the area. Exchange maps with a classmate and try to identify this new location.

3. **Observing** After doing the Explore activity on page 39, repeat the activity for a two-week period observing and recording the moon changes.

Critical Thinking

In your Journal, answer each of the following questions.

1. Suppose you knew there was a new moon last night. What kind of a moon would you expect to see tonight? Why?

Program Resources

Review and Assessment, pp. 5–10 L1
Performance Assessment, Ch. 1 L2
PASC
Alternate Assessment in the Science Classroom
Computer Test Bank L1

2. Study the topographic map of a hill below. Then, fill in the blanks in this sentence: The elevation at the top of the hill is between _____ and _____. Explain your answers.

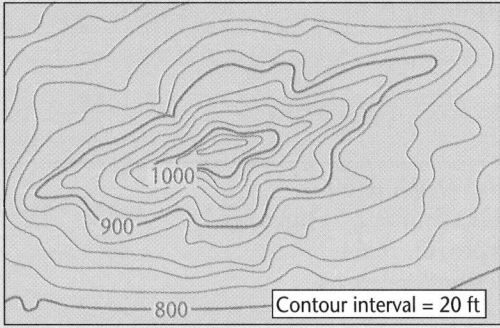

Contour interval = 20 ft

Problem Solving

Read the following problem and discuss your answers in a brief paragraph.

The North Star can be seen from most places in the United States in the night sky when you look north. At different latitudes, the North Star appears at different heights in the sky. The angle between the surface of Earth at your location and the position of the North Star is equal to your latitude. This angle can be found by using an instrument called a sextant. You can make a sextant with a protractor and straw. Hold or pin the straw so that it pivots at the center point of the protractor. Hold the protractor up close to your eyes, so its base is parallel to the ground. Sight along the straw to find the North Star. Keep the base of the protractor parallel to the ground.

1. Why do you think the base has to be horizontal?

2. What is your latitude?

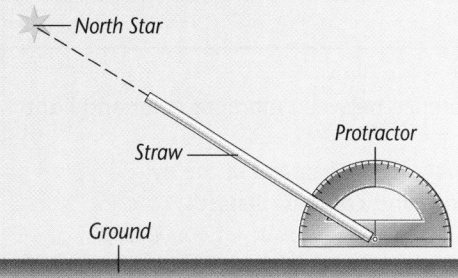

North Star

Straw

Protractor

Ground

CONNECTING IDEAS

Discuss each of the following in a brief paragraph.

1. Theme—Scale and Structure Suppose that you want to find the approximate elevation and location of a place. What kind of map would you use? Why?

2. Theme—Scale and Structure How could all latitudes in the continental United States be north and all longitudes be west?

3. Theme—Scale and Structure Two craters overlap. How can you determine which crater is the younger?

4. Science and Society What are two environmental problems caused by sanitary landfills?

5. Life Science Connection Name three factors that are important in determining where crops can be grown.

✔ Assessment

Portfolio Review the portfolio options that are provided throughout the chapter. Encourage students to select one product that demonstrates their best work for the chapter. Have students explain what they learned and why they chose this example for placement in their portfolios.

Additional portfolio options can be found in the following **Teacher Classroom Resources:**

Multicultural Connections, pp. 5, 6
Making Connections: Integrating Sciences, p. 5
Making Connections: Across the Curriculum, p. 5
Concept Mapping, p. 9
Critical Thinking/Problem Solving, p. 9
Take Home Activities, p. 6
Laboratory Manual, pp. 1–6
Performance Assessment P

Critical Thinking

1. The night after a new moon, a thin wedge is visible. The position of the moon in its revolution around Earth has changed so that a small part of its lighted side is visible.

2. 1040 feet and 1060 feet; With a 20 foot interval, the highest contour line on the map is at 1040. The small space in the center of the 1040 ft. contour line would then be under 1060 ft. in elevation because there is no 1060 ft. contour line.

Problem Solving

1. It has to be parallel to the ground because the angle to the North Star must be measured from a fixed reference line, in this case, the surface of Earth. If the measurement cannot be made on the ground, then it must be made from a parallel line above the ground.

2. Answers will vary with students' locations. Check a map for the latitude of the area.

Connecting Ideas

1. A topographic map would be used to find elevation and location. Topographic maps have both elevation as well as longitude and latitude shown on them.

2. The equator is to the south of the United States, thus, all latitudes in the U.S. are north latitude. The prime meridian is to the east, making all longitudes in the United States west longitude.

3. The younger crater overlaps the older crater. The younger crater's rim would be in the older crater.

4. Toxic chemicals and other hazardous wastes may seep into soil and groundwater. Hazardous gases may be produced as garbage decomposes.

5. Answers may include growing season, climate, type of soil, and landforms.

Chapter Organizer

SECTION	OBJECTIVES	ACTIVITIES & FEATURES
Chapter Opener		**Explore!**, p. 53
2-1 The Nature of Light (2 sessions, 1 block)	1. **Identify** how light travels. 2. **Distinguish** between objects that create light and those that only reflect light. 3. **Compare and contrast** opaque, translucent, and transparent materials. **National Content Standards: (5-8) UCP2, A1–2, G1–3**	**Find Out!**, p. 55 **Explore!**, p. 56 **Explore!**, p. 58
2-2 Reflection and Refraction (3 sessions, 1.5 blocks)	1. **Discuss** the different types of reflection. 2. **Describe** what happens to light during refraction. **National Content Standards: (5-8) UCP2, A1–2, G2**	**Find Out!**, p. 60 **Investigate 2-1:** pp. 62–63 **Explore!**, p. 64 **Skillbuilder:** p. 64
2-3 Color (3 sessions, 2 blocks)	1. **Examine** white (visible) light. 2. **Explain** the difference between pigment color and light color. 3. **Describe** the functions of the parts of the eye. 4. **Describe** how light and color are sensed. **National Content Standards: (5-8) UCP2, UCP5, A1–2, E1–2, F1, F5, G1–3**	**Explore!**, p. 66 **Find Out!**, p. 68 **Design Your Own Investigation 2-2:** pp. 72–73 **Science and Society**, p. 78 **Art Connection**, p. 79 **Technology Connection**, p. 80 **A Closer Look**, pp. 68–69 **Life Science Connection**, pp. 74–75

ACTIVITY MATERIALS

EXPLORE!

p. 53 flashlight, tape, drinking straw, construction paper, white paper

p. 56* 2 chalkboard erasers, flashlight

p. 58* flashlight, clay, white cardboard, piece of plate glass, piece of frosted glass, sheet of cellophane, book, waxed paper, sunglasses, tissue paper, piece of clothing such as your jacket, notebook, printed page

p. 64* opaque cup, water, pencil

p. 66* prism, white paper, flashlight, sunlight or light from a projector, crayons

INVESTIGATE!

pp. 62–63 4 pocket mirrors, flashlight, book

DESIGN YOUR OWN INVESTIGATION

pp. 72–73 scissors; glue, tape, or paper clips; poster board; art paper

FIND OUT!

p. 55* 3 index cards, drawing compass, light, metric ruler, clay

p. 60* aluminum foil

p. 68* 3 flashlights; green, blue, and red cellophane; 3 rubber bands; white paper

KEY TO TEACHING STRATEGIES

The following designations will help you decide which activities are appropriate for your students.

- **L1** Basic activities for all students
- **L2** Activities for average to above-average students
- **L3** Challenging activities for above-average students
- **LEP** Limited English Proficiency activities
- **COOP LEARN** Cooperative Learning activities for small group work
- **P** Student products that can be placed into a best-work portfolio
- Activities and resources recommended for block schedules

Need Materials? Call Science Kit (1-800-828-7777).

⏱ **OUT OF TIME?** We recommend that students do the activities with an asterisk.

Chapter 2 Light and Vision

TEACHER CLASSROOM RESOURCES

Student Masters	Transparencies
Study Guide, p. 10 **Making Connections: Integrating Sciences**, p. 7 **Making Connections: Across the Curriculum**, p. 7 **Making Connections: Technology and Society**, p. 7 **Laboratory Manual**, pp. 7-8, Producing a Spectrum	**Section Focus Transparency 4**
Study Guide, p. 11 **Concept Mapping**, p. 10 **Activity Masters**, Investigate 2-1, pp. 11–12 **Critical Thinking/Problem Solving**, p. 10 **Multicultural Connections**, p. 7 **Take Home Activities**, p. 7 **Science Discovery Activities**, 2-2, 2-3 **Laboratory Manual**, pp. 9–10, Refraction of Light	**Section Focus Transparency 5**
Study Guide, p. 12 **Critical Thinking/Problem Solving**, p. 5 **Activity Masters**, Design Your Own Investigation 2-2, pp. 13–14 **Multicultural Connections**, p. 8 **Science Discovery Activities**, 2-1 **Laboratory Manual**, pp. 11-14, Parts of the Eye	**Teaching Transparency 3**, Light Colors/Pigment Colors **Teaching Transparency 4**, The Eye **Section Focus Transparency 6**

ASSESSMENT RESOURCES	TEACHING & TECHNOLOGY
Review and Assessment, pp. 11–16 **Performance Assessment**, Ch. 2 **PASC*** **MindJogger Videoquiz** **Alternate Assessment in the Science Classroom** **Computer Test Bank**	**Spanish Resources** **Cooperative Learning Resource Guide** **Lab and Safety Skills** **Science Interactions, Course 1, CD-ROM** **Computer Competency Activities**

*Performance Assessment in the Science Classroom

NATIONAL GEOGRAPHIC TEACHER'S CORNER

National Geographic Society Products

To order the following products for use with this chapter, call the National Geographic Society at 1-800-368-2728:

- "The Sense of Sight," by Michael E. Long, November 1992.

National Geographic Society Products Available From Glencoe

To order the following products for use with this chapter, call the National Geographic Society at 1-800-368-2728:

- *Everyday Science Explained* (Book)
- *Messengers to the Brain* (Book)
- *Color: Light Fantastic* (Video)
- *Learning to See* (Video)
- *STV: Solar System* (Video)
- *Newton's Apple: Physical Sciences* (Laserdisc)

GLENCOE TECHNOLOGY

The following multimedia resources are available from Glencoe

Science and Technology Videodisc Series (STVS)
Physics
  Laser Eye Surgery
Physical Science CD-ROM

Teacher Classroom Resources

These Teacher Classroom Resources are examples of the materials in the Teacher Classroom Resources package.

TEACHING AIDS

Section Focus Transparencies

4 SECTION FOCUS TRANSPARENCY — Section 2-1

WHAT TIME IS IT?
Look at the photograph below. Although there aren't any clocks around, it is possible to infer at what time the photo was taken.

1. About what time of the day was the photograph taken? How do you know?
2. If the picture had been taken on a cloudy day, how would that affect your ability to answer the question above? Why?

L1

5 SECTION FOCUS TRANSPARENCY — Section 2-2

WHICH SIDE IS UP?
How do you think this photograph was made? If you guessed trick photography, try again. When the water in this lake is calm, a nearly perfect, upside-down image of the mountain reflects off of its glassy surface, much like a mirror.

1. Which side of the photograph is the reflected image? How can you tell?
2. Would the reflected image be the same if it were windy and the water were wavy? Why?

L1

6 SECTION FOCUS TRANSPARENCY — Section 2-3

A WORK OF ART
Great artists are masters at using color in their paintings. The colors they choose can create paintings of different moods. Using basic colors, an artist can mix paints to make whatever color he or she desires.

1. How many different colors are used in this painting?
2. What makes one color different from another?
3. What colors do you think the artist mixed to make the purple color in the flowers?

L1

Teaching Transparencies

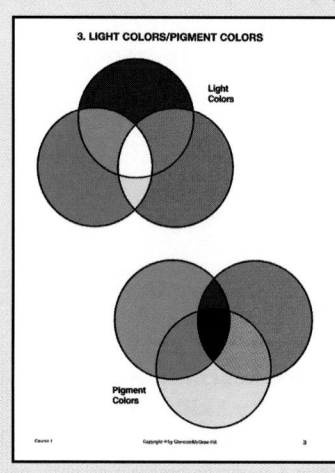

3. LIGHT COLORS/PIGMENT COLORS

Light Colors

Pigment Colors

L2

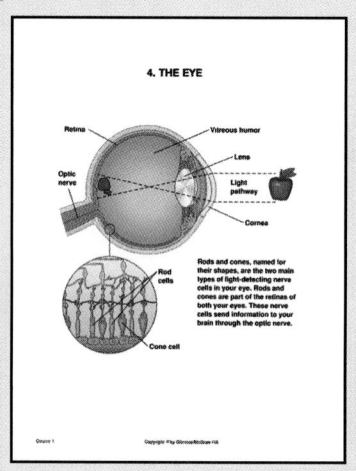

4. THE EYE

Retina · Vitreous humor · Lens · Optic nerve · Light pathway · Cornea

Rod cells

Cone cell

Rods and cones, named for their shapes, are the two main types of light-detecting nerve cells in your eye. Rods and cones are part of the retinas of both your eyes. These nerve cells send information to your brain through the optic nerve.

L2

HANDS-ON LEARNING

Science Discovery Activity*

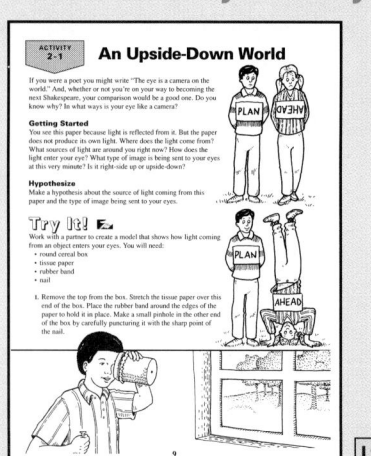

ACTIVITY 2-1 **An Upside-Down World**

If you were a poet you might write "The eye is a camera on the world." And, whether or not you're on your way to becoming the next Shakespeare, your comparison would be a good one. Do you know why? In what ways is your eye like a camera?

Getting Started
You see this paper because light is reflected from it. But the paper does not produce its own light. Where does the light come from? What sources of light are around you right now? How does the light enter your eye? What type of image is being sent to your eyes at this very minute? Is it right-side up or upside-down?

Hypothesize
Make a hypothesis about the source of light coming from this paper and the type of image being sent to your eyes.

Try It!
Work with a partner to create a model that shows how light coming from an object enters your eyes. You will need:
· round cereal box
· tissue paper
· rubber band
· nail

1. Remove the top from the box. Stretch the tissue paper over this end of the box. Place the rubber band around the edges of the paper to hold it in place. Make a small prehole in the other end of the box by carefully puncturing it with the sharp point of the nail.

9

L1

Laboratory Manual*

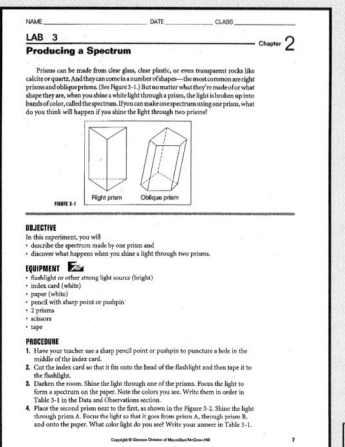

NAME _____ DATE _____ CLASS _____

LAB 3 — Chapter 2
Producing a Spectrum

Prisms can be made from clear glass, clear plastic, or even transparent rocks like calcite or quartz. And they can come in a number of shapes—the most common are right prisms and oblique prisms. (See Figure 3-1.) But no matter what they're made of or what shape they are, when you shine a white light through a prism, the light is broken up into bands of color, called the spectrum. If you can make one spectrum using one prism, what do you think will happen if you shine the light through two prisms?

Right prism   Oblique prism

FIGURE 3-1

OBJECTIVE
In this experiment, you will
· describe the spectrum made by one prism and
· discover what happens when you shine a light through two prisms.

EQUIPMENT
· flashlight or other strong light source (bright)
· index card (white)
· paper (white)
· pencil with sharp point or pushpin
· 2 prisms
· scissors
· tape

PROCEDURE
1. Have your teacher use a sharp pencil point or pushpin to puncture a hole in the middle of the index card.
2. Cut the index card so that it fits onto the head of the flashlight and then tape it to the flashlight.
3. Darken the room. Shine the light through one of the prisms. Focus the light to form a spectrum on the paper. Note the colors you see. Write them in order in Table 3-1 in the Data and Observations section.
4. Place the second prism next to the first, as shown in the Figure 3-2. Shine the light through prism A. Focus the light so that it goes from prism A, through prism B, and onto the paper. What color light do you see? Write your answer in Table 3-1.

7

L2

Take Home Activity

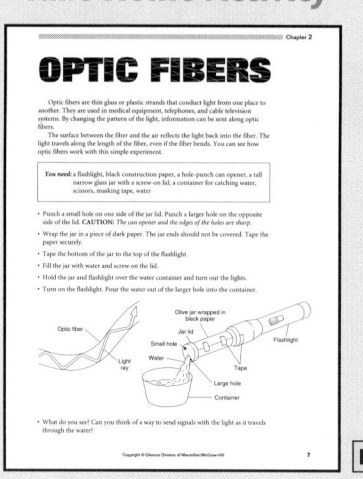

Chapter 2

OPTIC FIBERS

Optic fibers are thin glass or plastic strands that conduct light from one place to another. They are used in medical equipment, telephones, and cable television systems. By changing the pattern of the light, information can be sent along optic fibers.

The surface between the fiber and the air reflects the light back into the fiber. The light travels along the length of the fiber, even if the fiber bends. You can see how optic fibers work with this simple experiment.

You need: a flashlight, black construction paper, a hole-punch can opener, a tall narrow glass jar with a screw-on lid, a container for catching water, scissors, masking tape, water

· Punch a small hole on one side of the jar lid. Punch a larger hole on the opposite side of the lid. CAUTION: The can opener and the edges of the holes are sharp.
· Wrap the jar in a piece of dark paper. The jar ends should not be covered. Tape the paper securely.
· Tape the bottom of the jar to the top of the flashlight.
· Fill the jar with water and screw on the lid.
· Hold the jar and flashlight over the water container and turn out the lights.
· Turn on the flashlight. Pour the water out of the larger hole into the container.

Olive jar wrapped in black paper · Jar lid · Small hole · Tape · Optic fiber · Light ray · Water · Large hole · Flashlight · Container

· What do you see? Can you think of a way to send signals with the light as it travels through the water?

7

L1

Chapter 2 Light and Vision

Study Guide*

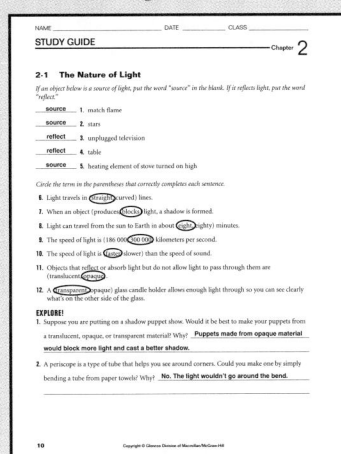

Concept Mapping

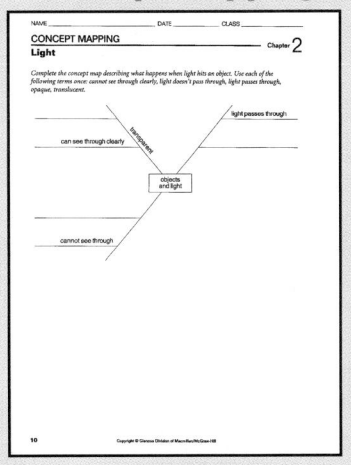

Critical Thinking/Problem Solving

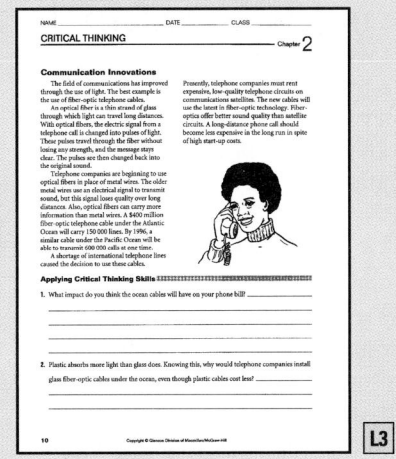

Integrating Sciences

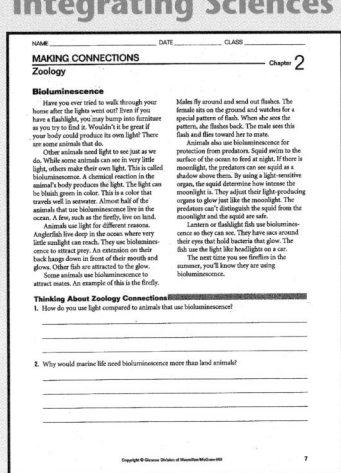

Across the Curriculum

Technology and Society

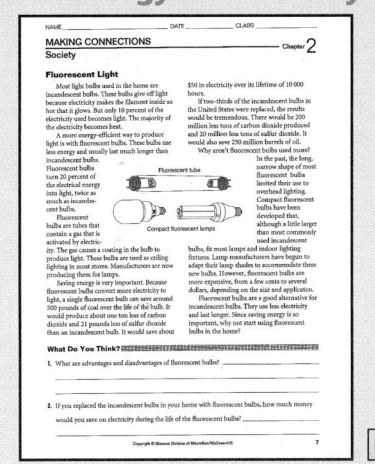

Multicultural Connection**

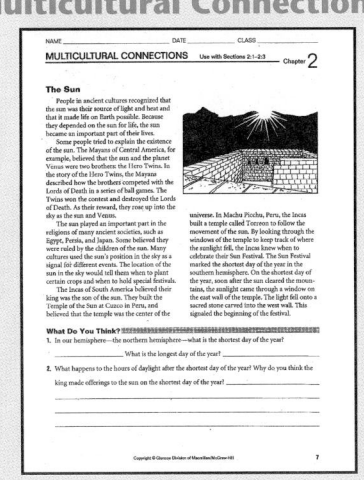

Performance Assessment

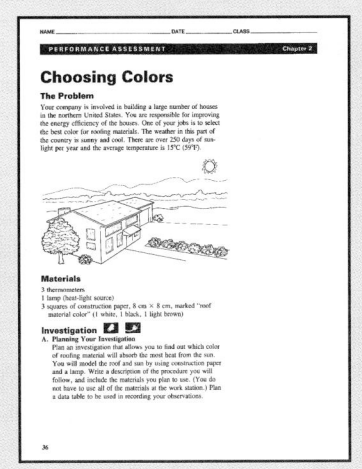

Review and Assessment

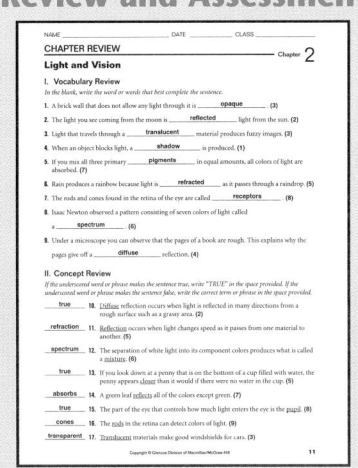

Light and Vision

THEME DEVELOPMENT

One theme that this chapter supports is systems and interactions. The properties of light interact with structures in our eyes so we see the world in a variety of ways. Light interacts with various objects to produce colors and reflections.

CHAPTER OVERVIEW

Students will learn how light allows us to see images. Students will compare and contrast objects that produce light and those that reflect it. Students will discover some properties of visible light and will explore the principles of light refraction and reflection.

Tying to Previous Knowledge

Have groups of students stand in different places in the room, with one group looking out the window. Ask each group to name the colors they see and to describe which objects they see casting shadows. Then ask what makes the colors and shadows visible. Students probably will suggest the sunlight or lightbulb shining on them. They may also suggest that shadows are produced when objects block the path of light.

INTRODUCING THE CHAPTER

Have students look at the photographs on page 52. Ask students to describe times when they have needed a flashlight.

00:00 OUT OF TIME?

If time does not permit teaching the entire chapter, use the Chapter Overview on this page, Reviewing Main Ideas at the end of the chapter, and the Chapter 2 audiocassette to point out the main ideas of the chapter.

CHAPTER 2
LIGHT and VISION

Did you ever wonder...

✓ **Why you can mix paints to get completely different colors?**

✓ **How long it takes for sunlight to travel to Earth?**

✓ **Why it's hard to distinguish different colors in moonlight?**

Science Journal

Before you begin to study about light and vision, think about these questions and answer them *in your Science Journal*. When you finish the chapter, compare your journal write-up with what you have learned.

Answers are on page 81.

*D*o you ever want a drink of water in the middle of the night? You grope for the flashlight with your hand in the dark room. Finally you find it, turn it on, and point the flashlight beam to find the door. The dresser is visible if you point the light beam at it, but as soon as you move the light toward the door, the dresser seems to disappear. The only objects you can see are the ones at which you point the flashlight beam.

Seeing any object requires light. Light may come from a flashlight, the light bulbs in your house, or from the sun. There are probably other sources of light in your home as well. Can you name them? How does light allow you to see the world around you? In this chapter, you'll learn about light and how it affects what you see.

▶ *In the activity on the next page, explore some characteristics of light and shadows.*

52

Learning Styles	Kinesthetic	Explore, pp. 53, 56, 64; Activity, p. 59; Investigate, pp. 72–73; Life Science Connection, pp. 74–75
	Visual-Spatial	Demonstration, p. 54; Activity, pp. 70, 73; Visual Learning, pp. 55, 57, 61, 67, 71; Find Out, pp. 55, 60, 68; Explore, pp. 58, 66; Investigate, pp. 62–63; A Closer Look, pp. 68–69
	Interpersonal	Across the Curriculum, p. 75; Activity, p. 77
	Intrapersonal	Science at Home, p. 81
LS	**Linguistic**	Multicultural Perspectives, p. 56; Visual Learning, pp. 59, 70; Activity, p. 65; Discussion, p. 67; Across the Curriculum, pp. 70, 74

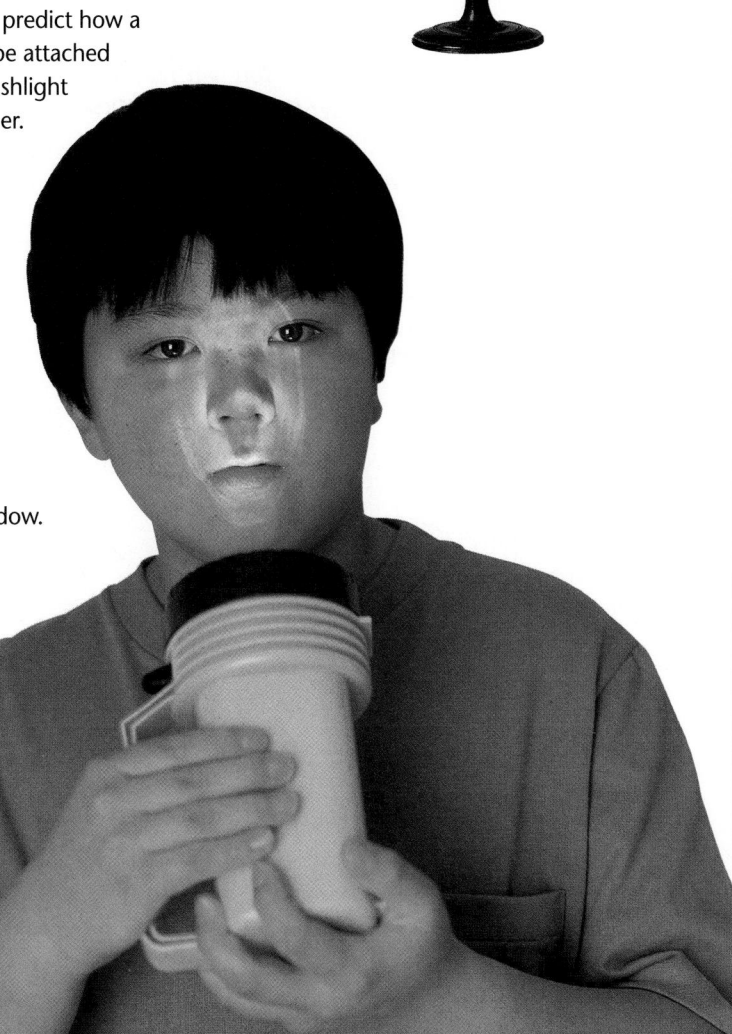

Explore! ACTIVITY

What do shadows tell you about light?

What To Do

1. Use a flashlight, tape, a drinking straw, construction paper, and white paper for this activity.

2. Cut a shape from construction paper and tape it to the end of a drinking straw. Tape the white paper to the chalkboard or wall. Dim the lights in the room.

3. *In your Journal*, draw a diagram to predict how a shadow will appear when the shape attached to the straw is an inch from the flashlight and when it is a foot from the paper.

4. Test your prediction.

5. Vary the distance of the shape between the flashlight and the paper.

6. How does the shadow change as the shape is moved closer to the paper?

7. *In your Journal*, draw a diagram to explain how the light is traveling from the flashlight to the paper in order for the shape to make a shadow.

Uncovering Preconceptions

Many students think that black is a color. However, true black is an absence of color, since a truly black object absorbs the light that strikes it and reflects no light back to the eye.

Explore!

What do shadows tell you about light?

Time needed 15 minutes

Materials flashlight, tape, drinking straw, construction paper, white paper

Thinking Processes recognizing cause and effect, observing and inferring

Purpose To observe how an object affects the path of light when placed in it.

Teaching the Activity

Science Journal Encourage students to sketch their predictions of how the shadow will appear at varying distances, and then sketch their results. Students should infer that the light is traveling in straight lines from the flashlight to the paper in order for the shape to make a shadow.

Expected Outcome

Students will observe that the farther an object is moved from the light source, the sharper the shadow will become.

Answer to Question

6. shadow becomes sharper

✔ Assessment

Content Working in small groups, have students demonstrate their knowledge of light and shadows by creating a cartoon that shows how distance from a light source affects a shadow. Use the Performance Task Assessment List for Cartoon/Comic Book in **PASC**, p. 61. **COOP LEARN** Assessment should be for all students. **L1**

PREPARATION

Planning the Lesson

Refer to the Chapter Organizer on pages 52A–D.

Concepts Developed

This section begins a study of the basic principles of light. Students will learn the difference between luminous and nonluminous objects. They will identify how light travels. Students also will compare and contrast opaque, translucent, and transparent materials.

1 MOTIVATE

Bellringer

Before presenting the lesson, display **Section Focus Transparency 4** on an overhead projector. Assign the accompanying **Focus Activity** worksheet. L1

LEP

Demonstration Select one object in the room, such as a large map. Have students look at it under ordinary lighting. Then pull the shades and have students look at the object as it is illuminated by several flashlights. Have students discuss what they saw. Students will observe that the amount of the object visible changes with the amount of light available. L1

2 TEACH

Tying to Previous Knowledge

Have students think about the difference between light and darkness. Ask them to discuss what they see when they turn on the lights in the morning and when they turn off the lights at night. Also ask them when they see shadows.

2-1 The Nature of Light

Section Objectives

- Identify how light travels.
- Distinguish between objects that create light and those that only reflect light.
- Compare and contrast opaque, translucent, and transparent materials.

Key Terms

reflection, opaque, transparent, translucent

Light and Reflection

Every day, all day long, light gives you information about the world around you. During the day, the sun is often your primary source of light. You also receive light from light bulbs, fire, and even from the television. Most things that you see are visible because light from somewhere else bounces off them. You can read this page because light is reflecting off it to your eyes. Light bouncing off something is **reflection**. You see your bed-room door when you shine a flashlight on it because the door reflects some of the light from the flashlight. You see your clothes in front of you in your closet because they reflect some of the light from your bedroom lamp directly back to your eye. You are unable to see these things when there is no source of light.

Exactly how does light travel from its source to objects and then to your eye?

Figure 2-1

Without light, there is no vision. You are able to see objects that are not sources of light because light is reflected from them.

B Some of that light reflects from the shirt directly to Anita's eyes. Her eyes send messages to her brain and she sees the shirt.

A Some light produced by the lamp falls upon Anita's shirt.

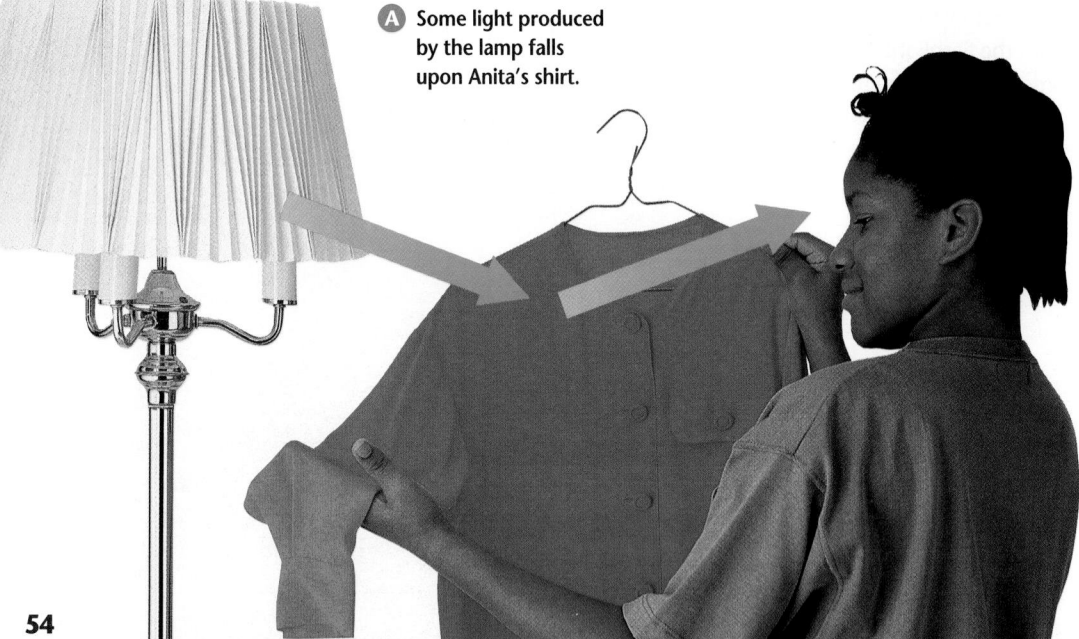

54

Program Resources

Study Guide, p. 10

Making Connections: Integrating Sciences, p. 7, Bioluminescence L2

Making Connections: Across the Curriculum, p. 7, In the Dark L2

Making Connections: Technology and Society, p. 7, Fluorescent Light L2

Laboratory Manual, pp. 7–8, Producing a Spectrum L2

Section Focus Transparency 4

How Light Travels

You saw from the Explore activity that when an object blocks light, a shadow is produced. The appearance of the shadow depends on the positions of the light source and the object blocking the light. When the object was closer to the light source, the shadow was large and fuzzy. When the object was closer to the paper, the shadow was smaller, darker, and sharper. But, a shadow was always produced. The light did not reach every part of the paper. Are shadows alone enough proof that light travels in straight lines? Look at the Find Out activity and **Figure 2-2** for more evidence.

Find Out! ACTIVITY

How does light travel?

It appears that light does not travel around corners. We can observe a beam of light to determine how it travels.

What To Do

1. Stack three index cards together. Carefully punch a hole through the middle of all three cards with the point of a compass.
2. Using small pieces of clay to support them, stand the cards about 2.5 cm apart, with the holes lined up.
3. Place a light bulb about 5 cm behind the hole in the back index card.

Conclude and Apply

1. What do you see when you look through the holes in the index cards?
2. Move the center card slightly out of line. What do you see now?
3. Based on what you have observed, *in your Journal*, describe how light travels.

Figure 2-2

Light travels in a straight line. When an object blocks the passage of light, the area behind the object is not lit—it is in shadow.

(A) The raised part of a sundial, the gnomon, blocks the sun's rays and casts a shadow on the sundial surface. Because Earth rotates, the position of the sundial in relation to the sun keeps changing as the day progresses.

(B) As the sundial's position changes, the angle of the shadow changes, marking the passing hours of the day.

55

Explore! ACTIVITY

What does a beam of light look like?

You can observe white light traveling from its source.

What To Do

1. Choose an area that is not brightly lit. Smack two chalky chalkboard erasers together.

2. Shine a flashlight into the cloud of chalk dust.

3. *In your Journal*, describe what the beam of light looks like.

4. Shine the flashlight in another area, away from the chalk dust, and observe the light beam. Why do you think you can see the beam of light better in the chalk dust?

5. *In your Journal*, suggest other properties of light that you observe.

When you observed the light beam in the chalk dust, you saw a straight beam of light. You could see the beam because it bounced or reflected off the chalk dust particles. It didn't matter how you turned the flashlight, the light beam always traveled in a straight line.

■ The Speed of Light

Whichever type of light source you have, it seems that light zips instantly from it to the object and back to you. It appears that you can see everything immediately. Light travels much faster than anything else we know of, but it does take some time for it to travel from place to place. How fast does light travel?

Throughout history, many people have tried to measure how fast light travels, but it wasn't until the late 1800s that scientists found that light travels at about 300 000 kilometers per second. That's fast enough to go from the sun to you in about eight minutes. It's so fast that a beam of light could cross your bedroom several million times before you can let go of the light switch.

Light at Different Speeds

Light travels through air at about 300 000 kilometers per second. Light travels through other materials at different speeds. The speed at which light travels through any material depends on the nature of the material. For example, light travels more slowly in water than in air. **Figure 2-3** illustrates the speed of light through several common materials. Using the information provided in the figure below, determine whether light travels faster through diamond or glass.

Figure 2-3

The speed of light is dependent upon the material through which it travels. Light travels faster through some materials than it does through others.

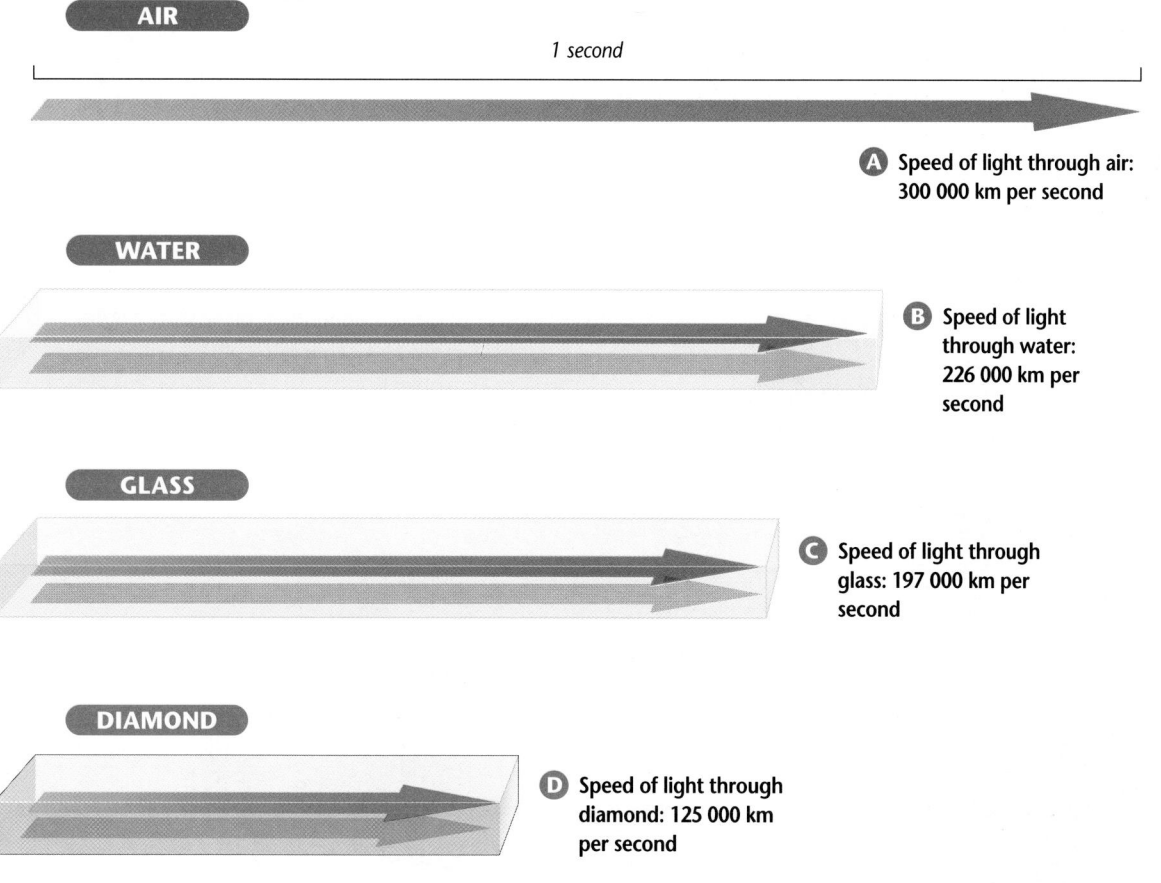

1 second

 A Speed of light through air: 300 000 km per second

B Speed of light through water: 226 000 km per second

C Speed of light through glass: 197 000 km per second

D Speed of light through diamond: 125 000 km per second

How Do We Know?

How was the speed of light first timed?

One of the most famous experiments to time light speed was performed in 1926 by an American, Albert Michelson, on Mount Wilson in California. Using two large mirrors 22 miles apart and a rotating mirror with eight sides, he was able to bounce light beams between the mirrors and make the most precise measurements of the speed of light ever made with mechanical methods.

Theme Connection The Explore and Find Out activities should show that light energy behaves in predictable ways. Point out that the basic nature of light rays is what makes our perception of color possible. It also accounts for the way mirrors work and the visual tricks of refraction. Stress that visible light, as well as other types of radiation, is actually a form of energy.

Visual Learning

Figure 2-3 To help students read and interpret graphs, ask the following questions: Through which material does light travel fastest? *air* Through which material does light travel most slowly? *diamond*

NATIONAL GEOGRAPHIC SOCIETY

 **Videodisc**

Newton's Apple: Physical Sciences

Sideview Mirror
Chapter 4, Side A
Comparing the paths of light rays

48324-49267

Meeting Individual Needs

Visually Impaired For students who have difficulty perceiving images, use a light source that produces heat. Explain that there is another way to demonstrate a shadow. Guide a student's hand close enough to a light source to feel the warmth. Have the student place a heavy piece of paper between the source and the hand. That would demonstrate shadow because the heat is less. L1

ENRICHMENT

Activity Provide students with old magazines. Have them cut out photographs that use shadows. Have them create a collage showing the various ways shadows are used in photography. Have students explain where the light source is in each photograph. L1

Explore!

What happens to light when it hits different objects?

TECH PREP

Time needed 10 minutes

Materials flashlight, lump of clay, and piece of white cardboard (for each student). Have objects available such as piece of plate glass, piece of frosted glass, sheet of cellophane, book, waxed paper, sunglasses, tissue paper, clothing, notebook, printed page

Thinking Processes observing and inferring, comparing and contrasting, recognizing cause and effect

Purpose To discover the differences among opaque, translucent, and transparent materials.

Preparation If materials are limited, have students work cooperatively in groups.

Teaching the Activity

Demonstration Shine a very strong light on a screen or large white paper. Hold up each object in front of the screen. **LEP**

Science Journal Suggest that students record their observations in chart form in their journals. In recording the characteristics of the three groups of objects, students may describe one group as impenetrable or dense, another group as cloudy or smoky, and the other group as clear. **L1**

Expected Outcomes

Students should report that opaque materials block light entirely; translucent materials allow some light to pass through; and transparent materials block almost no light.

✔ Assessment

Process Have students identify other classroom objects as transparent, translucent, or opaque. Students could then make posters showing what happens to light when it hits these objects. Use the Performance Task Assessment List for Poster in **PASC**, p. 73. **L1**

Opaque, Transparent, and Translucent

In the last three activities, you observed that light travels in straight lines. Several different things can happen when light hits an object. The nature of the object will determine what happens when light hits it. You've learned that light reflects off objects. The same objects also absorb some of the light that strikes them and can allow light to pass through. Collect several objects made of different materials for the Explore activity that follows to see what happens when light hits them.

Explore! ACTIVITY

What happens to light when it hits different objects?

Try shining light on different kinds of objects to see what happens to the light.

What To Do

1. Obtain a flashlight, a piece of clay, and a piece of white cardboard. Anchor the cardboard vertically on your desk using the clay.

2. Gather a variety of objects such as a piece of plate glass, a piece of frosted glass, cellophane, a book, waxed paper, sunglasses, tissue paper, an article of clothing, a notebook, and a printed page.

3. Select one object and place it in front of the cardboard. Shine the flashlight on the object and observe what you see on the white cardboard.

4. Record *in your Journal* what happened to the light as it hit each object.

5. Try sorting the objects into three groups depending on what the light did as it came into contact with each object.

Meeting Individual Needs

Learning Disabled You will need a page from a newspaper, a sheet of clear cellophane, a sheet of waxed paper, and a 15 cm × 15 cm piece of cardboard. Have the student put each of the materials over the newspaper one at a time and tell how well it can be read. Have the student label each item as opaque, translucent, or transparent. **LEP**

Figure 2-4 shows three different things that can happen when light hits an object. Light can pass directly through some objects. Other objects scatter light that passes through, while others completely block the light.

Objects that reflect or absorb light but do not allow light to pass through them are **opaque**.

When something allows enough light to pass through so that you can clearly see objects on the other side, it is **transparent**.

Any materials that let light through, but do not allow objects on the other side to be clearly seen are called **translucent**.

Whether the sun, a flashlight, or a candle is your source of light, you can see the objects around you. You can see yourself in a mirror, find your way with a flashlight, or see your family sitting at the dinner table because all of these objects reflect light back to your eyes. Light travels fast and in straight lines.

Figure 2-4

A This candle holder is opaque. It reflects or absorbs all of the light that hits it, allowing none to pass through. You can see light from the candle escaping from the open top, but not through the sides of this holder.

B This candle holder is translucent. It allows light through, but bends the light in so many different directions that all you see are fuzzy images. Are most things that you see opaque, transparent, or translucent?

C This clear glass candle holder is transparent. Like window glass, it allows light to travel directly through so that you clearly see what's on the other side of the glass.

check your UNDERSTANDING

1. Name two items not discussed in this section that you could see without reflected light.

2. The moon is about 387 000 kilometers from Earth. How long does it take light to get from the moon to Earth?

3. How do transparent, opaque, and translucent objects differ? Give an example of each.

4. **Apply** Imagine a star that is four light-years away. That is, it takes four years for light from the star to reach us. Do we see the star as it is now or as it was?

check your UNDERSTANDING

1. Examples of answers are fire in a fireplace, hot burners on a stove, luminous watch dials, or anything that creates its own light.

2. 1.29 seconds

3. Transparent objects, such as a window pane, allow most light that hits them to pass through. Translucent objects, such as melted margarine, allow only some light to pass through. Opaque objects, such as a book, allow no light to pass through.

4. as it was four years ago

3 ASSESS

Check for Understanding

To help students answer the Apply question in Check Your Understanding, ask them if the star exploded today, when would we see it? *in four years*

Reteach

Activity For students who have difficulty understanding the relation between the path of light and shadows, have a student stand in front of the classroom window with one arm stretched out to the side. Have the student raise and lower that arm while other students observe the change in the shadow cast. L1 LEP

Extension

Activity Have students who have mastered the concepts of light and shadow research how to set up a shadow theater. Then have them write and perform a short play using this technique. L3

4 CLOSE

Activity

Using connected light bulbs, have students put their hands close to the light source and then move slowly away. They should be able to tell you that the closer the light source and the reflective object are, the brighter the object will appear. Students will also notice a change in the shadow produced. LEP L1

PREPARATION

Planning the Lesson

Refer to the Chapter Organizer on pages 52A–D.

Concepts Developed

This section illustrates how a mirror works. It also focuses on the relationship between light and the texture of surfaces. Finally, an understanding of the refraction of light is developed.

1 MOTIVATE

Bellringer

Before presenting the lesson, display **Section Focus Transparency 5** on an overhead projector. Assign the accompanying **Focus Activity** worksheet. [L1]
[LEP]

Find Out!

How does light reflect on smooth and bumpy surfaces?

Time needed 5 minutes

Materials square of aluminum foil (about 15 cm × 15 cm)

Thinking Processes comparing and contrasting, observing and inferring

Purpose To demonstrate the difference between regular and diffuse reflection.

Teaching the Activity

Discussion To help students understand what they see in the crumpled foil, use Figure 2-5. Remind students that the eye sees only light rays directed at it. Draw an example on the board of how the eye perceives direct and diffused lines.

Reflection and Refraction

Section Objectives

- Discuss the different types of reflection.
- Describe what happens to light during refraction.

Key Terms

refraction

When Light Bounces and Bends

We've discussed the nature of light—light travels in straight lines; light can be reflected; and light can travel at different speeds through different materials. In this section, you'll take a closer look at what happens when light bounces or reflects and learn about another property of light. In the Find Out activity you'll learn more about reflection.

Find Out! ACTIVITY

How does light reflect on smooth and bumpy surfaces?

The type of surface reflecting the light is important when you are trying to see yourself. You can compare how light reflects off different surfaces.

What To Do

1. Observe your reflection in a piece of smooth aluminum foil.
2. Crumple the foil, spread it out, and observe your reflection again.

Conclude and Apply

1. Which piece of foil reflected more clearly?
2. How does the reflection on the smooth foil differ from the reflection on the crumpled foil?
3. What do you think happens to light that is reflected from any rough surface?

Figure 2-5

Ⓐ When light reflected from an object reaches a mirror, the smooth surface of the mirror reflects light back to your eyes in orderly, straight lines. If you look at a mirror, such as the mirror-tiled surface of the building on the far right of this photo, you see a reflected image of everything facing the mirror.

Ⓑ When light reflected from an object reaches an uneven surface, the light scatters, bouncing every which way. If you look at the rough stone walls of the older building on the left side of the photo, you see no reflected image. Why?

60 Chapter 2 Light and Vision

Science Journal Have students record their responses to Conclude and Apply questions in their journals. [L1]

Expected Outcomes

Students should be able to explain the difference in reflection from a smooth and from a crumpled surface.

Conclude and Apply

1. the smooth piece

2. The reflection on the smooth foil is clear; the reflection on the crumpled foil is diffuse.

3. The reflection is diffused by the rough surface.

✔ Assessment

Content Have students make their own scientific drawing showing how light reflects off smooth and bumpy surfaces. Use the Performance Task Assessment List for Scientific Drawing in **PASC**, p. 55. [L1] [P]

Regular Reflection

If you were holding a mirror in place of this book, some of the light around you would reflect off you into the mirror, then bounce straight back to you. When a reflection is very clear and looks just like the object, it is called a regular reflection. Smooth surfaces, such as mirrors, produce regular reflections.

Diffuse Reflection

When you crumple a piece of foil, the foil changes from one mirror into thousands of tiny mirrors. Each surface reflects light, but the light bounces in many different directions. The crumpled foil produces a diffuse reflection.

Why can't you see yourself on the smooth surface of this page? If you were to look at this page under a microscope, you would see that the surface of this paper is extremely rough. Light reflecting off this page is also producing a diffuse reflection.

Figure 2-6

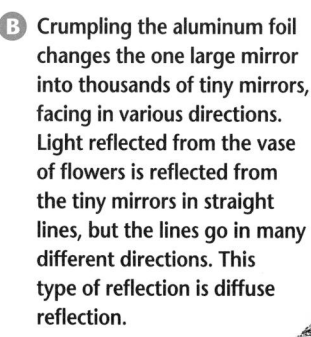

Ⓐ Smooth aluminum foil is a mirror. When light is reflected from the vase of flowers to the foil, the foil reflects the light to your eyes in orderly, straight lines. You see a reflected image of the vase and flowers. The reflection produced by the smooth foil is regular reflection.

Ⓑ Crumpling the aluminum foil changes the one large mirror into thousands of tiny mirrors, facing in various directions. Light reflected from the vase of flowers is reflected from the tiny mirrors in straight lines, but the lines go in many different directions. This type of reflection is diffuse reflection.

Ⓒ Some diffuse reflections produce no reflected image at all. When light reflected from the flowers hits the relatively rough white paper, the light scatters in many more directions and does not produce a reflected image.

2-2   Reflection and Refraction   **61**

INVESTIGATE!

2-1 Mirror Reflections

Planning the Activity

Time needed 20–25 minutes

Purpose To observe how mirrors reflect light.

Process Skills observing and inferring, predicting, using spatial relationships

Materials 4 pocket mirrors, flashlight, book

Preparation Have students bring square or rectangular mirrors to class for this activity.

Teaching the Activity

Troubleshooting Before darkening the classroom, you may wish to have partners read the procedure for using the flashlight and mirrors. Make sure that students conduct the investigation in a darkened room.

Process Reinforcement To get a preliminary feel for how mirrors reflect light, students could prop up two of the mirrors at an angle to each other. They could then shine the flashlight, so that it hits one mirror and reflects off the other, at different angles. You may want to have students form a hypothesis about how mirrors reflect light.

Science Journal Encourage students to write descriptions and draw diagrams as they record their observations in their journals. Then have students record their answers to the questions in their journals.

Mirror Reflections

When you see light reflected from a mirror, you see an entire object—your head, a car, or a chair, for example. You probably don't think about what happens when light reflects in a mirror. You can easily discover how mirrors reflect light in this activity.

Problem

How does light reflect in a mirror and how can this be used?

Materials

4 pocket mirrors   book
flashlight

What To Do

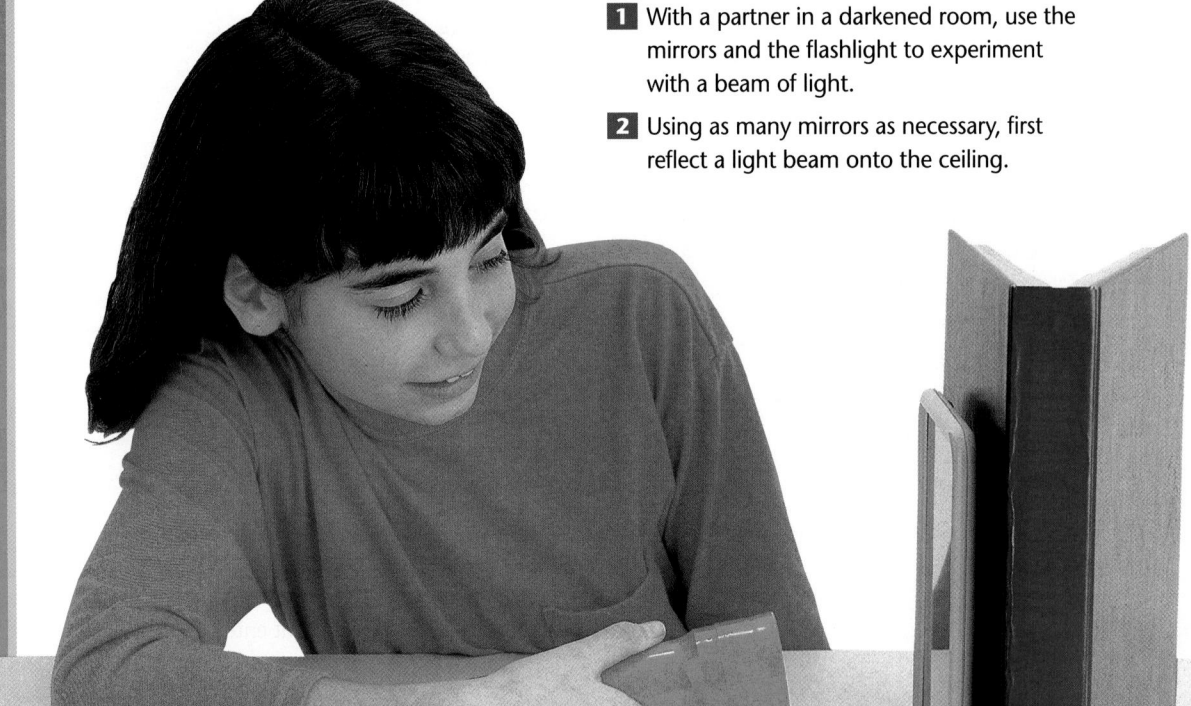

1 With a partner in a darkened room, use the mirrors and the flashlight to experiment with a beam of light.

2 Using as many mirrors as necessary, first reflect a light beam onto the ceiling.

Program Resources

Science Discovery Activities, 2–2, An Upside-Down World, 2–3, Waterscope Wonders

Activity Masters, pp. 11–12, Investigate 2-1

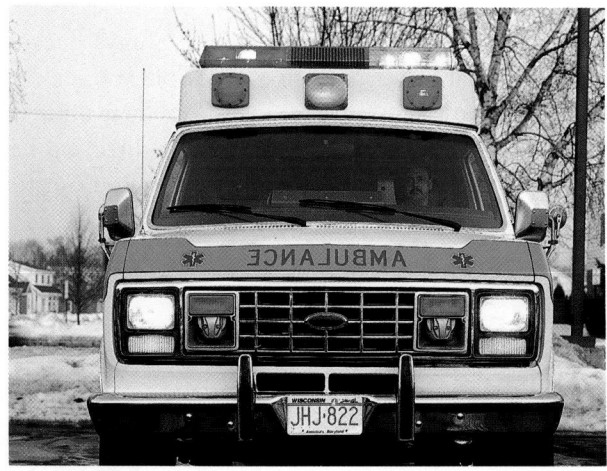

Why do you think that the front of some emergency vehicles, like this one, have backwards writing on the front?

3 Now, place a book upright on a desk.

4 Position the mirrors so that a light beam striking a mirror placed in front of the book is reflected to the back of the book.

5 Now, use your mirrors to reflect light into another room.

6 Using descriptions and diagrams, record the different positions of the mirrors, the flashlight, and the light beam for each trial *in your Journal.*

Analyzing

1. Make a statement that tells how light is reflected from a mirror.

2. What must you do to make the light beam change direction?

3. If you reversed the positions of the flashlight and the point at which the reflected beam strikes an object, how would the path of the light be affected?

Concluding and Applying

4. When might it be necessary to bounce light with mirrors? Name some situations in which a mirror would be more convenient than a light source.

5. *Predict* how you would arrange your flashlight and mirrors to get an image of the flashlight that would continue to be reflected from one mirror to the other. Write your prediction *in your Journal* and then try it.

6. **Going Further** Construct a periscope to see around corners or over fences using what you have learned in this activity. Draw a diagram to show how the light enters the periscope and reaches the viewer's eye.

✔ Assessment

Portfolio/Performance Have students place the diagram of the periscope that they made in Going Further (see reduced student text) in their portfolios. Then have groups of students write and perform a skit in which a mystery is solved using mirror reflections. Use the Performance Task Assessment List for Skit in **PASC**, p. 75. **P**

L1

Expected Outcomes

Students should observe that when light hits a mirror, it bounces off in straight lines. They may infer that the light hits a mirror and bounces off a mirror at the same angle.

Answer to Student Text Question Drivers looking into their rearview mirrors will be able to read the backwards writing.

Answers to Analyzing/ Concluding and Applying

1. Light is reflected from a mirror at the same angle which it strikes the mirror.

2. Cause the beam to be reflected from a mirror

3. No change in the light's path, but direction would be opposite.

4. Bathrooms, clothing stores, makeup counters, oldstyle signal system, telescopes, binoculars, some types of lasers. Anytime there is no straight-line path to where light is needed.

5. Place two mirrors parallel and facing one another. Put the flashlight between them.

6. Mirrors in a periscope face each other, are parallel, and are placed at 45° to the long axis of the periscope. Light enters the periscope, is reflected down the long axis, and then reflected to the user's eye.

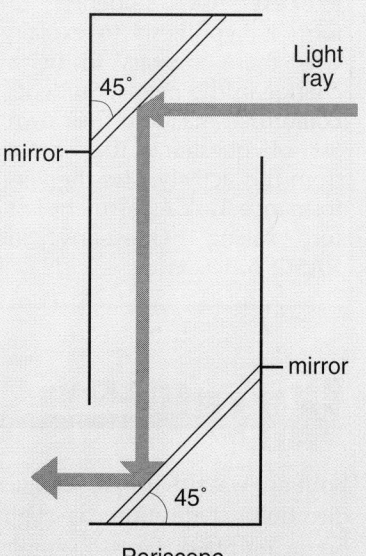

Light ray

45°

mirror

mirror

45°

Periscope

Teaching the Activity

Troubleshooting Do not fill the glass entirely.

Science Journal Students will record in their journals that the pencil appears broken as they move away from the cup. Light changes speed as it passes from water to air and from air to water. Students may infer that we see the effects of this change in the speed of light as a change in the pencil—which seems to bend. L1

Expected Outcome

Students will see that the pencil appears broken at the water line.

Answer to Questions

3. no difference
4. The pencil looks different, as though in two parts.

✔ Assessment

Process Ask students to form a hypothesis to explain why there appears to be a change in the pencil. Students could then make up their own list of questions that arose from this activity. Use the Performance Task Assessment List for Asking Questions in **PASC**, p. 19. L1

Refraction

We started out this chapter by talking about how light always travels in straight lines. Now you've made light change direction by bouncing it off objects—by reflecting it. This doesn't mean that we were wrong.

Light always does travel in straight lines, but you can change its direction.

Are there other ways to make light change direction besides reflection? You can find out for yourself in the next activity.

Explore! ACTIVITY

Why does light refract?

What To Do

1. Fill an opaque cup with water.

2. Place a pencil in the cup. The pencil should extend out of the water and the cup.

3. Stand directly over the cup and observe the pencil in the area where it leaves the water. What do you see?

4. Continue to observe the pencil and the water as you slowly back away from the cup. How does the pencil look now?

5. *In your Journal*, describe how the pencil appeared to change as you moved away from it.

SKILLBUILDER

Comparing and Contrasting
Compare and contrast reflection and refraction. Provide an example of each. If you need help, refer to the **Skill Handbook** on page 659.

In the Explore activity, you saw that your view of a pencil changed as you moved away from the water. We know that light changes speed as it moves through different substances. When light moves through water and then through air to your eyes it changes speed and direction.

Refraction is the process of bending light when it passes from one material to another. Imagine a toy truck being pushed on a hard surface. The right wheels encounter a rug and the truck suddenly slows and turns to the right, toward the slower wheels. When light rays pass from one substance to another and change speed, they turn or bend like the toy truck. The light rays reflected from the pencil bent as they moved

---

SKILLBUILDER

Both are ways that light changes direction. Reflection is light bouncing off surfaces. Example: a mirror. Refraction is light bending when it travels from one medium to another. Example: the pencil in water in the Explore. L1

Meeting Individual Needs

Visually Impaired Use flex straws to help visually impaired students understand why light refracts. Fill a cup of water almost full. Put in one straw. Cut off the straw at the bottom so that the flex stands at the water line. Flex the straw so that it no longer appears to bend where it enters the water.

Explain to students that the straws will feel the way the pencil in water looks. Have students compare by touch a straight straw and the bent straw.

from the water to the air. They changed direction by changing the angle of the straight line. The light did not travel in a curved pathway.

Light travels in straight lines. You can change its direction by reflecting it with mirrors or refracting it by passing it through different materials. The way light reflects and refracts determines how you see the world.

Figure 2-7

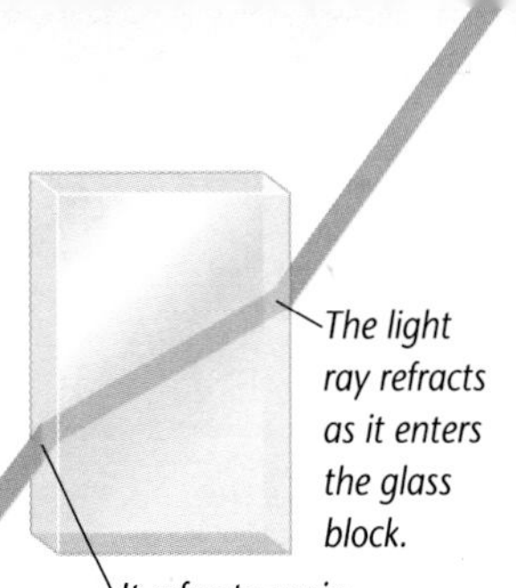

A The beam of light passing through this glass block refracts when it enters the glass and again when it leaves. Notice that the angle of the line in which the light travels has changed but the light still moves only in a straight line.

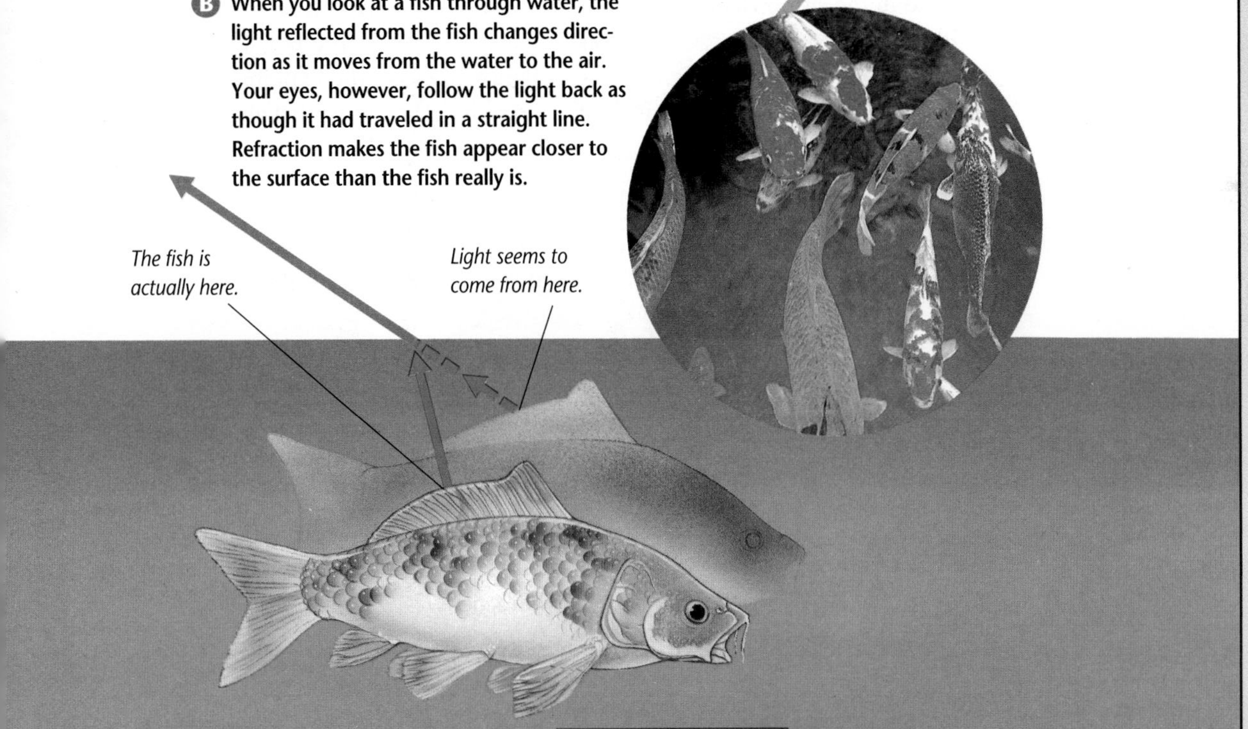

The light ray refracts as it enters the glass block.

It refracts again as it leaves the glass block.

B When you look at a fish through water, the light reflected from the fish changes direction as it moves from the water to the air. Your eyes, however, follow the light back as though it had traveled in a straight line. Refraction makes the fish appear closer to the surface than the fish really is.

The fish is actually here.

Light seems to come from here.

check your UNDERSTANDING

1. Draw diagrams showing how light is reflected from a dark, still pool of water and a rushing stream. Label the diagrams diffuse reflection and regular reflection.

2. Draw a diagram of light rays bending as they move from one material to another material.

3. Draw a diagram of how light would travel from a coin on the bottom of a swimming pool to your eyes if you were standing by the side of the pool.

4. **Apply** Draw a diagram showing how you might be able to see light from a room around a corner.

check your UNDERSTANDING

1, 2, 3, 4. Drawings should follow information in the section.

Program Resources

Take Home Activities, p. 7, Optic Fibers **L1**

Laboratory Manual, pp. 9–10, Refraction of Light **L2**

3 ASSESS

Check for Understanding

Assign small groups of students to work on questions 1 to 4. To help students answer the Apply question, point out that a light ray from another room would bounce off several walls before arriving at your eye. **COOP LEARN**

Reteach

Demonstration To demonstrate refraction, have students drop a washer into a glass of water and try to put a pencil through the hole in the washer. Explain that the light reflected from the washer bends away from the normal as it emerges from the water. Because the light ray that reaches your eye is lowered, the image of the washer appears higher in the water than the washer actually is. Students should see that refraction causes the washer to appear in a different place. **LEP** **L1**

Extension

Acquiring Information Have students research how the speed of light affects refraction. Have them look up the speed of light in air, water, and glass. Have students identify a pattern in the change in speed and direction of the light ray as it crosses a boundary. *an increase in speed; light bends away from an imaginary line perpendicular to the boundary; a decrease in speed; light bends toward the imaginary line* **L3**

4 CLOSE

Activity

Have each student write statements defining reflection and refraction and give an example of each. Students can exchange their work with partners. **L1**
COOP LEARN **P**

PREPARATION

Planning the Lesson

Refer to the Chapter Organizer on pages 52A–D.

Concepts Developed

The previous section focused on light reflection and refraction. This section introduces students to the visible light spectrum. White light is a mixture of all colors of light. Students will observe how different combinations of light and pigment determine the colors that we see. They will study the role our eye-brain system plays in enabling us to detect light, dark, and color.

1 MOTIVATE

Bellringer

Before presenting the lesson, display **Section Focus Transparency 6** on an overhead projector. Assign the accompanying **Focus Activity** worksheet. **L1**

LEP

Explore!

What colors are in sunlight or light from a light bulb?

Time needed 15 minutes

Materials prism; white paper; crayons; light source such as a flashlight, sunlight, or a slide projector

Thinking Processes observing and inferring, making models

Purpose To show that the colors of the spectrum will bend and separate when white light is passed through a prism.

Teaching the Activity

Demonstration You may

2-3 Color

Section Objectives

- Examine white (visible) light.
- Explain the difference between pigment color and light color.
- Describe the functions of the parts of the eye.
- Describe how light and color are sensed.

Key Terms

spectrum, retina, receptors, rods, cones

White Light

One morning, as you dress for school, you notice that your new green sweater does not look the same as it did in the store when you bought it. It's too late to change, so you wear it to school. Your classmates compliment you as soon as they see it. You look at your sweater and it looks different from the way it did at home! What's happening here? Shouldn't everything look the same when you see it in any light? Isn't all light white? Do the Explore activity to learn more about white light.

Explore! ACTIVITY

What colors are in sunlight or light from a light bulb?

What To Do

1. Place a prism between a light source and a piece of white paper.
2. Move the prism until you observe a spectrum of colors on the paper.
3. *In your Journal*, record the colors you observe. Make a sketch of what you see and label it.
4. What is white light made of?

Figure 2-8

White light is a mixture of all colors of light. When white light passes through a prism, different colors are refracted different amounts. The different amounts of refraction separate the white light into a spectrum.

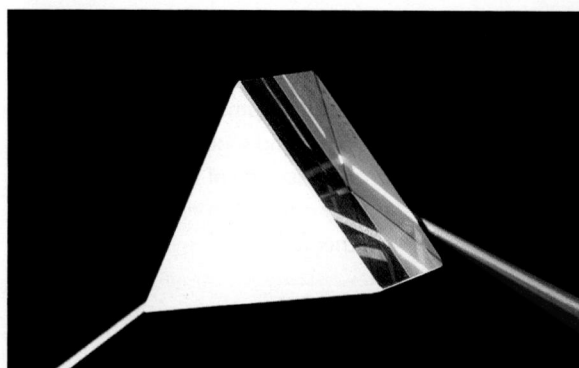

---

want to set up a prism and a light source so that the spectrum will be displayed fully on a large screen. **LEP**

Science Journal Have students describe what they see and have them suggest why light shows up as it does with a prism. *A prism breaks up white light into its basic components.* **L1**

Expected Outcome

Students should observe the seven basic colors of the spectrum in the ROY G BIV pattern—

red, orange, yellow, green, blue, indigo, violet.

Answer to Question

4. White light is made of many colors of light.

✔ Assessment

Oral Have students explain what white light is made of. Have them support their responses with observations. Use the Performance Task Assessment List for Making Observations and Inferences in **PASC**, p. 17. **L1**

A Mixture of Lights

In the 1600s, Isaac Newton first observed and explained how a prism affects light in the way you've just seen. He proposed that white light is a mixture of all colors. He also showed that different colors of light bend differently—refract—as they pass through a prism, so they emerge as separate bands of color called a **spectrum**. Let's see what the spectrum tells us about the makeup of light. When Newton looked at the spectrum, he observed a pattern that had roughly seven colors, as seen in **Figure 2-9**.

Do the activity on the next page to find out what happens when colors of light combine.

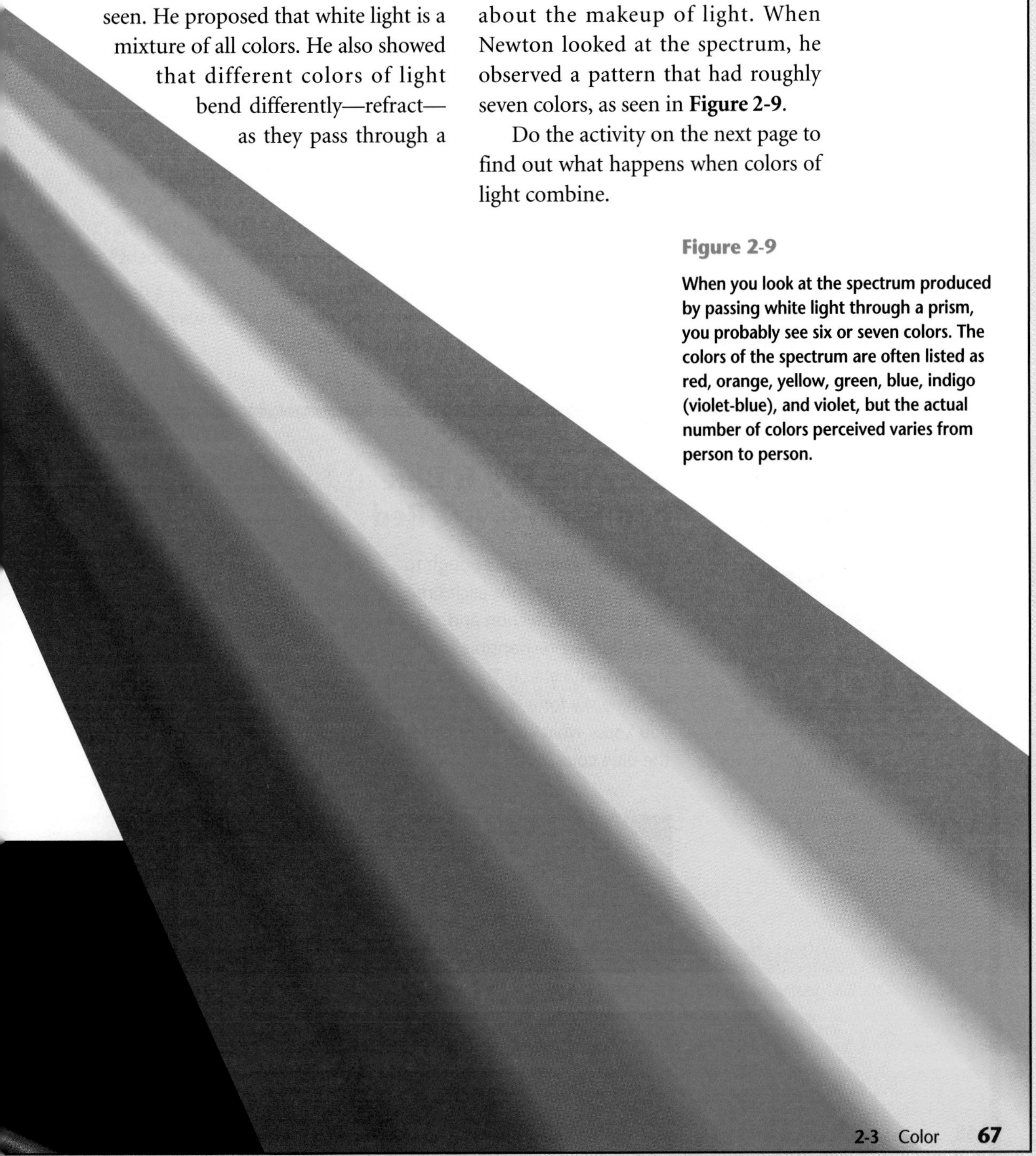

Figure 2-9

When you look at the spectrum produced by passing white light through a prism, you probably see six or seven colors. The colors of the spectrum are often listed as red, orange, yellow, green, blue, indigo (violet-blue), and violet, but the actual number of colors perceived varies from person to person.

Discussion Ask students how colored objects appear to them in a dark room. They may suggest that the objects appear to be black. Can students figure out the reason? *Colored objects in a dark room appear black because there is so little light for the objects to reflect.* L1

2 TEACH

Tying to Previous Knowledge

Ask students to describe sunlight. They will probably use the phrase *white light.* Point out that if you can see it, then it is part of the visible spectrum. Also point out that we do not really see sunlight. Instead we see the light reflected from objects on Earth.

Content Background

The colors in the light spectrum are different wavelengths. The shortest waves are at the violet end of the spectrum. They bend the most when passed through a prism. The longest are at the red end of the spectrum; they bend the least.

Visual Learning

Figure 2-8 Have students communicate the effect of refraction that occurs when white light passes through a prism. *Refraction separates white light into its component colors.*

Figure 2-9 Have students identify which color is refracted the most and the least as light passes through a prism. *Violet, red.*

Multicultural Perspectives

Symbolic Colors

Different colors have symbolic importance in many cultures. The Pueblo represent directions with color. White stands for the east, yellow usually stands for north, blue represents west, and red stands for south.

The Cherokee feel that colors represent important cultural ideas. Red means success, blue is trouble, black represents death, and white stands for happiness. Divide students into cooperative groups. Have each group do library research about the symbolic importance of color among Native American peoples. COOP LEARN L2

GLENCOE TECHNOLOGY

CD-ROM

Science Interactions, Course 1, CD-ROM

Chapter 2

How do lights mix?

Time needed 10 minutes

Materials three flashlights; three rubber bands; green, blue and red cellophane; white paper

Thinking Processes observing and inferring, comparing and contrasting, recognizing cause and effect, making models

Purpose To observe how the three primary colors of light combine to make other colors.

Teaching the Activity

You may wish to have students work in groups of three with each student holding a flashlight. COOP LEARN

Science Journal Have students use colored pencils or crayons to diagram how the lights mixed. L1

Expected Outcomes

Students will observe that green and blue combine to make cyan, green and red make yellow, blue and red make magenta, and white is the result of all three combined.

How do lights mix?

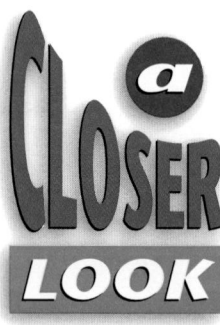

You know that white light is made up of the colors you observed with the prism. When those colors combine, white light appears.

What To Do

1. Obtain three flashlights; three rubber bands; green, red, and blue cellophane; and white paper.

2. Fold the green cellophane into a square several layers thick and place it over the lens of one flashlight. Hold it in place with a rubber band.

3. Do the same with the other colors of cellophane and the other flashlights.

4. In a darkened room, shine all three flashlights onto the white paper to make three circles that overlap.

Conclude and Apply

1. What do you see where green and blue overlap? Green and red? Blue and red? All three?

2. Move just one flashlight closer to and farther away from the paper. How do the colors change? Move each of the other two flashlights. What happens?

3. *In your Journal,* draw and label a diagram showing one of the ways the lights mixed.

Why the Sky Is Blue and Sunsets Are Red

Rainbows are rare enough to be a new delight each time we see one. Reflection and refraction are responsible for their colorful effect. Think about the blue sky for a moment. Do you know what accounts for the blue color?

Rayleigh Scattering

The different colors we see in the sky are not caused by refraction. An effect known as Rayleigh scattering scatters sunlight to produce the colors. Gas molecules in the atmosphere scatter blue light from the sun.

Purpose

A Closer Look expands students understanding of light by explaining how light interacts with the atmosphere to produce different colors of the sky at different times.

Content Background

The two bottom layers of Earth's atmosphere, the troposphere and the stratosphere, are dense with molecules of air, primarily oxygen and nitrogen. The white light of the sun passes through this maze of molecules. Blue light is scattered from air molecules. Near the surface of Earth, water molecules and other large particles, such as soot and dust, make the air more dense. The longer reddish wavelengths are scattered. Sunsets are red because sunlight late in the day passes through more of the dusty lower atmosphere and is therefore scattered more.

Mixing Colors with Lights

Were your results similar to what is shown in **Figure 2-10**? Blue plus green made a shade of blue called cyan, blue plus red made a shade of red called magenta, and red plus green made yellow. Where all three colors of light mixed, you got white. How did you get white with only three colors?

If you could combine the three colored lights, each with just the right brightness, you would be able to produce every color of the spectrum. Red, blue, and green light are called the primary colors of light.

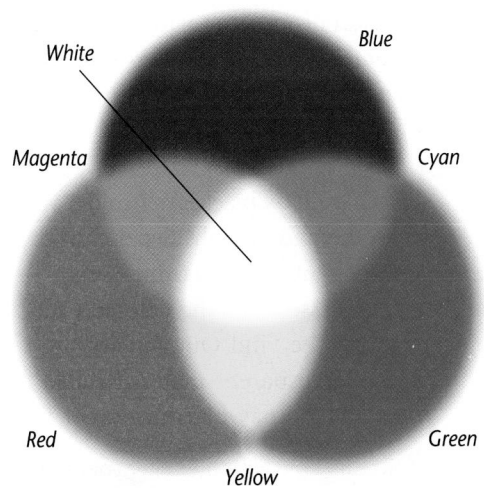

Figure 2-10

White light is produced when the three primary colors of light—red, green, and blue—are mixed.

Green and yellow scatter very little, and orange and red light scatter even less in the upper atmosphere. You can see this when the sun is high in the sky by comparing the sky nearer the sun to the sky farther away. The sky is whiter near the sun because you are looking at an area that contains all colors. The sky farther away has more scattered blue light, which is reflected to our eyes by dust particles and water droplets.

Sunrise and Sunset

At sunrise and sunset, light from the sun passes through more of the atmosphere close to Earth's surface. This layer of the lower atmosphere contains particles that scatter sunlight even more effectively. Blue and violet light and some of the green and yellow light scatter so well that only red and orange light are left to reach our eyes. If you look at the sky farther away from the sun, you can still see the scattered blue and violet light.

You Try It!

Make a simple model of the atmosphere. Fill a clear glass with water and add two to five drops of whole milk. Darken the room and shine a flashlight through the glass. What color do you see coming through the glass? Look at the light coming out of the top of the glass. It has been scattered (reflected away) by the milk particles. What color is it?

Find Out!

Conclude and Apply

1. cyan, yellow, magenta, white

2. The size and shape of the magenta, cyan, yellow, and white change. The intensity of the colors also changes, producing a variety of shades. The white may develop a pinkish tone. The size, shape, and intensity of the other overlapping colors change as each light source moves closer or farther away.

✔ Assessment

Performance Have groups of students use their observations about how lights mix to create lighting for a skit or play. Use the Performance Task Assessment List for Skit in **PASC**, p. 75. COOP LEARN L1

GLENCOE TECHNOLOGY

🔘 **Videodisc**

STVS: Physics

Disc 1, Side 1
Schlieren Photography (Ch. 10)

Fiber Optics on the Farm (Ch. 16)

Disc 1, Side 2
Laser Identification of Fibers (Ch. 11)

Chroma Key (Ch. 22)

Teaching Strategy

Have pairs of students make enlarged diagrams of what the atmosphere looks like as it scatters short blue wavelengths and as it scatters long red wavelengths. L2

Answers to

You Try It!

1. red and orange
2. blue

Going Further ▸

Students can work in small groups to find out how air pollution affects the colors seen in the sky. Have students make a list of types of air pollution. These should include auto exhaust and factory emissions. Have each group find color pictures of pollution and create a display that answers the question: What effect does pollution have on the color of the sky? Use the Performance Task Assessment List for Display in **PASC**, p. 63. COOP LEARN L1

Visual Learning

Figure 2-11 Have pairs of students use the figure to communicate their understanding of how filters affect the colors we see. L1

Teacher F.Y.I.

Discuss with students the meaning of the phrase "looking at the world through rose-colored glasses." Tell students this concept is based on reflection and transparency. It means that a person thinks about the world as if it were seen through a red filter. Everything would have a pleasant rosy or warm tint.

Inquiry Question **How can a colored light bulb be like a filter?** *The color of the glass of the bulb is the same color as the light it gives off. The colored glass absorbs the other colors of light given off by the filament.*

Across the Curriculum

Daily Life

Tell students that lighter colored materials absorb less light than darker materials. Have students apply this knowledge to answer these questions: **What color clothing should you wear on hot summer days if you want to be as cool as possible?** *lighter colors* **What color clothes should you wear on cold winter days in order to stay as warm as possible?** *darker colors* **What color clothes do you think would be best for people living in desert climates?** *white*

Light Filters

The color you see when light hits an object depends on which colors are reflected and which ones are absorbed. The paper on this page reflects all colors, so you see white. The ink on the page absorbs all colors, so you see it as black. Any colors you see in the photographs depend on the color of light reflected from them.

Sunglasses, like the colored cellophane in the Find Out activity, are a type of transparent material called a filter. Filters may be transparent for one or more colors of light. These colors pass through the filters, but the other colors are absorbed. The color of the filter is the color of light that passes through it.

Imagine it's a sunny day, so you put on your sunglasses. Your sunglasses absorb certain colors of light and allow other colors to pass through like the filters in **Figure 2-11**. Everything you observe through your sunglasses is seen in the colors of light that pass through your sunglasses.

Figure 2-11

These gym shoes and socks are seen under white light.

A White objects reflect all colors equally, so they appear to be the color of the light falling on them. The shoes and socks are viewed through a red filter.

B These shoes and socks are viewed through a blue filter. How do they appear now?

ENRICHMENT

TECH PREP

Activity Students can learn to separate and control variables by experimenting with different types of light and colorful clothes. The day before doing this activity, tell students to wear colorful clothes. Set up a fluorescent light bulb and an incandescent bulb. Turn off the room lighting. Have students stand in front of one light and then the other. Discuss how each type of light affects the colors of their clothes. LEP L1

Mixing Colors with Pigments

A blue that you see produced from mixing paints and a blue that you see from mixing lights may appear exactly the same to your eyes. However, two very different processes produced the same blue. Paints are examples of pigments—materials that absorb some colors and reflect others. You can make any pigment color by mixing different amounts of the three primary pigments of yellow, magenta, and cyan. A primary pigment's color depends on the color it reflects.

The key to understanding why mixing colored pigments is different from mixing colored lights lies in a very simple fact. When you mix lights, you are adding different light to the mixture. When you mix pigments, in a way, you are taking different colors away from the mixture. More colors of light are actually being absorbed, and not reflected for you to see. **Figure 2-10** shows what happens when you mix colored lights. Compare it with **Figure 2-12**, which shows what happens when you mix pigments. What do you see?

Mix colored lights, and the resulting colors are lighter because they're

Figure 2-12

The three primary colors of pigment appear black when they are mixed.

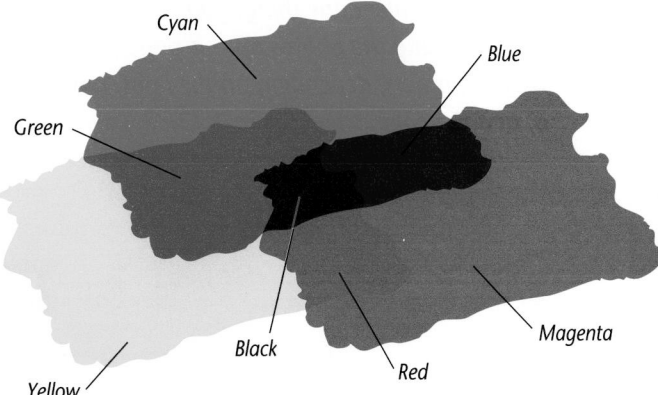

Cyan
Blue
Green
Magenta
Black
Red
Yellow

closer to white—there are more different colors in the mix. If you mix pigments, the resulting colors are closer to black because they absorbed more colors of light—fewer different colors are reflected to your eyes. But, just as you can dim and brighten three primary colored lights to produce any color, you can mix primary pigments in unequal amounts to produce any color. The results are the same—you can create the entire spectrum. All paint colors are produced by mixing only a small number of pigments.

DID YOU KNOW?

Our eyes are most sensitive to the color yellow-green. That's why the preferred color for new emergency vehicles, such as fire trucks, is yellow-green.

Figure 2-13

A variety of colors can be produced by mixing just two of the primary pigments. If you start with yellow and add magenta, a bit at a time, each addition will produce a new color.

Program Resources

Study Guide, p. 12
Critical Thinking/Problem Solving, p. 5, Flex Your Brain
Laboratory Manual, pp. 11–14, Parts of the Eye L2
Teaching Transparencies 3 and 4 L2
Section Focus Transparency 6

2-3 Color **71**

Uncovering Preconceptions
Ask students to name the three primary light colors. *red, blue, green* They may name the primary pigment colors—red, yellow, blue—instead. From their art education, students may have a working knowledge of color. Emphasize that the primary pigment colors are different from the primary light colors because light colors come from luminous objects, and pigment colors are reflected.

Visual Learning

Figure 2-12 To help students distinguish colors from light sources and reflected colors, have them compare and contrast Figure 2-12 with Figure 2-10. Point out the difference when all primary colors are mixed together.
Figure 2-13 Provide pigments and have groups demonstrate the effects of adding magenta to yellow, as shown in Figure 2-13. L1

Inquiry Questions How can red, blue, and green light make all the colors? *All colors are achieved by adding and mixing the primary colors at various degrees of brightness.* **Why are primary colors different in lights and paint?** *Primary light colors are produced by light sources. Primary pigment colors are the result of reflection and absorption.*

Flex Your Brain Use the Flex Your Brain to have students explore COLORS. L1

2-2 Seeing Colors

Preparation

Purpose To determine which colors and color combinations are most easily seen by the human eye.

Process Skills forming a hypothesis, observing and inferring, interpreting data, making and using tables, organizing information

Time Required 40 minutes— 10 for cutting out letters and organizing materials, 15 for testing and recording data, 15 to find percentages and answer questions

Materials See reduced student text.

Possible Hypothesis Students probably will not agree on the easiest colors to see at a distance, but groups should come up with one hypothesis to test for distance and one hypothesis to test for ease of seeing color combinations at a distance. Bright yellow-green is an easy color to see at a distance. Light colors against dark backgrounds are easy to see and visa versa. Students may pick exact colors to test.

The Experiment

Process Reinforcement Make certain that students set up their data tables so they can add up the number of times a color or color combination is chosen to figure percentages. Adding downward is a lot easier than adding sideways.

Possible Procedures Test all colors in the same place under the same lighting and from the same distance. Comparison tests done side by side (as seen in the photo) are more accurate than testing one color at a time.

Teaching Strategies

Troubleshooting Art paper and poster board of many colors will make this lab more interesting. To save paper, use smaller test samples at a shorter distance.

Seeing Colors

Think about the many things you see each day—flowers, the sky, the words and the photos on this page. Some things are easier to see than others, depending on the light, how far away from you it is, and the color.

Preparation

Problem

Which colors are easiest for human eyes to see? If you want the letter M to stand out from a given distance, does the color of the letter and the background affect your ability to see it?

Form a Hypothesis

Look at the photos on this page, then decide on a hypothesis predicting the easiest single color and combination of colors to see from a long distance.

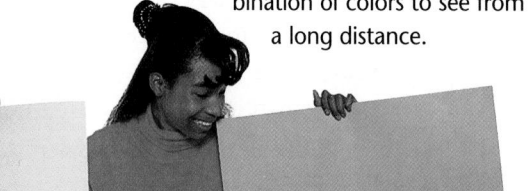

Objectives

- Observe what color is easiest for most people to identify from a long distance.
- Compare and contrast color combinations for ease of identifying letters at a distance.
- Demonstrate that letters are easier to read against certain color backgrounds.

Materials

scissors
glue, tape, or paper clips
posterboard and art paper of various colors including black and white

Safety Precautions

Be careful when using scissors.

Program Resources

Activity Masters, pp. 13–14, Design Your Own Investigation 2-2

Multicultural Connections, p. 8, Light Phenomena L1

Science Discovery Activities, 2-1, Color-Go-Round

DESIGN YOUR OWN
INVESTIGATION

Plan the Experiment

1. Look at the posterboard and paper. As a group decide what colors you will test for your hypothesis.

2. Discuss the best way to test your hypothesis, then write a procedure for the experiment. Make certain that colors are compared under the same conditions of distance and light.

3. Design a data table *in your Science Journal* or on a word-processing program.

4. Determining which color or colors are easiest to identify is a judgment. Most people will agree on the easiest color to see, but perhaps not every-one. To make your data more reliable, you need to test each color more than once. Ten people determining which color is easiest to identify gives stronger information than one person determining that color.

Check the Plan

1. Make certain your data table is designed to record each individual test.

2. Before you start the experiment, have your teacher approve your plan.

3. Carry out your experiment. Make observations and record your data.

Analyze and Conclude

1. **Interpret Data** Was your hypothesis supported by the data? Use your data to explain why or why not.

2. **Use Numbers** In each experiment, add up and record the number of times a specific color was chosen as the easiest to see from a long distance away. What percentage of the time was this color chosen? Percentages are figured by dividing the number of times chosen by the total number of trials. Then multiply by one hundred and add a percent sign. Use your calculator.

3. **Use Numbers** Find what percentage of the time the most easily seen combination of colors was chosen in the tests.

4. **Interpret Data** Using the percentages found in the questions above, write a conclusion about which colors and color combinations are easiest to see.

Going Further

If you were designing a sign to be seen easily at a distance, what colors would you choose? Use your data to explain your choice.

Discussion Tell students that not everyone sees colors exactly the same. That is why many people should compare the colors and color combinations. Some students may find that they have a form of color blindness.

Science Journal Based on the information they find from this experiment, have students write about what colors signs and costumes for stage productions should be.

Expected Outcome

Outcome will depend on what colors are available to test. Overall data from groups should agree on easiest color to see and easiest color combinations. In part of the test, yellow-green and light, bright colors against a dark background should be easiest to see.

Analyze and Conclude

1. Students may have predicted the wrong colors, but they should be able to determine which colors were easiest to see. Research shows that yellow-green is easy to see for most people.

2. Students should use their calculators to determine percentages. They should include this figure on their data table.

3. See answer #2.

4. This question depends on information in answer #3.

Going Further ▐▐▐▐▶

This depends on collected data. Students may suggest yellow-green against black.

✔ Assessment

Process Have students explore how our eyes perceive colors. Challenge students to make a color appear to change without changing it. From their research of color combinations, they should realize that the perception of a color can be changed by changing its background.

2-3 Color **73**

ENRICHMENT

Activity Have students use a hand lens to examine some colored comics. Direct them to look for red areas made of red (magenta) dots, blue areas made of blue (cyan) dots, and yellow areas made of yellow dots. Ask: **How are other colors made?** *green—blue and yellow dots; orange—red and yellow dots* [L1]

Life Science CONNECTION

Theme Connection The theme that this section supports is energy. The point is made that without the color spectrum of light rays, we would see the world in black, white, and shades of black.

Across the Curriculum

Social Studies

Tell students that during the 1960s, many young people wanted to go back to more natural ways of living. Part of their alternative life-styles involved weaving and dying their own cloth. Suggest that students find magazines from the late sixties and seventies such as Life and Time. What examples can students find of natural cloths and dyes used in a creative manner?

Connect to

Life Science

Cats' eyes have a vertical slit pupil. They also have a much greater percentage of rods in their retinas. Cats are predators; distinguishing movement and shape is more important than seeing colors.

Light and Pigments Together

Figure 2-14

A An object's color can change depending on the light you see it in. The walls of this experiment box, lit here by white light, appear blue, yellow, red, and white.

B Lit by red light, the same walls appear purple, orange, red, and pink.

The Investigate you just finished shows that the color of an object is determined by the light shining on it. Objects merely absorb or reflect certain colors of light or allow certain colors to pass through. Our idea of what something looks like depends upon the light we see it by.

Remember the sweater we talked about at the beginning of this section? The color looked different depending on where you were. The color of the light source determined the color you saw. Look at **Figure 2-14** to compare the same colors in different lights.

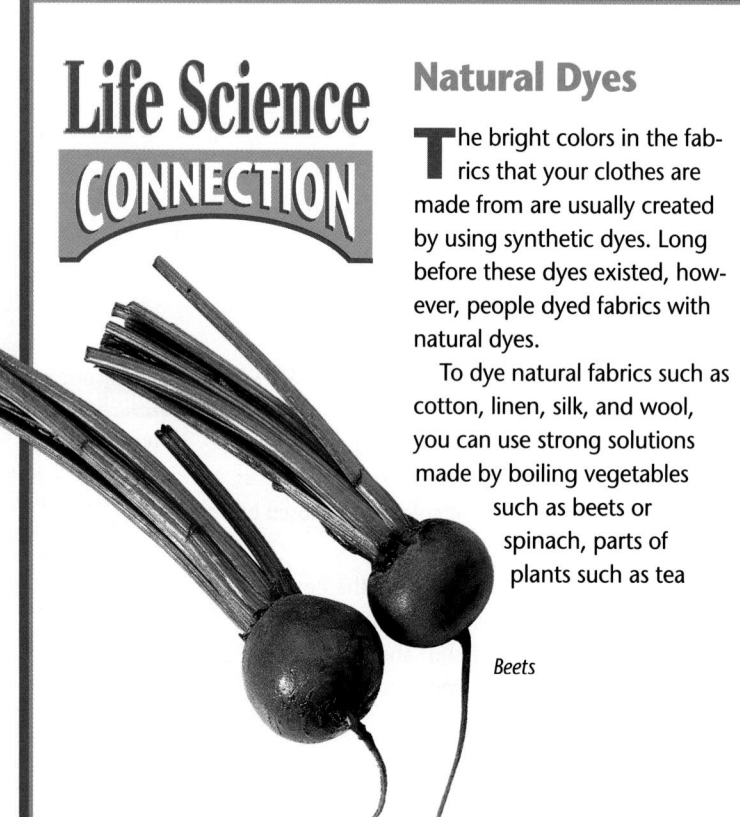

Life Science CONNECTION

Natural Dyes

The bright colors in the fabrics that your clothes are made from are usually created by using synthetic dyes. Long before these dyes existed, however, people dyed fabrics with natural dyes.

To dye natural fabrics such as cotton, linen, silk, and wool, you can use strong solutions made by boiling vegetables such as beets or spinach, parts of plants such as tea

Beets

leaves or wildflowers, and barks of various trees. The shades will vary depending on the strength of the dye solution, the soil in which the plant grew, whether fresh or dried plant parts were used, and the way the fabric was treated before it was dyed.

The colors from natural dyes are not as bright as those from chemical dyes. They are much softer—often described as "earth tones."

Here are some of the colors you can get from common plants:

reds: oregano leaves, tea leaves, beet roots, fruits of the lipstick tree

74   Chapter 2

Life Science CONNECTION

Purpose

The Life Science Connection further develops the concepts of color presented in this section by explaining how people have used colors occurring in the plant world to enhance their lives.

Content Background

A variety of chemical structures in plants react to the energy of light. These chemical structures result in the many colors seen in plants. A common chemical that reacts to

sunlight is chlorophyll, which produces the greens in plants. Because natural dyes are made from vegetable solutions, the colors produced come from the interaction of plant chemicals and sunlight that occurred when the plant was alive. Unlike most paint pigments, which are made from minerals, natural dyes fade fairly quickly, especially if exposed to sunlight. Mordants are chemical compounds that prolong the life of the color.

Your Eyes

Light enters your eye through an opening called the pupil, a black spot in the middle of the eye, that looks like a small hole. It is really the entrance to a fluid-filled chamber inside the eye.

The colored area around the pupil is the iris. This ring of tiny muscles usually contains some blue or brown pigment. The iris muscles control the size of the pupil according to the amount of light available.

In bright light, the iris muscles contract and make the pupil smaller. This limits the amount of light enter-ing the eye. In dim light, the iris muscles relax. The pupil enlarges, allowing more light to enter the eye. The light that enters your eye passes through the lens, a structure that focuses the light on the retina at the back of your eye. The lens changes its shape as your eyes focus on objects near and far away. The **retina** is a tissue that is sensitive to light. Your eyes can detect all kinds of colors and shapes, lights and shadows, because of the retina. Your eyes can be compared to two cameras, always taking pictures.

Connect to...
Life Science

Cats have excellent vision in dim light but they cannot distinguish colors very well. Investigate how a cat's eye might be different from your eye.

Across the Curriculum

Fine Arts

TECH PREP

Tell students how the Impressionist painters of the late 1800s took advantage of the way we see color. Instead of working with the outlines and details of objects, they made their images with rough patches of color. The Impressionists were especially interested in portraying different intensities of light. To do this, they developed a system of using complementary colors. Claude Monet and Pierre Renoir were two major Impressionists.

Postimpressionist Georges Seurat went even further. His technique was called pointillism. He used tiny dots of bright color instead of mixing colors on his palette. Up close to his painting, you can see each dot, but from a distance, the colors seem to merge.

Divide the class into groups. Have each group pick an Impressionist or Postimpressionist artist. Have students research their chosen artist to find out how that person used color. **L2** **COOP LEARN**

Onions

Spinach

yellow: barberry stems and roots, goldenrod blooms, saffron crocus blooms, onion skins
violet: hibiscus flowers, oregano leaves
blue: cornflowers, hollyhock flowers, wild indigo branches
green: onion skins, sorrel leaves, spinach leaves, dyer's broom tops
brown: hibiscus flowers, juniper berries, tea leaves
gray: blackberry shoots
black: barberry leaves, yellow dock roots

Before dyeing a fabric, it must be treated with a mordant, a dye-setting compound. Without a mordant, the fabric's color would easily wash or fade away. Alum, ammonia, and vinegar are mordants.

You Try It!

Make your own dye.

1. Use from one-half to one cup of leaves or flowers of the plants listed.
2. Place the plants in a plastic bag and crush them with your fingers.
3. Place the crushed plants in one cup of water and heat the water to a gentle boil. Simmer gently until most of the color has been removed from the plants.
4. Cool the mixture, filter it, and collect the liquid.
5. Cut two pieces of worn cotton fabric. Soak one piece in vinegar, wring it out, and allow it to dry.
6. Place both pieces of fabric into the dye. Stir until the fabric absorbs the color.
7. Wring out the fabric and allow them to dry. Then try washing them. Does the color come out?

GLENCOE TECHNOLOGY

 Videodisc

Science Interactions, Course 1 Integrated Science Videodisc

Lesson 1

Teaching Strategy

Tell students that production of synthetic dyes began in 1855. Ask why a synthetic dye might be more useful than a natural dye.

Troubleshooting

When doing the You Try It activity, be sure students do not use their bare hands for crushing. The oil on their fingers will affect the dye's action. If possible, have students use mortars and pestles to grind the leaves or flowers. Plastic bags make a fine, even breakdown of the material difficult, especially if it is dried.

Answers to
You Try It!

7. Both pieces faded, but the vinegar-soaked piece faded less.

Going Further ▐▐▐▐▶

Students can work in groups to find out how synthetic dyes are made and used. Assign research into azo dyes, vat dyes, sulfur dyes, reactive dyes, and disperse dyes. Encourage students to find examples of each type of dye either in photographs or actual swatches of cloth. Have students display their findings on a bulletin board. Use the Performance Task Assessment List for Bulletin Board in **PASC**, p. 59. **L2** **COOP LEARN**

How You See Color

The retinas in your eyes contain two different kinds of light-sensitive structures. These structures, or light **receptors**, respond to changes in light and color. **Rods** are receptors that are sensitive to light and dark. You see black-and-white images when the rods in your retina are stimulated by dim light. **Cones** are receptors that are sensitive to all the colors in the visible spectrum of light. You have three types of cones, referred to as red, green, and blue cones. Different colors of light will cause different combinations of cones to respond to them. The combination of these three types of cones lets you see the entire spectrum. You see color images when the cones in your retina are stimulated by color and bright light. Because color-blind people lack one or more kinds of these cones, they see colors differently, as shown in **Figure 2-16**.

■ Night Vision

Have you noticed that you don't see colors very well at night, or any time the light is dim? This is because cones, the color receptors in your retina, are not very sensitive to dim light. The light entering your eyes must be fairly bright before you can see colors. Rods

Figure 2-15

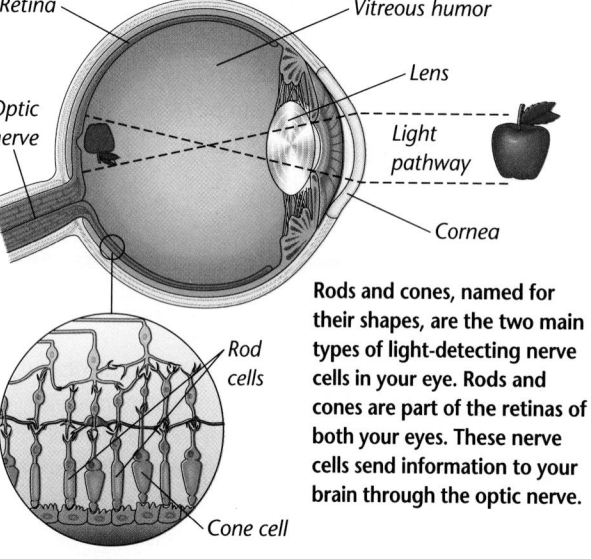

Rods and cones, named for their shapes, are the two main types of light-detecting nerve cells in your eye. Rods and cones are part of the retinas of both your eyes. These nerve cells send information to your brain through the optic nerve.

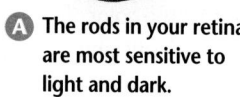

A The rods in your retina are most sensitive to light and dark.

B Red cones respond primarily to red and yellow light.

C Green cones respond primarily to green and yellow light.

D Blue cones react primarily to blue and violet light.

E Your brain works to combine all this information into a single image made up of both black-and-white and color images.

76 Chapter 2 Light and Vision

Three kinds of cones

In the middle 1960s, scientists showed for the first time that individual cone receptors of the eye react best to different light colors. Using a device that measures the amount of light absorbed by a tiny object, they found that single cone receptors absorbed more light of one color than light of other colors. They reasoned that if a cone absorbed light of a certain color, then that color would cause a reaction in the cone. By repeating the experiment on many different cones from the eye, they found that all cones can be classified into only three groups. One group absorbs blue light best, another absorbs green light best, and the third absorbs red. We now call these the blue, green, and red cones of the eye. These three kinds of cones are all we have and all we need to see all the colors of the spectrum.

are much more sensitive to dim light.

The rods and cones in the retinas of both your eyes send information to your brain through a large nerve at the back of the retina. Your brain interprets the information as images and color.

How important is light and color to your life? Think about the different things you look at every day. Imagine what your life would be like without your vision. In the next chapter, you will learn about sound and your sense of hearing.

Figure 2-16

Color-blind people lack one or more of the three kinds of cones and cannot distinguish all colors. Color patterns like these are used to test for color blindness.

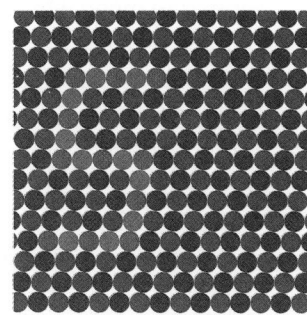

B People who cannot distinguish between red and green cannot see the 5 and 3 in the second pattern below.

A People who cannot distinguish the difference between blue and yellow cannot see the 5 in the first pattern to the right.

check your UNDERSTANDING

1. What colors make up white light?
2. What is the difference between red pigment and red light?
3. Name each part of the eye and describe its function in allowing you to read the words on this page.
4. What color would you see if all types of cones in your eyes were strongly stimulated?
5. **Apply** What color would you see if you mixed yellow and blue light? Why? Yellow and blue pigments? Why?

check your UNDERSTANDING

1. all colors, but at least green, blue, and red
2. Red pigment has no "redness" of its own; it reflects red light. Red light is actually red.
3. pupil—lets light into eye; iris—adjusts pupil size, according to amount of light; lens—focuses light on retina; rods—enable eye to see black text and white page; optic nerve—sends information eye receives to brain
4. white
5. White; yellow is a mixture of red and green. The three make white. Pigments: you see black since blue is a mixture of cyan and magenta, and yellow, cyan, and magenta combine to make black as in Figure 2-12.

3 ASSESS

Check for Understanding

To help students answer the Apply questions, review Figure 2-10 on page 69. Be sure students see the colors that result when primary colors of lights are combined. Then compare and contrast Figure 2-10 with Figure 2-12 on page 71.

Reteach

Demonstration Have students demonstrate how pigments interact to make colors. Obtain paper and some primary pigment paints from the art department. Have students work in groups and combine small amounts of various pigments to make other colors. For each color, have them describe which colors are reflected and which are absorbed. Be sure they mix all three primary pigments to produce black. Students should see that black pigment absorbs all light and reflects none. **L1**

LEP

Extension

Activity Have students who have mastered this section try to describe a multicolored object without referring to the specific colors involved. **L2**

4 CLOSE

Activity

Have students apply their knowledge of how lights mix. Use the colored flashlights from the Find Out activity on page 68. Have groups of students create their own light show by playing the lights in patterns around a darkened room. Give the groups ten minutes to come up with a pattern to display to the class.

LEP COOP LEARN

Science and Society

Purpose
Science and Society further develops Section 2-3 by giving examples of how the light and color we perceive affects the way we act.

Content Background
Sunlight stimulates the pineal gland, which produces melatonin. Lack of light impairs the control of the gland's function. Dr. Alfred Lewy was one of the first to treat the depression caused by SAD by using high levels of light to readjust the production of melatonin. Lewy suggests that all people would benefit by an early morning dose of sunlight.

Teaching Strategy Ask students to describe how they feel about the color of the classroom. Does it make them comfortable and relaxed? Or does it make them tense? Have students explain their reactions.

Activity
Have students work in pairs to examine various rooms in the school such as the study hall, cafeteria, library, and main office. Have them describe the color scheme in each room. Ask students to determine what effect the colors have on students, faculty, and staff. Then have each pair write a letter to the principal recommending changes and explaining how the changes could help the school.

Answers will vary depending on the restaurants and stores that students observed. Students should be able to relate colors to moods and emotions. They probably will say that the colors in a fast-food restaurant are intended to make people cheerful and eager to eat.

Science and Society

Light and Color in Our Lives

Most of the time, we just take light for granted. More and more, researchers are finding that light is necessary for our mental and physical well-being. Light can affect how we feel about our surroundings, how we feel about others, and how well we do our work.

Performance

While people respond differently, in general researchers have found that the brighter the room, the better the performance. Brightness generally seems to improve performance until the light reaches the point of glare. At this point, people tend to feel tired, irritable, or bored.

Sound Level

The brightness of light also affects the sound level of conversation. When people enter a brightly lit room, they tend to gather in larger groups, while dimmer rooms lend themselves to smaller groups and quieter conversations.

Health

During the dark hours of the night, the brains of animals and humans secrete a chemical, melatonin, which affects many of the body's functions, including temperature and mood. During the fall and winter, there are more hours of darkness and, therefore, more melatonin is produced. This causes some people to feel very tired and sometimes depressed. One form of this condition is known as Seasonal Affective Disorder (SAD).

Artificial Daylight

The symptoms of SAD are treated by having the patient sit under very bright light—about three times brighter than normal—for several hours early in the morning. This seems to change the body's rhythms and makes the patient feel better.

Color

Color also has important effects on the way we feel and on our actions. Colors like orange tend to make us feel more active and may affect the amount that we eat. Blue is a more calming color.

Science Journal
In your Science Journal, compare the mood created by the lighting and colors used in different stores or restaurants.

Going Further
Have student groups make up lists of phrases and sentences in which people use color to express an emotion. For example, "I really feel blue today". Ask them to discuss the emotions that colors seem to represent. **COOP LEARN** L1

Seeing Things Differently

I f you look closely at the petals of a rose, you will see that what appears from a distance to be a solid color is actually made up of many colors. The pink rose petals may have specks of red, blue, white, or yellow in them.

Impressionists

Early painters tried to copy their subjects exactly. They wanted the pink rose in the painting to look exactly like the pink rose on the bush. During the 1870s and 1880s, a group of artists decided that they were more interested in the impression of a rose than its exact look. As a result, they were called Impressionists.

These artists used color to reach their goals. They did not give objects firm outlines or fill in large areas of their canvasses with solid colors as others had done. Instead, they noticed that when they looked at a rose or a person, there were not distinct outlines, but a gradual blending of colors as one shade touched another. They showed the light as it affected colors, so that sometimes the same pink rose was pale, other times very vivid.

Monet

Claude Monet was one of the best-known Impressionist painters. He is famous for his ability to show how light played among the flowers in his garden. He painted many pictures of the same flowers in the same place,

but they were all different, depending on the angle of the sun, the time of day, and the season of the year. One of Monet's paintings entitled *Water Lilies* is shown here.

The paintings of the Impressionists are very valuable today, and people consider them some of the most beautiful art ever created. Although they do not contain every detail of their subject, the paintings do give an "impression" of the subject that includes the feelings that the artist had about the subject.

You Try It!

Look out the window or go outside and look at an area of trees, grass, and flowers. Do you see every detail in the landscape? Do you see every vein in the leaves or cracks in the tree bark? Or do you get an "impression" of the scene?

Going Further ⏵

Some students may want to know more about the Impressionist painters. They can write a letter of inquiry to art museums. One museum that has an excellent collection of Monet's paintings is:

The Museum of the Art Institute of Chicago
South Michigan Ave. and East Adams St.
Chicago, IL 62220

Use the Performance Task Assessment List for Letter in **PASC**, p. 67. **L2**

Purpose

Art Connection reinforces Section 2-3 by describing how a certain group of artists, the Impressionists, used their perception of light to create paintings.

Content Background

Impressionism is important in art history because it made a change from the traditional tonal mode of painting. In tonal painting, color was subordinated to light and dark patterns. Impressionist painters concentrated on chromatic richness and intensity, so shadows were portrayed as colors instead of shades of gray. The Impressionist painters were scientists in their own right because they realized how changes in light also affected a viewer's perception of an object. Although these artists lived long before many scientific discoveries about the nature of light, their work revealed a world of colored light. The artists' eyes discovered and their brushes proved that individual bits of color come together in the human brain to form a single image.

Teaching Strategy
Invite students to demonstrate the effects of sunlight on how artists depict the world. Have students work in pairs to paint the world. Assign each pair an object in the room. One student should try to render the object at one time of day. The partner should attempt it at another time. Have students use loose strokes of colored crayons to show what they see. Then compare the two drawings.
COOP LEARN **L1**

Answers to
You Try It!

Students should say they have a general impression, not a detailed view.

Technology Connection

Purpose

This Technology Connection relates infrared radiation to the energy spectrum that includes visible light, as is discussed in Section 2-3.

Content Background

Major contributions to the science of infrared astronomy were made beginning in the early sixties by the physicist, Frank Low. Originally, Low was working in the electronics industry on a superconducting bolometer. After Low published his paper on a germanium bolometer, a device in which changes in the conductivity of the germanium was directly proportional to the amount of energy entering the bolometer, astronomers began to seek Low out. Eventually Low went to work for the NRAO, the National Radio Astronomy Organization, where he continued to break ground in infrared observations of planets and stars using his bolometers.

Infrared radiation observation and research by Low and his colleagues led scientists to propose installing an infrared telescope on a satellite above Earth's blanketing atmosphere. NASA, in conjunction with the Netherlands and Great Britain, began a collaborative effort to launch the satellite. It was finally successfully launched in 1983. The infrared telescope scanned the sky for ten months and led to many new observations within the solar system and in faraway galaxies.

Teaching Strategy

A successor to Low's satellite was scheduled to be launched by the European Space Agency in 1993. Have students research to find out if this satellite was launched and what the goals for the satellite were. L2

Technology Connection

Infrared Astronomy

One of the newer techniques for learning about the universe is infrared astronomy. Visible light is part of an energy spectrum that includes X rays near the high energy end and radio waves near the low energy end. Infrared radiation has less energy than visible light but more than radio waves. Most infrared radiation coming from space is absorbed by water vapor and carbon dioxide in Earth's atmosphere, but certain wavelengths pass through and can be detected.

Problems

One problem with studying infrared radiation on Earth is that Earth, and the instruments themselves, emit infrared radiation. Infrared telescopes are set up at high altitudes to avoid detecting Earth's infrared radiation and to avoid interference from Earth's atmosphere.

Using infrared detectors in space can solve problems with Earth's radiation and atmospheric interference, and cooling the instruments to extremely low temperatures ends the problem of the equipment's own radiation.

Ms. Adriana Ocampo, a scientist working at the Jet Propulsion Laboratory, works with the Near-Infrared Mapping Spectrometer (NIMS), a device that detects infrared radiation from objects in space and maps the radiation using computer technology.

Ms. Ocampo's NIMS is being used within our own solar system by the space probe *Galileo*, which was launched in 1989. *Galileo* arrived at Jupiter in December of 1995, where it will remain through 1997. NIMS is being used to study the atmosphere and surface chemistry of Jupiter and its satellites.

Ms. Ocampo is a planetary geologist. She utilizes NIMS data to study the makeup and development over time of the surfaces of Jupiter's moons.

*inter*NET CONNECTION

Use the World Wide Web to find infrared images at NASA's Jet Propulsion Laboratory (JPL). How might such data on planetary atmospheres and temperatures be used?

*inter*NET CONNECTION

The home page for NASA's Jet Propulsion Laboratory can be found at **http://jpl.nasa.gov**

Going Further ▶

Have students investigate how infrared-sensitive materials are used in certain types of binoculars, scopes, and home security systems. Students should find out that these materials detect objects giving off infrared radiation and are effective even in total darkness. Students could then present their findings in a research report. Use the Performance Task Assessment List for Writing in Science in **PASC,** p. 87. L3 P

Science Journal

Review the statements below about the big ideas presented in this chapter, and answer the questions. Then, re-read your answers to the Did You Ever Wonder questions at the beginning of the chapter. *In your Science Journal,* write a paragraph about how your understanding of the big ideas in the chapter has changed.

2 When light moves from one kind of material to another, the light changes speed, which causes it to refract. *Explain how refraction affects what your eyes see when you observe an object that is underwater.*

3 White light (or visible light) is a combination of three main colors of light—red, blue, and green. How you see color depends on the combination of light and pigment. *What colors of light will yellow pigment reflect when viewed in white light?*

1 When light hits an object, the light is reflected, absorbed, or passes through the object. The object may reflect some colors, absorb others, and still let others pass through. *How do you know that a material is translucent?*

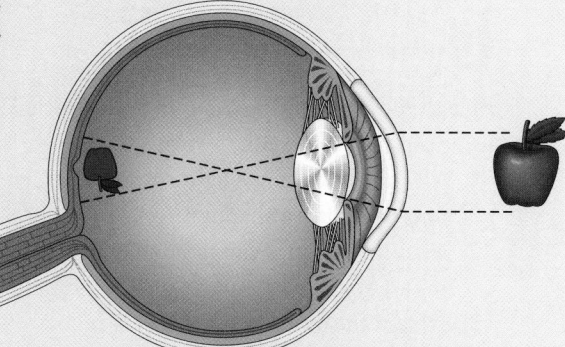

4 Light reflected from an object enters your eye and strikes the light-sensitive retina. Here rods react to light and dark and cones react to color. Your brain combines the information into one visual image of the object. *Why do some people see colors differently from other people?*

Science Journal

Did you ever wonder...

• The pigments reflect some of the primary colors and absorb others in the white light that shines on them. When pigments are mixed, different colors are absorbed and reflected. (p. 71)

• It takes about eight minutes for sunlight to travel to Earth. (p. 56)

• There is not enough light to stimulate the cones that respond to the color. (p. 76)

Science at Home

Students can try this optical illusion at home. In the early evening, have students hold a dime at arm's length and slowly move it toward them until the dime blots out the moon and record the distance from the dime to the eye. Repeat the procedure later when the moon looks smaller. **Now what distance is the dime from the eye?** *The same distance as before.* **What does this imply?** *An optical illusion made the moon appear larger at the horizon.* **L1**

Although visible light is all around us, to many students it will seem like an abstraction because it doesn't occupy space like a desk or a chair or another person. The strategy below will allow students to appreciate the reality of light and color.

Teaching Strategy

To give students additional opportunities to manipulate colors firsthand, ask students to write a description of a color, give it a name, and then invent it by mixing various paints. In effect, students will be using knowledge gained in the chapter to *predict* how various combinations of pigments produce new colors. Urge students to be creative. For example, they might name a color blue-jean-blue or redhead-orange. Using such real-life colors, students could match their invented color to the real thing. **L1**

Answers to Questions

1. Light passes through a translucent material when it hits it but objects are not clearly seen through translucent materials.

2. The path of light from the water is bent as it passes from the water into the air, but your eyes follow the light back as though it had traveled in a straight line, making the image underwater appear closer than it actually is.

3. When viewed in white light, yellow pigment reflects red and green light, and absorbs blue.

4. People who see colors differently probably lack one or more of the three kinds of cones.

GLENCOE TECHNOLOGY

 MindJogger Videoquiz

Chapter 2 Have students work in groups as they play the Videoquiz game to review key chapter concepts.

Using Key Science Terms

1. opaque
2. spectrum
3. translucent
4. refraction

Understanding Ideas

1. Light beams travel in straight lines and cannot bend around the object blocking the light.

2. When light hits an object it can pass through, be reflected, or be absorbed.

3. When light reflects from a smooth surface, it bounces off the surface in an orderly way and you see the reflected image as well as you see the object itself. This is called a regular reflection. When light hits a bumpy surface, it bounces off each tiny surface in straight lines in different directions producing a diffuse reflection.

4. Cones let you see things in color. There are three types of cones in the eye. Red cones respond to red and yellow light, green cones respond to green and yellow light, and blue cones respond to blue and violet light. Rods do not help you see color, but they are much more sensitive to light and let you see in light that is too dim for your cones.

5. The colors are produced by two different processes. When you mix lights to make colors, you are adding different light to the mixture. When you mix paints, more colors are actually being absorbed, and not reflected for you to see.

Developing Skills

1. See reduced student page for the completed concept map.

2. The light will travel until it is absorbed or reflected by an object.

Using Key Science Terms

cone	retina
opaque	rod
receptor	spectrum
reflection	translucent
refraction	transparent

An analogy is a relationship between two pairs of words generally written in the following manner: a:b::c:d. The symbol :: is read "as." For example, cat:animal::rose:plant is read "cat is to animal as rose is to plant." In the analogies that follow, a word is missing. Complete each analogy by providing the missing word from the list above.

1. window:wall::transparent: _____
2. notes:musical scale::colors: _____
3. cellophane:transparent::waxed paper:

   _____
4. mirror:reflection::prism: _____

Understanding Ideas

Answer the following questions in your Journal using complete sentences.

1. Why does blocking a light source with an object produce a shadow?
2. What three things can happen when light hits an object?
3. Explain the difference between a regular reflection and a diffuse reflection.
4. Describe the functions of cones and rods.
5. How do the processes of producing colors by mixing paints and mixing lights differ?

Developing Skills

Use your understanding of the concepts developed in this chapter to answer each of the following questions.

1. **Concept Mapping** Using the following events, complete the sequence concept map of how you see: *Light is focused on the retinas. Light enters the pupils of your eyes. Information is sent through a nerve to your brain. Your brain uses this information to make the images you see. The light passes through the lenses of your eyes.*

Initiating event

You turn a light on in a dark room.

Event 1

Light enters the pupils of your eyes.

Event 2

The light passes through the lenses of your eyes.

Event 3

Light is focused on the retina.

Event 4

Information is sent through a nerve to your brain.

Final outcome

Your brain uses this information to make the images you see.

2. **Observing and Inferring** Use the index cards and flashlight from the Find Out activity on page 55 and investigate how far a flashlight will shine. Dim the lights in the classroom. Increase the distance between the flashlight and the cards and then between the cards themselves. How far did the flashlight shine? Think about how far light can travel.

Critical Thinking

In your Journal, *answer each of the following questions.*

1. Imagine that you have one lamp with a small, bright bulb. You'd like to avoid a harsh glare on your paper. What might you do?

2. Sylvia did an experiment in which she shone different colored lights on an object that had been placed in front of white paper. Below are her data and observations. What color was the object? What else do you know about the object?

Data and Observations

Color of light	Color of object	Cast a shadow
yellow	yellow	yes
green	green	yes
red	red	yes

3. Explain how a stained glass window can absorb light, reflect light, and/or let light pass through. Based on your answer, tell what parts of the window are opaque, translucent, or transparent.

Problem Solving

Read the following problem and discuss your answers in a brief paragraph.

The new member of the drama club lighting crew was beside himself. The director of the play wanted two more spots of light on the set. He wanted more white light on the stage and a spot of yellow light on the backdrop. But only six lights were available. Two had permanent red filters in place, two had permanent blue filters, and two had permanent green filters. The crew member asked the lighting chief. She told him an easy way to solve the problem.

1. How would you make more white light?
2. How would you make a spot of yellow light on the backdrop?

CONNECTING IDEAS

Discuss each of the following in a brief paragraph.

1. **Theme—Systems and Interactions** How does a flashlight use reflection? Look closely at the parts of a flashlight. You may want to take it apart and examine them.

2. **Theme—Scale and Structure** Name two opaque things, two transparent things, and two translucent things in the room with you now.

3. **Theme—Systems and Interactions** What would absorb more sunlight—a white piece of plastic or a black piece of plastic?

4. **A Closer Look** Recall that a prism breaks up light into the colors of the spectrum. What do you think is acting like a prism in the case of a rainbow?

5. **Science and Society** Auditoriums and theaters often dim the lights before a performance begins. Why do you think this is done?

Critical Thinking

1. Use a translucent shade or a frosted bulb.

2. The object was white, and opaque or translucent.

3. Colored glass reflects the colors it is stained. For example, red stained glass reflects red. Red-colored glass is also transparent to red light allowing red light to pass through. If the glass is clear, it would be transparent to all colors. Some glass is translucent, allowing some diffused light to pass through. The lead between pieces of glass is opaque, allowing no light to pass through.

Problem Solving

Mix light from one red, one blue, and one green light to get white. Mix light from one red and one green to get yellow.

Connecting Ideas

1. The light from the bulb is reflected from and focused by a curved mirror behind the bulb.

2. Possibilities include: opaque—notebook, pencil; translucent—fingernails, fluorescent fixture covers; transparent—window glass, glass on a clock

3. The black piece, because it absorbs all colors of light. White reflects all colors.

4. The rain drops act like a prism.

5. The dimly lit theater will tend to reduce the conversation level so the performers on stage can be heard when the curtain goes up.

✔ Assessment

Portfolio Review the portfolio options that are provided throughout the chapter. Encourage students to select one product that demonstrates their best work for the chapter. Have students explain what they learned and why they chose this example for placement in their portfolios.

Additional portfolio options can be found in the following **Teacher Classroom Resources:**

Multicultural Connections, pp. 7, 8
Making Connections: Integrating Sciences, p. 7
Making Connections: Across the Curriculum, p. 7
Concept Mapping, p. 10
Critical Thinking/Problem Solving, p. 10
Take Home Activities, p. 7
Laboratory Manual, pp. 7–14
Performance Assessment P

Chapter Organizer

SECTION	OBJECTIVES	ACTIVITIES & FEATURES
Chapter Opener		**Explore!**, p. 85
3-1 Sources of Sound (2 sessions, .5 block)	1. **Recognize** that sounds are created by vibrations. 2. **Distinguish** between compression and rarefaction. 3. **Describe** the way sound travels through matter. National Content Standards: (5–8) UCP2–3, UCP5, A1–2, B3, G2	**Find Out!**, p. 87 **Find Out!**, p. 89 **Skillbuilder:** p. 91 **Life Science Connection,** pp. 90–91
3-2 Frequency and Pitch (3 sessions, 1.5 blocks)	1. **Use** the length or thickness of a vibrating object to predict whether its sound will be high or low. 2. **Describe** the relationship between pitch and frequency. 3. **Compare** the sound frequencies humans hear with the sound frequencies animals hear. National Content Standards: (5–8) UCP3, UCP5, A1–2, G1–3	**Explore!**, p. 94 **Find Out!**, p. 95 **Design Your Own Investigation 3-1:** pp. 98–99 **A Closer Look,** pp. 96–97
3-3 Music and Resonance (3 sessions, 1.5 blocks)	1. **Distinguish** between music and noise. 2. **Explain** how different musical instruments produce sounds of different quality. 3. **Describe** resonance. National Content Standards: (5–8) UCP1–2, A1, B3, E1–2, F2, F4–5, G1–3	**Explore!**, p. 102 **Investigate 3-2:** pp. 104–105 **Technology Connection,** p. 107 **Science and Society,** p. 108 **Teens in Science,** p. 109

ACTIVITY MATERIALS

EXPLORE!

p. 85* ruler
p. 94* metal ruler
p. 102 tuning fork, rubber mallet

INVESTIGATE!

pp. 104–105 2 tuning forks of different frequencies (256 Hz or higher); 1000-mL graduated cylinder (or bucket or pitcher about 30 cm deep); rubber mallet; plastic or glass tube 2.5 cm in diameter, about 45 cm long, open at both ends; metric ruler

DESIGN YOUR OWN INVESTIGATION

pp. 98–99* felt tip marker, test tubes with 2.5-cm diameter, test-tube rack, small graduated cylinder, water, metric ruler

FIND OUT!

p. 87* coiled spring, small piece of colored string
p. 89* metal coat hanger; string, about 1 m long; flexible wire, about 1 m long
p. 95* guitar, violin, or other stringed instrument

Need Materials? Call Science Kit (1-800-828-7777).

⏱ OUT OF TIME? We recommend that students do the activities with an asterisk.

KEY TO TEACHING STRATEGIES

The following designations will help you decide which activities are appropriate for your students.

L1 Basic activities for all students
L2 Activities for average to above-average students
L3 Challenging activities for above-average students
LEP Limited English Proficiency activities
COOP LEARN Cooperative Learning activities for small group work
P Student products that can be placed into a best-work portfolio
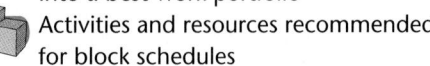 Activities and resources recommended for block schedules

Chapter 3 Sound and Hearing

TEACHER CLASSROOM RESOURCES

Student Masters	Transparencies
Study Guide, p. 13 **Concept Mapping**, p. 11 **Critical Thinking/Problem Solving**, pp. 5, 11 **Multicultural Connections**, p. 9 **Making Connections: Technology and Society**, p. 9 **Science Discovery Activities**, 3-1, 3-2 **Laboratory Manual**, pp. 15–16	**Teaching Transparency 5**, The Decibel Scale **Section Focus Transparency 7**
Study Guide, p. 14 **Take Home Activities**, p. 8 **Making Connections: Integrating Sciences**, p. 9 **Activity Masters**, Design Your Own Investigation 3-1, pp. 15–16	**Teaching Transparency 6**, The Ear **Section Focus Transparency 8**
Study Guide, p. 15 **How It Works**, p. 6 **Making Connections: Across the Curriculum**, p. 9 **Multicultural Connections**, p. 10 **Activity Masters**, Investigate 3-2, pp. 17–18 **Science Discovery Activities**, 3-3	**Section Focus Transparency 9**

ASSESSMENT RESOURCES	TEACHING & TECHNOLOGY
Review and Assessment, pp. 17–22 **Performance Assessment**, Ch. 3 **PASC*** **MindJogger Videoquiz** **Alternate Assessment in the Science Classroom** **Computer Test Bank**	**Spanish Resources** **Cooperative Learning Resource Guide** **Lab and Safety Skills** **Science Interactions, Course 1, CD-ROM** **Computer Competency Activities**

*Performance Assessment in the Science Classroom

NATIONAL GEOGRAPHIC TEACHER'S CORNER

National Geographic Society Products

To order the following products for use with this chapter, call the National Geographic Society at 1-800-368-2728:

- *Everyday Science Explained* (Book)
- *Messengers to the Brain* (Book)
- *How Loud Is Too Loud?* (Kids Network Curriculum Unit)
- *Listen! Hear!* (Video)

GLENCOE TECHNOLOGY

The following multimedia resources are available from Glencoe.

Science and Technology Videodisc Series (STVS)
Animals
  Songbird Study
  How Bats Hear
Human Biology
  Ear Implants
  Hearing by Touch

Glencoe Physical Science Interactive Videodisc
Waves and Sound

Physical Science CD-ROM

Teacher Classroom Resources

These Teacher Classroom Resources are examples of the materials in the Teacher Classroom Resources Package.

TEACHING AIDS

Section Focus Transparencies

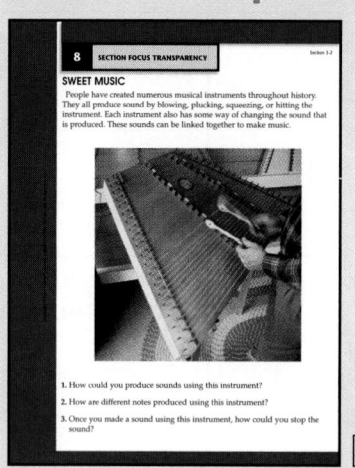

Teaching Transparencies

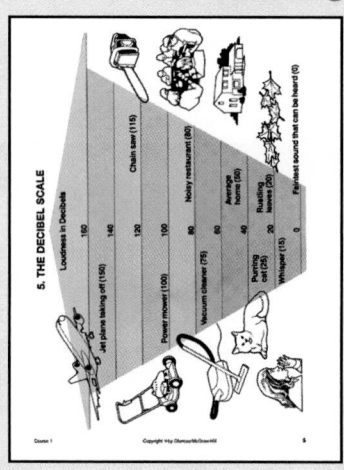

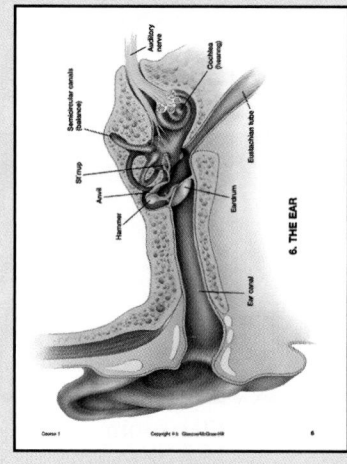

HANDS-ON LEARNING

Science Discovery Activity*

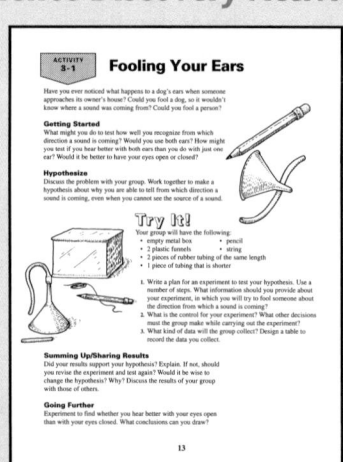

Laboratory Manual*

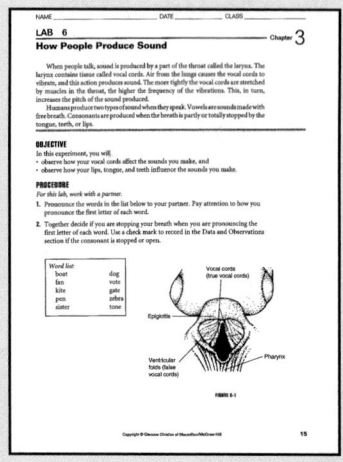

Take Home Activity

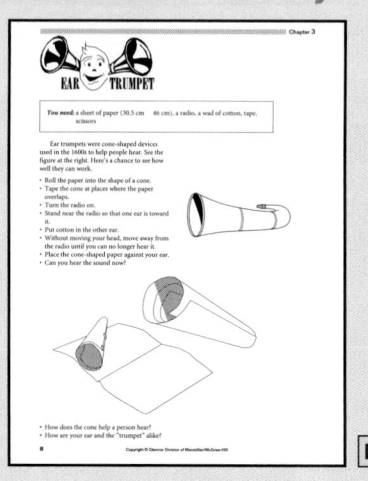

*There may be more than one activity for this chapter.

Chapter 3 Sound and Hearing

REVIEW AND REINFORCEMENT

Study Guide*

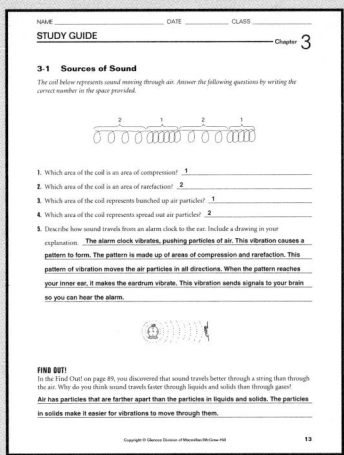

Concept Mapping

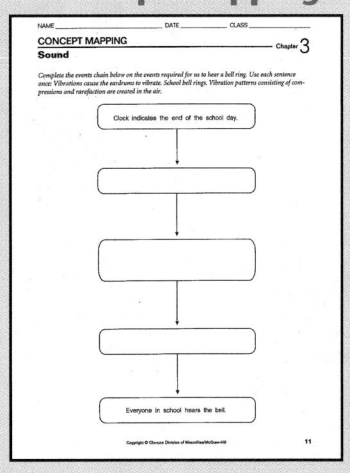

Critical Thinking/ Problem Solving

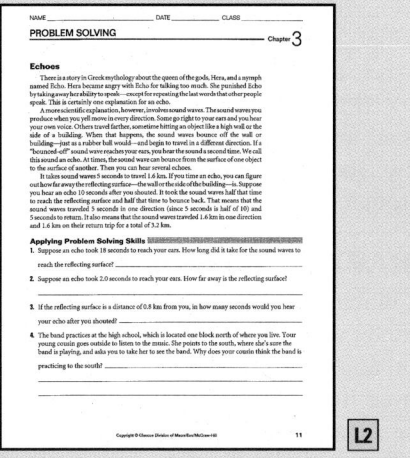

ENRICHMENT AND APPLICATION

Integrating Sciences

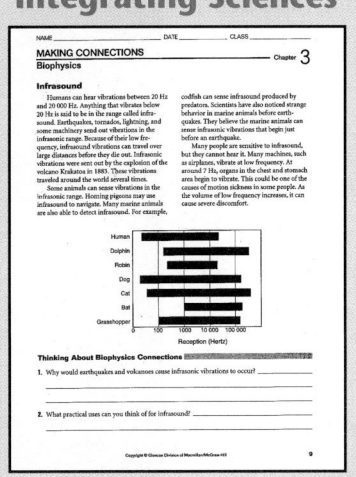

Across the Curriculum

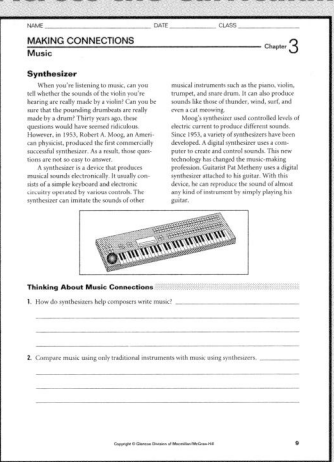

Technology and Society

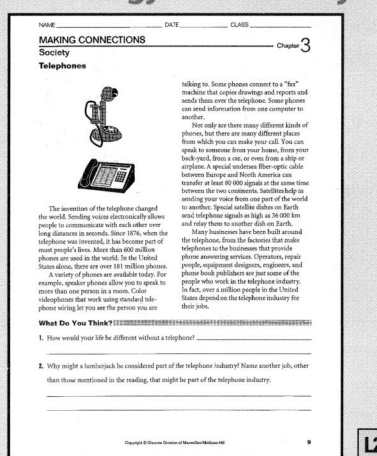

ASSESSMENT

Multicultural Connection**

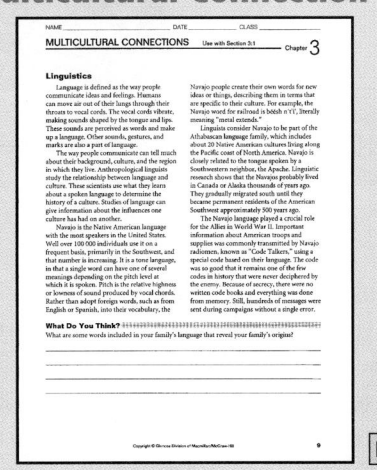

Performance Assessment

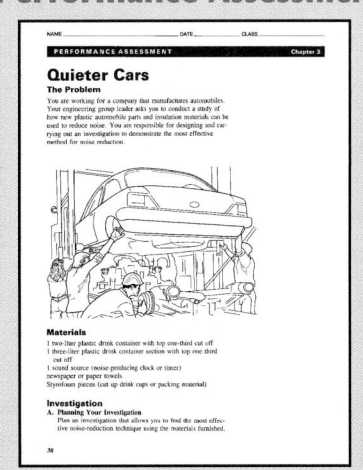

Review and Assessment

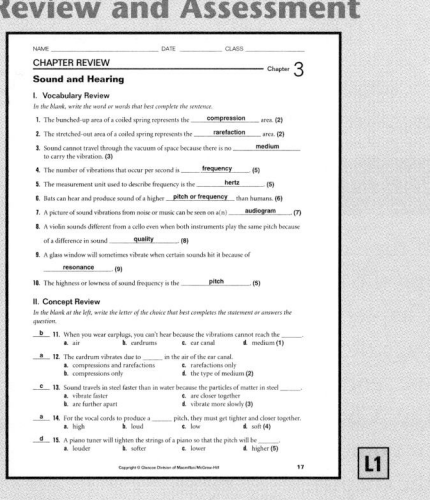

*One per section   **Two per chapter

Sound and Hearing

THEME DEVELOPMENT

One theme of this chapter is energy. Students explore sound energy, the product of vibration carried to our ears by air or another medium. They observe how sound energy travels through matter.

CHAPTER OVERVIEW

Vibrating objects produce sound through waves of compression and rarefaction. Sound only travels through a medium, which may be a solid, a liquid, or a gas. Different media conduct sound differently. Pitch is affected by the length, width, and tension of vibrating objects. The larger or looser the vibrating object, the lower the pitch. The rate of vibration is the frequency. Only certain frequencies are audible to humans. Other animals have the ability to hear some frequencies that are inaudible to humans.

Tying to Previous Knowledge

Show students a picture of a rainbow and a guitar. Point out that there are various colors on a rainbow as well as various notes on a guitar. Pluck each guitar string. Explain that the next chapter is about sound and lead a discussion about ways in which sound may be like light. Encourage them to see that both are energy. Colors of light are of different frequencies. Likewise, the notes on a guitar are of different frequencies.

00:00 OUT OF TIME?

If time does not permit teaching the entire chapter, use the Chapter Overview on this page, Reviewing Main Ideas at the end of the chapter, and the Chapter 3 audiocassette to point out the main ideas of the chapter.

Sound and Hearing

Did you ever wonder...

- ✓ If you could hear sounds on the moon?
- ✓ Why you make a sound when you blow into a soft drink bottle?
- ✓ Why the different strings on a stringed instrument produce different sounds?

Science Journal

Before you begin to study about sound and hearing, think about these questions and answer them *in your Science Journal.* When you finish the chapter, compare your journal write-up with what you have learned.

Answers are on page 110.

84

Today is moving day. Welcome to your new home! As you unpack, you hear a dog yapping next door. Someone turns a stereo up, and you hear your favorite song. A voice calls out, "Randy, please turn that down," and suddenly you can barely hear the music. Now you hear the whine of an electric saw, the pounding of a hammer, and the roar of traffic.

You've learned a bit about your new neighborhood before you've even had a chance to look around. You know that there is a dog next door, someone named Randy likes the same music you do, carpenters are working on your block, and a busy street is close by.

How did you discover all this? Simply by listening.

▶ *In this chapter, you'll find out what all sounds have in common, and how you are able to distinguish one sound from another.*

Learning Styles **LS**	Kinesthetic	Find Out, p. 87; Activity, pp. 88, 92; Multicultural Perspectives, p. 95; Investigate, pp. 104–105; Project, p. 110
	Visual-Spatial	Visual Learning, pp. 88, 97; Discussion, p. 103
	Intrapersonal	Activity, p. 86
	Logical-Mathematical	Visual Learning, pp. 90, 92; Across the Curriculum, p. 91
	Linguistic	Visual Learning, p. 100; Life Science Connection, pp. 90–91; Discussion, pp. 91, 101; Multicultural Perspectives, p. 92
	Auditory-Musical	Explore, pp. 85, 94, 102; Find Out, pp. 89, 95; Discussion, p. 92; Activity, pp. 93, 100, 103, 106; Demonstration, p. 96; Across the Curriculum, p. 96; A Closer Look, pp. 96–97; Investigate, pp. 98–99; Visual Learning, p. 103

Explore! ACTIVITY

How are sounds produced?

When you hear a sound, you are actually sensing the vibrations of your eardrums.

What To Do

1. Place a wooden ruler on a desk so that more than half of the ruler extends over the edge of the desk.

2. Use one hand to hold the ruler firmly against the desk. With the other hand, snap the free end of the ruler so it vibrates up and down.

3. *In your Journal*, describe what you see and hear.

ASSESSMENT PLANNER

PORTFOLIO
Refer to page 112 for suggested items that students might select for their portfolios.

PERFORMANCE ASSESSMENT
Process, pp. 85, 87, 99, 105
Skillbuilder, p. 91
Explore! Activities, pp. 85, 94, 102
Find Out! Activities, pp. 87, 89, 95
Investigate, pp. 98–99, 104–105

CONTENT ASSESSMENT
Oral, p. 102
Check Your Understanding, pp. 93, 100, 106
Reviewing Main Ideas, p. 110
Chapter Review, pp. 111–112

GROUP ASSESSMENT
Opportunities for group assessment occur with Cooperative Learning Strategies.

INTRODUCING THE CHAPTER

Have students examine the photograph on page 85 and suggest what sounds might be heard. Possible answers may include honking horns, engines, people talking.

Uncovering Preconceptions

Students may not connect sound with vibration. You may want to ask students if they feel any vibrations when they are exposed to very loud music.

Explore!

How are sounds produced?

Time needed 5 minutes

Materials ruler

Thinking Processes observing and inferring, comparing and contrasting, forming operational definitions, modeling

Purpose To observe that a vibration causes a sound.

Teaching Strategies

Troubleshooting Remind students to hold the ruler firmly to the desk. Students will find it easier to make the ruler vibrate if they strike or pluck it with an upward motion.

Expected Outcomes

The ruler's vibrations produce sound. The pitch of the shortened section is higher.

Answers to Questions

Students hear a tone from the ruler.

✔ Assessment

Process Have students make a list of other objects they could make vibrate to test how changing the size of the object affects the sound. Use the Performance Assessment Task List for Assessing a Whole Experiment and Planning the Next Experiment in **PASC**, p. 33. **L1** **P**

Planning the Lesson

Refer to the Chapter Organizer on pages 84A-D.

Concepts Developed

Chapter 2 described properties of light. In this section, students learn about sound. After learning that vibrations cause sound, students use Slinkies to visualize how sound travels through matter by compression and rarefaction. Students learn that sound travels only through a medium and they compare how well different media conduct sound. Students then contrast the speeds of sound and light.

1 MOTIVATE

Bellringer

 Before presenting the lesson, display **Section Focus Transparency 7** on an overhead projector. Assign the accompanying **Focus Activity** worksheet. L1
LEP

Activity Ask students to take out a pencil and paper. While they sit quietly for 60 seconds, ask them to listen and write down whatever they hear. Typical sounds might include the scratching of pencils, the hum of fluorescent lights, voices in the hallway, and traffic noises. When the minute is over, make a list of the sounds on the chalkboard. Ask students if they heard any sounds that they could not identify. Have them guess what might have produced each sound. L1

3-1 **Sources of Sound**

Section Objectives

■ Recognize that sounds are created by vibrations.

■ Distinguish between compression and rarefaction.

■ Describe the way sound travels through matter.

Key Terms

compression
rarefaction
medium

Figure 3-1

The harp, like all stringed instruments, makes sound when a player causes its taut strings to vibrate.

Vibrations Produce Sound

What sounds did you hear on your way to school this morning? The growling engine and squeaking brakes of a school bus? The roar of traffic as you walked along the street? A radio blaring from a neighbor's open window? Did you notice the sound of your own footsteps or the wind blowing past your ears? We hear so many different sounds all the time that we aren't usually aware of all of them. Sit quietly for a minute and listen to the sounds around you. What do you hear right now?

■ Sound and Matter

Your ears allow you to recognize many different sounds, but do you know what these sounds have in common? All sounds are produced by vibrating objects. When you did the Explore activity at the beginning of this chapter, you created a sound by making the ruler vibrate up and down rapidly. Vibrations are very quick, back-and-forth motions repeated over and over again. You could see the ruler vibrate, and those vibrations made a sound you could hear.

Here is an activity that will give you some clues about what sound is and how it travels through matter to your ears.

Program Resources

Study Guide, p. 13
Laboratory Manual, pp. 15–16, How People Produce Sound L2
Concept Mapping, p. 11, Sound L1
Critical Thinking/Problem Solving, p. 11, Echoes L2
Multicultural Connections, p. 9, Linguistics L1

Teaching Transparency 5 L2
Section Focus Transparency 7
Making Connections: Technology and Society, p. 9, Telephones L2

How does sound travel through matter?

Vibrations produce sound. But what is sound and how does it travel through matter to reach your ears?

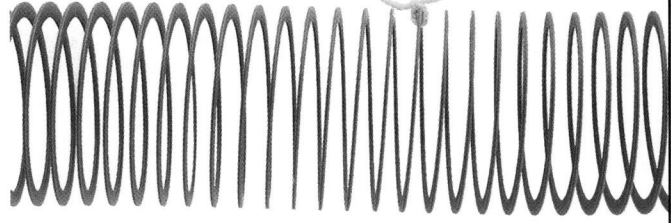

What To Do

1. Working with a partner, obtain a coiled spring and a small piece of colored string.

2. With your partner, put the coiled spring on the floor and stretch it to a length of at least 2 meters. Make sure all of the coils are about the same distance apart.

3. With one hand, squeeze together about 15 or 20 of the coils near you and observe what happens to the unsqueezed coils.

4. Release the bunched coils.

5. Quickly push your end of the coiled spring toward your partner, then pull it back to its original position.

6. Repeat the motion several times. Vary the speed.

7. Hold your end of the spring steady while your partner does the pushing.

8. Now, tie the colored string to one of the coils. Repeat the experiment and observe the string.

Conclude and Apply

1. *In your Journal*, draw what happens when you squeeze the coils.

2. Describe what happens when you push and pull the spring.

3. What happens to the string as the spring moves?

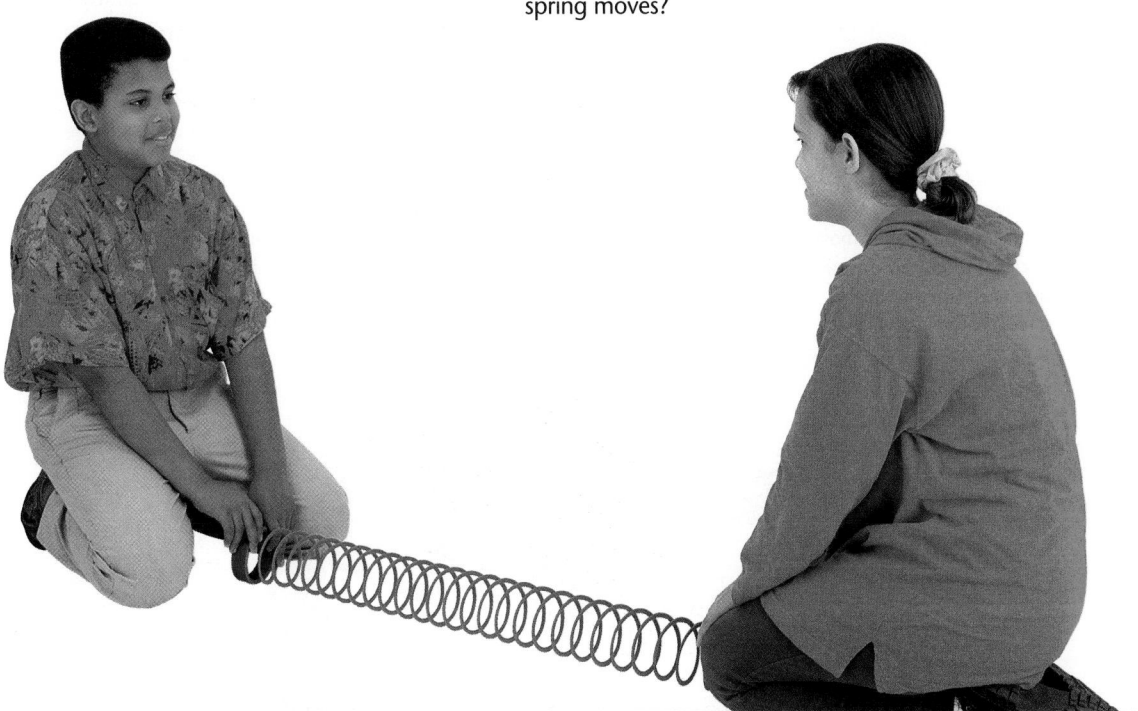

 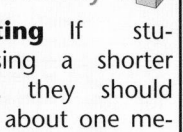

2 TEACH

Tying to Previous Knowledge

Have students recall what they learned about making observations in Chapter 1 and their observations of light in Chapter 2. In this section students will use their sense of hearing to observe how sound travels.

Theme Connection In this section the students see that a vibrating ruler creates a pattern of compressions and rarefactions. This supports the theme of stability and change because the pattern causes a change in the surrounding environment. In this instance, the compressions and rarefactions travel through air particles, causing the eardrum to change by vibrating.

Visual Learning

Figures 3-2 and 3-3 Refer to the figures and then have students try to visualize what it would be like if they could *see* sound. After students discuss the idea, have them write paragraphs in their journals in which they describe the aural characteristics of entering the classroom at the start of class. *When they enter, the classroom is quiet but it fills rapidly with people and sound.*

Connect to . . .

Life Science

Sounds are produced by pushing air through the air passages and nasal sacs, and causing certain structures, such as the nasal plug, to vibrate. The fatty melon, located on the forehead, focuses the sounds forward. Sounds are received by the melon and the lower jaw and channeled through the acoustic window to the inner ear. Information about the sound is sent to the brain. Diagrams should show the structures mentioned.

Figure 3-2

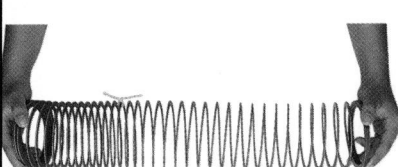

A When you pushed against the coiled spring, the coils near the string were close together and the other coils were farther apart.

B The pattern of squeezed and stretched coils was transmitted down the length of the spring.

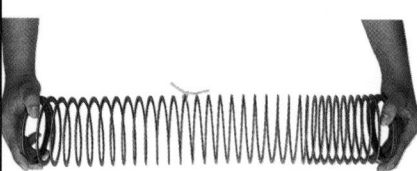

C The string did not travel with the pattern. The coils, unlike the pattern, did not travel the length of the spring. Each coil moved only slightly forward and then back to its original position.

In the Find Out activity, you created a pattern that moved down the length of the coiled spring. **Figure 3-2** shows how the pattern looks.

■ Compression and Rarefaction

Air is made up of particles so tiny you cannot see them. When everything is quiet, these air particles are about the same average distance apart, just like the coils of the stretched out spring.

A vibrating object pushes air particles just as your hand pushed the coiled spring. The vibrations create a pattern of bunched-up and spread-out particles that moves through the air. The pattern spreads out from the object in all directions. The part of the pattern with bunched-up particles is called the area of **compression**. The part with spread-out particles is called the area of **rarefaction**. **Figure 3-3** shows how a vibrating ruler affects the air particles around it.

Figure 3-3

A When the ruler vibrates upward or downward, it pushes the particles of air in front of its movement closer together—forming an area of compression.

B At the same time, the air particles on the opposite side of the ruler spread farther apart—forming an area of rarefaction.

C As the ruler vibrates up and down, it creates a pattern of compressions and rarefactions that travels through the air to your ear.

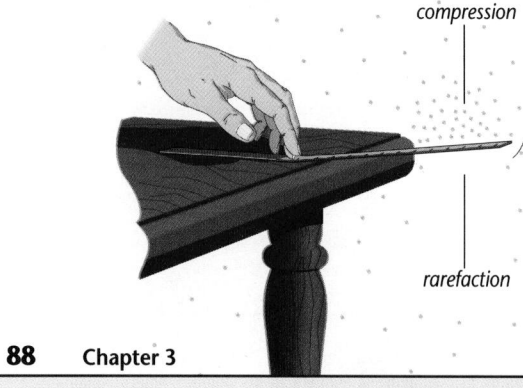

compression

rarefaction

88 Chapter 3

ENRICHMENT

Activity Use a finger or pen to tap on a desk, first very softly and then very loudly. Ask students what makes the difference between the taps. *The difference in the amount of compression and rarefaction; the louder tap was made by striking the table with greater energy, which in turn produced a greater amount of compression.*

Have students model this difference using their coiled springs. Have students model loudness by pushing on one end to set up waves in the coiled spring. Have students begin with a small push, which represents a soft, low-energy tap. Students should observe that this motion produces a small wave. Then have students use a large and sudden push, representing a loud, high-energy tap. They will observe that it produces a large wave.

Sound Requires a Medium

Vibrations produce sound by creating patterns of compression and rarefaction. Any solid, liquid, or gas that carries the pattern of sound is called the **medium** for that sound. For sound, the medium is the matter between the vibrating object and your eardrum. The medium conducts the sound to your ears. The sounds you hear almost always come to your ears through the medium of air, which is a gas.

Have you ever listened to sounds under water or put your ear against a wall or the ground to try to hear something more clearly? In the Find Out activity you can investigate how sound travels through other materials.

Connect to...
Life Science

Dolphins and porpoises make and analyze reflected sounds to communicate and locate prey. This is called echolocation. Make a diagram illustrating how dolphins produce and receive sounds.

Find Out! ACTIVITY

Can sound travel through string?

Air isn't the only kind of matter that sound can travel through.

What To Do

1. Tie about 50 cm of string to each end of a metal coat hanger.
2. Wrap the other ends of the string around each index finger.
3. Gently swing the hanger so the hook taps against a table or chair. Listen for the sound it makes.
4. Place your index fingers with the string attached in your ears and tap the hanger again.

Conclude and Apply

1. Which is a better conductor of sound, air or the string and your fingers?
2. Predict how a wire in place of the string would conduct sound.

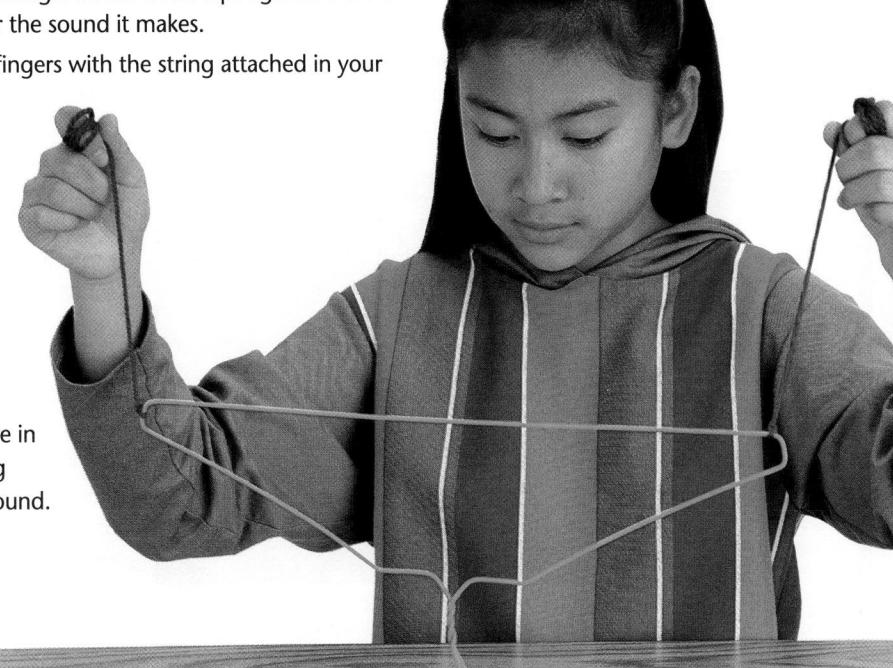

Find Out!

Can sound travel through string?

Time needed 15 minutes

TECH PREP **Materials** (per student) metal coat hanger or other light metal object, light string about 1 meter long, (thread or dental floss), flexible wire about 1 meter long

Thinking Processes thinking critically, observing and inferring, comparing and contrasting, recognizing cause and effect, modeling

Purpose To compare the ability of different materials to conduct sound.

Preparation You may wish to have students work in pairs.

Teaching the Activity

Ask students to keep quiet while performing this activity. They may wish to repeat it. Tell students to wrap the string around their fingers to medium tightness, rather than knotting it. **L1** **LEP**

Science Journal Students can record their observations in their journals. **L1**

Expected Outcomes

Solid objects are better conductors of sound than is air, and wire is a better conductor than string.

Conclude and Apply

1. the string and fingers
2. Sound would travel better through the wire.

✔ Assessment

Content Ask what the air, the string, and the wire have in common. Do they all conduct sound? *yes* Do some materials conduct sound better than others? *yes*

Meeting Individual Needs

Hearing Impaired Hearing-impaired students will not be able to respond to the Find Out as presented. All hearing-impaired students should be able to feel the vibrations conducted by the string or wire by placing their fingers in their ears, and they may be able to compare the abilities of the string and wire to conduct vibrations. Encourage these students to conduct as many observations as they may require.

Program Resources

Critical Thinking/Problem Solving, p. 5, Flex Your Brain
Science Discovery Activities, 3-1, Fooling Your Ears; 3-2, Stopping Sound

Flex Your Brain Use the Flex Your Brain activity to have students explore COMPRESSION and RAREFACTION.

Figure 3-4

The alarm goes off but no sound is heard. Explain what has happened.

As you saw in the Find Out activity, sound also travels through solids. Sound travels faster through liquids and solids than through gases.

One reason liquids and solids conduct sound better than air is that sound travels through them much faster than it travels through air. **Figure 3-5** shows the speed of sound through various materials.

■ Sound and a Vacuum

If you clapped your hands on the moon, you would make absolutely no sound. Why not? Because the moon has no atmosphere. There is no air through which sound can travel. There is no medium to carry the vibration patterns.

In a famous experiment shown in **Figure 3-4**, a clock's alarm is set to go off. The clock is placed on a piece of thick felt and covered with a glass dome. When the alarm rings, you hear it even though the sound is muffled by the felt and the glass cover. The experiment is repeated, but this time the air is pumped from the glass dome. The alarm goes off but you hear no sound. Why? There is no air inside the glass to carry the sound. Why is the piece of felt necessary in this experiment? Felt is used to muffle any sound that would be carried by the table.

Life Science CONNECTION

Sounds Are All Around You

Your ears are very sensitive instruments. They pick up many more sounds than you're aware of because your brain does such a good job of blocking out what you don't need to

hear. Unless you make a conscious effort to listen to all the sounds around you, you may have no idea that many of them are there.

Take a moment just to listen. What do you hear? What about the sounds outside? Imagine how hard it would be to concentrate in noisy places like the subway station pictured here, if you always noticed every sound around you.

Because your senses are constantly bombarded with all sorts of information, your brain needs a way to tune out some of it. Otherwise you couldn't concentrate.

Figure 3-5

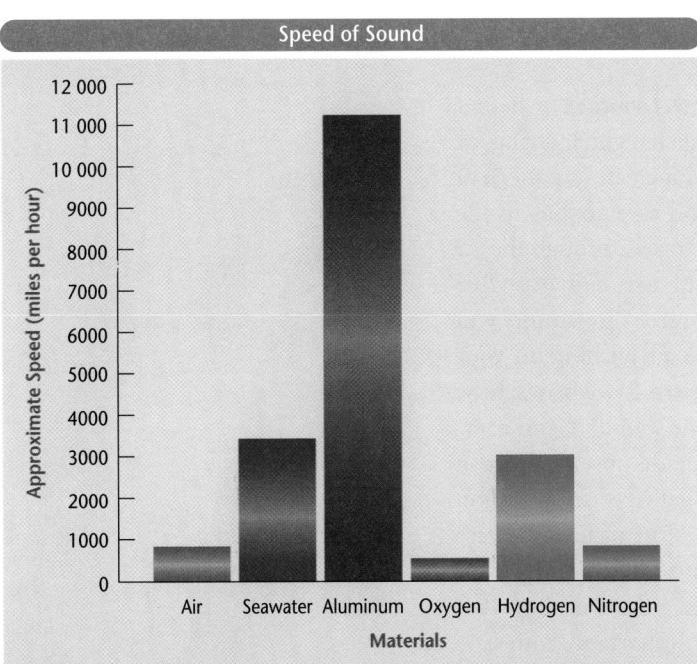

Speed of Sound

Approximate Speed (miles per hour) — Materials: Air, Seawater, Aluminum, Oxygen, Hydrogen, Nitrogen

Making and Using Graphs

Use the information in **Figure 3-5** to answer these questions. If you need help, refer to the **Skill Handbook** on page 657.

1. Which material conducts sound fastest?
2. Which material conducts sound most slowly?
3. What is the speed of sound through aluminum?
4. Air is made up of many gases, including oxygen and nitrogen. How can you relate this information to the sizes of the bars on the graph for these materials?

SKILLBUILDER

1. aluminum
2. oxygen
3. about 11 000 miles per hour
4. The size of the bar for air is about the same as the size of the bars for the gases contained in air. Sound travels at about the same speed through air and the materials that make up air. **L1**

Content Background

The speed of sound depends on the temperature and the medium through which the sound is traveling. Atoms are usually close together in solids, which is why solids transmit sound faster than air does.

Discussion Tell students that one of the world's few solo percussionists is a hearing-impaired New Zealander. Ask students how they think she is able to play. Help them understand that she "hears" vibrations with her legs and feet.

Across the Curriculum

Math

Tell students that sound travels through the air at about 330 m/s, while light crosses short distances almost instantaneously. Identify a location that is about 1 km away from your school. Ask students to tell how long it would take the sound of thunder to reach them if lightning struck that location. $\frac{1000\ m}{330\ m/s} = 3.0$ seconds Repeat the question for a location that is about 2, 3, 4, and 5 km from your school.

RAS

An area in your brain called the reticular activating system (RAS) sorts out all the information your senses provide. The RAS is a network of nerve cells deep within your brain stem at the top of your spinal cord. It helps you focus your attention on specific sounds while tuning out all the others.

The RAS regulates your level of awareness by screening the messages from your senses and passing on only what seems important or unusual. For instance, your RAS helps you ignore the sound of lockers slamming in the hallway to concentrate on what your

teacher says. And when a fire alarm goes off in that same hallway, your RAS automatically puts that message through.

Science Journal
Write a paragraph *in your Science Journal* that describes the sounds you hear in the morning when you wake up and the sounds you hear right before you fall asleep at night. What do these sounds tell you about your environment and the activities that are going on?

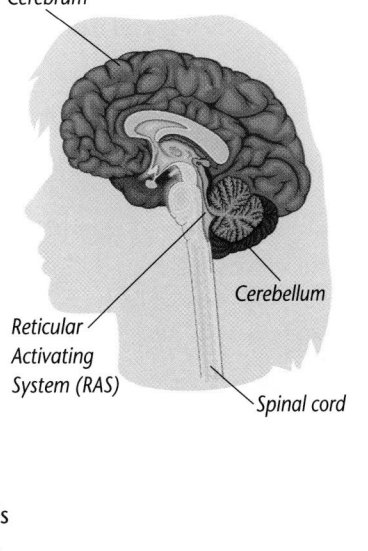

Cerebrum

Cerebellum

Reticular Activating System (RAS)

Spinal cord

top. Our sense of pitch depends on which part of the cochlea vibrates.

Teaching Strategies

Play music at a fairly loud volume and simultaneously read aloud the fourth paragraph describing the reticular activating system. Then ask students to explain what the RAS is and where it is located. Students will probably have trouble paying attention to both things at the same time. Ask them to describe their experience.

Science Journal

Answers will vary but students may say that the nighttime sounds indicate decreased activity of daytime animals and increased activity of species that are active during the night. Morning sounds may indicate increased activity of organisms active during the day such as birds.

Going Further ⑴⑴⑴➤

Have small groups develop tests of what stimuli the RAS passes on and what stimuli it filters out. Each group's test should present two stimuli and record which stimuli the subject perceived. After groups have conducted their own tests, have them exchange tests and compare the results. Use the Performance Task Assessment List for Designing an Experiment in **PASC**, p. 23.

L2 **P** **COOP LEARN**

Activity Have students line up and practice creating a "human wave" as seen at sports stadiums. Once they have perfected the wave, they can try creating waves of different speeds to simulate the different speeds at which sound moves through different types of matter. L1 LEP

Discussion Grip one end of a plastic ruler and wiggle it, rotating your hand at the wrist, as fast as you can. Ask students why it is not producing sound even though it is vibrating. Encourage students to understand that not all vibrations produce sound that we can hear. Have students discuss what kinds of vibrations might produce sound and, if appropriate, lead students to understand that sound consists of much faster vibrations than the ruler produced.

Visual Learning

Figure 3-6 When a friend shouts your name, how do you get the message? *Pattern of compressions and rarefactions reach eardrum →eardrum vibrates→causes hammer, anvil, and stirrup to vibrate→fluid in cochlea vibrates→receptor hairs in cochlea vibrate sending message to brain→brain analyzes and interprets message→you "hear" your name.*

How You Hear Sound

When an object vibrates, whether it's a vibrating ruler, a guitar string, or your vocal cords, it creates a pattern of compressions and rarefactions in the air. The pattern travels through the air particles to your ear, causing your eardrum to vibrate. The sound you hear is really the air pushing on your eardrums. **Figure 3-6** shows how vibrations become sound in your ears.

When you made observations in Chapter 1, you gathered information with your eyes. Most of us are aware that we depend on our sight to observe the world around us. But what you hear is also important. Your ears give you information about things you can't see. Remember moving day and how you found out about a possible new friend named Randy?

Figure 3-6

When a friend shouts your name, how do you get the message?

Ear canal

92

World Languages

There are over 3000 languages spoken in the world today, not including dialects. All of these languages have sound patterns created from the 20 to 60 sounds that human speech organs can make. Have student volunteers say a few words or sentences in different languages. Have the students listen for sounds they recognize.

If there are no students in the class who speak a foreign language, videotape short excerpts from newscasts on foreign-language stations. Many cable television systems have stations broadcasting in other languages such as Spanish and Japanese. Many foreign language films are available on videotape.

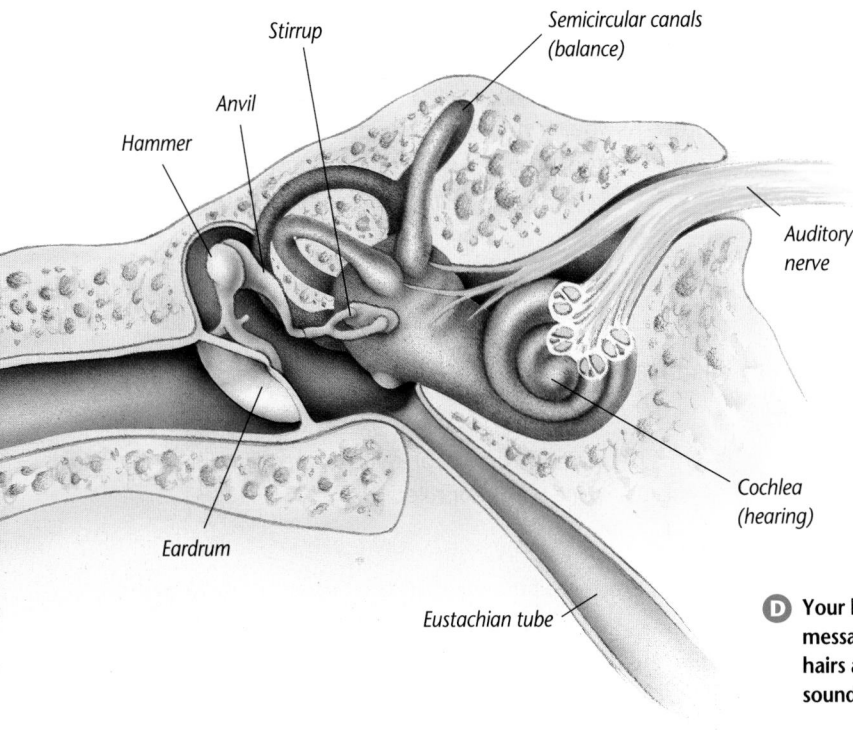

A The pattern of compressions and rarefactions created by your friend's vocal cords reaches your ear and causes your eardrum to vibrate.

B The vibrations pass from your eardrum to three tiny bones in your middle ear, the hammer, the anvil, and stirrup. Each bone vibrates in turn. The vibrations pass from the stirrup to the cochlea, a fluid-filled chamber.

Stirrup

Semicircular canals (balance)

Anvil

Hammer

C Tiny receptor hairs in the cochlea are attached to a nerve that sends messages to your brain. When sound vibrations passed on by the three bones cause the cochlea fluid to vibrate, these hairs bend back and forth, sending different sound messages to your brain.

Auditory nerve

Cochlea (hearing)

Eardrum

Eustachian tube

D Your brain analyzes the messages sent by the receptor hairs and interprets them as the sound of your name.

check your UNDERSTANDING

1. Why doesn't a ruler that is sitting by itself on your desk make a sound? What must happen to an object before it makes a sound?

2. Explain how to use a coiled spring to show the compressions and rarefactions that are created by a vibrating object.

3. Explain how a vibrating object that is some distance away from you causes your eardrums to vibrate.

4. **Apply** Imagine you are traveling in outer space. You see a nearby satellite explode when a large meteor crashes into it. Would you hear the boom? Why or why not?

check your UNDERSTANDING

1. It is not vibrating. It must be set into vibration to create a sound.

2. When you push one end, you cause some of the spring's coils to bunch together into areas of compression and some to spread apart in areas of rarefaction. The pattern of compressions and rarefactions moves down the length of the coiled spring.

3. It creates a pattern of compression and rarefaction in the particles of the air, which is carried to your ear.

4. No, because there are no air particles to provide a medium for the sound waves.

3 ASSESS

Check for Understanding

Discussion To help students answer the Apply question, ask them to describe sounds included in movie space battles. Ask them if they think the sounds are realistic and why they think so. Ask them why they think the filmmakers add these sounds. Use the Performance Task Assessment List for Making Observations and Inferences in **PASC**, p. 17.

Reteach

Demonstration Help students recognize how sound moves through air by setting off a chain of dominoes and making the pieces fall, one after another. Explain that this is another way to visualize how sound travels. **L1** **LEP**

Extension

Activity Have students compare and contrast the ability of different materials to conduct sound by extending the Find Out on page 89. Have them tie different metal objects to the string and test each one. They will see how lower-pitched sounds produced by larger objects are not easily heard through the air but travel well through the string. **L3**

4 CLOSE

Allow students to develop their ability to recognize cause and effect by having them tie two tin cans together with a long piece of string. Puncture a hole in the center of the bottom of the can and tie a knot so the string does not slip through the hole. Have one student talk into one end while another student listens at the other end. Ask students how the second student can hear the first student. *The voice causes the string to vibrate.* **L1**

PREPARATION

Planning the Lesson

Refer to the Chapter Organizer on pages 84A–D.

Concepts Developed

In this section, students learn that the pitch of a sound depends on its frequency. They will also learn how the length or width of a vibrating object affects its pitch by determining how fast it vibrates.

Explore!

Does the length of a vibrating object affect the sound it makes?

Time needed 5 minutes

Materials metal ruler

Thinking Processes observing and inferring, comparing and contrasting, recognizing cause and effect

Purpose To determine that greater length produces lower frequency vibration and lower pitch.

Science Journal Have students record their observations and their answers. [L1]

Teaching the Activity

Discussion After students experiment, extend the activity by having them slide the ruler inward and outward as they pluck it, so that they hear the pitch changing.

Expected Outcomes

Students observe that a long section of ruler vibrates more slowly and produces a lower tone, while a short length vibrates faster, producing a higher tone.

Answers to Questions

5. The sound is higher when the ruler is shorter. The sound is lower when the ruler is longer.

3-2 Frequency and Pitch

Section Objectives

- Use the length or thickness of a vibrating object to predict whether its sound will be high or low.
- Describe the relationship between pitch and frequency.
- Compare the sound frequencies humans hear with the sound frequencies animals hear.

Key Terms

frequency
hertz (Hz)
pitch

The Pitch of Sound

Have you ever made a sound by blowing across the top of a soft drink bottle? Did you notice that the sound changes as you drink more and more of the liquid inside the bottle? As you drink, what replaces the liquid in the bottle? Air.

When you blow across the top of the bottle, the air inside the bottle vibrates. The amount of air inside determines whether the sound will be higher or lower. Do the next activity to find out what happens if you change the length of a vibrating solid.

Explore! ACTIVITY

Does the length of a vibrating object affect the sound it makes?

At the beginning of this chapter you learned you could create a sound with your ruler.

What To Do

1. Extend exactly half the ruler's length beyond the edge of the desk.
2. Snap the free end and listen to the sound.
3. Vary the length of the ruler that extends over the desk and observe the sound changes.
4. *In your Journal*, record how the sound changes as you change the length of the ruler.
5. When is the sound higher? Lower?
6. When does the ruler vibrate faster? More slowly?

---

6. The ruler vibrates faster when it is shorter. The ruler vibrates more slowly when it is longer.

✔ Assessment

Performance Have small groups create musical instruments that operate on the short-long principle. Use the Performance Task Assessment List for Group Work in **PASC**, p. 97. [L1] COOP LEARN

Program Resources

Study Guide, p. 14
Teaching Transparency, 6 [L1]
Take Home Activities, p. 8, Ear Trumpet [L1]
Making Connections: Integrating Sciences, p. 9, Infrasound [L2]
Section Focus Transparency 8

In the Explore activity, you observed that a longer segment of ruler produces a lower sound than a shorter segment. **Figure 3-7A** and **B** shows the compressions and rarefactions your vibrating ruler makes in the air around it.

Figure 3-7C shows one cycle of the ruler. The number of times an object moves back and forth in one second is one way to keep track of an object's motion. The number of times an object vibrates in one second is called its **frequency**. We measure frequency with a unit called hertz, named after the German scientist Heinrich Hertz. One **hertz** (abbreviated Hz) is a frequency of one vibration per second or one cycle per second.

This next activity will show you another way to create high and low sounds.

Figure 3-7

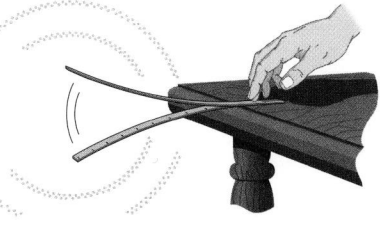

Ⓐ Because a long ruler vibrates slowly, it creates bands of compression and rarefaction that are farther apart. The result is a low sound.

Ⓑ A shorter ruler vibrates more rapidly than a longer one. Faster vibrations create compressions and rarefactions that are closer together. The sound is higher.

Ⓒ Each back-and-forth vibration of the ruler's motion is one cycle.

1 Cycle

Find Out! ACTIVITY

How do you make changes in sound on the strings of an instrument?

Take a close look at an acoustic guitar, violin, or other stringed instrument. Do the strings all look the same?

What To Do

1. Choose a stringed instrument and pluck the strings, one at a time.
2. Choose one string and pluck it. Push the string firmly against the fret board with your finger and pluck it again.
3. Loosen one string by turning the tuning peg. Pluck the string as you gradually tighten the peg.

Conclude and Apply

1. Which has a higher sound, a thick string or a thin string under the same tension? Why?
2. *In your Journal,* discuss two ways to change the sound a string makes.

95

Multicultural Perspectives

Musical Instruments

Musical instruments from all over the world apply the principles outlined in this section. Pan pipes from Greece and South America are played by blowing across tubes of different lengths. The pitch of African drums varies with their sizes. The finger harp consists of different lengths of thin metal or wood that are flicked with the fingers to produce tones. Striking a steel drum from the Caribbean in the center with sticks produces a low note due to the large area being set into vibration. Striking higher up toward the rim produces higher notes, as the area of vibration is more restricted. Have students work in small groups to create musical instruments. L1 **COOP LEARN**

■ Changing Pitch

Your ears recognize differences in sound frequencies as differences in pitch. **Pitch** refers to the highness or lowness of the sound you hear. When you hum along with your favorite music, you raise and lower the pitch of your own voice. If you listen carefully, you can hear that the pitch of your voice rises and falls even when you're just talking with friends.

As you discovered in the Find Out activity, thicker strings produce lower-pitched sounds. Changing the length of a string and changing the tension on the string will also change its pitch. **Figure 3-8** shows three ways to obtain different pitches on stringed instruments.

Figure 3-8

Thickness

A Because they are heavier, thicker strings vibrate more slowly than thinner strings vibrate. The slower the vibration, the lower the pitch.

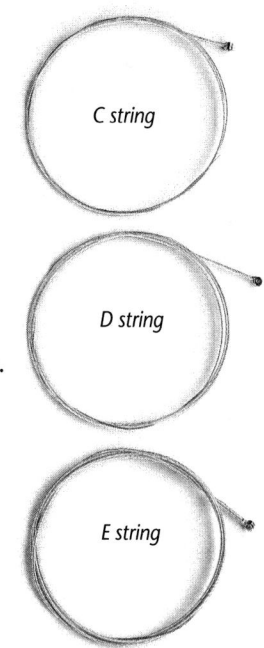

C string

D string

E string

Your Vocal Cords

What happens when you talk? How do you make those sounds? Just like other sounds, speech is produced by vibrations. When you speak or sing, the air you breathe out vibrates your vocal cords. The vocal cords are two thick folds of lip-shaped tissue that stretch across your larynx near the top of your windpipe.

Controlling Pitch

When you make a high-pitched sound, your muscles in your larynx stretch your vocal cords, which tightens them and brings them closer together. When you make a lower sound, your vocal cords relax a bit and move farther apart. You can control the pitch of your voice by tensing or relaxing your vocal cords. But changing the pitch of your voice rarely takes conscious effort because your brain adjusts your vocal cords automatically.

When you whisper, you form words with just your tongue and lips. Place your hand on your throat and say something out loud. Can you feel the vibrations? Now touch your throat and whisper. Your vocal cords should keep still.

Ventriloquists make their voices seem to come from someone else—often from a puppet or dummy.

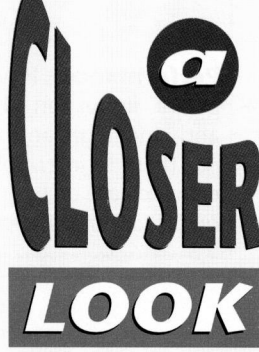

Purpose

The role of vocal cords in producing speech is an example of some principles of sound presented in Chapter 3. Section 3-1 presented the concept that vibrations produce sound. In Section 3-2, students found that the pitch of a sound can be altered by changing the tension of the vibrating object. The pitch of people's voices is controlled by the tensing and relaxing of their vocal cords.

Content Background

The tension of the vocal cords is the factor in voice production over which we have the least conscious control. Adult men have vocal cords approximately one-third longer than adult women. This accounts for the deeper male voice. The other key factors in voice production are breath control and resonance in the nasal passages. This resonance is controlled by the position of the soft palette in the roof of the mouth.

Length

B Holding a string down against the fretboard shortens the vibrating part of the string and raises the pitch of the note produced. The shorter the vibrating object, the faster the vibrations and the higher the pitch.

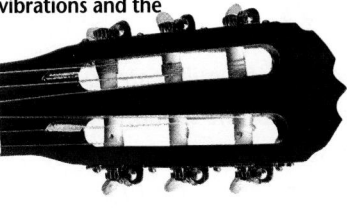

Tension

C Turning the tuning peg one way reduces the tension on the string. The vibrations slow down and the pitch gets lower. When you turn the peg the opposite way, the tension increases, the vibrations speed up, and the pitch gets higher.

Ventriloquists speak by moving only the tip of the tongue.

You Try It!

Practice these steps in front of a mirror, watching carefully for movement.

1 Bring your teeth together without tightening your jaw.

2 Part your lips slightly and smile a little.

3 Move your tongue to sound the vowels.

4 Next, try the consonants. Sounds like f, v, p, b, and m are tricky. With practice, you can learn to imitate these sounds with just your tongue.

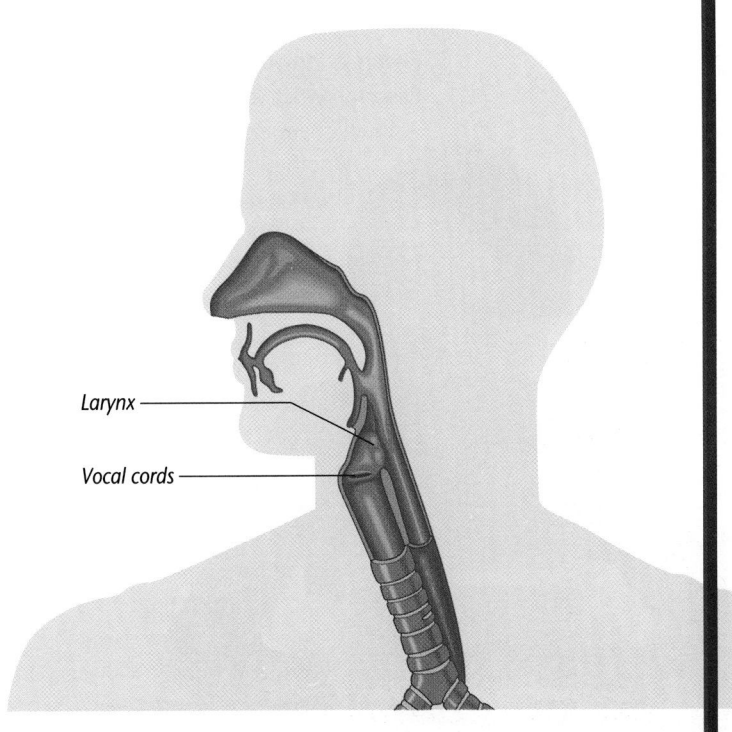

Larynx

Vocal cords

GLENCOE TECHNOLOGY

 Videodisc

The Infinite Voyage: The Great Dinosaur Hunt

Chapter 9
Communication Theories

The Infinite Voyage: The Dawn of Humankind

Chapter 3
Evolution of Speech: Reasons and Physiological Variants

Visual Learning

Figure 3-8 There are two ways of raising the pitch played by a guitar string. Pressing down on the fretboard effectively shortens the strings causing them to vibrate more quickly. Tightening the string speeds up the vibrations. Tightening and loosening pegs, however, is a painstaking and slow process. Fretboards are considered a great innovation in the history of stringed instruments. If possible, show students a photo of a Celtic harp, which has no fretboard.

Providing a large, even volume of breath while lifting the soft palette provides the loudest, most resonant, and farthest-carrying sound.

Teaching Strategies

The You Try It activity shows students how ventriloquists speak. Organize students into small groups to perform the activity. Ask each group to list the letters of the alphabet, noting whether each is easy or hard to produce with the teeth together and lips slightly apart. Then have students categorize the alphabet into two groups: those letters pronounced using the lips and those pronounced without them. **COOP LEARN**

Going Further ▐▐▐▐▐▐▐▶

Divide the class into small groups to research ventriloquism. Questions students should try to answer may include:

1. Do ventriloquists really "throw their voices" or is this an illusion?

2. What are the earliest known origins of ventriloquism and what purposes was it used for?

Have the groups present their findings as brief oral reports. Use the Performance Task Assessment List for Oral Presentation in **PASC,** p. 71. **L2** **COOP LEARN** **P**

DESIGN YOUR OWN INVESTIGATION

3-1 Length and Pitch

Preparation

Purpose To determine how the amount of water in a test tube affects pitch.

Process Skills making and using tables, observing, interpreting data, modeling, sequencing, predicting, forming hypotheses, comparing and contrasting, measuring in SI

Time Required 50 minutes

Materials See reduced student text.

Possible Hypotheses Students should hypothesize that the test tube with the most water will produce the lowest pitch or the one with the least water has the highest pitch.

The Experiment

Troubleshooting If you have limited equipment, different groups can test tubes of different sizes but make certain that the test tubes within each group are the same size.

Process Reinforcement Data tables should include two data columns for each tube tested—one for the amount of water in milliliters and one for the pitch relative to the tubes beside it. Students should be aware that more than one person should test the pitch. Not everyone is able to discern small changes in pitch.

Possible Procedures Using identical test tubes, fill them with varying volumes of water. One test tube should remain empty. Record the water volume in each tube. Blow across the top of each tube with the same amount of force. Record the relative pitch of the sound that results. Arranging the tubes from highest to lowest pitch will show the effect of water volume on pitch.

DESIGN YOUR OWN INVESTIGATION

Length and Pitch

You have learned that shortening the length of a guitar string speeds up the vibrations and raises the pitch of the sound. If you produce a sound by blowing across a test tube full of water, what happens to the pitch of that sound when you empty the tube to half full and blow across it?

Preparation

Problem
When air is blown across the top of a test tube, the column of air inside the tube vibrates. When water is added to the tube, what happens? How does this added water affect pitch?

Form a Hypothesis
Based on what you have learned about changing pitch, decide on a hypothesis for your group. Write it down.

Objectives
- Observe how pitch changes with varying amounts of water in the test tube.
- Conclude from your investigation how the length of the vibrated column of air affects pitch.

Materials
test tubes with an approximate diameter of 2.5 cm
test-tube rack
felt-tip marker
water
small graduated cylinder
small metric ruler

Safety Precautions

Be careful handling glass test tubes.

Program Resources

Activity Masters, pp. 15–16, Design Your Own Investigation 3-1

DESIGN YOUR OWN
INVESTIGATION

Plan the Experiment

1 Examine the materials and plan how your group will test the hypothesis. Write a step-by-step procedure.

2 *In your Science Journal,* draw diagrams of all the test tubes you use. Record your data on this diagram.

3 How will you measure the length of the column of air? Be sure to record the measurements on your diagram.

Check the Plan

1 Who will blow across the test tubes? Do the tubes need to be identical? Who will judge the sound? Do you need more than one person's opinion to judge the pitch?

2 Before you begin your experiment, make certain that your teacher approves your plan.

3 Carry out the experiment. Make observations and record your data on the diagrams *in your Science Journal.*

Analyze and Conclude

1. **Analyze Data** Which test tube produced the lowest pitch? How much water was in it?

2. Which test tube produced the highest pitch? How much water was in it?

3. **Compare and Contrast** Compare the pitches of all the bottles. How did the amount of water in the test tube affect the pitch of the vibrating column of air?

4. **Conclude** Make a statement about how the length of a vibrating column of air affects its pitch.

5. **Use Math** From your diagram, construct a bar graph relating column size to pitch.

6. What basic musical instrument does your test-tube instrument resemble?

Going Further

Can you create a musical scale by blowing across test tubes with varying amounts of water in them? How many test tubes will you need to create a musical scale of one octave?

Teaching Strategies

Discussion Tell students that musical instruments are based on vibrating columns of air of different lengths.

Science Journal Have students fill in the data tables in their journals. Students may want to investigate musical instruments based on the length of a column of air and write about them.

Expected Outcome

Tubes with more water produce higher pitches than tubes with less water.

Analyze and Conclude

1. the one with the most water. Students should record the amount of water in mL.

2. the one with the least water. Students should record the amount of water in mL.

3. Adding more water to the test tube raised the pitch.

4. The longer the vibrating column of air, the lower the pitch; or the shorter the vibrating column of air, the higher the pitch.

5. Students' graphs will vary but should show that as the length of the column of air increases, the pitch decreases.

6. a flute

Going Further |||||||▶

Students will need eight test tubes to make an octave.

✔ Assessment

Process With a pencil, tap the same tubes to construct a musical scale. Explain the unexpected difference between the two scales. When a tube is tapped, more water produces a lower pitch because the column of air is not vibrating nor is the water; the glass of the tube is vibrating. Striking an empty test tube gives the highest pitch. Touching the tube with fingers or putting water inside the tube slows down or deadens vibrations of the glass. More water slows the vibrations more and produces a lower pitch. This is what happens when a metal cymbal is struck and then touched to deaden the sound.

3 ASSESS

Check for Understanding

Discussion To help students answer the Apply question, ask them to describe the three ways a guitarist can change pitch.

Reteach

Activity Ask students to make percussion instruments out of tightly stretched leather or canvas stretched over different sized cans. By tapping each can with the eraser end of a pencil, they will detect that different sounds are produced depending on the volume of air. L1

Extension

Activity Experiment with sound by filling three 1-L bottles to different levels. Have a different student blow across each bottle's mouth. Have students explain why the bottle with less water has the lowest tone. *The column of air is vibrating, not the bottle and water.* L2

4 CLOSE

Demonstration

Challenge students to recognize cause and effect in musical instruments by showing them a slide trombone. Explain that the pitch is changed by moving the slide in and out. Ask students whether it makes a high or low note with the slide out. *Low, because the slide enlarges the column of air inside.* L1

Sound Frequencies

Have you ever seen a dog or cat perk up its ears as if it just heard something when you didn't hear anything at all? Dogs, cats, and other animals can hear sounds that humans can't hear. **Figure 3-9** shows the sound frequencies that humans and animals can hear.

As the figure shows, many animals hear a wider range of sound frequencies than humans. Hearing is important for an animal's survival. Finding food and detecting danger depends on a keen sense of hearing. Many animals have a keener sense of hearing than either sight or smell. The design of many animals' ears differs from that of humans. Bats have large outer ears that pick up more vibrations than small ears. Animals such as dogs and cats have movable outer ears which they move to locate sound without moving their heads.

There are many more sounds around us than our ears can hear. Not all the sounds we hear are pleasing to us. In the next section, you will learn how to distinguish between the sound of music and and the sound of noise.

Figure 3-9

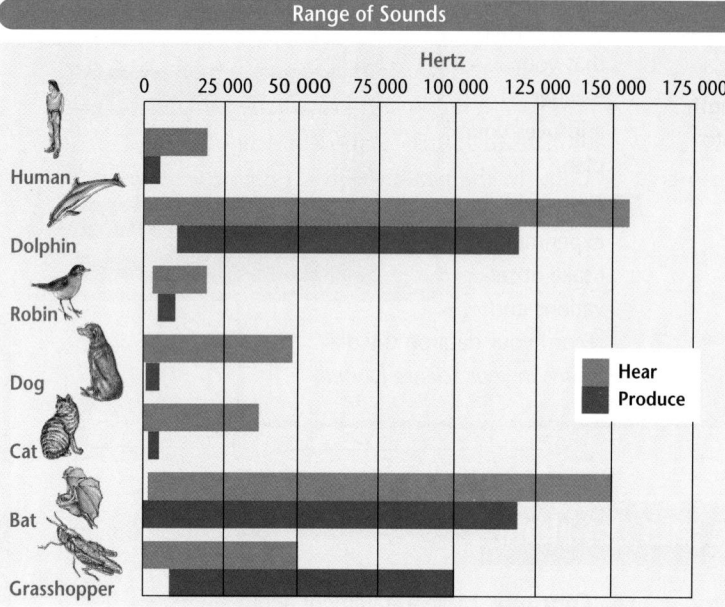

check your UNDERSTANDING

1. If you stretched out a thick rubber band and plucked it, how would the pitch of that sound compare with the pitch of a thin rubber band stretched to the same tension? Why?
2. Would a whistle with a sound frequency of 3000 Hz have a higher or lower pitch than a whistle whose frequency was 5000 Hz?

Explain your answer.
3. Use **Figure 3-9** to compare the abilities of humans and dolphins to hear different frequencies.
4. **Apply** Explain how an acoustic guitarist can play all the pitches of a ballad with just six strings.

Visual Learning

Figure 3-9 Ask students to describe situations where they observed animals being able to hear what people could not.

check your UNDERSTANDING

1. The thick band makes a lower note. It vibrates more slowly.
2. A 3000-Hz whistle has a lower pitch; there are fewer vibrations each second.
3. Dolphins can hear much higher pitches.
4. Guitarists can produce many notes from a string by varying its length and tightness.

3-3 Music and Resonance

What Is Music?

You've just gotten home from school. You throw down your books, turn on some music, and dance around the room as you sing along with your favorite song. Has anyone ever called your favorite music a jumble of noise? Have you ever heard music that sounds noisy and confusing to you? What is the difference between music and noise?

Both music and noise are sounds. We sometimes think of noise as unpleasant or annoying. But some noisy sounds, like falling rain or ocean waves, can also be pleasant.

You can make noise by tapping your pencil on the desk or speaking nonsense syllables. You can also use those sounds to create music. You could tap your pencil in rhythm or make up a melody for nonsense syllables. A sound that's considered noise in one situation might be music in another. The opposite is also true. A radio left on a music station overnight is noise to the person who's trying to sleep in the next room. Audiograms show a picture of sound vibrations. Using the audiograms in **Figure 3-10**, how would you describe the difference between the vibrations made by a noise, such as an electric saw, and the vibrations made by a musical instrument, such as the violin?

Section Objectives

- Distinguish between music and noise.
- Explain how different musical instruments produce sounds of different quality.
- Describe resonance.

Key Terms

resonance

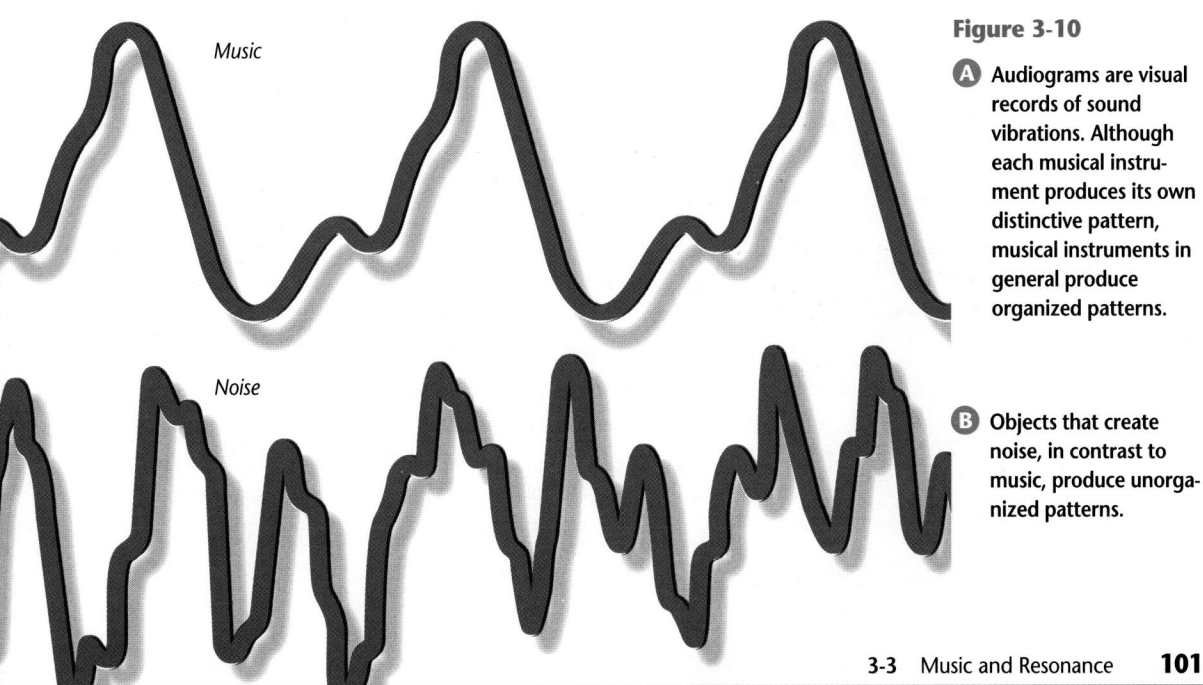

Music

Noise

Figure 3-10

 **A** Audiograms are visual records of sound vibrations. Although each musical instrument produces its own distinctive pattern, musical instruments in general produce organized patterns.

B Objects that create noise, in contrast to music, produce unorganized patterns.

Program Resources

Study Guide, p. 15

How It Works, p. 6, Is That Really My Voice? L1

Making Connections: Across the Curriculum, p. 9, Synthesizer L1

Multicultural Connections, p. 10, The Marimba L1

Section Focus Transparency 9

Visual Learning

Figure 3-10 Ask students what specific features of the top wave makes it "organized." If necessary, point out the repeated pattern of twin peaks, one much higher than the other. Ask students if they can see a similar pattern in the wave produced by noise. *This wave has no such regular pattern.*

Planning the Lesson

Refer to the Chapter Organizer on pages 84A-D.

Concepts Developed

The previous lesson discussed pitch as the result of frequency. In this section, students learn the difference between music and noise, identify the factors that affect sound quality of musical instruments, and experiment with resonance, the tendency of objects to vibrate at the same frequency as the source of sound.

1 MOTIVATE

Bellringer

Before presenting the lesson, display **Section Focus Transparency 9** on an overhead projector. Assign the accompanying **Focus Activity** worksheet. L1 LEP

Discussion Ask students if they know what an electric guitar sounds like *without* electric power. Students should recognize that an electric guitar makes very little sound without power. Ask students to contrast this with an acoustic guitar, like the one used for the Find Out in the previous section. Students should recognize that the acoustic guitar makes much more sound than the powerless electric guitar. Ask students to brainstorm reasons the two differ. Students should observe the contrast between the hollow body of the acoustic and solid body of the electric.

In the previous lessons, students learned that sound is caused by the vibration of an object. In this lesson, students learn that the vibration of one object can set another object vibrating, whether or not the two objects are touching.

Theme Connection The theme is stability and change. Resonance causes a change in another object by causing it to vibrate at the same frequency. Resonance is at the heart of the study of sound. All acoustic musical instruments, including the human voice, depend on resonance for their loudness and sound quality. The theme of energy transfer is implied in the study of resonance where one object causes another to vibrate at the same frequency.

Explore!

What is resonance?

 **Time needed** 5 minutes

Materials tuning fork, rubber mallet

TECH PREP **Thinking Processes** observing and inferring, comparing and contrasting, recognizing cause and effect, forming operational definitions

Purpose To recognize that resonance can amplify sounds.

Preparation Make sure students use wooden tables or desks if possible, with and without drawers.

Teaching the Activity

Discussion The most effective resonance is obtained from hollow wooden objects with thin walls. Have students place the tuning fork against the top of a desk or table with drawers or other space immediately beneath the top. Lead students to conclude that a table or desk with drawers results in the greatest resonance.

Sound Quality

Think back to the activity in which you played the guitar. The thinnest string on a guitar vibrates at about 330 Hz. You can play a note that vibrates at about 330 Hz on a

Figure 3-11

Ⓐ Stradivarius violins produce a quality of sound unmatched by any other violin.

Ⓑ Chemical analysis has shown the wood used in the violins had a higher than normal salt content. The wood also absorbed more varnish than usual.

Ⓒ The varnish contained minerals that made it extremely hard. All these factors contribute to the pleasing, rich tones produced when the violins are played.

clarinet, a cello, a piano, or a trumpet. You could even sing a note at this pitch with your voice. But, even though the pitch is the same, the quality of the sound will be different. An instrument's quality of sound depends on a variety of things—whether the sound is made by a vibrating string or a vibrating column of air, the material the instrument is made of (wood, metal, plastic, or a singer's vocal cords), the size of the instrument, and its shape. The way the vibrations are set into motion can also have an effect on sound quality. For example, strumming guitar strings with your fingers and plucking them with a plastic guitar pick produces sounds with different qualities.

Here is an activity that shows how the quality and the loudness of a sound can be changed. The activity uses a tuning fork, which is a metal object designed to vibrate at a particular frequency.

Explore! ACTIVITY

What is resonance?

What To Do

1. Hold a tuning fork by the stem.

2. Gently strike one of the fork's prongs with a rubber mallet. What do you hear?

3. Strike the tuning fork again. Stand the base of the stem on a table or desk top. What do you hear now?

4. *In your Journal*, tell how the sound changes when the fork base is held against the table.

Science Journal Have students answer the questions in their journals. **L1**

Expected Outcome

The tuning fork will sound louder when touching the desk.

Answers to Questions

2. A soft, musical tone.

3. The same tone, but louder.

✔ Assessment

Oral Show students pictures of musical instruments from reference books. Ask which they would expect to produce more resonance, for example, a grand piano or an upright piano? A violin or a cello? A flute or an oboe? **L1**

Resonance

The tuning fork in the Explore activity doesn't make a very loud sound all by itself. But what happens when you hold the stem of the tuning fork against the table? The sound gets louder. Can you explain why? You know that vibrating objects produce sound, and that sound can travel through solids such as the table as well as through air. **Figure 3-12** shows what happens when the tuning fork is placed against the table.

An acoustic guitar's sound doesn't come just from the vibrating string

making patterns of compression and rarefaction in the air. It also comes from the vibrations of the guitar body and the air inside it. When the string is attached to the guitar, the body of the instrument and the air inside it vibrate at the same frequency as the string. This tendency for an object to vibrate at the same frequency as another sound source is called **resonance**. Resonance means to resound, or to sound again. Resonance is what caused the sound of the vibrating tuning fork to get louder when you placed it against the table.

Figure 3-12

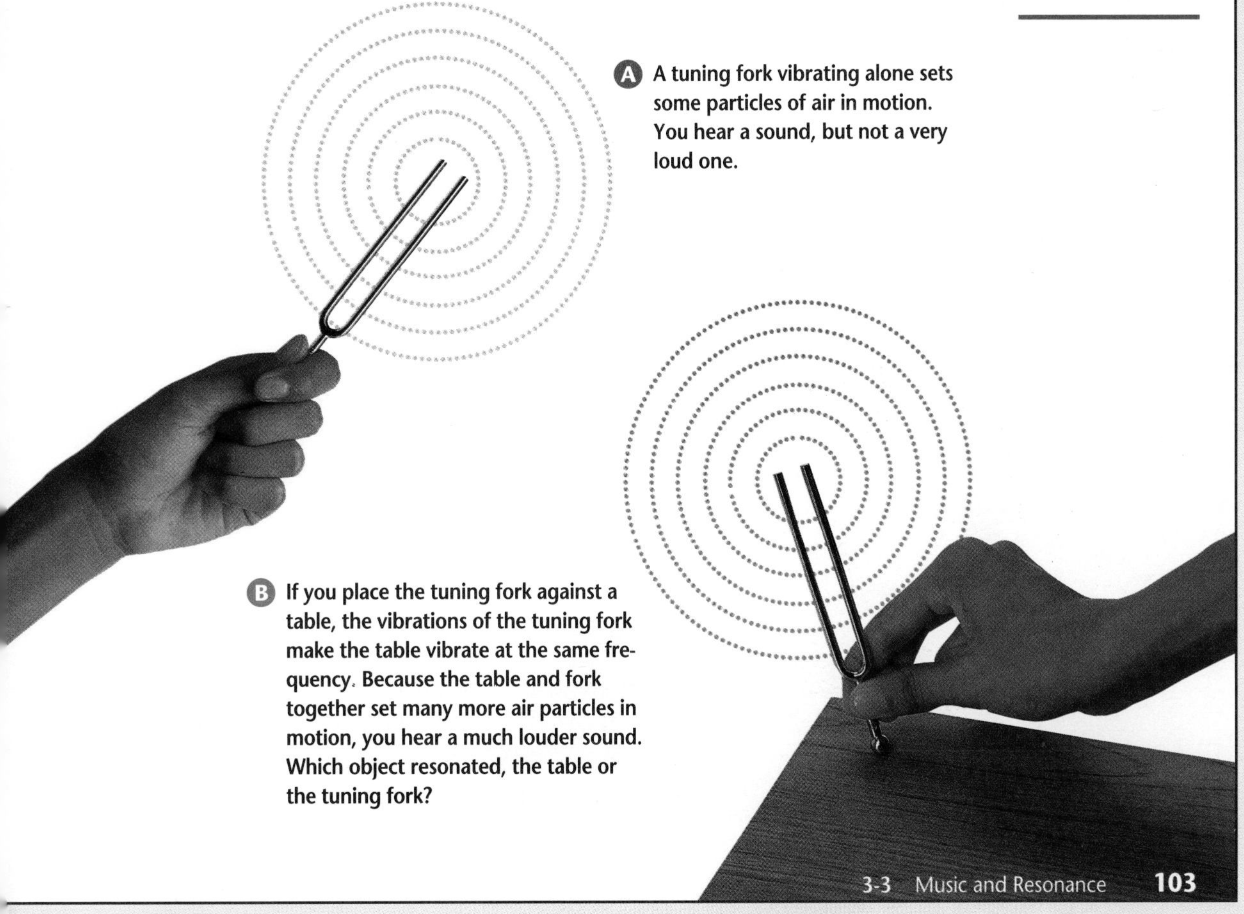

A A tuning fork vibrating alone sets some particles of air in motion. You hear a sound, but not a very loud one.

B If you place the tuning fork against a table, the vibrations of the tuning fork make the table vibrate at the same frequency. Because the table and fork together set many more air particles in motion, you hear a much louder sound. Which object resonated, the table or the tuning fork?

Content Background

For any object, the resonant frequency is the frequency at which it takes the least energy to make the object vibrate. Bridges are designed so that wind or traffic vibrations will not match their resonant frequency. If strong vibrations match the bridge's resonant frequency, the bridge may vibrate enough literally to tear it to pieces.

Activity Playing a kazoo demonstrates resonance in action in a fun way. Ask for volunteers to form a kazoo chorus. Have them begin by simply blowing through their kazoos. They might be surprised to discover that no sound is produced. Next have them hum a simple tune into the kazoo. Now they will notice that the kazoo gives a buzzing quality to their voices. Ask students if they can explain what happens. Lead them to discover that blowing air did not vibrate the membrane inside the kazoo, but humming did. If possible, take apart a kazoo. Otherwise it will be necessary to explain that there is a thin membrane inside that resonates in response to sound. L1

Visual Learning

Figure 3-12 **Which object resonated, the table or the tuning fork?** *The table resonated with the sound of the tuning fork.*

GLENCOE TECHNOLOGY

 Videodisc

Science Interactions, Course 1

Integrated Science Videodisc Lesson 1

ENRICHMENT

Discussion Help students recognize cause and effect by obtaining a diagram of the nasal passages and sinuses. Ask students what happens to the sound of people's voices when they have a very stuffed-up nose. Students should enjoy imitating a person talking with a stuffed-up nose. Ask students to suggest reasons for this change in sound. After students have made several suggestions, display the diagram and ask students to explain how resonance might play a part in this change in sound. Students should identify the sinuses as chambers in which air resonates to the frequency of the vocal cords; when the sinuses are blocked, this resonance is lost. L3

Planning the Activity

Time needed 30 minutes

Purpose To recognize the effect of a sound medium's size on its resonant frequency

Process Skills organizing information, making and using tables, observing and inferring, recognizing cause and effect, measuring in SI, predicting, modeling

Materials Any water container, at least 30 cm deep and with a mouth wide enough to permit measurement with a ruler from the surface of the water upward. If they are available, laboratory stands into which the pipes can be clamped will make the activity easier.

Teaching the Activity

Process Reinforcement Be sure students set up their data tables correctly. Tables should include columns with space to note the tuning fork frequency and the length of the column of air in the glass or plastic tube.

Possible Procedure Have students work in pairs. Distribute a higher-pitched and a lower-pitched tuning fork to each pair of students. Students will produce the most accurate measurement if they test each tuning fork three times and average the measurements.

Troubleshooting In raising and lowering the tube, students should be careful not to let the tube touch the container or the tuning fork. Tell students to study the illustration that shows the tips of the tuning fork being held close to the opening of the tube. Remind them to hold the tube as still as possible after determining the position of loudest resonance to allow accurate measurement.

Science Journal Have students create their data tables and then record their observations in their journals. They should also construct their graphs and answer the questions in their journals. L1

INVESTIGATE!

Length and Resonance

A tabletop resonates with the frequency of a vibrating tuning fork. The body of a guitar resonates with its vibrating strings. In this experiment, investigate the resonance of the air inside a glass tube.

Problem
Can you find the length of a tube of air that will resonate with a given sound frequency?

Materials
2 tuning forks of different frequencies (256 Hz or higher)
1 1000-mL graduated cylinder (or bucket or pitcher about 30 cm deep)
metric ruler

plastic or glass tube, 2.5 cm in diameter, about 45 cm long, open at both ends
rubber mallet
water

What To Do

1 Copy the data table *into your Journal.*

2 Find the number and the letters Hz on your tuning fork and record it under *Tuning Fork Frequency* in the data table.

3 Fill the graduated cylinder or bucket with water.

4 Hold one end of the tube while you place the other end partway into the cylinder or bucket of water (see photo **A**).

Meeting Individual Needs

Visually Impaired Allow visually impaired students to make relative measurements in Investigate 3-2. Provide them with wide strips of non-corrugated cardboard at least 2.5 by 20 cm. Once the point of loudest resonance is determined, have the students grasp the tube gently between thumb and forefinger and slide their hands down until their fingers touch the water. (A partner or clamp must hold the tube motionless.) A cardboard strip should then be placed against the tube so that students feel its end just touching the water. With their free hand, students then grasp the top of the strip where it meets the end of the tube and bend it to mark its length. Repeat with the second tuning fork. Compare the two strips to draw conclusions about the length of the air column and its resonant frequency.

A

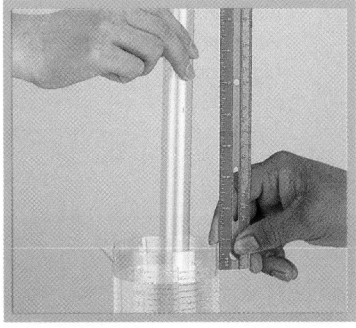
B

5 Have your partner strike the tuning fork with the mallet and hold the fork over the tube.

6 Raise or lower the tube in the water until the loudest sound is produced.

7 Have your partner *measure* the distance from the top of the tube to the water's surface (see photo **B**). Record the length in the table. This is the length of the column of air that resonates with the vibration of the tuning fork.

8 Repeat Steps 5–7 for the second tuning fork.

Sample data

Data and Observations	
Tuning Fork Frequency	Length of the Column of Air
The higher the frequency of the tuning fork,	
the shorter the length	of the column of air

Analyzing

1. *Interpret* your table to answer these questions. For which tuning fork is the length of the column of air longer? Which column of air resonates at the lower frequency?

2. How does the length of a column of air relate to its resonant frequency?

Concluding and Applying

3. Obtain a different frequency tuning fork by trading with another group. Look at its frequency and *predict* how the length of the column of air that resonates with this tuning fork will compare with your earlier trials. Record your prediction in your Journal. Repeat the experiment and see how your prediction compares with what you observe.

4. **Going Further** Have you ever heard an object in a room buzz when a certain note is played loudly on the radio? Explain what causes this to happen.

Expected Outcome

A longer air column has a lower resonant frequency than does a shorter air column.

Answers to Analyzing/ Concluding and Applying

1. The tuning fork with lowest frequency. The longest column.

2. The longer the tube, the lower its resonant frequency.

3. Predictions should follow the rule that the greater the length of tube, the lower its resonant frequency.

4. The musical note was the resonant frequency for that object.

✔ Assessment

Process Have students work as a class to collect and organize the data from this Investigate and create a display showing the frequency of the different tuning forks and the length of the air columns that resonate with them. Encourage students to use color to make the display easier to read and more visually interesting. Use the Performance Task Assessment List for Display in **PASC**, p. 63. L1 P

GLENCOE TECHNOLOGY

  CD-ROM

Science Interactions, Course 1, CD-ROM

Chapter 3

Program Resources

Activity Masters, pp. 17–18, Investigate 3-2

Science Discovery Activities, 3-3, Music to Your Ears

Check for Understanding

Discussion Remind students of the discussion of electric and acoustic guitars. Ask them to use what they now know about resonance to answer the Apply question.

Reteach

Activity Have students observe resonance by providing two tuning forks of the same frequency. Have one student strike a tuning fork. Have another student hold the second tuning fork near the first one. The second fork should start vibrating, too. Ask students why this happens. *The tuning forks have the same frequency.* Have students predict what would happen if the tuning forks were different frequencies. *The second one would not vibrate.*

COOP LEARN L1

Extension

Activity Allow students to recognize the effect of frequency on resonance by playing a note on a piano or acoustic guitar. (If a piano is used, press down on the right hand pedal, which frees the strings to vibrate.) Have students loudly sing the note you just played. Stifle the note as students continue to sing, then release that string. That string (and no others) should be vibrating by itself. Have students explain why the string continues to vibrate. LEP L2

4 CLOSE

Allow students to compare and contrast sounds by obtaining recordings of classical music, rock music, and noisy sound effects. Play selections and ask students to identify them as music or noise. LEP L1

Figure 3-13

A Each pipe in a pipe organ contains a column of air. The organ produces sound by forcing air through the pipes. The forced air causes the column of air in the pipes to vibrate. The vibrations produce the sound.

B The sound a pipe makes depends on its shape and size. Longer pipes produce the lowest notes, while shorter pipes produce higher notes.

In the Investigate, you found that the resonant frequency of an air column was related to its length. The type of material from which an object is made and the object's shape also affect the frequency at which an object resonates. Thick plate glass windows will vibrate as heavy trucks rumble by on the street. A certain note played on the piano may cause a chandelier to vibrate.

There are many sounds around us, with many different qualities. Sometimes people have different ideas about what makes a sound pleasant and what makes a sound unpleasant. What kinds of sounds do you like? Rain falling or rap music? What is your least favorite sound? Traffic noise or thunder? The next time you hear a sound that you really like, try to describe its characteristics.

check your UNDERSTANDING

1. Compare and contrast music and noise.
2. Why can you identify different musical instruments just by listening, even when they are playing the same pitch?
3. Using a guitar, how would you explain resonance?

4. **Apply** The body of an acoustic guitar is hollow, with an opening just under the strings. Explain why it is constructed this way. Hint: Remember that the strings are not the only part of the guitar that vibrates.

106 Chapter 3   Sound and Hearing

check your UNDERSTANDING

1. Musical sound is neatly organized into regular patterns. Noise consists of disorganized, unpatterned sounds.

2. Instruments have different sound qualities depending on their shape, size, and material.

3. Plucking the string causes it to vibrate. Energy from the vibrating string causes the guitar and the air inside to vibrate at the same frequency.

4. The body of the guitar and the air inside the guitar resonate with the strings. The hole allows the sound to exit.

Technology Connection

Active Noise Control

Technology Connection

How many noisy sound sources can you name? Loud music, jack-hammers, car mufflers, lawn mowers, jet engines, electric motors in fans and drills—and more. Passive methods of noise control try to reduce noise by surrounding the sound source with either foam insulation to absorb the noise or baffles to redirect it. Now there's a better idea. It's called active noise control.

How Does It Work?

Think of a noise source—say, an exhaust fan in the kitchen. Now picture its noise signature on an audiogram. There would be jagged hills and valleys in a pattern on the screen. Each noise source has its own unique noise signature.

Active noise control (ANC) devices use computer technology to produce something called anti-noise, whose noise signature is a mirror image of the noise's. Where the noise's pattern shows a hill, the anti-noise's pattern shows an identical but reversed valley. Where the noise has a valley, the anti-noise has a hill of exactly the same shape.

The ANC device analyzes the noise, predicts the signature of the anti-noise, and projects the anti-noise through speakers located near the fan. When the noise and the anti-noise reach your ear (or an oscilloscope) together, the result is a flat noise signature and no noise at all.

Factory workers, firefighters, pilots, construction workers, and many others are using ANC delivered by headphone to save their hearing and improve their efficiency on the job. Some new ANC devices have a built-in feedback loop so that if the noise changes, the anti-noise changes.

In one ANC car muffler design, a microphone near the exhaust pipe samples the engine sound. The computer chip in the ANC device produces the correct anti-noise, which is played by speakers mounted around the exhaust pipe. This design reduces engine noise 10 percent more than conventional mufflers. These ANC mufflers may be available on cars very soon.

Using Computers

Using graphics software, make a diagram using audiograms that explains how ANC technology works.

Purpose

The Technology Connection is a good follow-up to the discussion of noise in Section 3-3. Noise control methods based on the interaction of sounds that cancel each other out are being developed to cut unpleasant noises.

Teaching Strategy Stereo listeners occasionally notice 'dead spots' in a room—areas in which the music being played will suddenly sound muffled. These spots are produced by the same principles which make Active Noise Control a possibility. When two speakers are placed in a room, there may be points where the sound waves from one speaker travel just slightly farther than the sound waves from the other. Where the two waves overlap, they may be far enough out of sync that while one wave is at its crest, the other is at the bottom of a trough: as with Active Noise Control, the crest and trough two waves will cancel each other out. The music sounds muffled.

Demonstration

If possible, obtain an oscilloscope and hook a pair of speakers up to it. Set the scope to produce a steady tone; point out the crests and troughs to students. Clear the area in front of the speakers and angle the speakers inward. Invite the students to try to find dead spots. Point out to students that the purpose of Active Noise Control is to form a very large, very finely tailored dead spot, which covers up a specific noise over a large area.

Using Computers

Answers will vary but should show that only regular sounds can be canceled by ANC.

Going Further ▐▐▐▶

Ask students to assume that their local government wants to build a new airport near their community. Organize two teams to debate the pros and cons of the airport. Work with each team to help them brainstorm a list of reasons to support their position. The pro team should focus on economic benefits: new jobs, tax revenue, and improved transportation. The con team should focus on noise pollution and its consequences for the community.

Have several representatives from each team present the team's position and arguments, then solicit responses from individual students on the issue. Interested students may wish to research actual airport projects and their encounters with local opposition. The construction of Tokyo's airport in Japan is one of the world's leading examples. Use the Performance Task Assessment List for Venn Diagram and Pro/Con Issue in **PASC**, p. 95. **L1** **COOP LEARN**

Science and Society

Purpose

The beautiful tone of Louis Armstrong's trumpet or Midori's violin depends on the factors discussed in Section 3-3 that affect sound quality.

Content Background

Louis Armstrong is recognized as the leading trumpeter in jazz history. As a child, he followed the brass bands through the streets of New Orleans and came to know many of the pioneers of jazz. Although early on he played the trumpet in marching bands and on Mississippi riverboats, it was not until 1922 when he came into his own. That year, his hero, Joe "King" Oliver, sent for him to play second trumpet in a Chicago band. This led to a series of recordings with Oliver's Creole Jazz Band, including such classics as "Dipper Mouth Blues" and "Canal Street Blues." The records he made from 1925-1928 with his Hot Five and Hot Seven bands established his preeminence. From the 1930s on, Armstrong also gained fame as a film star, comedian, and bandleader.

Teaching Strategies

Have small groups of students develop a list of factors affecting the sound of a musical instrument, based on their reading of this feature. Point out that this list does not reflect the many possible variations that produce a particular sound. `COOP LEARN`

Answers to

You Try It!

Answers will vary. Students may have different ideas of how Armstrong's and Midori's music makes them feel.

Science and Society

Heart and Soul

Louis Armstrong, left; Midori, above

The sounds from instruments are the result of vibrations. A trumpet player blows into a mouthpiece, causing the air in the body of the trumpet to vibrate. The pitch and sound can be changed by varying the pressure of the player's lips on the mouthpiece, and by opening or closing three valves.

A violin player draws a bow across the strings of the violin to produce vibrations. The body of the violin vibrates at the same frequency. The strings produce different sounds. The player changes the pitch by pressing the strings against the neck of the violin, changing the length of the string that is vibrating.

Playing an instrument requires more than using the science of sound. It requires the skill of the player. Trumpeter Louis Armstrong and violinist Midori are recognized as musicians with very special talents.

Louis Armstrong

Louis Armstrong was born in 1900 in New Orleans—the city where jazz was born. He moved to Chicago, then on to New York where he became a major figure in jazz. In addition to playing the trumpet, he sometimes sang nonsense syllables, called scatting. Armstrong died in 1971, but he is still remembered worldwide by his nickname Satchmo and revered for what he could do with sound.

Midori

On her third birthday, Midori was given a violin half the size of an adult's. Midori practiced endlessly with her mother, a professional violinist in Osaka, Japan. Midori was invited to play in a summer festival at Aspen, Colorado when she was only eight years old. Her power, technique, and skill amazed everyone. Midori and her mother moved to New York City in 1982 so that Midori could study at the Juilliard School of Music. Since then, Midori has performed in many countries, including her native Japan.

You Try It!

Listen to a recording of Louis Armstrong and to one of Midori. How does the music make you feel?

Going Further ⫸

A number of your students probably play musical instruments. If so, have them bring their instruments and play them for the class. For each instrument, ask the student to describe how it makes its sound and how the pitch and quality of the sound may be varied. Lead the discussion by reminding students what they learned in the chapter concerning vibration as the source of sound, the means for changing the pitch of sound, and the factors affecting sound quality.

For example, a trumpet player may describe his or her vibrating lips as the source of sound, the valves as a means of changing the length of the column in order to change its pitch, and the metal as producing a hard, brassy sound. Use the Performance Task Assessment List for Group Work in **PASC**, p. 97.

Teens in SCIENCE

Making Waves—Sound Waves, That Is

West Virginian Torey Verts knows a lot about why things sound the way they do. She's a professional sound engineer. "When you listen to a record, you are hearing a lot more than your favorite band. Computers get a lot of use in the studio today. We can completely change a band's sound. For example, if the singer can't hit high notes, the engineer can turn a dial, and suddenly there's no problem. We can speed the music up or add special effects and synthesizers. Even though the sound engineer can do all these things, I don't think

musicians have much to worry about. After all, who wants to see a computer in concert?"

Think Fast

Torey, like many sound engineers, is a musician herself. "If I could, I would be up on the stage playing my guitar. That's why I love to work live concerts. Engineering lets me be a part of the sound. Of course, live concerts can be tough. If something goes wrong with the sound, you've got to fix it fast. You can't ask the audience to take a break while you find a loose connection." Torey has found that a good understanding of scientific principles can really help when you need to solve a problem.

"If you want to get involved in music, you've got to learn as much as you can

about the sciences," says Torey. "But you also need to know what sounds good. The best way to learn is to listen. Try to hear what it is you like about a song. What makes it sound good? What would make it sound better? And don't be afraid to listen to bad music either. Knowing what doesn't work is just as valuable as knowing what does."

You Try It!

Many radios, stereos, and CD players let you make adjustments to the sound that you hear. As you listen to a song, gently turn the treble knob as far to the right as it will go. Play the same song again and adjust the bass knob. What is the difference between these two adjustments? Reminder: When you have completed this assignment, be sure to return both knobs to their original positions.

Expand Your View **109**

Teens in SCIENCE

Have students read the page and look at the photos. The two teaching strategies below ask students to explain the basic mechanism of sound and media, and to explain the relationship among vibration, frequency, and resonant frequency. Have students use the terms *vibration, medium, frequency, music and noise,* and *resonance* as they review the photographs.

Teaching Strategies

Ask students to explain why their Find Out activity with the coiled spring was a good model for how sound moves through the air. Help students as required to explain that vibrating objects alternately compress and rarefy the air and send out a series of compression and rarefaction waves, transmitting the vibration to your ear. Other than the vibrational motion, the air molecules do not change place.

Answers to Questions

1. The air particles move closer together and farther apart in a regular pattern which moves outward from the vibrating object. The pattern matches the frequency of the vibrating object.

2. Space is a near vacuum. There is insufficient atmosphere to transmit sound waves.

3. Increasing the tension on a guitar string increases the pitch of the note it produces. Decreasing the tension produces the opposite effect.

4. The air inside the violin and the wooden body of the violin both resonate when a string is plucked.

GLENCOE TECHNOLOGY

MindJogger Videoquiz

Chapter 3 Have students work in groups as they play the Videoquiz game to review key chapter concepts.

Science Journal

Review the statements below about the big ideas presented in this chapter, and answer the questions. Then, re-read your answers to the Did You Ever Wonder questions at the beginning of the chapter. *In your Science Journal,* write a paragraph about how your understanding of the big ideas in the chapter has changed.

1 Sounds are created by vibrating objects. When an object vibrates, it creates a pattern of compressions and rarefactions in the particles of the air. *How do air particles in an area around a vibrating object move?*

2 We usually hear sounds that travel to our ears through the air. But any kind of matter can conduct sound. *Why couldn't an observer in space hear an explosion on a nearby satellite?*

3 Your ears recognize differences in sound frequencies as differences in pitch. Pitch refers to the highness or lowness of a sound. *How does changing the tension on a guitar string affect the pitch of the note it produces?*

4 The tendency for any object to vibrate at the same frequency as another sound source is called resonance. Resonance means to re-sound, or to sound again. *When a violin string is plucked, what resonates?*

Project

Gather a wide range of materials, such as empty cans, plastic sheeting, rubber bands, cigar boxes, dried beans, empty bottles, etc. Have students make musical instruments such as drums, rubber-band stringed instruments strung on cigar boxes or on nails hammered into plywood, maracas made of cans filled with dried beans, a bottle xylophone and spoons. L1 LEP

Science Journal

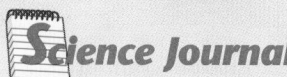

Did you ever wonder...

• Sounds cannot be heard on the moon because there is no air to carry them. (p. 90)

• The column of air in the bottle vibrates, producing sound. (p. 94)

• Changing the length, tension, or diameter of a string changes its resonant frequencies. (pp. 96-97)

Using Key Science Terms

compression	pitch
frequency	rarefaction
hertz (Hz)	resonance
medium	

Give the science term with a meaning opposite to that of the following phrases.

1. a squeezed together area
2. a spread out area
3. two sound sources vibrating at different frequencies.

For each set of terms below, explain the relationship that exists.

4. compression, vibration, rarefaction
5. frequency, pitch
6. pitch, compression, resonance

Understanding Ideas

Answer the following questions in your Journal using complete sentences.

1. What does the frequency or pitch of a sound source describe?
2. What creates the pattern of sound that travels to your ears when an object vibrates?
3. In what unit is frequency measured?
4. How does sound travel in air compared to water?
5. How does the pitch of a 4-inch column of air compare to the pitch of a 3-inch column of air?

Developing Skills

Use your understanding of the concepts developed in this chapter to answer each of the following questions.

1. **Concept Mapping** Using the following terms, complete the concept map of sound: *compressions, gas, liquid, medium, rarefactions, solid, vibrations.*

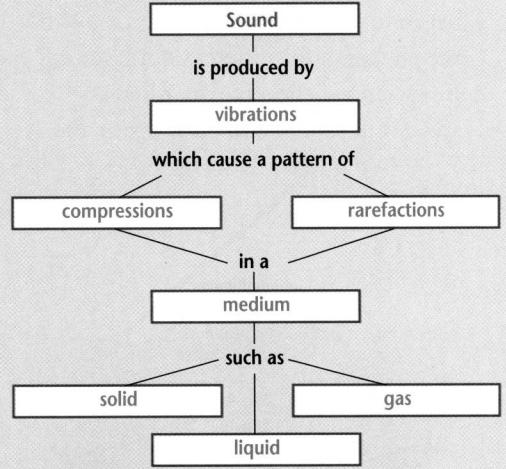

2. **Making and Using Graphs** Using **Figure 3-9** on page 100, arrange in order from narrowest range of frequencies heard to widest range of frequencies heard, the following: dog, dolphin, bat, human.
3. **Predicting** Repeat the Find Out activity on page 89 using a different material, such as thick rope or plastic string. Predict how the new material will conduct sound as compared to the string. Test your predictions.

Using Key Science Terms

1. rarefaction
2. compression
3. resonance
4. As sound vibrations travel through a medium, they create rarefactions and compressions.
5. As the frequency increases, we hear a higher pitch.
6. Compressions in one medium can result in compressions in a second medium. Each medium produces the same pitch, and the result is resonance.

Understanding Ideas

1. How many times the sound source vibrates every second determines the frequency or pitch.
2. Compression and rarefaction create the pattern of sound that travels to the ears.
3. Frequency is measured in hertz.
4. Sound travels faster in water than in air.
5. The pitch of a 4-inch column of air is lower than the pitch of a 3-inch column of air.

Developing Skills

1. See reduced student page for completed concept map.
2. human, dog, bat, dolphin
3. Students' answers will vary depending on the material used.

Program Resources

Review and Assessment, pp. 17–22
Performance Assessment, Ch. 3 [L2]
PASC
Alternate Assessment in the Science Classroom
Computer Test Bank [L1]

Critical Thinking

1. The air column in the trumpet is slightly longer and therefore the pitches were slightly lower.

2. Change the scene. In outer space, there is little or no air, so space is a poor conductor of sound.

Problem Solving

Students' answers will vary.

1. soda bottle, blowing over the bottle makes the air inside the bottle vibrate

2. fork, hit the tines of the fork which then vibrate

3. spoons, hitting bowls of spoons against one another makes them vibrate

4. a pan cover hit with a wooden spoon vibrates

5. a closed jar half full of beans, the glass and lid vibrate when the jar is shaken.

Connecting Ideas

1. Frequency describes how fast an object is vibrating, pitch is the sound a person hears. Greater or faster vibration produces higher sound or pitch.

2. The end of a stethoscope collects the sound and concentrates the vibrations. The air column in the stethoscope tubing vibrates at the same rate as the heart. The air column conducts the sound to the doctor's ears.

3. As you drink more soda, the air column in each bottle becomes longer and the sound from each bottle deepens.

4. Students' answers will vary. Air conditioners, fans, car mufflers, refrigerator motors.

5. The size of the "box" used in each stringed instrument provides for a different resonance from the instrument.

Critical Thinking

In your Journal, *answer each of the following questions.*

1. At the beginning of an outdoor band concert, Lee's trumpet was in tune. During intermission the trumpet was left out in the sun. The sun warmed up the metal, expanding it, and actually making the trumpet slightly larger. When Lee began to play after intermission, how did the pitches of the different notes sound?

2. In a science-fiction movie, when a space ship explodes in outer space, the

vibrations from the sound nearly destroy a nearby spaceship. If you were technical advisor for the movie, what would your advice be about this scene?

Problem Solving

Read the following problem and discuss your answers in a brief paragraph.

Your class is putting on a variety show, and you've been assigned to put together a "kitchen" band. That is, you must come up with instruments made from ordinary household items like pots and pans, glasses, silverware, and other common articles.

1. Suggest at least five different types of instruments for your band. Remember, they must be able to produce different pitched notes and play a melody.

2. How is each of the instruments played and where does the sound come from? What vibrates to produce the sound?

CONNECTING IDEAS

Discuss each of the following in a brief paragraph.

1. Theme—Stability and Change How are frequency and pitch related?

2. Theme—Systems and Interactions Explain how a stethoscope would help a doctor listen to a patient's heart.

3. Theme—Stability and Change You are drinking a soft drink out of a 10-ounce bottle. Every time you take a swallow, you blow across the top of your bottle to make a sound. Describe how the sound will change as you get closer and closer to finishing your drink.

4. Technology Connection Identify some noises in your neighborhood that might be reduced by ANC.

5. Science and Society A violin, a viola, and a cello are about the same shape, but don't sound the same. What could cause this?

✔ Assessment

Portfolio Review the portfolio options that are provided throughout the chapter. Encourage students to select one product that demonstrates their best work for the chapter. Have students explain what they learned and why they chose this example for placement into their portfolios.

Additional portfolio options can be found in the following **Teacher Classroom Resources:**

Making Connections: Integrating Sciences, p. 9

Multicultural Connections, pp. 9–10

Making Connections: Across the Curriculum, p. 9

Concept Mapping, p. 11

Critical Thinking/Problem Solving, p. 11

Take Home Activities, p. 8

Laboratory Manual, pp. 15–16

Performance Assessment P

Observing the World Around You

In this unit, you investigated how your senses function and how they are used to obtain information about your world. Your senses were used to observe patterns and features on Earth and in the sky.

You used light and sound as two primary means of obtaining information about your world. The color of the light and whether it is reflected from a surface or refracted by it enabled you to identify characteristics of objects.

Try the exercises and activity that follow—they will challenge you to use and apply some of the ideas you learned in this unit.

CONNECTING IDEAS

1. You may have seen flash photographs showing several people. The flash is bright white, so why do some of the people have red eyes and other people have eyes that appear normal?

2. You and your friends are in a swimming pool. While underwater, one of your friends taps on a metal railing. Why do you have trouble locating the sound source while you are underwater? How might the interaction of your other senses help you determine where the metal railing is located?

Exploring Further ACTIVITY

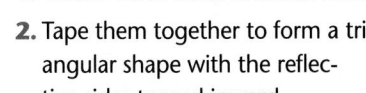

What makes all the patterns of light you see inside of a kaleidoscope?

What To Do

1. Obtain three small identical mirrors.

2. Tape them together to form a triangular shape with the reflective sides turned inward.

3. Tape a triangular piece of paper to one end of your kaleidoscope, and drop pieces of colored paper into it.

4. Look inside and describe what you see. What causes the pattern of color in your kaleidoscope?

Observing the World Around You

THEME DEVELOPMENT

In Unit 1, the theme of scale and structure was used in observations of patterns of landforms and patterns of objects in the sky. The theme of energy was explored with emphasis on light and sound energy. Senses were used to observe both stability and change and to observe interactions within Earth and sky systems.

Connections to Other Units

The concepts developed in Unit 1 will help students better observe and understand natural phenomena in the physical world and their interactions as explored in Unit 2.

Connecting Ideas
Answers

1. In flash photographs some people have eyes that appear normal while others have eyes that appear red because of the size of the pupil of the person's eyes at the time the photograph was taken. If the room is dark, the pupils are large and light from the flash reflects off the retina. The red reflected light makes the eyes appear red.

2. Sound travels faster in water than you can respond to it. Seeing your friend's movement may help.

Exploring Further

What makes the patterns of light you see inside a kaleidoscope?

Purpose To observe the patterns made by a kaleidoscope and infer what causes those patterns

Background Information

Objects in a kaleidoscope are reflected in the mirrors, and the images in the mirrors are again reflected. This produces a mosaic effect.

Materials If mirrors are unavailable, use aluminum foil glued to small pieces of cardboard.

Troubleshooting Make sure students securely tape their mirrors together, and that the mirrors form a triangle.

Answers to Questions

The pattern of color is caused by the reflection of the paper in the mirrors.

✔ Assessment

Process Ask students to explain in writing how their knowledge of a particular topic has developed during the study of this unit. Encourage students to include how any misconceptions they may have had were corrected. Use the Performance Task Assessment List for Writing in Science in **PASC**, p. 87. **L1** **P**

UNIT 2

Interactions in the Physical World

UNIT OVERVIEW

Chapter 4   Describing the Physical World

Chapter 5   Matter in Solution

Chapter 6   Acids, Bases, and Salts

UNIT FOCUS

In Unit 1, students learned about how they use their senses of sight and hearing to observe patterns and features on Earth and in the sky. As they study Unit 2, students will use these senses as well as smell and taste to identify various substances on Earth.

THEME DEVELOPMENT

The themes of stability and change and systems and interactions are evident in the descriptions of matter and its interactions in this unit. Many of the interactions within the systems in these chapters cause either physical or chemical change. However, not all the interactions cause changes. Some interactions with some kinds of matter are stable. Energy can cause change, and energy is a result of certain changes. The theme of scale and structure allows students to distinguish between properties of matter.

Connections to Other Units

The concepts developed in this unit can be related to those in Unit 1 where the various physical properties of light and sound waves were shown to interact with the senses to create color, shadow, pitch, and volume. Students will learn the difference between physical and chemical properties and how they affect the interaction of one type of matter with another.

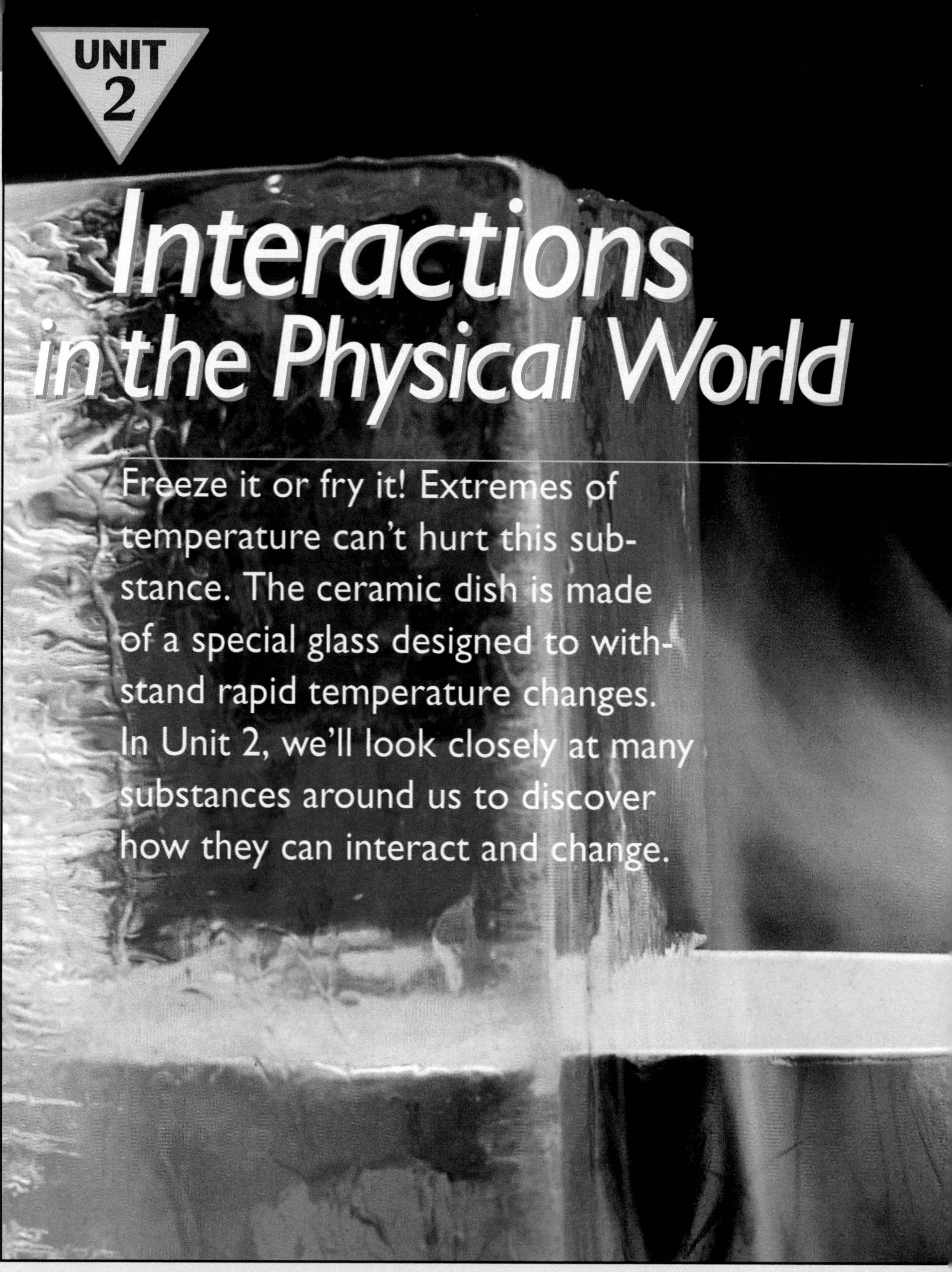

UNIT 2

Interactions in the Physical World

Freeze it or fry it! Extremes of temperature can't hurt this substance. The ceramic dish is made of a special glass designed to withstand rapid temperature changes. In Unit 2, we'll look closely at many substances around us to discover how they can interact and change.

GLENCOE TECHNOLOGY

 Videodisc

Use the *Science Interactions, Course 1* **Integrated Science Videodisc** lesson, *Pollution Detectives*, at the end of this unit. This videodisc allows students to apply concepts they've learned in the unit to the investigation of an environmental problem.

NATIONAL GEOGRAPHIC
try it!

Do all materials interact in the same way? Salt, a substance necessary for us to survive, is a result of sodium and chlorine combining chemically. Not all materials, however, will react with each other. Some materials will not even mix together. What do you think will happen when a sample of powdered drink mix is shaken vigorously with water, then with oil?

What To Do

1. Place some water into a jar and add some drink mix.

2. Close the jar and shake it rapidly. Stop shaking and observe what has happened to the drink mix and water.

3. Now, try the same thing using cooking oil instead of water. How did the oil and water differ?

4. Based on what you have observed, what predictions can you make about other mixtures?

115

Try It!

Purpose Students will observe that different substances interact differently.

Background Information

When forming solutions, materials mix evenly with materials that are similar. Have students describe other substances that they know of that mix and those that do not.

Materials jar with lid, beverage mix, oil, water

Answers to Questions

Student predictions may include that some substances mix easily, and others do not.

✔ Assessment

Oral Have students describe pairs of substances that mix and other pairs of substances that do not mix. Students can then make a poster showing substances that they identified. Use the Performance Task Assessment List for

Poster in **PASC,** p. 73.
COOP LEARN **L1**

Chapter Organizer

SECTION	OBJECTIVES	ACTIVITIES & FEATURES
Chapter Opener		**Explore!**, p. 117
4-1 Composition of Matter (2 sessions; 1 block)	1. **Differentiate** among substances, elements, compounds, and mixtures. 2. **Give examples** of heterogeneous and homogeneous mixtures. **National Content Standards:** (5–8) UCP1–3, A1–2, G1–2	**Find Out!**, p. 118 **Find Out!**, p. 122 **Design Your Own Investigation 4-1:** pp. 124–125 **A Closer Look**, pp. 120–121 **How It Works**, p. 145
4-2 Describing Matter (2 sessions; 1 block)	1. **Recognize** examples of physical properties. 2. **Measure** length, volume, and mass of different materials. 3. **Relate** density to mass and volume. **National Content Standards:** (5–8) UCP1, UCP3, A1, B1, G1–2	**Explore!**, p. 127 **Explore!**, p. 129 **Investigate 4-2:** pp. 132–133 **Science and Society**, pp. 146–147 **Art Connection**, p. 148
4-3 Physical and Chemical Changes (3 sessions; 1.5 blocks)	1. **Distinguish** between physical and chemical changes. 2. **Differentiate** between chemical and physical properties. **National Content Standards:** (5–8) UCP3, A2, B1, B3, E1, G2	**Find Out!**, p. 135 **Find Out!**, p. 136 **Skillbuilder:** p. 138
4-4 States of Matter (2 sessions; 1 block)	1. **Distinguish** among solids, liquids, and gases. 2. **Describe** physical changes relating to solids, liquids, and gases. **National Content Standards:** (5–8) UCP3, A1, B3, E1, F5, G1–3	**Explore!**, p. 139 **Explore!**, p. 140 **Find Out!**, p. 141 **Explore!**, p. 142 **Earth Science Connection**, pp. 142–143

ACTIVITY MATERIALS

EXPLORE!

p. 117 2 heatproof jars, lids, wooden splint, flashlight, matches
p. 127 pencil, button
p. 129* golf ball-sized rock, 500-mL beaker, water
p. 140* rock, fork, penny, paper clip, cotton ball, feather, sand
p. 142 2 balloons, jar, lid, rectangular container, alcohol, mirror, cardboard, flat dish

INVESTIGATE!

pp. 132–133 water, saturated saltwater mixture, rubbing alcohol, unknown (liquid) substance, 100-mL graduated cylinder, pan balance and set of masses, goggles

DESIGN YOUR OWN INVESTIGATION

pp. 124–125* rock salt, aluminum foil, baking soda, piece of granite, sugar water in a vial, copper wire, glass of lemonade, graphite, vinegar and oil salad dressing

FIND OUT!

p. 118 glass, microscope, slide, dropper
p. 122* 2 test tubes, iron and sulfur powder, microscope, slide, magnet, mortar and pestle
p. 135 can, ice, water, tongs, dish, hot plate
p. 136 goggles, old penny, ammonia, forceps, paper towels, jar with lid
p. 141* several see-through containers, food coloring, pitcher, water, measuring cup

Need Materials? Call Science Kit (1-800-828-7777).

⏱ **OUT OF TIME?** We recommend that students do the activities with an asterisk.

KEY TO TEACHING STRATEGIES

The following designations will help you decide which activities are appropriate for your students.

- **L1** Basic activities for all students
- **L2** Activities for average to above-average students
- **L3** Challenging activities for above-average students
- **LEP** Limited English Proficiency activities
- **COOP LEARN** Cooperative Learning activities for small group work
- **P** Student products that can be placed into a best-work portfolio
- Activities and resources recommended for block schedules

Chapter 4 Describing the Physical World

TEACHER CLASSROOM RESOURCES

Student Masters	Transparencies
Study Guide, p. 16 **Multicultural Connections,** p. 12 **Science Discovery Activities,** 4-1 **Activity Masters,** Design Your Own Investigation 4-1, pp. 19-20 **Critical Thinking/Problem Solving,** p. 5	**Teaching Transparency 7,** Classification of Matter **Section Focus Transparency 10**
Study Guide, p. 17 **Concept Mapping,** p. 12 **Making Connections: Technology and Society,** p. 11 **Science Discovery Activities,** 4-2 **Activity Masters,** Investigate 4-2, pp. 21-22 **Laboratory Manual,** pp. 17–20, Measurement and Graphing	**Teaching Transparency 8,** SI Units **Section Focus Transparency 11**
Study Guide, p. 18 **Take Home Activities,** p. 10 **Multicultural Connections,** p. 11 **Science Discovery Activities,** 4-3	**Section Focus Transparency 12**
Study Guide, p. 19 **Making Connections: Across the Curriculum,** p. 11 **Making Connections: Integrating Sciences,** p. 11 **Critical Thinking/Problem Solving,** p. 12 **How It Works,** p. 8 **Laboratory Manual,** pp. 21–24, Properties of Matter	**Section Focus Transparency 13**

ASSESSMENT RESOURCES	TEACHING & TECHNOLOGY
Review and Assessment, pp. 23-28 **Performance Assessment,** Ch. 4 **PASC*** **MindJogger Videoquiz** **Alternate Assessment in the Science Classroom** **Computer Test Bank**	**Spanish Resources** **Cooperative Learning Resource Guide** **Lab and Safety Skills** **Science Interactions, Course 1, CD-ROM** **Computer Competency Activities**

*Performance Assessment in the Science Classroom

NATIONAL GEOGRAPHIC TEACHER'S CORNER

National Geographic Society Products

To order the following products for use with this chapter, call the National Geographic Society at 1-800-368-2728:

- *Everyday Science Explained* (Book)
- *Solid, Liquid, Gas* (Video)

GLENCOE TECHNOLOGY

The following multimedia resources are available from Glencoe

Science and Technology Videodisc Series (STVS)
Chemistry
  Losing Weight by Design
Earth & Space
  Mapping with a Rifle
  Map Science
  Charting Air Space

The Infinite Voyage Series
Sail On, Voyager

National Geographic Society Series
STV: Solar System

Physical Science CD-ROM

Teacher Classroom Resources

These are key components of the classroom resources package.

TEACHING AIDS

Teaching Transparencies

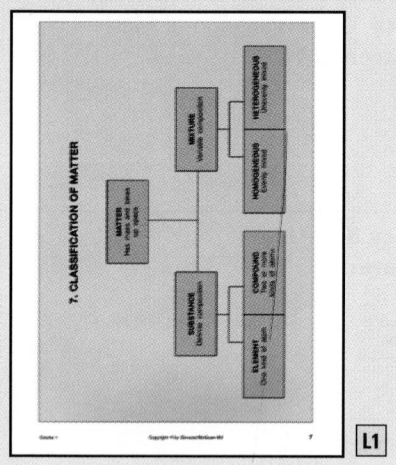

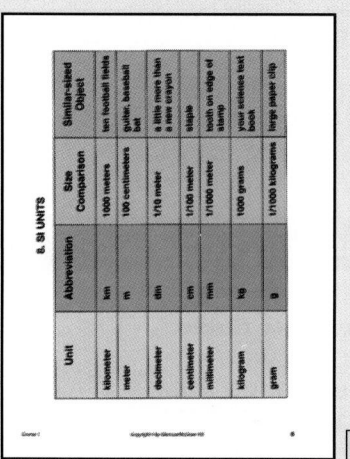

Section Focus Transparencies

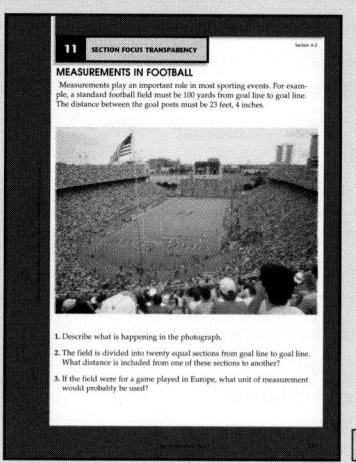

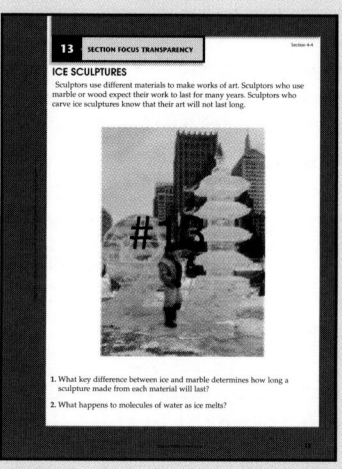

HANDS-ON LEARNING

Science Discovery Activity*

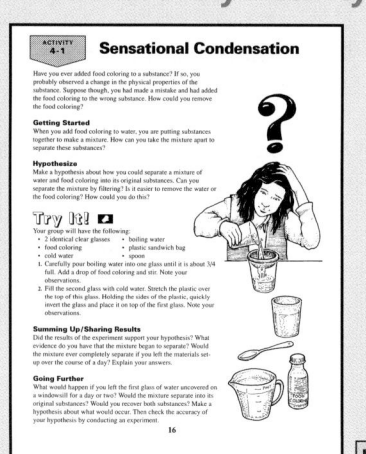

Laboratory Manual*

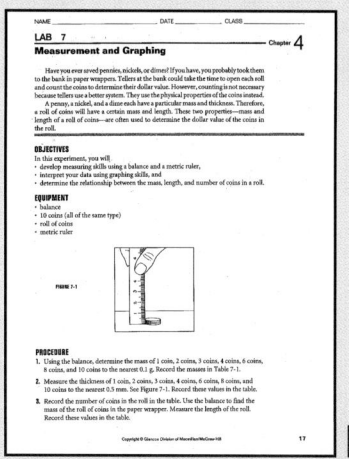

Take Home Activity

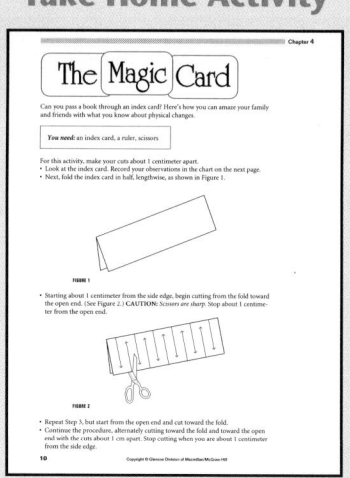

Chapter 4 Describing the Physical World

Study Guide*

Concept Mapping

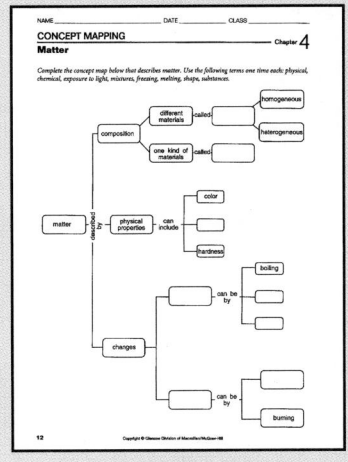

Critical Thinking/ Problem Solving

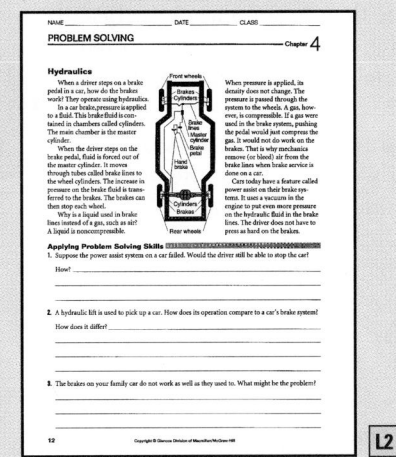

Integrating Sciences

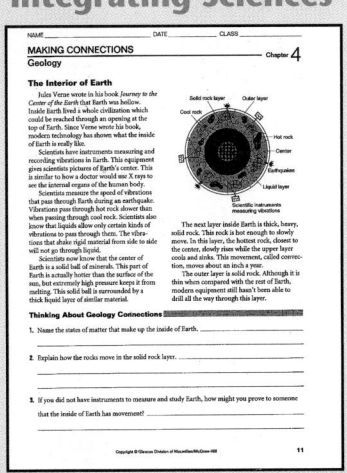

Across the Curriculum

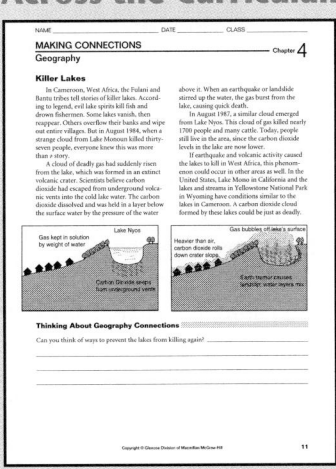

Technology and Society

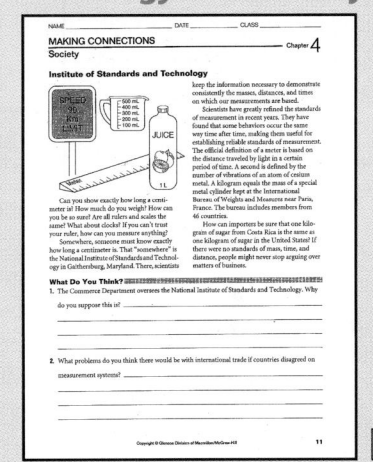

Multicultural Connection**

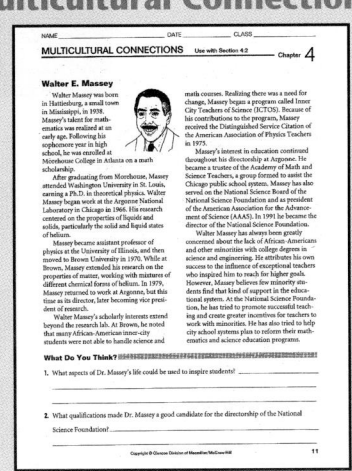

Performance Assessment

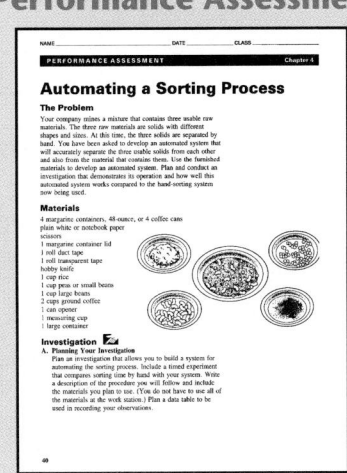

Review and Assessment

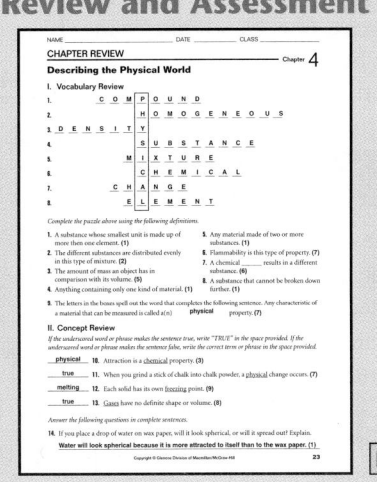

Describing the Physical World

CHAPTER 4

Describing the Physical World

THEME DEVELOPMENT

The themes supported by this chapter are scale and structure and stability and change. Students discover the size and structure of many materials and how these affect their interaction with one another. Students learn that apparently stable structures can change.

CHAPTER OVERVIEW

Students will distinguish elements, compounds, and mixtures, and compare and contrast heterogeneous and homogeneous mixtures. They will demonstrate how physical properties of matter are used to distinguish materials. Students will also be introduced to SI.

In addition, students will compare and contrast physical and chemical changes. Finally, they will learn how to identify the three basic states of matter.

Tying to Previous Knowledge

Show students an unbroken egg and ask in what ways you can make the egg change. *They may mention cracking, mixing, boiling, or frying.*

INTRODUCING THE CHAPTER

Have students look at the photographs on pages 116–117 and discuss how clearly they can see the buildings. Ask what they think might cause any changes in sharpness from front to back.

00:00 OUT OF TIME?

If time does not permit teaching the entire chapter, use the Chapter Overview on this page, Reviewing Main Ideas at the end of the chapter, and the Chapter 4 audiocassette to point out the main ideas of the chapter.

Did you ever wonder...

✓ Why cake batter pours, but cake crumbles?

✓ How you can smell someone's perfume after that person has left?

✓ Why water splashes, but chalk breaks?

✓ If you could get fresh water from the ocean?

Science Journal

Before you begin to study about describing the physical world, think about these questions and answer them *in your Science Journal*. When you finish the chapter, compare your journal write-up with what you have learned.

Answers are on page 149.

How the air feels, looks, and smells are clues as to what may be in the air that day. When the air is hazy and feels damp, it might mean that smoke, fog, and other matter are in the air. In and around some cities, those clues might mean that the air contains smog—a kind of air pollution.

Clues about how the air feels, looks, and smells, however, don't tell you exactly what air is. Is air the same all over the world? Is it made of one thing or a number of things? How do you begin to identify, describe, and classify the kinds of matter in the air around you?

You do some of these things already. This chapter will give you new ways of observing and classifying not only air, but many other materials in the physical world.

▶ *In the activity on the next page, explore a model of smog.*

116

Learning Styles	Kinesthetic	Explore, pp. 117, 127, 142; Activity, p. 119; Multicultural Perspectives, p. 130
	Visual-Spacial	Find Out, pp. 118, 122, 135, 136, 141; Visual Learning, pp. 120, 134, 143; A Closer Look, pp. 120-121; Investigate, pp. 124–125; Demonstration, p. 126
	Interpersonal	Activity, p. 134
	Intrapersonal	Activity, p. 140
	Logical-Mathematical	Demonstration, p. 128; Across the Curriculum, pp. 128, 130; Visual Learning, pp. 129, 130, 131; Explore, p. 129; Investigate, pp. 132–133
LS	Linguistic	Visual Learning, pp. 121, 126, 137; Activity, p. 121; Multicultural Perspectives, p. 125; Debate, p. 130; Across the Curriculum, p. 136

Explore! ACTIVITY

Can you make a model of smog?

What To Do

1. Place two closed, heat-proof jars in front of a dark background.

2. Leave one jar alone. Open the other jar.

3. With your teacher's help, drop a burning wooden splint into the open jar. Quickly close the lid so that the flame goes out and the jar fills with smoke.

4. Shine a flashlight through both jars. *In your Journal*, describe what happens to the light.

ASSESSMENT PLANNER

PORTFOLIO
Refer to page 151 for suggested items.

PERFORMANCE ASSESSMENT
Process, pp. 117, 118, 122, 125, 127, 129, 133, 136, 140, 141, 142
Skillbuilder, p. 138
Explore! Activities, pp. 117, 127, 129, 139, 140, 142
Find Out! Activities, pp. 118, 122, 135, 136, 141
Investigate, pp. 124–125, 132–133

CONTENT ASSESSMENT
Oral, p. 135
Check Your Understanding, pp. 126, 134, 138, 144
Reviewing Main Ideas, p. 149
Chapter Review, pp. 150–151

GROUP ASSESSMENT
Opportunities for group assessment occur with Cooperative Learning Strategies.

Uncovering Preconceptions

A common misconception is that liquid, solid, and gaseous water are three different substances. Explain that ice, liquid water, and water vapor are chemically identical but are in different physical states.

Explore!

Can you make a model of smog?

Time needed 10 minutes

Materials two heatproof jars with lids, matches, wooden splints, flashlight

Thinking Processes making models, observing and inferring, forming operational definitions

Purpose To infer that smog consists, in part, of smoke, or particles suspended in air

Teaching the Activity

Explain to students that smog is more than just smoke. It is a combination of smoke, particles given off by engines and factories, and tiny drops of moisture that trap smoke and particles in the air.

Science Journal Students should note that the light is partially blocked by the smoke but is not blocked by the air.

Expected Outcomes

Students will observe that the beam of light goes straight through the empty jar but looks hazy in the smoke-filled jar.

✔ Assessment

Process Have students apply their observations to model an explanation of the difference between the sunlight on clear and smoggy days. Students could prepare a slide show or photo essay showing some causes and effects of smog. Use the Performance Task Assessment List for Slide Show or Photo Essay in **PASC**, p. 77. [L1] **COOP LEARN** [P]

PREPARATION

Planning the Lesson

Refer to the Chapter Organizer on pages 116A–D.

Concepts Developed

Substances are made up of elements or compounds. These substances form the bases for all types of mixtures.

1 MOTIVATE

Bellringer

 Before presenting the lesson, display **Section Focus Transparency 10** on an overhead projector. Assign the accompanying **Focus Activity** worksheet. L1
LEP

Discussion To help students communicate, ask them what they think matter is made of. **How do they think it can be described?** Accept all reasonable answers at this time.

Find Out!

What are some characteristics of water?

Time needed 5 to 10 minutes

Materials glass of water, microscope slide, microscope, dropper

Thinking Processes observing and inferring, comparing and contrasting

Purpose To observe and identify properties of matter.

Preparation If you do not have a microscope, use a magnifying glass and a small square of glass or plexiglass.

Teaching the Activity

Troubleshooting Use distilled water if you can. Pollutants in tap water can give it color and odor.

4-1 Composition of Matter

Section Objectives
- Differentiate among substances, elements, compounds, and mixtures.
- Give examples of heterogeneous and homogeneous mixtures.

Key Terms
substance, mixture, heterogeneous mixture, homogeneous mixture, element, compound

How Can You Identify Substances?

The jars in the Explore activity were filled with air and smoke. The light that went through the jar filled with air came out in a clear beam. However, something in the smoke affected how the light came out. Air and smoke are two different kinds of matter.

Find Out! ACTIVITY

What are some characteristics of water?

What To Do

1. *In your Journal,* make some observations about a glass of plain water. What color is it? How does it smell? How does it feel?

2. Place a drop of water from the top of the glass on a microscope slide. Look at it under a microscope at low power.

3. Compare it with a drop taken from the bottom of the glass. Try taking a smaller drop from the middle of the glass. Is there any difference in how each drop looks under the microscope?

4. Compare drops of water from your glass with drops from a classmate's glass.

Conclude and Apply

1. What characteristics of water did you observe?

2. What can you conclude about the characteristics of all the drops of water in a glass of water?

One drop of water is exactly like every other drop of water. In fact, all pure water everywhere has the same characteristics. It is colorless, odorless, tasteless, feels wet, and is made of only one kind of material.

Another way to describe water is to say that water is a substance. A **substance** is anything that contains only one kind of material. Sugar is also a substance. It is made of only one kind of material and always has the same characteristics. It tastes sweet and dissolves when mixed with water. List other everyday things you see or use that are substances.

Science Journal Students can write their observations and the answers to Conclude and Apply in their journals. L1

Expected Outcomes

Students will notice that the water looks the same.

Conclude and Apply

1. Water is colorless, odorless, wet.
2. The characteristics of all drops are the same.

✔ Assessment

Process Students can work in small groups to write songs about the characteristics of water. Use the Classroom Assessment List for Song with Lyrics in **PASC**, p. 79. L1 **COOP LEARN** **P**

Identifying Mixtures

Is paper a substance? Do you think paper is made of only one kind of material? Look closely at the paper and slice of pita bread shown in **Figures 4-1B** and **4-1C**. Would you say bread is a substance? Why or why not?

Both paper and bread contain tiny bits of several different substances. Any material made of two or more substances in which the basic identity of each substance is not changed is called a **mixture**.

Think back to the two jars you used in the Explore activity at the beginning of this chapter. Did either of the jars contain a substance? You might think that the jar with only air in it contained a substance and the smoke-filled jar contained a mixture of air and smoke. Actually, air itself is a mixture, although you can't see the different substances that make up air.

In **Figure 4-1B**, were the different materials in pita bread mixed together evenly, or were they scattered throughout? A mixture in which the different substances are distributed unevenly is called a **heterogeneous mixture**. Both bread, chocolate milk, and the mixture of smoke and air are heterogeneous mixtures.

Suppose you make a mixture of salt and water. If you stirred the salt and water together thoroughly, would you be able to see the salt? Do you think that all parts of this mixture would taste the same? The salt is distributed evenly throughout the water. A mixture in which the different substances are distributed evenly throughout is called a **homogeneous mixture**. Salt water is a homogeneous mixture. What other homogeneous mixtures can you name?

Figure 4-1

Ⓐ The separate particles of tea and water in this glass of cold tea are too small to be seen and are evenly distributed. These properties make cold tea a homogeneous mixture.

Ⓑ If you look carefully at this pita bread you can see various sized particles of different substances scattered throughout. Pita bread is a heterogeneous mixture.

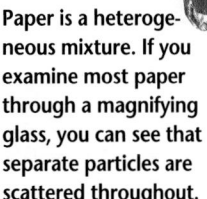

Ⓒ Paper is a heterogeneous mixture. If you examine most paper through a magnifying glass, you can see that separate particles are scattered throughout.

Ⓓ No matter how long you let powdered drink mix and water stand, the evenly distributed particles that make up this homogeneous mixture will not separate.

4-1 Composition of Matter **119**

2 TEACH

Tying to Previous Knowledge
Ask students whether they have ever watched a construction crew pour concrete. Ask them to describe the concrete. Point out that fresh concrete is a mixture of solids—portland cement, sand, rocks—and water.

Uncovering Preconceptions
Students may use the word *substance* to describe any kind of matter. Remind students that a substance contains only one kind of material.

Activity This activity allows the students to prepare and observe a mixture that will form concrete.

Materials needed are 10 g of crushed calcium carbonate tablets, 20 g of sand, 30 g of small, clean gravel, 5 g of water, paper cup, and stirrer.

Have students use a stirrer to mix the dry ingredients in the paper cup. Ask them to describe what they see. Then have them add the water and mix thoroughly with the stirrer. **Caution** students not to touch the wet concrete, as they could get chemical burns. Ask them to describe the appearance of the mixture. Allow the mixture to dry overnight. Before they check the cups, ask students what they think might have happened overnight. ⏹L1

Flex Your Brain Use the Flex Your Brain activity to have students explore MIXTURES.

Student Text Question
What other homogeneous mixtures can you name? *Answers may include cream, ice cream, lemonade, and coffee.*

Program Resources

Study Guide, p. 16
Teaching Transparency 7 ⏹L1
Multicultural Connections, p. 12, Luis Alvarez ⏹L1
Critical Thinking/Problem Solving, p. 5, Flex Your Brain
Section Focus Transparency 10

Meeting Individual Needs

Visually Impaired Use this activity to help students observe how matter forms mixtures.

Materials needed are a shallow pan, 7 or 8 small marbles, and 7 or 8 large marbles.

Have students feel the marbles to note the different sizes. Put the marbles in the pan and shake the pan gently. Have students feel the marbles and note how they are jumbled together.

GLENCOE TECHNOLOGY

 Videodisc

Science Interactions, Course 1
Integrated Science Videodisc
Lesson 2

Figure 4-2

If you stirred sand into a glass of water and poured the mixture into a filter-lined funnel, the water would slowly drip through. The sand would not. Many heterogeneous mixtures can be separated this way.

■ Separating Mixtures

When you make smoky air or salt water, you are putting substances together to make a mixture. Can you take mixtures apart to separate the substances?

Hand-separating the substances in some mixtures is one way to take them apart. What other methods can be used? Suppose you had a glass of sand and water. It would be difficult to separate the sand and water by hand. **Figure 4-2** demonstrates one method of mixture separation.

Can a filter be used to separate salt and water? Usually not. The particles that make up salt are small and are

Separating Mixtures

You know that you can use properties like size, color, or shape to separate substances in a mixture. You can use the property of boiling point by heating liquid mixtures and collecting each substance as its vapor changes back to a liquid.

Property of Attraction

A more difficult property to observe is the attraction that particles in a substance have for one another and for other substances. For example, when you place a drop of water on a piece of waxed paper, the water stays in a spherical shape. But when you drop the water onto newspaper or a paper towel, the water spreads out. How can this be explained? On the waxed paper, the particles in the water have a greater attraction for one another than they do for the waxed paper. On the

120 Chapter 4

distributed throughout the water. They would pass through the filter with the water. Some other method of separation is needed for most homogeneous mixtures. Such a method might involve changing any liquid part of the mixture into a vapor. The solid part would remain behind, as shown in **Figure 4-3**. Evaporation and boiling are two such methods.

You've seen how substances and mixtures are related and how they are different. And you've seen that there are different kinds of mixtures. You know that there is more than one way to separate a mixture. This knowledge can help you classify materials.

Figure 4-3

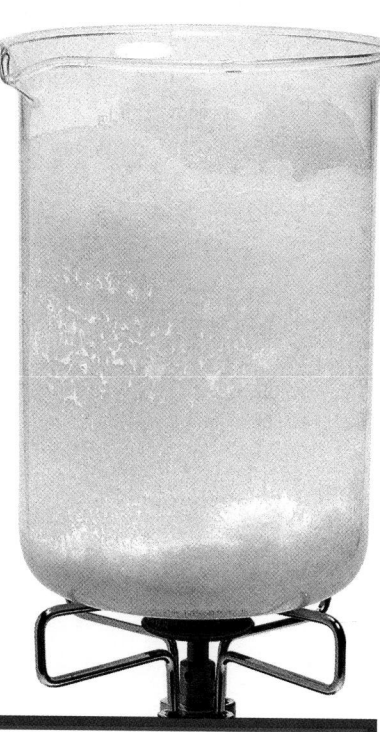

Salt stirred into a glass of water cannot be separated with a filter. But if you boil the mixture, the water will vaporize and rise into the air as steam. The salt will be left behind in the beaker. Boiling is one way to separate some homogeneous mixtures. What is another method?

paper towel, though, the water particles have a greater attraction for the paper.

This property of attraction is another method that can be used to separate mixtures.

You Try It!

1 Cut newspaper, paper towel, and filter paper into strips 2 cm wide by 8 cm long.

2 Tape one end of each strip to the middle of a pencil.

3 Dip a toothpick into green food coloring and make a line across the bottom of

each paper strip about 2 cm from the bottom. Allow the lines to dry.

4 Add 15 mL of water to a jar.

5 Place the pencil across the top of the jar so that just the tips of the paper strips contact the water. The strips should not touch the sides of the jar.

6 Wait 10 to 15 minutes and record your observations *in your Science Journal.* How would you use your observations to describe the attraction of the colored pigments for themselves and the paper?

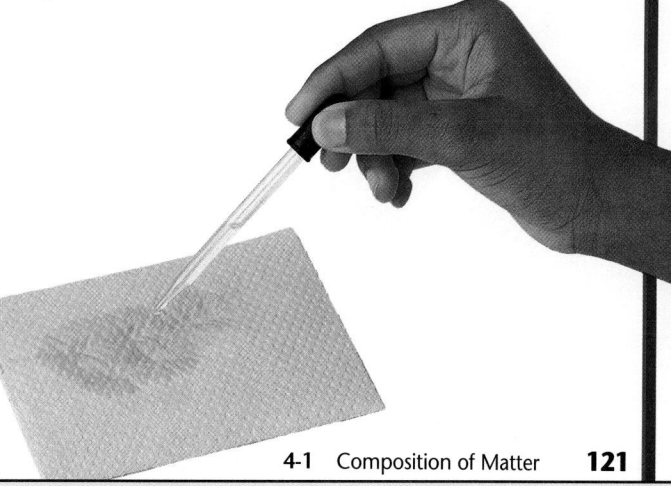

other when they contact different surfaces. Lay a paper towel over a sheet of waxed paper. Pour some water on the paper towel. Have students hypothesize what will happen. Students should find that water has beaded on the waxed paper. L1

Discussion

Before assigning the You Try It, ask students to predict what will happen to the food coloring. Accept all predictions.

Science Journal

The different pigments in the coloring separate, which indicates that their attraction for the paper is greater than their attraction for each other.

Going Further ▐▐▐▐▶

Have students work in small groups to discuss the idea that the properties of one substance allow it to interact in special

ways with another substance. Have them make two lists of common materials such as paper, water, glue, wood, etc. Then have students decide what properties items on the first list might have that allow them to react to items on the second list. Have groups present their lists to the class. Use the Performance Task Assessment List for Making Observations and Inferences in **PASC,** p. 17. L1 **COOP LEARN**

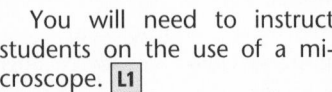

How is an element different from a mixture?

Time needed 15–20 minutes

Materials test tubes, iron powder, sulfur powder, glass slides, microscopes, small magnets, mortar and pestle

Thinking Processes observing and inferring, comparing and contrasting, classifying

Purpose To investigate the difference between an element and a mixture.

Preparation If microscopes are unavailable, hand-held magnifying glasses can be used instead.

Teaching the Activity

You will need to instruct students on the use of a microscope. L1

Science Journal Have students record the answers to the What To Do questions in their journals. L1

Expected Outcomes

Students should find that, even after they have ground up and mixed the iron and sulfur, they still retain their individual properties.

Answers to Questions

1. Iron powder is gray-black and shiny; solid sulfur is bright yellow and dull.

2. Iron crystals resemble needles; the sulfur powder will have more varied appearances depending on the kind of sulfur used and how it was ground or powdered.

3. Iron particles move toward the magnet.

4. Sulfur particles are not attracted by the magnet.

5. a mixture

6. You can still see the separate particles of iron and sulfur.

7. The iron particles in the mixture are attracted to the magnet, while the sulfur particles are not.

Conclude and Apply

1. Only sulfur particles remain.

2. Iron is magnetic; sulfur is not. Iron particles were separated by the magnet.

✔ Assessment

Process Have students compare and contrast how iron and sulfur particles behave in the presence of a magnet. Students can also write a paragraph describing how an element is different from a mixture. Use the Performance Task Assessment List for Writing in Science, **PASC**, p. 87. L1 P

Calcium, phosphorus, iron, potassium, and sodium are some of the elements your body needs to function properly. Create a chart that includes how each element is used by your body and a food source for each element.

What would you get if you took apart your bike? The wheels are made of rubber, the seat may be covered with plastic, and the frame may be made of alloys or metals. The lights may be covered with glass. With the right equipment, you could separate the alloys into the metals from which they are made. The same is true of the rubber, plastic, and other parts. Each is some combination of simpler materials.

■ Elements

Eventually, though, you would reach a point where you couldn't break down the parts into any simpler materials. At that point, you would have a collection of elements. An **element** is a substance that cannot be broken down further into simpler substances by ordinary physical or chemical means.

How can you demonstrate this? Let's do the next activity to find out.

Find Out! ACTIVITY

How is an element different from a mixture?

What To Do

1. Fill a small test tube about half full of iron powder. Fill another test tube with the same amount of sulfur powder. Record the physical appearance of each.

2. Take a few grains of iron, place the iron on a glass slide, and look at it under the microscope. Make another slide of sulfur particles and examine them under a microscope. Record *in your Journal* how the iron and sulfur are different.

3. Now hold a small magnet near the slide containing the iron. What do you observe? Clean the magnet and hold it near the sulfur particles. What do you observe now?

4. Empty both test tubes into a mortar. Take a pestle and carefully grind the two substances together until the contents look the same throughout. What have you just made?

5. Take a few grains of this mixture, put it on a glass slide, and look at it under a microscope. What do you observe? Remove the slide from the microscope and hold a small magnet near it. What happens?

6. Clean the magnet and repeat this step until no more particles are attracted from the slide. Examine the slide once again under the microscope.

Conclude and Apply

1. What do you see now?

2. What can you conclude from your observations?

Mixtures are made by combining two or more substances together in such a way that each keeps its own properties. Mixtures can physically be separated into simpler substances. If you mixed together the particles of iron and sulfur, you made a mixture. This mixture could be separated back into iron and sulfur by using a magnet.

Iron and sulfur particles, however, cannot be broken down any further, using either physical or chemical methods. If you were to continue grinding down samples of each into smaller and smaller particles, you would still be left with particles of iron and sulfur. Iron and sulfur are already in their simplest forms. Both iron and sulfur are examples of elements. Elements are known as the building blocks of matter.

■ Compounds

What would have happened if you had heated the iron and sulfur mixture? You would have made a new substance which would not look like the iron and sulfur mixture, nor iron or sulfur by itself. You would have made something totally new called iron sulfide. Heating can cause such chemical changes to occur.

Iron sulfide has properties different from those of either iron or sulfur. It is a new substance. A substance whose smallest unit is made up of more than one element is a **compound**. Iron sulfide is an example of a compound. What properties does a compound have that make it different from either an element or a mixture?

Figure 4-4

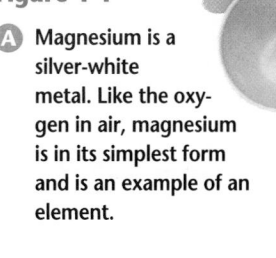

Ⓐ Magnesium is a silver-white metal. Like the oxygen in air, magnesium is in its simplest form and is an example of an element.

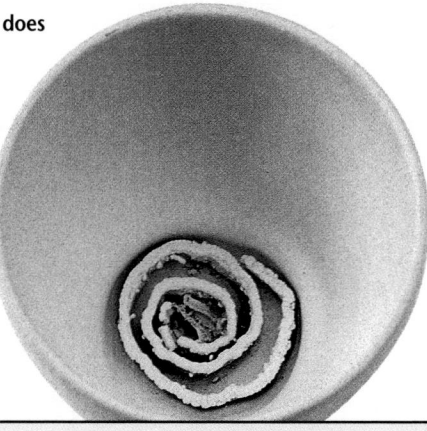

Ⓑ Heating magnesium in the presence of oxygen causes a change. The magnesium and oxygen combine chemically to make a new substance—magnesium oxide.

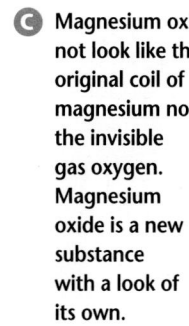

Ⓒ Magnesium oxide does not look like the original coil of magnesium nor the invisible gas oxygen. Magnesium oxide is a new substance with a look of its own.

Inquiry Question What properties of a compound make it different from either an element or a mixture? *Compounds are composed of more than one kind of element, but unlike mixtures, they cannot be separated by physical means. The elements in a compound do not retain their own properties.*

Life Science

ELEMENTS

calcium:
  for: strong bones and teeth, blood clotting, muscle and nerve activity;
  found in: milk, eggs, green leafy vegetables

phosphorus:
  for: strong bones and teeth, muscle contraction, stores energy;
  found in: cheese, meat, cereal

iron:
  for: carries oxygen in hemoglobin in red blood cells;
  found in: raisins, beans, spinach, eggs

potassium:
  for: balance of water in cells, nerve impulse conduction;
  found in: bananas, potatoes, nuts, meat

sodium:
  for: fluid balance in tissues,nerve impulse conduction;
  found in: meat, milk, cheese, salt,beets, carrots

Meeting Individual Needs

Learning Disabled To complete the Find Out on page 122, pair students with more able partners and have them divide the work. The learning disabled students can fill the test tubes, put powder on the slides, grind the substances, and use the magnet. Their partners can record each pair's findings. COOP LEARN

DESIGN YOUR OWN
INVESTIGATION

4-1 Elements, Compounds, Mixtures

Preparation

Purpose To classify materials as elements, compounds, or mixtures.

Process Skills observing, classifying, making and using tables, comparing and contrasting

Time Required 45–60 minutes

Materials You may want to have small amounts of each material ready for the students to observe instead of having them obtain their own.

Possible Hypotheses Students might say that elements will have their names on the periodic table, compounds have a double name, and mixtures can have more than one type of thing visible in them.

Plan the Experiment

Process Reinforcement You may wish to have students work cooperatively in designing their data tables.

Possible Procedures Thoroughly observe each object, and check each object's properties against the list of identifying characteristics. Record observations and classifications in the data table.

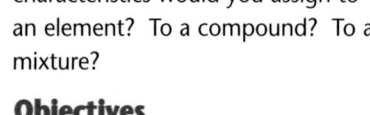

DESIGN YOUR OWN
INVESTIGATION

Elements, Compounds, Mixtures

Developing a system of classification helps turn a definition into a tool for solving problems. For example, you can classify vehicles as a car, pick-up truck, or van based on identifying characteristics. Can a similar system can be made to distinguish among elements, compounds, and mixtures?

Preparation

Problem
How can their differences help you distinguish among elements, compounds, and mixtures?

Form a Hypothesis
Find the definitions of elements, compounds, and mixtures from your text. If you were classifying objects based on these definitions, what characteristics would you assign to an element? To a compound? To a mixture?

Objectives
• Define element, compound, heterogeneous mixture, and homogeneous mixture.
• Develop a list of identifying characteristics based on the definitions.
• Classify an object as an element, a compound, a heterogeneous mixture, or a homogeneous mixture.

Possible Materials
small amount of rock salt
glass of lemonade
aluminum foil
baking soda
small piece of granite
copper wire
piece of graphite (carbon)
vinegar and oil salad dressing

Safety Precautions

Never eat, drink, or taste anything used in a laboratory experiment.

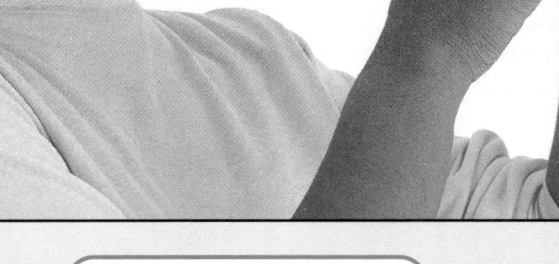

Program Resources

Activity Masters, pp. 19–20, Design Your Own Investigation 4-1

Science Discovery Activities, 4-1, Sensational Condensation

DESIGN YOUR OWN
INVESTIGATION

Plan the Experiment

1 Work as a group to choose objects and agree on a hypothesis. Record *in your Science Journal* the identifying characteristics that you will look for as you classify the objects.

2 Design a data table *in your Science Journal* to record the names of your test objects and the classifications you assign them.

Check the Plan

1 Do your identifying characteristics correspond to the definitions of substances and mixtures?

2 How will you keep track of your observations and explanations?

3 Before you begin, have the teacher check your plan and your list of objects.

4 Carry out the experiment.

Copper Vinegar and oil Aluminum Granite Graphite Rock salt Lemonade Baking soda

Analyze and Conclude

1. Observe and Infer If you know the name of a substance, how can you find out if it is an element?

2. Compare and Contrast How do compounds differ from mixtures?

3. Classify What homogeneous mixtures did you identify? How did you determine the difference between homogeneous and heterogeneous mixtures?

4. Classify Did your list of identifying characteristics help you to correctly classify the objects? How would you change your list if you were to repeat the experiment?

5. Make and Use Tables Make a table that lists the four kinds of substances and mixtures, their differences, and the classifications you made. Look in the Skill Handbook under Making Tables if you need help.

Going Further

Use your list of identifying characteristics to classify the contents of your refrigerator at home. Identify whether there are more substances or mixtures.

Multicultural Perspectives

Changing Hypotheses About Matter

Inform students that one of the earliest attempts to classify matter was made in Greece more than 2000 years ago. Ancient Greeks believed that all matter was made up of earth, fire, water, and air. Centuries later, scientists used electricity to break apart water into oxygen and hydrogen, showing that water is not an element. Have students research and report on the origins of modern chemistry in alchemy, which flourished in Alexandria, Egypt, from about 300 A.D. to 1600 A.D.

Teaching Strategies

Science Journal Have students create their data tables and write their answers to the questions in their journals.

Expected Outcome

Students should classify the materials as follows:
Elements—copper, aluminum, carbon
Compounds—rock salt, baking soda
Homogeneous mixtures—lemonade
Heterogeneous mixtures—granite, vinegar and oil salad dressing

Analyze and Conclude

1. Check to see if the name of the element is on the periodic table.

2. The different elements in a compound are chemically combined and cannot be separated by physical means. Moreover, unlike a mixture, a compound has different properties from the elements that combine to make it.

3. Lemonade was homogeneous. Homogeneous mixtures are uniform throughout so that you cannot tell there is more than one substance there. In heterogeneous mixtures, there is an uneven distribution of the materials that make up the mixture.

4. Answers will vary.

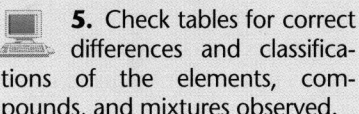 **5.** Check tables for correct differences and classifications of the elements, compounds, and mixtures observed.

✔ Assessment

Process Hold up a pencil. Ask a volunteer to classify the pencil as an element, a compound, or a mixture. It is a mixture made up of graphite, wood, a metal ring, and a rubber eraser. Working in small groups, students can brainstorm and classify as elements, compounds, or mixtures their own lists of everyday items. Use the Performance Task Assessment List for Group Work in **PASC,** p. 97.

Going Further ▌▌▌▌▶

Students might identify pizza, salad, salsa, and soda as mixtures; baking soda and water as compounds; and aluminum (foil) as an element. Students will probably identify more mixtures.

3 ASSESS

Check for Understanding

Ask students to identify some properties of salt and of sand. Point out that a property they share is not a good basis for separation.

Reteach

Demonstration Reinforce students' ability to differentiate homogeneous and heterogeneous mixtures. Materials needed are a large jar of water, spoon, sugar, and food coloring.

Fill the jar with water and then dissolve some sugar in the water. Ask students if the materials are evenly mixed. *yes* Then add some food coloring and do not stir. Ask if the materials are evenly mixed. *no* Students should recognize that when the materials are evenly mixed there is a homogeneous mixture. L1

Extension

Acquiring Information Have students who have mastered this section look up and define *colloids* or *suspensions*. L2

4 CLOSE

Ask students to identify examples of substances, elements, compounds, and heterogeneous and homogeneous mixtures in the classroom. Remind students that they should consider their own body fluids in identifying them. L1

In the Investigation you just completed, different elements, compounds, and mixtures were examined

Figure 4-5

A Carbon is an element that has three natural forms: diamond, graphite, and amorphous charcoal. Carbon has a unique ability to combine with itself and with other elements in various ways to form millions of different compounds. Sugar is a compound made of carbon, hydrogen, and oxygen.

B If you mix cinnamon and sugar, you have a mixture called cinnamon-sugar. This mixture can be separated back into cinnamon and sugar by physical means. How would you go about it?

and compared. Let's now summarize the differences between compounds and mixtures.

First, compounds cannot be separated by physical means. If you melted water when it was in the form of ice, the hydrogen and oxygen in it would not separate out—you would simply get liquid water.

Second, the substances that make up a compound do not keep their own properties. **Figure 4-5** demonstrates this property using sugar as an example. A mixture such as brass, on the other hand, still retains many of the physical properties of the elements that are used to make it—copper and zinc.

Third, the same compound always has the same composition. If you went from store to store buying samples of sugar and then took the time to break each down into its elements, you would always end up with the same amounts of carbon, hydrogen, and oxygen.

In the next section, you will learn how to describe different materials.

check your UNDERSTANDING

1. Differentiate between water and chocolate milk.
2. Give three examples of heterogeneous mixtures you use at school. Tell why they are heterogeneous.
3. Name three homogeneous mixtures you might eat, drink, or use at home. Explain your choices.
4. How can you distinguish between a piece of cotton and a piece of bronze?
5. State three properties of sugar that make it a compound.
6. **Apply** How might you separate a mixture of salt and fine sand?

check your UNDERSTANDING

1. Water is a compound; chocolate milk is a mixture of several substances.

2. Paper, pen, and pencil; made of two or more substances that are not evenly distributed.

3. Tea with sugar, salt water, dishwashing liquid: component substances are evenly distributed.

4. Two ways are by color and by density.

5. It cannot be separated by physical means; the elements do not keep their own properties; the proportion of elements is always the same.

6. Add water and stir until the salt disappears; pour through a filter to separate the sand; evaporate the water to separate the salt.

Describing Matter

Physical Properties

Suppose you are given the task of separating the materials in some wild bird food. You would observe the color, size, and shape of each type of grain or seed. You could describe some of the grains and seeds as flat, round, small, pointed, yellow, black, white, or striped. You could also use the characteristics of shape, size, and color to help you distinguish one kind of seed in the mixture from another. Notice that you can make such descriptions without changing the grains in any way.

When you use characteristics such as color, shape, and brittleness to describe an object or a material, you are naming some of its physical properties. Any characteristic of a material that can be observed or measured is a **physical property**. When you describe physical properties, the substances that make up the material are not changed. How are the physical properties of chalk different from those of an aluminum can?

Later in your studies, you will observe the color, brittleness, and hardness of some materials. But first, familiarize yourself with some of the most common measurements related to physical properties.

Section Objectives

- Recognize examples of physical properties.
- Measure length, volume, and mass of different materials.
- Relate density to mass and volume.

Key Terms

physical property
density

PREPARATION

Planning the Lesson

Refer to the Chapter Organizer on pages 116A–D.

Concepts Developed

This section explains how matter can be measured using SI. Recognizing physical properties contributes to students' understanding of how matter varies.

1 MOTIVATE

Bellringer

Before presenting the lesson, display **Section Focus Transparency 11** on an overhead projector. Assign the accompanying **Focus Activity** worksheet. L1
LEP

Student Text Question

How are the physical properties of chalk different from those of an aluminum can? *Chalk and an aluminum can differ in color, texture, strength, smell, and size.*

 Explore! ACTIVITY

Can you measure without a ruler?

What To Do

1. Walk across the front of your classroom. As you walk, line up the heel of one foot with the toe of the other foot.

2. *In your Journal,* record how many footsteps you took to walk across the classroom.

3. Compare this number with the number of steps your classmates used. You'll probably find that the number of footsteps used to measure your classroom differs from one student to another.

4. Now try to measure a pencil and a button using footsteps. How many footsteps do you think it is from your house to school?

Answers to Question

Answers will vary, depending on the size of the student's foot and the distances to be measured.

✔ Assessment

Process   Have students compare their results in Step 1 with results attained using one other nonstandard unit, such as the length of their bodies or a somersault. Students could then create a cartoon that shows which units are most reliable and why. Use the Performance Task Assessment List for Cartoon/Comic Book in **PASC,** p. 61. L1 P

 **Explore!**

Can you measure without a ruler?

TECH PREP

Time needed   10 minutes

Materials   pencil, button

Thinking Processes   observing, comparing and contrasting, forming operational definitions

Purpose   To observe that measurement techniques can vary.

Teaching the Activity

Science Journal   Before students do this activity, have them estimate how many footsteps each measurement will equal. Then have students record their results and compare estimates with actual measurements. L1

Expected Outcomes

The number of footsteps will vary from student to student.

Demonstration Guide students to infer that mass and volume are not co-dependent.

Display one kilogram of each of five materials of different densities, such as feathers, cotton balls, paper, sand, sugar, or flour. Do not tell students the mass of each. Ask students to order the materials from heaviest to lightest, and to explain how they arrived at their estimates.

2 TEACH

Tying to Previous Knowledge

Ask students how they can tell an item of clothing will fit without trying it on. Or ask how they know how much taller they are this year than last. They will probably point to sizes and measurements.

Theme Connection The theme is scale and structure. Any sample of matter has certain physical properties that can be measured. For example, kilometers are used to express measurements of large distances. Millimeters are used to express measurements of small objects.

Using the Table Explain to students that meters, divisions of meters, and multiples of meters measure length or distance. Divisions and multiples of grams measure mass. So a book could be measured in both meters and grams.

Across the Curriculum

Astronomy

Because stars are very far from Earth, scientists had to come up with a unit of measurement much larger than that used to measure distances on Earth. They based this unit on the speed of light. A light-year is how far a ray of light travels in one year—about 9 400 000 000 000 km. The star Proxima Centauri is about 4.3 light-years away from Earth. Have students figure out how many kilometers that is.

Table 4-1

Most of the world's countries use standard units of measurement. The standard units were designed to multiply and divide easily and to make both large and small quantities convenient to measure. This table shows the International System of Units, abbreviated SI.

SI Units			
Unit	Abbreviation	Size Comparison	Similar-sized Object
kilometer	km	1000 meters	ten football fields
meter	m	100 centimeters	guitar, baseball bat
decimeter	dm	1/10 meter	a little more than a new crayon
centimeter	cm	1/100 meter	staple
millimeter	mm	1/1000 meter	tooth on edge of stamp
kilogram	kg	1000 grams	your science textbook
gram	g	1/1000 kilogram	large paper clip

In the Explore activity, you measured objects in "footstep" units. Maybe this type of unit works well enough for some measurements, but it has some problems. For example, the measurements are not the same from one person to the next. In addition, very large and very small measurements are not easy to make with this unit.

People around the world need to be certain that their measurements are understood by others. They also want to be sure that a bolt of cloth in Delhi, India, will be measured in the same way as a bolt of cloth in Paris, France. Therefore, most of the world's countries use standard units of measurement. This standard measuring system is called the International System of Units, abbreviated SI, and is shown in **Table 4-1**.

■ Length

You may use a meterstick to measure the length of an object, such as a book. Just what does it mean to measure length? Is length the number of pages between the covers of a book? Or is it the number of minutes from the beginning to the end of a movie? In scientific measurement, length is the distance between two points. That distance could be the diameter of the period at the end of this sentence or the distance from Earth to the moon.

Study **Table 4-1**. What SI units of measure might you use to measure the length of a book? Would the book be changed when its length is measured? Is length a physical property of this object? Explain.

■ Volume

In addition to measuring the length of your book, you may also wish to use the meterstick to measure the book's width and height. If you then multiplied these three measurements, you would find out how much space the book occupied.

You would have found the volume of the book. Volume is the amount of space an object or a material occupies. Can you measure the length, width, or height of a substance like water? Liters and milliliters are the most common units used to express the volume of

---

Program Resources

Laboratory Manual, pp. 17–20, Measurement and Graphing L2
Study Guide, p. 17
Concept Mapping, p. 12, Matter L1
Teaching Transparency 8 L2
Section Focus Transparency 11

ENRICHMENT

Field Trip Have students visit the local supermarket. With the permission of the manager, have students inspect packaging that indicates volume and compare and contrast packages of different sizes or shapes that are supposed to contain the same volume. **Do the packages appear to have the right amount of mass? How can differences be accounted for?** *The density of the contents varies.* L1

water and similar substances. **Figure 4-6** demonstrates volume measurement of a sugar cube and water.

How much space do you think a small rock takes up? Can you use a ruler to help you find its volume?

Figure 4-6

A milliliter and a cubic centimeter are equivalent measures of volume—the amount of space an object or a material occupies.

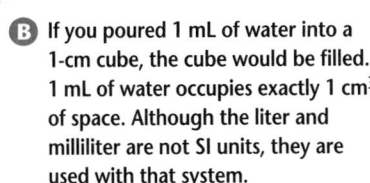

A The volume of a sugar cube can be found by multiplying its length times its width times its height. The cube's volume is 1 cm x 1 cm x 1 cm or 1 cubic centimeter, written 1 cm^3. Would you say that volume is a physical property? Why?

B If you poured 1 mL of water into a 1-cm cube, the cube would be filled. 1 mL of water occupies exactly 1 cm^3 of space. Although the liter and milliliter are not SI units, they are used with that system.

Explore! ACTIVITY

How can you measure the volume of a rock?

What To Do

1. Use a rock about the size of a golf ball.

2. Add 250 mL of water to a 500-mL beaker.

3. Carefully add the rock.

4. Record *in your Journal* what happens to the level of the water in the beaker. Is the change related to the volume of the rock?

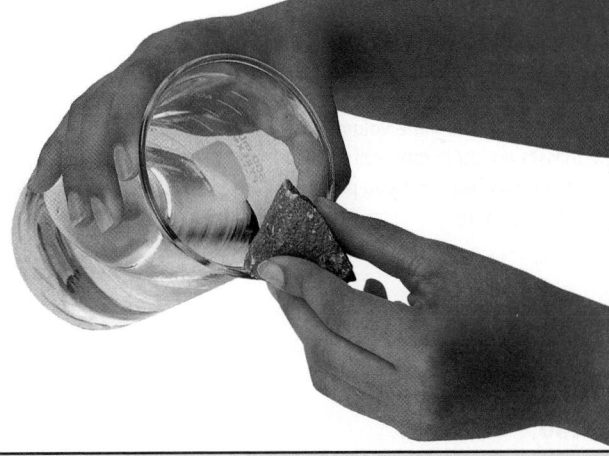

Explore!

How can you measure the volume of a rock?

Time needed 5 minutes

Materials golf ball-sized rock, 500-mL beaker, water

Thinking Processes observing and inferring, forming a hypothesis, forming operational definitions, measuring in SI, interpreting data, recognizing cause and effect

Purpose To measure volume by water displacement.

Teaching the Activity

Discussion Ask students whether two objects can occupy the same space at the same time. They will probably say no. Remind students that the amount of space taken up by each object is its volume.

Daily Life

Measurement of volume by water displacement can be used in cooking. For materials that are difficult to measure, such as peanut butter, this method is ideal. Half fill a large measuring cup with cold water and note the level. Add peanut butter until the water level rises by the amount indicated in the recipe.

Content Background

The metric system acts as a universal language for scientists. It was developed in 1791 and has been used in most European countries since then. The English system was standardized in the 15th century, but its units are not based on a common factor. The metric system uses a base of ten, which allows more systematic conversions.

Debate Have students compare and analyze. Point out that there are two ways to measure physical properties of matter—directly, by using a ruler or scale, or indirectly, by using mathematical formulas. Divide students into two groups. Have one group argue the merits of direct measurement while the other group argues the merits of indirect measurement. [L3]

COOP LEARN

Visual Learning

Figure 4-7 Help students use prior experiences to understand the relationship among mass, volume, and density.

Figure 4-8 Discuss the benefits of knowing an object's mass as well as its volume in order to calculate its size. **Which would feel heavier if you picked it up? Why?** *The golf ball, which has the greater mass.*

DID YOU KNOW?

A newborn African elephant is about 95 cm high and has a mass of about 120 kg. By the time the elephant reaches full size (at about age 20), it is 340 cm high and has a mass of about 5400 kg.

When you put the rock in the beaker, the water level went up. The two materials in the beaker took up more space than the water alone. The amount the water level went up tells you the volume of the rock. You can find the volume of the rock by subtracting the volume before the rock was added from the final volume.

■ Mass

Although you can now find the volume of a book, a rock, or a glass of water, can you tell how much material is in each one? Mass is the amount of matter in an object or a material. Look at the table tennis ball and golf ball in **Figures 4-7** and **4-8**. The golf ball has more mass than the table tennis ball.

If you study **Table 4-1** on page 128, you'll notice that kilogram is an SI unit of mass. Masses of small objects are measured in grams.

Like length and volume, mass is a physical property that is used to describe materials. Knowing the mass of an object could be useful in gathering more information about the object. How might knowing the mass and the volume of an object be useful?

Figure 4-8

This double pan balance clearly shows that although the table tennis ball and the golf ball appear to have about the same volume, their masses are quite different. Which would feel heavier if you picked them up? Why?

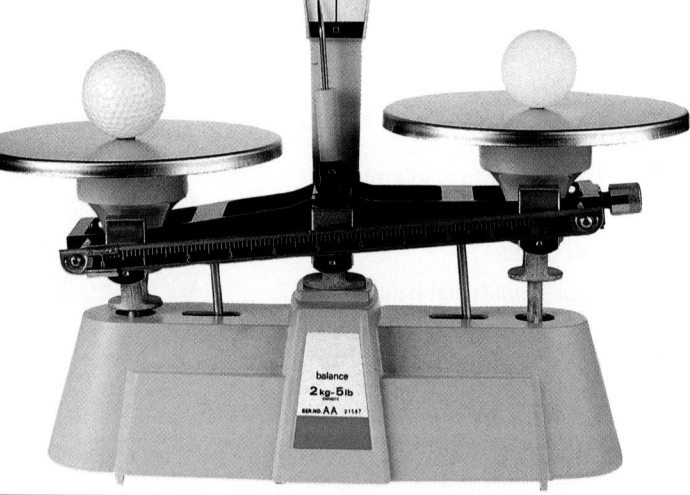

Figure 4-7

The golf ball contains more material than the table tennis ball. You might say the golf ball has more matter. Mass is the amount of matter in an object or in a material. The golf ball has more mass than the table tennis ball.

Inside a golf ball

Inside a table tennis ball

---

130 Chapter 4

Multicultural Perspectives

Counting with a Quipu

Most societies develop ways to make measurements and record information. The Incas of South America kept accounts using a quipu, a knotted string. By tying knots at varying distances on different colored wool yarn, the Incas could count and transmit information to others. Counting with the quipu was apparently based on the decimal system and included the use of zero. Writing in 1549, the Spanish historian Pedro Cienza de León recorded the quipu's precision. Incas explained to him that the quipu kept accounts "with such accuracy that not so much as a pair of sandals would be missing." Have students research and construct their own quipus and then demonstrate them to the class. [L2] **LEP**

■ Density

Which of the grocery bags shown in **Figure 4-9** would you rather carry? Density is another physical property used to describe materials. **Density** is the amount of mass an object or a material has compared to its volume. Density can be expressed as the mass of an object divided by its volume. Recall that grams (g) are units of mass, and cubic centimeters (cm^3) are units of volume. So one way density can be measured is in grams per cubic centimeter, written g/cm^3.

Suppose two identical bags were tightly closed, and you couldn't see inside them. Would you be able to tell which bag contained sand and which contained sugar? Surely you could determine the mass of the material in each bag. Perhaps you could guess the volume of each material. But suppose you knew the density of the material in each bag. Would you then be able to tell whether a particular bag contained sand or sugar?

Because sand and sugar have different densities, you could use this property to tell which material was in each bag. In the activity that follows, you will use density to identify a material.

Figure 4-9

Ⓐ These grocery bags are the same size and have equal volume. Both are filled to capacity. The bag of paper towels has much less mass than the bag of cans.

Ⓑ The amount of mass an object or a material has compared to its volume is a measure of its density. Which has the greater density— the bag of cans or the bag of towels?

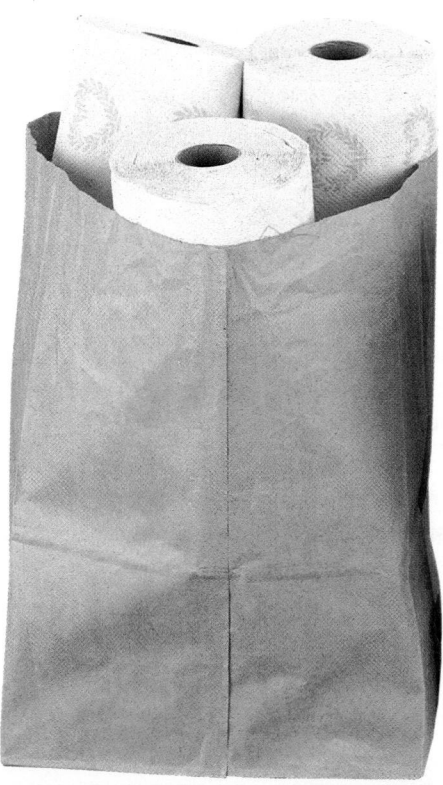

131

4-2 Using Density

Time needed 30 minutes

Materials See reduced student text.

Process Skills making and using tables, inferring, drawing conclusions, measuring in SI, interpreting data, comparing and contrasting, sampling and estimating

Purpose To identify an unknown substance using density data.

Preparation To prepare the saturated salt solution, dissolve 300 g of sodium chloride in 900 mL of distilled water. For the unknown, use any of the three known liquids. You may wish to give different unknowns to different students.

Teaching the Activity

Safety Students should wear goggles, which will prevent eye irritation. Be sure students do not ingest the alcohol or breathe its concentrated fumes. No open flames should be allowed during the activity.

You might want to do this as a group activity but challenge students to interpret data on their own. COOP LEARN

Student Journal Have students write their data tables as well as their observations and calculations in their journals. L1

 GLENCOE TECHNOLOGY

Videodisc

Science Interactions, Course 1
Integrated Science Videodisc

Lesson 1

INVESTIGATE!

Using Density

In this activity, you will find the density of three materials. You will use this information to help you identify an unknown material.

Problem
How can density be used to identify an unknown material?

Materials
water
rubbing alcohol
unknown (liquid) substance
100-mL graduated cylinder

saturated saltwater mixture
pan balance and set of masses
goggles

Safety Precautions

Avoid open flames.

What To Do

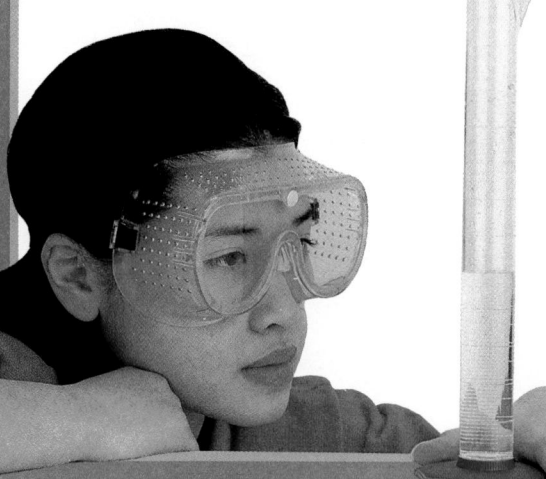

1 Copy the data table *into your Journal.*

2 Use the balance to measure the mass, in grams, of a clean, dry graduated cylinder (see photo **A**). Record the mass in your table.

3 Fill the cylinder with water to the 50-mL mark (see photo **B**).

4 *Measure* the mass of the filled cylinder and record it in your table under the heading *Total Mass* (see photo **C**). Then discard the water as directed by your teacher.

5 *Calculate* the mass of the water by subtracting the mass of the empty cylinder from the total mass. Record the result under the heading *Actual Mass.*

Program Resources

Activity Masters, pp. 21–22, Investigate 4-2

Making Connections: Technology and Society, p. 11, Institute of Standards and Technology L2

Science Discovery Activities, 4-2, Underwater Fire

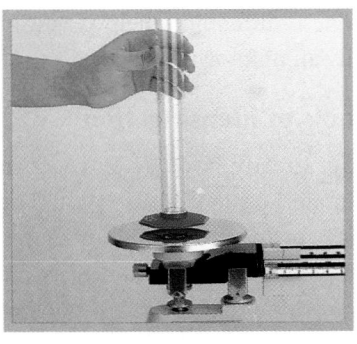

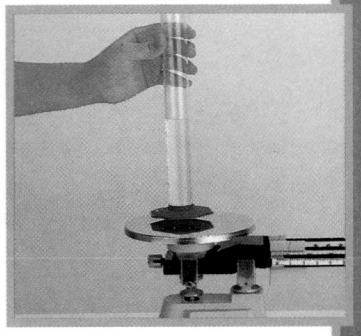

A   **B**   **C**

Sample data

Data and Observations

Material	Mass of Cylinder	Total Mass	Actual Mass	Volume	Density (g/cm³)
Water	117.59	168.31	50.72	50 mL	1.01
Salt water	117.59	176.69	59.10	50 mL	1.18
Alcohol	117.59	162.04	44.45	50 mL	0.89
Unknown	117.59	167.60	50.01	50 mL	1.00

6 Repeat Steps 3-5, first using the salt water, then the rubbing alcohol, and finally the unknown material. **CAUTION:** *Alcohol burns readily, and its fumes can be irritating. Wear goggles. Be sure that the room is well-ventilated, and there are no open flames.*

7 Record the data for each material.

Analyzing

1. **Calculate** the density for each material by dividing its actual mass by its volume. Round to two decimal places.

2. Which known material had the highest density?

Concluding and Applying

3. What was the unknown material?

4. How did finding the density of the unknown material help you identify it?

5. **Going Further** What other physical properties might you also look for and measure in identifying materials?

Expected Outcome

The unknown has a density that is equal to the density of one of the known solutions. Students will infer that the liquids are the same.

Answers to Analyzing/ Concluding and Applying

1. water, 1 g/cm³; rubbing alcohol, 0.89 g/cm³, saturated salt water, 1.18 g/cm³.

2. salt water

3. Answers will vary depending on the unknown used.

4. The density of the unknown liquid was the same as the density of one of the known liquids.

5. Color, boiling point, melting point, odor

✔ Assessment

Process Have students find the densities of other liquids and solids. To find the volume of a solid, students can use the water displacement method they used in the Explore on page 129. Use the Performance Task Assessment Lists for Carrying Out a Strategy and Collecting Data and for Using Math in Science in **PASC,** pp. 25 and 29. **L1**
COOP LEARN

Meeting Individual Needs

Learning Disabled   Some students may have difficulty understanding density. Have students blow up a balloon. Explain that the space in the balloon is its volume. Then have the student rest the balloon on a hand. The "feel" of the balloon is its mass. Put water into another balloon until it is the same size as the balloon blown up with air. Ask which balloon has more mass. *The water-filled balloon* Ask how the vol- umes compare. *They are the same.* Ask which has the greater density. *The water-filled balloon. Density is mass divided by volume.* **L1**

3 ASSESS

Check for Understanding

1. Have students identify the two physical properties that determine density. *mass and volume*

2. Have students answer questions 1 through 3 and discuss the Apply question.

Reteach

Activity Have students prepare two sets of flash cards—one set with definitions of words in this section, and the other set, with the words themselves. Students can then invent a card game based on matching words and definitions. L1

Extension

Activity Have students work in pairs to challenge other pairs to determine the identity of a mystery liquid by using its density. Each pair chooses two different liquids. They determine the density of each, and use one of the two liquids as the mystery liquid. Then they trade with another pair of students. Each pair finds the densities of the two known liquids and then determines the identity of the other liquid. L2 **COOP LEARN**

4 CLOSE

Have students identify, observe, and describe objects in the room using physical properties. L1

Figure 4-10

Seawater   Milk

Ⓐ The densities of milk and seawater are both 1.03 g per cm³. What tools of identification other than density could you use to identify these substances?

Ⓑ This bolt and piece of chalk are nearly the same length and volume. What tools of identification could you use to identify these substances?

Chalk

Bolt

Ⓒ This 1/2 kilogram of sugar and 1/2 kilogram of salt have the same mass and are the same color. What tools of identification could you use to identify these substances?

Salt   Sugar

Like sand and sugar, the water, salt water, and alcohol you used in the Investigate have different densities. These differences helped you to identify an unknown substance.

■ Tools of Identification

You already have many tools in your material identification kit. The physical properties of color, shape, length, volume, mass, and density are some of these tools or clues. Look at the objects pictured in **Figure 4-10**. Is knowing just the density enough to identify these objects? Is knowing only the length and volume enough information? What about mass and color? When you are trying to identify materials, remember to use every clue or tool you have available. Using only one or two tools alone may not be completely reliable. As you continue with this chapter, you will be adding more tools to your identification kit.

check your UNDERSTANDING

1. Choose an object or a substance that you use at home or at school. Describe it using at least three of the physical properties you learned about in this section.

2. How do length, volume, and mass differ from one another? What units are used to measure each?

3. What physical properties of a wooden block is its density related to? How could density be used to identify another sample of wood?

4. Apply How might you use physical properties to identify and separate broken glass and water?

134   Chapter 4   Describing the Physical World

check your UNDERSTANDING

1. Example: a pencil is a solid, yellow, and 19 cm long.

2. Length is a measure of distance (meters); volume is a measure of space (liters); mass is a measure of matter (kilograms).

3. Its mass and its volume. Determine the density of an unknown sample of wood and compare it with the known densities of different woods. The identity of the unknown sample can be determined, provided that two or more woods do not have the same density.

4. Although their colors may be similar, glass is a solid and won't pass through a filter; water will pass through a filter.

Physical and Chemical Changes

Physical Changes

Think about the different properties of chalk. Length, color, and brittleness are some of those properties.

Do substances change when you change their physical properties? Does chalk still remain chalk?

Figure 4-11

A When you break sticks of chalk in pieces, length and mass change. But the substance is still chalk.

B Even if you ground the pieces to powder, you would still have chalk. Breaking and grinding are called physical changes.

Section Objectives

- Distinguish between physical and chemical changes.
- Differentiate between chemical and physical properties.

Key Terms

physical change
chemical change
chemical property

 **ACTIVITY**

Do changes in physical properties affect substances?

What To Do

1. Put an empty can with no label on it into the freezer for use in a little while.

2. Remove one ice cube from the freezer and place it on a small dish. Watch what happens. Is the substance that forms the same as the substance that made up the ice cube?

3. Put the dish with this substance in it back in the freezer. Observe the dish after an hour. What changed? How does this new substance compare with an ice cube?

4. With your teacher's help, boil some water. Is the steam that goes into the

air the same substance as the water you started with?

5. Remove the can from the freezer. Using tongs, carefully hold the cold can near the steam. Record what happens *in your Journal.*

Conclude and Apply

1. How are ice and steam the same as the water you drink?

2. How are they different?

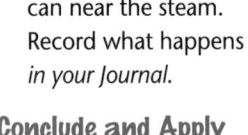 **Find Out!**

Do changes in physical properties affect substances?

Time needed   Several hours

Materials   clean, empty can; tongs; hot plate or Bunsen burner; small dish; ice cubes; water; pan or heat-resistant glassware

Thinking Processes   observing and inferring, comparing and contrasting

Purpose   To demonstrate that water remains the same substance after a physical change.

Preparation   Prepare ice cubes ahead of time.

Teaching the Activity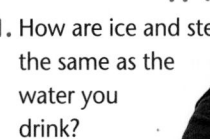

Science Journal   Have students elaborate on the second Conclude and Apply question in their journals. For example, students could compare the temperature of drinking water to that of steam or ice. L1

Planning the Lesson

Refer to the Chapter Organizer on pages 116A–D.

Concepts Developed

Chemical changes produce new substances with new physical and chemical properties. Physical changes do not.

1 MOTIVATE

Bellringer

Before presenting the lesson, display **Section Focus Transparency 12** on an overhead projector. Assign the accompanying **Focus Activity** worksheet. L1 LEP

Discussion   Have students describe the smell of burnt toast. Is the smell the result of materials that were not present before? *yes* Is the toast the same as it was earlier? *New materials are present. The toast is different.* L1

Expected Outcomes

Students will discover that water can exist in three physically different phases, or states—liquid, solid, and gas. L1

Conclude and Apply

1. All are made of the same substance, H_2O.

2. They have different physical properties, such as shape and volume.

✔ Assessment

Oral   Have students work in small groups to create a skit comparing and contrasting three states of water. Use the Performance Task Assessment List for Skit in **PASC**, p. 75. L1
COOP LEARN **LEP**

2 TEACH

Tying to Previous Knowledge

Have students identify changes that occur when a cake bakes. Ask them to discuss what clues (visual, tactile, olfactory) they use to determine the changes.

Find Out!

Can the identity of a substance be changed?

Time needed 40 minutes

Materials ammonia, old, discolored penny, safety goggles, jar and lid, forceps, paper towel

Thinking Processes observing and inferring, recognizing cause and effect

Purpose To demonstrate a chemical change.

Teaching the Activity

Safety Caution students that ammonia fumes can damage the lining of the respiratory system. Goggles prevent eye irritation and forceps prevent chemical burns. L1

Science Journal Have students write their observations in their journals. L1

Expected Outcomes

The penny will become shinier. The ammonia changes to a bluish color.

Conclude and Apply

1. The color of the solution changed from colorless to blue. The penny became shinier.

2. Yes. A chemical reaction took place.

✔ Assessment

Process Have students work in small groups to prepare questions about the changes they observed and how they could find out the cause. Use the Performance Task Assessment List for Asking Questions in **PASC**, p. 19. L1

COOP LEARN

Figure 4-12

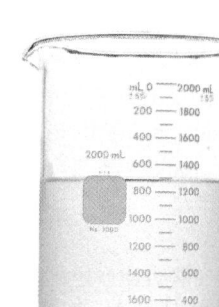

A When ice cubes melt, their shape changes.

B The water that made up the ice cubes is the same substance as the water that forms in the beaker. If you put the beaker of water in a freezer, what would happen to the water?

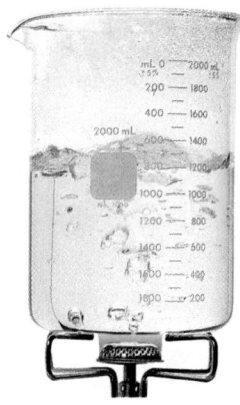

C If you boil the water, its shape and volume would again change, but the substance remains water. Boiling, melting, and freezing are physical changes.

Find Out! ACTIVITY

Can the identity of a substance be changed?

Wear safety goggles while you or others are doing this activity.

What To Do

1. Observe the physical properties of an old, discolored copper penny.

2. Place the penny in a small glass jar with a lid.

3. Add about 2 tablespoons of household ammonia and close the jar quickly. Observe the contents of the jar after one-half hour.

4. With forceps, carefully remove the penny and rinse it off with water. Dry it with a paper towel and examine it. Do you think the penny will be old-looking again if you let it sit for awhile?

Conclude and Apply

1. *In your Journal*, record what ways the properties of the penny and the ammonia changed.

2. Do you think the substances changed?

136 Chapter 4 Describing the Physical World

In the Find Out activity, some of the physical properties of the substance water changed, but the water itself was not changed. Changes in physical properties caused by melting, freezing, boiling, and breaking, for example, are physical changes. In a **physical change**, the physical properties of a substance may change, but the kind of substance does not change. You can change the physical properties of many substances. All of the changes will be physical changes and will not change the identity of the substance. But when a chemical change occurs, such as wood being burned, is the identity of the substance changed? Let's find out.

Program Resources

Take Home Activities, p. 10, The Magic Card LEP

Study Guide, p. 18

Multicultural Connections, p. 11, Walter E. Massey L1

Science Discovery Activities, 4-3, Cool Garbage

Section Focus Transparency 12

Across the Curriculum

History

Challenge students to describe the energy involved in chemical and physical changes. Guide them to infer that chemical and physical changes usually involve the input or output of energy, often in the form of heat. James Prescott Joule, an English physicist, used the income from a brewery he owned to support his scientific investigations of heat. Have student volunteers research Joule's contribution to our understanding of heat and the concept of conservation of energy.

Chemical Changes

In the Find Out activity, you saw that some of the copper in the penny was changed into a different substance. A change during which one of the substances in a material changes into a different substance is a **chemical change**. During a chemical change, the identity of a substance changes.

What clues did you have that a chemical change took place in the penny? Certainly, the change to the shiny copper color was one indication. What other clues can tell you that a chemical change has taken place? The smell of burnt toast or an automobile's exhaust fumes can be evidence that new substances have been formed. The smell is different from the smell of bread or gasoline. The foaming of fizzy tablets in a glass of water and the smell of ozone in the air after a thunderstorm are also signs that chemical changes have occurred. When a rocket blasts off, the light, sound, and smoke that accompany it are all clues that chemical changes are taking place.

Figure 4-13

Ⓐ Wood chips are made of certain substances.

Ⓑ When wood chips are set afire, a chemical change takes place.

Ⓒ Are the ashes and smoke that result still wood? Why?

How Do We Know?

Tests for Chemical Change

Sometimes the usual clues for identifying a chemical change do not work. Perhaps there is no smell or maybe you missed the flash of light. There is one way you can be sure a chemical change has taken place. If the substances after the change are different from the substances before the change, you can be certain the change was chemical.

There are several tests that tell you what substances are present in a material. One such test is a little like looking at the colors of a rainbow formed when the sun shines through drops of rain. This test involves using an instrument called a spectroscope to look at the colors produced by glowing substances.

When viewed through a spectroscope, each glowing substance produces its own personal rainbow. Sodium, for instance, produces a particular yellow light. No other substance produces the same color light as sodium.

If the light produced by a glowing substance has the same colors in it before and after a change, then the substances have not been changed. You could conclude that the change was physical. What would you conclude if the colors of light produced before and after a change are different?

GLENCOE TECHNOLOGY

 Videodisc

STVS: Chemistry
Disc 2, Side 1
Fire Safety Tests (Ch. 6)

Fire-Resistant Clothing (Ch. 7)

 Videodisc

The Infinite Voyage: The Future of the Past
Chapter 1
Preserving Frescoes in Florence

Chapter 4
The Statue of Liberty Restoration Project

Visual Learning

Figure 4-13 Ask students to suggest ways that the wood chips could undergo physical changes. *Possible answers: putting them in a plastic bag, chopping them into smaller chips* Then have students identify other chemical changes. *Possible answers: forest fire, campfire, cooking an egg, baking a cake* Then ask: Are the ashes and smoke that result from burning wood still wood? Why? *No. During a chemical change the substances in a material change into different substances.*

Meeting Individual Needs

Physically Challenged This activity involves measuring, mixing, and observing on a scale that is easier for some students to work with. Materials needed are one cup of tepid water, two teaspoons of sugar, one-quarter teaspoon of powdered dry yeast, one cup of flour, and two bowls.

Put a half cup of water in each bowl. To each bowl add a teaspoon of sugar and a half cup flour. Put the yeast in only one bowl. Stir the mixtures. Put the bowls in sunlight or over a radiator. After an hour, check the bowls. Help students relate that the yeast bowl is bubbly due to chemical changes caused by the yeast. The other bowl will look the same as it did before the timing began. LEP

SKILLBUILDER

The candle's mass and length change. The wick shortens. The wax melts and hardens. The first two changes are chemical. The melting and hardening of the wax is a physical change. L1

3 ASSESS

Check for Understanding

Have students discuss question 4 under Check Your Understanding. Have students write their ideas on the chalkboard in two columns, one marked *Chemical Change* and one marked *Physical Change*.

Reteach

Discussion Write the following on the chalkboard: *drying clothes on a clothesline; burning coal; digesting food.* Have students classify the changes as physical or chemical and give reasons for their classifications. L1

Extension

Activity Have students find out about the chemical changes that take place when photographic film is exposed to light. Students can prepare a bulletin board display. L3

4 CLOSE

Show students some paper clips and a pile of sawdust. Ask whether water or heat would be more likely to produce a chemical change in each. *Heat could chemically change sawdust. Water would rust the paper clips.* L1

Chemical Properties

Wood burns because there is something about it that makes it able to burn. This characteristic, called flammability, is a chemical property of wood. A **chemical property** is any characteristic that gives a substance the ability to undergo a chemical change. Why are fire doors not made out of wood?

Some substances undergo chemical change when they are exposed to light. The next time you visit a drugstore, look around. Notice that some vitamins, drugs, and other products that are sensitive to light are stored in containers that light can't get through. These substances are changed into other substances when light comes in contact with them. Hydrogen peroxide, when exposed to light, changes chemically into water and oxygen gas and is no longer useful for cleaning wounds.

SKILLBUILDER

Observing and Inferring
Observe a burning candle and record your observations. For example, you might note how the candle changes over time. What evidence do you observe that physical and chemical changes are taking place as the candle burns? If you need help, refer to the **Skill Handbook** on page 659.

Figure 4-14

A Most un-treated cloth is flammable— it has the chemical property of flammability, which means it burns easily.

B This playsuit has been badly burned in a test for flam-mability. Since 1953, The Flammable Fabrics Act has made it illegal to sell chil-dren's sleep-wear in the United States that burns easily. To comply with this law and keep people safe, manufacturers chemically treat clothing to reduce its flammability.

check your UNDERSTANDING

1. When you mix sugar in water, the sugar dis-appears. Explain why this is an example of a physical change rather than a chemical change.
2. How is flammability different from burning?
3. Why is light sensitivity considered a chemi-cal property rather than a physical property?
4. **Apply** Give one example of a physical change and one example of a chemical change that might occur when a meal is prepared.

check your UNDERSTANDING

1. The sugar simply dissolves in the water but it is still sugar. If left in an open container, the water evaporates, leaving behind the sugar.
2. Flammability is a chemical property; it de-scribes the ability of a substance to burn. Burning is a chemical change; it results in the formation of new substances.
3. Light sensitivity is the ability of a sub-stance to change into another substance when exposed to light.
4. Physical: food is cut into pieces. Chemical: when food is cooked some substances in the food may react to form new substances.

4-4 States of Matter

Solids, Liquids, and Gases

Can you see a way to group some objects together in the photo below? Perhaps you'd put the water, alcohol, and seawater in one group. What properties do sand, sugar, a penny, a nail, and a rock have in common? In which group would you put the air inside the jar? Practically everything you are likely to see or use can be classified as a solid, a liquid, or a gas. These terms refer to the three basic states of matter. Clearly solids, liquids, and gases have different properties. What properties can you use to identify solids, liquids, and gases? You will group objects in the following activity.

Section Objectives
- Distinguish among solids, liquids, and gases.
- Describe physical changes relating to solids, liquids, and gases.

Explore! ACTIVITY

How can objects be grouped?

What To Do

1. Look at the objects and materials pictured in the photograph below.

2. Can you see a way to group some objects together?

3. *In your Journal*, write down your suggested classifications and place each object under the appropriate heading.

4. Can you think of a different method of grouping these objects together?

PREPARATION

Planning the Lesson
Refer to the Chapter Organizer on pages 116A–D.

Concept Development
Matter always assumes one of three basic states—solid, liquid, or gas. Students will discover how these three differ.

1 MOTIVATE

Bellringer
Before presenting the lesson, display **Section Focus Transparency 13** on an overhead projector. Assign the accompanying **Focus Activity** worksheet. [L1] [LEP]

Discussion Ask students if they have ever heard of fuel line freeze-up. In winter, water condensation in the gas tank can freeze and block the fuel line. The engine will stall. How many states of matter do students think are involved in the process described above? Discuss with students how temperature affects the state of matter.

find other ways of sorting the items (by transparency/opacity, by the ability to hold a shape).

✔ Assessment

Performance Have students list examples of solids, liquids, and gases in things they might find around the house. *soap, juice, bubbles in soft drinks* Use the Performance Task Assessment List for Making Observations in **PASC**, p. 17. [L1]

Explore!

How can objects be grouped?

Time needed 15–20 minutes

Materials No special materials are required.

Thinking Processes classifying, comparing and contrasting, making and using tables

Purpose Students will examine the objects photographed and devise a system of classification by which they might be sorted.

Teaching the Activity

Science Journal Students could describe alternate classification systems, noting whether some more effectively sort all of the items than others. [L1]

Expected Outcomes

Students should recognize that the items can be grouped into solids, liquids, and gases. They will make a three-column table in their journals dividing the items by this criterion. They may

2 TEACH

Tying to Previous Knowledge

Ask students what liquids, solid, and gas are present in their bodies. Accept all reasonable responses. Students will probably list blood and tears as liquids, bones as a solid, and air in their lungs as a gas.

Theme Connection The themes are stability and change and energy. Ask students what happens if you heat an ice cube and a piece of steel. Students may say that the ice melts and the steel gets hot. Point out that if the steel is heated further, it will melt. When sufficient amounts of heat energy are added or subtracted from a substance, the substance will change its state.

Identifying Solids

You probably have a pretty clear idea of what a solid is. Certainly a rock is solid. But how would you describe a solid so that anyone would know what you mean? You might say that a solid is hard. Cotton balls and pillows are solids, too, yet they don't seem very hard, do they? The following activity explores the common properties of solids.

Explore! ACTIVITY

What do all solids have in common?

What To Do

1. Examine a small rock, a fork, a penny, a paper clip, a cotton ball, a feather, a grain of sand, and any other solid you can find.

2. List *in your Journal* as many physical properties of solids as you can. Do they seem to have any physical properties in common?

3. Imagine placing these objects in different containers such as a tray or a bowl. Do you think the shape or size of any solid will change during such activities?

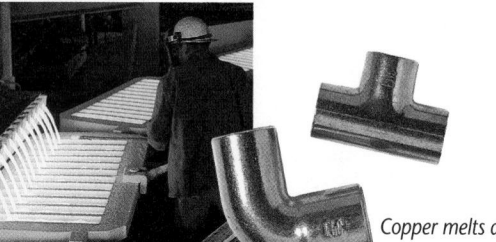
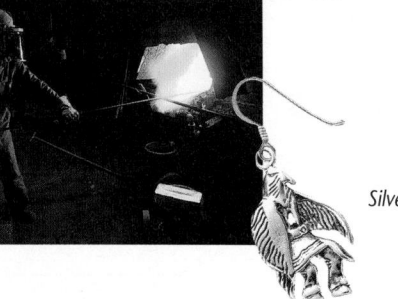

Figure 4-15

Every substance has its own melting point.

Paraffin melts between 50° C and 57° C.

Copper melts at 1083° C.

Silver melts at 961° C.

None of the solids changed in size or shape when you held it or put it in different containers. Any material that has a definite volume and a definite shape is a solid.

Melting is a physical change in which a solid becomes a liquid. An ice cube would melt in your hand but a penny wouldn't. Why do these two solids act so differently?

Room temperature is about 23°C. Ice melts at 0°C, which is lower than room temperature, while copper doesn't melt until its temperature reaches a little more than 1083°C. So, copper has a much higher melting point than ice. Each solid substance has its own melting point. Melting point is a physical property of solid substances.

Identifying Liquids

You know that when a solid melts it forms a liquid. Yet the properties of liquids are clearly different from the properties of solids. What properties do all liquids share that could help you identify them?

Find Out! ACTIVITY

What do all liquids have in common?

What To Do

1. Find several see-through containers having different shapes. The more unusual the shape the better.

2. Add some food coloring to a pitcher of water.

3. Use a measuring cup or graduated cylinder to pour the same amount of colored water into each container. Observe what happens to the shape of the water in each container.

4. Pour the water from one container back into the graduated cylinder. Did the volume change?

Conclude and Apply

1. *In your Journal*, record whether the shape of the colored water changed. When?

2. Do liquids have a definite volume? How do you know?

As you observed in the Find Out activity, liquids can be poured, and can change shape to fit the container they are in. Any matter that has a definite volume, but takes the shape of its container is a liquid.

Freezing is a physical change in which a liquid becomes a solid. Each liquid has its own freezing point. Water, for example, freezes at 0°C.

Mercury, which is a liquid at room temperature, won't freeze until the temperature is -38.87°C. Like the melting point of a solid, the freezing point of a liquid is a physical property.

Figure 4-16

When grape juice is in a container, it takes the shape of the container. Liquids do not have a definite shape the way solids do. This physical property can help you identify liquids. Any matter that has a definite volume, but takes the shape of its container, is a liquid.

DID YOU KNOW?

Solid carbon dioxide changes to carbon dioxide gas without first changing to a liquid. Because of this property, solid carbon dioxide, often called dry ice, is used to make smoke rise up from the stage in plays and concerts. The smoke is actually water vapor condensing as it comes into contact with cold carbon dioxide gas.

Find Out!

What do all liquids have in common?

Time needed 10 minutes

Materials several transparent bottles of different shapes, pitcher of water, food coloring, measuring cup

Thinking Processes observing, comparing and contrasting, drawing conclusions, forming operational definitions, classifying

Purpose To demonstrate that a specific volume of liquid can take different shapes.

Teaching the Activity

Troubleshooting Be sure the volume of water poured into each container is the same.

Science Journal Have students indicate how the shape of the container affected the *perception* of volume. In what ways might the owners of a restaurant exploit such perceptions? *They might use glasses with shapes that make a small amount of beverage seem like more.* L1

Expected Outcomes

The volume measured will not change. The liquid will take on the shape of the container.

Conclude and Apply

1. Yes, whenever the water was put into a different container.

2. Yes, because the volume of a liquid doesn't change.

✔ Assessment

Process Have students work in groups to discuss how to make wise consumer decisions in purchasing liquids. They should investigate which shapes of containers make it look like there is more liquid than others. Use the Performance Task Assessment List for Consumer Decision Making Study in **PASC**, p. 43. L1

COOP LEARN

Program Resources

Study Guide, p. 19

Laboratory Manual, pp. 21–24, Properties of Matter L3

How It Works, p. 8, How Does Cement Harden L2

Critical Thinking/Problem Solving, p. 12, Hydraulics L2

Making Connections: Integrating Sciences, p. 11, The Interior of Earth L2

Making Connections: Across the Curriculum, p. 11, Killer Lakes L2

Section Focus Transparency 13

What are some properties of gases?

Time needed 10 minutes

Materials two balloons of different shapes, clear jar with cap, cardboard, box, rubbing alcohol, mirror, small dish

Thinking Processes observing, drawing conclusions

Purpose To observe that gases take the shape of their containers but do not have a definite volume.

Teaching the Activity

Science Journal Have students record their answers to the questions. Students can also draw pictures illustrating their observations. L1

Expected Outcomes

Students will notice that gases take the shape of their container and do not have definite volumes.

Answers to Questions

1. Air takes the shape of each balloon.
2. Air takes the shape of its container.
3. The alcohol turns into a gas and evaporates. The breath condenses on the mirror.

✔ Assessment

Process Have students work in small groups to create posters about the properties of gases. Use the Performance Task Assessment List for Poster in **PASC**, p. 73. L1 **COOP LEARN**

Identifying Gases

Air is a mixture of gases that is all around you, but you can't see it. How can you tell when a substance is a gas?

Explore! ACTIVITY

What are some properties of gases?

What To Do

1. Blow up two differently shaped balloons. Record what happens *in your Journal*.

2. Cover a clear, glass jar with a cap. Cover a rectangular container with a piece of cardboard. What is the shape of the air occupying each container?

3. Put a drop or two of rubbing alcohol into a small, flat dish. What happens after a few minutes? Hold a mirror close to your mouth and gently breathe out onto the mirror. What do you observe?

Earth Science CONNECTION

Are Tin and Oxygen Liquids?

The state of a substance depends upon its temperature. We classify a substance as solid, liquid, or gas according to which state it is in at "room temperature" on Earth—23°C.

Water

You know that water freezes at 0°C and boils at 100°C. Water can exist naturally in the solid, liquid, and gaseous states on Earth. But what about other planets? What is "room temperature" on them? The surface temperature of a planet would determine in what state a substance would be.

If you lived on Mercury or Neptune, would substances exist in the same states as they do on Earth? On Mercury, surface temperatures may range from -193°C to 427°C just from nighttime to daytime. If water existed on Mercury, it would constantly be changing states.

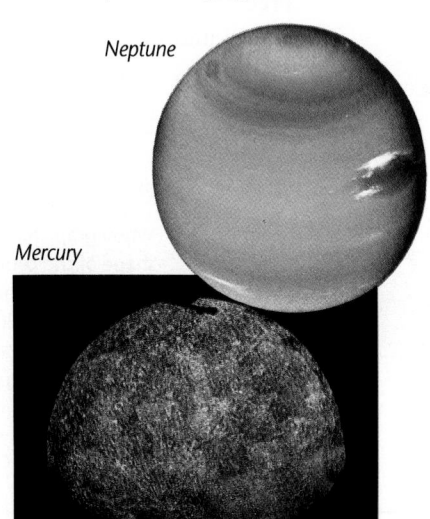

Neptune

Mercury

Earth Science CONNECTION

Purpose

The Earth Science Connection reinforces Section 4-4 by relating the states in which certain substances naturally exist to the range of temperatures on Earth and other planets.

Content Background

The differences between states of matter occur because of molecular arrangement. In solids, chemical bonds hold the molecules in relatively fixed positions. When the temperature of a solid increases, the atoms in its molecules move more quickly. This movement causes some solids to melt, or change to liquid. The molecules in gases are far apart and so move quickly in that state. Then, as they are slowed by lower temperatures, some gases can change to liquids. The change in states of matter also depends on pressure. For example, water boils at 100°C at sea level. As the altitude increases, the air pressure decreases. At 90 000 ft above sea level, the pressure is

When you blow into a balloon, the gases you breathe out go into the balloon. These gases take up space inside the balloon, so they have volume. Could you measure the volume using a graduated cylinder? What did you conclude from the Explore activity about the shape of gases? Matter that has no definite shape and no definite volume is a gas.

A gas spreads out to fill a container it's in, no matter how large the container. Remember what happened to the alcohol in the dish in the Explore activity? After a few minutes, it disappeared into the air. The liquid alcohol quickly turned into a gas. The alcohol evaporated. When you open a bottle

Figure 4-17

Ⓐ If you drop an antacid tablet into a glass of water, the sodium hydrogen carbonate in the tablet releases carbon dioxide gas bubbles.

Ⓑ The bubbles rise to the surface, break, and become part of the air in the room. Carbon dioxide, like all gases, has no definite shape and no definite volume. What is the main difference between a liquid and a gas?

Iron, Lead, and Gases

What about iron and lead? Look at the table. Is there anywhere on Earth where these substances would exist as liquids? How about on Mercury? As you can see, lead would sometimes be in the liquid state if it were found on Mercury.

Saturn

Earth

Substance	Melting Point,°C	Boiling Point,°C
Hydrogen	−259	−253
Iron	1536	2860
Lead	327	1740
Oxygen	−218	−183
Tin	232	2270

What about gases? We just assume that oxygen and carbon dioxide are gases. That's because average temperatures on Earth are above the boiling points of these substances. Remember that during the

Planet	Temperature
Mercury	−193 to 427°C
Earth	−88 to 58°C
Mars	−124 to -31°C
Saturn	−176°C
Neptune	−218°C

process of boiling, a liquid is changing to a gas. It doesn't have to be considered hot by humans living on Earth.

You Try It!

Look at the boiling points of the substances in the table above. Then look at the temperatures of the planets on the table to the left. On which planet(s) would tin be a liquid? On which planet(s) would oxygen be a liquid?

so low that water boils without any increase in heat.

Teaching Strategy
Based on the planet temperatures given in the table, have the students describe what the other planets might look like.

Answers to
You Try It!
1. Mercury
2. Mercury, Neptune

Going Further ⫸
Assign a different metal to each small group of students. Have each group research how its metal can be used in various ways. Have students relate the physical properties of the metal to the uses they come up with. Have each group prepare a booklet about its metal and then share its findings with the class, accompanied by actual examples or pictures. Encourage students who want to learn more about liquid

oxygen and the uses of liquid air to research experiments done by Louis Cailletet who liquefied oxygen in 1877 and Zygmunt Wroblewski who liquefied hydrogen a few years later. Use the Performance Task Assessment List for Booklet or Pamphlet in **PASC**, p. 57. L2 COOP LEARN P

Student Text Question

How can what you've learned in this chapter help explain how smoke, fog, and other substances in the air can reach areas thousands of meters from their source? *Particles in the air are not limited to a particular space. They can drift as air moves.*

3 ASSESS

Check for Understanding

Have students answer questions 1 through 3 and discuss question 4 under Check Your Understanding. Have students make drawings of the situations they suggest as answers.

Reteach

Activity Have students put together a bulletin board display about the three states of matter. They could use pictures from magazines or their own drawings. L1

Extension

Activity Have students who have mastered this section look up *sublimation*, which is the change from a solid directly to a gas. Have them discover examples of sublimation. L3

4 CLOSE

Have students infer what changes of state they could expect to see in the streets after a snowstorm. L1

Ice

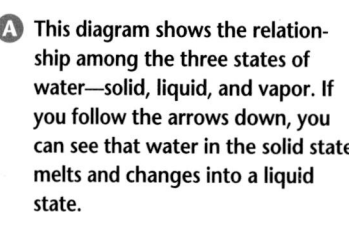

Figure 4-18

A This diagram shows the relationship among the three states of water—solid, liquid, and vapor. If you follow the arrows down, you can see that water in the solid state melts and changes into a liquid state.

Melting   Freezing

Liquid water

B As the liquid evaporates, the water changes to the gaseous state of water, which is water vapor.

Evaporation, boiling   Condensation

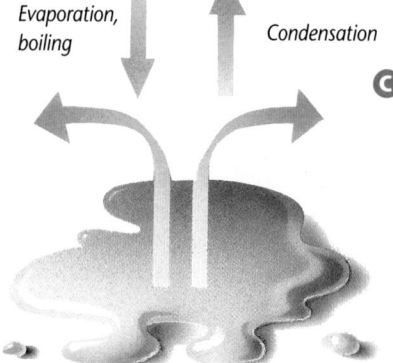

C If you follow the arrows up, you see that condensation changes water in the gaseous state into the liquid state. If liquid water freezes, it turns into ice.

---

of perfume, which contains alcohol, the odor soon becomes noticeable. The gas given off does not stay in the opened container. It spreads out to occupy the entire space available to it.

Can a gas become a liquid? What happens when you bring a cold can near steam? Some of the boiling water goes into the air as water vapor. The water forms liquid water when it touches the cold can. The water vapor condenses to form liquid water. Cooling speeds up condensation. Now think about what condensed on the mirror in the Explore activity. How do you know there is water vapor in the air you breathe out?

Figure 4-18 shows the relationship among solids, liquids, and gases. How can what you've learned in this chapter help explain how smoke, fog, and other mixtures in the air can reach areas thousands of meters from their source?

---

check your UNDERSTANDING

1. How are a brick, milk, and helium, which is sometimes used to fill balloons, different?
2. Is chocolate syrup a solid, a liquid, or a gas? Why do you say so?
3. What physical change occurs when you leave ice cream in a dish on the counter for a few minutes?
4. **Apply** Describe a place or a situation where you could find water as a solid, as a liquid, and a gas all at the same time.

---

check your UNDERSTANDING

1. A brick is a solid, so it has a definite shape and definite volume. Milk is a liquid, so it has a definite volume but takes the shape of its container. Helium is a gas, so it has no definite volume and no definite shape.

2. Liquid, because it has definite volume but takes the shape of its container

3. It changes to a liquid by melting.

4. In a refrigerator/freezer; ice in the freezer, water in the refrigerator, water vapor in the air contained in the refrigerator

HOW IT WORKS

Taking a Spin

Did you ever go into an amusement park ride that looked like a large, round room? You and the other riders stand with your back against the wall. Then the room begins to rotate, and you are pressed against the wall. When the room is spinning fast enough, the floor drops, and you are held against the wall.

What Is a Centrifuge?

In scientific and technical work, machines that work very much like this amusement park ride are often used to separate materials from mixtures. These machines are called centrifuges, and they work by separating materials in mixtures according to the density of the material. A centrifuge is pictured on this page.

How Are Centrifuges Used?

A good example of the use of a centrifuge is for the separation of the materials in blood. Blood is a heterogeneous mixture containing plasma, blood cells, and other materials. Since plasma and blood cells are used for different purposes, it is necessary to separate them. Blood is placed in small tubes that hang down from the part of the machine that spins. The tubes are mounted so that, as the machine begins to spin, the tubes pivot upward into a horizontal position. As the machine spins around, sometimes at thousands of spins per minute, the denser

materials move to the bottom of the tube, and the less dense materials stay toward the top of the tube. Because red and white corpuscles, plasma, platelets, and other materials in the blood have different densities, they will separate.

You Try It!

Other liquids can be separated by using the action of a centrifuge. For example, cream can be separated from milk. Which do you think has the greater density—cream or milk? Design an experiment to find out. You might be surprised.

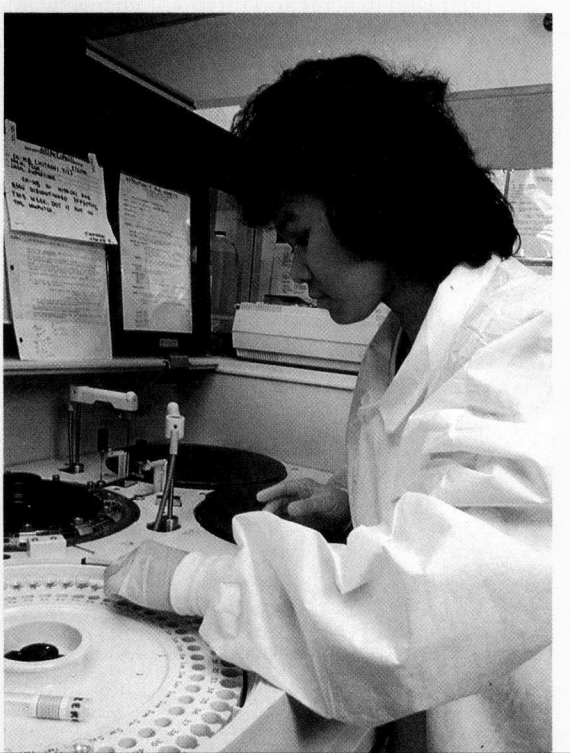

Going Further ▪▪▪▪▪▶

Have students find out how centrifuges are used in their own towns or cities. Students can work in pairs to write letters of inquiry to local doctors, hospitals, and industries asking what type of centrifuge is used, how it makes medical or industrial practice easier, and what separation process it replaces. Then have students share their information with the class. Other students work in small groups to explore the importance of the centrifuge. They can brainstorm about

what things could not be done or would be done very slowly if the centrifuge had not been invented. Use the Performance Task Assessment List for Group Work in **PASC**, p. 97. COOP LEARN L2

Purpose

How It Works expands Section 4-1 by explaining a common method of separating mixtures using a centrifuge. This method works because materials in the mixture have different densities, a property discussed in Section 4-2.

Content Background

Because the particles in mixtures vary in size, several types of centrifuges are needed for different purposes. The spin drier is a basic type of centrifuge that separates liquids from solids. Separative centrifuges operate at high speeds and can separate cream from milk and impurities from oil and clarify wine and beer. Zonal centrifuges can separate particles into various levels and densities. These machines are used for separation of subcellular particles such as nuclei, RNA, and DNA. Ultra centrifuges operate at extremely high speeds and are used to separate solids from very fine suspensions.

Teaching Strategies

After reading the selection, ask students to name other possible uses of centrifuges. Have the class discuss the applicability of a centrifuge for each suggestion.

Activity

Students who have not ridden the amusement park ride described in the selection can feel the effect of centripetal force by riding on a playground merry-go-round while someone else makes it go around very fast. L1

Answers to

You Try It!

Milk. Students may take the mass of equal volumes of milk and cream. The heavier milk has the greater density.

Science and Society

Purpose

Science and Society expands the discussion of SI units in Section 4-2 by looking at the two systems of measurement used in the United States, English and SI (metric).

Content Background

The history of metric use in the United States goes back over a century before the Metric Conversion Act. In 1866, the use of metric weights and measures was made legal. In 1875, the United States was the only English-speaking nation in the Diplomatic Conference on the Meter. The traditional measurements of the English system continued to be used for everyday purposes, but the metric system was favored by scientists.

International scientific advances led to a renewed debate about adopting the metric system. In the late sixties, the National Bureau of Standards carried out a study of metric use in the United States. The study showed that metrics were becoming more popular in the United States. Since the 1975 law, more and more industries have converted to metric use. Even though none of the 50 states have passed laws that require use of SI units, most of the states have laws that encourage it. Metric measures can now be seen side by side with English units. Road signs often give distances in kilometers. Gasoline pumps register in liters as well as gallons. Many soft drinks are packaged in liters.

Science and Society

Metrics for All?

In the United States, athletes compete on courses that are measured in meters, medicine is sold in milligrams and milliliters, and many automobile parts are measured in metric units. However, carpenters still buy lumber in feet and inches, and farmers measure their land in acres and their crops in bushels. Fabric is sold by the yard, and milk by the quart. Most highway signs give distances in miles and speed limits in miles per hour.

The Metric Conversion Act

In 1975, the Metric Conversion Act became law. The law states that the federal government will coordinate and plan the increasing use of the metric system on a voluntary basis. So, for the time being, we are living with two different systems.

The Metric System Controversy

For nearly 100 years, those favoring the metric system have argued for its widespread use in the United States, but opponents have argued just as vigorously against it. People from industry say that such a change would require them to replace or convert their machinery—a costly process. Those in favor say that machinery is often replaced anyway, and the cost would be a one-time expense that would produce lasting benefits. Trade would be easier with other countries, most of which use metrics.

Since the metric system is based on multiples of 10, calculations are much easier. Using metrics might reduce calculation errors and save time. Some people have estimated that up to two years could be cut out of traditional math courses in school because fractions and conversions would not be taught.

The System

The metric system, called SI from the French "Le Systeme Internationale d'Unités," is the standard system of measurement used worldwide. All SI units and their symbols are accepted and understood by the scientific community. In SI, each type of measurement has a base unit—meter (m) for length, kilogram (kg) for mass, etc. Look at the table on the following page to see these base units.

In the English system,

Comparing English and SI Units

1 mile

1 kilometer

(Mile is 1.6 times longer than kilometer)

1 yard

1 meter

(1.09 x longer than yard)

1qt

1 liter

(1.06 x larger than quart)

1 pound

1 kilogram

(weighs 2.2 pounds)

you have to remember that there are 12 inches in one foot, 3 feet in one yard, and 5280 feet in one mile. In SI units, there are prefixes that indicate which multiple of 10 should be used. You would say that there are 10 millimeters in a centimeter, 100 centimeters in a meter, and 1000 meters in a kilometer—all multiples of 10.

Metric Conversion

Many people resist the switch to metrics because they are not very familiar with metric units. They have grown up using feet, pounds, and gallons and feel more comfortable continuing to use them. Most people, however, do not really know very much about these units, especially about how they relate to each other. Do you know how many cubic inches are in a fluid ounce? How many ounces are in a pound, or inches in a mile? For that matter, did you know that there are two kinds of miles—the nautical mile and the statute mile?

SI Base Units

Measurement	Unit	Symbol
Length	Meter	m
Mass	Kilogram	kg
Time	Second	s
Electric current	Ampere	A
Temperature	Kelvin	K
Amount of substance	Mole	mol
Light intensity	Candela	cd

How many feet are in each? How many square feet make one acre?

Should We Convert?

Do you think the government should pass a law requiring that the United States convert completely to SI units by a given time? Would the advantages of adopting the metric system outweigh the possible disadvantages?

USING MATH

Look at the diagram comparing SI and English units. Which SI unit would you use to measure each of the following?

a. a large carton of milk

b. your height

c. your mass

d. the length of your arm

e. distance across a state

Content Background

Some conversion factors in the English system are shown below. These conversions are good examples of the complicated relationships between English units.

1 fluid ounce = 1.805 cubic inches

16 ounces = 1 pound

63 360 inches = 1 mile

1 nautical mile (no longer in official U.S. use) = 6080 feet

1 statute mile = 5280 feet

43 560 square feet = 1 acre

Teaching Strategy
Conduct a class debate of the questions posed in the last paragraph of this excursion.

Activity

After students have selected the SI measurements for the Using Math activity, have them actually measure the items in SI units. You will need to provide students with a metric ruler, a metric scale, a large milk carton and a graduated beaker or graduated cylinder. **L1** **LEP**

Enrichment

Students can work in small groups to better understand the need for standards of measurement. Have each group devise a new system of length measurement. Units might include school supplies such as paper, pencils, and erasers and furniture parts such as desktops, chair backs, and seats. Have groups exchange tables of measure for their systems and try to measure objects in the room with these systems. Discuss how confusing the measuring got. **L2**
COOP LEARN

USING MATH

a. liter

b. meter

c. kilogram (here the measurement will be greater than ten for your students.)

d. centimeter

e. kilometer

Going Further ▸

Have students write to their congressional representatives explaining why they would or would not support a complete conversion by the U.S. to the metric system. Letters to senators should be mailed to the Senate Office Building, Washington D.C. 20510, and letters to representatives should be mailed to the Cannon House Office Building, Washington, D.C. 20510. Use the Performance Task Assessment List for Letter in **PASC**, p. 67.

Art Connection

Purpose

Art Connection expands Section 4-2, in which students learn that color is a physical property. This excursion also shows how objects and movement can be represented in terms of color.

Content Background

Alma Woodsey Thomas was the first African-American woman to have a solo exhibition at the Whitney Museum of American Art (1972). During her years of teaching, she was an important force in the Washington arts scene. She organized clubs and art lectures for her students, established art galleries in the public schools, and helped found the Barnett-Aden Gallery, one of the first galleries in Washington devoted to modern art. Thomas's use of color dominated the entire canvas. Her broad strokes of color suggested images instead of limiting them to specific forms.

Teaching Strategy

Have pairs of students read the article together. Then ask students to discuss what Thomas probably felt about the use of color. They will probably say that she used mosaics of color patches to create abstract images of nature. She saw nature revealed as color. Students might also want to find examples of Alma Thomas's paintings so that they can better understand color field painting. Students could describe her use of color in each painting.

Going Further ⫸

Have students look at pictures by Alma Thomas as well as other color field painters. Suggest that they look at work by Nassos Daphis, Helen Frankenthaler, Ellsworth Kelly, and Frank Stella. Then students can work in small groups to describe the vi-

Art Connection

Alma Woodsey Thomas—Color Field Painter

Alma Thomas was an artist who achieved prominence in the mainstream art community. She worked in the modern tradition of Color Field painting.

Education

She was born in 1891 in Columbus, Georgia. Because her aunts were teachers, she decided at an early age that teaching could be her way to a better life, too. Her family moved to Washington, D.C., in 1907. In 1924, she was the first graduate of the new art department at Howard University.

Thomas taught art in the Washington schools for 35 years. During that time, she earned an M.A. at Teachers College of Columbia University. During her teaching career, she exhibited realistic paintings in shows of African American artists. In the 1950s, she took painting classes at American University and became interested in color and abstract art.

Color and Abstract Art

By 1959, Thomas's paintings had become abstract. By 1964, she had discovered a way to create an image through small dabs of paint laid edge to edge across the painting's surface. In *Iris, Tulips, Jonquils, and Crocuses*, the color bands move vertically and horizontally across the canvas to represent a breeze moving over a sunlit spring garden. In

Autumn Leaves Fluttering in the Wind (shown at right), rust-colored patches move in patterns like those of swirling autumn leaves. The glimpses of blue, yellow, and green between the patches represent the sky and land.

Thomas's paintings are mosaic patches of color that she said, "represent my communion with nature." She wrote, "Color is life. Light reveals to us the spirit and living soul of the world through colors."

What Do You Think?

Alma Thomas wrote that she was "intrigued with the changing colors of nature as the seasons progress." Describe how you would paint a natural scene using the Color Field painting style.

sual impact of color field paintings. Suggest that students think of the painting as a detail of some larger image. Also suggest that students think of how color field painters are using the two dimensions of the canvas instead of trying to create an illusion of three dimensions. Use the Performance Task Assessment List for Writer's Guide to Nonfiction in **PASC,** p. 85. **COOP LEARN** L2

Science Journal

Review the statements below about the big ideas presented in this chapter, and think about each question. Then, re-read your answers to the Did You Ever Wonder questions at the beginning of the chapter. *In your Science Journal*, write a paragraph about how your understanding of the big ideas in the chapter has changed or expanded.

1 A substance is made of only one kind of material. A mixture consists of two or more substances, each with its own identity. *Are the materials shown substances or mixtures?*

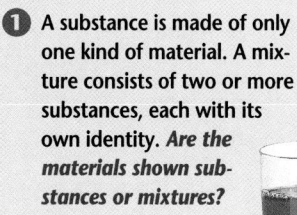

2 An element is a substance that cannot be broken down further into simpler substances by ordinary physical or chemical means. A compound is made of two or more elements that are chemically combined, and always has the same chemical composition. *Does a chemical compound have the same physical properties as the elements that make it up?*

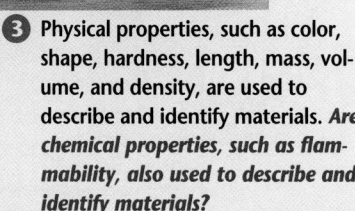

3 Physical properties, such as color, shape, hardness, length, mass, volume, and density, are used to describe and identify materials. *Are chemical properties, such as flammability, also used to describe and identify materials?*

4 A substance is the same after a physical change. *Is a substance different after a chemical change?*

Science Journal

Did you ever wonder...

• Cake batter is a liquid. After the chemical changes brought about by baking, the cake becomes a solid. (p. 137)

• Its particles are dispersed in the air and remain a while before drifting. (pp. 143–144)

• A liquid can change its shape; a solid cannot, so it breaks. (pp. 140–141)

• Evaporate salt water and then condense the vapor back into fresh water. (pp. 120–121)

Project

Have students keep a list of physical and chemical changes they see each day for several days. Make sure they distinguish between the two types of change. Have students check their lists to eliminate duplicates. Then divide the total number of changes by the number of days observed. This will give the number of physical and chemical changes that happen on an average day. Use the Performance Task Assessment List for Analyzing the Data in **PASC,** p. 27. L1

Have students study the pictures and read the captions. Then have students expand their understanding of the main ideas by inventing or finding new examples of the ideas. **COOP LEARN**

Teaching Strategies

Divide the class into four groups. Assign each group a photograph or diagram. Have each group come up with another idea that illustrates the concept involved. They should then draw a picture or find a photograph from a magazine that relates to their idea. They should write a caption that explains what is happening. These can be put together in book form or on a poster for display. Sample Ideas:

1. Raisin bran and milk form a mixture.

2. A person on a scale is measuring a physical property.

3. Snow or ice thaw to form puddles of water.

4. A nail left outside will rust.

Answers to Questions

1. Soft drink–mixture; sugar–substance

2. A chemical compound does not have the same physical properties as the elements that make it up.

3. Yes

4. A substance is different after a chemical change.

GLENCOE TECHNOLOGY

MindJogger Videoquiz

Chapter 4 Have students work in groups as they play the Videoquiz game to review key chapter concepts.

Using Key Science Terms

Answers will vary. Possible answers include: chemical change–wood burning, combustion, tarnishing; chemical property–flammability, light sensitivity; compound–salt (sodium chloride), sugar (sucrose); density–grams per cubic centimeter, kg/L; element–copper, silver, tin; homogeneous mixture–salt water, clear tea; heterogeneous mixture–bread, chocolate milk, smog; physical change–boiling, melting, freezing; physical property–length, density, color; substance–sugar, water

Understanding Ideas

1. Answers will vary. Examples include filtering, evaporation, and hand separation.

2. Its mass and volume

3. A. A substance is different after a chemical change; a substance is the same after a physical change.

B. In a heterogeneous mixture, the different substances are distributed unevenly; in a homogeneous mixture, they are distributed evenly.

C. A physical property is any characteristics of a material that can be observed or measured; a chemical property is any characteristic that gives a substance the ability to undergo a chemical change.

D. A substance contains only one kind of material; a mixture is made of two or more substances.

4. A solid has a definite volume and definite shape; a liquid has a definite volume but no definite shape; a gas has no definite volume and no definite shape.

Developing Skills

1. See reduced student text.

2. A. 40 pebbles; B. Approximately 55 g

Using Key Science Terms

Give two examples of each of the following:

chemical change	heterogeneous
chemical property	mixture
compound	mixture
density	physical change
element	physical property
homogeneous	substance
mixture	

Understanding Ideas

Answer the following questions in your Journal using complete sentences.

1. Name two ways of separating mixtures.

2. What do you need to know about a material to determine its density?

3. a. How do a chemical change and a physical change differ?

b. How do a heterogeneous mixture and a homogeneous mixture differ?

c. How do a physical property and a chemical property differ?

d. How do a mixture and a substance differ?

4. How do the properties of solids, liquids, and gases differ?

Developing Skills

Use your understanding of the concepts developed in this chapter to answer each of the following questions.

1. Concept Mapping Using the following terms, complete the concept map of

---

classification below: *compounds, elements, heterogeneous, mixtures*

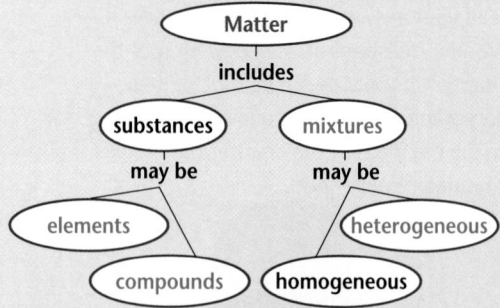

2. Making and Using Graphs The following graph shows the mass of a given number of pebbles. Use the graph to: a) estimate the number of pebbles in a sample that has a mass of 20 g; b) estimate the mass of 110 pebbles.

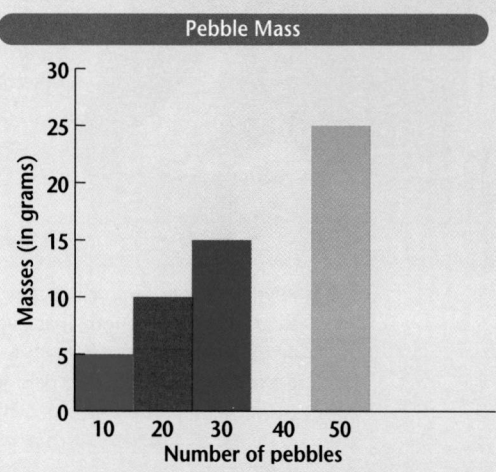

3. Observing and Inferring You fill a pan up to the top with cold water, which you plan to heat to make tea. Before the water comes to a boil, you notice that a small amount of water has spilled over the top of the pan. How did this happen?

---

3. As water becomes warmer, its density decreases slightly, and the water expands. When the water in the pan expanded, a small amount flowed over the top.

4. 1 700 000 g/m³

Program Resources

Review and Assessment, pp. 23–28 L1
Performance Assessment, Ch. 4 L2
PASC
Alternative Assessment in the Science Classroom
Computer Test Bank L1

4. Measuring in SI An object has a density of 1700 kg/m^3. What is its density in units of g/m^3?

Critical Thinking

In your Journal, *answer each of the following questions.*

1. How might you determine whether a sample of matter is a substance or a mixture?

2. The density of steel is greater than the density of water, so a solid bar of steel sinks when it is placed in water. What can you infer about a steel ocean liner, knowing that it floats on water?

3. You have a 200 cm^3 glass of milk. What is the mass of the milk?

Problem Solving

Read the following paragraph and discuss your answers in a brief paragraph.

You are walking along the beach, when you find a small, shiny chunk that looks like silver. You would like to determine whether the chunk is actually pure silver or just some other silvery-colored material.

1. Assume that you have available a sample of silver having the same mass as your chunk. Using only materials you can find in your kitchen, how can you use density to see if the material might be silver?

2. Assume your density test shows that your chunk might be silver. With the help of a chemist, how might you use chemical properties to further test your chunk to see if it is silver?

CONNECTING IDEAS

Discuss each of the following in a brief paragraph.

1. **Theme—Systems and Interactions** List at least two physical properties that can be observed using each of your five senses.

2. **Theme—Stability and Change** You know that milk is stored in the refrigerator rather than in the cupboard. Explain why this is so.

3. **Science and Society** Give three arguments in favor of the United States converting completely to the SI system, and three arguments against it. Which arguments do you feel are stronger?

4. **Earth Science Connection** Imagine that a new planet has been discovered farther from the sun than Pluto. Would you expect the new planet to have an atmosphere in which you could breathe? Explain.

5. **How It Works** Explain how centrifuges can be used to separate heterogeneous mixtures, such as blood.

Assessment

Portfolio Review the portfolio options that are provided throughout the chapter. Encourage students to select one product that demonstrates their best work for the chapter. Have students explain what they learned and why they chose this example for placement into their portfolios.

Additional portfolio options can be found in the following **Teacher Classroom Resources:**

Making Connections: Integrating Sciences, p. 11

Multicultural Connections, pp. 11–12

Making Connections: Across the Curriculum, p. 11

Concept Mapping, p. 12

Critical Thinking/Problem Solving, p. 12

Take Home Activities, p. 10

Laboratory Manual, pp. 17–24

Performance Assessment P

Critical Thinking

1. Mixtures can be separated by mechanical means.

2. The ship has less mass per unit volume than a steel bar. Because it contains air, its total density is less than the density of water.

3. 200 cm^3 × 1.03 g/cm^3 = 206 g

Problem Solving

1. Put a small glass in a bowl, filling the glass to the rim with water, then putting the material sample into the full glass. Measure the volume of the water that spilled into the bowl. This is the volume of the material. Simply divide the mass by its volume to find its density.

2. Silver is one of the least reactive metals. You might try to test your chunk with substances that will react with other metals, but not with silver. For example, copper sulfate solution will react with many metals, but not with silver. Chemical testing will help you to determine whether or not your chunk is silver.

Connecting Ideas

1. Answers may include: sight, color and shape; hearing, pitch and loudness; touch, temperature, and texture.

2. Milk undergoes a chemical change which causes it to spoil. At a lower temperature, milk spoils more slowly.

3. Suggested answers for: 1. Using the same measurements as other countries; 2. Easier to multiply and divide; 3. Eliminates conversion. Against: 1. Change many measuring devices; 2. Would have to learn the SI system; 3. Former measurements need conversion.

4. No; the temperature would be too cold for oxygen to exist as a gas.

5. Centrifuges separate heterogeneous mixtures by using the force of spinning motion to move denser materials to the bottom.

Chapter Organizer

SECTION	OBJECTIVES	ACTIVITIES & FEATURES
Chapter Opener		**Explore!**, p. 153
5-1 Types of Solutions (3 sessions; 2 blocks)	1. **Classify** solutions. 2. **Define** and **identify** solutes and solvents. 3. **Describe** three factors that affect the rates at which solids and gases dissolve in liquids. **National Content Standards: (5–8) UCP1, UCP3, A1–2, F1, G1–2**	**Find Out!**, p. 154 **Find Out!**, p. 155 **Explore!**, p. 156 **Investigate 5-1:** pp. 158–159 **Find Out!**, p. 160 **Find Out!**, p. 161 **Find Out!**, p. 163 **Skillbuilder:** p. 164 **A Closer Look,** pp. 162–163
5-2 Solubility and Concentration (3 sessions; 1.5 blocks)	1. **Describe** how solubility varies for different solutes and for the same solute at different temperatures. 2. **Interpret** solubility graphs. 3. **Compare** and **contrast** saturated and unsaturated solutions. 4. **Infer** solution concentrations. **National Content Standards: (5–8) UCP1, A1–2, F1, G1–2**	**Explore!**, p. 165 **Skillbuilder:** p. 167 **Design Your Own Investigation 5-2:** pp. 168–169 **Life Science Connection,** pp. 170–171 **Science and Society,** pp. 178–179 **Economics Connection,** p. 180
5-3 Colloids and Suspensions (2 sessions; 1 block)	1. **Distinguish** between a colloid and a suspension. 2. **Recognize** at least two mixtures that are colloids. **National Content Standards: (5–8) UCP3, A1, F2, F4, G1–3**	**Explore!**, p. 173 **Explore!**, p. 175 **Literature Connection,** p. 177

ACTIVITY MATERIALS

EXPLORE!

p. 153 sugar, salt, flour, corn meal, rice, 5 glasses, water, stirrer, teaspoon
p. 156* 1 package soft drink mix, water, clear pitcher, flashlight, stirrer, straw
p. 165* 2 beakers, salt, sugar, stirrer
p. 173* gelatin, transparent bowl, flashlight
p. 175* glass, muddy water, stirrer

INVESTIGATE!

pp. 158–159 safety goggles, Epsom salt (magnesium sulfate), graduated cylinder, large beaker, spoon or stirring rod, 2 small beakers; thick, water-absorbent string

DESIGN YOUR OWN INVESTIGATION

pp. 168–169 distilled water, 10-mL

graduated cylinder, large test tube, table sugar (sucrose), test-tube holder, laboratory balance, thermometer in a two-hole stopper, copper wire stirrer, safety goggles, beaker, hot plate, scoop for sugar, oven mitt

FIND OUT!

p. 154* powdered sugar, mortar and pestle, peppermint oil, castile soap, dropper, scoops, calcium carbonate, dishcloth, soil, corn syrup, toothbrush, scale
p. 155 beakers, sand, salt, funnels, ring stands with rings, filter paper, balance
p. 160* distilled water, grinding instrument, stopwatch, cm graph paper, stirrer, sugar cubes, glasses
p. 161 4 glasses, stirrer, sugar, teaspoon
p. 163 soft drink, balloons, tape

KEY TO TEACHING STRATEGIES

The following designations will help you decide which activities are appropriate for your students.

L1 Basic activities for all students
L2 Activities for average to above-average students
L3 Challenging activities for above-average students
LEP Limited English Proficiency activities
COOP LEARN Cooperative Learning activities for small group work
P Student products that can be placed into a best-work portfolio
Activities and resources recommended for block schedules

Need Materials? Call Science Kit (1-800-828-7777).

⏱ OUT OF TIME? We recommend that students do the activities with an asterisk.

Chapter 5 Matter in Solution

TEACHER CLASSROOM RESOURCES

Student Masters	Transparencies
Study Guide, p. 20 **Multicultural Connections**, p. 13 **Activity Masters**, Investigate 5-1, pp. 23–24 **Critical Thinking/Problem Solving**, p. 5 **Making Connections: Across the Curriculum**, p. 13 **How It Works**, p. 9 **Concept Mapping**, p. 13 **Making Connections: Technology and Society**, p. 13 **Science Discovery Activities**, 5-1, 5-2 **Laboratory Manual**, pp. 25–28, Solutions	**Teaching Transparency 9,** The Solution Process **Section Focus Transparency 14**
Study Guide, p. 21 **Activity Masters**, Design Your Own Investigation 5-2, pp. 25–26 **Making Connections: Integrating Sciences**, p. 13 **Laboratory Manual**, pp. 29–32, Solubility; pp. 33–36, Densities of Solutions	**Teaching Transparency 10,** Solubility Graph **Section Focus Transparency 15**
Study Guide, p. 22 **Critical Thinking/Problem Solving**, pp. 5, 13 **Multicultural Connections**, p. 14 **Take Home Activities**, p. 12 **Science Discovery Activities**, 5-3	**Section Focus Transparency 16**

ASSESSMENT RESOURCES	TEACHING & TECHNOLOGY
Review and Assessment, pp. 29–34 **Performance Assessment**, Ch. 5 **PASC*** **MindJogger Videoquiz** **Alternate Assessment in the Science Classroom** **Computer Test Bank**	**Spanish Resources** **Cooperative Learning Resource Guide** **Lab and Safety Skills** **Science Interactions, Course 1, CD-ROM** **Computer Competency Activities**

*Performance Assessment in the Science Classroom

NATIONAL GEOGRAPHIC TEACHER'S CORNER

National Geographic Society Products Available from Glencoe

To order the following products for use with this chapter, call the National Geographic Society at 1-800-368-2728:

National Geographic Society Series
STV: Biodiversity
STV: Atmosphere

GLENCOE TECHNOLOGY

The following multimedia resources are available from Glencoe:

Science and Technology Videodisc Series (STVS)
Ecology
  Evaluating Artificial Reefs
  Aquaculture

Glencoe Physical Science Interactive Videodisc
Behavior of Gases

Physical Science CD-ROM

Teacher Classroom Resources

These are key components of the classroom resources package.

TEACHING AIDS

Section Focus Transparencies

Teaching Transparencies

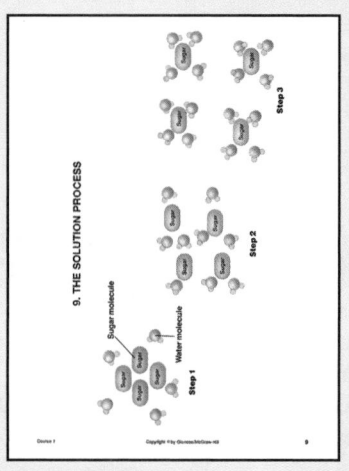

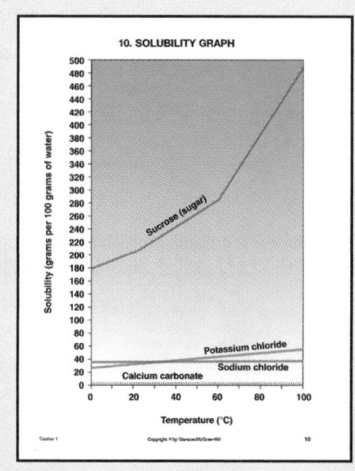

HANDS-ON LEARNING

Science Discovery Activity*

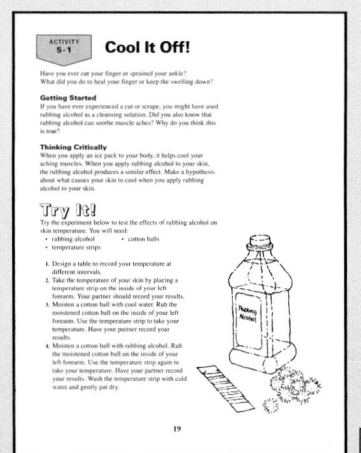

Laboratory Manual*

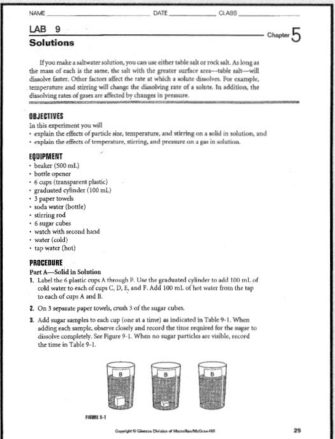

Take Home Activity

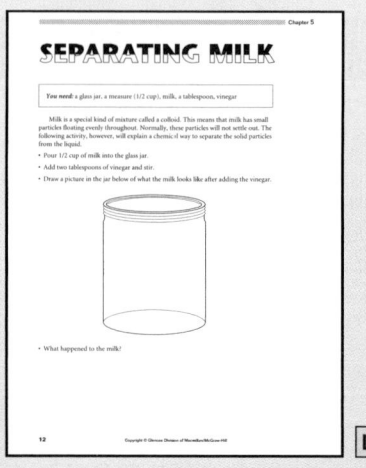

Chapter 5 Matter in Solution

Study Guide*

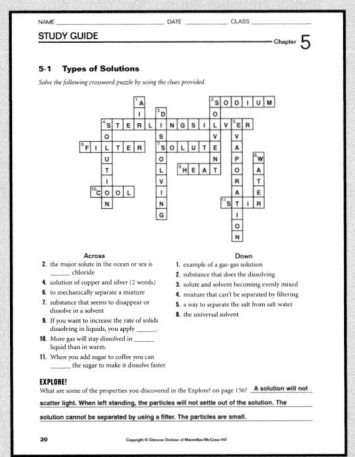

Concept Mapping

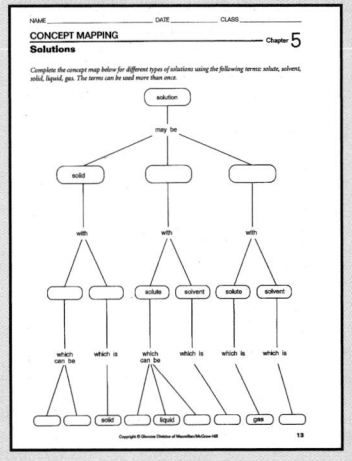

Critical Thinking/ Problem Solving

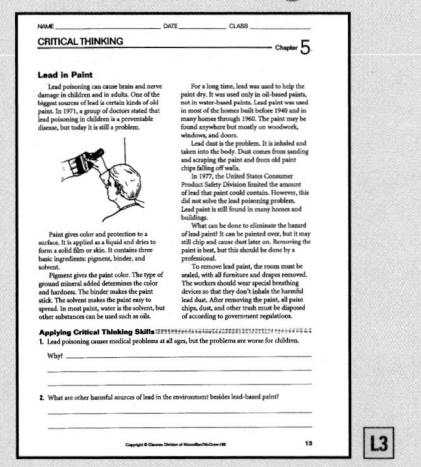

Integrating Sciences

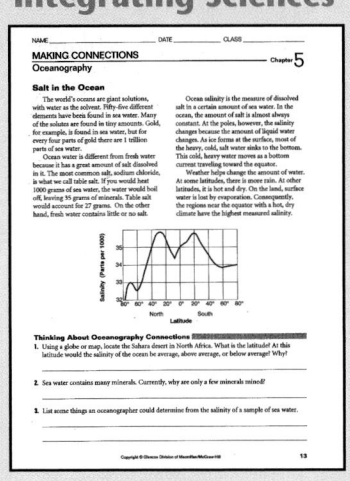

Across the Curriculum

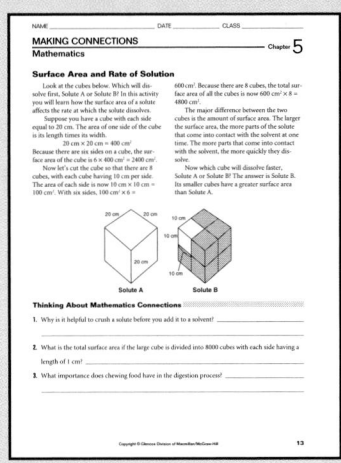

Technology and Society

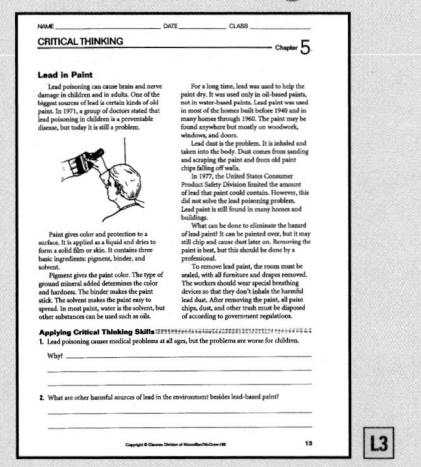

Multicultural Connection**

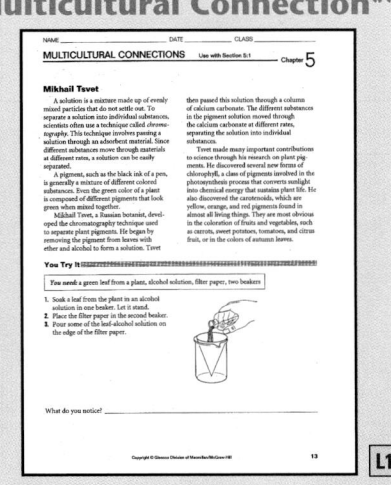

Performance Assessment

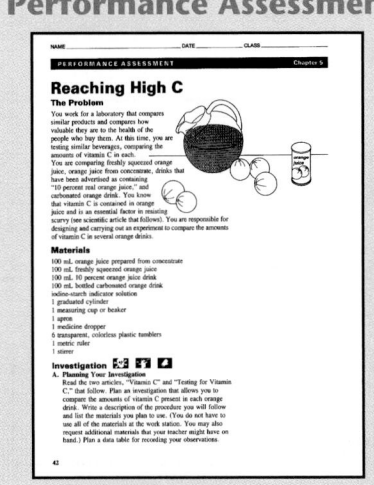

Review and Assessment

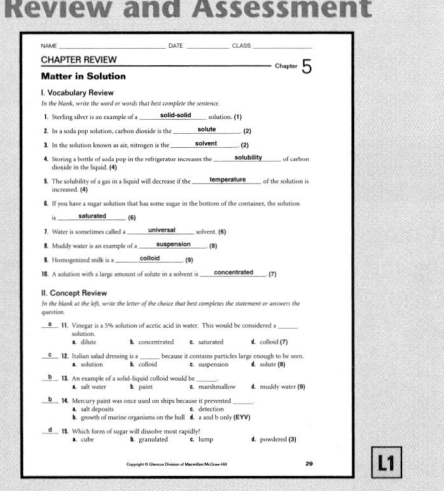

*One per section   **Two per chapter

CHAPTER 5

MATTER in SOLUTION

Matter in Solution

THEME DEVELOPMENT

The major differences between solutions and mixtures involve changes that take place when a solute dissolves and how stable those changes are. The theme of systems and interactions is revealed by the behavior of solutions as systems whose individual particles interact to produce large-scale behavior.

CHAPTER OVERVIEW

In this chapter, students will study qualitative and quantitative aspects of matter in solution. Finally, students investigate the differences between true solutions and other mixtures.

Tying to Previous Knowledge

Display two beakers, one filled with tap water and the other with salt water. First ask students to identify which is which. Then add known amounts of salt to both beakers until the solutions are saturated. Encourage students to develop an explanation for what happened. *Water can hold only so much salt, so the salt water, being closer to the limit, dissolved less salt.*

INTRODUCING THE CHAPTER

Have students read the paragraph on page 152. Ask students to think of other mixtures that they have eaten.

Did you ever wonder...

✓ Where the sugar goes when it dissolves in a glass of lemonade?

✓ Why you can add more sugar to a cup of hot tea than to a glass of iced tea?

✓ Why you don't have to shake milk before you drink it?

Science Journal

Before you begin to study about matter in solution, think about these questions and answer them *in your Science Journal.*

Answers are on page 181.

t's a cold morning, and you've just gone into the kitchen for breakfast.

As you eat, changes are going on. Your instant oatmeal is getting soggier in your bowl. Bits of powdered hot chocolate settle out in your cup, but they vanish again when you stir them with your spoon. Your apple juice, however, remains unchanged.

From a scientist's view, you are eating three mixtures made with water. Yet as you have probably guessed by now, not all mixtures look or act the same. Why is your apple juice clear, while your hot chocolate is cloudy? Why doesn't your oatmeal disappear in water like your hot chocolate? This chapter will help you answer questions such as these.

▶ **In the activity on the next page, explore materials that seem to disappear in water.**

152   *Salt*   *Sugar*   *Corn meal*   *Flour*

If time does not permit teaching the entire chapter, use the Chapter Overview on this page, Reviewing Main Ideas at the end of the chapter, and the Chapter 5 audiocassette to point out the main ideas of the chapter.

Learning Styles		
Kinesthetic	Find Out, pp. 154, 163; Activity, p. 166	
Visual-Spatial	Explore, pp. 153, 156, 165, 173, 175; Demonstration, pp. 154, 164, 172, 174; Find Out, pp. 155, 161; Investigate, pp. 158–159; Visual Learning, p. 161	
Interpersonal	Activity, pp. 157, 166; Science at Home, p. 181	
Intrapersonal	Activity, p. 171	
Logical-Mathematical	Activity, p. 159; Find Out, p. 160; Across the Curriculum, p. 161; Demonstration, p. 165; Visual Learning, pp. 167, 172; Investigate, pp. 168–169	
LS **Linguistic**	Visual Learning, pp. 157, 171; Across the Curriculum, p. 157; Discussion, p. 166; Life Science Connection, pp. 170–171	

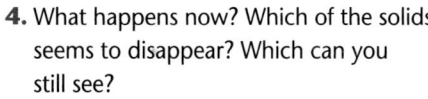

Explore! ACTIVITY

What materials seem to disappear in water?

What To Do

1. Add a teaspoon of sugar, salt, flour, corn meal, and rice to separate glasses of water.
2. Observe and record what happens *in your Journal*.
3. Stir the contents of each glass and wait for a few minutes more.
4. What happens now? Which of the solids seems to disappear? Which can you still see?

Rice

Uncovering Preconceptions

Students may think that vigorous stirring can cause more of a solute to dissolve. Actually, stirring only affects the *rate* of solubility.

Explore!

What materials seem to disappear in water?

Time needed 15 minutes

Materials teaspoon, sugar, salt, flour, corn meal, rice, five glasses of water, stirring rods

Thinking Processes observing, comparing and contrasting, recognizing cause and effect

Purpose To compare the solubility of different substances.

Preparation Make sure each glass contains the same amount of water at the same temperature.

Teaching the Activity

Science Journal Have students note in their journals what is the same and different about the way each substance interacts with water. L1

Expected Outcomes

Students should find that the sugar and salt disappear, flour and corn meal turn the water cloudy, and the rice just gets wet.

Answers to Questions

4. The corn meal, flour, and rice start to settle at the bottom; salt and sugar; corn meal, flour, and rice

✔ Assessment

Process To compare and contrast the structures of the soluble and insoluble materials, have students examine the five substances with a hand lens. Students can record their observations in their journals. Use the Performance Task Assessment List for Making Observations and Inferences in **PASC**, p. 17. L1

5-1 # Types of Solutions

PREPARATION

Planning the Lesson

Refer to the Chapter Organizer on pages 152A–D.

Concepts Developed

This section allows students to discover qualitative properties of solutions. The distinction between a solution and other mixtures is explored.

1 MOTIVATE

Bellringer

 Before presenting the lesson, display **Section Focus Transparency 14** on an overhead projector. Assign the accompanying **Focus Activity** worksheet. [L1]
LEP

Demonstration Show two beakers, one with tap water, the other with a mixture of water and acid. (pH 4) Which water appears more "dangerous?" *They look the same.* Identify the acidified water as having a similar acid content to acid rain, which is a *solution* in which acids are dissolved in rainwater. [L1]

Find Out!

How can you make a homemade cleanser?

TECH PREP

Time needed 20 minutes

Materials mortar and pestle, scale, scoops, 8 g powdered sugar, 2 drops peppermint oil, 3 g castile soap, 22 g calcium carbonate, corn syrup, dirt, dishcloth, toothbrush, water, dropper

Thinking Process observing and inferring, designing an experiment, measuring in SI, modeling

Purpose To observe the ingredients in a mixture.

Section Objectives

- Classify solutions.
- Define and identify solutes and solvents.
- Describe three factors that affect the rates at which solids and gases dissolve in liquids.

Key Terms

solution
solute
solvent

What Is a Solution?

Think back to the cold-weather breakfast at the beginning of this chapter. The hot chocolate, the oatmeal, and the apple juice were all examples of mixtures, which you read about in Chapter 4. Each was a mixture with water. In the Explore activity, you saw how sugar and salt seem to disappear in water, while flour, corn meal, and rice do not.

Shampoo, toothpaste, and detergent all depend on water to make them work, too. When you clean your hair, teeth, or clothes, you are trying to remove bits of dirt. One way to do this is to mix the dirt with something that will remove it, such as a mixture of soap and water. How can you demonstrate this? The following Find Out activity should help.

Find Out! ACTIVITY

How can you make a homemade cleanser?

What To Do

1. Place 8 g of powdered sugar in a mortar.
2. Add two drops of peppermint oil and 3 g of castile soap.
3. Mix these together with a pestle.
4. Add 22 g of calcium carbonate, and mix thoroughly.
5. Finally, add corn syrup until a paste is produced. This is now a homemade cleanser.
6. Carefully stain a small section of material, such as a dishcloth, with dirt.
7. Apply some of your homemade cleanser to an old toothbrush, along with a few drops of water. Try scrubbing the stain. Record what happens *in your Journal.*

Conclude and Apply

1. Wash away the cleanser with a little water. What happens now?
2. Explain *in your Journal* why you think this is happening.

Teaching the Activity

Discussion Have students focus on the properties of the ingredients of the mixture.

Science Journal Have students record their observations in their journals. [L1]

Expected Outcome

The cleanser retains its cleaning ability.

Conclude and Apply

1. The dirt stain disappears.

2. Stain dissolves in the mixture.

✔ Assessment

Performance Discuss how the mixture would be changed if any of the ingredients were eliminated. Would it still work as a cleanser? Have students design an experiment to test such mixtures. Use the Performance Task Assessment List for Designing an Experiment in **PASC**, p. 23.
COOP LEARN [L1]

■ Separation of Solutions

If you could separate the ingredients from your cleanser, you would find that their physical appearance has been changed upon mixing, but not their individual properties. Such a material is a mixture, as is shown in **Figure 5-1**. Mixtures may be separated by mechanical means. Can all mixtures be separated as easily as this?

Figure 5-1

Everyday cleansers such as shampoo and bar soap are mixtures. They work in much the same way as the cleanser you made in the Find Out activity.

Find Out! ACTIVITY

Can all mixtures be separated by mechanical means?

What To Do

1. Fill two 400-mL beakers about halfway with water.
2. Add 10 g of sand to the first beaker and 10 g of table salt to the second beaker. Stir both mixtures well.
3. Clamp a funnel to each of two ring stands and fold a piece of filter paper to fit in each.
4. Place an empty beaker underneath each funnel.
5. Slowly pour some of your mixture of sand and water into one funnel.
6. Now, pour some of your mixture of salt and water into the other funnel. Record your observations *in your Journal.*

Conclude and Apply

1. Which mixture leaves something behind on the filter paper?
2. Which mixture does not?
3. Which mixture can be separated by filtering, a mechanical means?

■ Properties of Solutions

Sand and salt form mixtures with water, yet they cannot be separated in the same way. While sand can be separated from water by filtering, salt cannot. Why do you think this is? Certain mixtures, such as salt and water, are called solutions and cannot be separated by filtration. Have you ever made instant soft drink on a hot day and watched the crystals disappear in water as you stirred them? Clear soft drink is an example of a solution. What properties does a solution have that make it different from other kinds of mixtures?

5-1   Types of Solutions   **155**

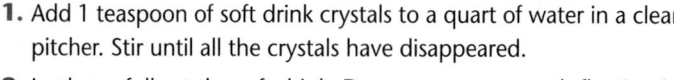

Explore!

What are the properties of a solution?

Time needed 15 minutes

Materials 1 package unsweetened soft drink mix, 1 quart water in a clear pitcher, flashlight, straw, stirrer

Thinking Processes comparing and contrasting, forming operational definitions

Purpose To observe the properties of a solution.

Preparation Use a flavor of soft drink that will produce a clear solution.

Teaching the Activity

Discussion Encourage students to be alert for differences of any type in the soft drink solution. Talk about uniformity. Ask what happened to the crystals. Is there still evidence of their existence?

Science Journal Have students record their initial and follow-up observations in their journals. Encourage them to record the date and time of each observation. L1

Expected Outcomes

The crystals will disappear completely; the soft drink will not change overnight. Students can conclude that the particles are evenly mixed.

Answers to Questions

2. no

3. The soft drink is clear; no crystals can be seen.

4. They taste the same.

5. Yes; no particles settled out.

✔ Assessment

Process In the Find Out activity on page 154, students should have observed that the cleanser ingredients change appearance but retain other original physical properties. The soap still cleans; the sugar is still sweet; the peppermint oil is still minty. In their journals, have students contrast the cleanser to a solution. *The main difference is that particles are evenly mixed in a solution.* P L1

Explore! ACTIVITY

What are the properties of a solution?

What To Do

1. Add 1 teaspoon of soft drink crystals to a quart of water in a clear pitcher. Stir until all the crystals have disappeared.

2. Look carefully at the soft drink. Do you see any crystals floating in the pitcher? Record your observations *in your Journal*.

3. Darken the lights in the room and shine a flashlight through the pitcher. What do you observe?

4. Use a straw to remove a small amount of soft drink from the top of the pitcher. Taste it. Now carefully use the straw to remove a small amount of soft drink from the bottom of the pitcher. Taste it. How do the two samples compare?

5. Cover the pitcher and let it sit undisturbed overnight. Look at the soft drink again the next day. Does it look the same?

Figure 5-2

The sea is one of the world's largest solutions. Sodium chloride is the major solute and water is the solvent.

When you make a soft drink, you mix together two kinds of materials, soft drink crystal and water. Because they are so small and are mixed evenly throughout the water, you don't see the single particles of soft drink. Also, a sample from the top will have the same number of dissolved particles as one drawn from the bottom and so will taste as sweet. After waiting, neither material will settle out. Any mixture made up of tiny particles that are evenly mixed and do not settle out is called a **solution**.

A solution is made up of two types of materials, one of which may seem to disappear in the other. Remember the soft drink crystal disappearing in the water? Any substance that seems to disappear, or dissolve, is called a **solute**. The substance in which the solute dissolves is called a **solvent**. Generally, the substance present in the largest amount is the solvent. In your soft drink, what is the solute? What is the solvent? Because water can dissolve so many different solutes to form solutions, chemists often call it a universal solvent.

Program Resources

Study Guide, p. 20 L1

Laboratory Manual, pp. 25–28, Solutions L3

Multicultural Connections, p. 13, Mikhail Tsvet L1

Teaching Transparency 9 L2

Section Focus Transparency 14

Meeting Individual Needs

Visually Impaired Set up samples of sugar/salt/flour/starch/cotton and water mixtures for examination. Direct visually impaired students to take a pinch of each mixture between thumb and forefinger and test it for texture, viscosity, and uniformity. Encourage them to draw conclusions about the solubility of each substance. The sense of taste may also be employed.

Types of Solutions

Solutions are important to all living things, as shown in **Figure 5-3**. Water carries dissolved nutrients to all parts of a plant. The ocean is a vast water solution of minerals and dissolved gases from Earth's crust. Medicines are often solutions of different chemicals. Some of your body fluids, such as urine and saliva, are water solutions.

Solutions may be mixtures of two or more solids, liquids, or gases, or any one of these in another. Some types of solutions are shown in **Figure 5-3**. Typically, the solute is named first, followed by the solvent. For example, if the solute is a gas and the solvent is a liquid, they form a gas-liquid solution. What type of solution is lemonade that is made with powdered drink mix? **Figure 5-3** shows examples of different solutions.

You've separated sand and water by filtering, so you know this mixture is not a solution. Solutions cannot be separated by filtering because the dissolved particles are too small. How can solutions be separated? In the following Investigate, we will examine solutions and evaporation.

Figure 5-3

Ⓐ The air you breathe is a gas-gas solution of oxygen and other gases dissolved in nitrogen.

Ⓑ This salt water aquarium holds a solid-liquid solution of sodium chloride and water.

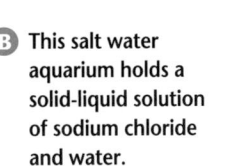

Ⓒ Sterling silver is a solid-solid solution of 7.5 percent copper and 92.5 percent silver.

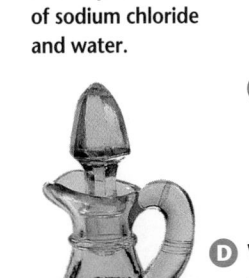

Ⓓ Vinegar is a liquid-liquid solution, made up of 5 percent acetic acid and 95 percent water.

Ⓔ What can you observe about this soft drink that indicates that it is a gas-liquid solution?

Composition of Solutions		
	Type of Solution	Examples
Gas solution	gas-gas	air
Liquid solution	gas-liquid liquid-liquid solid-liquid	soft drink vinegar salt water
Solid solution	solid-solid	brass sterling silver

5-1 Types of Solutions **157**

INVESTIGATE!

5-1 Evaporation and Solutions

Planning the Activity

Time needed 20 minutes to set up

Purpose To determine if solutions can be separated by evaporation.

Process Skills observing and inferring, recognizing cause and effect, forming a hypothesis, making and using tables, measuring in SI, predicting, interpreting data, modeling

Materials Cotton knitting or crochet yarn can be used for string.

Teaching the Activity

Discussion Before beginning this activity, ask students to review what they have learned about the properties of a solution. What methods for separating solutions have been tried in previous activities? What were the results of these efforts?

Process Reinforcement Be sure that students set up their data tables correctly. There should be a column for the date of each observation. Under the column labeled "Observations," there should be enough space to note observations of the beakers and of the string. If students are having difficulty, you may wish to go over the mechanics of making and using tables.

Possible Hypotheses Because the solute and solvent in a solution are "inseparable," students might expect that they will evaporate "together" and neither will be left behind. This is a reasonable, but incorrect, assumption.

Possible Procedures Place beakers where they can be observed but not knocked over or disturbed throughout the experiment. Also, assign students to check that the string ends remain submerged as the liquid level goes down.

Science Journal Have students record their daily observations and conclusions in their journals. L1

Evaporation and Solutions

You've seen that solutions cannot be separated by letting them stand or by filtering. In this activity, you'll try to separate a solution using evaporation.

Problem

Can solutions be separated by evaporation?

Materials

safety goggles
Epsom salt (magnesium sulfate)
large beaker (400 mL)
thick, water-absorbent string

water
graduated cylinder (100 mL)
2 small beakers (250 mL)
spoon or stirring rod

Safety Precautions

What To Do

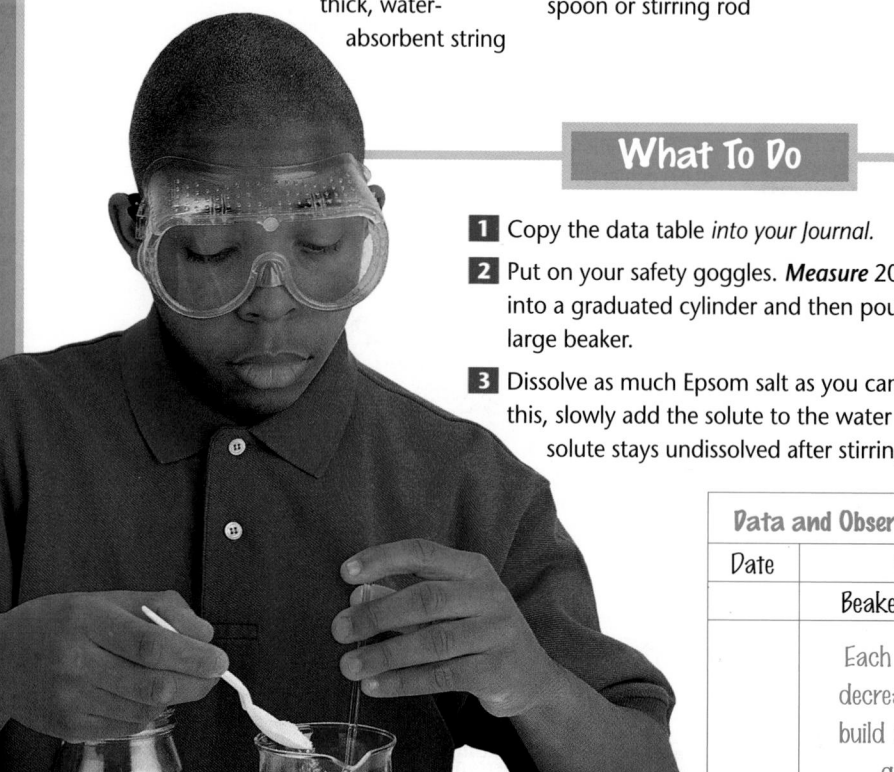

1 Copy the data table *into your Journal.*

2 Put on your safety goggles. *Measure* 200 mL of water into a graduated cylinder and then pour this into the large beaker.

3 Dissolve as much Epsom salt as you can in the water. To do this, slowly add the solute to the water until some of the solute stays undissolved after stirring.

Sample data

Data and Observations		
Date	Observations	
	Beakers	String
	Each day, water level decreases and crystals build up on string in a greater amount	

Program Resources

Activity Masters, pp. 23–24, Investigate 5-1

Making Connections: Technology and Society, p. 13, Smog L2

Science Discovery Activities, 5-1, Cool It Off! 5-2, Candy-ade

Meeting Individual Needs

Learning Disabled Have students make checklists before beginning an activity that require students to:

• obtain all equipment and materials they will need, including "extras" such as pens

• write the procedure, step by step

• write what they expect the experiment to show

• make tables for collecting data

A

4 Fill the two small beakers with the Epsom-salt solution. Place them side by side about 10 cm apart. Drape the string between the beakers with the ends of the string submerged in the solutions. It should be set up as in the illustration (see photo **A**). The string should sag slightly between the beakers. Let the setup stand undisturbed for several days.

5 *Observe* the setup every few days and record your observations in your data table.

Analyzing

1. What happened to the water level in the beakers? Where did the water go?

2. What happened on the string between the beakers?

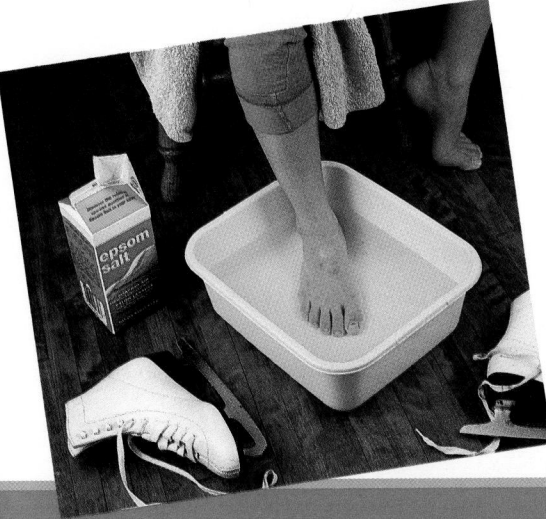

Concluding and Applying

3. *Predict* the effect of the following changes in the outcome of this Investigation.
   a. You dissolved only half as much Epsom salt in the water.
   b. No string was placed between the beakers.

4. Which part of a solid-liquid solution evaporated in the activity, the solute or the solvent? Which part was left behind?

5. ~~Going Further~~ *Infer* why evaporation can't be used to separate gas-gas solutions?

Epsom salt and hot water make up a solution that has long been a home remedy for reducing inflammation and easing aching muscles.

Expected Outcome

Students will observe that the Epsom salt is left behind. This shows that, while solutions cannot be separated by certain other mechanical means, some can be separated by evaporation.

Answers to Analyzing/ Concluding and Applying

1. The water level went down. The water evaporated into the air.

2. It dripped liquid and formed a white crust hanging down from the string and building up under it.

3. a. There would be less crust on the string.

b. The crust would be left in the beakers.

4. solvent; solute

5. A liquid turns to gas when it evaporates. In a gas-gas solution, the solvent is already gas, so further evaporation cannot occur.

✔ Assessment

Performance Have students imagine that they live in a coastal city that is experiencing a water shortage. Ask students to write a letter to the editor of their local newspaper giving their opinion on whether or not the city should try extracting salt from seawater to make drinking water. Students' letters must discuss evaporation and solutions. Use the Performance Task Assessment List for Letter in **PASC**, p. 67. **P** **L1**

ENRICHMENT

Activity Materials needed are wide-mouth 1-gallon jars and distilled water. Have students measure the amount of particulate air pollution in your area. Set jars outside, filled with about 1 inch of distilled water. Leave the jars exposed for 30 days. Then bring the jars in and heat to evaporate the remaining water. Have students calculate the total amount of dustfall per unit area (e.g., square mile). **L3**

Dissolving

Figure 5-4

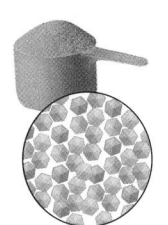  

(A) If you take a scoop of soft drink crystals ...

(B) and add them to water, the crystals break down.

(C) The crystal particles become evenly mixed with the water particles.

Remember the last Explore activity? When you added the soft drink crystals to water, the crystals seemed to disappear. You can taste the soft drink, so you know the materials from the crystals are still there. Where did they go? Just how do the particles of solid and liquid mix together in the first place? How does the process of dissolving work?

Figure 5-4 shows how dissolving occurs. How do you think this happens? Earlier you learned that a solution is made up of tiny particles. Does the particle size of a solid affect how fast it can dissolve in a liquid?

Find Out! ACTIVITY

How does particle size affect dissolving?

What To Do

1. Pour 100 mL of distilled water into each of two containers.

2. Grind three sugar cubes into a fine powder. Place this powder on a sheet of folded paper for easy pouring.

3. Have a partner place three whole sugar cubes into one container of water at the exact same time that you add the three powdered cubes to the other container. Immediately start timing. Stir both solutions.

4. *In your Journal,* record the times when the powdered sugar has dissolved and when the three cubes have dissolved. Which took longer to dissolve, the powder or the cubes?

5. Grind up three more sugar cubes.

Spread the powder as thinly and as evenly as possible onto a sheet of centimeter graph paper (see photo).

6. Count and record how many squares the powder covers. This number is the surface area, in cm^2, of the part of the paper that is covered.

7. The surface area of the powder is six times that of the area covered. What is the surface area of the three whole sugar cubes? How do these two numbers compare?

Conclude and Apply

1. Which particles seem to dissolve faster—smaller or larger ones?

2. Which particles have a greater total surface area—smaller or larger ones?

A larger surface area lets more solid solute come in contact with more solvent. Grinding or breaking up a solid solute increases the surface area of the solute, as you can see in **Figure 5-5.** Thus, dissolving happens more quickly when a solid solute is broken into smaller pieces. What other factors affect the rate of dissolving?

Figure 5-5

Ⓐ Each side of a 10-cm cube has a surface area of 100 cm². The total surface area for the cube is 600 cm².

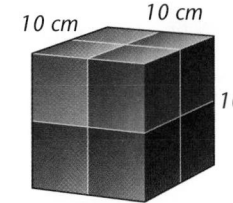

10 cm 10 cm

10 cm

Total surface area =
(10 cm × 10 cm) ×
6 sides = 600 cm²

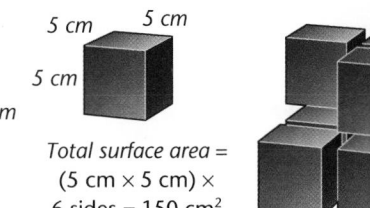

5 cm 5 cm

5 cm

Total surface area =
(5 cm × 5 cm) ×
6 sides = 150 cm²

Total surface area =
8 blocks × 150 cm² = 1200 cm²

Ⓑ When the cube is broken down into eight 5-cm cubes, the total surface area increases to 1200 cm². Unbroken, the cube had 6 surfaces. How many surfaces are there now?

 Find Out! ACTIVITY

How do stirring (or shaking) and temperature affect dissolving?

What To Do

1. Fill two glasses with water.

2. Add 3 teaspoons of sugar to the first glass and let it sit.

3. Add 3 teaspoons of sugar to a second glass and stir the contents rapidly.

4. Stop stirring and let the second glass sit. Record your observations *in your Journal.*

5. Fill a third glass with hot water and a fourth one with ice water. Add 3 teaspoons of sugar to each. Do not stir.

6. In which glass does the sugar seem to dissolve more quickly?

7. Now, stir the contents of each glass. What do you observe?

Conclude and Apply

1. How does shaking or stirring affect dissolving?

2. How does temperature affect dissolving?

5-1 Types of Solutions **161**

Find Out!

How do stirring (or shaking) and temperature affect dissolving?

Time needed 10 minutes

Materials four glasses, sugar, teaspoon, stirring rod, water, ice, hot water

Thinking Processes observing and inferring, comparing and contrasting, recognizing cause and effect, concept mapping, interpreting data

Purpose To observe how motion and temperature affect solubility.

Preparation Make sure that each glass contains the same amount of water.

Teaching the Activity

Discussion Have students predict what will happen before each experiment.

Science Journal Have students compare and contrast the effects of stirring and temperature on the rate of dissolving. **L1**

Figure 5-6

A If a solute, such as dye, is added to a solvent, such as water, the dye will slowly dissolve and mix with the water particles.

B Stirring speeds up the dissolving process.

You have now found out that sugar dissolves faster in hot, rather than cold water. **Figure 5-6** shows how the solute and solvent mix together faster when stirred or shaken. Solids are not the only substances that can be dissolved in liquids. As you already know, a carbonated soft drink is an example of a gas dissolved in a liquid. Shaking or stirring an opened bottle of soft drink causes it to spurt out as more gas particles, exposed to the surface, freely escape. Shaking or stirring slows down dissolving for a gas in a liquid, but speeds up dissolving for a solid in a liquid. How does temperature affect the dissolving of a gas in a liquid?

Can We Change the Freezing Point?

Water freezes at 32°F (0°C). However, this doesn't always suit our needs. Ice may be great for hockey, but it can be disaster for someone walking on a frozen sidewalk. So, we alter our environment to suit ourselves.

Antifreeze

We use antifreeze to melt ice on car windshields, and to prevent ice from forming in the water used to cool automobile engines. Water takes up more volume when it freezes. If water freezes in a car engine, it can expand enough to crack the engine!

Lowering the Freezing Point

Experiments show that by adding certain substances to water, we may lower the solution's freezing point. The freezing point depends upon the number of particles dissolved in a liquid. When you add

Find Out! ACTIVITY

How does temperature affect dissolving of a gas in a liquid?

What To Do

1. Carefully open a chilled bottle of carbonated soft drink.

2. Cover the opening with a balloon and secure it with tape.

3. Shake the drink. **CAUTION:** *Don't point the bottle at anyone.* Record what happens *in your Journal.*

4. Repeat the procedure with an unchilled bottle of soft drink of the same size, brand, and flavor. Contrast the results from the cold and warm drinks.

Conclude and Apply

Assuming both bottles had an equal amount of gas before they were opened, which bottle had the greater amount of dissolved gas in it after shaking?

antifreeze to the water in the radiator of your car, the freezing point of the water in your radiator goes down to a lower temperature.

Salt

Salt is a substance that lowers water's freezing point and is used to melt snow on roads and sidewalks. Like commercial antifreeze, the more salt that is added, the lower the freezing point of the substance.

You Try It!

1. Fill three small unbreakable containers with equal amounts of cool tap water.

2. Dissolve one tablespoon of salt in one container and

two tablespoons of salt in another. Add nothing to the third container of water.

3. Put the three containers of water in your freezer and leave them there. Check them every hour. Which of the contents froze first? Second? Last? Explain your observations *in your Journal.*

5-1 Types of Solutions **163**

Answer to
You Try It!

Plain water; water with one tablespoon of salt; water with two tablespoons of salt; substances added to water lower its freezing point, and different amounts of a substance lower it different amounts.

Going Further ⅢⅢ➤

Have students work in small groups to explore another property of some liquid-liquid solutions. Have them combine 100 mL of water

with another 100 mL of water in a 200-mL graduated cylinder. Then have them combine 100 mL of water with 100 mL of ethanol in another 200-mL graduated cylinder. Compare the results. Students should find that the alcohol-water solution yields a smaller volume (about 5 or 6 mL less) than the water-water solution. This is because the particles fit closer together in the alcohol-water solution. COOP LEARN L2

Find Out!

How does temperature affect dissolving of a gas in a liquid?

Time needed 10 minutes

Materials chilled carbonated soft drink in a bottle, unchilled carbonated soft drink in a bottle, balloons, masking or duct tape

Thinking Processes observing and inferring, comparing and contrasting, making and using graphs

Purpose To test the relationship between temperature and the solubility of a gas in a liquid.

Teaching the Activity

Safety To avoid eye injuries, a bottle of carbonated liquid should never be pointed at anyone, and goggles should be worn.

Discussion The undissolved gas in the liquid causes the balloons to inflate. The more undissolved gas there is, the more the balloons will inflate.

Science Journal Have students record their observations and conclusions in their journals. L1

Expected Outcomes

The warm liquid had more undissolved gas, so it inflated the balloon more.

Conclude and Apply

the cold soft drink

✔ Assessment

Content Ask students to compare and contrast the effects of temperature on the solubility of gases and solids. Students can record their responses in paragraph form or as a chart. Select the appropriate Performance Task Assessment List in **PASC.** P L1

The media center is colder because of air conditioning; therefore, the water in the tank will probably be colder and dissolve more oxygen gas for more fish. L1

3 ASSESS

Check for Understanding

Discuss the Apply question with students. Then ask them what the technicians would have found if the bottle contained a solution. *no particles floating in the bottle*

Reteach

Demonstration To demonstrate the effect of temperature on dissolving, place egg coloring tablets in glasses of hot and cold vinegar. To demonstrate the effect of stirring, place egg coloring tablets in two glasses of cold vinegar. Stir the vinegar in one glass. To demonstrate the effect of surface area, place a whole tablet in a glass of vinegar and a crushed tablet in an equal amount of vinegar.

Extension

Acquiring Information Have students investigate the effect of boiling and freezing on the separation of a solute from its solvent. L3

4 CLOSE

Discussion

Tell students that chlorine tablets were added to a swimming pool. Ask them what type of solution this makes in the pool *solid-liquid* and to identify the solute and the solvent. *chlorine tablets; water* Then ask what would happen to the solution if the pool heater were turned on. *Chlorine tablets dissolve more quickly.* L1

Using Observations to Form a Hypothesis

You are helping set up one fish bowl in your warm, sunny classroom and another in the air-conditioned media center. Both tanks are the same size and contain the same amount of water. Based on the amount of oxygen that can dissolve in the water, to which tank would you be able to add more fish? Explain why. If you need help, refer to the **Skill Handbook** on page 663.

Figure 5-7

A The amount of a gas that can be dissolved in a liquid decreases as the temperature of the liquid increases.

B Fish take the oxygen they need to live from the water. What might happen if the temperature of the water in this fish tank increases?

In the last activity, you found that the warm soft drink inflated the balloon more. This happened because when the bottles were opened and the pressure inside them was released, more gas escaped from the warm soft drink than from the cold soft drink. More carbon dioxide gas can dissolve in cold water than in warm water. The same is true for all gases. More gas can dissolve in cooler solvents.

The amount of a gas in a solvent has a very important consequence to fish and other aquatic animals. As Earth's atmosphere warms, the ocean's surfaces warm as well. What happens to the amount of oxygen dissolved in the water as the water is warmed? What effect might this have on fish, which require oxygen in order to live?

All solutions are mixtures, but not all mixtures are solutions. A solution is made up of tiny particles that are evenly mixed and do not settle out on standing. Unlike other types of mixtures, solutions cannot be separated by mechanical means. The rate of dissolving solute in a solvent is influenced by particle size, stirring or shaking, and temperature. As you have seen, however, gas solutes are affected by these actions differently than solid solutes in the same liquid solvents.

check your UNDERSTANDING

1. Air, vinegar, and sterling silver are three solutions. Identify which kind of solution each is. Name the solute and solvent in each.
2. You sprinkle some powdered sugar on a warm, moist, freshly baked cake. A while later you notice that the powdered sugar has disappeared. What happened to it?
3. A soup recipe calls for bouillon powder or bouillon cubes. Use of which ingredient would speed up the making of the soup?

Why does this occur?
4. **Apply** A laboratory receives a bottle of red liquid for analysis. After it sits overnight, bits of red powder are found on the bottom of the bottle. When the lab technicians shine a light through it, the top of the beam appears pink, while the bottom half looks dark red. Small particles are floating throughout the bottle. Does the bottle contain a solution? Explain your answer.

check your UNDERSTANDING

1. Air—gas-gas solution; oxygen and other gases, solutes; nitrogen, solvent. Vinegar—liquid-liquid solution; acetic acid, solute; water, solvent. Sterling silver—solid-solid solution; copper, solute; silver, solvent.
2. The sugar particles dissolved (or seemed to disappear) in the moisture of the cake.

3. Bouillon powder because smaller solute particles dissolve faster.
4. No. A solution is a mixture of tiny particles that are evenly mixed and will not settle out. The laboratory found that the particles were large enough to be seen with a flashlight, were not evenly mixed, and settled out.

5-2 Solubility and Concentration

Reaching the Limit

Most of the solutions you know about have water as the solvent. Water is the best solvent known because more substances can dissolve in water than in any other liquid. Do you think that two different solutes can dissolve in water in the same amounts? Let's find out.

Do sugar and salt dissolve in water in the same amounts?

What To Do

1. Put 100 mL of water in each of two 400-mL beakers.

2. Add 50 g of table salt to one beaker and 50 g of sugar to the other beaker.

3. Stir both at the same rate. Compare and contrast the results in your Journal. How much of each solute dissolved?

Not all solutes can dissolve in water in the same amounts. You have just seen that more sugar than salt can dissolve in the same amount of water. Look at **Figure 5-8**. Only 16.3 g of the chemical calcium chromate can be dissolved in 100 g of water at room temperature before no more will go in, while the same amount of water can dissolve 120 g of chromium (III) sulfate!

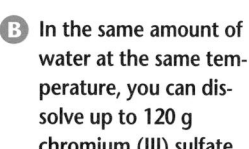

Figure 5-8

Ⓐ You can dissolve up to 16.3 g calcium chromate in 100 g of water at room temperature before the solute stops dissolving.

Calcium chromate

Ⓑ In the same amount of water at the same temperature, you can dissolve up to 120 g chromium (III) sulfate.

Chromium (III) sulfate

Section Objectives

- Describe how solubility varies for different solutes and for the same solute at different temperatures.
- Interpret solubility graphs.
- Compare and contrast saturated and unsaturated solutions.
- Infer solution concentrations.

Key Terms

solubility, saturated, unsaturated, concentrated, dilute

PREPARATION

Planning the Lesson

Refer to the Chapter Organizer on pages 152A–D.

Concepts Developed

Students discover solubility in water and temperature's effect. The concepts of concentrated and dilute are also introduced.

1 MOTIVATE

Bellringer

 Before presenting the lesson, display **Section Focus Transparency 15** on an overhead projector. Assign the accompanying **Focus Activity** worksheet. L1

LEP

Demonstrate Unseen by you, have students slowly add and stir a "secret" compound (see Figure 5-9) into 100 mL of water in a beaker until no more will dissolve. Have them label the beaker with the amount of substance (g) dissolved in it. Now announce that you can "magically" identify the secret compound. *The solubility graph on page 167 tells you how much of a compound dissolves at room temperature (20°C).*

Answer to Question

3. All the sugar dissolved in the water, but some salt settled to the bottom.

✔ **Assessment**

Process Have students list factors affecting solubility. Ask them whether any of these variables could account for the differences they observed in this activity. Use the Performance Task Assessment List for Analyzing the Data in **PASC,** p. 27. L1

Explore!

Do sugar and salt dissolve in water in the same amounts?

Time needed 10 minutes

Materials two 400-mL beakers, 200 mL water, 50 g table salt, 50 g sugar, stirring rod

Thinking Processes observing and inferring, comparing and contrasting, measuring in SI

Purpose To compare the solubility of substances.

Teaching the Activity

Troubleshooting Stress that students handle and stir both substances in the same manner.

Science Journal Have students predict their results and then record their observations in their journals. L1

Expected Outcome

More sugar than salt dissolves.

2 TEACH

Tying to Previous Knowledge

Have students recall the properties of solutions from Section 5-1. Ask them to name the three factors that affect how quickly a solute dissolves in a solvent. *particle size, temperature, and shaking or stirring*

Discussion Earlier, students learned that temperature can affect the *rate* at which substances dissolve. Here, stress that temperature increases the *amount* that can be dissolved as well (for the solutes shown). Have students speculate on why some solutes are soluble and why some are relatively insoluble. [L1]

Activity You will need various solvents such as turpentine, paint thinner, nail polish remover; various solutes such as sugar, salt, and paint.

In a well-ventilated place, have small groups of students compare the properties of the various solvents to water. How well do they dissolve each of the solutes? **CAUTION:** Make sure adequate ventilation is provided and students wear goggles when using these solvents. [L1]
[COOP LEARN]

GLENCOE TECHNOLOGY

  Software

Computer Competency Activities

Chapter 5

 Videodisc

Science Interactions, Course 1 Integrated Science Videodisc

Lesson 2

Solubility

The amount of a substance that can dissolve in 100 g of solvent at a given temperature is called **solubility**. **Figure 5-9** records the solubility of various solutes in water at room temperature.

What is the solubility of potassium chloride at room temperature? Thirty-four grams of potassium chloride will dissolve in 100 g of water at room temperature before no more will go in. Its solubility in water at room temperature is 34.0 g/100 g water.

Substances that seem to dissolve in a liquid at a certain temperature to form a solution are said to be soluble. Those that do not seem to dissolve are called insoluble. Solubility tables help you learn how much of a solid will dissolve in a solvent at a given temperature.

In the salt solution that you just made, you started out with 50 g of salt, but not all of it dissolved. According to **Figure 5-9**, only 36.0 g of the salt dissolved in 100 g of water. What do you think would happen if you poured even more salt into the solution? In solid-liquid solutions, when no more solid solute will dissolve, the extra solute settles out. The solution becomes saturated because no more solute will dissolve in the solvent. A **saturated** solution is one that has dissolved all the solute it can hold at a given temperature.

Figure 5-9

This table shows the amount of various substances that will dissolve in 100 g of water at room temperature.

Solubility of Various Substances		
Substances	Uses	Solubility
Barium sulfate	X rays	0.00025 g/100 g water
Calcium carbonate	chalk	0.0015 g/100 g water
Lithium carbonate	ceramics	1.3 g/100 g water
Potassium chloride	light salt	34.0 g/100 g water
Sodium chloride	table salt	36.0 g/100 g water
Sucrose	sugar	204.0 g/100 g water

Potassium chloride is used as a bleaching agent in some photograph processing procedures.

Lithium carbonate is a component of glazes for ceramics. Glazing prevents the item from absorbing liquids.

Sucrose is the most common of sugars and the least expensive. Its role as a sweetening agent in cookies is only one of its many uses.

166 Chapter 5 Matter in Solution

ENRICHMENT

Activity Have students investigate the connection between solubility and cleaning ability. Materials needed are an oil-based paint, turpentine, a water-based paint, and paint thinner.

Have students test the solubility of both types of paint in both solvents. Then have them apply two samples of each type of paint to a clean, dry surface and let dry. Then have them try to remove the samples with each solvent to find which removes each type of paint better.

Have students explain their results. *Each substance was removed more successfully by the solvent with which it was more soluble.* **CAUTION:** *Make sure adequate ventilation is provided and students wear goggles when using these solvents.*

Use the Performance Task Assessment List for Carrying Out a Strategy and Collecting Data in **PASC,** p. 25. [L2]

Solubility Graphs

Solubility graphs can help you compare solubilities of the same substance at different temperatures. For example, look at **Figure 5-10**. The solubilities of four substances are drawn in this graph. Temperature is plotted on the horizontal axis, and solubility is plotted along the vertical axis.

Look again at the graph in **Figure 5-10**. Of the substances shown on the graph, which is the least soluble at 60°C? What happens to the solubility of a given solute as the temperature rises?

If you raise the temperature of the water, you can add more potassium chloride to the solution. What do you think would happen to a sugar solution if you heated it? Would more or less sugar solute be able to be dissolved at a higher temperature? The next activity will show you.

Figure 5-10

This solubility graph tells how much of a substance can dissolve in 100 grams of water at a given temperature.

A Find the sucrose line and follow it up. The graph shows that more and more sucrose can be dissolved in the same amount of water as the water temperature increases.

B The chart shows that calcium carbonate is practically insoluble at any temperature.

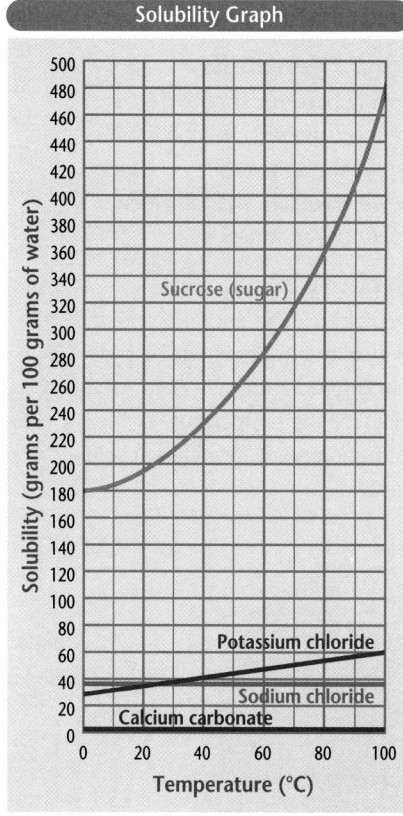

Solubility Graph

C Using the chart, you can predict how much of a substance can be dissolved in 100 g of water at a specific temperature. For example, at 60°C about 285 g of sucrose can be dissolved.

D How much sodium chloride can be dissolved at 80°C? How much potassium chloride can be dissolved at this temperature?

5-2 Saturating a Solution at Different Temperatures

Preparation

Purpose To observe how temperature affects solubility and to construct a solubility graph.

Process Skills making and using graphs, interpreting data, measuring in SI, observing and inferring, recognizing cause and effect, making and using tables, comparing and contrasting, modeling, sampling and estimating

Time Needed 30 minutes

Materials Prepare the two-hole stopper with thermometer and copper wire stirrer in advance.

Safety Precautions Solutions will be hot when the experiment is completed.

Possible Hypothesis The students might predict that more sugar will dissolve in warmer water.

The Experiment

Process Reinforcement Be sure students copy their data tables correctly. A space for the calculation of the amount of sugar that can be added to 100 g of water is an additional possible category in the table.

Possible Procedures The students have their choice of changing the amount of sugar or the amount of water. In either experiment, heat the water to about 80°C with stirring and then allow it to cool. As the dissolved sugar solution cools, note the temperature at which sugar crystals form. This temperature is the saturation temperature. Repeat the test for each different solution.

Saturating a Solution at Different Temperatures

A solvent can dissolve only a certain amount of solute before it is saturated. However, changing the temperature changes the situation. The point of saturation is different for different temperatures.

Preparation

Problem
How does the solubility of table sugar in water change at different temperatures?

Form a Hypothesis
Form a hypothesis about whether there is a change in solubility for sugar when temperature is changed. How will solubility change when temperature changes?

Objectives
- Predict how the saturation point of a solution will change with differing temperatures.
- Measure the temperature at the saturation point for different solutions.

Materials
a two-hole stopper containing a thermometer and a copper wire stirrer
large test tube
test-tube holder
distilled water
table sugar
graduated cylinder
laboratory balance
hot plate
beaker of water
safely goggles
oven mitt

Safety Precautions

Be careful when using the hot plate and when inserting the stopper into the test tube.

Sample data

Data and Observations			
Grams of Sugar	mL of Water	Saturation Temperature (°C)	Grams of Sugar Per 100 g of Water
28.7	10	60°C	287
28.7	12	40°C	239
28.7	14	20°C	205

Program Resources

Study Guide, p. 21 [L1]
Laboratory Manual, pp. 29–32, Solubility [L3]; pp. 33–36, Densities of Solutions
Activity Masters, pp. 25–26, Design Your Own Investigation 5-2
Teaching Transparency 10
Making Connections: Integrating Sciences, p. 13, Salt in the Ocean [L2]
Section Focus Transparency 15

DESIGN YOUR OWN
INVESTIGATION

Plan the Experiment

1 You will need to find the temperature at the saturation point for different solutions. The saturation point is reached when the solution cools enough that crystals of solute begin to form in the solution. If you begin with 28.7 g of sugar and 10 mL of water and run one test, how would you change the solution for another test? Would you add 2 mL of water? Would you add 2 g of sugar? In your group, decide how you will change the solution for each of three trials.

2 Copy the data table into your Science Journal and record the second and third solutions your group has agreed to test.

3 Set up the equipment as shown in the photo. How will you dissolve the sugar? You should not need to heat the solution to

more than 80°C. How will you find the saturation point? Watch the temperature closely as the solution cools and you look for crystals.

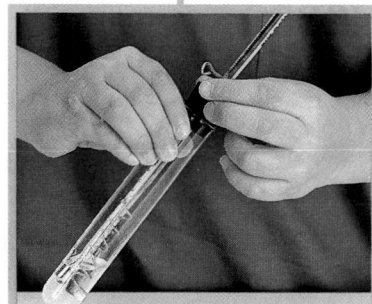

Check the Plan

1 What will stay constant in the three trials? What will change?

2 Will you stir the contents of the tube while it heats? While it cools?

3 Before you start the experiment, have your teacher approve your plan.

4 Carry out your experiment. Make observations and complete your data table *in your Science Journal.*

Analyze and Conclude

1. Use Math Calculate the grams of sugar per 100 g of water at each saturation temperature. Use the following formula and insert the appropriate numbers for grams of sugar and milliliters of water from your data table.

Mass of sugar = $\dfrac{? \text{ g sugar}}{? \text{ mL water}} \times 100 \text{ mL of water}$

2. Measure in SI What were the three saturation temperatures?

3. Compare and Contrast How did the mass of sugar that dissolved in 100 g of water change as the temperature changed?

4. Hypothesize Explain how your hypothesis was supported or disproved.

5. Use Graphs Graph the solubility versus temperature for the sugar-water solution.

Going Further

Predict the solubility of sugar at 0°C, the freezing point of water, and 100°C, the boiling point of water.

5-2 Solubility and Concentration **169**

Meeting Individual Needs

If students have difficulty understanding the formula in Step 7 above, restate it as a proportion:

$$\frac{\text{dissolved sugar}}{100 \text{ mL water}} = \frac{28.7 \text{ g sugar}}{x \text{ mL water}}$$

To find the amount of sugar that dissolves in 100 mL, have students solve the proportion for "dissolved sugar."

Teaching Strategies

Troubleshooting Stress observation for this activity. Students need to look closely to see when crystals form. Suggest that one student watch for crystals while a partner keeps track of and records temperature.

Discussion Discuss why adding water or sugar to the test tube changes the saturation temperature. It changes the ratio of sugar to water in the test tube, in effect making a new solution that has a different saturation temperature.

Science Journal Have students record their observations in their journals. They should also construct their graphs and answer the questions in their journals.

Expected Outcome

The amount of sugar that dissolves in water increases with temperature. The precise relationship between temperature and solubility can be represented on a graph. Students will recognize that this graph can be used to infer information about solubility at other temperatures.

Analyze and Conclude

1. Check calculations. The first answer will be 287 g of sugar at about 60°C. If they change the amount of water, the answers will be 239 g and 205 g. If they change the amount of sugar, the answers will be 307 g and 327 g.

2. If the amount of water is changed, the temperatures are 60°C, 40°C, and 20°C. If the amount of sugar is changed, the temperatures are 60°C, 65°C, and 73°C.

3. Less sugar could dissolve at lower temperatures.

4. Answers will vary.

5. Graphs should approximate the curve for sucrose in Figure 5-10 on page 167.

Going Further ⁞⁞⁞⁞⁞▶
179 g, 487g

Uncovering Preconceptions

To make a solubility graph (as in Investigate) students may ask why they couldn't just keep adding solute to water at a constant temperature until the solution became saturated. This approach would work, but it would be difficult to maintain constant temperature. To get around this difficulty, water is heated so it can hold "extra" solute. As the water cools, students can observe the saturation temperature, the exact temperature at which solute comes out of solution.

Theme Connection The theme this section supports is systems and interactions as it shows that a solution is a system composed of a solute and a solvent. As these interact, the solute continues to dissolve until the solution becomes saturated. These systems can also be described as concentrated or dilute, depending on the amount of solute compared to the amount of solvent.

Inquiry Question Why does air, a solution, feel much more humid on a hot day than on a cool day even if the relative humidity on both days is the same? *On a hot day the air can hold much more water than on a cool day.*

Connect to...
Earth Science

In a limestone cave, calcium carbonate is being dissolved and deposited. Give a presentation about two different speleothems. Include a drawing and a description of how each one is formed.

You have just seen that the solubility of sugar increases with an increase in temperature. That is, the amount of sugar that dissolves in water increases as the temperature increases. A cold solvent will usually hold less solute than a hot solvent will hold. As the temperature of the solvent rises, more solute will dissolve in it.

■ Unsaturated vs. Saturated Solutions

When you added 50 g of sugar to water in the Explore activity, all the sugar dissolved in the water. A solution that can hold more solute at a given temperature is an **unsaturated** solution. An unsaturated solution has room for more solute particles. Each time a saturated solution is heated to a higher temperature, it may become unsaturated. The term unsaturated is qualitative. It doesn't give you an exact amount. In an unsaturated solution, more solute can be dissolved in the solvent. As soon as the solution becomes saturated at a given temperature, no more solute can be dissolved at that temperature.

Suppose you make a saturated sugar solution at 60°C and then let it cool to room temperature. Part of the solute will become solid again. Why do you think this happens? Most saturated solutions behave in a similar way when cooled.

Life Science CONNECTION

Keeping the Balance

When the human body is working properly, the concentration of fluids inside its cells is the same as the concentration of fluids that surround them. For example, sodium chloride makes up slightly less than one percent of blood cells and the fluid around them. If the concentration of sodium chloride is higher outside the cells than inside, fluids ooze out

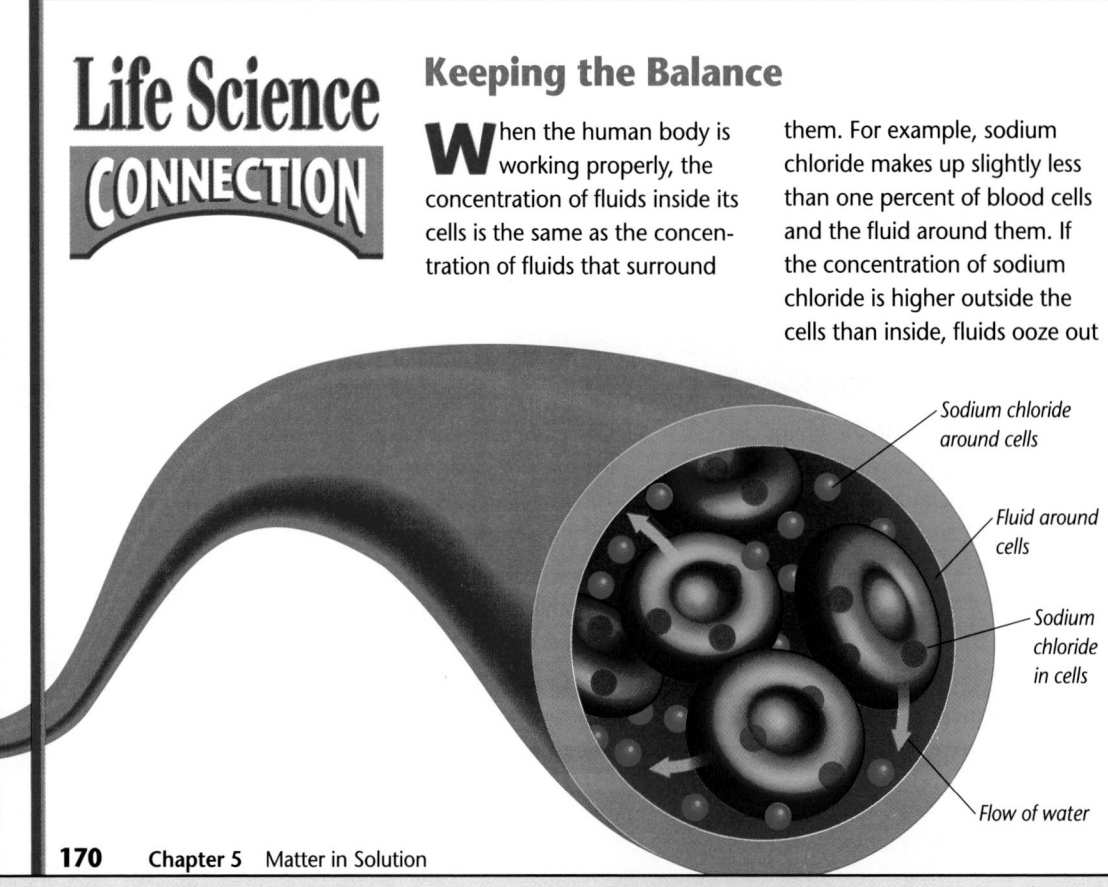

Sodium chloride around cells

Fluid around cells

Sodium chloride in cells

Flow of water

Purpose

The Life Science Connection extends Section 5-2 by explaining the importance of a balanced sodium chloride concentration inside and outside of blood cells.

Content Background

The process by which a semi-permeable membrane allows solvent molecules from a dilute solution to pass through to a more concentrated solution is called osmosis. Osmotic pressure is an unbalanced concentra-

tion that triggers osmosis. Two solutions of equal concentration, and thus equal pressure, are called isotonic. Where concentrations differ, the more concentrated solution is called hypertonic and the less concentrated, hypotonic.

Teaching Strategy

Have students consider how the body perspires during exercise. Then show them one or two commercially available products that athletes drink during and after ex-

Concentration

You've already discovered that solubility is the number of grams of solute that will dissolve in 100 grams of solvent. Sometimes you do not need to know exactly how much solute is dissolved. All you may need to know is that one solution contains more solute than another solution contains.

A **concentrated** solution has a large amount of solute in a solvent. A **dilute** solution has a small amount of solute in a solvent. The terms, concentrated solution and dilute solution, can be used daily to describe such

Figure 5-11

You add two tablespoons of chocolate to your milk; your friend adds six tablespoons. Which drink is more dilute? Which is more concentrated? Why?

Visual Learning

Figure 5-11 Discuss with students how the taste of the two glasses of chocolate milk would compare. **Which drink is more dilute? Which is more concentrated? Why?** *Your drink is more dilute because it has a smaller amount of solute. Your friend's drink is more concentrated because it has a larger amount of solute.*

Activity You will need two cans of the same type of soda, one at room temperature and the other well chilled.

With the permission of their parents, have students perform a taste test at home to decide which tastes "fizzier" and, therefore, has more carbon dioxide gas in solution. Have students share their findings with the class and relate them to what they have learned about temperature and solubility. *The cold soft drink tastes more fizzy. Therefore, the solubility of a gas must decrease with temperature.* **L1**

GLENCOE TECHNOLOGY

 Videodisc

STVS: Chemistry
Disc 2, Side 1
Uses for Super Slurper (Ch. 15)

Cloud Chemistry (Ch. 22)

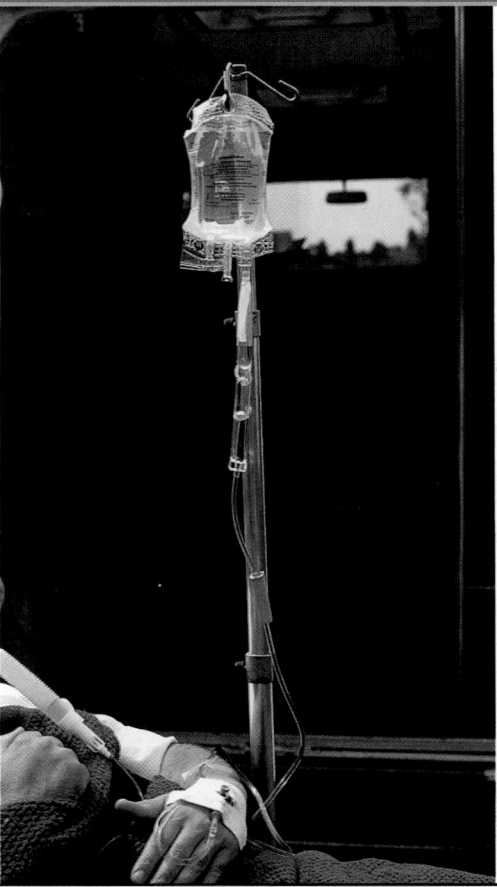

of the cells and into the surrounding fluid and reduce the sodium chloride concentration there. This change could cause the cells to become short of water, or dehydrated.

On the other hand, if the sodium chloride concentration outside the cells is lower than the concentration inside the cells, the cells can become flooded with fluids. This problem can become severe enough to cause cells to burst.

Intravenous Fluids

Doctors attempt to balance fluid concentrations by giving patients fluids intravenously, as shown in the photograph. In this procedure, a bag containing the proper fluids is hung above the patient and allowed to drain through a tube into the patient's vein and through the bloodstream. In this way, cells are surrounded by fluids containing the proper concentrations of substances.

What Do You Think?

How would a patient's body react if it received too much sodium chloride intravenously?

ercise. Read the ingredients label and point out that some of these are solutions or compounds that the body loses through perspiration. Lead students to conclude these products restore proper fluid concentrations by replacing the sodium chloride.

Answer to
What Do You Think?
The cells will become dehydrated as they lose liquid to the surrounding fluid.

Going Further ⁞⁞⁞⁞▶
Have students discuss the role blood plays in transporting oxygen throughout the body. Once the lungs transfer oxygen to the blood, how might body temperature affect the oxygen concentration in the blood?
Have students work in small groups to research the composition of blood and what happens to fluids in the body during hemorrhaging. How does this relate to

concentration levels in the blood? Then have students share their findings, accompanied by diagrams, with the class. Use the Performance Task Assessment List for Group Work in **PASC**, p. 97. **COOP LEARN**

3 ASSESS

Check for Understanding

Before assigning the Check Your Understanding questions, pose these problems: A saturated solution contains 75 g of substance X dissolved in 100 g of water at 50°C. Give the answer that is most likely.

- At 60°C, will the solution be saturated or unsaturated? *unsaturated*
- At 60°C, will more substance X dissolve in the solution? *yes*
- At 60°C, how will the concentration of the solution change if more substance X is added? *It will increase.*
- At 40°C, will the original solution be saturated or unsaturated? *saturated*

Reteach

Demonstration Show a pictorial version of a solubility graph. In diagram form, show several beakers, all labeled "saturated solutions" containing 100 g water. In the first beaker, show 200 g sugar dissolved in the water at 20°C; in the second beaker, show 210 g dissolved at 30°C. At other temperatures, show separate beakers with different amounts of sugar dissolved. L1

Extension

Acquiring Information Have students who have mastered the concepts in this section investigate what happens when two solutes are dissolved in the same solvent. For example, what

Figure 5-12

One way to describe solution concentrations precisely is to state the percentage of solute by volume of solution.

A One drink label states that the percentage of juice by volume is 100 percent.

B The other drink contains 10 percent juice and 90 percent water. Which of the two drinks is more concentrated?

common items as medications, cleaning products, tea, coffee, lemonade, soup, and even chocolate milk.

Concentrated and dilute are relative terms, like large and small. But there are ways you can describe solution concentrations precisely. Have you ever read the label on a juice box to see how much actual juice is in there? Look at **Figure 5-12**. How much juice are you actually drinking when you purchase such products?

What have you discovered about solubility in this section? Solubility of a solution is the amount of solute that can dissolve in 100 g of solvent at a given temperature. The solubility of a solid solute in a solvent generally increases with temperature. Solubility tables and graphs help you predict how much of a solute will dissolve in a solvent at a given temperature. Solvents that hold as much solute as they can contain are saturated, while those that hold less solute than they can are unsaturated. Concentrated solutions contain more solute in a solvent than equal amounts of dilute solutions.

Are there any other kinds of mixtures that are not solutions? Yes, and in the next section you will read about two.

check your UNDERSTANDING

1. Suppose you want to make super-sweet lemonade. You stir 2, 3, 4, or more teaspoons of sugar into a cup of lemonade, and it all disappears. But eventually you add another teaspoon of sugar, and it no longer dissolves. Why?
2. Look back at **Figure 5-10**. How much sugar would have to be dissolved in 100 g of water at 30°C to form a saturated solution?

How much sugar would have to be added to form an unsaturated solution at the same temperature?
3. **Apply** You add 1 teaspoon of lemon juice to a cup of water, and a friend adds 4 teaspoons of lemon juice to another cup of water. Into whose cup would more spoons of sugar need to be added to make the lemonade sweet? Explain.

will happen if sugar is added to a solution that is already saturated with salt? L3

4 CLOSE

Have students summarize their knowledge of solubility by brainstorming how each of these factors affects how a substance dissolves: type of solute, type of solvent, temperature, particle size, color, stirring, concentration, taste. L1

check your UNDERSTANDING

1. It means that the solubility limit of sugar in lemonade has been reached and the solution has become saturated.
2. 210 g; anything less than 210 g
3. The friend's; it is more concentrated, and would need more sugar to make it sweet.

 # Colloids and Suspensions

Nonsolutions

No matter what the concentration of a solution, the particles are always too small to be seen. Because of this, gas solutions and liquid solutions are always transparent. They may be colored, like coffee, tea, or apple juice, but they are always transparent.

Think about things in a kitchen that seem to be transparent. Water, maple syrup, glass, and gelatin desserts come to mind. They are all mixtures, and they all look transparent. Are they all solutions? Let's find out.

Section Objectives
- Distinguish between a colloid and a suspension.
- Recognize at least two mixtures that are colloids.

Key Terms
colloid
suspension

 **ACTIVITY**

Are all transparent mixtures solutions?

What To Do

1. Make a gelatin dessert by following the package directions. Use a transparent bowl.

2. When the gelatin has set, look at it from the side. Record *in your Journal* whether it is transparent.

3. Shine a beam of light from a flashlight through the dessert. Is the gelatin still transparent? Is gelatin a solution?

When you mix water with another material, the combination meets part of the definition of a solution. It is a mixture. You can tell by looking at it whether it also satisfies a second part of the definition. Its particles are too small to see. If you're not sure, shine a light through the mixture, as you did when you made gelatin. True liquid

solutions will be clear. Nonsolutions will not. A solution will not show the path of the beam of light. Think back to when you made a soft drink in the first section and shined a flashlight through the transparent pitcher. Did you see any particles? Solution particles are too small to block the light from going straight through.

 Explore!

Are all transparent mixtures solutions?

Time needed 15 minutes

Materials gelatin dessert mix, transparent bowl, flashlight

Thinking Processes comparing and contrasting, classifying, forming operational definitions, predicting

Purpose To observe the properties of a colloid.

Preparation Prior to class, prepare the gelatin.

Teaching the Activity

Demonstration In contrast, you might want to shine the flashlight through a solution to show that it remains transparent.

Science Journal Have students note their ob-

PREPARATION

Planning the Lesson

Refer to the Chapter Organizer on pages 152A–D.

Concepts Developed

Sections 5-1 and 5-2 focused on solutions. In this section, the differences among solutions, colloids, and suspensions are explored.

1 MOTIVATE

Bellringer

Before presenting the lesson, display **Section Focus Transparency 16** on an overhead projector. Assign the accompanying **Focus Activity** worksheet. L1 LEP

---

servations and conclusions in their journals. L1

Expected Outcome

The gelatin will not be transparent when light is shined through it. Students should conclude that gelatin is not clear like a solution.

Answers to Questions

3. no; no.

✔ Assessment

Oral Ask students to describe other qualities of solutions. Have them predict whether the gelatin will share these properties. Use the Performance Task Assessment List for Formulating a Hypothesis in **PASC**, p. 21. L1

174 Chapter 5 Matter in Solution

Demonstration You will need a beaker of "clean" pond water and a flashlight. Have students describe the appearance of the water in normal room light *fairly clear* Then shine a flashlight through it and have students describe its appearance. *cloudy* Discuss why this difference occurs. *The water contains particles that are small enough to appear transparent, but large enough to be seen when a light is shined through them.* L1

2 TEACH

Tying to Previous Knowledge

Have students review the properties of a solution and list them on the board. Be sure to include uniformity as a major property as well as transparency and the size of solute particles.

Theme Connection The theme that this section supports is stability and change. The composition of solutions and colloids is stable, but the composition of a suspension will change as the particles settle.

Visual Learning

Figure 5-13 Have students recall their observations from the Explore activity on page 173 as they look at Figure 5-13. **Which is a solution?** *the saltwater*

How Do We Know?

Scientists in the time of Zsigmondy could not explain why colloids didn't settle out like suspensions. Zsigmondy's microscope showed *structural* differences between colloids and other mixtures. This example brings to light a more general concept of chemistry: Microscopic form can be used to explain macroscopic function.

A nonsolution, however, will show a clearly defined beam of light. Look at **Figure 5-13**.

Figure 5-13

(A) The particles in a solution are too small to scatter light. If you shine a light through a mixture and the light shines straight through, the mixture is a solution.

(B) The container directly in front of the light is filled with salt water and the other with gelatin dessert.

(C) The gelatin dessert particles scatter the light, but the salt water does not. Which is a solution?

How Do We Know?

The Ultramicroscope

Richard Zsigmondy, an Austrian chemist, worked at a glass factory making both clear and colored glass. Colored glass is made when different substances are mixed in melted clear glass. Zsigmondy wanted to study how particles behaved in his glass. In 1902, he developed a device that he called an ultramicroscope, shown in the figure. An ultramicroscope focuses on a colloid at right angles to the light source. The background is dark. Only the light scattered from colloid particles enters the microscope. The colloid particles can be seen not as particles with definite outlines, but as small sparkles. The device enabled Zsigmondy to see that particles in colloids are constantly moving in random, zigzag paths. This is one reason they do not settle out. In 1925, Zsigmondy received a Nobel Prize in Chemistry for his work on colloids.

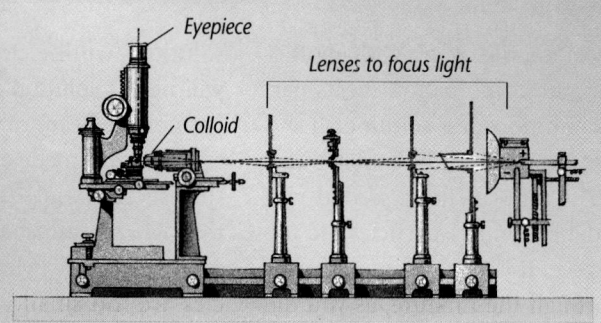

Eyepiece

Lenses to focus light

Colloid

■ Colloids

Not all liquid mixtures are solutions. Milk is a mixture of water, fats, proteins, and other substances. It is a mixture but not a solution. Milk is a colloid. A **colloid** is a mixture that, like a solution, does not settle out. Unlike the small particles in a solution, however, the particles in a colloid are large enough to scatter light. That is why milk looks white.

Dirty air may be a colloid, too. Its particles scatter light. You have seen the scattering of light by dust particles when a beam of sunlight shines into a dark room through a slit in the curtain. The dust particles, many of them too small to be seen, look like bright speckles as the light is scattered by them.

Some mixtures, such as muddy water, are neither solutions nor colloids. What are they called?

Meeting Individual Needs

Visually Impaired To help visually impaired students understand how a beam of light shining through a colloid is visible when viewed at an angle of 90°, use a box or other simple object that has a right angle. Guide each student's hand along one edge of the box to represent the beam of light. Then use a perpendicular edge to represent the line of sight of the viewer.

Program Resources

Study Guide, p. 22 L1
Critical Thinking/Problem Solving, p. 5, Flex Your Brain; p. 13, Lead in Paint
Multicultural Connections, p. 14 L1
Take Home Activities, p. 12, Separating Milk L1
Science Discovery Activities, 5-3, A Foggy Situation
Section Focus Transparency 16

Explore! ACTIVITY

When is a mixture not a solution or a colloid?

What To Do

1. Fill a glass with muddy water.
2. Let the glass sit for 30 minutes. Record your observations *in your Journal*.
3. Stir the contents carefully. What do you see now?
4. Wait a few minutes and then examine the glass again. Is the color the same throughout the glass?
5. Now, let the glass sit overnight. What do you observe the next day? Is muddy water a solution?

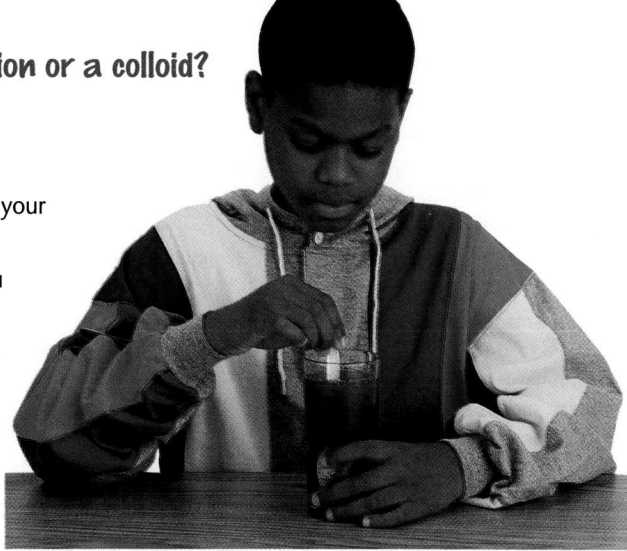

■ Suspensions

Muddy water is a suspension. A **suspension** is a mixture containing a liquid in which visible settling occurs. The particles of solute are larger than the particles of solvent. The force of gravity causes these particles to settle out.

As in a colloid, the particles in a suspension are not homogeneous. In the last activity, you noticed that the muddy water turned different shades of brown, even after stirring. A suspension is different from a solution because its particles are not evenly mixed and are large enough to settle out.

Figure 5-14 classifies mixtures according to four basic properties.

Figure 5-14

You can classify most mixtures by testing for the four properties listed in the chart.

Solutions, Colloids, and Suspensions			
Description	Solutions	Colloids	Suspensions
Scatter light	No	Yes	Yes
Settle upon standing	No	No	Yes
Can be separated using filter paper	No	No	Yes
Sizes of particles	Small	Medium	Large

A How would you classify a mixture that scattered light but did not settle upon standing and could not be separated using filter paper?

B If you could see the particles in a mixture without using a microscope, how would you classify it?

Explore!

When is a mixture not a solution or a colloid?

Time needed 30 minutes

Materials muddy water, glass, stirrer

Thinking Processes comparing and contrasting, classifying, recognizing cause and effect, forming operational definitions

Purpose To investigate the properties of a suspension.

Preparation Chocolate syrup and water can be substituted for muddy water.

Teaching the Activity

Demonstration Put a solution and a colloid next to the muddy water for contrast. Note that they do not settle. Stress the role of gravity in the settling process.

3 ASSESS

Check for Understanding

After students complete the Apply question, ask them what they expect would happen to the bits of pulp if the glass of juice were in space and why. *The bits of pulp would not settle; gravity would not pull them down.*

Reteach

Concept Mapping To help students organize data, guide them in developing a hierarchy including solutions, colloids, and suspensions based on how well particles "disappear." Solutions are at the top because particles appear to vanish completely in them. Colloids are next, since particles seem to disappear in them but can be viewed indirectly when they scatter light. Suspensions are at the bottom since the particles in them are easy to see as they settle or are separated through a filter. L1

Extension

Activity Have students who have mastered the concepts in this section investigate a chemical property of colloids and suspensions by testing whether the particles in milk or muddy water make the boiling point of these liquids different from the boiling point of clear water. L3

4 CLOSE

Activity

Have students make a Venn diagram to present the information in Figure 5-14. L1

Figure 5-15

A Bronze—made from copper and tin—is a solid-solid solution commonly used to make ornamental objects, such as this ceremonial vessel from China.

B Fog is an example of a colloid that contains liquid particles mixed in a gas. Its states of matter are described as liquid in gas.

C Dust in the air is a type of colloid.

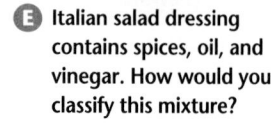

D Paint is a colloid made of solid pigment particles in solvents. How would you describe its states of matter?

E Italian salad dressing contains spices, oil, and vinegar. How would you classify this mixture?

Particles in some mixtures are large enough to interfere with the movement of light through the liquid. These mixtures are not classified as solutions, but they may be suspensions or colloids. If the particles in a mixture settle out of the system, the system is a suspension. If particles of a cloudy or light-scattering system do not settle out, the system is a colloid.

Homogeneous mixtures make up a large percentage of the materials in your life. The air you breathe, the water you drink, and even the steel that was used to build your school—each is a homogeneous mixture. Suspensions and colloids are not homogeneous mixtures. Although materials may seem very different, many of them have a great deal in common.

check your UNDERSTANDING

1. List two examples of colloids.
2. Identify mayonnaise as a colloid or a suspension. Explain your answer.
3. **Apply** A glass of orange juice sits on the kitchen counter. After an hour, bits of pulp settle out on the bottom of the glass. What type of mixture does this represent?

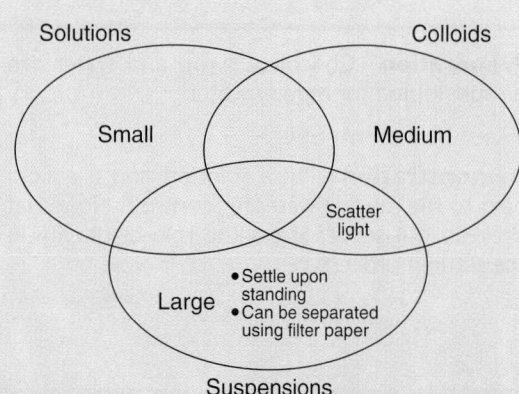

check your UNDERSTANDING

1. Answers may include gelatin dessert, milk, smoke, fog.
2. Colloid. Its particles are too small to be seen and do not settle out.
3. suspension

Literature Connection

Down to the Sea Again

One of the world's largest solutions—the ocean—has inspired exploration throughout recorded history. Scientists and explorers have written volumes filled with ocean facts, and they continue to examine the vast solution that makes up almost three-fourths of Earth's surface.

The sea fascinates creative writers and storytellers as well. They have passed along their sea legends from generation to generation. To storytellers, the ocean may be a setting for adventure or a symbol with spiritual meaning, not a scientific problem to be solved.

Herman Melville's *Moby Dick* is a famous example of a story dealing with the seafarer's way of life. Other writers who have used the sea to express themselves include Walt Whitman, Stephen Crane, and William Shakespeare.

The poem "Sea Fever" is yet another example of sea-inspired writing. The poem was written by John Masefield, England's poet laureate from 1930 to 1967, who became a sailor when he was 13 years old. When you read the poem, you'll realize that the poet draws mainly from the visual and emotional parts of his experience.

As you read the poem, think of ways a scientist might talk about the things the poet describes. Do you think of the sea as Earth's vast solution? Or do you see it as a place of romance?

I must go down
to the seas
again, To the
lonely sea and
the sky.
And all I ask is a
tall ship and a
star to steer
her by.
And the wheel's
kick and the
wind's song
and the white
sail's shaking.
And a gray mist
on the sea's face, and a gray dawn breaking.

I must go down to the seas again, for the call
of the running tide
Is a wild call and a clear call that may not
be denied;
And all I ask is a windy day with the white
clouds flying,
And the flung spray and the blown spume,
and the sea-gulls crying.

I must go down to the seas again to the
vagrant gypsy life,
To the gull's way and the whale's say where the
wind's like a whetted knife;
And all I ask is a merry yarn from a laughing
fellow-rover,
And quiet sleep and a sweet dream when
the long trick's over.

Science Journal

In your Science Journal, write a short poem about a special place. Describe the sights, sounds, smells, tastes, and emotions you experienced there.

Going Further ⑄

Have students work in small groups to research sea explorers and their discoveries from one of the following countries: Portugal, Spain, The Netherlands, England, France, and the United States. Then have each group prepare journals as if it had accompanied one of the explorers researched. Use the Performance Task Assessment List for Writer's Guide to Fiction in **PASC,** p. 83.

COOP LEARN **L2**

Literature Connection

Purpose

The Literature Connection discusses how the ocean, an enormous solution to scientists, has stimulated creativity in writers and poets.

Content Background

Many writers have also spent time at sea as sailors and have used the knowledge gained in their writings. Herman Melville, an American author (1819–1891), not only sailed, but deserted one ship and participated in a mutiny. His sea experiences provided ideas for such books as *Typee, Moby Dick,* and *Redburn.* Jack London, another American author (1876–1916), worked both as a fish patroller and as an oyster pirate. His lifestyle inspired him to become a writer. He is known for many great books such as *The Sea Wolf.* Richard Henry Dana, Jr., an American lawyer (1815–1882), tried sailing to improve his health. His *Two Years Before the Mast* is a classic book. John Masefield, an English poet (1878–1967), was a naval cadet officer as a youth. The word "trick" in the last line of the poem is the time a sailor spends at the pilot's wheel steering the ship.

Teaching Strategy

To help students develop their communication skills, have each student bring a brief poem or essay relating to the sea to class. Read them in small groups and discuss the emotions they bring forth. **COOP LEARN**

Activity

Have students list five properties of the ocean that they feel inspired so many writers. Do these properties describe the physical characteristics of a solution? Can any other solution evoke such emotion as the oceans can?

Science *and* Society

Purpose

Science and Society applies concepts of solutes and solvents presented in Sections 5-1 and 5-2 by discussing the concentration of pollutants and trash in the ocean.

Content Background

During the last century, people thought ocean resources were limitless and self-sustaining. Today we know this to be false. Numerous forms of pollution are harmful to the ocean's environment. Materials released into the water irresponsibly can interfere with the ecosystems found in the water by settling and forming undesirable deposits. Sometimes thermal pollution occurs. Radioactivity can be released. Some materials that are broken down naturally under normal circumstances, can build up to dangerous levels. Agriculture can compound problems by yielding pesticides and other harmful chemicals. Urban areas contribute large amounts of toxic wastes and, often, poorly treated sewage. Methods of controlling ocean pollution include governmental legislation, research into waste treatment and waste recycling, and chemical treatment of water.

Science *and* Society

Cleaning Up the Oceans

You already know that the oceans are loaded with salt. But have you ever thought about other substances that are in seawater? Oceanographers and other scientists are very interested in knowing more about the contents of the oceans. Some of these experts are simply curious while others want to know more about the vast amounts of chemicals dumped or washed into the oceans from industrial and hospital waste.

Chemical Waste

For example, some people have dumped radioactive hospital wastes—materials from testing procedures and cobalt treatments—into the oceans. Now, concerned citizens want to know if these wastes will leak from their containers and mix in the seawater. Will they harm sea plants and animals? Will contaminated seawater affect humans? Is there some way to prevent such pollution? Mercury is another dangerous industrial waste that has been dumped into seawater. In the past, shipbuilders used paint that contained mercury because the mercury prevented the growth of marine plants and animals on the hulls of ships.

Years ago, some people thought mercury could be dumped on the ocean floor without harming anyone. They figured that because the mercury would not dissolve easily in the seawater, it could be safely deposited there.

However, scientists eventually learned that chemical processes in the ocean changed the mercury so that it could get into fish. People eating the fish could get mercury poisoning, which can cause very serious health problems. The United States government now prohibits the dumping of wastes that contain mercury, and the use of mercury compounds in paint.

Thinking that a chemical spill or the dumping of small amounts of toxic waste wouldn't make much difference in the vast ocean, people today have made the oceans even more complex by adding such things as sewage, oil, and other pollutants. However, ocean pollution does not go unnoticed, and many countries are trying to control it through their own laws and through agreements with other countries.

Solid Waste

In addition to chemicals, tons of solid materials are deposited into the oceans. Plastics thrown into oceans pose problems because they do not easily decompose or dissolve in seawater. That means they may clutter the oceans for years, sometimes harming wildlife and frequently destroying the natural beauty of the sea.

One biologist walked 1.5 miles along the beach of a Pacific island to see how much trash he could find. He found hundreds of pieces of garbage, including 74 bottle tops, 25 shoes, 6 light bulbs, toys, cigarette lighters, and a football. The amazing thing about his discoveries was that the island was at least 3000 miles from any continent.

He knew there must be an enormous amount of trash floating in the ocean if he was able to find so much on an uninhabited island. He was so concerned about what he had found there that he wrote to a scientific journal to tell his story.

Cleaning Up the Problem

What can we do about this severe problem? The United States Congress has passed laws to control some dumping of industrial

wastes into the oceans. Scientists are trying to develop ways to control environmental damage caused by accidental chemical spills, such as oil spills. In fact, some researchers have released oil-eating organisms into the ocean to break up oil spills. Others have attempted to contain oil spills by surrounding them with barriers.

Who Is Involved?

Individuals and organizations everywhere worry that pollution has gotten out of control. They're working to repair the environment by talking to their congressional representatives and even by picking up trash on beaches.

What Do You Think?

You and a partner are in charge of a campaign to clean up the oceans. Think of ways that individuals and organizations can help and explain how you will promote your ideas.

Going Further |||||▶

Arrange a field trip to visit a local stream, lake, river, or ocean beach that needs some cleanup. Upon arrival, have students note the general appearance of the area. Then have them put on gloves and pick up trash and place it in sacks for proper disposal. After returning, have students write how they felt about the area when they arrived, as they worked, and when they had finished cleaning up. **COOP LEARN** **L1**

Teaching Strategy Show students pictures of the Exxon Valdez oil spill and describe various methods used to contain and clean up the oil. Have students share anything they have read about this oil spill. Then discuss whether or not long-term problems will likely result from the spill and, if so, what they might be.

Activity

Invite a representative of your local government to discuss water pollution problems in your town, county, or state. Then have students work in small groups and make posters that can be used to make the rest of the school or community aware of these problems and how they can be dealt with. Use the Performance Task Assessment List for Poster in **PASC,** p. 73. **COOP LEARN** **L1**

Research

Have students work in groups to research different environmental groups. Discuss what each group represents and how each tries to reach its goals. Do the groups act responsibly? Have groups share their findings with the class and select an organization the class would like to join.

Possible groups to consider are:

Greenpeace International
Earthfirst!
Sierra Club
National Geographic Society
Friends of the Earth
Cousteau Society
National Audubon Society

COOP LEARN **L2**

Answers to
What Do You Think?

Groups can hold fund-raising benefits and use the proceeds for an advertising campaign or donate the money to environmental organizations. Individuals and groups can write to members of Congress or ask qualified speakers to educate the public at community meetings.

economics connection

Purpose

The Economics Connection applies the information about concentrations in Section 5-2 by discussing what happens to a body of water in which the salt concentration is constantly increasing.

Content Background

There are several types of salts. Normal salts contain no ions that will cause an acid or base reaction in a solution. Acid salts contain some hydrogen ions, which can act as an acid in a reaction. Basic salts contain hydroxide ions. These salts are called simple salts. So-called double and complex salts in solution will usually yield metal ions.

Salt comprises about two thirds of the dissolved solids in ocean water, or about 2.8 percent by weight. The composition of the Great Salt Lake is about 15 percent by weight. The concentration of the Dead Sea in the Middle East is double that of the Great Salt Lake.

Salt is often found in the form of salt domes on land. Large deposits are found in Louisiana and Texas in the United States and in several European countries. Many other countries produce salt by seawater evaporation.

Teaching Strategy

Historically, salt was a common medium of exchange. It was often the main type of good traded along ancient caravan routes. Have students list the reasons for the importance and value of salt in early societies. Consider the difficulty of travel and transportation and the location of major sources of production. L1

economics connection

So You Think the Ocean Is Salty?

What do you think happens when rivers flow into a sea that doesn't drain, but instead loses a lot of its water through evaporation?

You may have guessed that the concentration of salt in the sea increases with time.

The Great Salt Lake

The Great Salt Lake of Utah is a well-known salt sea that has grown so salty that it has become a tourist attraction and a source of various natural resources.

Gathering Salt

Businesspeople once sold salt taken from the lake. At first, dried salt was found on the shore and hauled away. The problem with that method, however, was removing the mud gathered along with the salt.

Later, some ambitious folks boiled the salty water in large containers until they were left with salt they could sell. Unfortunately, they were also left with other chemicals that had been dissolved in the water. However, salt was not always easy to come by in those days, so slightly contaminated salt was more acceptable then than it would be today.

In some cases, salt is all too easy to collect at the Great Salt Lake. Swimmers, for example, are attracted to the Great Salt Lake because they can float on its surface with little effort. But when they return to shore, the water evaporates from their skin, leaving them covered with salt crystals.

What Do You Think?

Why is the concentration of salt water increasing in the Great Salt Lake?

Answer to
What Do You Think?

It gets little fresh water while at the same time evaporation continues to cause a higher concentration of salt.

Going Further ▐▐▐▐▐▶

Have students work in pairs to research other bodies of water in the world that have high concentrations of salt. In addition, have students compare salt concentrations in the four oceans. Are there differences? What might account for this? Students can present their findings in the form of posters or pamphlets. Select the appropriate Performance Task Assessment List in PASC. COOP LEARN L2

Science Journal

Review the statements below about the big ideas presented in this chapter, and answer the questions. Then, re-read your answers to the Did You Ever Wonder questions at the beginning of the chapter. *In your Science Journal*, write a paragraph about how your understanding of the big ideas in the chapter has changed.

1 Solutes dissolve in solvents to form solutions. Solutions may be made up of solids, liquids, and gases. *What properties do solute particles demonstrate?*

2 Solutes dissolve in solvents at different rates. *What factors affect the rate of dissolving?*

3 Solubility is the amount of solute that can dissolve in 100 g of solvent at a given temperature. *How is a concentrated solution different from a dilute solution?*

4 Colloids and suspensions are two other types of mixtures. *How do colloids and suspensions differ from solutions?*

Use the diagrams to review the main ideas presented in the chapter.

Teaching Strategy

Have students describe the dissolving process from a solute's perspective.

• The solute begins in a "clump" with other solute particles. What happens when they are added to water? *The clump is split up until water surrounds each particle. The solute particles become evenly spaced throughout the water. Individual particles are so small they are invisible.*

• The solvent is heated. How does heating allow more solute to dissolve? *Heating spreads out the water particles so more solute particles can fit between them. Heating also increases the energy and motion of the water particles, so they move more rapidly to surround solute particles.*

Answers to Questions

1. The particles mix evenly and are too small to be seen; the particles will not settle out.

2. The rate of dissolving is affected by particle size, shaking or stirring, and temperature.

3. A concentrated solution has a large amount of solute in solvent; a dilute solution has a small amount of solute in solvent.

4. Colloids differ from solutions in their ability to scatter light. Suspensions are different from solutions because they have large particles that will settle out.

Science Journal

Did you ever wonder...
• The sugar crystals are broken down to such a small particle size and mixed so evenly in the water that they disappear and become part of a solution. (p. 160)
• The solubility of sugar in water increases as temperature rises. More sugar will dissolve per unit volume in hot than in iced tea. (p. 167)
• Milk is a colloid, a mixture that does not settle out. (p. 174)

Science at Home

Have students prepare samples of colorless soda at temperatures ranging from near-freezing to near-boiling in glasses. Have family members rank the liquids based on how fizzy they appear. Then, have students rank the liquids based on their temperatures. Compare the two rankings. The colder liquids should all rank near the top according to fizziness. They contain more dissolved carbon dioxide gas, which causes the fizziness. **L1**

GLENCOE TECHNOLOGY

 MindJogger Videoquiz

Chapter 5 Have students work in groups as they play the Videoquiz game to review key chapter concepts.

Using Key Science Terms

1. *Suspension* does not belong because the other three terms involve true solutions and a suspension is not a solution.

2. *Solution* does not belong because it does not refer to the amount of solute in a solvent.

3. *Colloid* is the only term that does not refer to how much solute can dissolve in a solvent.

4. *Solvent* is the only term that is not a type of mixture.

Understanding Ideas

1. a) Usually the solubility increases as the temperature increases.

b) Usually the solubility decreases as the temperature increases.

2. stirring, increasing solvent temperature, decreasing solute particle size

3. dilute, unsaturated

4. There would be no visible change because the sodium chloride solution is unsaturated over the temperature range.

5. Particle size is smallest in a solution, larger in a colloid, and largest in a suspension.

Developing Skills

1. See reduced student page for completed concept map.

2. First solution: 1.2 kg or 1200 g are dissolved in 5 liters or 5000 g of water.

$$\frac{1200 \text{ g salt}}{5000 \text{ g water}} = \frac{0.24 \text{ g salt}}{1 \text{ g water}}$$

Second solution: 56 g are dissolved in 200 mL or 200 g of water.

$$\frac{56 \text{ g salt}}{200 \text{ g water}} = \frac{0.28 \text{ g salt}}{1 \text{ g water}}$$

The second solution is more concentrated.

3. The cube on the right may have been made with cold water; the cube on the left may have been made with hot water. More gas can dissolve in cold water; cold water may have dissolved air that gets trapped as the water

U sing Key Science Terms

colloid	solute
concentrated	solution
dilute	solvent
saturated	suspension
solubility	unsaturated

For each set of terms below, choose the one term that does not belong and explain why it does not belong.

1. solute, solvent, solution, suspension

2. unsaturated, saturated, concentrated, solution

3. colloid, dilute, solubility, concentrated

4. solution, suspension, solvent, colloid

U nderstanding Ideas

Answer the following questions in your Journal using complete sentences.

1. How does an increase in temperature usually affect the solubility of:
a) a solid in a liquid?
b) a gas in a liquid?

2. What are some methods that you might use to increase the rate of dissolving of a solid?

3. Five grams of sucrose are dissolved in 100 g of water. Which of the Key Science Terms can be used to describe the resulting solution?

4. A solution of 30 g of sodium chloride in 100 g of water is cooled from 90°C to 10°C. What visible change would you notice in the solution?

5. Compare the size of particles in a solution, suspension, and colloid.

D eveloping Skills

Use your understanding of the concepts developed in this chapter to answer each of the following questions.

1. Concept Mapping Fill in the concept map using the following terms: *gelatin, muddy water, salt water, solution, suspension.*

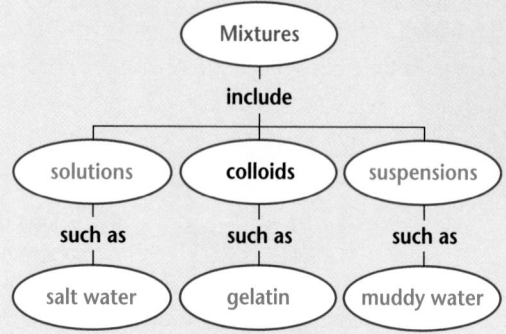

```
           Mixtures
              |
           include
    ┌─────────┼─────────┐
 solutions  colloids  suspensions
    |          |          |
 such as    such as    such as
    |          |          |
salt water  gelatin   muddy water
```

2. Measuring in SI Two salt solutions are prepared. In the first solution, 1.2 kg of salt are dissolved in 5 liters of water. In the second solution, 56 g of salt are dissolved in 200 mL of water. Which solution is more concentrated?

3. Observing and Inferring Gas trapped in ice can make it look cloudy. Which of these ice cubes was probably made with cold water? Which was probably made with hot water? Explain.

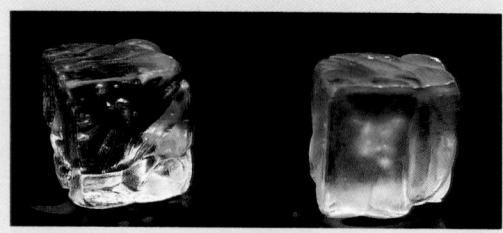

freezes, causing cloudiness. Less gas can dissolve in hot water; any dissolved air was driven out of the hot water as it was heated. Less trapped gas produces clearer ice.

4. 5°C: calcium carbonate, potassium chloride, sodium chloride, sucrose

75°C: calcium carbonate, sodium chloride, potassium chloride, sucrose

Program Resources

Review and Assessment, pp. 29–34 [L1]
Performance Assessment, Ch. 5 [L2]
PASC
Alternate Assessment in the Science Classroom
Computer Test Bank [L1]

4. **Sequencing** Use **Figure 5-10** to list the four substances in order from least soluble in water to most soluble, at 5°C and at 75°C.

Critical Thinking

In your Journal, answer each of the following questions.

1. White light shining through a prism separates into a spectrum of colors. Can a colloid act as a prism? Explain.

2. Oil and vinegar do not mix. Yet, when you pour a small amount of each in a bottle and then shake the bottle vigorously, you can make a mixture of salad dressing that will stay together for some time. Explain why you think this is possible, using the concepts learned in this chapter.

3. A large bottle of fabric softener states it contains enough softener to soften 100 loads of laundry. A different brand in a smaller bottle also states it contains enough to soften 100 loads of laundry. Explain how this can be.

Problem Solving

Read the following paragraph and discuss your answers in a brief paragraph.

On a rafting trip, your friend asks you if you want a cracker. Your mouth is dry, but you eat it anyway. The cracker has absolutely no taste, and you struggle to swallow it. Then you take a long, gulping drink of water. Now you eat another cracker, and it tastes delicious!

At home later that afternoon your mother gives you a couple of crispy cookies for a snack. They taste okay, but with a glass of cold juice, they taste even better.

Oh, no! Broccoli for dinner! You'll eat it because you know it's good for your body, but you also know that if you eat it quickly, you can hardly taste the broccoli. The longer you chew it, the more you can taste it.

Think about the three sets of facts just presented. Try to draw a conclusion from them. Hint: Remember what this chapter is about.

Critical Thinking

1. No; light can travel through a prism. A colloid scatters light but does not allow it to travel through its particles.

2. By shaking the two, you make the particles very small so they form a suspension. After a while, the influence of gravity brings the smaller particles into contact and they separate back into oil and vinegar. This is why salad dressings need to be shaken before being used.

3. The softener in the smaller bottle is probably more concentrated, and less is needed per load.

Problem Solving

The crackers were tasteless until they mixed with water. The same thing happened with the cookies and juice. The broccoli was deliberately kept out of a liquid, saliva, as much as possible so it would not be tasted. An appropriate conclusion is that the flavor-producing chemicals in food can be tasted only when they are dissolved in a liquid.

Connecting Ideas

1. This is a physical change because no chemical reaction has taken place, no new substance has been created, and the solute could be recovered if the solvent were evaporated away.

2. Yes, there could be solid-solid and gas-gas solutions, neither of which involves a liquid. There could also be solid-liquid, liquid-liquid, and gas-liquid solutions that involve some liquid other than water.

3. Salt is corrosive and would destroy a car's radiator.

4. As the Great Salt Lake loses more and more of its water through evaporation, salt is left behind, increasing the concentration in the lake.

CONNECTING IDEAS

Discuss each of the following in a brief paragraph.

1. **Theme—Stability and Change** Powdered drink mix forms a solution in water. Is this a physical change or a chemical change? Explain.

2. **Theme—Interactions and Systems** The moon has no water on it. Could there be solutions on the moon? Explain your answer.

3. **A Closer Look** Salt is cheaper than antifreeze.

Why do you suppose we don't put salt in the car radiator, instead of antifreeze?

4. **Economics Connection** Why is the concentration of salt in the Great Salt Lake so high?

✔ Assessment

Portfolio Review the portfolio options that are provided throughout the chapter. Encourage students to select one product that demonstrates their best work for the chapter. Have students explain what they learned and why they chose this example for placement in their portfolios.

Additional portfolio options can be found in the following **Teacher Classroom Resources:**

Multicultural Connections, pp. 13, 14

Making Connections: Integrating Sciences, p. 13

Making Connections: Across the Curriculum, p. 13

Concept Mapping, p. 13

Critical Thinking/Problem Solving, p. 13

Take Home Activities, p. 12

Laboratory Manual, pp. 25–32

Performance Assessment P

Chapter Organizer

SECTION	OBJECTIVES	ACTIVITIES & FEATURES
Chapter Opener		**Explore!,** p. 185
6-1 Properties and Uses of Acids (3 sessions; 1.5 blocks)	1. **Describe** the properties of acids. 2. **Name and compare** some common acids and their uses. National Content Standards: (5–8) UCP1, A1, F2–3, F5, G1	**Find Out!,** p. 186 **Explore!,** p. 189 **Earth Science Connection,** pp. 188–189 **Science and Society,** pp. 207–208 **SciFacts,** p. 209
6-2 Properties and Uses of Bases (2 sessions; 1.5 blocks)	1. **Describe** the properties of a base. 2. **Name and compare** some common bases and their uses. National Content Standards: (5–8) UCP1, A1–2, G1	**Find Out!,** p. 192 **Explore!,** p. 193 **Design Your Own Investigation 6-1:** pp. 194–195 **Find Out!,** p. 196 **Skillbuilder:** p. 197
6-3 An Acid or a Base? (2 sessions; 1 block)	1. **Analyze** a pH reading and tell what it means. 2. **Explain** what an indicator shows about acids and bases. National Content Standards: (5–8) UCP1, UCP3, A1–2, F1, G1–2	**Skillbuilder:** p. 199 **Investigate 6-2:** pp. 200–201 **Literature Connection,** p. 208
6-4 Salts (1 session; .5 block)	1. **Observe** a neutralization reaction. 2. **Explain** how salts form. National Content Standards: (5–8) UCP1–2, F1–3, F5, G1, G3	**Find Out!,** p. 203 **A Closer Look,** pp. 204–205

ACTIVITY MATERIALS

EXPLORE!

p. 189* any household container with ingredient list

p. 193* any household container with ingredient list

INVESTIGATE!

pp. 200–201* safety goggles; apron; test tubes; test-tube rack; 100-mL graduated cylinder; red cabbage juice indicator; grease pencil; 7 dropping bottles with: household ammonia, colorless carbonated soft drink, baking soda solution, sodium hydroxide solution, hydrochloric acid solution, distilled water, white vinegar

DESIGN YOUR OWN INVESTIGATION

pp. 194–195* lemon juice, household ammonia, vinegar, cola, baking soda, table salt, test tubes, test-tube rack, stir-ring rods, distilled water, red and blue litmus paper, orange slices, deodorant, antacid tablet

FIND OUT!

p. 186* marble chips, 250-mL beaker, dilute solution of sulfuric acid (vinegar or soda water will also suffice), laboratory balance (or other exact scale), stirrer, filter paper, towel

p. 192* baking soda, laundry detergent, cornstarch, sugar, salt, clear plastic cups, paper, magnifying glass, tablespoon, labels, water, stirrer, 100-mL graduated cylinder

p. 196 glass jar, water, Epsom salt, household ammonia, teaspoon

p. 203* ammonia, white vinegar, test tubes, litmus paper, dropper, stirrer, 10-mL graduated cylinder

KEY TO TEACHING STRATEGIES

The following designations will help you decide which activities are appropriate for your students.

- **L1** Basic activities for all students
- **L2** Activities for average to above-average students
- **L3** Challenging activities for above-average students
- **LEP** Limited English Proficiency activities
- **COOP LEARN** Cooperative Learning activities for small group work
- **P** Student products that can be placed into a best-work portfolio
- Activities and resources recommended for block schedules

Need Materials? Call Science Kit (1-800-828-7777).

⏱ **OUT OF TIME?** We recommend that students do the activities with an asterisk.

Chapter 6 Acids, Bases, Salts

TEACHER CLASSROOM RESOURCES

Student Masters	Transparencies
Study Guide, p. 23 **Critical Thinking/Problem Solving,** p. 14 **How It Works,** p. 10 **Making Connections: Technology and Society,** p. 15 **Multicultural Connections,** p. 15 **Making Connections: Integrating Sciences,** p. 15	**Teaching Transparency 12,** Acid Rain **Section Focus Transparency 17**
Study Guide, p. 24 **Activity Masters,** Design Your Own Investigation 6-1, pp. 27–28 **Making Connections: Across the Curriculum,** p. 15 **Laboratory Manual,** pp. 37–40, Acids, Bases, and Indicators	**Section Focus Transparency 18**
Study Guide, p. 25 **Activity Masters,** Investigate 6-2, pp. 29–30 **Science Discovery Activities,** 6-1, 6-2, 6-3	**Teaching Transparency 11,** The pH Scale **Section Focus Transparency 19**
Study Guide, p. 26 **Multicultural Connections,** p. 16 **Concept Mapping,** p. 14 **Laboratory Manual,** pp. 41–42, Neutralization **Take Home Activities,** p. 13	**Section Focus Transparency 20**

ASSESSMENT RESOURCES	TEACHING & TECHNOLOGY
Review and Assessment, pp. 35–40 **Performance Assessment,** Ch. 6 **PASC*** **MindJogger Videoquiz** **Alternate Assessment in the Science Classroom** **Computer Test Bank**	**Spanish Resources** **Cooperative Learning Resource Guide** **Lab and Safety Skills** **Science Interactions, Course 1, CD-ROM** **Computer Competency Activities**

*Performance Assessment in the Science Classroom

NATIONAL GEOGRAPHIC TEACHER'S CORNER

National Geographic Society Products

To order the following products for use with this chapter, call the National Geographic Society at 1-800-368-2728:

- *Acid Rain* (Kids Network Curriculum Unit)

GLENCOE TECHNOLOGY

The following multimedia resources are available from Glencoe:

Science and Technology Videodisc Series (STVS)
Ecology
  Acid Rain and Plants
  Fish and Acid Rain

National Geographic Society Series
STV: Human Body Volume 1
GTV: Planetary Manager

Physical Science CD-ROM

Teacher Classroom Resources

These are key components of the classroom resources package.

TEACHING AIDS

Teaching Transparencies

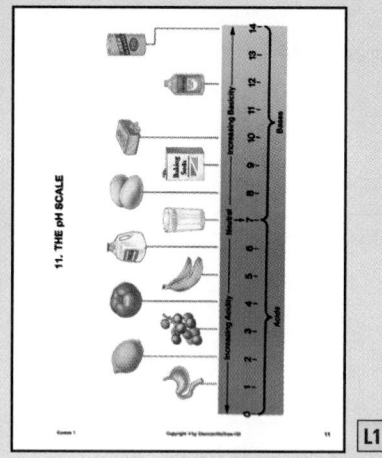

Section Focus Transparencies

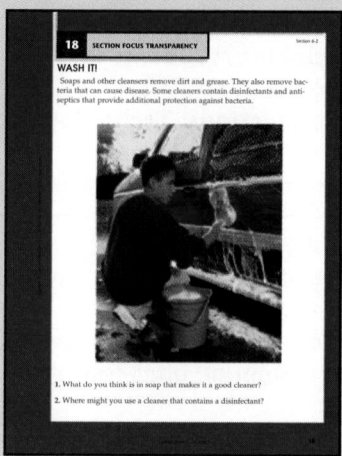

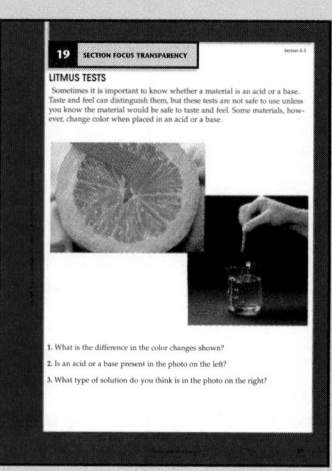

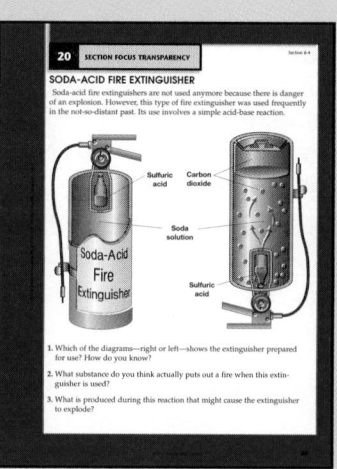

HANDS-ON LEARNING

Science Discovery Activity*

Laboratory Manual*

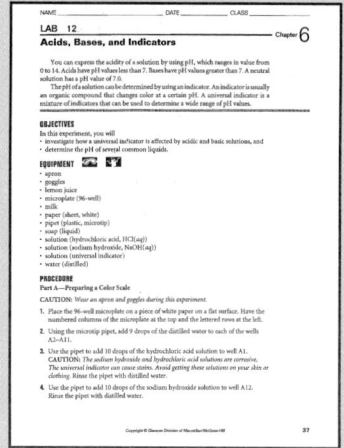

Take Home Activity

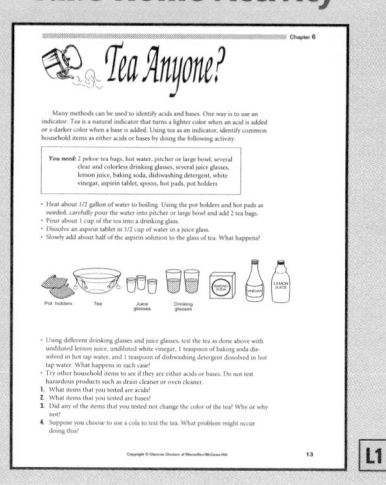

Chapter 6 Acids, Bases, and Salts

REVIEW AND REINFORCEMENT

Study Guide*

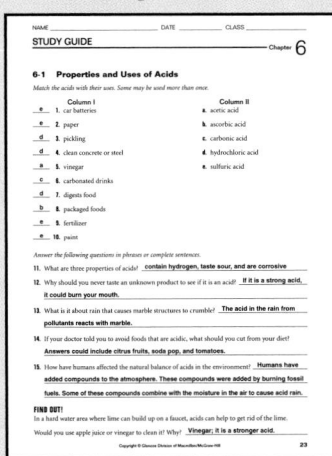

Concept Mapping

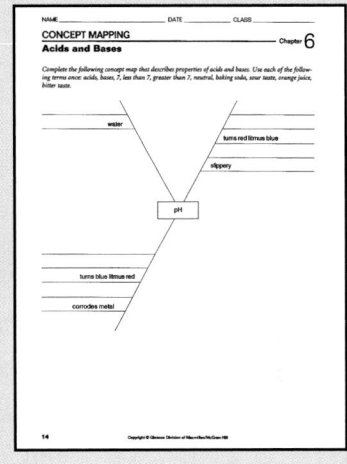

Critical Thinking/ Problem Solving

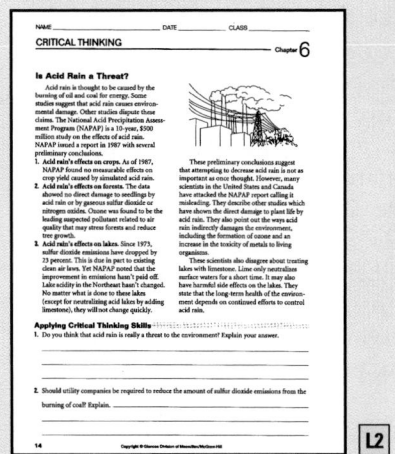

ENRICHMENT AND APPLICATION

Integrating Sciences

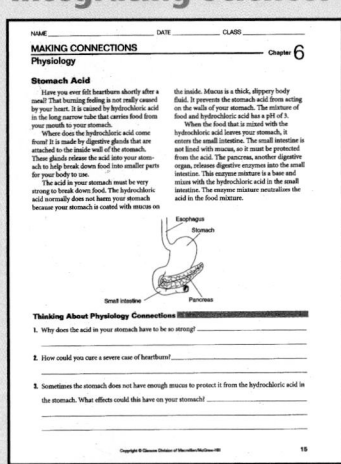

Across the Curriculum

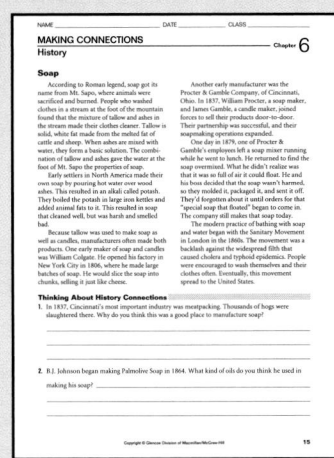

Technology and Society

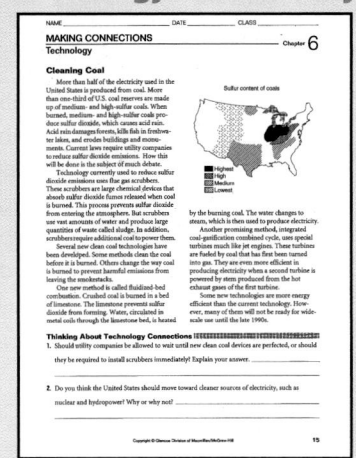

ASSESSMENT

Multicultural Connection**

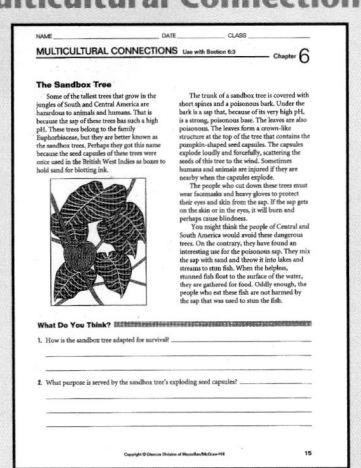

Performance Assessment

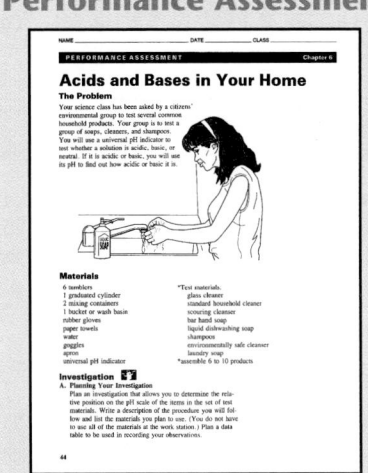

Review and Assessment

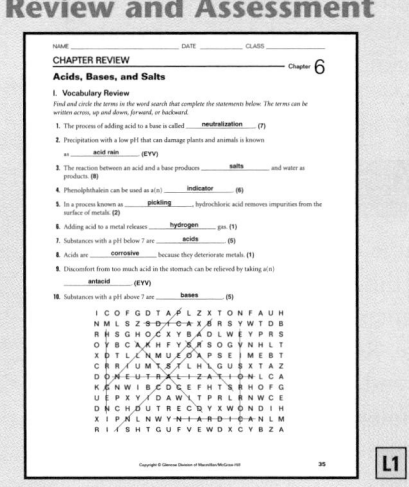

Acids, Bases, and Salts

THEME DEVELOPMENT

The theme that this chapter supports is systems and interactions. pH gives chemists perspective for understanding how acids and bases, some of the most reactive compounds on Earth, interact and change in highly predictable ways. Whether a substance may be an acid or a base is determined by observing its interaction with litmus paper.

CHAPTER OVERVIEW

In this chapter, students will study acids, bases, and salts. Acids and bases neutralize each other, forming a salt and water. Litmus paper can be used to identify acids and bases. The acidity of a substance is measured using pH.

INTRODUCING THE CHAPTER

Help students recognize cause and effect by telling them that the stone structures in the photographs on these pages are hundreds of years old, yet they have deteriorated more in the last half-century than in their entire previous history. Ask students to hypothesize why this is so.

Uncovering Preconceptions

Students may think rainwater qualifies as "acid rain" if it is at all acidic. Actually, natural rainwater is acidic, measuring as low as pH 5.0.

00:00 OUT OF TIME?

If time does not permit teaching the entire chapter, use the Chapter Overview on this page, Reviewing Main Ideas at the end of the chapter, and the Chapter 6 audiocassette to point out the main ideas of the chapter.

Acids, Bases, & Salts

Did you ever wonder...

✓ Why lemons taste sour?

✓ Why a skull-and-crossbones symbol is on a can of drain cleaner?

✓ How an antacid can settle your upset stomach?

 **Science Journal**

Before you begin to study about acids, bases, and salts, think about these questions and answer them *in your Science Journal.* When you finish the chapter, compare your journal write-up with what you have learned.

Answers are on page 210.

You might think that what is made from stone will last forever. If you were to examine the stone used in structures, it would give you clues about what has happened in the environment since the stone was placed there. You might notice that many famous buildings and landmarks are gradually being worn away. How does this happen?

In this chapter, you will find out about the substances—acids, bases, and salts—that control many chemical changes around us. Understanding the properties of these substances can help you understand some problems of our environment and what some possible solutions are.

▶ *In the activity on the next page, explore some structures that may have been damaged by these chemical changes.*

Quebec City, Quebec, Canada

184

Learning Styles	Kinesthetic	Find Out, pp. 186, 192; Activity, pp. 196, 197, 199
	Visual-Spatial	Explore, p. 185; Demonstration, pp. 186, 193, 198, 204; Visual Learning, pp. 187, 192, 196; Activity, pp. 188, 190, 191, 198; Earth Science Connection, pp. 188–189; Investigate, pp. 194–195, 200–201; Find Out, p. 203; Science at Home, p. 210
	Interpersonal	Activity, p. 188
	Logical-Mathematical	Activity, pp. 197, 202; Across the Curriculum, p. 199
LS	**Linguistic**	Multicultural Perspectives, pp. 187, 194; Explore, pp. 189, 193; Activity, pp. 191, 206; Visual Learning, pp. 199, 205; A Closer Look, pp. 204–205

Explore! ACTIVITY

Is stone deteriorating in your neighborhood?

What things in your neighborhood are made of stone? Can you see the effects that certain chemicals in the environment have on stone?

What To Do

On the next nice day, explore the area around your home or school.

1. Examine structures such as buildings, sidewalks, and streets. Examine the gravestones in a cemetery if one is nearby.

2. Look for examples of deteriorating rock. What is evidence of deterioration of rock?

3. *In your Journal*, explain what you think could be causing the deterioration.

Worn stone columns at the Acropolis in Athens, Greece

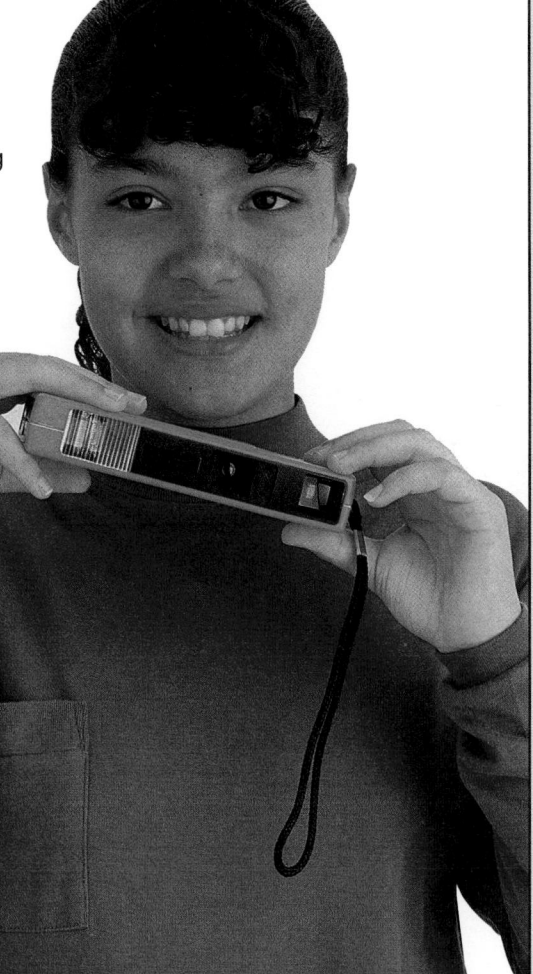

ASSESSMENT PLANNER

PORTFOLIO
Refer to page 212 for suggested items that students might select for their portfolios.

PERFORMANCE ASSESSMENT
Process, pp. 185, 186, 193, 195, 201
Skillbuilder, pp. 197, 199
Explore! Activities pp. 185, 189, 193
Find Out! Activities pp. 186, 192, 196, 203
Investigate! pp. 194–195, 200–201

CONTENT ASSESSMENT
Oral, p. 196
Check Your Understanding, pp. 191, 197, 202, 206
Reviewing Main Ideas, p. 210
Chapter Review, pp. 211–212

GROUP ASSESSMENT
Opportunities for group assessment occur with Cooperative Learning Strategies.

Explore!

Is stone deteriorating in your neighborhood?

Time needed 90 minutes

Thinking Processes observing and inferring, recognizing cause and effect

Purpose To observe the effects of chemicals in the environment on stone.

Preparation Consult a map before the trip begins. A neighborhood guidebook may list structures that are old or noteworthy.

Teaching the Activity

Discussion Have students compare and contrast the condition of stone and concrete on some new structures with that of older structures. **L1**

Troubleshooting If there is not enough time for a field trip, have students analyze pictures or slides of crumbling stone structures.

Science Journal Have students record their observations in their journals. **L1**

Expected Outcome

Students should observe some degree of deterioration caused by chemicals in the environment.

Answer to Question

Possible answers include acid rain and air pollution.

✔ **Assessment**

Process Have small groups of students record their observations and inferences on posters. Posters should include sketches with written annotations describing the stone's appearance and possible causes of deterioration on damaged stone. Use the Performance Task Assessment List for a Poster in **PASC**, p. 73. **L1** **COOP LEARN**

PREPARATION

Planning the Lesson

Refer to the Chapter Organizer on pages 184A–D.

Concepts Developed

In this section students will discover the role acids play in everyday life and in industry. Students also will investigate properties of acids.

1 MOTIVATE

Bellringer

 Before presenting the lesson, display **Section Focus Transparency 17** on an overhead projector. Assign the accompanying **Focus Activity** worksheet. L1
LEP

Demonstration Wearing goggles and a lab apron, put a magnesium strip on the petri dish on the overhead projector. Add a drop of hydrochloric acid. As the class sees hydrogen gas bubbles forming, point out that they are observing a chemical reaction. L1

Find Out!

What effect does a substance known as an acid have on rock?

Time needed 20 minutes

Materials marble chips, 250-mL beaker, 0.1 *M* sulfuric acid, filter paper, laboratory balance, towel, stirrer

Preparation To prepare acid, add 10 mL of concentrated sulfuric acid to 1 L of water.

Thinking Processes thinking critically, observing and inferring, recognizing cause and effect, measuring in SI

Purpose To observe the reactivity of an acid.

6-1 Properties and Uses of Acids

Section Objectives
- Describe the properties of acids.
- Name and compare some common acids and their uses.

Key Terms
acids

Acids in the Environment

In the chapter opener, you saw how the rock of sidewalks, streets, gravestones, and perhaps even structures in your neighborhood can deteriorate. Do this simple activity to discover one reason why rocks wear away.

Find Out! ACTIVITY

What effect does a substance known as an acid have on rock?

The effects of acid on certain types of rock can be easily observed.

What To Do

1. Mass about 5.0 g of marble chips and place them in a 250-mL beaker.
2. Next, add 50 mL of a very dilute solution of sulfuric acid. **CAUTION:** *Sulfuric acid is poisonous and can burn the skin. You may also use vinegar or club soda, but the reaction time will be much slower.*
3. Observe the mixture for several minutes.
4. Stir the mixture vigorously. Then let it sit until it stops bubbling.
5. Pour the mixture through a filter paper placed in a funnel.
6. Rinse and dry the marble chips and mass them again.

Conclude and Apply

1. Did the mass of the rocks change?
2. *In your Journal,* explain your answer to Question 1.

You've just discovered that acid can have a harmful effect on marble. Marble is a rock commonly used in buildings and sculptures. When acids are present in the air, they react with stone and certain other materials and can damage the environment. How do acids get into the air?

186 Chapter 6 Acids, Bases, and Salts

Teaching the Activity

Discussion Have students predict what the acid might do to the chips. Ask students if they have seen crumbling on buildings. Ask them to hypothesize the cause of the damage. L3

Science Journal Students should record their observations in their journals. L1

Expected Outcome

The marble chips lose mass. Students may conclude that the acid damaged the chips.

Conclude and Apply

The acid causes some of the marble to become part of the solution.

✔ Assessment

Process Have students hypothesize the effect of other dilute acids. Then students can repeat the activity with other acids and compare and contrast their results. Use the Performance Task Assessment List for Formulating a Hypothesis in **PASC**, p. 21. L1

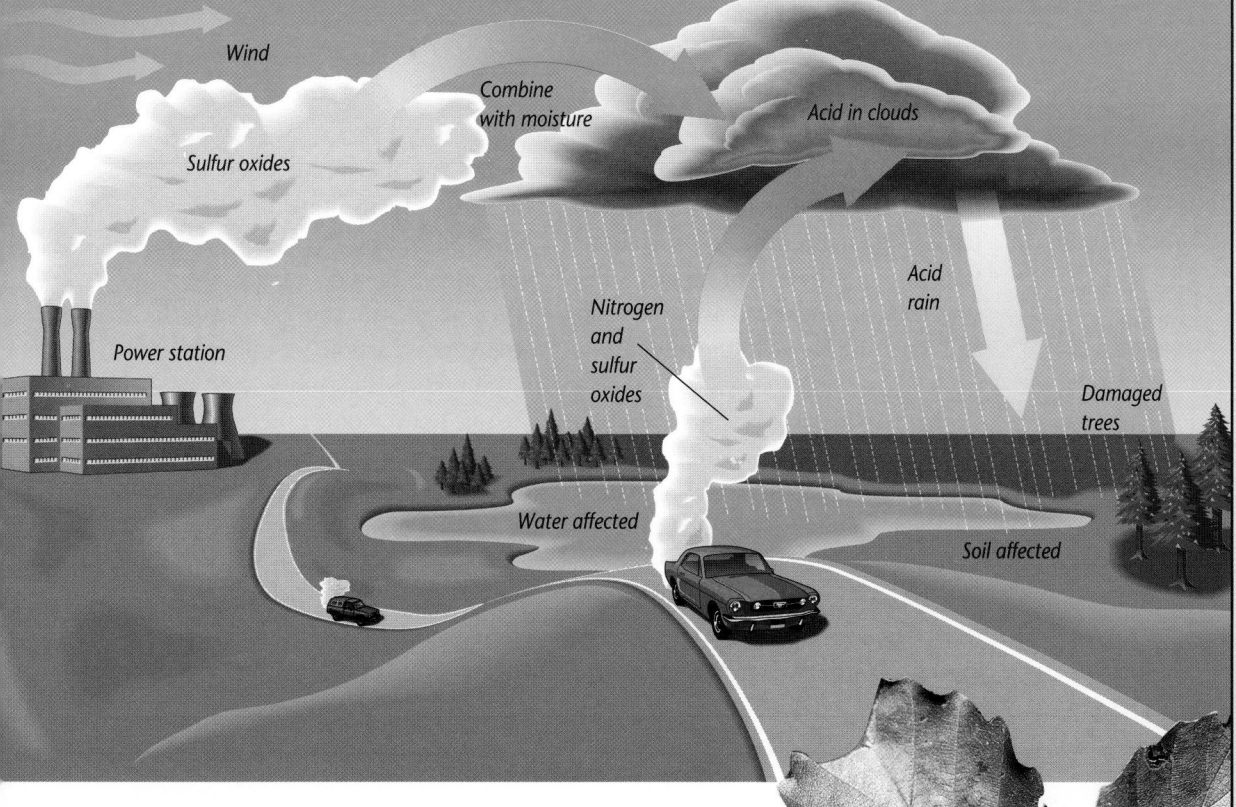

Wind

Combine
with moisture

Sulfur oxides

Acid in clouds

Power station

Acid
rain

Nitrogen
and
sulfur
oxides

Damaged
trees

Water affected

Soil affected

Figure 6-1

A Nitrogen oxides and sulfur oxides released as waste from traffic, factories, and power stations rise into the air. When these compounds combine with water in the air, they dissolve into nitric acid and sulfuric acid. These acids become part of raindrops, which fall to Earth as acid rain.

B Acid rain damages plants, kills fish, and erodes building materials. The destruction is not always local. Clouds of acid-rain drops can be carried by the wind and fall as acid rain hundreds of kilometers from their source.

■ Acid Pollution

Volcanic eruption and other natural phenomena add pollutants to our atmosphere, causing rain to be acidic. However, since humans started affecting the environment, the natural balance of this pollution cycle has been changed by compounds added to the atmosphere by human activities.

Examine **Figure 6-1**. You can see how compounds produced by

the burning of fossil fuels, such as coal or petroleum products, combine with moisture in the air. The result of this is acid rain, which can fall great distances from its source.

Acid rain may make the water in lakes unfit for fish to live. It can harm and kill plants, as can be seen by the leaf in **Figure 6-1**. Even hard building materials, such as marble, can be seriously damaged.

Tying to Previous Knowledge

Discuss the properties of solutions studied in Chapter 5. Point out that acids are special kinds of solutions that are very reactive. Discuss the concepts of *concentrated* and *dilute* as they relate to acids.

Theme Connection Stability and change is an important theme in this chapter. Ask students how acid formed in the atmosphere changes the stability of substances on Earth when it falls as acid rain.

Visual Learning

Figure 6-1 Help students recognize cause and effect as they study the illustration and read the captions. Certain nonmetallic oxides dissolving in water in the air form acids that eventually fall to Earth in rain. Help students compare and contrast this to their observations in the Find Out activity on page 186.

Content Background

Even in the most unpolluted circumstances, natural rainwater is not pure H_2O. As the rain falls through the atmosphere, it dissolves significant quantities of oxygen, nitrogen, and carbon dioxide. Carbon dioxide gives rainwater a slightly acidic character that is not harmful to humans. During electrical storms, lightning can cause oxygen, nitrogen, and water vapor to react to form significant amounts of nitric acid (HNO_3).

Meeting Individual Needs

Visually Impaired Have students develop their observation skills by using their sense of touch to identify the effect of the acid on the marble chips. Students should hold well-rinsed and dry acid-treated chips in one hand, and untreated chips in the other. Have students describe the differences in weight and texture. Then place one acid-treated chip into a group of untreated ones and have students locate it by touch.

Multicultural Perspectives

Fighting Scurvy

For centuries, scurvy was known as the "scourge of the navy," killing many sailors during long voyages. Native North Americans in 1534 successfully treated sick crewmen from a ship captained by Jacques Cartier. Have students research the cause of scurvy and identify the role of acids in its treatments.

GLENCOE TECHNOLOGY

 CD-ROM

Science Interactions, Course 1, CD-ROM

Chapter 6

Activity Have students check with local and state agencies about the composition and pH of soil in the area. They should find out how local growers treat the soil and with what types of chemicals. Where does the runoff from the land go? Sources of information include:

4-H organizations

County Agricultural Extension Offices

State and Federal Departments of Agriculture

Local growers [L3] **COOP LEARN**

Activity A well-known story states that a dentist placed a tooth in a glass of cola and found that its acid ingredients disintegrated the tooth overnight. While this tale may be apocryphal, students can compare and contrast the "disintegration" properties of soft drinks. Have them place various objects (clay, fingernail clippings, hair, plastic, stone, pencil lead, pencil eraser, bone, and so on) in various soft drinks for various lengths of time. The effect on a piece of raw hamburger is dramatic! Have them write a report on their findings.

Students may wish to compare the effects of different colas as well. [L1] **LEP**

Connect to

Earth Science

Locations of limestone caves in your state may be obtained by contacting the state Department of Natural Resources.

Connect to...
Earth Science

Surface water is usually slightly acidic. It will dissolve exposed limestone that is at or near the surface. Make a poster that shows the location of limestone caves in your state.

Properties of Acids

If you had a glass of orange juice this morning, you remember the tart taste in your mouth. A sour or tart taste is one property of an acid. The sour taste of foods such as citrus fruits and tomatoes is due to the presence of weak acids.

Many acids are extremely reactive. Most acids contain the element hydrogen. By reacting with some metals, these acids seem to destroy or corrode the metal and release hydrogen as a gas. Have you seen a car battery that is corroded where cables are attached? Corrosion is a chemical change that may occur when acid and metal come in contact with each other. Some acids react more strongly with metals than do others. You could rank acids on the basis of their strength of reaction.

You know that most acids in food are safe to eat. Some other acids, however, are strong and can damage body tissues. That's why taste should never be used as a way to test for the presence of an acid. Acids are part of your everyday life. Most common **acids** are compounds that contain hydrogen, taste sour, and are corrosive.

Earth Science CONNECTION

Turning Rocks to Soil

Do you remember seeing something gray, green, or brownish covering bare rocks or the limbs of trees? This scaly-looking substance is really a group of organisms called lichens.

What Are Lichens?

Lichens are made up of fungi and algae that have formed a relationship helpful to both of them—the algae provide food through photosynthesis, and the fungi provide protection. There are three basic groups of lichen: crusty or flaky, papery or leafy, and stalked or branching.

Some lichens appear to grow out of solid rock. If you look with a microscope, however, you can see that tiny threadlike growths anchor the lichen to the rock surface.

Rocks into Soil

Rock-growing lichens play an important role in converting rocks into soil. Lichens produce a dilute, acidic solution that slowly dissolves the minerals. Soon cracks appear in the rock and the threads of the lichens dig deeper into the rock. The cracks fill with water that freezes and melts, making the cracks bigger.

After a long time, the rocks break apart into smaller pieces, and eventually become soil.

Purpose

The Earth Science Connection extends Section 6-1 by describing how the acid solution produced by lichens starts the process of breaking down rocks into soil. It also introduces the use of lichens as an acid-base indicator.

Content Background

Lichens reduce rock to soil in several ways. The acids secreted by the lichens may dissolve minerals in the rock. This allows water and then larger plants to enter the rock and weather it. Some lichens have a unique way of weathering their rock homes. The lichens expand and attach to the rock when wet. As they dry, the lichens contract, pulling tiny particles off the host rock. Eventually small amounts of soil are formed on the rock. This soil will then support larger plants whose roots will continue the weathering action.

Acids Around You

The acids in acid rain, in car batteries, and in citrus fruits are just three examples of the acids around you. There are others that affect your life every day. Some are probably quite familiar to you.

Explore! ACTIVITY

What are some common acids?

If you look around your home, you will discover that many familiar items contain acids.

What To Do

1. With an adult's help, check ingredients on labels to see if acids are listed.

2. Make a list of the acids that are in your home and how they are used.

3. Bring your list to school and compare it with your classmates' lists.

4. Make a class list of all the products containing acid that are used in the home.

Once this process begins, plants can also grow in the cracks and speed up the breakup of the rocks.

When the rocks break apart, new soil with new minerals forms and more plants begin to grow. Soil that is worn out from over-use or eroded by wind or water is made healthier with the addition of new soil from broken rocks.

Other Uses

In addition to breaking up rocks, lichen acids are also useful ingredients in perfume making. They are used to make all the different ingredients in perfume mingle together to

make a pleasant smell. Lichen extracts are also used to manufacture antibiotics, medicines, puddings, and fabric stiffeners.

You Try It!

You can use lichens to test whether a solution is acidic or basic.

Gather some lichens from rocks and trees. Then dip one into an acidic solution, such as white vinegar, and another into a basic solution, such as household ammonia. What do you observe?

Teaching Strategy

Have students examine rocks near their homes and school for lichen colonies. Ask them to compare the different forms and shapes. Have students share their findings.

Answers to

You Try It!

The lichen placed in the acid turns red, and the lichen placed in the base turns blue.

Going Further ▸▸▸▸

Lichens break down rock into soil, but this is not the only means of soil formation. Have students research the various natural ways in which soil is formed. Have them discuss these methods and then hypothesize how the methods might be affected by air pollution. Use the Classroom Assessment List for Formulating a Hypothesis in **PASC**, p. 21. L2 **COOP LEARN**

Explore!

What are some common acids?

Time needed 30 minutes

Materials any household container that has an ingredient list

Thinking Processes practicing scientific methods, observing, organizing information, making and using tables, classifying, comparing and contrasting

Purpose To identify common household products that contain acids.

Teaching the Activity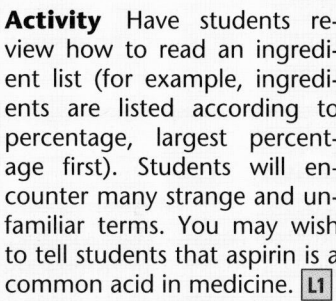

Activity Have students review how to read an ingredient list (for example, ingredients are listed according to percentage, largest percentage first). Students will encounter many strange and unfamiliar terms. You may wish to tell students that aspirin is a common acid in medicine. L1

Activity Students can use reference books to look up the functions of various acids that they find and share them with the class. For example, phosphoric acid adds tartness to soft drinks. L2

Science Journal Encourage students to write about this activity. They might invent categories such as: "most unlikely substance containing acid" or "acid with most difficult name to pronounce." L1

Expected Outcome

Students will identify a variety of different acids found in common products.

✓ Assessment

Performance Have students write newspaper articles about acids—both useful and dangerous—in their lives. Use the Performance Task Assessment List for Newspaper Article in **PASC**, p. 69.

189

Industrial Acids

You may have been surprised to learn how many acids you use every day. You probably found that vinegar contains acetic acid and soft drinks contain carbonic acid and phosphoric acid. Besides aspirin, how many acids did you find that are important to your health? Ascorbic acid is often added to packaged foods to help them stay fresh longer. There are some acids that you may depend on that you never directly use. These industrial acids are used in the manufacture of several important products.

How Do We Know?

Hydrogen in Acids

One of the key properties of acids is that most of them contain hydrogen. But no one has ever seen a single acid particle. How do we know acids contain hydrogen? Thousands of reactions of acids have been studied, and in almost every case hydrogen has either been produced or was a key element in the reaction.

■ Hydrochloric Acid

Because some acids react with certain metals, acids play an important role in industry. One acid that is both commonly and industrially used is hydrochloric acid. Industrial-strength hydrochloric acid is a colorless liquid that gives off strong fumes. The fumes not only can burn your skin but can also harm your lungs and eyes. The fumes react with water on the surfaces of these organs, causing severe burns. Crude hydrochloric acid, or muriatic acid as it is commonly called, is used in industry in a process known as *pickling*. In this process, impurities are removed from metal surfaces by dipping the metals in hydrochloric acid. Muriatic acid is also used to clean concrete and excess mortar from brick.

You might be surprised to learn that hydrochloric acid also helps you digest the hamburger you may have eaten for lunch. Hydrochloric acid in the stomach helps break down food so that it can be further digested. The stomach is protected from this stomach acid by a coating of mucus that keeps the acid away from the stomach lining. Hydrochloric acid can make a hole in metal or a cotton cloth, yet it does not necessarily harm the lining of your stomach.

Figure 6-2

Hydrochloric acid is both a common acid and an industrial acid. Crude hydrochloric acid, or muriatic acid as it is commonly called, is used in industry to clean the surfaces of materials such as concrete or steel.

Figure 6-3

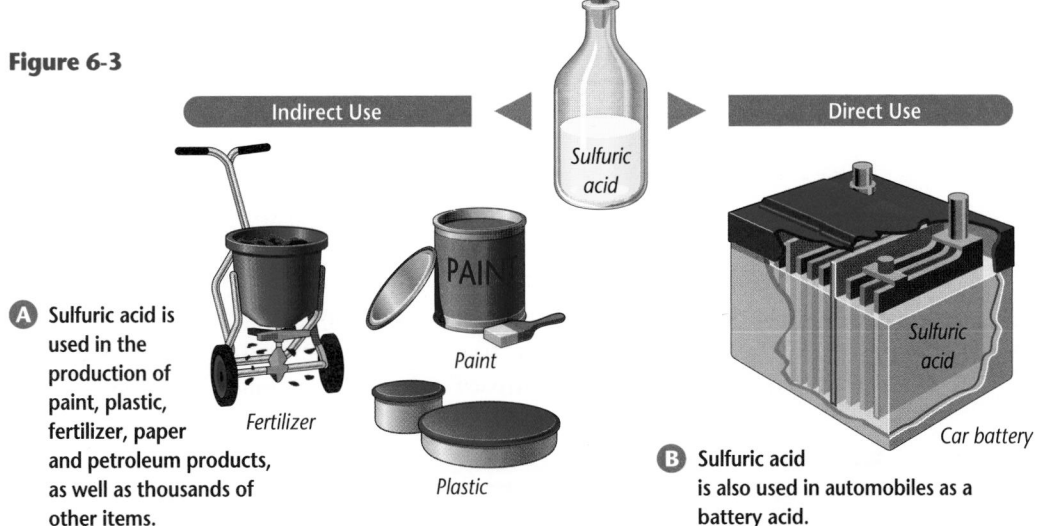

Indirect Use ◀ Sulfuric acid ▶ Direct Use

A Sulfuric acid is used in the production of paint, plastic, fertilizer, paper and petroleum products, as well as thousands of other items.

Fertilizer

Paint

Plastic

B Sulfuric acid is also used in automobiles as a battery acid.

Car battery

Sulfuric acid

■ Sulfuric Acid

You learned about the effect of dilute sulfuric acid on marble at the beginning of this chapter. Sulfuric acid is another industrial acid commonly used in automobile batteries and is often called battery acid. Concentrated sulfuric acid is a thick, syrupy liquid that can also cause severe burns. Sulfuric acid is one of the most widely used chemicals in the world. Examples of its uses are shown in **Figure 6-3**. It is used in the production of metals, fertilizers, plastics, paper, and petroleum products, as well as thousands of other items. The amount of sulfuric acid a country uses is a measure of how economically advanced the country is. Over 30 billion kilograms of sulfuric acid are produced every year in the United States. Half of it is used by industries in this country, and the rest is exported.

You have seen what a useful group of chemicals acids can be. They can help or threaten your health. Weak acids in food keep you healthy, but acid rain threatens the environment. In the next section, you will learn about another group of chemicals that can also affect your environment.

DID YOU KNOW?

In the late 1800s, gold was discovered in western North America. Mistaken for gold in many areas was a mineral called fool's gold. An acid would dissolve most of the fool's gold but leave the real gold unchanged. Thus, *acid test* came to stand for a test that reveals the genuine article.

check your UNDERSTANDING

1. How are all acids alike? How do acids differ from one another?
2. From the list of acids developed by your class, name the two you use most often and tell how you use them.
3. **Apply** You want to store lemon slices in the refrigerator. Will you store them in a plastic bag or in aluminum foil? Explain your answer.

---

check your UNDERSTANDING

1. Acids all taste sour and most contain hydrogen; they differ in reactivity.
2. Sample answer: acetic acid in vinegar and citric acid in juice
3. Plastic bag; the lemon would react with the aluminum foil.

---

3 ASSESS

Check for Understanding

Have students write *true* or *false* for each statement and write a sentence or two to explain their answers.

Acids—
- have a sour taste. (T)
- should never be touched. (F)
- are common in foods. (T)
- have a distinctive smell. (F)
- exist in your body. (T)
- can be strong or weak. (T)
- are used in batteries. (T)
- can cause pollution. (T)
- react only with metals. (F)
- have no practical uses. (F)

Reteach

Discussion Some students may not grasp how one acid can destroy metal while the same amount of another acid can be a food ingredient. Compare acid strength to calories in food. Some acids are weak; some are strong. In the same way, an ounce of some foods has more calories than an ounce of other foods. L1

Extension

Activity Have students who have already mastered the concepts in this section test the corrosiveness of acids such as those in soda pop, dissolved aspirin, and grapefruit juice on such metals as iron, steel, stainless steel, and copper. L2

4 CLOSE

Activity

Have students suppose that they are given a mystery substance to identify. They may ask twenty questions to determine whether the substance is an acid. Have them devise questions to arrive at a definitive answer. Use the Performance Task Assessment List for Asking Questions in **PASC**, p. 19. L1

6-2

PREPARATION

Planning the Lesson

Refer to the Chapter Organizer on pages 184A–D.

Concepts Developed

This section introduces students to bases by exploring their properties and identifying a few common bases. The use of an acid-base indicator is introduced. Industrial applications are pointed out to emphasize the everyday use of bases.

Visual Learning

Figure 6-4 To understand the sequence of events when bases help clean, have students study the drawing as they read the captions. Then call on students' experiences to reinforce the concepts described in the captions.

Find Out!

Which materials are bases?

Time needed 15 minutes

Materials baking soda, laundry detergent, cornstarch, table sugar, table salt, clear plastic cups, tablespoons, magnifying glasses, 100-mL graduated cylinder, water, paper, labels, stirrer

Thinking Processes inferring, comparing and contrasting, forming operational definitions, measuring in SI

Purpose To classify materials as bases, using tactile criteria.

Preparation Use a white powdered laundry detergent.

Teaching the Activity

Troubleshooting Fingers may remain slippery for other tests if hands are not adequately washed between tests.

 6-2

Properties and Uses of Bases

Section Objectives
■ Describe the properties of a base.
■ Name and compare some common bases and their uses.

Key Terms
bases

Properties of Bases

You probably think of the word *base* as part of the game played on a baseball diamond. But bases, like acids, are an important group of chemical compounds that you use every day.

You can find out more about bases by doing the following activity.

Find Out! ACTIVITY

Which materials are bases?

Bases have properties that you can observe.

What To Do

1. Measure 1 tablespoon each of baking soda, laundry detergent, cornstarch, sugar, and salt.

2. Place each material on a piece of paper.

3. Using a magnifying glass, examine each sample and describe it.

4. Rub each material between your fingers. **CAUTION:** *These materials are safe to touch. Do not test any unknown material by touching it.*

5. Label each cup with the name of a material.

6. In each cup, add each material to 300 mL of water and stir.

7. Now, touch each liquid with your fingertips. Rub your fingers together. Be sure to wash your fingers after touching each liquid.

Conclude and Apply

1. Can you identify each material by looking at it?

2. Can you identify, when dry, these materials by the way they feel?

3. Do all the liquids feel the same?

Most undissolved bases are solids. Although there are materials, such as oil, that are not bases but are slippery, when a base is dissolved in water, it feels slippery because it reacts with the oil on your skin. Which of the powders in the Find Out activity may have been bases? Bases also have a bitter taste, but strong bases, just like strong acids, can burn the skin. Never use taste or touch as a way to test for the presence of an *unknown* base.

---

quately washed between tests.

Science Journal Have students record their observations in their journals. **L1**

Expected Outcome

The baking soda and detergent solutions will feel slippery; the others do not.

Conclude and Apply

1. Student answers will vary depending on how familiar students are with the products.

2. No, cornstarch and baking soda will feel similar, as will salt and sugar.

3. The baking soda and detergent feel slippery; the others do not.

✔ Assessment

Performance Have small groups create a poster about common uses of bases in everyday life. Use the Performance Task Assessment List for Poster in **PASC**, p. 73. **L1** **COOP LEARN**

Bases Around You

When you washed your face this morning, you probably used a product that has a base in it. Laundry detergents and shampoos may also contain bases. Bases help clean because of a very interesting property that is explained in **Figure 6-4**. Therefore, the soap can remove the grease or dirt from your skin, clothes, or hair and then be rinsed away by water. **Bases**, then, are compounds that taste bitter, are usually solids, and feel slippery when dissolved in water. Like acids, bases can be weak and not very reactive, or they can be strong and violently reactive.

You have seen that many products in your home contain acids. Explore how bases can also be found in many household products, including soaps, detergents, and shampoos.

Figure 6-4

Bases help clean because of a very interesting property.

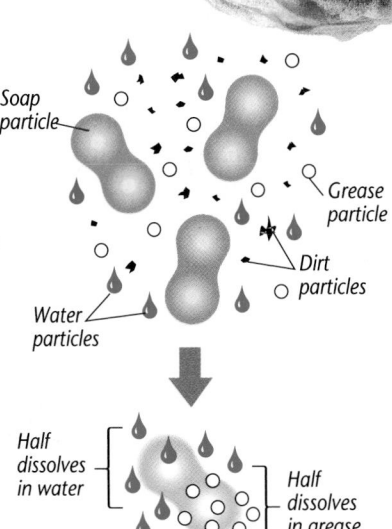

A One end of a soap, detergent, or shampoo particle is soluble in grease, while the other end is soluble in water.

B The soap can pull the grease or dirt away from your skin, clothes, or hair and then be rinsed away by water.

Soap particle
Grease particle
Dirt particles
Water particles
Half dissolves in water
Half dissolves in grease
Dirt is washed away

Explore! ACTIVITY

What bases can you find around you?

To identify a base, one clue to look for is the word hydroxide in the name of any chemical.

What To Do

1. Look at the common bases and the ways they are used that are listed in the table.
2. Many items in your home contain bases. With the help of an adult, find out what these items are and how they are used.

Some Common Bases and Their Uses	
Name	**Where Found**
Aluminum hydroxide	Deodorants, antacids
Ammonium hydroxide	Household cleaner (ammonia water)
Calcium hydroxide	Manufacture of mortar and plaster
Magnesium hydroxide	Laxatives, antacids
Sodium hydroxide	Drain cleaner

Acids and bases have several characteristics, such as reactivity, in common. How can you identify which substances are acids and which substances are bases? This Investigate will show you.

6-2 Properties and Uses of Bases **193**

Explore!

What bases can you find around you?

Time needed 30 minutes

Materials any household container that has an ingredient list

Thinking Processes observing, interpreting data, making and using tables

Purpose To classify common household items as bases.

Teaching the Activity

Troubleshooting Students are more likely to find bases in products that are kept in the garage, under the sink, and in the bathroom; stay away from most edible products unless they are baked goods. **L2**

Science Journal Just as students wrote about common acids in their journals, have them enter the knowledge gained about common bases in their journals as well. **L1**

6-1 Identifying Acids and Bases

Preparation

Purpose To classify materials as acids or bases by using litmus paper.

Process Skills observing and inferring, classifying, interpreting data, practicing scientific methods, predicting, making and using tables

Time Required one class period

Materials Having the materials already in prelabeled test tubes would minimize spillage and help with time.

Preparation For a 5% baking soda solution, dissolve 50 g baking soda in 950 mL of water.

Safety Precautions Make sure that the students avoid getting the materials on their skin. If they do, they should rinse the affected area in cold water.

Possible Hypotheses Students may hypothesize that lemon juice, vinegar, cola, and orange slices will turn the paper red, but ammonia, baking soda, and deodorant will turn it blue. They may hypothesize that table salt will have no effect.

The Experiment

Process Reinforcement Be sure students set up their data tables correctly. There should be space to check each item as turning litmus paper blue or red or no color change.

Possible Procedures Obtain six strips of each color of litmus paper. Dissolve in water each solid to be tested. Test each liquid using red and blue litmus paper, and record results in the table.

Identifying Acids and Bases

Acids and bases are used for different purposes. Their uses often correspond to the way they react to various substances. Litmus paper, which is red in the presence of an acid and blue in the presence of a base, can be used to identify acids and bases.

Preparation

Problem
How can acids and bases be identified?

Form a Hypothesis
As a group, form a hypothesis that will help you determine which substances are acids and which are bases.

Objective
• Identify acids and bases based on their reactions with litmus paper.

Possible Materials
test-tube rack
six test tubes
household ammonia
cola
table salt
lemon juice
vinegar
baking soda
orange slices
deodorant
piece of antacid tablet
red and blue litmus paper
stirring rods
distilled water

Safety Precautions

Goggles and apron should be worn at all times when using these weak acids and bases. Do not allow the substances to contact your skin.

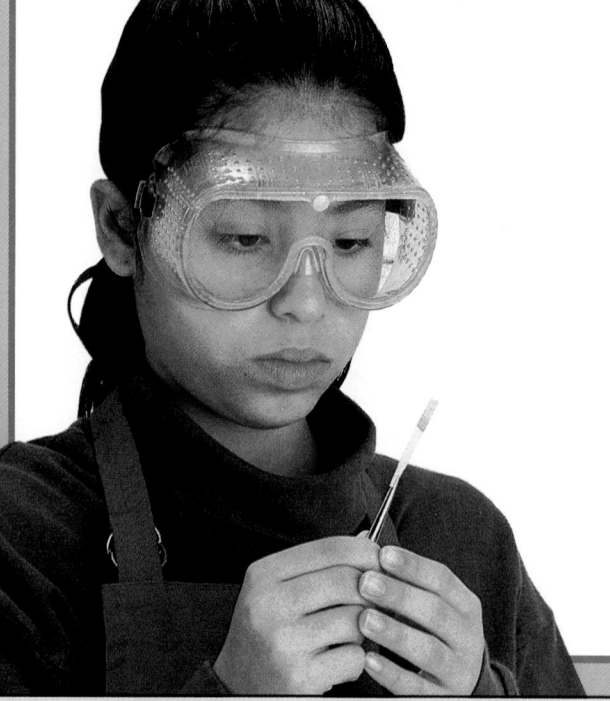

Multicultural Perspectives

Concrete
Concrete is a substance that changed the world. With it, civilizations built bridges, viaducts, roads, and buildings that have stood for thousands of years. The Egyptians were the inventors of this base-containing substance. The formula included calcium carbonate (from limestone), clay, sand, and gravel. Later on, the Romans improved on the original Egyptian formula by adding volcanic ash to the mix.

Have students research and then prepare a poster comparing and contrasting the composition of concrete used today and that of the Egyptians. In addition, have them research the effect concrete had on Egyptian life and how it was first used. L2

DESIGN YOUR OWN
INVESTIGATION

Plan the Experiment

1 Within your group, choose six of the available substances to test.

2 What procedure will you follow to test each substance? Will you use both kinds of litmus paper? Is it important to use clean stirring rods?

3 Water is a neutral liquid; it is neither acidic nor basic. How will you use it as you test the solids?

4 How will you keep track of your findings? In your Science Journal, design a table or chart to use during your experiment.

Check the Plan

1 Have another group read your plan after they have finished writing their own. Do they understand what you plan to do?

2 Before you begin, have your teacher approve your plan.

3 Carry out the experiment.

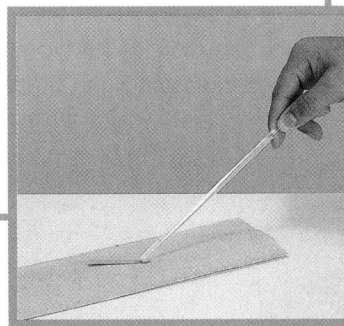

Analyze and Conclude

1. Observe What changes did you observe for acids? For bases?

2. Infer Which substances did you infer were acids? Bases?

3. Analyze How effective was your procedure? Are there things you would change if you were to do it again?

4. Conclude List any substances that showed no change with the litmus test. What can you determine about them?

Going Further

Predict how the bases you identified in the Explore activity on page 193 would affect litmus paper.

Meeting Individual Needs

Learning Disabled Students might benefit from a mnemonic device to help them remember some of the properties of bases. Have students generate a mnemonic of their own.
Bitter **A**nd **S**lippery **I**n **C**leaners
Have them also generate a mnemonic for litmus tests, such as aci**D**
re**D**

Base
Blue

Program Resources

Study Guide, p. 24
Laboratory Manual, pp. 37–40, Acids, Bases, and Indicators **L3**
Activity Masters, pp. 27–28, Design Your Own Investigation 6-1
Making Connections: Across the Curriculum, p. 15, Soap **L2**
Section Focus Transparency 18

Teaching Strategies

Troubleshooting Point out that students should check "no effect" only in the event that neither color litmus paper changes its color. All acids and bases will fail to change at least one color of litmus paper.

Discussion Help students classify these materials by having students think of each as having three possible hidden identities: acid, base, or neutral. The litmus paper allows each true identity to be known.

Expected Outcome

Acids will turn blue litmus paper red and leave red litmus paper unchanged. Bases turn red litmus paper blue and leave blue litmus paper unchanged. Neutral substances leave both papers unchanged.

Analyze and Conclude

1. Acids turned blue litmus red. Bases turned red litmus paper blue.

2. Lemon juice, vinegar, cola, and the orange slice were acids. Ammonia, baking soda, the antacid tablet and the deodorant were bases.

3. Answers will vary.

4. Salt and distilled water showed no color change. Both are neutral—neither acidic or basic.

✔ Assessment

Process Have students working in small groups design an experiment to determine the acidity of ponds, lakes, streams, rivers, and puddles in their area. Use the Performance Task Assessment List for Designing an Experiment in **PASC**, p. 23. **L1**
COOP LEARN

Going Further ⟫⟫⟫

Acids would turn blue litmus red. Bases would turn red litmus blue.

Bases and Your Health

You have learned that several acids, such as ascorbic acid, vitamin C, are important to your health. As with acids, bases are important to your health and well-being. Blood and many other body fluids are mildly basic; that is, they contain a base. Your body would not function properly without the correct balance of acids and bases. For example, antacid tablets, which are a mild base, will reduce excess stomach acid and also help maintain the acid-base balance. The Find Out activity that follows shows you how to make a common antacid.

Find Out! ACTIVITY

How can you make an antacid with a base?

Household ammonia contains ammonium hydroxide. The chemical name for Epsom salt is magnesium sulfate. When these two compounds react, they produce magnesium hydroxide. Magnesium hydroxide is a base found in a product called milk of magnesia.

What To Do

1. Fill a glass jar half full of water.
2. Stir in 1 teaspoon of Epsom salt into the water.
3. Pour 2 teaspoons of household ammonia into the jar. Do not stir! Let this solution stand for five minutes. **CAUTION:** *DO NOT taste this solution. DO NOT let anyone else taste this solution.*

Conclude and Apply

1. What do you observe happening?
2. What do you think the white milky substance is?
3. Why do you think it is called *milk* of magnesia?

You have read how a base can react with an acid in your stomach and how other bases are needed for your health. Bases are also useful in industry to make many products that we use daily.

Industrial Bases

The most widely used base is ammonia gas dissolved in water. Pure ammonia gas has a distinctive and very irritating odor. You may know this base as household ammonia, a cleaner. It is especially good for cleaning windows. It is also used to manufacture fertilizers, medicines, plastics, refrigerants, and dyes.

Calcium hydroxide, commonly called lime, is often used on lawns and gardens where the soil is too acidic. One effect of the amount of acid in soil is shown in **Figure 6-5**. Different plants need different amounts of acid in the soil. Calcium hydroxide reduces the acidity of the soil.

Sodium hydroxide, or lye, is a very strong base. Because a great amount of thermal energy is released when sodium hydroxide dissolves in water, it is used as a drain cleaner and an oven cleaner. Sodium hydroxide is very dangerous. Anyone using it must wear gloves and eye protection and must avoid inhaling the fumes.

The next time you use any of the products mentioned in this section, stop and think about what properties the product has and why it has them.

SKILLBUILDER

Comparing and Contrasting

List ways that acids and bases are similar. Then list the ways in which they are different. If you need help, refer to the **Skill Handbook** on page 659.

Figure 6-5

Hydrangeas produce blue flowers in acidic soil and pink flowers in basic soil. How are hydrangeas like the litmus paper you used in the Investigation?

check your UNDERSTANDING

1. Dishwater feels slippery. What can you infer about the detergent used to wash dishes?
2. How is the structure of a base related to its cleaning properties?
3. **Apply** State three different uses for common bases and identify the specific base used.

check your UNDERSTANDING

1. It must be a base.
2. For some bases, one end of the molecule is soluble in oils and the other end is soluble in water.
3. Possible answers include household cleaning (ammonia water), antacids (magnesium hydroxide), and drain cleaner (sodium hydroxide).

PREPARATION

6-3

6-3 # An Acid or a Base?

PREPARATION

Planning the Lesson

Refer to the Chapter Organizer on pages 184A–D.

Concepts Developed

This section develops the idea of acids and bases as part of a continuum that runs from strongly acidic to strongly basic. pH, a numeric way of representing how acidic or basic a substance is, is introduced. Indicators can be used to estimate pH.

1 MOTIVATE

Bellringer

 Before presenting the lesson, display **Section Focus Transparency 19** on an overhead projector. Assign the accompanying **Focus Activity** worksheet. L1

LEP

Demonstration Wearing goggles, put several drops of phenolphthalein in 100 mL of dilute acid. Stir thoroughly. Then begin adding base dropwise to the solution. Suddenly, when the solution reaches the basic pH range, the presence of phenolphthalein will make it turn bright pink, indicating a base is present. Explain that the phenolphthalein color change works much in the same way that litmus paper works. L1

2 TEACH

Tying to Previous Knowledge

Review the properties of acids and bases. Ask students if they think a scale that measures how acidic or basic a solution is would be useful. Then tell them they will learn about such a scale in this section.

Section Objectives
- Analyze a pH reading and tell what it means.
- Explain what an indicator shows about acids and bases.

Key Terms
pH
indicator

The Acid-Base Balance

You have learned that acids and bases are important in keeping your body functioning properly.

■ In Your Body

You could never digest your food without the hydrochloric acid in your stomach. During digestion, the food moves from the stomach to the small intestine. Bile, which is made by the liver to help digest food, is then added to the acidic food mixture, adjusting the acidity of the mixture. If the mixture is too acidic, disorders such as ulcers can result.

On the other hand, blood is basic (that is, it contains a base), and in order for food nutrients to be safely absorbed by the blood, they, too, must be basic. If the acid-base balance in your body becomes unbalanced, you could become seriously ill.

Figure 6-6

The pH of a solution indicates how acidic it is. The pH scale is a series of numbers used to measure pH. The scale starts with 0 to indicate the most acidic solution and ends with 14, which indicates the most basic solution. The number 7 on the scale identifies neutral solutions that are neither acidic nor basic.

| Strong acid pH 0 | Stomach contents pH 1 | Lemon pH 2 | Soft drink pH 3 | Tomato pH 4 | Coffee pH 5 | Milk pH 6 |

0 1 2 3 4 5 6

More acidic

Program Resources

Study Guide, p. 25
Teaching Transparency 11, The pH Scale L1
Section Focus Transparency 19

ENRICHMENT

Activity Have students fill several small containers with pond water. Add a different dose of ammonia to each container but leave one as a control. Repeat with dilute hydrochloric acid and another set of containers. After several weeks, have students record the growth they observe inside each container. Have them compare algae growth in the set of acidic containers and in the set of basic containers. L3

■ The pH Scale

It is important to maintain the balance between acids and bases in swimming pools and in tanks for tropical fish. You need to control the acidity of the water. To control the acidity, you need to adjust the pH by adding either acids or bases.

What is pH? **pH** is a measure that shows the acidity of a solution. The pH scale shown in **Figure 6-6** is used to measure pH. Its values range from 0 to 14.

Looking at **Figure 6-6**, you can see that solutions with a pH value less than 7 are acidic. The lower the value, the more acidic the solution. This means that a solution with a pH value of 1 is very acidic. Solutions with a pH value greater than 7 are basic. The higher the pH number, the more basic the solution. The number 7 on the pH scale represents a neutral solution that is neither acidic nor basic. Pure water has a pH value of 7.

Any material that can be put into solution can be tested to find its pH. You can see that a variety of common items, from foods to batteries, are acidic, and a variety of items, from cleaners to baking soda, are basic.

How can we find the pH of an acid or a base? The following Investigate will show you one way.

Making and Using Graphs

Find the pH of the following materials: rainwater, apples, club soda, seawater, drain cleaner, distilled water, and milk of magnesia. Now use a bar graph to plot the pH of the materials against the pH scale in **Figure 6-6**. If you need help, refer to the **Skill Handbook** on page 657.

Visual Learning

Figure 6-6 Have students give comparisons such as, *Milk is more acidic than eggs but more basic than tomatoes.*

Inquiry Question If 100 mL of solution that has a pH of 5 is mixed with 100 mL of solution that has a pH of 3, will the resulting solution have a pH of 8? If not, estimate what the pH of the combined solution will be. *A pH of 4 would be a good estimate.*

GLENCOE TECHNOLOGY

Software

Computer Competency Activities

Chapter 6

Across the Curriculum

Math

A solution of pH 1 has *10 times* more hydrogen ions than does an equal volume of a solution of pH 2, 100 times more hydrogen ions than a solution of pH 3, and so on. Have students complete this table. **L2**

pH	H⁺ ions	pH	H⁺ ions
6	50	3	50 000
5	500	2	500 000
4	5000	1	5 000 000

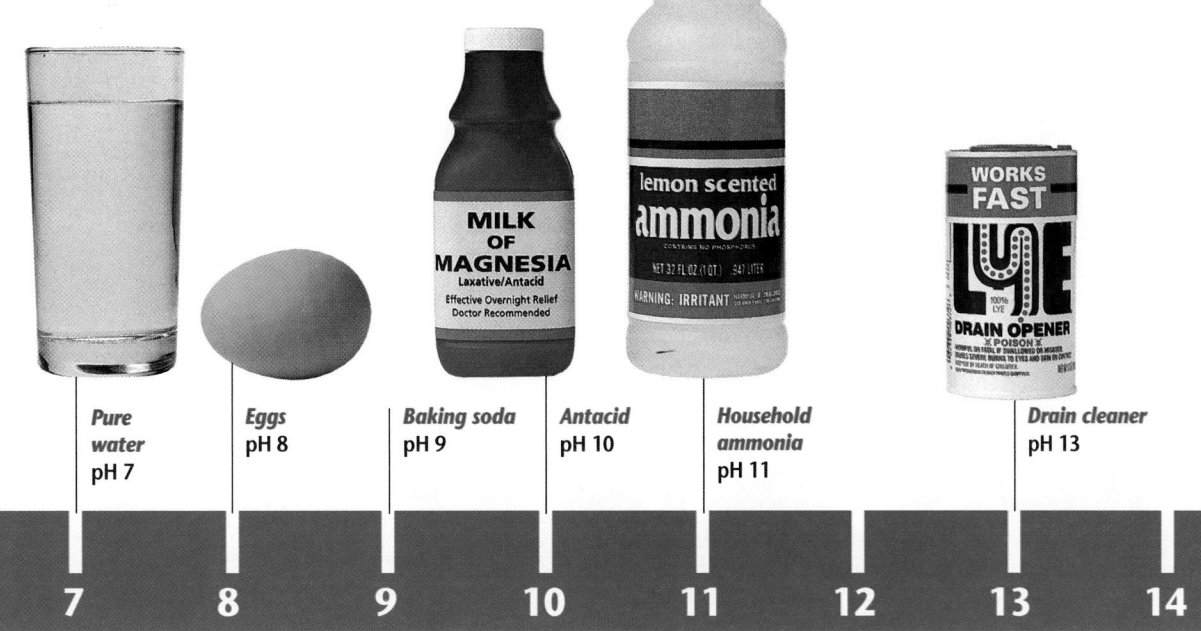

Pure water pH 7 | *Eggs* pH 8 | *Baking soda* pH 9 | *Antacid* pH 10 | *Household ammonia* pH 11 | *Drain cleaner* pH 13

7 8 9 10 11 12 13 14

Neutral   **More basic**

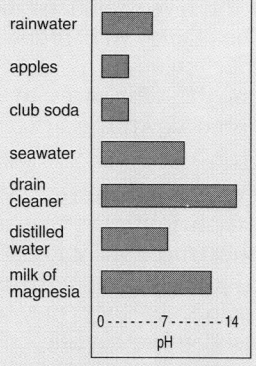

rainwater
apples
club soda
seawater
drain cleaner
distilled water
milk of magnesia

0 ---- 7 ---- 14
pH

ENRICHMENT

Activity Explain that human taste buds can detect four different tastes—sour, bitter, salty, and sweet. Acids and bases account for two of the four tastes.

Prepare samples such as honey, vinegar, salt solution, lemon juice, unsweetened chocolate, and so on. Blindfold students and present them with the samples. Have them rate each sample as "sour," "bitter," "salty," or "sweet." On the basis of taste, have them predict which foods might contain acids and which ones might contain bases. When they take off the blindfolds, have them check to see how accurate their ratings were.

Note: Before starting, check whether students have food allergies. Also point out that these tests are only safe because common foodstuffs selected by responsible adults are being used. **L2**

6-2 Finding pH

Planning the Activity

Time needed 30 minutes

Purpose To classify substances on the basis of their relative acid and base strengths by using cabbage juice as an indicator.

Process Skills observing and inferring, comparing and contrasting, measuring in SI, making and using tables, interpreting data, predicting, classifying

Materials See reduced student text.

Preparation To make the indicator, cut red cabbage into quarters and grate into a bowl. Add one to two cups of water and let stand. When the water turns a deep red remove the cabbage and pour the solution into a glass jar through a strainer. For a 5% HCl solution, dissolve 50 g of concentrated hydrochloric acid in 950 mL of water. For a 2% sodium hydroxide solution, dissolve 20 g of sodium hydroxide in 980 mL of water.

Teaching the Activity

Safety Cabbage juice can stain clothing. If any juice splatters on students' clothing, have them use cold water to rinse the stain immediately. You might also wish to keep a commercial stain stick handy and rub the stick on the stain after it has been rinsed. This will make it more likely that the stain will be removed when the clothing is laundered.

Process Reinforcement Students should create data tables that will allow them to record a predicted pH, the color of the indicator, and the relative pH of each of the seven solutions they test. The table will probably have four columns, with the names of the substances to be tested listed down the first column. L1

Possible Hypotheses Students may hypothesize that different substances will cause the cabbage juice to change to different colors, or they may go further and make specific color predictions for each of the different substances.

Troubleshooting With litmus paper, precise shade of color was not important. Here, shade is important. Have students familiarize themselves with the various colors (see table) beforehand so they have a good idea of what to look for.

Possible Procedures Students may record their predictions at any point in the procedure, as long as they do so before adding the substances to be tested to the cabbage juice.

Discussion Help students recognize the role of

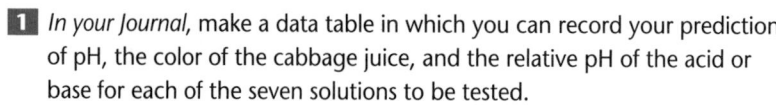

Finding pH

Certain substances change color when pH changes. A common substance of this type found in nature is red cabbage juice. In this Investigate, you will test the pH of household liquids using red cabbage juice.

Problem

How can cabbage juice indicate the relative pH of acids and bases?

Materials

safety goggles	apron
7 test tubes	test-tube rack
100-mL graduated cylinder	red cabbage juice grease pencil

7 dropping bottles with:

household ammonia	colorless carbonated soft drink
baking soda solution	sodium hydroxide solution
hydrochloric acid solution	distilled water
white vinegar	

Safety Precautions

Use caution when working with acids and bases. Wear lab aprons and goggles.

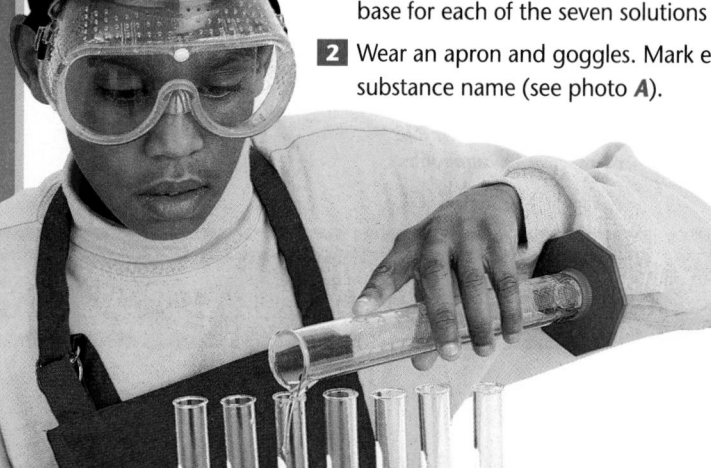

What To Do

1 *In your Journal*, make a data table in which you can record your prediction of pH, the color of the cabbage juice, and the relative pH of the acid or base for each of the seven solutions to be tested.

2 Wear an apron and goggles. Mark each test tube with the substance name (see photo **A**).

3 Fill each test tube with 15 mL of the cabbage juice.

4 Use the following table of colors to predict the relative pH of the test solutions. Record these predictions in your table.

---

constants and variables in an experiment by stressing that the colors they see result from the interaction of the red cabbage juice and the substances tested. Ask the following questions. What remains constant in this experiment? What is the variable? *The constants are the 15 mL of cabbage juice per test tube and the five-drop measure of the different substances. The variable is the different substances.* Ask if students used another indicator, what would remain constant? *The relative acidity of the substances.*

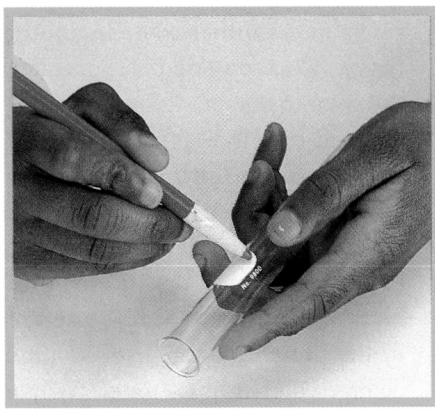

A

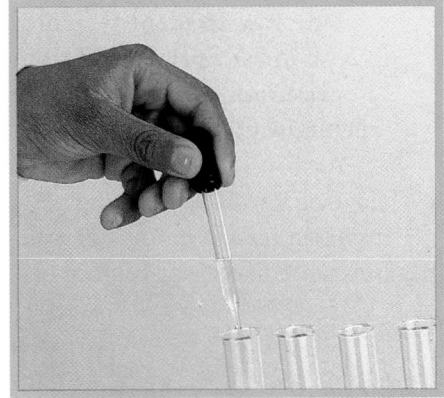

B

5 Add 5 drops of each test solution to the test tube labeled with its name (see photo **B**). **CAUTION:** *If you spill any liquids on your skin, rinse the area immediately with water. Alert your teacher if any liquid is spilled in the work area.*

6 *Observe* any color changes of the cabbage juice. In your data table, record the color and relative pH of each solution.

Cabbage Juice Color	Relative pH
bright red	strong acid
red	medium acid
reddish purple	weak acid
purple	neutral
blue green	weak base
green	medium base
yellow	strong base

Analyzing

1. *Classify* which test solutions were acids and which were bases.
2. Which base was weakest?
3. Which acid was strongest?
4. *Infer* why distilled water didn't change the color of the cabbage juice.

Concluding and Applying

5. How do your predictions *compare* with the results?
6. How does cabbage juice indicate the relative strength of acids and bases?
7. Going Further *Predict* how other substances at home would react with the cabbage juice. Ask an adult to help you test your predictions.

Program Resources

Activity Masters, pp. 29–30, Investigate 6-2
Science Discovery Activities, 6-1, Secret Message; 6-2, Healthy Hair; 6-3, Investigating Indicators

Meeting Individual Needs

Physically Challenged Students can be helped to observe more accurately if they preview the activity so they know what hues to expect from the indicator. When the class does the experiment, they can brief lab partners on what colors to expect from various pH's and how vivid those colors should appear. Have students use crayons or paints to prepare a chart to compare with the test tubes.
COOP LEARN

Science Journal Have students record their data predictions on their table and answer the questions in their journals. **L1**

Expected Outcome

The following substances will turn the indicator the colors indicated: distilled water, purple; white vinegar, red; ammonia, green; soft drink, red; baking soda, blue-green; sodium hydroxide, yellow; hydrochloric acid, bright red.

Answers to Analyzing/ Concluding and Applying

1. acids: soft drink, hydrochloric acid, vinegar; bases: ammonia, baking soda, sodium hydroxide
2. baking soda
3. hydrochloric acid
4. Water is neutral.
5. Answers will vary.
6. The more yellow the cabbage juice turns, the more basic the substance being tested is; the brighter red the cabbage juice turns, the more acidic the substance being tested is.
7. Answers will vary.

✔ Assessment

Process Have students work in small groups to create a color-coded poster of a pH scale using the colors they observed in the cabbage juice tests. Students can include their own drawings or pictures from magazines as examples of substances with various pH ratings. Use the Performance Task Assessment List for Poster in **PASC**, p. 73. **L1**
COOP LEARN

GLENCOE TECHNOLOGY

 Videodisc

Science Interactions, Course 1 Integrated Science Videodisc

Lesson 2

Check for Understanding

As they answer the Apply question, have students think not in terms of which substances are poison, but rather, which substances are harmful if handled or swallowed.

Reteach

Activity Give students strips of universal indicator paper that have been dipped in various acidic or basic solutions. Have them sequence the strips from strongly acidic to strongly basic. L1

Extension

Activity Materials needed are lemon juice, beakers, and an acid-base indicator.

Students can test the pH of different dilutions of an acidic solution such as lemon juice. With an indicator, they find the pH of the undiluted juice. Then they dilute the juice by mixing 1 part lemon juice and 1 part water, 2 parts water, 3 parts water, and so on. Ask what ratio of dilution will make the solution neutral. Basic? *A very high dilution will make it approach, but not reach, neutral; it is impossible to make it basic using water.* L3

4 CLOSE

Activity

Have students collect water from various sources in the immediate vicinity such as rainwater, different tap waters, river and seawater, pond water, bottled water, and puddle water. Have them use an indicator to test and compare the pH of each sample. L1

Indicators and pH

One way to identify acids and bases is to use a pH meter. This meter uses electrical measurements to read out precise pH values.

As you saw in the Investigate, another way of determining pH is by using an indicator. An **indicator** is a substance that is one color in one pH range and another color in another range. To determine pH, you match the color to that of a known color.

Acid-base indicators come in two varieties—a solution or indicator paper. You already know about one of the most common indicators—litmus paper. Remember that litmus turns blue in the presence of a base and red in the presence of an acid. Would it change color in pure water? Why or why not?

Another common acid-base indicator is called phenolphthalein. Phenolphthalein is colorless in an acid, but it turns bright pink in the presence of a base. Phenolphthalein is also used to treat paper strips.

Another indicator that involves several color changes is universal indicator, shown in **Figure 6-7**.

For living things to grow and be healthy, there must be a balance in acids and bases. In the next section, you'll discover a way in which acid rain damage might be temporarily repaired.

Figure 6-7

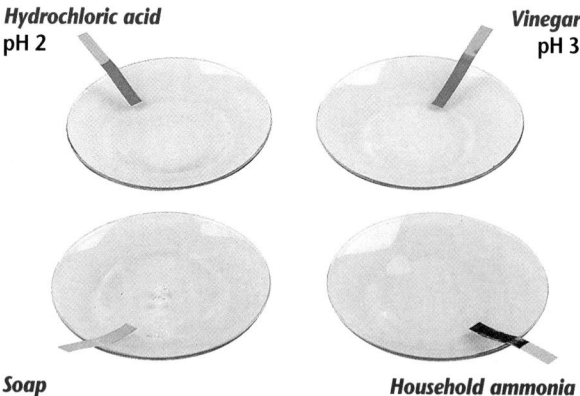

Hydrochloric acid
pH 2

Vinegar
pH 3

Soap
pH 8

Household ammonia
pH 11

A The acidity of a solution can affect the color of certain dyes found in nature. Various dyes react at different pHs. A combination of indicators is used to make a universal indicator paper.

B Starting with pink for the strongest acids, the paper goes through various color stages. Green indicates pH 7 and dark purple indicates the strongest bases.

check your UNDERSTANDING

1. Describe two common acid-base indicators. Which colors will you observe in an acidic solution? In a basic solution?

 2. Arrange the following list of solutions in order from most acidic to most basic.
rainwater, pH 5.8
club soda, pH 3.0
seawater, pH 8.0
drain cleaner, pH 13.0
distilled water, pH 7.0

3. **Apply** Which of the substances mentioned in Question 2 should have a poison symbol on its label? Explain why.

check your UNDERSTANDING

1. Litmus: An acid will turn blue litmus red, and a base will turn red litmus blue. Phenolphthalein: Colorless in the presence of an acid and pink in the presence of a base.

2. Club soda, rainwater, distilled water, seawater, drain cleaner

3. Drain cleaner; it is such a strong base that it will burn your stomach if swallowed.

6-4 Salts

Neutralization

You've probably heard commercials on television that say, "Are you bothered by an upset stomach? Try our product to neutralize excess stomach acid. You'll feel better fast!"

Would the pH of such a product be less than or greater than 7? For you to feel better, the pH would have to be greater than 7 because only a base can neutralize an acid.

Neutralization is the chemical reaction that occurs between an acid and a base. During neutralization, acidic and basic properties are canceled, or neutralized. For example, you've read how lime is used to neutralize acidic soil and how you can use an antacid to neutralize excess acid in your stomach.

■ Products of Neutralization

A **salt**, another type of compound, is formed as part of neutralization. Water is also formed. This reaction can be written as follows:

acid plus base produces a salt plus water

The same reaction can also be expressed using the names of specific chemicals:

hydrochloric acid plus sodium hydroxide produces sodium chloride plus water

To see how neutralization occurs, do the following Find Out activity.

Section Objectives
■ Observe a neutralization reaction.
■ Explain how salts form.

Key Terms
neutralization
salt

Find Out! ACTIVITY

How does neutralization take place?

What To Do

1. Put 10 mL of household ammonia in one test tube and 10 mL of white vinegar in another.

2. Test each with litmus paper. What do you observe? Which liquid is the acid? Which liquid is the base?

3. Add ammonia, drop by drop, to the test tube of vinegar, stirring with a stirring rod after each drop.

4. After adding 5 drops, test your solution with litmus paper.

5. Then, test with litmus for each additional drop.

Conclude and Apply

1. *In your Journal*, describe what change you observe in the litmus.

2. Record how many drops of ammonia it took to get a neutral solution.

3. Explain how you know the solution is neutral.

6-4

PREPARATION

Planning the Lesson

Refer to the Chapter Organizer on pages 184A–D.

Concepts Developed

Students learn that when an acid and base react, they neutralize each other and a salt is formed.

the solution with litmus paper after *each* additional drop of ammonia is added.

Science Journal Have students write their impressions of what took place in the activity. Have them speculate about the nature of acids and bases and how they are able to transform each other. L1

Expected Outcome

The solution will reach a neutral pH. The students may conclude that a base can be used to make an acidic solution neutral.

Conclude and Apply

1. When neutral, the solution will not change the color of the red or blue litmus paper.

2. Answers will vary.

3. The solution would not change either color of litmus paper.

Find Out!

How does neutralization take place?

Time Needed 10 minutes

Materials ammonia, white vinegar, 10-mL graduated cylinder, litmus paper, dropper, 2 test tubes, stirring rod

Thinking Processes thinking critically, observing and inferring, recognizing cause and effect, interpreting data

Purpose To observe the neutralization process.

Teaching the Activity

Discussion Talk about what *neutralize* means. Here, acids and bases are used to cancel the effect of one another.

Safety Provide adequate ventilation and warn students that both reactants have strong fumes.

Troubleshooting Stress that, after the fifth drop of ammonia is added, students must test

✔ Assessment

Content Show students a color chart for universal indicator paper. Ask them to predict what would happen if the neutral solution were tested with universal indicator paper. *The neutral solution should have a pH of 7, so the paper would turn green.* L1

Before presenting the lesson, display **Section Focus Transparency 20** on an overhead projector. Assign the accompanying **Focus Activity** worksheet. L1

LEP

Demonstration This activity will allow students to see that a salt is a product of neutralization.

Crush two antacid tablets in warm water. Add a few drops of universal indicator. Add drops of acid until the solution is neutral (see indicator label). Have the class verify that the solution is now neutral. To prove that a salt is formed during neutralization, slowly evaporate the solution. Students should be able to observe the formation of a salt the next day.

2 TEACH

Tying to Previous Knowledge

Review the pH scale and talk about acids and bases in relative terms. Ask the students how they could make an acid solution (a) more or (b) less acidic. (Remind them of the antacid they made in Section 2.) Tell them that in this section they will learn how this is done. L1

Salts

We usually think of salt as just table salt, sodium chloride. Many different salts, however, are formed in neutralization reactions. You come in contact with some of these salts every day but may not have realized that these compounds are salts. One example is calcium carbonate, or chalk. This salt is also used in the manufacturing of paint and rubber tires. Another salt, potassium nitrate, is also known as saltpeter and is used to make fertilizer and explosives.

Like acids and bases, salts are important in industry. **Figure 6-8**

shows some salts and gives examples of how they are used. For example, table salt is important in food preparation. It's used in the curing of hams and bacon and in the production of lunch meats such as bologna, sausage, and wieners.

Many salts are also useful as raw materials. Chlorine, a chemical used to purify water, is obtained from sodium chloride. Other salts are used in the production of rubber, water softeners, chemicals, paints, fertilizers, and many other products you use every day.

Food Pioneer

You have learned in this chapter that salts are used in the preparation of foods. Pioneering food chemist Dr. Lloyd A. Hall developed new methods of using salts in preparing and preserving foods.

Hall was born in Elgin, Illinois in 1894, and he developed an interest in chemistry during high school. In 1925, Hall became director of research at Griffith Laboratories, where he studied the use of salts in meat curing.

Preserving Foods

Hall developed a method of combining sodium with nitrate and nitrite so that the sodium could preserve the meat before the nitrogen-containing salts could penetrate the meat and cause disintegration. The new technique was called flash-drying, and the resulting crystals

Dr. Lloyd A. Hall

Purpose

A Closer Look extends the concepts presented in Section 6-4 to the properties and uses of salts in the food industry.

Content Background

From ancient times, food preservation required the use of salt. In the corning process, meat first cured in a brine of salt, sugar and saltpeter. Later the meat was cooked as needed. The drying process involved rubbing meat with a mixture of salt, spice, and sugar and then covering it with

brine for several weeks. Removed from the brine, the meat hung from a ceiling beam until dried. Pepper rubbed into the meat added flavor and, as a bonus, also acted as an insect repellent.

Teaching Strategies

Have students take an inventory at home or at a grocery store and list the various ingredients found in beef jerky and smoked sausage. Have them identify which ingredients act as preservatives. Then have them list food items found in their house-

Figure 6-8

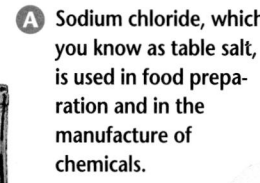

A Sodium chloride, which you know as table salt, is used in food preparation and in the manufacture of chemicals.

B Baking soda, sometimes called sodium bicarbonate, is the salt sodium hydrogen carbonate. This salt is part of the extinguishing material in one type of fire extinguisher, and is also commonly used in food preparation.

C This iron chromate salt is a yellow pigment used to color ceramics, glass, and enamels.

D Potassium permanganate is used in tanning leather. This salt is also used to purify water.

were more effective than any meat-curing salts used before. Hall's work had a major impact on the meat industry.

Spices had been used for many years to preserve food. Hall found that many spices were actually contaminating the foods they were supposed to preserve. The spices were often infested with molds, yeasts, and bacteria. Hall developed a method to sterilize these spices without ruining their appearance, quality, and flavor. His method was also used for preparing medicines, medical supplies, and cosmetics.

Fats and oils often spoiled, or became rancid, when components of the fat interacted with the oxygen in the air. Many antioxidants, agents that prevent spoiling, would not dissolve in fat, so they could not effectively mix with the product. Hall developed a fat-soluble antioxidant mixture that was 99.64 percent sodium chloride. The remaining ingredients were the key to preserving the fats and oils.

Hall served as a science adviser in two wars. He helped solve the problem of keeping food for the military fresh and healthful. He was the first African American to be on the board of directors of the American Institute of Chemists.

What Do You Think?

If you've ever gone on a camping or hiking trip, you may have wished you could take along some of your favorite foods. What might a food chemist invent to keep those foods from spoiling along the way?

6-4 Salts **205**

Check for Understanding

Assign questions 1–3 of Check Your Understanding. After students have completed the Apply question, have them look for cake recipes and locate the acid ingredient in each. Baking powder has acid already in it, usually in the form of tartaric acid, calcium acid phosphate, or sodium aluminum sulfate.

Reteach

Analogy To describe a neutralization reaction, use the analogy of balancing a seesaw. You start out unbalanced (i.e., acid), on one end of the seesaw. To balance the seesaw, you must add an equal amount of base to the other side. **LEP** **L1**

Extension

Activity Have students give the names of the salts that would form in these neutralizations.

a. Nitric acid and potassium hydroxide *Potassium nitrate*

b. Carbonic acid and calcium hydroxide *Calcium carbonate*

c. Hydrochloric acid and ammonium hydroxide *Ammonium chloride* **L3**

4 CLOSE

Have students design an experiment by imagining they have returned in a time machine to the Middle Ages, a time when salt was a rare and precious commodity. They have bragged to the king that they can manufacture salt. Have them describe what materials and equipment they would need to perform this feat. **L1**

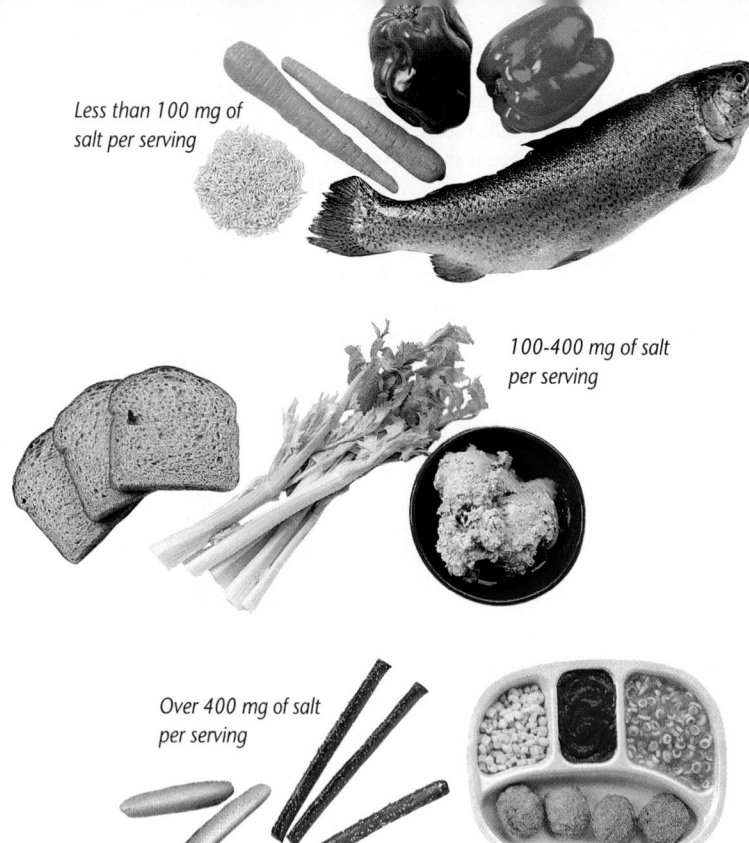

Less than 100 mg of salt per serving

100-400 mg of salt per serving

Over 400 mg of salt per serving

Figure 6-9

Ⓐ People need only a limited amount of sodium chloride to maintain the body's chemical balance. You can get all the sodium you need in unsalted foods.

Ⓑ Some foods naturally have more sodium than others. Some experts think the amount of sodium consumed should be limited to 2000 mg a day.

Sodium chloride is important in your diet and the diet of animals. Limited amounts of salt are needed to maintain the body's chemical balance. When salt is lost through perspiration and excretion, it must be replaced if you are to stay healthy. Workers and athletes laboring under hot conditions often take special drinks to replace the salt they perspire away. Animals such as deer will travel miles to get to a salt lick.

In the beginning of this chapter, you read about the damage acid rain causes to monuments, buildings, and the environment. Now you know how these reactions take place. You also know that acids and bases make it possible for living things to function, grow, and be healthy. The important thing is to keep a balance between these two types of compounds. As you look about your world, be aware of the effects of acids and bases.

check your UNDERSTANDING

1. Hydrochloric acid, hydrogen chloride, can be neutralized by calcium hydroxide. What salt is produced by this reaction?

2. A solution turns phenolphthalein pink. What would you observe if the solution were tested with litmus?

3. **Apply** Baking soda is used to make a cake rise, but baking soda will not react unless an acid is present. Which of these would be a good acid to use: vinegar, hydrochloric acid, lemon juice, or water? Explain your answer.

check your UNDERSTANDING

1. calcium chloride

2. The litmus paper would be blue because a substance that turns phenolphthalein pink is a base.

3. Lemon juice; it is acid that is not dangerous to eat undiluted and has a better taste for a cake than does vinegar.

Program Resources

Study Guide, p. 26
Laboratory Manual, pp. 41–42, Neutralization **L3**
Multicultural Connections, p. 16, Salt
Concept Mapping, p. 14, Acids and Bases
Take Home Activities, p. 13 **L1**
Section Focus Transparency 20

Science and Society

Acid Rain

Cars, airplanes, home heaters, most industrial machines, and giant electric generators are all powered by fuels that come from coal and oil. These fuels have been very important in making our world industrialized and comfortable. But they also produce harmful pollution. When coal and oil are burned for energy, they give off gases into the atmosphere. When these gases combine with water in the clouds, the result is acid rain.

Like any acid solution, acid rain affects the substances on which it falls.

Fish, crops, forests, and the surfaces of sculptures and buildings (like the Parthenon, built in ancient Greece, pictured on p. 208) can all be damaged by acid rain.

It's Not a New Problem

You may be surprised to learn that the term acid rain was first used in 1872 by Robert Angus Smith, who wrote about polluted air and rain in and near large cities in England. It was a century later, though, that people began to regard acid rain as a serious problem. Today, acid

rain is a problem that all industrialized countries have to consider. For example, the effects of acid rain on the fertile fields of Mexico are just as harmful as the effects of smog on the people of Mexico City.

Its Effects

There are different kinds of acid rain. Sulfuric acid and nitric acid are produced when certain sulfur and nitrogen compounds combine with rainwater. Even when highly diluted, these acids are harmful to the environment. The more concentrated the acids, the more serious is the problem.

When acid rain falls to Earth, the soil and water become more acidic. Among the first organisms to suffer are plankton. Plankton live in the water and are eaten by small fish, which are eaten by larger fish, and on and on through the food chain.

Science Journal

While answers will vary, most students should conclude that it is a serious environmental threat.

Going Further ⅢⅢ➤
Have students research current positions of different public interest groups and industry-sponsored groups regarding the economic issues behind the acid rain problem. Have each student

present his/her group's position in the form of a poster, letter to the editor, or speech to a congressional committee investigating acid rain. Use the Performance Task Assessment List for Letter, Poster, or Oral Presentation in **PASC,** pp. 67, 73, or 71. L2 COOP LEARN P

Science and Society

Purpose
Science and Society reinforces the discussion of acids in Section 6-1 by explaining what happens when certain chemicals dissolve in atmospheric moisture to form acid rain.

Content Background
Acid rain is just one form of acid deposition. Dry acid deposition also occurs on a regular basis. Both depositions have pH values lower than 5.6.

Acid rain results from the release of sulfur oxides, mostly by coal-fired plants, and from nitrogen oxides in auto emissions. Acid rain is the first global-scale problem caused by burning fossil fuels. Some of our pollution ends up in Canada, and we get some of theirs. North American, European, and Asian pollution finds its way to the Arctic Circle. Controlling acid rain necessitates each country first confronting its own problem and then acting together with others.

Teaching Strategies
Have students debate the importance of the acid rain issue. One group should support the position of industry and one group the position of environmentalists. L2
COOP LEARN

Activity
Have students make lists of family activities that consume power, such as driving motor vehicles, operating appliances, using lawn mowers, and so on. Have them consider how and to what degree they could reduce the amount of energy they personally consume. Have students share ideas and make a class list of ways students will work to conserve energy. L1 COOP LEARN

Literature Connection

Purpose
This Literature Connection, "In Times of Silver Rain," balances the unfavorable effects of acid rain that students learned about in this chapter by presenting one poet's lyric images of rain.

Content Background
Langston Hughes (1902–1967) is celebrated as the leader of the "Harlem Renaissance," the 1920s outpouring of music, art, and literature from New York's Harlem section. Hughes was raised by his maternal grandmother in his hometown of Joplin, Missouri. He attended Columbia University for a year in 1921, and then set off to see the world. To this end, he shipped out as a merchant seaman, worked in a Paris nightclub, and bused tables in Washington, DC. His break came when poet Vachel Lindsay ate at the restaurant and Hughes slipped some of his poems under his plate. Impressed, Lindsay helped launch Hughes's career. In addition to poetry, Hughes published biographies, children's stories, song lyrics, and articles.

Teaching Strategies
To capture the sound of Hughes's imagery, suggest that students read the poem aloud with a partner. Then, student pairs might work together to make collages or paintings that capture the poem's mood and imagery. **L1** **COOP LEARN**

Discussion
Have students read the poem on their own. Then explain that Hughes was writing in the 1920s, before the effects of acid rain were known. Have students discuss how Hughes's poem might be different if he had written it today. **L1**

When the plankton are killed, the fish that depend upon them for food soon die.

Soil and land plants can also suffer from the effects of acid rain. Acid rain can dissolve important mineral nutrients and make the soil highly acidic. Some plants are adapted to normally acidic soil, but others are not. These are the ones that suffer from acid rain.

Possible Solutions
Science and industry are constantly searching for cleaner, more efficient fuels. Harnessing the energy in the sun, the wind, and the atom as an alternative to coal and oil is a major goal. Industry is also hard at work to find ways to use coal and oil without causing harmful by-products.

Literature Connection

I n this chapter, you've read about acid rain and the unfavorable effects it has on the environment. To get a different view of rain, read the poem, "In Times of Silver Rain" by Langston Hughes

(Bontemps, Arna Wendell, ed. *Golden Slippers: An Anthology of Negro Poetry for Young Readers*).

To Langston Hughes, rain is filled with beautiful images that inspire poetry. His image of "silver rain" evokes the idea of a shimmering cascade of raindrops—each like a silvered mirror reflecting the effect of rain on the plants growing on the plain.

"In Times of Silver Rain"

What Do You Think?
What images does rain evoke for you?

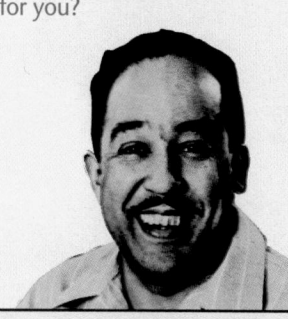

Answers to
What Do You Think?
Some students may agree with Hughes's life-affirming images of rain, while others may suggest dreary and depressing images to capture rain's less cheerful side.

Going Further ⫸
Have students work in small groups to prepare a display about the Harlem Renaissance. Break the class into three groups. Have one group research the art; another the music, and the third the literature. Notable artists of the movement include James Chapin and Romare Bearden. Musicians include composers Duke Ellington and Fats Waller, blues singers Ethel Waters and Bessie Smith, and entertainers Josephine Baker, Florence Mills, and Bill Robinson. Some of the writers were Countee Cullen, Claude McKay, Zora Neale Hurston, Jean Toomer, and Arna Bontemps. Use the Performance Task Assessment List for Display in **PASC,** p. 63. **L2** **COOP LEARN**

How bad is battery pollution?

INSIDE A BATTERY

Alkaline manganese cell

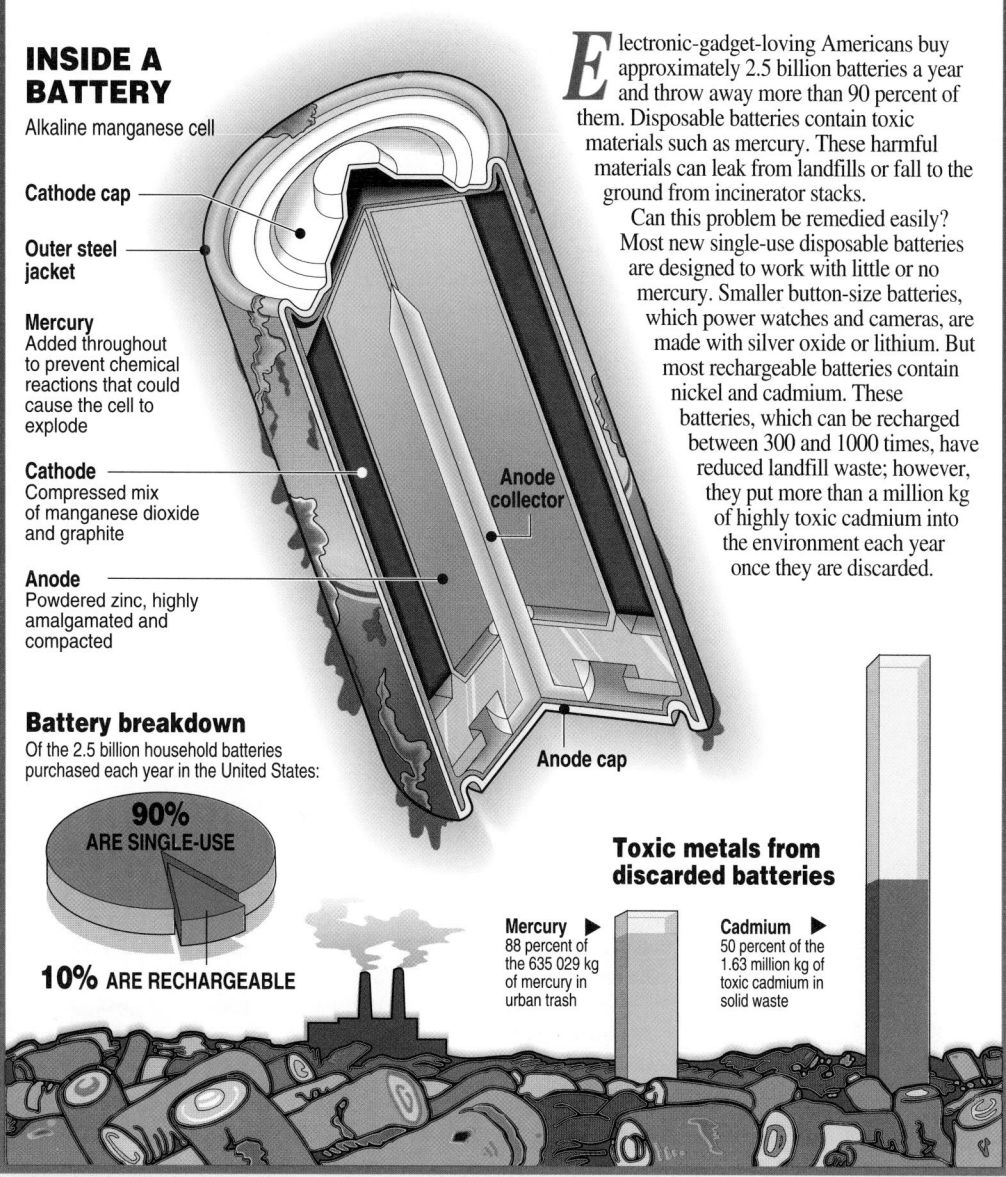

Cathode cap

Outer steel jacket

Mercury
Added throughout to prevent chemical reactions that could cause the cell to explode

Cathode
Compressed mix of manganese dioxide and graphite

Anode
Powdered zinc, highly amalgamated and compacted

Anode collector

Anode cap

Electronic-gadget-loving Americans buy approximately 2.5 billion batteries a year and throw away more than 90 percent of them. Disposable batteries contain toxic materials such as mercury. These harmful materials can leak from landfills or fall to the ground from incinerator stacks.

Can this problem be remedied easily? Most new single-use disposable batteries are designed to work with little or no mercury. Smaller button-size batteries, which power watches and cameras, are made with silver oxide or lithium. But most rechargeable batteries contain nickel and cadmium. These batteries, which can be recharged between 300 and 1000 times, have reduced landfill waste; however, they put more than a million kg of highly toxic cadmium into the environment each year once they are discarded.

Battery breakdown

Of the 2.5 billion household batteries purchased each year in the United States:

90% ARE SINGLE-USE

10% ARE RECHARGEABLE

Toxic metals from discarded batteries

Mercury ▶
88 percent of the 635 029 kg of mercury in urban trash

Cadmium ▶
50 percent of the 1.63 million kg of toxic cadmium in solid waste

Science Journal

In your Science Journal, plan a public service brochure aimed at encouraging consumers to make use of rechargeable batteries and write a rough draft.

NATIONAL GEOGRAPHIC
SCIFACTS

How bad is battery pollution?

Purpose

In Section 6-1, students learned about pollution caused by acids in the environment. These SciFacts introduce students to another environmental hazard: disposable batteries.

Content Background

Primary or disposable batteries depend on the storage of internal chemicals to produce energy. When the chemicals are depleted, the battery is "dead." Secondary or rechargeable batteries transform one type of chemical into another. When the transformation is complete, the battery must be recharged in order to be used again. The recharging process involves restoring the chemicals to their original state.

Research

Divide students into two groups. Have one group research other uses for mercury, and the second group research uses for cadmium. For instance, mercury is used in thermometers, insecticides, and medicines; cadmium is used for industrial purposes to plate steel and other metals. Students' research should include environmental and health problems associated with mercury and cadmium contamination. Have the two groups share the results of their research.

Science Journal

Students' brochures will vary, but should contain information concerning the environmental damage caused by disposable batteries, as well as points in favor of using rechargeable batteries.

Use the diagrams to review the main ideas presented in the chapter.

Teaching Strategies

Have students make acid and base "profile" cards. The paper color should correspond to the colors of the red cabbage indicator. Divide the class up into seven teams. Each team should use all available resources to make cards corresponding to the sample card shown below.

Profile of a BASE
Name: ANTACID
Active Chemical: Sodium hydrogen carbonate Red Cabbage Indicator Color: Green pH: 9.5 Category: Medium Base Feel: Slippery Color: White Taste: Bitter Hazardous to Touch: No Poisonous: No Practical Use: Medicine

Substances might include: sulfuric acid (pH 0), washing soda (pH 12), saliva (pH 6.2–7.4), milk (pH 6.3–6.6), cooking oil (pH 7), grapefruit juice (pH 2.5), hydrochloric acid (pH 0), bottled water (pH will vary). **L1** **COOP LEARN**

Answers to Questions

1. Answers might include vinegar, citrus fruits, and tomatoes.

2. Answers might include ammonia cleaners and drain and oven cleaners.

3. These tissues can be burned.

4. Answers might include sodium chloride as table salt and calcium chloride, which is used to melt snow.

GLENCOE TECHNOLOGY

MindJogger Videoquiz

Chapter 6 Have students work in groups as they play the Videoquiz game to review key chapter concepts.

Science Journal

Review the statements below about the big ideas presented in this chapter, and answer the questions. Then, re-read your answers to the Did You Ever Wonder questions at the beginning of the chapter. *In your Science Journal*, write a paragraph about how your understanding of the big ideas in the chapter has changed.

1 Acids are compounds that taste sour, may release hydrogen gas in reactions with active metals, and turn blue litmus paper red. *What common foods contain acids?*

2 Bases taste bitter, feel slippery, and turn red litmus paper blue. *What common cleaning products contain bases?*

3 Acids and bases can be very reactive. *What can happen to skin and lung tissues if they are in contact with strong acids or strong bases?*

4 In a neutralization reaction, an acid reacts with a base to produce a salt and water. *Name some common salts and their uses.*

Science at Home

Baking soda is a base that reacts with acid and produces gas to make cakes rise. Students can demonstrate this by making two small cakes, both using baking soda. Add lemon juice, vinegar, or cream of tartar to ingredients of the first cake. Use no acid in the second cake. After baking, compare how much each cake rose. The first cake should be much lighter. Do not use baking powder because it already contains an acid ingredient. **L1** **LEP**

Science Journal

Did you ever wonder...

• A sour taste is one property of an acid. Citrus fruits contain acid. (p. 188)

• Drain cleaner contains a very strong base, sodium hydroxide. It reacts vigorously with many things, including human tissue, which it can burn. (pp. 192-193)

• The stomach contains hydrochloric acid. An upset stomach is caused by an excess of acid. An antacid is a base that neutralizes the excess acid. (p. 196)

Using Key Science Terms

acid   neutralization
base   pH
indicator   salt

Each of the statements below contains a word or words that makes it wrong. Rewrite the sentence, replacing the incorrect word or words with a term that uses the correct science word from above.

1. An acid turns litmus paper blue.
2. A base has a sour taste and reacts with aluminum.
3. An acid and a base react, forming an indicator.
4. A measure of acidity is called neutralization.
5. A substance used to test for an acid is called an acid test.
6. A salt is always produced in a decomposition reaction.

Understanding Ideas

Answer the following questions in your Journal using complete sentences.

1. Explain how hydrochloric acid in your body is important to your life and health.
2. How might the burning of coal in one area of the country affect the environment in another area?
3. What would be a good indication that a grapefruit contains an acid?
4. Name five common acids found in your home and/or school.
5. What is the most commonly used base and what is it used for?

Developing Skills

Use your understanding of the concepts developed in this chapter to answer each of the following questions.

1. **Concept Mapping** Complete the concept map of acids and bases.

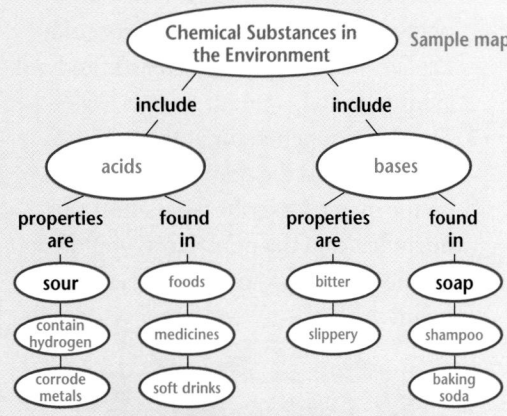

Sample map

2. **Comparing and Contrasting** Repeat the Find Out activity on page 186. Use pieces of limestone or concrete rather than using pieces of marble. Compare the results with those made in the original activity.
3. **Separating and Controlling Variables** In the Find Out activity on page 203 why is it necessary to stir after each drop of ammonia is added to the vinegar?
4. **Observing and Inferring** After doing the Find Out activity on page 192, use litmus paper to identify each substance as an acid or a base. How do these results compare with those from the original activity? Which of the substances were neither acidic nor basic?

Using Key Science Terms

1. *A* base turns litmus paper blue.
2. *An acid* has a sour taste and reacts with aluminum.
3. An acid and a base react, forming *a salt.*
4. A measure of acidity is called *pH*.
5. A substance used to test for an acid is called an *indicator*.
6. A salt is always produced in a *neutralization* reaction.

Understanding Ideas

1. Food in the stomach needs to be broken down so it can be digested and used by the body. The hydrochloric acid in the stomach is necessary for breaking down this food.
2. Burning coal releases sulfur dioxide, which reacts with water in the atmosphere to form sulfurous acid. This acid is carried on the wind and falls as acid rain hundreds of miles away, polluting lakes and threatening wildlife.
3. Grapefruit has a sour taste.
4. Answers will vary but may include acetic acid (vinegar), ascorbic acid (lemon juice), sulfuric acid (battery acid), and carbonic and phosphoric acids (in soft drinks).
5. Ammonia in water; it is used as a cleaning agent and fertilizer.

Developing Skills

1. See reduced student page for concept map.
2. The limestone or concrete will also lose mass. Students may conclude that acids damage these materials also.
3. The acid and base must come in contact with each other for the reaction to take place. If the mixture were not stirred, some materials might remain unreacted and give a false reading with the litmus paper.

Program Resources

Review and Assessment, pp. 35–40 [L1]
Performance Assessment, Ch. 6 [L2]
PASC
Alternate Assessment in the Science Classroom
Computer Test Bank [L1]

4. Acids will turn blue litmus paper red and leave red litmus paper unchanged. Bases turn red litmus paper blue and leave blue litmus paper unchanged. Neutral substances leave both papers unchanged. Those materials determined to be bases in the activity are confirmed by litmus. The other substances are neutral.

Critical Thinking

1. The soil is mildly basic.

2. Aluminum is not a suitable container for orange juice, which is acidic, because metals may react with acids.

3. The water is acidic and could react with the mortar in the pool's wall and the metal in the pump and filter. Add a base, such as lime, to the water.

Problem Solving

Hydrangeas must be a natural indicator, like red cabbage juice, that turn color in acidic or basic soil. Manuel and Sheila's soil must be opposite with respect to pH. To get blue hydrangeas, Manuel should adjust his soil so its pH matches that of Sheila's.

Connecting Ideas

1. Tomato sauce is acidic and could react with aluminum, which would spoil its flavor.

2. For virtually all landforms, acid rain weathers rock and damages plant and animal life.

3. Sodium hydroxide dissolves completely in water. It cannot be separated by filters and will not scatter light. It is a homogeneous mixture or solution.

4. Possible answers include reducing the amount of coal used to produce electricity and scrubbing sulfur dioxide out of smokestack gases after coal has been burned.

Critical Thinking

In your Journal, *answer each of the following questions.*

1. The leaves on the red cabbage in your garden are blue-green in color. What does this indicate about the soil?

2. The inside of a paper carton that your orange juice came in has a silvery color. Explain why it is probably plastic and not aluminum foil.

3. The chlorine generator at the pool has gone wild, and the pool's pH is as shown on the meter. Describe what effect this might have on the pool. How would you get the water back to where it is safe to swim?

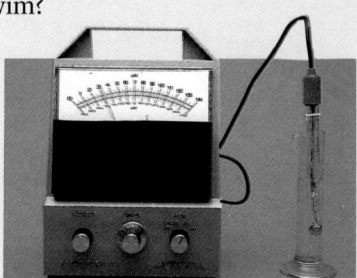

Problem Solving

Read the following problem and discuss your answers in a brief paragraph.

Sheila Jones grows beautiful blue hydrangeas. Her neighbor, Manuel Ortiz, wants to grow some just like them. Sheila gives several roots to Manuel.

When Manuel's flowers bloom, they aren't blue, but pink! The person at the nursery tells Manuel to check his soil.

What do you think is the reason that Manuel's blue hydrangeas are pink? Why are Sheila's hydrangeas blue? What should Manuel do to get blue hydrangeas?

CONNECTING IDEAS

Discuss each of the following in a brief paragraph.

1. Theme—Systems and Interactions A recipe for spaghetti sauce states that it must be cooked in a nonaluminum pot. What is the reason for this instruction?

2. Theme—Stability and

Change Predict the effect of acid rain on various types of landforms.

3. Theme—Systems and Interactions You dissolve sodium hydroxide in water. Is this mixture heterogeneous or homogeneous? Is the mixture a solution?

4. Science and Society What steps can be taken to reduce the amount of acid-producing substances in the air?

✔ Assessment

Portfolio Review the portfolio options that are provided throughout the chapter. Encourage students to select one product that demonstrates their best work for the chapter. Have students explain what they learned and why they chose this example for placement into their portfolios.

Additional portfolio options can be found in the following:

Teacher Classroom Resources:

Making Connections: Integrating Sciences, p. 15
Multicultural Connections, pp. 15–16
Making Connections: Across the Curriculum, p. 15
Concept Mapping, p. 14
Critical Thinking/Problem Solving, p. 14
Take Home Activities, p. 13
Laboratory Manual, pp. 37–42
Performance Assessment P

Interactions in the Physical World

In this unit, you investigated how substances on Earth differ. You observed how some materials dissolve in others, forming solutions, while others do not.

You can describe an iron nail using physical properties, such as its size and mass, and chemical properties, such as its ability to react with oxygen to form rust.

Physical and chemical properties can also be used to identify substances as acids, bases, and salts.

Try the exercises and activity that follow—they will challenge you to use and apply some of the ideas you learned in this unit.

CONNECTING IDEAS

1. Ocean water is very salty. You may have accidentally tasted some while swimming, or noticed that salt crystals formed on your body as you sat in the sun to dry after swimming. How does the salt appear on your skin?

2. Suppose you are trying to decide which vinegar to buy. One brand is much cheaper than the other brand. You wonder if the cheaper brand has been diluted with water. If this is true, the acidity of the cheaper, diluted vinegar should be lower than the more expensive brand. How could you design an experiment to see if the acidity was the same for both brands?

Exploring Further ACTIVITY

What hazardous materials can be found around your home?

What To Do

1. Make a card file of dangerous materials around your home.

2. Identify acids, bases, and other toxic solutions and materials.

3. Include the danger posed by each substance and the steps to be taken if the material is accidentally spilled or taken into your body.

4. Use this activity to compile a first aid file for the home.

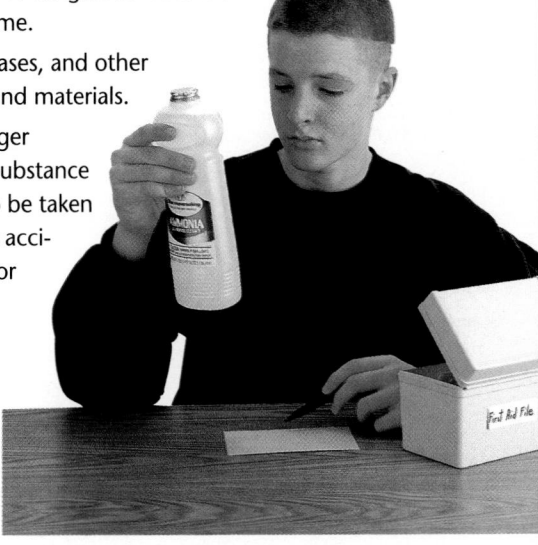

Interactions in the Physical World

THEME DEVELOPMENT

The themes developed in Unit 2 were scale and structure, systems and interactions, stability and change, and energy. The theme of scale and structure helped students distinguish between the physical and chemical properties of matter. The more they understand about materials in the world that are stable and those that change, the better they are able to see patterns of change in the environment, and make valid predictions.

Connection to Other Units

Students continued to use their senses to observe and classify, as they learned in Unit 1, while conducting a variety of investigations of the physical world in Unit 2. The investigations of substances and their interactions in Unit 2 will help students understand the more complex interactions explored in Unit 3.

Connecting Ideas

Answers

1. The salt was dissolved in the water. When the water evaporated, the remaining salt appears white and crystalline on the skin.

2. Students suggestions may vary but the brands could be tested with pH paper.

Exploring Further

What hazardous materials can be found around your home?

Purpose To apply and communicate knowledge of the properties of materials.

Background Information

Many people are surprised to discover how many hazardous materials are in everyday use at home. Most cleaning agents are basic and some, such as drain cleaner, can be toxic. Acids are more-commonly found in the garage. Additionally, many pesticides and almost all rodent killers are very toxic.

Materials Index cards

Troubleshooting Students can often find first aid information on product packaging.

✔ Assessment

Poster Have students work in small groups to write newspaper articles about the dangers in many common household materials and first aid measures that can be taken in case of accidents. Use the Performance Task Assessment List for Newspaper Article in **PASC**, p. 69. L1

COOP LEARN P

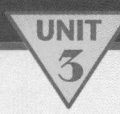

UNIT 3

Interactions in the Living World

UNIT OVERVIEW

UNIT FOCUS

In Unit 3, students extend their understanding of systems and interactions to living things, from the smallest to the largest. Students gain an appreciation of the similarities and differences among living organisms.

THEME DEVELOPMENT

Systems and interactions is a major theme developed in Unit 3. All living things share the same basic need for food. The relationship among organisms—those that produce food through photosynthesis and those that need to obtain food from other sources—is seen. Scale and structure also is a theme in this unit. Students will compare organisms, regardless of size or complexity.

Connections to Other Units

This unit builds on the observation skills developed in previous units. It also provides information that will help students bridge to the next unit. Unit 4 covers concepts dealing with water, erosion, and soil formation. It will directly relate to the identification of those water areas on Earth where specific water life-forms may be found. The formation of soil and erosion will explain where minerals present in soil originally came from.

UNIT 3

Interactions in the Living World

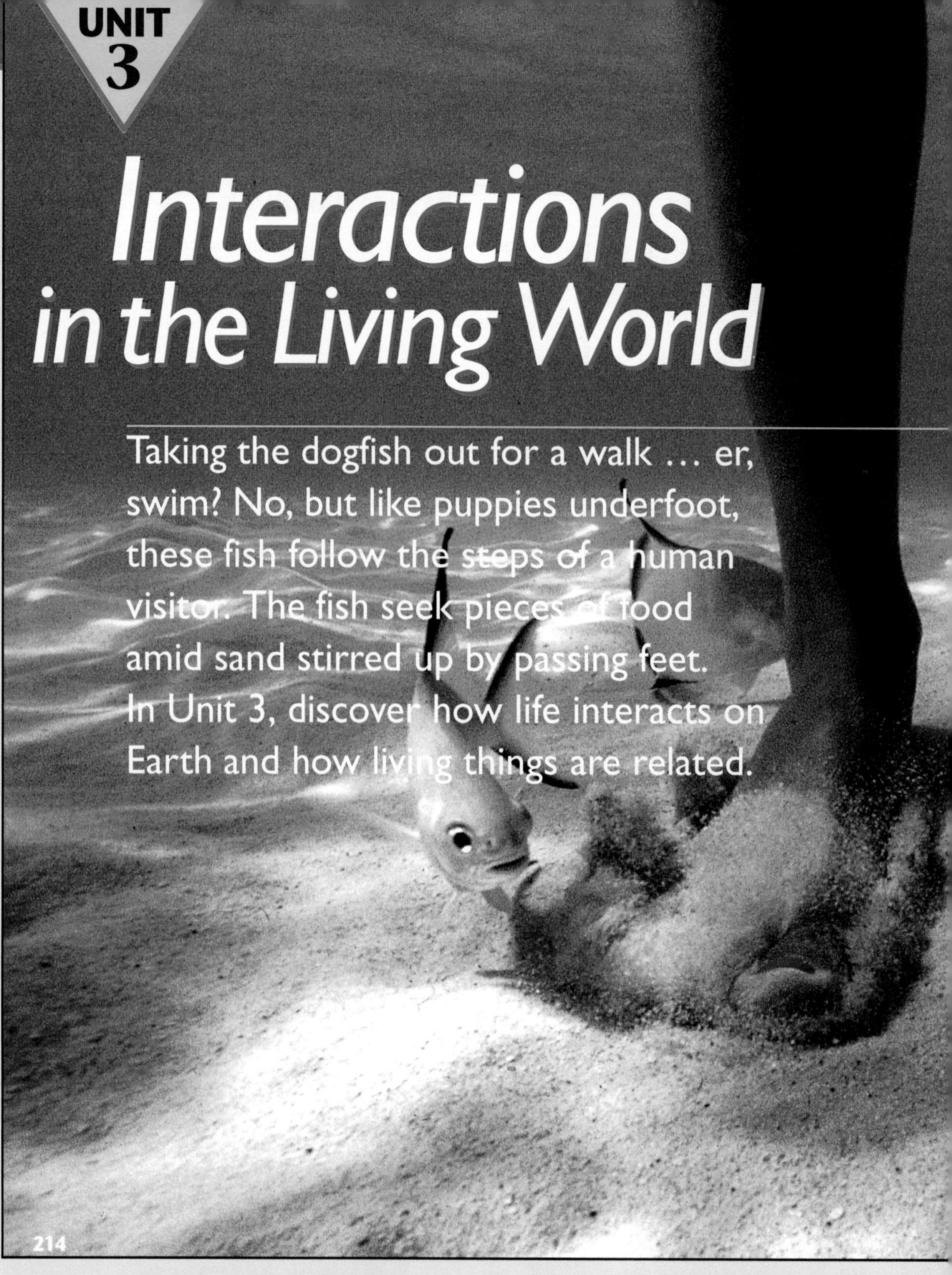

Taking the dogfish out for a walk … er, swim? No, but like puppies underfoot, these fish follow the steps of a human visitor. The fish seek pieces of food amid sand stirred up by passing feet. In Unit 3, discover how life interacts on Earth and how living things are related.

214

GLENCOE TECHNOLOGY

Videodisc

Use the *Science Interactions, Course 1* **Integrated Science Videodisc** lesson, *Ecosystems*, with Chapter 11 of this unit. This videodisc reinforces and enhances the concepts of ecosystems and ecological balance.

NATIONAL GEOGRAPHIC
try it!

What do we really know about the living things around us? What do they look like? How do they move? Do they all eat the same way? Do they all make sounds? Take time to observe a living animal closely, very closely.

What To Do

1. With a partner, spend 15 minutes watching a pet gerbil or other organism available in your classroom. Describe the organism's surroundings. What does the gerbil do while you watch it? Record each observation that you make *in your Science Journal*.

2. *In your Science Journal,* write a definition for the word *life* based on your observations. Is *life* easy to define? Compare your observations with those of your classmates.

3. Try to find somethng nonliving that fits a classmate's definition. Try to find a living thing that is excluded by the definition. Is your own definition complete and accurate?

215

Discussion Some questions you may want to ask your students are:

1. Define the word *life* without using the words *living, alive,* or *life* in your definition. Students may suggest descriptions such as "breathing," "moving," and so on. Most will emphasize familiar animals in their attempts, neglecting plants and very simple animal forms.

2. Other than color, how might plants and animals differ? How are they alike? Students will probably name some similarities and differences but will miss many, such as movement and methods of reproduction, which will be explored in this unit.

3. What is the role of the gases in our atmosphere in helping to maintain life on Earth? Most students will suggest that oxygen is necessary.

The answers to these questions will help you establish what misconceptions your students may have.

Try It!

Purpose Students will develop a working definition of life through observing living organisms.

Background Information

For larger organisms, most people can agree on whether or not an organism is alive. However, defining life becomes increasingly complex as scientific methods extend our powers of observation and discovery.

Materials gerbils or other small animals for students to observe

✔ Assessment

Oral Have students work in small groups to compare and discuss their definitions of life. Then have them refine their definitions as they observe each other. Use the Performance Task Assessment List for Group Work in **PASC**, p. 97. **COOP LEARN** L1

Chapter Organizer

SECTION	OBJECTIVES	ACTIVITIES & FEATURES
Chapter Opener		**Explore!**, p. 217
7-1 What Is the Living World? (2 sessions; 1.5 blocks)	1. **Determine** the characteristics of living things. 2. **Apply** the characteristics of living things to determine if something is alive or not. **National Content Standards: (5–8) UCP1, A1, C1, C5, G1, G2**	**Find Out!**, p. 218 **Design Your Own Investigation 7-1:** pp. 222–223 **Geography Connection**, p. 239 **Science and Society**, pp. 240–241
7-2 Classification (2 sessions; 1 block)	1. **Recognize** how a classification system allows scientists to communicate information. 2. **Describe** the levels of the system used to classify organisms. 3. **Explain** the characteristics that make up the five kingdoms of organisms. **National Content Standards: (5–8) UCP1, UCP2, A2, G1, G3**	**Explore!**, p. 225 **Explore!**, p. 226 **A Closer Look**, pp. 230–231 **How It Works**, p. 242
7-3 Modern Classification (4 sessions; 2 blocks)	1. **Demonstrate** that a classification system can show how organisms are related. 2. **Identify** the traits scientists use to classify organisms. **National Content Standards: (5–8) UCP1, UCP2, A2, C1, C5, F2, F4–5, G1**	**Find Out!**, p. 233 **Explore!**, p. 234 **Investigate 7-2:** pp. 236–237 **Chemistry Connection**, pp. 234–235

ACTIVITY MATERIALS

EXPLORE!

p. 217 candle, matches, flowering plant
p. 225* blank 3 × 5 pieces of paper, assorted audiotapes
p. 226 assorted tree leaves
p. 234* collection of animal pictures

INVESTIGATION!

pp. 236–237 paper and pencil

DESIGN YOUR OWN INVESTIGATION

pp. 222–223 1 package dry yeast, table sugar, 4 test tubes, water, dropper, microscope, microscope slides, coverslips, flat-ended toothpicks, salt, baking soda, labels, measuring spoons

FIND OUT!

p. 218* mustard seeds, gravel, 2 jars, marker, water
p. 233* No special materials are required.

KEY TO TEACHING STRATEGIES

The following designations will help you decide which activities are appropriate for your students.

L1	Basic activities for all students
L2	Activities for average to above-average students
L3	Challenging activities for above-average students
LEP	Limited English Proficiency activities
COOP LEARN	Cooperative Learning activities for small group work
P	Student products that can be placed into a best-work portfolio
	Activities and resources recommended for block schedules

Need Materials? Call Science Kit (1-800-828-7777).

⏱ **OUT OF TIME?** We recommend that students do the activities with an asterisk.

Chapter 7 Describing the Living World

TEACHER CLASSROOM RESOURCES

Student Masters	Transparencies
Study Guide, p. 27 **Critical Thinking/Problem Solving,** p. 15 **Multicultural Connections,** p. 18 **Making Connections: Integrating Sciences,** p. 17 **Activity Masters,** Investigate 7-1, pp. 31–32 **Making Connections: Technology and Society,** p. 17	**Section Focus Transparency 21**
Study Guide, p. 28 **Concept Mapping,** p. 15 **Multicultural Connections,** p. 17 **Making Connections: Across the Curriculum,** p. 17 **Take Home Activities,** p. 15 **Laboratory Manual,** pp. 43–46, Classification	**Teaching Transparency 13,** Life's Five Kingdoms **Teaching Transparency 14,** Classification Pyramid **Section Focus Transparency 22**
Study Guide, p. 29 **Flex Your Brain,** p. 5 **Activity Masters,** Investigate 7-2, pp. 33–34 **Science Discovery Activities,** 7-1, 7-2, 7-3	**Section Focus Transparency 23**

ASSESSMENT RESOURCES	TEACHING & TECHNOLOGY
Review and Assessment, pp. 41–46 **Performance Assessment,** Ch. 7 **PASC*** **MindJogger Videoquiz** **Alternate Assessment in the Science Classroom** **Computer Test Bank**	**Spanish Resources** **Cooperative Learning Resource Guide** **Lab and Safety Skills** **Science Interactions, Course 1, CD-ROM** **Computer Competency Activities**

*Performance Assessment in the Science Classroom

NATIONAL GEOGRAPHIC TEACHER'S CORNER

Index to National Geographic Magazine

The following articles may be used for research relating to this chapter:

- "Life Without Light," by Ian R. MacDonald and Charles Fisher, October 1996.
- "Dead or Alive: The Endangered Species Act," by Douglas H. Chadwick, March 1995.
- "Rebirth of a Deep-Sea Vent," by Richard A. Lutz and Rachel M. Haymon, November 1994.

National Geographic Society Products Available From Glencoe

To order the following products for use with this chapter, contact your local Glencoe sales representative or call Glencoe at 1-800-334-7344:

- *STV: Biodiversity* (Video)
- *Eye on the Environment: Pollution* (Poster)
- *Eye on the Environment: Vanishing Wildlife* (Poster)

Additional National Geographic Society Products

To order the following products for use with this chapter, call the National Geographic Society at 1-800-368-2728:

- *Earth's Endangered Environments* (CD-ROM)
- *Animal Classes Series* (5 Videos)
- *The Diversity of Life* (Video)
- *Pollution: World at Risk* (Video)

Teacher Classroom Resources

These are key components of the classroom resources package.

TEACHING AIDS

Section Focus Transparencies

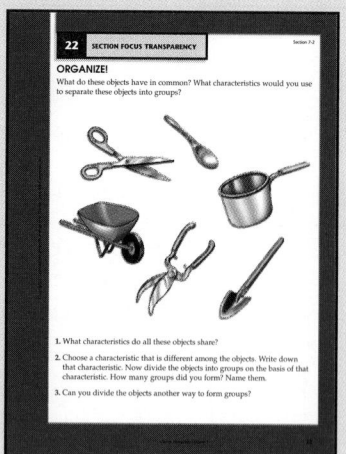

Teaching Transparencies

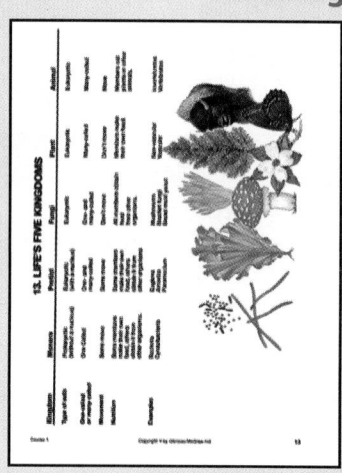

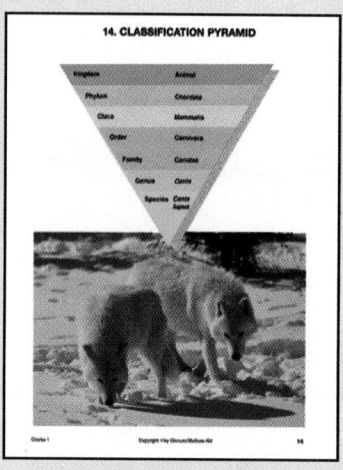

HANDS-ON LEARNING

Science Discovery Activity*

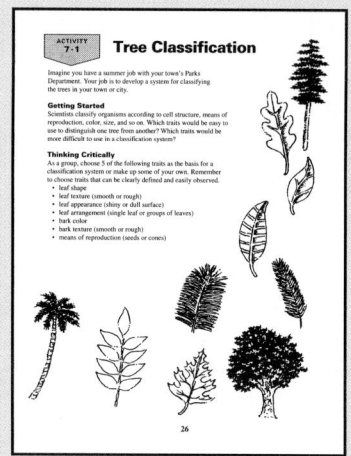

Laboratory Manual*

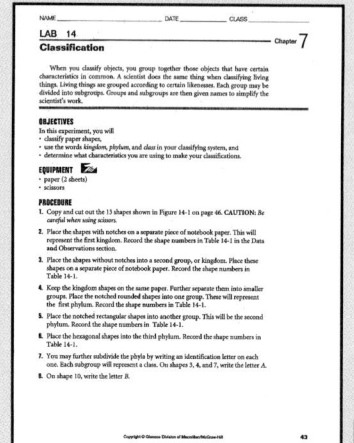

Take Home Activity

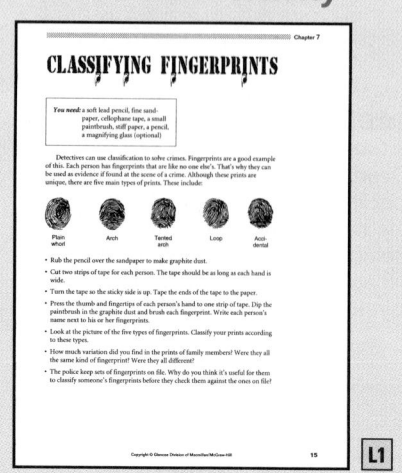

*There may be more than one activity for this chapter.

Chapter 7 Describing the Living World

Study Guide*

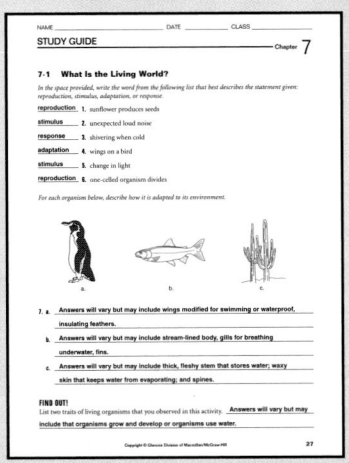

Concept Mapping

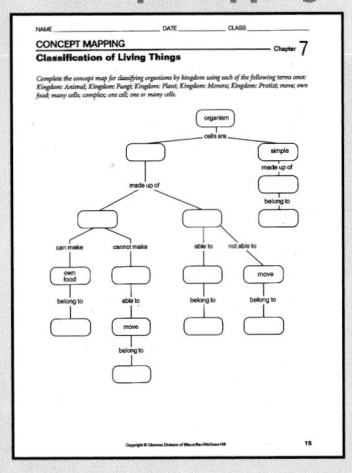

Critical Thinking/ Problem Solving

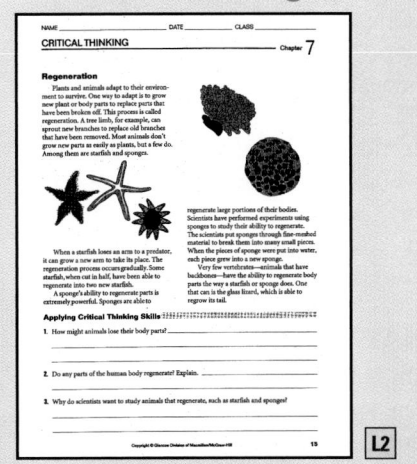

Integrating Sciences

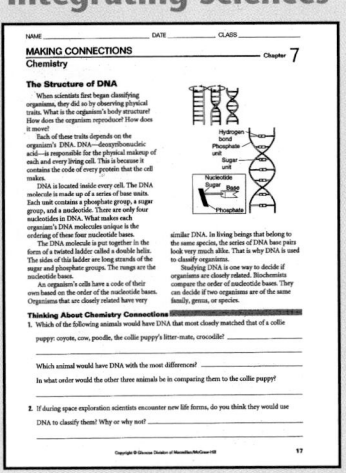

Across the Curriculum

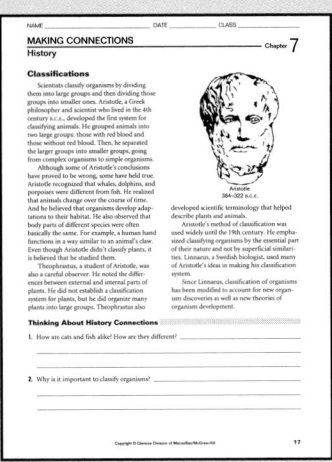

Technology and Society

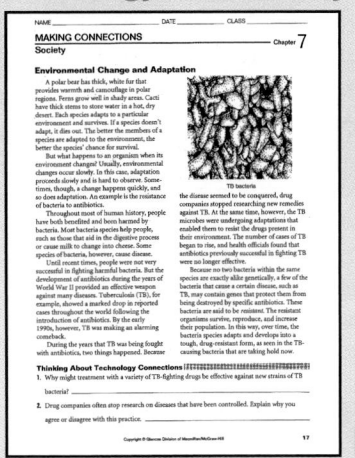

Multicultural Connection**

Performance Assessment

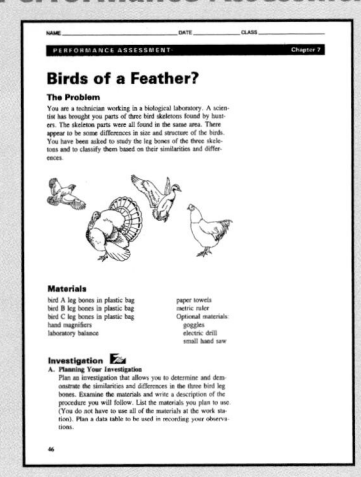

Review and Assessment

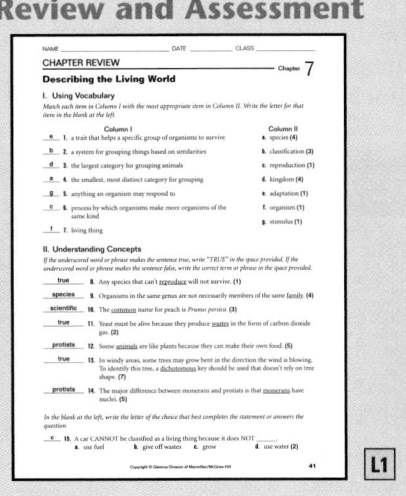

Describing the Living World

THEME DEVELOPMENT

One theme of this chapter is systems and interactions. The Linnaean system of binomial nomenclature is a system of classification for living things, based on body structure and interactions with the environment. This system applies to all organisms regardless of size or scale.

CHAPTER OVERVIEW

In this chapter, students will learn six traits that distinguish living from nonliving things. Then, they will be introduced to the categories of the Linnaean system. Students will apply their knowledge of classification to specific organisms and use Linnaean categories to understand relatedness among organisms and name them.

Tying to Previous Knowledge

To reinforce the concept of classification, students can name some of the terms and concepts they use to classify parts of Earth, such as mountains and plateaus.

INTRODUCING THE CHAPTER

Have students look at the chapter opener photograph. Students can think about what makes a zebra a living thing. Then they can brainstorm lists of things that are alive and that are not alive.

OUT OF TIME?

If time does not permit teaching the entire chapter, use the Chapter Overview on this page, Reviewing Main Ideas at the end of the chapter, and the Chapter 7 audiocassette to point out the main ideas of the chapter.

Describing the Living WORLD

Did you ever wonder...

✓ **What makes something alive?**

✓ **Why names are so important?**

✓ **How or if you and your cat or dog are related?**

Science Journal

Before you begin to study about the features of living things and how they are classified, think about these questions and answer them *in your Science Journal*. When you finish the chapter, compare your journal write-up with what you have learned.

Answers are on page 243.

A s you look at the zebras across the bottom of the page, you can imagine the thundering sound that they make. Can you imagine the vibration that you would feel if they were real? There is no doubt that you are looking at a picture of life in action here. Is everything around you alive? What are some things that you look for when you decide something is alive or not?

▶ *In the following activity, observe some objects and begin to think more about what life is like.*

216

Learning Styles	Kinesthetic	Investigate, pp. 222–223
	Visual-Spatial	Find Out, pp. 218, 223; Visual Learning, pp. 220, 224, 229, 231; Demonstration, p. 221; Activity, pp. 230, 233, 238; Explore, p. 234; Project, p. 243
	Interpersonal	Explore, p. 225; A Closer Look, pp. 230–231
	Intrapersonal	Activity, p. 232
	Logical-Mathematical	Activity, p. 223
LS	Linguistic	Explore, pp. 217, 226; Discussion, pp. 219, 224, 227, 229; Visual Learning, p. 219; Research, pp. 221, 231; Debate, p. 221; Activity, p. 226; Across the Curriculum, pp. 228, 230; Multicultural Perspectives, pp. 229, 237

Explore! ACTIVITY

What are some signs of life?

What To Do

1. Carefully observe a lighted, dripping candle for up to five minutes. Do not touch the flame or move the candle. *In your Journal*, list some adjectives that you would use to describe any lifelike characteristics of the candle.

2. Now observe a flowering plant. *In your Journal*, list adjectives for lifelike characteristics of the plant. Then do the same for the zebras or an animal you might have in the classroom.

3. With a partner, discuss which of these objects is alive. Write a conclusion statement that both of you agree on. Be prepared to defend your statement.

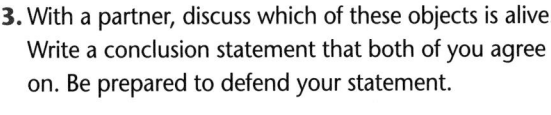 **ASSESSMENT PLANNER**

PORTFOLIO
Refer to p. 245 for suggested items that students might select for their portfolios.

PERFORMANCE ASSESSMENT
Process, pp. 218, 223, 225, 233
Explore! Activities, pp. 217, 225, 226, 234
Find Out! Activities, pp. 218, 233
Investigate, pp. 222–223, 236–237

CONTENT ASSESSMENT
Oral, p. 235
Check Your Understanding, pp. 224, 232, 238
Reviewing Main Ideas, p. 243
Chapter Review, pp. 244–245

GROUP ASSESSMENT
Opportunities for group assessment occur with Cooperative Learning Strategies.

 Explore!

What are some signs of life?

Time needed 15–20 minutes

Materials candle, match, flowering plant

Thinking Processes observing and inferring, comparing and contrasting, forming a hypothesis, classifying, forming operational definitions

Purpose To observe and make inferences about the characteristics of living things.

Teaching the Activity

Safety Caution students not to touch the lighted candle or hot wax. Use large-based candles so they can not be accidentally knocked over.

Discussion Before beginning the activity, students can brainstorm some characteristics of living things.

Science Journal Suggest that students write a description of the candle after 1, 3, and then 5 minutes. **L1**

Expected Outcome
Students will recognize that the plant and the animal are alive but the candle is not.

Answers to Questions
1. bright, drippy, flickering, warm, changing
2. The plant springs back when touched; zebras move, make noise, and breathe.
3. Accept all reasonable answers. Students will probably conclude that all show some signs of life.

✔ Assessment

Content Have small groups list characteristics of living things, then name nonliving things that exhibit any characteristic(s) of living things. Use the Performance Task Assessment List for Group Work **PASC**, p. 97. **L1** **COOP LEARN**

PREPARATION

Planning the Lesson

Refer to the Chapter Organizer on pages 216A–D.

Concepts Developed

Students will explore the six traits of living organisms that distinguish them from nonliving things.

Find Out!

What are some differences between living and nonliving things?

Time needed 15 minutes the first day, 10 minutes the second day

Materials mustard seeds, gravel, 2 jars, water, marker

Thinking Processes observing, forming a hypothesis, interpreting data, comparing and contrasting, classifying, recognizing cause and effect

Purpose To compare living and nonliving things.

Teaching the Activity

Discussion Ask students how they know the gravel will not grow. Growth is a characteristic only of living things. Encourage students to define *nonliving* as lacking in the traits described in the text.

Science Journal Suggest that students record their observations in a two-column chart in their journals. L1

Expected Outcomes

The mustard seeds will sprout but the gravel will not.

What To Do

2. Mustard seeds and the gravel are small, yellow, dry, and appear lifeless. There is no way observation can tell if either is alive. Students will predict that seeds will germinate and the gravel will not.

7-1 What Is the Living World?

Section Objectives

- Determine the characteristics of living things.
- Apply the characteristics of living things to determine if something is alive or not.

Key Terms

organism,
reproduction,
stimulus,
adaptation

Is It Alive?

If you visit an aquarium, or if you ever stop to watch tropical fish in a pet store, you might see some fish like the ones in the photograph that opens this unit. Look again at the photograph of the coral reef on page 214. If you were face-to-face with a fish, you'd know you were looking at a living thing. But what about the water and the sand? Or the coral? Are any of these things living?

What do living things do that nonliving things don't do? This next activity will help you identify some characteristics or traits you can use to distinguish living and nonliving things.

Find Out! ACTIVITY

What are some differences between living and nonliving things?

What To Do

1. From your teacher, obtain some mustard seeds and some gravel pieces about the same size as the seeds.

2. Describe the seeds. Can you tell if they are alive? Describe the gravel. Do you really know that the gravel isn't alive?

3. Mark 2 separate jars as A and B. Put the seeds in jar A and the gravel in jar B.

4. Soak both the seeds and the gravel for 24 hours in small, but equal amounts of water.

5. What do you predict will happen?

6. Observe the soaked seeds and the gravel for the next two days.

Conclude and Apply

1. What can you conclude about whether the seeds or gravel are alive? What signs of life do either show?

2. Based on your observations, what would you say are some differences between living and nonliving things?

Conclude and Apply

1. Some of the seeds begin to germinate in 24 hours; more by 48; but not the gravel. Most will conclude that the seeds are alive, but the gravel is not.

2. Living things appear to grow with the help of water. Accept all reasonable answers.

✔ Assessment

Process Have students compare and contrast characteristics of gravel and grains. Students can create a simple chart to present the similarities and differences. Use the Performance Task Assessment List for Making Observations and Inferences in **PASC**, p. 17. L1

Organisms

As you look around your classroom, you'll see people talking to one another, reading, or scratching an itch. Maybe it's lunchtime and your class eats lunch in the classroom, so people are drinking and eating. A plant in the window is drooped over from lack of water. A guppy in the aquarium has just laid more eggs. You and your classmates, the plant, and the guppies are all organisms. An **organism** is a living thing. In this chapter, you will learn about different organisms. Does every organism seem alive all the time? As you saw with the seeds in the Find Out activity, sometimes, it isn't so easy to tell. Seeds don't appear alive until they start to grow. Neither do the barnacles, such as those in **Figure 7-1**. Read about these organisms, then turn the page to learn about the characteristics shared by all organisms.

Figure 7-1

As adults, barnacles resemble rocks. They stay in one place and go for long periods showing no outward signs of being alive.

A An adult barnacle attaches itself to some object under water and stays there for the rest of its life. The object can be a rock, another organism, a pier, or the bottom of a ship.

B Once attached to something, a hard shell forms around the adult barnacle. An attached barnacle opens part of its shell periodically for feeding. Only then does it appear alive because it extends its legs from its shell and sweeps food into its mouth.

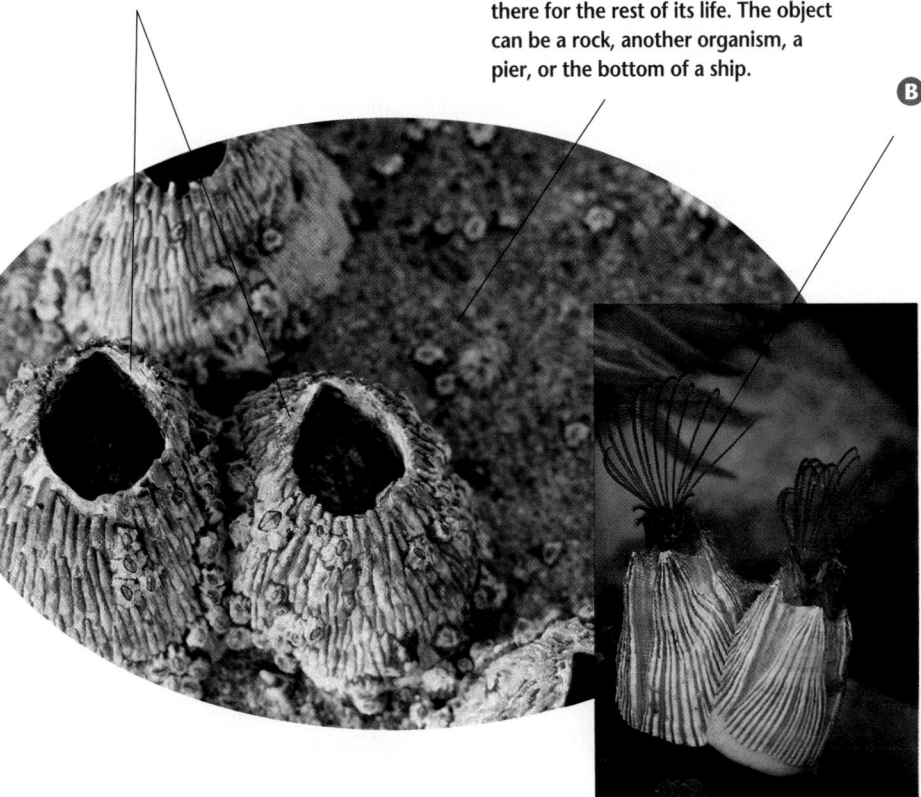

7-1 What Is the Living World? **219**

Program Resources

Study Guide, p. 27
Critical Thinking/Problem Solving, p. 15, Regeneration L2
Multicultural Connections, p. 18, The Yangtze L1
Making Connections: Integrating Sciences, p. 17, The Structure of DNA L2
Section Focus Transparency 21

Visual Learning

Figure 7-1 Ask a volunteer to read aloud the information given as other students study the photographs of barnacles. Students can list the ways that barnacles are like living things and list the ways that barnacles are like nonliving things.

Uncovering Preconceptions
Students may think that the difference between living and nonliving things is obvious. Ask students how they tell the difference. Accept all answers, and then explain that this chapter will show that the difference is not always obvious.

1 MOTIVATE
Bellringer
Before presenting the lesson, display **Section Focus Transparency 21** on an overhead projector. Assign the accompanying **Focus Activity** worksheet. L1 LEP

Discussion Students compare and contrast living and nonliving things. Obtain a videotape of the classic Boris Karloff *Frankenstein.* Show students the sequence in which Dr. Frankenstein melodramatically brings his monster to life using the electrical power of lightning. (If the film is not available, have students take turns telling the familiar story.) Ask students how the doctor knew his creature lived. *It moved.* Tell students that this section describes the six traits used to determine if something is alive or not.

2 TEACH

Tying to Previous Knowledge
In the previous two chapters, students have classified parts of the physical world. Review briefly the classifications—physical and chemical; elements and alloys—they have studied and then explain that this section describes ways to classify the living world. Have students suggest possible categories for living things.

Traits of Living Things

Organisms, from mustard seeds to elephants, have traits that set them apart from nonliving things. Think about each trait described here. Think about how you yourself fit the definition of organism as you study each of these.

Figure 7-2

■ Organisms Are Made Up of Cells

Cells are the basic units of all organisms in which life functions take place. Some organisms, such as bacteria, are made up of just one cell. Larger organisms, such as cats, birds, bees, and trees, are made up of billions of cells. Large animal organisms also usually have many different kinds of cells. You have bone cells, nerve cells, and muscle cells. Plants contain a variety of different types of cells as well.

Most cells are very small. You'll find that you can't see most cells without the help of a microscope.

Ⓐ Organisms are made up of cells. In many-celled organisms, cells may be highly specialized, like these blood cells.

■ Organisms Use Water and Food and Produce Wastes

Every organism is made up of a great deal of water. For example, your body is about two-thirds water. Most of this water is found inside cells where chemical reactions take place. There, energy is released from food compounds. As a result, you are able

to move, think, grow, and produce new cells. Water also carries away most waste products that are made in cells.

Ⓑ Organisms need water and food to develop and live.

■ Organisms Reproduce

If organisms did not reproduce, how would all the forms of life on Earth continue to exist? **Reproduction** is the process by which organisms make more organisms of the same kind. As you can see in **Figure 7-2C**, a single sunflower produces many seeds. Each seed is capable of producing a flower that produces more seeds. Reproduction ensures that a particular type of organism survives.

Ⓒ Organisms reproduce, making more of the same kinds of living things.

 Organisms grow and develop.

Organisms Grow and Develop

So long as they are supplied with food for energy, organisms grow. Small elephants develop into large elephants. Skinny saplings grow up to become large trees. Some organisms grow in terms of the number of individuals rather than becoming larger in size. One-celled organisms usually grow to a certain size and then reproduce, resulting in an increased number of organisms.

Organisms Respond

When a cat hears the sound of a can opener, it comes running into the kitchen expecting to be fed. When you hear an unexpected loud noise, do you jump? When it gets cold, do you shiver?

 Organisms respond to their environment.

When you do any of these things, you are responding to changes in your environment. Anything an organism responds to is a **stimulus**. A stimulus produces a response—a change in behavior on the part of an organism.

Stimuli are received through the senses. Stimuli can be in the form of changes in light, temperature, sound, touch, or taste. Chemicals also act as stimuli.

Organisms Are Adapted to Their Environment

What is there about a fish that enables it to survive in water but not on land? Why does a cactus survive in the desert? Any trait of an organism that helps it survive in its environment is an **adaptation**. Fish have adaptations called gills that enable them to remove oxygen from water. Fish can't remove oxygen from air as you do. A cactus has special tissues that are adapted to hold water. The better an organism is adapted, the better its chances for surviving long enough to reproduce. Adaptations are inherited. They are not temporary adjustments that an organism makes to changes going on around it.

In the Investigate that follows, think about whether common yeast shows any or all of the characteristics of life you've learned about here.

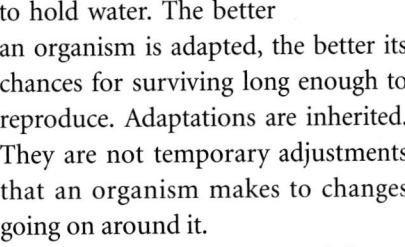 Organisms such as cacti are adapted for retaining water.

Content Background

Adaptation occurs through natural selection. In reproduction, random changes occur in the genes of a population that control an organism's growth. Some of these changes, or mutations, result in variations that make an organism more likely to survive and reproduce. Any variation favoring survival and reproduction is passed on to new generations. In the competition for survival, organisms that lack this new variation are less likely to survive. As these organisms die out, most members of the species finally have the new variation.

Research Have students conduct library research on the *Viking* lander missions to Mars and write an essay describing the experiments that the lander performed to search for signs of life on Mars. The essays should describe when the missions were launched from Earth, when they reached Mars, the experiments conducted, equipment used, and results. Use the Performance Task Assessment List for Writing in Science in **PASC**, p. 87. L3

Demonstration To help students recognize the need for six traits to distinguish living from nonliving things, crumple a piece of paper and place it in a large ashtray. Making sure that students can see the paper, set it on fire with a match and let it burn to ash. Then tell students that they watched fire grow and use and release energy. Ask students if fire is alive. *no* Help students conclude that nonliving things may have one or two traits on the list but that only living things have all six. L2

7-1 Living or Nonliving?

Preparation

Purpose To determine whether a substance is living or nonliving.

Process Skills comparing and contrasting, making and using tables, practicing scientific methods, observing, forming a hypothesis, separating and controlling variables, interpreting data

Time Required 45–50 minutes

Materials See reduced student text. To make sugar water, use the ratio of one teaspoon sugar to one cup water. Dry yeast can be found in the refrigerated section at the supermarket.

Possible Hypotheses Groups may hypothesize that yeast will show traits of a living organism, and that they will be able to show that yeast is living.

The Experiment

Process Reinforcement Reinforce making and using tables by referring students to Making and Using Tables on page 656 in the Skill Handbook. It is important to reinforce the need for separating and controlling variables in this experiment. Students can read about this in the Skill Handbook on page 665.

Possible Procedures Divide the class into as many groups as you have available microscopes. Stagger the times when students start the experiment to allow groups to use the same microscopes. Students should record observations every 5 minutes. Students should wait 15 minutes for changes in the yeast tube. Before and after the experiment, students should take a drop from each test tube, put it on a clean slide, and cover it with a coverslip to observe under the microscope. Students should draw what they see.

Living or Nonliving?

In this section, you have learned about the traits of living things. During this investigation, you will observe how substances react to food. Use your knowledge about living things to help you prove which substances are living.

Preparation

Problem
How can you prove that something is living or nonliving?

Form a Hypothesis
As a group, decide on a statement or prediction about whether yeast, baking soda, and salt are living or not. Record your hypothesis.

Objectives
• Observe the changes over time when something is fed.
• Compare the reactions of test items to the traits of living organisms.

Possible Materials
microscope
coverslips
4 test tubes each with 20 drops of sugar water
1 package dry yeast
salt (NaCl)
baking soda (NaHCO$_3$)
measuring spoons
labels

Safety Precautions

Dispose of the yeast as directed by your teacher.

Meeting Individual Needs

Physically Challenged Some students with physical limitations may not be able to manipulate the materials used in this activity. Allow those students to observe and record the information gathered during this investigation.

Program Resources

Activity Masters, pp. 31–32, Design Your Own Investigation 7-1

Making Connections: Technology and Society, p. 17, Environmental Change and Adaptation L1

DESIGN YOUR OWN
INVESTIGATION

Plan the Experiment

1 As a group, decide how you will test your hypothesis. Design the experiment to prove if any of the three test items are alive. Write down the steps of your experiment.

2 What are the variables? What is the control? Before you begin, label the tubes for your experiment and record what each one will contain. Make certain that you use the same amount of each substance in the test. In this experiment, sugar water is the food.

3 Make a data table. In the table, list the traits of living things that you will observe.

4 Immediately after you put each test item into the sugar water, observe a sample of each under the microscope. At the end of the experiment, observe the samples again and compare.

Check the Plan

1 Investigations like this one take time. Plan to take observations every 10 minutes for 40 minutes. After every observation, record what you saw.

2 Before you start your experiment, make certain your teacher approves your plan.

3 Carry out the experiment.

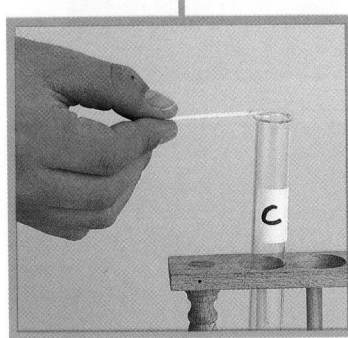

Analyze and Conclude

1. **Observe** How did test tubes A, B, C, and D change over the time you observed them?

2. **Infer** What was in the test tubes that changed? Was the change an indication of a life process occurring?

3. **Conclude** What waste product was produced in the experiment? What do you think it is? Why did it occur?

4. **Scientific Illustration** Make drawings of what you see under the microscope at the beginning and at the end of the experiment for each tube.

5. **Collect Data** In your data table, check off each trait that applies to each item tested.

6. **Conclude** Use your data to conclude whether each test item is living or nonliving.

Going Further

Design an experiment in which you test for each of the characteristics of life as described on pages 220 and 221.

ENRICHMENT

Activity To develop students' ability to interpret data, students can perform the Investigate (but without microscopic examination), using water, vinegar, and baking soda. All test tubes receive 20 drops of water. In addition, test tube B gets 20 drops of vinegar; tube C gets 4 toothpick ends of baking soda; and tube D gets 20 drops of vinegar and 4 toothpick ends of baking soda. Students will observe the bubbling reaction of tube D. Ask them if this indicates that baking soda is alive and, if not, how else they can explain the reaction. *The bubbles are the result of an acid-base chemical reaction.*

Students may need assistance in connecting this activity with concepts learned in Chapter 6 on Acids, Bases, and Salts. ☐L3

Teaching Strategies

Troubleshooting Check the date of the yeast. The fresher the yeast, the better the results of the experiment.

Science Journal Students should record observations and drawings in their science journals.

Expected Outcome

Students should recognize that only the yeast exhibits traits of living things.

Analyze and Conclude

1. The test tube with the yeast bubbled up as the yeast fed on the sugar. The other substances did not react.

2. Yeast; the change showed that yeast was feeding on sugar. This was an indication of digesting food, a life process.

3. A gas bubbled up. Carbon dioxide was produced by the breakdown of food.

4. Students should record their observations of what is in each tube at the beginning and at the end of the experiment. They should see the yeast cells.

5. Traits mentioned: organisms are made up of cells; use water and food and produce wastes; reproduce, grow, respond; and are adapted to their environment. The yeast broke down food and produced waste, and grew. The salt and baking soda should not have shown any of these traits.

6. The hypothesis that predicted that yeast is alive was supported because yeast showed traits of living things while the other substances did not.

✔ Assessment

Process Have students repeat the experiment with plain water instead of sugar water in all the test tubes. Use the Performance Task Assessment List for Designing an Experiment in **PASC**, p. 23.

Going Further ⫸

Check student designs for correct use of variables and scientific methods.

7-1   What Is the Living World?   **223**

Visual Learning

Figure 7-3 As students study the figure, have them compare the needs of living things for food and water with their own needs.

3 ASSESS

Check for Understanding

Show students a fish, small animal, or plant and a rock. Ask them which one is alive. *the animal or plant* Ask students to explain, using the six traits of living things, how they know it is alive.

Reteach

Concept Mapping Draw a concept map on the chalkboard, with "living thing" in the center and the six traits in circles around and connected to it. Students can give examples of both living and nonliving things that illustrate each trait. `L1`

Extension

Scientific Writing Have students write a science fiction story describing scientists who visit another planet and try to apply the six traits to determine whether or not a green blob is a living thing. Use the Performance Task Assessment List for Writer's Guide to Fiction in PASC, p. 83. `L3`

4 CLOSE

Discussion

Hold up a piece of paper and ask how it compares to a tree. Why is the piece of paper nonliving while its source, a tree, is living? Ask the same question regarding a cotton ball and a piece of cotton cloth.

Needs of Living Things

■ Food

Living things change the substances they take in as food. In the Investigate, you may have observed that the test tube that contained yeast, sugar, and water, produced bubbles. To stay alive, most organisms break down sugar and energy is released. In breaking down sugar, carbon dioxide and water are produced in a process called cellular respiration. The bubbles that you saw are bubbles of carbon dioxide.

■ Water

"I'm dying of thirst!" How often have you heard that statement? What do you suppose would have happened if you had not supplied water for the yeast in the Investigate? No matter how much food is available, unless the food contains water or other water is available, an organism will not survive long.

Living organisms come in many sizes, shapes, and colors. In the next sections, you will learn why and how traits are used to classify living things.

Figure 7-3

Ⓐ Like every organism, this barn swallow needs water to live. Water in food, the bird's cells, or from a pond or stream satisfies this need.

Ⓑ Growth and development take energy. The fuel that supplies that energy is food. All organisms need food. This starfish eats mussels and clams. This food supplies the starfish with energy.

check your UNDERSTANDING

1. Name a trait of living things that each of the following nonliving objects appears to possess. Then explain why they are, nonetheless, not alive.
   a. a gasoline engine
   b. a crystal of salt
   c. a robot
2. Give two examples of organisms responding to a stimulus in their environment.
3. How is an adaptation different from a response to a stimulus?
4. **Apply** A pigeon pecks at bread crumbs in a park. How does the pigeon show the characteristics of living things? Does the pigeon show all the characteristics? If it does not, can you still say it is alive?

check your UNDERSTANDING

1. a. A gasoline engine consumes fuel. b. A crystal of salt grows. c. A robot responds to its environment. Each lacks other traits.
2. Dogs bark and trees lose leaves.
3. Adaptation involves a change in an entire species; responses are individual and not lasting.
4. Pigeon moves, shows need for food, response to environment, and adaptation to environment. If numerous pigeons are present, it is obvious that reproduction takes place in the population.

Classification 7-2

Order Out of Confusion

You rush into the record store, eager to buy the latest recording by your favorite group. You head for the first table and begin looking for it. Something must be wrong! CDs are mixed with audiotapes. Rock is mixed with reggae; country with classical; rap is right next to opera. Nothing about this place is in order! How can you find anything?

Section Objectives
- Recognize how a classification system allows scientists to communicate information.
- Describe the levels of the system used to classify organisms.
- Explain the characteristics that make up the five kingdoms of organisms.

Key Terms

classification
kingdom
phylum
genus
species

ACTIVITY

How do categories help you find what you want?

What To Do

1. Obtain a stack of blank 3 × 5 pieces of paper for category labels from your teacher.

2. Obtain several types of audiotapes or CDs.

3. With a partner, sort the tapes or CDs and develop categories for them.

4. Write each category on a separate piece of paper. Stack each matching audio-tape or CD by its label.

5. Did you and your partner always agree on the categories?

6. How many ways did you find to classify the audiotapes or CDs?

Stores make it easy to find what you want by organizing recordings according to whether they are CDs or audiotapes, by musical style, and by artist. This organization is a form of classification. **Classification** is any system used to group ideas, information, or objects based on their similarities.

Your method of classifying in the Explore activity may have differed from your partner's. Of course, nature isn't as easy to organize as a store. But classifying living things helps scientists organize their knowledge so that they can communicate about specific organisms.

7-2 Classification **225**

How do categories help you find what you want?

Time needed 10–15 minutes

Materials blank 3 × 5 pieces of paper, assorted audiotapes or CDs

Thinking Processes organizing information, classifying, modeling

Purpose To experiment with classification by groups and subgroups.

Teaching the Activity

Assign partners to work together. Partners can each develop their own categories and then compare. **COOP LEARN**

Discussion Ask students to discuss their favorite kind of music or performers. Lead a brief discussion of categories students might use, such as type of music (jazz, classical, rock), performer, or composer. Suggest a numbered list or outline format for depicting the categories.

7-2

PREPARATION

Planning the Lesson

Refer to the Chapter Organizer on pages 216A–D.

Concepts Developed

Students will study the binomial classification system developed by Linnaeus and explore the features of organisms that belong to each of the five major kingdoms.

Science Journal Suggest students list their own categories as well as their partner's categories. **L1**

Expected Outcomes

Students will develop different systems of categorizing a stack of audiotapes.

Answers to Questions

5. Most will answer that there were at least two opinions.

6. Most will have at least two ways to classify the cassettes—type of music, vocal or instrumental, types of instruments, volume are possibilities; CD or cassette.

✔ **Assessment**

Process Have students compare their categories with the categories of at least two other classmates. Ask students to describe situations in which each method of categorizing would work better than any other method. Use the Performance Task Assessment List for Making and Using a Classification System in **PASC**, p. 49. **L1**

7-2 Classification **225**

1 MOTIVATE

Bellringer

Before presenting the lesson, display **Section Focus Transparency 22** on an overhead projector. Assign the accompanying **Focus Activity** worksheet. L1

LEP

Activity Ask three volunteers to each place an object on a table where all students can see them. Ask students to name traits the objects have in common. Traits may reflect shape, color, material, or purpose. Explain that scientists use similar approaches to classify the immense number of organisms on Earth. L1 LEP

2 TEACH

Tying to Previous Knowledge

In the previous section, students distinguished between living and nonliving things using six traits. Ask students how they might distinguish between different kinds of living things by their traits. Encourage them to suggest using such traits as color, shape, size, and habits in order to distinguish among living things.

Explore!

How would you name things?

Time needed 15 minutes

Materials assorted leaves from different plants

Thinking Processes organizing information, classifying

Purpose To classify and name living things.

Preparation Sort leaves into envelopes and distribute or make separate piles and have student pairs take one from each pile. **COOP LEARN**

Teaching the Activity

Encourage students to use words such as *round, oval, long,* and *pointed* to describe leaf shape. Point out other traits such as leaf edge, vein pattern, color, texture, and shininess. Have students find two traits shared by at least two leaves. L1

Science Journal Have students sketch and/or describe the leaves in their journals as they define each leaf group. L1

Expected Outcomes

Students identify traits by which leaves may be classified and develop names that permit identification.

A Scientific Way to Classify

What kind of bird do you picture when you hear the name *robin*? If you live in the United States, you may think first of the bird shown in **Figure 7-4**. But the name *robin* means something different to people living in England. They call the bird in **Figure 7-5**, robin. And the name means something else again to people in China, who call the bird in **Figure 7-6** robin. You can see that these robins aren't the same bird at all!

Scientists have a way to classify and name the huge numbers of organisms found on Earth to help keep order. Try this method yourself in the following activity.

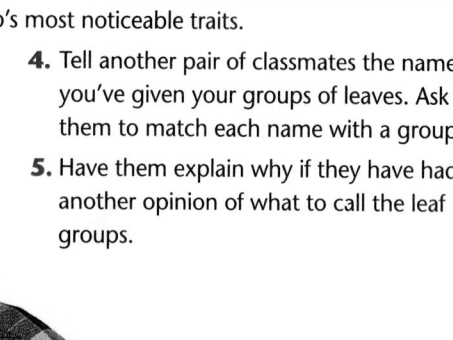
Figure 7-4

Figure 7-5

Explore! ACTIVITY

How would you name things?

What To Do

1. Obtain an assortment of tree leaves from your teacher.
2. Work with a partner to determine traits of the leaves that enable you to classify them into two or more groups.
3. After you have classified them, write a name on an index card that describes each group's most noticeable traits.
4. Tell another pair of classmates the names you've given your groups of leaves. Ask them to match each name with a group.
5. Have them explain why if they have had another opinion of what to call the leaf groups.

Answers to Questions

Students should be prepared with reasonable explanations for the names they choose.

Assessment

Performance Each student pair can create a poster that explains their leaf groups. Provide art supplies such as colored paper, paint, markers, and yarn. Posters can then be displayed in the classroom. Use the Performance Task Assessment List for Poster in **PASC**, p. 73. P L1

Common Names

People use common names, such as *robin*, for organisms that they are familiar with. Usually, that's all right because people from the same region know what organism everyone is referring to. But scientists around the world would run into trouble if they relied on common names. As in the case of the robin, they have to be able to communicate accurately with each other even if they don't speak each other's language.

A Good Reason for Classification

Organisms are classified according to their traits and given names that distinguish them from other kinds of organisms. Such a system enables scientists around the world to know they are communicating about the same organism. This is important, especially when there is a need to identify a disease-causing organism where use of a wrong scientific name could result in misdiagnosis or death.

The classification system that scientists rely on gives a unique name to each kind of known organism. The name identifies the organism and enables scientists to know how it is related to other organisms that are similar to it. Although all three birds in **Figure 7-4** through **7-6** each has the common name robin, each one has a different scientific name. A scientific name distinguishes each robin from all other kinds of robins and all other kinds of birds.

In the 1700s, a classification and naming system was developed by a Swedish scientist named Carolus Linnaeus. Linnaeus originally began by grouping all organisms into two large categories: animals and plants. These categories are called kingdoms. A **kingdom** is the most general and the largest group of organisms in the classification system. Today, scientists recognize three other kingdoms in addition to plants and animals: monerans, protists, and fungi. Turn the page to learn about each of these kingdoms.

Figure 7-6

Connect to...

Earth Science

Organisms aren't the only things with a classification system. Find out what the Mohs' mineral scale is used for.

How Do We Know?

Linnaeus and the Early History of Classification

Carolus Linnaeus established the modern system of classification in the 1700s. At an early age, Linnaeus's friends and relatives noted his love for the natural world, especially botany (the study of plants).

In 1728, Linnaeus attended medical school for a short time. There, he met a botanist who persuaded him to continue his interest in plants. Later, Linnaeus became a professor of botany, and directed a major scientific expedition to study plant life in the Arctic.

It was from these early studies that Linnaeus got the idea to establish a system of classification. The 1700s were important for the field of botany. Many of the world's plants were being described and studied by scientists at this time. Linnaeus established his system to help students of botany quickly put these newly described plants into categories.

Daily Life

Help students recognize that they use informal classification of plants and animals every day. For example, most students quickly learn to recognize poison ivy and poison oak if they live in an area where they are widespread. Similarly, they distinguish among flies, yellowjackets, bees, mosquitoes, horseflies, and deerflies. Have students share how they classify dogs, cats, birds, food, and so on. L1

Content Background

Nearly 2000 years before Linnaeus began his work, the Greek philosopher Aristotle developed a system to group living things. He divided all organisms into two kingdoms: plants and animals. He then divided the animal kingdom into groups depending on where the animals lived—on land, in water, or in the air. The plant kingdom was also divided into three groups. Eventually, it was obvious that there were too many exceptions. Frogs spend time on land *and* in water: to which category would they belong? Yet Aristotle's system was used for nearly 2000 years until it was replaced by the work of Linnaeus.

Uncovering Preconceptions

Many students may think there are only two kingdoms: plants and animals. Spend extra time discussing the other three kingdoms. Monerans are one-celled organisms lacking a nucleus, or central part that controls the cell. Protists can be one-celled or many-celled and have a nucleus. Fungi may look plantlike, but are not green and they obtain food by absorbing it from other organisms. Plants are complex, many-celled organisms that make their own food from sunlight. Animals are complex, many-celled organisms that eat other organisms.

Life's Five Kingdoms

Linnaeus's system consisted of only two kingdoms into which he grouped all organisms. As scientists studied the characteristics of more and more different kinds of organisms, they realized that more kingdoms were needed to classify accurately. Since the 1950s, living things have been grouped into five kingdoms.

On these two pages are but a few examples of the members of each kingdom. The characteristics of each kingdom are also given.

Figure 7-7

■ Life's Diversity

You can see from the examples that each kingdom contains much variety. During your lifetime, you may see only a few of these organisms, but each is important in the living world. Each organism plays a role which you will learn more about in Chapters 12 and 16, which both discuss relationships in the living world.

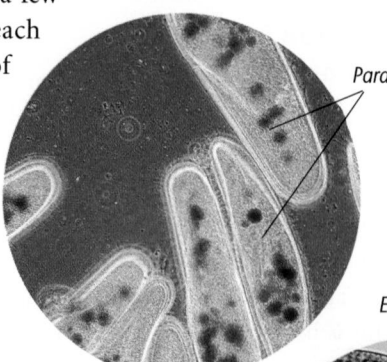
Paramecia

Euglena

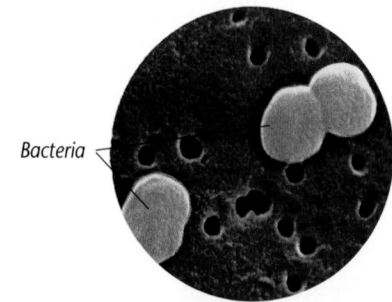
Bacteria

Moneran Kingdom

Representatives of the group of one-celled organisms known as monerans have been present on Earth for about 3.5 billion years. The single cell of the moneran has a very simple organization and does not include a nucleus. About 1800 species of monerans have been named. Monerans are grouped into bacteria and cyanobacteria.

Cyanobacteria

Protist Kingdom

A The cells of protists, such as paramecium and Euglena, are more complex and organized than monerans. Protists have a nucleus that controls the cell's activities. Protists have lived on Earth for about one billion years. About 38 000 species have been identified.

B Protists are a varied group. Most protists are one-celled, but some are many-celled. Some protists, like the paramecia, swim by moving surface hairs; others whip around with a single hair-like structure. Some protists are animal-like, some others are plant-like and can make their own food, while some are parasites. You will learn about monerans and protists in Chapter 8.

228 Chapter 7 Describing the Living World

Program Resources

Making Connections: Across the Curriculum, p. 17, Classifications L2

Teaching Transparency 14 L2

Laboratory Manual, pp. 43–46, Classification L1

Take Home Activities, p. 15, Classifying Fingerprints L1

Mushroom

Yellow morel

Roquefort cheese

Fungus Kingdom

A Fungi are either one-celled or many-celled. In Chapter 8, you'll learn how fungi obtain food by decomposing other organisms.

B Various types of mushrooms and yeast are examples of fungi. Even the blue part of Roquefort cheese is the result of a fungus.

Plant Kingdom

A There are at least a quarter of a million known plant species. Scientists suspect that many more species will be discovered, especially in tropical rain forests. Plants are many-celled organisms, yet their ancestors were probably one-celled green algae—members of the protist kingdom.

B The oldest land plant fossils are about 400 million years old. In Chapter 10, you'll learn how plants make their own food by using light from the sun.

Plant fossil

A rain forest in Olympic National Park, Washington

Animal Kingdom

A Humans and butterflies are just a few of the groups of organisms classified as animals. Animals are many-celled organisms with complex body structures. In Chapter 9, you'll learn how animals must consume other organisms to survive.

Human and butterfly

B There is evidence that animals first appeared on Earth about 700 million years ago. Today, there are about one million known animal species. Some animals, such as certain worms, are so small they can be seen only with the help of a microscope. Others, such as the blue whale are the largest organisms on Earth.

 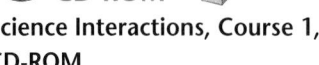

Beyond the Five Kingdoms

Activity To reinforce students' classifying skills, write the Linnaean categories on separate sheets of cardboard or stiff paper. Thumbtack them to a bulletin board in jumbled order and have students rearrange them in correct order from kingdom down through species. L1

Across the Curriculum

Language Arts

Discuss how the common names of plants and animals can be misleading. For example, starfish are not fish and prairie dogs are more closely related to squirrels than to dogs. Have students suggest other names that could be misleading, such as sea horse and elephant seal. L2

SKILLBUILDER

Students should produce a pie graph with areas in the following proportions: about 70% animals, 20% plants, 7% fungi, nearly 3% protists, and not even 1% monera. The portion of the pie graph for monerans should be so small that it is only a labeled line.

Beyond the Five Kingdoms

SKILLBUILDER

Making and Using Graphs

Make a pie graph to show the number of species in each kingdom. If you need help, refer to the **Skill Handbook** on page 657.

Kingdom	Number of Species
Monera	1800
Protist	38 000
Fungi	100 000
Plant	285 000
Animal	1 million

Classification doesn't stop with five kingdoms however. Organisms within kingdoms are grouped into smaller and smaller categories.

■ The Subgroups

The kingdom is still the largest category into which living things are grouped, just as it was in Linnaeus's time. The plant kingdom is separated into subgroups called divisions. The fungi, monera, protist, and animal kingdoms are separated into subgroups called phyla. (The singular is phylum.) If you compare your search for a song in the record store to the animal kingdom, the store itself would be the kingdom and it would have two phyla—the audiotape section and the CD section.

Each **phylum** is separated into still smaller groups called classes. A class can be compared to the category of music you look under for your song—for example, rock music. Each class is further separated into groups

A History of Plant Classification

Plant classification is a system that is used to group and identify plants found worldwide. When did humans first start classifying plants?

The art on both this page and the next are examples of uses and classification of plants by people of different cultures. The image to the immediate right is a page from an Islamic botany book. The image on the next page is a detail from a mural, painted by Diego Rivera, depicting the preparation of herbal medicines.

A Long History

Plant classification has been practiced for thousands of years in one form or another. People have classified plants based on their uses. For example, some plants were used for food, others for making cloth and dyes. Still others, called herbs, were used for medicine.

Herbals

Around 300 B.C.E., Theophrastus, a Greek, devised a system to classify plants. Theophrastus wrote a book called an *herbal* in which he identified and described plants. To assist him, he hired a number of traveling students to

Purpose

This excursion provides an historical view of classifying organisms introduced in this section.

Content Background

Careers in botany require a college degree in botany or a related biological field, with many positions also requiring an advanced degree. Botany is divided into many specialties, including plant pathology (the study of plant diseases) and paleobotany (botanical fossils). The largest number of botanists are employed as teachers and researchers. Another large group works for government or international agencies such as the United Nations. Oil and chemical companies also employ botanists to help in the search for oil and in the development of fertilizers and herbicides.

Teaching Strategy

Show students a sample of coffee beans

called orders. An order can be likened to the alphabetical sections you find under rock music. Each order is separated into families. A family can be compared to all the audiotapes of a particular rock group.

Each family is further separated into subgroups called genera. (The singular is genus.) A genus can be compared to the particular cassette that has the song you want. Finally, each **genus** is made up of the smallest categories of all—**species**. A species might be compared to the song itself.

Figure 7-8

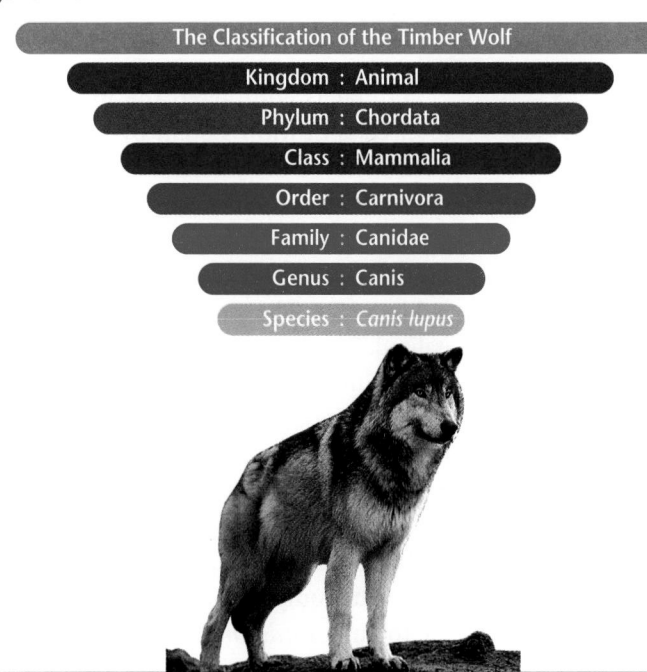

The Classification of the Timber Wolf
Kingdom : Animal
Phylum : Chordata
Class : Mammalia
Order : Carnivora
Family : Canidae
Genus : Canis
Species : *Canis lupus*

collect and observe plants in other places. As a result, he was able to include descriptions of over 500 species of plants.

After Theophrastus, there was little change in plant classification for nearly 2000 years. Classification remained mostly based on how plants could be used, instead of how they looked.

Finally, in 1735, Linnaeus published *System of Nature*, in which he described hundreds of plants and gave each a scientific name. Many of these names are still in use today.

You Try It!

Working with a partner, make an herbal identifying plants found in your area. Describe and draw each type of plant. Make up common names for each of your plants and label them. Find out how some of these plants may have been used.

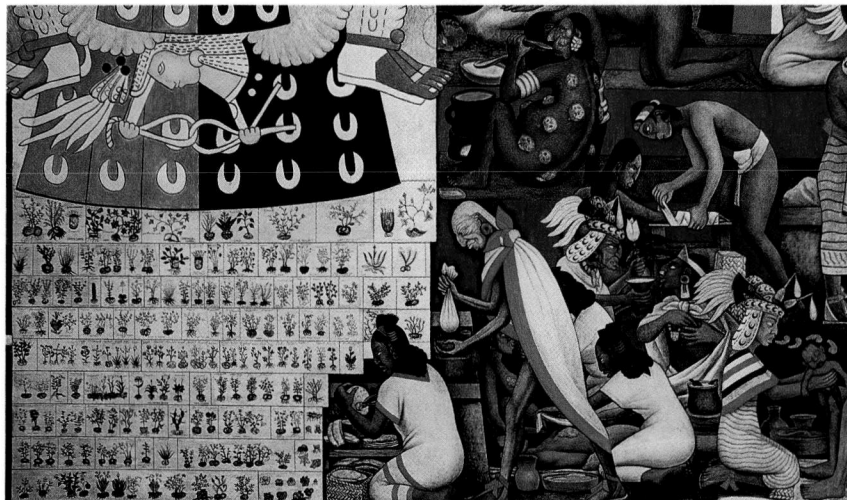

231

and of loose tea. Point out that both come from plants and both are used to make hot drinks. Ask students if they can assume that the plants are closely related. Coffee, known as *Coffea arabica,* of the family Rubiacae, originated in Africa. Tea, known as *Camellia sinensis,* of the family Theaceae, originated in Southeast Asia. The drink coffee is made from the roasted beans of the plant, while tea is made from the dried leaves.

Answer to
You Try It!

Common names will reflect color, shape, location, and other features.

Going Further ▸

Divide the class into small groups. Have each group choose a food, object, or medicine that they believe is made from plants. Have the groups use the library to find out if their chosen object comes from plants. If not, they should choose a new object and

try again. Have each group prepare an oral presentation covering the scientific name of the plant, where it grows, and what products are made from it. Students also can prepare posters showing the structure of the plant and what parts of it are used commercially. L1 **COOP LEARN** P

The questions in Check Your Understanding focus on the purpose and principles of Linnaean classification. Use the Apply question to review why Latin names are used, how they fit into the seven Linnaean categories, and why organisms are given the particular Latin names they have.

Reteach

Discussion Help students relate scientific names to their own names. Their last names tell about the family group they belong to, while their first names distinguish them from other family members. In scientific names, the first name tells what group (genus) the organism belongs to and the second name distinguishes it from other group members. **LEP** **L1**

Extension

Activity To apply their classifying skills, have students work in groups to create a matching game. Students should write descriptions of organisms that offer clues about what kingdom they belong to. Have groups exchange descriptions and try to identify the kingdom of each. Or have the class play the game in two teams. You read the descriptions written by one team to the other team and each correct answer wins one point. The team with the most points at the end wins. **COOP LEARN** **L3**

4 CLOSE

Activity

Ask students to use their research skills to find the scientific names of common plants and animals in a dictionary. Have each student draw a picture of one plant or animal and label it with its scientific name. Display the drawings on a bulletin board in the classroom. **LEP** **L1**

A Scientific Way to Name

Linnaeus completed the multilevel classification system by giving each kind of organism a two-part scientific name. The first part of the scientific name is the genus of the organism, such as *Turdus* for the robin in **Figure 7-4**. The genus is always capitalized. The second part of the scientific name describes the organism more specifically and often tells something specific about the organism. For example, what might *migratorius* tell you about a bird? It tells you that the bird migrates, or moves from place to place. The complete specific name for the robin is *Turdus migratorius*. Other examples are given in **Figure 7-9**. A scientific name identifies a specific organism and only that organism.

Linnaeus used Latin words to name organisms because it was learned and understood by most scientists at that time. Scientists continue to use Latin names today and use Linnaeus's system to name any new organisms they discover. Some organisms have latinized names of the person who identified them, or for the

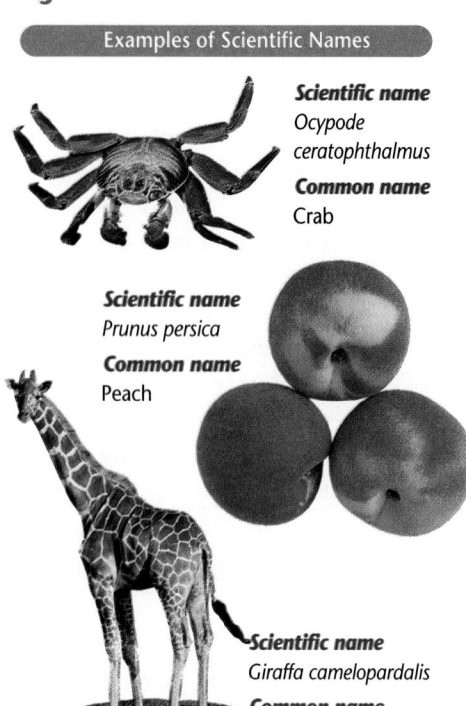

Figure 7-9

Examples of Scientific Names

Scientific name
Ocypode ceratophthalmus
Common name
Crab

Scientific name
Prunus persica
Common name
Peach

Scientific name
Giraffa camelopardalis
Common name
Giraffe

location in which they were found.

In the next section, you will learn that classification continues to change and now involves more than how an organism appears.

check your UNDERSTANDING

1. The scientific name for a particular large cat is *Panthera leo*. Why would a scientist want to use this name when communicating about the organism rather than just "large cat"?
2. Which level of classification would hold the largest number of organisms? Which would hold the fewest?
3. **Apply** Suppose a scientist discovered a new organism and named it *Panthera migratorius*. List everything that the name tells you about the organism.

check your UNDERSTANDING

1. "Large cat" could describe many different animals, from lions to leopards, and cause confusion.
2. A kingdom holds the most organisms; a species holds the fewest.
3. Answers should mention a large cat that migrates.

 **Modern Classification**

Organisms Are Related

You've seen how classifying can help scientists distinguish different species of organisms, but classifying is also important for showing how organisms are related. For example, if there is a new kind of disease-causing bacterium, it is important to know what it has in common with known bacteria. A treatment or medicine that controls or kills a known kind of disease-causing bacteria might also work on a closely related bacterium.

In the Find Out activity that follows, find out how classifications can show where relationships exist between organisms.

Section Objectives
- Demonstrate that a classification system can show how organisms are related.
- Identify the traits scientists use to classify organisms.

Find Out! ACTIVITY

How closely related are three common animals?

What To Do

1. Look at the table below, beginning with the kingdom level, and compare the cat, leopard, and deer at each level.

2. *In your Journal*, note where the animals become different from each other. Then answer the questions that follow.

Conclude and Apply

1. At what level are domestic cats and deer different?

2. At what level are domestic cats and leopards different?

3. Animals closely related to the deer might be found at what levels?

4. Comment on the sentence: "The more levels two organisms both belong to, the more closely they are related."

Group	Domestic Cat	Leopard	Deer
Kingdom	Animalia	Animalia	Animalia
Phylum	Chordata	Chordata	Chordata
Class	Mammalia	Mammalia	Mammalia
Order	Carnivora	Carnivora	Artiodactyla
Family	Felidae	Felidae	Cervidae
Genus	Felis	Panthera	Odocoileus
Species	*Felis cattus*	*Panthera pardus*	*Odocoileus virginianus*

Find Out!

How closely related are three common animals?

Time needed 5–10 minutes

Thinking Processes organizing information, classifying, comparing and contrasting

Purpose To observe and infer that animals with similar classifications are closely related.

Teaching the Activity

Discussion Have students compare the leop-ard, domestic cat, and deer. L1

Troubleshooting Ask students to read aloud the category names for each animal and correct their pronunciation as needed. LEP

Science Journal Students may want to copy the table and answer the questions in their jour-nals. L1

Expected Outcome

Students recognize that the domestic cat and

PREPARATION

Planning the Lesson

Refer to the Chapter Organiz-er on pages 216A–D.

Concepts Developed

Students will discover how scientists use traits to classify or-ganisms and determine their re-latedness.

1 MOTIVATE

Bellringer

Before presenting the lesson, display **Section Focus Trans-parency 23** on an overhead pro-jector. Assign the accompanying **Focus Activity** worksheet. L1 LEP

Activity Have students iden-tify similar and different traits in two animals. L1 LEP

leopard are more closely relat-ed than the domestic cat and deer.

Conclude and Apply

1. at order level

2. at genus level

3. at order, family and genus levels

4. The more levels two organ-isms share, the more features they share and the more closely related they are.

✔ Assessment

Process In pairs, students can predict animals that are related at the genus level to those in the chart. Then have students find the Latin names of the animals. Use the Perfor-mance Task Assessment List for Formulating a Hypothesis in **PASC**, p. 21. L1
COOP LEARN

Time needed 15 minutes

Materials pictures of animals

Thinking Processes organizing information, classifying, thinking critically, comparing and contrasting, interpreting scientific illustrations

Purpose To identify traits useful for classification.

Preparation Distribute five to six pictures to groups of students. Or paste each picture on posterboard and display them.

Teaching the Activity

Discussion Ask students to give other examples of animals that have the same coloring, but have very little in common, for example, killer whales, penguins, skunks, and zebras. Do the same for animals that live in similar environments, for example, frogs and fish. L1

Science Journal Ask students to list other obvious traits that would not be useful for classifying animals. L1

Expected Outcomes

Students should recognize that color and location are not useful traits for classification.

Answers to Questions

1. Answers will vary. Students will recognize that color and location are not useful ways to classify. Neither is fur versus feathers.

What Traits Show Relationships?

You expect a media store to have audiotapes and CDs organized into understandable categories. In the same way, scientists need a classification system based on useful traits—traits that will help scientists see how organisms are related. What characteristics do you think would be useful?

Explore! ACTIVITY

Do obvious traits always show relationships?

What To Do

1. Collect a set of animal pictures. First, classify the animals according to color. What animals have ended up in the same group? How much do these animals have in common?

2. Now group the animals according to where they live: for example, in water, on land, underground, and so on.

3. *In your Journal*, explain if color or location are useful ways to classify.

Modern Tools for Classification

Chemistry CONNECTION

Sometimes the relationship between two different organisms is so close that it is hard to classify them based on body traits you can see or on fossil evidence alone. In such cases, scientists may look inside organisms and examine the life processes that go on.

Chemical Clues

The giant panda is an example of how an organism may be reclassified because of

Brown bear

Giant panda

Raccoon

Chemistry CONNECTION

Purpose

The Chemistry Connection describes how using traits to determine the relatedness of species can lead to reclassification.

Content Background

One of the more recent approaches in animal classification involves studying an animal's genetic material, or its DNA. All living organisms contain DNA, which contains a code that determines how an organism will look, and to a degree, how it

will act. Because DNA is slightly different from organism to organism, scientists use it to find out which organisms are related.

In one method, DNA taken from the cells of one organism is joined together with DNA from another organism. The more closely related the two organisms are, the better the DNA joins together, or matches. Scientists have used this method, called DNA hybridization, to reclassify certain animals. For example, panda bears were thought to be related to raccoons

How useful were color and location as classifying traits? If scientists classified organisms just by color or location, what do you think they would say about black bears, black widow spiders, and black birds?

As you've seen, this classification system also groups very different organisms together so it isn't a very useful approach.

Information from the Past

Scientists also study the traits of fossils, the ancestors of today's organisms. Fossils show how organisms that lived millions of years ago are related to organisms that live today. Fossil evidence has demonstrated that horses and donkeys are more closely related than horses and goats because horses and donkeys have more ancestors in common than do horses and goats.

Using a Key

A variety of methods are used to identify related organisms. As you have learned, many are based on appearance. One tool, called a *dichotomous key* is used to identify organisms. There are keys for plants, mushrooms, fish, butterflies, and every other kind of organism. In the Investigate on the next page, learn how to identify two kinds of birds by using a key.

new evidence. Giant pandas were thought to be bears. In the 1980s, pandas were reclassified as raccoons, animals that share some of the panda's physical traits. But later studies of their body structures and especially of the chemical composition of their cells, showed that giant pandas were more closely related to bears. Modern chemistry techniques can separate the DNA, the genetic or inherited code, of one organism from another. Similar DNA codes indicate that two organisms are related.

As you can see, classification is not written in stone, so to speak. Scientists continue to develop finer tools for determining the shared and the unique characteristics of organisms. As more is learned, new ways to classify living things are accepted.

Birds of a Feather

To refine classifications of some organisms, scientists also use tools that help them look at the microscopic details of organisms. Dr. Roxie Laybourne of the National Museum of Natural History in Washington, D.C. uses an electron microscope for bird classifications. For over 30 years, Dr. Laybourne has been studying the structure, shape, and coloring of bird feathers to find out clues that show how birds are related. The electron microscope allows Dr. Laybourne to see the small details of bird feathers. "I'm studying what it's like on the inside of the barbules, the smallest 'hairs' on the feathers," says Dr. Laybourne. "With the electron microscope, I can see lots of spots inside the barbules. This is another good way to tell one kind of feather from another, and one species of bird from another."

What Do You Think?

What benefits are there to using a combination of classification tools rather than just one?

7-3 Modern Classification **235**

Explore!

✔ **Assessment**

Oral Have student pairs identify other traits for classifying animals and regroup the pictures according to those traits. Use the Performance Task Assessment List for Making and Using a Classification System in **PASC**, p. 49. [L1]

2 TEACH

Tying to Previous Knowledge

In the previous section, students learned about Linnaean system of classification and what kinds of organisms are grouped in the five kingdoms. Tell students that in this section, they will learn how scientists figure out how to classify organisms.

Theme Connection The theme of systems and interactions is supported in this section. Students learn how to apply a system of classification to individual organisms. In the Find Out activity on page 233, students compare classification of various animals. In the Explore activity on page 234, students attempt to identify useful traits for a classification system.

Inquiry Questions Ask students to think about how scientists classify fossil animals. **What traits of living creatures must these scientists do without?** *Answers may include body coverings, color, and methods of reproduction.* **What traits do they think scientists use?** *Answers may include the shape of bones or other hard body parts.*

and not bears, because of some physical similarities. Scientists have determined that pandas really are bears and not related to raccoons.

Teaching Strategy

Divide the class into six-person teams and each team into two three-person groups. Have each group develop a list of traits for each of the people in the other three-person group in their team. Then have the group decide which two of the three people in the other group are more alike and report their results to the class.

COOP LEARN [L1]

Answers to

What Do You Think?

Students' answers will probably state a combination of methods.

Going Further ▸▸▸▸▸

Have students research the classification of pandas. Students should write a brief essay describing the controversy and explaining why pandas have been reclassified twice. To extend this activity, have students research the classification of the horseshoe crab, which was once believed to be related to other crabs but has since been shown, through studies of its blood, to be more closely related to spiders than to other crabs. Use the Performance Task Assessment List for Investigating an Issue Controversy in **PASC**, p. 65. [L3]

7-2 Using a Key

Planning the Activity

Time needed 10 minutes

Purpose To use a key in classifying organisms.

Process Skills organizing information, making and using tables, thinking critically, comparing and contrasting

Materials No special materials or preparation are required for this activity.

Teaching the Activity

Demonstration To help students understand how the key works, illustrate steps 1–3 on the chalkboard in the form of a flowchart, with the instructions in boxes connected by lines. Each box should contain a question such as **"Crest on head?"** and the words *yes* and *no*. Lines connect *yes* and *no* to the next appropriate box.

Process Reinforcement Be sure students set up their data tables correctly. The side of the table should include labels that correspond to the photographs in the student text.

Possible Hypotheses Students may hypothesize that the birds pictured are not very closely related. Therefore, they will expect them to belong to different genera.

Possible Procedures Remind students to study the bird they are identifying as they make their choices. Walk around the room as students work. If they have the wrong answer, direct them to try again and check their choices.

Using a Key

In this activity, you will learn to use a key to identify jay birds. A key is a step-by-step guide for identifying organisms that requires that you make a choice between two statements at each step until a name is reached for an organism.

Problem

How is a key used to identify jays?

Materials

paper and pencil

What To Do

1 Look at the two jays pictured on the next page.

2 Begin with Step 1 of the Key to Jays of North America. Select one statement that is true about that bird and follow the direction it takes you in. Follow each succeeding step until you identify the bird by its common and scientific names. Use the key to classify the bird labeled **A**.

3 *In your Journal*, make a data table like the one shown. Write the common name and scientific name for the jay.

4 Now use the same procedure to classify the species of jay labeled **B**.

Sample data

Data and Observations		
Jay	Scientific Name	Common Name
A	Cyanocitta cristata	Blue jay
B	Cyanocorax yncas	Green jay

Meeting Individual Needs

Visually Impaired Visually impaired students will not be able to do the Investigate as written. To give them practice with classification, have one or more students help you write detailed descriptions of the animals in the pictures. You or a fully sighted student can describe a group of animals to visually impaired students. Have these students suggest characteristics that can be used for classification.

Program Resources

Study Guide, p. 29

Activity Masters, pp. 33–34, Investigate 7-2

Science Discovery Activities, 7-1, Tree Classification; 7-2; Winging It; 7-3, Name that Fossil

Section Focus Transparency 23

A

B

Key to Jays of North America

1a If the jay has a crest on the head, go to Step 2.

1b If the jay has no crest, go to Step 3.

2a If the jay's crest and upper body are mostly blue, it is a blue jay, *Cyanocitta cristata*.

2b If the jay's crest and upper body are brown or gray, it is a stellar's jay, *Cyanocitta stelleri*.

3a If the jay is mostly blue, go to Step 4.

3b If the jay has little or no blue, go to Step 6.

4a If the jay has a white throat, outlined in blue, it is a scrub jay, *Aphelocoma coerulescens*.

4b If the throat is not white, go to step 5.

5a If the jay has a dark eye mask and gray breast, it is a gray-breasted jay, *Aphelocoma ultramarinus*.

5b If the jay has no eye mask and has a gray breast, it is a pinyon jay, *Gymnorhinus cyanocephalus*.

6a If the jay is mostly gray and has black and white head markings, it is a gray jay. *Perisoreus canadensis*.

6b If the jay is not gray, go to Step 7.

7a If the jay has a brilliant green body with some blue on the head, it is a green jay, *Cyanocorax yncas*.

7b If the jay has a plain brown body, it is a brown jay, *Cyanocorax moria*.

Analyzing

1. Using the key, how many species of jay can you *infer* are in North America?

2. How many genera can be identified with this key?

Concluding and Applying

3. How do you know that this key doesn't contain all the species of jays in the world?

4. Why wouldn't you be successful in identifying a robin using this key?

5. ~~Going Further~~ Why wouldn't it be a good idea to begin in the middle of a key, instead of with the first step?

Science Journal Have students record their data and responses to the questions in their journals. ⃞L1

Expected Outcomes

Students will use the key to identify the birds and be able to draw conclusions about bird species and genera.

Answers to Analyzing/ Concluding and Applying

1. Eight different species

2. Five different genera

3. The title of the key specifies that it is a key to the jays of North America.

4. The key identifies jays, not robins.

5. The first pair of descriptions are the most general. The following descriptions are more specific. Correct identification depends on making the correct choice between the first pair of descriptions.

✔ Assessment

Content Ask students to list the traits they used to identify the two birds. Use the Performance Task Assessment List for Analyzing the Data in **PASC**, p. 27. ⃞L1

Multicultural Perspectives

Six Thousand Ways to Say "Camel"

Research into classification systems of other cultures has shown that people who live according to ancient, traditional ways often have extremely detailed systems of classifying organisms. The bedouin of North Africa, for example, have nearly 6000 descriptive words for *camel*, versus perhaps six words used by Europeans or Americans. The Hanunóo of the southern Philippines identify 1625 different plant forms grouped into 890 categories. A Western botanist would identify the same collection as consisting of 1100 species and 650 genera. The two broadest categories of Hanunóo classification are things that cannot be named and things that can be named. Have students consider what items they would put in those two categories.

3 ASSESS

Check Your Understanding

Extend question 1 by asking students to name the species and genus for both animals. *Panthera* is the genus they share; *pardus* and *tigris* are their distinct species. Use the Apply question to review all of the traits that scientists use in classification.

Reteach

Classifying Demonstrate how to use a key by making a transparency of a key. Show students photographs of an organism that can be identified using the key. Have them go through the identification steps aloud. **L1**

Extension

Classifying Have students select four to five common objects and write a key that would allow a person who had never seen them before to identify what they are. The key may be written in list form or as a flowchart. **L3**

4 CLOSE

Activity

To reinforce students' classification skills, obtain photographs of several plants and animals and learn the classification of each. Then show students the photographs, two at a time, and ask them to guess how many categories, from kingdom through species, the organisms share. **L1**

Kingdom : Animalia

Phylum : Chordata

Class : Mammalia

Order : Carnivora

Family : Ursidae

Genus : Ursus

Species : *Ursus arctos*

■ In Summary

With numerous levels in the classification system in use, how do scientists know which group an organism belongs to? You have seen that an organism's traits are useful in determining this. However, your idea of a trait should now be larger than it was when you started this chapter.

Now you know that scientists may look at all the organism's traits. These include cell structure, methods of reproduction, methods of obtaining food, body structure, body coverings (hair, fur, feathers), color, size, and so on. By examining the traits an organism has in common with others and the traits that make it unique, scientists can place the organism into the appropriate kingdom, phylum, class, order, family, genus, and species.

Figure 7-9

As you have learned, the classification of an organism, such as the brown bear shown here, is based on more than merely looking at its coat color. In identifying and naming an organism, its relationships to other members of its kingdoms are revealed.

check your UNDERSTANDING

1. The scientific name for the leopard is *Panthera pardus*. The scientific name for the tiger is *Panthera tigris*. Explain how their names indicate that leopards and tigers are related.

2. Why is height not a good trait to use in classifying trees?

3. **Apply** You are given an unknown organism to classify. How will you go about identifying and naming it?

check your UNDERSTANDING

1. The scientific names show that both animals belong to the same genus, *Panthera*, the closest relationship two different species can have.

2. The heights of trees can change with their age.

3. Identify the organism's traits and compare the traits to those of other organisms. You might use a key to identify and name an organism. If it is an organism that no one has ever identified before, you may give it a two-part name based on its relationship to similar organisms, where it was found, or the name of the person who found it.

Geography Connection

Hydrothermal Vents

Can you imagine strange places at the bottom of the ocean where water comes out of Earth at temperatures of more than 300° F and where many unusual organisms thrive? Such places are called hydrothermal vents.

Hydrothermal Vents

There are many cracks in the ocean floor. So, what causes hydrothermal vents? Cold water (34°–37° F) sinks down into the cracks, where it comes into contact with hot rocks in Earth that heat the water to great temperatures. The hot water dissolves chemicals found in the rocks and flows back into the ocean floor. The vents are most common in the deepest parts of the ocean, as deep as 20 000 feet.

Who Lives There?

Groups of unusual animals such as those in the photograph on this page seem to thrive near these vents. Usually, at such depths, lone animals survive on debris or on any prey with which they come in contact. Therefore, scientists were indeed surprised when, in 1977, off the coast of South America, they found a large community of animals living near a hydrothermal vent.

Scientists were even more surprised that such communities could exist when the chemicals flowing from the vents usually would be deadly to most organisms. It since has been discovered that in the bodies of the organisms that live near the vents, bacteria are present that convert the harmful chemicals into food and energy for the larger organisms.

The organisms found in vent communities are numerous and unique. They include giant clusters of tube worms that grow to lengths of ten feet, huge white clams, and jellyfish-like animals called siphonophores.

You Try It!

Find out more about one of the organisms discussed in this article or choose another organism found in the deep sea to research. What special adaptations does the organism have for its life in the deep?

Geography Connection

Purpose
One of the traits of an organism described in Section 7-1 is its adaptation to an environment. The Geography Connection describes an unusual environment to which living things have become adapted.

Content Background
Hydrothermal vents were discovered by the research ship *Knorr*, from Woods Hole Oceanographic Institute. The scientists went to the eastern Pacific Ocean, where a previous team had measured unusually warm water temperatures on the seabed 1.5 miles deep. They surveyed 10 miles of seafloor with a camera-mounted "sled." When photos revealed the warm springs, two scientists descended in the U.S. Navy's *Alvin* submarine. They found up to 2.5 gallons of hot water pouring out of vents every second and a dark "smoke" of minerals in the water. Huge white clams, brown mussels, and white crabs clung to the rocks on the dark bottom.

Teaching Strategy
Ask students why scientists were so surprised by the hydrothermal vent communities. Lead students to see that the absence of sunlight means that there is normally little plant or animal life on the bottom.

Answer to
You Try It!
Students can check encyclopedias under Oceans, Bathypelagic Zone, and Abyssal Zone and the card catalog under Hydrothermal Vent. Have them present their results in captioned drawings that show the animal's adaptations and behaviors.

Going Further ⫸
Divide students into cooperative groups and assign each group a different aspect of vents: geology, environment, living things, and scientific exploration. Have the groups do library research to obtain information. Tape together sheets of paper to create a large drawing area and have the groups work together to draw a cutaway illustration of a vent community. In addition to showing geology, life, and exploration, have students write information they learned on large index cards and tape them to appropriate places on the illustration.

For information, students should consult:

Cone, Joseph. *Fire Under the Sea: The Discovery and Exploration of Hot Springs in the Sea Floor.* New York: Morrow, 1991.

Fodor, R.V. *The Strange World of Undersea Vents.* Hillside, NJ: Enslow, 1991.

Use the Performance Task Assessment List for Group Work in **PASC,** p. 97. [L2] **COOP LEARN**

Science and Society

Purpose

Several traits of organisms cited in Section 7-1 include a need for food and water and adaptation to an environment. Reasons cited in this excursion for species becoming endangered include pollution of water and destruction of food sources and environments.

Content Background

Extinction is an ongoing process of life. New species evolve and existing species are becoming extinct at a steady rate. In addition, however, there have been mass extinctions of species every 26 to 30 million years since the beginning of the fossil record. The most famous is the extinction of the dinosaur for reasons that are still being debated. The next mass extinction may be caused by humans, unless present trends change. Excessive hunting has been a major cause in nearly all 46 modern extinctions of large land animals and of 88 bird species. The passenger pigeon is one of the most famous examples. In 1850, they represented 40 percent of all birds in North America; 50 years later, they had vanished. The destruction of habitat has been the cause of fewer extinctions so far, but is of the greatest concern for the future. Tropical forests, for example, support half of Earth's 5 to 10 million species in about 7 percent of the total land area. From one quarter to one third of tropical forests are already gone, and an additional 75 000 square miles are cut down every year. Unless this rate declines, the forests will be gone in 200 years.

EXPAND
your view

Science and Society

Endangered Species

What do the giant panda, the brown pelican, and the pincushion cactus have in common? Each of these organisms is an endangered species. Endangered species are those organisms that are in danger of becoming extinct. When a species becomes extinct, no members of that species are any longer found on Earth.

Illegal Hunting

Many more organisms have become extinct during the time of Earth's existence than are found today. Usually, extinction is a natural process that may or may not occur over a long period of time. Dinosaurs are probably the best known extinct animals.

Today, however, many species become endangered as the result of the activities of human beings. Humans kill animals for their fur, tusks, and other body parts, as well as for food and sport. Many endangered animals are now protected by laws, but poachers, people who hunt illegally, disobey the laws and kill and trap thousands of endangered species yearly.

The African elephant and the black rhino are examples of animals that are endangered because of poaching. Poachers illegally kill the elephants for their ivory tusks. Black rhino horns are collected for questionable medical reasons.

Habitat Destruction

Besides the demand for products like ivory, another cause for the increasing number of endangered and extinct species is the destruction of the places where organisms live. As the human population increases, so does the need for places for humans to live. The areas where plants and animals live are cleared to make room for housing, industry, roads, and farming for this increased population. When land is cleared, most of the plants are killed. With habitats destroyed, both food sources and territories are disrupted for

Pincushion cactus

organisms such as the giant panda and pincushion cactus.

As a result of the destruction of land, some species of organisms are becoming extinct before they are identified. This is especially true as the largely unexplored rain forests throughout the world are cleared to make room for ranching and other development.

Giant panda

Pollution

Pollution also plays an increasing role in destroying organisms, such as the brown pelican. As water becomes polluted, oxygen and nutrients that are necessary to keep fish on which shore birds depend, plants, and other inhabitants alive disappear. It isn't long before all wildlife are affected.

It is too late to save species that are already extinct, but what is being done to save those that currently are endangered?

National and international laws have been passed to protect endangered species and stop the sale of products from the species. More than 100 nations have signed the Convention on International Trade in Endangered Species (CITES) agreement. Among other things, the agreement makes it illegal to trade furs and skins of endangered species and bans the

trading of ivory from elephant tusks. It also regulates the trading of live animals and birds for pets.

To protect endangered species in the United States, the first Endangered Species Act was passed in 1973. The act makes it illegal to harm an organism on the Endangered Species list. Also, the law lets the government label certain areas as critical for wildlife. These are areas that species need in order to survive. The government is not allowed to disturb any lands that are identified as critical for wildlife.

Zoos and wildlife preserves are helping to increase the population of many endangered species. Through breeding programs, many of these efforts have been successful. However, in some cases, as with the giant panda, breeding in places other than the animal's natural surroundings is not always easy or successful.

Even with these efforts and more, the fight to save endangered species is far from over. You are likely to see many more species become extinct in your lifetime.

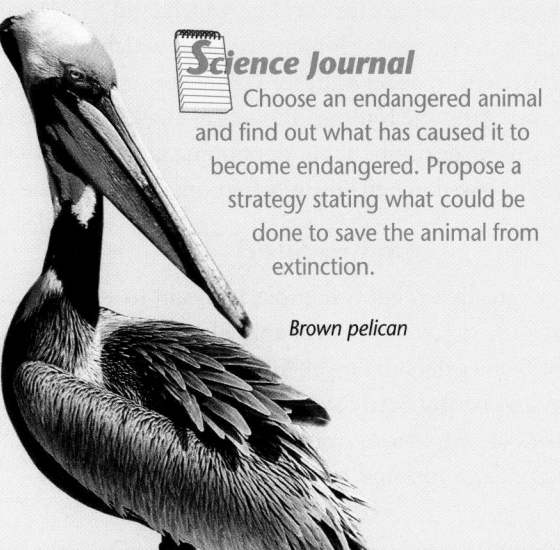

Science Journal
Choose an endangered animal and find out what has caused it to become endangered. Propose a strategy stating what could be done to save the animal from extinction.

Brown pelican

Going Further ▐▐▐▐▶

Form two teams to represent opposing sides in a debate: those concerned about the destruction of species and global warming resulting from deforestation, and those who need to cut down the forests to live. Have the teams brainstorm arguments to support their positions, then stage a debate in which three speakers from each team present the arguments and both teams get one chance to rebut the arguments of the other. Close the debate by suggesting that solutions to the problem will require cooperation between the two sides and ask students for suggestions. Suggestions may include grants from wealthy nations to improve farming methods, investment by wealthy nations in enterprises which provide people with other ways to make a living, and the creation of protected national parks in the third world. Use the Performance Task Assessment List for Oral Presentation in **PASC,** p. 71. `COOP LEARN`
`L2`

HOW IT WORKS

Purpose

This excursion looks at efforts to inventory and save endangered species of animals and plants.

Content Background

Preserving seeds in seed banks may prevent the extinction of plants that could be valuable at some future date. Ideally, seeds in seed banks are stored under controlled conditions. Every so often, depending on the type of seed, the seeds are planted, allowed to go through their life cycle and new seeds are harvested. While seeds that are stored for long periods of time may be viable, some environmentalists question the young plants' ability to survive. Others think that old varieties, which contain invaluable genetic material, have the ability to survive in changed environmental conditions.

Teaching Strategy

Have students suggest reasons why it would not be desirable for the genetic content of plant species to be lost. If possible, show students pictures of the great variety of potatoes grown in the Andes and tell them how each variety grows best under certain climatic conditions. Have students debate the value of spending government funds on seed banks. **L1**

For more information on seed banks write to:

Native Seeds/Search
2509 N. Campbell Ave. #325
Tucson, AZ 85719

Southern Exposure
P.O. Box 170
Earlysville, VA 22936

HOW IT WORKS

Maintaining Diversity

Many species of organisms are rapidly becoming extinct. But how many? And how rapidly?

U.S. Biological Survey

The government is mounting a survey the goal of which is to inventory each plant and animal species in the United States. About 1700 biologists from seven federal agencies (including the Fish and Wildlife Service, the National Park Service, and the Bureau of Land Management) will search wetlands, forests, deserts, bayous, shorelines and mountains.

The result will be a complete picture of the nation's ecosystems. When the survey is complete, it can be used as a biological base for deciding which land can be developed for human uses and which land needs to be set aside for wildlife preserves. After the National Biological Survey is complete, decisions about developing land will be based on reliable information.

Saving Seed

You already know of projects meant to save endangered species of animals, such as the bald eagle. But do you know about seed savings banks? Seed Savers Exchange is a network of people who swap seeds that are becoming more and more difficult to find.

Their goal is to make sure that these species such as the the Scarlet Runner bean do not become extinct. The bean's seeds are pictured on this page.

Most gardeners today buy seeds from seed companies. Seed companies find it profitable to sell mostly hybrid seeds, which come from plants created by crossing two parent varieties. Hybrids tend to be healthy (because of hybrid vigor), and they can be designed to have just the qualities we want.

Before commercial seed companies, gardeners saved seeds from their best plants each year to plant the next spring. They also traded seeds among themselves. The exchange is carrying on that tradition.

Seed Savers Exchange and other seed savings banks know that their mission needs to be carried out now, before it's too late. In 1900, over 7000 varieties of apple trees could be found in the United States. Today there are fewer than 1000 varieties left.

inter**NET** CONNECTION

Information on biodiversity and habitat preservation projects is available from the National Biological Service on the World Wide Web. Locate a project based near you and find out how your class can contribute.

inter**NET** CONNECTION

The home page for the National Biological Service is **http://www.nbs.org**

Going Further ⟩⟩⟩

Have students work in groups to research what can be done to help prevent plant extinctions. One resource students may wish to read is *Rain Forest in Your Kitchen: The Hidden Connection Between Extinction and Your Supermarket* by Martin Teitel. Students can create a booklet, pamphlet, or poster presenting several different things people can do to help prevent plant extinctions. Select the appropriate Performance Task Assessment List in **PASC.** **L2** **COOP LEARN**
P

Science Journal

Review the statements below about the big ideas presented in this chapter, and answer the questions. Then, re-read your answers to the Did You Ever Wonder questions at the beginning of the chapter. *In your Science Journal*, write a paragraph about how your understanding of the big ideas in the chapter has changed.

1 Organisms are made of cells, need water and food, grow, reproduce, respond to stimuli, and adapt to their environments. *Do all organisms always show all of these features all the time? Explain your answer.*

| Kingdom : Animal |
| Phylum : Chordata |
| Class : Mammalia |
| Order : Carnivora |
| Family : Canidae |
| Genus : Canis |
| Species : *Canis lupus* |

2 Animals are classified scientifically by kingdom, phylum, class, order, family, genus, and species. Plants are grouped into divisions instead of phyla. *Suggest reasons why someday scientists might devise other kingdoms in the classification system.*

3 Classifying organisms helps scientists understand how they are related. *How can things like fossils and DNA show relationships among organisms?*

chapter 7
REVIEWING MAIN IDEAS

The suggestions below will provide students with an opportunity to compare and contrast classification levels used in the Linnaean system.

Teaching Strategies

• Ask students to suggest three tests, based on what they learned in Section 7-1, that they could apply to an object in order to determine if it is living. *Possible answers may include observing it to see if it uses food and water, to see if it grows, to see if it reproduces, or to see if it responds to its environment.*

• Ask students questions that test their understanding of how the levels of the Linnaean system are related. Each question should have the following format: **"Two organisms belong to the same family. Are they part of the same phylum?"** *yes*

Answers to Questions

1. All organisms do not always show all the features of life. Reasons will vary. Students may site plants, which require time to observe movement and reproduction; difficulty of seeing cells. Accept all reasonable answers.

2. Accept all reasonable answers. Scientists may use finer chemical differences to separate groups. Some may say that monerans and protists may be regrouped because they are so varied.

3. Fossils may show structural similarities among ancient groups; DNA will reveal chemical similarities.

Science Journal

Did you ever wonder...

• Six traits: cellular structure, a need for food and water, growth, reproduction, response to the environment, and adaptation. (pp. 220–221)

• Names can be used to distinguish organisms and to show how organisms are related. (p. 227)

• You are both members of the kingdom *Animalia*, the phylum *Chordata*, and the class *Mammalia*. (pp. 231, 233)

Project

Have students select an organism found at home: a pet (other than a cat, which is classified in the text), houseplant, insect, or common household bacteria. Have students observe it closely and write a detailed description of its traits. Have students find the scientific name of the organism by using encyclopedias and field guides at the library. L2

GLENCOE TECHNOLOGY

 MindJogger Videoquiz

Chapter 7 Have students work in groups as they play the Videoquiz game to review key chapter concepts.

Using Key Science Terms

1. Stimulus; genus and species are most specific classification subgroups; also the groups used to identify an organism

2. organism; in classification kingdom is the largest group; phylum is a major subgroup into which a kingdom is divided.

3. Changes in temperature, touch, and sound

4. dogs, people, plants, bacteria

5. Monera, Protist, Fungus, Plant, Animal

6. Subgroup of genus; choices may be any given in chapter.

Understanding Ideas

1. Living things are made up of cells, need water and food, grow, reproduce, respond to their environment, and are adapted to their environments.

2. Scientists use classification to make it easier to study, name, and communicate.

3. A kingdom includes many different phyla.

4. They share more traits with each other.

5. Identification involves naming a specific organism. Classification comes first and places an organism in a broad category after which it can be named.

Developing Skills

1. See reduced student text.

2. Overlap for cat and leopard: Animalia, Chordata, Mammalia, Carnivora, and Felidae. Overlap for cat and deer: Animalia, Chordata, Mammalia.

3. A field guide is usually devoted to one group. They are not usually keys, but give much of the same information.

Critical Thinking

1. Information in yellow pages is grouped according to large categories, which are subdivided into more refined categories.

Using Key Science Terms

adaptation	phylum
classification	reproduction
genus	species
kingdom	stimulus
organism	

For each set of terms below, choose the one term that does not belong and explain how the two remaining terms are related.

1. genus, species, stimulus

2. kingdom, organism, phylum

Give an example of each of the following.

3. stimulus

4. organism

5. kingdom

6. species

Understanding Ideas

Answer the following questions in your Journal using complete sentences.

1. List the characteristics that make living and nonliving things different from one another.

2. Why is a classification system helpful?

3. How is a phylum different from a kingdom?

4. The more levels two organisms both belong to in the classification system, the more closely they are related. Why?

5. Why is identifying an organism different from classifying the same organism?

Developing Skills

Use your understanding of the concepts developed in this chapter to answer each of the following questions.

1. Concept Mapping Complete the concept map on kingdoms shown here.

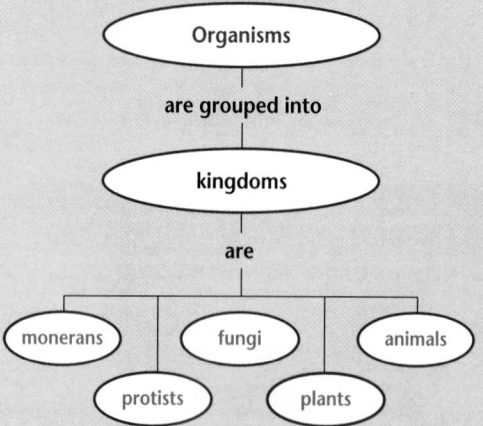

2. Comparing and Contrasting After doing the Find Out activity on page 233 use the classification scheme of the domestic cat and a leopard to construct a Venn diagram. Draw two overlapping circles. Label one circle for each animal named in the table. In the overlapping area, write the scientific names of the groups to which both animals belong. Write the scientific names that are not shared by both animals within each animal's circle, outside of the overlapping section. Make another Venn diagram using the classification scheme of the domestic cat and a deer. Now use the Venn diagram to help you answer the Conclude and Apply questions at the end of the activity.

---

2. Answers will vary but should note decrease in size of kingdom from animals through monerans; that plants and animals grow largest and have most complex structure whereas monerans and protists are smallest and simplest; and that monerans and protists are oldest.

3. They belong to same genus but different species.

Program Resources

Review and Assessment, pp. 41–46 [L1]
Performance Assessment, Ch. 7 [L2]
PASC
Alternate Assessment in the Science Classroom
Computer Test Bank [L1]

---

3. Classifying After doing the Investigate activity on page 236, look at a field guide and see how these guides are used to identify organisms.

Critical Thinking

In your Journal, *answer each of the following questions.*

1. How is a yellow pages phone book similar to scientific classification?

2. Study the pie graph below. Then use information from the chapter to write a paragraph comparing the five kingdoms in terms of size.

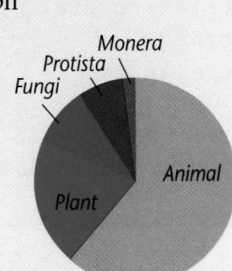

3. The scientific name for present-day humans is *Homo sapiens.* At what level are today's humans related to their ancestors, *Homo habilis?* At what level are we different from *Homo habilis?*

Problem Solving

Read the following problem and discuss your answers in a brief paragraph.

You are traveling through a rain forest. You see a plant that you think has never been described and classified.

1. What could you do to determine if you have discovered a new species?

2. Why would it be important to know if a plant that is used for making a medicine is or is not a new species?

3. What traits would you look for in order to classify and identify the plant?

Problem Solving

1. Observe its traits and compare them to all known species to which it may be related.

2. A new species may offer new and important information about life, or it may be a possible food crop or may even have medicinal uses.

3. size, color, structure, habitat, method of reproduction

Connecting Ideas

1. Sight gives us information about how things look. Hearing gives us information about the sounds made around us. Taste gives us information about whether things are edible. Smell tells us about objects around us without having to look at them. Touch tells us about conditions in the world and the positions of our bodies.

2. Answers will vary, but could include items such as land and water, or mountains, plains, valleys, and plateaus.

3. A person's last name, like an organism's genus name, tells what group he or she belongs to. A person's first name identifies him or her as an individual, but the scientific name of an organism does not refer to individual organisms. Scientific names could not work for people because they classify only to the species level; they do not distinguish different members of the same species.

4. Many endangered species, some of which are unknown, are not covered by the laws, and the laws can be difficult to enforce.

CONNECTING IDEAS

Discuss each of the following in a brief paragraph.

1. Theme—Systems and Interactions Explain how senses transmit information about the physical world in which we live.

2. Theme—Scale and Structure Refer to Chapter 1 and create a classification system for different landforms.

3. Theme—Systems and Interactions In what ways does a person's two-part name compare with an organism's scientific name? Would it be possible to name individuals in the same way we name organisms? Why or why not? How would this show relationships?

4. Science and Society If there are laws protecting endangered species, why are species still becoming extinct?

✔ Assessment

Portfolio Review the portfolio options that are provided throughout the chapter. Encourage students to select one product that demonstrates their best work for the chapter. Have students explain what they learned and why they chose this example for placement in their portfolios.

Additional portfolio options can be found in the following **Teacher Classroom Resources:**

Multicultural Connections, pp. 17, 18
Making Connections: Integrating Sciences, p. 17
Making Connections: Across the Curriculum, p. 17
Concept Mapping, p. 15
Critical Thinking/Problem Solving, p. 15
Take Home Activities, p. 15
Laboratory Manual, pp. 43–46
Performance Assessment P

Chapter Organizer

SECTION	OBJECTIVES	ACTIVITIES & FEATURES
Chapter Opener		Explore!, p. 247
8-1 The Microscopic World (3 sessions; 1.5 blocks)	1. **Classify** organisms as being one-celled or many-celled. 2. **Recognize** the difficulty of determining whether or not viruses are living. 3. **Conclude** that a one-celled organism is capable of carrying out as many life functions as a many-celled organism. **National Content Standards: (5–8) UCP1–2, UCP5, A1–2, C1, F1, F5, G1–3**	Find Out!, p. 248 Investigate 8-1: pp. 252–253 Science and Society, pp. 269–270
8-2 Monerans and Protists (2 sessions; 1 block)	1. **Describe** the major characteristics and activities of monerans and protists. 2. **Describe** the way monerans and protists affect other living things. **National Content Standards: (5–8) UCP1, UCP5, A1, C1, F1–2, F5, G1**	Find Out!, p. 255 Skillbuilder: p. 258 Explore!, p. 259 Explore!, p. 260 Find Out!, p. 261 A Closer Look, pp. 256–257 Geography Connection, p. 272
8-3 Fungus Kingdom (3 sessions; 2 blocks)	1. **Describe** the major characteristics and activities of fungi. 2. **Describe** the way fungi affect other living things. **National Content Standards: (5–8) UCP1, A1–2, C1, E1, F1–3, F5, G1**	Explore!, p. 263 Design Your Own Investigation 8-2: pp. 266–267 Explore!, p. 268 Chemistry Connection, pp. 264–265 Teens in Science, p. 271

ACTIVITY MATERIALS

EXPLORE!

p. 247 dry yeast, warm tap water, microscope slide, microscope

p. 259 wet mount slide of pond water, microscope, wet mount slides of tap water

p. 260* diatomite, slide, coverslip, water, microscope, scrapings from the inside walls of a fishtank

p. 263 store-bought mushroom, hand lens

p. 268 bread mold, hand lens, slide, coverslip, water, microscope

INVESTIGATE!

pp. 252–253 3.7 cm × 0.7 cm bolt; 2 nuts to fit bolt; 2 pieces #22-gauge wire, 14 cm long; polystyrene ball, 4.5 cm in diameter; pipe cleaners, cut in 2-cm lengths

DESIGN YOUR OWN INVESTIGATION

pp. 266–267 peaches, apples, or oranges, kitchen knife or fork, water, paper towels, plastic bags, paper plate

FIND OUT!

p. 248* wet mount slide of Euglena, wet mount slide of planarian, flashlight, microscope

p. 255* spoiled milk, unspoiled milk, slide, microscope, methylene blue stain, coverslips

p. 261 wet mount slides of *amoeba* and *paramecium*, labels, microscope

KEY TO TEACHING STRATEGIES

The following designations will help you decide which activities are appropriate for your students.

L1	Basic activities for all students
L2	Activities for average to above-average students
L3	Challenging activities for above-average students
LEP	Limited English Proficiency activities
COOP LEARN	Cooperative Learning activities for small group work
P	Student products that can be placed into a best-work portfolio
	Activities and resources recommended for block schedules

Need Materials? Call Science Kit (1-800-828-7777).

⏱ **OUT OF TIME?** We recommend that students do the activities with an asterisk.

Chapter 8 Viruses and Simple Organisms

TEACHER CLASSROOM RESOURCES

Student Masters	Transparencies
Study Guide, p. 30 **Activity Masters**, Investigate 8-1, pp. 35–36 **Multicultural Connections**, p. 20 **Making Connections: Integrating Sciences**, p. 19 **Making Connections: Technology and Society**, p. 19 **Critical Thinking/Problem Solving**, p. 5	**Teaching Transparency 16**, Protists **Section Focus Transparency 24**
Study Guide, p. 31 **Critical Thinking/Problem Solving**, p. 16 **How It Works**, p. 12 **Multicultural Connections**, p. 19 **Science Discovery Activities**, 8-2, 8-3 **Laboratory Manual**, pp. 47–48, Life in Pond Water **Laboratory Manual**, pp. 49–50, Shapes of Bacteria **Laboratory Manual**, pp. 51–54, Monerans	**Teaching Transparency 15**, The Microscope **Section Focus Transparency 25**
Study Guide, p. 32 **Take Home Activities**, p. 16 **Making Connections: Across the Curriculum**, p. 19 **Concept Mapping**, p. 16 **Activity Masters**, Design Your Own Investigation 8-2, pp. 37–38 **Science Discovery Activities**, 8-1	**Section Focus Transparency 26**

ASSESSMENT RESOURCES	TEACHING & TECHNOLOGY
Review and Assessment, pp. 47–52 **Performance Assessment**, Ch. 8 **PASC*** **MindJogger Videoquiz** **Alternate Assessment in the Science Classroom** **Computer Test Bank**	**Spanish Resources** **Cooperative Learning Resource Guide** **Lab and Safety Skills** **Science Interactions, Course 1, CD-ROM** **Computer Competency Activities**

*Performance Assessment in the Science Classroom

NATIONAL GEOGRAPHIC TEACHER'S CORNER

Index to National Geographic Magazine	National Geographic Society Products Available From Glencoe	Additional National Geographic Society Products
The following articles may be used for research relating to this chapter: • "Viruses: On the Edge of Life, On the Edge of Death," by Peter Jaret, July 1994. • "Bacteria: Teaching Old Bugs New Tricks," by Thomas Y. Canby, August 1993. • "The Disease Detectives," by Peter Jaret, January 1991.	To order the following products for use with this chapter, contact your local Glencoe sales representative or call Glencoe at 1-800-334-7344: • *Newton's Apple Life Sciences,* "Mold." (Videodisc)	To order the following products for use with this chapter, call the National Geographic Society at 1-800-368-2728: • *Everyday Science Explained* (Book) • *Bacteria* (Video) • *Protists: Threshold of Life* (Video) • *Virus!* (Video)

Teacher Classroom Resources

These Teacher Classroom Resources are examples of the materials in the Teacher Classroom Resources package.

TEACHING AIDS

Section Focus Transparencies

Teaching Transparencies

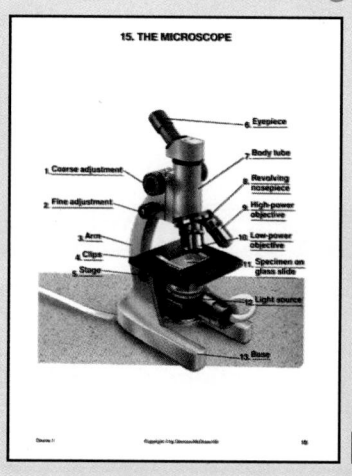

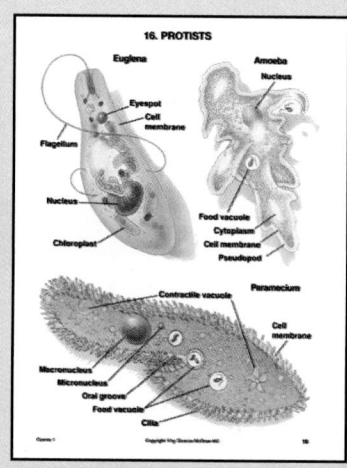

HANDS-ON LEARNING

Science Discovery Activity*

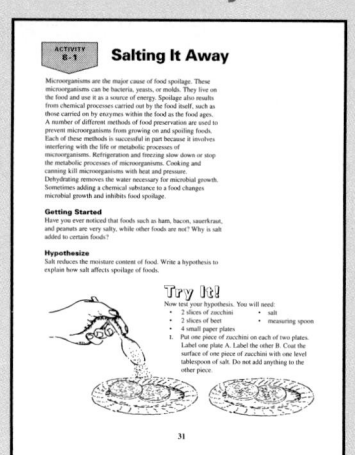

Laboratory Manual*

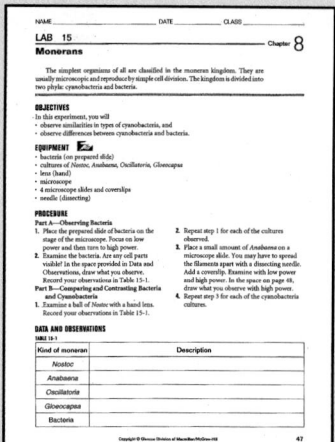

Take Home Activity

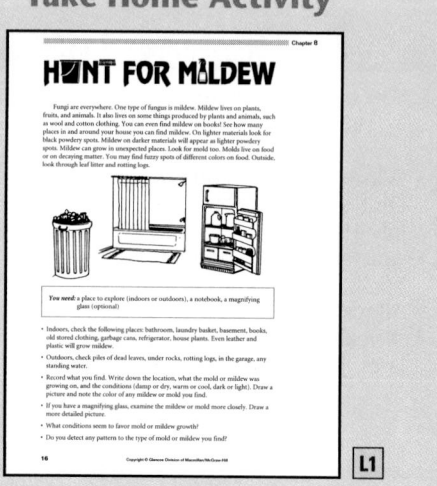

Chapter 8 Viruses and Simple Organisms

REVIEW AND REINFORCEMENT

Study Guide*

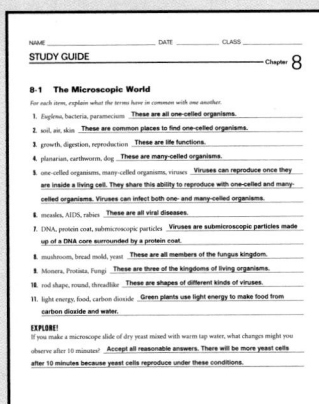

Concept Mapping

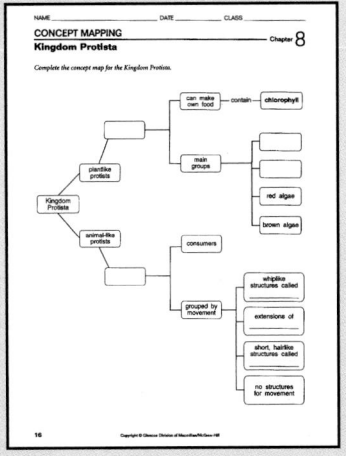

Critical Thinking/ Problem Solving

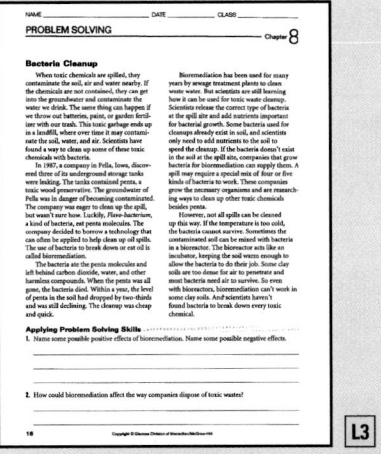

ENRICHMENT AND APPLICATION

Integrating Sciences

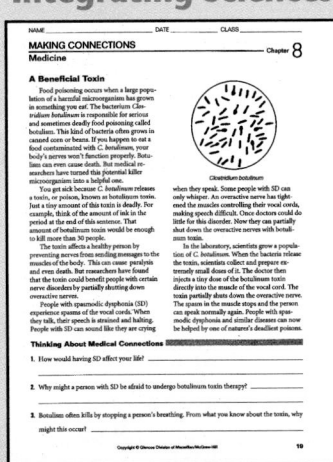

Across the Curriculum

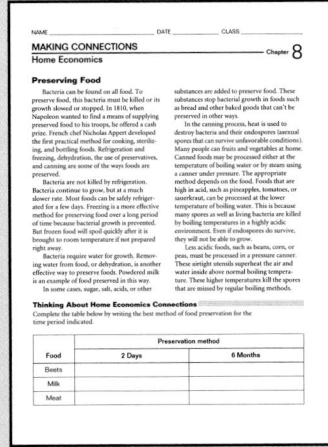

Technology and Society

ASSESSMENT

Multicultural Connection**

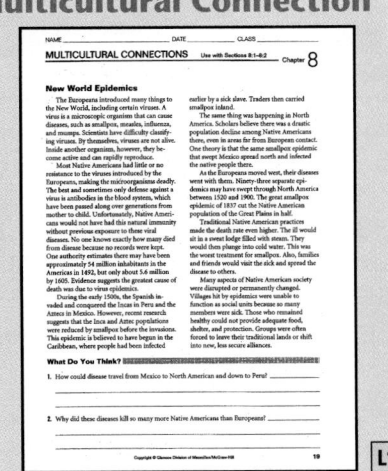

Performance Assessment

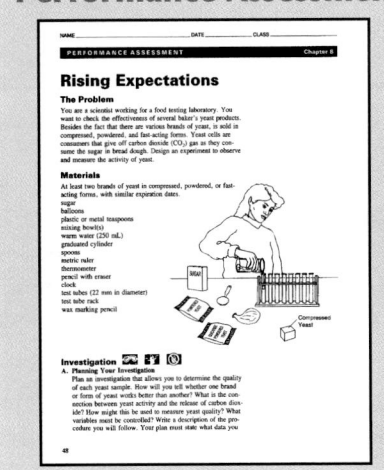

Review and Assessment

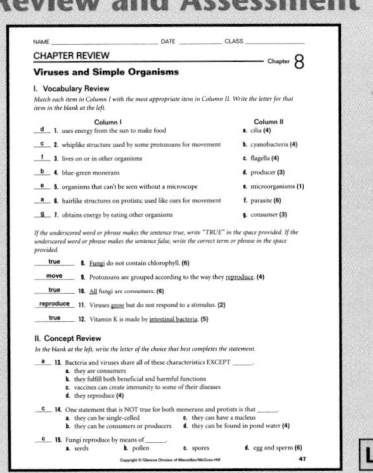

*One per section   **Two per chapter

Viruses and Simple Organisms

THEME DEVELOPMENT

The theme is scale and structure. Section 8-1 explores some of the differences and similarities between one-celled and many-celled organisms. The scale and structure of monerans relative to other organisms and how the structure of cyanobacteria enables them to colonize are explored.

CHAPTER OVERVIEW

Students will discover that one-celled organisms carry out the same life functions as many-celled organisms. Students also explore the nature of viruses.

Tying to Previous Knowledge

Have students brainstorm a list of animals. Then divide the class into small groups and have them devise classification systems.

INTRODUCING THE CHAPTER

Have students discuss the experience of opening refrigerator containers of forgotten foods. Have them describe how the foods looked and smelled, and suggest what was going on.

VIRUSES and Simple Organisms

Did you ever wonder...

✓ **What a virus looks like?**

✓ **Why milk in a refrigerator doesn't usually sour, but milk left out does?**

✓ **Whether all bacteria and fungi are harmful?**

Science Journal

Before you begin to study about viruses and simple organisms, think about these questions and answer them *in your Science Journal.* When you have finished the chapter, compare your journal write-up with what you have learned.

Answers are on page 273.

Helping to clean out the refrigerator on a Saturday morning is probably not your idea of fun. There are all those "mystery" containers. On top of that, when you open them, the mystery continues. Was it cottage cheese, or was it spaghetti sauce? What was once food is now fuzzy and may be pink or green, and definitely smelly. You share your world with many microscopic organisms. Some cause disease and some protect you from disease. Others decompose food and piles of leaves. Some live on your skin and others live in oceans and lakes and produce the oxygen you breathe. You can't live with many of them, and you stay alive because of others.

▶ *In the activity that follows, take a look at yeast, one member of this very important group.*

246

⏲ OUT OF TIME?

If time does not permit teaching the entire chapter, use the Chapter Overview on this page, Reviewing Main Ideas at the end of the chapter, and the Chapter 8 audiocassette to point out the main ideas of the chapter.

Learning Styles	**Kinesthetic**	Activity, pp. 249, 254; Investigate, pp. 252-253, 266-267; Explore, p. 263
	Visual-Spatial	Explore, pp. 247, 259, 260, 268; Find Out, pp. 248, 255, 261; Activity, pp. 250, 256, 262; Visual Learning, pp. 251, 254, 257, 258, 263, 265; Demonstration, p. 263
	Interpersonal	Multicultural Perspectives, p. 251; Activity, p. 259; Chemistry Connection, pp. 264-265
LS	**Intrapersonal**	Science at Home, p. 273
	Linguistic	Across the Curriculum, p. 251; Research, pp. 253, 261; A Closer Look, pp. 256-257; Visual Learning, p. 259; Activity, p. 268

Explore! ACTIVITY

What does yeast look like?

What To Do

1. Examine dry yeast with a hand lens.

2. Mix a small amount of yeast with cold tap water in a small dish.

3. In a second dish, mix very warm, but not steaming hot, tap water with a small amount of yeast. Add a small amount of sugar to the mixture.

4. Place a very small drop of the first mixture on a microscope slide. Add a cover slip.

5. Examine it first under low power, then under high power. *In your Journal*, describe and draw some yeast cells. Then examine a drop of the second mixture.

6. Wait 5 minutes and look at the yeast in both dishes again.

7. What changes do you see? Describe them *in your Journal.*

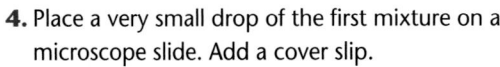

THE ORIGINAL ALL NATURAL YEAST

ACTIVE DRY YEAST — NET WT. 7 g. (¼ OZ.)

ACTIVE DRY YEAST — NET WT. 7 g. (¼ OZ.)

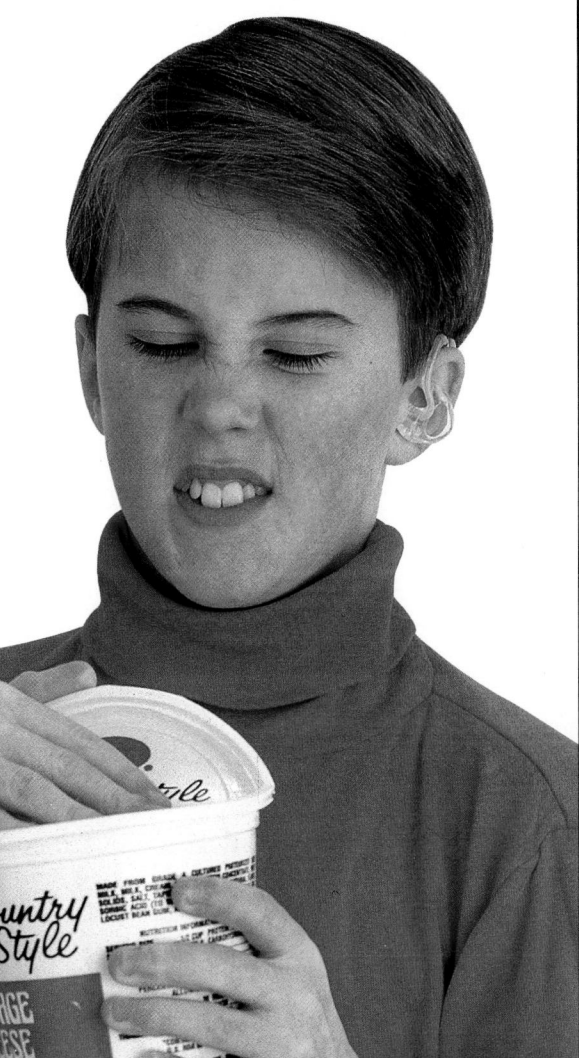

Explore!

What does yeast look like?

Time needed 60 minutes

Materials dry yeast, warm tap water, microscope and slides

Thinking Processes observing and inferring, comparing and contrasting

Purpose To observe that yeast grows in warm water.

Teaching the Activity

Troubleshooting For the yeast to grow, the water must be warm, not hot. In Step 2, have students test the water on their wrists.

Science Journal Have students include before and after sketches of the yeast in their journals. L1

Expected Outcomes

Students will see that yeast are one-celled organisms. Students also should observe that the yeast grow faster in warm water with sugar added.

Answers to Questions

7. Yeast in the dish with sugar showed more growth and some bubbles.

✔ Assessment

Process Ask students to carry out the same activity, using cold tap water. Have students predict what they expect to see under the microscope, then have them describe and draw their results. Use the Performance Task Assessment List for Carrying Out a Strategy and Collecting Data in **PASC**, p. 25. L1

ASSESSMENT PLANNER

PORTFOLIO
Refer to page 275 for suggested items that students might select for their portfolios.

PERFORMANCE ASSESSMENT
Process, pp. 247, 255
Skillbuilder, p. 258
Explore! Activities, pp. 247, 259, 260, 263, 268
Find Out! Activities, pp. 248, 255, 261
Investigate, pp. 252-253, 266-267

CONTENT ASSESSMENT
Oral, pp. 261, 263
Check Your Understanding, pp. 254, 262, 268
Reviewing Main Ideas, p. 273
Chapter Review, p. 274-275

GROUP ASSESSMENT
Opportunities for group assessment occur with Cooperative Learning Strategies.

PREPARATION

Planning the Lesson

Refer to the Chapter Organizer on pages 246A-D.

Concepts Developed

In this section, students will determine if an organism is one- or many-celled and note that all organisms carry out the same life functions. They will also explore viruses.

Find Out!

How do two organisms compare?

Time needed 30 minutes

Materials wet-mount slides of *Euglena* and planarian, penlight flashlights, microscopes

Thinking Processes observing and inferring, comparing and contrasting, recognizing cause and effect, interpreting scientific illustrations.

Purpose To identify common characteristics of complex and simple organisms.

Preparation Label all slides clearly to avoid confusion.

Teaching the Activity

Discussion Ask students what they think a one-celled organism can and cannot do. Remind students that planaria are very small, but made up of many cells.

Science Journal Have students record their descriptions and draw *Euglena* and planarian in their journals. L1

Expected Outcome

Students should conclude that both one-celled and many-celled organisms are active and respond to stimuli such as light.

8-1 The Microscopic World

Section Objectives

- Classify organisms as being one-celled or many-celled.
- Recognize the difficulty of determining whether or not viruses are living.
- Conclude that a one-celled organism is capable of carrying out as many life functions as a many-celled organism.

Key Terms

microorganisms
producer
consumer
viruses

Simple and Complex Organisms

Many of the foods you eat, such as hamburgers and salads, come from large, complex organisms such as cattle and plants. But you might be surprised to know that many of the foods you eat are produced with the help of simple organisms such as bacteria and fungi. "Simple" does not mean that these organisms are unimportant or are capable of doing only one or two things. The yeast you looked at in the Explore activity on page 247 use sugar to obtain energy for themselves in a process that is anything but simple. Many organisms are called *simple* only because they are small. Yet their life processes are as complex as those in larger organisms such as a dog, a flatworm, and you. In the following Find Out activity, observe how some so-called simple and complex organisms compare.

Find Out! ACTIVITY

How do two organisms compare?

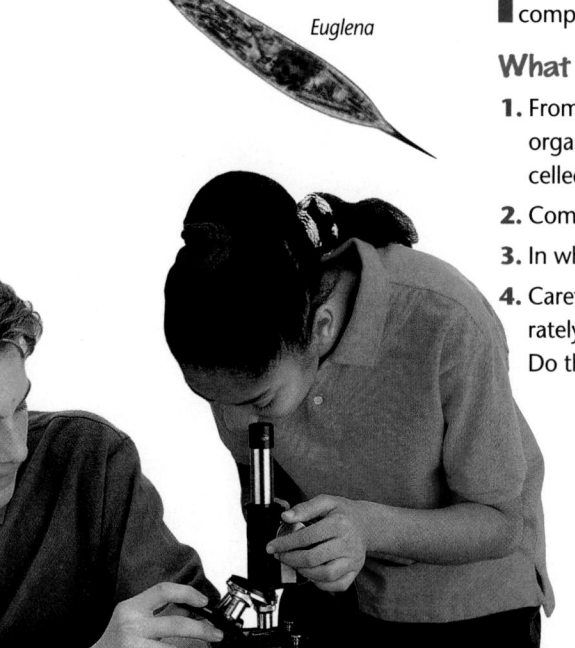

Euglena

Planarian

In spite of differences in size, simple and complex organisms share many things in common.

What To Do

1. From your teacher, obtain a slide of a living one-celled organism called *Euglena* and another slide of a many-celled organism called a planarian.
2. Compare the sizes of the two organisms.
3. In what ways do the organisms resemble each other?
4. Carefully observe the activities of each organism separately. Aim the beam of a penlight flashlight at the slide. Do the organisms respond in any way to the light?

Conclude and Apply

1. Is *Euglena* limited in what it can do because it is only one cell?
2. Does having many cells enable the planarian to do more than *Euglena*?
3. *In your Journal*, write a statement comparing the terms *simple* and *complex* based on your observations.

Conclude and Apply

1. *Euglena* is not limited because it consists of only one cell; it can carry out all life functions.
2. If one part of the planarian is damaged, the other cells can still survive.
3. Accept all reasonable comparisons. Most should say that it would be difficult to tell the difference between the two based on using these live organisms. Some may say the use of the two terms doesn't mean much.

✔ Assessment

Content Ask students to write a paragraph or make a table comparing and contrasting the characterisitics of *Euglena* and the planarian they observed. Use the Performance Task Assessment List for Making Observations and Inferences in **PASC,** p. 17. L1 P

One-Celled Organisms

If you were fortunate enough to be able to look at live *Euglena* in the Find Out activity, you probably noticed that it is a very lively organism. Just because an organism is one-celled doesn't mean that it is dull and uninteresting. The living world contains an enormous variety of one-celled organisms. Like many other living things, one-celled organisms respond to stimuli such as changes in light, heat, or chemicals in their environment. They may move toward or away from a stimulus. One-celled organisms, like the ones in **Figure 8-1**, carry out all life functions. They move, grow, consume food, release energy from food, produce waste products, and reproduce.

of soil, your skin, other animals, or food left too long in the refrigerator to find them. They live in fresh and salt water. They live in you and on you. Many keep you healthy. A few of them make you ill, but most are beneficial. Some are used to make food and others break down food. It's hard to see them, but they are everywhere.

Figure 8-1

One-celled organisms perform all of life's functions. Movement is one of these activities.

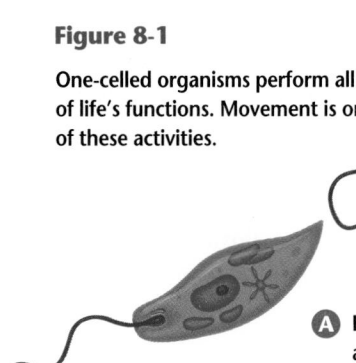
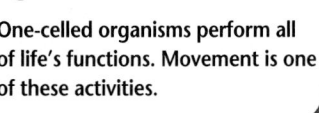

A By alternately contracting and stretching their bodies, Euglenas move themselves along.

B Euglenas also have a thin whiplike structure that helps them to move.

Paramecium

C *Paramecium* rolls rapidly through water, sweeping food into its mouth with the help of numerous hairlike projections that cover its body.

■ Where are they found?

As hard as you might try, it is difficult to find a place where one-celled organisms do not live. You only have to check the air you breathe, a handful

■ How can you see them?

Organisms such as bacteria, protists, and fungi that are too small to be seen with the unaided eye are called **microorganisms**. You already know that living things are called organisms. *Micro-* means *small*, so *micro*organisms are very small organisms.

To see the thousands of one-celled and the small, many-celled organisms that will be studied in this chapter, you will have to look through a microscope, as you did to see the yeast and the *Euglena*.

Meeting Individual Needs

Visually Impaired Assist visually impaired students in completing the Find Out activity on page 248. Pair each visually impaired student with a fully sighted partner. Instruct the partner to complete the microscope steps, describing what is occurring in the slides. **COOP LEARN**

Program Resources

Study Guide, p. 30
Making Connections: Integrating Sciences, p. 19, A Beneficial Toxin L1
Making Connections: Technology and Society, p. 19, Who Should Be Allowed to Manufacture AZT? L1
Critical Thinking/Problem Solving, p. 5, Flex Your Brain
Section Focus Transparency 24

📖 Before presenting the lesson, display **Section Focus Transparency 24** on an overhead projector. Assign the accompanying **Focus Activity** worksheet. L1 LEP

Activity Have a student break an egg into a dish, taking care not to rupture the yolk. Provide students with hand lenses so they can examine the yolk sac. As they observe it, point out that the yolk is a single cell. Explain that a cell can range in size from microscopic to huge. To reinforce this, point out the size of an ostrich egg. L1

2 TEACH

Tying to Previous Knowledge

Begin by asking students to brainstorm characteristics of cells. Have a volunteer write a list on the chalkboard. Then ask students to recall the six characteristics of living things that they used in Chapter 7 to determine if something is alive or not. Have a volunteer list the characterisitics on the chalkboard.

Theme Connection The theme that this section develops is scale and structure. In this section, students see that one-celled and many-celled organisms carry out the same life processes but on different scales. Students also learn how the structures of viruses affect their ability to penetrate cells. This becomes evident when the students study Figure 8-3.

Flex Your Brain Use the Flex Your Brain activity to have students explore SIMPLE ORGANISMS. L1

Many-Celled Organisms

How do one-celled and many-celled organisms compare? Does a many-celled organism have an advantage over a one-celled organism? Many-celled organisms are usually larger than one-celled organisms and of course, have more cells.

The cells in larger organisms also are more specialized. That means that certain cells perform only one type of job. For instance, your bone cells form bone tissue. Bone cells are not like the cells that line your digestive tract that absorb food.

In addition, if one part of a many-celled organism is injured, the organism may be able to repair itself and will probably be able to survive. If you break a leg, your leg heals and you go on living. On the other hand, if the membrane of a one-celled organism is punctured, the organism dies. It would seem that being many-celled might have some advantages.

■ Classifying Simple Organisms

As you learned in Chapter 7, there are five kingdoms of living organisms. In this chapter, you will look more closely at characteristics and examples of three of those kingdoms—Monera, Protista, and Fungi. **Table 8-1** below compares these three kingdoms and illustrates a member of each kingdom. Besides structures, you'll also notice that the organisms are compared in terms of how they either produce or obtain energy. A **producer**, such as a green plant, uses light energy to make food from carbon dioxide and water. A **consumer**, such as an animal or a fungus, is unable to make its own food.

Table 8-1

Characteristics of the Monera, Protista, and Fungi Kingdoms			
Characteristic	Monera	Protista	Fungi
One-celled?	all	most	some
Many-celled?	no	some	most
Has a nucleus?	no	yes	yes
Producer/Consumer?	both	both	consumer
Examples	bacteria cyanobacteria	amoeba *Euglena* algae	mushroom bread mold yeast

Monera: rod-shaped *Bacillus* bacteria

Protista: *Euglena*

Fungi: mushroom

Viruses

You may have noticed that none of the kingdoms includes viruses. **Viruses** are submicroscopic particles made up of a DNA or an RNA center, which is surrounded by a protein coat. DNA is the substance found in cells that determines the characteristics of organisms. RNA is a substance that cells use to make proteins.

Viruses come in a variety of shapes and sizes. You will note that scientists classify viruses by shape as described in **Figure 8-2**.

Why do you think viruses are not included in any kingdom? It has been observed that viruses cannot grow, respond to a stimulus, or break down food to release energy. However, when a virus enters a living cell, it does use materials in the cell to reproduce. The ability to reproduce is one of the few things viruses share in common with living organisms. What do some viruses look like? Work with a model in the Investigate on the next page to find out.

Figure 8-2

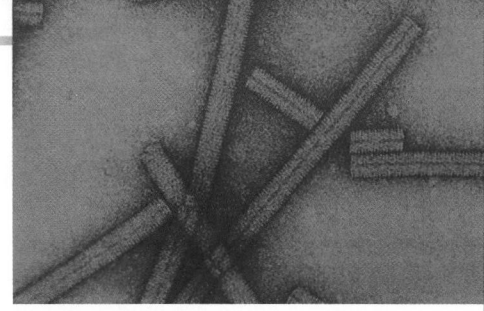

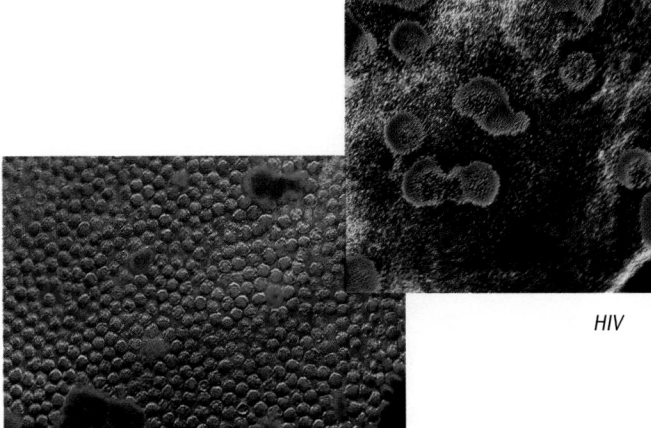

Ⓐ The existence of viruses has been known since the late 1890s when the the tobacco mosaic virus was discovered. Although scientists had proof of the existence of viruses, they had no way to see them until 1939, when the first electron microscopes were available.

Tobacco mosaic virus

HIV

Polio virus

Ⓑ Viruses come in a variety of shapes and sizes. Along with the thin, rod shape of the tobacco mosaic virus, there are viruses that are round or shaped like spiked balls, or thin threads. HIV, which causes AIDS, is round.

Visual Learning

TECH PREP

Figure 8-2 Explore with students the different shapes and sizes of the viruses in Figure 8-2. Inform students that in the 1980s, the smallpox virus became the first infectious agent to be eliminated from the human population. The virus is now held only in two laboratories around the world: at the Centers for Disease Control in Atlanta and in Russia. Recently, scientists debated whether to destroy the last remaining smallpox virus or to continue to hold it. Ask students what arguments might be offered in favor of saving it. **L3**

Across the Curriculum

Language Arts

Have students look up the Latin meaning of *virus* and explain why it was given that name. *It means "poison," which probably refers to its effects on the human body.* **L1**

Daily Life

Some of the diseases caused by viruses include rabies, polio, diarrhea, yellow fever, measles, mumps, chicken pox, shingles, respiratory diseases, warts, hepatitis, rubella (German measles), and multiple sclerosis. Students can research how each of these viral diseases is controlled or prevented. **L2**

How Do We Know?

How do new viruses form?

How new viruses form was a mystery until 1969, when Nobel Prize winner Max Delbrük experimented by injecting two different types of viruses into one type of bacterial cell. The viruses responded rapidly and new viruses erupted, destroying the bacterial cells. Delbrük examined the new viruses and found that they were similar to the two he first injected. He also found a third type of virus that had traits that were a combination of the original two. Delbrük concluded that when two types of viruses enter a cell, DNA is exchanged, and new viruses form.

GLENCOE TECHNOLOGY

 Videotape

The Secret of Life
Use the videotape *Nothing to Sneeze At: Viruses* to explore how the immune system reacts to the influenza virus.

Multicultural Perspectives

Early Vaccines

During the late 19th century, Japanese bacteriologists made great progress in developing serums to prevent diseases prevalent at the time. In 1889, Shibasaburo Kitasato, a Japanese bacteriologist, produced the first pure culture of the tetanus bacillus. His achievement finally made it possible to create a vaccine against tetanus. He was also one of the discoverers of the bacterium responsible for bubonic plague, and helped to produce a serum used to combat plague epidemics.

Have interested students investigate the early history of vaccine production and vaccinations. Students can then work together to create a display to communicate their findings. Use the Performance Task Assessment List for Display in **PASC**, p. 63. **L3** **COOP LEARN**

Shapes of Viruses

Planning the Activity

Time needed 30 to 60 minutes

Purpose To compare and contrast viruses with each other and with cells.

Process Skills interpreting scientific illustrations, making and using tables, classifying, observing and inferring, comparing and contrasting, making models

Materials See reduced student text.

Preparation Be sure to have extra polystyrene balls on hand.

Teaching the Activity

Demonstration Before students begin working, demonstrate how to twist the wire tightly.

Troubleshooting Urge students to work slowly and carefully, referring frequently to the illustrations.

Process Reinforcement Ask students why scientists might make models of viruses. Be sure they understand that scientists use models of viruses to better understand their behavior and to aid other scientists in recognizing a particular virus. Students should recognize that viruses are extremely small, only a tiny fraction of the size of a cell and smaller than the smallest bacteria. As students refer to the illustrations in making their models, urge them to keep the true size of a virus in mind.

Possible Procedures Students may work individually or in pairs. If time is short, one student may construct model A, and the other model B.

Discussion Have students recall that although viruses have different shapes, all viruses consist of a strand of genetic material coated with protein.

Science Journal Have students answer the questions in their journals. Some students may want to draw a diagram of their model in their journals for reference. [L1]

Viruses all contain similar structures and materials, yet they differ greatly in shape. In this activity, you can observe and make models of some viruses.

Problem
How can you make a model of a virus?

Materials
3.7 cm × 0.7 cm bolt
2 pieces #22-gauge wire, 14-cm long pipe cleaners, cut in 2-cm lengths
2 nuts to fit bolt
polystyrene ball, 4.5 cm in diameter

Safety Precautions

Be careful when working with wire.

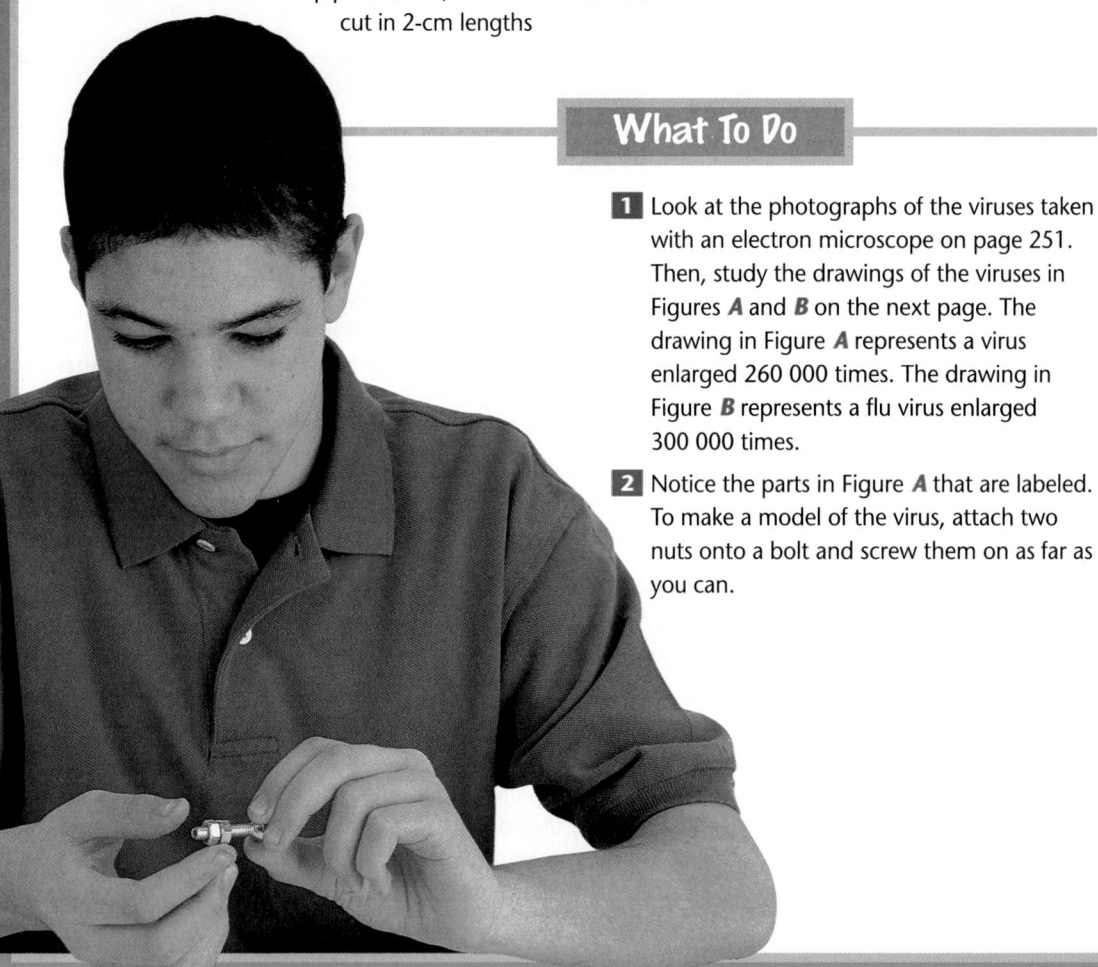

252

What To Do

1 Look at the photographs of the viruses taken with an electron microscope on page 251. Then, study the drawings of the viruses in Figures *A* and *B* on the next page. The drawing in Figure *A* represents a virus enlarged 260 000 times. The drawing in Figure *B* represents a flu virus enlarged 300 000 times.

2 Notice the parts in Figure *A* that are labeled. To make a model of the virus, attach two nuts onto a bolt and screw them on as far as you can.

Program Resources

Activity Masters, pp. 35–36, Investigate 8-1

Across the Curriculum

Math

Explain to students that total magnification is derived by multiplying the eyepiece power (usually 10×) by the objective magnification (10×, 30×, 40×, or 43×). Then ask: What is the total magnification of your microscope if the eyepiece is 10×, and the high power objective is 40×? *The total magnification on high power is 400x.*

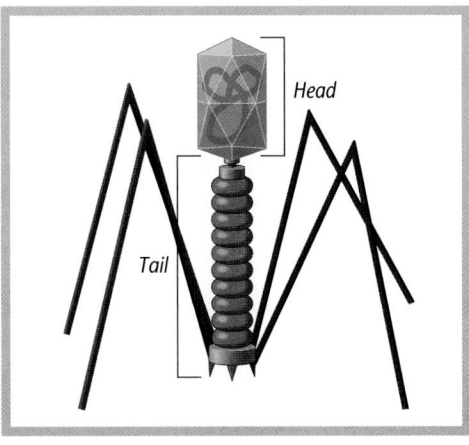

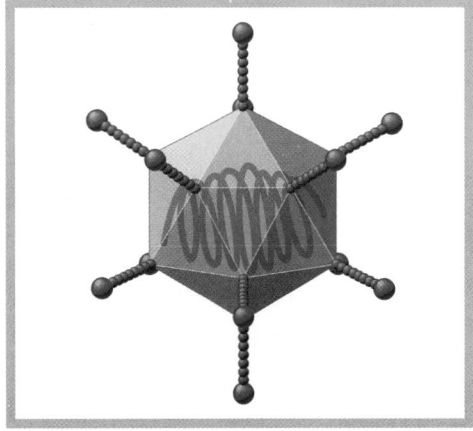

A   B

3 Twist the wires around the bolt near the bottom. Make the wire as tight as you can. Fold the wire ends and bend them so that they look similar to the drawing.

4 Use the polystyrene ball and pipe cleaners to make a model of the flu virus in Figure **B**.

Analyzing

1. *Compare* your models with Figures **A** and **B**. How are they alike?

2. *Contrast* the two viruses. How are they different from each other?

3. What is there about the structure of a virus that seems to help it get inside your cells?

Concluding and Applying

4. How does a virus differ from a cell?

5. **Going Further** What kinds of illnesses have you heard of that are caused by viruses? *Make a table* of at least three diseases caused by viruses that you have researched in library references. In the table, list the disease, what it affects, and where it is found. You may want to use a computer to make your table.

Expected Outcomes

Students will observe that viruses and cells have very different structures.

Answers to Analyzing/ Concluding and Applying

1. Model A seems to have a head, neck, and tail with appendagelike fibers. Model B is sphere-like with spikes coming from it but no other parts.

2. They differ in shape: A has a head, neck, and tail with appendages; B has only a head.

3. Students may say that a virus enters a person's cells by using its shape to attach to the cell membrane.

4. Cells have cytoplasm, cell membranes, and other structures; a virus has only DNA or RNA with a coat around it.

 5. Answers may include flu, AIDS, colds, and chicken pox. Student tables can show three diseases listed down the left of the table and rows across the top that indicate what the disease affects (i.e., chicken pox affects skin) and where the disease is found (i.e. worldwide). Other possible columns would be age-affected; how cured.

✔ Assessment

Performance Have students look at the illustration of a virus on page 254, Figure 8-3-A. Ask them to locate the genetic material (DNA or RNA) in the virus. Have students determine how they could indicate the location of this genetic material on the models they made. Use the Performance Task Assessment List for Model in **PASC**, p. 51. L1 P

GLENCOE TECHNOLOGY

 Software

Computer Competency Activities

Chapter 8

3 ASSESS

Check for Understanding

1. Students can make a chart or poster comparing and contrasting cells and viruses, working individually or in small groups. L1

2. Urge students to review the Investigate activity before responding to the Apply question in Check Your Understanding.

Reteach

Illustrate Have pairs of students write and illustrate a pamphlet showing how a virus infects a cell and how the transmission of viruses can be slowed or prevented. L1 P

Extension

Research Have students research the AIDS virus. They can collect articles from newspapers, books, and magazines. Why are scientists finding this disease so difficult to cure? As a group, have students discuss ways to help society cope with the disease. L3

4 CLOSE

Activity

Students can summarize what they have learned by making drawings or models of one-celled organisms to accompany their models of viruses. Label and display the drawings and models around the classroom. LEP L1

How Viruses Behave—or Misbehave

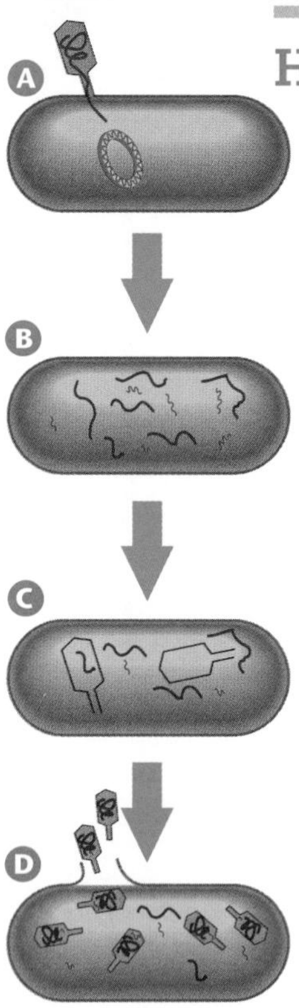

Figure 8-3 indicates that once inside a cell, a virus reproduces, turning the cell into a virus factory. In doing so, the cell may be destroyed. Destruction of cells by some viruses is the key to viral diseases such as AIDS, which destroys certain types of white blood cells. Other viruses, such as the ones that cause cold sores, may enter the body and stay for years before reproducing.

Viruses cause disease in plants and animals. Some of the viral diseases you are familiar with are influenza, measles, chicken pox, mumps, polio, rabies, and AIDS.

Figure 8-3

A A virus attaches itself to a host cell and injects its DNA into the cell.

B The virus DNA takes over the cell and uses the cell's energy and other materials to duplicate itself.

C New viruses form from the duplicated virus DNA.

D The new viruses burst from the cell and go on to infect other cells. The burst cell is destroyed.

How do viruses get from one place to another? Influenza or flu viruses are spread through coughing and sneezing. Other viruses are spread by contact with objects that are contaminated with virus particles.

■ Helpful Viruses

Scientists have begun to take advantage of a virus's ability to enter diseased cells and change the cell for the better in a process called gene therapy.

In this section, you learned that all organisms—whether they are one-celled or many-celled—carry out the same basic life functions. Viruses, while they carry out the life process of reproduction, are not considered organisms because they don't grow, eat, produce food, or respond to stimuli. Viruses can, however, reproduce when inside a living cell. As you learned, viruses come in a variety of shapes and may be classified into groups according to their shapes. In the next section, you will learn about specific simple organisms.

check your UNDERSTANDING

1. Place each of the following organisms into one of the three kingdoms discussed in this section:
   Organism A: many-celled; is a consumer
   Organism B: one-celled; has no nucleus
   Organism C: one-celled; has a nucleus

2. On what basis are monerans, protists, and fungi classified?

3. Apply If you found an unknown but inactive structure in a test tube, what steps could you take to determine whether or not it is a virus?

check your UNDERSTANDING

1. Organism A is probably a fungus, but could be a protist. Organism B is a moneran. Organism C is probably a protist, but could be a fungus.

2. Monerans, protists, and fungi are classified according to their structure and how they produce or use energy.

3. Students may say they would examine the structure to see if it resembled any known virus. They might check to see if it contained DNA or RNA with a coat around it. They could inject the possible virus into bacteria to see whether it is able to reproduce itself.

Monerans and Protists

Monerans

Have you ever accidentally tasted milk that was left out of the refrigerator too long? Milk kept at room temperature spoils and tastes bitter, but milk kept in a refrigerator spoils only if it is left in there for a very long period of time. What is in milk that makes it spoil?

Find Out! ACTIVITY

What makes milk spoil?

Are there any differences between room-temperature milk and milk that is refrigerated?

What To Do

1. Observe a sample of refrigerated milk and a sample of spoiled milk. Describe differences *in your Journal.*

2. Label two microscope slides, one A and one B.

3. Place a very small drop of refrigerated milk on slide A.

4. Add a very small drop of methylene blue stain to the milk. Then place a coverslip on the sample.

5. Prepare slide B for the spoiled milk in the same way.

6. Observe each slide under a microscope.

Conclude and Apply

1. What did you observe about the spoiled milk that you did not observe in the refrigerated milk? What do you think caused differences between the two samples?

Where do you think the organisms come from that cause food to spoil? What conditions are needed for these organisms to grow? Monerans called *bacteria*, are present in small numbers in any carton of milk you take off a supermarket shelf, even though milk has been pasteurized. Pasteurization is the process of heating food to kill harmful bacteria. But pasteurization doesn't kill all the bacteria. Keeping food cooled slows down the rate of reproduction of bacteria in foods.

Section Objectives

- Describe the major characteristics and activities of monerans and protists.
- Describe the way monerans and protists affect other living things.

Key Terms

flagella
cilia

Connect to...

Physics

Not all microorganisms can be killed by boiling. The food industry treats canned food by pressure. Find out what happens inside a pressure cooker to ensure safety in foods.

8-2 Monerans and Protists **255**

Planning the Lesson

Refer to the Chapter Organizer on pages 246A–D.

Concepts Developed

In this section, students will learn about the characteristics and activities of monerans and protists.

Connect to . . .

Physics

Under pressure, the temperatures of liquids can be raised higher than they would be at normal atmospheric pressure. As a result, in a pressure cooker, harmful and non-harmful organisms can be killed off.

1 MOTIVATE

Bellringer

Before presenting the lesson, display **Section Focus Transparency 25** on an overhead projector. Assign the accompanying **Focus Activity** worksheet. L1
LEP

Expected Outcome

Spoiled milk has more bacteria.

Conclude and Apply

The refrigerated milk contains less bacteria. A higher temperature caused the bacteria to reproduce more in the non-refrigerated milk samples.

✔ Assessment

Process Have students design an experiment to determine how long it takes milk to spoil at different temperatures. Allow time for students to share their designs. Use the Performance Task Assessment List for Designing an Experiment in **PASC**, p. 23. L1 P

Find Out!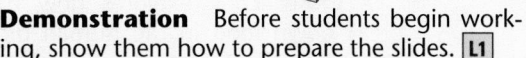

What makes milk spoil?

Time needed   25–30 minutes

Materials   milk, spoiled milk, microscope and slides, methylene blue stain, coverslips

Thinking Processes   observing and inferring, comparing and contrasting.

Purpose   To demonstrate the effect bacteria has on milk.

Teaching the Activity

Demonstration   Before students begin working, show them how to prepare the slides. L1

Science Journal   Students should record their observations in their journals. L1

Solution Preparation
Methylene blue stain   Dissolve 1.5 g methylene blue in 100 mL of ethyl alcohol. Dilute by adding 10 mL of solution to 90 mL of water.

2 TEACH

Tying to Previous Knowledge

Divide the class into three equal groups. Have each group review the names and general traits of one of the first three kingdoms of life forms they learned in Chapter 7. Have each group give the report to the class, working from their notes. COOP LEARN L1

Theme Connection The themes that this section supports are scale and structure and stability and change. In this section, students will explore the scale and structure of monerans and protists relative to other organisms. Monerans are instruments of change wherever they live. Protists are indicators of change in that when conditions change, protists will be affected.

Content Background

Bacteria are so small that about 50 000 bacteria could fit on the head of a pin. Bacteria have different shapes. A rod-shaped bacterium is called a *bacillus,* a round bacterium is called a *coccus,* and a bacterium shaped like a corkscrew is called a *spirillum.*

Characteristics of Monerans

There are two main groups of monerans—bacteria and blue-green bacteria, also called cyanobacteria. sometimes categorized by their forms which may be rodlike, round, and spiral.

■ Bacteria

Bacteria are one-celled organisms that have no nucleus. They have, instead, a single circular DNA molecule. Bacteria have an outer boundary called a cell wall. They are found in every possible place on Earth, from snow to boiling springs, to the lining of your intestine and on your skin.

Bacteria have been classified and named like other organisms. They are

■ Cyanobacteria

Cyanobacteria are monerans that contain green and blue pigments, which is why they are also called blue-green bacteria. They live in fresh water, salt water, and soil. Cyanobacteria are producers. They contain chlorophyll, a green substance that enables them to make food from carbon dioxide in the presence of sunlight and also produce oxygen, which you breathe.

Ecosystems Inside and Out

In Chapter 11, you will study about ecosystems. Ecosystems are made up of groups of organisms interacting with each other and with their environment. Usually ecosystems are thought of as being rainforests or deserts, but you are part of a very important ecosystem that you cannot even see. This ecosystem, in your intestines, plays a role in keeping you healthy. At birth, your digestive system is free of all bacteria. Early on however, certain helpful bacteria begin to live and reproduce on your skin and in your intestines.

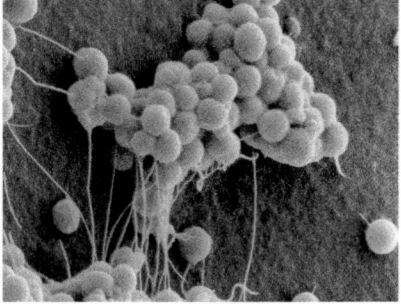
Bacteria on the surface of human skin

Intestinal Bacteria

Some types of bacteria, such as *Escherichia coli,* help you digest your food by breaking down some of the large molecules of food that your body is unable to break down on its own. In this way, an internal

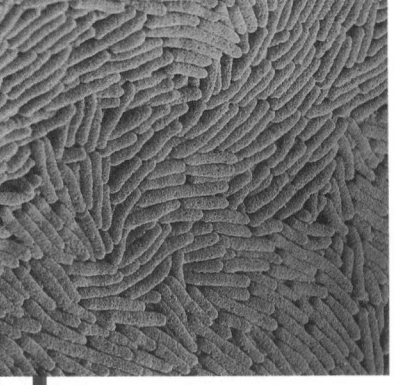
Escherichia coli bacteria is a common bacterium that inhabits your intestine.

Purpose

A Closer Look explores how bacteria break down food in the human digestive system. This feature explains the beneficial relationships between the human body and bacteria, which students may think of only as germs.

Content Background

Just as eliminating an organism from an ecosystem in the outer world can create problems, eliminating an organism from the ecosystem of the body also can create problems. For example, eliminating organism A can allow organism B, normally consumed by the organism A, to multiply too quickly. Many health care professionals today urge people to avoid taking unnecessary amounts of antibiotics since overuse of antibiotics may disturb the body's natural balance of bacteria.

Teaching Strategy

Have students work in groups of two or

How Do Monerans Affect Other Living Things?

Monerans are both helpful and harmful to other organisms much larger than themselves. The word bacteria may make you remember a sore throat or illness that was caused by these microorganisms. Perhaps it is unfortunate that most publicity given to bacteria is negative, because only a few bacteria cause illness. Most bacteria are important for their helpful qualities.

■ Bacteria as Recyclers

Besides protecting your body from certain diseases, many bacteria digest and recycle materials. Spoiled food is spoiled only because it is no longer useful to you. However, it is very useful to bacteria. With the help of bacteria, materials are broken down to simpler substances and become available for use by other living things. You will learn more about the importance of organisms that recycle materials when you study ecology in Chapters 11 and 16.

Figure 8-4

When the temperature of milk is controlled, and certain bacteria are added to milk, the milk changes into yogurt. Read about yogurt cultures on a yogurt carton.

Visual Learning

Figure 8-4 Have students read the ingredient list on several different brands of yogurt. Then have a yogurt tasting session. Have students research how yogurt is made. An encyclopedic-type cookbook will explain.

GLENCOE TECHNOLOGY

 **Videodisc**

STVS: Plants & Simple Organisms

Disc 4, Side 1
Using Bacteria to Detect Carcinogens (Ch. 7)

Bacterial Miners (Ch. 9)

Bacterial Waste Treatment (Ch. 10)

New Uses for Algae (Ch. 12)

ecosystem forms. You provide food and a place to live for the bacteria, and the bacteria kill toxic bacteria that might invade your body.

Some bacteria on your skin thrive on sweat and other products produced by your skin. They also help destroy certain harmful bacteria that are constantly trying to invade the human organism.

Making Vitamins

Besides digestion, bacteria have other important functions in your body, such as making vitamins. For instance, a normal diet does not contain as much vitamin K as your body needs. Fortunately, the bacteria in your intestine can make extra vitamin K for you. Without enough vitamin K, your body cannot form blood clots to stop you from bleeding when you cut yourself. Vitamin B$_{12}$ is also made by bacteria. Your body uses B$_{12}$ to make red blood cells. If you didn't have enough vitamin B$_{12}$, you might suffer from anemia. Maintenance of your bodily ecosystem, then, is important for keeping you healthy.

What Do You Think?

Antibiotics are used to treat bacterial infections. What do

Lactobacillus acidophilus is a bacterium normally found in the mouth and intestines.

you think happens to the bacteria normally found in your intestines when you take antibiotics for a long period of time?

GLENCOE TECHNOLOGY

 CD-ROM

Science Interactions, Course 1, CD-ROM

Chapter 8

---

three to read and discuss this feature. Ask students to identify four ways that bacteria keep humans healthy. Have students make illustrated posters that portray these benefits. Use the Performance Task Assessment List for Poster in **PASC**, p. 73.
COOP LEARN L1

Answers to
What Do You Think?

1. If the antibiotic is specific enough, the infectious bacteria should be controlled while the beneficial bacteria are left alone. Broad-spectrum antibiotics might kill both types of bacteria, leaving you free from infection but unable to digest your food properly for a while.

Going Further ▅▅▅▶

Divide the class into groups to research the benefits to humans of other bacteria and organisms. Each group should identify the name of the bacteria or organism, the state of the food (fresh or spoiled) that they eat, and the useful by-product. Assign: bacteria used in the production of cheese, wine and vinegar, penicillin, vitamin B$_{12}$, vitamin K, bread dough, or bacteria used in soil enrichment and to decompose organic waste. Have groups share their findings with the class. **COOP LEARN**
L2

Content Background

All cyanobacteria have phycocyanin, a blue pigment, and bacteriochlorophyll, a green pigment. Cyanobacteria were once classified as plants because of their photosynthetic pigments.

Visual Learning

Figures 8-5, 8-6, and 8-7 Ask students to examine the three figures on this page and explain what they show about bacteria. *Although some bacteria are harmful to humans, humans use other bacteria for their own benefit.* Begin a class list of harmful bacteria and beneficial bacteria. L1

SKILLBUILDER

Because the bacteria grew through many generations and suddenly died, they probably have used up the food supply. It is also possible that waste products have built up. L1

Across the Curriculum

Math

Students can work cooperatively to determine whether they would rather have $10 every 20 minutes for 13 hours or one penny doubled every 20 minutes for 13 hours. *The penny is by far the better choice. When calculated arithmetically, it yields $400; exponentially, $5 billion. The $10, in contrast, yields $390.* Explain that this is called exponential growth. **COOP LEARN** L2

■ Bacteria and Food

Using Observations to Form a Hypothesis

You are a researcher in a laboratory. The bacteria in an experiment you have been working on have reproduced in large numbers. Then, all of a sudden, most of the bacteria began to die. You need to find out what caused the deaths of the bacteria. What hypothesis can you make about what is happening to the bacteria? If you need help, refer to the **Skill Handbook** on page 663.

Have you ever eaten bacteria for lunch? As you can see from **Figures 8-6** and **8-7**, many foods are made with the help of bacteria. Bacteria break down substances in milk to produce different types of cheeses. They are also the important factor in making yogurt and sauerkraut.

■ Industrial Uses

Figure 8-5

Industries other than the food industry also rely on bacteria. Microbiologists have found ways to put bacteria to use in manufacturing medicines, cleansers, adhesives, and other products. Bacteria that digest oil are used to clean up oil spills around the world. Bacterial cultures are also now used in products that unclog drains.

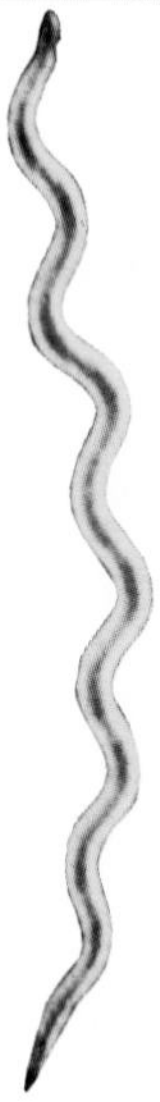

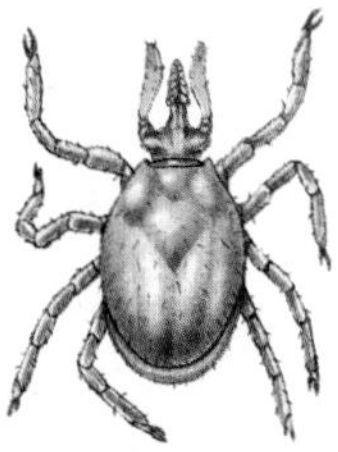

■ Disease-Causing Bacteria

Lyme disease is caused by the spiral bacterium, *Borrelia burgdorferi*, carried by several kinds of ticks in woodlands, brushy areas, and coastal grasslands. These ticks are also carried by white-tailed deer.

Although many bacteria are helpful rather than harmful, you have probably had firsthand experience with bacteria that cause disease. Ear infections, tooth decay, and strep throat, for example, are caused by bacteria. Other bacteria that cause diphtheria and whooping cough in humans have had significant impact on world history. Acne infections result when bacteria use oils produced by cells as a source of food.

258   Chapter 8   Viruses and Simple Organisms

Figure 8-6

Part of the preparation process of olives that you buy in the store involves treating the olives in a bacterial solution.

■ Cyanobacteria

Cyanobacteria provide food and oxygen for organisms in lakes and ponds. However, too many cyanobacteria in a pond can be harmful. Have you ever seen a pond covered with smelly green slime?

Fertilizers washed into ponds from nearby farm fields supply nutrients for the cyanobacteria in the pond. The cyanobacteria then reproduce rapidly. The nutrients get used up and large numbers of cyanobacteria begin to die. Other bacteria in the pond begin to feed on their remains. As the bacteria consume the dead cyanobacteria, they use up the oxygen in the water. In extreme cases, reduced oxygen causes fish and other organisms in the pond to die. The pond develops foul odors and a slimy appearance as organisms begin to decompose.

Figure 8-7

Bacteria break down substances in milk to produce cheese. The flavor and texture of the cheese depends mostly on the kinds of bacteria present. The holes in this Swiss cheese are the result of a species of bacteria that gives off carbon dioxide.

Program Resources

Study Guide, p. 31
Critical Thinking/Problem Solving, p. 16, Bacteria Cleanup L3
How It Works, p. 12, How Do Disinfectants Work? L2
Multicultural Connections, p. 19, New World Epidemics L1
Section Focus Transparency 25

Laboratory Manual, pp. 51–54, Life in Pond Water L2; pp. 49–50, Shapes of Bacteria L3; pp. 47–48, Monerans L2
Teaching Transparency 15 L2
Science Discovery Activities, 8-2, To Peel or Not to Peel; 8-3, Mushroom Designs

---

Protists

You might say that protists are nature's variety show. That is because there are so many different groups of them, but they do all share some things in common. Protists need moist surroundings. Some live in damp soil, rotting logs, or the bodies of other organisms. Most, however, live in oceans, ponds, swamps, or other bodies of water. The protists are a large group of organisms with many different shapes and characteristics. **Figure 8-8** shows the variety and importance of forms in the protist kingdom.

Explore! ACTIVITY

What lives in pond water?

What To Do

1. Make a wet-mount slide of a drop of pond water.

2. Observe the slide under both low and high power of the microscope.

3. Compare this slide with a wet mount of aged tap water.

4. *In your Journal*, describe what you see in both slides.

5. What did you find living in pond water that you didn't find in tap water?

Figure 8-8

Kingdom Protista contains a variety of organisms. While varied, they share some characteristics—all have cells with one or more nuclei. Scientists hypothesize that protists are the ancestors of modern fungi, plants, and animals. They are classified as plantlike, animal-like, and funguslike.

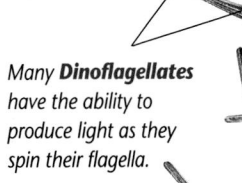

Many **Dinoflagellates** have the ability to produce light as they spin their flagella.

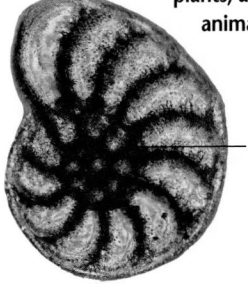

Foraminiferans secrete fingerlike projections that contain calcium and harden into a chalky, many-chambered shell. When foraminiferans die, they sink to the bottom of the ocean.

Radiolarians capture food by extending fingerlike projections, or false feet.

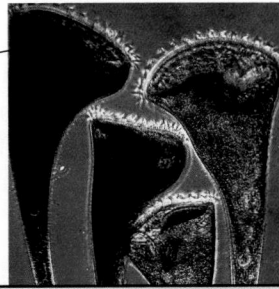

Stentors have fine hairlike projections that help them move.

Volvox colonies contain chlorophyll and produce oxygen. They are representative of green algae.

8-2   Monerans and Protists   **259**

ENRICHMENT

Activity Have small groups of students create a flip book for one type of protist, such as flagellates, ciliates, amoebas, and so forth. On each page of the flip book they should show one stage of the organism's movement. When the pages are bound and flipped, the protist should appear to move. The more pages showing a smaller segment of the movement, the more interesting the book. Use the Performance Task Assessment List for Group Work in **PASC**, p. 97. **L3**

Explore!

What lives in pond water?

Time needed 15-30 minutes

Materials fresh pond water, tap water, slides, microscopes

Thinking Processes observing and inferring, comparing and contrasting

Purpose To compare and contrast organisms in pond water and tap water.

Teaching the Activity

Demonstration Show students how to prepare the slides.

Troubleshooting Use water from an established aquarium if pond water isn't available.

Science Journal Have students record their observation in their journals. **L1**

Expected Outcomes

Students will observe a great many protists in pond water, but not in tap water.

Answer to Question

The slide of pond water has various living organisms; there are none in the tap water.

✓ Assessment

Performance Have students make a poster of some of the protists they observed in the pond water. Use the Performance Task Assessment List for Poster in **PASC**, p. 73. **L1**
COOP LEARN **P**

Visual Learning

Figure 8-8 Students can use an unabridged dictionary to find out background information on the five protist examples in the figure, including the language the name is derived from. They can make a bulletin board display. Suggest that students label each of the organisms and write a brief paragraph highlighting its characteristics. **L1**

Plantlike Protists

Plantlike protists are known as algae. All algae can make their own food because they contain chlorophyll. However, not all algae are green because some have other pigments that cover up the green color of their chlorophyll. **Figure 8-9** shows examples of brown algae, also known as brown seaweed, and red algae.

There are six main groups of algae. Each group has its own characteristics. You can observe the characteristics of one group—diatoms—in the following Explore activity. Diatomite is a sandy soil made up of shells left by one type of protist, diatoms, that died about 20 million years ago.

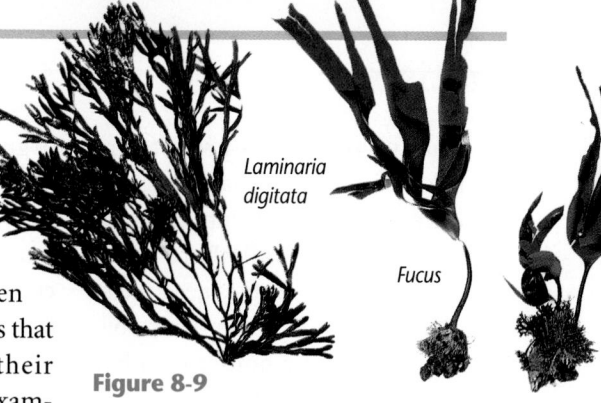

Laminaria digitata

Fucus

Figure 8-9

Each type of algae carries pigments in its cells. The various pigments absorb and reflect different wavelengths of light, which makes each type of algae appear to be a distinctive color. Nongreen algae include red, brown, and golden-brown algae.

Rhodymenia

Chondrus crispus

Explore! ACTIVITY

What is a diatom?

What To Do

1. Look at a sample of diatomite on a slide under low and then high power using a microscope.

2. Then, scrape the inside wall of a fish tank and make a slide. You should be able to see diatom shells in the diatomite and living diatoms in the fish tank scrapings.

3. How many different shapes of shells can you find?

4. Why do you think diatom shells survive?

The diatoms and shells you looked at in the Explore activity were from two different places—salt water and fresh water. In salt water, diatoms form an important part of plankton—the organisms on which whales and many types of fish feed. Diatom shells contain silica.

Animal-like Protists

Some protists are animal-like in that they are consumers. They do not contain chlorophyll. Animal-like protists are called protozoans. They are classified into groups according to the way they move. In the following activity, compare two different groups of protozoans.

Find Out! ACTIVITY

What characteristics do different protozoans have?

What To Do

1. Make two wet-mount slides, one of a live amoeba and the other of live paramecia.

2. Observe each organism under low power, then under high power.

Conclude and Apply

1. *In your Journal*, describe how each protozoan moves. Describe how it feeds or avoids obstacles.

2. What characteristics of the protozoans that you looked at would you use to classify them into separate groups?

■ Types of Movement

The amoeba that you observed moves about by forming false feet. Look at **Figure 8-10** to learn more about false feet.

The paramecium belongs to the group of protozoans that have short, hairlike structures called **cilia** all over their bodies. They use the cilia like oars to move and to sweep food into their mouths. Other protozoans have whip-like structures called **flagella** for moving through their watery surroundings. A fourth group of protozoans, such as those that cause malaria, have no structures for movement. They are parasites, living on other organisms and feeding on their tissues.

Figure 8-10

A To feed, the amoeba extends its false feet on either side of food, such as a bacterium.

B The false feet flow over and around the food, until it is completely enclosed.

C The food vacuole floats within the cytoplasm. Strong enzymes are used to digest the food.

D Undigested food is forced out of the cell as the vacuole explodes outward.

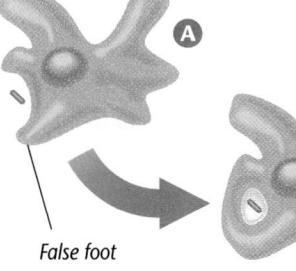
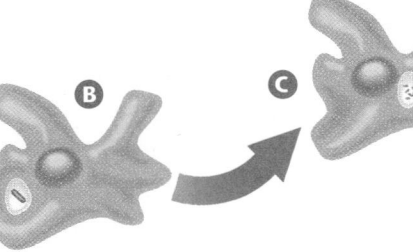

False foot

261

Find Out!

What characteristics do different protozoans have?

Time needed 30–45 minutes

Materials viable amoeba and paramecia, slides, microscopes

Thinking Processes observing and inferring, comparing and contrasting, classifying

Purpose To classify amoeba and paramecia according to their characteristics, including movement and appearance.

Preparation Be sure to have sufficient quantities of viable amoeba and paramecia to make extra slides, if needed.

Teaching the Activity

Demonstration Before students begin working, review the procedure for making the wet mounts.

Troubleshooting Remind students to label each slide. LEP

Science Journal Suggest that students record their observations in a two-column table in their journals. L1

Expected Outcomes

Students will observe that protozoans differ in their appearance and means of locomotion.

Conclude and Apply

1. Amoeba will recoil from many obstacles; paramecia will use their cilia to "roll" away from obstacles.

2. appearance and locomotion

✓ Assessment

Oral Ask students to suppose they have a dozen slides of different protozoans. Have them describe how they would try to classify them into different groups. Students should prepare an oral statement, including a list of characteristics one might use. Use the Performance Task Assessment List for Oral Presentation in **PASC**, p. 71. L1

Content Background

Of the 100 000 or so protozoans, only a few cause disease, but the illnesses they do cause can be severe or even fatal. Malaria, sleeping sickness, and dysentery are diseases caused by protozoans. Malaria still results in the death of over a million people every year throughout the world.

3 ASSESS

Check for Understanding

Have students answer the following questions in small groups. How are monerans alike? How are they different? How are protists alike? How are they different? Pairs of students can answer the Check Your Understanding questions.

Reteach

Concept Mapping Help students make concept maps of the classification of monerans and protists. L1 P

Extension

Classifying Ask students to research diseases that are caused by bacteria. Have them classify the diseases into two groups—those which can usually be combatted by the body's natural immune system and those which usually need to be treated with antibiotics. L3

4 CLOSE

Activity

Provide students with a page of diagrams and characteristics that students can identify as either "moneran" or "protist." LEP L1

How Do Protists Affect Other Living Things?

Many protists are useful to other organisms. For example, algae, the protists that contain chlorophyll, are helpful to just about everyone, including you. Not only do they produce food in the presence of sunlight, but at the same time, they also produce most of the oxygen that Earth's organisms depend upon.

Funguslike protists, such as the slime mold in **Figure 8-11** are probably little noticed by most people. However, slime molds are recyclers that help break down leaves and fallen trees.

■ Disease-Causing Protists

Not all protists are helpful to other organisms. One type, *Plasmodium*, causes malaria, which kills more people on Earth each year than any other disease. **Figure 8-12** shows the protist *Trypanosoma*, another disease-causing protist.

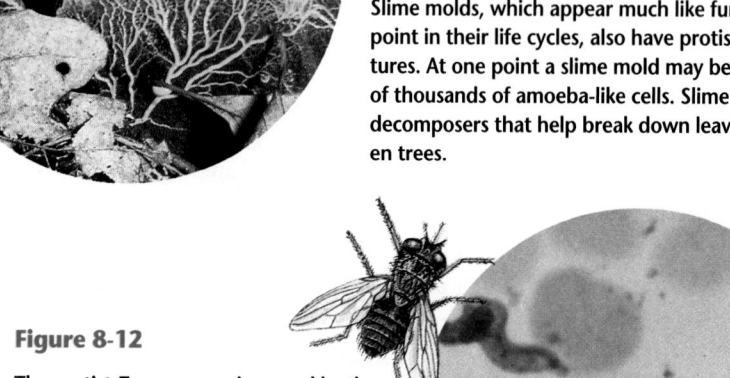

Figure 8-11

Slime molds, which appear much like fungi at some point in their life cycles, also have protist-like features. At one point a slime mold may be made up of thousands of amoeba-like cells. Slime molds are decomposers that help break down leaves and fallen trees.

Figure 8-12

The protist *Trypanosoma* is spread by the bloodsucking tsetse fly in Africa. When a trypanosome enters the bloodstream, it causes African sleeping sickness. The symptoms of this disease are fever, swollen glands, and extreme sleepiness.

check your UNDERSTANDING

1. Why are plantlike protists important in the world?
2. How are animal-like protists classified?
3. How are monerans different from protists?
4. **Apply** Defend the statement: Monerans are not simple organisms.

check your UNDERSTANDING

1. Plantlike protists are an important food source for many organisms and produce most of the world's oxygen.
2. Animal-like protists are classified according to the way they move.
3. Monerans are one-celled organisms with no nucleus. Protists may be one-celled or many-celled, but they have a nucleus.
4. Answers will vary. Students may argue that since monerans carry out all the activities of any other living organism, it is incorrect to call them simple.

Fungus Kingdom

Fungi

Have you seen any fungi lately? None? You may be surprised to learn that one-fifth of all the different kinds of organisms on Earth are fungi. Hundreds of these fungi—most of them molds—live in the soil under your feet, and in your home!

Become acquainted with one common member of Kingdom Fungi through a trip to the grocery store.

Section Objectives

- Describe the major characteristics and activities of fungi.
- Describe the way fungi affect other living things.

Key Terms

parasite

Explore! ACTIVITY

What can you learn about fungi by observation?

What To Do

1. Look at a mushroom that you might buy in a grocery store.

2. Describe its size.

3. Carefully pull the cap off the stalk and lay it aside.

4. Use your fingers to pull the stalk apart lengthwise. Continue gently pulling the stalk apart.

5. Now, look at the underside of the cap. Look at all the thin membranes. Using a hand lens, look at one of the membranes. What can you see?

6. *In your Journal,* describe how the mushroom is constructed.

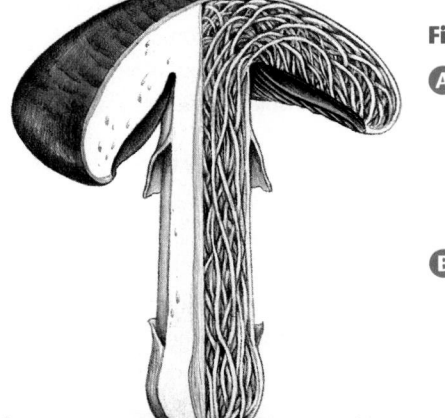

Figure 8-13

Ⓐ Fungi do not have the specialized tissues and organs of plants that you will learn about in Chapter 10. The body of a fungus is usually a mass of many-celled, threadlike tubes. None of the cells contains chlorophyll.

Ⓑ Fungi were once classified as plants. Based on your observations, would you accept this classification? In what ways do fungi differ from plants? In what ways do they remind you of plants?

PREPARATION

Planning the Lesson

Refer to the Chapter Organizer on pages 246A–D.

Concepts Developed

In this section, students will learn to distinguish the characteristics and activities of fungi.

1 MOTIVATE

Bellringer

Before presenting the lesson, display **Section Focus Transparency 26** on an overhead projector. Assign the accompanying **Focus Activity** worksheet. L1

LEP

Visual Learning

Figure 8-13 Based on your observations, would you accept this classification? *No, because fungi do not contain chlorophyll.* **In what ways do fungi differ from plants?** *They lack chlorophyll and can't make food.*

---

like parts and do not contain chlorophyll.

Answers to Questions

Mushrooms are many-celled. Students may see many spores.

✔ Assessment

Oral Have students relate what they learned about fungi by observing the mushroom. Use the Performance Task Assessment List for Oral Presentation in **PASC**, p. 71. L1

Explore!

What can you learn about fungi by observation?

Time needed 30 minutes

Materials mushrooms from the grocery store, hand lens

Thinking Processes observing and inferring, practicing scientific methods

Purpose To distinguish how fungi differ from plants.

Teaching the Activity

Demonstration Show students how to carefully pull the stalk apart lengthwise. Set up a spore print demonstration using an open cap face down on a white sheet of paper.

Science Journal Have students sketch or describe their observations in their journals. L1

Expected Outcomes

Students will observe that fungi have thread-

2 TEACH

Tying to Previous Knowledge

Have students identify the five kingdoms (Moneran, Protist, Fungus, Plant, Animal) as you list them on the chalkboard. Then help students recall what they learned about the Fungus Kingdom in Chapter 7.

Theme Connection The themes that this chapter supports are scale and structure and stability and change. Fungi can be both useful and harmful to humans, as the information on these pages shows.

Uncovering Preconceptions

Many students will think that all mushrooms with the exception of those purchased in a grocery store are poisonous. Point out that only a few species are poisonous, but it is difficult for even experts to tell some of the poisonous species from similar non-poisonous ones.

NATIONAL GEOGRAPHIC SOCIETY

Videodisc

Newton's Apple: Life Sciences

Mold

Chapter 3, Side A

Molds are fungi

31146-31738

Uses of mold

34875-35703

How Do Fungi Affect Other Living Things?

Figure 8-14

Fungi live in moist shady areas. Bracket fungi often grow on trees in the woods.

People find fungi useful in many ways. The fungi called yeast make bread and pizza crust rise by producing gases. Other fungi give some cheeses, such as blue cheese, very different flavors that people like. Many people enjoy mushrooms on pizza and in salads and other dishes. One kind of fungus, called *Penicillium*, produces penicillin that doctors prescribe for patients with diseases caused by certain kinds of bacteria. If you've ever seen the green, powdery fungus growing on rotting oranges and old bread, then you're already familiar with what *Penicillium* looks like.

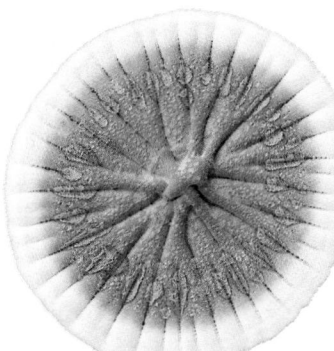

Figure 8-15

The green growth in this photograph is *Penicillium*, from which the antibiotic penicillin is derived.

Chemistry CONNECTION

Secrets in the Soil

Being a soil scientist may not seem exciting, but in what other field can a person take frequent trips to scoop up a pocketful of soil?

Soil scientists travel all over the globe, from the arctic to equatorial jungles, collecting samples of the local soil fungi. Most of the soil fungi in the world are as yet undiscovered, scientists judge, so it's vital to collect samples from environments that are rapidly disappearing, such as tropical rain forests.

Why Collect Soil Fungi?

Soil fungi have many uses. One fungus native to Japan is being added to forest soil in the eastern United States because the fungus kills gypsy-moth caterpillars, which are serious leaf-eating pests.

Another soil fungus, found originally in the Canary Islands, contains a drug that suppresses the immune response in humans. Patients who receive transplanted organs may be given this drug to prevent transplant rejection.

Nature's Control

A new kind of cockroach trap, marketed as the country's first "biopesticide," contains a fungus that eats through the hardened, outer shell of the cockroach, eventually killing the insects.

Chemistry CONNECTION

Purpose

Chemistry Connection reinforces Section 8-3 by describing how fungi grow and consume organic materials. Even though fungi are among the less obvious organisms on Earth, they have great impact on other life-forms.

Content Background

In the late 1920s, Alexander Fleming noticed a mold on a culture of staphylococcus bacteria. He saw that a ring of bacteria around the mold specks had been killed. The mold, which he named *penicillium*, killed some bacteria but not others. It was not until World War II that other researchers, trying to find a drug to combat war wounds, turned back to Fleming's work. They succeeded in producing the drug penicillin in quantity, saving thousands of lives. To this day, penicillin is a powerful and useful antibiotic, but some disease organisms have developed a resistance to it.

■ Fungi as Recyclers

You may have seen fungi growing on an old tree lying on a forest floor. If so, you've seen an example of fungi in their most important job. Fungi are able to break down, or decompose, organic material. Food scraps, clothing, dead plants, and animals are all made of carbon-containing organic material. Fungi cannot make their own food. They do not contain chlorophyll. They are consumers. Along with many bacteria, fungi decompose organic materials and recycle these materials back to the soil. These materials are then used by plants to grow. Like many bacteria, fungi help rid Earth of mountains of waste.

Some fungi, such as the ones that cause athlete's foot and ringworm, are parasites. **Parasites** are organisms that live on or in other living things and feed on them. Fungi that are parasites cause some of the most damaging diseases in plants. Wheat rust and corn smut are two fungi that can destroy food crops.

How do fungi grow best? The Investigate on the following pages will help you find an answer.

Figure 8-16

Corn smut is a damaging fungus that attacks corn plants.

Across the Curriculum

Math

The stinkhorn fungus *Dictyophora*, found in tropical Brazil, is one of the world's fastest-growing organisms, growing at a rate of 2.5 centimeters every five minutes. If its full height is 10 centimeters, how long does it take to mature? *20 minutes* L2

Visual Learning

Figures 8-15 and 8-16 Have students read the captions, then contrast the different effects of the two fungi.

Content Background

Bacteria and fungi are decomposers that play a vital role in the recycling of materials in nature. Gardeners and farmers use these microorganisms to make compost from waste vegetable matter. Have students research how compost is made and discuss the possible benefits of using compost rather than chemical fertilizers. L2

Fungi are also a big part of a field called bioremediation. Bioremediation uses living organisms to clean up wastes. Other methods, such as incineration and disposal in landfills, don't provide permanent solutions to hazardous wastes. But when you feed certain hazardous wastes to the right soil fungi, the end result is just carbon dioxide (and a bigger population of fungi). The hazardous wastes—such as PCBs, pesticides, dioxins, coal tars, and heavy fuels—are neutralized with no unwanted byproducts.

Fungi to the Rescue!

Four kinds of native fungi are presently being used to remove a pollutant from a reservoir in California's San Joaquin Valley. The pollutant, selenium, has accumulated in the reservoir because of intensive irrigation of farmland. Selenium is present in levels high enough to kill birds nesting around the reservoir. Since the Kesterton National Wildlife Refuge is nearby, a nontoxic solution is critical. The fungi ingest the selenium, convert it from a water-soluble form to a gaseous form, then "burp" the selenium into the atmosphere. The gaseous form attaches to airborne particulates and drifts away. Tests have shown the fungi can reduce selenium levels by 60 to 70 percent within three years.

You Try It!

Ask five people these two questions: What is the best-known antibiotic in the world? (penicillin) What was penicillin developed from? (fungus) How many of your subjects knew both answers?

Teaching Strategy

Have students work in pairs to read the essay and then identify 5 people for each of them to interview. After completing their interviews, ask them to make a graph or table that summarizes the results of their 10 interviews. Select the appropriate Performance Task Assessment List in **PASC.** COOP LEARN
L2

Going Further ▐▐▐▐▐▶

Students can research illnesses caused by fungi, such as athlete's foot. Have students work in groups to create booklets showing how people can prevent illnesses caused by fungi. Students can contact local physicians.

8-2 The Work of Fungi

Preparation

Purpose To design and carry out an experiment that tests the growth of mold on fruit.

Process Skills forming a hypothesis, designing an experiment, collecting and analyzing data, separating and controlling variables, modeling, comparing and contrasting, recognizing cause and effect, observing and inferring

Time Required 60 minutes to set up the experiment; 4 days for observation

Materials Additional materials that might be useful are a thermometer for monitoring and controlling temperature and a cardboard box to create a dark environment.

Safety Precautions Be sure students wash their hands before and after working with molds. **Caution:** Students who are allergic to molds should not handle them.

Possible Hypotheses Some students might hypothesize that mold will grow better in dark conditions or in damp conditions. Some might think both darkness and dampness are needed for mold to grow.

Plan the Experiment

Process Reinforcement Be sure that students develop a hypothesis that can be tested with a single experiment, using materials available in the class. Remind students to identify the controls and variables. Encourage students to create a data table that will allow them to collect and record their data.

Possible Procedures Several possible procedures include washing fruit with soap and water; washing fruit with plain water; rubbing with a disinfectant; making a cut in the fruit; putting fruit in a brown bag or a sealable plastic bag; putting fruit in a dark area or in a lighted area; putting fruit in the refrigerator or in a warm, dark place. Accept all reasonable approaches.

The Work of Fungi

Have you ever seen mold grow on fruit? When conditions are right, mold can cover and penetrate a fruit with hundreds of thousands of tiny threadlike branches called hyphae. The cells of hyphae release substances that break down organic materials in the fruit. Then bacteria move in, causing spoilage.

Preparation

Problem
Under what conditions does a mold grow best on fruit?

Form a Hypothesis
Has anyone in your group ever seen mold on fruit? As a group, discuss the conditions under which the mold grew. Then form a hypothesis that can be tested in your experiment.

Objectives
- Design an experiment using several variables to promote mold formation on fruit.
- Compare and contrast conditions that promote mold formation.
- Infer how certain conditions interact to promote or prevent the growth of mold.

Possible Materials
peaches, apples, or oranges
paper towels
plastic bags
kitchen knife or fork
paper plate
water

Safety Precautions
    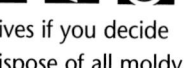

Be careful with knives if you decide to cut any fruit. Dispose of all moldy products as directed by your teacher.

Program Resources

Study Guide, p. 32
Take Home Activities, p. 16 [L1]
Making Connections: Across the Curriculum, p. 19, Preserving Food [L1]
Concept Mapping, p. 16
Activity Masters, pp. 37–38, Design Your Own Investigation 8-2
Science Discovery Activities, 8-1
Section Focus Transparency 26

DESIGN YOUR OWN
INVESTIGATION

Plan the Experiment

1 Examine the materials provided. Decide how you will use them in your experiment.

2 Design a procedure to test your hypothesis. Write down what you will do at each step.

3 How will you record your data? If you need a table, design one now in your Science Journal.

Check the Plan

1 List the conditions that you think will promote mold formation on fruit. Which condition will you test?

2 If you are testing more than one condition, have you allowed for a control in your experiment? What is the control?

3 Make sure your teacher approves your experiment before you proceed.

4 Carry out your experiment. Record your observations.

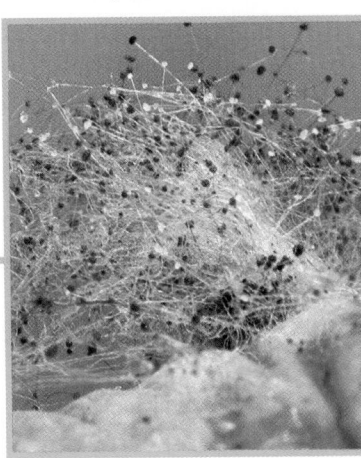

How is the fungus growing on this bread different from the mold you observed growing on fruit?

Analyze and Conclude

1. **Compare and Contrast** Which conditions promoted mold formation? Which conditions prevented mold formation?

2. **Recognize Cause and Effect** Was your hypothesis supported? If not, explain why it might still be right.

3. **Observe and Infer** On which fruits did mold grow the easiest? Suggest reasons for this.

4. **Infer** Suggest a good use for the moldy fruit at the end of this experiment instead of throwing it into the trash. Give a reason for your suggestion.

5. **Interpret Data** Based on the outcome of your experiment, what steps would you take to prevent oranges from molding quickly at home?

Going Further

You may have formed several hypotheses while designing this experiment. Test one of these hypotheses or design a new experiment based on an observation made while conducting this one.

Multicultural Perspectives

Ancient Medical Practices

Some time before 2000 B.C., Egyptians developed one of the world's first sophisticated medical practices. Egyptian healers were so famous that people traveled to them from all around the Middle East. For example, the Egyptians were known to have put moldy bread on wounds. What can students infer about the action of the fungus? Connect this activity with Alexander Fleming's discovery of penicillin and its action on bacteria.

Have students find out about other "modern" treatments used by the Egyptians and create a class chart of their findings. L2

Teaching Strategies

Troubleshooting Encourage students to use a control.

Science Journal Have students record their observations in their journals. They should also create their data tables and answer the questions in their journals.

Expected Outcome

Students will observe that molds grow better in damp, dark conditions.

Analyze and Conclude

1. Damp, warm, dark conditions promoted mold. Dry, cool, light conditions prevented mold. Mold may also have been prevented by certain wrapping materials.

2. A hypothesis might not be supported yet still not be wrong. Further data must be collected before the hypothesis can be evaluated. For example, a student who hypothesizes that dark conditions promote mold may not see any growth if the fruits are very dry. Additional data using less dry fruits in dark and bright conditions would need to be collected before deciding whether or not darkness promotes mold.

3. The soft-skinned fruits, such as peaches, seemed to mold more easily than hard-skinned fruits, such as apples. Students may observe that mold grows easier on fruits that are damaged.

4. Moldy fruit can be added to a composting bin where the molds in the fruit will help to break down other organic matter.

5. Answers may vary but should include suggestions on keeping the oranges cool and dry.

✔ Assessment

Performance Ask students to identify the controls and variables they used in this experiment.

Going Further ▐▐▐▐▶

Students may wish to design experiments that will test the effects of temperature, light, humidity, acidity, or other factors on mold growth.

How Do Fungi Meet Their Life Needs?

As you saw in the Investigate, moisture is an important factor in the growth of fungi. Under the right conditions, fungal threads can penetrate fabric, leather, fur, wood, paint, and even some plastics, to obtain the nutrients and energy they require. They do not make their own food. Can you think of a way to prevent fungal growth?

You can take a look at the reproductive parts of a fungus in the following Explore activity.

Explore! ACTIVITY

How do fungi reproduce?

What To Do

1. Examine some black or green mold on bread using a hand lens.
2. *In your Journal,* describe what you see.
3. Remove a small bit of mold and make a wet mount of the fungus.
4. Observe the mold under low power with a microscope.
5. What do you see all over the slide? What do you think these are?

A fungus begins life as a spore. In the Explore activity, you observed spore cases containing millions of these powdery reproductive cells.

You may now be thinking that even the simplest organism is very complex. The organisms you have studied so far are very different from one another in some ways. And yet they are all alike in the kinds of activities they carry out to stay alive!

check your UNDERSTANDING

1. How do fungi differ from cyanobacteria in how they obtain energy?
2. Explain why fungi are capable of reproducing in large numbers.
3. How do the moisture requirements of fungi compare with those of bacteria for life?
4. **Apply** Why doesn't a mushroom farmer have to worry about how much sunlight his or her crop gets?

4 CLOSE

Activity

Ask students to write newspaper articles on what the world would be like without fungi. Use the Performance Task Assessment List for Newspaper Article in **PASC**, p. 69. **L1**

check your UNDERSTANDING

1. Fungi are consumers and cyanobacteria are producers.
2. Their spore cases contain millions of spores, or reproductive cells.
3. Both require moisture for growth.
4. Mushrooms and other fungi do not use sunlight to make their own food.

Science and Society

Using Viruses To Fight Disease

One of the most amazing processes in the human body is the immune response system. Immunity is our body's built-in defense against disease.

When a disease-causing bacterium or virus enters your body through breaks in your skin or through moist membranes, your immune system mobilizes a series of defenses against these foreign particles. White blood cells traveling through the circulatory system surround and engulf most intruding microorganisms.

Antibodies and Antigens

In other instances, white blood cells produce substances called antibodies. Antibodies are proteins that deactivate and destroy particular microorganisms. Antibodies are produced in response to proteins called antigens that are located on the surfaces of bacteria and viruses. When your body produces antibodies to particular antigens, you are said to be immune to those microorganisms.

Immune for How Long?

Immunity can last a few months or years, as with the flu virus, or it can last an entire lifetime, as with the polio virus. Unfortunately, immunity against one type of antigen or virus does not protect you against all others, even though they may be similar. This is one reason why you catch a cold more than once. Your colds may seem like the same disease, but each is caused by a different virus.

Active Immunity

When your body produces antibodies in response to a particular antigen, it is called active immunity.

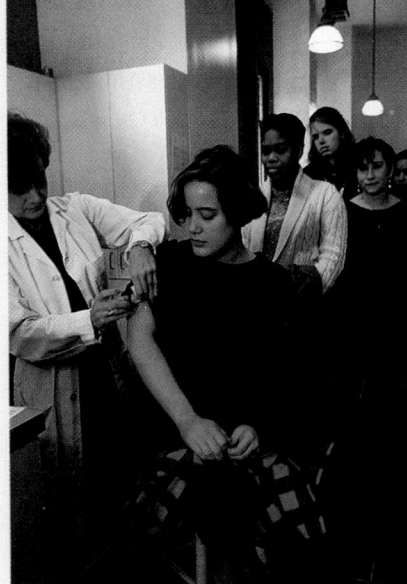

Active immunity can either be natural or artificial. When your body produces antibodies as a result of contact with a disease-causing microorganism, this is a natural process. But your body can also produce antibodies after injection with a vaccine. A vaccine is a solution of dead or weakened bacteria or viruses which, when injected into the body, causes an immune response. Vaccination is an artificial process.

Science and Society

Purpose
In Section 8-1, students learn that certain viruses cause disease. Science and Society builds on this information by describing how viruses can be used to fight disease.

Content Background
In 1986, the U.S. Department of Agriculture granted the world's first license to market a living organism produced by genetic engineering, a virus used as a vaccine to prevent a disease in pigs. This would not have been possible without Pasteur's work more than a century earlier.

The development of vaccines, like the discovery of penicillin, was an accident. In 1879, while working on a way to stop the spread of anthrax, Louis Pasteur discovered that bacteria could be weakened. They failed to cause disease but still conveyed immunity. Two years later, Pasteur demonstrated that injecting someone with a weakened strain of the anthrax bacteria brought immunization against the disease. He called this method of creating immunity vaccination, to honor Edward Jenner's use of cowpox (vaccinia) to prevent smallpox. Working with others, Pasteur developed vaccines useful in battling many of the other major diseases of his day—cholera, tuberculosis, tetanus, diphtheria, and rabies.

Teaching Strategy Have students read the article and role-play the activities of bacteria or viruses, and the parts of the body's immune system.

Going Further ⟩⟩⟩⟩⟩➤

TECH PREP

Have students research the two types of polio immunizations, IPV (inactive polio virus) and OPV (oral polio virus). Students should investigate why the live virus in OPV provides lifetime protection while the inactive virus in IPV may not. Have them present their findings as a pamphlet designed to provide public health information. Use the Performance Task Assessment List for Booklet or Pamphlet in **PASC**, p. 57. L2 P

Content Background

AIDS is believed to be caused by the Human Immunodeficiency virus known as HIV. This virus belongs to a class of viruses known as retro viruses. After infection with the HIV virus, a person can remain without symptoms for many years, 7-8 being the average. Scientists now know that during what was previously called a dormant period, the HIV virus is active in the lymphatic system.

It is debatable if HIV itself makes people sick. In later stages of infection it does produce illness as others viruses do when a person is infected with them. HIV does cause infected people to be vulnerable to opportunistic infections. It does this by destroying T-cells and making monocytes, a type of white blood cell, nonfunctional. T-cells and monocytes are at the hub of the human immune response and are responsible for all secondary immune responses such as the production of antibodies.

Although scientists around the world are expanding their understanding of HIV, a cure or immunization is not yet available. A vaccine made from inactive viruses is still in the early stages of development. Many researchers believe that the development of vaccine is more likely than a cure.

Teaching Strategy
After students have finished reading the article, divide the class into teams. Ask them to debate whether vaccines and drugs that have not been through clinical trials should be made available to people suffering from potentially fatal diseases.

Genetically-Engineered Bacteria

By studying the DNA and RNA of particular bacteria and viruses, scientists have found that they can change the characteristics of microorganisms using genetic engineering techniques. RNA, like DNA, determines the characteristics of an organism. These altered microorganisms, when packaged in vaccines, are incapable of causing disease. But they do trigger immune responses against unaltered forms of the same microorganisms. The flu vaccine is an example of a vaccine made with an altered form of the virus.

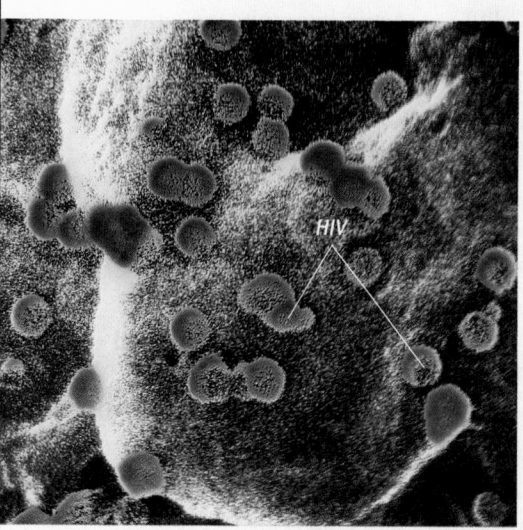
A white blood cell infected with HIV

No Simple Solution

But scientists have also found that it is not so easy to develop vaccines against other viruses. HIV which causes AIDS is a good example of this. Dr. Flossie Wong-Staal of the University of California at San Diego has been working on a vaccine that contains just the outer protein coat of the HIV virus and not the virus's DNA. She is hoping that these modified viruses will act as decoys to stimulate an immune response.

This technique may not necessarily work with HIV. Scientists know that HIV mutates, or changes form, often. One type of vaccine would not necessarily be effective against other forms of the virus. HIV is an example of a retrovirus. Scientists have never produced a human vaccine against a retrovirus.

Scientists are hoping that new vaccine research using genetic engineering techniques may help produce an HIV

Dr. Flossie Wong-Staal

vaccine. In one study, scientists removed a piece of DNA that controls the production of antigens from HIV. The piece of DNA was then injected into an insect's cell, and the cell began to produce large amounts of HIV antigen. Scientists hope the artificial antigens will stimulate an immune response to the HIV virus.

Science Journal
New drugs are often tested first on animals, such as mice. The best test, though, is to inject a healthy person with the virus to observe if later the body protects itself with an immune response.

People who have a disease often volunteer to test a new drug or vaccine. *In your Science Journal* write a paragraph describing your opinion of human testing.

Going Further ▌▌▌▌▌►
A country may require international certificates of vaccination against yellow fever and cholera. Many countries require other vaccines as well. Have students select a country and find out what vaccines are required for visitors. Students can also research what vaccinations are required for entry into the United States. Local public health offices should be able to provide information. **L1**

COOP LEARN

Science Journal
Accept all logical responses. Some students may think human testing could lead to unnecessary deaths or diseases. Others may think that if a person already has the disease, he or she has nothing to lose.

When Are Gibberellins Too Much of a Good Thing?

Scientists know that gibberellic acid, produced by a fungus, is just one of many substances that can dramatically affect plant growth. For example, researchers have observed that some citrus trees treated with large amounts of gibberellins grow six times faster than usual and suffer damage. But scientists have also learned to treat plants with smaller amounts of gibberellins to cause faster growth without damage.

When Audrey Cruz, a Filipino student in the accelerated studies program at Whitney Young High School in Chicago, learned about gibberellins, she knew there must be some way to determine how much gibberellic acid would best benefit plants.

Designing an Experiment

Audrey planted beans and obtained gibberellins from a local plant store.

When the bean plants were about 5 inches tall, she sprayed most of them with the gibberellins. To learn the effects of different amounts of the hormones, she used different concentrations of gibberellins. Her "control plants" received no gibberellins.

From the experiment, Audrey learned that the lowest concentration of gibberellic acid (5 parts per million) caused some growth, but not as much as caused by more concentrated acid (10 parts per million). Yet, plants died soon after they were treated with the highest concentration (20 parts per million).

New Data/New Questions

Audrey says she learned some interesting lessons from her project. She showed that too much gibberellic acid can be bad for the plant, as she had suspected. But she learned some unexpected lessons.

For example, after careful analysis of her results, she believed that the gibberellins in higher concentrations probably killed some of the bean plants. But she wondered if other factors could also have been involved. Had she given the plants too much water? Had something else affected the plants' growth?

Audrey learned something important about doing scientific experiments—to use caution in interpreting her results.

What Do You Think?

By questioning her own experimental technique, Audrey demonstrated the need for scientific objectivity. If you were to do Audrey's experiment, what would you do the same? What would you do differently?

Going Further ▌▌▌▌▌▶
Students can work with partners to find out what fruits and vegetables they eat that have been treated with gibberellic acid. They might also want to learn if they should apply gibberellic acid to plants in their homes and gardens. For information, students can contact these agencies:
Cornell Cooperative Extension
1425 Old Country Road
Plainview, N.Y. 11803
(516) 454-0900

New York Botanical Garden
Southern Blvd. and 200th Street
Bronx, N.Y. 10458
(718) 817-8700

Purpose
This feature extends the material in Section 8-3 by giving further information about the way fungi affect other living things.

Content Background
Gibberellin is applied to dwarf plants to stimulate them to grow to normal size. Minute applications transform bush beans to pole beans and dwarf corn to normal corn. Perhaps the most widespread horticultural application is in grape production. Gibberellin is now regularly applied in the cultivation of a number of grape varieties, such as "Thompson seedless," to increase berry size. In Japan, gibberellic acid is used to induce seedlessness in certain grape varieties. It is also applied to citrus trees, mainly to improve market acceptability of the fruits. The hormone is applied by spray when the fruit is blossoming.

Plants suffering from *bakanae*, the "foolish seedling disease," are conspicuous not only because of their extreme height but also because of their pale, spindly appearance.

Teaching Strategies Students can work in pairs to read the selection, alternating paragraphs. Then, have students discuss how they might modify Audrey's experiment to control the variables mentioned in the final two paragraphs. When students have read the selection, they might like to work in teams to reproduce Audrey's experiment.
`COOP LEARN`

Answers to
What Do You Think?

Answers may vary but should include controlling the amount of water given to the plants as well as observation of other factors affecting plant growth.

Geography Connection

Purpose
This Geography Connection adds to Section 8-2 by showing how bacteria can be beneficial to humanity. In that section, students learned how bacteria help digest and recycle materials in dead organisms and are used in the production of medicines, cleansers, and adhesives. Here, they find out how bacteria are used to help clean up oil spills and toxic wastes.

Content Background
Oil coming into contact with water forms a frothy mass that smothers living organisms in a toxic sauce. Although bacteria greatly assist in the cleanup procedures on the beaches, "the technology for cleaning up oil lags far behind the technology for carrying it out," and the bacteria are not used to clean up animals. Scientists believe that it will take seven to ten years after an oil spill before the shores regain their natural balance. Even then, however, some rare species of plants and animals may be gone.

Teaching Strategies
Have small groups of students read the selection together and list some things they can do personally to reduce our dependence on oil. **COOP LEARN**

Answers to
What Do You Think?
Answers will vary. Some students might suggest that using bacteria to fight pollution could result in toxic waste products from the bacteria.

Geography Connection

Bacteria to the Rescue

In 1989, the Exxon Valdez spilled millions of gallons of oil on the Alaskan coastline. The oil soaked as much as two feet deep into the beaches. When nothing else cleaned up the spill, scientists enlisted the help of bacteria. These microorganisms, living in the soil and water, feed on hydrocarbons, a principal ingredient of oil. The scientists sprayed fertilizer on the beaches to stimulate the bacteria to grow. Within two weeks, the oily beaches that had been sprayed were much cleaner than those that had not. The number of bacteria had tripled, and they were "gobbling up" the oil. Now a similar process has been implemented to get bacteria to feed on and break up clogs in household drains.

Other Useful Bacteria
Many types of bacteria help solve our pollution problems. For example, some species of bacteria can metabolize sulfur, which occurs in large amounts in some kinds of coal. When this high-sulfur coal burns, it releases sulfur into the atmosphere, causing acid rain. However, when sulfur-eating bacteria are mixed into piles of high-sulfur coal, they change it into a much cleaner fuel.

Ongoing Problems
Medical wastes with radioactive ingredients are usually buried in toxic dumps. When these wastes decay, they produce radioactive methane gas. This gas can seep out and pollute the air. Research scientists have begun studying bacteria that change methane into water. When fertilized with nitrogen and phosphorous, these bacteria work two or three times faster to digest the dangerous methane gas.

What Do You Think?
If scientists can develop bacteria to metabolize pollution and waste, chemical companies might be able to eliminate waste treatment plants. Dump sites might be able to speed up the degradation of nontoxic waste, so that it converts to soil more quickly. Toxic waste might be changed into less harmful compounds.

Solutions to problems sometimes can cause new problems. Can you think of any new problems that might result from using bacteria to fight pollution?

Going Further ⅢⅢⅢ➡
Have students play the role of a major oil company and design an advertising campaign to reassure the public that oil spills can be prevented or cleaned up and the environment restored. Suggest that students include a variety of media, such as newspaper, radio, and television ads, posters, and endorsements by scientists and public figures. Students may find it helpful to read some of the material oil companies have issued after large tanker spills. **COOP LEARN** **L2**

chapter 8
REVIEWING MAIN IDEAS

Science Journal

Review the statements below about the big ideas presented in this chapter, and answer the questions. Then, re-read your answers to the Did You Ever Wonder questions at the beginning of the chapter. *In your Science Journal*, write a paragraph about how your understanding of the big ideas in the chapter has changed.

1 Monera, Protista, and Fungi are the kingdoms of microscopic and simple organisms. *Why can you say that these organisms are anything but simple?*

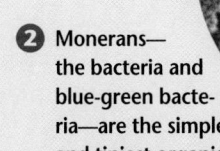

2 Monerans—the bacteria and blue-green bacteria—are the simplest and tiniest organisms on Earth. They are found almost everywhere. *How do monerans that are producers differ from those that are consumers?*

3 Protists are organisms with complex cells that are either plantlike or animal-like. *How do protists differ from monerans?*

4 Fungi help decompose dead organisms and recycle the materials of which these organisms were made. *Explain why fungi are important.*

5 Most scientists consider that viruses are nonliving, noncellular structures. *What do they lack that living organisms have?*

Have students look at the five pictures on this page and read the statements to review the main ideas of this chapter.

Teaching Strategy

Divide the class into four groups. Assign each group one of the pictures, numbers 2–5. Tell each group they are going to impersonate the organisms in their picture. They are to create a brief skit illustrating an average day in the life of the organism. Have them write a script and assign roles to everyone. Suggest that students create simple costumes, use commonplace props, and use sound effects. As an introduction to the performance, read the caption to the first picture. Use the Performance Task Assessment List for skit in **PASC**, p. 75.

COOP LEARN **L1**

Answers to Questions

1. Although these organisms are small, they carry out all the activities of other life forms: they grow, reproduce, obtain energy, and respond to stimuli.

2. Monerans that are producers use sunlight to make their own food. Monerans that are consumers obtain energy by consuming organic material.

3. Protists have a nucleus and monerans do not.

4. Fungi break down organic material so that other organisms can use the materials.

5. They lack a nucleus and do not perform the functions that cells perform.

Science Journal

Did you ever wonder...

• Viruses come in a variety of shapes and sizes—rod-shaped, round, shaped like spiked balls, or shaped like thin threads. (p. 251)
• The room temperature apparently enables the bacteria to be more active. Chilling milk will slow bacterial growth. (p. 255)
• Not all bacteria and fungi are harmful. Bacteria protect the body from certain diseases. Fungi are used to produce drugs. (pp. 258, 264)

🏠 Science at Home

Have students cover a small piece of bread or fruit on a plate for a week. During the course of the week, students should make daily entries in a log describing what they see. At the end of the week, have students add the decomposing material to a home or community compost heap. Use the Performance Task Assessment List for Science Journal in **PASC**, p. 103. **L1**

GLENCOE TECHNOLOGY

MindJogger Videoquiz

Chapter 8 Have students work in groups as they play the Videoquiz game to review key chapter concepts.

Using Key Science Terms

1. Cilia and flagella are hairlike projections (short and large respectively) used to classify protists. Both enable organisms to move.

2. Bacteria and cyanobacteria are the two groups of organisms that make up the Moneran Kingdom.

3. Viruses are considered parasites because they harm the cell or organism they attack.

4. Producers are organisms that make their own food; consumers need to take in food to obtain energy.

5. Microorganisms are very small organisms; viruses are very small organism-like structures.

6. Living things are consumers or producers. Parasites are a type of consumer.

Understanding Ideas

1. Many-celled organisms have different types of cells that are specialized to carry out different functions.

2. A moneran does not have a nucleus; a protist has a nucleus.

3. Students may say that they would look for the organism's method of locomotion, as that is the way scientists classify protozoans.

4. Like plants and animals, fungi need oxygen to carry out respiration. Unlike plants, fungi do not contain chlorophyll; therefore, they cannot make their own food. Also, they do not have the specialized tissues and organs of plants.

5. They are found in wet or damp environments because they require water to grow.

6. Algae may be green in color and have chlorophyll.

7. The protists with no structures for movement live on and in other organisms.

U sing Key Science Terms

cilia	parasite
consumer	producer
flagella	virus
microorganism	

For each set of terms below, explain a relationship that exists.

1. cilia—flagella
2. bacteria—cyanobacteria
3. parasite—virus
4. producer—consumer
5. microorganism—virus
6. parasite—producer

U nderstanding Ideas

Answer the following questions in your Journal using complete sentences.

1. How do many-celled organisms differ from one-celled organisms?
2. What is the main difference in cell structure between a moneran and a protist?
3. What would you look for in order to classify protists in a sample of pond water?
4. How are fungi similar to plants? How are they similar to animals?
5. Why do you suppose fungi are more often successful in wet or humid environments?
6. Why are algae described as being plantlike?
7. Which protist group would be described as parasites? Explain your choice.

Developing Skills

1. See reduced student page for completed concept map.

2. Diatoms, Radiolarians, and Foraminiferans—because they produce hard shells of either silica or calcium.

3. Accept all logical experimental designs. Students should find that conditions that encouraged mold on fruit will be about the same as on wood or paint—moisture and warmth.

D eveloping Skills

Use your understanding of the concepts developed in this chapter to answer each of the following questions.

1. Concept Mapping Complete the concept map for simple organisms.

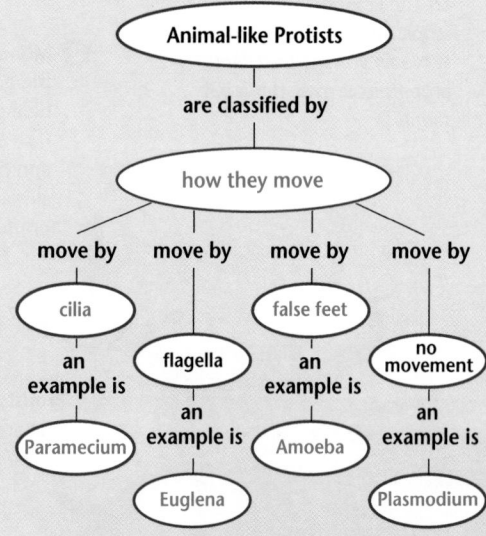

2. Recognizing Cause and Effect Most fossils are formed by organisms that have some sort of hard covering or contain hard substances. Which protists probably have left fossils? Explain your choice.

3. Comparing and Contrasting Think about places where you have seen mold growing other than on food. After doing the Investigation on page 266, design an experiment to find out what conditions are necessary for mold to grow on a piece of untreated wood and in a jar of paint. Were the conditions the same as those for growing mold on food?

Program Resources

Review and Assessment, pp. 47–52 L1
Performance Assessment Ch. 8 L2
PASC
Alternate Assessment in the Science Classroom
Computer Test Bank L1

Critical Thinking

In your Journal, *answer each of the following questions.*

1. Study the volvox in **Figure 8-8**. Why are these organisms considered to be complex? What do they contribute to the environment?
2. How is it an advantage to monerans, protists, and fungi, to be able to respond to changes in their environment?
3. How can cyanobacteria be both beneficial and harmful to a lake?
4. A scientist discovered a new kind of organism growing on a rotting log. The organism was using the log as a food source and, at the same time, helping to decompose it. How would the scientist determine in which kingdom the new organism should be placed?
5. Use your knowledge of virus reproduction to explain why it is difficult to get rid of viruses.

Problem Solving

Read the following problem and discuss your answers in a brief paragraph.

Imagine that a new kind of virus wipes out all bacteria on Earth.

1. What disadvantages for humans would there be in a world without bacteria? What advantages?
2. Would there be any way humans could live in a world without bacteria? Explain your reasons. What other organisms might adapt to fill the role of missing bacteria?

CONNECTING IDEAS

Discuss each of the following in a brief paragraph.

1. **Theme—Stability and Change** Large chalk beds of forminiferans, as much as much as 1000 feet thick, exist in Mississippi and Georgia. Why can you conclude that these areas were covered by ocean at one time?

2. **Theme—Systems and Interactions** How do bacteria and fungi contribute to the needs of all living things?
3. **Theme—Systems and Interactions** How does a one-celled organism differ from a single cell that is part of a many-celled organism?
4. **Science and Society**

How is the action of healthy white blood cells similar to that of the amoeba in Figure 8-10?
5. **Chemistry Connection** Explain reasons why control of insect pests might be better accomplished with organisms such as fungi than with strong, laboratory-made pesticides.

Critical Thinking

1. Algae are complex because they perform complex activities, such as making food and producing oxygen.

2. Because they are so small, these organisms are extremely vulnerable to changes in their environments, so they need to be able to adjust to changes in order to survive.

3. They are beneficial because they provide food and oxygen to organisms in lakes and ponds. They are harmful when the pond is covered with a green slime, preventing sunlight from penetrating water. When they die, decomposition begins.

4. By determining if it is one-celled or many-celled and if it has a nucleus.

5. Because new viruses enter cells, the disease spreads quickly, making it very hard to eradicate.

Problem Solving

1. Humans could not benefit from foods made by bacteria. Also, we could not digest plant foods efficiently. An advantage would be that many illnesses would not exist.

2. Accept all reasonable answers. Illnesses caused by bacteria would not be a problem anymore. More fungi might fill the role of bacteria in decomposition.

Connecting Ideas

1. Foraminiferans are found in the ocean today.

2. By breaking down and decomposing other organisms, thereby releasing materials.

3. The one-celled organism performs all life functions; a cell from an organism cannot exist on its own.

4. White blood cells move with amoeba-like action to engulf bacteria.

5. Accept all reasonable answers. It may reduce the use of pesticides.

✔ Assessment

Portfolio Review the portfolio options that are provided throughout the chapter. Encourage students to select one product that demonstrates their best work for the chapter. Have students explain what they learned and why they chose this example for placement in their portfolios.

Additional portfolio options can be found in the following **Teacher Classroom Resources:**

Multicultural Connections, pp. 19, 20

Making Connections: Integrating Sciences, p. 19

Making Connections: Across the Curriculum, p. 19

Concept Mapping, p. 16

Critical Thinking/Problem Solving, p. 16

Take Home Activities, p. 16

Laboratory Manual, pp. 47–54

Performance Assessment P

Chapter Organizer

SECTION	OBJECTIVES	ACTIVITIES & FEATURES
Chapter Opener		**Explore!**, p. 277
9-1 What Is an Animal? (3 sessions; 1.5 blocks)	1. **Describe** the characteristics all animals have in common. 2. **Classify** different animals by some of their characteristics. **National Content Standards:** (5–8) UCP1–2, UCP5, A1–2, B3, C1, C3, G1	**Explore!**, p. 279 **Find Out!**, p. 280 **Design Your Own Investigation 9-1:** pp. 282–283 **Skillbuilder:** p. 285 **Physics Connection**, pp. 286–287
9-2 Reproduction and Development (4 sessions; 2 blocks)	1. **Distinguish** between sexual and asexual reproduction. 2. **Trace** the stages of complete and incomplete metamorphosis. **National Content Standards:** (5–8) UCP3, UCP5, A1, C1–2	**Explore!**, p. 289 **Find Out!**, p. 293 **Teens in Science**, p. 306 **SciFacts**, p. 304
9-3 Adaptations for Survival (3 sessions; 1 block)	1. **Explain** how adaptations allow animals to survive on Earth. 2. **Give examples** of some animal adaptations. **National Content Standards:** (5–8) UCP4–5, A1, C1, C3, C5, F5, G1–3	**Explore!**, p. 296 **Investigate 9-2:** pp. 300–301 **A Closer Look**, pp. 298–299 **History Connection**, p. 305 **Literature Connection**, p. 305

ACTIVITY MATERIALS

EXPLORE!

p. 277 optional tape recorder

p. 279* beaker or small bowl, snail, goldfish or guppy

p. 289* raw chicken egg, shallow bowl or dish, hand lens

p. 296 No special materials are required.

INVESTIGATE!

pp. 300–301* hand lens, toothpick, vinegar, flashlight, live earthworms, shallow pan, paper towels, 500-mL beaker, water, cotton swab

DESIGN YOUR OWN INVESTIGATION

pp. 282–283* 20 mealworms; pan balance; plastic storage box; 20 g bran flakes; 20 g dry oatmeal; 20 g sugar-coated corn flakes; 20 g broken, unsalted wheat crackers; cheesecloth; gram masses; hand lens; forceps; water

FIND OUT!

p. 280 No special materials are required.

p. 293 vial of food and fruit flies in different stages of development

KEY TO TEACHING STRATEGIES

The following designations will help you decide which activities are appropriate for your students.

L1 Basic activities for all students

L2 Activities for average to above-average students

L3 Challenging activities for above-average students

LEP Limited English Proficiency activities

COOP LEARN Cooperative Learning activities for small group work

P Student products that can be placed into a best-work portfolio

Activities and resources recommended for block schedules

Need Materials? Call Science Kit (1-800-828-7777).

00:00 **OUT OF TIME?** We recommend that students do the activities with an asterisk.

Chapter 9 Animal Life

TEACHER CLASSROOM RESOURCES

Student Masters	Transparencies
Study Guide, p. 33 **Multicultural Connections,** pp. 21, 22 **Making Connections: Integrating Sciences,** p. 21 **Making Connections: Across the Curriculum,** p. 21 **Take Home Activities,** p. 17 **Critical Thinking/Problem Solving,** p. 5 **Making Connections: Technology and Society,** p. 21 **Activity Masters,** Investigate 9-1, pp. 39–40 **Science Discovery Activities,** 9-1 **Laboratory Manual,** pp. 55–56, Vertebrates	**Section Focus Transparency 27**
Study Guide, p. 34 **Concept Mapping,** p. 17 **Critical Thinking/Problem Solving,** p. 17 **How It Works,** p. 13 **Science Discovery Activities,** 9-2	**Teaching Transparency 18,** Frog Life Cycle **Section Focus Transparency 28**
Study Guide, p. 35 **Activity Masters,** Investigate 9-2, pp. 41–42 **Science Discovery Activities,** 9-3	**Teaching Transparency 17,** Invertebrates/Vertebrates **Section Focus Transparency 29**

ASSESSMENT RESOURCES	TEACHING & TECHNOLOGY
Review and Assessment, pp. 53–58 **Performance Assessment,** Ch. 9 **PASC*** **MindJogger Videoquiz** **Alternate Assessment In the Science Classroom** **Computer Test Bank**	**Spanish Resources** **Cooperative Learning Resource Guide** **Lab and Safety Skills** **Science Interactions, Course 1, CD-ROM** **Computer Competency Activities**

*Performance Assessment in the Science Classroom

NATIONAL GEOGRAPHIC TEACHER'S CORNER

National Geographic Society Products

To order the following products for use with this chapter, call the National Geographic Society at 1-800-368-2728:

- "Sanctuary: U.S. National Wildlife Refuges," by Douglas H. Chadwick, October 1996.
- "Koalas," by Oliver Payne, April 1995.
- "Emperors of the Ice," by Glenn Oeland, March 1996.
- "Remote World of the Harpy Eagle," by Neil Rettig, February 1995.
- "Animals at Play," by Stuart L. Brown, December 1994.

National Geographic Society Products Available From Glencoe

To order the following products for use with this chapter, call the National Geographic Society at 1-800-368-2728:

- *Animal Classes Series* (5 Videos)
- *The Great Cover-Up: Animal Camouflage* (Video)

GLENCOE TECHNOLOGY

The following multimedia resources are available from Glencoe:

The Secret of Life
In the Land of Milk and Money: Biotechnology

Science and Technology Videodisc Series (STVS)
Animals
   Jellyfish

Life Science CD-ROM

Teacher Classroom Resources

These are key components of the classroom resources package.

TEACHING AIDS

Section Focus Transparencies

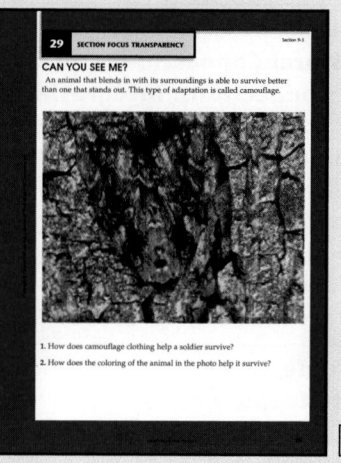

Teaching Transparencies

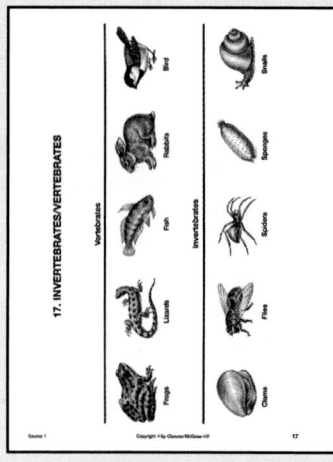

HANDS-ON LEARNING

Science Discovery Activity*

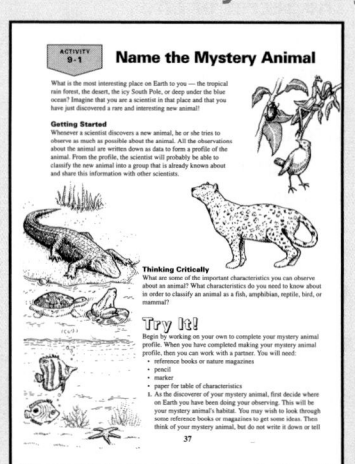

Laboratory Manual*

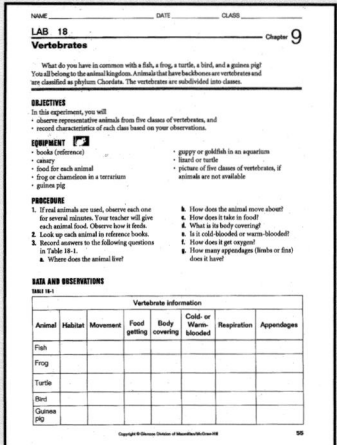

Take Home Activity

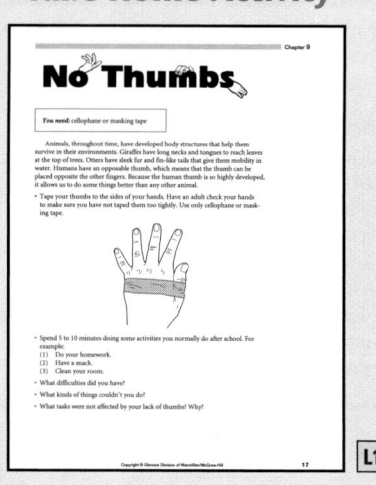

*There may be more than one activity for this chapter.

Chapter 9 Animal Life

REVIEW AND REINFORCEMENT

Study Guide*

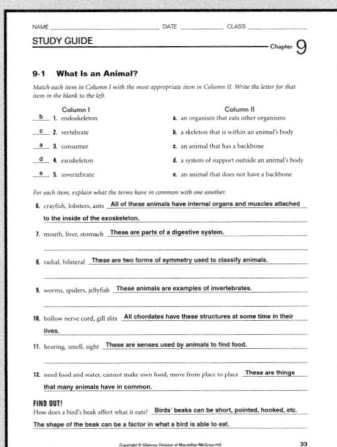

Concept Mapping

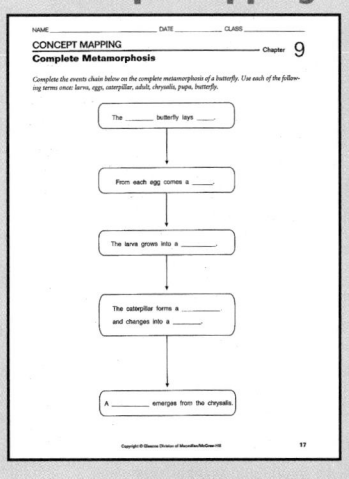

Critical Thinking/ Problem Solving

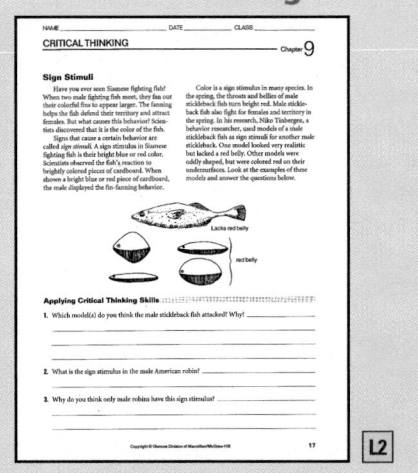

ENRICHMENT AND APPLICATION

Integrating Sciences

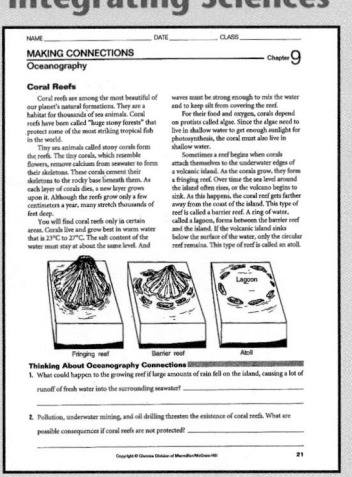

Across the Curriculum

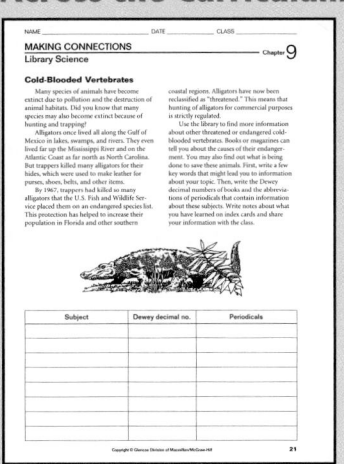

Technology and Society

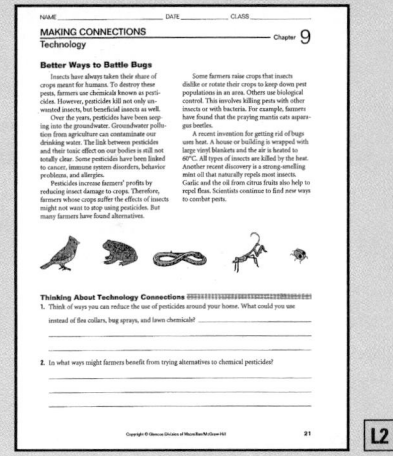

ASSESSMENT

Multicultural Connection**

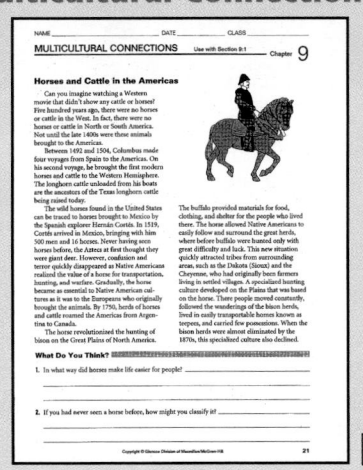

Performance Assessment

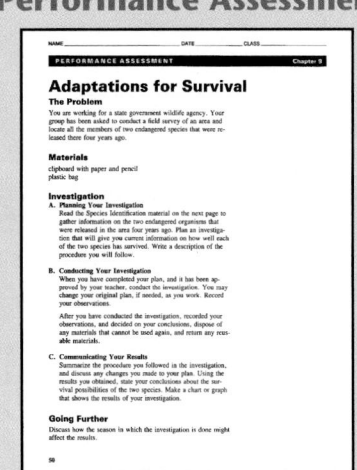

Review and Assessment

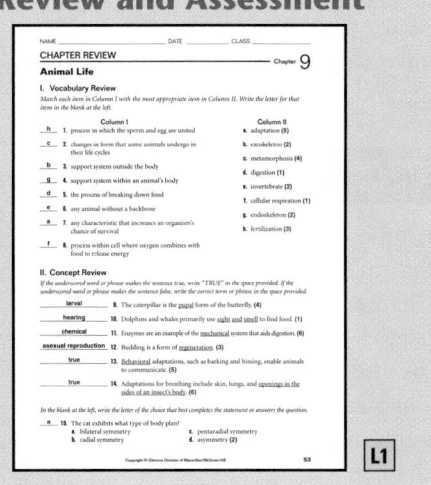

Animal Life

THEME DEVELOPMENT

The themes that this chapter supports are systems and inter-actions and scale and structure. Members of the animal kingdom share common characteristics. Some of these characteristics include methods of obtaining food and ways of perceiving and re-sponding to the environment. All animals are consumers. Thus, they have adaptations that en-able them to find food.

CHAPTER OVERVIEW

In this chapter, students will study the characteristics that de-fine all members of the animal kingdom. They also will learn how scientists classify animals. Students will be introduced to the concepts of sexual and asex-ual reproduction and animal de-velopment. Finally, students will explore adaptations that help different species survive.

Tying to Previous Knowledge

Arrange students in small groups and have them discuss some of the characteristics and behaviors of different kinds of familiar animals. Generate a class discussion about each of the characteristics and behaviors. L1

INTRODUCING THE CHAPTER

Have students look at the photograph on page 277. Ask students to brainstorm different types of animals, then discuss which animals would make good pets and which would not.

00:00 OUT OF TIME?

If time does not permit teach-ing the entire chapter, use the Chapter Overview on this page, Reviewing Main Ideas at the end of the chapter, and the Chapter 9 audiocassette to point out the main ideas of the chapter.

CHAPTER 9

Animal Life

Did you ever wonder...

✓ How snakes and earth-worms differ?

✓ Whether frogs and butterflies have anything in common?

✓ How you locate food?

Science Journal

Before you begin to study about animals, think about these questions and answer them *in your Science Journal.* When you fin-ish the chapter, compare your journal write-up with what you have learned.

Answers are on page 307.

---

f you were given two minutes to list all of the animals that share your life, what examples would you give? Rats and rabbits? Snakes and squirrels? Cats and crickets? Bees, birds, and bats? Animals are found in big cities, small towns, and out in the country. You may think raccoons and deer are found only in the woods, yet an amazing number of them show up in suburban backyards and city parks. Countless other animals may share your home with you. You might say that you are surrounded by animals! And don't forget that people are animals, too. How long is your list now?

▶ **In the activity on the next page, explore some traits that are unique to an animal that lives near you or with you.**

276

What do animals around you do?

Both cats and dogs are domesticated or tame animals that live together with humans. Other animals, such as squirrels, ants, and snakes, are not domesticated. How does living with or apart from humans affect the behaviors of animals?

What To Do

1. Choose an animal that you see every day. It may be domesticated or not. Without interfering with its activities, observe the animal closely.

2. *In your Journal,* describe what the animal looks like, and where and how it lives.

3. How does the animal behave when it doesn't seem to be aware of you? Describe how it responds to you.

4. List what it eats and how it obtains its food.

5. If possible, record any sounds your animal makes.

6. *In your Journal,* use your observations to suggest why you think this organism is classified as an animal.

ASSESSMENT PLANNER

PORTFOLIO
Refer to page 309 for suggested items that students might select for their portfolios.

PERFORMANCE ASSESSMENT
Process, pp. 283, 296, 301
Skillbuilder, p. 285
Explore! Activities, pp. 277, 279, 289, 296
Find Out! Activities, pp. 280, 293
Investigate, pp. 282–283, 300–301

CONTENT ASSESSMENT
Oral, p. 279
Check Your Understanding, pp. 288, 295, 303
Reviewing Main Ideas, p. 307
Chapter Review, pp. 308–309

GROUP ASSESSMENT
Opportunities for group assessment occur with Cooperative Learning Strategies.

Explore!

What do animals around you do?

Time needed   2 days

Materials   tape recorder, an animal

Thinking Processes   observing, classifying, forming operational definitions

Purpose   To observe animal behaviors and characteristics.

Teaching the Activity

Provide data sheets that list activity questions. **L1**

Troubleshooting   Emphasize that animals should not be teased or mistreated.

Science Journal   Have students draw a picture of the animal and its habitat. Ask students to note how the animal uses its environment. **L1** **LEP**

Expected Outcome

Students will observe and describe their animal and its habits and behaviors.

Answers to Questions

Answers will vary depending on the animal chosen for study. Example: birds rest in trees or on the ground, have feathers, are found in groups or alone, and communicate through song or call.

✔ **Assessment**

Performance   Have students work in small groups. Give each group a small bowl with a goldfish and ask them what features of this organism characterize it as an animal. Sprinkle a small amount of fish food into the bowl to observe the reaction. Use the Performance Task Assessment List for Group Work in **PASC**, p. 97. **L1** **COOP LEARN**

PREPARATION

Planning the Lesson

Refer to the Chapter Organizer on pages 276A–D.

Concepts Developed

Students will identify the characteristics that all animals have in common. For example, they cannot make their own food, so they must eat. These characteristics can be used to classify animals.

1 MOTIVATE

Bellringer

Before presenting the lesson, display **Section Focus Transparency 27** on an overhead projector. Assign the accompanying **Focus Activity** worksheet. L1
LEP

Activity Help students identify and compare animal characteristics by showing a film about animals, like *All Things Animal* (Barr Productions). Then, have students list characteristics of the animals they saw. Ask students to look for characteristics that animals share. L1

2 TEACH

Visual Learning

Figure 9-1 Use the figure to help students compare and contrast characteristics of animals. Using captions, students can develop a concept of "adaptation." Using the heads land, air, and water, students can make a table to brainstorm adaptations animals have.

 **9-1** # What Is an Animal?

Section Objectives

- Describe the characteristics all animals have in common.
- Classify different animals by some of their characteristics.

Key Terms

invertebrate
vertebrate
endoskeleton
exoskeleton

Kingdom of Animals

Have you ever thought about all the groups that you are a part of? You are part of a family, a class in school, and a neighborhood. Maybe you belong to the school band or basketball team. As a member of each group, you have things in common with other members of each group. For example, you are probably about the same age as the other students in your class and you live in the same city or town.

■ You Are an Animal

While doing the Explore activity on page 277, you may have noticed that you and the animal you observed share some characteristics. You may both move around on legs. You may

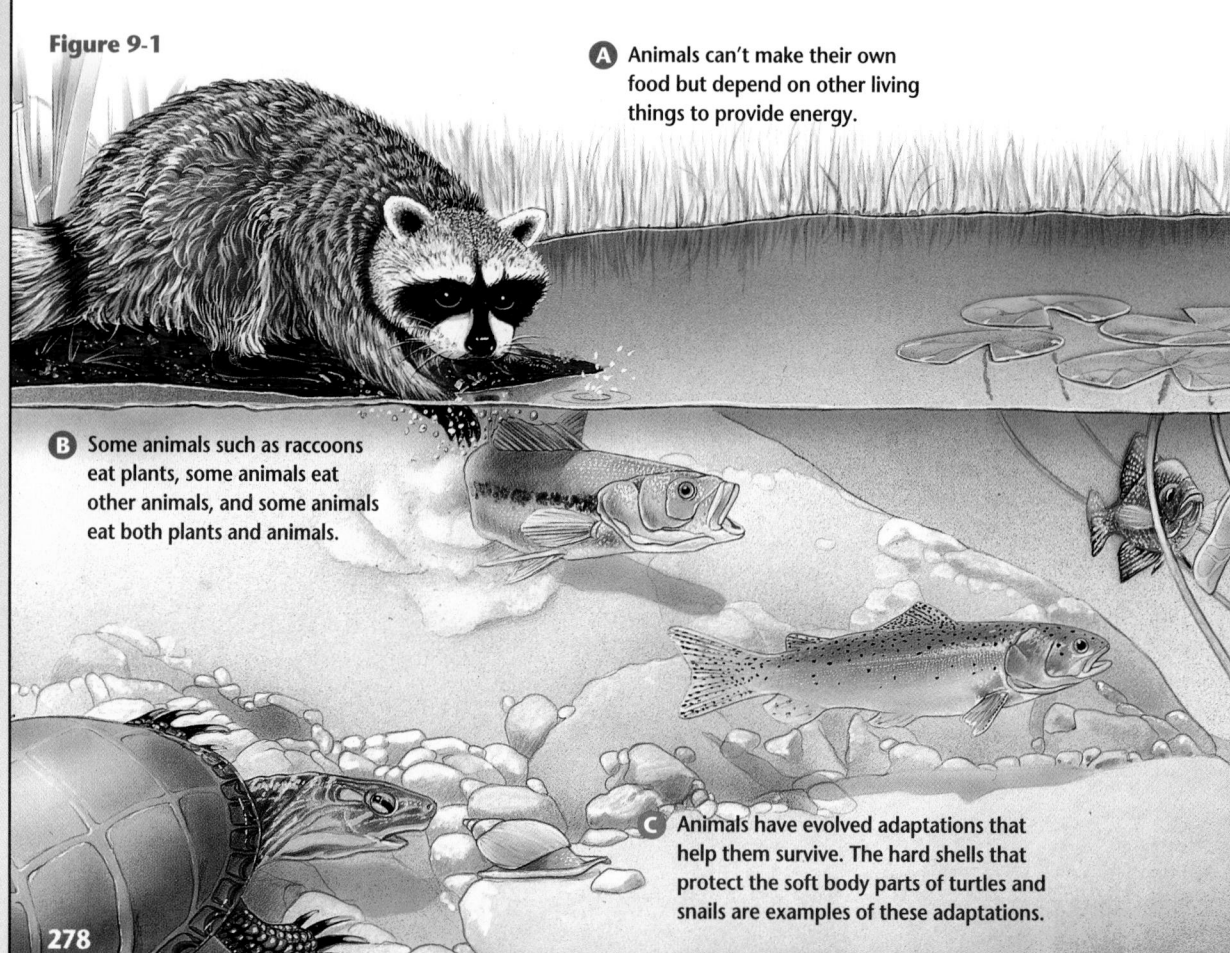

Figure 9-1

A Animals can't make their own food but depend on other living things to provide energy.

B Some animals such as raccoons eat plants, some animals eat other animals, and some animals eat both plants and animals.

C Animals have evolved adaptations that help them survive. The hard shells that protect the soft body parts of turtles and snails are examples of these adaptations.

278

Program Resources

Study Guide, p. 33
Laboratory Manual, pp. 55–56, Vertebrates L2
Multicultural Connections p. 21, Horses and Cattle in the New World L1
Making Connections: Integrating Sciences, p. 21, Coral Reefs L2
Section Focus Transparency 27

ENRICHMENT

Activity Have students classify animals by preparing a bulletin board with pictures of animals representing the various phyla. This can be an ongoing project. As students learn more about animal characteristics, reproduction, and adaptive behavior, this new information can be added to the display. Research will be needed for some phyla. L2 **COOP LEARN**

both have eyes to see with, ears to hear with, and teeth for chewing. Look at the variety of organisms in **Figure 9-1** and ask if you have some of the same adaptations described there. Then, observe some other live organisms in the Explore activity that follows.

ACTIVITY

What are the characteristics of animals?

What To Do

1. Obtain a beaker with a fish and a snail in it.

2. Compare how the two animals move. Does movement give an animal an advantage?

3. *In your Journal,* propose a definition for "animal" based on your observations. Why do you and a snail or fish share the animal kingdom?

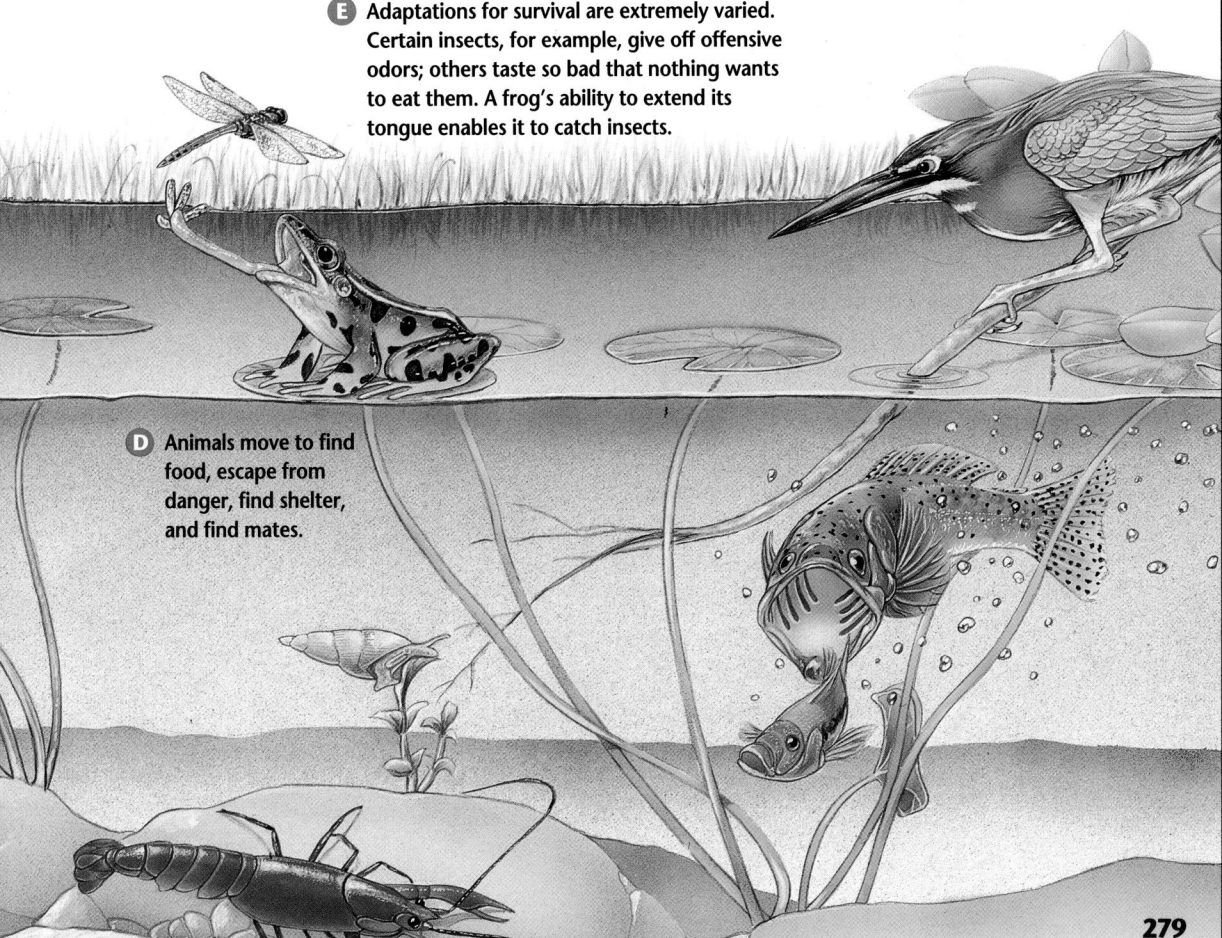

E Adaptations for survival are extremely varied. Certain insects, for example, give off offensive odors; others taste so bad that nothing wants to eat them. A frog's ability to extend its tongue enables it to catch insects.

D Animals move to find food, escape from danger, find shelter, and find mates.

279

Explore!

What are the characteristics of animals?

Time needed 10 minutes

Materials beaker or small dish, snail, goldfish or guppy

Thinking Processes observing, comparing and contrasting

Purpose To compare and contrast how snails and fish move to identify traits that characterize them as animals.

Teaching the Activity

Troubleshooting Make sure that students disturb the fish as little as possible so that it behaves normally.

Discussion Have students brainstorm ways they are similar to and different from snails and fish. List students' responses on the chalkboard.

Science Journal Students should include that animals are living things that move. They may also include traits such as take in food, have head and tail ends, respond to stimuli. **L1**

Expected Outcome

Students will observe that both snails and fish move.

Answers to Questions

2. The fish swims quickly; the snail moves slowly. Movement is related to food getting, seeking shelter, taking care of young.

3. Humans, snails, and fish share the animal kingdom because all eat, reproduce, move, and respond to their environment.

Assessment

Oral Give each student two animals on which to make an oral presentation to the class. Have them include characteristics that the two animals share, characteristics that set them apart, and characteristics that the student has in common with the animals. Use the Performance Task Assessment List for Oral Presentation in **PASC**, p. 71. **L1**

Multicultural Perspectives

Animal Legends

Many Native Americans placed great importance on animals. Pueblo farmers in the parched Southwest thought that the rattlesnake could cause rain to fall. In the spring, the Pueblo captured live rattlers, and during the rain ceremony, stroked the snakes with feathers. After the ceremony, they returned the snakes gently to the ground, where it was hoped they would bring rain.

Interested students can research other animal legends of Native Americans and then share their findings with the class. **L2**

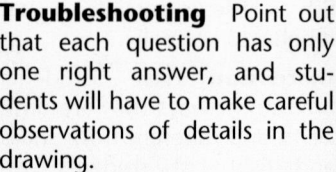

Find Out!

How useful is a bird's beak?

Time needed 15 minutes

Materials No special materials are required; it can be done independently.

Thinking Processes observing and inferring, interpreting scientific illustrations, comparing and contrasting

Purpose To interpret scientific illustrations as a way of recognizing that variations in animals' characteristics are related to lifestyle.

Teaching the Activity

Troubleshooting Point out that each question has only one right answer, and students will have to make careful observations of details in the drawing.

Science Journal Have students make their own comparisons of the bird beak pictures in their journals. L1

Expected Outcomes

Students will compare bird beaks and determine how they limit the type of food the bird feeds on.

Conclude and Apply

1. heron
2. sparrow
3. black skimmer
4. osprey
5. pelican
6. Accept all supported opinions.

✔ Assessment

Performance Students can find pictures of birds other than those shown in the activity. Each student can draw pictures of two birds and write descriptions relating characteristics of the bird's beaks to the types of foods they eat. Small groups can then create booklets about birds' beaks. Use the Performance Task Assessment List for Booklet or Pamphlet in **PASC,** p. 57. L1

COOP LEARN P

Consumers and Producers

Food in some form provides the energy all organisms need to carry out their life processes. But, recall from Chapter 8 that animals can't make their own food. Green plants, on the other hand are producers. Producers make their own food. You will learn more in Chapter 10 about the process whereby plants use water, carbon dioxide, and energy from the sun to produce food.

In contrast to plants, animals are consumers. Consumers are organisms that are unable to make their own food. They must consume other organisms in order to obtain energy. Consumers are adapted in specific ways for getting the energy and nutrients they need. In the following Find Out activity, see how one feature, the beak, affects the survival of one type of consumer, namely birds.

Find Out! ACTIVITY

How useful is a bird's beak?

Each species of bird is uniquely adapted for obtaining the food it requires.

What To Do

Observing the beaks to the right, and using the description below, match the bird to the type of food it eats.

Conclude and Apply

1. Which bird has a long, sharply pointed bill to spear fish?
2. Which bird's bill is short and thick, enabling it to crush seeds?
3. Which bird's lower bill enables it to skim food from the water's surface?
4. The powerful, sharp, hooked beak on this bird enables it to tear flesh.
5. This bird has a pouch under its lower jaw in which it collects fish.
6. *In your Journal,* write a statement explaining how an adaptation such as a beak affects survival in birds.

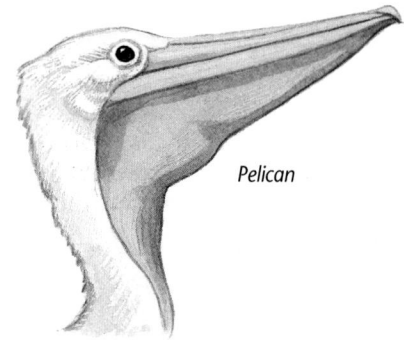
Pelican

Heron

Sparrow

Osprey

Black skimmer

Finding Food

As you learned in the Find Out activity, birds have unique body parts adapted for obtaining food. You are probably more familiar than you realize with adaptations that enable animals to feed.

Many animals depend heavily on their senses to locate food. How do you find food at a mall? What senses do you use? Is it possible that you smell food first and then see the restaurant? Maybe you hear the clatter of dishes? The important words here are *smell*, *see*, and *hear*.

■ Sight and Smell

Humans have fairly well-developed senses of sight, smell, and hearing, but other members of the animal kingdom are even more dependent on these senses for survival than you are. Hunting animals, such as the coyote or wolf, will often pick up the scent of prey. They will follow the scent until they find the animal they are tracking.

The expression "eagle-eyed" gives a clue to an animal that depends on a very sharp sense of sight. For birds of prey that hunt, such as eagles, owls, and hawks, excellent vision is a necessary adaptation for the survival of their species.

■ Hearing

Not all organisms rely on sight or smell for survival however. In Chapter 3, you learned that sound travels through solids, liquids, and gases. That means that you might be able to hear under almost any condition. Scientists have found that hearing is particularly well-developed in many marine animals. Dolphins and whales use echolocation to determine where an object is located.

In the following Investigate, find out what attracts a mealworm to food.

Figure 9-2

Ⓐ The mountain bluebird's large eyes allow the bird to keep a look-out for danger and for insects to eat.

Ⓑ A mountain bluebird searches for food by hovering in the air and watching for insects on the ground. When it sees an insect, the bird swoops down and uses its strong beak to catch and hold its prey.

9-1 What Is an Animal? **281**

Program Resources

Critical Thinking/Problem Solving, p. 5, Flex Your Brain

Making Connections: Technology and Society, p. 21, Better Ways to Battle Bugs ⎣L2⎦

ENRICHMENT

Activity Students can observe animal behavior by making bird feeders from milk cartons and hanging them outside the classroom windows. Have students place birdseed in the feeders and watch as birds come to eat. Have students keep a daily record of how different birds act when they feed. ⎣L1⎦ **COOP LEARN**

Discussion Have students brainstorm animals other than birds (soil dwellers, those that are nocturnal, and so on) and the types of foods they consume to obtain energy.

Content Background

Large eyes located on the sides of the head in most birds allow birds to see a much wider area than other animals, including people. The eyes focus independently of each other so a bird can see two different images at the same time. Many predatory birds such as owls and eagles can focus straight ahead with both eyes focusing on the same image, as humans do. This allows the bird to judge depth and distance and helps it capture prey. Most birds can see in color, which helps them find food.

Activity Snakes do not have external ears, but they can hear. To help students experience how snakes hear, you will need a rubber mallet and a tuning fork.

Have students strike a tuning fork with a rubber mallet, then hold the stem of the tuning fork against the bones of the skull directly behind the ear. The sound will be transmitted through the bones of the head to the ear. Tell students that snakes also hear by bone conduction. ⎣L2⎦ **LEP**

Flex Your Brain Use the Flex Your Brain activity to have students explore USES OF SENSES. ⎣L1⎦

Videodisc

GTV: Planetary Manager

Animal

47447
Animal

47490

9-1 What Is an Animal? **281**

9-1 Picky Eaters

Preparation

Purpose To design a controlled experiment to determine the food preferences of mealworms.

Process Skills observing, measuring in SI, making and using tables, designing an experiment, forming a hypothesis, separating and controlling variables, observing and inferring, interpreting data, predicting, recognizing cause and effect, sampling and estimating

Time Required 20 minutes to set up; several minutes each day for four more days

Materials Mealworms can be obtained from a pet store. Be sure mealworms are active and not about to pupate, or you may not see results. Other types of flaked cereals may be used, but do not mix sugar-coated cereals with plain cereals.

Safety Precautions Students should wash their hands after handling mealworms. Caution students not to put cereal in their mouths.

Possible Hypotheses Students may hypothesize that mealworms will show food preferences, or they may go one step further by hypothesizing that mealworms will show a preference for a particular cereal.

Plan the Experiment

Process Reinforcement Be sure students set up their data tables correctly. The test foods should be listed down the side, and the days (day 1, day 2, etc.) across the top. In the data tables, students will record the number of mealworms observed each day at each food source. If students are having trouble, refer them to Making and Using Tables on page 656 in the Skill Handbook.

Picky Eaters

Are there certain foods that you do not like to eat? Maybe you don't like spinach because of its taste or texture. Other members of your family may not like corn. Do animals, like people, have food preferences? In the following activity, you will observe how mealworms respond to different kinds of food.

Preparation

Problem
How can you determine food preferences of mealworms?

Form a Hypothesis
As a group, discuss the factors that might influence the mealworms' preference for one food over another, such as moistness or odor. Agree upon these factors, then form a hypothesis about the food that mealworms prefer.

Objectives
• Design an experiment that tests the mealworms' responses to various types of foods.
• Compare the mealworms' responses to different foods.
• Infer why mealworms prefer some foods over others.

Possible Materials
20 mealworms
plastic storage box
pan balance
cheesecloth
gram masses
hand lens
forceps
20 g bran flakes
20 g dry oatmeal
20 g sugar-coated corn flakes
20 g broken, unsalted wheat crackers
water

Safety Precautions

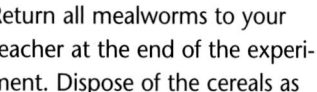

Return all mealworms to your teacher at the end of the experiment. Dispose of the cereals as directed by your teacher.

DESIGN YOUR OWN
INVESTIGATION

Plan the Experiment

1 Examine the materials provided. Which materials will you use? How will you use them?

2 Agree upon a way to test your hypothesis. Write down what you will do at each step.

3 Assign tasks to members of the group.

4 Design a table for recording your data. Decide when data will be measured and recorded.

Check the Plan

1 Determine where you will put the different foods in the plastic box. How will you calculate how much food the mealworms have eaten?

2 Decide where to place the mealworms in the box. Be sure to discuss how to move the mealworms back to their starting place so that the experiment can be repeated several times.

3 Before you start the experiment, have your teacher approve your plan.

4 Carry out your experiment. Complete your data table *in your Science Journal.*

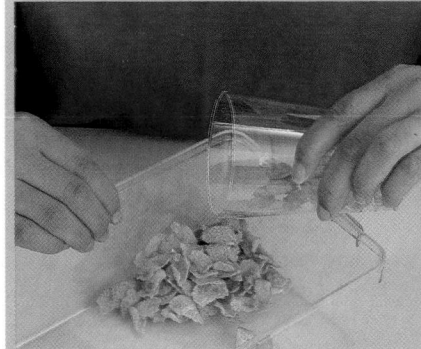

Analyze and Conclude

 1. Use Numbers Calculate the total number of mealworms that preferred each food.

2. Use Numbers Calculate the mealworms' daily average of food intake for each food type.

3. Conclude Which type of food did most mealworms prefer? Did this support your hypothesis? Explain.

4. Infer Infer why the mealworms were attracted to a particular food.

Going Further

Predict whether mealworms might be attracted to other food choices. Design an experiment to test this prediction. Test the prediction.

Program Resources

Activity Masters, pp. 39–40, Design Your Own Investigation 9-1

Science Discovery Activities, 9-1, Name the Mystery Animal

Possible Procedures Place different food types in separate corners of the box. Food types should not touch one another. Moisten food materials. Place mealworms in the center of the box equidistant from each food type. Students should count the number of mealworms in each food location daily.

Teaching Strategies

Troubleshooting Remind students to handle the mealworms carefully. Keep boxes in a warm (not hot) location. Do not get cereal soggy. If mealworms do not respond, replace them with more active mealworms.

Science Journal Have students record their hypotheses, procedures, and results in their journals.

Expected Outcome

Students will observe that mealworms do have food preferences.

Analyze and Conclude

1. Answers will vary.

2. Answers will vary.

3. The food most mealworms preferred was cooked or moistened oatmeal. Answers will depend on students' hypotheses. Help students to understand that an unsupported hypothesis is not wrong. There are no right or wrong hypotheses, only supported or unsupported ones.

4. Answers will vary.

✔ Assessment

Process Have students working in cooperative groups design an experiment to determine food preferences of fellow students. Tell them they are cafeteria managers, and it is their job to find out what will sell in the cafeteria and what will not. Once they have tabulated their results, have the groups of students present their findings to the school dietician and ask him or her to talk to the class about how lunch menus are determined.

Going Further ⑆

Answers will vary, but are likely to include other forms of cereals and crackers.

Figure 9-3 Frogs have complete digestive systems, possessing a mouth where food enters, and an anus where wastes pass out of the body. This type of body plan is sometimes referred to as a tube within a tube. Have students trace the path of food through the frog's digestive system. Point out that the liver is not actually part of the digestive tube (no food passes through it).

Figure 9-4 Students can model how a starfish's tube feet work. Have students fill medicine droppers with water, squeeze out a few drops of water, then place their fingers over the end of the dropper as they let go of the bulb. They will feel a slight suction. The suction pressure that each tube foot exerts is greater than what students feel. A starfish has numerous tube feet, thereby creating enormous suction pressure.

GLENCOE TECHNOLOGY

Videodisc

The Infinite Voyage: Secrets from a Frozen World

Chapter 1
The Southern Ocean—A Rich Marine Ecosystem

Chapter 2
Krill: The Vital Link of the Food Chain

How Animals Digest Food

Figure 9-3

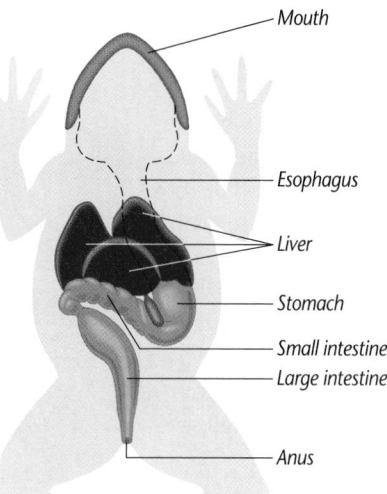

Mouth
Esophagus
Liver
Stomach
Small intestine
Large intestine
Anus

A Food enters through the mouth. The frog's tongue enables it to grab whole insects.

B Food passes through a series of organs in the digestive system. Enzymes in the stomach and small intestine break the food down chemically. The digested food is absorbed and used throughout the body.

C Undigested materials are expelled as waste through the anus.

Locating food is one thing, but once an animal has the food it needs, it has to break it down into substances it can use. This is done through the process of digestion. Most many-celled animals have digestive systems. In a digestive system, food is broken down mechanically and chemically as it passes through different parts of a specialized body tube. These animals have two body openings, a mouth and an anus, and food moves in only one direction. **Figure 9-3** describes how one animal, the frog, digests food.

What happens when an animal doesn't have a complex digestive system like the frog's? Some organisms, such as sponges, feed by filtering water that flows through the pores in its body. **Figure 9-4** shows how another organism, the starfish, feeds.

Figure 9-4

A Starfish use their tube feet to obtain food, such as clams, mussels, and worms. Here, the starfish attaches its tube feet to both halves of a mussel shell and pulls until the shell opens.

B The starfish pushes its stomach out through its mouth and spreads the stomach over the food. Enzymes secreted by the stomach turn the food into a soupy liquid, which is taken into the stomach.

C Then, the starfish pulls its stomach back into its body and digestion continues.

284 Chapter 9 Animal Life

ENRICHMENT

Activity Students can practice classifying skills by making a mobile of the different animals they have read about in this section. Beneath each animal, students should write the animal's name and the fact they find most interesting about it. When students have finished this section, have them rearrange the order of animals on their mobiles to reflect what they have learned about the classification of animals. L1

Program Resources

Multicultural Connections, p. 22, The Center of Sami Life L1

Making Connections: Across the Curriculum, p. 21, Cold-Blooded Vertebrates L3

Take Home Activities, p. 17, No Thumbs L1

How Scientists Classify Animals

As you can see, there are many different features to consider when looking at animals. Scientists use these different characteristics to place animals into groups.

■ Invertebrates and Vertebrates

Try this simple activity. Reach around and run your fingers down the middle of your back from neck to waist. Describe what you feel.

When you ran your hand down your back, you felt a backbone. Biologists use the characteristic of having or not having a backbone as a way to separate members of the animal kingdom into two groups. Most animals in the living world don't have backbones. These animals belong to an

enormously large group known as the invertebrates. An **invertebrate** is any animal that doesn't have a backbone. Worms, clams, jellyfish, flies, and spiders are examples of invertebrates.

You belong to the smaller group of animals that do have backbones. You are a **vertebrate**, an animal that has a backbone. **Figure 9-5** shows examples of different groups of vertebrates and invertebrates.

SKILLBUILDER

Making and Using Graphs

This bar graph gives you some idea of the relative number of species of invertebrates and vertebrates that biologists have identified. How many species of invertebrates are shown by the graph? Which group of animals is most abundant? Which group has the fewest number? If you need help, refer to the **Skill Handbook** on page 657.

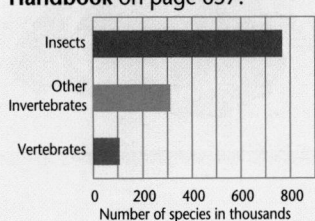

Number of species in thousands

Figure 9-5

Vertebrates

| Frogs | Lizards | Fish | Rabbits | Birds |

Invertebrates

| Clams | Flies | Spiders | Sponges | Snails |

Content Background

The classification of animals has changed over the years, as more details about animal characteristics are learned. A good source of information on classification is *Five Kingdoms* by Margulis and Schwartz (Freeman, 1988). The book is written at an adult level, but the scenes that place each group into their environment can be used to answer some student questions.

Uncovering Preconceptions
Many students do not think of invertebrates as animals. Using several nature magazines from the library, have students classify any animals in articles as vertebrates or invertebrates.

Visual Learning

Figure 9-5   What groups of vertebrates are shown? *Introduce students to the group names of the examples shown: birds, mammals, fish, reptiles, amphibians* **What do spiders and clams have in common?** *They are invertebrates.*

SKILLBUILDER

1. About 1025
2. Insects
3. Vertebrates L1

 TECHNOLOGY

 CD-ROM

Science Interactions, Course 1, CD-ROM

Chapter 9

ENRICHMENT

Activity Have students demonstrate good posture. Have them stand with their heads high, but not pushed back. The shoulders should be held back just far enough to allow ease of breathing. Arrange students in pairs to practice holding their backbones correctly. Tell them to try to imagine that a weighted line hangs down from beside their ears to just in front of their ankles. LEP L1

ENRICHMENT

Research Help students recognize cause and effect by pointing out that the most serious complication of a fractured backbone is injury to the spinal column, which can result in paralysis. Have students find out what first aid measures are recommended when it is suspected that a person has fractured his or her backbone. Should the person be moved? They may wish to contact their local rescue squad for information. L2

Chordates are classified together because of the presence of a notochord, a rod-shaped structure composed of cells that supports the body of lower chordates. In higher vertebrates, the notochord is present in the embryo but is later replaced by the vertebral column, or spine.

Inquiry Questions Many sea stars, sea urchins, and sand dollars have pentaradial symmetry. **What does *penta*- mean?** *It means five.* **Which form of symmetry do you think allows an animal to move more efficiently on land? bilateral Why?** *It allows animals to organize their movement in a straight line and in a certain direction.*

Figure 9-6

Organisms have one of three basic body plans.

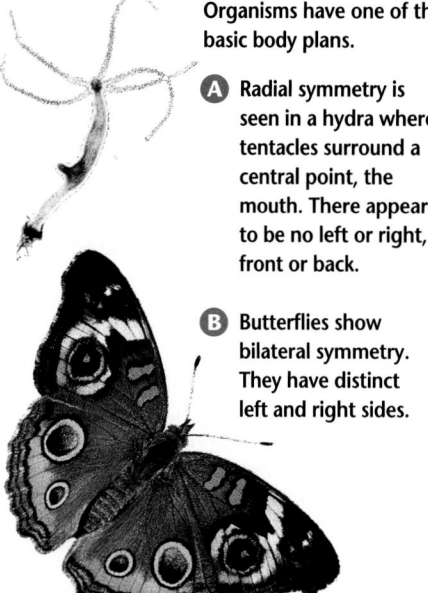

A Radial symmetry is seen in a hydra where tentacles surround a central point, the mouth. There appears to be no left or right, front or back.

C Organisms, such as sponges, which show no recognizable symmetry, are described as asymmetrical.

B Butterflies show bilateral symmetry. They have distinct left and right sides.

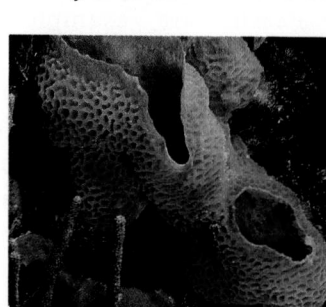

Vertebrates belong to a group called the chordates. All chordates have a hollow nerve cord and gill-like openings at some time during their life. Fish, amphibians, reptiles, birds, and mammals all have this feature and are therefore classified together.

■ Body Plans

Biologists also look at each animal's basic body plan, or how body parts are arranged, when describing and classifying an organism. Body plan is the animal's symmetry. In **Figure 9-6**, you can learn about different body plans.

Physics
CONNECTION

Creature Features

Building models of animals can help you understand their adaptations. How can you combine serious science with fun arts and crafts? Just ask biomechanics

researcher Mimi Koehl from the University of California at Berkeley.

Biomechanics

Biomechanics is the field of science in which scientists ask how living organisms are affected by the laws of physics. By studying the physical environment an animal lives in, scientists can begin to explain why animals are shaped in a certain way and why they behave as they do. For example, the shapes of fish and

Malaysian flying frog

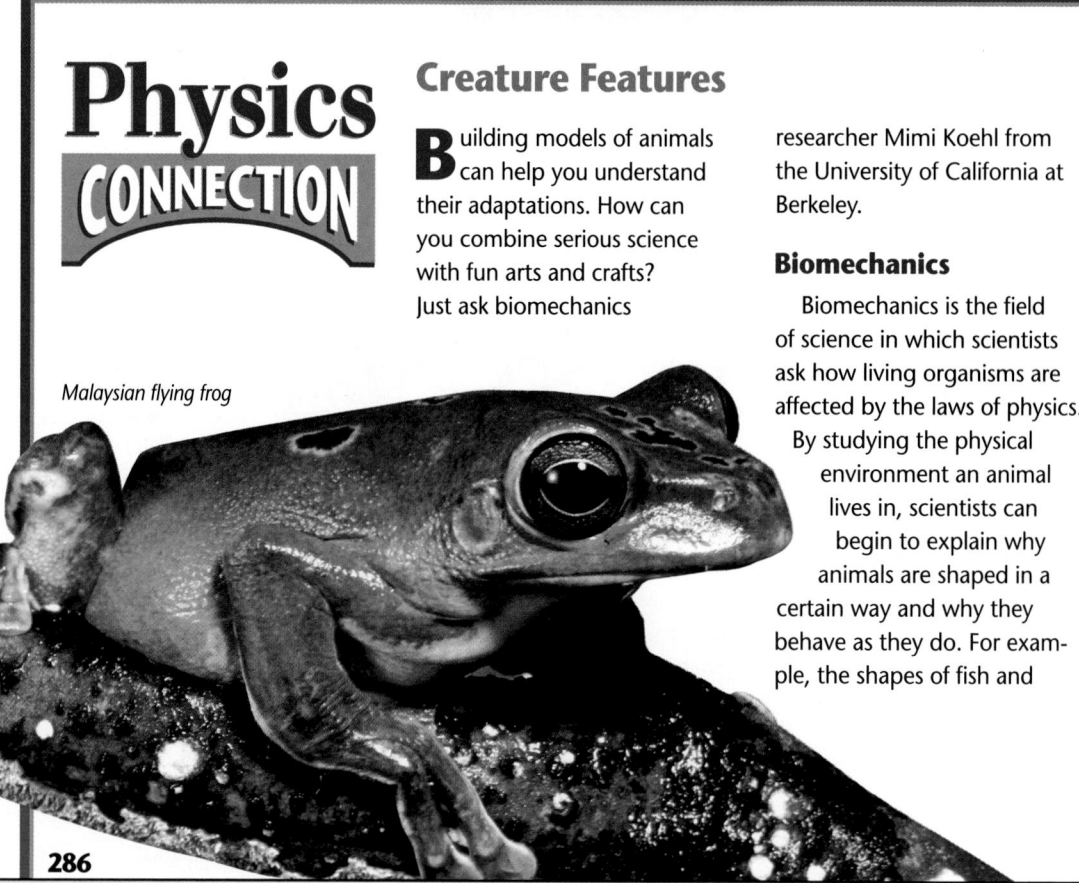

286

Physics
CONNECTION

Purpose
Physics Connection describes how two scientists have used models to study physical adaptations of animals. It also helps students understand that adaptations are based in the physical and chemical characteristics of organisms.

Content Background
Wind tunnels have enabled biomechanics researchers to measure the action of air as it moves against an object. Wind tunnels

show how much resistance a given object meets as it moves through the air. Ask students to identify some human products that might be tested in a wind tunnel. Answers might include airplanes or cars.

Like flying squirrels and flying fish, flying frogs do not truly fly as birds do. The flying frog's body is designed to produce resistance, rather than reduce it. This slows the frog's velocity as it falls from a high place, allowing it to glide and giving it the ap-

■ Skeletons: A Means of Support

Have you ever watched a tall building under construction? Steel beams provide the support for the walls, floors, and roof. Animals also have support systems called a skeleton. Remember when you felt your backbone? The backbone is part of your internal support system. But skeletons may be internal or external.

■ Internal and External Skeletons

Figure 9-7 shows an example of an **endoskeleton**, a skeleton that is within an animal's body. All vertebrates have endoskeletons. Endoskeletons, such as

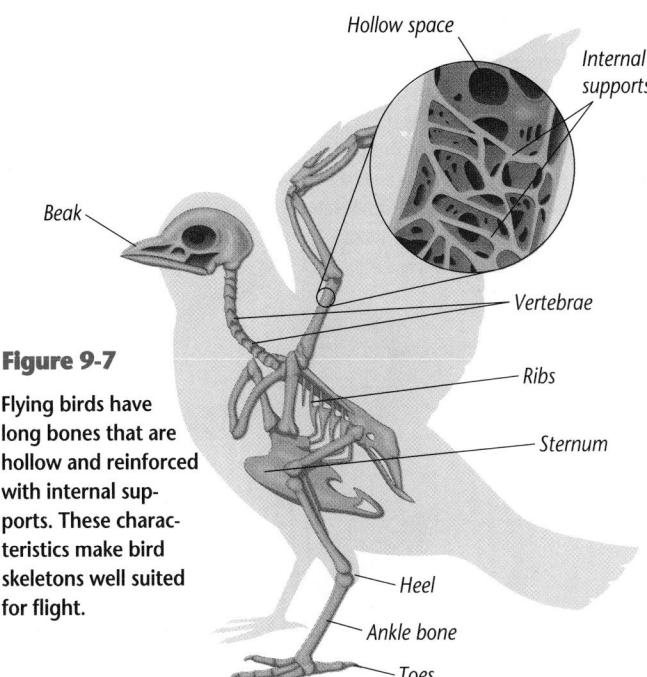

Hollow space

Internal supports

Beak

Vertebrae

Ribs

Sternum

Heel

Ankle bone

Toes

Figure 9-7

Flying birds have long bones that are hollow and reinforced with internal supports. These characteristics make bird skeletons well suited for flight.

Content Background

Most animals are symmetrical in some way. Only a few can be described as asymmetrical. Many asymmetrical animals are sessile (attached at the base, or without any distinct projecting support). Animals with radial symmetry do not move efficiently. Either they are sessile or they float or crawl. Most animals with bilateral symmetry move headfirst.

Visual Learning

Figure 9-7 Reinforce the meanings of the root words *endo-* (meaning internal or inside) and *exo-* (meaning external or outside). Students can classify a variety of skeleton models or photographs; crab, crayfish, insect, cat, rat according to *exoskeleton* and *endoskeleton* categories.

Across the Curriculum

Language Arts

Have students look up the prefix *zoo-* and some words that include it. This prefix refers to animals. The branch of biology that deals with the study of animals is zoology. Scientists who study animals are zoologists. Ask students what a zoological park is. *a zoo*
L2

dolphins allow these animals to swim through water with little resistance.

But explaining animal structure is not always so easy, especially with live animals or animals that have become extinct.

Flying Frogs

Dr. Koehl has worked with biologist Sharon Emerson of the University of Utah to explain the structure of a flying amphibian—the Malaysian flying frog. Dr. Emerson has studied this tiny frog in its native habitat. Dr. Emerson wanted to know whether this frog's interesting adaptations, such as large webbed feet, had anything to

do with the frog's gymnastic abilities.

Koehl and Emerson made several realistic models. They also made individual legs, "hands," and feet of nonflying frogs.

To investigate whether the frog's shape is related to its acrobatic abilities, Koehl and Emerson tested their model frogs in a small wind tunnel. The wind tunnel allowed the researchers to investigate how wind resistance affects the frog as it flies through the air. They also replaced parts of the model with the interchangeable nonflying-frog parts to see how these affect wind resistance.

From the experiments, Koehl and Emerson observed that the frog's adaptations, such as its webbed feet, helped it control how it moved through the air much like a parachutist controls direction during a fall. They concluded that the flying frog's shape is not so much beneficial to flying as it is to how it controls its fall.

You Try It!

Go to the library and learn about the adaptations of some of your favorite animals. Read about their behaviors in their natural environments. How are your animal's adaptations related to its behavior?

pearance of flight. (See *Discover*, December 1991)

Teaching Strategy

Have students make a model to observe how webbed feet might affect wind resistance. Have students loosely hold a fork upside down about one and a half feet away from a fan. Ask them to observe whether the wind affects the fork. Then have students repeat the experiment with tissue paper wrapped around the tines of

the fork. Have students determine how the addition of the tissue "webbing" is like the webbed feet of the Malaysian flying frog, and how this webbing helps it "fly." L1

Answers to You Try It!

There are many examples of adaptations that are related to behavior. Students can investigate an animal with a similar adaptation, such as a flying squirrel.

Going Further ▪▪▪▪▶

Ask students to choose an adaptation of one of their favorite animals. Challenge them to make a model of this adaptation to demonstrate how this feature works in the animal's environment. Students may wish to use fans, aquariums, or earth to represent the specific environment. Use the Performance Task Assessment List for Models in **PASC**, p. 51. L1

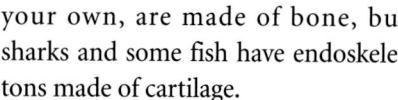

Figure 9-8 Infer how the crab and the snake move. What role does the skeleton play in each animal's movement? *Snakes lack legs and move by twisting their bodies in a wave-like pattern. Muscles are attached to the internal skeleton. Crabs have an exoskeleton to which muscles are attached. They use their legs for movement.*

3 ASSESS

Check for Understanding

1. Play a game to guess an animal's identity by writing clues about its food-gathering methods, body parts and senses, and classification. Write the name of the animal on the front of an index card and clues on the back. Students can play the game in pairs or teams. Use the Performance Task Assessment List for Group Work in **PASC**, p. 97.
`COOP LEARN`

2. Ask students to make a chart showing how invertebrates are different from vertebrates.

Reteach

Activity Have students make a concept map showing the steps scientists can use to classify a new animal. *Scientists first determine if the animal is a vertebrate or invertebrate. Next, they determine the animal's symmetry. Last, they compare the animal's characteristics to those of other organisms.* `L1`

Extension

Discussion In the last 300 years, at least 300 vertebrate animals have become extinct. By comparison, estimates show that dinosaur species died off at the rate of about one per 1000 years. Have students explain what causes animals to become extinct today and what can be done to help preserve them. `L3`

Figure 9-8

A *Endoskeletons* All vertebrates have an internal skeleton, also called an endoskeleton. Cells that make up endoskeletons are constantly renewed. Muscles are attached to the outside of the endoskeleton, giving the outside of most vertebrate bodies a soft texture.

B *Exoskeletons* The shell of a crab is its exoskeleton. This outside skeleton protects the animal's internal organs from loss of fluids and supports its body. The crab sheds its exoskeleton every few weeks as it grows, and a new one develops underneath. The internal organs and the muscles of crayfish, lobsters, shrimp, crabs, ants, bees, and beetles are attached to the inside of the exoskeleton.

check your UNDERSTANDING

1. An adult sponge is an animal that doesn't move from place to place. It's attached to one spot on the ocean floor. Why can you still consider it an animal?
2. Classify each of the animals listed as vertebrate or invertebrate. Which of the animals has an endoskeleton? Which of them has an exoskeleton?
    bee   mouse   robin   ant   deer
3. Apply If you found a fossil of an organism with a backbone, what characteristics would you know about the organism?

---

your own, are made of bone, but sharks and some fish have endoskeletons made of cartilage.

Some animals have an internal skeleton that is neither bone nor cartilage. Animals such as jellyfish and earthworms maintain their form because of fluids contained under pressure in compartments in their bodies.

Many invertebrates such as the crab in **Figure 9-8B**, have an **exoskeleton**, a support system on the outside of the body.

In this section, you've thought about characteristics that separate animals from other groups of living things. You also learned that animals are classified in different ways based on body structure adaptations. In the next section, learn how different animals reproduce.

4 CLOSE

Have each student draw the outline of a familiar animal. Based on what they learned in this section, what words could they write in the outline to describe the animal? Student descriptions should include endoskeleton or exoskeleton, vertebrate or invertebrate, and the organism's symmetry. `L1`

check your UNDERSTANDING

1. It is an animal because it is a consumer and digests its food.
2. Bee and ant are invertebrates. The rest are vertebrates. All vertebrates have endoskeletons; the bee and ant have exoskeletons.
3. It is an animal, a consumer, a vertebrate, a chordate, and has an endoskeleton.

9-2 Reproduction and Development

How Animals Reproduce

Have you ever stopped at a pet shop to look at puppies, rabbits, and kittens? By the time these animals get to the store, they're already several weeks old and changing rapidly. As you laugh at their actions, you're not thinking about the important processes of reproduction that took place to make these young animals.

Reproduction in animals is either sexual or asexual. Each process is characterized by certain adaptations. For example, eggs are an important adaptation in reproduction for many animal groups. In the Explore activity below, observe an egg to identify features that help chickens and other birds to survive as a species.

Section Objectives

■ Distinguish between sexual and asexual reproduction.
■ Trace the stages of complete and incomplete metamorphosis.

Key Terms

fertilization
regeneration
metamorphosis

Explore! ACTIVITY

What's in a bird egg?

What To Do

Take a close look at a chicken egg to learn something about the typical parts of a bird egg.

1. Open a chicken egg into a shallow bowl.

2. Use a hand lens to take a close look at the shell. How do you suppose a shell is helpful to the developing bird?

3. With the help of the diagram of the egg, identify the parts inside the egg.

4. *In your Journal,* infer a function for each part you see in the egg.

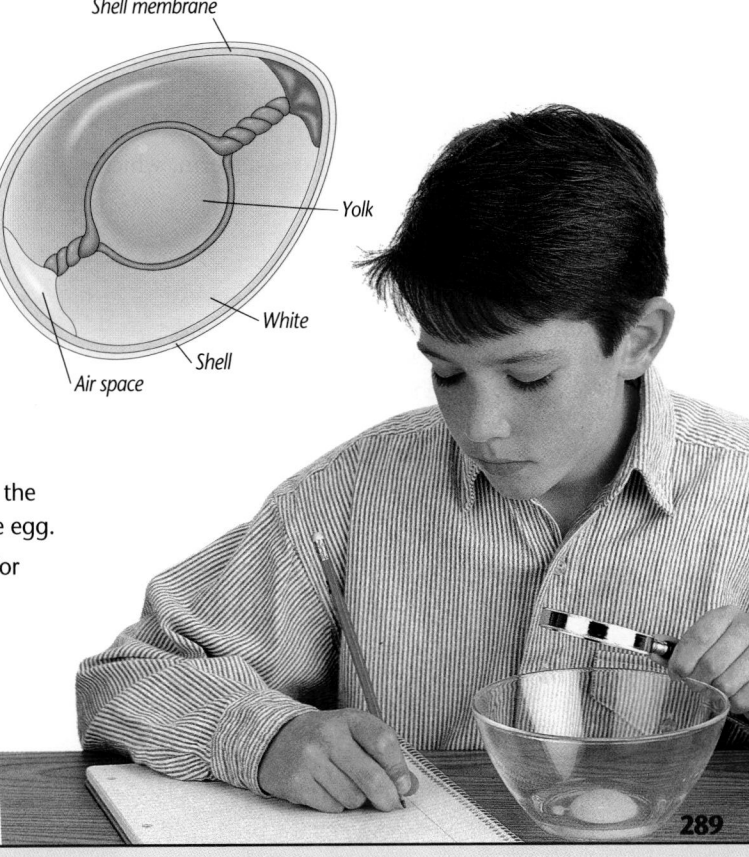

Shell membrane
Yolk
White
Shell
Air space

289

Explore!

What's in a bird egg?

Time needed 25 minutes

Materials one raw egg per student group, shallow bowl or dish, hand lens

Thinking processes observing and inferring, interpreting scientific diagrams

Purpose To observe structures involved with early animal development.

Teaching the Activity

Safety Be sure students wash their hands after handling eggs.

Troubleshooting Students should carefully crack the eggs open so that the yolk is not broken. If eggs are not fertilized, students will not be able to see developing embryos.

Science Journal Make sure students include the following: shell membrane, yolk, white, shell. They may not be able to see the air space

2 TEACH

Tying to Previous Knowledge

To reinforce classifying skills, arrange students in small groups to make posters showing how scientists classify animals, which they learned in the previous section. Tell them to leave about one-quarter of the poster blank. When everyone has finished working, have each group share its poster with the class in a brief oral presentation. Explain to students they will now learn two other ways that animals can be classified, by how they reproduce and how they develop. **What animals can you think of that need a lot of care after they are born?** *mammals, such as humans, foxes, and kittens, and birds, such as penguins, robins, and cardinals* **Which ones are independent very quickly?** *Students can cite examples of fish, amphibians, and reptiles.* You can have students complete the posters when they have finished this section. L2 CCOP LEARN P

Connect to

Physics

Decibel: a unit of intensity of sound, expressed as loudness, rated on a scale with zero (0) being inaudible, to beyond 120, where sound can inflict pain.

Teacher F.Y.I.

Explain that virtually all bony fishes can regenerate amputated fins, but cartilaginous fishes cannot. Ask students to name some cartilaginous fishes. *rays and sharks*

Connect to...

Physics

Many animals have evolved behaviors that promote sexual reproduction. The coqui frog in Puerto Rico makes a sound like its name (ko-*kee*) to attract a mate. The sound can be as loud as 108 decibels. Find out what a decibel is and why 108 of these would be deafening.

Sexual Reproduction

Animal groups that reproduce sexually have separate male and female individuals. In sexual reproduction, sperm from a male unite with one or more eggs produced by a female in a process called **fertilization**. In animal groups, fertilization may be external or internal.

■ External Fertilization

Reproduction results in a generation of new individuals that are like their parents. Different animal species may produce one or many new individuals at one time. For starfish, sea urchins, fish, frogs, and other water-inhabiting animals, huge numbers of eggs and sperm are usually produced at one time. These organisms are dependent on water to carry out fertilization. During external fertilization, the female releases eggs into the water. The male then releases sperm, which swim to the eggs. In each instance, all of the eggs are usually fertilized.

■ Internal Fertilization

During internal fertilization, a female produces one or more eggs that are kept in the body. The male then deposits sperm in a fluid into the female's body. The sperm swim to and unite with the eggs. Organisms that have fertilized eggs that undergo development in the female's body generally produce smaller numbers of offspring. Organisms that undergo internal fertilization are not usually restricted to living in or near water to accomplish fertilization.

Egg

Egg-producing ovaries

Figure 9-9

(A) *External Fertilization* Among organisms that reproduce through external fertilization, water is a key factor. Female fish release large numbers of eggs into the water. The male fish release large numbers of sperm that swim to the eggs and fertilize them.

(B) *Internal Fertilization* During internal fertilization, eggs remain in the female. While less dependent on water for fertilization than fish or other aquatic animals, sperm from the male moves in a fluid through the female reproductive system toward the egg. Once fertilized, an egg develops in the female, as with humans, or may be laid with a protective shell, as with alligators and ducks.

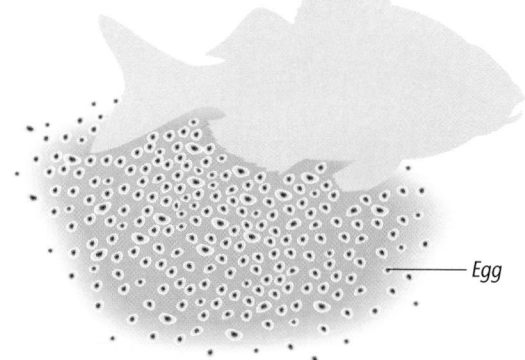

Egg

ENRICHMENT

Research Have students use reference materials to research and report on gestation periods in mammals and the average number of offspring born to different species. Some gestation periods are shown below. L1

hamster	16 days	cow	281 days
rabbit	31 days	horse	336 days
dog	61 days	giraffe	442 days
cat	63 days	whale	450 days

Program Resources

Study Guide, p. 34
Critical Thinking/Problem Solving, p. 17, Sign Stimuli L2
Teaching Transparency 18
Section Focus Transparency 28

Asexual Reproduction

Asexual reproduction is the production of a new organism from just one parent. No eggs or sperm are exchanged. Among animals, asexual reproduction occurs in many invertebrates, such as the aphids in **Figure 9-10**.

Hydra and sponges can reproduce by an asexual process called budding. During budding in hydra, as shown in **Figure 9-11**, one or more new individuals form on the parent organism. At some point, the new organism breaks off from the parent. The new organism has the same characteristics as the parent.

■ Regeneration

A few animals form whole new body parts by regeneration. **Regeneration** occurs when an animal regrows a missing part or regrows from only a portion of the original body. Regeneration may also be a form of asexual reproduction when whole new individuals develop. Sponge growers cut large sponges into smaller pieces and throw the pieces back into the ocean. Each separate piece grows into a larger sponge in as little as eighteen months.

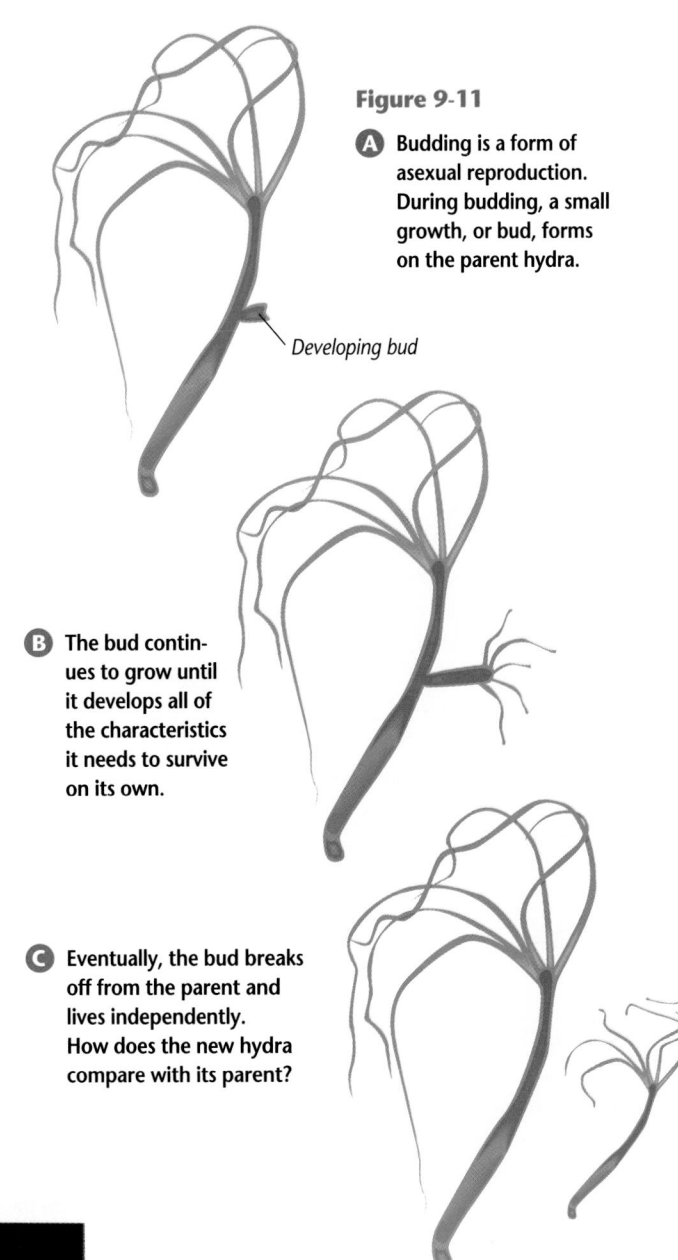

Figure 9-11

Ⓐ Budding is a form of asexual reproduction. During budding, a small growth, or bud, forms on the parent hydra.

Developing bud

Ⓑ The bud continues to grow until it develops all of the characteristics it needs to survive on its own.

Ⓒ Eventually, the bud breaks off from the parent and lives independently. How does the new hydra compare with its parent?

Figure 9-10

Aphids are insect pests commonly found feeding on rose bushes. Aphids reproduce almost continuously by asexual means, and for most of the year, produce only female offspring.

Uncovering Preconceptions
Some students might not think of sponges as animals because they are sessile. Explain that sponges carry on all the processes of life, including reproduction, just as other animals do.

Discussion Tell students the hydra was named for a mythical giant water monster with nine heads. Hercules, a Greek hero, was supposed to slay the monster. But each time he cut off one head, Hydra grew two more to take its place. Ask: How is the real hydra like the mythical one? *A real hydra can grow new organisms by budding, asexual reproduction.*

GLENCOE TECHNOLOGY

 Videodisc

STVS: Animals

Disc 5, Side 2
Studying Sharks (Ch. 3)

Nerve Regeneration in Garfish (Ch. 4)

Raising Super Fish (Ch. 7)

Sea Turtle Mystery (Ch. 8)

Alligator Courtship (Ch. 9)

Visual Learning

Figure 9-11 Have a volunteer read aloud the captions as students study the figure. **How does the new hydra compare with its parent?** *The new hydra is identical to the parent except for size.* **Do you think the new hydra will continue to grow? Why?** *The new hydra will grow to about the size of the parent and eventually bud itself.*

Across the Curriculum

Health

Point out that humans undergo a process similar to regeneration in which damaged organs grow larger to compensate for the loss. If three-quarters of the human liver is removed, for example, the remaining portion enlarges to a mass equivalent to the original organ. Recently, surgeons removed part of a mother's liver and transplanted it to her daughter, where it grew large enough to function properly. Have students find out about other human organs (such as skin) that regenerate. **L2**

Animal Development

Development of a fertilized egg into an adult varies in the animal world. Animals such as birds and mammals have young that look very similar to the adult form of their species, so changes during their development are not too dramatic. For example, kittens have many of the characteristics that an adult cat has. But many young animals at first look nothing like their parents. What happens to these animals as they progress to adulthood?

■ Metamorphosis

During development, many members of the animal kingdom undergo extensive changes in form. The changes in form that organisms undergo in their life cycles are called **metamorphosis**. *Meta-* means after or beyond and *morpho* means form. One familiar example is the frog in **Figure 9-12**. On this page and the next, you can see the changes that frogs undergo. Other organisms change form completely as they grow from egg to adult. Two types of metamorphosis are known in insect life cycles—complete and incomplete. In the Find Out activity that follows, take a look at the stages of metamorphosis that occur during the development of fruit flies.

Figure 9-12

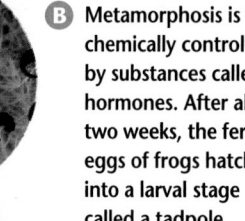

A The female frog lays up to 3000 eggs, which are fertilized by the male's sperm after the egg mass has been laid in water.

B Metamorphosis is chemically controlled by substances called hormones. After about two weeks, the fertilized eggs of frogs hatch into a larval stage called a tadpole.

C The tadpole lives in water, has a long tail, and uses gills to take in oxygen from the water.

Four weeks old

D As the tadpole grows, its tail begins to be absorbed. Legs form and lungs develop.

Six weeks old

How does a fruit fly change as it develops?

Changes in form are easy to observe in fruit flies because they go through their life cycle in a short period of time.

What To Do

You will observe different stages of development in live fruit flies.

1. From your teacher, obtain a vial containing food and fruit flies in different stages of development.

2. Using a hand lens and without opening the vial, observe the fruit flies.

3. Use the diagrams below as a guide to identify different stages of development.

4. Record your observations *in your Journal.*

5. Repeat your observations every day for two weeks. Record changes that you see.

Conclude and Apply

1. How many stages did you observe? Describe all the stages of development that you observed.

2. Draw each stage and arrange them in the order in which you think development occurs.

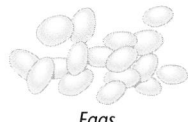
Eggs

Larvae

Pupae

Adult

E The young frog's tail continues to be absorbed and the frog begins spending more time on land than in the water.

Nine weeks old

Adult frog

F A mature frog has no tail, lives on land, breathes air, and eats insects rather than algae.

Twelve weeks old

9-2 Reproduction and Development **293**

How does a fruit fly change as it develops?

Time needed 15 minutes the first day; a few minutes each day for two weeks

Materials vials containing food and fruit flies in different stages of metamorphosis

Thinking Processes observing and inferring, interpreting data, sequencing, interpreting scientific illustrations, comparing and contrasting

Purpose To observe the stages of metamorphosis of the fruit fly.

Preparation Obtain a vial of fruit flies and food for each student group. Arrange for an appropriate place to store the vials. **COOP LEARN**

Teaching the Activity

Distribute the vials and explain that the different-looking creatures are all fruit flies in varying stages of metamorphosis.

Science Journal Have students research the California medfly problem of the 1980s. Have them find out which stage in the life cycle of the fruit fly caused damage to crops. **L1**

Expected Outcome

Students will observe the stages of development in live fruit flies and infer that fruit flies undergo complete metamorphosis.

Conclude and Apply

1. Four stages. Students should include eggs, larvae, pupae, and adult.

2. Drawings should reflect stages shown on page 293.

✔ Assessment

Content Have students research other animals that go through metamorphosis and draw pictures of the developmental stages for each animal they choose. Use the Performance Task Assessment List for Scientific Drawing in **PASC**, p. 55. **L1** **P**

Meeting Individual Needs

Visually Impaired Assist visually impaired students in observing the metamorphosis of the fruit fly by doing the following:

• Assign a partner to each visually impaired student.

• Have students together obtain the food and fruit flies from you.

• Give student pairs an enlarged copy of the diagram of fruit fly metamorphosis.

• Direct them to compare what they see on the blowup to what they see in the vial. Provide strong magnifying glasses.

• Both students should respond to the questions. **COOP LEARN**

STAGE 1 Egg

STAGE 2 Larva

STAGE 3 Pupa

Figure 9-13

A **Complete Metamorphosis** begins about seven days after an egg is fertilized. The caterpillar egg develops into a larva which chews the egg case open and stretches out, as in Stage 2.

B Once hatched, a caterpillar eats several times its body weight in one day. As it grows, the larva molts—its exoskeleton splits and a new one forms. Just before its last molt, the mature larva attaches itself to a twig or leaf to prepare for the pupa stage.

C The third stage in the metamorphosis is the pupa. The caterpillar spins a covering from a silken thread it produces. The larva now enters the pupa stage. During this time, chemicals called hormones cause vast changes to take place that result in a complete change in form inside the pupa.

■ Complete and Incomplete Metamorphosis

Insects exhibit two different kinds of metamorphosis. The distinct and different stages of development that you saw in the Find Out activity tell you that fruit flies undergo complete metamorphosis. Butterflies also undergo complete metamorphosis. **Figure 9-13** across the top of this page and the next shows how completely different the adult stage in a butterfly is from the larval stage.

Other animals change form too, but the changes are less distinct. In contrast to complete metamorphosis, incomplete insect metamorphosis involves three stages. As you can see in **Figure 9-14**, grasshoppers undergo incomplete metamorphosis. First, eggs hatch into nymphs, which look very similar to the adult, only smaller. Periodically, nymphs molt and grow larger. Once wings develop, the grasshoppers are mature.

Figure 9-14 **STAGE 1** Eggs

A **Incomplete Metamorphosis** begins as the grasshopper egg hatches and a nymph (as in B) emerges. The nymph is an immature form that looks like an adult grasshopper, only much smaller.

STAGE 2 Nymph

B The nymph, which has undeveloped reproductive organs and lacks wings, molts five or six times as it grows and develops.

STAGE 3 Adult

C It takes 40 to 60 days for a nymph to become a mature, winged, adult grasshopper that can reproduce.

294 Chapter 9 Animal Life

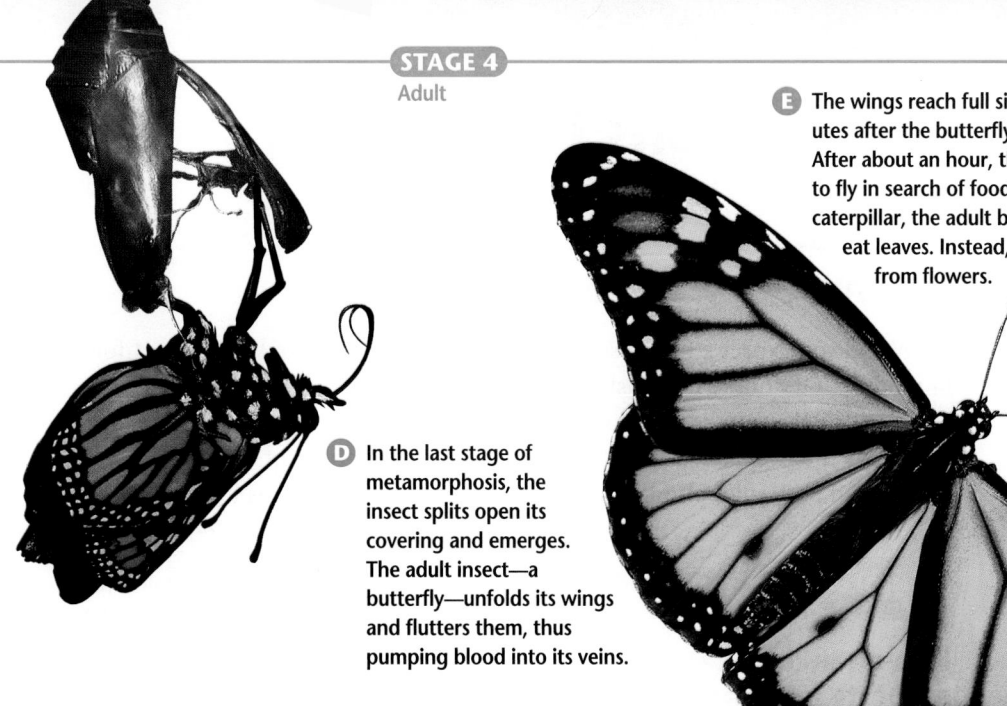

D In the last stage of metamorphosis, the insect splits open its covering and emerges. The adult insect—a butterfly—unfolds its wings and flutters them, thus pumping blood into its veins.

E The wings reach full size about 30 minutes after the butterfly has emerged. After about an hour, the insect is ready to fly in search of food. Unlike the caterpillar, the adult butterfly does not eat leaves. Instead, it drinks nectar from flowers.

■ Survival During Development

Animals have evolved with a variety of adaptations that enable young to survive to adulthood. Some of these adaptations involve reproducing in large numbers, shells, or internal development. Eggs and larvae are common sources of food for other animals. As a result, these stages are frequently destroyed by changes in the environment, such as drying or lack of food. How do these species survive? Animals that develop through the process of metamorphosis seem to reproduce in large numbers. In contrast, birds, and reptiles reproduce in smaller numbers. These species have adapted ways to protect their young as they develop outside the female's body. After fertilization, a protective shell forms around an egg. The shell keeps the egg from drying out.

Most mammals develop within the female's body. Internal development provides protection for the embryo.

check your UNDERSTANDING

1. Compare sexual and asexual reproduction. Is there an advantage to sexual reproduction?
2. Compare and contrast complete and incomplete metamorphosis.
3. **Apply** In what way does an organism that reproduces by internal fertilization have an advantage over one that reproduces by external fertilization?

check your UNDERSTANDING

1. Sexual reproduction—two parents involved; asexual—only one parent involved. More protected offspring may survive.
2. Both involve changes in form; complete—four completely different forms; incomplete—three stages–nymph looks similar to adult.
3. Female animals that reproduce by internal fertilization release eggs that stay within their bodies, and produce smaller numbers of offspring. In animals that have external fertilization, greater numbers of eggs are produced, but comparatively few offspring survive to adulthood.

3 ASSESS

Check for Understanding

Before students answer the Check Your Understanding questions, have them work in small groups to make a chart with two columns, labeled *Internal Fertilization* and *External Fertilization*. Have students list as many characteristics of each method of reproduction as they can. Use the Performance Task Assessment List for Group Work in **PASC**, p. 97. **COOP LEARN**

Reteach

Activity Help students infer the difference between sexual and asexual reproduction. Bring in a photograph and photo copies of a familiar scene. Have students create their own drawings from the photograph. Display drawings and copies. How are the copies like asexual reproduction? *They are identical with the photo, the parent.* How are the drawings like sexual reproduction? *Each is similar to the parent yet different.* **L1**

Extension

Activity Have students create a display with pictures, photographs, clay, and other found objects to illustrate metamorphosis. Use the Performance Task Assessment List for Display in **PASC**, p. 63. **L2** **P**

4 CLOSE

To reinforce sequencing skills, give students pictures of the fertilized eggs, larva, pupa, caterpillar, cocoon, and adult moth or butterfly and have them place the pictures in correct order. **L1**

PREPARATION

Planning the Lesson

Refer to the Chapter Organizer on pages 276A–D.

Concepts Developed

In Section 9-2, students learned that some animals reproduce sexually; others, asexually. Now students will see how adaptations have enabled animals to survive.

1 MOTIVATE

Bellringer

Before presenting the lesson, display **Section Focus Transparency 29** on an overhead projector. Assign the accompanying **Focus Activity** worksheet. L1

LEP

Explore!

Does an animal's environment tell you something about its adaptations?

Time needed 20 minutes

Thinking Processes observing and inferring, comparing and contrasting, predicting

Purpose To observe different environments and infer the types of adaptations necessary for survival in each environment.

Teaching the Activity

Science Journal Ask students whether water and food would be plentiful in each of the environments. L1

Expected Outcome

Students will infer how environment can influence physical adaptations in species.

9-3 Adaptations for Survival

9-3

Section Objectives

■ Explain how adaptations allow animals to survive on Earth.

■ Give examples of some animal adaptations.

Key Terms

cellular respiration
metabolism

Surviving Where You Are

Would you expect a fish to survive in a forest? Of course not. Fish have adaptations that make them better suited for life in water. Birds, bears, beetles, and butterflies also have characteristics that make them suited for their environment. Consider the characteristics of each of the environments and the adaptations of animals in each area in the following Explore activity.

Explore! ACTIVITY

Does an animal's environment tell you something about its adaptations?

What To Do

Study the pictures showing the different environments where organisms live.

1. *In your Journal,* make a chart with the headings *Tundra, Desert, Grassland,* and *Rain Forest.*

2. Under each heading, write a description of the area from what you see in the pictures.

3. Brainstorm a list of the kinds of adaptations animals would probably have to have to survive in each of these environments? Include this list in your chart.

Desert in Alamo Canyon, Arizona

Grasslands of southern Brazil

Alaskan tundra in fall

Rain forest in Venezuela

Answers to Questions

Answers will vary. *Tundra:* carnivorous, herbivorous, thick fur or fat for warmth, hibernation, feet or paws adapted for moving across snow or ice. *Desert:* nocturnal, thick skin or fur to withstand extreme temperatures, adaptations for finding or retaining water. *Grassland:* carnivorous, herbivorous, living in herds for protection, relatively fast running speeds to escape predators. *Rain forest:* carnivorous, herbivorous, adaptations for living in trees, waterproof fur, feathers, or skin.

✔ Assessment

Process Have students choose one of the environments from the activity and predict the kinds of animals that might live there and what those animals might eat. Students can then use reference materials to check their predictions. Select the appropriate Performance Task Assessment List in **PASC.** L1

Physical Adaptations

It's fairly easy to conclude that an organism living in a rain forest might have some adaptations that differ from an organism that lives in the arctic tundra. An adaptation is any characteristic that increases an organism's chances for survival. Adaptations may be structural, like bird beaks. Others are harder to detect. For instance, cows have symbiotic bacteria that produce the enzyme that allows them to digest cellulose. Even behaviors can be considered adaptations.

■ Body Structure Adaptations

Remember the Find Out activity you did that involved bird beaks? The beak is only one adaptation that birds have. The activity could have been done using the birds' feet as well, as **Figure 9-15** shows.

Bird feet are an example of an adaptation that involves the animal's body structure. An anteater's tongue and the teeth of a deer are also body structure adaptations. Beaks, teeth, feet, and tongues are all body structures. Are there any adaptations you can think of inside the body?

■ Internal Body Adaptations

Internal adaptations may be structural and chemical. One internal adaptation found in animals is the organ used to obtain oxygen. Most organisms require oxygen for respiration. **Cellular respiration** occurs inside the body's cells when oxygen combines with digested food to release energy from chemical bonds in food.

Figure 9-15

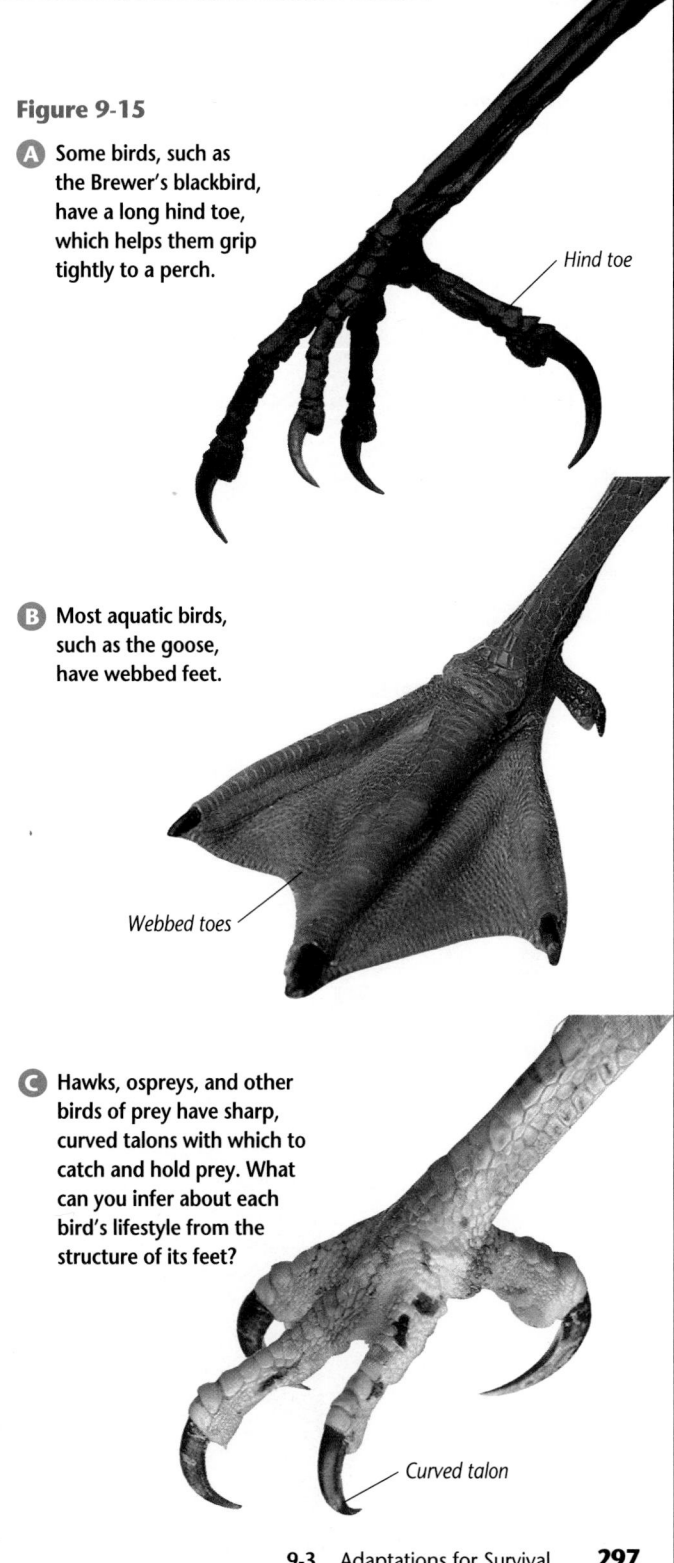

A Some birds, such as the Brewer's blackbird, have a long hind toe, which helps them grip tightly to a perch.

Hind toe

B Most aquatic birds, such as the goose, have webbed feet.

Webbed toes

C Hawks, ospreys, and other birds of prey have sharp, curved talons with which to catch and hold prey. What can you infer about each bird's lifestyle from the structure of its feet?

Curved talon

9-3 Adaptations for Survival **297**

┌─ **Program Resources** ─┐

Study Guide, p. 35
Teaching Transparency 17 [L2]
Section Focus Transparency 29

Visual Learning

Figure 9-15 What can you infer about each bird's lifestyle from the structure of its feet? *Birds with long hind toes spend much time perching in trees and bushes. Birds with webbed feet spend most of their time in water. Birds with curved talons catch and eat other animals.*

Activity Help students relate physical adaptations to environment by having them imagine that they had to survive without "modern conveniences." What problems would they encounter and how would they solve them? Have students describe in words or drawings where they would sleep and how they would get food. [L1]

2 TEACH

Tying to Previous Knowledge

Students are familiar with pets' behaviors. Ask them to relate instances of feeding behavior and evidence of communication with other animals and with humans. Have students explain what they think causes these behaviors. *Communication, for example, allows animals to find food, defend against enemies, and reproduce successfully.*

Activity Arrange students in pairs. Have one student keep time and count while the other holds his or her arms straight out and flaps. See how many times a person can flap in one minute and how long it takes for the person to tire. Have students switch tasks and repeat the activity. Ask students why our chest muscles would get tired more readily than other muscles, such as those in our legs. *Our leg muscles would be more developed because we use them to walk.* Why would birds have well-developed chest muscles? *They use these muscles for flying.* Why are certain parts of an animal's body adapted in different ways, as with our leg muscles? *Such adaptations help the animal to survive better.* [L1] **COOP LEARN** **LEP**

Student Text Question
Are there any adaptations you can think of inside the body? *Internal adaptations can include differing skeletal systems, hearts with different numbers or chambers, digestive systems, and so on.*

Figure 9-16

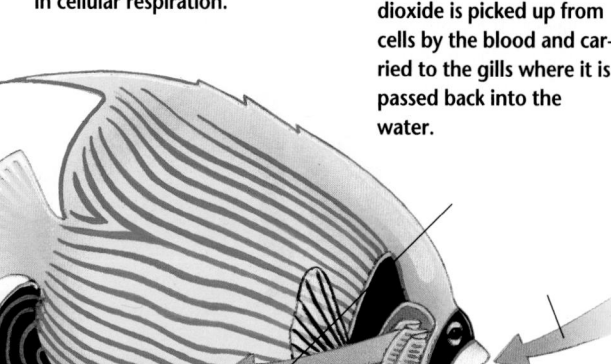

A As water passes over the gills, capillaries within them absorb oxygen dissolved in the water. The blood carries the oxygen to body cells, where it is used in cellular respiration.

B One of the waste products of cellular respiration is carbon dioxide. Carbon dioxide is picked up from cells by the blood and carried to the gills where it is passed back into the water.

Gills

■ Adaptations for Breathing

In the animal kingdom, there is a tremendous variety of ways in which organisms take in oxygen for cellular respiration. **Figure 9-16** shows a fish with gills, which it uses to remove oxygen from water. Insects have small openings along the sides of their bodies through which air containing oxygen enters. Most amphibians have lungs but they can also obtain oxygen through their skin. And of course, humans obtain oxygen through air breathed into lungs.

Cellular respiration is one of the chemical changes that goes on in an organism. The total of all of the chemical changes that take place in an

Konrad Lorenz

Konrad Lorenz was a famous scientist who used his sense of sight to observe animals. Lorenz was a founder of the science of ethology—the study of animal behavior.

Lorenz thought that in order to learn about animal

Konrad Lorenz

behavior, an animal must be observed in its natural environment. Before Lorenz, scientists had mainly studied animal behavior in laboratories.

The Importance of Observation

Lorenz devoted much of his time to watching colonies of birds, including geese. From these observations came one of his most important discoveries. In 1935, Lorenz concluded that if a mother goose is not present when her baby geese hatch, they will consider the first moving object they see to be their mother.

298

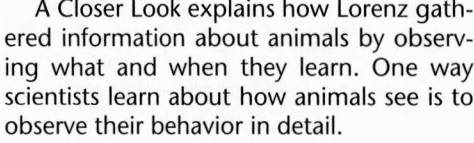

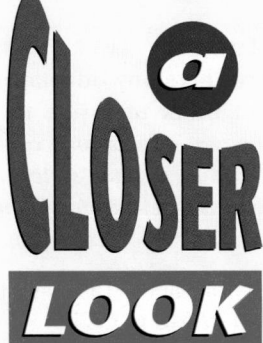

Purpose

A Closer Look explains how Lorenz gathered information about animals by observing what and when they learn. One way scientists learn about how animals see is to observe their behavior in detail.

Content Background

Lorenz's discovery of imprinting contributed to our understanding of how behavioral patterns evolve within a species. His findings were influential in the develop-

ment of learning theory. They suggest that animals are genetically programmed to learn specific information crucial for survival. Lorenz applied his theories to humans as members of social groups. In his controversial book, *On Aggression* (1966), Lorenz argued that aggression, like imprinting, is inborn and serves some useful functions. Unlike imprinting, however, aggression can be modified to play a more socially useful role.

organism is its **metabolism**. During the metabolism of food, heat is released. The faster food is broken down, the greater the amount of heat energy released. Why do you suppose you get warm when you exercise? Could this be related to increased metabolism from increased muscle action?

■ Human Adaptations for Respiration

Not all organisms adapt to the same conditions in the same way. People who live in high mountain areas such as in the Andes of South America where there is less atmosphere have been found to have larger hearts that circulate blood more quickly. They also have more red blood cells to carry oxygen than people who live at sea level. But, some people who live in similar conditions in Tibet do not have these adaptations.

■ Behavior Adaptations

Animals demonstrate many kinds of behavioral adaptations. A bluejay squawks or shrieks if a cat comes too close to its nest site. Some bugs quickly skitter away when a light shines on them. The set of responses an organism exhibits to changes in its environment is its behavior. In the Investigate that follows, you will see how an earthworm reacts to several changes in its surroundings.

Will Anyone Do?

Lorenz found that if the moving object happened to be himself, the baby geese would follow him as if he were their mother. He found this to be true with ducks, too, as in the photograph below. Lorenz called this behavior *imprinting.* Imprinting is an animal instinct in which the animal becomes attached to another organism at a critical time soon after birth or hatching. Imprinting is important because in order for baby animals to survive, they must recognize a mother who will feed and protect them.

Lorenz continued his work in animal behavior and, in 1973, was awarded the Nobel prize for his work. Lorenz died in 1989.

You Try It!

Choose an animal to observe for a week and keep a diary of the animal's activities. What conclusions can you make about the animal's behavior from your observations?

9-2 Earthworm Behavior

Planning the Activity

Time needed 30–60 minutes

Purpose To examine the external structures of the earthworm and note its responses to various stimuli.

Process Skills making and using tables, observing and inferring, recognizing cause and effect, designing an experiment

Materials See reduced student text. Obtain worms from local bait shop or collect locally.

Preparation Distribute materials before distributing the worms.

Teaching the Activity

Safety Caution students to handle the earthworms gently and **absolutely avoid** getting vinegar on the worms.

Process Reinforcement Encourage students to use their observations to infer how earthworms behave in their natural habitat. Students might infer that earthworms come out at night, seek out moist soil, and try to escape if touched.

Discussion If the worms are not on top of the soil when students open the container, ask students what might have driven the worms under the soil. *light* Discuss whether earthworms have "eyes" to respond to light. Ask students what they might do to get worms to come to the surface. *Place the container in the dark for a few minutes.* Direct students to consider this when shining the flashlight on the worms. In both cases, the worms will move away from the light. L2

Science Journal Before doing the experiment, students should think about what kind of senses earthworms might have and predict how the earthworms will react to each stimuli in the experiment. Have students record their predictions in their journals. L1

INVESTIGATE!

Earthworm Behavior

How can you determine the effect some conditions have on an earthworm? In this activity, you will observe some earthworm characteristics and infer how they enable the earthworms to survive.

Problem

How is an earthworm adapted to live in soil?

Materials

hand lens	toothpick
vinegar	flashlight
live earthworms in slightly moist soil	shallow pan
	paper towels
500 mL beaker	cotton swab
	water

Safety Precautions

Earthworms can be released into the soil when you have completed your investigation. Earthworms are valuable because they aerate and mix soil as they move through it.

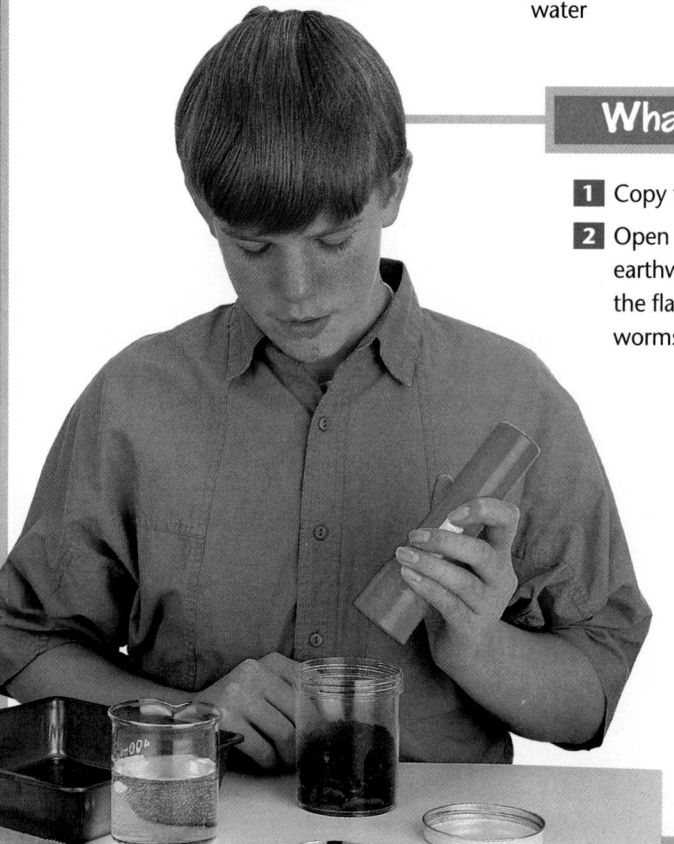

What To Do

1 Copy the data table *into your Journal.*

2 Open the container to be sure some of the earthworms are on the top of the soil. Shine the flashlight on the worms. Record how the worms react.

Sample data

Data and Observations

Condition	Response
Light	Responses may
Fingers	vary with
Touch-front	the health
Touch-back	of the worms.
Vinegar	

Program Resources

Activity Masters, pp. 41–42, Investigate 9-2

Science Discovery Activities, 9-3, Something to Flap Over

ENRICHMENT

Activity Explain that governments issue special postage stamps to honor famous people, national events, and endangered animals such as the rhino or leopard. Ask students to select one animal from this chapter and design a stamp honoring it. In addition to drawing a picture of the animal, students should include part of its habitat and the animal's name. Then ask how they could use this stamp to help the animal it depicts. L2 P

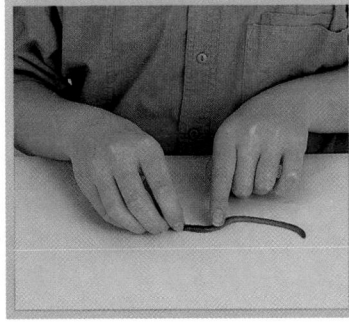

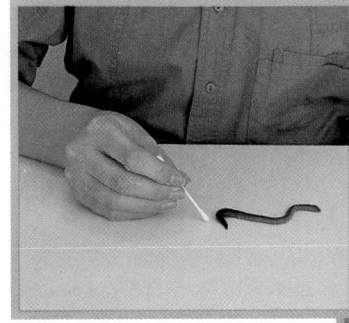

A **B** **C**

3 Moisten your hands and remove an earthworm from the container. **CAUTION:** *Use care when working with live animals.* Keep your hands moist while working with the earthworm. Hold the worm gently between your thumb and forefinger. Observe its movements and record them in the table.

4 Rub your fingers gently along the body. With a hand lens, *observe* the small hair-like bristles that you feel.

5 With the toothpick, gently touch the worm on the front and back ends. Record your observations.

6 Dip the cotton swab in vinegar. Place it in front of the worm on a wet paper towel. *Do not touch the worm with the vinegar.* Record what you observe.

Head end

Ringlike segments

All earthworms are made up of many segments. They have no legs, and it is sometimes difficult to tell which end is the head. The head is the end closest to the thick, heavy band on the worm's body.

Analyzing

1. What happens when light is shined on the earthworms?

2. How does the earthworm react to touch?

3. How does the earthworm react to the vinegar?

Concluding and Applying

4. *Infer* how the earthworm's reaction to light is an adaptation for living in soil.

5. How are the bristles an adaptation for living in soil?

6. ~~Going Further~~ *Design an experiment* to find out how the earthworm reacts to different temperatures.

Activity Oil on birds' feathers helps keep them water-repellent, a very important adaptation. To show how oil waterproofs feathers, cut two 6-cm squares of construction paper. Have a volunteer cover one with petroleum jelly and leave the other one plain. Drop six drops of water on each square. Have students describe what they observe. [L1]

[LEP]

Content Background

In territorial species, only individuals with territories mate. This helps ensure an adequate food supply for offspring. It also passes on the genes of the best adapted individuals to the next generation.

Visual Learning

Figure 9-17 How does this behavior differ from a dog greeting someone familiar? *The dog's tail might wag. The bark would not be threatening, and the dog would not maintain eye contact.*

Theme Connection Animals' protective adaptations provide further examples of the theme of scale and structure. Ask students to study Figures 9-17 and 9-18. **When a dog feels threatened, does it look larger?** *Yes, the hair on its back stands up, and it holds its tail up high.* **How can an animal's coloring or skin pattern protect it from predators or help it hunt?** *By allowing the animal to blend into the environment and remain undetected by predators or prey.*

As you review your results from the Investigate activity, you should notice that the bristles are a structural adaptation. The responses the earthworm made to light, touch, and vinegar are behavioral adaptations. Just as body structure can aid survival, different kinds of behavior can also determine if an organism will survive.

Figure 9-17

An animal may defend its territory to protect its food supply or a mate. Defending its territory, a dog may growl, bark, and bare its teeth. How does this behavior differ from a dog greeting someone familiar?

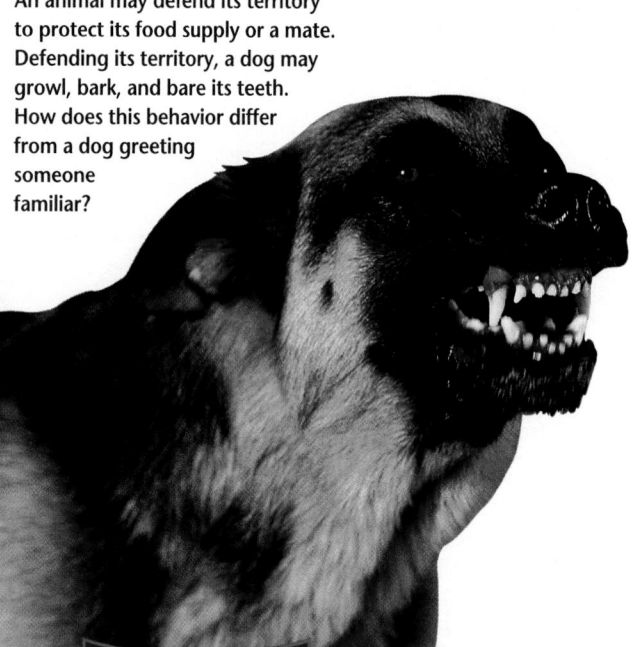

How Do We Know?

Do some animals rely on one another's behavior?

Some behaviors are helpful to more than one species of animal. For example, the frog-eating bats in parts of Central and South America hunt frogs at night. Researchers know that the bats use sound to find the frogs. But how do the bats distinguish between poisonous and nonpoisonous frogs?

A Batty Experiment

The researchers set up two speakers and a frog-eating bat in a large cage. One speaker broadcast the mating call of a poisonous frog. The second speaker played the mating call of a nonpoisonous frog. The bat always flew toward the sound from the nonpoisonous frog. The researchers concluded that not only did the bats use sound to find the frogs, but the bats used the sound to distinguish poisonous and nonpoisonous frogs. These bats show a feeding behavior that is dependent on the mating call behavior of frogs.

■ Mating Behaviors

Many animals take part in activities that attract members of the opposite sex. These behaviors also ensure that both animals are ready to mate at the same time. The peacock spreads his feathers to attract the peahen. Some insects release chemicals that attract members of the opposite sex. Other insects rub their legs or wings together to produce sounds that attract mates.

Many behavioral adaptations are ways for animals to communicate. In addition to marking their territory and finding mates, animals communicate to warn of danger, give directions for finding food, and maintain social order within a group.

■ Protective Behaviors

Look at **Figure 9-18**. Can you see any animals? Some adaptations protect animals from being seen as they hunt or from being attacked and eaten by other animals. Camouflage is an adaptation that allows an animal to

---

Multicultural Perspectives

Courtship Around the World

People in all cultures exhibit courtship rituals. Colombians, for example, have a great number of courtship dances. On the Caribbean coast, people dance the *bullerengue, lumbalu,* and the circular *cumbia.* The national dance, called the *bambuco,* came from the Andean zone. Waving kerchiefs, male and female partners mime a courtship of pursuing and flirting. This same dance is done with a waltz step in southern Colombia. Ask students if they think any of the dances that they know are courtship rituals. Have students investigate courtship dances in other cultures. [L1]

hide by blending into its surroundings. The stripes of a tiger or a zebra are one type of camouflage. A chameleon can turn different shades of green and brown as it moves through vegetation. A mouse or a snake can stay safely hidden when they lie motionless in leaves on a forest floor, as seen in **Figure 9-18**.

You have learned about several adaptations animals have that allow them to survive in their environment. Some of these adaptations are physical, such as specific beaks, feet, or claws. Other adaptations are behaviors that help animals communicate or find food. Without these adaptations, animal groups would not survive.

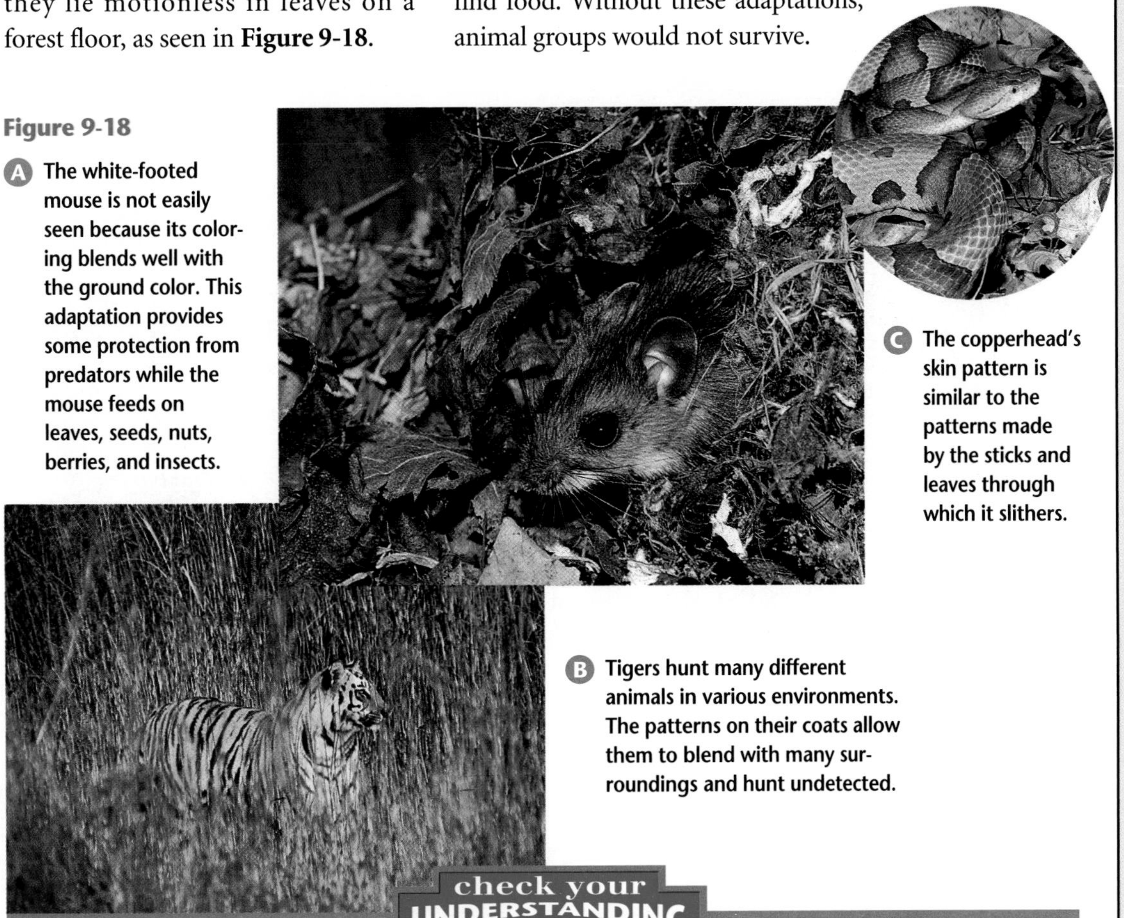

Figure 9-18

A The white-footed mouse is not easily seen because its coloring blends well with the ground color. This adaptation provides some protection from predators while the mouse feeds on leaves, seeds, nuts, berries, and insects.

C The copperhead's skin pattern is similar to the patterns made by the sticks and leaves through which it slithers.

B Tigers hunt many different animals in various environments. The patterns on their coats allow them to blend with many surroundings and hunt undetected.

check your UNDERSTANDING

1. How do these adaptations increase the chance for the animal or the animal's species to survive?
   a. eyesight of an owl
   b. whiskers of a cat
   c. courtship dance of a bird
   d. blending into environment

2. Give an example of an adaptation in body structure and explain how it helps the organism survive.

3. **Apply** When a honeybee locates a source of nectar, it flies back to the hive and performs a special dance. What type of adaptation is this? How is it useful for the bees?

check your UNDERSTANDING

1. **a.** The owl has a better chance of catching food. **b.** The cat's whiskers help it sense narrow spaces in which it might get stuck. **c.** The bird will find a mate and produce young. **d.** can't be seen by predators or prey

2. Possible answers include short, strong beak that cracks seeds; large ears which gather sound waves.

3. The dance is a behavioral adaptation that communicates the location of food to the whole hive. It helps maintain an adequate food supply for the hive.

3 ASSESS

Check for Understanding

1. Have students create a children's storybook explaining the animal adaptations they learned about. Have students include an overview of physical adaptation and animal behavior with specific examples. Students can exchange books and write reviews of each other's work. L2 P

2. Have students work in small groups to discuss the kinds of information bees would have to communicate to enable other bees to locate a source of nectar. Have groups develop mime skits or dances to convey information about how to locate a special area of the school, such as the library or cafeteria. Use the Performance Task Assessment List for Group Work in **PASC**, p. 97. COOP LEARN L2

Reteach

Activity To help students practice classifying behaviors, have them make a chart with the headings *Territorial Behavior*, *Courtship Behavior*, and *Protective Behavior*, and have them write examples of each of the behaviors under the headings. L1 P

Extension

Activity Students can research some of the different techniques scientists have used to study the language-learning abilities of primates, for example, Ameslan chimps and the symbol language Yerkish, developed for apes. L3 P

4 CLOSE

Reinforce classifying skills. Write the word *communication* on the board. Ask students to brainstorm ways that animals communicate. Then classify these methods as: sounds, visual displays, or chemicals. Have students demonstrate the second method by playing a brief game of charades. L1

Can male animals give birth?

Purpose

The ability of the male sea horse to give birth makes it unique in the animal kingdom. These SciFacts give students an added perspective to the information on reproduction provided in Section 9-2.

Content Background

Scientists estimate that about two dozen species of sea horses exist around the world, mainly in warm, coastal waters. Sea horses range in size from 4 cm to 30 cm. The female sea horse lays up to several hundred eggs at a time in her partner's pouch. During labor, the male sea horse expels his young through a tiny opening in his pouch. Each miniature sea horse is approximately 1 cm long.

Activity

Show students pictures or slides of sea horses and other aquatic animals in their natural habitats. Make a chart on the board and have students list the similarities and differences between the animals shown. Be sure that habitat is considered as well as physical appearance.

Science Journal

Assess student poems based on creativity and a basic understanding of the sea horse.

NATIONAL GEOGRAPHIC SciFacts

Can male animals give birth?

Unlike most fathers in the animal kingdom, the male sea horse is the one who gets pregnant and gives birth. After a female deposits eggs in the male's brood pouch, he fertilizes them and gives the developing embryos protection and nourishment. About 21 days later, he expels his young in a long and exhausting labor. Depending on the species, the offspring will number from as few as four to as many as a world-record 1572.

Sea horses mate for life and reinforce their bonds with ritual "greetings." Each morning the female visits her pregnant partner. Both sexes brighten in color and then twirl around a sea grass shoot, holding on to it with their tails.

Various species of the genus *Hippocampus* are found in the coastal waters of six continents. A legal global trade of 20 million sea horses annually–primarily for medicines, aquariums, curios, and food–could lead to the widespread collapse of sea horse populations.

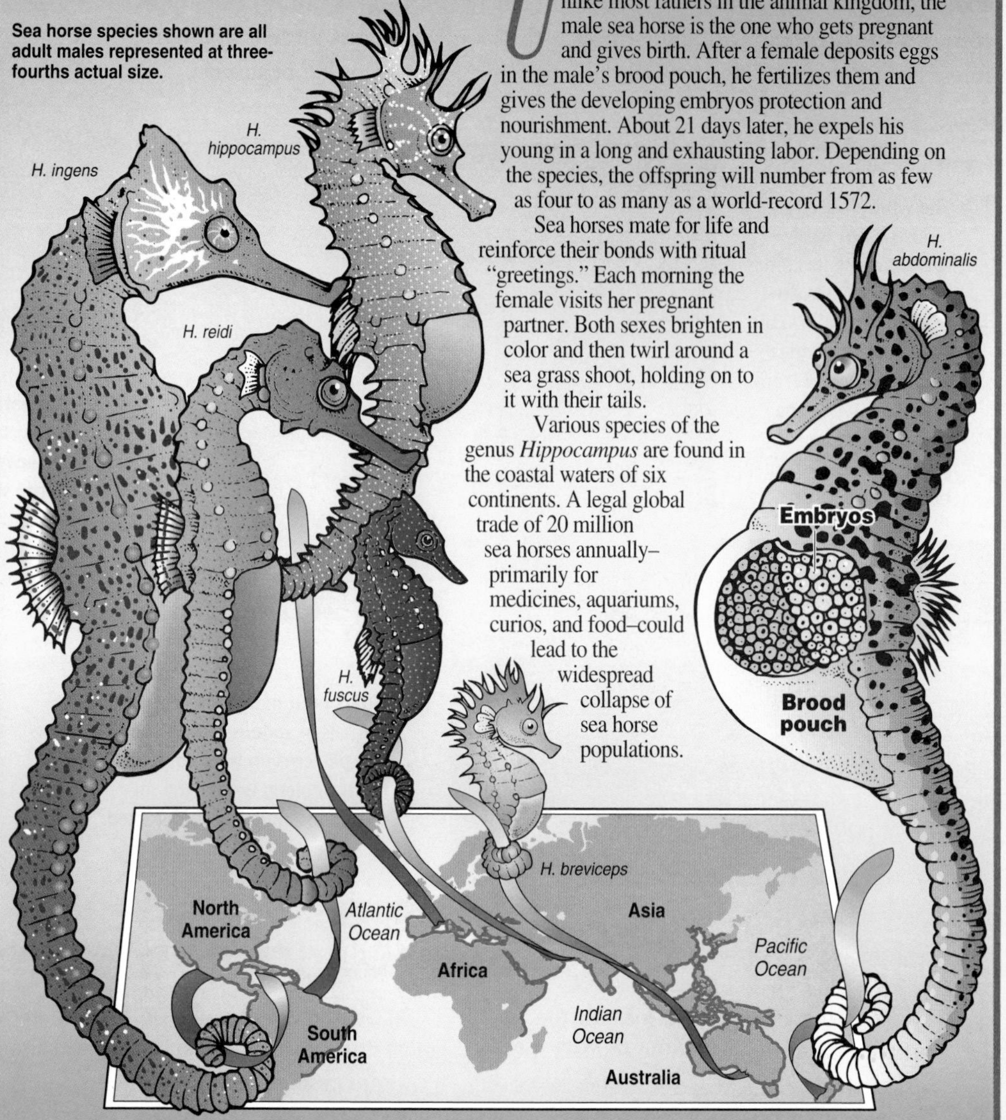

Sea horse species shown are all adult males represented at three-fourths actual size.

H. ingens

H. hippocampus

H. reidi

H. fuscus

H. abdominalis

Embryos

Brood pouch

H. breviceps

North America

Atlantic Ocean

Asia

Pacific Ocean

Africa

Indian Ocean

South America

Australia

Science Journal

Think about why very small animals are important to life. *In your Science Journal,* write a poem about sea horses or another small animal after researching information about that animal.

HISTORY CONNECTION

Domesticating Animals

Paintings in caves show that animals were tamed, or domesticated, thousands of years ago. Once people settled into communities and learned how to plant crops, they also learned to use animals as a source of meat, milk, and wool, and to tame them to perform tasks.

Using Tame Animals

The first animal to be domesticated was the dog. People bred them for jobs, such as guarding and hunting.

Later, people saw advantages in using tame sheep and goats to supply meat, milk, and wool. Cattle, pigs, and donkeys were soon domesticated, too. The Anasazi, prehistoric Native Americans of the American Southwest, domesticated turkeys for food and clothing. They used their feathers to make robes and blankets.

What Do You Think?

What can you tell about the people who made the drawings shown in the picture? What can you tell about the animals?

HISTORY CONNECTION

Purpose
History Connection extends the discussion of adaptations in Section 9-3, focusing on the domestication and breeding of certain species.

Teaching Strategy Lead students in a discussion of the evolution of domesticated animals. Ask students to identify any "wild" behaviors that remain. Students may mention the startle response; fear reactions; running in fear of loud noises and sudden movement; growling, hissing, and baring of teeth for defense; stalking of prey.

Answers to
What Do You Think?

You can tell that the people were farmers or herders and that they had domesticated animals.

Literature Connection

Do Birds Have Knees? Do Ladybugs Sneeze?

Sometimes poets use absurd images to convey ideas. Find and read the poem "A Love Song" by poet Raymond Richard Patterson (Adoff, Arnold, ed. *Black Out Loud: An Anthology of* *Modern Poems by Black Americans*).

What do you think is the answer to the poet's initial question? How did your knowledge of animals help you determine the answer to the questions?

Science Journal
Think of other silly images of animals. *In your Science Journal* write and illustrate a picture book to share with a younger child. Use a question like those in the poem on each page of your book.

Literature Connection

Purpose
Literature Connection extends Section 9-1, in which students learn about characteristics of animals.

Content Background
Although most of the questions can be answered in the negative, two could have "yes" answers. Snakes do not have legs, so one would not expect them to have hips. However, pythons are among the few snakes that have rudimentary pelvic (hip) bones. Penguins do not have arms, but the bones in a bird's wing are similar in many ways to the bones in a person's arms.

Teaching Strategy Ask students to answer the questions in the title of this excursion. Have students give reasons for their answers. *Birds have knees, although they do not have kneecaps as humans do. Ladybugs do not sneeze, because a ladybug has no nose.*

Going Further ⑊⑊⑊▶
Have student pairs select an animal and report to the class when it was domesticated, by whom, for what desirable traits, how it has been bred over the years to enhance these traits, and any lost traits or weaknesses that have resulted from this breeding. Have students choose a whole class of animals or a specific animal. Possibilities are: cats, beef cattle, dairy cattle, chickens or turkeys, dogs, horses, fishes, sheep, minks, rabbits, or laboratory rats. Use the Performance Task Assessment List for Oral Presentation in **PASC,** p. 71.
⌊L2⌋

Teens in SCIENCE

Purpose
Teens in Science extends Section 9-2, in which students learn about regeneration. This excursion describes Jessica Knight's experiments with regeneration in flatworms.

Content Background
Planarians are flatworms, members of phylum Platyhelminthes. Many experiments have been done to investigate regeneration in planarians. When a planarian is cut into pieces, the head end regenerates the rest of the body more quickly than any other piece can.

Teaching Strategy
Have students discuss Jessica's experiments. Some students may wish to try some of these experiments. Point out that the processes described are more complicated than they sound. All equipment must be very clean and scalpels used to make cuts must be very sharp.

What Do You Think?
Accept all answers that can be supported logically. Students will probably say that the research will be helpful.

Teens in SCIENCE

Jessica Knight

Jessica Knight, a 16-year-old student at the North Carolina School of Science and Mathematics, wanted to know more about planarians. She wondered what would happen to planarians if they were exposed to various amounts of light.

Researching a Question

By reading about experiments conducted on planarians, Jessica learned that a fragment of planarian about the size of a period on this page could regrow into a

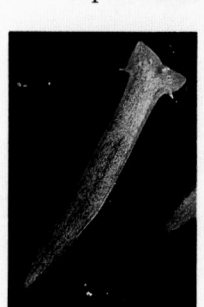

Two-headed planarian

complete planarian within two weeks. She also learned that, when exposed to too much light, cancerous cells grow in some planarians. Maybe, she thought, planarians could teach us something about the changes caused in cells by ultraviolet light, an agent which is thought to cause cancer.

The Experiment

For her experiments, Jessica chose *Dugesia tigrina*, a brown planarian that has a primitive nervous system.

Using a dissecting microscope for close observation, Jessica sliced them apart in a variety of ways. Then she watched how they regenerated missing parts. "Depending on how you cut them," Jessica says, "you can come up with two heads on one end, or a tail on each end, or head on each end, or a head coming out the side."

Collecting the Data

After weeks of observing and videotaping regeneration in planarians, Jessica was ready to study the effects of various amounts of ultraviolet light on regeneration and cancer growth. She planned

to use a light meter and calculations to determine how much light the planarians actually receive.

What levels of light affect regeneration? How does light affect regeneration? What levels cause tumors? How much light kills the planarians? Jessica believes she will have answers to these questions when she has completed her experiment. She thinks that damage caused by the ultraviolet light will be comparable to damage ultraviolet light does to human skin.

What Do You Think?
Do you think planarian research could someday show scientists how to treat wounds in humans?

Going Further ▸
Have students use reference books to find out more about planarians and their structure. Have students relate the simple structure to the worms' ability to regenerate. L2

Science Journal

Review the statements below about the big ideas presented in this chapter, and answer the questions. Then, re-read your answers to the Did You Ever Wonder questions at the beginning of the chapter. *In your Science Journal*, write a paragraph about how your understanding of the big ideas in the chapter has changed.

1 A diet of flying insects satisfies this frog's need for energy. *How does this activity characterize it as an animal?*

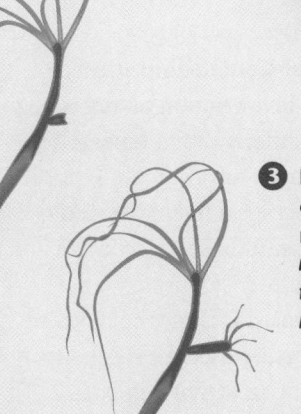

2 An animal has one of two main kinds of support—inside the body and outside the body. *What are these two types of support called, and what are some examples of animals having each type?*

3 Different animal groups exhibit sexual or asexual reproduction. *What is the major difference between these two forms of reproduction?*

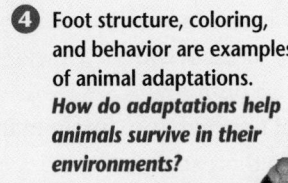

4 Foot structure, coloring, and behavior are examples of animal adaptations. *How do adaptations help animals survive in their environments?*

Science Journal

Did you ever wonder...

• Snakes have a backbone; earthworms do not. (pp. 285–286)

• Both frogs and butterflies undergo metamorphosis as they grow from egg to adult. (pp. 292–294)

• People locate food by sound, sight, and odors. (p. 281)

Project

Have students classify animals by constructing a large poster or bulletin board display. Have them divide the main heading, *Animals*, into two divisions, *Invertebrates* and *Vertebrates*. Under each subheading, students should include names and pictures of representative animals with descriptions of characteristics of each. Select the appropriate Performance Task Assessment List in PASC. **COOP LEARN** [L1] [P]

chapter 9
REVIEWING MAIN IDEAS

Have students work in groups to illustrate the main ideas of the chapter.

Teaching Strategy

Divide the class into four groups. Assign one of the main ideas to each group. Have each group make a collection of pictures that illustrate their topic. **COOP LEARN** [L1]

1. animals running, flying, swimming; animals stalking, chasing, or holding prey; and animals grazing on plants

2. animal bones, drawings of skeletons, pictures of lobsters or crabs, or the exoskeletons of these animals

3. baby animals, including photos that indicate parental care

4. adaptations, such as a giraffe's long neck, an elephant's trunk, a crab's claws, or a grasshopper's long back leg

Answers to Questions

1. The frog is a consumer. It must use plants or other animals as food. Being a consumer is a characteristic of animals.

2. Exoskeletons–crabs, lobsters, and insects; endoskeletons–fish, snakes, humans

3. Organisms produced by asexual means are produced by one parent; those produced sexually are produced by two separate parents.

4. Accept all reasonable explanations. Adaptations for individual environments help organisms by providing means for obtaining food; by providing conditions condusive to successful reproduction; by providing protection from severe changes in temperature.

GLENCOE TECHNOLOGY

 MindJogger Videoquiz

Chapter 9 Have students work in groups as they play the Videoquiz game to review key chapter concepts.

Using Key Science Terms

1. a. *metabolism*–all the chemical activities an organism undergoes; *cellular respiration*–one aspect of metabolism–the release of energy from food.

b. *metamorphosis*–series of developmental changes in certain organisms; *regeneration*–the repair or replacement of body parts.

c. *endoskeleton*–an internal skeleton; *exoskeleton*–an outer hardened, shell-like skeleton.

d. *vertebrate*–an animal with a spinal column; *invertebrate*–an organism lacking a backbone.

2. *exo-*: outer (Greek); *endo-*: within (Greek). These roots help you to understand that an exoskeleton would be on the outside of the body and an endoskeleton would be within the body.

3. cellular respiration

Understanding Ideas

1. Plants are producers and make food from which they release energy; animals are consumers, incapable of making food; obtain energy by consuming or eating other organisms.

2. A bird has an endoskeleton; a bee has an exoskeleton.

3. egg, larva, pupa, adult

4. Humans have bilateral symmetry because their left and right sides reflect one another.

5. It is the uniting of egg and sperm to produce a new individual.

Developing Skills

1. See reduced student page for completed concept map.

2. Graph should show following divisions: Known Insects, 84%; Arachnids, 13%; Crustaceans 3%.

3. Mealworms should show a preference for the apple or molasses over the aspartame.

4. Students should be able to observe that if temperature is too low or too high, fruit flies

---

U sing Key Science Terms

cellular respiration	metabolism
endoskeleton	metamorphosis
exoskeleton	regeneration
fertilization	vertebrate
invertebrate	

1. What is the relationship between the following pairs of terms?
metabolism and cellular respiration
metamorphosis and regeneration
endoskeleton and exoskeleton
vertebrate and invertebrate

2. Using a dictionary, research the meaning of the roots *endo-* and *exo-* and explain how knowing those meanings helps you understand the meaning of the terms *endoskeleton* and *exoskeleton*.

3. What term in the list above is used to explain how organisms release energy?

U nderstanding Ideas

Answer the following questions in your Journal *using complete sentences.*

1. What is the difference between plants and animals in terms of how they obtain energy?

2. How is the skeleton of a bird different from the skeleton of a bee?

3. What are the stages of complete metamorphosis in a butterfly?

4. Explain what type of symmetry humans have.

5. What does fertilization accomplish in a sexually reproducing organism?

D eveloping Skills

Use your understanding of the concepts developed in this chapter to answer each of the following questions.

1. Concept Mapping Complete the concept map using these terms: *one parent, sexual, asexual, two parents.*

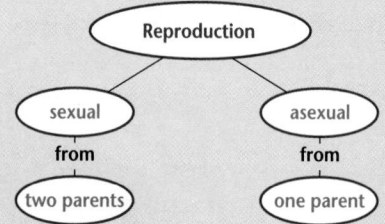

2. Making and Using Graphs Make a pie graph of the species of arthropods.

Class of Arthropod	# of species
Arachnids	100 000
Crustaceans	25 000
Insects (known)	700 000

3. Comparing and Contrasting Repeat the mealworm investigation on pages 282 and 283 using different food types. Use aspartame (a sugar substitute), pieces of apple, or molasses. Compare the results of this activity with those made in the original activity.

4. Interpreting Data Change the type of food or temperature used for raising fruit flies in the Find Out activity on page 293 to see how development of the larvae might be affected.

---

will not develop. Warmer temperatures will be more conducive to development than cooler temperatures. Foods high in sugars (fruits) will probably be most beneficial to development.

Critical Thinking

In your Journal, *answer each of the following questions.*

1. Compare and contrast the advantages and disadvantages of sexual and asexual reproduction.

2. Why do animals that develop inside the parent produce fewer eggs than those animals that develop externally?

3. What characteristic puts humans in the same group as snakes but in a different group from worms?

4. Give a specific example of how behavior can increase an animal's chance for survival.

5. Suppose a bird hatched with a beak that was shaped differently from beaks of other birds in the same species. Would this be helpful or harmful? Explain.

Problem Solving

Read the following problem and discuss your answers in a brief paragraph.

1. Suppose a female fish lays 100 000 eggs in a season. The male's sperm can fertilize 60 percent of the eggs. How many fertilized eggs are there?

2. After fertilization, 90 of every 100 of the fertilized eggs are eaten by other animals. Now how many eggs are left?

3. An additional 20 percent of the remaining eggs are destroyed when the water temperature gets too cold. How many eggs actually hatch?

4. If only half of those fish actually live to be adults, calculate the total number that survive.

CONNECTING IDEAS

Discuss each of the following in a brief paragraph.

1. **Theme—Stability and Change** Explain how a keen sense of sight is an adaptation for survival.

2. **Theme—Scale and Structure** Discuss whether complete metamorphosis is an advantage over incomplete metamorphosis?

3. **Theme—Systems and Interactions** Why is a tumbleweed not an animal even though it moves from place to place?

4. **A Closer Look** Kittens are born with their eyes shut and do not see their mothers for the first few days. How might imprinting take place between cats and their kittens?

5. **Physics Connection** How can experimenting with models of animals help you understand their adaptations?

✔ Assessment

Portfolio Review the portfolio options that are provided throughout the chapter. Encourage students to select one product that demonstrates their best work for the chapter. Have students explain what they learned and why they chose this example for placement in their portfolios.

Additional portfolio options can be found in the following **Teacher Classroom Resources:**

Multicultural Connections, pp. 21, 22

Making Connections: Integrating Sciences, p. 21

Making Connections: Across the Curriculum, p. 21

Concept Mapping, p. 17

Critical Thinking/Problem Solving, p. 17

Take Home Activities, p. 17

Laboratory Manual, pp. 55–56

Performance Assessment [P]

Critical Thinking

1. Sexual reproduction requires two parents, while asexual reproduction requires only one. An advantage of asexual reproduction is that it is faster than sexual reproduction. Sexual reproduction can combine characteristics from both parents, leading to greater variety.

2. Fertilized eggs that are developed internally have a better chance to survive and grow to adult organisms.

3. the presence of a backbone

4. Possible answers include: a dog barking can deter competitors for food; making loud noises can alert other animals to the presence of a predator and help them to escape.

5. The different shape would be helpful if it enabled the bird to obtain more food or protect itself from enemies. It would be harmful if it prevented the bird from doing these things.

Problem Solving

1. $100\ 000 \times 0.60 = 60\ 000$

2. $60\ 000 - 54\ 000 = 6000$

3. $6000 - 1200 = 4800$

4. $4800 \div 2 = 2400$

Connecting Ideas

1. Keen sight can help an animal locate food and avoid enemies. Both actions improve chances of survival.

2. Some of the stages of complete metamorphosis do not depend on having a food source as do the intermediate stages of incomplete metamorphosis.

3. A tumbleweed is not a consumer.

4. Accept all reasonable answers. Smell might be a common choice.

5. Some students may say that zoos help keep endangered species alive. Other students may say that wild animals should never be captured.

Chapter Organizer

SECTION	OBJECTIVES	ACTIVITIES & FEATURES
Chapter Opener		**Explore!**, p. 311
10-1 What Is a Plant? (2 sessions; 1 block)	1. **List** the traits of plants. 2. **Describe** the structures and functions of roots, stems, and leaves. **National Content Standards: (5–8) UCP1, UCP5, A1, C1, G1–2**	**Explore!**, p. 313 **How It Works,** p. 339
10-2 Classifying Plants (2 sessions; 1 block)	1. **Compare** and **contrast** vascular and nonvascular plants. 2. **Compare** and **contrast** plants that produce seeds in cones with those that produce seeds in fruits. **National Content Standards: (5–8) UCP1, UCP5, A1, C1, G3**	**Find Out!**, p. 318 **Explore!**, p. 320 **History Connection**, p. 340 **Literature Connection,** p. 340
10-3 Plant Reproduction (3 sessions; 1.5 blocks)	1. **Trace** the stages in the life cycle of a moss. 2. **Describe** the structure and function of a flower. 3. **List** methods of seed dispersal. **National Content Standards: (5–8) UCP3, UCP5, A1–2, C1–2, G1**	**Find Out!**, p. 322 **Find Out!**, p. 325 **Design Your Own Investigation 10-1:** pp. 328–329 **A Closer Look,** pp. 326–327
10-4 Plant Processes (2 sessions; 1 block)	1. **Explain** the role of stomata in gas exchange in plants. 2. **Compare** photosynthesis and respiration. **National Content Standards: (5–8) UCP1, UCP3, UCP5, A1, B3, C1, F2, F4–5, G1, G3**	**Find Out!**, p. 331 **Skillbuilder:** p. 332 **Investigate 10-2:** pp. 334–335 **Chemistry Connection**, pp. 332–333 **Science and Society,** pp. 337–338

ACTIVITY MATERIALS

EXPLORE!

p. 311 items for salad, such as lettuce, carrots, beans, peppers, cheese; salad dressing; mixing bowl; paring knife; plates; utensils

p. 313 a variety of plants

p. 320* unshelled peanuts, hand lens

INVESTIGATE!

pp. 334–335* lettuce, dish, water, coverslip, microscope, microscope slide, salt solution, forceps, paper towel, pencil

DESIGN YOUR OWN INVESTIGATION

pp. 328–329* 6 paper cups, 12 bean seeds, 12 radish seeds, 12 watermelon seeds, paper towels, water

FIND OUT!

p. 318* tall jar, water, blue food coloring, stirrer, fresh celery stalk with leaves, scalpel or sharp knife

p. 322* cutting from philodendron plant, glass, water, scissors

p. 325* flower, black paper, scalpel, hand lens

p. 331* 2 plastic bags, with twist-ties; 2 potted seedlings; petroleum jelly; water; label or felt marker; measuring cup

KEY TO TEACHING STRATEGIES

The following designations will help you decide which activities are appropriate for your students.

L1 Basic activities for all students

L2 Activities for average to above-average students

L3 Challenging activities for above-average students

LEP Limited English Proficiency activities

COOP LEARN Cooperative Learning activities for small group work

P Student products that can be placed into a best-work portfolio

Activities and resources recommended for block schedules

Need Materials? Call Science Kit (1-800-828-7777).

[00:00] **OUT OF TIME?** We recommend that students do the activities with an asterisk.

Chapter 10 Plant Life

TEACHER CLASSROOM RESOURCES

Student Masters	Transparencies
Study Guide, p. 36 **Critical Thinking/Problem Solving**, p. 18 **Multicultural Connections**, p. 23 **Making Connections: Across the Curriculum**, p. 23 **Science Discovery Activities**, 10-1	**Section Focus Transparency 30**
Study Guide, p. 37 **Multicultural Connections**, p. 24 **Making Connections: Integrating Sciences**, p. 23 **Science Discovery Activities**, 10-2 **Laboratory Manual**, pp. 57–58, Vascular and Nonvascular Plants	**Section Focus Transparency 31**
Study Guide, p. 38 **Take Home Activities**, p. 18 **Critical Thinking/Problem Solving**, p. 5 **How It Works**, p. 14 **Making Connections: Technology and Society**, p. 23 **Activity Masters**, Design Your Own Investigation 10-1, pp. 43–44 **Science Discovery Activities**, 10-3	**Teaching Transparency 19**, Parts of a Flower **Teaching Transparency 20**, Monocot and Dicot **Section Focus Transparency 32**
Study Guide, p. 39 **Concept Mapping**, p. 18 **Activity Masters**, Investigate 10-2, pp. 45–46 **Laboratory Manual**, pp. 59–60, Plant Respiration **Laboratory Manual**, pp. 61–62, Plant Growth	**Section Focus Transparency 33**

ASSESSMENT RESOURCES	TEACHING & TECHNOLOGY
Review and Assessment, pp. 59–64 **Performance Assessment**, Ch. 10 **PASC*** **MindJogger Videoquiz** **Alternate Assessment in the Science Classroom** **Computer Test Bank**	**Spanish Resources** **Cooperative Learning Resource Guide** **Lab and Safety Skills** **Science Interactions, Course 1, CD-ROM** **Computer Competency Activities**

*Performance Assessment in the Science Classroom

NATIONAL GEOGRAPHIC TEACHER'S CORNER

Index to National Geographic Magazine

The following articles may be used for research relating to this chapter:

- "The Gift of Gardening," by William S. Ellis, May 1992.

National Geographic Society Products Available From Glencoe

To order the following products for use with this chapter, contact your local Glencoe sales representative or call Glencoe at 1-800-334-7344:

- *STV: Plants* (Videodisc)
- *Newton's Apple Life Sciences,* "Plant Growth" and "Mold." (Videodisc)

Additional National Geographic Society Products

To order the following products for use with this chapter, call the National Geographic Society at 1-800-368-2728:

- *Kingdom of Plants Series* (5 Videos)
- *Photosynthesis: Life Energy* (Video)
- *Pollination* (Video)

Teacher Classroom Resources

These are key components of the classroom resources package.

TEACHING AIDS

Teaching Transparencies

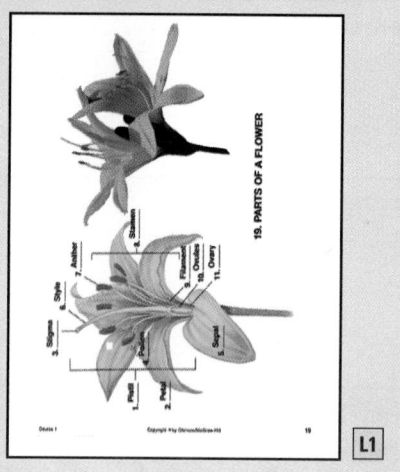

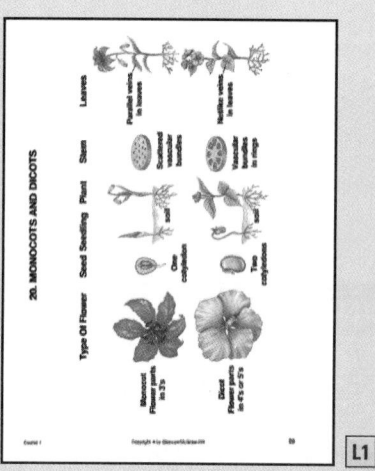

Section Focus Transparencies

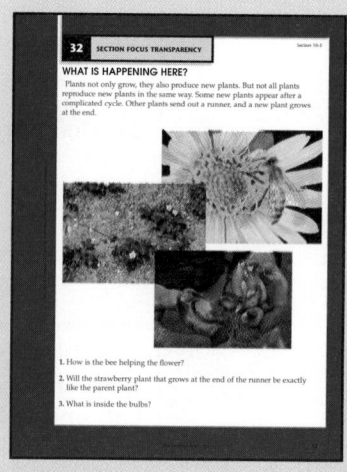

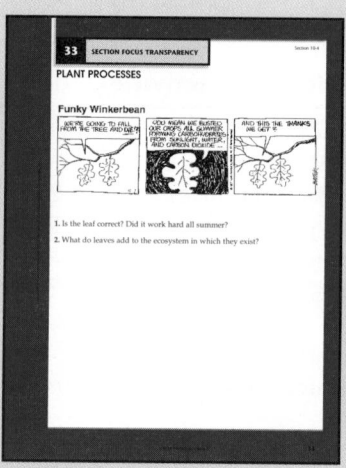

HANDS-ON LEARNING

Science Discovery Activity*

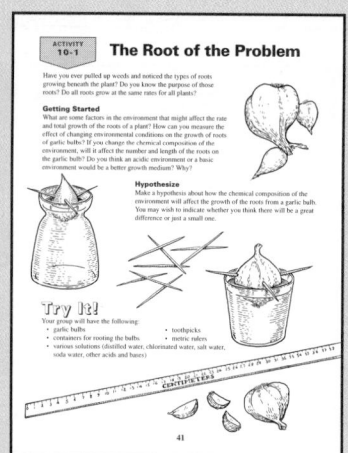

Laboratory Manual*

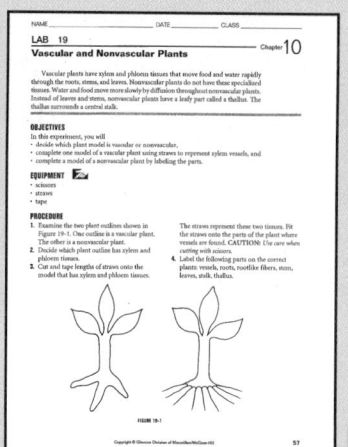

Take Home Activity

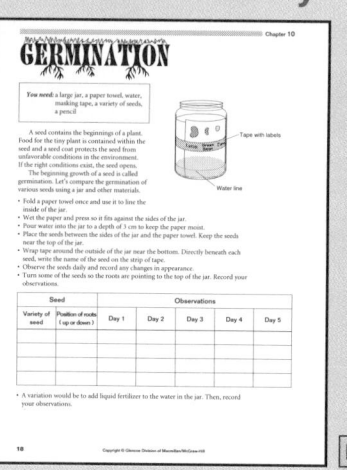

*There may be more than one activity for this chapter.

Chapter 10 Plant Life

Study Guide*

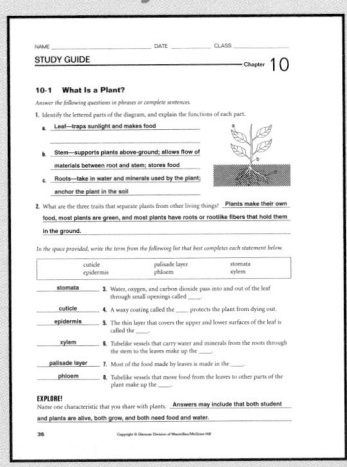

Concept Mapping

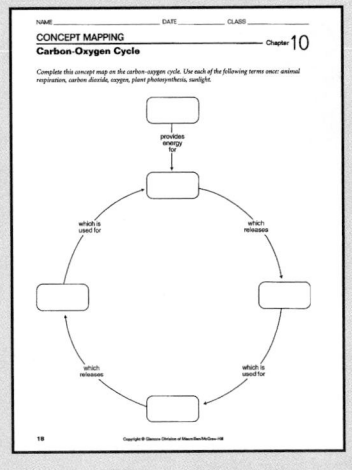

Critical Thinking/ Problem Solving

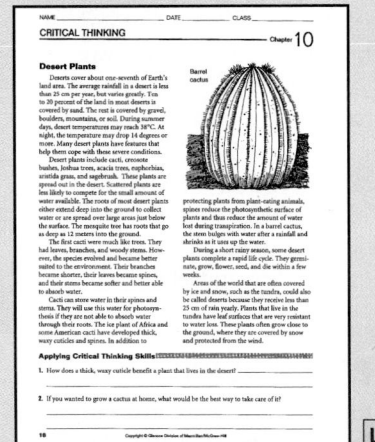

Integrating Sciences

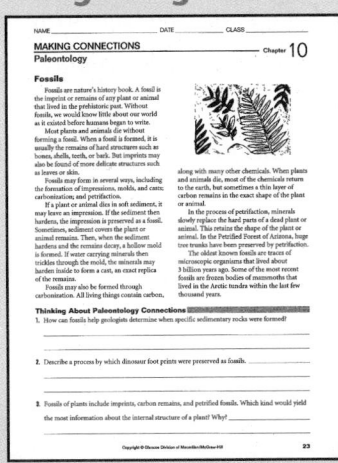

Across the Curriculum

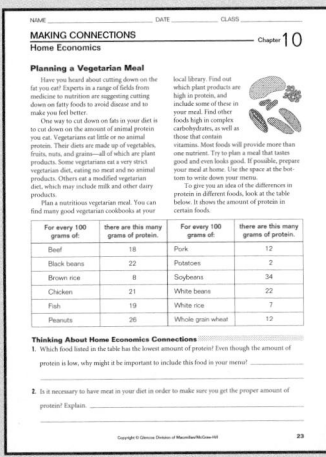

Technology and Society

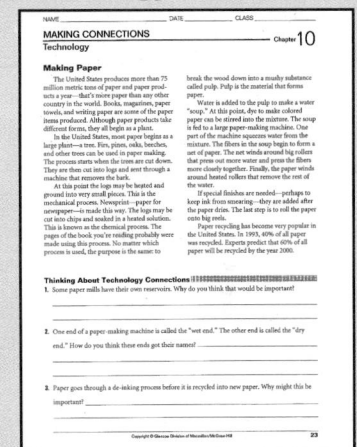

Multicultural Connection**

Performance Assessment

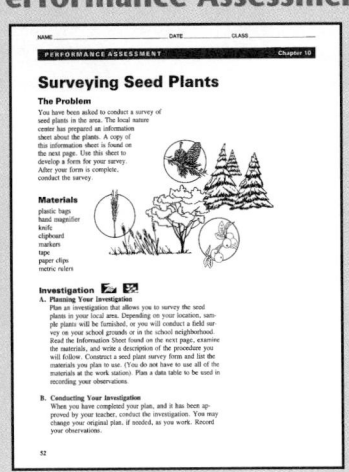

Review and Assessment

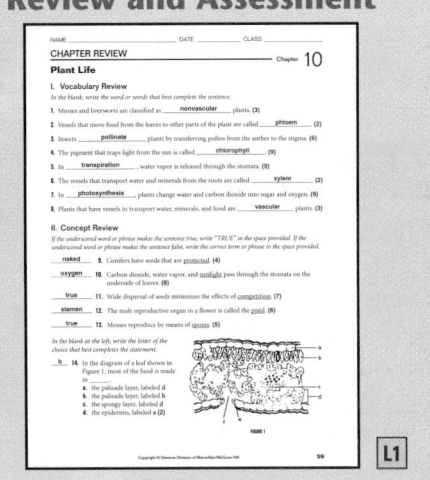

*One per section   **Two per chapter

Plant Life

THEME DEVELOPMENT

The themes that this chapter supports are scale and structure, systems and interactions, and stability and change. Students learn a system of categories based on plant structure. They explore interactions within plants and between the environment and plants. Students will appreciate the role plants play in affecting stability and change in Earth's environment.

CHAPTER OVERVIEW

In this chapter, students will study the characteristics that define plants. Students will study the structure and function of roots, stems, and leaves. They will also learn about plant classification, reproduction and seed formation, and life processes.

Tying to Previous Knowledge

Divide the class into four groups and assign each a plant type: mosses, ferns, cone-bearing plants, and fruit-bearing or flowering plants. Have each group write a description of that plant type. **COOP LEARN**

INTRODUCING THE CHAPTER

Have students name plants that grow in their area or identify them by distinctive characteristics: appearance, smell, leaf shape, flowers, and so on. Ask what is similar or different about the plants listed.

```
00:00  OUT OF TIME?
```

If time does not permit teaching the entire chapter, use the Chapter Overview on this page, Reviewing Main Ideas at the end of the chapter, and the Chapter 10 audiocassette to point out the main ideas of the chapter.

Plant Life

Did you ever wonder...

✓ Why leaves of many plants turn bright colors in the fall?

✓ What roots are for?

✓ Why insects are attracted to flowers?

Science Journal

Before you begin to study about plants, think about these questions and answer them *in your Science Journal.* When you finish the chapter, compare your journal write-up with what you have learned.

When was the last time you and your friends got together for a movie? After reading the newspaper ads, you chose the show. You paid for your ticket, rushed to the concession stand, and then looked for several seats in a row. The lights went out, and you enjoyed two hours of laughter, tears, or thrills.

Unless the movie title was The Eggplant that Ate Chicago, the last thing on your mind was "plants." But think about all the ways plants were involved. Trees provided the material for the newspaper, ticket, popcorn box, and soft drink cup. Popcorn—first used by Native Americans—is the seed of a plant. The flavoring in your drink came from plant parts. Even your clothes or the seat covers may be made of cotton grown in India. In this chapter, you will find out about plants and their roles in your life.

▶ *In the activity on the next page, explore some of the characteristics that make a plant, a plant.*

Learning Styles	**Kinesthetic**	Find Out, pp. 318, 331
	Visual-Spatial	Activity, pp. 312, 321, 326, 336; Explore, pp. 313, 320; Visual Learning, pp. 314, 324, 327; Discussion, p. 316; Find Out, pp. 322, 325; Demonstration, p. 326; A Closer Look, pp. 326–327; Investigate, pp. 328–329, 334–335; Chemistry Connection, pp. 332–333; Science at Home, p. 341
	Interpersonal	Explore, p. 311; Activity, p. 312; Across the Curriculum, p. 319
LS	**Linguistic**	Visual Learning, pp, 313, 317, 333; Demonstration, pp. 315, 323; Multicultural Perspectives, pp. 320, 324, 329; Activity, pp. 324, 325, 330; Across the Curriculum, pp. 327, 333; Discussion, p. 332

ACTIVITY

What's for lunch?

What To Do

1. With your classmates, bring in items to make a salad. Many different items can be included, such as carrots, lettuce, beans, peppers, and cheese.

2. Before you put your salad together, identify the items that are plants. What makes them plants?

3. How are the plant items in your salad different from foods that aren't plants? List the differences *in your Journal*.

4. Were some salad ingredients made from plants? What about the salad oil or dressing?

311

Explore!

What's for lunch?

Time needed 15 to 20 minutes

Materials salad ingredients, salad dressing, mixing bowl, plates, paring knife, utensils

TECH PREP

Thinking Processes observing, comparing and contrasting

Purpose To distinguish plants from nonplant foods.

Preparation Have students bring in a variety of ingredients or prepared salads. As an alternative, students can make "paper" salads using pictures cut from magazines.

Teaching the Activity

Discussion Have students offer as much information as they can about each ingredient.

Safety Remind students to cut away from themselves.

Science Journal Have students write their findings in their journals. L1

Expected Outcome

Students will distinguish plant and nonplant foods.

Answers to Questions

2. Students should identify plant foods as growing in the ground, being mostly green, having roots, stems and leaves, etc.

3. Unlike plants, nonplant items aren't green and lack roots, stems, and leaves. Mushrooms and eggs are nonplant items.

4. Salad ingredients made from plants include lettuce, tomatoes, carrots. Salad oil or dressing contains vegetable oils.

✔ Assessment

Content Invite students to "create" a new salad. Have students list and describe the ingredients, and then classify them as plants or nonplant foods. Use the Performance Task Assessment List for Invention in **PASC**, p. 45. L1

ASSESSMENT PLANNER

PORTFOLIO
Refer to page 343 for suggested items that students might select for their portfolios.

PERFORMANCE ASSESSMENT
Process, pp. 313, 318, 322, 335
Skillbuilder, p. 332
Explore! Activities, pp. 311, 313, 320
Find Out! Activities, pp. 318, 322, 325, 331
Investigate, pp. 328–329, 334–335

CONTENT ASSESSMENT
Oral, p. 331
Check Your Understanding, pp. 315, 321, 330, 336
Reviewing Main Ideas, p. 341
Chapter Review, pp. 342–343

GROUP ASSESSMENT
Opportunities for group assessment occur with Cooperative Learning Strategies.

10-1

PREPARATION

Planning the Lesson

Refer to the Chapter Organizer on pages 310A–D.

Concepts Developed

Students identify the traits of plants that distinguish them from other living things and make observations of plant structure, including roots, stems, and leaves.

1 MOTIVATE

Bellringer

 Before presenting the lesson, display **Section Focus Transparency 30** on an overhead projector. Assign the accompanying **Focus Activity** worksheet. L1

LEP

Activity This activity will enable students to identify the importance of plants in their lives.

You will need a pencil, book, cotton garment, fresh vegetable, piece of bread, can of coffee, sneaker, and rubber eraser.

Ask students what these objects have in common. List students' ideas on the chalkboard. Accept all suggestions while seeking the response that all of them come from plants. Help students to identify the plants involved. Lead them to understand that plants are essential to our lives. L1

NATIONAL GEOGRAPHIC SOCIETY

 Videodisc

STV: Plants

What Is a Plant?
Unit 1, Side 1
What Is a Plant? (In its entirety)

01094-28709

What Is a Plant?

Section Objectives
- List the traits of plants.
- Describe the structures and functions of roots, stems, and leaves.

Key Terms
xylem
phloem

What Makes a Plant a Plant?

Have you ever taken a walk in the woods or in a park with trees and flowers? Maybe someone you know has a garden. Your neighbor may grow herbs in a kitchen window. Most certainly you've seen weeds growing at the sides of roads. Trees, flowers, vegetables, herbs, and weeds are all plants.

Nearly all animal life depends on plants. If there were no more plants, animals would not go on living for very long. You know that you are an animal. But do you know what makes you different from a plant?

Remember that all organisms can be grouped according to their traits. In the previous chapter, you learned that all animals share certain traits. Plants, too, have traits in common. These traits separate plants from the other kinds of living things.

Figure 10-1

Plant types differ widely from one another, but most plants share the following three characteristics:

A Plants don't depend on other organisms for their food. Instead, they make their own food.

B Plants are usually green.

C Plants usually don't move around. Most plants have roots or rootlike structures that hold them in the ground. There are several different kinds of roots shown in the photographs on these two pages.

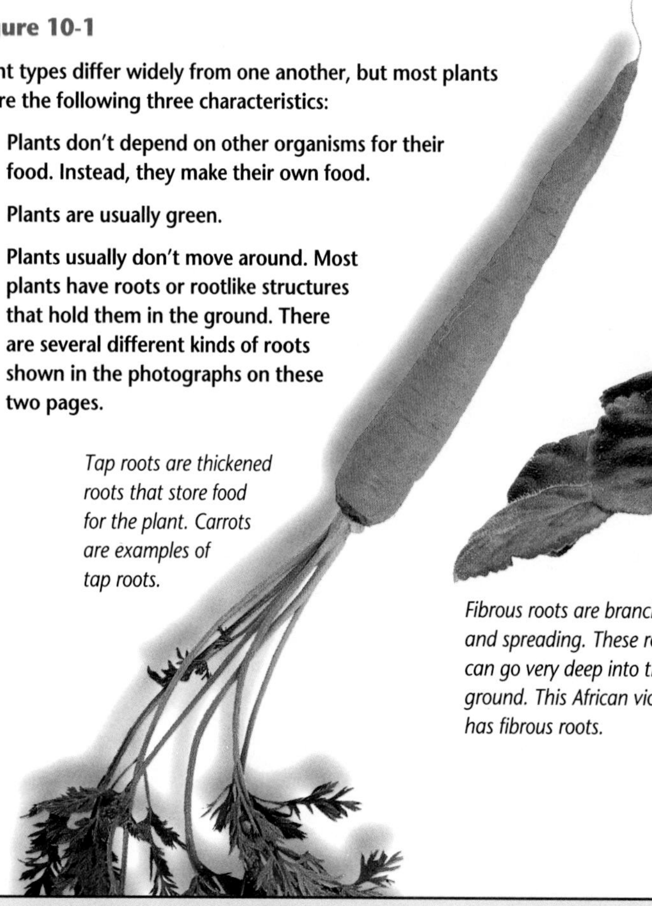

Tap roots are thickened roots that store food for the plant. Carrots are examples of tap roots.

Fibrous roots are branched and spreading. These roots can go very deep into the ground. This African violet has fibrous roots.

┌─────────────────────────────────────┐
Program Resources

Study Guide, p. 36
Critical Thinking/Problem Solving,
p. 18, Desert Plants L2
Section Focus Transparency 30
└─────────────────────────────────────┘

┌─────────────────────────────────────┐
Meeting Individual Needs

Visually Impaired For the Explore activity on page 313, allow visually impaired students to use their sense of touch on the plants in order to make observations of root, stem, and leaf. Pair the students with a fully-sighted student to observe and record their tactile and olfactory observations. **COOP LEARN**
└─────────────────────────────────────┘

Plant Structure

If you were asked to draw a plant, you'd probably draw one similar to the plants shown in **Figure 10-1.** Most plants you're familiar with have roots, stems, and leaves. Take a look at these structures in the following activity.

 ACTIVITY

Are all roots, stems, and leaves alike?

What To Do

1. Obtain several plants from your teacher. Look at the plants carefully. Can you identify the roots of each plant?

2. How are all the roots alike? How are they different?

3. Compare and contrast the stems and leaves of the plants, too. Record your observations *in your Journal.*

You can see from the Explore activity that roots, stems, and leaves are not all alike. Yet each structure of the plant has a specific function that helps keep the plant alive.

■ **Roots**

Suppose you took a walk to observe plants in your neighborhood. You might see trees, potted plants in windows, and dandelions in sidewalk

Plants, like this cactus, that live where there is little rainfall have compact root systems that are close to the surface of the ground. These roots quickly absorb water from dew or rain.

cracks. But you actually saw only about half of each plant! You saw only the parts of the plants that are aboveground. You probably did not see any roots. Most plant roots are below the surface of the ground. The root systems of some plants are as large or larger than the rest of the plant. Why must root systems be so large?

Roots have two important functions. All the water and minerals used by a plant enter the plant through its roots. Roots also anchor the plant in the soil. Without roots, a plant could be blown away by wind or washed away by water. Sometimes roots also store food. When you eat carrots or beets, you are eating roots with stored food.

313

2 TEACH

Tying to Previous Knowledge

In the previous chapter, students learned about the characteristics of animals. Introduce the lesson by having students suggest three to five ways plants differ from animals. Record their responses on the chalkboard.

Visual Learning

Figure 10-1 Have students list in their journals the three traits shared by most plants. Discuss how these traits separate plants from other kinds of living things. Then, as students examine the different kinds of roots shown in the figure, ask: **What kind of root is a carrot?** *tap root, a root with stored food* L1

Expected Outcome

Students will observe that plant structures take many different forms.

Answers to Questions

2. Roots grow at the bottom of plants. Some are systems of long, thin fibers; others are single, thick growths.

3. Stems are generally stiff. Some are thick, some thin, some woody. Most leaves are green.

✔ Assessment

Process The roots, stems, and leaves of different plants can appear quite different. Ask pairs of students to analyze their observations and infer why certain plants appear so different, yet their structures serve the same purpose. Have students form hypotheses based on their inferences. Use the Performance Task Assessment List for Formulating a Hypothesis in **PASC,** p. 21.
COOP LEARN L1

Explore!

Are all roots, stems, and leaves alike?

Time needed 5 to 15 minutes

Materials a variety of plants

Thinking Processes observing, organizing information, classifying, comparing and contrasting

Purpose To observe and identify plant structural differences.

Preparation Have available at least one plant for each group of students. Groups can swap plants. COOP LEARN

Teaching the Activity

Science Journal Ask students to record their observations on three separate pages in their journals—one each for roots, stems, and leaves—with a column on each page for similarities and a column for differences. Encourage students to illustrate their observations. L1

Figure 10-2

Stems carry out several functions. They support the above-ground parts of a plant.

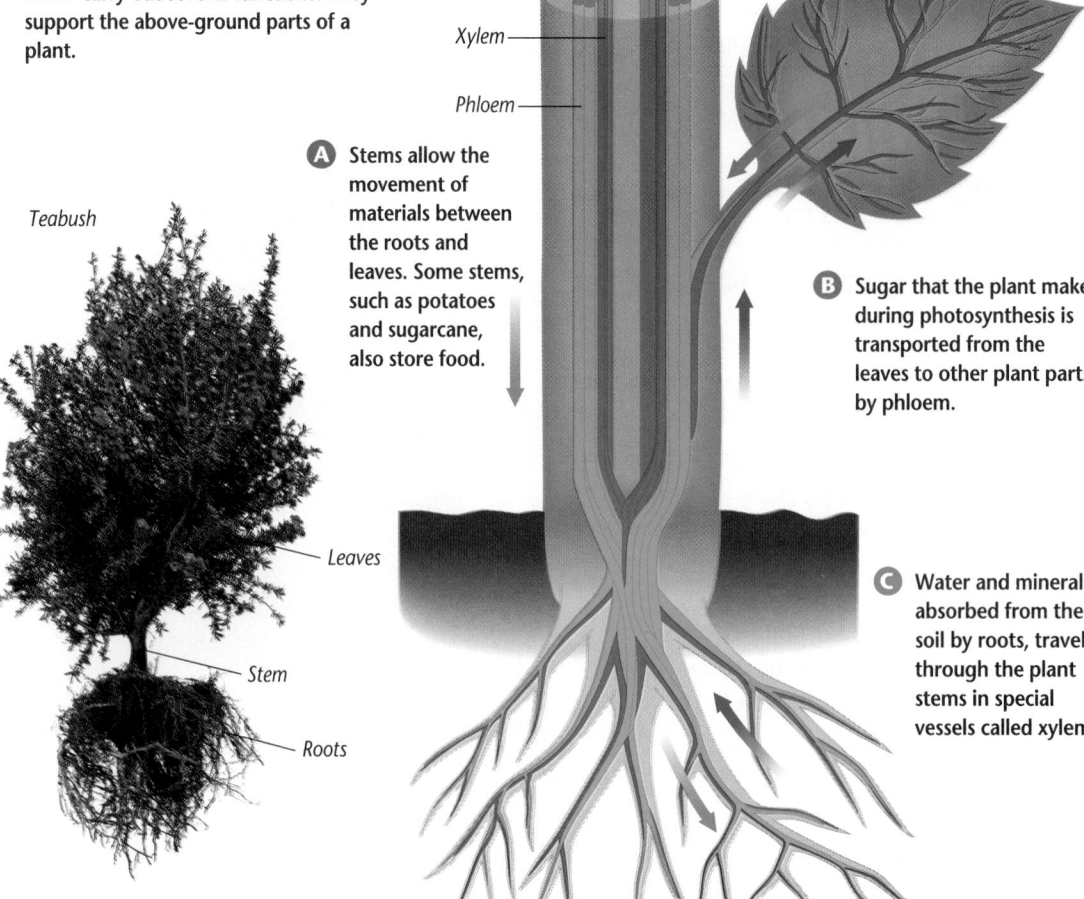

A Stems allow the movement of materials between the roots and leaves. Some stems, such as potatoes and sugarcane, also store food.

Teabush

Leaves

Stem

Roots

Xylem

Phloem

B Sugar that the plant makes during photosynthesis is transported from the leaves to other plant parts by phloem.

C Water and minerals absorbed from the soil by roots, travel through the plant stems in special vessels called xylem.

■ Stems

Stems have several jobs in keeping a plant healthy. They support the plant, store food, and allow for the movement of materials through vessels called xylem. **Xylem** is made of tubelike vessels that transport water and minerals up from the roots through the stem to the leaves of a plant. Xylem is dead and is what we call wood. The second type of vessel—phloem—is alive. **Phloem** is made of tubelike vessels that move food from the leaves to other parts of the plant. **Figure 10-2** shows where xylem and phloem are located in a teabush.

■ Leaves

Did you notice different kinds of leaves in the Explore activity? Leaves come in all shapes and sizes. A cactus has sharp spines, while a holly's leaves are dark, shiny, and prickly. One pine's needles are long and thin, yet another's are short and thick. No matter what shape and size leaves are,

Figure 10-3

Leaves are protected—top and bottom—by a thin layer, called the epidermis. A waxy coating called the cuticle sometimes covers the epidermis and protects the plant from drying out.

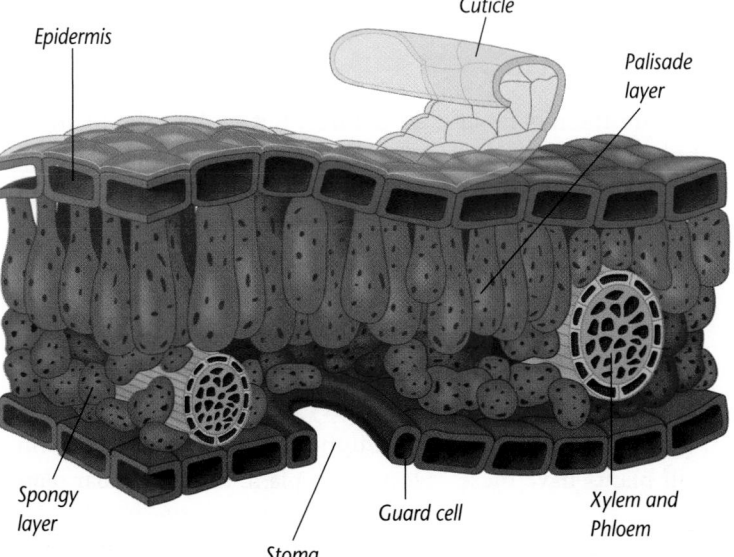

Epidermis

Cuticle

Palisade layer

Spongy layer

Stoma

Guard cell

Xylem and Phloem

Ⓐ Two different layers are located between the upper and lower layers of the epidermis. The palisade layer, where most food is made, is just below the upper layer of the epidermis. A spongy layer is located between the palisade layer and lower layer of the epidermis.

Ⓒ The spongy layer contains many air spaces as well as the xylem and phloem that transport water, minerals, and food to and from the leaves.

Ⓑ Materials needed by the plant, such as water, oxygen, and carbon dioxide, pass in and out of the leaf through small openings called stomata. Guard cells around each stoma control the size of these openings.

they are the plant's organs for trapping sunlight and making food. You can see the structure of a typical leaf in **Figure 10-3**. Notice the jobs of the different parts of the leaf as you study the figure.

In this section, you read about the traits that set plants apart from other organisms. You've also seen some of the structures many plants have. In the next section, you will find out how to use these structures to group plants.

check your UNDERSTANDING

1. List the traits of plants.
2. What are the main functions of roots, stems, and leaves?
3. Which tissue transports water in a plant?
4. How does carbon dioxide enter a plant?
5. **Apply** Why would a cactus usually have a thick, waxy coating covering the stem just like some leaves do?

check your UNDERSTANDING

1. Plants don't depend on other organisms for nutrients. They make their own food. Most plants are green. Plants usually don't move from location to location. Most plants have roots, or rootlike fibers, that hold them in the ground.
2. Roots anchor plants and absorb water and minerals. Stems support the plant, store food, and contain xylem and phloem for transporting materials. Leaves trap light and make food.
3. xylem
4. through the stomata, openings in the leaf
5. to prevent water from escaping

---

3 ASSESS

Check for Understanding

Assign questions 1–4 and the Apply question from Check Your Understanding. Extend the Apply question by asking students to explain why cacti must conserve water.

Reteach

Discussion Discuss students' descriptions of fruit or flower-bearing plants completed in the Chapter Opener activity, Tying to Previous Knowledge. Ask students to identify correct features and features that may be missing. Have students help complete an illustration on the chalkboard showing roots, stems, leaves, xylem, and phloem. L1

Extension

Discussion Ask students to explain how plants get what they need *from* the environment without damaging it. Have students write a brief report describing how adaptations such as roots, stems, and leaves solve this problem. Use the Performance Task Assessment List for Writing in Science in **PASC**, p. 87. L3

4 CLOSE

Demonstration

Have students identify the leaves, roots, and stems of various types of plants and give a one-sentence description of the function of each. Descriptions should reveal students' knowledge of plant structure. L1

10-2

PREPARATION

Planning the Lesson

Refer to the Chapter Organizer on pages 310A–D.

Concepts Developed

In this section, students learn the purpose of plant vascular tissue (xylem and phloem) and of seeds. They are introduced to plants that lack roots, stems, and leaves, and learn to classify plants as nonvascular, vascular without seeds, or vascular with seeds.

1 MOTIVATE

Bellringer

Before presenting the lesson, display **Section Focus Transparency 31** on an overhead projector. Assign the accompanying **Focus Activity** worksheet. L1

LEP

Discussion Ask students to look closely at their own wrists and describe what they see. Students will see, under the skin, a network of faint blue lines and should be able to identify these as veins, or at least as blood vessels. Ask students what is the purpose of these vessels. Ask them if they think plants have a similar system. *Student responses will vary.* Tell students that they will perform an experiment in this section that will show that although plants may not have veins as many animals do, they do have structures that move water throughout the plant.

10-2 Classifying Plants

Section Objectives

■ Compare and contrast vascular and nonvascular plants.

■ Compare and contrast plants that produce seeds in cones with those that produce seeds in fruits.

Key Terms

nonvascular plant
vascular plant

Nonvascular Plants

When you went to the movies at the beginning of this chapter, you discovered one way you could classify plants—by their usefulness. Scientists, however, determine which groups to place plants in by observing their structures.

In the last section, you looked at and described some typical plants. Each had roots, stems, and leaves. You may think that all plants have these structures, but look at the plants in **Figure 10-4**. They don't have the structures you might associate with plants. The mosses and liverworts in **Figure 10-4** belong to a group called nonvascular plants.

Recall from the last section that xylem and phloem are tubelike vessels that carry water, minerals, and food throughout the roots, stems, and leaves of a plant. A **nonvascular plant**

Figure 10-4

Mosses and liverworts are examples of nonvascular plants. Notice that they don't have flowers or cones, which means that they cannot produce seeds. Instead, mosses and liverworts reproduce by spores. You will learn more about this method of plant reproduction later in this chapter.

Program Resources

Study Guide, p. 37
Laboratory Manual, pp. 57–58, Vascular and Nonvascular Plants L2
Multicultural Connections, p. 24, Sacred Blue Corn L1
Section Focus Transparency 31
Making Connections: Integrating Sciences, p. 23, Fossils L2
Science Discovery Activity, 10-2

Meeting Individual Needs

Learning Disabled Some students will find it helpful to keep a vocabulary journal of key terms that they can refer back to at any time during the chapter. Each time a new term is introduced, they should record it in their journal. From this lesson, they might record the terms *nonvascular plant* and *vascular plant*. LEP

is a plant that lacks tubelike vessels to transport water, minerals, and food. Nonvascular plants also lack roots, stems, and leaves. They do have rootlike fibers—stalks that look like stems—and leaflike green growths. Study the traits of the mosses and liverworts in **Figure 10-4**. How do mosses and liverworts reproduce? Where do they live?

Mosses and liverworts are often the first plants to grow in areas that have been ravaged by fire. They also grow on newly formed rocks such as those found in lava beds. As the plants grow, their rootlike fibers move into small cracks in the rocks' surfaces. Mosses release chemicals that actually begin to break down the rocks. As these plants grow and die, the decaying plant material adds nutrients to the newly formed soil. Eventually, other plants are able to survive in the same area.

A Because nonvascular plants aren't able to transport water efficiently, they must live in moist areas.

B Mosses and liverworts are often found growing on tree trunks, on rocks, or next to streams. How does the lack of transport vessels affect the height of nonvascular plants?

2 TEACH

Tying to Previous Knowledge

Ask students to list ways water is distributed from a reservoir or from a well. Most students will suggest a system of pipes and conduits. Tell students that vascular plants also have systems to distribute water.

Theme Connection Students learn a classification system for plants based on structure. In the Find Out, students learn to identify xylem, which is part of the structure of vascular plants. The theme of scale and structure is supported as students use the structure of seeds for classification.

Uncovering Preconceptions Before reading this section, students may think that all plants have roots, stems, leaves, and vascular tissue. Using a clump of moss, you can show them that plants come in many different forms. Letting them gently pull apart moss will enable them to see major differences between large rooted plants and these simpler organisms.

GLENCOE TECHNOLOGY

 Videodisc

Science Interactions, Course 1 Integrated Science Videodisc

Lesson 3

Inquiry Question What conclusion can you draw about a part of a forest where the ground was thickly covered with moss? *Moss plants are nonvascular, thus they can grow only where water is readily available. Therefore, the ground must be moist most of the time.*

Visual Learning

Figure 10-4 Ask students where they would expect to find moss. *areas of moisture, such as deep forests or along streams* In what parts of the country would they expect not to find moss? *dry, desert regions* **How does the lack of transport vessels affect the height of nonvascular plants?** *They can't grow very tall.*

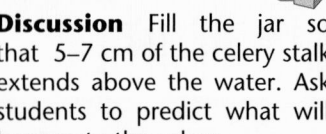

Seedless Vascular Plants

Suppose you are a contestant on a game show. The final question asks you, "If plants without xylem and phloem are called nonvascular, what are plants with these structures called?" You would be correct if your answer was "Vascular." Xylem and phloem make up a plant's vascular system. Thus, a **vascular plant** is a plant with xylem and phloem.

Find Out! ACTIVITY

Which tissue is xylem?

Many of the plants you are familiar with are vascular plants. These plants include vegetables, such as broccoli, corn, carrots, and celery. How can you observe the vascular tissue in a piece of celery?

What To Do

1. Pour water into a tall jar until it is 3/4 full.

2. Add 25 drops of blue food coloring and stir.

3. Use a scalpel or sharp knife to slice about 1/4" off the end of a fresh stalk of celery. **CAUTION:** *Use care when working with a scalpel.* Make sure the celery has its leaves attached.

4. Place the celery into the colored water. Leave the celery undisturbed for 48 hours. After this time, remove the celery from the jar.

5. Use a scalpel or sharp knife to cut horizontally through the celery about 1/3 of the way from the end of the stalk.

6. Observe the cut ends of the celery.

7. Holding the celery in both hands, snap the stalk in two. Pull apart the pieces. This will expose stringlike structures within the celery. Observe these stringlike structures.

8. Make a cut through one of the small stems that attaches the leaves to the main celery stalk. Observe the cut ends of the stem.

Conclude and Apply

1. Explain *in your Journal* how you know which tissue is xylem.

2. Where is the xylem located?

3. What evidence do you have that the colored water moved from the main stalk (stem) of the celery to the leaves?

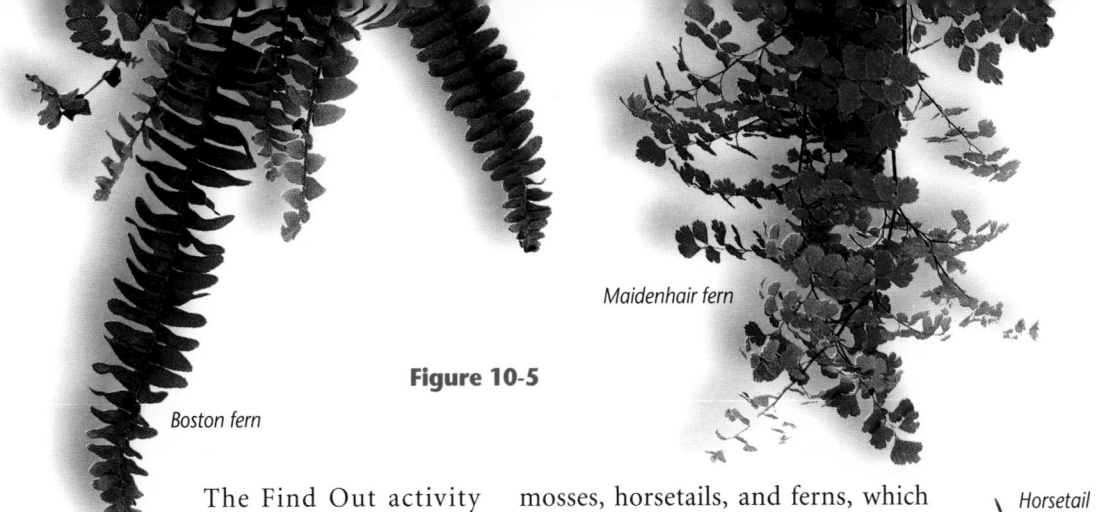

Maidenhair fern

Figure 10-5

Boston fern

Horsetail

The Find Out activity illustrates why vascular plants can grow taller and thicker than nonvascular plants. The vascular tissues carry water and minerals to all parts of the plant. Vascular plants can also live in drier areas.

Scientists divide vascular plants into groups according to how they reproduce. Vascular plants without flowers or cones produce spores instead of seeds. These seedless vascular plants include club mosses, spike mosses, horsetails, and ferns, which you can see in **Figure 10-5**. Ferns are the most numerous seedless vascular plants. While some tropical tree ferns may grow to be 5 meters tall, ancient tree ferns that lived about 300 million years ago grew as high as 25 meters, or as tall as a six-story building. These ferns and other plants were the raw materials from which coal and other fossil fuels formed.

Wood fern

Silver fern

Classifying Plants **319**

Content Background

Some mosses have water-conducting cells, but they are not organized as vascular tissue. Mosses, like amphibians, evolved from water-dwelling organisms and still need water for fertilization. Mosses have rhizoids that hold them to the ground but that do not conduct water as true roots do.

GLENCOE TECHNOLOGY

 CD-ROM

Science Interactions, Course 1, CD-ROM

Chapter 10

NATIONAL GEOGRAPHIC SOCIETY

 Videodisc

STV: Plants

How Plants Are Used
Unit 3, Side 2
How Plants Are Used

01134-26590

What Is a Flower?
Unit 2, Side 1
What Is a Flower?

28945-53624

Across the Curriculum

 **TECH PREP**

Geography

Obtain several field guides to trees, shrubs, and other plants such as the *Golden Guides,* Peterson guides, or Petrides guides. Ask pairs of students to look through a guide and select a particular plant. Provide students with outline maps of North America or have them trace a map of North America from an atlas. Help students find the guide's state-by-state (or province-by-province) range information on their plants and have them color in the areas where the plants grow. **COOP LEARN** **L2**

Seed Plants

By far the largest group of plants on Earth consists of the seed plants. Over 235 000 species of seed plants have been discovered. Like ferns, seed plants are vascular—they have vascular tissue. Seed plants are different from both nonvascular plants and seedless vascular plants because they have roots, stems, and leaves and grow from seeds. As you do the following Explore activity, you will discover what's in a seed.

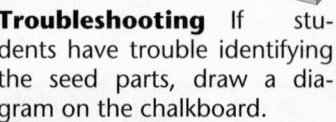

Explore! ACTIVITY

What's in a seed?

What To Do

1. Use a peanut that's still in its shell. Open the shell to expose the seeds.

2. Take the reddish-brown covering off one of the seeds. Carefully pull apart the two halves of the seed. Examine the halves with a hand lens. Which part of the seed is the young plant?

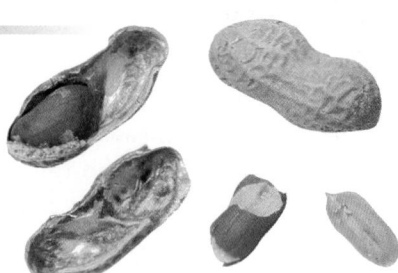

3. Find the parts that you think would become stem, leaves, or roots. *In your Journal,* summarize what you think the rest of the seed is used for.

Figure 10-6

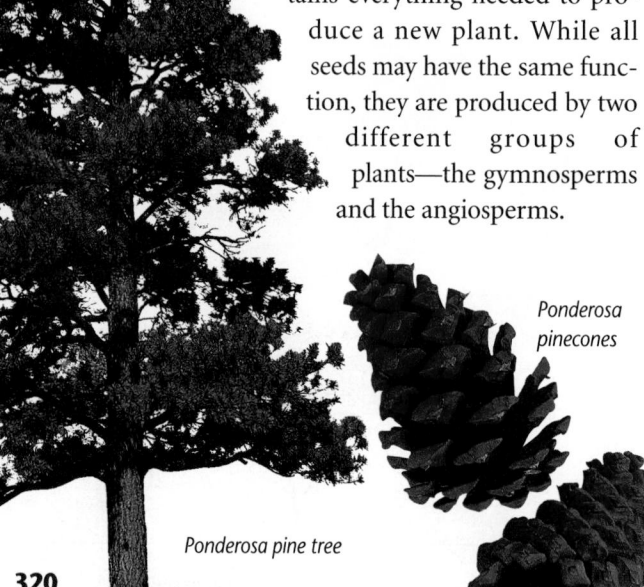

Ponderosa pinecones

Ponderosa pine tree

320

You just saw that a seed contains an undeveloped plant and stored food. Each seed contains everything needed to produce a new plant. While all seeds may have the same function, they are produced by two different groups of plants—the gymnosperms and the angiosperms.

■ Gymnosperms

Both the oldest trees and the tallest trees alive today are gymnosperms. Gymnosperms are vascular plants that produce their seeds on cones. The word gymnosperm means "naked seed" and is very descriptive because the seeds of these plants are not protected.

Figure 10-6 shows an example of a gymnosperm. You may know most of the plants in the gymnosperm group by the name evergreen because they remain green throughout the year. You are probably most familiar with the different types of evergreens, such as the pines, firs, and spruces.

Angiosperms

One of the best parts of a summer picnic is biting into a cold, juicy watermelon. Watermelons and many of the other foods you eat are examples of angiosperms. An angiosperm is a vascular plant in which the seed is enclosed and protected inside a fruit. You will learn more about how seeds form in both angiosperms and gymnosperms in the next section.

Angiosperms are also known as flowering plants. The variety of flowering plants seems endless. Stately oaks and graceful dogwoods, delicate rice and hearty corn plants, colorful bird-of-paradise and white yucca flowers are a few examples.

You've just read about the major classification of plants into nonvascular plants, vascular plants that have no seeds, and vascular plants that produce seeds. In the next section, you'll discover how each of these plants reproduces.

Figure 10-7

Orchid

Watermelon slice

Bird-of-Paradise

Chrysanthemum

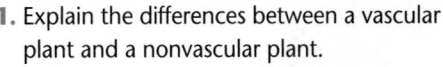

check your UNDERSTANDING

1. Explain the differences between a vascular plant and a nonvascular plant.
2. Why is a pine tree placed in a different group from a cherry tree?
3. **Apply** You notice some beautiful flowers in a field. There are yellow flowers, white flowers, and purple flowers. What can you tell about the kinds of plants they are part of?

10-2 Classifying Plants **321**

---

check your UNDERSTANDING

1. Vascular plants have xylem and phloem tubelike vessels to transport water and other materials. Nonvascular plants lack such tubelike vessels.
2. Pine trees produce seeds in cones. Cherry trees are flowering trees that produce seeds protected by fruit.
3. They are angiosperms that have seeds and vascular tubelike vessels.

---

3 ASSESS

Check for Understanding

Use the Check Your Understanding questions to gauge students' grasp of the traits of vascular and nonvascular plants, and seed-producing and seedless vascular plants. Ask students to identify additional traits and to explain their answers.

Reteach

Activity Have students help you make a classification table on the chalkboard. The major headings are *Vascular* and *Nonvascular* plants. Under *Vascular* are the subheadings *seed-producing* and *spore-producing*. Under *seed-producing* are *gymnosperm* and *angiosperm* subsubheadings. Ask students to provide examples of each type. L1 COOP LEARN

Extension

Activity Have students who have mastered the section concepts repeat the Find Out activity on page 318 using carrots, houseplant cuttings, and moss. After 24 hours, students should observe and record any changes in the plant and try to draw conclusions about the plant's structure. L3

4 CLOSE

Activity

Display a flower, moss, an evergreen cone, an orange, and a fern frond or a picture of each. Have students identify the sample that comes from (1) a non-vascular plant, (2) a seedless, vascular plant, (3) a gymnosperm, and (4) an angiosperm. *moss, fern frond, cone, flower, orange* L1

PREPARATION

Planning the Lesson

Refer to the Chapter Organizer on pages 310A–D.

Concepts Developed

Students learn about asexual and sexual reproduction in plants, including reproduction by spores, cones, and flowers. They identify the structures used by plants for sexual reproduction. Students also explore key factors in seed development and dispersal.

1 MOTIVATE

Bellringer

Before presenting the lesson, display **Section Focus Transparency 32** on an overhead projector. Assign the accompanying **Focus Activity** worksheet. [L1] [LEP]

Section Objectives

- Trace the stages in the life cycle of a moss.
- Describe the structure and function of a flower.
- List methods of seed dispersal.

Key Terms

pollination

10-3 Plant Reproduction

Plants from Plants

What is the first picture that comes to mind when you hear the word *nursery?* Most of us usually associate babies or young children with that word. However, sometimes the word is associated with plants. A plant nursery is where you can go to buy plants. In most cases, the plants you buy are young. You might walk through large greenhouses full of young plants as you decide which ones to buy. Where do all the plants come from?

Find Out! ACTIVITY

How can you grow new plants?

Sometimes new plants can grow from plant parts. Try this activity to see how you can grow a new plant.

What To Do

1. Take a cutting from a philodendron plant. Include part of the stem and one or two leaves.
2. Place one end of the stem in water.
3. Observe the cutting for a week or two.

Conclude and Apply

1. What happens to the cutting after several weeks?
2. How is what you have observed similar to what happens when you plant a seed? Record your observations *in your Journal.*

Find Out!

How can you grow new plants?

Time needed 15 minutes for setup; daily observation time for 2 weeks

Materials philodendron plant, water, glass, scissors

Thinking Processes comparing and contrasting, recognizing cause and effect, separating and controlling variables, observing and inferring

Purpose To observe asexual reproduction in plants.

Preparation This activity can be performed as a demonstration or by groups. You may need several plants to provide enough cuttings. [COOP LEARN]

Teaching the Activity

Troubleshooting
Place cuttings in sunlight to promote rooting. Check water level each day.

Science Journal Students can record their observations in a chart. [L1]

Expected Outcome

The cutting will develop roots in water.

Conclude and Apply

1. It grows roots.
2. Both produce a new plant.

✔ Assessment

Process Have students work in small groups to compare their findings. Ask students to analyze why the results differed, taking all variables into consideration. [L1] [COOP LEARN]

Asexual Reproduction

Some angiosperms can reproduce from their roots, stems, or leaves. Reproduction of new plants from roots, stems, or leaves is called vegetative reproduction. The new plants that result are identical to the parent. **Figure 10-8** shows several plants that reproduce by vegetative reproduction. Which of these plants develops new plants from the roots and which develops new plants from the stems?

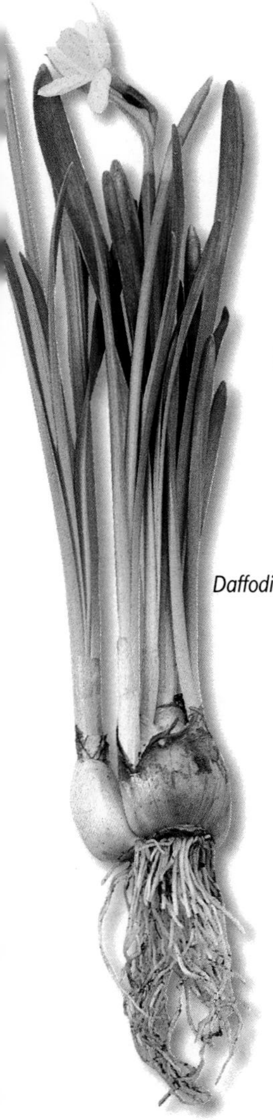

Daffodil

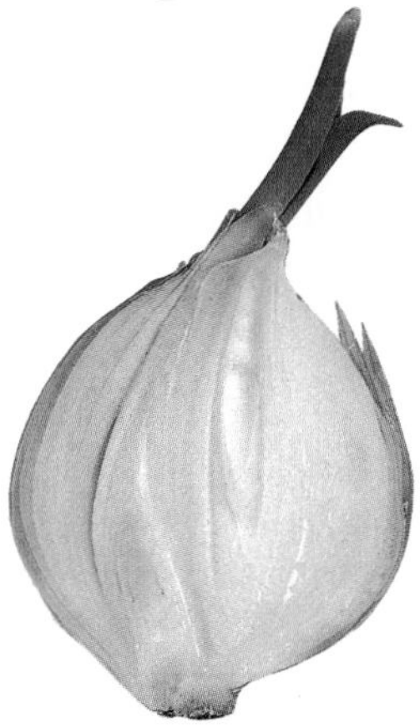

Onion

Figure 10-8

(A) Onions, daffodils, and potatoes all reproduce from the stem of the plant. Onions and daffodils produce bulbs, the part of the stem from which new plants originate. The potato itself is a thickened stem.

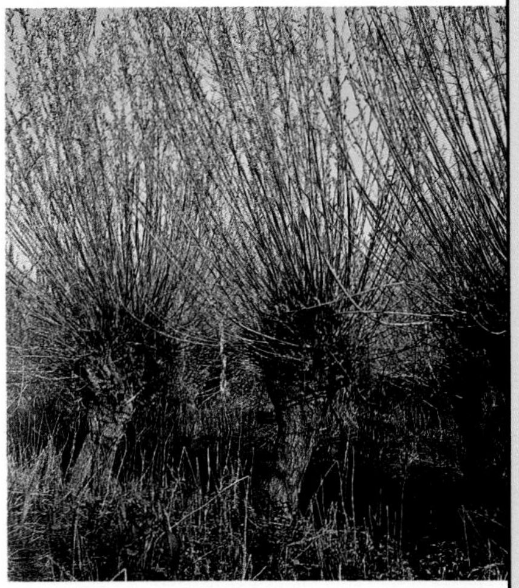

Willow tree

(B) The roots of the sweet potato, blackberry, or willow develop shoots that grow into separate plants.

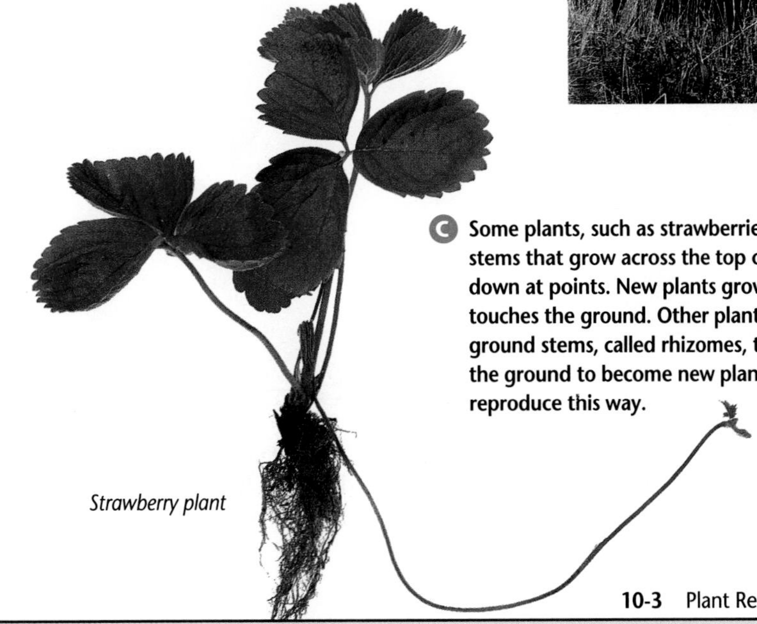

Strawberry plant

(C) Some plants, such as strawberries, grow runners—stems that grow across the top of the soil and touch down at points. New plants grow where the runner touches the ground. Other plants develop underground stems, called rhizomes, that push through the ground to become new plants. Lawn grasses can reproduce this way.

Demonstration Bring in a spider plant with baby plants. Ask students to explain how they think these plants are produced. After introducing the term, *asexual reproduction,* have students suggest other plants that reproduce in a similar manner. *strawberries by runners* L1

2 TEACH

Tying to Previous Knowledge

In the previous section, students classified plants partly by their ability to produce seeds. Tell students that in this section, they will learn in detail that plants reproduce asexually and sexually by seeds or spores.

Theme Connection One theme that this section supports is stability and change. Students learn how interactions with the environment play a role in plant reproduction.

Student Text Question **Which of these plants develops new plants from the roots and which develops new plants from the stems?** *From the roots: willow. From the stems: onions, daffodils, potatoes, strawberries.*

Flex Your Brain Use the Flex Your Brain activity to have students explore FLOWERS. L1

NATIONAL GEOGRAPHIC SOCIETY

Videodisc

STV: Plants

What Is a Seed?
Unit 4, Side 3
What Is a Seed?

00952-26224

Program Resources

Study Guide, p. 38
Teaching Transparencies 19 L1 ; **20** L1
Take Home Activities, p. 18, Germination L1
Section Focus Transparency 32
How It Works, p. 14, Velcro L1
Making Connections: Technology and Society, p. 23, Making Paper L2
Critical Thinking/Problem Solving, p. 5, Flex Your Brain

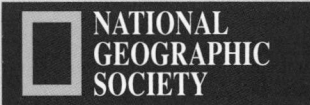

Sexual Reproduction

Recall from Chapter 9 that animals have both sexual and asexual reproduction. In sexual reproduction, the new organism develops from two parents. Sexual reproduction in plants involves the production of either spores or seeds.

■ Reproduction by Spores

Nonvascular plants, such as the mosses, and seedless vascular plants, such as ferns, reproduce from spores.

Follow the life cycle of a moss in **Figure 10-9** to see how it reproduces.

■ Seeds from Flowers

Instead of producing spores during sexual reproduction, many plants produce seeds. Pine trees and other gymnosperms produce seeds on cones. Angiosperms produce seeds within flowers. In the following activity, you'll study the parts of a flower to see where seeds are produced.

Figure 10-9

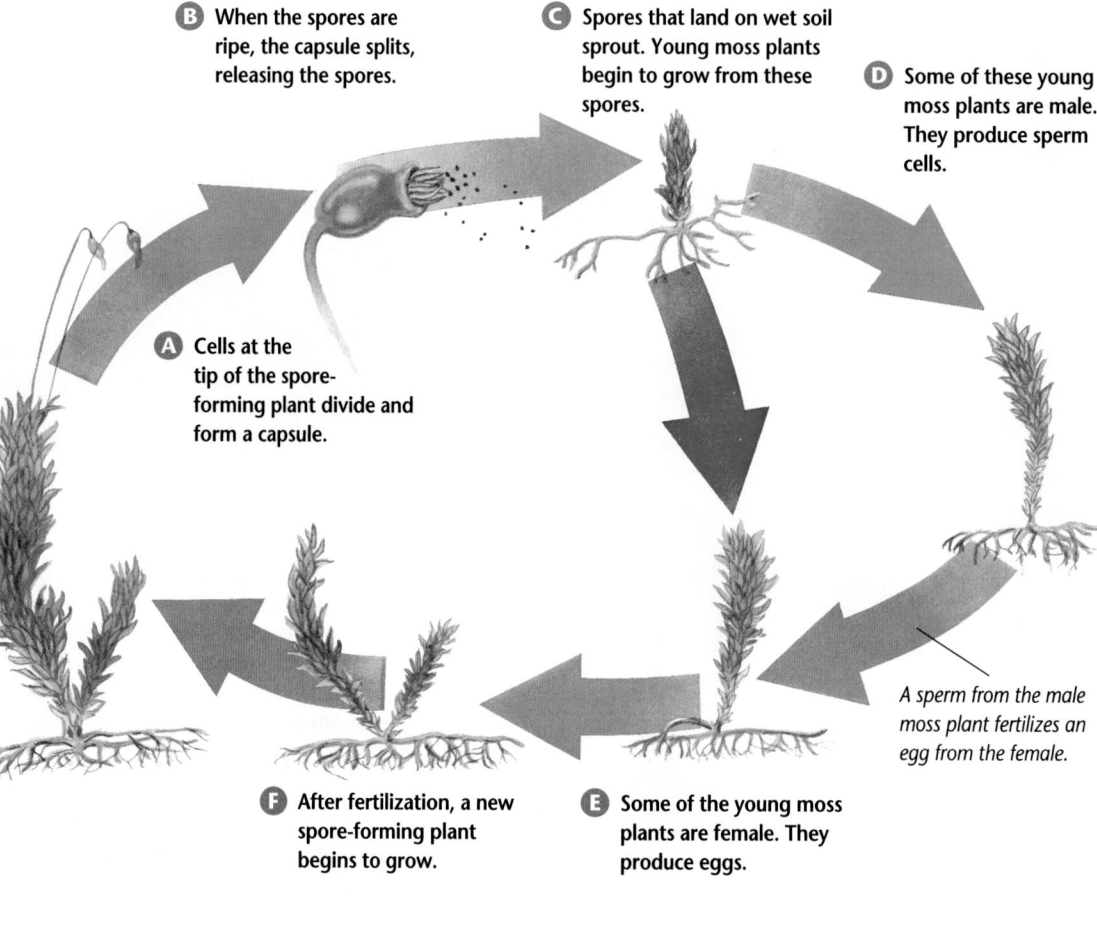

B When the spores are ripe, the capsule splits, releasing the spores.

C Spores that land on wet soil sprout. Young moss plants begin to grow from these spores.

D Some of these young moss plants are male. They produce sperm cells.

A Cells at the tip of the spore-forming plant divide and form a capsule.

A sperm from the male moss plant fertilizes an egg from the female.

F After fertilization, a new spore-forming plant begins to grow.

E Some of the young moss plants are female. They produce eggs.

Multicultural Perspectives

Land of the Ripe Rice Ears

Rice is the dominant food crop in much of Asia. The average Japanese eats about 84 kg of rice per year, compared to only about 3 kg for Americans. Rice originated in India about 3000 B.C.E. and spread both east and west, reaching Japan about 250 B.C.E. Rice is an annual grass that grows 4 feet high. Rice seeds are sown in prepared beds; when seedlings are 25–50 days old, they are transplanted into flooded paddies to mature.

One name for Japan is *Mizuho-no-kuni*, which means Land of the Ripe Rice Ears. Have students research other uses of rice in Japan. *drink (rice wine), clothing (sandals, rain capes, and rain hats made of rice straw), writing paper, and building materials.* L2

Find Out! ACTIVITY

What are the parts of a flower?

Flowers are more than pretty things for us to enjoy. A flower allows a plant to carry out the important job of reproduction.

What To Do

1. Examine the flower your teacher will give you. Use the illustration to see how the parts are arranged.

2. Remove the outer row of leaflike parts called the sepals. The structures inside the sepals are petals. Remove the petals.

3. Next, locate the stamens, the thin stalklike structures with expanded tops. Remove the stamens.

4. Look at the stamens with a hand lens. Observe the top part called the anther and the stalk called the filament.

5. Tap the anther against a piece of black paper to knock out the pollen

grains. Examine the pollen grains with a hand lens.

6. The structure that remains is the pistil. The stigma is at the top. The stalklike part is the style. The ovary is the swollen base of the pistil. Use a scalpel to cut across the ovary. **CAUTION:** *Always be careful with sharp instruments.* Use a hand lens to look at the inside of the ovary.

Conclude and Apply

1. What functions might the petals have?

2. How is the stigma adapted for trapping pollen grains?

3. How might pollen travel to the stigma?

4. *In your Journal,* diagram and label the parts of a flower.

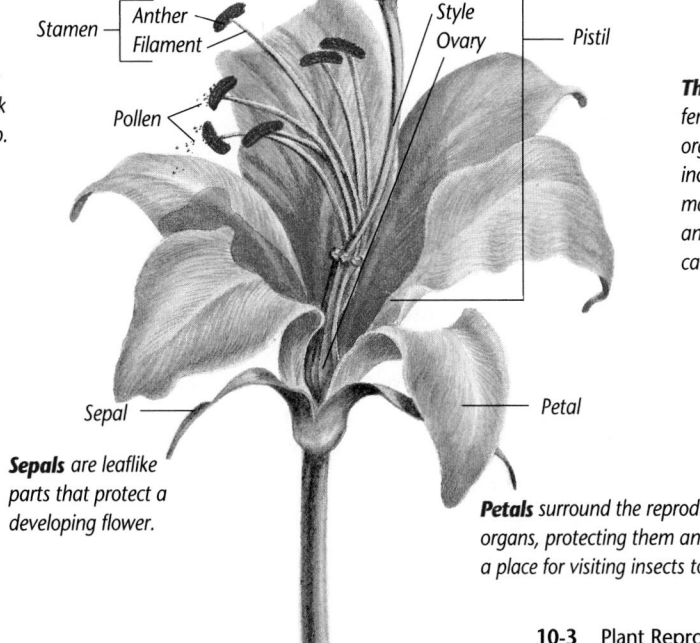

Stamens are the male reproductive organs. Each stamen has a slender stalk with a thick anther on top. Pollen grains form in the anther. Sperm develop in the pollen grains.

Stamen — Anther
Filament

Pollen

Stigma
Style
Ovary

Pistil

The pistil is the female reproductive organ. The pistil includes a sticky stigma, a stalklike style, and a swollen base called the **ovary.**

Sepal

Sepals are leaflike parts that protect a developing flower.

Petal

Petals surround the reproductive organs, protecting them and providing a place for visiting insects to land.

10-3 Plant Reproduction **325**

Find Out!

What are the parts of a flower?

Time needed 15 to 20 minutes

Materials large flower with prominent reproductive organs such as a tulip, gladiolus, or day lily; black paper; hand lens; scalpel

Thinking Processes observing and inferring, comparing and contrasting, interpret scientific illustrations

Purpose To observe and diagram the parts of a flower.

Preparation Students may work in groups of two or three. A nursery or flower shop may be willing to donate flowers. COOP LEARN

Teaching the Activity

Encourage students to take time with their observations.

Safety Students who are allergic to flowers shouldn't touch them during this activity.

Science Journal After students diagram and label the parts of a flower, have them explain the function of each part. L2

Expected Outcome

Students become familiar with the parts of a flower.

Conclude and Apply

1. to attract insects and birds, leading to pollination, to protect young pistils and stamens

2. It is sticky.

3. It could be blown by the wind or carried on the body of an insect or bird.

4. Make sure students include male and female reproductive organs, petals, and sepals.

✔ Assessment

Content Give students a diagram of a flower without labels. Have students label the parts of the flower and orally explain the function of each part. P L1

ENRICHMENT

Activity So efficient are pines in their reproductive methods that it is quite possible for a pine tree in Scotland to be pollinated by another in Norway. Ask students to research how this is possible, and then make a map showing the path the pollen takes. L3
COOP LEARN

Meeting Individual Needs

Visually Impaired Help visually impaired students identify the parts of a flower in the Find Out activity by pairing them with fully sighted students. Have students observe by touching the various structures of the flower, with the fully sighted student naming each one. For the same activity, have students examine a pine cone carefully for shape, size, texture, and stickiness.

Demonstration To compare and contrast methods of plant reproduction, obtain a piece of moss, a cone, and a flower. Show students the materials and ask them which one reproduces with spores. *the moss* Ask which of the three materials reproduce using pollen. *the cone and flower* Ask which of these depends primarily on the wind for pollination. *the cone* Ask which of the two contains a pistil and a stamen. *the flower*

Activity Have students trace the sequence of steps in the process of pollination. Students can make a flow chart or a drawing to illustrate the process. [L1]

How Do We Know?

Students gain detailed knowledge of the role of insects in the pollination of flowers. For more information, students can read Hamilton, David. *Flowers.* New York, Arcade, 1990.

Inquiry Question What are the ways plants depend on the environment to assist with reproduction? *pollination by insects and wind, seed dispersal by wind, water, and other organisms, and the need for water for germination*

How Do We Know?

Pollinating Flowers

Scientists used movie cameras to show how pollination happens. They took pictures showing a bumblebee climbing into a flower. As the bee climbs in, you can see an anther full of pollen coming down on its back. As the bee leaves, flecks of pollen grains are sticking to its body. These grains are left on the stigmas of other flowers it visits.

What did you notice when you looked inside the ovary? You should have seen tiny structures that resemble seeds. However, these structures are not actually seeds. When the eggs inside these structures are fertilized, seeds will then develop. Each structure produces one seed.

■ Seed Development

Before a seed can develop in a seed plant, a pollen grain must be transferred from the male to the female. In angiosperms, wind, insects, birds, or other animals may transfer pollen from the anther to the stigma. When pollen lands on the stigma, it forms a tube that grows down through the style and into the ovary. Sperm from the pollen grain travel through the tube and unite with the egg contained within the small structures inside the ovary. The transfer of pollen grains from the stamen to the stigma is called **pollination**.

The shape, size, and color of flowers are important factors in how they are pollinated. For example, large, brightly colored flowers may attract insects that pollinate the flowers. Night-blooming flowers may have

---

Advertising in the Real World

Wanted: Insect, bird, bat, or other animal to carry pollen from male to female plant. No experience necessary, but must be able to follow instructions. Benefits competitive."

An ad like this might work for plants if insects and other animals could read. But plants have better ways of attracting willing workers. The color of the blossom may attract a particular pollinator. About 80 percent of all flowers are pollinated by insects, and the rest by wind, birds, and mammals. Which

insect or mammal comes to a flower depends on what the flower looks and smells like. For example, hummingbirds are attracted to big, showy, red flowers like the hibiscus for their nectar.

Of Birds and Bees

Bees sometimes go to red flowers, too. That's because of the scent they put out, but not because of the red color. Bees see mostly blue, lavender, and yellow colors, so those are the colors of flowers to which bees travel. White attracts flies or night-flying insects and bats, depending on the scent.

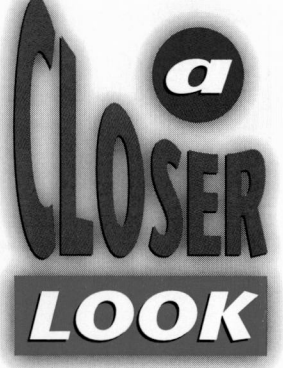

Purpose

A Closer Look extends the information presented in Section 10-3 on plant reproduction by describing some plant adaptations that promote pollination.

Content Background

In evolution, any adaptation that improves an organism's chances of surviving and reproducing will tend to be passed to offspring. Over time, more successful survivors and reproducers prosper in the com-

petition while less successful ones become extinct. An evolutionary relationship that has occurred between specific plants and animals is coevolution. Coevolution occurs when two species of organisms evolve structures and behaviors in response to changes in each other over a long period of time. The pollination of flowers by animals is the most common example of coevolution.

strong scents that attract pollinators such as bats. And those flowers that are pollinated by the wind may be pale in color or white, have small petals, or maybe no petals at all. Can you think of any wind-pollinated flowers?

Figure 10-10 shows the parts of a seed. The seed contains an embryo plant, food for the embryo, and a protective seed coat. Remember the peanut you examined earlier in this chapter? You could identify the young plant and its food. In the following activity, you'll investigate the relationship between soaking the seed coat and the time it takes the seed to begin to grow.

Figure 10-10

Each seed is a complete growing environment for the embryo plant inside. The seed provides food until the young plant can produce its own, and the seed coat provides protection until the new plant gains size and strength.

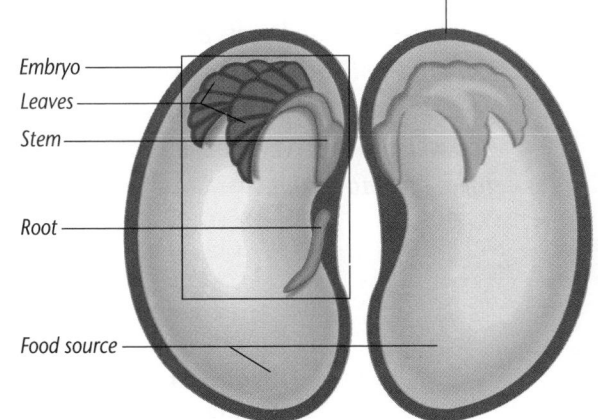

Seed coat

Embryo
Leaves
Stem
Root
Food source

Most birds have poorly developed senses of smell. That's why hummingbirds are attracted to color. Bright red flowers are attractive to hummingbirds.

Once a pollinator is attracted to a flower, the appearance of the flower sometimes contributes to the pollination process. On the blossom, there might be a cluster of dots or a color pattern visible to insects.

You Try It!

Design your own flower that would attract a particular type of pollinator. Use color or scent or both. Mark the flower with patterns that might help lead the pollinator to the pollen. Be sure to include the stamen (male) and stigma (female) parts so pollination can take place. You may either draw your flower or make a model flower from colored tissue paper. How might a flower pollinated by the wind differ?

10-3 Plant Reproduction **327**

10-1 They're All Wet

Preparation

Purpose To determine the effect of water on the speed of seed germination.

Process Skills predicting, sampling and estimating, organizing information, making and using tables and graphs, designing an experiment, observing, forming a hypothesis, separating and controlling variables, interpreting data

Time Required 25 minutes for two days, brief weekly observations for three weeks

Materials See reduced student page.

Possible Hypotheses Students may hypothesize that soaking will decrease the time it takes for seeds to sprout. Others may go a step further and hypothesize that a certain amount of soaking time will decrease the time it takes for seeds to sprout. Each group should reach a concensus about their hypothesis.

Plan the Experiment

Process Reinforcement Help students devise ways to separate and control variables (soaking times, temperature, light) that could affect the outcome of their experiment. Review the mechanics of making and using data tables and/or graphs before students begin.

Possible Procedures Divide each seed type into two samples. Using tap water, soak one sample for an hour and the other sample for 24 hours. Remove seeds from the water and place them between paper towels. Label clearly. Record the amount of time it takes for each group of seeds to begin sprouting, and the amount of time it takes for 25 percent, 50 percent, and 75 percent of the seeds in each group to sprout. A control group could be unsoaked seeds.

DESIGN YOUR OWN
INVESTIGATION

They're All Wet

You know that seed coats protect seeds, but a plant embryo cannot grow until this coat breaks open. In this activity, you will soak seeds in water to observe what effect this has on the seeds' coats. You will also observe whether soaking the seeds affects the time it takes them to sprout.

Preparation

Problem
How does soaking affect the time it takes seeds to sprout?

Form a Hypothesis
As a group, form a hypothesis about what might happen to seeds that are soaked in water for various lengths of time.

Objectives
- Predict the effect soaking has on the time it takes seeds to sprout.
- Infer what function water plays in the seeds' ability to sprout.

Materials
6 small cups (paper or plastic)
12 radish seeds
12 watermelon seeds
12 bean seeds
paper towels

Program Resources

Activity Masters, pp. 43–44, Design Your Own Investigation 10-1
Science Discovery Activities, 10-3, No Place to Grow

Meeting Individual Needs

Physically Challenged Physically challenged students may require special assistance during the Investigate. Have another student assist in preparing the seeds. The physically challenged student might record the data tables and the team's observations.

DESIGN YOUR OWN
INVESTIGATION

Plan the Experiment

1 Examine the materials provided by your teacher. Then design an experiment that uses these materials to test the effects of different soaking times on seeds.

2 Plan a data table in your Science Journal for recording your observations.

3 Because your test may last several weeks, assign daily tasks to all members of the group. Who will observe the seeds each day? Who will record the observations?

Check the Plan
Discuss and decide upon the following points and write them down.

1 Have you allowed for a control in your experiment? What is it?

2 How long will you conduct your test? How will you observe the sprouting seeds without injuring the seeds?

3 Make sure your teacher approves your experiment before you proceed.

4 Carry out your experiment. Record your observations.

Analyze and Conclude

1. **Compare and Contrast** Which seeds sprouted first? Last?

2. **Infer** What can you infer about the types of seeds and the times it took for them to sprout?

3. **Separate and Control Variables** Why did you soak the seeds for different amounts of time?

4. **Interpret Data** Infer what function water played in this experiment.

5. **Draw a Conclusion** How does soaking time affect the time it takes for a seed to begin growing? Did your observations support your hypothesis?

Going Further
Predict what would happen if you used tea or lemon juice as a soaking solution.

Multicultural Perspectives

The Worldly Potato
Few plants have been as important to the world as the potato, which is a tuber or food storage unit for the plant rather than a root. Potatoes are from South America, where they have grown for at least 13 000 years. The ancient Inca not only grew them, they created a portable, freeze-dried product from them by crushing them to press out the water and then freezing them at night. The Spanish brought the potato to Spain in the 1500s, and from there it spread throughout Europe and to China, Japan, and North America.

The potato had a great impact on every place it reached. It was a nutritious food that was easy to grow. Have students find out what the nutritional value of the potato is compared to rice and wheat. **L1**

Teaching Strategies

Discussion Explain to students that they will be testing the effect of a single variable—the length of time a seed soaks in water—on the sprouting of three different kinds of seeds.

Science Journal Have students record their data table, observations, and graph or chart of their results in their journals.

Expected Outcome
Students will observe that the seeds soaked longest germinated first.

Analyze and Conclude

1. The seeds that were soaked for the longest time began to grow first. If students used unsoaked seeds as a control, these seeds would sprout last.

2. Students will observe that the different types of seeds have different seed coats, which affects the time it takes for the seeds to sprout.

3. The seeds were soaked for different amounts of time to see whether the length of soaking time affected when a seed would begin to grow.

4. Water was absorbed by seed tissues, which swelled and broke the seed coat.

5. Soaking seeds reduces germination time. Answers will vary depending on hypotheses.

✔ Assessment

Performance Have students use clay to make models of a seed before and after it has sprouted. Students can work with a partner or in their groups.

Going Further ‖‖‖‖▶
Both tea and lemon juice are acids. They might weaken the seed coats faster than water.

GLENCOE TECHNOLOGY

 Software

Computer Competency Activities

Chapter 10

Check For Understanding

1. Have students, working in pairs, describe the reproductive cycles of spore and seed-producing plants, then prepare reports with diagrams. [L1] [P]

2. Assign the Check Your Understanding and Apply questions.

Reteach

Diagram To help students master the terms and methods of plant reproduction, draw four concept maps on the chalkboard, one each for asexual, spore, gymnosperm, and angiosperm reproduction. Key ideas, such as "flower" or "cone," are drawn in circles surrounding and connected to the appropriate main concept in the center, for example, gymnosperm—cone. [L1] [LEP]

Extension

Activity Have students make scientific drawings to describe their observations in the two Find Out activities. For the Find Out activity on page 322, have students provide three drawings of the philodendron cutting: still on the parent plant, newly placed in the glass of water, and in the glass with roots showing. For the Find Out activity on page 325, students can draw a cutaway view of the flower and identify its structures with labels. [L2]

4 CLOSE

Activity

Ask each student to prepare an outline of plant reproduction, using traditional outlining techniques—Roman numerals, capital letters, Arabic numerals, lowercase letters. Remind students that they can use heads from the text, but that they will also have to identify key parts of the text and summarize the material. [L2] [P]

Figure 10-11

Fruits and seeds are dispersed by animals, wind, water, and sometimes people.

A Fleshy fruits, such as oranges and tomatoes, are filled with water and sugar. Animals are attracted to these fruits, eat them, and may spit out the seeds or disperse them in their wastes.

C Sticker or burr type fruits, like those of the common thistle, may stick to animals or the clothing of people and be carried far away from the original plant.

B Winged "helicopters" of the maple and silky dandelion seeds are carried away by the gentlest breezes.

D Many fruits and seeds of plants growing near water, such as the coconut, contain air chambers that allow them to float and be carried for miles in water.

■ Seed Dispersal

Imagine what would happen if all seeds began to grow close to the parent plant. The young plants would compete with the parent plant, and with each other, for light, water, soil, and nutrients. The dispersing of seeds away from the parent plant helps reduce the competition for these resources and gives each plant a better chance of survival.

You and even your pets disperse seeds. Small seeds may stick to your shoes. Hooked seeds may stick to your dog's fur or to your clothes. In the next section you will learn how plants get the energy they need to produce flowers and seeds.

check your UNDERSTANDING

1. Trace the life cycle of a moss.

2. How does the seed of an angiosperm develop?

3. What is the function of flowers?

4. List two ways seeds can be dispersed.

5. Apply Suppose you're eating a piece of watermelon, and you spit out some of the small, black structures contained in the fruit. What part of the plant are you eating? How are you helping the plant reproduce?

check your UNDERSTANDING

1. Spores are produced by a spore-producing plant. A sex-cell-producing plant grows from the spores and produces sperm and eggs. These unite to form a new spore-producing plant.

2. Pollen grains land on the stigma. A pollen tube grows down from the stigma to the ovary. Sperm travel through the pollen tube and unite with the egg to form a seed. An embryo plant then develops within a seed.

3. reproduction

4. wind, seeds floating on water

5. You are eating the fruit, which was once the ovary. You are dispersing its seeds.

Transpiration

Remember the last time you took part in a physically active game? It may have been aerobics in gym class, a soccer game, or simply chasing your dog around the yard. Think about how your body reacted. Your face was probably flushed, and you breathed hard. You're used to the idea that people and other animals breathe. But the idea that gas exchange also happens in plants may seem a little strange. The next activity will allow you to observe this exchange indirectly.

Section Objectives

- Explain the role of stomata in gas exchange in plants.
- Compare photosynthesis and respiration.

Key Terms

transpiration
photosynthesis
chlorophyll

Find Out! ACTIVITY

Where does the water come from?

Who ever heard of a sweaty plant? Yet, plants do release water vapor as a part of the gas exchange that they undergo. This activity will give you a clue how.

What To Do

1. Obtain two plastic bags, two potted seedlings, petroleum jelly, and water from your teacher.

2. Pour the same amount of water into the pots of both plants.

3. Label one bag "petroleum jelly" and the other bag "no petroleum jelly."

4. Put one seedling in the bag labeled "no petroleum jelly." Seal the bag and put it in a sunny window.

5. Rub petroleum jelly on the bottom of all the leaves on the second plant. Put this seedling in the other bag. Seal the bag and place it next to the first plant.

6. Wait several hours or until the next day and observe the bags.

Conclude and Apply

1. In which bag did water droplets collect?

2. *In your Journal,* explain where the water came from.

3. What did the petroleum jelly prevent from happening?

Find Out!

Where does the water come from?

Time needed 40 minutes on the first day, 15 minutes the second day

Materials two plastic bags with twist-ties, two potted seedlings (young bean plants grown from seed are ideal), petroleum jelly, labels or felt markers, measuring cup, water

Thinking Processes thinking critically, observing and inferring, comparing and contrasting, recognizing cause and effect

Purpose To observe which part of the plant releases vapor in the air.

Preparation This activity can be performed as a demonstration or by groups of students.
COOP LEARN

Teaching the Activity

Rub petroleum jelly on the *bottom* of the leaves where most stomata are. For best results, wait 24 hours before observing the bags.

Planning the Lesson

Refer to the Chapter Organizer on pages 310A–D.

Concepts Developed

Students learn about the plant process of transpiration and the relationship between photosynthesis and respiration.

1 MOTIVATE

Bellringer

Before presenting the lesson, display **Section Focus Transparency 33** on an overhead projector. Assign the accompanying **Focus Activity** worksheet. [L1] **LEP**

Science Journal Students can record their observations and answer the questions in their journals. [L1]

Expected Outcome

Students will observe that plants release water vapor.

Conclude and Apply

1. in the bag without petroleum jelly

2. It condenses from water vapor that passes out of the plant

3. It prevents water vapor from escaping from the plant through its stomata.

✔ Assessment

Oral Ask students to infer what might happen if a person had a layer of petroleum jelly over her skin while playing soccer. *She would be unable to perspire, leading to overheating.* Use the Performance Task Assessment List for Making Observations and Inferences in **PASC**, p. 17. [L1]

Discussion Ask students to imagine that they are in a spacecraft that has crash-landed on the moon. Have them help you make a list of things they will need to survive. The list should include air, food, and water. Explain that plants have related needs, which students will learn about in this section.

2 TEACH

Tying to Previous Knowledge

Remind students of the traits of living things described in Chapter 7. Ask them to suggest other important plant traits besides reproduction. Students should identify cellular structure, growth, needing food and water, response, and adaptation. This section explores basic processes that are characteristic of living plants.

Theme Connection The themes that this section supports are systems and interactions. Students learn about basic plant interactions with their environment through transpiration and photosynthesis. Transpiration is observed when the students perform the Find Out activity on page 331.

Discussion Ask students to suggest what purpose the plastic bags and petroleum jelly serve in the Find Out activity. Encourage students to see that the plastic bags produce an airtight space, while petroleum jelly prevents gas exchange through stomata.

Interpreting Scientific Illustrations

Stomata are located on leaves. Look at **Figure 10-12**. Where are the stomata located on this leaf? What is one reason for this location? In pond lilies, the leaves float on the water. Where would you expect the stomata to be on pond lily leaves? For help refer to the **Skill Handbook**, page 667.

How does gas exchange take place in plants? Study **Figure 10-12**. It shows how gases move into and out of leaves, and how water vapor is given off by the stomata as these gases are exchanged. Water is absorbed by roots and is transferred up to the leaves through xylem vessels. Once in the leaves, most of the water evaporates. The water vapor is released from inside the leaf through the stomata. The loss of water vapor through the stomata of a leaf is called **transpiration**. Plants lose large amounts of water every day through the process of transpiration. This water is a major source of the water vapor in air.

Think about the Find Out activity you just completed. Water formed inside one bag. It formed when water vapor that transpired from the plant collected on the inside of the bag. The other bag had little or no water. Recall that you rubbed petroleum jelly on the bottom of this plant's leaves. Based on what you know about the structure of a leaf, what did the petroleum jelly do? Where do you think most of the stomata are located on a leaf?

In Living Color

In the fall, the leaves of many shrubs and trees change from their familiar green to shades of red, yellow, orange, and brown. This happens because such plants are breaking down chlorophyll in colder weather. The lack of chlorophyll shows us other pigments present in the leaf that we can't normally see. Some of these pigments transfer energy from sunlight to chlorophyll.

Chromatography

You can reveal the pigments in a leaf by a technique called chromatography. This technique uses a solvent that causes the pigments to separate. The pigments travel up a piece of paper, each pigment stopping at a different place. The more soluble a pigment is, the farther up the paper it travels.

What To Do

1. Obtain a piece of filter paper at least 15 cm long.
2. Use a pencil to mark two X's 2 cm from the bottom and 1.5 cm from the sides of the filter paper. Use the Figure as a guide.
3. Use a dropper to add pigment to the paper strip between the two X's you

Purpose

This Chemistry Connection discusses pigments in plants and gives students a technique for separating the pigments—chromatography.

Content Background

Thousands of pigments are found in plants. The basic one is chlorophyll, which is green; then, xanthophyll, which is yellow; and the anthocyanins, which range from pale pink to red to a rich purple.

To prepare solution: Bring 150 mL of water to a boil. Place entire package of spinach into the boiling water. Bring to a boil again. After several minutes, remove the spinach from the water and squeeze out all excess water. Transfer the boiled spinach to a 400 mL beaker containing 80 mL of ethyl alcohol. Heat the beaker until the alcohol begins to bubble, about 30 seconds. CAUTION—Do not heat alcohol solution with an open flame such as a Bun-

Figure 10-12

Plants release water vapor through openings in the leaves called stomata. Each stoma is surrounded by two guard cells that regulate the size of the opening.

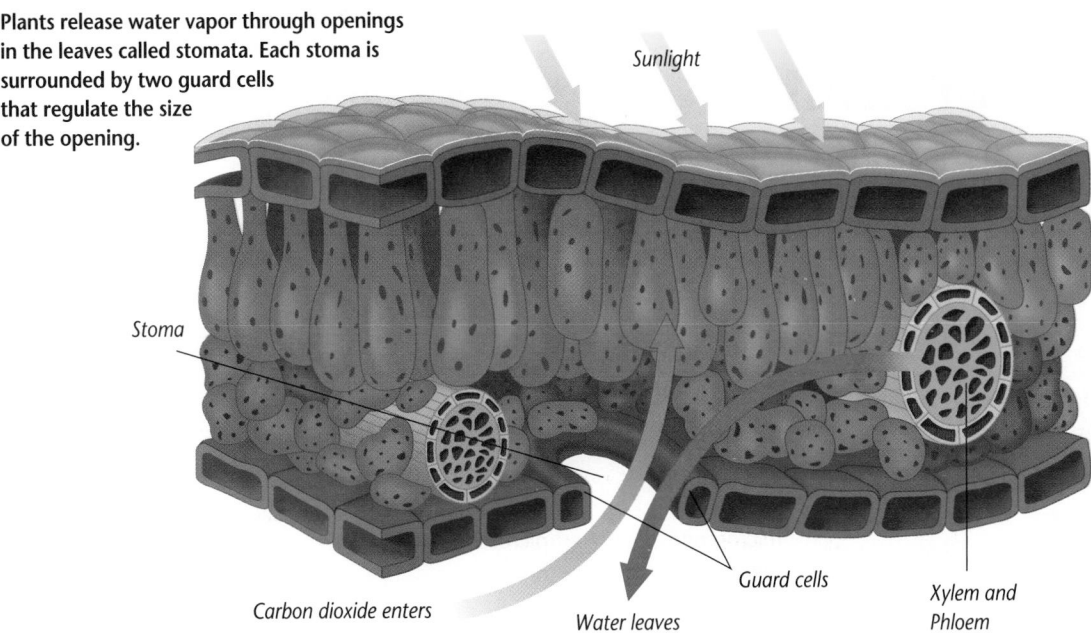

Sunlight

Stoma

Carbon dioxide enters

Water leaves

Guard cells

Xylem and Phloem

SKILLBUILDER

1. on the lower surface

2. To conserve water, for gas exchange; Students may also say to keep dust and dirt out of the opening.

3. on the upper surface L1

Visual Learning

Figure 10-12 Have students explain how gases move into and out of leaves and how water vapor is given off by the stomata as these gases are exchanged. L1

Across the Curriculum

Daily Life

Inform students that people worry about the loss of tropical rain forests partly because plants use carbon dioxide, the major gas involved in the "greenhouse effect" that scientists believe will raise global temperatures by 1° to 5°C over the next 50 years. Use of fossil fuels has resulted in increased levels of carbon dioxide in the atmosphere. Development worldwide has also shrunk forests. One proposed remedy to the loss of tropical rain forests is planting trees to use carbon dioxide and produce more oxygen. Have students research other proposed solutions to the problem. L2

---

made. Allow the spot of pigment to dry.

4. Continue to add pigment to the paper until you have a dark spot, about 20 drops.

5. Using the Figure as a guide, roll the end of the filter paper around a straw. Fasten the filter paper around the straw with a paper clip.

6. Remove the paper and straw assembly from the jar. Add solvent to a height of 0.5 cm in the jar.

7. Rest the straw across the top of the jar, with the paper dangling into the solvent. Make sure the bottom end of the paper strip *just* touches the solvent.

8. Do not shake or move the jar for at least 15 minutes.

9. Remove the paper strip from the jar. This is your chromatogram.

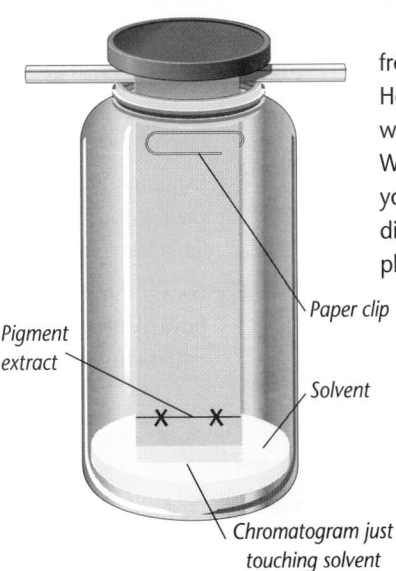

Paper clip

Pigment extract

Solvent

X X

Chromatogram just touching solvent

10. Notice the different bands of color on the chromatogram. Each color band is a different pigment.

What Do You Think?

The solution you tested came from a green plant—spinach. How many different pigments were in the solution you tested? What colors were they? How do you think your results would differ if you used a different plant solution?

---

sen burner. Use a hot plate. Allow to cool. Reheat it several more times. Squash the spinach with a glass rod. Reheat and squash until the alcohol solution becomes a dark green color. Enough pigment is now available for the entire class. Use acetone or petroleum ether as the solvent.

Teaching Strategy

This activity can be performed by groups of three or four. Make sure students adjust the paper length so that the solvent

does not touch the pigment dot. Have students create a data table to record their results. **COOP LEARN** L2

Answers to

What Do You Think?

4 pigments: carotene (orange), xanthophyll (yellow), chlorophyll a (bright green), chlorophyll b (dull green). The colors would likely be different, depending on the plant solution used. Student chromatograms may not clearly show all four pigments.

Going Further ▓▓▓▓▶

Does sunlight have an effect on leaf color? Have students, working in groups, design and carry out an experiment to compare the effects of keeping the leaves of a plant in sunlight and in darkness. Students should include a plan for performing chromatography experiments to compare pigments in the two sets of leaves. Students can record their plans, observations, and results on posters to share with the class. **COOP LEARN** L3

10-2 Stomata

Planning the Activity

Time needed 30 minutes

Purpose To observe the function of stomata.

Process Skills organizing information, making and using tables, observing and inferring, comparing and contrasting, recognizing cause and effect, practicing scientific methods, forming operational definitions

Materials See reduced student text.

Preparation Soak the lettuce in water for up to an hour before the start of class. To prepare the salt solution, dissolve one teaspoon of table salt in 100 mL of tap water.

Teaching the Activity

Safety Remind students to handle the microscope carefully.

Process Reinforcement To reinforce students' skills of making and using tables, review where data and observations will be recorded in the table.

Discussion Ask students whether they can quench their thirst by drinking salt water. Explain that in living tissues, water moves from high-concentration areas to low-concentration areas. Salt water has a lower concentration of water than fresh water because salt has displaced some of the water. Thus, drinking a salt solution tends to draw water from your body's tissues.

Science Journal Suggest that students write a few sentences interpreting their data table in their journals. L1

INVESTIGATE!

Stomata

You learned that stomata are openings through which oxygen, carbon dioxide, and water pass. In this activity, you will learn what stomata look like and how they work.

Problem

How do stomata work?

Materials

lettuce	dish
water	coverslip
microscope	microscope slide
salt solution	forceps
paper towel	pencil

Safety Precautions

Use care handling the microscope.

What To Do

1 Copy the data table *into your Journal.*

2 From a dish of water containing lettuce leaves, choose a lettuce leaf that is stiff from absorbing the water.

3 Bend the leaf back and use the forceps to strip off some of the transparent tissue covering the leaf. This is the epidermis (see photo **A**).

4 Prepare a wet mount of a small section of this tissue (see photo **B**).

5 Examine the specimen under low and then high power of the microscope (see photo **C**). Draw and label the leaf section in your data table.

6 Locate the stomata. Count how many are present and how many are open. Record these numbers in your data table.

Program Resources

Study Guide, p. 39

Laboratory Manual, pp. 59–60, Plant Respiration L2 ; pp. 61–62, Plant Growth L3

Concept Mapping, p. 18, Carbon—Oxygen Cycle L1

Activity Masters, pp. 45–46, Investigate 10–2

Section Focus Transparency 33

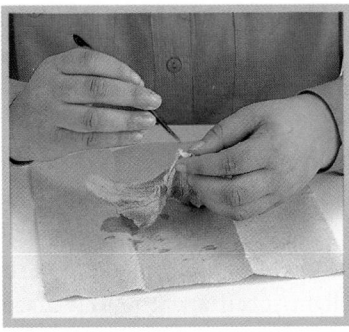

A

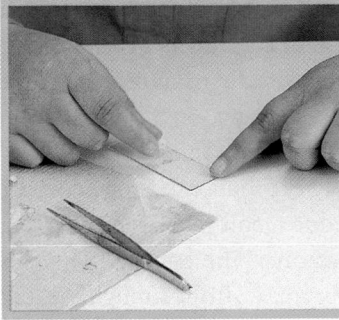

B

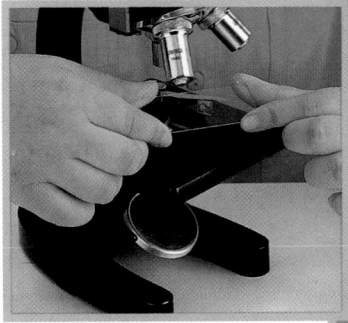

C

7 Place a paper towel at the edge of the coverslip and draw out the water. Using a dropper, add a few drops of salt solution at the edge of the coverslip. The salt solution will spread out beneath the coverslip.

8 Examine the preparation under low and then high power of the microscope. Draw and label the leaf section in your data table.

9 Repeat step 6.

Sample data

Data and Observations		
	Water Mount	Salt Solution
Number of Stomata	Number of stomata will vary.	
Number of Open Stomata	Spacing will vary.	
Drawing of Leaf Section	Stomata open	Stomata closed

Analyzing

1. *Describe* the guard cells around a stoma.

2. How many stomata did you see in each leaf preparation?

3. *Calculate* the percentage of the stomata open in water and in salt water. Which type of water had a higher percentage of open stomata? Which had a lower percentage of open stomata?

Concluding and Applying

4. *Infer* why the lettuce leaf became stiff in water.

5. *Infer* why more stomata were closed in the salt solution.

6. ~~Going Further~~ *Predict* what would happen if you soaked the lettuce in a stronger salt solution. Would more or fewer stomata close?

Expected Outcome

Students learn that stomata open and close based on how much water is contained in plant tissues.

Answers to Analyzing/ Concluding and Applying

1. Guard cells look almost like thick lips.

2. The number of stomata should be about the same.

3. The number of stomata present in the water mount would be approximately 12, in the salt solution approximately 10. The number of stomata open in the water mount could be approximately 10, in the salt solution approximately 2. In the water mount most stomata should be open. In the salt solution all stomata should be closed.

4. Tissues in the leaf absorbed water.

5. Water diffused *from* guard cells *to* salt solution. They lost water, so they closed.

6. Going Further: More concentrated salt solutions may cause more stomata to close.

✔ Assessment

Process Invite students to compare and contrast how other vegetables react to being soaked in plain water. Which vegetables would cooks want to soak in plain water? Why? *Cooks would want to soak vegetables eaten raw, such as carrots, celery, and lettuce to make them crisper; and watery vegetables such as eggplant to make them firmer when cooked.* Use the Performance Task Assessment List for Making Observations and Inferences in **PASC**, p. 17. L1

Meeting Individual Needs

Physically Challenged Standard desks and laboratory tables may be poorly positioned for wheelchair-bound students to use in the Find Out and Investigate activities. Provide low tables or a short length of board that can be clamped across the handles of the wheelchair and serve as a work space.

Meeting Individual Needs

Visually Impaired You can familiarize visually impaired students with the structure of stomata by preparing in advance two enlarged models from clay, showing the guard cells in the closed, flattened position and in the open, turgid position. Touching the models will help students understand how the guard cells control the movement of gases in and out of the stomata.

Use the Check Your Understanding questions to ask students for complete definitions of transpiration and photosynthesis. Extend the Apply question by asking students to give examples of producer-consumer dependence.

Reteach

Tell students that stomata can be compared to their own lungs, which take in oxygen and release carbon dioxide. Ask students how they can prove that their lungs release water vapor. Students may mention seeing their breath on a cold day or breathing mist on a mirror. L1

Extension

Activity Have students pour 25 mL of bromothymol blue into a beaker and note the original color. Then have them place a straw into the solution and blow into it. Students should note the color changes to green. Ask students what they have added to the solution by blowing through the straw. *carbon dioxide, to which the solution reacts by turning green* Have students add *Elodea* plants to the beaker and place under a bright light for 24 hours. Record the color change after this period of time. Ask students to explain their observations. *Students should infer that photosynthesis in the plants removed carbon dioxide from the solutions, changing the color from green back to the blue.* L3

4 CLOSE

Discussion

Ask students to imagine that they want to grow plants on the moon. Have them brainstorm a list of the basic necessities and explain the purpose of each. L1

Photosynthesis

You know that animals are consumers. Almost all consumers depend on plants, either directly or indirectly, for food. Almost all plants, on the other hand, do not depend on other organisms for food. They produce their own.

Plants produce food in a series of chemical reactions. The energy for these reactions comes from sunlight. **Photosynthesis** is the process in which plants use light to produce food. During photosynthesis, plants use sunlight to change water and carbon dioxide into sugar and oxygen.

Chlorophyll is a green pigment in plants that traps the light from the sun. Plants are green because chlorophyll absorbs the blue, violet, and red parts of the light spectrum. It reflects green light, so plants appear green to us.

Let's look at what happens to the products formed during photosynthesis. Some of the sugars formed are used by the plant for its own life processes, such as growth. Some sugar is stored. When you eat carrots or potatoes, you are eating stored food.

Some of the oxygen that forms during photosynthesis is used by the plant itself. Some oxygen passes out of the leaves through the stomata. This oxygen may eventually be used by other organisms for respiration. The carbon dioxide produced by other organisms during respiration is used by plants to make food during photosynthesis. In this way, photosynthesis and respiration are linked in a never-ending cycle that provides energy to all living things.

Figure 10-13

The processes of photosynthesis and respiration are linked. Some of the end products of each process are the starting products for the other process.

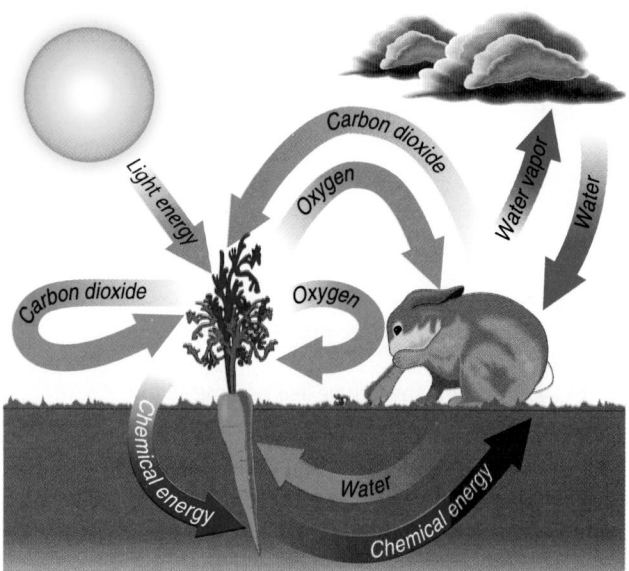

check your UNDERSTANDING

1. What role do stomata play in gas exchange in plants?
2. What are the starting and end products of photosynthesis and respiration?
3. What is the role of the green pigment chlorophyll in photosynthesis?
4. **Apply** How is all life on Earth dependent on sunlight?

check your UNDERSTANDING

1. Carbon dioxide, oxygen, and water vapor pass in and out of a leaf through stomata.
2. Photosynthesis uses carbon dioxide, water, and light energy to form sugar and oxygen. In respiration, organisms break down sugar and oxygen to water and carbon dioxide, and release energy.
3. Chlorophyll traps light energy from the sun, which is needed to help change water and carbon dioxide into sugar and oxygen.
4. Plants require light for photosynthesis; animals, as consumers, depend on plants either directly or indirectly for food.

Science and Society

Green Plants vs. Industrial Growth

Like animals, plants can be harmed by the actions of humans. Laws are sometimes used to protect endangered species—animals or plants whose existence is threatened. In what ways are plants threatened, and by whom?

Most people have heard about the threat to tropical rain forests. In Central and South American countries, these diverse forests are being cleared to create new agricultural land for growing populations and to provide lumber. Many South American people depend on these cleared lands for growing crops, even though the soil is not very rich.

Global Warming

Scientists hypothesize that continued destruction of the rain forests will contribute to a condition known as global warming.

This map shows surface temperatures on Earth. The warmer areas are shown in red, while the cooler areas appear in blue.

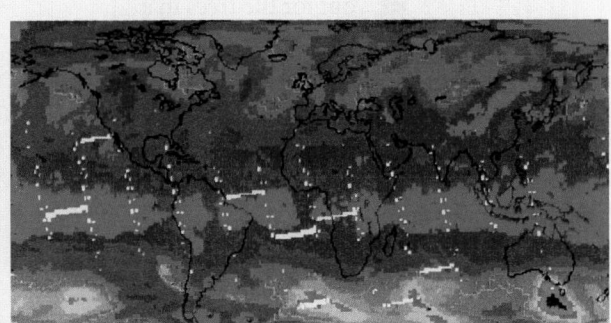

Global warming is a rise in average atmospheric temperatures. When certain gases, like carbon dioxide, build up in Earth's atmosphere, they trap heat that normally escapes through the atmosphere. This raises temperatures. Carbon dioxide is produced when the trees are burned as people clear the rain forests. Adding to this is the loss of the trees that use carbon dioxide in the process of photosynthesis. Earth's plants play a major role in regulating carbon dioxide levels in the atmosphere.

Untapped Resources

The loss of rain forests and other forested areas is harmful in other ways, too. You may wonder why it would matter to us if some obscure plant species ceased to exist. So far, only five percent of the plant species in the world have been analyzed for their potential as medicines. Plants that hold cures for diseases could disappear before we ever get a

Science and Society

Purpose
To survive, plants must be able to carry on the processes discussed in Section 10-4. This excursion focuses on how human activity in industrialized areas interferes with these processes and affects plant survival.

Content Background
Acid rain is rain of higher than normal acidity formed when burning coal, oil, and gas, release sulfur and nitrogen oxides into the air. These compounds combine with water vapor to produce sulfuric acid and nitric acid. They are then carried in wind and deposited in rain and snow. Acid rain, which is most severe in southeastern Canada, the northeastern U.S., and western Europe, has been associated with the death of lake fish and forest trees, changes in soil chemistry, and deterioration of building exteriors. But due to the difficulty of isolating variables in nature, it is not clear to what extent acid rain is responsible for these effects. Many government and industry leaders want stronger evidence before passing laws that will require spending hundreds of millions of dollars on pollution control.

Teaching Strategy Ask students to name the ways human activity threatens plant species and the impact that the destruction of plant life has on humans.

Demonstration

Construct a simple greenhouse from a cardboard box with a sheet of glass over the top. Cut a hole in the side of the cardboard large enough to let one hand enter the box, and seal it with paper and masking tape. Set the box in sunlight for 30 minutes. Then remove the covering from the hole and have students put their hands inside. They should find that the interior of the box has grown very warm. Explain that the glass acts as a heat trap that lets the light and heat of the sun into the box but does not let the heat leave. Tell students that this is the basis for the greenhouse effect, which a majority of scientists believe is warming the Earth. Air pollutants, including methane and carbon dioxide, trap and reflect the heat of the sun in our atmosphere and are expected to boost average global temperatures 2° to 5C° over the next 50 years, with consequences that are very difficult to predict. You should, in fairness to the scientific community, also point out that not all scientists agree with this hypothesis.

Answers to
You Try It!

Trees are available to schools free of charge from the Forestry Service.

chance to find them. Rain forests are particularly important because they contain nearly half of the world's species of plants and animals. This diversity is important in maintaining Earth's fragile ecological balance.

Threats from Industry

Wild plant species face additional threats in industrialized parts of the world. Here, threats come from the introduction of exotic species, diseases, insect pests, and water and air pollution. Environmental changes that affect plants usually affect all living creatures. Some plants will not grow where air pollution exists. If the air is not good enough for them, is it good enough for us?

Before the Industrial Revolution in England, most peppered moths were white. Their white color enabled them to blend in with the lichen-covered bark of the trees. As the Industrial Revolution came into full swing, soot killed the lichens and blackened the trees. Then black peppered moths became dominant. After Britain passed laws to clean

the air, lichens returned, and white peppered moths again became common. The lichen served as an indicator of environmental health. An environmental change sufficient to cause the black peppered moth to be more common must have been both widespread and long-lasting.

Environmental Monitors

Plants are in touch with their environment and are an indicator of the quality of our environment. In fact some people think plants can feel pain and pleasure the way animals can. No scientific evidence exists for this. But plants do send and receive electrical signals within their tissues. Scientists have attached electrodes to leaves and shown that changes in the quality or amount of air, light, or water in a plant's environment make differences within the plant's voltages.

An **urban forester** is someone who looks out for the trees in a city. Fighting tree diseases such as oak wilt and educating the public about them are part of the job. An urban forester studies botany and chemistry and attends forestry school. Local, state, and federal governments hire foresters.

Going Further ⅢⅢⅢ▶

Divide students into three groups for a debate on acid rain. Have one group represent utilities and heavy industries, which generate 37 percent of all air pollution and are alleged to be the major sources of acid rain. Have a second group represent American environmentalists who want to force these companies to spend hundreds of millions of dollars to reduce emissions that they believe cause acid rain. Have a third group represent Canadian fishers who believe that acid rain

produced in the United States is polluting Canadian rivers and lakes and killing fish.

Work with each group to develop scientific, economic, and political arguments for their positions. After the groups have completed their research, moderate a debate in which each group appoints two representatives to present its arguments and have one chance to rebut the other teams' arguments. Use the Performance Task Assessment List for Investigating an Issue Controversy in **PASC**, p. 65. **L2** **COOP LEARN**

HOW IT WORKS

The Soil-Less Garden

Growing plants in nutrient solutions in tanks instead of in soil is called hydroponics. It comes from Latin words that mean to work the water. Hydroponics was first developed in California in 1929, by Dr. W. F. Gerische.

Growing plants in greenhouses in nutrient solutions has advantages. If soil is unsuitable, or at a premium in an area, or if there are diseases in the soil, hydroponics is a good alternative. It works with a wide variety of plants, including tomatoes, lettuce, and carnations.

Variations on a Theme

Two methods of hydroponic gardening are commonly used. In the original method, plants grow in a shallow, watertight container that holds the nutrient solution. Wood fiber, peat, or some other growing medium is supported by a wire framework a few centimeters above the surface of the liquid. Seedlings are set into the growing medium with their roots in the nutrients held in the shallow container. The nutrient solution must have air circulated through it, and must have its pH checked regularly. The solution needs to be completely changed every 10 to 14 days.

A more popular method uses a layer of sand or gravel along with the nutrient solution. The sand or gravel, which is the growing medium, is held in individual pots or in rectangular containers. The nutrient solution may be fed from the top or pumped up from the bottom and allowed to drain back for reuse.

One difficulty with hydroponics is that nutrient solutions must be checked and adjusted daily. The nutrients used are essentially the same ones found in fertile soil, or in fertilizers. However, great care must be taken that the nutrients remain in the proper concentration and are aerated so the plants can take them in.

Inert compound

Plastic mesh

Nutrient solution

What Do You Think?

Do you think vegetables grown in a hydroponic garden would taste any different from those grown in soil? What plants might not be suited to hydroponic growing?

Going Further ⑄⑄⑄➤
Divide the class into groups to construct simple hydroponic gardens for the classroom. Have each group select a different plant to grow. Begin by having students research hydroponics.

Bourgeois, Paulette. *The Amazing Dirt Book.* Reading, MA: Addison-Wesley, 1990.

Bridwell, Raymond, *Hydroponic Gardening.* Santa Barbara, CA: Woodbridge Press, 1990.

Follow the plans in these books or the illustration shown here to build a hydroponic structure in a plastic container. Obtain seedlings of the plants and change the solution every two weeks.

COOP LEARN | **L1** | **LEP**

HOW IT WORKS

Purpose

In Section 10-1, the typical plant is described as one with roots that anchor it to the soil and absorb nutrients from the soil. This excursion describes how certain plants may grow without soil, using a method known as hydroponics.

Content Background

One of the first practical uses of hydroponics was during World War II, when the United States had military bases on Pacific islands that lacked fertile soil or suffered from soil-borne plant diseases. When these islands could not be regularly supplied with fresh vegetables, the soldiers used hydroponics to grow their own. In the future, hydroponics may find a new use in space. For space stations, long-distance spacecraft, and colonies on other planets, hydroponics will offer a means to grow plants that will provide both air and food for human beings.

Teaching Strategy Ask students to identify the labeled parts of the hydroponic system and explain what purposes they serve: *the inert compound supports the plant, the plastic mesh supports the inert compound, and the nutrient solution provides nutrients and water for the plant.* Then ask students to name the advantages and disadvantages of hydroponics. *Can grow plants in infertile places; need to control nutrient solution.*

Answers to

What Do You Think?

Plants lacking extensive root systems, such as cacti or mosses, might not be suitable.

HISTORY CONNECTION

Purpose
This History Connection discusses ancient and current uses of medicinal plants.

Content Background
For thousands of years, garlic has been eaten to guard against plagues, respiratory problems, colds, inflammations in the mouth, ear infections, and high blood pressure. Herbalists in ancient Rome used marjoram to make people feel calm, relieve bruises, and treat eye diseases. Both Chinese and Native-American traditional medicine use ginseng root to treat senility, diabetes, mental illness, anemia, fever, headache, and stomach ailments. Eucalyptus is used in commercial medicines to clear mucus from the nose and lungs and, in traditional practice, eucalyptus oil is considered an antiseptic, astringent, and stimulant. Chlorophyll or its derivatives are added to gum and mints. Chewing fresh parsley helps cure bad breath.

Teaching Strategy
Ask students to suggest how the development of Linnaeus's scientific classifications has helped in the use of medicinal plants.

Purpose
This literature connection expands the chapter concepts through the story "Sunkissed: An Indian Legend."

Content Background
Some Native Americans did not have a written language, although they sometimes recorded legends in pictographs. To pass on beliefs and values necessary for the survival of the tribe, peo-

---

HISTORY CONNECTION
Ancient Medicinal Plants

Around 4000 years ago, a Chinese emperor put together a book that described more than 300 medicinal plants. Early Sumerians and Egyptians used plants for healing. Later the Greeks and Romans also provided additional information about medicinal plants.

During the Middle Ages, monks in Europe studied and translated ancient texts about healing herbs. Every monastery had a Physick Garden for growing the herbs. Such gardens later became common at castles, courts, and hospitals.

By the 13th century, there was a system of classifying plants. New books were written giving herbal prescriptions for illnesses. Eventually, scientists learned to isolate the healing ingredient and to make more of it.

Early forms of drugs, such as aspirin and insulin, can be traced to medicinal plants used by Native Americans. In the 1940s and 1950s, chemist Percy Julian, shown in the photo, developed drugs from chemicals in soybeans. His synthetic cortisone is used by arthritis sufferers today.

What Do You Think?
If you were a scientist, how much attention would you pay to ancient beliefs?

Literature Connection
Sunkissed Flowers

You learned that plants need sun in order to live and grow. "Sunkissed: An Indian Legend," as told by Alberto and Patricia de La Fuente (Peña, Sylvia Cavazos, ed. TUN-TA-CA-TUN), illustrates the importance of the sun to all living things—especially to the survival of plants.

A legend is a story that has been passed along from earlier times. It may be partly true or simply an imaginative way of explaining some of the things that took place in nature. Obtain a copy of "Sunkissed: An Indian Legend" and read its explanation of how one kind of flower, the Margarita, was changed for all time by a special kiss from the sun.

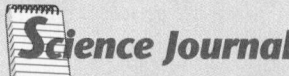

Science Journal
In your Science Journal write a short legend that tells why a certain kind of flower is like it is today. Just choose a flower and let your imagination take over!

---

ple learned the tribal legends and communicated them orally to the following generation.

Some Native Americans believe that each creature in nature has a power by which it maintains itself and affects others. Each tribe has a different name for this power. Many tribes also recognize a primary power that is the source of all life. In this legend, that power is the sun, called "Tonatiuh," the King of Light.

Teaching Strategy
After students read the legend on their own, have the class work together

to act it out. You may wish to videotape it to watch later.

Science Journal
Accept all logical stories. Students should use correct content regarding plants in their legends.

Science Journal

Review the statements below about the big ideas presented in this chapter, and answer the questions. Then, re-read your answers to the Did You Ever Wonder questions at the beginning of the chapter. *In your Science Journal*, write a paragraph about how your understanding of the big ideas in the chapter has changed.

1 Plants are either vascular or nonvascular. Nonvascular plants and some vascular plants, such as the fern, reproduce by spores. Most vascular plants, such as the poppy, reproduce by seeds. *How do the plants in the photographs reproduce?*

2 Cones are the reproductive organs of gymnosperms. Flowers are the reproductive organs of angiosperms. *What are the parts of the flower shown here?*

3 Transpiration, respiration, and photosynthesis are processes carried on by all plants. Photosynthesis and respiration are plant processes that provide the food and energy needed for plant growth. Transpiration is water loss from the leaves of plants. *How are photosynthesis, respiration, and transpiration related?*

Science Journal

Did you ever wonder...

• Chlorophyll is a green pigment in plants. Plants are green because chlorophyll reflects green light. Plants produce little or no chlorophyll in cold weather, making other pigments in the leaves visible. (pp. 332, 336)

• All water and minerals used by plants enter through the roots. Roots also anchor plants in the soil. (p. 313)

• Insects are attracted by the flowers' color, scent, and nectar. (p. 326)

Science at Home

Have students obtain seeds for an herb or spice that would be useful for cooking at home. Have them plant two herb gardens in large, flat pans or pots according to the directions on the seed package. Students can test the reaction of the herbs to light by placing one garden in direct sunlight and the other in shade and then observing the differences in growth. **L1**

chapter 10
REVIEWING MAIN IDEAS

Ask students to read the page and study the three figures. Then have them review their knowledge of plants by developing a table and by drawing detailed, scientific illustrations of vascular and nonvascular plants and of angiosperms and gymnosperms. **LEP**

Teaching Strategies

• Have students create a table with the two main heads *Vascular* and *Non-vascular* and two subheads under *Vascular*, *spore-producing* and *seed-producing* plants. Ask students to provide an example for each category.

• Organize students into groups of 3 or 4. Have them make cutaway drawings of the reproductive organs of an angiosperm or gymnosperm. Students should label all of the important structures, from roots to reproductive organs. Have students come up with some statements that show they understand the differences between gymnosperms and angiosperms. **COOP LEARN**

Answers to Questions

1. Liverworts reproduce by spores. Flowering plants reproduce by seeds.

2. Pistil (stigma, style, ovary), stamen (anther, filament), sepal, petal.

3. Photosynthesis and respiration are complementary processes—the products of one provide the reactants for the other. As plants carry out respiration, they produce water, which leaves the plant by transpiration.

GLENCOE TECHNOLOGY

 **MindJogger Videoquiz**

Chapter 10 Have students work in groups as they play the Videoquiz game to review key chapter concepts.

Using Key Science Terms

1. xylem
2. nonvascular plant
3. pollination
4. chlorophyll
5. photosynthesis
6. transpiration
7. vascular plant
8. phloem

Understanding Ideas

1. The palisade layer made most of the food.

2. Nonvascular plants would be found living in moist areas because these plants can't transport water within their bodies.

3. Angiosperms are plants in which the seed is enclosed and protected inside a fruit. Gymnosperms produce unprotected seeds on cones.

4. Pollen may be transferred by wind, insects, birds, or other animals.

5. Some of the glucose, which is produced during photosynthesis, is used by the plant for its own life processes and some of it is stored. Some of the oxygen that forms passes out of the leaves through the stomata.

Developing Skills

1. See reduced student page.

2. Students should be able to identify the leaves and root or stem of the young plant and recognize that the rest of the seed is food for the plant.

3. Students should observe more water droplets in the bag in the sun and infer that plants lose more water on a sunny day.

Critical Thinking

1. Pollen grains from the male cones stick in the fluid produced by the female cones.

2. Spore-producing plants grow only after eggs and sperm (which are produced by sex-cell-producing plants) unite.

Using Key Science Terms

chlorophyll   pollination
nonvascular plant   transpiration
phloem   vascular plant
photosynthesis   xylem

Each of the following sentences is false. Make the sentence true by replacing the italicized word with a word from the list above.

1. *Phloem* is made up of vessels that transport water and minerals throughout a plant.

2. A *vascular* plant lacks xylem and phloem.

3. The process by which pollen grains move from the stamen to the ovules is *transpiration.*

4. *Xylem* is the green pigment in plants that traps light.

5. The process in which plants use light to produce food is *transpiration.*

6. The loss of water vapor through the stomata of a leaf is called *pollination.*

7. A *nonvascular* plant has xylem and phloem.

8. *Chlorophyll* are vessels that transport food throughout a plant.

Understanding Ideas

Answer the following questions in your Journal using complete sentences.

1. Why is the palisade layer in leaves important?

2. Where might you find nonvascular plants living? Why?

3. Explain the difference between an angiosperm and a gymnosperm.

4. List some ways pollen may be transferred.

5. What happens to the products formed during photosynthesis?

Developing Skills

Use your understanding of the concepts developed in this chapter to answer each of the following questions.

1. **Concept Mapping** Complete the concept map of the life cycle of a moss.

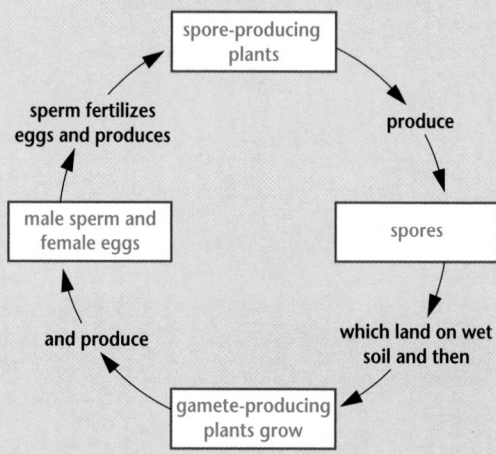

2. **Observing** Repeat the Explore activity on page 320, using a lima bean and a corn seed. Identify the young plant, the stored food, and the seed coat in each seed.

3. **Observing and Inferring** After doing the Explore activity on page 331, observe how sunlight affects transpiration. Water two potted seedlings and cover and seal each of them in a plastic bag. Put one plant in the sunlight and a box over the other plant. After two days compare the amount of water inside the bags. How does sunlight affect transpiration?

3. the waxy cuticle; the stomata; the epidermis.

4. Color would not attract animals that are active at night. A strong scent guides animals to the plant.

5. Birds with long, narrow beaks, such as hummingbirds, feed from flowers. The beak's shape is an adaptation that enables it to reach nectar.

6. The surface area of the plant with many large leaves is greater and includes more stomata. This plant will lose more water through transpiration.

Program Resources

Review and Assessment, pp. 59–64 L1
Performance Assessment, Ch. 10 L2
PASC
Alternate Assessment in the Science Classroom
Computer Test Bank L1

Critical Thinking

In your Journal, *answer each of the following questions.*

1. What is the function of the sticky fluid that is produced on pine cones?
2. In nonvascular plants, why are the spore-producing plants dependent on the gamete-producing plants?
3. What two features of a leaf help prevent water loss?
4. Why do flowers that are pollinated at night often have a strong scent?
5. Some birds pollinate when they feed on nectar from flowers. Look at the birds shown in the picture. Which one probably feeds on nectar? How can you tell?

6. Which plant do you think loses more water through transpiration, a plant with few, small, leaves, or one with many, larger, leaves? Explain your answer.
7. Describe how the structure of seeds helps plants reproduce.

Problem Solving

Read the following problem and discuss your answers in a brief paragraph.

Dawn went with her grandmother to buy seeds to plant in the garden. She noticed that the package of zinnia seeds she wanted to buy said that the seeds were 95 percent viable. Her grandmother explained that viable meant living. She said that 95 out of 100 seeds would sprout and grow.

They purchased the seeds and headed home to plant them around the border of their garden.

How could Dawn and her grandmother find out if the seeds they planted were viable? List the steps you would take to determine if a seed is viable.

CONNECTING IDEAS

Discuss each of the following in a brief paragraph.

1. **Theme—Scale and Structure and Systems and Interactions** What are some traits and processes that plants share with animals?

2. **Theme—Systems and Interactions** Why is photosynthesis important for maintaining Earth's present atmosphere?

3. **Theme—Systems and Interactions** List three ways plants affect your life.

4. **How It Works** Compare and contrast two methods of hydroponic gardening.

5. **Science and Society** Why is the preservation of the Amazon rain forest so important?

✔ Assessment

Portfolio Review the portfolio options that are provided throughout the chapter. Encourage students to select one product that demonstrates their best work for the chapter. Have students explain what they learned and why they chose this example for placement in their portfolios.

Additional portfolio options can be found in the following **Teacher Classroom Resources:**

7. Seeds contain the embryo or young future plant as well as stored food for the developing embryo.

Problem Solving

If the seeds Dawn and her grandmother planted sprout and grow, they are viable. Dawn could soak 100 seeds and count the number that sprout. If 95 or more sprout, the statement on the seed package is true.

Connecting Ideas

1. Both plants and animals have sexual and asexual methods of reproduction. They both use respiration to release energy from food.

2. Photosynthesis uses carbon dioxide from the atmosphere to produce food. In the process, it releases large amounts of oxygen that other living things use.

3. Possible answers: many homes are heated with fossil fuels, you eat plants or plant parts; your oxygen is a product of photosynthesis.

4. One method of hydroponic gardening is the original method in which plants are grown in a shallow, watertight container containing a nutrient solution, and seedlings are set with roots in the solution which must have air circulated through it. The solution needs to be changed every 10 to 14 days. Another more popular method uses a layer of sand or gravel with the nutrient solution. The sand or gravel, which is the growing medium is in individual pots or rectangular containers. The nutrient solution is fed from the top or pumped from the bottom and allowed to drain for reuse.

5. The preservation of the Amazon forest is important to prevent the disappearance of plant species, global warming, and air pollution.

Chapter Organizer

SECTION	OBJECTIVES	ACTIVITIES & FEATURES
Chapter Opener		**Explore!**, p. 345
11-1 What Is an Ecosystem? (2 sessions; 1 block)	**1. Distinguish** between populations and communities. **2. Distinguish** between habitats and niches. **3. Describe** the structure of an ecosystem. **National Content Standards: (5–8) UCP1–2, UCP5, C4, G1–2**	**Explore!**, p. 346 **Find Out!**, p. 349
11-2 Organisms in Their Environments (3 sessions; 1.5 blocks)	**1. Describe** a food chain and its relationship to a food web. **2. Explain** how natural cycles are important in the environment. **National Content Standards: (5–8) UCP1–2, A1, B3, C4, F2**	**Explore!**, p. 352 **Investigate 11-1:** pp. 354–355 **Skillbuilder:** p. 357 **Chemistry Connection,** pp. 358–359 **Science and Society,** pp. 372–373
11-3 How Limiting Factors Affect Organisms (3 sessions; 1 block)	**1. Identify** some limiting factors. **2. Describe** adaptations of organisms to limiting factors. **National Content Standards: (5–8) UCP3–5, A1–2, C1, C4–5, F2, F4–5, G1–2**	**Design Your Own Investigation 11-2:** pp. 362–363 **A Closer Look,** pp. 368–369 **Teens in Science,** p. 371

ACTIVITY MATERIALS

EXPLORE!

p. 345 No special materials are required.

p. 346* No special materials are required.

p. 352* No special materials are required.

INVESTIGATE!

pp. 354–355* water, bowl, forceps, glass slide, coverslip, light microscope, magnifying glass, owl pellet, cardboard, glue

DESIGN YOUR OWN INVESTIGATION

pp. 362–363* small paper cups, water, hand lens, plastic wrap, labels, cotton swabs, mold source, sugarless dry cereal, dry macaroni, dry potato flakes, dry crackers without salt

FIND OUT!

p. 349* ruler, white drawing paper or tracing paper, crayons or markers

KEY TO TEACHING STRATEGIES

The following designations will help you decide which activities are appropriate for your students.

L1 Basic activities for all students

L2 Activities for average to above-average students

L3 Challenging activities for above-average students

LEP Limited English Proficiency activities

COOP LEARN Cooperative Learning activities for small group work

P Student products that can be placed into a best-work portfolio

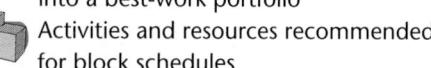 Activities and resources recommended for block schedules

Need Materials? Call Science Kit (1-800-828-7777).

00:00 OUT OF TIME? We recommend that students do the activities with an asterisk.

Chapter 11 Ecology

TEACHER CLASSROOM RESOURCES

Student Masters	Transparencies
Study Guide, p. 40 **Concept Mapping,** p. 19 **Take Home Activities,** p. 19 **Multicultural Connections,** pp. 25, 26 **Making Connections: Across the Curriculum,** p. 25 **Making Connections: Technology and Society,** p. 25 **Science Discovery Activities,** 11-1	**Teaching Transparency 22,** Organisms and Their Environment **Section Focus Transparency 34**
Study Guide, p. 41 **Critical Thinking/Problem Solving,** pp. 15, 24 **Activity Masters,** Investigate 11-1, pp. 47–48 **Science Discovery Activities,** 11-2	**Teaching Transparency 21,** Cycles **Section Focus Transparency 35**
Study Guide, p. 42 **Making Connections: Integrating Sciences,** p. 25 **Activity Masters,** Design Your Own Investigation 11-2, pp. 49–50 **Science Discovery Activities,** 11-3 **Laboratory Manual,** pp. 63–66, Physical Conditions and Behavior **Laboratory Manual,** pp. 67–70, Living Space	**Section Focus Transparency 36**

ASSESSMENT RESOURCES	TEACHING & TECHNOLOGY
Review and Assessment, pp. 65–70 **Performance Assessment,** Ch. 11 **PASC*** **MindJogger Videoquiz** **Alternate Assessment in the Science Classroom** **Computer Test Bank**	**Spanish Resources** **Cooperative Learning Resource Guide** **Lab and Safety Skills** **Science Interactions, Course 1, CD-ROM** **Computer Competency Activities**

*Performance Assessment in the Science Classroom

NATIONAL GEOGRAPHIC TEACHER'S CORNER

Index to National Geographic Magazine

The following articles may be used for research relating to this chapter:

• *Water: The Power, Promise, and Turmoil of North America's Fresh Water,* A Special Edition, November 1993.

National Geographic Society Products Available From Glencoe

To order the following products for use with this chapter, contact your local Glencoe sales representative or call Glencoe at 1-800-334-7344:

• *STV: Water* (Videodisc)
• *STV: Habitats* (Videodisc)

Additional National Geographic Society Products

To order the following products for use with this chapter, call the National Geographic Society at 1-800-368-2728:

• *The Curious Naturalist* (Book)
• *A Swamp Ecosystem* (Video)
• *An Ecosystem: A Struggle for Survival* (Video)
• *Life in the Sea Series,* "Web of Life" (Video)
• *Old-Growth Forest: An Ecosystem* (Video)
• *Pond-Life Food Web* (Video)

Teacher Classroom Resources

These are key components of the classroom resources package.

TEACHING AIDS

Section Focus Transparencies

Teaching Transparencies

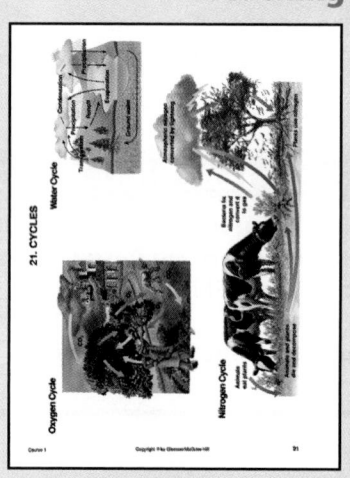

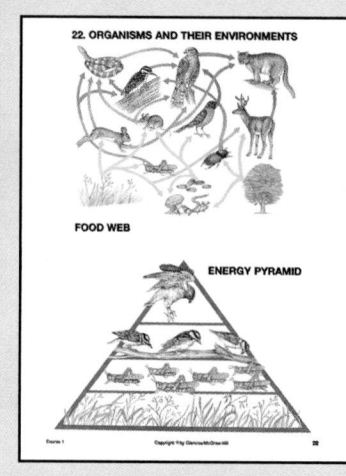

HANDS-ON LEARNING

Science Discovery Activity*

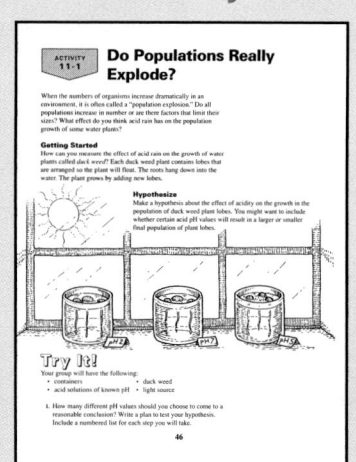

Laboratory Manual*

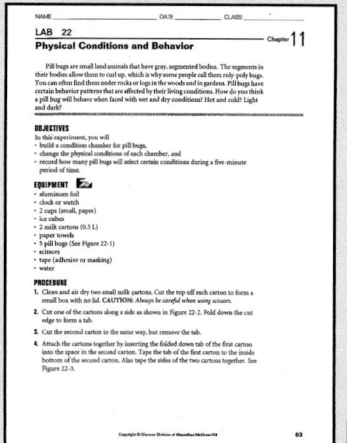

Take Home Activity

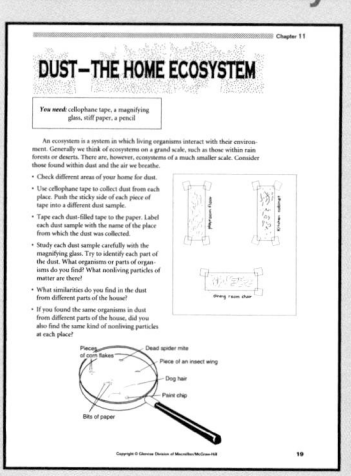

*There may be more than one activity for this chapter.

Chapter 11 Ecology

REVIEW AND REINFORCEMENT

Study Guide*

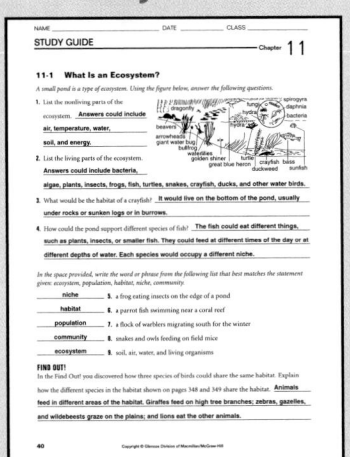

Concept Mapping

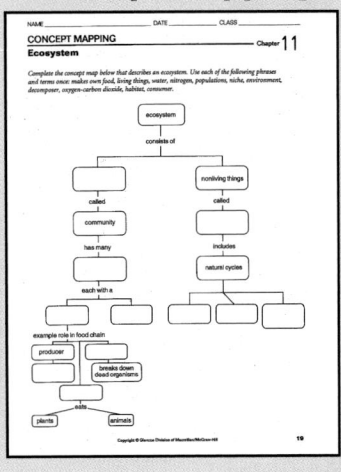

Critical Thinking/ Problem Solving

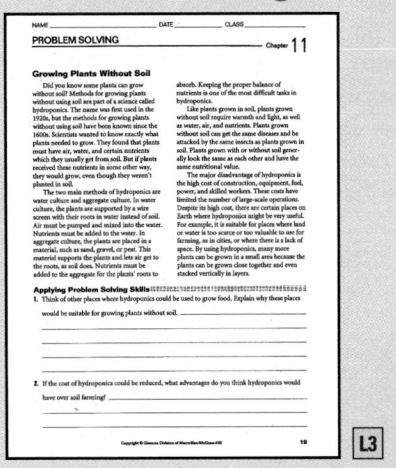

ENRICHMENT AND APPLICATION

Integrating Sciences

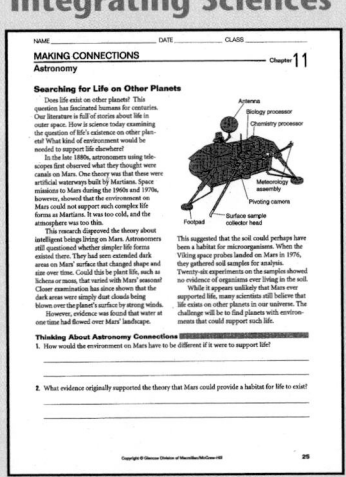

Across the Curriculum

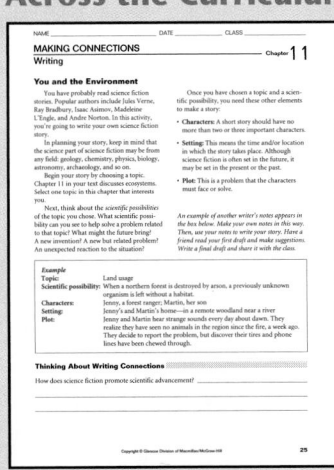

Technology and Society

Technology and Society content

ASSESSMENT

Multicultural Connection**

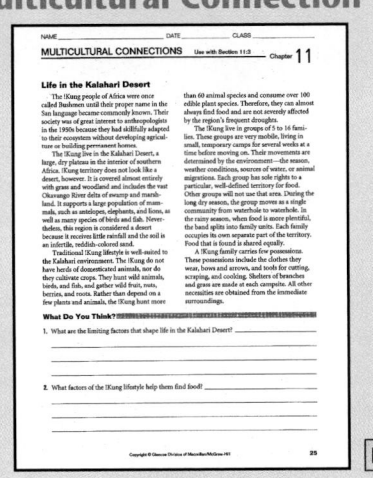

Performance Assessment

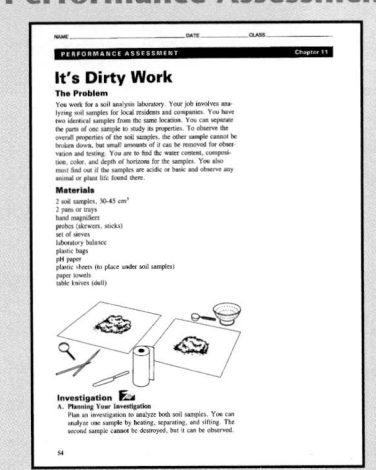

Review and Assessment

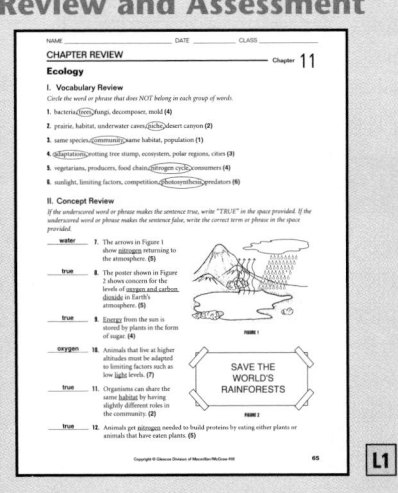

CHAPTER 11

ECOLOGY

Ecology

THEME DEVELOPMENT

In an ecosystem, organisms interact with members of their own population and with other members of their community, as well as with the nonliving aspects of the environment. Natural cycles provide an example of systems that maintain a balance of resources in the environment. The theme of energy is evident in the food web in an ecosystem.

CHAPTER OVERVIEW

In this chapter, students will discover how plants and animals interact with each other as well as with nonliving components in an ecosystem. Energy flows from producers to consumers to decomposers. This flow of energy is illustrated as simple food chains and then as more complex food webs.

Students also will explore limiting factors.

Tying to Previous Knowledge

Ask students what they ate for dinner last night. Discuss where these foods come from. Which were producers? (Chapter 10) Which were consumers? (Chapter 9)

INTRODUCING THE CHAPTER

Have students examine the photographs. Ask them to find examples of producers and consumers. Ask students how the animals and plants benefit by living together.

OUT OF TIME?

If time does not permit teaching the entire chapter, use the Chapter Overview on this page, Reviewing Main Ideas at the end of the chapter, and the Chapter 11 audiocassette to point out the main ideas of the chapter.

Did you ever wonder...

✓ **What's going on when bread gets moldy?**

✓ **Why our oxygen supply doesn't get used up?**

✓ **Why penguins don't live in your neighborhood?**

Science Journal

Before you begin to study about ecology, think about these questions and answer them *in your Science Journal.* When you finish the chapter, compare your journal write-up with what you have learned.

344

Have you been to the zoo recently? If so, you know that modern zoos keep animals in areas that resemble their natural surroundings. Just as it would in nature, the zoo giraffe finds its food in tall trees that grow in a grassland, while zebras and wildebeests graze on the grass.

▶ *How can different types of living things survive in the same area? How do these living things interact with the nonliving world around them? Start exploring the interaction of living things with the activity on the next page.*

Learning Styles	Kinesthetic	Investigate. pp. 354–355, 362–363; Chemistry Connection, pp. 358–359
	Visual-Spatial	Explore, p. 345; Discussion, pp. 346, 353; Visual Learning, pp. 348, 356, 360, 366; Find Out, p. 349; Activity, pp. 351, 357, 358, 359, 360; Demonstration, p. 358
	Interpersonal	Multicultural Perspectives, p. 357
	Logical-Mathematical	Across the Curriculum, p. 350; Using Math, p. 349; Activity, p. 367
	Linguistic	Explore, pp. 346, 352; Visual Learning, pp. 350, 353, 369, 370; Multicultural Perspectives, pp. 350, 353, 367; Activity, pp. 353, 356; Across the Curriculum, pp. 356, 365; Discussion, pp. 361, 365, 368, 370; Research, p. 364; A Closer Look, pp. 368–369; Science at Home, p. 374

LS

Explore! ACTIVITY

What would you discover on a neighborhood safari?

What animals and plants live and grow in your neighborhood? How might these living things interact? Go on a neighborhood safari or expedition with two friends and find out.

What To Do

1. Choose a small area near your school to study local plants and animals. Try to find out where they live, how they get what they need to live, and how they are influenced by the nonliving parts of their surroundings.

2. *In your Journal,* write your observations or make drawings of what you see.

3. Compare your findings with those of other groups.

ASSESSMENT PLANNER

PORTFOLIO
Refer to p. 376 for suggested items that students might select for their portfolios.

PERFORMANCE ASSESSMENT
Process, pp. 346, 352, 363
Skillbuilder, p. 357
Explore! Activities, pp. 345, 346, 352
Find Out! Activity, p. 349
Investigate, pp. 354–355, 362–363

CONTENT ASSESSMENT
Oral, p. 345
Check Your Understanding pp. 351, 360, 370
Reviewing Main Ideas, p. 374
Chapter Review, pp. 375–376

GROUP ASSESSMENT
Opportunities for group assessment occur with Cooperative Learning Strategies.

Uncovering Preconceptions

Students may believe that any changes to the environment are destructive to all living things. Some human-made and natural disasters may be beneficial to some organisms. Ask students if they can think of any organisms that might benefit from the clearing of a forest. *For example, decomposers (page 353) benefit from the death of animals, while some animals are better adapted to open fields.*

Explore!

What would you discover on a neighborhood safari?

Time needed 60 to 120 minutes

Thinking Processes observing, comparing and contrasting, practicing scientific methods, classifying

Purpose To observe interactions in an ecosystem.

Preparation This activity will be most satisfying to students if conducted at a time of the year when animals and plants are abundant.

Teaching the Activity

Troubleshooting Help students choose ecosystems with a variety of easily identifiable organisms.

Science Journal Have each group member record the group's observations. **L1**

Expected Outcomes

Students should be able to identify several populations of animals and plants as well as several nonliving parts in their environment.

✔ Assessment

Oral Have students relate how living things interacted in the area they observed. Use the Performance Task Assessment List for Making Observations and Inferences in **PASC**, p. 17. **L1**

PREPARATION

Planning the Lesson

Refer to the Chapter Organizer on pages 344A–D.

Concepts Developed

This section teaches students that living and nonliving parts of Earth interact. The knowledge they acquired when studying the living world in Chapter 7, animal species in Chapter 9, and the plant kingdom in Chapter 10 is integrated into the study of population and communities and the ways in which they influence each other.

1 MOTIVATE

Bellringer

Before presenting the lesson, display **Section Focus Transparency 34** on an overhead projector. Assign the accompanying **Focus Activity** worksheet. [L1]
[LEP]

Discussion Display pictures of animal and plant populations on the bulletin board. Ask students to describe how different populations interact to make a community. Then point out that communities interact with nonliving things to form ecosystems. [L1]

Explore!

In how many different surroundings do you carry out your everyday activities?

Time needed 20 to 25 minutes

Thinking Processes observing and inferring, recognizing cause and effect, classifying, predicting

Purpose To observe and infer how one interacts with one's surroundings.

Section Objectives

- Distinguish between populations and communities.
- Distinguish between habitats and niches.
- Describe the structure of an ecosystem.

Key Terms

habitat, population, community, niche, ecosystem

Living Things and Their Natural Surroundings

You may sometimes move through your daily activities without thinking about your surroundings. In school, you spend much of your day in rooms interacting with classmates and teachers. After-school activities bring you in contact with different groups of friends and acquaintances. Your location may shift from school to an outdoor playing field, local shopping center, or the busy sidewalk that leads you home. You may not notice the yellow tulips in a pot on a neighbor's porch or the footprints left by a raccoon as it ran through a muddy flowerbed.

Each day you move into and out of many different surroundings. Explore your surroundings in the activity that follows.

Explore! ACTIVITY

In how many different surroundings do you carry out your everyday activities?

Sometimes we're not even aware of all the places we visit in a day or of the living things that surround us in those places.

What To Do

1. For one day, *in your Journal* record the different places in which you work, play, and live. Record the plants, animals, or other organisms you interact with. List the non-living parts of each place too, such as air, noise, sunlight, or artificial light.

2. Be sure to explain how you interact with each living and nonliving part of your surroundings. For example, did you drink any water? Eat a piece of fruit? Ride a bicycle?

3. Next, think of how separate places might be related or connected to each other. How might something that happens in one place have an effect on another place?

346 Chapter 11 Ecology

Teaching the Activity

Science Journal To help students in recognizing cause and effect, choose an activity, such as sweeping the sidewalk, and have students write a journal entry about interacting with their surroundings. [L1]

Expected Outcome

Journal entries should indicate a variety of surroundings and parts of each with which students interacted.

Answers to Questions

One example is that if you have an especially active day at school, you may want to have more of a snack than usual when you get home.

✔ Assessment

Process Have students list which parts of their environment affect them and which parts they affect. Use the Performance Task Assessment List for Making Observations and Inferences in **PASC,** p. 17. [L1] [P]

■ Environment and Habitat

As you filled out your Journal, you became aware of the many different living things that you interact with each day. No matter where you are, you interact with plants, animals, and other organisms. But you also interact with nonliving things. You breathe in oxygen, feel the warmth of the sun, or brace yourself against the force of the wind. You are surrounded by your environment. The environment consists of everything around an organism.

Think back to the Journal you kept. You may have been in several different environments. School, a soccer field, a movie theater, a bus or subway car, a grocery store, or a crowded beach are some examples. Now think of the place where you live. The particular place where an organism lives is its **habitat**. Your habitat might be your neighborhood or local community. A starfish may live in an underwater cave. A rattlesnake's habitat may be a canyon in the desert.

Figure 11-1

A Organisms live in many different kinds of habitats. The eagle's habitat, for example, includes both land and the air.

B The coyote lives in a variety of land habitats including deserts, mountains, and prairies.

C The mallard's habitat is this small pond and the grassy area near the water's edge.

D The sea is the habitat of the octopus.

347

---

2 TEACH

Tying to Previous Knowledge

Relate the study of animals in Chapter 9 to that of plants in Chapter 10 by talking about seed dispersal by consumers. For example, in the East African savanna, acacia tree fruit contains a seed which may be eaten and destroyed by beetle larvae. But when the acacia fruit is eaten by grazing animals, the acids of their digestive systems kill the larvae. The seeds pass out of the animals in their feces. Thus the seeds can be spread to germinate elsewhere in the savanna.

Theme Connection As students investigate the surroundings in which they and other organisms carry out their everyday activities, they should develop an appreciation for the way in which living and nonliving parts of the environment interact to form ecosystems. This theme of systems and interaction becomes evident as the students do the Explore activity on page 346.

GLENCOE TECHNOLOGY

Videodisc

STVS: Plants and Simple Organisms

Disc 4, Side 1
Bacteria in the Clouds (Ch. 4)

Bacterial Control of Mosquitoes (Ch. 11)

Simple Forms of Life in the Antarctic (Ch. 16)

Program Resources

Making Connections: Across the Curriculum, p. 25, You and the Environment **L3**

Making Connections: Technology and Society, p. 25, No-Till Farming **L1**

Teaching Transparency 22 **L1**

Meeting Individual Needs

Gifted Have students research and prepare a report on the artificial reefs that have been built to provide habitats for organisms that live in the sea. Ask students to construct a model from papier-mâché or other materials, or draw a sketch of the major components of these reefs. **L3**

Populations and Communities

Figure 11-2 shows some of the animals that live on the grasslands of Africa. Each individual animal within each group is a member of a population. A **population** is a group of individuals of the same species that live in an area at the same time.

Choose one of the environments you recorded in the Explore activity. What animal and plant populations live there?

As you study the different populations in **Figure 11-2**, notice that they interact with each other. Together, these populations form a community. A **community** is made up of all the populations that live and interact with each other in an area.

Think of the community in which you live. What are five populations that live and interact with you? Describe the relationships you have with each other.

How do different populations share the same habitat? This next activity will show you.

Figure 11-2

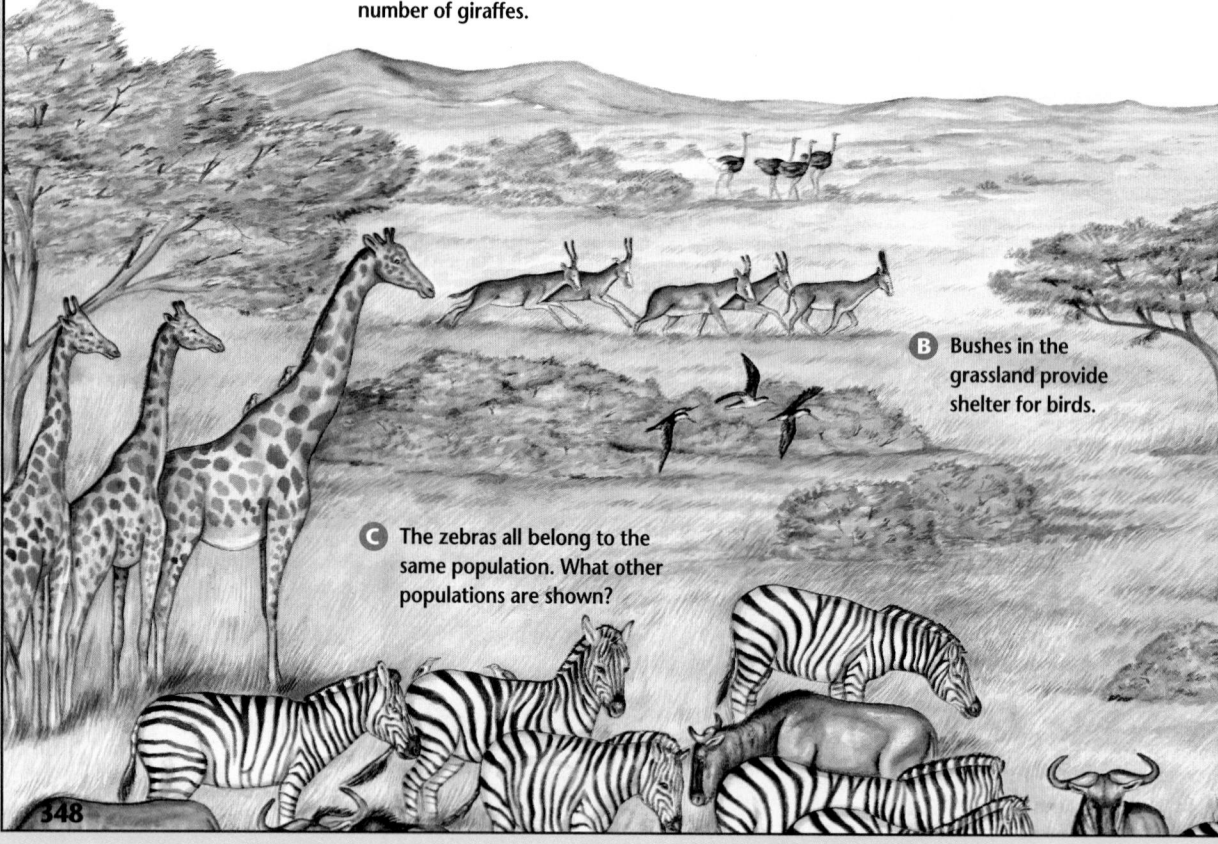

A In this natural habitat, large herds of zebra and wildebeest can be found next to a smaller number of giraffes.

B Bushes in the grassland provide shelter for birds.

C The zebras all belong to the same population. What other populations are shown?

Can three different bird species share the same habitat?

How can different animals share the space in which they live and feed? This activity will help you understand.

What To Do

1. Copy the diagram of the tree *into your Journal*.

2. Develop a key to show each of the three species of warblers.

3. Use your key to fill in the diagram of the tree to show where each species of bird spends most of its time feeding. The following observations will help you. The Cape May warbler feeds in areas 1a and 2a. The Bay-breasted warbler feeds in areas 3a, 3b, 3c, and 4c. The Myrtle warbler feeds in areas 3b, 3c, 5c, and 6c.

Conclude and Apply

1. In which parts of the tree does each warbler feed? Which birds share some parts of the tree?

2. How can three bird species feed in the same tree?

Bay-breasted warbler

Cape May warbler

Myrtle warbler

A B C B A

D Lions eat gazelles and other animals of the plain. When gazelles flee, they warn others of danger.

E The plant and animal populations shown here live and interact with one another. Together, they form a community.

ENRICHMENT

Using Math in Science Give students experience in estimating populations. Acquire a picture with a large population of a single species of plants or animals that is fairly evenly distributed throughout the picture (for example, a photo of bats on a cave ceiling). Divide students into three groups. Make one copy of the picture for each group. Have one group of students estimate the population. Have a second group count the actual number. Have the third group divide the photo into a gridwork of squares, count the population in a square, and multiply that number by the total number of squares. Have students silently record their numbers and the time it took to obtain them. Then have groups compare answers.

COOP LEARN L2

Can three different bird species share the same habitat?

Time needed 15 to 20 minutes

Materials ruler, white drawing paper or tracing paper, crayons or markers

Thinking Process interpreting scientific illustrations, observing and inferring, making models, interpreting data

Purpose To interpret an illustration portraying relationships between populations in a community.

Teaching the Activity

Make sure students understand how to create a key for their diagram. Suggest that students use a different color or pattern to represent each type of bird.

Science Journal Suggest that students use tracing paper to copy the diagram of the tree and tape it into their journals. Check that students label their keys. L1

Expected Outcome

Students should recognize that the three bird species are able to feed in the same tree because they have divided up the feeding areas within the habitat.

Conclude and Apply

1. Parts of the tree where each warbler feeds will be illustrated in journal sketches. The Bay-breasted warbler and the Myrtle warbler share some parts of the tree.

2. They feed in different areas of the tree.

✔ Assessment

Content Ask students to write a brief explanation of the advantages the three bird species gain by sharing the tree. *Sharing the tree ensures food and allows each to adapt to gathering food sources and quickly locating food.* Use the Performance Task Assessment List for Writing in Science in **PASC,** p. 87. L1 P

Niches

Visual Learning

Figure 11-3 Have volunteers describe for the class the different niches that make up the habitat shown in Figure 11-3. Next, ask each student to write a paragraph explaining how all the various organisms shown in the figure can exist together in the same community. L1 P

Content Background

Very few living things live alone. Nearly every living thing depends upon other living things.

The word *ecology* comes from the Greek word *oikos* meaning "house" or place to live. Ecologists study how organisms act together and how they are adapted to their environments.

Across the Curriculum

Math

Use fish and game magazines to find graphs that show changes in animal population size. Make transparencies of the graphs and display them on the overhead projector. Have students interpret the graphs and write brief reports explaining any population trends. L2

In the Find Out activity you saw how feeding areas within a habitat could be divided so that several species could share the same living space.

If you observe other communities in nature, you will find the same thing. Many populations can live in the same area because each species fills a specific role in the community. The role of an organism within its community is the organism's **niche**. A barn cat, for example, fills a certain role in a farm community by eating mice and rodents. An organism's habitat is part of its niche. What an organism eats, when it eats, and where it eats are also part of its niche. Look back at the Find Out activity. You could describe part of the Cape May warbler's niche by saying that the Cape May warbler feeds at the top outer branches of a tree. The way an organism reproduces and raises its young are part of its niche, too. **Figure 11-3** gives other examples of niches.

Figure 11-3

A Zebras feed on the tall, coarse grasses, wildebeests feed on the leafy center layer, and the small gazelles feed on the tender new shoots.

C Lions also share this community but unlike the other animals mentioned on this page that eat plants, lions eat other animals.

B Hippos spend the hot days in the water and graze on grass at night. Gazelles, zebras, and wildebeests graze during the day.

D No two plants or animals meet their needs in exactly the same way. Each species fills a particular niche within the habitat. In this way, species can exist together in the same community.

Multicultural Perspectives

An Ancient View of Earth

Concern about the well-being of Earth's varied ecosystems has increased in recent decades, but the concept is not new. Beginning about 2500 B.C.E. Chinese scientists developed a view of the entire universe as one huge organism. Scientists kept thorough records of natural events, including the weather and star movements. Natural disasters were interpreted as symptoms of the emperor or his officials failing in their duties and thus making the organism ill. Have students research and share the current view of the role of humans in environmental disasters facing some ecosystems today. L1

Elements of an Ecosystem

You've learned that organisms interact with each other in communities. They also interact with nonliving things in their environment. These relationships form an ecosystem. An **ecosystem** is a community of organisms interacting with one another and with the environment. A rotting tree stump is a small ecosystem, as you can see in **Figure 11-4**.

Cities, redwood forests, polar regions, and oceans are examples of larger ecosystems. How would you describe your ecosystem? As you continue through this chapter you'll learn about how organisms interact in an ecosystem. You'll find out about how the roles different organisms play all contribute to the ecosystem's changes and stability.

Figure 11-4

A rotting tree stump is a small ecosystem. Here insects, bacteria, and other organisms all interact with one another.

Rotting tree trunks are often attacked by wood-boring beetles.

Bracket fungi and moss may grow on the cool dark side of the trunk.

Leaf litter—leaves, twigs, and bark—provide food and shelter for slugs, snails, beetles, and flies.

check your UNDERSTANDING

1. Give an example of a population of plants.
2. What is the difference between a population and a community?
3. What is the difference between a niche and a habitat?

4. **Apply** Describe the niche you fill in your community. Identify how this niche benefits two other populations that share your community.

check your UNDERSTANDING

1. Sample answers include pine trees in a forest, grass in a field, or dandelions in a yard.
2. A population is a group of organisms belonging to the same species that live in the same area at the same time. A community involves different populations of organisms and their interactions.

3. A habitat is just one part of a niche; a niche also involves what an organism eats, how an organism reproduces, and so on.
4. Niche would be that of an omnivorous middle-school student. You can help populations by watering the lawn or houseplants, filling a bird feeder, feeding the dog.

3 ASSESS

Check for Understanding

Have students answer Check Your Understanding questions 1–4. After students share the niches they named in response to the Apply question, ask them to name their community and ecosystem. *Ecosystem may be urban community or may be local school district.*

Reteach

Discussion Have students describe the relationships between population, community, and ecosystem. Point out that populations interact to make a community. Stress that communities and nonliving things make up the ecosystem. L1

Extension

Acquiring Information Have students who have mastered the concepts in this section find out how wind and temperature interact to produce the windchill factor. The windchill is an example of how two environmental factors interact. L3

4 CLOSE

Activity

Select a habitat close to the school such as a nearby tree or pond. Have students observe it for several minutes. Ask the class what populations they found there and what niche each species occupied. Discuss the nonliving components of the ecosystem and how the populations affect and are affected by them. L1

PREPARATION

Planning the Lesson

Refer to the Chapter Organizer on pages 344A–D.

Concepts Developed

In this section, students explore the cycle of energy from the sun to producers, consumers, and decomposers within ecosystems. Water and nutrient cycles are also presented.

1 MOTIVATE

Bellringer

Before presenting the lesson, display **Section Focus Transparency 35** on an overhead projector. Assign the accompanying **Focus Activity** worksheet. L1

LEP

Explore!

How do organisms get their food?

Time needed 10 to 15 minutes

Thinking Process classifying, making and using tables

Purpose To classify organisms as either producers or consumers of food.

Teaching the Activity

Discussion Ask students why producers can make their own food while consumers cannot. *Producers have the ability to photosynthesize, or convert sunlight to energy.*

Science Journal Suggest that students record their observations in a two-column chart. They should label one column *Producers,* and label the other, *Consumers.* L1

Activity You could extend this activity by introducing the terms *herbivore* (plant eater), *carnivore* (meat eater), and *omnivore* (plant and meat

eater) and having students further classify the organisms they identified as consumers. L2

Expected Outcomes

Students should have little trouble distinguishing these organisms as either plants (producers) or animals (consumers).

Answers to Questions

Producers: cactus, fern, moss, grass, geranium, tree

Organisms in Their Environments

Section Objectives

- Describe a food chain and its relationship to a food web.
- Explain how natural cycles are important in the environment.

Key Terms

decomposer
food chain
food web

Food Producers and Consumers

In Chapter 9, you learned that all animals consume other organisms for food. These animals may consume other animals, plants, or plants and animals. In Chapter 10, you saw that plants produce their own food. Think back to the organisms at the beginning of this chapter. Which organisms are consumers? Which are producers? Producing or consuming food is one of the major niches an organism fills. It's one of the major ways in which organisms interact within their environments.

Explore! ACTIVITY

How do organisms get their food?

What To Do

1. Below is a list of organisms that you might find if you took a walk around your school.

2. Study the list and classify the organisms into two groups: those that can make their own food and those that cannot make their own food.

cactus	fern	earthworm
gerbil	student	geranium
moss	bird	fish
grass	ant	butterfly
spider	squirrel	tree

3. Which organisms did you classify as producers? Which did you classify as consumers? Answer the questions and record your observations *in your Journal.*

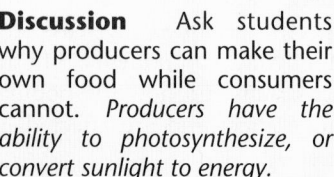

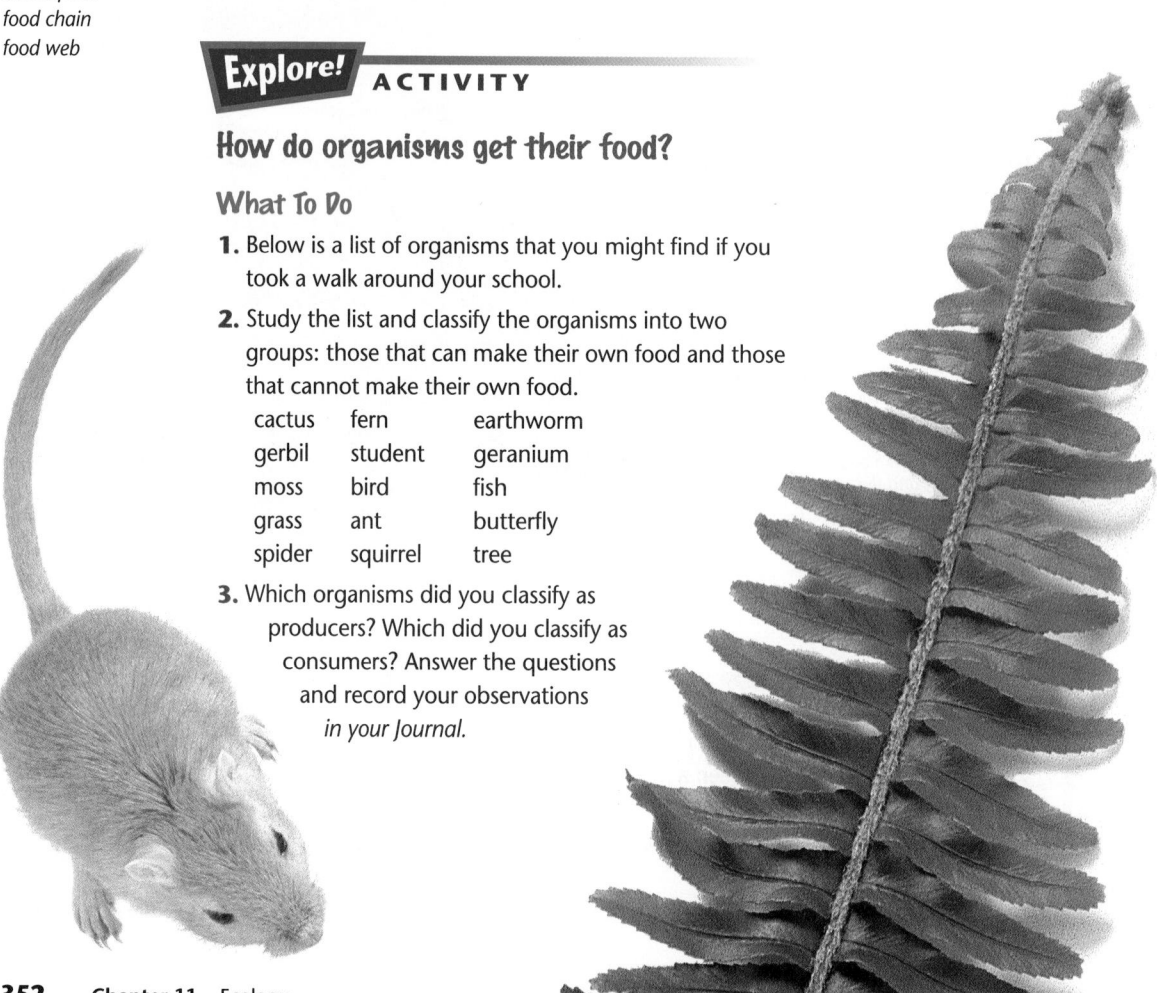

Consumers: earthworm, gerbil, student, fish, bird, butterfly, ant, spider, squirrel

✔ Assessment

Process Have students list organisms in or around their homes that are producers and organisms that are consumers. Use the Performance Task Assessment List for Science Journal in **PASC**, p. 103. L1 P

Decomposers

Now look at the organisms you classified in the Explore activity as consumers. Some consumers, such as gerbils, eat only plants. Other consumers, such as spiders, eat only animals. Still other consumers eat both plants and animals. Which type of consumer are you? Are all humans the same type of consumer?

In the first section, you learned that each organism has its own niche. One kind of consumer, called a decomposer, has an especially important niche within the community. A **decomposer** is an organism that gets its food by breaking down dead organisms into nutrients. As a result, nutrients within dead organisms are recycled back into the environment. Study **Figure 11-5** to learn more about decomposers.

Have you ever seen mold on a rotting log or on an old tomato? These organisms, called fungi, are decomposers. Most bacteria are decomposers, too.

Producing, consuming, and decomposing to obtain food are ways organisms interact. Do the next Investigate to see some of the relationships in an ecosystem.

Figure 11-5

A Perhaps you've seen mold on a rotting log or on a spoiled fruit or vegetable. That mold and these mushrooms are members of the fungus kingdom. Most fungi are decomposers—they get their food by breaking down dead organisms into nutrients.

 B Many types of bacteria are also decomposers. In compost heaps, bacteria break down dead plant material into the nutrients that living plants need.

11-2 Organisms in Their Environments **353**

INVESTIGATE!

11-1 What Do Owls Eat?

Planning the Activity

Time needed 45 to 60 minutes

Purpose To interpret data to learn about an ecosystem.

Process Skills observing and inferring, classifying, designing an experiment, interpreting data, estimating, interpreting scientific illustrations

Materials Owl pellets can be obtained from a biological supply house. Students may choose additional materials for preparing their exhibits.

Preparation Organize students into groups or have them choose their groups. Distribute materials to each group.

Teaching the Activity

Safety Make sure students wash their hands after handling the owl pellets and their contents. Caution students not to touch their faces or mouths while examining the pellets.

Troubleshooting Give the students hints as to what they should look for when making observations. If possible, provide pictures or models of the skeletons of small birds and mammals to aid students in identifying the contents of their pellet.

Process Reinforcement Be sure that each group develops a plan for examining, identifying, and displaying the contents of the pellet. Each member of the group should be given an opportunity to participate.

Possible Hypotheses Students may hypothesize that owls are consumers. Some students may hypothesize the owls are carnivores or omnivores. Still others may go further to form hypotheses about the specific prey of owls.

What Do Owls Eat?

Owl pellets are made of indigestible things an owl has swallowed, including fur and bones. These pellets form in an owl's stomach and then the owl coughs them up. Examining an owl pellet can tell you much about what is going on in a small part of the owl's ecosystem.

Problem
What role do owls play in their ecosystem?

Materials

water	bowl
forceps	glass slide
coverslip	light microscope
magnifying glass	owl pellet
cardboard	glue

Safety Precautions

Use care when handling microscope slides and coverslips. Dispose of all materials properly.

What To Do

1 With your group design a way to investigate what an owl pellet is made of and what its contents are. You should make a display of the contents of the owl pellet. After your plan has been approved by your teacher, carry it out.

2 *In your Journal*, write a short summary of your design and of what you found the contents of the owl pellet to be. Use the table on page 355 to help you identify the contents of the owl pellet.

Possible Procedures As students take apart the pellets, they should be careful to keep each item in the pellet intact to aid in identifying the contents. To allow each student to participate, groups may choose to have each member examine each item silently, using a magnifying glass or microscope as needed. Once each student has had a chance to look at an item, members of a group can discuss the identity of the item. Groups and individuals may consult pictures and models to aid in identifying the contents of their pellets. **L1**

Discussion Ask students what an investigator who was able to examine the contents of their stomachs that day might find. What would those contents tell the investigator about the environment in which the students live?

Science Journal Have students record their plans and results in their journals. **L1**

Barn owls live on all continents except Antarctica. They are known everywhere as the farmer's friend because they destroy harmful rodents that live in barns and eat grain.

Owl Pellet Contents

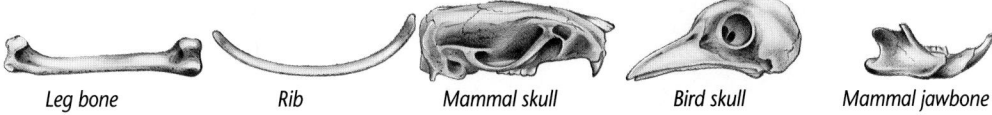

Leg bone   Rib   Mammal skull   Bird skull   Mammal jawbone

Analyzing

1. What made up the outside of the owl pellet?

2. What did you see inside the pellet? How many of each kind of thing were there?

3. What role does an owl play in its ecosystem? Is it a producer or a consumer? How do you know?

4. Describe the niche of an owl. Include where the owl lives, when it feeds, and what it eats. Describe the niche of an owl's prey. How are the two similar? How do they differ?

Concluding and Applying

5. If one owl pellet is produced each day, *estimate* the number of organisms eaten by the owl in a single day. Estimate the number of organisms an owl needs to eat to survive for one year.

6. Going Further *Design an experiment* to figure out what might happen to the population of owls if there were a sudden explosion in the population of mice.

Program Resources

Activity Masters, pp. 47–48, Investigate 11–1

Science Discovery Activities, 11-2, Is It Balanced

Expected Outcomes

Students should realize that an owl occupies the niche of a predator in its ecosystem, preying on mice and other birds; that an owl is a consumer, not a producer; and that an owl is responsible for keeping down the population of mammals that many humans consider to be harmful.

Answers to Analyzing/ Concluding and Applying

1. Feathers and/or fur from the animals that have been eaten by the owl may make up the exterior of the pellet.

2. Bones. The number of bones observed will vary.

3. It is a predator, a consumer. You can find the remains of what it consumed in the pellet.

4. An owl is usually a night-hunting predator that lives in a nest in trees or other structures. The prey is often small, herbivorous mammals that live in or on the ground. The prey may be other birds or, in some cases, lizards. The niches are almost completely different except that the habitat overlaps.

5. Answers will vary depending on the number of skulls or other unique bones found. There may be as many as 5-6 skulls in a pellet. That would indicate 5-6 animals per day, resulting in the consumption of 1800 or more prey animals per year.

6. Answers will vary, but should include looking for signs of an increase in the owl population.

✓ Assessment

Process Have students imagine that they are able to obtain pellets of owls from many different environments on different continents of the Earth. Then have them work in groups to design a plan for examining and identifying the contents of the different pellets. Ask them how they could use the information they gathered from the pellets to learn about the different environments in which the owls live. Use the Performance Task Assessment List for Designing an Experiment in **PASC,** p. 23. L1
COOP LEARN

Energy Flow in an Ecosystem

Figure 11-6

A Plants use sunlight to produce food in the form of sugar. Plants store the food they make in their roots, leaves, and stems.

B A mouse eats the plants. Some of the food in the plants is stored in the mouse's body. The mouse's body changes some of the food into energy to run, eat, and breathe.

C The snake eats the mouse. The mouse supplies energy for the snake to live.

D Hawks eat a variety of small animals including snakes. The hawk's body uses the energy gained from the snake to carry out its life processes. This flow of food energy from one organism to another continues throughout the food chain.

Whether you observe a small part of an ecosystem, such as in the last activity, or an entire large ecosystem, such as the ocean, you'll find feeding relationships among the organisms. Food is required for the life processes of every organism, whether it is a producer, a consumer, or a decomposer. How do producers, consumers, and decomposers interact?

Remember that through the process of photosynthesis, plants use light energy from the sun to make food. This chemical reaction changes water, carbon dioxide, and light into sugar and oxygen. The sugar is food that can be stored and used later by the plants.

When animals—consumers—eat plants, the energy in the plants is passed to the animals. These animals are then eaten by other animals, which in turn may also be eaten. Each time, food energy passes from one animal to the next. Energy is also passed on when decomposers break down dead organisms. Each organism in this relationship is like a link in a chain. A **food chain** is a model of how the energy in food is passed from organism to organism in an ecosystem.

Figure 11-6 shows the elements of a simple food chain. Plants produce food for themselves which is stored in roots, stems, and leaves. Animals eat plants, and the animals are, in turn, eaten by other animals. In this way energy from the sun is distributed throughout an ecosystem.

Food Webs

A feeding relationship in a single food chain is simple. However, most organisms get their food from more than one source. For example, a bear eats fish, berries, honey, and insects. An owl eats different kinds of rodents and snakes. Sometimes one food source can provide food for many different organisms. For example, grass is eaten by rabbits, cattle, deer, and horses. Thus, an organism can belong to several different food chains. When related food chains are combined, a food web is formed. A **food web** is the combination of all the overlapping food chains in an ecosystem. **Figure 11-7** is an example of a food web.

Consumers with varied diets—in other words, those that belong to a fairly large food web—have a better chance of survival than those with limited diets. If something happens to disturb one supply of food, the consumer can obtain food from another food chain in the web.

Sequencing

Study the list of organisms below. Decide in which sequence (or order) they should be placed to make the most likely food chain. If you need help, refer to the **Skill Handbook** on page 653.

bass	person
insect larvae	crayfish

Figure 11-7

A food web is a model of the overlapping food chains in an area. Follow the food chains that the seed-eating bird belongs to in this web. Are grasshoppers important to hawks in this food web? Why or why not?

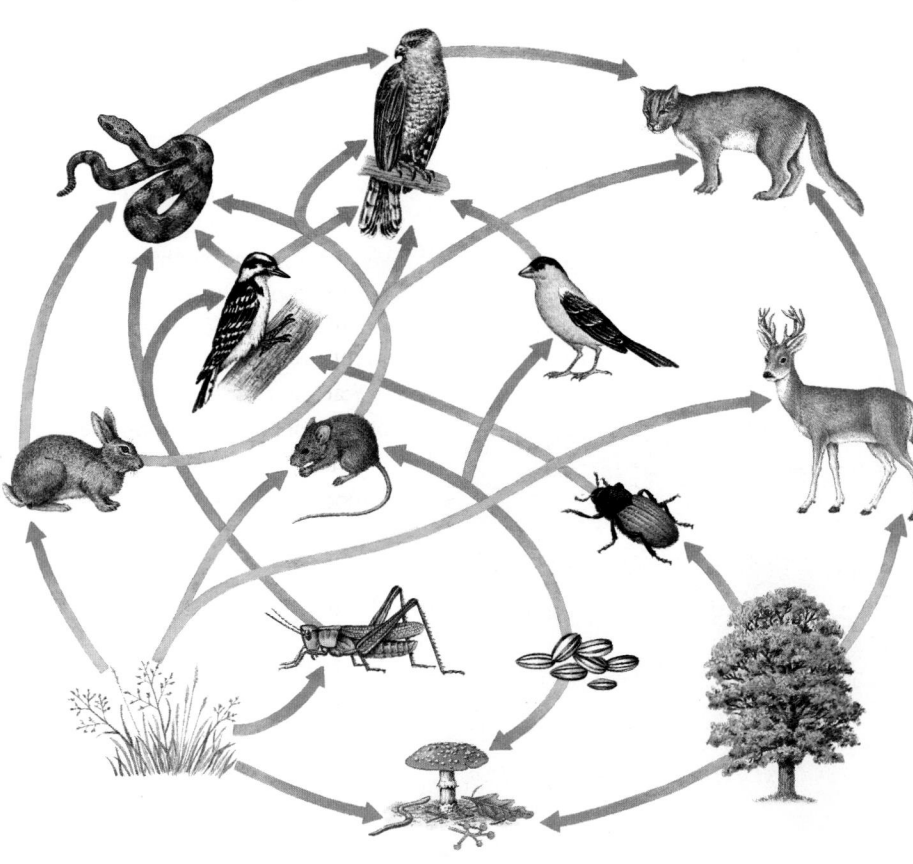

Inquiry Question Can you think of another advantage that consumers of varied diets have over those with limited diets? *When one food supply runs low a consumer of a varied diet can always switch to another. Consumers with limited diets cannot.*

Activity To reinforce students' sequencing skills, have them describe or diagram how organisms may change places in the food chain throughout their lives. *Tadpoles eat plants whereas frogs and toads eat animals. Many insect larvae also eat producers and then become meat-eaters as adults.* L1

Uncovering Preconceptions Failure to understand the complex interactions within some food webs has led to the misuse of pest control methods over the years. For example, people think the grasshopper is a voracious pest that destroys crops. Yet at least one grasshopper, the lubber, eats only broad-leaved plants, such as the sunflower, which allows more sunlight and moisture to reach the grass roots.

SKILLBUILDER

1. insect larvae
2. crayfish
3. bass
4. person L1

Flex Your Brain Have students use the Flex Your Brain activity to Explore ECOSYSTEMS. L1

Multicultural Perspectives

Ecosystems Around the World

Bilingual or multicultural students may have difficulty recognizing the animal and plant populations mentioned and pictured in the text. Whenever possible, encourage them to share information about food chains and ecosystems in their native countries. **LEP**

Program Resources

Teaching Transparency 21 L2
Critical Thinking/Problem Solving, p. 5, Flex Your Brain

Demonstration Help students formulate a model of the water cycle by demonstrating evaporation and condensation in the water cycle by boiling water in a teakettle. Hold a cold glass plate over the steam and the steam will condense.

Activity Have students make a chart that shows at what ocean depths you would expect to find producers, consumers, and decomposers. Students should explain their chart in a brief paragraph. L2 P

depth	type
surface	producer, consumer
middle range	consumer
bottom	decomposer, consumer

Producers would be found near the surface where sunlight is available for photosynthesis. Consumers could be found at any depth, but plant eaters would feed near the surface. Decomposers could be found near the ocean floor where organic debris settles.

Connect to . . .
Physics

Because matter can be neither created nor destroyed there is a limited amount of matter which must be used for all things. If materials on Earth were not recycled, all of the available matter would soon be consumed. Living things could not reproduce, maintain body systems, or survive. Therefore, recycling of matter is essential to life on Earth.

Connect to...
Physics

Scientists know that matter can be neither created nor destroyed. Use this information to explain why recycling of materials is essential to life on Earth.

Natural Cycles

Organisms need a constant supply of energy to live. The sun provides the energy that flows through most of the food chains in our world.

Organisms also have other needs for survival, including water and nutrients. These needs are met from a limited supply of Earth's natural resources. Unlike energy from the sun, these resources aren't continually replaced. Instead, they are constantly being used and recycled. Without the recycling of materials, organisms would quickly run out of the water and nutrients they need.

■ Water Cycle

Energy from the sun causes water in Earth's oceans to evaporate. Hot air rises and carries water vapor up into the cooler atmosphere. Here water vapor forms clouds. Water falls from the clouds as rain or snow. Plants absorb water from the soil. Animals drink water or obtain it by eating plants. Both plants and animals use or store some water and return the rest to the environment. Plants release water through their leaves. Animals release water with waste products. Water evaporates, and the cycle continues.

Oxygen Makers

Earth's atmosphere provides some essential ingredients for life: oxygen, carbon dioxide, nitrogen, and water vapor. Earth's original atmosphere consisted of ammonia, carbon dioxide, carbon monoxide, hydro-

gen, methane, nitrogen, sulfur dioxide, and water vapor, but very little oxygen. Earth's current atmosphere is about 21 percent oxygen. So where did all the oxygen come from?

You learned that oxygen is a product of photosynthesis—the process by which plants and some bacteria and algae absorb carbon dioxide from the air, combine it with water in the presence of sunlight to convert it to sugar, and then release oxygen into the air. The oxygen content of Earth's atmosphere began to increase about 2.3 billion years ago, when cyanobacteria and algae started growing in the ocean. In a way, we owe

Purpose
The Chemistry Connection extends the discussion of the role of plants in the oxygen-carbon dioxide cycle, which is presented in this section. Students also have the opportunity to observe the production of oxygen by plants as a by-product of photosynthesis.

Content Background
The simplified equation for photosynthesis is:

$$6CO_2 + 12H_2O + \text{light energy} \rightarrow C_6H_{12}O_6 + 6O_2 + 6H_2O$$

Whereas land plants utilize the free carbon dioxide in the atmosphere, aquatic plants use carbon dioxide dissolved as a gas or as carbonates in the water. The rate of photosynthesis in land plants is greater when there is an increase in the amount of carbon dioxide. Earth's early atmosphere, therefore, may have allowed plants to grow faster and produce more oxygen.

Nitrogen Cycle

Nitrogen is an element used by most organisms to build proteins and other body chemicals. While most of the atmosphere is made of nitrogen, organisms can't use this form of the element. Instead, plankton in water and bacteria in the soil and in the roots of bean and pea plants change the gas into a form that can be taken in and used by plants. Animals get nitrogen by eating either plants or animals that have eaten plants. The nitrogen is returned to the soil when animals release waste products or when dead organisms decay. **Figure 11-8** shows the nitrogen cycle.

Figure 11-8

In the nitrogen cycle, nitrogen passes from the atmosphere to living things, then back again to the atmosphere.

Nitrogen is changed to forms plants can use by lightning

Bacteria convert nitrogen back to gas

Bacteria change nitrogen into forms plants can use

Bacteria

Content Background

Oxygen and carbon make up much of the body's carbohydrates, proteins, and fats. They are also involved in many chemical reactions. The oxygen-carbon dioxide cycle is driven by photosynthesis and respiration.

Nitrogen fixation is one of the major processes of the nitrogen cycle. In nitrogen fixation, bacteria that live on the roots of legumes (alfalfa, clover, peas, and beans) convert nitrogen into ammonium compounds. Lightning also plays a role in nitrogen fixation.

Activity To reinforce students' ability to interpret scientific illustrations, give them photo-copied diagrams of the water, oxygen-carbon dioxide, and nitrogen cycles without labels. Then have students fill in the labels. L1

NATIONAL GEOGRAPHIC SOCIETY

Videodisc

STV: Atmosphere What Is the Atmosphere?

Unit 1, Side 1
Atmospheric Gases

03109-07851

our existence to bacteria and algae.

The ability of plants to convert carbon dioxide into oxygen is one reason why people are so concerned about our vanishing forests. So far, scientists have not found a way to make photosynthesis occur outside living organisms.

You Try It!

You can observe the results of photosynthesis.

Materials

small live plant
deep basin or pan
tall, narrow glass container
    or test tube
water hose or a bucket full
    of water

What To Do

1. Take your materials to a sunny location.
2. Fill the basin with water.
3. Fill the glass container to the brim.
4. Completely cover the open end of the container or test tube with your thumb.
5. Place the container upside down in the basin and remove your hand, without

letting air inside.

6. Thread a runner of the plant into the container while keeping the open end of the container completely submerged. Don't cut the runner from the plant!
7. Leave the setup in the sun several hours. Record what you observe when you return. How might it be related to photosynthesis?

Teaching Strategy

Have students examine the effects of sunlight on the amount of oxygen produced. Place one setup in bright sunlight, one setup in a shaded area, and cover one with a box to eliminate any sunlight. After several hours, have students compare the pockets of air in the top of the containers and draw conclusions.

Going Further ▥▥▥▶

Ask students to write a story from the viewpoint of one of the nonliving components of an ecosystem, air. Have them include the exchange of gases, the presence of sunshine, and the protection afforded by the ozone layer. Suggestions include how they are affected when rain forests are chopped down, what the destruction of the ozone layer would do to them, and how they function differently during the

day and at night. Use the Performance Task Assessment List for Writer's Guide to Fiction in **PASC**, p. 83. L2 P

3 ASSESS

Check for Understanding

1. Have students diagram a food web for organisms in your area. Discuss the importance of food webs and how human activities or natural disasters disrupt food webs.

2. Have students share the food chains they constructed. Identify the producer(s) which appeared most often.

Reteach

Activity To help students formulate a model of the inter-relationships in food chains, play the food web game. List producers, consumers, and decomposers on separate index cards. Give a card to each student. Give each producer a ball of colored yarn. Have producers roll the ball to their primary consumers, who roll it to each subsequent consumer until each chain is complete. This creates a food web. L1

Extension

Activity Have students use field guides and reference books to construct a food web. Let students choose an animal and find out what the animal eats and what eats the animal. L3

4 CLOSE

Activity

Use the classroom terrarium as an ecosystem. Have students identify producers (mosses and liverworts) and consumers (salamanders, turtles, lizards). Point out that the decomposers are bacteria and fungi. L1

■ Oxygen-Carbon Dioxide Cycle

At this moment, plants, animals, and most other organisms are removing oxygen from the atmosphere. Why hasn't Earth's oxygen supply been used up? The answer is that oxygen is recycled.

Remember from Chapter 10 that during respiration, plants and almost

Figure 11-9

A During respiration, plants and almost all other organisms take in oxygen and release carbon dioxide into the air. During photosynthesis, plants take in carbon dioxide from the air and release oxygen.

B Together the processes of photosynthesis and respiration continually recycle oxygen and carbon dioxide.

all other organisms take in oxygen and release carbon dioxide. Carbon dioxide also enters the atmosphere when decomposers break down dead organisms. However, during photosynthesis, plants take in carbon dioxide from the air and release oxygen.

Lately, many people have become concerned with another way in which carbon dioxide is released. Sometimes dead organisms do not decay. After millions of years, they turn into fossil fuels such as coal. When the fossil fuels are burned, carbon dioxide is released. The use of fossil fuels has greatly increased since the 1800s. Thus, the amount of carbon dioxide in the atmosphere has also increased. Continued use of fossil fuels is changing the delicate balance of Earth's atmosphere.

Another human activity that can affect the oxygen-carbon dioxide cycle is cutting down Earth's rain forests. This can result in less oxygen being produced and released into the air because there are fewer plants to carry out photosynthesis.

Natural recycling that occurs on Earth ensures that living things will be able to obtain the materials needed for life.

check your UNDERSTANDING

1. Explain the relationship between a food chain and a food web.

2. Is it possible to find a food chain that includes only a producer and decomposer? Explain your answer.

3. Describe what might happen if the water cycle was interrupted.

4. Apply Identify a meat or fish product that you've recently eaten. Construct a food chain that shows the feeding relationships that preceded your eating the product. Place yourself at one end of the food chain.

check your UNDERSTANDING

1. A food chain is one line of food transfer from organism to organism. A food web consists of all the food chains that overlap in an ecosystem.

2. Yes, a producer may die from a disease and decay.

3. Eventually all of the water would be used. Oceans and rivers would dry out. All organisms would die.

4. Possible answer:

sun → aquatic plant → fish → student

How Limiting Factors Affect Organisms

Requirements for Life

At the zoo, you'll notice that animals are displayed in different environments. Some animals, such as some penguins, are housed behind glass, where the temperatures can be kept cool during hot weather. Others, such as rattlesnakes, are found in hot, dry display areas. Still others, such as

bats, are literally kept in the dark. Why do all these different animals have such different requirements?

Plants and other living things also are found growing in different conditions. Why might this be? Do the activity that follows to begin exploring what factors might limit growth.

Section Objectives
- Identify some limiting factors.
- Describe adaptations of organisms to limiting factors.

Key Terms
limiting factor

Figure 11-10

A Many modern zoos build enclosures that closely resemble the natural environments of the animals. Jungle-like environments with plenty of places to climb make primates feel at home.

B The natural environment of this bamboo viper snake is hot and dry. Here the snake is kept in a glass enclosure with air that is just the right temperature and has just the right moisture content to keep the snake healthy.

C Birds are kept in roomy aviaries filled with plants that would grow in the bird's natural habitat.

D Animals, such as polar bears, that need a certain temperature to survive may be kept in glass enclosures where temperature can be easily monitored.

Program Resources

Study Guide, p. 42
Section Focus Transparency 36

PREPARATION

Planning the Lesson
Refer to the Chapter Organizer on pages 344A–D.

Concepts Developed
Now that the students understand relationships between living and nonliving parts of an ecosystem, they are introduced to limiting factors. These factors are the conditions that determine the survival of an organism, population, or species in its environment. Biotic and abiotic limits are shown, as well as adaptations that some plants and animals have made in order to survive.

1 MOTIVATE

Bellringer
Before presenting the lesson, display **Section Focus Transparency 36** on an overhead projector. Assign the accompanying **Focus Activity** worksheet. L1
LEP

Discussion Discuss your region and the limiting factors there; talk about climate, geography, and soil. Ask students how the plants and animals in your region have adapted to your environment.

11-2 How Do Molds Grow?

Preparation

Purpose To observe and infer how a limiting factor affects the growth of a population.

Process Skills Observing, inferring, recognizing cause and effect, using variables, interpreting data, forming operational definitions

Time Required Begin this activity on a Monday or Tuesday of a five-day week so students can make daily observations. 10 minutes to plan; 30 minutes to set up; 5 minutes for observations on the next three days; 15 minutes to analyze on fourth day; 20 minutes to answer Analyze and Conclude questions

Materials Mold source can be obtained from a biological supply house. Any pasta or bread product may be used in the experiment.

Possible Hypothesis Students may predict that water is a limiting factor for the growth of molds. If students predict that a different factor limits the growth of molds, they need to design their investigation to test that factor.

The Experiment

Process Reinforcement Be sure students set up their data tables correctly. Test foods down the side and days of the experiment across the top is an effective design. Students can use one table to estimate the rate of mold growth and use another table to record actual data and illustrate growth.

Possible Procedures Prepare two cups for each food. Add the same amount of material to each cup and label each. Add just enough water with the spray mister to moisten food. Rub a moist cotton swab across the dish of growing mold and then rub the cotton swab across the surface of the food. Do this for all six cups. Observe the cups for the next 4 days. Observations should be recorded immediately.

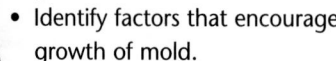

How Do Molds Grow?

Molds are fungi that can feed on just about anything. Think about where molds grow in our environment. Do they grow everywhere or are there factors that limit the growth of molds?

Preparation

Problem

What basic factor limits the growth of molds?

Form a Hypothesis

As a group make a hypothesis about what factor seems most important in encouraging the growth of mold.

Objectives

- Identify factors that encourage growth of mold.
- Evaluate data.
- Determine which factor is a strong limiting factor to the growth of mold.

Possible Materials

6 small paper cups
hand lens
labels
mold source (teacher supplies)
spray bottle of water
plastic wrap
cotton swabs
dry potato flakes
dry macaroni
sugarless, dry cereal
other dry saltless, sugarless food

Safety Precautions

After transferring the mold source, wash your hands thoroughly. All surfaces in the experiment in touch with microorganisms should also be washed thoroughly. Do not inhale, taste, or touch material from the mold source. If you have a mold allergy, do not handle the mold.

DESIGN YOUR OWN
INVESTIGATION

Plan the Experiment

1 Examine the materials provided and decide how you will use them to test the group's hypothesis.

2 What is the limiting factor that you are testing? How will it be introduced in the experiment?

3 A small amount of food at the bottom of each cup is enough to feed mold. To introduce mold into each cup, rub a moist cotton swab across the dish of growing mold, then rub the cotton swab across the surface of the food in each cup. Try to put the same amount in each cup.

4 Mold grows over a period of days. Checking mold growth day by day should be taken into account in making your plan and your data table.

Check the Plan

1 How will you keep the environment in your experimental cups from change or contamination?

2 Where will you keep your experiment? Is it a neutral environment? Are the conditions for all the cups the same?

3 How long do you think it will take the mold to grow? How often will you check the experiment? Make certain that each observation is recorded in your data table.

4 Before your begin the experiment, have it approved by your teacher.

5 Carry out your experiment and record your data.

Analyze and Conclude

1. **Compare and Contrast** In which cups did you see evidence of mold growth? In which cups was there no mold growth?

2. **Infer** Determine whether there is a factor that limits the growth of mold.

3. **Interpret Data** Did mold grow faster on one particular food?

4. **Draw a Conclusion** If you wanted to package food to sell, what is one way you could prevent mold from spoiling your product?

Going Further

Moisture is a limiting factor in mold growth. What other factors can you test to see if they limit mold growth?

Tying to Previous Knowledge

Review the characteristics and requirements of life from Chapter 7, with emphasis on adaptation. There are limiting factors in your area which should be recognizable by students. Mountainous or desert conditions provide obvious examples; bird migration or hibernation or competition that results from human encroachment into natural habitats are other possible factors.

Theme Connection The stability of many plant and animal populations is maintained only within a certain range of conditions. Rattlesnakes, sharks, marigolds, and cacti are each adapted to a particular environment and will die if there is a sudden change in the living or nonliving components.

Activity Different kinds of grasses grow well in different regions of North America. Have students analyze limiting factors to find out the variety of grass most suited for lawns in your area. Suggest that students visit a local nursery or lawn care center. Have them find out the soil conditions, watering practices, and fertilization requirements for each variety of grass. L2

GLENCOE TECHNOLOGY

Videodisc

The Infinite Voyage: Secrets from a Frozen World

Chapter 1
The Southern Ocean—A Rich Marine Ecosystem

Chapter 2
Krill: The Vital Link of the Food Chain

Chapter 6
The Antarctic Peninsula: Pack Ice and Life Cycles

Figure 11-11

Ⓐ Many types of penguins require a cold, wet habitat and could not survive in the hot, dry habitat that is home to the pinto chuckwalla.

Ⓑ Lack of moisture is a limiting factor for orchids, which need a lot of moisture to live, but not for cacti, which are adapted to thrive in a hot, dry habitat.

364

■ Limiting Factors

As you worked on the Investigate, you found that mold grew better on certain materials and under certain conditions. What might have happened if you had placed the materials in the refrigerator? Mold would still have grown, but much more slowly.

Environmental factors help determine whether an organism can live in the environment. Think about a shark. Will a shark survive in the ocean or a lake? You know that most sharks live in the ocean. Sharks are saltwater organisms, and most species of sharks can't survive in a freshwater lake.

Now think about plants. Cacti grow in dry areas such as deserts. Too much water will kill them. A **limiting factor** is any condition that influences the growth or survival of an organism or species.

Limiting factors can be nonliving, environmental conditions, such as temperature, wind, chemicals in the soil, amounts of light and water, as well as pollution in the water or air. Limiting factors can also be the relationships between living organisms in a community. You'll see what kinds of relationships can be limiting as you continue to read the chapter.

Program Resources

Laboratory Manual, pp. 63–67, Physical Conditions and Behavior L2 ; pp. 67–70, Living Space L3

Making Connections: Integrating Sciences, p. 25, Searching for Life on Other Planets L2

ENRICHMENT

Research Ask students to use the library to research the introduction of Russian thistle, starlings, English sparrows, gypsy moths, and kudzu into the United States. Or have students read and report on the work of John Garcia, who studies the interactions of animals and their environment in California. Have students give an oral presentation. Use the Performance Task Assessment List for Oral Presentation in **PASC,** p. 71. L2

Nonliving Limiting Factors

You saw that molds need water to grow. What kinds of limiting factors affect other living things?

■ Temperature

Every organism has a set of conditions that are best for survival. For example, most organisms can survive if the temperature falls within a certain range. For example, some fish in a freshwater aquarium carry on life processes in water that is room temperature (22°C). They can live in water that's 5.5 degrees warmer or colder. If the water's temperature goes beyond that 11-degree range for too long, the fish die.

■ Sunlight

The amount of direct sunlight an ecosystem receives is another limiting factor. Go to a greenhouse or a store that sells plants. A tag with instructions for proper care is often included with each plant. Bright, direct light kills some plants. Other plants thrive in it. Study **Figure 11-12** to learn more about sunlight as a limiting factor.

■ Rainfall

The amount of rainfall in an environment is also a limiting factor. Generally, each ecosystem has an average level of rainfall. Many factors work together to determine that level. If rainfall doesn't meet that level over a long period of time, plants die. Can you infer what happens to the animals, including humans, in the ecosystem?

Figure 11-12

For many plants, sunlight can be a limiting factor.

Ⓐ Bright, direct sunlight kills some plants. Other plants thrive on it. Impatiens grow best when planted in shady areas.

Ⓑ If you want petunias to thrive, you must plant them in sunny to partly sunny areas.

Ⓒ Zinnias grow best in full sun.

Inquiry Questions How do animals endure the heat in some deserts? *They stay in their burrows during the day and come out at night.* **Why do most tropical rain forest animals live in trees?** *More food sources are available in the lighted area of the canopy.*

Discussion Have students explore whether any place on the rain forest floor would ever have thick vegetation. *Any place where sunlight reaches the forest floor, such as a clearing, or along a riverbank, might have thick vegetation.*

Across the Curriculum

Geography

Have students find out about the nonliving limiting factors in the area in which they live. Direct them to use a globe or world map to determine the latitude, longitude, and altitude of their school. Have them research the climate, including the length, amount of rainfall, and average temperature of each of the seasons and the number of days of sunshine, as well as soil conditions, pollutants, erosion, and human land development and use. Students can then write reports on their findings. Use the Performance Task Assessment List for Writing in Science in **PASC**, p. 87. **L2**

Figure 11-13

More than half of the animal and plant species in the world live in the rain forest ecosystem. A great many species have adaptations that help them deal with limiting factors in the rain forest ecosystem.

The harpy eagle soars high above the upper canopy hunting for monkeys and sloths for food.

Spider monkeys eat mostly nuts and fruit from the forest.

Spider monkeys live in the upper canopy. They use their tails to swing from branch to branch.

The scarlet macaw uses its strong beak to eat nuts and seeds.

Sloths use their long curved claws as hooks to help them move within the trees or to hang upside down.

Boas, which grow to giant size in the rain forest, stalk tree frogs and iguanas for food.

Climbing plants and vines grow and twine around the trees and bushes in the understory.

The largest animals in the ecosystem live on the forest floor. This South American jaguar's coat helps it blend into the habitat and hide from its enemies.

Ocelots hunt birds and monkeys in the lower canopy and rabbits and small deer on the forest floor.

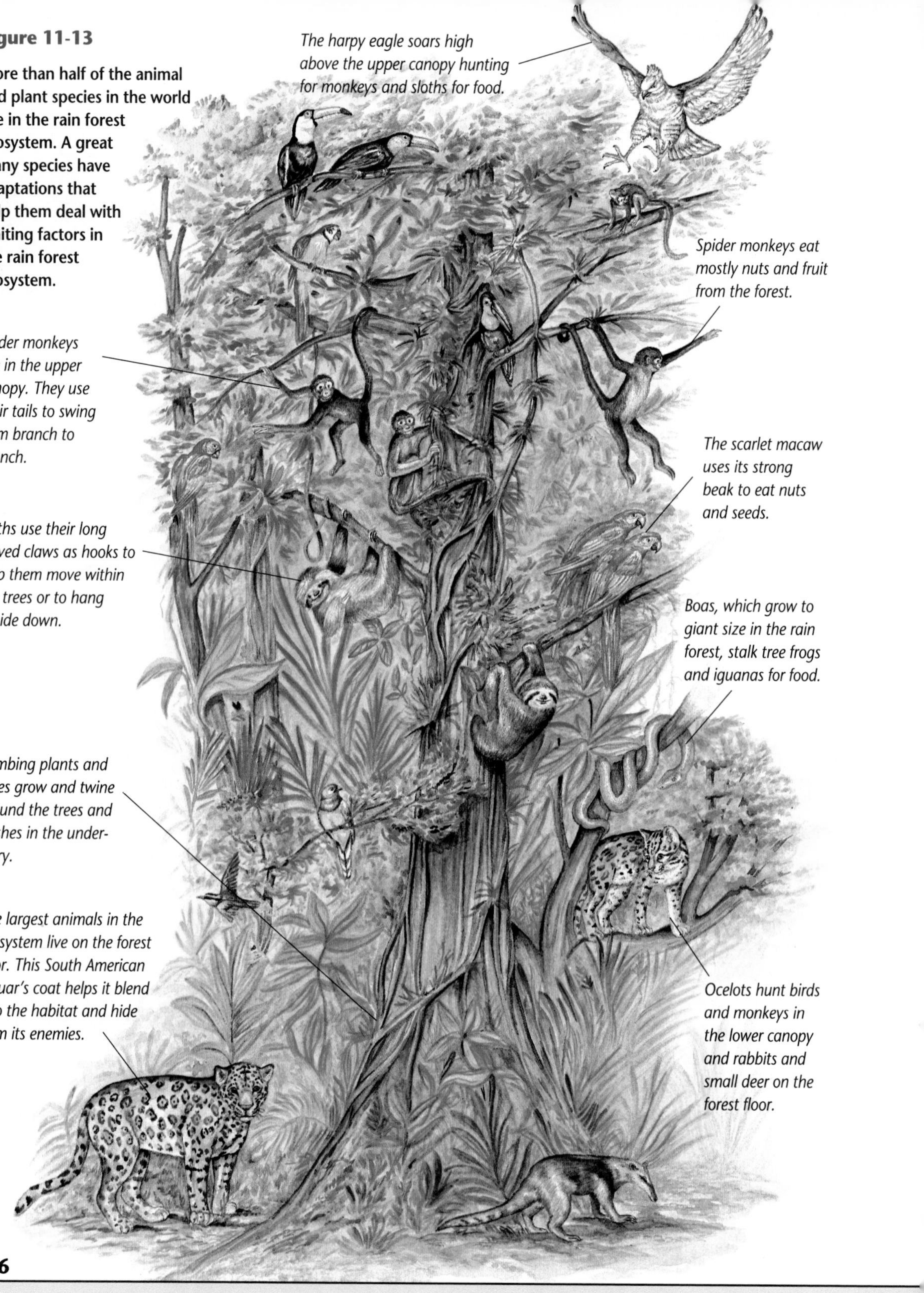

366

Meeting Individual Needs

Visually Impaired Pair visually impaired students with fully sighted students. Have students develop and tell a story with a setting based in a particular ecosystem. The community and the nonliving components should be central to the story. **COOP LEARN**

Other Organisms as Limiting Factors

You've learned about some non-living limiting factors. Now, find out how relationships between living organisms can be limiting factors.

■ Competition

Recall that in Section 11-1 you learned that each organism fills a niche in its community. Competition results when two organisms try to fill the same niche. The organisms compete for food, shelter, water, and other needs until one organism is forced to leave the area or dies.

One example of competition as a limiting factor is what happens to wildlife as humans move into an area. Humans build houses, cut down trees, pave roads—activities that greatly change the environment. Many animals are forced to live elsewhere. **Figure 11-13** shows other examples of competition.

■ Predator-Prey Relationship

A close look at a food chain can provide a clue to another behavior-linked limiting factor. Animals that catch and eat other animals are called predators. See **Figure 11-14** for one example of predators. The animals predators eat are called prey. The predator-prey relationship has an effect on the size of populations of both predators and prey. How many predator-prey relationships can you find in **Figure 11-13**? Usually the numbers of predators and prey within a community will stay about the

Figure 11-14

Lions hunt in groups. Working together, they are able to kill prey larger than themselves, such as wildebeests and zebras. Cooperative hunting also allows each individual lion to expend less energy in the hunt.

same. But look at **Figure 11-15** to see what happens when the size of one population changes. You'll notice that as the mouse (prey) population rises, the number of owls (predator) also rises. At some point the large owl population will eat so many mice that only a few will be left. Without enough mice to eat, the owl population will decrease. What do you think will happen next to the size of each population?

Figure 11-15

Owls are predators of mice. The chart shows how the populations of mice and owls change over several years. Notice that as the number of mice (prey) rises, the number of owls (predator) also rises.

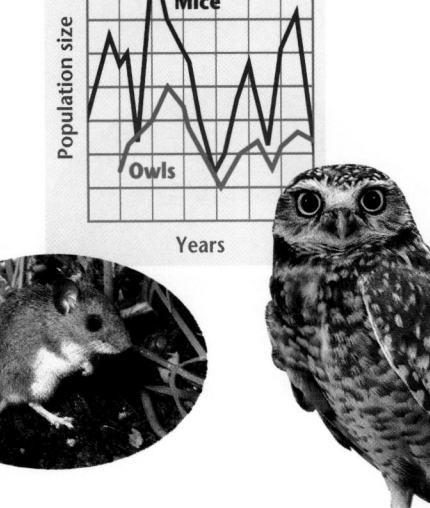

Multicultural Perspectives

George Washington Carver

George Washington Carver (1864–1943) conducted important research to extend and diversify the food chain. He found new ways to use peanuts, sweet potatoes, and pecans. Peanuts have nodules on their roots in which nitrogen-fixing bacteria can operate. These bacteria help replenish the soil by re-turning nitrogen to it. Born a slave in Missouri, Carver attended school in Iowa and served as director of agricultural research at Tuskegee Institute in Alabama.

Have student groups research and record the ways in which the peanut can be used as a food, such as cooking oil and peanut butter. Then have the groups share their findings and develop a list. **COOP LEARN** **P**

How Organisms Are Adapted

The success of an organism within its environment depends on how well it's adapted to that environment. An adaptation is a characteristic that increases the chance of an organism to survive in its environment. **Figure 11-16** shows one example of a plant adapted to a certain environment. Look back at **Figure 11-13** and see how many adaptations to rain forest conditions you can spot.

■ **Plant Adaptations**

Lack of rainfall in a desert is a limiting factor for many types of plants.

Why do cacti grow successfully in a desert environment? You'll recall from Chapter 10 that a cactus plant is adapted in two ways. First, its extensive root system quickly absorbs any water that may fall during a rainstorm. Second, its stems and leaves prevent the loss of water. The waxy covering called the cuticle prevents water loss through the stem. The leaves of the cactus are long, sharp spines. Because there is less leaf surface area, less water is lost through the spines than would be with other types of leaves.

Discussion Have students discuss and give examples of the kinds of adaptations different organisms have made to the ecosystems in which they live.

Content Background

The biosphere of Earth is divided into biomes, which are large geographic areas that have similar climates and climax communities. The major biomes are the tundra, northern coniferous forests, deciduous forests, grasslands, tropical rain forests, and deserts.

Succession is a gradual change in a community over time. Succession is discussed in more detail in Chapter 16.

Communities tend to alter the area in which they live in such a way as to make it less favorable for themselves and more favorable for other communities. There are two main types of competition. *Intraspecific competition* occurs between organisms of the same species. *Interspecific competition* occurs between organisms of different species.

Teacher F.Y.I.

Owls hunt mainly at night. In the dark, they can see from 10 to 100 times better than humans. Have students predict the owl's hearing capabilities. *They have superb hearing.*

How Wide Is Life's Comfort Zone?

You probably adjust the temperature of your home by using heaters and fans or air conditioners. How much heat and cold could you endure if you had to?

Many life-forms can't adjust the temperature where they live. How much heat and cold can living beings endure?

Cold

In Antarctica's McMurdo Sound, the water temperature ranges from -1.4°C to -2.15°C. Yet fish live and thrive in McMurdo Sound. Most of them are a kind of perchlike fish. How do they keep from freezing? Their bodies have evolved the

Antarctic fish

ability to produce compounds that act like strong antifreeze. Most other fish freeze when their body fluids cool to about -0.8°C, but McMurdo Sound's perchlike fish freeze only when their temperature goes down to about -2.2°C. McMurdo Sound very rarely gets that cold, so these fish have the habitat largely to themselves.

Heat

The prizewinner so far for surviving high temperatures is archaebacteria, a kind of microorganism that can grow at temperatures up to 110°C. Archaebacteria inhabit the waters near undersea volcanic vents.

Purpose

A Closer Look extends the discussion of how organisms are adapted to their environment. Students will learn about the ability of some species to adapt to a range of environmental conditions.

Content Background

The newly discovered archaebacteria are the oldest known monerans, or single-celled organisms that lack a nucleus. Archaebacteria cannot live in the presence of oxygen.

Although they seem to be rare, they are found in extremely hot, acidic, and alkaline environments like hot springs, cattle feed lots, and salt marshes. Although archaebacteria and eubacteria, or the true bacteria, both consist of single-celled organisms without a nucleus, they are not very closely related. Their cell walls and cell membranes differ in composition. Eubacteria are widespread organisms that can be found in almost every environment on Earth, inclu-

Adaptations to Low Light

Plants in rain forests are adapted to different limiting factors. Lower levels of the rain forest receive very little light. Any nutrients in the topsoil are quickly absorbed by the roots of tall trees, so the soil has few nutrients. Plants called epiphytes grow high up on the branches of the taller trees. They grow in the top layers of the forest and get water from the air and nutrients from decaying plant matter near their roots. In addition, growing high in the branches of a tree also exposes the plant to greater amounts of sunlight.

Figure 11-16

Ⓐ The silversword plant grows high in the craters of extinct volcanoes in Hawaii. During the day it's exposed to strong ultraviolet light. At night it's exposed to extreme cold.

Ⓑ The leaves of the plants have small hair-like projections that serve to protect the plant both from the cold and from the ultraviolet light.

Other Challenges

The high temperature isn't the only challenge these microbes have mastered. When volcanic gases dissolve in seawater, they make it very acidic. Some archaebacteria live in water with a pH of 1 or less—more acidic than stomach acid.

Other archaebacteria prefer water with a pH of 11.5—nearly as basic as the ammonia in household cleaners.

Some of these can also survive in solutions that are 36% salt. (For comparison, Atlantic and Pacific Ocean seawater is about 3% salt, and the Great Salt Lake is about 25% salt.)

That's not all; some archaebacteria even grow in the very deepest parts of the sea, at tremendous pressures.

From the cold ocean depths to the heat of volcanic vents—in strong acids or bases or salts—even the most extreme habitats can contain life.

What Do You Think?

Consider what kind of life-support suit you would need to visit Antarctic perchlike fish or archaebacteria where they live. Describe the extremes such a suit would have to protect you from.

In what other places—places that we now think to be uninhabitable—might we someday find living organisms?

Lake bed stained by the growth of Archaebacteria. **369**

ding hot springs, polar ice caps, and the guts of animals. However, they cannot withstand very high temperatures and acidic conditions. Most eubacteria also require oxygen.

Teaching Strategy

Before students read this feature, ask them to form hypotheses about the highest and lowest temperatures at which life can exist. Ask them to write their hypotheses in their journals and discuss them in small groups. Students can revise their hypotheses, if necessary, after reading and discussing this feature with the class. [L1]

Going Further ▶

Human beings, like bacteria, have also adapted to a wide range of conditions. Have students, working in small groups, select one group of people who have adapted to extreme conditions such as heat, cold, altitude, or drought. Have each group create a poster or bulletin board that illustrates and explains some of the adaptations people have developed to survive in an extreme environment. Select the appropriate Performance Task Assessment List in **PASC.** **COOP LEARN** [L2]

Visual Learning

Figure 11-17 In their journals, have students make a list of ways the camel is adapted to its environment. L1

3 ASSESS

Check for Understanding

Assign pairs of students to answer Check Your Understanding questions 1–4. As an introduction to the Apply question, ask students if they have ever gone on an overnight camping trip. Ask those who have what they brought with them. Discuss how these items protected them from nonliving and living limiting factors.

Reteach

Review Review common human activities with the students and ask them to identify limiting factors in them. Breathing, eating, drinking, and wearing clothes are a few behaviors that reveal limiting factors. L1

Extension

Activity Have students who have mastered the concepts in this section plant a cactus dish garden, paying particular attention to how requirements for soil, water, and light differ from those of other plants. Have students chart or graph the progress of the garden. L3 P

4 CLOSE

Discussion

Talk about living and nonliving limiting factors that apply to the students in the class. Ask students to discuss how they could survive in areas that they are not adapted to, such as the polar regions, the ocean, or the desert. L1

Figure 11-17

The camel's body is well adapted to its desert habitat.

Camels have two long toes on each foot. The two toes are connected by a broad cushion-like pad. The pad allows the camel to walk on sand without sinking, much like snow-shoes allow a person to walk on snow.

Camels can shut their nostrils and lips tightly to keep out blowing sand.

Thick eyebrows shade the camel's eyes from the bright desert sun and long lashes help keep out blowing sand.

Camel's knees have tough, leathery pads of skin, which cushion the impact and protect the camel from the burning hot sand when the camel kneels to rest.

■ Animal Adaptations

Just as plants are adapted to various environments, animals are also adapted. In **Figure 11-17** you can see how a camel is adapted to life in a desert environment.

Animals that live in very high mountain areas of the world also have adaptations to several limiting factors. For example, as you climb to higher elevations, the temperature drops and the amount of oxygen in the atmosphere decreases. Animals that live in this environment, such as the llama are protected against the cold and high altitudes.

If you were an astronaut orbiting Earth, you'd be able to see the white polar caps, the blue oceans, the brown deserts, and the green grasslands. Even at that distance you'd see that Earth is home to a variety of ecosystems. But only a closer look shows you how much the characteristics of each ecosystem influence the lives of the organisms that live there.

check your UNDERSTANDING

1. What effect would placing a plant in a closed box have on its growth? What limiting factor(s) is (are) involved?
2. Why would competition among organisms increase when resources are limited?
3. Would the adaptations of a cactus help or harm the plant if it were placed in a wet environment? Why?
4. **Apply** Limiting factors exist for humans as well as other organisms. Suppose you were on a camping trip in a desert. Make a list of the limiting factors you'd need to deal with. Identify the equipment you'd need to protect yourself from these limiting factors.

check your UNDERSTANDING

1. The plant's growth would be slowed or stopped. Light and possibly temperature are the limiting factors.
2. The same number of organisms would be competing for a smaller supply of resources.
3. These adaptations would harm the plant since they are designed to help the plant get more water.
4. Possible limiting factors include little water, cold nights, and hot days. Possible equipment includes canteens, blanket, water, and an umbrella to shade out the sun.

Teens in SCIENCE

One for All, and All for Trees

Have you heard about The Tree Musketeers? No, they don't carry swords or fight bad guys. But in their home town of El Segundo, California, these young people are real heroes.

Planting Trees

The Tree Musketeers began when several girls met in a Brownie Scout troop. "We were studying ecology," explains one of the founding members, Sabrina Alimahomed. "We learned how trees help clean the air and create rain. Since our state has an air quality problem and a drought, we felt that we could help by planting some trees."

Sabrina explains how the organization works. "The Tree Musketeers teaches people how to plant and care for trees," she says. "We got started by planting one tree, and the organization just grew from there. When we first heard about pollution, many of us felt frightened. We'd all assumed that because we were children, we didn't have to worry about things like the environment yet. But now we know that children need to care about the planet, too. Grown-ups always tell us that we can do anything if we set our minds to it. So, we decided to save the world."

There's no doubt about how valuable trees are to the environment. Trees not only help produce the oxygen we breathe, they also help prevent soil erosion, and control noise pollution by absorbing sound.

Saving Ecosystems

Over the years, more than 300 children have participated in The Tree Musketeers. Hundreds of trees have been planted.

Sabrina continued, "We hope that our work in The Tree Musketeers will continue to help both children and adults realize that when it comes to saving our planet, every person makes a difference."

Science Journal

Gather a clear plastic bag and some string. On a hot, sunny day, tie a small plastic bag around a few leaves hanging on a low tree branch. Leave the bag for a few hours. Now measure the amount of moisture that has collected on the inside of the bag. Record your results and observations *in your Journal.*

Teens in SCIENCE

Purpose

Teens in Science extends the discussion of limiting factors in Section 11-3 by showing how these can be changed by planting trees.

Content Background

Trees are defined as perennial woody plants with a single main stem from which branches and stems extend to form a crown. If they have broad leaves that are shed at the end of the growing season, they are called *deciduous.* *Coniferous* trees have needles that are shed less frequently. Trees are a source of wood, food, resins, rubber, quinine, turpentine, and cellulose. Tree rings can be used to measure the age of a tree, and often, they reflect climatic conditions of the past. Narrow rings can be an indication of drought or low temperature, while broad rings indicate years of favorable growing conditions.

Teaching Strategy

Have students tie bags around leaves in parts of the tree that are not exposed to strong sunlight and compare the amount of moisture that collects with that from the sunny part. Try the experiment with other plants such as evergreens and cacti.

Discussion

Ask students the name of your official state tree. Discuss whether there are any of these trees in your community today or if there were any of them in the past.

Going Further ▐▐▐▐▶

Have students get involved in a tree planting project in the community, if appropriate. Decide where the need is greatest in your area—whether to control soil erosion or noise pollution, produce more oxygen or rain, provide a habitat for local wildlife, or generally improve the beauty of the community. Have interested students write to The Tree Musketeers or any other national or local environmental group for more information on how to get started. L2

Science Journal

Students may want to try to drain the water collected into a graduated cylinder for measurement. If not enough water has collected they can make an estimate.

Science and Society

Purpose

Section 11-2 explains food chains and food webs. This Science and Society describes how changes in Chesapeake Bay caused by pollution have affected the food web in this ecosystem.

Content Background

English colonists settled in the Chesapeake Bay area in the seventeenth century and fishers, crabbers, and oyster hunters have made their living there ever since. It is estimated that 70 percent of the fish of the Atlantic coastal waters of North America spend at least part of their lives in the Chesapeake Bay. The state of Virginia harvested 693 million pounds of fish in 1990 from Chesapeake Bay—the largest catch of any state in the continental United States.

• The decay of algae that form the thick curtains is caused by bacteria that then consume even more oxygen than the algae. In spite of this, some species have actually flourished in this changed environment—bluefish and menhaden, for example.

• DDT and other chlorinated hydrocarbons have been replaced by insecticides containing compounds that break down more quickly into nontoxic forms. Biological control agents, or predator insects, may prove to be a better long-term substitute as pests become resistant to insecticides.

Science and Society — Saving Chesapeake Bay

You may have studied estuaries in geography or in a science class. An estuary forms where a river enters the sea, creating a mixture of fresh water and salt water. Chesapeake Bay is the largest estuary in the United States. The bay stretches 185 miles from the north where the Susquehanna River enters, to the south between Cape Charles and Cape Henry in southeastern Virginia.

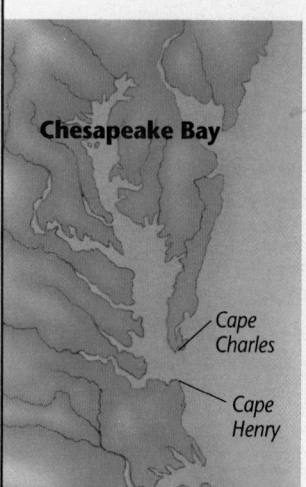

Chesapeake Bay

Cape Charles

Cape Henry

Location of Chesapeake Bay

Life in Chesapeake Bay

Chesapeake Bay is probably best known for the enormous number of crabs and oysters pulled from its waters. In addition to the large amount of aquatic life, the bay is host to many other animals.

The bay and its 7000 miles of shoreline are home to many water birds, including Canada geese, green-backed heron, snowy egrets, and the largest population of ospreys found in the United States.

Pollution Threatens the Bay

Forty-six major rivers and hundreds of tributaries carry fresh water to Chesapeake Bay. But the "fresh" water is really far from fresh. It's full of chemical runoff (contaminated water) from factories, sewage plants, and farms. These chemicals can cause diseases in fish and shellfish.

The Threat from Algae

But that's only one of the problems. Where water circulation is low—in narrow parts of rivers and estuaries—an overabundance of plant nutrients can cause problems, too. When runoff contains nitrates and phosphates, such as those found in fertilizers, the algae in the water

Osprey

feed on them and reproduce more rapidly. Curtains of these plantlike protists have been known to grow as much as three feet thick! This keeps the sun from reaching any oxygen-producing algae below.

When the nutrients in the water have been used up, the algae die and begin to decay. Rather than solving the problem, the decomposing algae make the situation even worse. The decaying process uses up the oxygen in the water, and the lack of oxygen makes it hard for fish and other aquatic organisms to survive.

DDT: Still a Threat

Polluted waters take a toll on water birds, too. From the 1950s to the early 1970s, runoff containing the pesticide DDT caused widespread contamination. Organisms can't eliminate DDT from their systems. The chemical becomes more concentrated in their body tissues with each step up the food chain.

Here's how DDT moved through one food chain in Chesapeake Bay. Runoff contaminated plants and organisms in the water. Fish ate these plants and organisms, or absorbed DDT directly through their skins. Birds ate the contaminated fish. When ospreys ate fish containing DDT, they

Blue crab

produced eggs with shells so thin that the chicks were crushed beneath their nesting parents. Because of this, the osprey's breeding population fell to about 1000 pairs. However, since the United States banned DDT in 1972, the osprey population in the bay has doubled.

Clean Up Underway

Efforts by the states of Maryland, Virginia, and Pennsylvania to clean up the bay have concentrated on wastewater treatment and regulation of development near its waters. Treating sewage more thoroughly than required by law, limiting construction and the use of concrete and asphalt, and finding ways to keep soil and chemicals out of the water may help save the bay.

inter**NET**
CONNECTION

The Environmental Protection Agency is involved in federal cleanup efforts at Chesapeake Bay. Use the World Wide Web to find information on the cleanup program that might also have an application in your area.

Teaching Strategy To model an estuary, fill a large clear glass baking dish with water about halfway to the top. Mix food coloring into a heavily salted (4–6 teaspoons) cup of water. Pour the salt water gently into one end of the dish. Students should see that the salt water sinks to the bottom of the dish. Explain that in an estuary, the mixture of salt and fresh water is not uniform. The force of the river flow and the tides of the ocean produce regions of varying salinity. Most freshwater fish can live only in fresh water. Contamination of this water may not be cleansed rapidly by the ocean tides.

inter**NET**
CONNECTION

To find information on the cleanup of the Chesapeake Bay, contact the Environmental Protection Agency at **http://www.epa.gov/r3chespk**

Going Further ▸

Major port cities such as New York, London, Hong Kong, and Calcutta are situated on estuaries. On a world map, identify estuaries and major cities that are located on them. Divide the class into small groups and assign each group one of the cities or estuaries. Have students research any destruction to the ecosystem that has occurred there as well as any efforts that are being made to clean it up. Students may find that many of these areas have more fish in them today than they did

ten or twenty years ago, while others are getting worse. Have the groups compare notes to see what factors seem to work in cleaning up this fragile type of ecosystem in different areas of the world. Use the Performance Task Assessment List for Group Work in **PASC**, p. 97. L1 **COOP LEARN**

Have students look at the four illustrations on this page. Ask them to describe details that support the main ideas of the chapter found in the statement for each illustration.

Teaching Strategies

Divide the class into four groups. Have students construct and label a three-dimensional model that incorporates and highlights a main idea of the chapter. Ideas include:

1. Re-create a woodlands scene inside a cardboard box. Include producers such as the grass and trees, and consumers such as the deer and raccoon. Nonliving components of the scene could include a stream of water, soil, and a rotting log.

2. Construct a model of a food chain. Predator-prey relationships, producers, consumers, and decomposers can be included in a more complex food web.

Answers to Questions

1. Conditions would include stagnant and wet. Organisms would include fish, frogs, algae, water lilies, insects, and so on.

2. Sun → Grass → Cow → Hamburger

3. Because without recycling materials would be used up and not available any more.

4. See Figure 11-17 on page 370 for a complete description.

GLENCOE TECHNOLOGY

MindJogger Videoquiz

Chapter 11 Have students work in groups as they play the Videoquiz game to review key chapter concepts.

Science Journal

Review the statements below about the big ideas presented in this chapter, and answer the questions. Then, re-read your answers to the Did You Ever Wonder questions at the beginning of the chapter. *In your Science Journal,* write a paragraph about how your understanding of the big ideas in the chapter has changed.

1 A community of organisms interacting with the nonliving environment is an ecosystem. *Describe conditions and organisms in a pond ecosystem.*

2 A food chain shows how food energy flows through an ecosystem. *Describe the food chain that leads to a hamburger on your plate.*

3 Many materials, such as water, oxygen, carbon dioxide, and nitrogen, are constantly being recycled through the environment. *Why is it necessary for water, oxygen, and nitrogen to be recycled?*

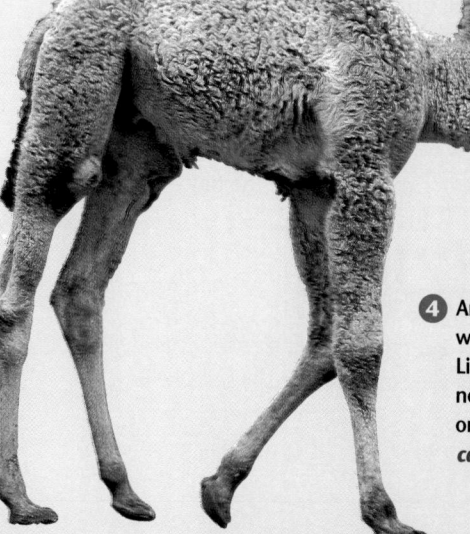

4 An organism's survival is related to how well it is adapted to the environment. Limiting factors, which may be living or nonliving, influence the survival of an organism or a species. *Describe how a camel is adapted to its environment.*

Science at Home

Have students write about their role in a food web and illustrate it. Ask them to make note of everything they eat during one 24-hour period. Have them find out where these food products came from. Are they from producers, consumers, or decomposers? What were their native habitats like? In the case of the consumers, from what did they obtain their food? Use the Performance Task Assessment List for Writing in Science in **PASC,** p. 87. **L1** **P**

Science Journal

Did you ever wonder...

• Molds are decomposers. These multicellular organisms break down dead organic matter such as bread for their food. (p. 353)

• Our oxygen is recycled. Although life forms use up oxygen during respiration, plants produce oxygen as a waste product of photosynthesis. (p. 360)

• Many species of penguins are adapted to survive in a habitat that has cold weather. (p. 364)

Using Key Science Terms

community	habitat
decomposer	limiting factor
ecosystem	niche
food chain	population
food web	

For each set of terms below, explain the relationship that exists.

1. niche—habitat
2. population—community
3. food chain—decomposer
4. food chain—food web
5. limiting factor—ecosystem

Understanding Ideas

Answer the following questions in your Journal using complete sentences.

1. What is the relationship between a population, community, and ecosystem?
2. Explain why producers and decomposers are important.
3. What is the source of the energy that flows through most of the food chains on Earth?
4. List three examples of limiting factors.
5. Give an example of how a living thing might be adapted to survive despite limiting factors in the environment.

Developing Skills

Use your understanding of the concepts developed in this chapter to answer each of the following questions.

1. **Concept Mapping** Using the following terms complete the concept map of limiting factors: *adaptations, competition, light, nonliving, predator-prey, rainfall.*

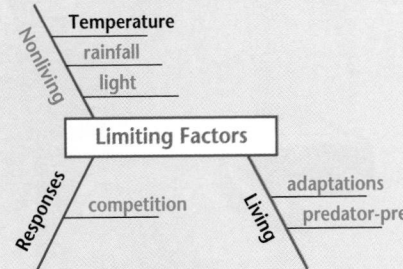

2. **Sequencing** Look at the food web on page 357. Follow the food chains of a snake and a hawk.

3. **Separating and Interpreting Data** Change the amount of water or temperature used for growing mold in the Investigate on pages 362-363 to see how growth might be affected.

Critical Thinking

In your Journal, *answer each of the following questions.*

1. Give examples of situations when too much or too little of a material becomes a limiting factor.
2. Explain why natural cycles are important in the environment.

Program Resources

Review and Assessment, pp. 65–70 [L1]
Performance Assessment, Ch. 11 [L2]
PASC
Alternate Assessment in the Science Classroom
Computer Test Bank [L1]

Developing Skills

1. See reduced student page for completed concept map.
2. grass → grasshopper → snake → hawk; grass → mouse → hawk
3. Responses should indicate that the molds that are denied light and exposed to extremes in temperatures are not as healthy as the control molds and some may even die.

Using Key Science Terms

1. Niche is the role an organism fills in a community and includes what it eats and where it lives. Habitat is its home and, therefore, part of its niche.

2. Population is the total number of individuals of a species in an area; community is all of the populations that interact with each other in that area.

3. Decomposers cause the decay of dead organisms; they are part of a food chain, often considered the "end" of the chain.

4. A food chain shows how food energy flows through a community; a food web is the combination of all the interlocking food chains in the community.

5. A limiting factor is a condition within an ecosystem—either living or nonliving—that influences the growth or survival of an organism or species.

Understanding Ideas

1. Populations interact to form a community. Communities of organisms interact with one another and with the environment making up an ecosystem.

2. Producers make the food for all organisms on Earth. Decomposers break down dead organisms into nutrients that producers and consumers can use again.

3. The sun provides the energy that flows through most of the food chains on Earth.

4. Answers will vary but may include temperature, wind, chemicals in the soil, amounts of light and water, food, and shelter.

5. Examples will vary. Cacti have a thick waxy cuticle and small needle-like leaves to conserve moisture.

Critical Thinking

1. Accept all reasonable answers. One possible answer: Marigolds will turn yellow and die if too much water is provided.

2. Earth has a limited supply of resources. If they were not recycled, we would run out of these resources.

3. The two organisms would compete with each other for the exact same role and resources. There wouldn't be enough to meet the needs of both.

4. A cow is a plant-eating consumer that lives in open fields.

5. Accept all reasonable answers. The removal of bark beetles might cause a decline in the skunk population or an increase in the tree population.

Problem Solving

1. present population of Animal A—100; year 1–200; year 2–400; year 3–800; year 4–1600

2. present population of Animal B—50; year 1–150; year 2–450; year 3–1350; year 4–4050

3. present population of Animal C—25; year 1–100; year 2–400; year 3–1600; year 4–6400

4. at present can support 10 000 animals; after year 1–9000; year 2–8100; year 3–7290; year 4–6561

5. in the fourth year

Connecting Ideas

1. A food web shows feeding relationships between organisms.

2. Sunlight supplies energy that producers use to make food.

3. All ecosystems must provide food, water, and a survivable climate. They'll differ in types of plants and animals.

4. Poisonous chemicals and pollutants enter the body of an organism that is consumed by other animals.

3. Explain why two species cannot occupy the same niche at the same time.

4. Describe the niche of a cow.

5. What effect would humans have if they tried to kill all the bark beetles in this community with a chemical?

Problem Solving

Read the following problem and discuss your answers in a brief paragraph.

Large areas of the rain forest are being destroyed each year. This loss of habitat means fewer animals can live in the remaining area. You are a biologist trying to predict how long the habitat can support three animal populations.

1. Animal A currently has a population of 100. Each year its population doubles. How large will its population be after 1 year, 2 years, 3 years, and 4 years?

2. Animal B has a population of 50. Each year its population triples. How large will its population be after 1 year, 2 years, 3 years, and 4 years?

3. Animal C has a population of 25. Each year its population quadruples. What is its population after 1 year, 2 years, 3 years, and 4 years?

4. Right now the area in the rain forest that you are studying can provide food for 10 000 animals. Because of continued habitat destruction, that number is reduced by 10 percent a year. Find out how many animals the habitat can feed after 1 year, 2 years, 3 years, and 4 years.

5. In which year will the total population outstrip the food supply?

CONNECTING IDEAS

Discuss each of the following in a brief paragraph.

1. Theme—Energy How does a food web show the relationship between organisms in an ecosystem?

2. Theme—Energy What role does sunlight play in an ecosystem on Earth?

3. A Closer Look Describe ecosystems that you are a part of. How are they alike and different?

4. Science and Society How are food webs related to the spread of poisonous chemicals and other pollutants in an ecosystem?

✔ **Assessment**

Portfolio Review the portfolio options that are provided throughout the chapter. Encourage students to select one product that demonstrates their best work for the chapter. Have students explain what they learned and why they chose this example for placement in their portfolios.

Additional portfolio options can be found in the following **Teacher Classroom Resources:**

Multicultural Connections, pp. 27, 28

Making Connections: Integrating Sciences, p. 25

Making Connections: Across the Curriculum, p. 25

Concept Mapping, p. 19

Critical Thinking/Problem Solving, p. 24

Take Home Activities, p. 19

Laboratory Manual, pp. 63–70

Performance Assessment P

Interactions In the Living World

In this unit, you classified organisms as plants or animals using characteristics such as the ability to move about freely and search for food and water.

You learned how plants manufacture their own food using sunlight and that animals then rely on this stored food in plants to survive. You also saw how plants and animals form an important part of the oxygen-carbon dioxide cycle.

Try the exercises and activity that follow—they will challenge you to use and apply some of the ideas you learned in this unit.

CONNECTING IDEAS

1. Trace the pathway of water from soil, through a vascular plant to its leaves, and back again to the soil. Explain what processes occurred in which plant parts, both inside and outside the plant. Relate the processes inside and outside the plant to their effects on the ecology of the plant's habitat.

2. Obtain samples of a wide variety of objects from your teacher. Design a classification scheme that will enable you to classify living and nonliving things. Decide if other groupings are needed to further classify the living things and the nonliving things.

Exploring Further   ACTIVITY

How Might Salt Be a Limiting Factor?

What To Do
Design an experiment that will determine if radish seedlings will grow in salt. Be sure to keep all factors except the amount of salt constant. The amount of salt given to the plants will be the only variable in your experiment. Record your design *in your Journal.*

Interactions in the Living World

THEME DEVELOPMENT

All living things are structurally similar. They also are similar in terms of their basic needs such as food, energy, reproduction, and response to the environment. Those concepts were developed in this unit as students were introduced to the variety of living things present on Earth and the technique used for classifying them.

Connections to Other Units

Living things are composed of chemicals. Thus, living and non-living matter is composed of atoms, molecules, and compounds. Thus Unit 3 is closely related to Unit 2. The relationship between living things and the systems moving on Earth's surface are explored in the following unit, Unit 4.

Connecting Ideas
Answers

1. Intake of water from the soil takes place via root hairs. Water loss from the leaf occurs through open stomata. Students must be aware of the water cycle and how the different states of water are involved.

2. A clue to the difference between living (or once living) and nonliving would be the microscopic organization (cellular organization) of each item.

Exploring Further

How might salt be a limiting factor?
Purpose To investigate the effect of salt on the germination and growth of radish seedlings.

Background Information
Radish seedlings will sprout in about 3–4 days after seeds are soaked in water overnight. Use sand or vermiculite in plastic cups as the material for planting the seedlings. Provide different concentrations of saltwater to be used in watering each cup of plants. Have students mark all plant cups with the saltwater concentration being used. Make sure that all plants receive the same volume of saltwater each time watering takes place.

✔ Assessment

Oral Have students make an oral presentation about the experimental designs and results. Use the Performance Task Assessment List Oral Presentation in **PASC**, p. 71. **COOP LEARN** **L1**

UNIT 4

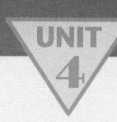

Changing Systems

UNIT OVERVIEW

UNIT FOCUS

In Unit 2, students studied relationships and interactions in the physical world. In Unit 3, this was extended to interactions and relationships in the living world. In Unit 4, students extend their understanding to systems and the changes that occur in systems.

THEME DEVELOPMENT

Themes developed in Unit 4 are energy, stability and change, and systems and interactions. How time, distance, and displacement are related provides the basis for understanding the changes that take place when motion occurs. Another important focus of this unit is how mechanical and chemical energy interact with matter to erode the surfaces of Earth, and thus create changes in the environment.

Connections to Other Units

The concept of motion, both on and near Earth, is strongly related to the next unit. In Unit 5, movement of Earth and Earth materials during earthquakes and volcanoes is described. Also, movement of the moon as part of the Earth-moon system and the effect of this motion on Earth's oceans is explained.

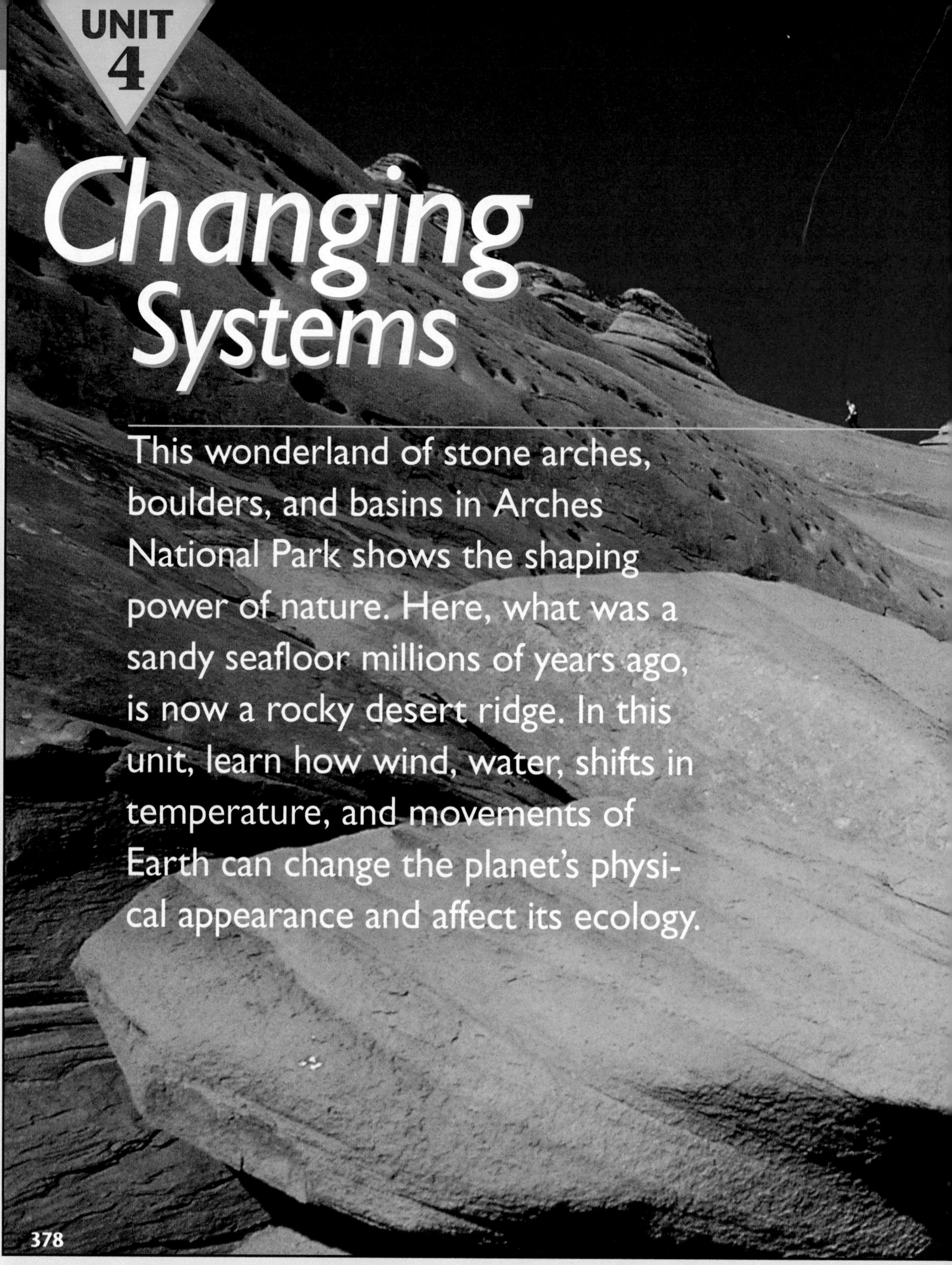

UNIT 4

Changing Systems

This wonderland of stone arches, boulders, and basins in Arches National Park shows the shaping power of nature. Here, what was a sandy seafloor millions of years ago, is now a rocky desert ridge. In this unit, learn how wind, water, shifts in temperature, and movements of Earth can change the planet's physical appearance and affect its ecology.

378

GLENCOE TECHNOLOGY

Videodisc

Use the *Science Interactions, Course 1* **Integrated Science Videodisc** lesson, *Motion,* with Chapters 12 and 13 to reinforce and enhance the concepts of velocity and acceleration.
Use the **Integrated Science Videodisc** lesson, *Ecosystems,* with Chapter 16. This videodisc reinforces and enhances the concepts of ecosystems and ecological balance.

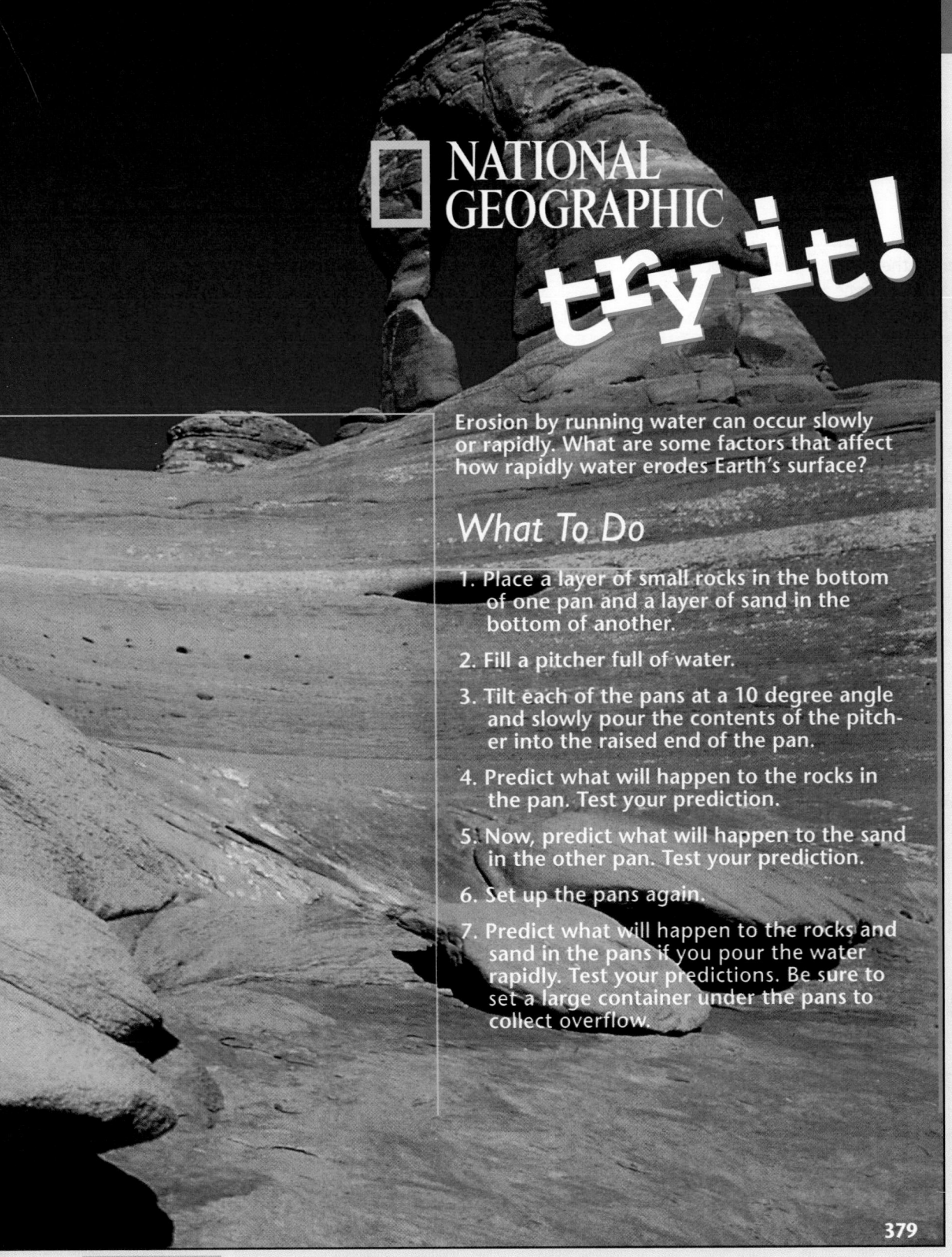

NATIONAL GEOGRAPHIC

try it!

Erosion by running water can occur slowly or rapidly. What are some factors that affect how rapidly water erodes Earth's surface?

What To Do

1. Place a layer of small rocks in the bottom of one pan and a layer of sand in the bottom of another.

2. Fill a pitcher full of water.

3. Tilt each of the pans at a 10 degree angle and slowly pour the contents of the pitcher into the raised end of the pan.

4. Predict what will happen to the rocks in the pan. Test your prediction.

5. Now, predict what will happen to the sand in the other pan. Test your prediction.

6. Set up the pans again.

7. Predict what will happen to the rocks and sand in the pans if you pour the water rapidly. Test your predictions. Be sure to set a large container under the pans to collect overflow.

379

Try It!

Chapter Organizer

SECTION	OBJECTIVES	ACTIVITIES & FEATURES
Chapter Opener		Explore!, p. 381
12-1 Position, Distance, and Speed (3 sessions; 1.5 blocks)	1. **Specify** the position of an object. 2. **Find** the distance along a path. 3. **Determine** the average speed of a moving object. **National Content Standards: (5–8) UCP2, UCP3, A1, B2, G1**	Explore!, p. 383 **Design Your Own Investigation 12-1:** pp. 388–389 **Life Science Connection,** pp. 384–385 **History Connection,** pp. 405–406 **Art Connection,** p. 408
12-2 Velocity (3 sessions; 1.5 blocks)	1. **Distinguish** between displacement and distance. 2. **Find** the average velocity of a moving object. 3. **Distinguish** between velocity and speed. 4. **Use** the concept of relative velocity. **National Content Standards: (5–8) UCP2–3, B2, G2**	Skillbuilder: p. 392 Explore!, p. 393 **A Closer Look,** 394–395
12-3 Acceleration (2 sessions; 1 block)	1. **Distinguish** between velocity and acceleration. 2. **Determine** acceleration from velocity change and time. **National Content Standards: (5–8) UCP2–3, A1, B2**	Investigate 12-2: pp. 398–399
12-4 Motion Along Curves (1 session; .5 block)	1. **Distinguish** between a displacement and a distance along a curved path. 2. **Find** the average velocity for motion along a curved path. 3. **Recognize** situations for which there is an acceleration, called the centripetal acceleration, even when the speed is constant. **National Content Standards: (5–8) UCP2–3, B2, E1, G1, G3**	Find Out!, p. 401 Explore!, p. 403 **Technology Connection,** p. 407

ACTIVITY MATERIALS

EXPLORE!

p. 381* lined, graph, and blank paper; pencil

p. 383* overhead projector, transparency of House of Terror map; copies of transparency for students

p. 393* battery-powered toy car, paper

p. 403* test tube with stopper, water, record turntable, masking tape

INVESTIGATE!

pp. 398–399 protractor, 10- to 12-cm length of string, button

DESIGN YOUR OWN INVESTIGATION

pp. 388–389* meterstick, stopwatch, masking tape

FIND OUT!

p. 401* protractor, metric ruler

KEY TO TEACHING STRATEGIES

The following designations will help you decide which activities are appropriate for your students.

- **L1** Basic activities for all students
- **L2** Activities for average to above-average students
- **L3** Challenging activities for above-average students
- **LEP** Limited English Proficiency activities
- **COOP LEARN** Cooperative Learning activities for small group work
- **P** Student products that can be placed into a best-work portfolio
- Activities and resources recommended for block schedules

Need Materials? Call Science Kit (1-800-828-7777).

[00:00] OUT OF TIME? We recommend that students do the activities with an asterisk.

TEACHER CLASSROOM RESOURCES

Student Masters	Transparencies
Study Guide, p. 43 **Critical Thinking/Problem Solving**, p. 5 **Take Home Activities**, p. 21 **How It Works**, p. 15 **Making Connections: Integrating Sciences**, p. 27 **Activity Masters**, Design Your Own Investigation 12-1, pp. 51–52	**Teaching Transparency 24**, Distance-Time Graph **Section Focus Transparency 37**
Study Guide, p. 44 **Concept Mapping**, p. 20 **Making Connections: Across the Curriculum**, p. 27 **Making Connections: Technology and Society**, p. 27 **Critical Thinking/Problem Solving**, pp. 5, 20 **Science Discovery Activities**, 12-1	**Teaching Transparency 23**, Displacement Map **Section Focus Transparency 38**
Study Guide, p. 45 **Multicultural Connections**, p. 28 **Science Discovery Activities**, 12-2, 12-3 **Laboratory Manual**, pp. 71–74, Speed and Acceleration **Activity Masters**, Investigate 12-2, pp. 53–54	**Section Focus Transparency 39**
Study Guide, p. 46 **Multicultural Connections**, p. 27	**Section Focus Transparency 40**

ASSESSMENT RESOURCES	TEACHING & TECHNOLOGY
Review and Assessment, pp. 71-76 **Performance Assessment**, Ch. 12 **PASC*** **MindJogger Videoquiz** **Alternate Assessment In the Science Classroom** **Computer Test Bank**	**Spanish Resources** **Cooperative Learning Resource Guide** **Lab and Safety Skills** **Science Interactions, Course 1, CD-ROM** **Computer Competency Activities**

*Performance Assessment in the Science Classroom

NATIONAL GEOGRAPHIC TEACHER'S CORNER

National Geographic Society Products

To order the following products for use with this chapter, call the National Geographic Society at 1-800-368-2728:

- Newton's Apple Physical Sciences (Videodisc)
  STV: Restless Earth

National Geographic Society Products Available From Glencoe

To order the following products for use with this chapter, call the National Geographic Society at 1-800-368-2728:

- Everyday Science Explained (Book)

GLENCOE TECHNOLOGY

The following multimedia resources are available from Glencoe.

Science and Technology Videodisc Series (STVS)
Physics
  Laminar Flow over Airplane Wings

Glencoe Science Interactions Interactive Videodisc
Motion

Physical Science CD-ROM

Teacher Classroom Resources

These are key components of the classroom resources package.

TEACHING AIDS

Teaching Transparencies

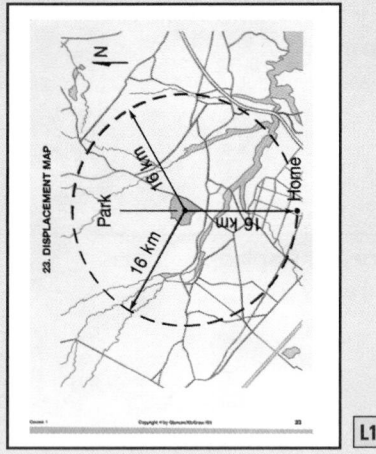

Section Focus Transparencies

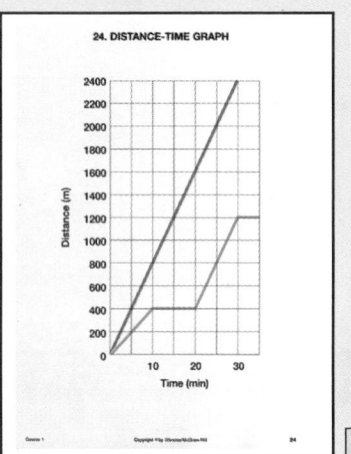

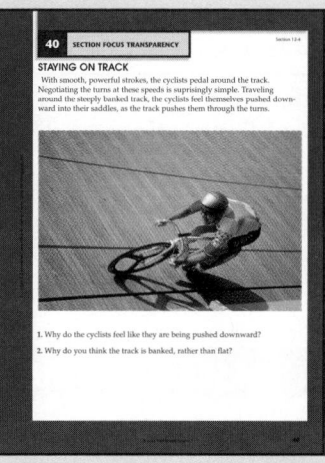

HANDS-ON LEARNING

Science Discovery Activity*

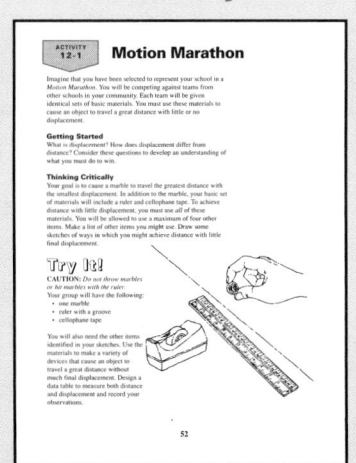

Laboratory Manual*

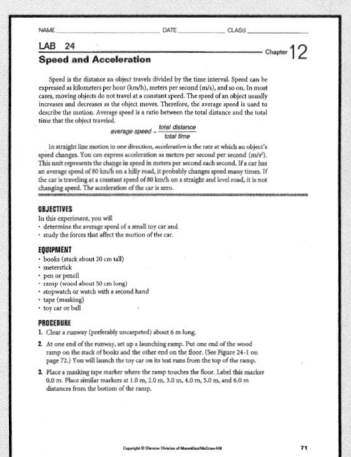

Take Home Activity

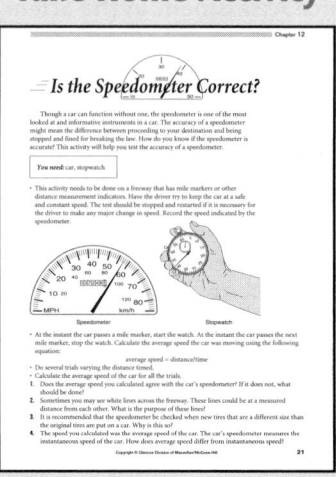

*There may be more than one activity for this chapter.

Chapter 12 Motion

REVIEW AND REINFORCEMENT

Study Guide*

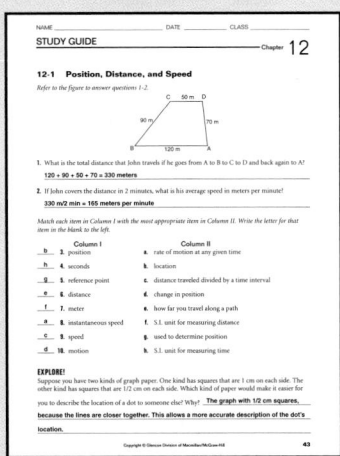

L1

Concept Mapping

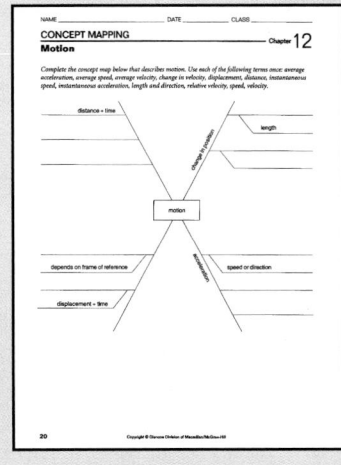

L1

Critical Thinking/ Problem Solving

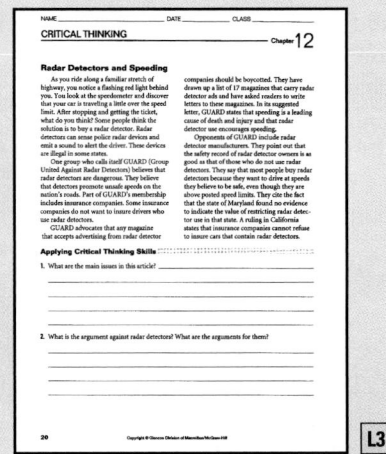

L3

ENRICHMENT AND APPLICATION

Integrating Sciences

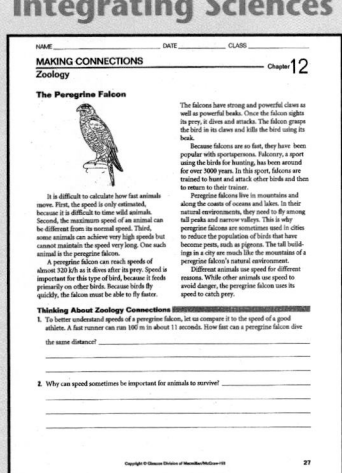

L2

Across the Curriculum

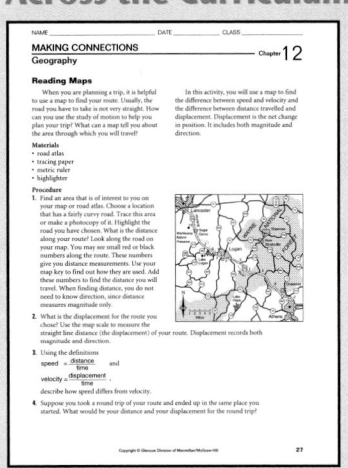

L2

Technology and Society

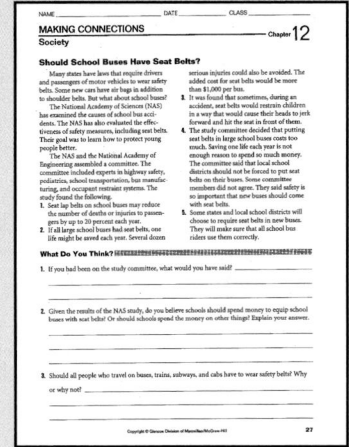

L2

Multicultural Connection**

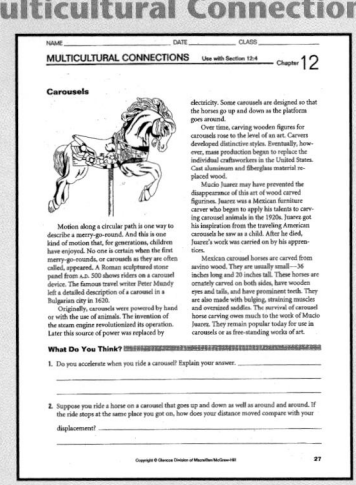

L1

ASSESSMENT

Performance Assessment

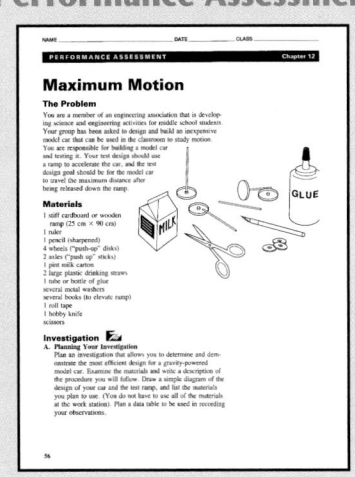

L2

Review and Assessment

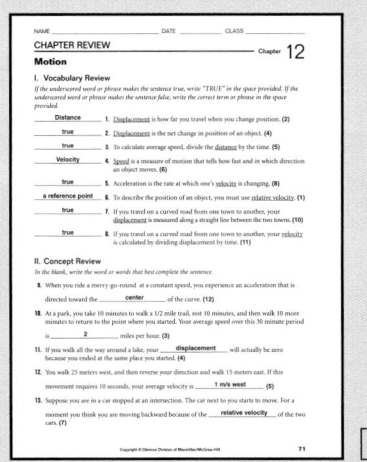

L1

Motion

THEME DEVELOPMENT

The term *motion* itself implies change. The position of an object and the change in its position over time are described in this chapter. The theme of systems and interactions is illustrated by looking at objects and their motion from different viewpoints.

CHAPTER OVERVIEW

In this chapter, students will learn to describe an object's position relative to a reference point. Students will learn that motion is any change in position. By measuring the distance an object moves or its displacement over time, students will calculate the object's speed and velocity. Finally, students will learn that objects accelerate by changing speed or direction.

Tying to Previous Knowledge

Have students point out any motion that they can see—the hands of a wall clock or the movement of a classmate. Ask them to recall implied motion such as the revolution of the moon or the movement of light through a prism.

INTRODUCING THE CHAPTER

Have students infer which objects are in motion in the picture of the amusement rides on pages 380–381. Ask them to describe other motion they saw when they last went to a fair.

00:00 OUT OF TIME?

If time does not permit teaching the entire chapter, use the Chapter Overview on this page, Reviewing Main Ideas at the end of the chapter, and the Chapter 12 audiocassette to point out the main ideas of the chapter.

MOTION

Did you ever wonder...

✓ Why you have a sensation of moving backward as you sit in a car while one next to you is pulling away from a traffic light?

✓ Why your stomach feels funny when you're on a roller coaster?

✓ Why you feel pulled outward when riding around a curve in a bus?

Science Journal

Before you begin your study of motion, think about these questions and answer them *in your Science Journal.* When you finish the chapter, compare your journal write-up with what you have learned.

Answers are on page 409.

380

*T*he first visit to the amusement park each summer is the best. This year, the high point of the visit is the new sky ride. Did you notice that as the chairs of the sky ride start forward, you feel pushed backward into your seat? Did you also notice that as the ride stopped for new passengers, you slid forward a little? Do you remember how the roller coaster went faster and faster as it went downhill? These are some of the effects of motion that we'll be exploring in this chapter.

▶ *In the activity on the next page, explore some of the ways you could describe a position to someone else.*

Learning Styles	Kinesthetic	Explore, pp. 381, 393; Activity, pp. 382, 384, 402
	Visual-Spatial	Explore, pp. 383, 403; Visual Learning, pp. 385, 390, 392, 397, 403; Multicultural Perspectives, p. 387; Across the curriculum, p. 387; Demonstration, pp. 392, 396, 400; Activity, pp. 400, 404
	Logical-Mathematical	Life Science Connection, pp. 384–385; Across the Curriculum, pp. 385, 397; Visual Learning, p. 386; Activity, pp. 387, 389, 393, 395, 404; Investigate, pp. 388–389, 398–399; Discussion, p. 391; A Closer Look, pp. 394–395; Find Out, p. 401; Project, p. 409
LS	**Linguistic**	Visual Learning, pp. 382, 385; Discussion, p. 390; Research, p. 392

Explore! ACTIVITY

How would you tell someone where you are?

Almost every day you have to tell someone where somebody or something is. How do you go about it?

What To Do

1. With your pencil, make a dot somewhere along the top edge of a sheet of unlined paper. Describe *in your Journal* the location of the dot on your paper.

2. Put a second dot on the same sheet of paper.

3. In your own words, describe its position and compare the location of the first dot with that of the second.

4. Ask a friend to read only your journal description and from the description, try to place two dots on another piece of paper to look like yours.

5. Repeat this exercise with lined paper and then with graph paper.

6. Does your ability to tell someone where the dots are improve each time? Why do you think this is so?

 # Position, Distance, and Speed

PREPARATION

Planning the Lesson

Refer to the Chapter Organizer on pages 380A–D.

Concepts Developed

In this section, students learn how to describe objects in terms of their position, direction of movement, and speed. They learn how to locate an object, measure the distance it travels, and find its average speed.

1 MOTIVATE

Bellringer

 Before presenting the lesson, display **Section Focus Transparency 37** on an overhead projector. Assign the accompanying **Focus Activity** worksheet. L1 LEP

Activity To help students communicate their observations, ask one student to walk, hop, or run from one side of the classroom to the other. Then ask the other students to classify motion in as many ways as they can. Answers may include the type of motion, the speed of the motion, and the direction of the motion. L1 LEP

Visual Learning

Figure 12-1 Can you describe the position of the merry-go-round without referring to any other object in the photograph? *No, to describe position, you must have a reference point.*

Section Objectives

■ Specify the position of an object.
■ Find the distance along a path.
■ Determine the average speed of a moving object.

Key Terms

position
distance
average speed

Position and Motion

One of the best things about going to an amusement park is going there with friends. Sometimes, however, not everyone can arrive together—you need to decide where and when people are to gather so that no one gets left out. As an example, you might specify at 2:00 P.M. by the front gate, or perhaps by the merry-go-round at noon.

■ Position

In the Explore activity, you tried to describe the location or position of a dot on a sheet of paper. If you described the position of the dot by saying how far it was from the top and side edges of the paper, you were using points along the edges of the paper as reference points. The **position** of an object must always be described by comparing it to a reference point. By the time you did the exercise with graph paper, all you had to do to describe your position to a friend was to say something like "over five squares from the left edge and down six squares from the top."

Where you are located is your position compared to a reference

Figure 12-1

To describe the position of something, you must have a reference point. This amusement park is located just outside the center of Calgary, Canada. Many of the food booths are just in front of the large Ferris wheel. The large Ferris wheel is to the left of the smaller Ferris wheel. Find the merry-go-round in the photo. Can you describe its position without referring to any other object in the photo?

Program Resources

Study Guide, p. 43
Teaching Transparency 24 L1
Take Home Activities, p. 21, Is the Speedometer Correct? L1
Critical Thinking/Problem Solving, p. 5, Flex Your Brain
Section Focus Transparency 37

---

point. A reference point might be the base of a Ferris wheel during your trip to the amusement park. A reference point could also be in front of a certain store in a mall or by the door of a particular building on a farm. The main thing is that everyone needs to know what the particular reference point is in order to locate his or her position.

■ Motion

What if you walk from the water slide to the tilt-a-whirl? Doesn't your position change? When you change your position you experience motion.

You could describe your change in position to a friend in terms of the number of steps you took walking from the water slide to the tilt-a-whirl. For your friend to be able to take the same walk, however, you would also have to say in what direction you walked. For example, you could say that you walked 75 steps north and then 100 steps west. The direction of motion can be described using compass directions or using the number of degrees in a circle. Let's explore how you might describe both the direction and distance of motion by using just a map.

Explore! ACTIVITY

How can you use a map to help describe motion?

What To Do

1. Using the figure shown, determine what distance and in what direction you must travel along the paths to go from the Tunnel of Love to the House of Terror.

2. What direction describes the path you would take when moving from the Ferris wheel to the Tunnel of Love?

3. Where do you end up if you travel directly west 50 meters from the Ferris wheel?

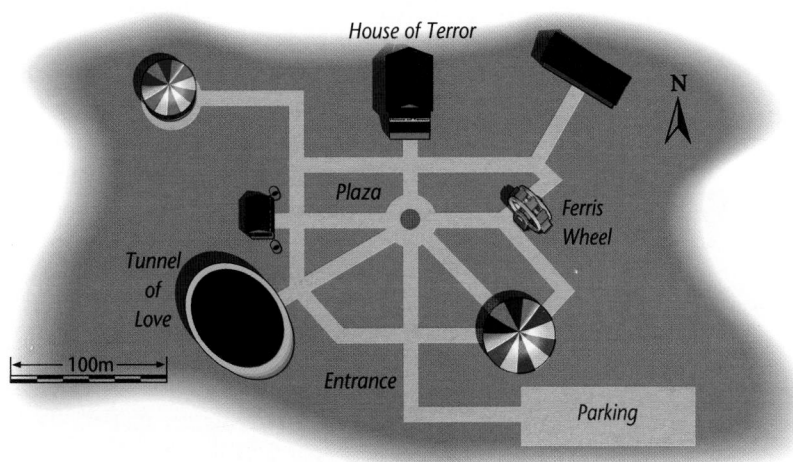

12-1   Position, Distance, and Speed   **383**

Explore!

TECH PREP

How can you use a map to describe motion?

Time needed   10 to 15 minutes

Materials   overhead projector, transparency and student copies of House of Terror map

Thinking Processes   observing and inferring, measuring in SI, sequencing

Purpose   To describe motion by measuring distance and identifying direction.

Preparation   If you do not have an overhead projector sketch the map on the chalkboard.

Teaching the Activity

Activity   Have students recognize spatial relationships by asking a volunteer to trace the path from the Tunnel of Love to the House of Terror. Then review with them how to use the map scale to find actual distance. L1 **COOP LEARN**

Tying to Previous Knowledge

Ask students to give examples of activities they or other organisms perform that involve motion. *getting out of bed, walking from the classroom to the cafeteria, running, and flying* Inanimate objects show motion as well. Have students name such objects. *cars, buses, ocean waves*

Flex Your Brain   Have students use the Flex Your Brain activity to explore MOTION.

Science Journal   Students may describe the routes they chose by referring to reference points in the amusement park or by using directional and distance designations. L1

Expected Outcome

Using the map, students should be able to measure the distance between two points and describe the distance and direction of motion needed to get from one place to another.

Answers to Questions

From Tunnel of Love to House of Terror go 100 m northeast then 38 m north. From the Ferris Wheel to the Tunnel of Love you would travel southwest. If you travel 50 m west from the Ferris Wheel you would end up at the plaza.

✔ Assessment

Process   Help students practice recognizing spatial relationships by drawing maps showing different routes through the amusement park. Have students describe the distance and direction of motion using the rides as reference points. Then have students make posters that include the maps and descriptions of motion. Use the Classroom Assessment List for Poster in **PASC**, p. 73. **COOP LEARN** P L1

Distance Along a Path

When you walk through the amusement park, you change your position with every step you take. In doing so, you travel a certain distance. **Distance** is how far you travel along a path while you change your position. You've seen that a distance can be measured using just about any units—the number of steps, or the number of city blocks, for example. In science, we usually use SI units. The SI unit for measuring distance is the meter, abbreviated m.

Walkers and joggers wear a small device to record the distance that they travel. If you measured your distance as you walked between rides, food booths, and shows, you might find that you had covered several thousand meters (1000 meters = 0.6 mi).

Figure 12-2

If you wanted a snack at the amusement park, you might walk 250 m from the merry-go-round to the snack bar. Walking back along the same route, you would travel another 250 m to return to your starting point. The distance along the path between those two points is 250 m; the total distance to the snack bar and back is 500 m.

Life Science CONNECTION

How Fast Do Animals Move?

All animals have specialized structures that allow them to move in different ways and at different speeds.

Animals can move on land, on water, under water, and through the air. They can move their whole bodies or just some of their parts. Animals can walk, run, leap, jump, climb, and dig. Most birds can run, walk, fly, glide, soar, swim, or waddle on land. Snakes can slither. Squids and scallops can move by jet propulsion.

For vertebrates, animals with backbones, movement requires muscles attached to movable bones, and energy. The faster an animal moves, the more energy it uses. For invertebrates, animals without backbones, movement requires

384 Chapter 12 Motion

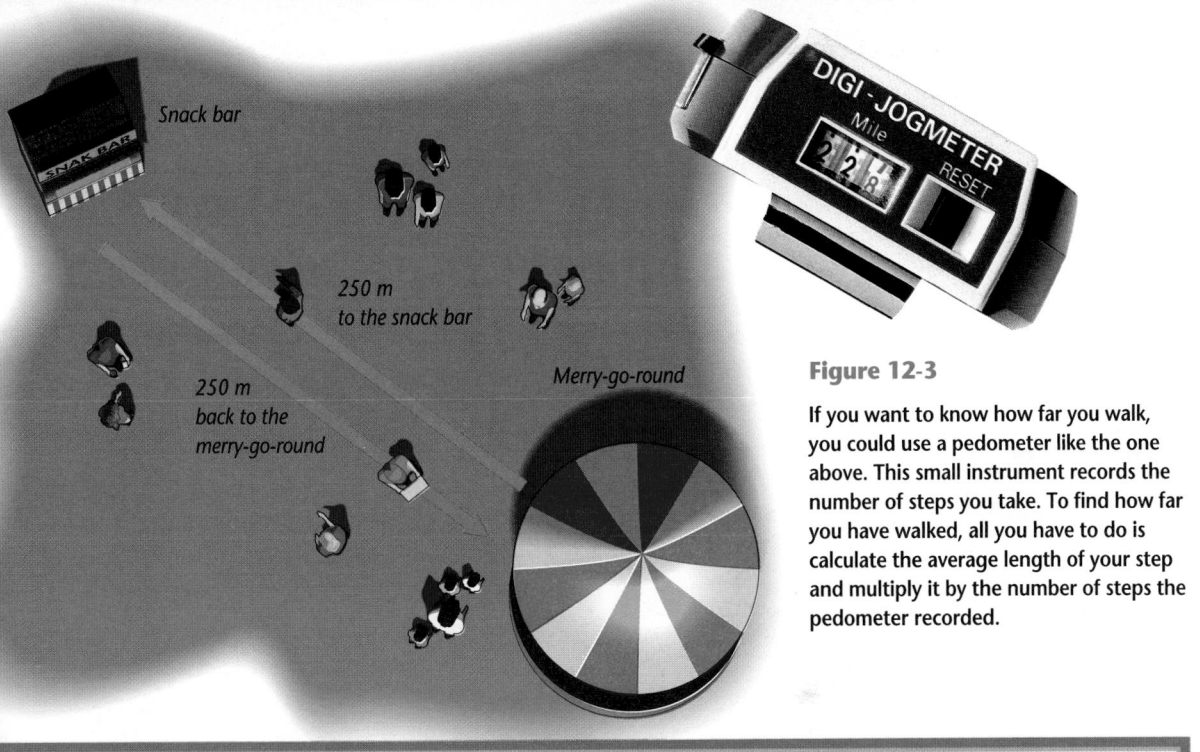

Snack bar

250 m
to the snack bar

250 m
back to the
merry-go-round

Merry-go-round

DIGI - JOGMETER
Mile
RESET

Figure 12-3

If you want to know how far you walk, you could use a pedometer like the one above. This small instrument records the number of steps you take. To find how far you have walked, all you have to do is calculate the average length of your step and multiply it by the number of steps the pedometer recorded.

Across the Curriculum

Mathematics

Ask students to work in pairs to find the fastest men's and women's times for running events. Have them make a double line graph of their findings by plotting the distance on the horizontal axis and the speed on the vertical axis. One line should represent men and the other women. Ask students to interpret their graphs by identifying any trends shown and inferring possible reasons for them. **L2** **COOP LEARN**

Visual Learning

Figure 12-2 Have students trace the path on the diagram. Then have them trace alternate paths and estimate if each path is longer or shorter than 500 m.

Figure 12-3 Help students begin to analyze words and word meanings by having them discuss the word pedometer. *Ped-* is the combining form of having a foot and *-meter* is the combining form of measure.

GLENCOE TECHNOLOGY

CD-ROM

Science Interactions, Course 1, CD-ROM

Chapter 12

other adaptations, such as waving hairs, wings, siphons, and sometimes many legs.

Does it surprise you that, generally, the larger the animal, the faster it can move? Does it also surprise you that swimming is the most energy-efficient way for an animal to move? Flying requires more energy, because the bird must overcome gravity. What animals do you think can move with the highest speeds?

The chart provides the fastest speeds ever measured for a variety of animals. Keep in mind that the greatest speed that an animal is capable of can only last a short amount of time. For every animal, there is a preferred speed that is usually far

less than its top speed.

USING MATH

Which animal has the fastest top speed? The slowest? How many times around a circular track could a rabbit run for every one circuit of a turtle?

What is your school record for the 50 or 100 meter dash? What is this speed in meters per second? How does this compare with the human top speed? What is your best speed in the 50 or 100 meter dash? How does this compare with the fastest human?

Speeds of Animals					
Mammals		**Birds**		**Amphibians**	
cheetah	26.7 m/s	vulture	17.0 m/s	frog	1.5 m/s
racehorse	19.1 m/s	ostrich (running)	18.0 m/s	**Fish**	
blue whale	18.0 m/s	penguin		tuna	20.0 m/s
dog	16.0 m/s	(swimming)	3.5 m/s	flying fish	10.0 m/s
rabbit	15.7 m/s	duck		salmon	3.0 m/s
cat	13.4 m/s	(swimming)	0.7 m/s	**Invertebrates**	
human	12.0 m/s	**Reptiles**		locust	4.5 m/s
squirrel	2.0 m/s	lizard	6.7 m/s	dragonfly	22.0 m/s
		turtle	2.0 m/s	ant	0.03 m/s

Teaching Strategy

Have students compare their own running speeds by checking with their physical education teachers for school records. These teachers may also be able to give them current age-group records for the state, country, and world. Then conduct a 50- or 100-meter dash for your class. Have students carefully measure the distance run and time the event. Check for any student who might have health considerations.

Answers to
Using Math
1. The cheetah
2. The ant
3. 8
4. Answers will vary with schools and individuals.

Going Further ⫸

Have students work in small groups to research and then compare and contrast other speeds. Topics could include the speeds of animals such as the frigate bird, the diving speed of the peregrine falcon as well as events and speed records set in the special and handicapped olympics. Ask each group to present their findings to the class. Use the Performance Task Assessment List for Oral Presentation in **PASC,** p. 71. **L2** **COOP LEARN**

Content Background

Speed is the rate of change in position. Instantaneous speed is the speed by which an object changes position at a given instant; it can be measured by a speedometer or radar gun, which aims microwave pulses at a moving object. Average speed requires less sophisticated equipment since you must measure only the total distance the object traveled and the total time it took.

Inquiry Question How could you measure the average speed of a car you are traveling in if all you had was a watch and no odometer? *You would have to divide a measurement such as the number of city blocks traveled or telephone poles passed by the number of minutes or hours traveled.*

Visual Learning

Figure 12-4 Have students trace the path of the roller coaster and predict where and how the speed changes. Then, as they read the caption, call on their experiences to reinforce the concepts described. Have students describe their own experiences riding a roller coaster, riding a bicycle, a skate board, or roller skates. Ask students what factors make the roller coaster speed up or slow down? *Going downhill, the force of gravity makes the roller coaster speed up. While going uphill, the force of gravity slows the roller coaster.*

If you know that the roller coaster takes 3 minutes to complete its course and the tracks measure 1170 meters from start to finish, what is the average speed of the roller coaster? *The average speed is 390 m per minute (1170/3 minutes).*

How Fast Is Fast?

Figure 12-4

The distance traveled by the roller coaster in a certain time is its average speed.

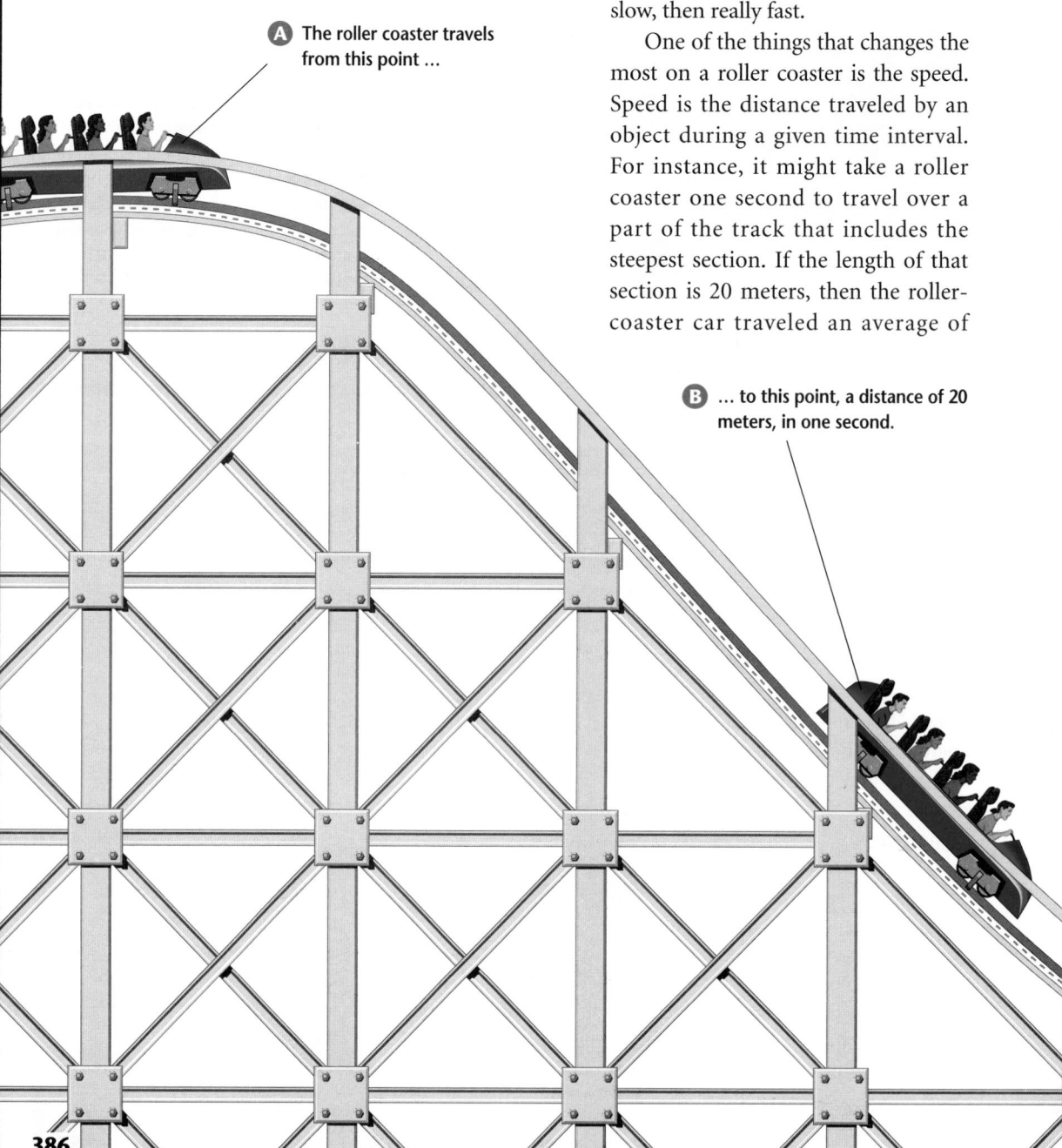

A The roller coaster travels from this point ...

B ... to this point, a distance of 20 meters, in one second.

386

Perhaps one of the most exciting rides you recall from your trip to the amusement park is the roller coaster. Up, down, around—first fast, then slow, then really fast.

One of the things that changes the most on a roller coaster is the speed. Speed is the distance traveled by an object during a given time interval. For instance, it might take a roller coaster one second to travel over a part of the track that includes the steepest section. If the length of that section is 20 meters, then the roller-coaster car traveled an average of

---

Meeting Individual Needs

Visually Impaired Encourage students to discuss ways motion can be determined through nonvisual means. Students may mention hearing objects move, feeling objects such as a mosquito move on you, and so on. Help students practice their observation skills by moving a sandpaper-covered block along a strip of sandpaper. Have students locate the starting place and put their fingers on it. Have them begin counting seconds (you may need to work with them on timing), or let them count ticks on a metronome while you move the block. Direct them to stop counting when they hear the movement stop. Let students locate the block and measure the distance with a meterstick. They should then be able to determine the average speed of the object. **L1** **LEP**

20 meters per second over that section. Walking around the amusement park might give you an average speed of around 1 meter per second, if you're not in much of a hurry.

■ Average and Instantaneous Speed

Suppose that in one hour, while walking around the amusement park, you run to catch up with some friends, stop to talk, buy some cotton candy, and watch the reptile show. Like most moving things, you do not maintain a constant speed during your walk. Even though your motion isn't constant, it is possible to find your average speed.

The **average speed** is found by dividing the total distance traveled by the total time required to travel the distance. That is,

$$\text{average speed} = \frac{\text{total distance}}{\text{time interval}}$$

In the next activity, you will find your average walking speed.

D The roller coaster doesn't travel at a constant speed all of the time that it goes around the tracks. Sometimes it goes faster, sometimes more slowly. If you know that the roller coaster takes 3 minutes to complete its course and the tracks measure 1170 meters from start to finish, what is the average speed of the roller coaster?

C The instantaneous speed traveled at this point of the ride is 15 meters per second.

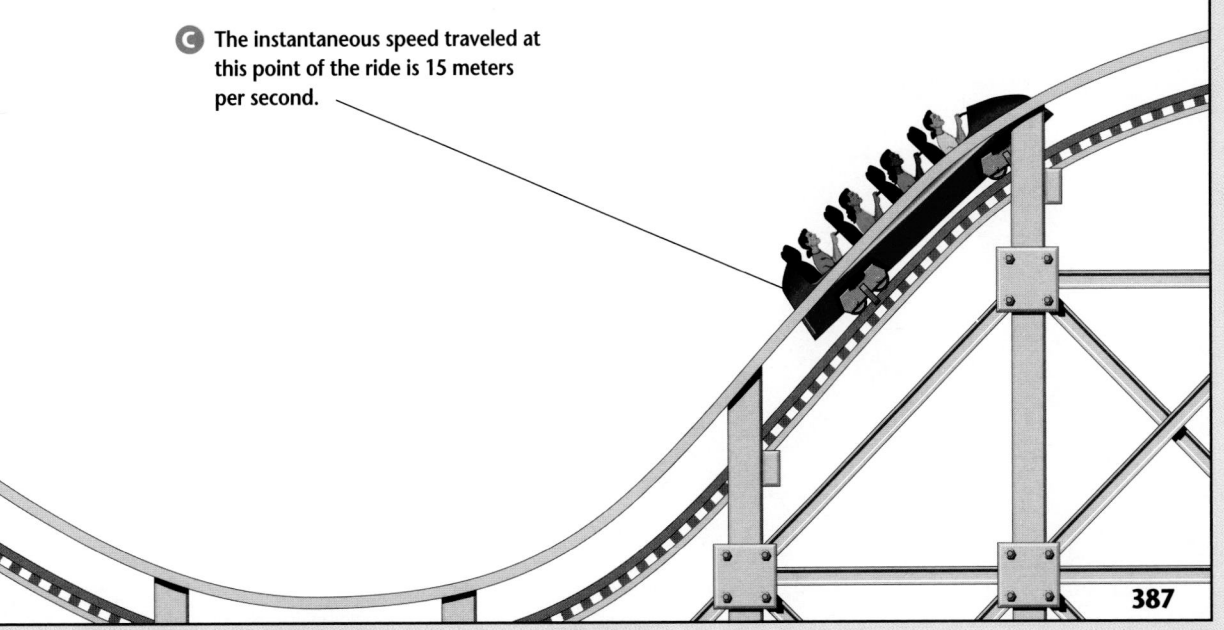

387

12-1 Average Walking Speed

Preparation

Purpose To develop a scientific experiment to find out if average walking speed is affected by the type of foot gear worn.

Process Skills making and using tables, interpreting data, designing an experiment, measuring in SI, predicting, making and using graphs, forming operational definitions, separating and controlling variables, comparing and contrasting, observing

Time Required 1 full class period

Materials Each group should test similar foot gear. Because girls and boys tend to wear some different types of shoes and boots, you may want to group students by gender so they can test similar foot gear.

Possible Hypotheses Students may hypothesize that running or walking shoes will give everyone the fastest average walking speed.

The Experiment

Process Reinforcement Be sure students set up their data tables correctly. Students could make a separate table for each student listing foot gear worn down the side and time, distance, and speed across the top.

Possible Procedures Mark off a distance and take turns walking quickly in different foot gear. Time and record data. Figure average speed for each foot gear and record that in the table.

Teaching Strategies

Troubleshooting Some students will have trouble figuring out averages. Average speed = distance (m)/speed (sec).

Discussion Have students discuss how they could determine if stride length affects walking speed. If students wish to determine this in this experiment, stride length can be measured and recorded in data tables.

Average Walking Speed

When you walk, do you feel more comfortable walking barefoot, in sandals, or in athletic shoes designed for walking or running? Have you ever wondered what your average speed is when you walk?

Preparation

Problem

Can the type of foot gear you wear increase your average walking speed?

Form a Hypothesis

As a group, form a hypothesis predicting what foot gear will increase average walking speed for each individual in your group.

Objectives

• Measure speed and find averages for each individual.
• Compare foot gear for walking speed.
• Graph and interpret your data.

Materials

meterstick
masking tape
stopwatch
shoes, boots, and sandals

DESIGN YOUR OWN
INVESTIGATION

Plan the Experiment

1 This is a group activity. Each group should test its hypothesis by designing a test procedure. Write it out step by step.

2 Speed is measured in meters per second (m/s). To find average speed, you need to know the distance traveled and the length of time each individual walked.

3 A walkway of a definite size is needed to conduct tests of the hypotheses. How long will your test track be?

4 How many trials will you conduct for each individual wearing a particular type of shoe? Testing someone 5 times is more reliable than testing once.

5 What foot gear will each individual wear? Should all the individu-als tested by one group wear similar foot gear? Make certain that shoes, boots, and sandals are free from mud and dirt.

6 Design data tables *in your Science Journal* or on a spreadsheet for recording your data.

Check the Plan

1 What is your control in this test?

2 What are your variables?

3 Who will collect the data?

4 Make certain your teacher approves your plan before you proceed.

5 Carry out the experiment, make observations, and record the data.

Pete Kain swims, cycles, and runs as a triathlete. If you knew the time and the distance he covers in each event, would you be able to find Pete's average speed as a triathlete? How?

Analyze and Conclude

1. Analyze Was your hypothesis supported by the data? Use your data to explain why or why not.

2. Interpret What foot gear produced the fastest average speeds in your group?

3. Analyze Use all the class data to make bar graphs of individual speeds in each type of foot gear. This can be done on a spreadsheet. When you have a graph for each type of foot gear, analyze each one to find out who was the fastest walker in all the different shoes. Was there a foot gear that caused some individual to walk slower while others walked faster?

4. Infer What would happen to your average walking speed if your running shoes had wet mud on them or they were worn down?

Going Further

If your have a suitable track, test which brand of running shoe increases average speed the most.

Science Journal Students should draw their data tables in their Science Journal and record their data.

Expected Outcome

Students will gather data, learn that more trials give better data, and determine which foot gear gives the fastest average walking speed for each individual, for the group, and for the class.

Analyze and Conclude

1. Answers may vary, but students should find that running and walking shoes provide the highest average walking speeds. Students should indicate the number of individuals in their group that had the fastest average walking speed in walking or running shoes or whatever foot gear worked best for them.

2. See #1.

3. Make graphs of the average speeds for all the class. Sandals should have a separate graph from boots and so on. Each individual should have a separate bar on each graph of a foot gear he or she wore.

4. Worn down or muddy running shoes do not provide the friction force that clean, new running shoes do. Therefore old or muddy shoes make you go slower.

Going Further ⅢⅢⅢ▶

This is a good experiment for nice weather. Remember that shoes should be of approximately equal age and cleanness.

✔ Assessment

Process Reinforce students' data interpretation skills by having them explain their bar graphs to other groups. Then have them think of other ways to graph or show their data, such as using line graphs or using symbols to show slow, medium, or fast average walking speeds. Use the Performance Task Assessment List for Graph from Data in **PASC,** p. 39.

ENRICHMENT

Activity Have students test their experimental skills by finding the average speed of a pet or a neighbor's pet using the method learned in Investigate 12-1. Then have students research the average speeds of other animals such as ants, turtles, horses, and ostriches. Have students organize their findings in a bar graph. L1

Program Resources

Activity Masters, pp. 51–52, Design Your Own Investigation 12-1

How It Works, p. 15, Speedometers L2

Making Connections: Integrating Sciences, p. 27, The Peregrine Falcon L2

Figure 12-5

Turtle   Human   Cougar

| 0.5 meters/second | 5.0 meters/second | 15 meters/second |

The numbers under each figure indicate the speed of that runner at the instant the photo was taken. Are these photographs a record of average or instantaneous speeds?

As you saw in the Investigate activity, the speed of an object at any one instant may not be the same as the object's average speed. Instantaneous speed is the rate of motion at any given instant. A car's speedometer shows instantaneous speed. In the fable of the tortoise and the hare, the hare's instantaneous speed could be much greater than the tortoise's speed. Which animal had the faster average speed? How do you know?

Walking around an amusement park, riding a roller coaster, driving a car, riding a skateboard, even orbiting planets—all involve a motion during an interval of time. For each motion, we can determine the object's average speed over the path taken by dividing the total distance traveled by the time interval.

The average speed of a moving object tells you how rapidly it travels but it does not tell which way the object goes. In the next section, you will learn how to combine speed and direction to describe an object's motion.

check your UNDERSTANDING

1. In your own words, describe what is meant by (a) position and (b) distance.
2. How could you measure the distance of your path from home to school?
3. List the information you need to calculate the average speed of a car traveling from your house to school.
4. **Apply** Florence Griffith Joyner set a world record by running 200 m in 21.56 s. What was her average speed?

12-2 Velocity

Displacement and Velocity

Have you ever ridden on a merry-go-round or carousel? Usually, a ride on the carousel involves 15 or 20 turns in a circular path. You may ride for three or four minutes, but you get on and off at the same place. Despite all of the trips around in a circle, you really haven't gone anywhere at all.

■ Displacement

Displacement is the net change in position of an object. A round trip to the store, like the round trips on the carousel, produces a displacement of zero because you end up where you started.

Displacement is described by both a distance and a direction. Distance is the length of the path only. Look at the map on page 392. To say that you live 16 kilometers from the amusement park is not necessarily helpful to another person. Sixteen kilometers could be anywhere in a circle of radius 16 kilometers from the amusement park. To make it more understandable and useful to others, you need to specify which way. As shown by the map, home is 16 kilometers south of the amusement park.

Suppose you drove to and from the amusement park. The odometer of the car measures the distance the car has moved. As with walking, the round trip measures a distance of 32 kilometers. Your displacement is still zero. Is there a way to describe motion that takes displacement into account?

Look at the map on page 392.

Figure 12-6

The round trip around a Ferris wheel produces a displacement of zero because you get off the ride at the same point that you got on.

391

Section Objectives

- Distinguish between displacement and distance.
- Find the average velocity of a moving object.
- Distinguish between velocity and speed.
- Use the concept of relative velocity.

Key Terms

displacement
average velocity
relative velocity

Program Resources

Study Guide, p. 44
Concept Mapping, p. 20, Motion L1
Critical Thinking/Problem Solving, p. 20, Radar Detectors and Speeding L3
Teaching Transparency 23 L1
Section Focus Transparency 38

PREPARATION

Planning the Lesson

Refer to the Chapter Organizer on pages 380A–D.

Concepts Developed

Students will compare and contrast average speed with average velocity and distance with displacement. They will be introduced to the concept of relative velocity, in which an object's apparent velocity depends on the observer's frame of reference.

1 MOTIVATE

Bellringer

 Before presenting the lesson, display **Section Focus Transparency 38** on an overhead projector. Assign the accompanying **Focus Activity** worksheet. L1
LEP

Discusssion To help students infer the relationship between the observer's point of view and an object's relative velocity, have students imagine that they are flying in an airplane traveling at 800 km/h, and one of them gets up and walks toward the front of the airplane at 1 m/s. **Ask how fast that person would appear to be moving to the other passengers on the plane.** *1 m/s* **Then ask how fast that person would appear to be moving if he or she could be seen by an observer on the ground.** *at the speed of the plane plus 1 m/s* Discuss what might account for this difference. L1

Tying to Previous Knowledge

If students were riding in a car, they could use the odometer to measure the distance and the speedometer to measure speed. These readings would not help them to know the direction in which they were traveling or how far they were from their starting point. Displacement and average velocity, concepts presented in this section, would help them determine this.

Visual Learning

Figure 12-7 To help students improve their spatial relationships, ask them to identify other locations such as 16 kilometers north of the amusement park.

Demonstration You will need a globe and masking tape. Put a small piece of masking tape somewhere on the equator. Have students imagine that the tape is a person. Spin the globe. Ask students how fast the person would think he or she was going. Since Earth is the frame of reference for the person, that person would not think he or she was moving at all. Compare this to how rapidly the person seems, to the students, to be spinning around in a circle. Stop the spinning globe in the same position it was in when you started. Ask students what the displacement of the person was. The answer is 0, since the starting and ending points are the same. This should be contrasted with the distance traveled, which was several rotations of Earth. [L1]

Flex Your Brain Use the Flex Your Brain Activity to have students explore VELOCITY.

Braking distance increases 4.6 m.

Figure 12-7

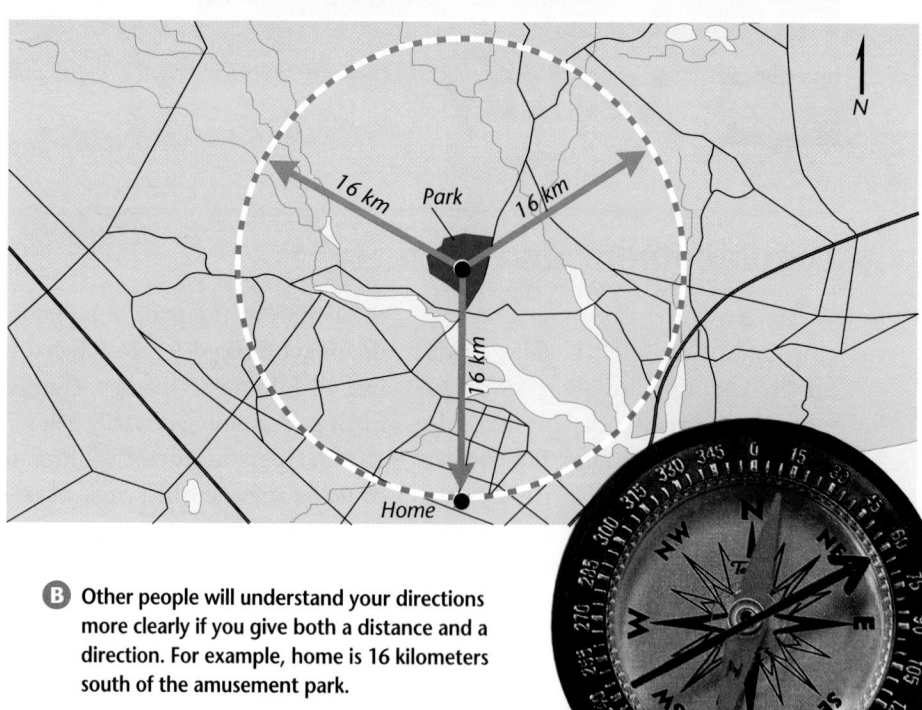

A To say that you live 16 kilometers from the amusement park is not necessarily helpful to another person. Sixteen kilometers could be anywhere in a circle of radius 16 kilometers from the amusement park.

16 km Park 16 km

16 km

Home

N

B Other people will understand your directions more clearly if you give both a distance and a direction. For example, home is 16 kilometers south of the amusement park.

SKILLBUILDER

Making and Using Graphs

The table gives the stopping distances for different speeds. Graph the points with the speed in m/s along the horizontal axis and the stopping distance in meters along the vertical axis. Connect the points with a smooth line.

Using the graph, determine how much braking distance increases when the speed increases from 4 m/s to 8 m/s. If you need help, refer to the **Skill Handbook** on page 657.

Speed (m/s)	Braking Distance (m)
2	1.0
4	2.3
6	4.2
8	6.9
10	10.0

■ Average Velocity

A measure of motion that tells you how fast and which way an object moves is velocity. You learned in the first section that the average speed of an object depends on the distance traveled within a certain time. Now that you know the difference between distance and displacement, do you think there could also be a measure of how fast an object moved over a certain displacement? There is! The measure of such motion is called **average velocity**.

Calculating an average velocity is similar to calculating an average speed. We are concerned here with the total displacement, which includes distance and direction rather than the total distance traveled. To find average velocity:

$$\text{average velocity} = \frac{\text{total displacement}}{\text{time interval}}$$

■ Relative Velocity

Have you ever sat in a car at a traffic light and felt you were moving backward? When you looked at other objects outside the car you realized that it was the car next to you moving forward.

392 Chapter 12 Motion

Program Resources

Critical Thinking/Problem Solving, p. 5, Flex Your Brain
Science Discovery Activities, 12-1, Motion Marathon

ENRICHMENT

Research Help students better understand cause and effect relationships by having them research the term *escape velocity*, the speed an object must attain to travel away from Earth. **Ask what would happen to a rocket that did not reach the escape velocity shortly after takeoff.** *It would orbit or fall back to Earth.* [L2]

The same effect can happen at the amusement park on a double roller coaster with a double track. While waiting for your car to move, the car in the roller coaster next to yours starts moving. You think that you are moving in the opposite direction.

This is an example of **relative velocity**. Relative velocity is the velocity of one object determined from the view, or frame of reference, of another object. Either one or both of the objects may be moving relative to some third object.

Explore! ACTIVITY

What is relative velocity?

What To Do

1. Send a battery-powered toy car along the length of a sheet of paper.

2. Observe the motion of the car.

3. Can you predict what will happen if you send the car along the paper while a friend pulls the paper on the table in the direction of the motion of the car? Try it!

4. Describe the car's motion.

5. Now, observe the motion of the car when the sheet is moved in a direction opposite to that of the car's. What do you see?

In the Explore, the car moved over the paper and over your desktop. The car's motion over both the paper and desk top was the same as long as the paper didn't move. As soon as you made the paper move, however, the car's motion over the desktop was not the same as its motion over the paper. When you describe the motion of the car or any object, you have to compare it to something else—in this case, either the desktop or the paper. The motion that you describe depends on your frame of reference, or what you are comparing the car's motion to. We usually use Earth as our frame of reference. For example, you speak of your motion to and from school in terms of the distance and direction you travel on the surface of Earth. We don't always use Earth as our frame of reference, however.

Program Resources

Making Connections: Across the Curriculum, p. 27, Reading Maps L2

Making Connections: Technology and Society, p. 27, Should School Buses Have Seat Belts? L2

ENRICHMENT

Activity Have students make a gently sloping ramp by propping a board on a textbook. Tell students to mark off a 5-m distance from the bottom of the ramp. Measure the length of the ramp. Let a marble roll down, and time it. Roll the marble again and measure its travel time to the 5-m mark. Calculate the average velocity in each case. Compute the difference in average velocity between the two cases. L2
LEP

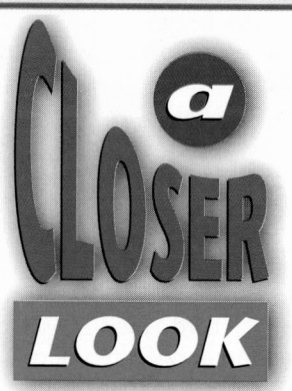

Purpose

A Closer Look reinforces Section 12-2 by having students consider the velocity of Earth relative to different frames of reference.

Content Background

Because Earth is not a perfect sphere, the text uses the circumference of Earth at the equator. The polar radius is about 19 km less than the equatorial one.

Earth actually takes 365.26 days to complete one orbit around the sun, which is why we add one day to the calendar every four years.

Teaching Strategies

Have students use everyday situations to analyze the concept of relative velocity. One possibility might be how the velocity of a merry-go-round horse would appear to someone on a neighboring horse compared to how it would appear to someone standing outside the merry-go-round.

394 Chapter 12 Motion

Figure 12-8

A An astronaut working outside a space shuttle would use the shuttle as her frame of reference. If the astronaut and shuttle are moving in the same direction around Earth at the same orbit speed, then their relative velocity would be zero. In this case, the astronaut would say that she is motionless compared to the space shuttle.

B On the ground, however, NASA controllers would describe both the astronaut's and the shuttle's orbit speed as about 40 000 kilometers per hour, using Earth as their frame of reference.

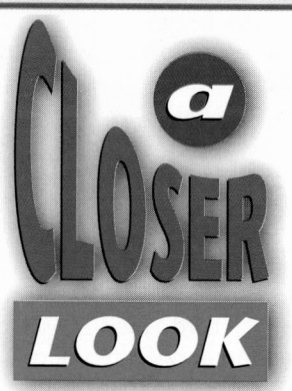

Around and Around We Go

Although we generally express our velocity relative to Earth, we can get some interesting information if we consider our velocity relative to other locations. For example, you know that Earth rotates once every 24 hours. When you stand still, your velocity relative to Earth is zero, yet you are moving as Earth rotates. If a friend were suspended above the North Pole in a spaceship and looking down at Earth, what would your friend observe your velocity to be?

The circumference of Earth at the equator is about 25 000 miles (40 000 km). This means that if you were standing at a

The photograph to the right was taken with a camera pointed at the North Star and with the shutter left open about four hours. The white lines are tracks made by stars as Earth rotated beneath them.

394 Chapter 12 Motion

USING MATH

1. about 584 040 000 miles (934 464 000 km)

2. about 1 600 110 miles (2 560 175 km)

3. about 66 671 miles/hour

4. The rotation of Earth, the revolution of Earth around the sun, and even the speed at which the entire solar system is moving is necessary in order to determine where an object will be when you are traveling great distances over a long period of time through space.

Going Further ⑊⑊⑊➤

Have students work in pairs to calculate speeds and distances traveled by other bodies in the solar system. The moon has a radius of about 1,086 miles and rotates once every 27.32 days. Have students use this information to determine the circumference of the moon and its rotational speed in miles per hour. *About 6796 miles; about 10 miles/hour*

Use the Performance Task Assessment List for Analyzing the Data in **PASC**, p. 27. **L3**
COOP LEARN

An astronaut making repairs to an object in space would use a different reference point to determine her velocity than a person tracking on the ground. Her only frame of reference is the shuttle.

check your UNDERSTANDING

1. A man walks 2 km due west, then turns around and walks 3 km due east. What distance does the man walk? What is the man's displacement?
2. Explain whether it is possible for a car's average velocity to be zero even if the car traveled a distance of 200 km.
3. An airplane flying toward the west has a speed of 200 km/h. What is the plane's velocity?
4. A student decides to have some fun on a moving walkway at the airport. The walkway moves to the north with a speed of 2 m/s, and the student walks at the same speed in the same direction. What velocity would a person standing next to the walkway measure to describe the student's motion?
5. **Apply** If the same student sees a friend standing next to the walkway and wants to remain at rest relative to that friend, at what velocity (speed and direction) should he or she walk?

point on the equator, you would move 25 000 miles in 24 hours as Earth rotates. That's about 1040 mi/hr (1674 km/hr). Your friend in the spaceship would observe you moving in a circle at that speed.

What is the speed of Earth as it revolves around the sun? Let's find out!

 USING MATH

Use your math skills to figure out how fast Earth would appear to be moving if you observed it from the sun. Although Earth's path or orbit around the sun is not a perfect circle, we can get a rough idea of our speed assuming that it is.

1 The circumference of a circle can be calculated using the equation $c = 2\pi r$ where c = circumference, r = the radius of the circle and $\pi = 3.14$. The average distance from Earth to the sun is about 93 000 000 miles (149 700 000 km). This is the radius of the circle or orbit that Earth travels around the sun. Calculate the circumference of the circle.

2 Earth travels this distance, c, in one year (365 days). How far does Earth travel in one day?

3 What is the speed of Earth in miles/hour?

4 Why do you think it might become more important to use relative velocity as we increase our travel in space?

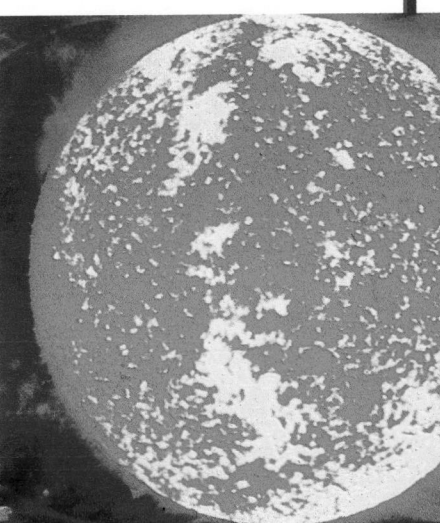

12-2 Velocity Along a Straight Line **395**

check your UNDERSTANDING

1. 5 km; 1 km due east
2. Yes; the average velocity of a car is its displacement divided by time. If the car traveled from point A to B and then back to A, its displacement is 0, and 0/time = 0.
3. 200 km/h west.
4. 4 m/s north
5. 2 m/s south, or opposite to the direction of the walkway's motion

3 ASSESS

Check for Understanding

To answer the Apply question, it is important that students realize that moving in the same direction and at the same speed as another object will make the relative velocity 0.

Reteach

Activity To strengthen students' skill at comparing and contrasting average speed and velocity, have a student walk to make a complete circle. Then ask the student to run in the same circle. Ask him or her whether walking or running produced a greater average speed. Then ask which produced the greater average velocity. Lead students to see that the average speed was greater when running because the same distance was covered in a shorter time. But the average velocity was the same because the displacement was zero each time. **LEP** **L1**

Extension

Discussion Ask students who have already mastered the section concepts to compare the speed and velocity of Earth and the moon as each completes one orbit. **L3**

4 CLOSE

Activity

Students will practice measuring average velocity of a partially filled balloon. Materials needed are a meterstick, a stopwatch, and a balloon. Blow up the balloon but do not tie the end. Let the balloon go. Have a student time the interval from when the balloon is released until it comes to rest. Have students measure the distance along a straight line from the point that the balloon was let go to the spot that it landed. Divide the balloon's displacement by the time elapsed to determine its average velocity. **COOP LEARN** **L2**

PREPARATION

Planning the Lesson

Refer to the Chapter Organizer on pages 380A–D.

Concepts Developed

So far the chapter has primarily dealt with average or constant speed and velocity. Acceleration occurs whenever the velocity of an object changes. The change can involve either a difference in speed or in direction of the object. The measurement of average acceleration must take into account how much the velocity of the object changed during a given time interval.

1 MOTIVATE

Bellringer

🕯️ Before presenting the lesson, display **Section Focus Transparency 39** on an overhead projector. Assign the accompanying **Focus Activity** worksheet. L1

LEP ▢▢

Demonstration Help students learn to test assumptions by having them observe how an object's velocity can change.

You will need a windup toy car. Wind up the car and place it on a desktop or the floor. Ask students to note the points at which the car accelerates. While most will note that the car speeds up until it reaches a constant speed, not all will be aware that the car also accelerates as it slows down. L1

12-3 **Acceleration**

Section Objectives

- Distinguish between velocity and acceleration.
- Determine acceleration from velocity change and time.

Key Terms

acceleration
average acceleration

Figure 12-9

Changing Velocity

Have you ever enjoyed an absolutely smooth ride? Surely not at the amusement park, where half of the fun comes from sharp turns, moving up and down, and spinning around in circles.

When the velocity of an object changes, the object accelerates. You often do not sense your motion when you are moving at a constant velocity. You do, however, sense motion when you accelerate. Anyone who has ridden on a roller coaster knows about acceleration. When the car takes off quickly from rest, you feel as though you are being pushed back in your seat. When the car's brakes are suddenly applied, you may feel as if you are moving forward. Even in the family car, you move forward if the car stops suddenly. These are everyday examples of acceleration.

When you are moving along in one direction at a constant speed, but then turn quickly to the right or left you feel a push outward. This is another example of acceleration. Since velocity involves both speed and direction, to change your velocity requires that you change either one or the other. Thus, acceleration can be produced by changing your speed (how much) or by changing your direction of travel (which way), or both.

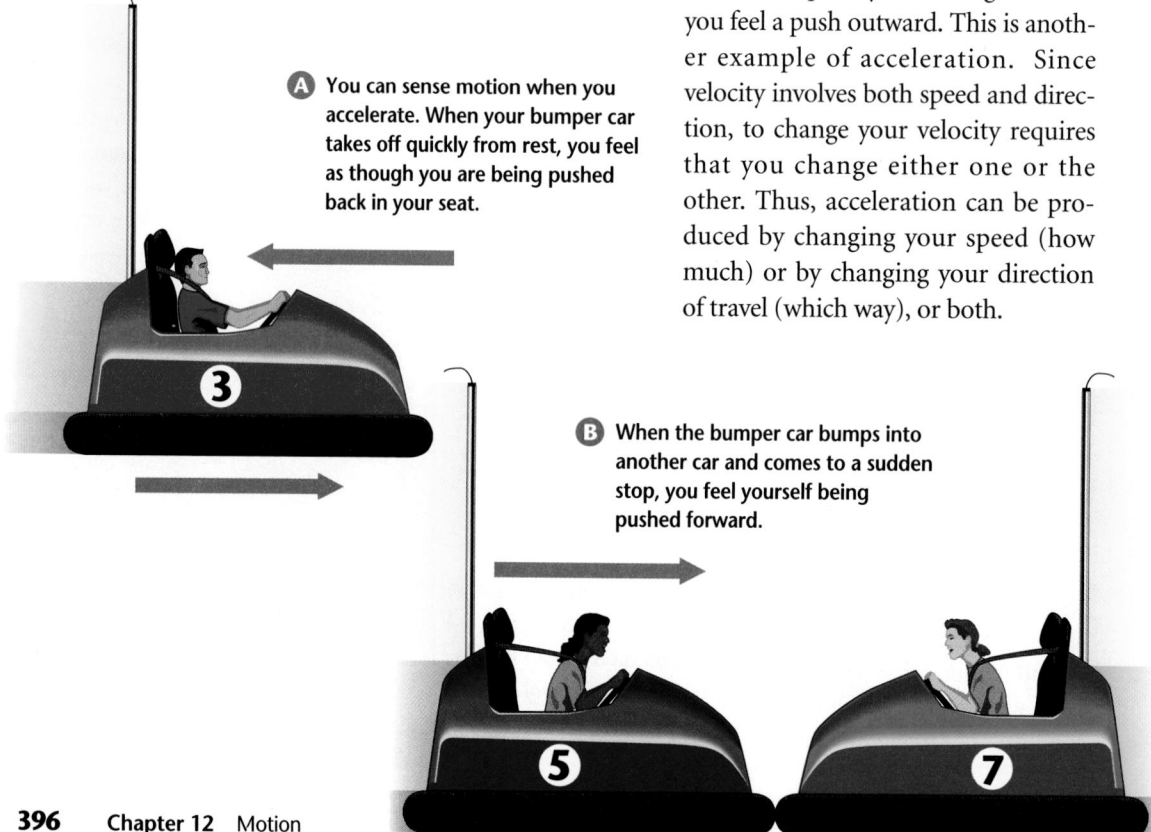

A You can sense motion when you accelerate. When your bumper car takes off quickly from rest, you feel as though you are being pushed back in your seat.

B When the bumper car bumps into another car and comes to a sudden stop, you feel yourself being pushed forward.

396 Chapter 12 Motion

⟨ Program Resources ⟩

Study Guide, p. 45
Laboratory Manual, pp. 71–74, Speed and Acceleration L3
Multicultural Connections, p. 28, Ellison Onizuka L1
Section Focus Transparency 39

You often use the word acceleration in conversation when you speak of something speeding up. Did you know you would be equally correct to refer to an object that is slowing down as accelerating? Astronauts undergo large accelerations when they take off from Earth. This is due to their large change in velocity as they move from Earth to their position in orbit around Earth. For this reason, astronauts have comfortably padded seats. When the shuttle and the astronauts return to Earth, they need to slow down. This slowdown is referred to as a negative acceleration.

Acceleration is the rate at which velocity is changing. Because velocity can change by changing either the speed or direction of motion, you can accelerate by changing speed, direction, or both. For an example, suppose you are on a water slide. Both your speed *and* acceleration change as you move over different parts of the slide.

Figure 12-10

Suppose you are on a water slide and speeding up as you move. Your speed changes by 1 meter per second for every second that you move. Your acceleration is 1 meter per second per second, or 1 m/s/s.

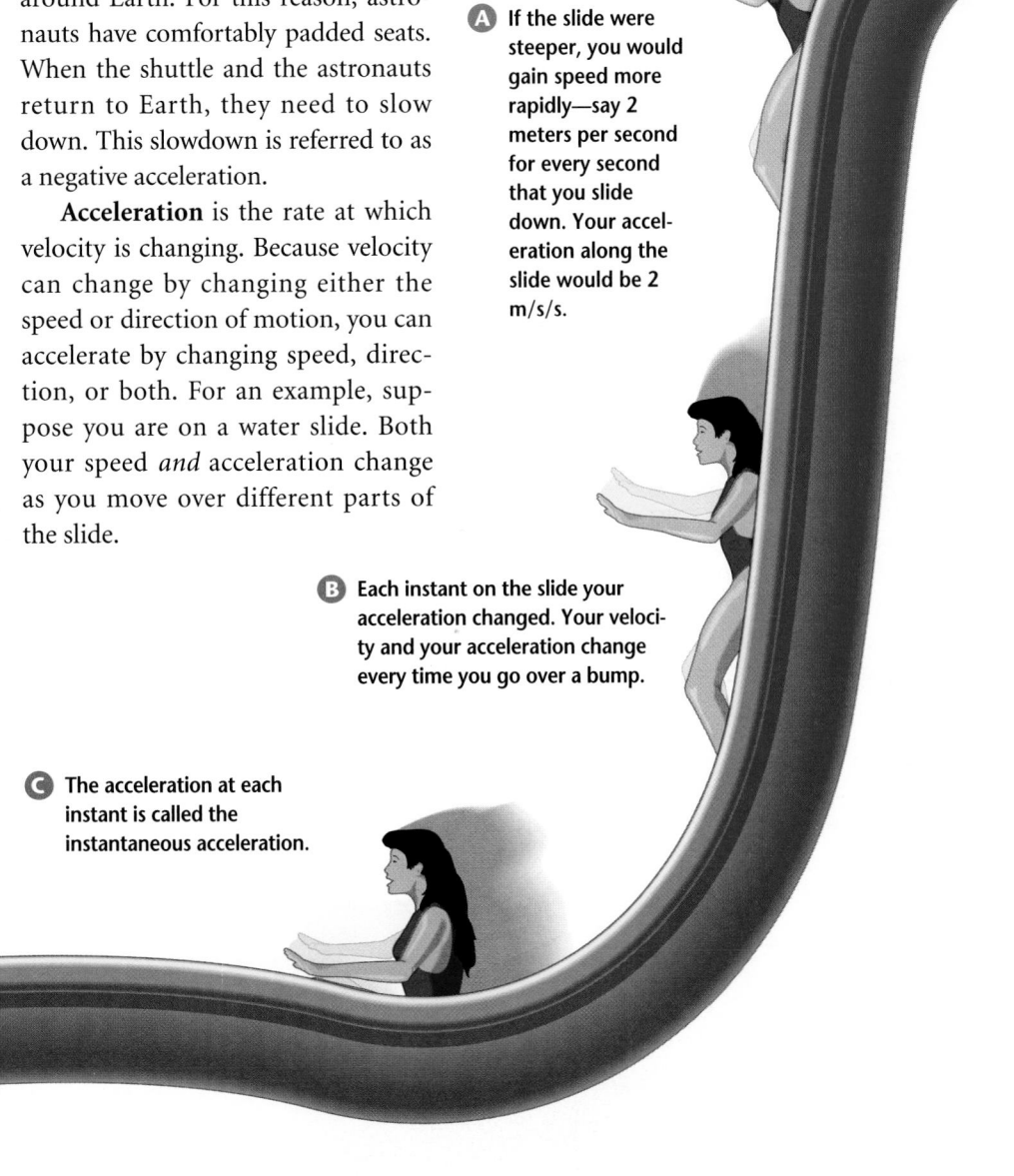

A If the slide were steeper, you would gain speed more rapidly—say 2 meters per second for every second that you slide down. Your acceleration along the slide would be 2 m/s/s.

B Each instant on the slide your acceleration changed. Your velocity and your acceleration change every time you go over a bump.

C The acceleration at each instant is called the instantaneous acceleration.

12-3 Acceleration **397**

2 TEACH

Tying to Previous Knowledge

Ask students how they know that they are moving when they are traveling in a car or elevator at a constant velocity (same speed and direction). **What clues do they use to note a change in velocity?** *They may compare their movement to that of objects moving faster or slower than themselves. Or they may just feel the change, such as the upward or downward movement of an elevator.*

Visual Learning

Figure 12-9 Students who haven't ridden bumper cars can relate to the illustration by recalling similar experiences in other moving vehicles such as cars or buses. **Figure 12-10** Help students understand that acceleration can refer to a decrease in speed as well as an increase in speed. Then ask students to list all the ways the girl accelerates. *The girl accelerates as her speed increases, as her direction changes and the slide curves upward and downward, and as she slows down when entering the water.*

Theme Connection The theme that this section supports is stability and change. From the time an object starts moving until it stops, it does not maintain a constant velocity. Acceleration, another facet of stability and change, is a change in velocity that occurs when an object changes its speed or direction.

Across the Curriculum

Health and Safety

Point out to students that when a car traveling about 50 km/h collides head-on with something solid, the car suddenly slows down. Any passenger not wearing a seat belt continues to move forward at the same speed the car was traveling, slamming into the dashboard or front seat at 50 km/h. A person wearing a seat belt becomes part of the car. As the car slows down, so does the passenger. **L1** **LEP**

12-3 Acceleration **397**

12-2 Instantaneous Acceleration

Planning the Activity

Time needed 15–20 minutes

Purpose To use a laboratory technique for measuring instantaneous acceleration.

Process Skills making and using tables, practicing scientific methods, observing and inferring, interpreting data, predicting, measuring in SI, forming operational definitions, recognizing cause and effect, observing, modeling

Materials protractor, string 10–12 cm long, button

Preparation Make sure that the object used to weigh down the string is heavy enough to pull the string taut and that the string is thin and smooth to allow it to move freely as it swings alongside the protractor. You may find that a key or 1/2" or 5/8" washer works better than a button with some types of string.

Teaching the Activity

Process Reinforcement Students should create data tables that will allow them to record several readings on the protractor while they are speeding up and slowing down. L1

Possible Hypotheses Students may hypothesize that the string will change directions as velocity changes.

Possible Procedures Some students may prefer to work in groups of three, with one student holding the protractor, another student reading the angle while running along with the first, and the third student recording the data.

Troubleshooting Depending on the protractors your students use, they may find that a 25 to 30 cm-long string works better.

Have students work in pairs, one to hold the protractor and one to read the angle. Tell students that the angle should be read as quickly as possible during their partners' movements, as students will quickly reach relatively constant velocities.

Instantaneous Acceleration

Have you ever wondered how quickly you can accelerate? You can make an accelerometer that will allow you to measure the instantaneous acceleration of moving objects.

Problem

How can you measure instantaneous acceleration?

Materials

protractor
10 to 12 cm length of string
heavy button

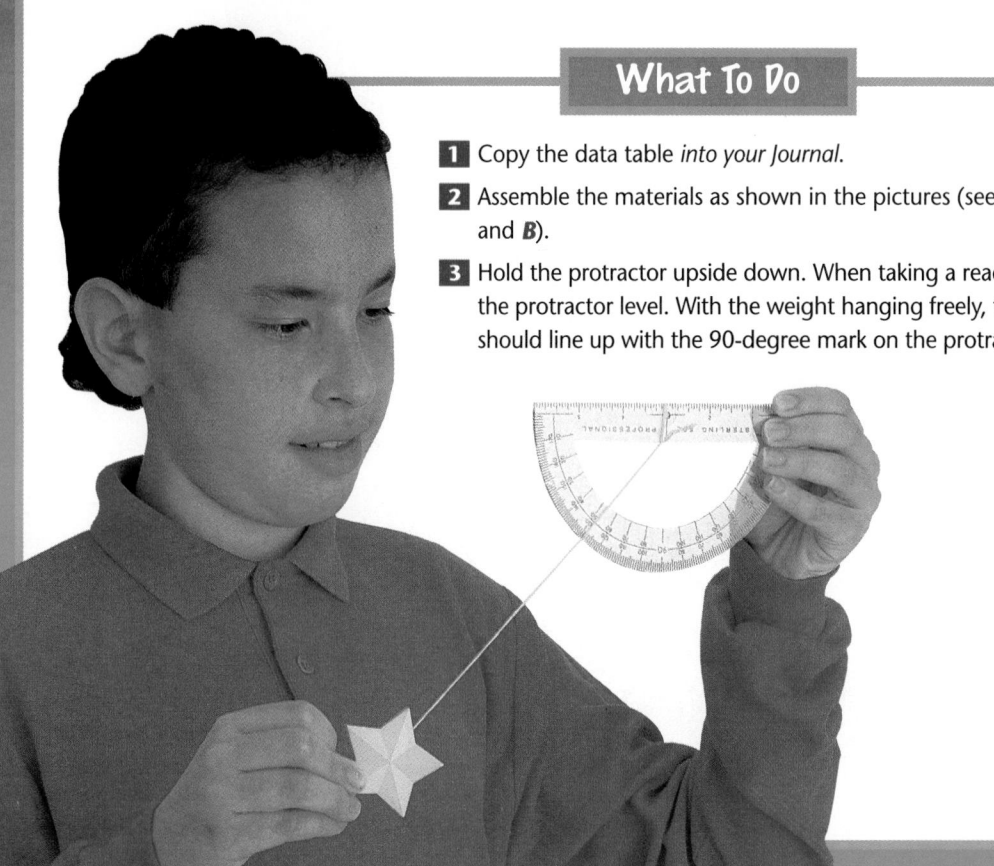

What To Do

1 Copy the data table *into your Journal*.

2 Assemble the materials as shown in the pictures (see photos **A** and **B**).

3 Hold the protractor upside down. When taking a reading, hold the protractor level. With the weight hanging freely, the string should line up with the 90-degree mark on the protractor.

Program Resources

Activity Masters, pp. 53–54, Investigate 12-2
Science Discovery Activities, 12-2, Acceler-Skate; 12-3, Flingball

---

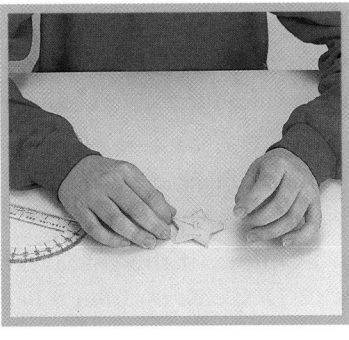
A

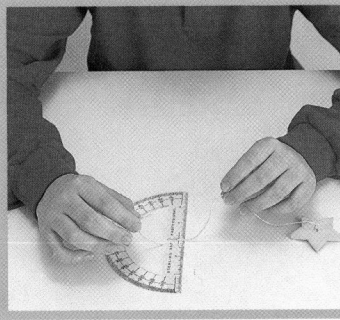
B

Conversion Chart	
Moving Object	Acceleration m/s/s
90°	0
80°	1.7
70°	3.6
60°	5.7
50°	8.2
45°	9.8
40°	12
30°	17
20°	27
10°	56
0°	—

4 Hold the accelerometer at arm's length in front of your face with the numbers facing you. Quickly move the accelerometer to one side. Observe the maximum angle of the string measured by the protractor. In what direction does the string move? What can you infer about the direction of acceleration? Try moving the accelerometer quickly to the other side. What does this tell you?

5 Use the conversion chart to convert the angle reading on the accelerometer to an acceleration in meters per second per second.

6 Hold the accelerometer level and begin to run. Have a friend run with you and read the angle. Enter this data in the table.

Sample data

Data and Observations	
Moving Object	Acceleration m/s/s
75°	2.63
50°	8.22

Analyzing

1. In what direction does the string move in comparison with the direction you move when you speed up? Describe.

2. How did the string behave as you slowed down?

Concluding and Applying

3. Describe the position of the string while you were moving with constant velocity.

4. **Going Further** Predict whether or not loose objects in an accelerating car would tend to move in the same direction of the string, or in the opposite direction.

Science Journal Have students record their data on their data tables and answer the questions in their journals. L1

Expected Outcomes

The acceleration measurements will likely vary widely, although increases should be seen. Students should find that the string swings away from the direction of motion as they speed up and toward the direction of motion as they slow down.

Answers to Analyzing/Concluding and Applying

1. The string moves in the opposite direction, away from the direction of motion.

2. The string moved toward the direction of motion.

3. The string hangs straight down.

4. They would move in the same direction as the accelerometer string.

✔ **Assessment**

Process Have students devise a laboratory technique for measuring instantaneous acceleration for upward and downward motion. Use the Performance Task Assessment List for Designing an Experiment in **PASC**, p. 23. COOP LEARN L1

GLENCOE TECHNOLOGY

 Videodisc

Science Interactions, Course 1, Integrated Science Videodisc

Lesson 4

Meeting Individual Needs

Physically Challenged Adapt the activity in one of these ways so that students in wheelchairs can participate.

1. The student can hold the accelerometer in one hand while another student pushes the chair.

2. Attach a meterstick to the wheelchair so that it protrudes forward. Attach the protractor to the stick so the string hangs down at eye level, thus freeing both hands to achieve greater acceleration. L1

Check for Understanding

Ask students to explain what their answer to the Apply question means. *The car's speed increases by 1.6 m/s every second that it accelerates.* Use the Performance Task Assessment List for Group Work in **PASC,** p. 97.

Reteach

Activity Direct a volunteer to begin walking at a constant velocity. Have the other students, one at a time, call out a different command that, when followed, will accelerate the volunteer's motion. These should include commands to increase speed, decrease speed, and change direction. The type of commands students call out should indicate their understanding of the different ways acceleration can occur to change velocity. **L1** **COOP LEARN**

Extension

Activity Attach a hard-boiled egg to the center of a skateboard with a small piece of clay. Thrust the skateboard against a wall about 2 m away. Secure another egg with rubber bands and repeat the crash. Have students compare the acceleration of the skateboard and its effect on the egg during each trial. **L2** **LEP**

4 CLOSE

Demonstration

Throw a ball up into the air and catch it. Ask students to analyze the ways the ball accelerated. *Some possible answers: It increased velocity when it was tossed; it was accelerating (down) during its entire flight as well as when you caught it.* **L1**

Figure 12-11

A If a go-cart on a straight, level track speeds up from 0 to 3 m/s in 8 seconds, the car's average acceleration is:

Average acceleration = change in velocity/time interval

$$= \frac{3/m/s - 0 \; m/s}{8 \; s}$$

= 0.375 m/s/s along the track

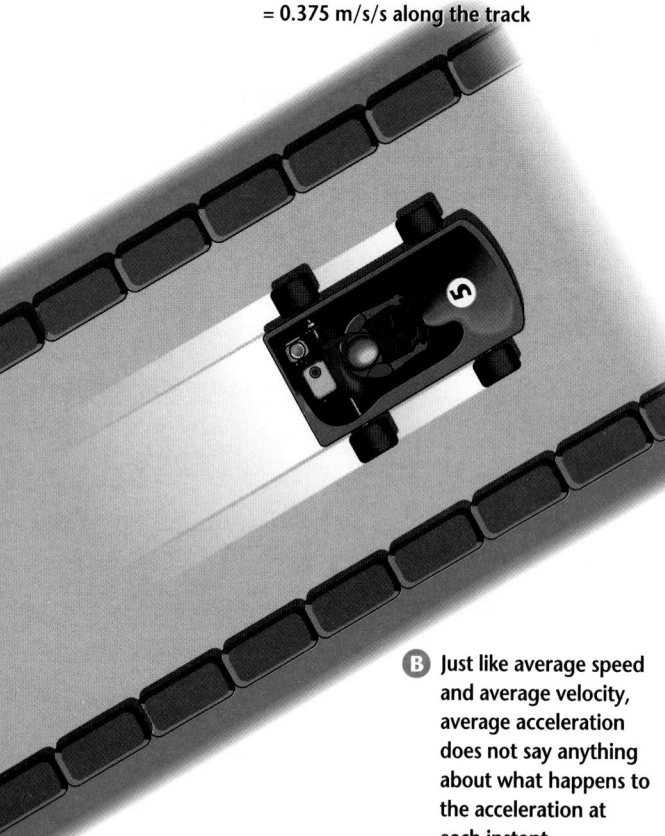

B Just like average speed and average velocity, average acceleration does not say anything about what happens to the acceleration at each instant.

■ Average Acceleration

Calculating average acceleration is similar to finding average velocity. **Average acceleration** is the change in the velocity divided by the time interval during which the change occurs:

Average acceleration = change in velocity/time interval with a certain direction

For example, if the go-cart in **Figure 12-11** speeds up from 0 to 3 m/s in 8 seconds, the go-cart's average acceleration is:

Average acceleration equals the change in velocity divided by the time interval (3m/s – 0 m/s)/8 s = 0.375 m/s/s along the track

Every day of your life, you move in a variety of different ways and experience acceleration of some kind. Whenever you slow down, speed up, turn a corner, or move in a circle, you are accelerating. Smooth rides at a constant velocity can be interrupted by a change in velocity. When observing the different motions around you, try to describe them in terms of position, distance, displacement, velocity, and acceleration.

check your UNDERSTANDING

1. A car moves with a constant velocity of 15 m/s north. What is the car's acceleration?
2. What must happen to the velocity of an object when the object is accelerating?
3. Explain how it is possible for an object to be accelerating if it is moving with constant speed.
4. **Apply** Calculate the average acceleration of a car that increases its velocity along a straight line from 10 m/s to 21 m/s in 7 seconds.

check your UNDERSTANDING

1. 0 m/s/s
2. Its velocity must change.
3. Its direction must be changing.
4. 1.6 m/s/s

12-4 Motion Along Curves

Changes in Position: Displacement

A long roller coaster has many curves and bends. Obviously, if you were in a hurry to get from one place to another, a path like the one described by the tracks of the roller coaster would not be the quickest way to go! You can easily see that a straight line is a shorter route.

Recall that you find displacement along a straight line by subtracting the starting position from the finishing position and noting direction of the motion. But how do you find displacement between two points along a curved line, a winding road for example?

Find Out! ACTIVITY

How is displacement along a curved path measured?

Look closely at the drawing. It shows both the actual path and the displacement of a train along an amusement park track. What is the train's displacement for this trip?

What To Do

1. Use the scale at the bottom of the diagram to determine the length of the displacement.

2. Use a protractor to measure the direction. Record your measurements *in your Journal.*

Conclude and Apply

1. What is the average velocity if the train completes the trip in 3 hours? Divide the total displacement by the time interval.

2. Is the distance along a curved path always greater than the displacement between the start and end points?

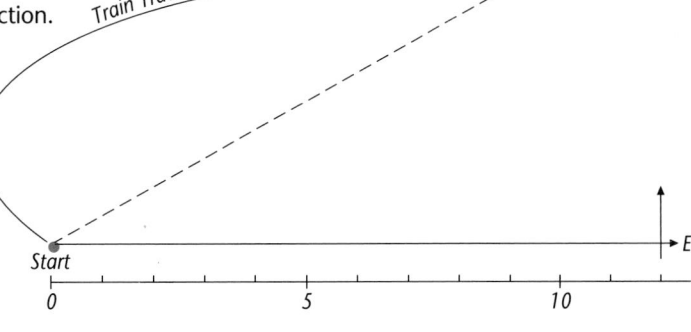

Section Objectives

- Distinguish between a displacement and a distance along a curved path.
- Find the average velocity for motion along a curved path.
- Recognize situations for which there is an acceleration, called the centripetal acceleration, even when the speed is constant.

Key Terms

centripetal acceleration

PREPARATION

Planning the Lesson

Refer to the Chapter Organizer on pages 380A–D.

Concepts Developed

This section describes motion along curves, which always involves acceleration. This knowledge is useful in understanding physics on Earth, and the orbits of satellites and planets.

Science Journal Ask students to explain why displacement along a curved path is less than distance actually traveled. Students should refer to the method used to measure displacement. L1

Expected Outcomes

Students should be able to measure displacement along a curved path and calculate average velocity.

Conclude and Apply

1. 3 km/h, 30 degrees north of east
2. yes

✔ Assessment

Performance Ask students to draw cartoon maps of three different routes. The first route should have a displacement of zero, the second route, a displacement equal to the distance traveled, the third route, a displacement less than the distance traveled. *If the route begins and ends in the same place, the displacement is zero. If the route is a straight line, the displacement is equal to the distance traveled. If the route follows a curved path, the displacement is less than the distance travelled.* Use the Performance Task Assessment List for Cartoon/Comic Book in **PASC,** p. 61. **COOP LEARN** P L1

Find Out!

How is displacement along a curved path measured?

Time needed   5–10 minutes

Materials   metric ruler, protractor

Thinking Processes   thinking critically, measuring in SI, forming operational definitions

Purpose   To measure displacement and calculate the average velocity along a curved path.

Preparation   Students may notice that one unit on the scale represents 1 km. Just measuring the displacement allows them quickly to find the actual length by changing centimeters to kilometers.

Teaching the Activity

Troubleshooting   Some students may need help in properly placing the protractor on the diagram to measure the direction.

 Before presenting the lesson, display **Section Focus Transparency 40** on an overhead projector. Assign the accompanying **Focus Activity** worksheet. L1

LEP

Activity This activity will allow students to observe acceleration by changing direction. You will need small rubber balls.

Direct students to put their arms straight out in front of them, palms up, and to put the balls in their palms. Tell them to run forward. The balls should roll toward the students since the direction of acceleration is forward. Have them repeat the experiment, this time walking forward and turning sharply left. The balls will appear to roll off their hands to the right. The balls, in fact, stay on their original path. Their velocity will not change unless a force acts upon them. L1 LEP

2 TEACH

Tying to Previous Knowledge

In their study of acceleration in a straight line in the previous section, students recalled that they are forced backward when a car accelerates forward. They should also be familiar with how their bodies are thrown to the left when the car turns to the right, and vice versa.

Theme Connection The motion of an object along a curved path involves stability and change. Even if it is maintaining a constant speed, it is undergoing a change, centripetal acceleration.

You learned in the Find Out that displacement along a curved path is measured along a straight line connecting the start and end points. If you travel on a curved road from one town to another, your displacement will be the distance, measured along a straight line, between the two towns. The actual distance you traveled along the curved road will be greater than the displacement.

Figure 12-12

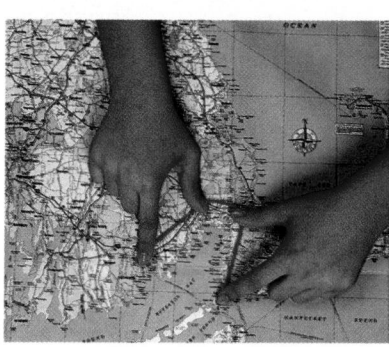

A Imagine planning an automobile trip from Falmouth to New Bedford, Massachusetts. The only route possible is a curved and hilly highway around Buzzards Bay.

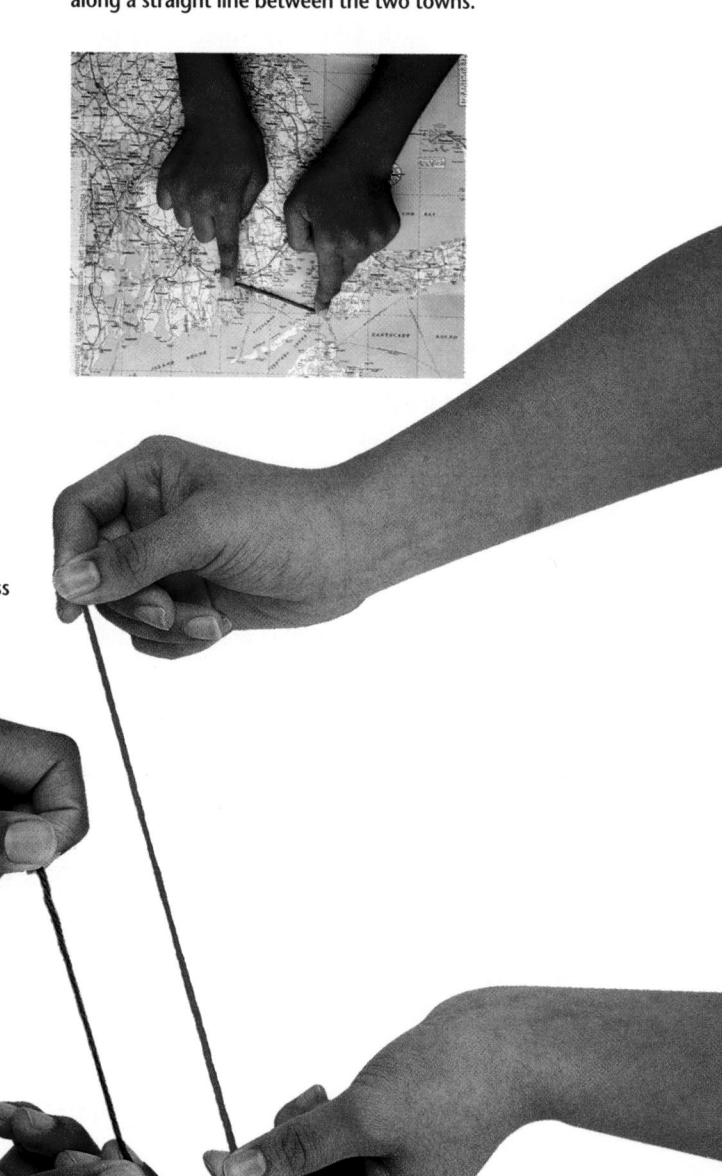

B Once you arrive at New Bedford, your displacement is the distance measured along a straight line between the two towns.

C Will the actual distance you travel be greater than or less than your displacement?

402

Program Resources

Study Guide, p. 46
Multicultural Connections, p. 27, Carousels L1
Integrated Science Videodisc, Motion
Section Focus Transparency 40

Meeting Individual Needs

Learning Disabled Encourage these students to act as volunteers. The forces involved in acceleration can be felt and experienced and will supplement the more intellectual concepts described in the text and class discussions. Also pair these students with peer helpers to complete any necessary calculations. **COOP LEARN**

Centripetal Acceleration

If you were traveling on a curve, in which direction would you accelerate? In the next exercise you will explore the direction of acceleration along a curved path.

You are already familiar with the effects of acceleration. When a car speeds up, you feel pushed backward. As an elevator starts upward, you feel pushed downward. By such experience, you know you are accelerating in a direction opposite to that in which you feel pushed. Therefore, in circular motion, since the push feels outward, acceleration must be inward. A device that shows direction inward when moving in a circle is the bubble accelerometer. In the next activity you will see how the device directly shows acceleration.

Connect to...
Life Science

Centrifuges spin materials in a circle. They are used to separate blood cells from plasma, cream from milk, and for concentrating viruses. Make a drawing of a centrifuge and explain how it works.

Connect to
Life Science
Answer: *Drawings will vary based upon the type of centrifuge that is drawn. As the centrifuge spins, heavier materials migrate to the outer end of the spinning container (test tube, capillary tube, etc.) and the lighter materials stay on top.*

Explore! ACTIVITY

What is the direction of acceleration of an object moving along a circular path?

What To Do

1. Fill a test tube with water. There should be 1 cm air space at the top after a stopper has been inserted into the test tube.

2. When you turn the test tube on its side, you will have a bubble accelerometer. Hold the test tube level and accelerate it to the right. Which way did the bubble move relative to the acceleration? Now lay the test tube on the radius of a record turntable. After attaching the test tube securely with masking tape, start the turntable rotating at its slowest speed. Which way does the bubble move? Could you use this bubble accelerometer to detect acceleration for other objects that can move in circular paths?

Bubble

Test tube with water and air bubble

Turntable

Explore!

3 ASSESS

Check for Understanding

Have students explain how the bubble accelerometer worked in the Explore activity. The water appears to move outward, forcing the bubble inward.

Reteach

Activity Have students observe the movement of lettuce leaves in a salad spinner. What is the purpose of the spinner? (*to remove water from the lettuce*) What happens to the water? (*the water moves outward through the holes in the spinner*) L1

Extension

Activity Have students use the turntable from the Explore. Place a small lightweight object such as a foam ball on the turntable and spin it at the slow speed. The object will fall off the edge of the turntable. Ask students to predict what would happen if the turntable were rotated at a higher speed. The object should fall off with a greater velocity. This shows that the faster the velocity around a curved path, the greater is the centripetal acceleration. LEP L2

4 CLOSE

Activity

Students will need a 2-m length of string, wooden bead (or any object heavy enough to weigh down the string).

Have students tie the bead on the end of the string and swing in a circular motion. **CAUTION:** *Be sure object is firmly attached to the string and that students are wearing goggles and are not close to other students while doing this activity.* Ask students in which direction the centripetal acceleration is. They should guess that it is directed along the string toward their hand. You may want to have one student do this activity at a time. L1

In the Explore, the inward motion of the bubble shows that the acceleration is directed toward the center of the turntable. The direction of the acceleration of an object moving along a circular path is toward the center of the circle and is called **centripetal acceleration**.

Figure 12-13

On a tilt-a-whirl, you spin around in a small circular path. At the same time, your small cart is sweeping in a larger circular path. At times, the acceleration would result in your being flung against the side of the cart so that you could be accelerated the same as the cart.

You've had a busy day at the amusement park. You have learned some amazing things about objects that travel in circular paths. Now you're hungry. You've really worked up an appetite. To get to the ice cream stand as quickly as possible, you would take the straightest path. You calculate the displacement in the same way, whether the path traveled is straight or curved. Remember that displacement and not the distance along any path traveled is used to calculate average velocity. If you ride in a roller coaster at a constant speed on a curved track, you are continually undergoing centripetal acceleration. You are always accelerating toward the center of the curve or circle that you are traveling around.

As a satellite or space shuttle travels in its orbit around Earth, even when it has a constant speed, it is accelerating toward Earth. You will find out what the source of this acceleration is in the next chapter.

check your UNDERSTANDING

1. If you travel on a curved road from one town to another, your displacement is always different from your actual distance traveled. One is always greater. Explain how you would measure each and why one is greater. Draw a diagram to help you with your explanation.
2. Explain what you would need to know to find the average velocity for motion along a curved path.
3. Why is an object accelerating when it is moving in a circle at constant speed?
4. **Apply** Could you use a bubble accelerometer to study the centripetal acceleration of a moving car? Explain how you would do it.

check your UNDERSTANDING

1. The shortest distance between two points is a straight line. This is also the displacement. Since the road is curved, your travel distance is greater.
2. total displacement and total time
3. Acceleration involves both speed and direction. When an object moves in a circle, its direction is constantly changing, so it is accelerating.
4. Yes. Tape the bubble accelerometer to a flat place in the car; watch the direction in which the bubble moves. The bubble moves in the direction of centripetal acceleration.

HISTORY CONNECTION

History of Time

How precisely did people know the time before watches and electric clocks were invented? How did people organize their days before agreeing on a system?

When people first began to farm and build towns, about 10,000 years ago, the most important times to know were the seasons of the year. These were easy to measure by counting the days since the sun was highest in the sky in summer, or since the river flooded in the spring. People soon realized that the seasons repeated themselves about every 365 days.

As life in towns required cooperation among many different people, it became important to know the time of day. The position of the sun in the sky gave a clue, and this could be measured using the first clock, a shadow clock or sundial as shown in the figure. About 4000 years ago in Egypt, an early shadow clock was a post in the ground. If you stood looking at the post toward north, the post would cast a shadow toward the west, to your left, as the sun rose in the morning. At sundown, the shadow would be cast toward the east, to your right, as the sun set in the west. At any point in the day, the shadow would fall on a different position on the ground. This position told the time. During the day, the shadow's tip would move

Teaching Strategies Have students compare and contrast digital and analog clocks, discussing the advantages and disadvantages of each. Students may feel that digital clocks are more precise, easier to read, and quicker to learn. On the other hand, analog clocks may show the passage of time better. Talk about common terms such as a quarter or half hour.

Activity
To complete the You Try It, have small groups set up shadow clocks in the school yard. If that is not possible, have students make a class shadow clock by securing the stick to the floor close to a south-facing window using masking tape.
COOP LEARN

HISTORY CONNECTION

Purpose
The History Connection extends Section 12-1 by presenting methods of measuring time that relied on the regular motion of various objects.

Content Background
Tell students that the piece of metal that stands in the center of a sundial is called a gnomon or style. The face of the sundial is called the dial face or plane. The earliest description of a sundial was written by a Chaldean astronomer named Berossus, in about 300 B.C. Daylight was divided into twelve equal sections called temporary hours because they changed with the seasons. A standard length for an hour didn't come until mechanical clocks were invented.

Clocks require a stable mechanical oscillator for their operation. The earliest clocks used a heavy weight attached to a string and were first built in the 9th century. Around 1500 coiled springs were introduced into clock making allowing for lighter-weight timepieces. Christiann Huygens, a Dutch scientist, built the first pendulum clock about 1656.

Electric clocks were built in the late 19th century. They have an electric motor synchronized with the frequency of alternating current.

Quartz clocks use the vibrations of a quartz crystal for more precise time keeping and were introduced in 1929. Atomic clocks which measure time by the oscillations of atoms or molecules were first built in 1948.

EXPAND
your view

Going Further ⫸

Have students work in small groups to build sand or water clocks. Have them use two plastic, wide-mouth containers for each clock. To make a sand clock, students should fasten the containers together at their bottoms and put a small hole through the bottoms. To make a water clock, students should stack the containers and put a small hole in the bottom of the top container and a matching hole in the top of the bottom container. Holes should be small so sand or water can drip at a slow but steady rate. Depending on the size of the containers, have students mark the level of sand or water at regular intervals. Use the Performance Task Assessment List for Invention in **PASC**, p. 45. L1

from left to right in an arc to the north.

The shadow clock was only useful during the day, and people began to invent new clocks that would work at night too. One of these was the water clock. If you made a water jug with a small hole in the bottom, the water would slowly leak out and the water level would go down. The level of the water in the jug would tell you how much time had passed since the jug was filled.

The water clock worked fine during the summer, but not when the water froze during winter. This problem was solved by substituting fine sand for water. By

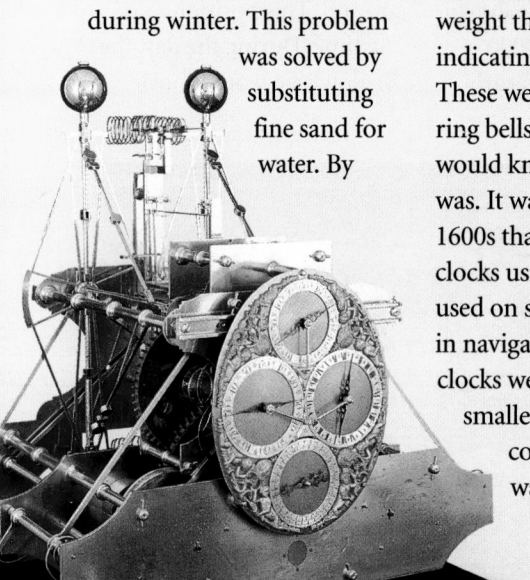

about the year 800, the sand clock became the hourglass, a bottle with two compartments separated by a partition with a small hole in it. When the sand all fell through the hole to the bottom compartment, someone quickly turned it over so that the sand would flow the other way.

The water and sand clocks were inconvenient because someone always had to be there to refill the water or turn the hourglass over.

By the 1300s, accurate mechanical clocks were invented using a falling weight that turned an axle, indicating the time in hours. These were used in towns to ring bells so that everyone would know what hour it was. It was not until the 1600s that precise portable clocks using springs were used on sailing ships to help in navigation. These ships' clocks were later made much smaller so that everyone could carry a precise watch.

You Try It!

Build a shadow clock and measure its accuracy. Place a straight stick in the ground. At 8:00 A.M., place a small stone at the tip of the stick's shadow. At every hour after 8:00, place another stone on the shadow's tip. On another day, choose five times during the day and use your shadow clock to measure the time as accurately as possible. Compare the shadow-clock times with the actual time measured with a clock. What are the differences between the shadow clock times and the actual times? How much error would you make in estimating the time using the shadow clock?

Check your shadow clock a few times for the next few weeks. Do the shadow tips still match perfectly with your stones at selected times? Explain why.

EXPAND
your view

Technology
Connection

The Technology of Thrills

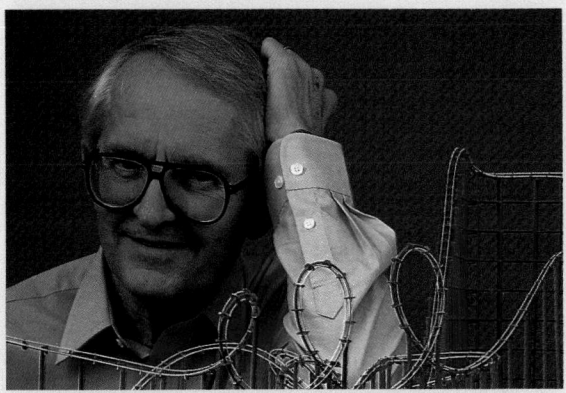

Ron Toomer gets motion sickness and hates to ride on roller coasters. He especially hates to ride on the Magnum XL-200 in Sandusky, Ohio, the world's largest roller coaster. The Magnum reaches a speed of 75 miles per hour and drops 210 feet on one of its downhill runs. When the cars reach the bottom, the riders are pressed into their seats with about as much force as astronauts taking off in the space shuttle. Ron Toomer is the engineer who designed the Magnum and 80 other roller coasters all over the world. He and the other engineers who design roller coasters have two important thoughts in mind when they begin to plan a new ride—how to make it safe for the riders, and how to produce as many thrills as possible.

Most roller coasters are now built from thick steel tubes. A team of engineers uses computers to calculate how strong the track has to be to stand up to the weight, speed,

and forces of the cars as they climb, coast down, and go into loops upside down. For every foot of track, the computers calculate the car's velocity and its forces upward, downward, and sideways. New coaster cars are made with wheels above, below, and inside the tracks to keep the car from flying off into space when it goes upside down through corkscrew loops. After the roller coaster is built, computers monitor the speed of each car and apply several different brakes if the car begins to move too fast. The steel tracks and supports are checked for cracks with X-ray machines, and padded steel lap bars and belts keep the riders safely in their seats. For every new thrill built into a roller coaster, new safety features are added.

If Ron Toomer thinks of a new loop or twist that he would find terrifying, he knows that it will be popular with the public and his own kids. His fun comes from using his knowledge of mechanical engineering to create the most terrifying but safest roller coaster rides he can think of.

 Using Computers

Take a survey of your classmates and friends to find out how many have ridden on a roller coaster, and how many of them would ride on one again. Ask what the roller coaster fans like about the rides. Using a spreadsheet, record your data and graph the results.

Going Further ⅢⅢⅢ➡
Have groups of students research the history of the roller coaster and present their findings in an oral presentation to the class or as a poster. Use the Performance Task Assessment List for Oral Presentation in **PASC,** p. 71. [L2] [P]

Technology
Connection

Purpose
The Technology Connection extends Sections 12-3 and 12-4 by describing how the concepts of acceleration and motion along curves are used to design roller coasters.

Content Background
Tell students that mechanical engineering is the study of design, construction, operation, and maintenance of machines for industry and everyday life.

Ask students to describe some of the symptoms and causes of motion sickness. *Nausea, vomiting, and dizziness are symptoms. Motion sickness can be caused by movement in a car, boat, airplane, etc.*

Teaching Strategies Help students formulate models that help explain roller coaster design by having them use straight and curved track from a model railroad or race-car track to demonstrate the design necessary to keep a roller-coaster car from flying off the track. Build a straight section of track that ends in a sharp curve. Push a car so that it travels the length of the track. Raise the straight end of the track slightly and repeat. Keep raising the track slightly until the car flies off the track. Then if possible, bank the curved section so that the outside edge is higher than the inside edge and repeat the raising of the straight end of track. [L2]

 **Using Computers**

Students who like the ride will say that it is fun or exciting. Students who do not like the ride will say that it is frightening or that it makes them sick.

Art
Connection

Art
Connection

How Do You Paint Motion?

Purpose
The Art Connection reinforces Section 12-1 by presenting a style of painting that portrays motion.

Content Background
Point out to students that the Futurist movement was not just an artistic period, but a philosophy of life. It was begun in 1909 in Italy by the poet Filippo Marinetti who wanted to destroy the culture of the past and replace it with a new art form and poetry based on dynamic movement and modern urban living. The Futurist movement grew out of the more famous Cubist school of art. Besides Balla, other Futurist painters included Umberto Boccioni, Joseph Stella, and Raymond Duchamp-Villon and his brother Marcel Duchamp.

Balla's *Dynamism of a Dog on a Leash* was inspired by the multiple-exposure photographs of movement by Eadweard Muybridge.

Teaching Strategies
Have pairs of students formulate models of motion by producing a series of pictures which can be flipped through to illustrate motion. One student can pose for the pictures while the other draws simple stick figures. Have students decide upon a movement to draw. It can be a simple dance step or a sports activity. Students may choose to show this same movement in one picture as they complete the What Do You Think activity. Then have the students change roles and create a second series of sketches. L2 COOP LEARN P

Answers to
What Do You Think?

1. Students may feel that the

I talian painter Giacomo Balla painted *Dynamism of a Dog on a Leash* in 1912. He and his artistic friends at the time called themselves "futurists." They were tired of paintings that only showed objects and people at one instant in time like a snapshot. They believed that the future of the modern world would be full of machines in motion and they wanted a new way to paint moving objects that showed the excitement of their motion.

In this painting, which is shown on the right, the dachshund dog and its fashionable lady are shown with their feet in all possible positions as the dog trots down the street on its daily walk. Looking at the picture, you may get the idea that motion produces a blur to the eye. Single instants of time blend into one another because the eye and the brain cannot keep up with rapidly changing positions.

This painting was probably inspired by the first primitive movies that were made a few years earlier. These were a series of photographs of moving animals and people taken quickly, one after the other. Looking at one picture and then the next by flipping through the photographs lets you imagine the motion taking place. Balla painted the

moving dog as if many of these photographs were added together in one place. The effect is to show changing positions over time in a single painting. This kind of painting was very unusual in 1912.

What Do You Think?

Do you think Balla was successful in capturing the rapid motions of the dachshund in this painting? Can you think of other ways that a painting or drawing can show motion? Draw a picture using one of these ideas.

drawing is just a blur and does not show the dog covering any distance. Others may feel that the drawing looks just as if the dachshund and lady are walking.

2. One possible way to imply motion is to draw something in the picture that is known by the viewer to be moving, such as a bird in the air.

Going Further ▐▐▐▐▶
Have students work in small groups to study paintings of other periods and see the way movement was shown in them. Include paintings by

the Cubists and Expressionists. Have the groups explore other art forms, such as sculpture and architecture, that were created during those periods. If possible, take your students on a field trip to an art museum. Then have each student select one piece of art that he or she sees as having movement. Ask students to write short reports explaining how the artists depicted motion in these works. Use the Performance Task Assessment List for Writing in Science in **PASC**, page 87. L2 COOP LEARN P

Science Journal

Review the statements below about the big ideas presented in this chapter, and answer the questions. Then, re-read your answers to the Did You Ever Wonder questions at the beginning of the chapter. *In your Science Journal*, write a paragraph about how your understanding of the big ideas in the chapter has changed.

1 Average speed is the total distance traveled divided by the total time required to travel the distance. *Who will win a race, the cyclist with the greatest instantaneous or average speed?*

2 Average velocity is a quantity giving the total displacement divided by the time interval. It includes direction. *What factors must sailboats take into consideration about the wind in order to win a race?*

3 Acceleration is a change in velocity that occurs over time. *If the girl on the right accelerates down the slide from 0 to 5 meters per second in 2 seconds, what is her average acceleration?*

4 When an object travels along a curved path at a constant speed, its direction of motion is constantly changing. *What kind of acceleration is the object experiencing?*

Science Journal

Did you ever wonder...

• Even though your car is not moving, you think you are moving in the opposite direction relative to the other car. (pp. 392–393)

• Your body senses a change in motion when you accelerate. The roller coaster's acceleration causes the funny feeling in your stomach. (pp. 396–397)

• You are accelerating inward so your body feels pushed outward. (p. 403)

Project

Ask students to collect newspaper and magazine headlines and photos that describe motion. The sports page should be a good source of materials, as will science pages and weather reports. Build a collage of the pictures and headlines. Build a second collage of headlines that use the terms of motion but do not imply physical movement, such as changes in attitude, rise and fall of the stock market, and acceleration of negotiations. Use the Performance Task Assessment List for Display in **PASC**, p. 63. **L1**

Students will participate in a contest to find the fastest attainable speeds and greatest acceleration for animals, humans, and machines. They research and write a short report about one of the other main ideas.

Teaching Strategies

Have students pick a category or assign all of them. The categories for the contest can include:

1. Fastest average speed for men and women in various races, such as the 100 meter, 200 meter, 400 meter, 1500 meter, 3000 meter, and marathons.

2. Fastest average speed for non-human animals. You can subdivide this category into land, air, and marine animals.

3. Fastest average speed for machines. You can also subdivide this category into land, air, and water speeds.

Students must be able to verify their answers by quoting the source of their information.

In a separate activity, students should write a short report about one of the other main ideas. For example, for average velocity, they could write about the way sailboats achieve an average velocity; acceleration, might include information about the astronauts aboard the space shuttle; centripetal acceleration, could have students research the curved tracks used for roller coasters or race cars.

Answers to Questions

1. average speed
2. speed and direction
3. 2.5 m/s^2
4. centripetal acceleration

GLENCOE TECHNOLOGY

 **MindJogger Videoquiz**

Chapter 12 Have students work in groups as they play the Video-quiz game to review key chapter concepts.

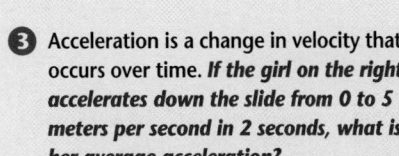

Using Key Science Terms

1. average acceleration
2. average speed
3. average velocity or relative velocity
4. position
5. centripetal acceleration

Understanding Ideas

1. Velocity always includes direction.
2. Displacement is a measure of the straight-line distance and direction from your starting point. If you travel two km north and then two km south, or if you travel along a curved path, your displacement from your starting point will be less than the actual distance you traveled.
3. Determine if it is changing position.
4. An odometer measures distance.
5. Yes. It has a negative acceleration.

Developing Skills

1. Note starting odometer reading → Note time of departure → Ride to grandmother's house → Note time of arrival and calculate travel time → Calculate average speed, distance traveled divided by time of travel.

2.
Event (m)	Average speed (m/s)
100	10.14
200	10.14
400	9.24
800	7.80
1500	7.15

The average speeds for the 100-m and 200-m races are the same. For the other races, average speeds become slower as the races become longer.

3. Answers will vary. The average speed should be between 7.80 m/s and 7.15 m/s. A guess of 7.60 gives a time of 2:11.6. The actual record is 2:13.9 or 7.47 m/s.

4. by 15 m

Using Key Science Terms

acceleration   displacement
average acceleration   distance
average speed   position
average velocity   relative velocity
centripetal acceleration

Each phrase below describes a science term from the list. Write the term that matches the phrase describing it.

1. total velocity change divided by total time
2. ten miles per hour
3. ten miles per hour south
4. comparing your location to the location of a known point
5. acceleration of a point on the edge of a spinning compact disc.

Understanding Ideas

Answer the following questions in your Journal using complete sentences.

1. Why is the term "10 miles per hour" not a description of velocity?
2. How can the amount of your displacement be less than the distance you travel?
3. How can you determine if an object is accelerating?
4. What does an odometer measure?
5. If an object is slowing down, could you say that it is accelerating?

Developing Skills

Use your understanding of the concepts developed in this chapter to answer each of the following questions.

1. **Concept Mapping** You travel in a car to your grandmother's house. You want to know your average speed for the trip. Create an events chain to show how you would find your average speed.
2. **Making and Using Tables** The following are men's American record times for five running events: 100 m–9.85 s; 200 m–19.73 s; 400 m–43.29 s; 800 m–1 min 42.60 s; 1500 m–3 min 29.77 s. Use these data to make a table listing the event and average speed of each event. Do you notice any pattern in the table?
3. **Interpreting Data and Predicting** Use your results from the previous exercise to predict the average speed for the 1000 m race. Then use your prediction to estimate the record time.
4. **Making and Using Graphs** The data in the illustration below give the stopping distances for different car speeds. Plot these points on a graph with the speed in m/s along the horizontal axis and the stopping distance in meters along the vertical axis. Connect the points with a smooth line. Using your graph, determine how much the stopping distance

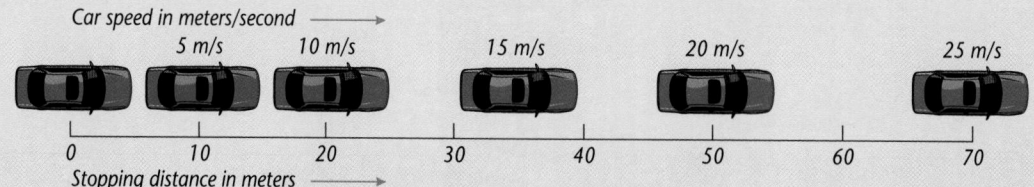

Car speed in meters/second ⟶

5 m/s   10 m/s   15 m/s   20 m/s   25 m/s

0   10   20   30   40   50   60   70

Stopping distance in meters ⟶

Program Resources

Review and Assessment, pp. 71-76 L1
Performance Assessment, Ch. 12 L2
PASC
Alternate Assessment in the Science Classroom
Computer Test Bank L1

increases when the speed increases from 15 m/s to 20 m/s.

Critical Thinking

In your Journal, *answer the following questions.*

1. When does the distance along a path equal the displacement?
2. If two friends riding bicycles together both have the same velocity relative to Earth, what is their velocity relative to each other?
3. How might you determine the speed of your car if the speedometer broke?
4. When you enter some tollways, you are given a toll card that tells the location and time you entered. When you exit, the time on the toll card can be checked. Could a driver ever be issued a speeding ticket with this information? Explain.

Problem Solving

Read the following problem and discuss your answers in a brief paragraph.

Your friend lives in another city in your state and wants to visit you. Your friend just called and wants to know the shortest highway route between your two cities. All you have is your state map and a piece of string. Can you determine the shortest route with just the map and the string? Can you determine the displacement? Try it.

Obtain a map of your state and select another city on the map. Using a piece of string, how could you find the shortest highway route from your city to the one you picked? How would you determine the displacement involved?

CONNECTING IDEAS

Discuss each of the following in a brief paragraph.

1. **Theme—Stability and Change** Cruise control in a car keeps the car's speed constant. Does this necessarily keep the car's velocity constant?
2. **Theme—Systems and Interactions** If you are on a train or a plane with the window shades down, how do you know that you are moving? What could you do to find out if you are moving or not?
3. **A Closer Look** Who moves faster as Earth rotates, you or a person at the equator? Explain.
4. **History Connection** Compare and contrast three different methods of measuring time. What are the advantages and the disadvantages of each method?
5. **Technology Connection** In which position(s) on a roller coaster track do you experience centripetal acceleration?

✔ Assessment

Portfolio Review the portfolio options that are provided throughout the chapter. Encourage students to select one product that demonstrates their best work for the chapter. Have students explain what they learned and why they chose this example for placement into their portfolios.

Additional portfolio options can be found in the following **Teacher Classroom Resources:**

Making Connections: Integrating Sciences, p. 27

Multicultural Connections, pp. 27–28

Making Connections: Across the Curriculum, p. 27

Concept Mapping, p. 20

Critical Thinking/Problem Solving, p. 20

Take Home Activities, p. 21

Laboratory Manual, pp. 71–74

Performance Assessment P

Critical Thinking

1. When the path traveled is in one direction in an exact straight line.
2. Zero
3. Use the odometer and a watch. Find the time it takes to travel one mile. Divide the distance by the time to determine the speed of your car.
4. Yes, if you know the total distance and the total time interval and there is a given speed limit, it is easy to determine if you traveled faster than the existing speed limit.

Problem Solving

Yes; for each route, place the string along the highway(s) between the two cities, exactly following all curves. Mark the length of each route on the string and compare to determine the shortest route. To determine displacement, place the string in a straight line connecting the two cities.

Connecting Ideas

1. No, you could be changing direction all the time.
2. You could not be sure unless you could interact with something in the outside world. You could only feel movement if you experienced a change in velocity. You have no other frame of reference. You could pull up the window shade and look for relative motion of the plane/train compared to trees, building, etc.
3. A person at the equator, because he or she travels a greater distance during one rotation, while the time of rotation, 24 hours, is the same.
4. Examples: clock, hourglass, water clock, and movement of sun, moon, and stars. Advantages and disadvantages are degree of accuracy, precision, convenience, and consistency.
5. At any point where the track is curved

Chapter Organizer

SECTION	OBJECTIVES	ACTIVITIES & FEATURES
Chapter Opener		Explore!, p. 413
13-1 Falling Bodies (3 sessions, 1.5 blocks)	1. **Calculate** the acceleration of a falling object given measurements of its position at various times. 2. **Describe** the motion of an object as it falls freely toward Earth. National Content Standards: (5–8) UCP2–3, A1–2, B2, G1–2	Find Out!, p. 415 **Investigate 13-1:** pp. 416–417 **Skillbuilder:** p. 421 **A Closer Look,** pp. 418–419
13-2 Projectile Motion (1 session, .5 block)	1. **Describe** the horizontal motion of a projectile. 2. **Describe** the vertical motion of a projectile. 3. **Explain** how vertical and horizontal motions of projectiles are independent. National Content Standards: (5–8) UCP2–3, A1, B2	Find Out!, p. 423 **History Connection,** p. 438
13-3 Circular Orbits of Satellites (2 sessions, 1 block)	1. **Describe** how a satellite is a projectile in free-fall. 2. **Connect** weightlessness to free-fall. 3. **Explain** satellite motion in terms of relative velocity. National Content Standards: (5–8) UCP2–3, B2, G1	Find Out!, p. 427 **Life Science Connection,** pp. 426–427 **Science and Society,** pp. 436–437
13-4 The Motion of a Pendulum (2 sessions, 1.5 blocks)	1. **Define** the period of a pendulum. 2. **Define** frequency, as it relates to periodic motion. 3. **Describe** the relationship between the period of a pendulum, the mass of a bob, the length of the pendulum, and the amplitude of the motion. National Content Standards: (5–8) UCP2–3, A1–2, B2, G1, G3	Design Your Own Investigation 13-2: pp. 432–433 **How It Works,** p. 435

ACTIVITY MATERIALS

EXPLORE!
p. 413* modeling clay, pencil

INVESTIGATE!
pp. 416–417 ruler

DESIGN YOUR OWN INVESTIGATION
pp. 432–433 80-cm piece of string, 8 metal washers, masking tape, ruler, meterstick, seconds timer

FIND OUT!
p. 415* string; heavy bolts or nuts or other object of that size
p. 423* meterstick, 2 coins
p. 427* No special materials required.

KEY TO TEACHING STRATEGIES
The following designations will help you decide which activities are appropriate for your students.

- **L1** Basic activities for all students
- **L2** Activities for average to above-average students
- **L3** Challenging activities for above-average students
- **LEP** Limited English Proficiency activities
- **COOP LEARN** Cooperative Learning activities for small group work
- **P** Student products that can be placed into a best-work portfolio
- Activities and resources recommended for block schedules

Need Materials? Call Science Kit (1-800-828-7777).

⏱ OUT OF TIME? We recommend that students do the activities with an asterisk.

Chapter 13 Motion Near Earth

TEACHER CLASSROOM RESOURCES

Student Masters	Transparencies
Study Guide, p. 47 **Concept Mapping**, p. 21 **Multicultural Connections**, p. 29 **Making Connections: Across the Curriculum**, p. 29 **Making Connections: Technology and Society**, p. 29 **Activity Masters**, Investigate 13-1, pp. 55–56 **Science Discovery Activities**, 13–1 **Laboratory Manual**, pp. 75–76, Speed of Falling Objects	**Teaching Transparency 25**, Strobe of Ball Falling **Section Focus Transparency 41**
Study Guide, p. 48 **Science Discovery Activities**, 13–2 **Laboratory Manual**, pp. 77–80, Projectile Motion	**Teaching Transparency 26**, Projectile Motion **Section Focus Transparency 42**
Study Guide, p. 49 **Critical Thinking/Problem Solving**, pp. 5, 21 **Multicultural Connections**, p. 30 **Take Home Activities**, p. 22 **Making Connections: Integrating Sciences**, p. 29 **How It Works**, p.16 **Science Discovery Activities**, 13–3	**Section Focus Transparency 43**
Study Guide, p. 50 **Activity Masters**, Design Your Own Investigation 13-2, pp. 57–58	**Section Focus Transparency 44**

ASSESSMENT RESOURCES	TEACHING & TECHNOLOGY
Review and Assessment, pp. 77–82 **Performance Assessment**, Ch. 13 **PASC*** **MindJogger Videoquiz** **Alternate Assessment in the Science Classroom** **Computer Test Bank**	**Spanish Resources** **Cooperative Learning Resource Guide** **Lab and Safety Skills** **Science Interactions, Course 1, CD-ROM** **Computer Competency Activities**

*Performance Assessment in the Science Classroom

NATIONAL GEOGRAPHIC TEACHER'S CORNER

Index to National Geographic Magazine	National Geographic Society Products Available From Glencoe	Additional National Geographic Society Products
The following articles may be used for research relating to this chapter: • "Searching for the Secrets of Gravity," by John Boslough, May 1989. • "Satellites That Serve Us," by Thomas Y. Canby, September 1983.	To order the following products for use with this chapter, contact your local Glencoe sales representative or call Glencoe at 1-800-334-7344: • *Newton's Apple Physical Sciences* (Videodisc) • *An Introduction to Physical Science, Part I* (2 Filmstrips)	To order the following products for use with this chapter, call the National Geographic Society at 1-800-368-2728: • *Everyday Science Explained* (Book)

Teacher Classroom Resources

These are key components of the classroom resources package.

TEACHING AIDS

Teaching Transparencies

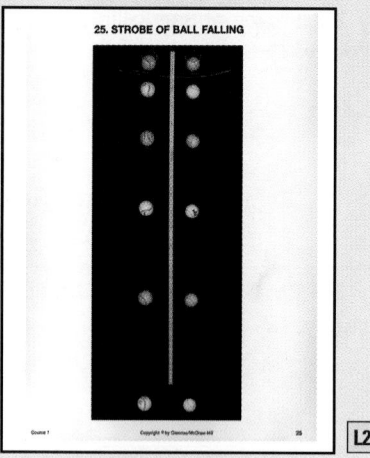

Section Focus Transparencies

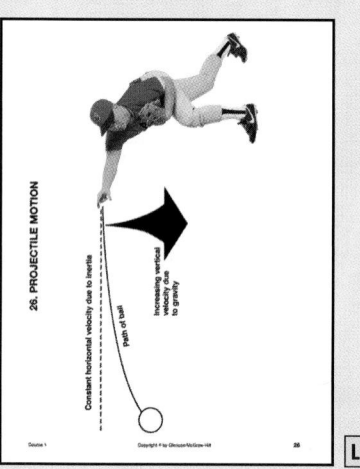

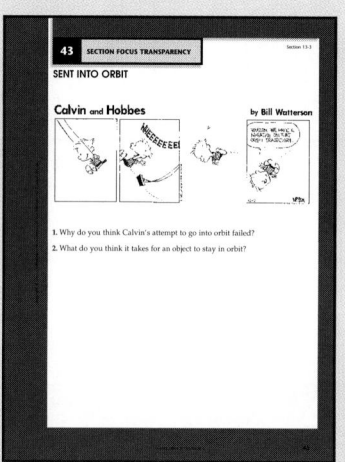

HANDS-ON LEARNING

Science Discovery Activity*

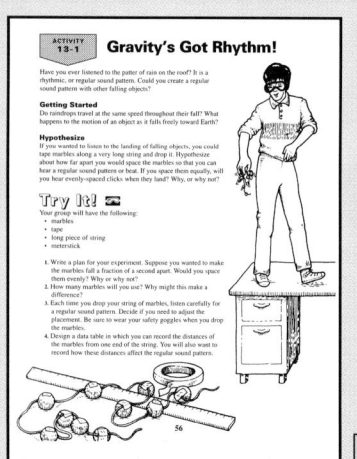

Laboratory Manual*

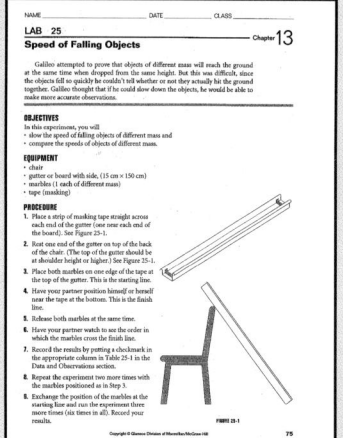

Take Home Activity

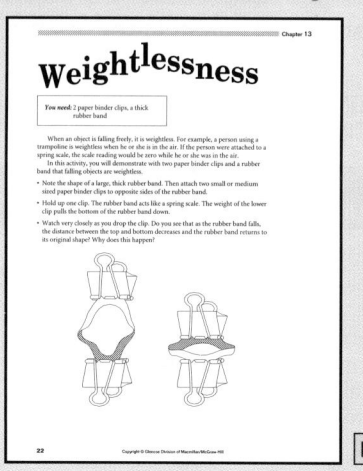

Chapter 13 Motion Near Earth

Study Guide*

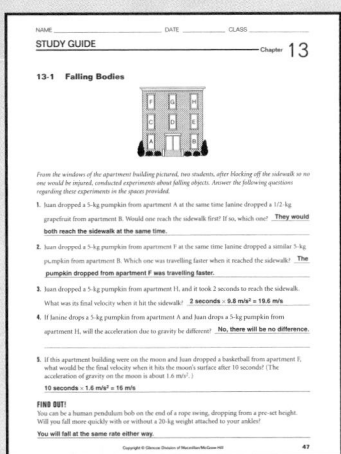

Concept Mapping

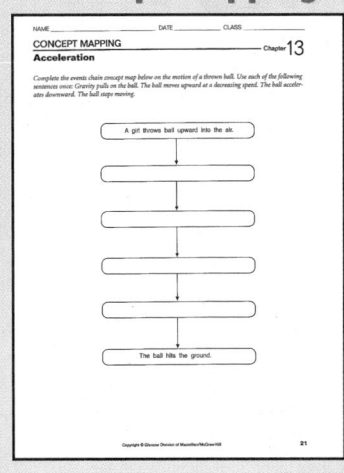

Critical Thinking/ Problem Solving

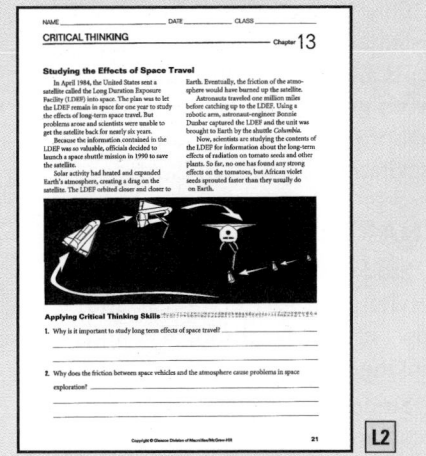

Integrating Sciences

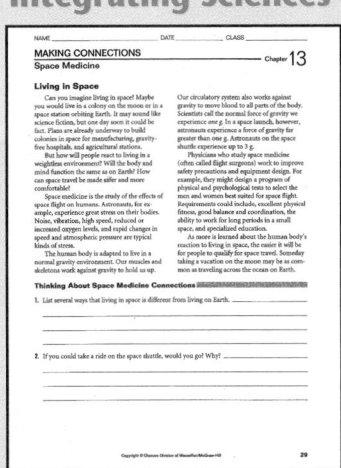

Across the Curriculum

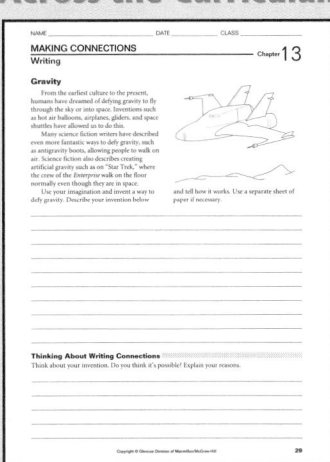

Technology and Society

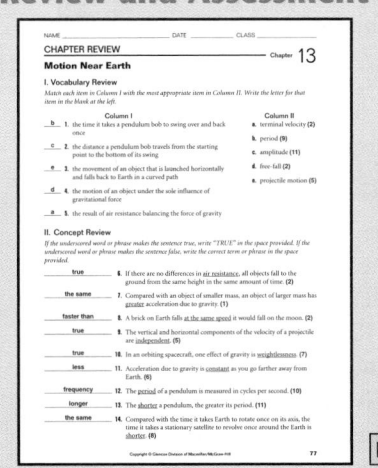

Multicultural Connection**

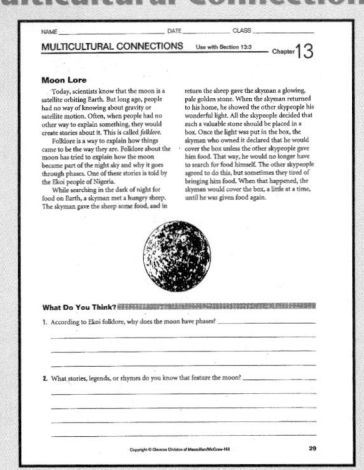

Performance Assessment

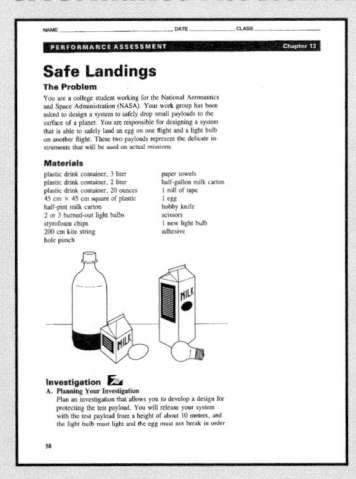

Review and Assessment

*One per section   **Two per chapter

CHAPTER 13

Motion Near Earth

THEME DEVELOPMENT

One theme of this chapter is systems and interaction. Students analyze the interaction between gravity and falling objects that produces acceleration in the gravitational system of Earth.

CHAPTER OVERVIEW

Students will discover what gravitational acceleration is and how it affects stationary and moving objects.

Students will investigate projectile motion and the motion of a pendulum. They will verify that only pendulum length—not bob weight or amplitude—determines a pendulum's frequency.

Tying to Previous Knowledge

Ask if a moving object's velocity tends to remain the same. Students will recall that, friction excepted, an object tends to keep moving at the same velocity.

INTRODUCING THE CHAPTER

Have students examine the photo on page 412 and describe what will happen next. Tell students that this chapter explains why and how objects fall.

Uncovering Preconceptions

Students may believe that a heavier object falls faster than a light object.

MOTION NEAR EARTH

Did you ever wonder...

✓ **Whether heavy objects fall to Earth more quickly than light objects?**

✓ **How satellites can stay in orbit?**

✓ **Why astronauts float around in the space shuttle?**

Science Journal

Before you begin to study about motion near Earth, think about these questions and answer them *in your Science Journal.* When you finish the chapter, compare your journal write-up with what you have learned.

Answers are on page 439.

Humans live only on Earth. We seldom stray far from its surface. You may have flown in an airplane, but that's still close to Earth. Some humans have gone to the moon. That may seem a long way from Earth, but compared with the distance to the other planets or the stars, it's not far at all.

Because we have spent nearly all of our time on Earth's surface, almost all our experience with motion has to do with the way objects move near or on Earth. When you shoot a basketball at a basket, you know the ball will fall back down. When you run down a basketball court, you know you are continually pushing off Earth and dropping back to it with each step.

▶ *How does motion close to Earth differ from what it would be in space? In this chapter, you'll explore motion on Earth and get an idea of how you would move elsewhere.*

412

00:00 OUT OF TIME?

If time does not permit teaching the entire chapter, use the Chapter Overview on this page, Reviewing Main Ideas at the end of the chapter, and the Chapter 13 audiocassette to point out the main ideas of the chapter.

Learning Styles		
Kinesthetic	Explore, p. 413; A Closer Look, pp. 418–419; Find Out, p. 423; Visual Learning, p. 425; Project, p. 439	
Visual-Spatial	Demonstration, pp. 414, 418, 420, 422, 430, 434; Find Out, p. 415; Visual Learning, pp. 419, 420, 424, 434; Activity, p. 424; Discussion, p. 425	
Logical-Mathematical	Investigate, pp. 416–417, 432–433; Across the Curriculum, p. 420; Activity, pp. 421, 423, 429, 434; Visual Learning, pp. 426, 430; Find Out, p. 427	
Linguistic	Visual Learning, pp. 415, 422; Life Science Connection, pp. 426–427; Across the Curriculum, pp. 428, 431; Multicultural Perspectives, p. 428; Activity, p. 429	

LS

Explore! ACTIVITY

How do things fall?

You learned about motion in Chapter 12. Does the size or mass of an object affect its fall?

What To Do

1. Make two balls out of modeling clay, one about 1.5 cm in diameter and the other about 4 cm in diameter.

2. Determine which ball is heavy and which one is light.

3. *In your Journal*, predict which ball will fall more quickly when both balls are dropped from the same height at the same time.

4. Drop them at the same time from the same height.

5. Describe *in your Journal* how the balls landed.

6. Attach one of the clay balls to each end of a pencil. Hold this object high and drop it. Repeat the activity, varying the position of the balls.

7. Record *in your Journal* how the object fell and landed each time you dropped it.

ASSESSMENT PLANNER

PORTFOLIO
Refer to page 441 for suggested items that students might select for their portfolios.

PERFORMANCE ASSESSMENT
Process, pp. 413, 417, 423
Skillbuilder, p. 421
Explore! Activity, p. 413
Find Out! Activities, pp. 415, 423, 427
Investigate, pp. 416–417, 432–433

CONTENT ASSESSMENT
Oral, p. 427
Check Your Understanding, pp. 421, 424, 429, 434
Reviewing Main Ideas, p. 439
Chapter Review, pp. 440–441

GROUP ASSESSMENT
Opportunities for group assessment occur with Cooperative Learning Strategies.

Explore!

How do things fall?

Time needed 15 minutes

Materials modeling clay, pencil

Thinking Processes observing and inferring, comparing and contrasting, measuring in SI, predicting, modeling, recognizing cause and effect, classifying, forming operational definitions

Purpose To determine if a heavier object falls faster than a lighter object.

Teaching the Activity

Have students work in pairs, taking turns dropping and observing from near floor level. [L1] **COOP LEARN**

Troubleshooting The clay balls will fly off the pencil and will need to be replaced before each drop.

Science Journal Have students record their observations and answers in their student journals.

Expected Outcome

Students observe that objects fall at the same rate.

Answers to Questions

Both balls hit at the same time when dropped together. When attached to a pencil, they still fall at the same rate; the pencil does not change orientation during the fall.

✔ Assessment

Process Invite students to explain whether their observations support their predictions. Students can make a cartoon comparing or contrasting their predictions. Use the Performance Task Assessment List for Cartoon/Comic Book in **PASC**, p. 61. [L1]

PREPARATION

Planning the Lesson

Refer to the Chapter Organizer on pages 412A–D.

Concepts Developed

The previous chapter discussed motion and its characteristics. In the first section of this chapter, students learn about motion caused by the acceleration of gravity both on Earth and on the moon.

1 MOTIVATE

Bellringer

 Before presenting the lesson, display **Section Focus Transparency 41** on an overhead projector. Assign the accompanying **Focus Activity** worksheet. [L1]

[LEP]

Demonstration Invite students to predict. Fill a small balloon with water and tie it. Hold the balloon 2 centimeters above a tabletop and ask students what will happen when you let go. Students should say that it will fall. Ask students if they think the balloon will break. Then drop the balloon (which will not break). Then hold the balloon a meter above the tabletop. Ask students if they think the balloon will break when dropped from this height. Most students may believe that the balloon will break. Ask them to form a hypothesis about why the balloon would break when dropped from a higher point. Students should say that the balloon will be going faster and will hit the tabletop harder. You should test this before doing it in front of the class. [L1]

Student Text Question
What always brings you back to Earth when you jump? *gravity*

Section Objectives

- Calculate the acceleration of a falling object given measurements of its position at various times.
- Describe the motion of an object as it falls freely toward Earth.

Key Terms

acceleration due to gravity

Describing How Things Fall

You experience falling objects every day. Raindrops or snowflakes may have fallen on you, and you may have seen leaves, seeds, or twigs fall from trees. When you jump, you always come back to Earth. What always brings you back to Earth when you jump? In this section, you will describe the motion of falling objects and learn that there is something similar about the motions of all falling objects, including yourself.

Figure 13-1

A Although many doubt the accuracy of the legend that Galileo dropped objects from the top of the Leaning Tower of Pisa to find which would hit the ground first, Galileo did experiment with falling objects.

414 Chapter 13

■ When Mass Varies

More than 2300 years ago, the Greek philosopher Aristotle was interested in the question—do heavy objects fall in the same amount of time as lighter objects? Aristotle, like others of his time, thought that Earth was the center of the universe and that objects with more mass would naturally rush more quickly to the center of the universe. Aristotle would have said that a large stone would fall more quickly than a small pebble.

In the early 1600s, the Italian scientist Galileo used experimentation to answer the question about falling objects. According to legend, Galileo dropped objects of different masses from the top of the Leaning Tower of Pisa to find out which would hit the ground first. See **Figure 13-1** to find out Galileo's conclusions.

B From his many experiments, Galileo concluded that all objects should fall from the same height in the same time.

Program Resources

Study Guide, p. 47
Laboratory Manual, pp. 75–76, Speed of Falling Objects [L2]
Concept Mapping, p. 21, Acceleration [L1]
Multicultural Connections, p. 29, Moon Lore [L1]
Teaching Transparency 25 [L2]
Section Focus Transparency 41

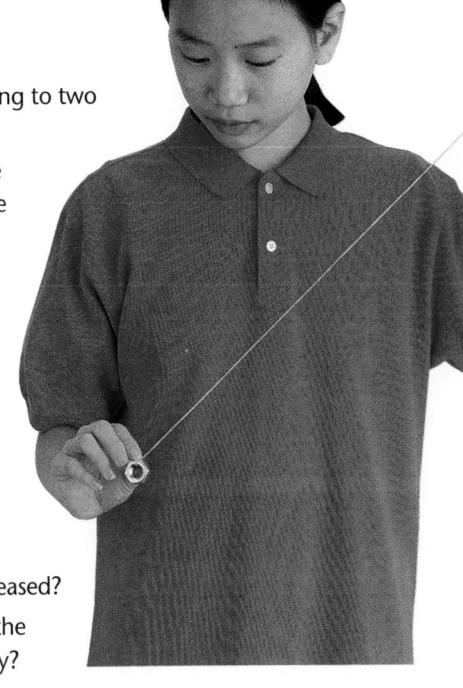

Find Out! ACTIVITY

Do heavier objects fall more quickly?

There are several ways to find out whether heavy objects fall more quickly than light objects. One way is to observe the motion of a pendulum.

What To Do

1. Make a pendulum by tying a 1 m length of string to two heavy bolts, nuts, or similar objects.

2. Hold the end of the string with one hand while you pull the objects, the pendulum bob, to one side about 25 cm with the other hand.

3. Release the bob and observe it closely.

4. Construct another pendulum the same length, using only one bolt, nut, or similar object.

5. Have a partner hold one pendulum while you hold the other. Start the pendulums swinging at the same time by pulling the bobs back about 25 cm and releasing them.

Conclude and Apply

1. What happened when the pendulums were released?

2. Which pendulum bob reached the bottom of the swing first? Did the heavy bob fall more quickly?

C Leaves float and drift when they fall because of air resistance.

As you saw in the Find Out activity, both bobs appear to reach the bottom of the swing at the same time. But you, like Galileo, couldn't really be certain whether the objects reached the bottom at the same time because they were moving too quickly. In the next activity, you have an opportunity to observe two balls of different masses, a baseball and a tennis ball, falling from the same height.

13-1 Falling Bodies **415**

Find Out!

Do heavier objects fall more quickly?

Time needed 15 minutes

Materials two 3-foot lengths of string; two or three bolts, nuts, or other heavy objects; and a light object such as a small piece of wood

TECH PREP

Thinking Processes observing and inferring, comparing and contrasting, recognizing cause and effect, modeling, measuring in SI

Purpose To learn, by comparing and contrasting the behavior of light and heavy bobs, that the swing of a pendulum is independent of the mass of its bob.

Preparation Students can more readily compare and contrast the movement of the pendulums if the first pendulum has one bolt for a bob and the second pendulum has two bolts. Students can see more clearly that the mass at the end of the string is doubled.

Visual Learning

Figure 13-1 Have students compare their procedures and observations in the Explore on page 413 with Galileo's.

Inquiry Question: What do you think would happen if you dropped a feather and a clay ball from the top of the Leaning Tower of Pisa? Why? *The clay ball would probably reach Earth before the feather because air resistance and air currents could slow the feather.*

Teaching the Activity

Troubleshooting Caution students holding the upper end of the string to keep it as still as possible. Have students pull the pendulum bob no more than about 30 cm to the side and release it; raising it farther disturbs the swing.

Caution students to raise the bob the same distance each time they try this.
COOP LEARN

Science Journal Have students write their responses to the Conclude and Apply questions in their journals. **L1**

Expected Outcome

Students observe that a pendulum set in motion continues to swing, and that the weight of the bob makes no difference in its speed.

Conclude and Apply

1. The light and heavy bobs swung at the same speed when the pendulums were released.

2. Both pendulums arrived at the same time.

✔ Assessment

Performance Have students make an invention consisting of two or more pendulums to show what they observed in this activity. Use the Performance Task Assessment List for Invention in **PASC**, p. 45. **P** **L1**

13-1 Acceleration of Falling Objects

Planning the Activity

Time needed 20 minutes

Purpose To calculate the acceleration of gravity.

Process Skills observing and inferring, making and using tables, making and using graphs, sampling and estimating, sequencing, interpreting scientific illustrations, observing, interpreting data, measuring in SI

Preparation Before starting, confirm that students copied the data table correctly.

Teaching the Activity

Process Reinforcement Guide students as they complete the data table by going over the following information. The *Position* column gives the position of the ball as shown by the metric ruler. The *Distance Fallen* column shows the *change* in position: the position in the previous image subtracted from current position. *Average Velocity* is the distance fallen divided by 0.1 second. L1

Troubleshooting Help students determine how to set up the calculations.

Science Journal Have students write their responses to the Analyzing/Concluding and Applying questions. L1

Acceleration of Falling Objects

Two falling balls were recorded by strobe photography. Six photographs were taken, each 1/10 of a second apart. Even though the balls are of different masses, they are falling at the same rate. Figure 13-2 is a drawing made from that photograph. At one side of the figure is a two-meter measuring stick. As each ball falls, there seems to be more distance between each succeeding image. You learned in Chapter 12 that distance traveled in a given time is a measure of speed. Therefore, if the balls are moving greater distances in the same period of time, they are speeding up or accelerating. Use Figure 13-2 to discover the acceleration of these two falling objects.

Figure 13-2

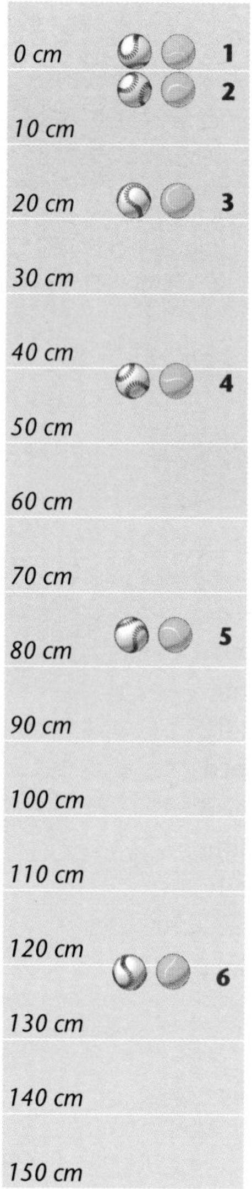

Problem

What is the acceleration of falling objects?

Materials

ruler   data table   **Figure 13-2**

What To Do

1 Copy the data table *into your Journal*.

2 Using the ruler as a guide, record the position of the first image of one ball as it starts to fall. Always *measure* the ball position from the same point on the ball.

Sample data

Data and Observations				
Image	Position (cm)	Distance Fallen (cm)	Average Velocity (m/s)	Time (s)
1	0			
2	5	5	0.5	0.10
3	20	15	1.5	0.20
4	44	24	2.4	0.30
5	78	34	3.4	0.40
6	123	45	4.5	0.50

Meeting Individual Needs

Learning Disabled Learning-disabled students may have trouble with the mathematics required by the Investigate but may be able to make accurate measurements of position. Team learning-disabled students with one or more other students. Make the learning-disabled student responsible for measurements. You may also wish to spend additional time with learning-disabled students, reviewing the fundamental concepts of velocity and acceleration and using the familiar example of a car's or bicycle's motion to explain the difference. **COOP LEARN**

3 Now, record the position of the second image.

4 *Calculate* the distance from the first to the second image by subtracting the position of the first image from the position of the second image. Record this distance in the table under Image 2.

5 The images were taken 1/10 of a second apart. Use this information to *calculate* the average velocity of the balls between the first and second image in meters per second by dividing the distance fallen (in meters) by 1/10 second.

6 Record the position of the rest of the images. Make sure you always measure to the leading edge of the ball.

7 Find the distance between each pair of images and the average velocity. Fill these in for the rest of the images.

8 The last column of the table shows the exact time the balls' velocity reached the average velocity. To calculate these times, we assumed the clock started at the time of the first image and that the ball reached average velocity halfway in time between any two images.

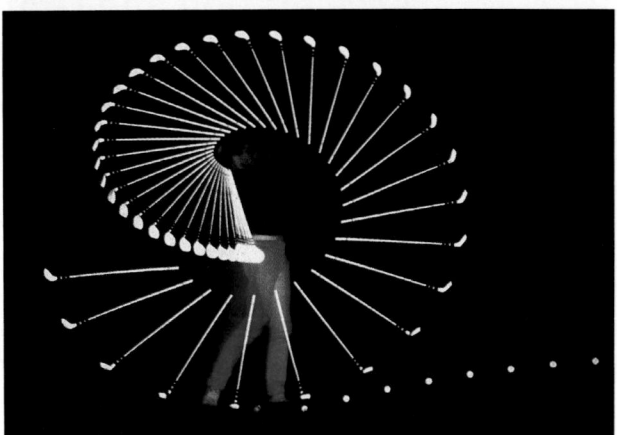

Harold E. Edgerton (1903-1990) developed strobe photography, which enables a photographer to take multi-image photographs of moving subjects. Each image in this photo was taken only a fraction of a second before the next.

Analyzing

1. Did the velocity of the balls change as they fell? How do you know?

2. Make a bar graph of distance fallen versus time. *Infer* what the graph tells about the balls' positions as they fell.

Concluding and Applying

3. How much did the average velocity increase between the second and third images? Between the third and the fourth images?

4. ~~Going Further~~ *Calculate* the balls' acceleration. Find the acceleration between the image at 0.10 s and the image at 0.30 s by dividing the increase in velocity by the time interval in seconds. What was your result?

Program Resources

Activity Masters, pp. 55–56, Investigate 13-1

Making Connections: Across the Curriculum, p. 29, Gravity L2

Making Connections: Technology and Society, p. 29, Space Law L2

Science Discovery Activities, 13-1, Gravity's Got Rhythm!

Expected Outcome

Students learn that gravity accelerates objects at 9.8 m/s^2.

Answers to Analyzing/ Concluding and Applying

1. Yes; the average velocity increased from Image 2 to 5. There was more space between images.

2. The distance the balls fell was greater during each time interval, therefore their positions were further and further from the start after each time interval.

3. As measured in the sample data: 1.5 m/s between the second and third and 2.4 m/s between the third and fourth images

4. As measured in the sample data: The velocity increased 2.9 m/s between 0.10 and 0.40. The acceleration was:

$$\frac{2.9 \text{ m/s}}{0.3 \text{ s}} = 9.7 \text{ m/s}^2.$$

✔ Assessment

Process Have students work in small groups to use the results they calculated. Ask students to draw a scientific diagram similar to figure 13-2 for a ball that is dropped from a height of 500 cm. Use the Performance Task Assessment List for Scientific Drawing in **PASC,** p. 55.
COOP LEARN P L1

In the previous chapter, students learned to distinguish between velocity and acceleration. Ask students to explain the differences between velocity and acceleration. Students will use both concepts to analyze how gravity affects objects' motion in this section.

Demonstration This activity will allow students to observe the constant acceleration of gravity. Materials needed are 2-m lengths of string and metal washers.

Securely attach a metal washer at each of the following positions along a 2-m length of string: 0 cm, 5 cm, 20 cm, 45 cm, 80 cm, 125 cm, and 180 cm. Hold the string by the end nearest to the 180 cm position over a metal wastebasket or pie plate. Ask students to predict what they will hear when the string is dropped. Release the string. Students should hear regular intervals between the clicks of the washers striking metal. Ask students to explain how this activity demonstrates the constant acceleration of gravity. *The constant time interval between clicks indicates that each washer, which had to fall a longer distance than the one beneath it, covered that distance in the same time, due to the constant acceleration of gravity.*

Acceleration and Gravity

You're probably familiar with the acceleration of a car. When a car changes speed or direction, we say that the car is accelerating. A car can speed up when you press on the accelerator. What causes the balls in the Investigate to speed up? As you will learn later, acceleration requires force. The force that produces the acceleration in the balls has been given the name gravity.

In the Investigate, when you measured the **acceleration due to gravity**, your result should have been close to 9.8 m/s/s, which can be written as 9.8 m/s^2. You saw that the acceleration was the same for two different objects of different size and mass. The acceleration due to gravity, which is given the symbol g, is the same for any free falling objects. **Figure 13-3A** and **B** compares the acceleration of an apple and a watermelon.

Free falling objects accelerate very quickly. Imagine that you climbed to the top of a tall building and dropped a pumpkin. **Figure 13-3C** shows how the velocity of that pumpkin would change because of acceleration due to gravity.

Terminal Velocity: Falling in Air

The term *free-fall* means that the only force acting on the falling body is gravity. An object in free-fall accelerates at a rate of 9.8 meters per second squared. However, for objects falling near Earth, the faster the object goes, the greater the air resistance.

If you were to put your hand out of a car window with your palm forward when the car is moving slowly, the force of air would not be very great. When the car is moving faster, the force would be much greater.

How does air resistance affect a falling object? In most cases not much if the object falls only a short way. But in a long fall, the effect can be very great.

Purpose

A Closer Look extends the discussion of the motion of a falling object by explaining the effect of air resistance.

Content Background

Air resistance is important in daily life. Automotive designers boost gas mileage by giving cars a shape with less air resistance. Some trucks have curved plastic shields to guide air around the trailer. Air resistance is also important in space flight. The space shuttle relies on air resistance to slow down from orbital speed to landing speed.

Teaching Strategy

Challenge students to think critically by asking them what the text means by "the forces acting on the diver are balanced." Have students describe the forces (gravity and air resistance) and why air resistance increases. *The faster you go, the more air friction there is.* Have students look at the picture and explain how these skydivers are

Figure 13-3

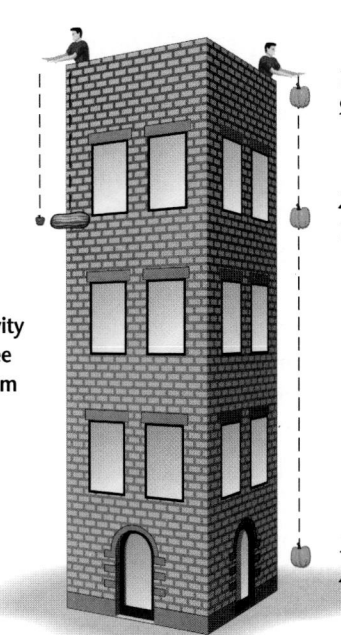

A An apple and a watermelon simultaneously dropped from a building would accelerate toward Earth at the same rate and reach the ground at the same time.

B Acceleration due to gravity (*g*) is the same for all free falling objects falling from the same height.

$$g = 9.8 \text{ m/s}^2$$

1 sec.
9.8 m/s

2 sec.
19.6 m/s

3 sec.
28.4 m/s

C When first released, the pumpkin's velocity is zero. After one second, the pumpkin is falling at 9.8 m/s. After two seconds, its velocity is 19.6 m/s. If the pumpkin were still falling after 10 seconds, it would be traveling about 98 m/s. That's about 353 km per hour, or 219 miles per hour!

Balanced Forces

For example, when sky divers fall toward the ground, they are falling through air, not a vacuum. The faster they fall, the greater the air resistance. When the divers' speed reaches about 240 kilometers per hour, they find that they are no longer accelerating. They are falling at a constant velocity. We therefore infer that the air resistance must balance the force of gravity. This means the forces acting on the diver are balanced. If there is no acceleration, the direction and speed of the diver remain constant. This constant velocity is called terminal velocity. Terminal velocity is not the same for all objects—it depends on the mass, size, and shape of the falling object. Generally, larger objects have a greater amount of air resistance on them. This is why, after a parachute opens, a diver falls at a velocity that is slow enough to allow a safe landing.

You Try It!

Show that the terminal velocity is not the same for all objects. Drop different objects—such as a flat sheet of notebook paper, a sheet of notebook paper folded in quarters, and a sheet of notebook paper crumpled into a small ball—from a height of about two meters. Do they fall at the same rate? If not, which falls fastest? Slowest? How can you explain your observations?

---

using terminal velocity. *They are maneuvering their bodies to increase velocity until all divers reach the same altitude. Then they adjust to the same body position to maintain the same velocity.*

Answers to
You Try It!

1. no
2. the crumpled ball of paper
3. the flat sheet of paper

4. Larger objects have more air resistance.

Going Further ▎▎▎▎▎▎▶

Have students work in small groups to experiment with parachutes, air resistance, and falling rate. Cut cloth into different-sized squares. Provide each group with one of each size cloth, string, and a small weight. Have groups design and conduct an experiment to determine the relation between the area of the cloth in cm² and the time it takes the parachute to fall 2 m.

Each experiment should measure the cloths, build parachutes, drop them from a fixed height, and time their fall. Ask each group to display its data in a table and a line graph. Have groups compare results and discuss any differences. Use the Performance Task Assessment Lists for Designing an Experiment, Carrying Out a Strategy and Collecting Data, and Graph from Data in **PASC**, pp. 23, 25, and 39. **L1** **P**
COOP LEARN

Acceleration on the Moon

Have you ever seen films of astronauts on the moon? If you have, you may have wondered why the astronauts seemed to float as they move from place to place. Why are astronauts on the moon able to jump up so high and then appear to float down?

The acceleration of gravity on the moon is less than the acceleration of gravity on Earth. Jumping astronauts on the moon reach a greater height before the acceleration of gravity slows them to a stop and causes them to fall back to the surface.

If you had the opportunity to play basketball on the moon, even Michael Jordan would be amazed at what you could do. Slam-dunking the ball would be simple because the lower acceleration due to gravity on the moon would allow you to jump higher.

Figure 13-4

If, while visiting the moon, you drop a brick and a beach ball simultaneously, both would reach the moon surface at the same time. All objects on the moon, regardless of mass, accelerate downward at the same rate. But that rate is less than the 9.8 m/s² rate at which objects fall near Earth. On the moon, the acceleration due to gravity is about 1.6 m/s², or about 1/6 that of Earth.

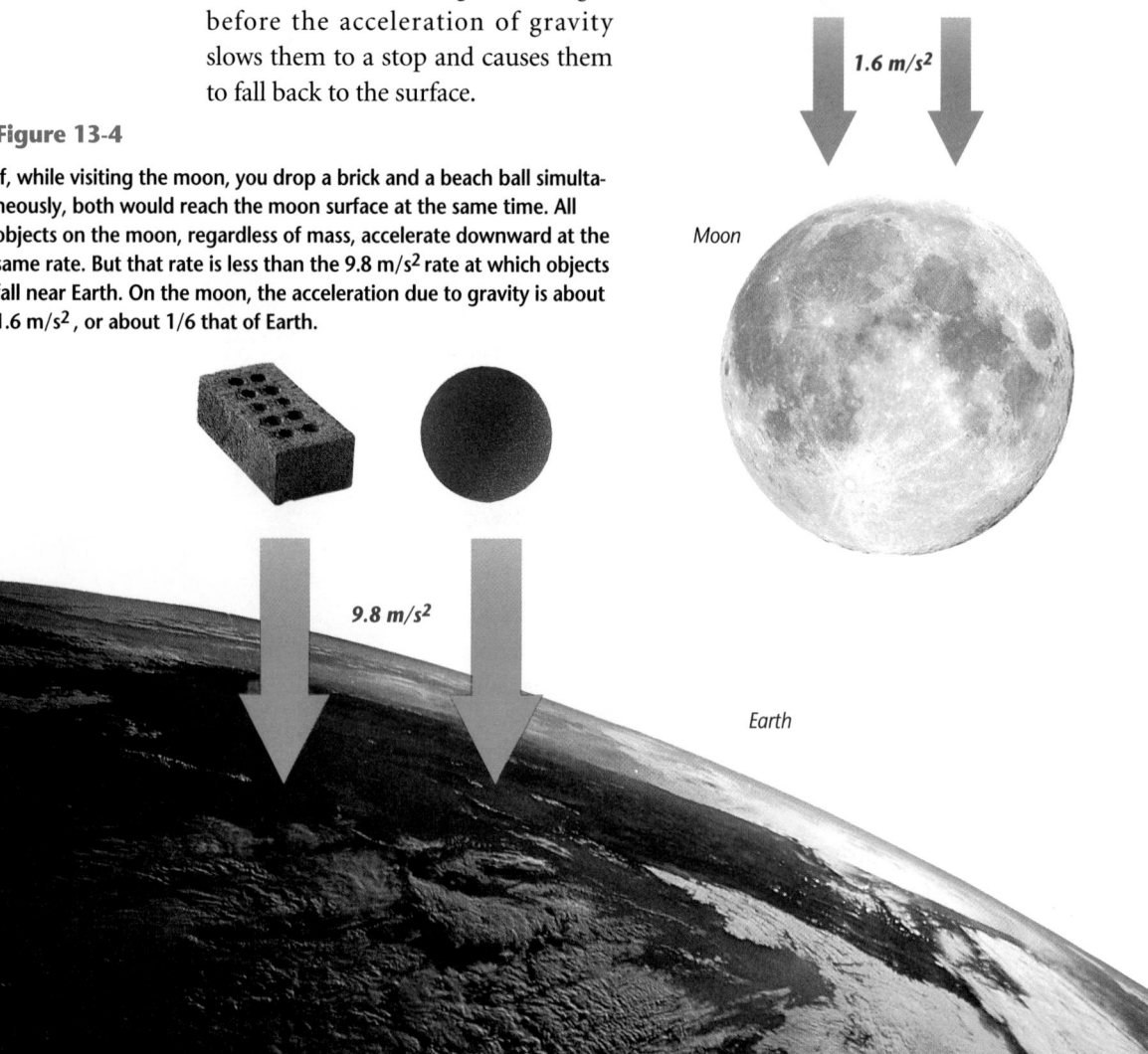

1.6 m/s²

Moon

9.8 m/s²

Earth

Figure 13-5

One important thing that determines how hard you must push off the ground to get moving when riding on a skateboard is gravity.

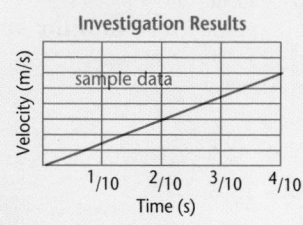
We experience gravity in our everyday activities. Gravity is important even as you walk or run.

Without air, heavy objects, light objects, and objects in between all fall with an acceleration equal to *g*. You now know at what rate of acceleration dropped objects fall at or near Earth's surface. What about when something is actually thrown, like when you throw a ball to the catcher to get a player out at home plate? Does this forward motion affect the way the object falls? In the next section, you will explore the kinds of motion that occur when an object is thrown.

check your UNDERSTANDING

1. Calculate the speed, in meters per second, of an object that has been falling from a building for eight seconds.
2. Explain in your own words how falling objects accelerate.
3. How could you show that the acceleration due to gravity of an object is a constant?
4. **Apply** Suppose an astronaut on the moon dropped a piece of rock from a height of 2 m and measured the time it took to fall to the moon's surface. Explain why a rock dropped from the same height on Earth would reach the ground more quickly than the rock dropped on the moon.

13-1 Falling Bodies **421**

check your UNDERSTANDING

1. 8×9.8 m/s = 78.4 meters/second
2. Gravity exerts a constant force on the object, making its velocity constantly increase.
3. Measure the velocity at several different times, and calculate the acceleration during each time interval. The accelerations should be the same.
4. Earth's gravity accelerates the rock at a higher rate than the moon's gravity.

13-2

PREPARATION

Planning the Lesson

Refer to the Chapter Organizer on pages 412A–D.

Concepts Developed

In this lesson, students learn that projectile motion has both a vertical (falling) component and a horizontal (forward travel) component that act independently of one another, so that a projectile falls at the same rate as an object that is not moving forward while falling.

1 MOTIVATE

Bellringer

 Before presenting the lesson, display **Section Focus Transparency 42** on an overhead projector. Assign the accompanying **Focus Activity** worksheet. [L1]
LEP

Demonstration Hold a small rubber ball in one hand and release it so that it falls straight down into your other hand. Then have one student catch the ball when you toss it. Ask students to describe the two motions. *vertical, horizontal.* Ask students if they think the ball was falling in both cases. Students should conclude that a thrown ball is still falling as it is moving forward. [L1]

Visual Learning

Figure 13-6 Introduce the idea that it is almost impossible to throw a ball perfectly horizontally. Even a minimal amount of upward thrust will increase the time it takes a ball to fall to Earth. **Does the forward motion of the ball affect its falling speed?** *no*

---

422 Chapter 13 Motion Near Earth

Projectile Motion

Section Objectives

- Describe the horizontal motion of a projectile.
- Describe the vertical motion of a projectile.
- Explain how vertical and horizontal motions of projectiles are independent.

Key Terms

projectile motion

The Motion of Projectiles

Everyone has, at one time or another, experienced projectile motion. What happens when you throw a stone or ball? The object will follow some curve through the air until it falls back down to Earth. Objects that are launched forward into the air, such as rockets, bullets, and satellites, are called projectiles. But even common objects, such as a soccer ball or baseball are projectiles. An object that is launched forward (horizontally) and then falls back to Earth is said to have **projectile motion.** How does such a thrown object move?

Figure 13-6

A Holding a basketball up to your chest, you fire the ball horizontally as hard as you can. The ball goes speeding away from your fingertips.

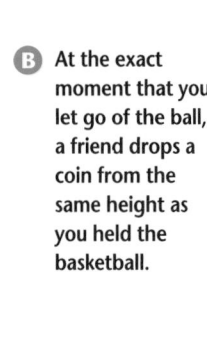

B At the exact moment that you let go of the ball, a friend drops a coin from the same height as you held the basketball.

C Both the coin and ball started from the same height. Both should fall at the same rate, but the basketball is moving forward as it falls. Does the forward motion of the ball affect the time it takes to fall?

422 Chapter 13 Motion Near Earth

Program Resources

Study Guide, p. 48
Laboratory Manual, pp. 77–80, Projectile Motion **COOP LEARN**
Teaching Transparency 26 [L1]
Science Discovery Activities, 13-2, Ready! Aim! Launch!
Section Focus Transparency 42

Meeting Individual Needs

Visually Impaired Visually impaired students can interpret the results of the Find Out by listening for the sounds of the coins hitting the floor. Students with normal vision may also find it interesting to perform the activity once with eyes closed in order to hear whether the coins make two separate clicks or one combined click as they reach the floor. [L1]

Find Out! ACTIVITY

Does forward motion affect falling speed?

You can investigate the motion of projectiles by doing the following activity with a meterstick and two coins.

What To Do

1. Lay a meterstick on a table with approximately 20 cm extending over the edge of the table.

2. Place one coin next to the meterstick and close to the edge of the table.

3. Place a second coin 15 to 20 cm further up the meterstick. Both coins should be just a few millimeters from the stick.

4. Holding your finger on the meterstick at the edge of the table, quickly swing the meterstick so it pivots and pushes both coins off the table at the same time.

Conclude and Apply

1. Which coin flew off the table faster?

2. Which coin hit the floor first?

3. How did the motion of the coins differ as they were pushed off the table?

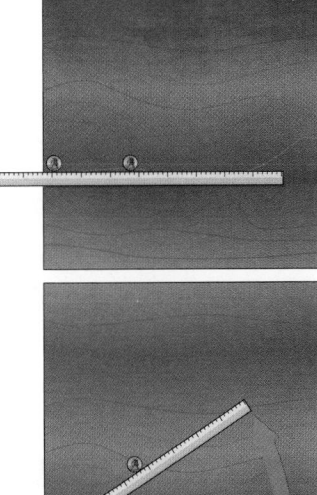

In the Find Out activity, both coins should have hit the floor at the same time, but it's difficult to see. The motion of the coins may have been too quick to let you analyze exactly what happened. Once again, we can use strobe photography to observe motion more accurately. The strobe photograph in **Figure 13-7** allows us to examine a similar experiment done with two golf balls.

As you can see, both balls are falling at the same rate. This means that a basketball fired from your hands horizontally will hit the ground at the same time as a coin dropped by your friend. How would you explain this?

Figure 13-7

A This strobe photograph shows the path of two golf balls. The red ball was dropped and the yellow one was fired horizontally.

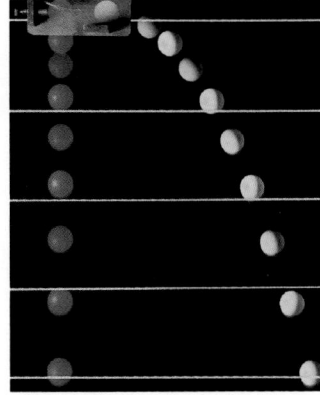

B The dropped ball fell straight downward. The fired ball moved horizontally as it fell.

C Both balls hit the ground at the same time. Both balls had a downward acceleration of 9.8 m/s². The forward motion of the yellow ball did not affect the rate at which it fell.

13-2 Projectile Motion **423**

---

2 TEACH

Find Out!

Does forward motion affect falling speed?

Time needed 15 minutes

Materials Meterstick and two coins

Thinking Processes observing and inferring, comparing and contrasting, formulating a hypothesis

Purpose To demonstrate that objects moving horizontally fall at the same rate as objects dropped vertically.

Preparation This activity can be performed by groups. Designate one student in each group to place the coins, one to handle the meterstick, and one or more to observe the motion of the coins. **COOP LEARN**

Teaching the Activity

Troubleshooting The meterstick is pivoted with a finger on it at the point where it crosses the table.

Discussion Have students perform the activity two or three times. Then ask them to describe the coins' motion.

Science Journal Have students make a diagram of their investigation. **L1**

Expected Outcome

Students should observe that the coin nearest the edge falls nearly vertically while the other coin moves in an arc, but that both coins hit the floor at the same time.

Conclude and Apply

1. The one pushed farthest.
2. neither one
3. See Expected Outcome above.

✔ Assessment

Process Challenge students to form a hypothesis to explain their results. Use the Performance Task Assessment List for Formulating a Hypothesis in **PASC**, p. 21. **L1**

ENRICHMENT

Activity Ask students to make a ramp from a plastic rule or half a cardboard tube from a roll of paper towels. Have them set up the ramp 5 cm from the edge of a table and experiment with setting the ramp at different angles and rolling ball bearings down it and onto the floor. Ask students to assume that it takes 0.5 second for the ball bearings to reach the floor from the top of the table. For each position of the ramp, have students measure from the edge of the table to the point of impact and then calculate the speed at which the ball bearing was moving when it left the table. The speed is calculated by dividing the distance by 0.5 second. **L3** **COOP LEARN**

13-2 Projectile Motion **423**

3 ASSESS

Check for Understanding

For the Apply question, have students draw diagrams showing the path of a horizontally thrown dart and the path for hitting the target.

Reteach

Demonstration Have students communicate by describing the motion of two projectiles. Rest a steel ball bearing at the edge of a table. Make a ramp from half a cardboard paper towel tube. Roll a second ball down the ramp so that it strikes the first ball. Have students observe that both balls hit the floor at the same time regardless of the paths they take. Ask students to explain why this happens. **LEP** **L1**

Extension

Activity Challenge students to recognize spatial relationships and to think critically. Have students demonstrate the variations in the paths of a projectile. Materials needed are 1.5-m flexible hose and 30-cm stick.

Tie an end section of the hose to the stick so that the stick holds it straight and rigid. Have students prop up the stick, hold the other end of the hose high, and drop a ball bearing into the high end. The ball bearing will accelerate down the length of the hose and come out the other end at an upward angle. Students can try different heights for the upper end and different angles for the stick, keeping a table of height, angle, and ball bearing's impact point. **L3** **P**

Horizontal and Vertical Components

We've been talking about projectile motion almost as if it were two separate motions. There's motion across the ground that we've been calling horizontal motion, and there's falling motion that we've been calling vertical motion.

The horizontal part of a velocity is the horizontal component. Similarly, the vertical part of a velocity is called the vertical component. Every motion has both horizontal and vertical components, but each component acts as if the other were not present. We say that the vertical motion is independent of the horizontal motion. The vertical component has an acceleration equal to 9.8 m/s². The horizontal component has a constant velocity.

So far, you've only looked at the motion of projectiles fired horizontally. However, projectiles that are launched up into the air at an angle, like the volleyball in **Figure 13-8**, behave the same way.

Understanding the motion of projectiles was an important step in our quest to put objects into orbit around Earth. In the next section, you'll learn more about these objects and how they stay in orbit.

Figure 13-8

Ⓐ If the path of a volleyball were recorded by strobe photography, the images would be taken at equal intervals. This illustration shows how the ball would travel.

Ⓑ The ball slows down as it rises and then speeds up as it falls from its highest point. The images remain the same distance apart horizontally.

30 cm 30 cm 30 cm 30 cm 30 cm

check your UNDERSTANDING

1. Describe three examples of a projectile not mentioned in this section.
2. Which object would hit the ground first, a baseball launched perfectly horizontally off the bat of a professional baseball player or a baseball dropped from the same height as the player's bat?
3. Explain how the motion of a basketball after a jump shot can be separated into vertical and horizontal components.
4. **Apply** Draw a diagram to show why a dart player has to aim above the target to hit the bull's-eye. Show the dart's path and the two kinds of motion involved.

424 Chapter 13 Motion Near Earth

4 CLOSE

Give three students who are about the same height identical rubber balls. Ask one to drop the ball from shoulder height, one to throw the ball horizontally, and one to throw the ball upward, all simultaneously. Ask students to explain the motion of the three balls. *The balls dropped and thrown horizontally should hit the floor together; the ball thrown upward should hit later.* **LEP** **L1**

check your UNDERSTANDING

1. Examples: a baseball, arrow, bullet.
2. Both land at the same time.
3. The basketball moves both horizontally and vertically; a strobe photo would show both movements.
4. Diagrams should show the dart falling during horizontal travel

13-3 Circular Orbits of Satellites

Newton's Prediction of Satellite Motion

Isaac Newton was born in 1642, the same year that Galileo died. Newton described, even back then, how a satellite could be put into orbit. It took another 270 years to develop the technology to send an artificial satellite into orbit. This satellite was *Sputnik I*, sent up by the Soviet Union on October 4, 1957.

■ How Projectiles Orbit Earth

Imagine Earth as perfectly round with no mountains or hills or air to slow or stop a projectile. Now, suppose you horizontally fire a projectile, perhaps a rifle bullet. The projectile moves horizontally and falls to the ground. But Newton had an interesting thought. Imagine that you throw

something, or fire it, with enough velocity that it travels many kilometers. As it falls, it falls just enough to follow the curvature of Earth. It would fall until it circled Earth. This

Sputnik I

is shown in **Figure 13-9**. **Figure 13-10** shows how to calculate how fast a projectile would have to travel horizontally to orbit Earth.

Figure 13-9

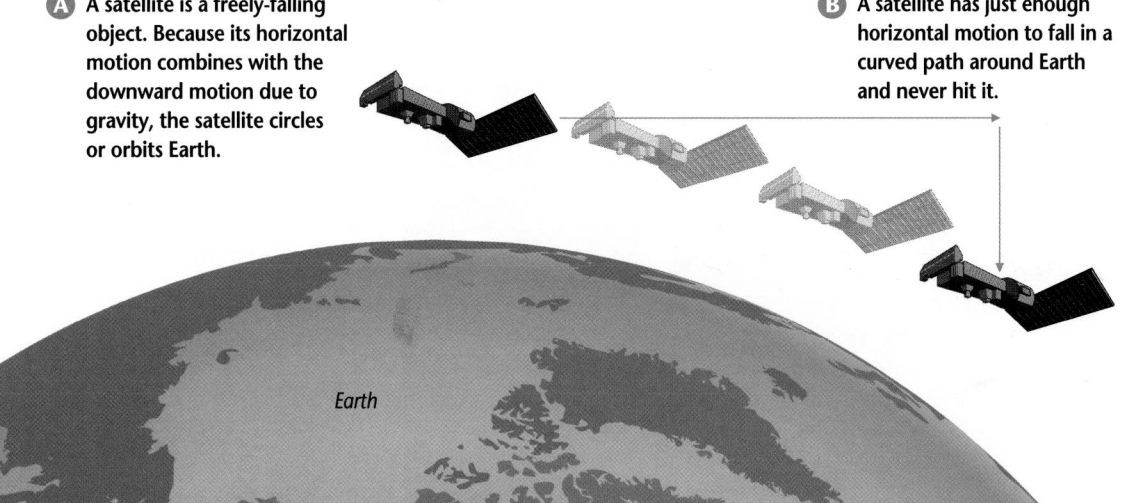

A A satellite is a freely-falling object. Because its horizontal motion combines with the downward motion due to gravity, the satellite circles or orbits Earth.

B A satellite has just enough horizontal motion to fall in a curved path around Earth and never hit it.

Earth

Program Resources

Study Guide, p. 49
Critical Thinking/Problem Solving, p. 21, Studying the Effects of Space Travel L2
Multicultural Connections, p. 30, Japanese Technology L1
Take Home Activities, p. 22, Weightlessness L1
Section Focus Transparency 43

Critical Thinking/Problem Solving, p. 5, Flex Your Brain
How It Works, p. 16, Training for Weightlessness L1
Making Connections: Integrating Sciences, p. 29, Living in Space L2
Science Discovery Activities, 13-3

---

2 TEACH

Tying to Previous Knowledge

In the previous section, students learned about projectile motion. In this section, students learn that putting a satellite into orbit is an extension of projectile motion. Ask students to describe their experiences with observing launches on TV or perhaps in person at Cape Canaveral.

Visual Learning

Figure 13-10 Have students work in groups to solve the problem presented. Suggest roles for group members, such as explainer, recorder, and math checker. Tell students that they may refer to the calculations provided as necessary. **COOP LEARN**

Content Background

"Weightlessness" is a term of convenience. Wherever there is mass, there is gravity and weight. In orbit the masses interacting are very small, so weights are nearly unobservable.

GLENCOE TECHNOLOGY

Videodisc

The Infinite Voyage: Sail On, Voyager

Introduction

Chapter 2
Technical Design and Capabilities

Figure 13-10

How fast would the satellite have to travel to continuously fall around Earth without hitting it?

A The satellite begins its vertical movement with a speed of 0 m/sec and accelerates downward at 7.8 m/s². At the end of 1 second, its average downward change in velocity for that second is 7.8 m/s divided by 2, or 3.9 m/s. So the satellite will fall 3.9 m during that second.

B Scientists have calculated that a projectile must travel 7.9 km horizontally in that second for its path to match the curve of Earth. Since there are 3600 seconds in one hour, this velocity is equal to

$$\frac{7.9 \text{ km}}{\text{s}} \times \frac{3\ 600 \text{ s}}{1 \text{ h}} = \frac{28\ 440 \text{ km}}{\text{h}}$$

A satellite must travel at 28 440 km per hour to continuously fall around Earth at the Equator.

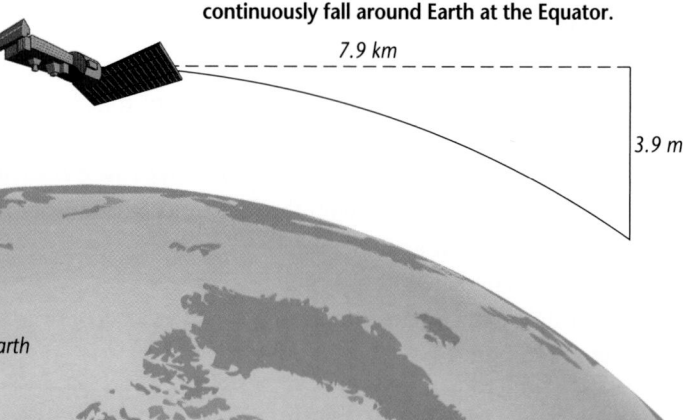

7.9 km

3.9 m

Earth

Life Science CONNECTION

Biological Effects of Weightlessness

On Earth, the force of gravity acts on everyone and everything. We call this downward force weight and measure it with a scale. Over time, the human body has adapted to the effects of gravity in this environment.

In an orbiting spacecraft, gravity still affects everything. Because everything is falling freely, however, the effect of gravity is different. If an astronaut stands on a scale in the spacecraft, the scale reads zero. For this reason, the astronaut is said to be weightless.

How do you think the human body would be affected by long periods of weightlessness? What happens when astronauts return to Earth's environment?

Areas of Concern

Our sense of balance depends upon gravity, but

Life Science CONNECTION

Purpose

This Life Science Connection describes the effect of weightlessness on people and the concern this creates for astronauts.

Content Background

When the body first becomes weightless, the fluid in the inner ear, which provides our sense of balance, is no longer held down by gravity and sloshes around, producing dizziness and nausea. The impact is short-lived, however, because the body's supply of blood, lymph, and other fluids rapidly redistributes upward from the lower body where gravity normally concentrates it. The higher fluid pressure overwhelms the inner ear's ability to sense motion changes.

Teaching Strategies

Have students work in small groups to develop a table listing the effects of weightlessness and possible remedies. Compare the tables and have students

Find Out! ACTIVITY

How did Sputnik I stay in orbit?

Sputnik I orbited about 516 km above Earth's surface. The satellite traveled at a horizontal velocity of 7.6 km/s. In each 7.6 km, it must fall toward Earth 4.18 m to stay in its orbit.

What To Do

1. Determine Sputnik's average downward velocity for one second.

2. Calculate the downward velocity at the end of that second.

Conclude and Apply

1. What is the relationship of average downward velocity to distance fallen in one second?

2. What is the acceleration due to gravity at this altitude?

The higher the altitude of a satellite's orbit, the lower the horizontal velocity required. This is because acceleration due to gravity is less further away from Earth. Perhaps you've seen pictures of astronauts floating inside a space shuttle as it orbits Earth. Why does this happen?

luckily, becoming disoriented in orbit has not become a major problem. Some astronauts did experience brief periods of nausea and disorientation, but these effects were short-lived.

All body functions are carefully monitored during a flight.

The only factor that has shown a significant change is the pulse rate, which often increases during lift-off and space walks. In all cases, however, the rate has returned to normal in a very short time.

Returning to Earth

Upon return to Earth, some astronauts feel faint when they first stand up. Some Soviet cosmonauts found it difficult to adjust to the effects of gravity on Earth after 17 days in orbit. For several days, their arms, legs, and head felt as if they were very heavy. They also seemed to have less blood and some changes in the walls of their veins. The reasons are not yet clearly understood.

Other observed effects of weightlessness are the loss of calcium in the bones and the loss of body mass. So far, astronauts have quickly regained lost body mass after returning to Earth.

With every returning space flight crew, doctors are learning more and more about how the human body reacts to long periods of weightlessness.

What Do You Think?

After several months in space, the heart, muscles, and bones will adapt to the conditions of free-fall. What problems could arise for astronauts when they return to Earth after several months in space?

Find Out!

TECH PREP

How did Sputnik 1 stay in orbit?

Time needed
5 minutes

Thinking
Processes comparing and contrasting, recognizing cause and effect

Purpose To calculate varying acceleration of gravity at different altitudes.

Teaching the Activity

Work through the sample problem in Figure 13-10 using the acceleration due to gravity at Earth's surface, 9.8 m/s². Have students apply this formula to the problem. **L2**

Science Journal Have students make their calculations in their journals.

Expected Outcomes

Students should calculate the average and final earthward velocity of *Sputnik* during one second of orbit and the acceleration of gravity at an altitude of 516 km. **L2**

Answers to Questions
1. 4.18 m/s
2. 2 × 4.18 m/s = 8.36 m/s.

Conclude and Apply
1. They are the same value.
2. 8.36 m/s²

✔ Assessment

Oral Have students formulate additional questions about satellites and their orbits. Use the Performance Task Assessment List for Asking Questions in **PASC**, p. 19. **L1**

---

name body functions that are not affected by weightlessness. List their suggestions on the chalkboard. **COOP LEARN**

Answers to
What Do You Think?

Without gravity, muscles and bones rapidly lose strength. Returning astronauts could be too weak to move, their hearts too weak to pump blood, and their bones too brittle to support them on Earth.

Going Further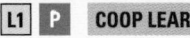

Divide the class into small groups and have each group conduct research on how human beings may live in space without suffering ill effects. Have each group focus on a specific activity in space: a flight to Mars, living in a colony on the moon, or living in a space station. Have the groups prepare oral presentations with visuals describing the problems life in space presents and how we hope to solve them.

For information, students can write to:

NASA Headquarters
Washington, DC 20546

Use the Performance Task Assessment List for Oral Presentation in **PASC**, p. 71. **L1** **P** **COOP LEARN**

Weightlessness

Connect to...
Life Science

How could astronauts exercise in space while experiencing weightlessness? Research several isometric exercises and tell why they would work on a spacecraft.

Why do we say that astronauts are weightless when in orbit? Have you ever jumped on a trampoline? **Figure 13-11** explains what happens when you jump on a trampoline.

In an elevator that starts downward rapidly, you may have a feeling that you suddenly weigh less. The floor of the elevator does not seem to push on your feet. When the roller coaster you're riding goes over the top of a hill at high speed, you feel lifted up and floating free—weightless.

These experiences occur when you are falling freely. Whatever the direction of your velocity, you are experiencing acceleration toward Earth in a free-fall. This is what we mean by **weightlessness**. If you could stand on a bathroom scale during this time, it would register zero.

Figure 13-11

From the time your feet leave the trampoline to the time at which they touch it again, you are experiencing weightlessness.

■ Free-fall in Orbit

The satellites in orbit around Earth are in free-fall and that is why everything appears to be weightless aboard a satellite. The satellite and everything in it, including the astronauts, are all falling toward Earth with exactly the same acceleration. Satellites fall toward Earth and never really escape Earth's gravitational pull. The acceleration due to gravity is less for higher orbits. But such satellites still fall toward Earth, with some acceleration, even though it is less than 9.8 m/s^2. Have you ever looked up at the sky on a clear night and seen a bright spot moving across the sky? This may have been a satellite. Some satellites appear to move across the sky, and others do not. Why do some satellites appear to remain stationary?

428

The Stationary Satellite

Imagine a satellite that always appears in the same position. A **stationary satellite** is placed in orbit around Earth at just the right altitude and just the right velocity so that it moves around its orbit once per day. It moves in the same direction as Earth is rotating. Relative to a person on Earth, this satellite has no motion.

Stationary satellites like the one shown in **Figure 13-12** are often used as communication satellites to relay radio messages from one city to others. For example, a stationary satellite placed over Lake George, Uganda could receive radio communications from Bombay and relay them to London. Stationary satellites are used to beam television transmissions of events live across the world.

You have learned in this section that satellites are constantly falling toward Earth as they orbit. You will find that the study of other falling bodies, such as a swinging pendulum, will help you understand much more about motion in many different areas of science.

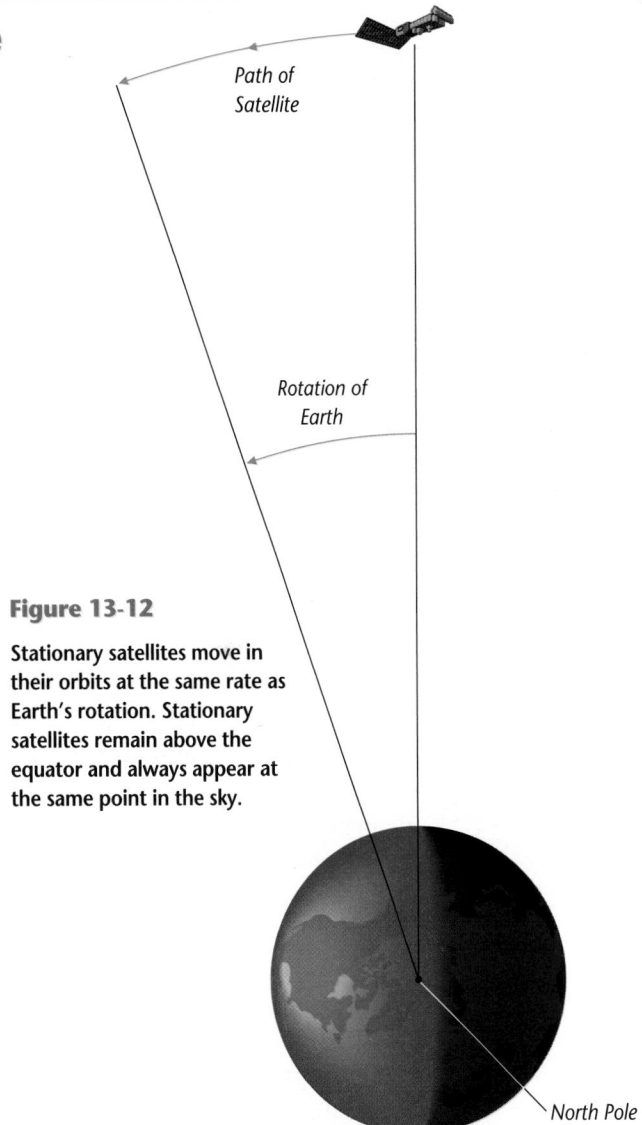

Path of Satellite

Rotation of Earth

North Pole

Figure 13-12

Stationary satellites move in their orbits at the same rate as Earth's rotation. Stationary satellites remain above the equator and always appear at the same point in the sky.

check your UNDERSTANDING

1. Explain how a satellite stays in orbit.
2. Imagine that you brought a bathroom scale into an elevator and stood on it. As you went down, what would you predict would happen to the weight it registers for you when the elevator first starts to move?

Explain your answer.
3. Why would a stationary satellite always appear in the same location to you if you could see it from Earth?
4. **Apply** Could an astronaut in the space shuttle exercise by lifting weights? Explain.

check your UNDERSTANDING

1. The satellite's horizontal velocity must be just great enough so that, in one second, it falls toward Earth the same distance as Earth's curvature deviates from the horizontal.
2. Less weight. You and the elevator are both accelerating toward Earth.
3. The satellite and Earth's surface, where you

are standing, are moving at the same relative velocity.
4. No, they are weightless in free-fall.

Check for Understanding

Discussion Use the questions to review the major points of the section: how projectile motion explains orbits; weight and free-fall; and the motion of satellites. Use the Apply question to let students apply their knowledge of free-fall to new areas. You may wish to have students speculate on other things astronauts can't do while weightless.

Reteach

Demonstration Students may find it hard to grasp how projectile motion can lead to orbital motion. Draw a circle on the chalkboard representing Earth. At the top of the circle, draw a cannon firing. Show the path of projectiles fired with increasing force: the path becomes flatter and the impact points move away from the cannon until the curve matches Earth's curve. **LEP** **L1**

Extension

Activity Have students calculate the acceleration of gravity experienced by an object that stays in orbit by falling 3.64 m toward Earth in 1 second. *7.28 m/s²* **L3**

4 CLOSE

Activity

Challenge students to compare and contrast the movement of a projectile and satellite. Have them turn to Figure 13-8 showing the path of a projectile. Note that the volleyball (projectile) accelerates as it falls toward Earth. Remind students that an object in orbit falls toward Earth and ask them if this means an orbiting object speeds up. Have students debate the question, then explain that the object does not increase in speed.

PREPARATION

Planning the Lesson

Refer to the Chapter Organizer on pages 412A–D.

Concepts Developed

Students describe the motion of a pendulum by period and frequency. Students experiment to determine the factors that affect a pendulum's period.

1 MOTIVATE

Bellringer

Before presenting the lesson, display **Section Focus Transparency 44** on an overhead projector. Assign the accompanying **Focus Activity** worksheet. L1

LEP

Demonstration Make two pendulums from string and small objects; one should be half the length of the other and have a heavier bob. Set both pendulums swinging. Ask students how they are different. *Responses should include string length, object size, and speed of swing.* Tell students that they will learn what makes pendulums behave as they do. L1

Visual Learning

Figure 13-13 A pendulum's frequency is the number of cycles the pendulum completes in 1 second. Ask students what is the frequency of this pendulum? *It completes 1/2 cycle in 1 second; the frequency is 1/2 cycle per second or 1/2 Hz.*

Figure 13-14 Help students understand how amplitude is measured: *horizontally from the bob's starting point to a vertical line at the bottom of the swing.*

Section Objectives

- Define the period of a pendulum.
- Define frequency, as it relates to periodic motion.
- Describe the relationship between the period of a pendulum, the mass of a bob, the length of the pendulum, and the amplitude of the motion.

Key Terms

period

Periodic Motion

Galileo is said to have watched a chandelier in a church slowly moving back and forth like a pendulum one day. He began to see that a pendulum's behavior would tell him something about how things fall on Earth. Earlier in this chapter you used a pendulum. You found that whatever the mass of the pendulum bob, the bob fell and reached the bottom of its swing in the same amount of time. Repeated motions such as those of a pendulum are called periodic motions. Let's begin to study the periodic motion of a pendulum by first observing and then defining the experimental variables.

■ Period

When you release a pendulum bob, it moves past the bottom and back up again to a position where it briefly stops. Then it reverses direction and swings back over to where you had let it go in the first place. The pendulum continues to swing back and forth. The time for the pendulum bob to swing over and back once is called the **period** of the pendulum. **Figure 13-13** illustrates the period of a pendulum.

■ Frequency

It is sometimes easier to use the number of times an object moves back and forth in one second, rather than how long it takes for one motion over and back. The number of times an object moves back and forth in 1 second is called frequency. We talked about the frequency of sound in

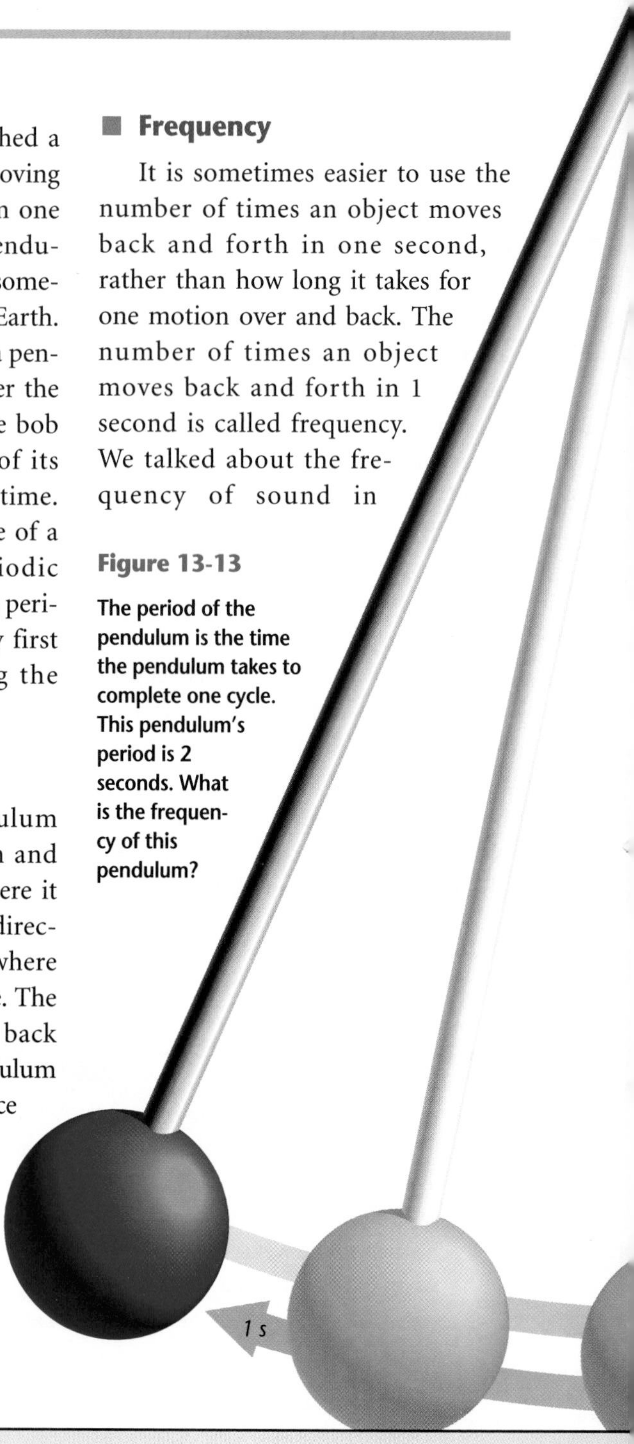

Figure 13-13

The period of the pendulum is the time the pendulum takes to complete one cycle. This pendulum's period is 2 seconds. What is the frequency of this pendulum?

1 s

Program Resources

Study Guide, p. 50
Section Focus Transparency 44

Chapter 3. The process of moving over and back once is called a cycle. Frequency then is measured in cycles per second. Recall that a cycle per second is given the name hertz, abbreviated Hz. What is the frequency of a playground swing that swings back and forth once every 3 seconds? What is the frequency of a swinging door that swings back and forth once every 2 seconds?

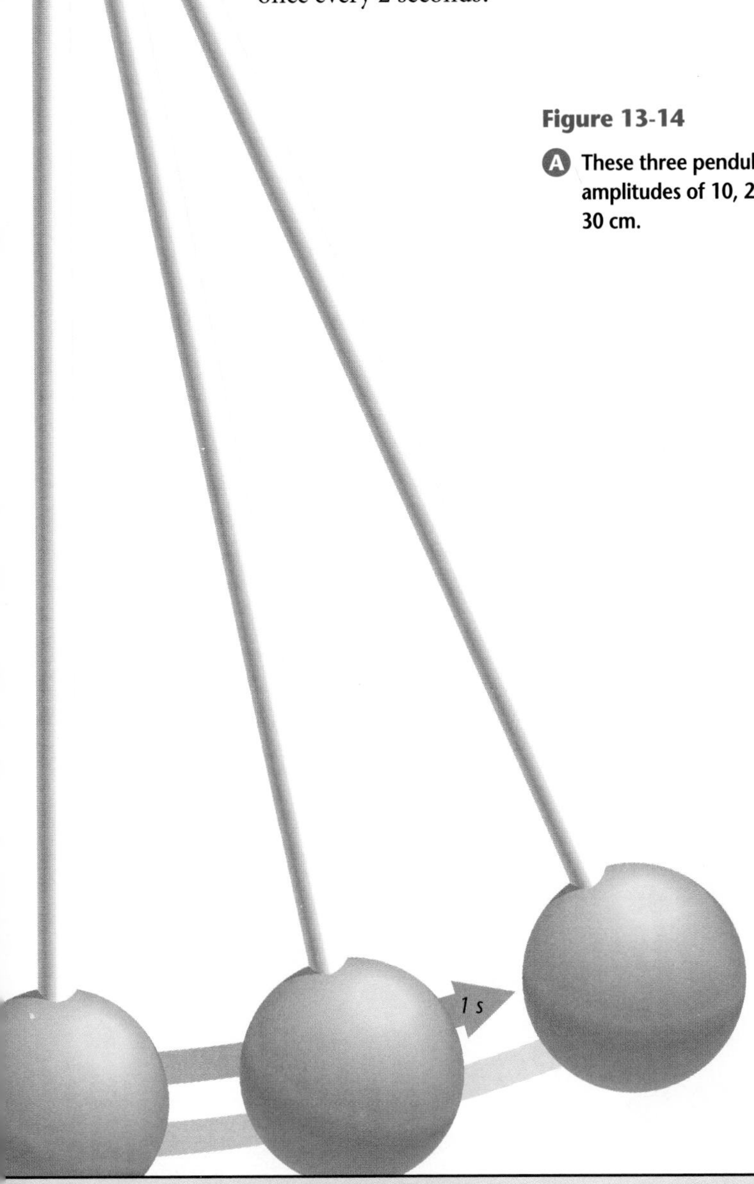

■ Amplitude

As a pendulum moves toward its lowest point, note the distance the pendulum bob has traveled from the starting point to the bottom of the swing. This distance is called the amplitude. **Figure 13-14** illustrates three different amplitudes.

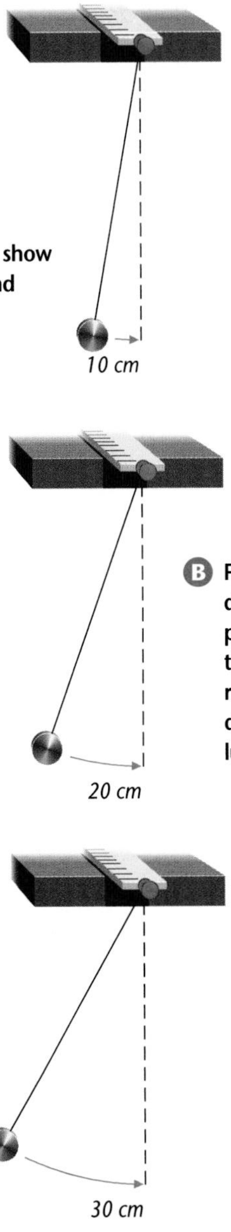

Figure 13-14

Ⓐ These three pendulums show amplitudes of 10, 20, and 30 cm.

10 cm

Ⓑ For relatively small distances, does pulling the bob farther back before you release it increase or decrease the pendulum's amplitude?

20 cm

1 s

30 cm

13-4 The Motion of a Pendulum **431**

2 TEACH

Tying to Previous Knowledge

In Section 13-1, students used a pendulum to explore gravity's acceleration. Students will learn what determines a pendulum's frequency. You might begin this lesson by asking students to describe pendulums they have seen, for example, on grandfather clocks.

Theme Connection The theme of stability and change is supported by the factors that affect the swing of a pendulum: period, frequency, and amplitude. For example, changing the length of the pendulum changes the time for swings and the pendulum period.

Student Text Questions
What is the frequency of a playground swing that swings back and forth once every three seconds? *1/3 = 0.33 Hz* **What is the frequency of a swinging door that swings back and forth once every two seconds?** *1/2 = 0.5 Hz*

Across the Curriculum

History

Inform students that pendulums revolutionized timekeeping. Within 30 years after Galileo died, pendulums reduced the average error of the best timepieces from 15 minutes to 10 seconds per day. Have students do research to learn more about scientific and technological insights that occurred in Europe during the Renaissance.

GLENCOE TECHNOLOGY

  CD-ROM

Science Interactions, Course 1, CD-ROM
Chapter 13

Meeting Individual Needs

Visually Impaired Visually impaired students who have difficulty observing pendulum motion can draw conclusions about pendulums from a metronome. Metronomes are upside-down pendulums powered by clockwork, and they produce a loud "tock" at the extreme ends of each swing. The speed of the swing is controlled by the distance of the bob from the pivot. Have visually impaired students compare the period of the metronome at various positions of the bob (measured relatively rather than numerically), with a fully sighted student timing 10-second intervals. Students will find that the farther the bob is from the pivot, the longer the period. **COOP LEARN** **L1**

13-2 The Period of a Pendulum

Preparation

Purpose To identify the factors that determine the period of a pendulum.

Process Skills making and using tables, making and using graphs, observing, separating and controlling variables, interpreting data, measuring in SI, recognizing cause and effect, predicting, designing an experiment, forming operational definitions

Time Required 50 minutes

Materials The tables to which the rulers will be taped need to be at least three feet high.

Possible Hypotheses Some students may hypothesize that increasing the pendulum bob's mass will lengthen the pendulum's period. The investigation will show that the bob's mass has no effect on the pendulum's period.

Plan the Experiment

Process Reinforcement Make sure that students set up their data tables correctly, testing only one variable—amplitude, mass, or length—at a time.

Possible Procedures Changing only one variable per trial, test amplitudes of 10 cm, 20 cm, 30 cm, and 40 cm; lengths of 30 cm, 40 cm, 50 cm, and 60 cm; and bob masses of 1 washer, 2 washers, 4 washers, and 8 washers.

Teaching Strategies

Troubleshooting Remind students not to throw the pendulum into its swing, but rather to let it drop and let the bob take over the operation.

Science Journal Have students record their answers to the Analyze and Conclude questions in their journals.

DESIGN YOUR OWN
INVESTIGATION

The Period of a Pendulum

A classic example of periodic motion is a pendulum in a grandfather clock. How does a pendulum help a clock keep time? What variables affect a pendulum's motion?

Preparation

Problem
What affects the period of a pendulum?

Form a Hypothesis
Think about the length of the pendulum, the bob mass, and amplitude (pull-back distance) of the pendulum. How do you think changing these variables will affect the period of the pendulum?

Objectives
• Separate and control variables in an experiment.
• Predict the effect of different tests on the period of a pendulum.

Materials
ruler
string
masking tape
meterstick
metal washers
seconds timer

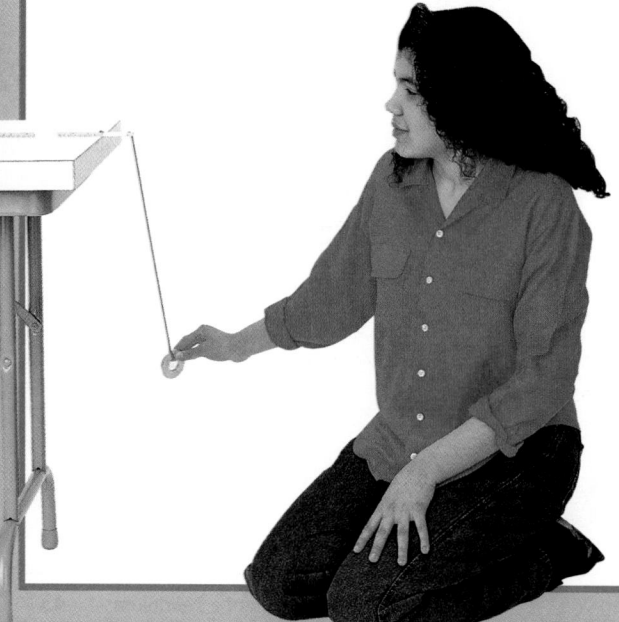

DESIGN YOUR OWN
INVESTIGATION

Plan the Experiment

1 Using the photo as a guide, explain how you will make your pendulum.

2 How will you measure the different variables?

3 Decide how you will vary each trial. The sample data tables will help guide you in your testing. Be sure you change only one variable in each trial.

Check the Plan

1 Prepare data tables *in your Science Journal* that are specific to your tests.

2 Have your teacher check your plan before you begin your experiment.

Sample data

Data and Observations

Length = __50__ cm   Mass __2__ washers

Pullback Distance (cm)	Time for 10 Swings (s)	Pendulum Period (s)
10	14	1.4
20	14	1.4
30	14	1.4

Sample data

Amplitude = __10__ cm   Length = __50__ cm

Pendulum Bob Weight (Number of Washers)	Time for 10 Swings (s)	Pendulum Period (s)
1	14	1.4
2	14	1.4
4	14	1.4

Sample data

Amplitude __20__ cm   Mass __2__ washers

Pendulum Length	Time for 10 Swings (s)	Pendulum Period (s)
30 cm	11	1.1
40 cm	12	1.2
60 cm	15	1.5

Analyze and Conclude

1. Compare Summarize the results of your experiment and compare them with your hypothesis.

2. Explain Looking back at your comparison in the last question, explain the effect of changing the pendulum's bob mass, amplitude, and length on the pendulum period.

3. Use Numbers Draw a graph plotting the pendulum's period for the different string length. Using your graph, predict the pendulum's period for a string of 100 cm. What is the relationship between the change in length of string and the change in period?

Going Further

If you were building a pendulum clock, how would you build it to make sure the clock would be as accurate as possible?

Expected Outcome

Students prove that length is the only factor affecting period, and that the period increases with length.

Analyze and Conclude

1. Answers will vary.

2. Changing length affected the period; neither changing mass nor amplitude had an effect.

3. Answers will vary depending on graphs, but should be about 2.0 s; the amount of increase will vary.

Going Further ⅢⅢⅢ➤

Answers will vary but should include the idea that the length of the pendulum affects its period.

✔ Assessment

Performance Have students work in cooperative groups to create the Pendulum Bowling Game. Have groups play each other's games. Students can then discuss which games worked best and why. Use the Performance Task Assessment List for Group Work in **PASC**, p. 97.

Program Resources

Activity Masters, pp. 57–58, Design Your Own Investigation 13-2

3 ASSESS

Check for Understanding

Use the Check Your Understanding questions to make sure students understand the difference between period and frequency and that pendulum length affects both. Accept all reasonable answers.

Reteach

Activity Use Data Tables A-C from Investigate on page 433. Fill in the *Time for 10 Swings* column from the Sample Data. Have students divide each number by 10 and fill in the *Pendulum Period* column. For each data table, ask students if all of the numbers in the *Pendulum Period* column are the same or different. Only Table C contains different numbers, indicating that only length affects the period. Ask students what this means. **LEP** **L1**

Extension

Activity Have students use the apparatus from Investigate 13-2 to determine the pendulum length that produces a 0.5-second and 1-second period. **L2**

4 CLOSE

Demonstration

Repeat the demonstration in Motivate: swinging unequal pendulums. Ask students to apply the knowledge they gained in Investigate 13-2 to identify what affects how pendulums act. *the length of string* **L1**

Figure 13-15

The steps shown model the process a scientist might use to discover what affects the period of a pendulum.

Changing Variables

In the Investigate, you studied the effects of mass, amplitude, and length on the period of a pendulum. These three factors that you tested are called

- State the problem as a question.
  What affects the period of a pendulum?

- Decide which variables might make a difference.
  length, mass, amplitude

- Choose a test variable. Keep all the other variables constant. Do several trials, changing the test variable in every trial. Measure the effects of the changes.
  Constant: length, mass
  Varies: amplitude

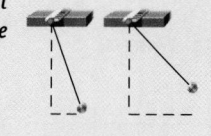

- Choose a different variable to test and repeat the above step.
  Constant: mass, amplitude
  Varies: length

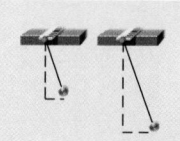

- Choose a different variable to test and repeat the above step.
  Constant: amplitude, length
  Varies: mass

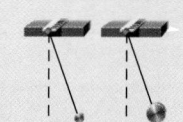

- Analyze the results and decide which variable or variables make a difference.

independent variables. The factor that depended on the value of the independent variable, the period, is called the dependent variable. An experiment is a procedure in which you hold all independent variables constant except one. **Figure 13-15** on the left shows one way to conduct an experiment.

■ Drawing Conclusions

Changing one independent variable at a time allows you to draw conclusions about the effect of that variable on what you're measuring, the dependent variable. If you had varied both length and mass at the same time, you would not know why the period changed. Experiments like these are used to discover the causes of the behavior of things you observe.

There are many other kinds of oscillations and vibrations that you will learn about later in this book. What you now know about the pendulum will help you understand their periods, frequencies, and amplitudes.

---

check your UNDERSTANDING

1. What's the difference between the frequency and the period of a pendulum?
2. Explain how you measure the frequency of the periodic motion of a pendulum. What units is frequency measured in?
3. If you wanted to change the period of a pendulum, which of the three variables you tested would you change?
4. **Apply** Give three examples of other objects or events that undergo periodic motion. How might you go about testing what affects the periodic motion of a guitar string?

---

check your UNDERSTANDING

1. Frequency is number of swings per second; period is time needed for one complete swing.
2. Number of swings per second is measured in cycles per second or Hertz (Hz).
3. length of string
4. Guitar strings, earthquakes, water waves. Fret string to change its length and listen to pitch.

HOW IT WORKS The Metronome

HOW IT WORKS

A metronome is a device that can be used to beat exact time for a musician. It works on the principle of the pendulum, which you have learned swings back and forth at a regular rate.

Adjusting the Tempo

As the rod in the metronome goes back and forth, it makes a ticking sound. The tempo, or rate of ticking, is adjusted by sliding the movable weight up or down the rod. Moving the weight away from the pivot produces a slower tempo. Moving the weight closer to the pivot produces a faster tempo. How is this similar to the rate of a pendulum?

A scale on the rod, or behind it, shows the number of swings per minute. For example, if you want one tick per second, you set the movable weight at 60. Where would you set it to get two ticks per second? How many ticks per second would you get if it were set at 90?

The First Metronome

The first metronome was probably made in 1815 in Amsterdam by an inventor named Dietrich Winkel. However, he did not get a patent for the device. In 1816, a German mechanic, Johann Maelzel, patented a similar device. Winkel went to court to get the patent rights. He won the court battle, but Maelzel's device was already widely used. In fact, many pieces of printed music have a notation, such as MM120, to indicate the tempo at which

the composer wanted the music to be played. The MM stands for Maelzel metronome, and the 120 indicates the number of beats per minute.

You Try It!

Listen to several pieces of music. Using a second hand on a watch or a clock, determine at which number a metronome would have been set when these musical selections were played.

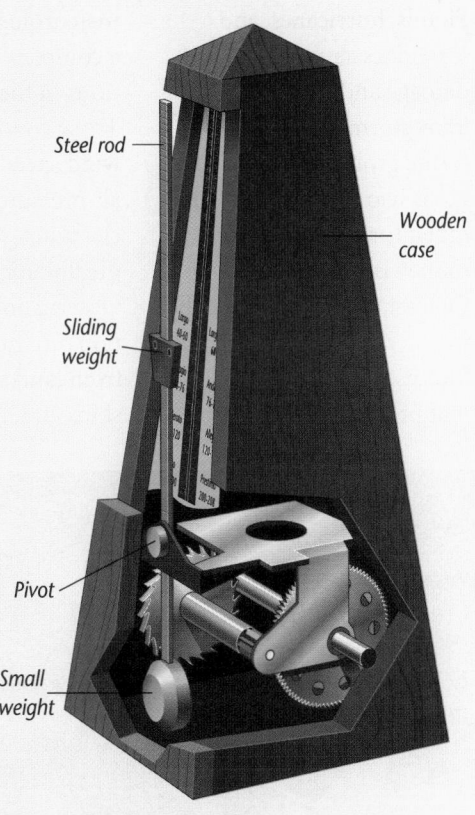

Steel rod

Wooden case

Sliding weight

Pivot

Small weight

Going Further ⫸

Have students work in small groups to build a pusiloge, a device invented by Santorio Santorio, a friend of Galileo, to measure pulse rate. Each group will need a cardboard paper towel tube, a laboratory stand with a clamp at right angles to it, a piece of string, small weight, paper, pencil, and tape. Have students attach one end of the string to the weight and tape one end to the outside of the tube in the middle. Place the tube over the laboratory clamp so that it is parallel to the

table. Turn the tube so that it winds up the string and lifts the weight. Have students start the weight swinging. By winding the weight up and down, they can change the period of the pendulum's swing. Play a recording of music and have students adjust their pendulums so that the pendulum's swing is the same as the beat of the music. Use the Performance Task Assessment List for Model in **PASC,** p. 51. **COOP LEARN** **L2**

Purpose

A metronome is a device that is used to beat exact time in music. Its operation is based on the principle of the pendulum, which is discussed in Section 13-4.

Content Background

A metronome is a pendulum that swings, not from its end, but from a point several inches in from the end. A heavier, fixed weight swings at the short end of the rod, while a lighter movable weight is positioned on the longer end to regulate the speed of the swing. Energy from a tightened coil spring is fed to the swinging rod by the kind of toothed escapement found in all mechanical clocks. The escapement also produces the loud ticking sound that makes the metronome useful. In addition to mechanical metronomes, musicians now use electronic ones that can be set digitally to produce an exact number of ticks per second.

Teaching Strategies
Allow students to compare and contrast. Demonstrate both for students, showing how the number of ticks per second is set.

Discussion

Challenge students to infer why musicians use metronomes. Explain as needed that composers can use metronome settings to tell musicians exactly how slow or fast a piece of music is to be played rather than letting each musician determine his or her own speed. Both music students and professional musicians also use metronomes as they practice, to improve their ability to keep a steady beat.

Science and Society

Purpose
The role of satellites in studying and predicting Earth's weather is one application of satellites, which are discussed in Section 13-3.

Content Background
Weather is a complex phenomenon that involves five basic elements: air pressure, temperature, wind, moisture, and precipitation. There are thousands of weather satellites that orbit Earth and give weather information several times a day. It is impossible to obtain that information in other ways. These satellites obtain information about the five basic weather elements. Because weather satellites are far above Earth they are able to give a "bigger picture." They can provide pictures of clouds that cover large areas of Earth's surface. Positions of clouds indicate areas of low pressure that precede systems and fronts. Satellites also carry sensitive instruments that can record data about temperature and humidity far above Earth's surface.

Teaching Strategies
Invite students to observe a weather map obtained from a daily newspaper. Use an overhead projector to project the map for the class and point out the major features of the map, such as high- and low-pressure areas, cloudiness, precipitation, and temperature readings. Ask students to explain each of the features and what they think it means. Point out that weather forecasting is based on identifying systems of weather, such as cloudless high-pressure areas or rainy low-pressure areas, measuring the direction and speed at which they are moving across Earth's surface, and predicting when the weather system will reach a given area.

Science and Society

Weather Satellites

Can you think of any natural occurrence that affects people's lives as much as the weather does? Farmers depend on the weather for good crops. Floods, hurricanes, and tornadoes cause terrible damage and loss of life. Snowstorms often cause traffic problems.

It's no wonder, then, that much time, effort, and money is put into trying to forecast, or predict, what the weather will be like. Accurate forecasts could mean that problems caused by the weather would be less frustrating, damaging, and deadly.

Collecting Data
Meteorologists, people who study the weather, have many tools and instruments such as those shown in the picture to help them. To make weather predictions, meteorologists need to have a complete picture of conditions in the atmosphere. These conditions include wind speed and direction, air pressure and how it is changing, temperature, precipitation, and humidity. Observations of these conditions are made regularly from land stations, from ships and buoys at sea, and from airplanes and balloons in the sky. As you can imagine, gathering all the data from all parts of the world presents a problem.

In 1960, the United States government put into orbit *Tiros I*, the first artificial satellite equipped to take pictures of Earth's weather in detail. *Tiros III*, launched in 1961, was the first satellite to discover a hurricane over the Atlantic Ocean.

More recent developments include an advanced series of satellites called the *Tiros-N*. Besides observing weather conditions, these satellites collect data about infrared radiation in Earth's atmosphere.

There are two kinds of weather satellites. One kind orbits Earth in a low orbit—800 to 1400 kilometers—that passes over the poles. Because Earth rotates, low-orbit satellites pass over a different area on each orbit. The other kind of weather satellite has a very high orbit—36 000 kilometers.

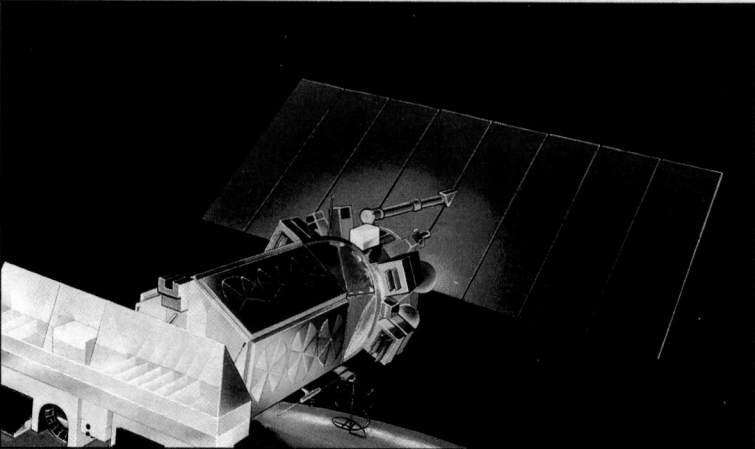

Tiros-N satellite

cloud patterns that occur during storms. The photo at the left shows one such cloud pattern. Besides taking pictures, satellites continually monitor the temperature of the water near the surface of the oceans.

Satellites collect information from their own instruments and relay it to ground stations for analysis. But they do more. The satellites also collect information from the thousands of land and water stations on Earth's surface. Instruments at these stations operate automatically, and the data they collect are transmitted by radio signal to a satellite. From there, the information is relayed to ground stations and analyzed by computers.

With this data-collecting system and high-speed computers to analyze the data, scientists continue to learn about the interaction of the air with land and the oceans. Researchers also

learn about how the atmosphere gains and loses heat by radiation. In the future, patterns of weather behavior will be better understood, and long-range weather prediction may become more reliable.

Satellite photograph of a hurricane in the Gulf of Mexico

At this height, the satellite goes around Earth in the same time it takes Earth to make one turn on its axis. In other words, the satellite is always over the same spot. Such satellites are so high, the pictures they take cover a large part of Earth's surface.

Weather Patterns

What kind of information can satellites gather? Satellites keep track of worldwide weather patterns. Weather patterns for big storms, such as hurricanes, can be identified and tracked so people can be warned.

Worldwide conditions such as snow and ice cover are easy to see on photos taken by satellites. Satellites also take pictures of the

*inter*NET CONNECTION

Use the World Wide Web to find out more about the American Meteorological Society and the topics they study. Explain which topics interest you and why.

Discussion

Ask students to explain why weather satellites are put into two different orbits. Explain as needed that low, polar orbits allow satellites to make in-depth studies of constantly changing portions of Earth's surface, while geostationary satellites give forecasters a stable view of larger weather systems and groups of systems. Have students refer to the article on pages 436–437 and name the kinds of data that satellites gather. List their answers on the chalkboard, including major weather patterns, cloud formations, ocean surface temperature, and data transmitted by ground stations.

Activity

Have students research the GOES (Geostationary Operational Environmental Satellite) weather satelllites, including the failure of GOES6 and plans to launch GOESI and GOESJ. Information on U.S. weather satellites and launch plans can be obtained from NASA, Meteorology, 2101 NASA Road 1, Houston, TX, 77058-3696.

*inter*NET CONNECTION

The home page for the American Meteorological Society can be found at
http://www.ametsoc.org

Going Further ⫸

Divide students into five-member groups to create weather maps showing the change in weather over a one-week period. Distribute eight copies of a black-and-white political map of your region or the U.S. (without state names, if possible) to each group. Make each member of the group responsible for watching one day's televised weather report or obtaining a newspaper weather map. Based on that resource, students should draw their own weather maps for that day, showing high- and low-pressure areas, areas of cloudiness and precipitation, and temperature readings at various standard locations. Display completed weather maps on a bulletin board. Use the Performance Task Assessment List for Bulletin Board in **PASC,** p. 59. **COOP LEARN** **L2** **P**

HISTORY CONNECTION

Purpose

The History Connection provides further information about Galileo's influence on the study of motion and the advancement of science.

Content Background

Galileo's early interest was in the science of motion. When Galileo heard about the Dutch invention of the telescope, he improved on it. Galileo created a 30-power telescope which he used to discover the mountains on the moon, new stars, four of Jupiter's moons, sunspots, and the rings of Saturn. Galileo's troubles with the Vatican began in 1615, when he publicly defended Copernicus. His problems culminated in 1632, after he published a book presenting arguments for the Copernican model; Galileo was imprisoned and the book burned. Under house arrest, Galileo secretly wrote another book on motion and astronomy, which was smuggled out of Italy.

Teaching Strategies
Challenge students to think critically. Ask them why they think the religious authorities of his day condemned Galileo's ideas about the sun-centered model of the universe. Explain that, until that time, people believed that a motionless Earth was at the center of the universe, and that heaven lay beyond the stars. Galileo's work and Copernicus's model denied this idea and, thus, seemed to attack the church.

Answers to
You Try It!

The steeper the slant, the higher the average speed, because the steeper slant provides an acceleration closer to that of a free fall.

HISTORY CONNECTION

Galileo

Galileo Galilei (1564–1642) began his career as a medical student, but soon gave up medicine for mathematics and the physical sciences. Early in his career, he studied and wrote about the motion of falling objects. His interests changed, though, when he heard about the invention of the telescope. He made one and used it to study objects in the sky. He observed the cratered surface of the moon, the "arms" of Saturn, the moons of Jupiter, the phases of Venus, and many stars that had not been seen before. Galileo found evidence that supported the Copernican (sun-centered) model of the solar system and thus cast doubt on the Ptolemaic (Earth-centered) model.

He published his observations and conclusions. Because his ideas went against the teachings of his religion at that time, he was arrested and forced to deny his ideas.

Despite these setbacks, he continued his studies on motion. The results of his studies were published in a book entitled *Dialogues Concerning Two New Sciences*.

Galileo's studies of falling bodies, motion on an inclined plane, and projectile motion all helped give a better understanding of motion. Even though he seemed to understand what is now called inertia, his writings never included the idea of inertia. The same is true of acceleration. He wrote about it but didn't connect it with forces.

A replica of Galileo's telescope

You Try It!

Repeat one of Galileo's experiments. On a smooth, level floor, prop one end of a 30-cm ruler up about 2 cm. Let a marble roll down the ruler. After it gets to the floor, see how far it rolls in two seconds. Do this three times and find the average distance. Then use $v = d/t$ to find the average speed. Change the slant of the ruler and repeat. What is the speed now? How would you explain the difference, if any?

Going Further ⅢⅢⅢ➡

Obtain a copy of Bertolt Brecht's *Galileo*, a play describing Galileo's work and its importance, his conflict with the church, and his old age. Copies may be available at your library:

Brecht, Bertolt. *Seven Plays*. New York: Grove Press, 1961.

Brecht, Bertolt. *Galileo*. New York: Grove Weidenfeld, 1966.

Divide the class into groups, and select short scenes from the play that convey the action. As-sign each group a scene to read. Have the groups read the scenes aloud in the proper sequence. When the reading is finished, ask students for their comments and reactions. Students could then create their own skit about Galileo. Use the Performance Task Assessment List for Skit in **PASC**, p. 75. **COOP LEARN** L1 P

chapter 13
REVIEWING MAIN IDEAS

Science Journal

Review the statements below about the big ideas presented in this chapter, and answer the questions. Then, re-read your answers to the Did You Ever Wonder questions at the beginning of the chapter. *In your Science Journal,* write a paragraph about how your understanding of the big ideas in the chapter has changed.

1 On or near Earth, the acceleration due to gravity is a constant 9.8 m/s². *If an elephant and a mouse fell freely off a high wire at the same time, which would reach the net first?*

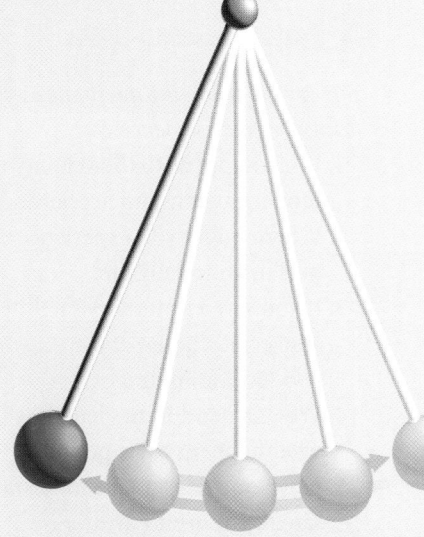

2 Projectiles have both vertical and horizontal movement. These objects move independently horizontally forward at a constant velocity and vertically downward at an increasing velocity because of the acceleration due to gravity. *What motions do gymnasts and baseballs have in common?*

3 Satellites stay in orbit around Earth because of their horizontal velocity and the acceleration due to gravity. *How does a satellite go to a higher orbit?*

Earth

4 The movement of a pendulum as it swings back and forth provides an opportunity to examine periodic motion and the variables affecting it. Pendulum length, not its mass or amplitude, determines the time it takes to swing back and forth once. *How does increasing the length of a pendulum affect its period?*

The strategies below ask students to recall how they calculated the acceleration of gravity; to compare gravitational acceleration to the tendency of an object to move at a continuous velocity; and to make predictions based on their knowledge of motion near Earth.

Teaching Strategies

Test students' understanding of the chapter by asking the following questions. What would happen if a satellite was fired into orbit at a speed greater than that needed to maintain the orbit? *It would move outward from Earth, following a curved path.* Are astronauts in space still weightless when they fire their rocket to move them around? *No, the rocket accelerates them, creating weight.* How would a pendulum behave in orbit? *There can be no pendulum motion in orbit because of weightlessness.* If gravity gets weaker the farther you are from Earth's center, would a pendulum on top of the world's highest mountain swing faster or more slowly than one at sea level? *more slowly*
L1

Answers to Questions

1. They would reach the net at the same time.

2. Both move forward at a constant velocity and accelerate downward at an increasing velocity.

3. Increase horizontal velocity

4. It increases the period.

Science Journal

Did you ever wonder...

• Heavy and light objects hit the ground at the same time if released at the same time, assuming no air resistance. (pp. 414–415)

• Satellites stay in orbit because their combined forward and falling motions make a path that matches the curvature of Earth. (pp. 425–426)

• Because they are freely falling around Earth, they are experiencing weightlessness. (p. 428)

Project

Have students build a kinetic sculpture in which a rolling marble tips balances, rings bells, spins water wheels, and leaps over gaps to demonstrate gravity's acceleration. Pathways can be built from cardboard bent into U-shaped channels, taped together, and mounted on cardboard legs. Students may wish to compete in groups to build the most interesting sculpture and explain the forces involved. **COOP LEARN** **L1**

GLENCOE TECHNOLOGY

 MindJogger Videoquiz

Chapter 13 Have students work in groups as they play the Videoquiz game to review key chapter concepts.

Using Key Science Terms

1. projectile motion
2. acceleration due to gravity
3. stationary satellite
4. period
5. weightlessness

Understanding Ideas

1. The velocity increases or decreases uniformly.

2. Students' answers will vary. Examples may include an elevator starting to move downward and a car going over the top of a hill.

3. It would still be directly over your house.

4. The period increases.

5. It takes one day.

6. Strobe photography is used to analyze objects in motion.

Developing Skills

1. See student page for completed concept map.

2. A satellite close to Earth must travel 7.9 km/s to stay in orbit. Sputnik travelled at a velocity of 7.6 km/s at a height of 516 km. A satellite orbiting at an altitude of 350 km would need to travel between 7.6 km/s and 7.9 km/s; 7.7 km/s is a good estimate.

3. At position 2, the rocket has enough velocity to match the acceleration due to gravity, so it goes into orbit. At position 3, the rocket gains velocity from rocket firing and moves away from Earth.

Critical Thinking

1. $9.8 \text{ m/s}^2 \times 3 \text{ s} = 29.4 \text{ m/s}$

2. The tennis ball thrown upward will be moving faster because it started falling from a greater height.

3. The ball reaches maximum height at the halfway point of its flight, so it should land near the 10-yard line.

U sing Key Science Terms

acceleration due to gravity	projectile motion
period	stationary satellite
	weightlessness

Each phrase below describes a science term from the list. Write the term that matches the phrase describing it.

1. throwing a ball
2. 9.8 m/s² at Earth's surface
3. appears not to move in the sky
4. seconds per cycle
5. condition of objects in free-fall

U nderstanding Ideas

Answer the following questions in your Journal using complete sentences.

1. What happens to the velocity of an object moving at constant acceleration?

2. Where might you experience weightlessness in your daily life?

3. If you saw a stationary satellite directly above your house at 6 A.M., where would you see it at midnight?

4. When a pendulum's length is increased, what happens to its period?

5. How long does it take for a stationary satellite to orbit Earth?

6. What method of photography is often used when analyzing objects in motion?

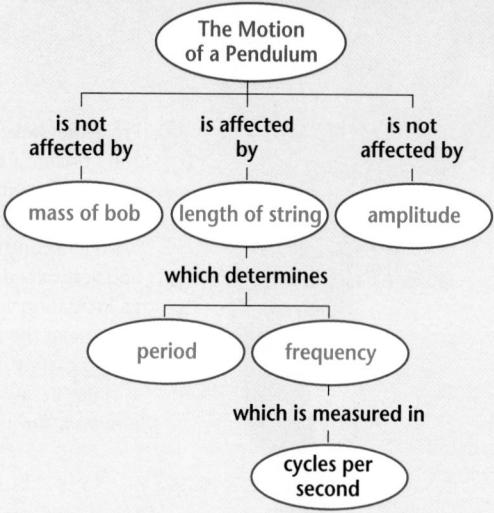

D eveloping Skills

Use your understanding of the concepts developed in this chapter to answer each of the following questions.

1. **Concept Mapping** Using the following terms, complete the concept map of the motion of a pendulum: *amplitude, frequency, length of string, mass of bob, period.*

```
        The Motion
        of a Pendulum
   ┌──────────┼──────────┐
is not      is affected   is not
affected by    by       affected by
   │           │           │
mass of bob  length of   amplitude
             string
              │
        which determines
          ┌───────┴───────┐
        period        frequency
                          │
                    which is measured in
                          │
                    cycles per
                    second
```

2. **Predicting** Use information from the Find Out activity on page 427 to predict the velocity of a satellite that orbits Earth at an altitude of 350 km.

3. **Interpreting Scientific Illustrations** In the diagram on the left, a rocket is fired from Earth in position 1. Describe what happens to its velocity in position 2. What do you think has to happen for it to go into position 3?

Program Resources

Review and Assessment, pp. 77–82 L1
Performance Assessment, Ch. 13 L2
PASC
Alternate Assessment in the Science Classroom
Computer Test Bank L1

Critical Thinking

In your Journal, *answer each of the following questions.*

1. A flower pot accidentally falls from the balcony of a 12th-floor apartment. You see it smash on the ground 3 seconds later. What is its velocity just before it hits the ground?

2. You have two tennis balls. You allow one to drop straight to the ground. The other you toss 5 meters straight up into the air and then let it fall to the ground. Which ball is moving faster when it hits the ground? Explain why.

3. A punter punts the football at the 50-yard line, and the ball reaches its maximum height at the 30-yard line. Where will the ball land?

Problem Solving

Read the following problem and discuss your answers in a brief paragraph.

You are part of a space expedition that has just landed on an unexplored planet. One of your first assignments is to determine the acceleration due to gravity on this planet. You are in charge of the crew being sent to the surface.

1. What materials would you have your crew assemble in order to determine the acceleration due to gravity on this planet?

2. List the steps in the procedure that you would have the crew follow to calculate the acceleration due to gravity on this planet.

Problem Solving

1. Materials: camera, strobe light, meterstick, object to drop.

2. Open shutter of camera in darkness and drop the object in camera's view next to meterstick while strobe light is firing every 1/10 second. On the developed photo, measure the distance traveled between intervals and calculate average velocity. Then subtract average velocity in first interval from average velocity in second interval to get change of velocity. Divide by number of seconds between flashes to get acceleration in units of meters per second per second.

Connecting Ideas

1. Some animals have greater muscular strength compared to body weight than humans; those animals can jump higher.

2. Yes; during liftoff, it increases as the shuttle's acceleration adds to gravity's acceleration. When the rocket switches off and the shuttle begins orbiting, the reading falls to zero.

3. Acceleration due to gravity is less on the moon, so the pendulum swings more slowly. Its period is longer than on Earth.

4. Students' answers will vary but should center on the concept that a pendulum bob's mass makes no difference in its period. Therefore, the mass or material of the sliding bob doesn't affect the period.

5. Students responses will vary but should include the information that if two objects have the same mass, the object with larger surface area usually falls more slowly and reaches terminal velocity faster. An object's mass is less important in considering terminal velocity than its size or shape.

6. Weather satellites provide information on large scale and far away weather events which can't be viewed in their entirety from the ground.

CONNECTING IDEAS

Discuss each of the following in a brief paragraph.

1. **Theme—Systems and Interactions** Why can some animals on Earth jump higher than humans, even though they are affected by the same gravity?

2. **Theme—Systems and Interactions** Do you think an accelerometer would work on a space shuttle? How do you think the accelerometer readings would change during and after lift-off, and finally in orbit?

3. **Theme—Stability and Change** Would the period of a pendulum be the same on the moon as it is on Earth? If not, why would it be different?

4. **How It Works** You and a friend are comparing metronomes. Yours has a sliding weight made of copper. Your friend's has a brass sliding weight. How does this affect the periods of the metronomes?

5. **A Closer Look** What is the relationship between a falling object's size, shape, and mass and the object's terminal velocity?

6. **Science and Society** How do weather satellites help weather forecasters?

✔ Assessment

Portfolio Review the portfolio options that are provided throughout the chapter. Encourage students to select one product that demonstrates their best work for the chapter. Have students explain what they learned and why they chose this example for placement into their portfolios.

Additional portfolio options can be found in the following **Teacher Classroom Resources:**

Making Connections: Integrating Sciences, p. 29

Multicultural Connections, pp. 29–30

Making Connections: Across the Curriculum, p. 29

Concept Mapping, p. 21

Critical Thinking/Problem Solving, p. 21

Take Home Activities, p. 22

Laboratory Manual, pp. 75–80

Performance Assessment P

Chapter Organizer

SECTION	OBJECTIVES	ACTIVITIES & FEATURES
Chapter Opener		*Explore!*, p. 443
14-1 Water Recycling (2 sessions; 1 block)	**1. Describe** how water moves through the hydrologic cycle. **2. Identify** the processes involved in the hydrologic cycle. **3. Demonstrate** runoff. National Content Standards: (5–8) UCP3, B2, G1–2	*Find Out!*, p. 444 *Skillbuilder:* p. 446 **Chemistry Connection,** pp. 446–447 **Economics Connection,** pp. 466–467 **Teens in Science,** p. 468
14-2 Streams and Rivers (3 sessions; 1.5 blocks)	**1. Describe** how and why streams form. **2. Discuss** characteristics of streams on steeply sloped land and streams on gently sloped land. National Content Standards: (5–8) UCP3, A1, B2	*Explore!*, p. 449 *Investigate 14-1:* pp. 452–453 **Science and Society,** pp. 464–465
14-3 Groundwater in Action (3 sessions; 1 block)	**1. Explain** how soil and rocks can be porous and permeable. **2. Describe** groundwater, aquifers, and the water table. **3. Explain** how groundwater is obtained from a well. National Content Standards: (5–8) UCP1–3, A1, B2, F2–3, F5, G1	*Explore!*, p. 455 *Design Your Own Investigation 14-2:* pp. 458–459 *Skillbuilder:* p. 463 **A Closer Look,** pp. 456–457

ACTIVITY MATERIALS

EXPLORE!

p. 443 waxed paper, cardboard, water, paper towels

p. 449* sand, stream table, 2- or 3-cm block of wood, water, sprinkling can, metric ruler

p. 455* sand, tub or stream table, water

INVESTIGATE!

pp. 452–453 2 pails, plastic hose, 2 screw clamps, stream table, sand, blocks of wood

DESIGN YOUR OWN INVESTIGATION

pp. 458–459 watch with second hand, 4 500-mL beakers, water, metric ruler, permanent marker, 25-mL graduated cylinder, potting soil, clay, sand, gravel

FIND OUT!

p. 444 large beaker, small beaker, water, plastic wrap, marble, lamp or several hours of direct sunlight, rubber band

KEY TO TEACHING STRATEGIES

The following designations will help you decide which activities are appropriate for your students.

L1 Basic activities for all students

L2 Activities for average to above-average students

L3 Challenging activities for above-average students

LEP Limited English Proficiency activities

COOP LEARN Cooperative Learning activities for small group work

P Student products that can be placed into a best-work portfolio

Activities and resources recommended for block schedules

Need Materials? Call Science Kit (1-800-828-7777).

00:00 **OUT OF TIME?** We recommend that students do the activities with an asterisk.

Chapter 14 Moving Water

TEACHER CLASSROOM RESOURCES

Student Masters	Transparencies
Study Guide, p. 51 **Concept Mapping,** p. 22 **Multicultural Connections,** p. 32 **Take Home Activities,** p. 23 **How It Works,** p. 17 **Making Connections: Integrating Sciences,** p. 31 **Science Discovery Activities,** 14-1 **Laboratory Manual,** pp. 81–82, The Hydrologic Cycle	**Teaching Transparency 27,** Water Cycle **Section Focus Transparency 45**
Study Guide, p. 52 **Multicultural Connections,** p. 31 **Making Connections: Across the Curriculum,** p. 31 **Activity Masters,** Investigate 14-1, pp. 59–60 **Science Discovery Activities,** 14-2 **Laboratory Manual,** pp. 83–86, Stream Patterns	**Teaching Transparency 28,** Distribution of Water on Earth **Section Focus Transparency 46**
Study Guide, p. 53 **Activity Masters,** Design Your Own Investigation 14-2, pp. 61–62 **Critical Thinking/Problem Solving,** pp. 5, 22 **Making Connections: Technology and Society,** p. 31 **Science Discovery Activities,** 14-3 **Laboratory Manual,** pp. 87–90, Permeability	**Section Focus Transparency 47**

ASSESSMENT RESOURCES	TEACHING & TECHNOLOGY
Review and Assessment, pp. 83–88 **Performance Assessment,** Ch. 14 **PASC*** **MindJogger Videoquiz** **Alternate Assessment in the Science Classroom** **Computer Test Bank**	**Spanish Resources** **Cooperative Learning Resource Guide** **Lab and Safety Skills** **Science Interactions, Course 1, CD-ROM** **Computer Competency Activities**

*Performance Assessment in the Science Classroom

NATIONAL GEOGRAPHIC TEACHER'S CORNER

Index to National Geographic Magazine

The following articles may be used for research relating to this chapter:

- "Our Polluted Runoff," by John G. Mitchell, February 1996.
- "The Great Flood of '93," by Alan Mairson, January 1994.
- *Water: The Power, Promise, and Turmoil of North America's Fresh Water,* A Special Edition, November 1993.
- "Ogallala Aquifer: Wellspring of the High Plains," by Erla Zwingle, March 1993.

- "The Colorado: A River Drained Dry," by Jim Carrier, June 1991.

National Geographic Society Products Available From Glencoe

To order the following products for use with this chapter, contact your local Glencoe sales representative or call Glencoe at 1-800-334-7344:

- *STV: Water* (Videodisc)

Additional National Geographic Society Products

To order the following products for use with this chapter, call the National Geographic Society at 1-800-368-2728:

- *Flood: Wrestling With the Mississippi* (Book)
- *What's in Our Water?* (Kids Network Curriculum Unit)
- *Fresh Water: Resource at Risk* (Video)
- *The Power of Water* (Video)
- *Water: A Precious Resource* (Video)

Teacher Classroom Resources

These are key components of the classroom resources package.

TEACHING AIDS

Section Focus Transparencies

45 SECTION FOCUS TRANSPARENCY Section 14-1

SLIP SLIDING AWAY!

Sinkholes, like this one in Florida, can be a hazard in populated areas. Sinkholes form in areas where limestone is common.

1. How do you think the sinkhole was formed?
2. How could you prevent your house from sliding into a sinkhole?

L1

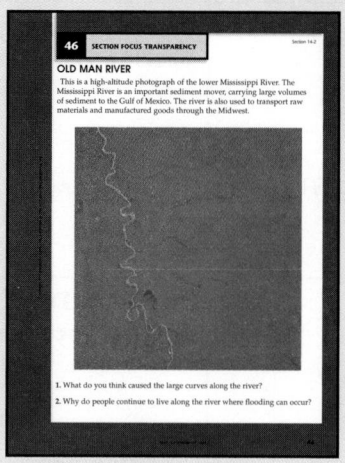

46 SECTION FOCUS TRANSPARENCY Section 14-2

OLD MAN RIVER

This is a high-altitude photograph of the lower Mississippi River. The Mississippi River is an important sediment mover, carrying large volumes of sediment to the Gulf of Mexico. The river is also used to transport raw materials and manufactured goods through the Midwest.

1. What do you think caused the large curves along the river?
2. Why do people continue to live along the river where flooding can occur?

L1

47 SECTION FOCUS TRANSPARENCY Section 14-3

THE LEANING TOWER

The Leaning Tower of Pisa, in Italy, has been a famous tourist attraction for many years. But what caused it to lean? The tower's tilt accelerated as the city grew and needed more water. Engineers are working to stabilize the tower.

1. What has happened to the ground in the area around the Leaning Tower?
2. How is this connected to water use?

L1

Teaching Transparencies

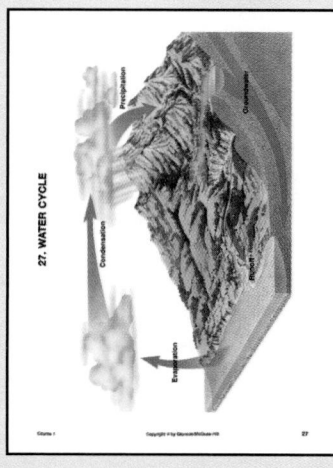

27. WATER CYCLE

L2

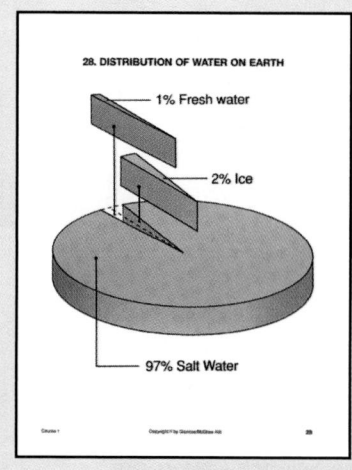

28. DISTRIBUTION OF WATER ON EARTH

1% Fresh water

2% Ice

97% Salt Water

L1

HANDS-ON LEARNING

Science Discovery Activity*

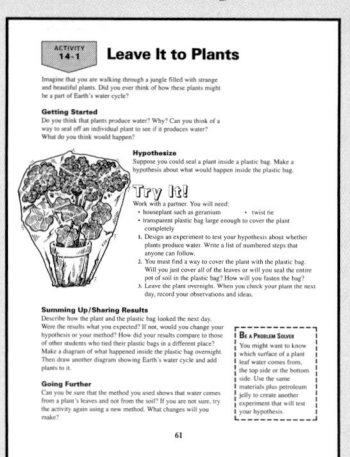

ACTIVITY 14-1 Leave It to Plants

Imagine that you are walking through a jungle filled with strange and beautiful plants. Did you ever think of how these plants might be a part of Earth's water cycle?

Getting Started
Do you think that plants produce water? Why? Can you think of a way to seal off an individual plant to see if it produces water? What do you think would happen?

Hypothesize
Suppose you could seal a plant inside a plastic bag. Make a hypothesis about what would happen inside the plastic bag.

Try It!
Work with a partner. You will need:
• houseplant such as a geranium
• transparent plastic bag large enough to cover the plant completely
• twist tie

1. Design an experiment to test your hypothesis about whether plants produce water. Write a list of numbered steps that anyone can follow.
2. You must find a way to cover the plant with the plastic bag. Will you just cover all of the leaves or will you seal the entire pot of soil in the plastic bag? How will you fasten the bag?
3. Leave the plant overnight. When you check your plant the next day, record your observations and ideas.

Summing Up/Sharing Results
Describe how the plant and the plastic bag looked the next day. Were the results what you expected? If not, would you change your hypothesis or your method? How did your results compare to those of other students who test their plastic bag in a different place? Make a diagram of what happened inside the plastic bag overnight. Then draw another diagram showing Earth's water cycle and add plants to it.

Going Further
Can you be sure that the method you used shows that water comes from a plant's leaves and not from the soil? If you are not sure, try the activity again using a new method. What changes will you make?

BE A PROBLEM SOLVER
You might want to know which surface of a plant leaf water comes from. Use the same materials plus petroleum jelly to create another experiment that will test your hypothesis.

61

L1

Laboratory Manual*

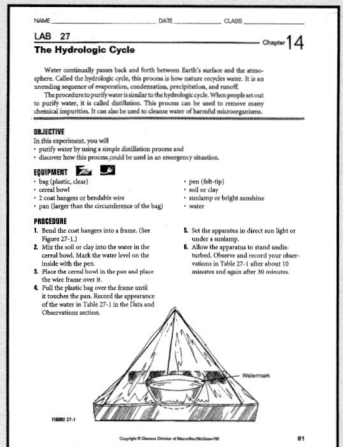

NAME _____ DATE _____ CLASS _____

LAB 27

The Hydrologic Cycle   Chapter 14

Water continually passes back and forth between Earth's surface and the atmosphere. Called the hydrologic cycle, this process is how nature recycles water. It is an unending sequence of evaporation, condensation, precipitation, and runoff.

The procedure to purify water is similar to the hydrologic cycle. When people set out to purify water, it is called distillation. This process can be used to remove many chemical impurities. It can also be used to cleanse water of harmful microorganisms.

OBJECTIVE
In this experiment, you will
• purify water by using a simple distillation process and
• discover how this process could be used in an emergency situation.

EQUIPMENT
• bag (plastic, clear)
• cereal bowl
• 2 coat hangers or bendable wire
• pen (larger than the circumference of the bag)
• pen (felt-tip)
• soil or clay
• sunlamp or bright sunshine
• water

PROCEDURE
1. Bend the coat hangers into a frame. (See Figure 27-1.)
2. Mix the soil or clay into the water in the cereal bowl. Mark the water level on the inside with the pen.
3. Place the cereal bowl in the pan and place the wire frame over it.
4. Pull the plastic bag over the frame until it touches the pen. Record the appearance of the water in Table 27-1 in the Data and Observations section.
5. Set the apparatus in direct sun light or under a sunlamp.
6. Allow the apparatus to stand undisturbed. Observe and record your observations in Table 27-1 after about 10 minutes and again after 30 minutes.

FIGURE 27-1

#1

L2

Take Home Activity

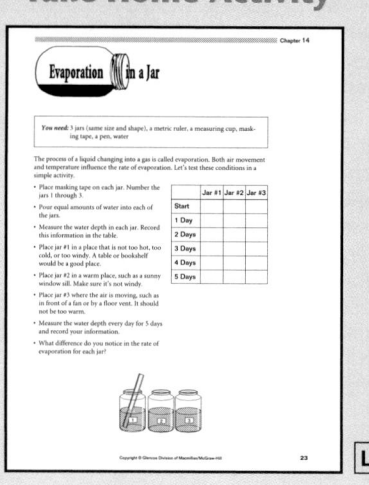

Evaporation in a Jar   Chapter 14

You need: 5 jars (same size and shape), a metric ruler, a measuring cup, masking tape, a pen, water

The process of a liquid changing into a gas is called evaporation. Both air movement and temperature influence the rate of evaporation. Let's test these conditions in a simple activity.

• Place masking tape on each jar. Number the jars 1 through 5.
• Pour equal amounts of water into each of the jars.
• Measure the water depth in each jar. Record this information in the table.
• Place jar #1 in a place that is not too hot, too cold, or too windy. A table or bookshelf would be a good place.
• Place jar #2 in a warm place, such as a sunny window sill. Make sure it's not windy.
• Place jar #3 where the air is moving, such as in front of a fan or by a floor vent. It should not be too warm.
• Measure the water depth every day for 5 days and record your information.
• What difference do you notice in the rate of evaporation for each jar?

	Jar #1	Jar #2	Jar #3
Start			
1 Day			
2 Days			
3 Days			
4 Days			
5 Days			

23

L1

Chapter 14 Moving Water

Study Guide*

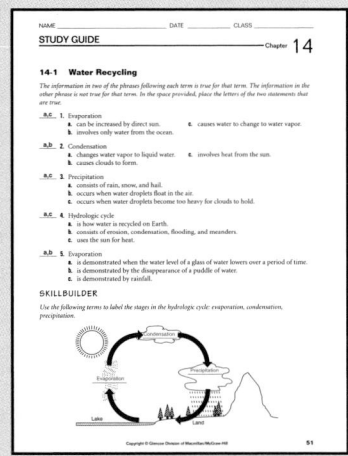

Concept Mapping

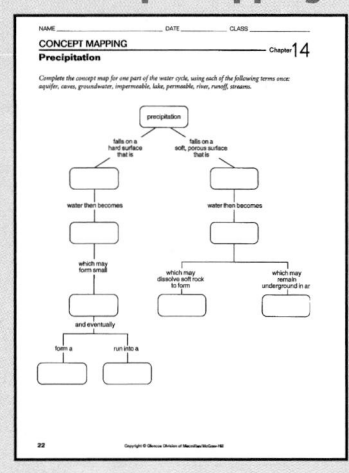

Critical Thinking/ Problem Solving

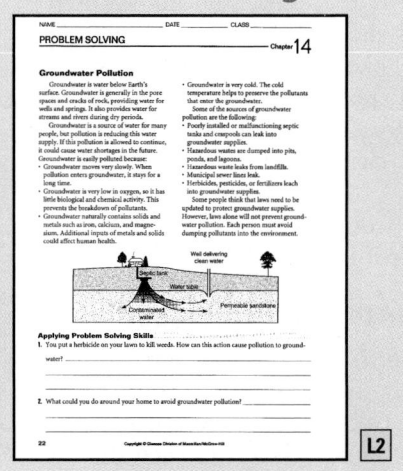

Integrating Sciences

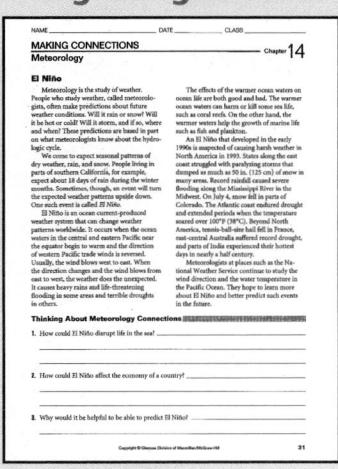

Across the Curriculum

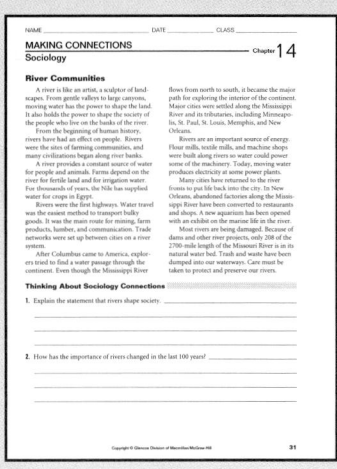

Technology and Society

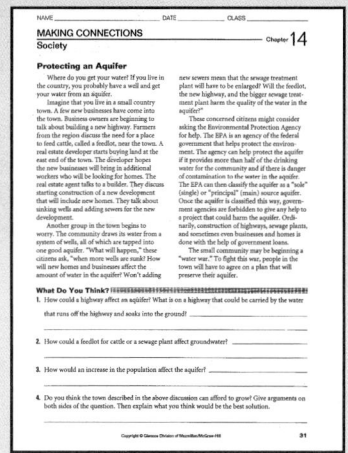

Multicultural Connection**

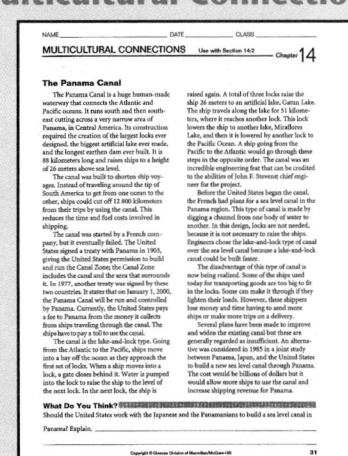

Performance Assessment

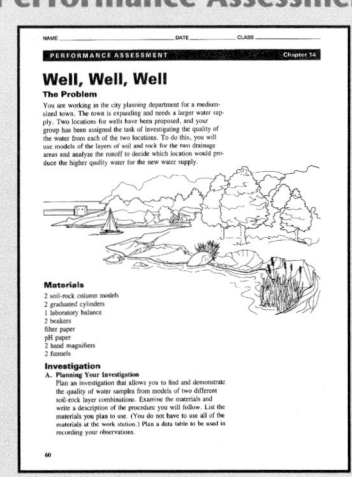

Review and Assessment

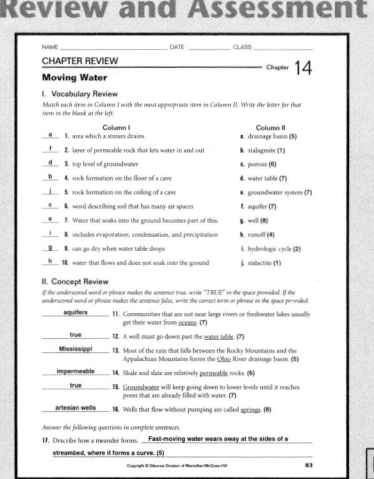

*One per section   **Two per chapter

Chapter 14 Moving Water **442D**

Moving Water

THEME DEVELOPMENT

Using the theme of systems and interactions, students learn how the various forms of water are linked in a hydrologic system that recycles the moisture of Earth. The theme of stability and change is used to describe how streams and rivers are formed and the differences in their strength and flow.

CHAPTER OVERVIEW

In this chapter, students will study the basic processes of the water cycle and how runoff functions in the cycle. Then students will trace the development of rivers and streams from runoff. Finally, the idea of an aquifer is introduced.

Tying to Previous Knowledge

Ask students if they think of water as in motion or as standing still. Explain that, because of gravity, water is almost constantly in motion.

INTRODUCING THE CHAPTER

Have students look at the photograph and then discuss bodies of water they have seen which are similar.

Uncovering Preconceptions

Many people take the availability of drinkable water for granted. But water resources depend on both the amount of rainfall in the area and the type of porous soil.

00:00 OUT OF TIME?

If time does not permit teaching the entire chapter, use the Chapter Overview on this page, Reviewing Main Ideas at the end of the chapter, and the Chapter 14 audiocassette to point out the main ideas of the chapter.

Moving Water

Did you ever wonder...

✓ **What happens to rain after it falls?**

✓ **How rivers form?**

✓ **Where well water comes from?**

📝 Science Journal

Before you begin to study about moving water, think about these questions and answer them *in your Science Journal.* When you finish the chapter, compare your journal entry with what you have learned.

Answers are on page 469.

Have you ever seen a raging, roaring stream like the one shown in the photograph below? Its rushing waters could take you on the most exciting ride of your life!

What if you could follow a tiny droplet of water? You might race with it down a swift stream, or you could follow a droplet as it lazily drifts down a gentle, quiet river. Eventually you might follow the droplet as it soaks into the ground. You might even find the droplet has made its way to a lake or to the ocean.

▶ *In the activity on the next page, explore how surfaces on Earth can affect what happens to water.*

442

Learning Styles	Kinesthetic	Explore, pp. 443, 449, 455; Find Out, p. 444
	Visual-Spatial	Visual Learning, pp. 448, 451, 460; Discussion, p. 450; Investigate, pp. 452–453; Design Your Own Investigation, pp. 458–459; Activity, p. 454; Demonstration, p. 456; Project, p. 469
	Interpersonal	Activity, pp. 445, 448
	Logical-Mathematical	Demonstration, p. 461; Across the Curriculum, p. 461
LS	**Linguistic**	Discussion, pp. 444, 446, 448, 456, 462; Visual Learning, pp. 445, 447, 454, 457, 462; Chemistry Connection, pp. 446–447; Across the Curriculum, pp. 447, 450, 451, 457; Multicultural Perspectives, p. 453; Activity, pp. 454, 463; A Closer Look, pp. 456–457; Research, p. 460

How do Earth's surfaces affect what happens to water?

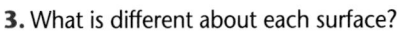

Have you ever wondered why water flows the way it does? What happens to the rainwater that falls on grassy areas compared to rainwater that falls on a parking lot?

What To Do

1. Put waxed paper over some cardboard. Slowly pour some water onto the paper and tilt the cardboard.

2. Repeat the procedure, this time using paper towels. *In your Journal* describe what happens to the water in each case.

3. What is different about each surface?

4. How do these differences affect the way water moves?

5. What kinds of surfaces on Earth cause water to move in the ways seen in this activity?

6. *In your Journal* describe some examples that you have seen of how different surfaces affect the way water moves.

ASSESSMENT PLANNER

PORTFOLIO
Refer to page 471 for suggested items that students might select for their portfolios.

PERFORMANCE ASSESSMENT
Process, pp. 443, 455
Skillbuilder, pp. 446, 463
Explore! Activities, pp. 443, 449, 455
Find Out! Activity, p. 444
Investigate, pp. 452–453, 458–459

CONTENT ASSESSMENT
Oral, p. 453
Check Your Understanding, pp. 448, 454, 463
Reviewing Main Ideas, p. 469
Chapter Review, pp. 470–471

GROUP ASSESSMENT
Opportunities for group assessment occur with Cooperative Learning Strategies.

Explore!

How do Earth's surfaces affect what happens to water?

Time needed 10 minutes

Materials water, cardboard, waxed paper, paper towel

Thinking Processes observing and inferring, modeling, comparing and contrasting, classifying, recognizing cause and effect

Purpose To observe that surfaces have an effect on the flow of water.

Teaching the Activity

Discussion Have students discuss how the activity relates to flowing water in their area such as after a rainstorm. L1

Science Journal Have students describe any difference in water flow.

Expected Outcomes

Students will notice that water runs off the waxed paper quickly and easily, but the paper towels absorb some of the water. Students should realize that different surfaces affect water differently.

Answers to Questions

2. The waxed paper allowed water to flow freely. The paper towel absorbed some of the water.

3. The waxed paper is hard and slick. The paper towel is soft and porous.

4. The water runs off the waxed paper. It is absorbed by the paper towel.

5. Clay or rock surfaces would act like waxed paper. Sandy soils and grassy areas would act like paper towels.

✓ Assessment

Process Have students repeat the activity using other types and layers of paper. Help them understand how Earth's surfaces affect what happens to water. Use the Performance Task Assessment List for Group Work in **PASC,** p. 97. L1

COOP LEARN

PREPARATION

Planning the Lesson

Refer to the Chapter Organizer on pages 442A–D.

Concepts Developed

In this section, students will learn that water moves through several processes.

1 MOTIVATE

Bellringer

Before presenting the lesson, display **Section Focus Transparency 45** on an overhead projector. Assign the accompanying **Focus Activity** worksheet. L1
LEP

Discussion Ask students to discuss what happens to the water Earth receives. They should conclude that some evaporates, some is absorbed by plants, some filters into the soil, some fills ponds and lakes, but most is runoff that eventually flows to the ocean. L1

Find Out!

How does water cycle through the environment?

Time needed 5 minutes to prepare, several hours until effect

Materials large beaker or jar, small beaker, plastic wrap, marble, rubber band, lamp

Thinking Processes measuring in SI, modeling, observing and inferring

Purpose To observe how water evaporates under heat and then condenses to form precipitation

Teaching the Activity

Troubleshooting Be sure the rubber band fits snugly.

Water Recycling

Water Cycle

Gray clouds roll in from the horizon, lightning flashes, thunder booms, and there you are—drenched by a sudden summer downpour. An hour later, the clouds have rolled past, the sky is bright blue, and puddles of rainwater are shimmering in the sunlight. Wait several hours more, and the puddles have disappeared. Where did the water go?

Section Objectives
- Describe how water moves through the hydrologic cycle.
- Identify the processes involved in the hydrologic cycle.
- Demonstrate runoff.

Key Terms
hydrologic cycle
runoff

Find Out! ACTIVITY

How does water cycle through the environment?

What To Do

1. Obtain a large beaker, a small beaker, some plastic wrap, a marble, a rubber band, and a lamp.

2. Pour 2 cm of water into the large beaker. Then, place the small beaker upright in the center of the large beaker.

3. Cover the opening of the large beaker loosely with plastic wrap. Seal the wrap with the rubber band.

4. Put the marble in the middle of the plastic wrap. What do you think the marble does?

5. Place the beaker under the lamp or in direct sunlight for several hours.

Conclude and Apply

1. *In your Journal* describe what occurred after the beaker sat for several hours.

2. How does this activity help show what happens to water on Earth?

Science Journal Have students record their answers to question 4 and the Conclude and Apply questions in their journals. L1

Expected Outcomes

The water in the large container evaporates, condenses on the plastic wrap, and should fall into the smaller container.

Conclude and Apply
1. See expected outcomes.
2. It shows the hydrologic cycle.

✓ Assessment

Process Have students discuss other situations where they have observed precipitation or condensation and then have them infer where the liquid came from. Use the Performance Task Assessment List for Making Observations and Inferences in **PASC**, p. 17. COOP LEARN L1

In the Find Out you saw how the liquid water went into the air as water vapor, a gas. You learned about liquids and gases in Chapter 4. Rainwater in puddles does the same thing. Did you see the water become vapor?

Let's follow what happens to a drop of water as it moves in a cycle from being vapor in the air to being liquid on Earth and from Earth back into the air just as the rainwater in the puddles. It is called the **hydrologic cycle**.

Figure 14-1

Hydrologic Cycle

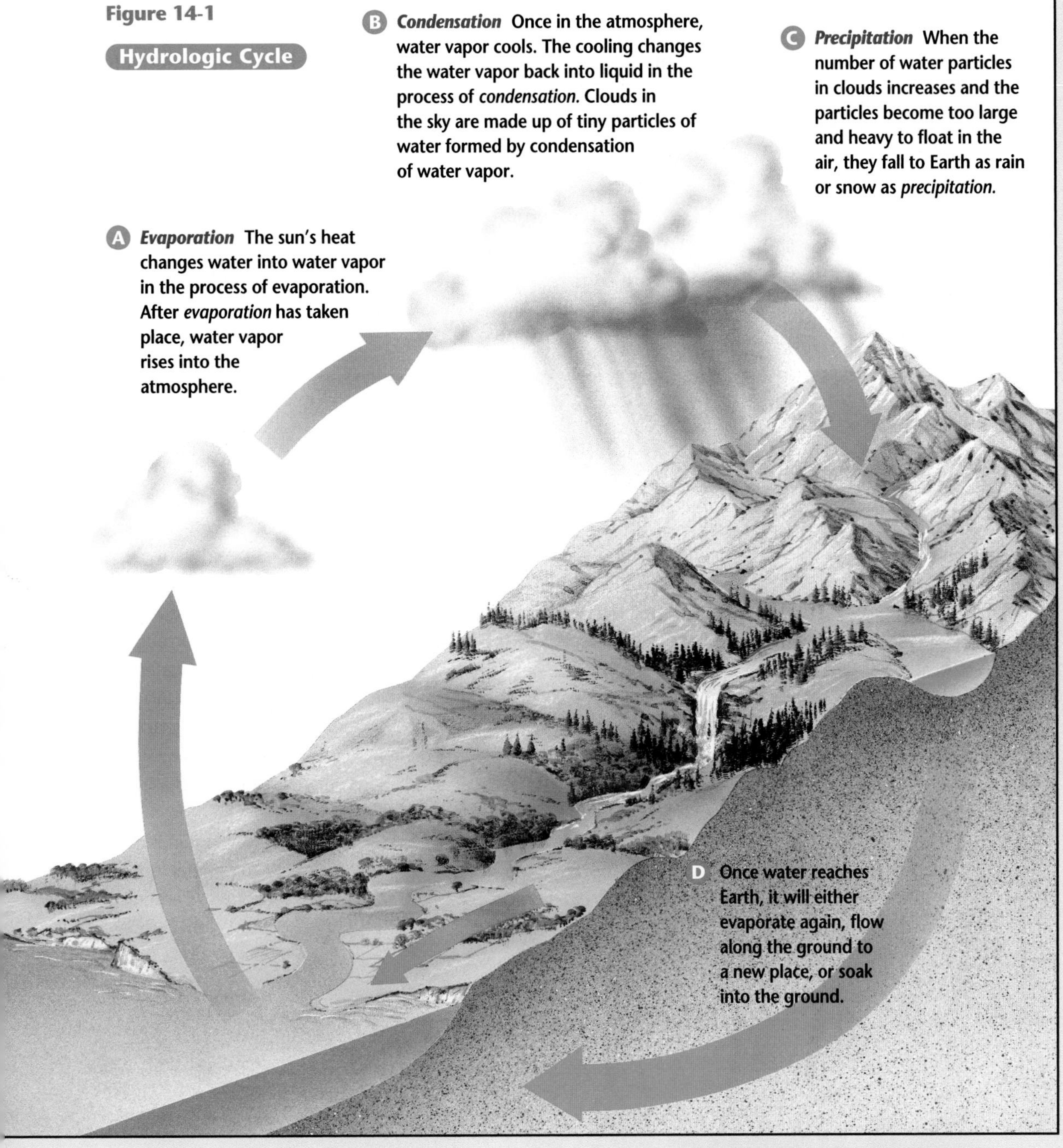

B *Condensation* Once in the atmosphere, water vapor cools. The cooling changes the water vapor back into liquid in the process of *condensation*. Clouds in the sky are made up of tiny particles of water formed by condensation of water vapor.

C *Precipitation* When the number of water particles in clouds increases and the particles become too large and heavy to float in the air, they fall to Earth as rain or snow as *precipitation*.

A *Evaporation* The sun's heat changes water into water vapor in the process of evaporation. After *evaporation* has taken place, water vapor rises into the atmosphere.

D Once water reaches Earth, it will either evaporate again, flow along the ground to a new place, or soak into the ground.

Program Resources

Study Guide, p. 51
Laboratory Manual, pp. 81–82, The Hydrologic Cycle L2
Concept Mapping, p. 22, Precipitation L1
Multicultural Connections, p. 32, Summer Monsoons L1
Teaching Transparency 27 L1
Section Focus Transparency 45

ENRICHMENT

Activity Divide the class into groups. Have each group make a list of places in their homes where evaporation or condensation takes place. They may name steam coming from boiling water or the frost that forms in a freezer. Challenge students to make a hypothesis that explains why these things happen. **COOP LEARN** L2

SKILLBUILDER

Rain falls to Earth. Water evaporates. Water vapor is forced upward and condenses. Condensation forms clouds. Water falls again as rain, snow, or hail. [L1]

Inquiry Question What do you think will happen to the amount of runoff from an area when an asphalt parking lot is built on top of the soil? *Runoff increases because the asphalt cannot absorb water.*

Discussion Grasses have many shallow roots that spread laterally across soil. A pine tree has a few long roots. Ask students to describe which plants would better reduce runoff and help prevent soil erosion. *They should conclude that grass roots can soak up more surface water and thus better prevent erosion.* [L2]

NATIONAL GEOGRAPHIC SOCIETY

Videodisc

STV: Water
Water Quality

Unit 1, Side 1
Great Lakes Watershed

02516-05710

Crop Irrigation

21935-24744

Runoff

Most people have spilled a glass of water sometime during their life. The glass topples over, and the water spills.

Sequencing

Sequence the events in the hydrologic cycle beginning with rain falling to Earth. If you need help, refer to the **Skill Handbook** on page 653.

If the water falls on a carpet, it quickly soaks in. If it spills on a tile floor, it forms a puddle. If the floor is uneven, the liquid soon flows toward the lowest spot.

Isn't this similar to what happened in the Explore activity? The water soaked into the paper towel but ran off the waxed paper.

This is also what happens to rainwater on Earth. Water soaks into the ground in some places but flows on the top of other surfaces. Water that flows and does not soak in is called **runoff**. Water that runs off will move along the ground and will eventually enter a stream.

What determines whether rainwater soaks into the ground or runs off? One factor is the ground itself. If the land is hard and smooth, like the waxed paper in your experiment or like rock in the real world, water will likely run off.

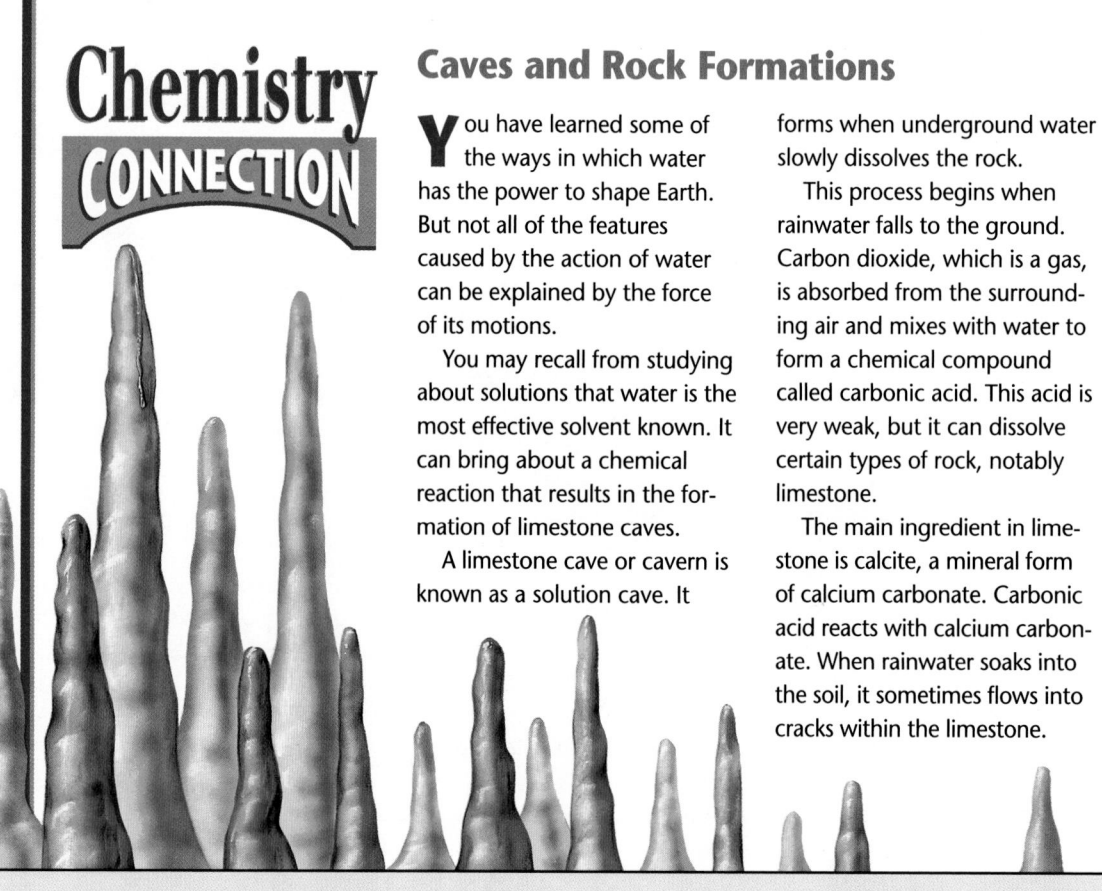

Chemistry CONNECTION

Caves and Rock Formations

You have learned some of the ways in which water has the power to shape Earth. But not all of the features caused by the action of water can be explained by the force of its motions.

You may recall from studying about solutions that water is the most effective solvent known. It can bring about a chemical reaction that results in the formation of limestone caves.

A limestone cave or cavern is known as a solution cave. It forms when underground water slowly dissolves the rock.

This process begins when rainwater falls to the ground. Carbon dioxide, which is a gas, is absorbed from the surrounding air and mixes with water to form a chemical compound called carbonic acid. This acid is very weak, but it can dissolve certain types of rock, notably limestone.

The main ingredient in limestone is calcite, a mineral form of calcium carbonate. Carbonic acid reacts with calcium carbonate. When rainwater soaks into the soil, it sometimes flows into cracks within the limestone.

Chemistry CONNECTION

Purpose

The chemistry connection extends this section's discussion of how water shapes Earth by describing how rainwater interacts with rock beneath Earth's surface.

Content Background

Another common feature associated with cave formation is a sinkhole. Like caves, sinkholes are created by the action of carbonic acid on limestone. Sinkholes are depressions in the ground visible at the surface that are caused when groundwater dissolves limestone beneath the depression, causing the ground to collapse. Some large sinkholes are caused by the collapse of cave ceilings. These large holes can collect runoff water and become lakes. Sinkholes are common in areas where there are large limestone deposits. In Europe, sinkholes are called karsts. This term is also used in the United States to describe topography containing caves and sinkholes. Sinkholes are common in Florida,

Figure 14-2

During the summer of 1993 flooding on the Mississippi River disrupted thousands of lives. Called one of the worst U.S. natural disasters of all times, the flooding caused an estimated 13 billion dollars in damages and killed 25 people.

A This sequence of photos show the flooding of McBride, Missouri. A levee on the Mississippi River 14 miles away from the town broke at 2 A.M. The first photo was taken at 10 A.M. as residents evacuated their homes.

B The second photo was taken at 5 P.M. A boat patrols to make sure that everyone has left as the water rises.

What other factors affect whether rain soaks in or runs off? You know from experience that sometimes rains are fast, hard driving, and heavy. Light rain falling over several hours will probably have time to soak into the ground while heavy rain may run off because it doesn't have time to soak in.

C At 4 P.M. on the next day, 38 hours after the levee broke, this photo was taken, showing the water at its highest level in the town.

The carbonic acid slowly dissolves the surrounding rock, creating underground holes called caverns. Sometimes the ground above caves in, creating a sinkhole.

Have you ever visited one of these caverns? If so, you have probably seen rock formations hanging from the ceiling that look almost like giant icicles. These rock formations are called stalactites. Stalagmites are rock formations that rise from the floor.

A single stalactite may take thousands of years to form. When underground water dissolves limestone, it absorbs calcite. The stalactite begins as a single drop of water clinging to the roof of the cave. As the drop of water evaporates, it loses some carbon dioxide. When that happens, calcite is deposited on the end of the stalactite. The stalactite grows as other drops of water cling to the outside of the stalactite.

Every so often, a drop of water will fall on the ground, depositing some calcite on the cave floor. After many years, the calcite will form a stalagmite, which looks like an upside-down stalactite.

inter*NET* CONNECTION

Information on the history and structure of Mammoth Cave is available from the National Park Service on the World Wide Web. Find out about the cave and describe an area in it that you would like to visit.

14-1 Water Recycling **447**

Visual Learning

Figure 14-3 Have students make a concept map that shows factors that determine how much water will soak into the ground and how much water will run off. Help students get started by asking: **Do you think the amount of moisture already in the soil affects runoff? Why?** *Yes. Dry soil will soak up a lot of water and reduce runoff, whereas wet soil will not.*

3 ASSESS

Check for Understanding

To answer the Apply question, be sure students know the meaning of the word *recycle*.

Reteach

Activity Have students work in groups to observe how different surfaces affect runoff. Materials needed are a paint tray half full of soil, a piece of cardboard, a clump of grass, and a beaker of water. Have students design and execute an experiment using the cardboard to simulate a paved street and the clump of grass to simulate a field. **LEP** **COOP LEARN**

Extension

Activity Have students find out what happens to runoff in their community. They should contact the community public works department and then prepare a map showing the location of storm sewer drains in the area of their school. **L3**

4 CLOSE

Discussion

Challenge students to infer what might cause droplets of water to form on the outside of a cold soda can. *The cooler temperature of the soda cools the air next to the can and the water vapor condenses to form small drops of water.* **L1**

Figure 14-3

A Several factors determine how much water will soak into the ground and how much will run off. One factor is the ground itself. Water will run off of hard rocky land and soak into soft soil. Another factor is the slope of the land. Gravity pulls water down steep slopes.

B Gentle rolling slopes and flat areas usually hold water until it either evaporates or sinks into the ground. The number of plants also affects runoff. Plants and their roots act like a sponge to soak up and hold water. Do you think the amount of moisture already in the soil affects runoff? Why?

Another factor that affects the amount of runoff is the slope of the land. You probably know that water flows downhill. This downward flow is due to Earth's gravity. Gently rolling slopes and flat areas usually hold water in place until it can evaporate or sink into the ground. Steep slopes, however, do not hold the water, and it runs off.

Do you think plants can affect runoff? Just like water running off a table, water on Earth tends to run off smooth surfaces. However, plants and their roots act like a sponge to soak up and hold water.

As you have read, there are many factors that affect runoff. A hard rain that falls on sloping, barren ground will probably run off. But a slow steady rain that falls on a level, grass-covered lawn will probably soak in and not become runoff. In the following sections, you'll learn what happens to water that soaks into the ground.

check your UNDERSTANDING

1. What supplies the energy for the hydrologic cycle?
2. What processes are involved in the hydrologic cycle?
3. How can runoff be decreased?
4. **Apply** Use the hydrologic cycle to explain why water can be described as recycled.

Program Resources

Take Home Activities, p. 23, Evaporation in a Jar **L1**

How It Works, p. 17, A City Water System **L2**

Making Connections: Integrating Sciences, p. 31, El Niño **L2**

Science Discovery Activities, 14-1, Leave It to Plants

check your UNDERSTANDING

1. the sun
2. evaporation, condensation, precipitation
3. Vegetation helps to decrease runoff.
4. The form of water is changed in the hydrologic cycle so that the water is continually reused.

 ## Streams and Rivers

Stream Development

Think of streams that flow near your home or streams you may have visited. Some streams are narrow, noisy, have steep sides, and flow swiftly.

These streams may form white-water rapids or waterfalls. Other streams are wide and slow-moving. Why do streams develop differently?

Section Objectives

- Describe how and why streams form.
- Discuss characteristics of streams on steeply sloped land and streams on gently sloped land.

Key Terms

meander
drainage basin

Explore! ACTIVITY

What happens to rainwater that runs off?

Remember the storm you were caught in at the beginning of the last section? You never imagined that one storm could bring so much rain in so little time. Not all the rain collected in puddles, however. Where did all the rest of the water go?

What To Do

1. Place sand in a stream table to a depth of 4 cm, but leave one end of the table empty. Put a block under the end of the table that is full of sand so that it is lifted 2 or 3 cm.

2. Use a sprinkling can to sprinkle water in the sand on the upper side of the stream table. The sprinkled water will be like rain falling on Earth.

3. *In your Journal* record what happens.

4. Explain how the water forms streams in the sand? Where did the water settle?

Explore!

What happens to rainwater that runs off?

Time needed 5 minutes

Materials stream table (see illustration on page 449), sand, block of wood, sprinkling can, metric ruler

Thinking Processes thinking critically, observing and inferring, making models

Purpose To demonstrate how streams form from runoff.

Preparation Use the illustration on page 449 as a guideline. Set up the stream table ahead of time. You can substitute bricks for wood blocks. Test the process to see how much water to sprinkle.

Teaching the Activity

Demonstration Before using the stream table, tilt a cookie sheet over a sink and sprinkle water on the uplifted end. Students can observe and describe the water's path.

PREPARATION

Planning the Lesson

Refer to the Chapter Organizer on pages 442A–D.

Concepts Developed

Students learned about runoff in the previous section. In this section they will investigate where much of that water goes.

1 MOTIVATE

Bellringer

Before presenting the lesson, display **Section Focus Transparency 46** on an overhead projector. Assign the accompanying **Focus Activity** worksheet. **L1**
LEP

Science Journal Before doing Step 2, have students write in their journals a prediction of what will happen. Then invite them to check their predictions.

Expected Outcome

The sprinkled water will tend to form into channels that run down to the lower end of the table.

Answers to Questions

See Expected Outcome.

✔ Assessment

Performance Invite students to work in small groups to re-create and extend their model as a science fair display. Use the Classroom Assessment List for Science Fair Display in **PASC**, p. 53. **L1** **COOP LEARN**
LEP

Discussion Use a map of the United States and Figure 14-6. Have students locate the Mississippi drainage basin on the map. Have them locate and identify some of the tributaries of the Mississippi. L1

2 TEACH

Tying to Previous Knowledge

Ask students to describe any brooks, streams, or rivers they have seen. Encourage them to comment on the size, the color of the water, and the slope of the surrounding land.

Theme Connection The theme that this section supports is systems and interactions. Streams and rivers form systems. People use these interconnected waterways for travel, fresh drinking water, and irrigation.

Content Background

Streams are classified as young, mature, or old. These names come from what geologists call the geomorphologic age of the land on which a stream flows. This is based on the principle that mountains wear down over time. Thus, steep land becomes less steep over time, as water wears it down. Mountainous areas are generally geomorphologically young, so streams flowing on them are labeled young. Rapidly flowing streams on steep slopes have much energy and erode the stream bottom more than the sides. A mature stream curves down a gradual slope and erodes its sides. It flows less swiftly through its valley. This stream will begin to meander. An old stream flows very slowly through the floodplain it has carved. The erosion of the sides causes major changes in its meanders.

As you saw in the previous Explore gravity causes water to flow downhill until it reaches the lowest point possible. Water flowing within a smaller channel is generally called a stream. Water in a larger channel is generally called a river.

Small streams eventually join together to form a larger stream. That larger stream will join with other large streams to form a river. Where do the rivers go? Do all rivers and streams have the same characteristics?

Streams that flow through steeply sloped areas run swiftly downhill. Such a stream may carve a narrow, steep, V-shaped valley because the running water wears away the stream bottom more than its sides. Streams on steep slopes may also form areas of white-water rapids or tumble over waterfalls.

Figure 14-4

Yellowstone National Park

A Streams that flow through steeply-sloped areas flow swiftly and fairly straight downhill. Because the rushing water wears away the stream bottom more than its sides, these streams often carve a narrow V-shaped valley into the land.

B The rapid flow of water often forms areas of white-water rapids and the stream may tumble over a waterfall on its path to a lower area of land.

450 Chapter 14 Moving Water

Program Resources

Study Guide, p. 52
Laboratory Manual, pp. 83–86, Stream Patterns L2
Multicultural Connections, p. 31, The Panama Canal L1
Section Focus Transparency 46

Across the Curriculum

Literature

The great Mississippi River formed the background for many of Mark Twain's stories. Have students read the chapter called "Sounding" in *Life on the Mississippi*. Discuss how the chapter relates to what they have learned about rivers.

A stream from a steep slope may eventually reach land that slopes very little. Or a stream may begin in such an area. A stream moving along a gradual slope flows much more slowly. Its valley is wide and low. The water has started to wear away the sides of the streambed, developing curves and bends in its path.

The curves in a river form because the speed of the water varies depending on the width of the stream channel. Water in wide, shallow areas of a stream is slowed down by the friction created with the bottom of the river. In deep areas, less water comes in contact with the bottom, so less friction results. Therefore, deep water can flow faster.

You can see that the river gently turns and curves as it moves along the gentle slope. This faster-moving water wears away the sides of the streambed where it flows more quickly, forming curves. A curve that forms in this way is a **meander**.

The broad, flat valley formed by a river on a gentle slope is called a floodplain. When the stream floods because of heavy runoff, it often covers part or all of its floodplain.

During the next Investigate you'll discover why streams have different characteristics.

Figure 14-5

The Yellowstone river, Yellowstone National Park

A Streams moving through flat or gentle slopes flow less rapidly than steep-sloped streams and flow in curves and bends.

B Curves in a river are formed when channels of deep water, free of the slowing force of friction from the stream bottom, flow faster than the rest of the water. This faster-moving water wears away the sides of the streambed, forming curves in the stream path.

Across the Curriculum

History

Inform students that the changing path of the lower portion of the Rio Grande has, in the past, created an international problem between the United States and Mexico. The international boundary between Mexico and the United States was originally defined as the main channel of the river; however, over the years it has become harder and harder to define the main channel of the river. In order to avoid continued problems, the border between the United States and Mexico was defined by a series of fixed points, on the basis of latitude and longitude. Ask students to research the Treaty of Guadalupe Hidalgo, which ended the Mexican War, to find out more about U.S./Mexican border problems affected by rivers.

Content Background

Because of its age, the Mississippi River has carved out a floodplain that covers over 39 000 square kilometers. Floodwaters can rise as much as 15 to 17 meters above the river's lowest stage. When floods occur on the Ohio and Missouri, heavy floods also happen along the Mississippi. This is because both rivers drain into the Mississippi.

GLENCOE TECHNOLOGY

CD-ROM

Science Interactions, Course 1 CD-ROM

Chapter 14

Program Resources

Teaching Transparency 28 L1
Making Connections: Across the Curriculum, p. 31, River Communities L2

Visual Learning

Figures 14-4 and 14-5 Have students use the figures to compare and contrast the two types of streams shown. Then, in their journals, have students contrast streams that flow through steeply sloped areas and streams that move through flat or gentle slopes.

14-1 Differences in Streams

Planning the Activity

Time needed 10–15 minutes

Purpose To demonstrate the effect of slope on the characteristics of a stream.

Process Skills thinking critically, observing and inferring, comparing and contrasting, recognizing cause and effect, making models, predicting, designing an experiment, separating and controlling variables

Materials 2 pails, plastic hose, 2 screw clamps, stream table, sand, blocks of wood

Preparation Have the stream table and materials set up in advance (see illustration below). You may want to have a large spoon or spatula ready to smooth out the damp sand.

Teaching the Activity

Process Reinforcement Be sure that students model two different types of streams. Check to see that they use the blocks to vary the height of the sand end. **L1**

Possible Hypotheses Students should hypothesize that slope affects the width and flow of a stream channel.

Troubleshooting Be sure the tube from the supply pail is secure. The tube that carries excess water out of the stream channel may become clogged with sand. To prevent this, be sure the stream channel is deep enough near the tube opening that the tube is above the sand. But also be sure the water is able to reach the tube.

Possible Procedures See the illustration below.

Science Journal In their journals, have students compare and contrast the two channels and how they are formed. Students can generalize their observations to include hypotheses about how their experiment relates to the formation of streams and rivers. **L1**

Differences in Streams

You've seen how streams form, but do streams have different characteristics? During this Investigate you will make your own models of streams.

Problem

What factors do you think control stream characteristics? Think about what you know about streams. How would you make your own streams? How can you control the flow?

Materials

2 pails	stream table
plastic hose	sand
2 screw clamps	blocks of wood

Safety Precautions

Wear an apron to protect your clothing.

What To Do

Work with your group and plan ways to set up your stream table to form different streams. Show your plan to your teacher. If you are advised to revise your plan, be sure to check with your teacher again before you begin. Carry out your plan keeping in mind:

1 When you set up the stream table, dampen the sand.

2 By using a screw clamp on the supply hose you can adjust the flow of water.

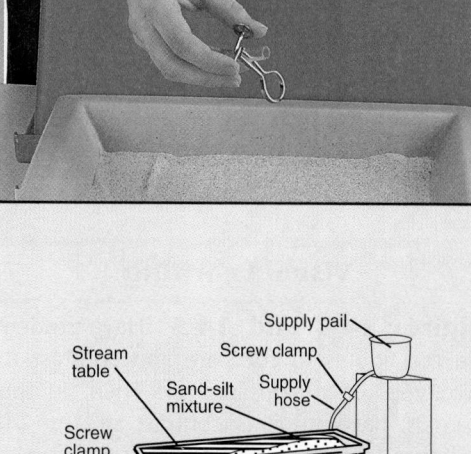

Supply pail
Screw clamp
Stream table
Sand-silt mixture
Supply hose
Screw clamp
Exit hose
Block of wood
Catch pail

Program Resources

Activity Masters, pp. 59–60 Investigate, 14-1

Science Discovery Activities, 14-2, How Does Your River Run?

The broad, heavy bulk of the steamboat restricts its use to slow-moving, meandering rivers such as the Mississippi.

The light, narrow frame of a kayak makes it easy to guide in fast moving white water rapids, such as these on the Colorado River.

3 Do not make the reservoir end higher than the other end of the stream table.

4 Smooth out the sloping sand from the previous channel before forming another one.

Analyzing

1. How could the flow of water be increased?

2 How could the flow of water from the supply pail be slowed down?

3. Describe the channel that was formed when the sand end was high.

4. Describe the channel that was formed when the sand end was lower.

Concluding and Applying

5. *Compare and contrast* the two types of stream channels.

6. *Determine the cause* of the differences between the two channels you made.

7. ~~Going Further~~ What kind of stream channels would you expect to form on plains? What kind form in mountainous areas?

Stream patterns will form. The steeper the slope, the narrower and faster the stream.

Answers to Analyzing/ Concluding and Applying

1. by loosening the screw clamp

2. by tightening the screw clamp

3. narrow and swift

4. broad and slow

5. The channel that formed when the sand end was high was narrower and swifter than the channel that formed when the sand end was lower.

6. The difference is caused by the change in slope.

7. Broad channels form on plains. Narrow channels form in mountainous areas.

✔ Assessment

Oral Have students assess their experiment by identifying and explaining what their models show about stream channels. Students can suggest ways to extend the experiment, such as by showing the effect of drought or flooding on a channel. Use the Performance Task Assessment List for Assessing the Whole Experiment and Planning the Next Experiment in **PASC**, p. 33. L1 **COOP LEARN**

NATIONAL GEOGRAPHIC SOCIETY

 Videodisc

STV: Water

Nile River, Egypt

52524

Colorado River, Grand Canyon; Arizona

52605

Multicultural Perspectives

The Nile

Inform students that ancient Egyptians made good use of the floodplain created by the Nile River. The Nile rose every year to flood stage. It left behind a deposit of rich soil for plentiful harvests. The regular annual flood also gave birth to the idea of a 365-day year. This was based on the average period between rises over fifty years.

Have students research the Aswan High Dam. Constructed in the 1960s, it has affected Egyptian life, food supplies, and environment. Invite students to present their findings in a poster or oral report. **COOP LEARN** L1

3 ASSESS

Check for Understanding

To help students with the Apply question, remind them that fast-flowing water tends to run in a fairly straight path. Ask students where fast-flowing water is usually found. *steep slopes*

Reteach

Activity Invite students to organize information. Have them write a description of what they think the bed of a mountain stream looks like. *narrow, with rocks and pebbles at its bottom*

Have students then write a description of the bed of a large, mature river. *The image should be of a wide bed with a sandy or muddy bottom, the result of wearing away of both bottom and sides.* **L1**

Extension

Activity To understand another very important use of rivers, have students research the names of some of Earth's rivers that form boundaries between states and countries. Ask them to make a table listing the name of the river and the states and/or countries that it divides. **L3**

4 CLOSE

Activity

Challenge students to formulate a model. Have them draw an imaginary river system. They should link several fast-flowing

Draining the Land

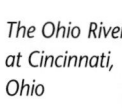

The Ohio River at Cincinnati, Ohio

Figure 14-6

Most of the rain that falls between the Rocky Mountains and the Appalachians forms the Mississippi River drainage basin. The rain drains into small streams and rivers that eventually flow into the Missouri or Ohio Rivers. In turn, these large rivers flow into the Mississippi River.

Mississippi River drainage basin

The Missouri River at Kansas City, Missouri

The Mississippi River at St. Louis, Missouri

The water in streams and rivers comes from rain or melted snow—that is, runoff.

All the water in a land area eventually flows down, or drains, into one stream. The area that a stream drains is called a **drainage basin**. Each stream has its own drainage basin.

The drainage basin of a large river usually includes the drainage basins of smaller rivers and streams. The largest drainage basin in the United States is the Mississippi River drainage basin. The Mississippi Basin is formed from most of the rain that falls between the eastern slope of the Rocky Mountains and the western slopes of the Appalachians.

All the streams and rivers in a major drainage basin form a river system. The Mississippi River system drains about one-third of the United States.

check your UNDERSTANDING

1. What causes streams to form and flow downhill?
2. Where are slow-moving streams most likely to be found?
3. Does water flow faster in wide or narrow channels?
4. **Apply** Why don't meanders form in streams on steep slopes?

454 **Chapter 14** Moving Water

streams from mountains to two or three slower, meandering rivers. These should flow into a large, highly curved river on a plain. Have students label the types of slopes and the types of rivers. **L1**

check your UNDERSTANDING

1. Streams are formed by runoff water and run downhill because of the force of gravity.
2. Slow-moving streams are most likely to be found in valleys and plains.
3. Water flows faster in narrow channels.
4. The water in a stream on a steep slope is flowing too quickly to form gentle curves.

 **Groundwater in Action**

Into the Ground

In the previous sections we saw that some rainwater forms runoff and some rainwater goes into streams and rivers. Some of this rain water, however, soaks into the ground and seems to disappear. Where did the water go?

Section Objectives

■ Explain how soil and rocks can be porous and permeable.

■ Describe groundwater, aquifers, and the water table.

■ Explain how groundwater is obtained from a well.

Key Terms

groundwater
aquifer
water table

Explore! ACTIVITY

How can the water level in the ground be changed?

What To Do

1. Fill a tub or stream table with sand and level it out. Pour water into the tub until the water is almost to the top of the sand.

2. Make a shallow hole in the sand so that you can see the water at the bottom of the hole. *In your Journal*, describe what this hole might represent.

3. Now, add more water to the sand. What happens to the level of the water in the hole? How can you change the level of the water?

As you observed, water that enters the sand moves from one place to another. There must be a lot of space between the fragments of sand. A drop of water that soaks into such ground would just seem to disappear because the ground has so much space within it.

Just watering a plant shows you how quickly water may soak into soil. Like most soils, the soil in which most houseplants are potted is made up of many tiny fragments. Some fragments may be sand-sized, some larger, and others smaller. The spaces among the fragments are called pores.

■ Permeability

Water that soaks into the ground collects in the pores and becomes part of the **groundwater**. In fact, it becomes part of a groundwater system.

PREPARATION

Planning the Lesson

Refer to the Chapter Organizer on pages 442A–D.

Concepts Developed

In this section, students will analyze how water enters soil and rock layers and forms groundwater reserves. Water accumulates underground and forms the water table. Students will also discuss other sources of water, such as aquifers and springs. They will explain how wells are used to get water from an aquifer.

1 MOTIVATE

Bellringer

Before presenting the lesson, display **Section Focus Transparency 47** on an overhead projector. Assign the accompanying **Focus Activity** worksheet. L1

LEP

Expected Outcome

The water level in the hole rises when more water is poured into the sand.

Answer to Question

3. The water level rises. Pour in more water or remove some with an eyedropper.

✔ Assessment

Process Have students make a scientific drawing of their model. You may wish to place the completed drawing in their portfolio. Use the Classroom Assessment List for Scientific Drawing in **PASC**, p. 55. L1 P

Explore!

How can the water level in the ground be changed?

Time needed 5 minutes

Materials tub, sand, water

Thinking Processes making models, predicting, interpreting scientific illustrations

Purpose To model how groundwater levels change.

Preparation Set up the sand and water ahead of time.

Teaching the Activity

Troubleshooting When adding more water to the sand, students should try to not disturb the hole they made. They may have to firm the sides of the hole.

Science Journal Answers will vary. The hole might represent a pond, lake, or well. L1

2 TEACH

Tying to Previous Knowledge

Challenge students to infer from where they get water for their homes. They may say from the faucet. Then ask where the water from the faucet originates. If they say the water company, ask where the water company gets it. Continue to query students until they recognize that the water they drink comes from one of Earth's many supplies of fresh water: rivers, lakes, reservoirs, aquifers, and springs.

Theme Connection The water we use comes from a system connected to the hydrologic cycle. Other parts of the cycle involve the movement of water through the air and over the land. The groundwater system is the movement of water through Earth.

Demonstration Model how water reacts to an impermeable material. You will need a piece of slate or a ceramic plate.

Pour some water onto the slate or plate. Have students observe what happens to the water. *It does not go anywhere.* Have students contrast what happened to the water in the sand from Explore and the water on the plate.

Figure 14-7

Various kinds of soil allow water to pass through at different rates.

A Peat soaks up and holds water.

B Water drains easily through light sandy soil.

C Clay soil is dense and stops water from draining through.

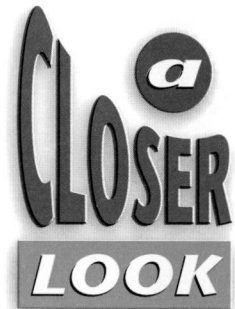

Hot Springs and Geysers

In some locations beneath Earth's surface, underground water comes in contact with hot rock. When that happens, the water heats up. If the water then makes its way to the surface, a hot spring is formed. Usually, hot springs bubble gently and are only a few degrees warmer than the surrounding air.

In some cases, however, the underground water heats up so that it bursts violently through Earth's surface. These hot springs are called geysers.

Geysers exist in locations where there are vast underground passageways for the water to travel through. Usually, all of these connecting tunnels, as shown in the illustration, lead to a single opening on the surface.

The groundwater, from its contact with hot, underground rock, is heated to very high temperatures, causing it to expand to fill these tunnels.

Pohuto and Prince of Wales Feathers geysers in New Zealand

456

Purpose

A Closer Look extends Section 14-3 by showing what can happen to groundwater that becomes heated, forming geysers.

Content Background

The thermal energy that heats underground water comes from pockets of magma close to Earth's surface. Yellowstone National Park has several types of features resulting from heated groundwater. Steam and other gases forced up through cracks in the rocks are called fumaroles. Sometimes the emerging steam brings acids with it to form a mud pot. The acids dissolve rock around the fissure and make a pool of bubbling mud. A hot spring's groundwater has unrestricted flow to the surface. A geyser forms when water is heated in a complex maze of underground passageways. When pressure in the passageways is reduced because the water has expanded and flowed out of the opening,

Water that stays above the ground becomes part of a river system. A groundwater system is similar to a river system even though it lies within the ground. However, instead of having stream channels that connect different parts of a drainage basin, a groundwater system may have connecting pores that water can move through.

Soil or rock that has many connecting pores is said to be permeable. Water can pass through such ground materials easily. Soil or rock that has few or very small pores is less permeable. Water can't pass through it as easily. Some materials, such as clay, shale, and slate, have very small pores or no pores at all. Because water can't pass through these materials, they are impermeable.

How quickly water seeps into the ground depends on the permeability of that ground. Do you think water would seep quickly into hard-packed soil? How would permeability affect groundwater?

To understand how impermeable materials affect groundwater, think about a raincoat. You know that a raincoat is designed to keep rainwater from getting through. So the raincoat is impermeable to water. But other kinds of clothing might let some or all of the rainwater through.

In the next Investigation, you'll discover how different soils affect permeability.

Visual Learning

Figure 14-7 Have students order the various soils by permeability. *sandy soil, peat, clay soil*

Across the Curriculum

Daily Life

Guide students to observe that, after a normal rain, puddles form on concrete or other hard surfaces. But the water is absorbed by grassy areas and ordinary soil.

NATIONAL GEOGRAPHIC SOCIETY

Videodisc

STV: Water Water Conservation

Unit 2, Side 1
Ogallala Aquifer

17877-21929

Preventing Groundwater Contamination

31262-33972

Aquifer showing layers, artwork

52537

Groundwater coming to surface as a spring

52539

This expanding water forces some of the water on top out of the ground, taking the pressure off the remaining water. The remaining water boils quickly, with much of it turning to steam. The steam shoots out of the opening like steam out of a teakettle, forcing the remaining water out with it. Once the geyser erupts, groundwater begins to refill the passageways, and the process begins again.

What Do You Think?

Some geysers follow a regular schedule. Old Faithful Geyser in Yellowstone National Park, for instance, erupts an average of once every 65 minutes. Can you explain why this happens?

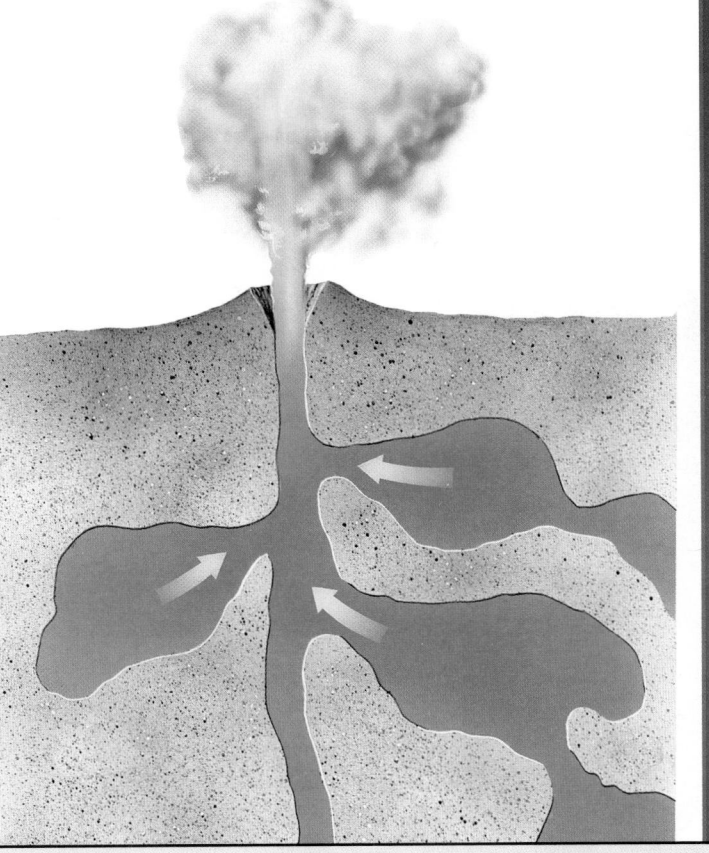

the water boils rapidly and much of it turns to steam. Water and steam are forced upwards in a sudden surge.

Answers to
What Do You Think?

After the geyser erupts, it takes 65 minutes for new water to seep back into the system, become heated, and erupt again.

Going Further ⫸

Have students find pictures of geysers and hot springs in magazines and books. Ask what they notice about the colors and formations in the area. What is responsible? Also ask if they think that life exists in these hot-water systems. *Algae, bacteria, and even some insects live in many hot springs. The life in hot springs and the minerals that surface through the action of the water cause bright colors and rock formations.*

Students can write to:
Yellowstone Association
P.O. Box 117
Yellowstone National Park, WY 82910
Use the Performance Task Assessment List for Letter in **PASC**, p. 67. **COOP LEARN**
L1

14-2 Ground Permeability

Preparation

TECH PREP

Purpose To predict, test, compare, and contrast the permeability of various soils.

Process Skills observing and inferring, comparing and contrasting, recognizing cause and effect, forming operational definitions, making and using tables, measuring in SI, predicting, interpreting data, separating and controlling variables, classifying, sequencing, modeling

Time Required one hour or more

Materials See student activity. Potting soil, clay, sand, and gravel can be obtained from a gardening store.

Possible Hypotheses Students may hypothesize that the size of the soils' connecting pores determines how fast water seeps into different soils.

The Experiment

Process Reinforcement Data tables should include columns with the following headings: Soil Combinations, Observations of Soil, Prediction, and Amount of Time to Sink In.

Possible Procedures Use the metric ruler and permanent marker to measure and mark the outside of each beaker 3 cm from the bottom. Fill each beaker with one of the test "soils." The test soils are any combination of the materials that are available to the students. After observing and recording the texture and color of the test soils, use the watch to determine how much time it takes for the water in the beaker to permeate the test soils to the 3-cm mark.

Ground Permeability

There are many different kinds of soils. Soils that have a lot of connecting pores are characterized as permeable. Permeability affects how fast water can seep into the soil and flow through the ground.

Preparation

Problem
What factors determine how fast water seeps into different soils?

Form a Hypothesis
As a group, discuss what factors might influence different soils' permeability. Agree upon these factors, then make a hypothesis that can be tested in your investigation.

Objectives
• Predict and compare the permeability of different soils.
• Measure the time it takes for water to seep into different soils.
• Infer why some soils are more permeable than others.

Possible Materials
watch with second hand
25-mL graduated cylinder
four 600-mL beakers
water
metric ruler
permanent markers
potting soil, clay, sand, gravel

Program Resources

Study Guide, p. 53
Laboratory Manual, pp. 87–90, Permeability [L3]
Activity Masters, pp. 61–62, Design Your Own Investigation 14-2
Science Discovery Activities, 14-3, Putting Water on the Table
Section Focus Transparency 47

DESIGN YOUR OWN
INVESTIGATION

Plan the Experiment

1 Examine the materials provided by your teacher. Decide which materials you will use and how you will use them in your experiment.

2 Design a procedure to test your hypothesis. Write down what you will do at each step of your test.

3 Design a table for recording data.

Check the Plan

1 Review how you will combine the different materials to create "test soils." Will your combina-

tions create different-colored and -textured test soils?

2 Have you determined how you will measure the water as it seeps into the soil?

3 Before you start the experiment, have your teacher approve your plan.

4 Carry out your experiment. Complete your data table *in your Science Journal* or on a computer spreadsheet.

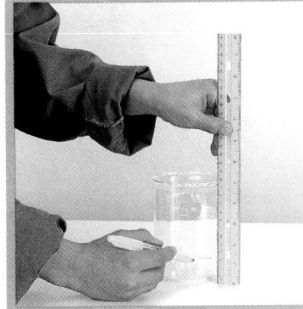

Analyze and Conclude

1. Measure Measure the time it takes for the water in the beaker to permeate the soil. Compare your observations with your hypothesis and your predictions, and record these in your data table.

2. Interpret Data Which of the soils was least permeable? Most permeable? How did you tell?

3. Infer Infer why some soils are more permeable than others.

Going Further

Explain how permeability affects groundwater flow. Be sure to discuss runoff in your answer.

Meeting Individual Needs

Learning Disabled You will need a large jar or beaker, water, pebbles, and rocks. Some students may have difficulty understanding the idea of permeability. Fill the beaker with pebbles and rocks. Have students note the spaces in between; these are pores. Slowly fill the jar with water. Have students observe how the water flows around the rocks and pebbles and fills the spaces in between. **L1**

Teaching Strategies

Troubleshooting Look over students' data tables to make sure they have included each factor that they will observe. Also, make certain students' test soils include pure samples of sand, clay, potting soil, and gravel in addition to mixtures of these materials.

Science Journal In their journals, have students describe which observations were most helpful in predicting soil permeability. **L1**

Expected Outcome

The gravel and sand will be more permeable than potting soil or clay.

Analyze and Conclude

1. Observations will vary.

2. The beaker containing gravel was the most permeable. Water soaked fastest into this soil. The beaker containing the clay was the least permeable. Water took the longest time to soak into this soil.

3. Less permeable soils may have smaller pores or they might be more compacted. More permeable soils may have larger pore spaces and be less compacted.

✔ Assessment

Performance Have students working in cooperative groups discuss the observations that they made. One group member can serve as the recorder. Then students can create a bulletin board on the permeability of various soils. Use the Performance Task Assessment List for Bulletin Board in **PASC**, p. 59 **COOP LEARN** **L1**

Going Further ⸻➤

Water that falls on Earth's surface can become groundwater quicker in some places, depending on the permeability of the ground. Water soaks quickly into the permeable soils, thus reducing the amount of runoff.

Figure 14-8

A Permeable, porous rocks, such as sandstone, allow water to pass through.

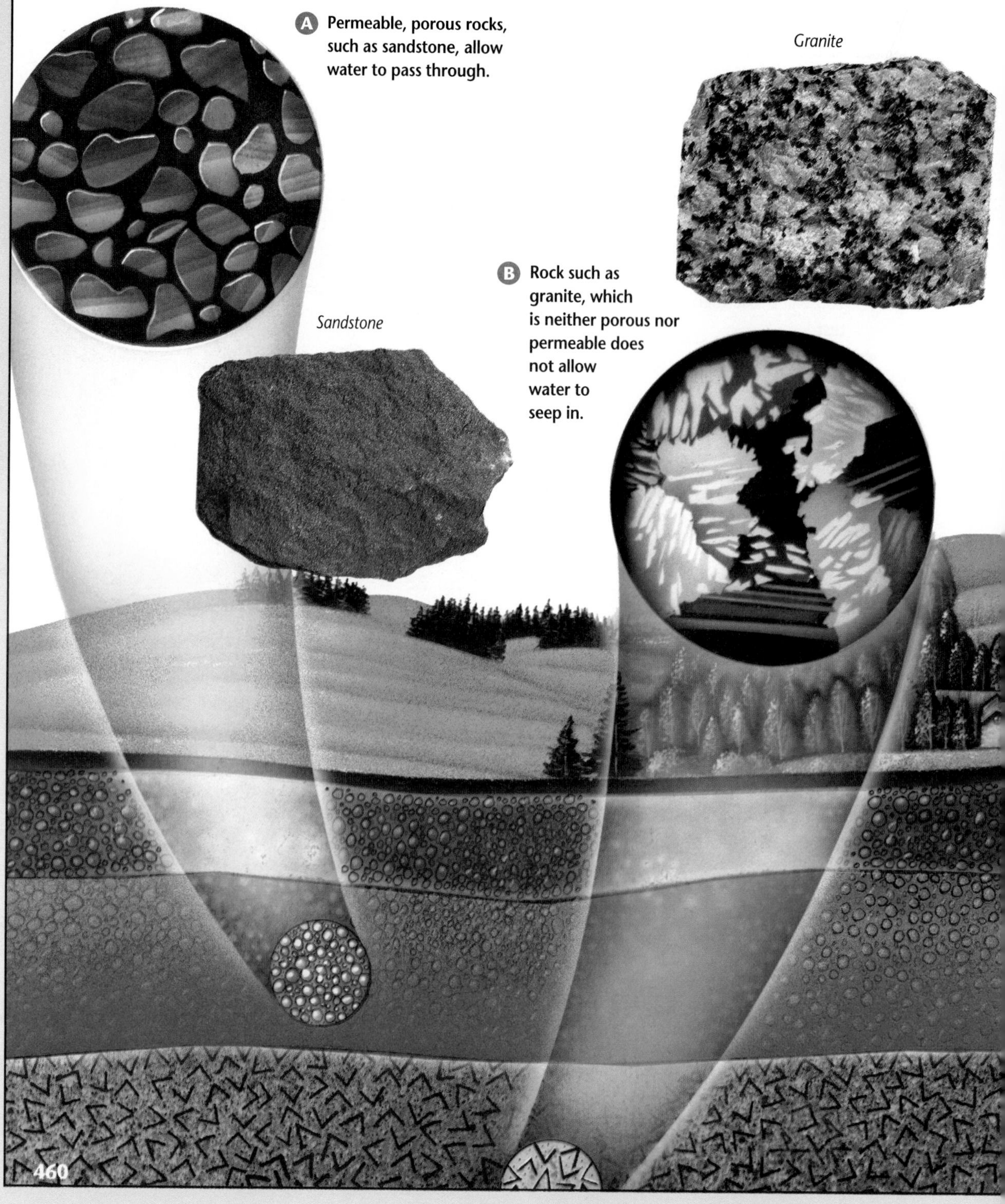

Granite

Sandstone

B Rock such as granite, which is neither porous nor permeable does not allow water to seep in.

460

■ Aquifers

How deep into the ground do you suppose groundwater can go? That depends on the permeability of the soil and rock. Groundwater will keep going down to lower levels until it reaches pores that are already filled with water. This water is resting on a layer of impermeable rock. When this happens, the impermeable rock acts like a dam, and the water can't move down any deeper. So the water begins to fill up the pores in the rocks above the impermeable rock. A layer of permeable soil or rock that allows water to move in and out freely is called an **aquifer**. Soils that contain sand or gravel and rocks like sandstones and limestones are often aquifers.

Why do you suppose aquifers are important to people? Aquifers are sources of water for many communities. In fact, if you do not live near a large river or a large freshwater lake, the chances are good that you get your water from an aquifer. Where does the water that you drink come from?

C When water moving through a permeable layer of rock reaches an impermeable layer, the water's downward motion is stopped. Once the pores to the permeable layer are filled, the water may move sideways through an aquifer—a permeable layer of soil or rock that allows water to move in and out freely. Soils that contain sand or gravel and permeable rocks are often aquifers.

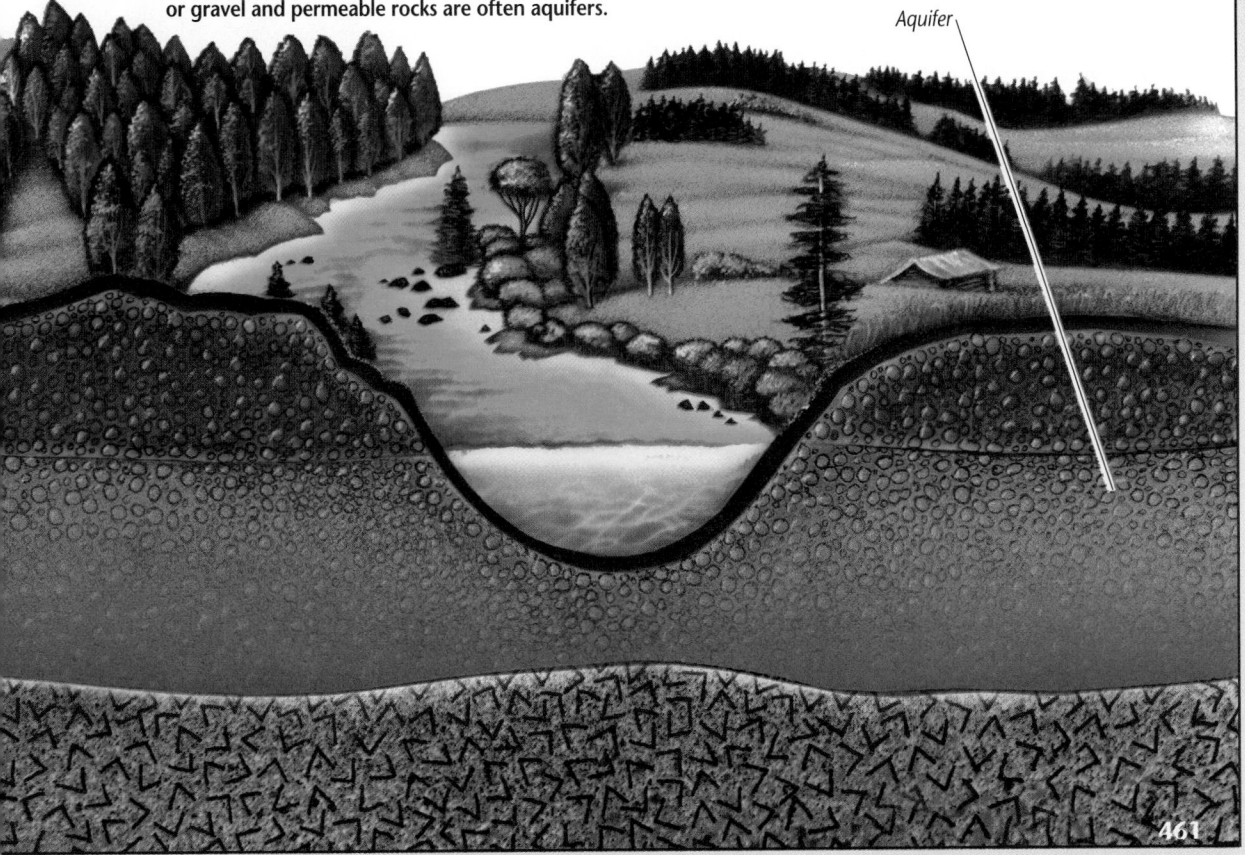

Aquifer

461

Wells and Springs

The water level in an aquifer may change from season to season. Recall the Explore activity you did at the beginning of this section. You discovered you could change the level of water in the aquifer you made. Knowing the level of water in the ground is important to many people because they get their drinking water from groundwater. Water wells are drilled down into the aquifers. Water from an aquifer flows into a well and then is pumped back up to the surface.

A well, as seen below, must go down past the water table to reach water. What is the water table? The **water table** is the top of the level where groundwater has collected in the ground. If the well is far enough below the water table, the well should provide a reliable source of cool drinking water in every season of the year.

During dry seasons, a well might dry up because the water table drops. The water table may also drop if too many wells are drilled in an area. In this instance, more water is taken out of the ground than can be replaced by rain. Unlike wells, most streams and

Figure 14-9

A When water from the aquifer flows into this well, it is pumped through pipes to the surface to provide drinking water. Why is knowing the depth of the water table during both normal and dry seasons important when digging a well?

B When water travels from a higher to a lower level and becomes trapped between two layers of impermeable rock, pressure builds up as the water attempts to reach the water table. If a well is drilled at that point, the pressure will push the water to the surface without pumping. Wells that flow without pumping are called artesian wells.

Aquifer

Water flows from an artesian well.

Drawing water from a well in Egypt.

462 Chapter 14 Moving Water

Figure 14-10

Wherever the water table meets Earth's surface, groundwater may flow out of the rock or soil as seen in the photo on the left. These places where the water table is exposed are called springs. The photo above is of Mammoth hot springs in Yellowstone National Park.

SKILLBUILDER

The added amount of water being drawn from the wells causes the water table to drop. Nature's remedy would be enough rain to replenish the water table. **L1**

rivers do not run dry in dry weather. One reason is that streams and rivers are usually lower in elevation than the surrounding land.

In some places, the water table meets Earth's surface. Groundwater simply flows out of the rock or soil at these places. Springs can be found on hillsides or any other place where the water table is exposed at the surface. Springs can often be used as a source of water.

You have taken a long journey through the hydrologic cycle. Rainwater can evaporate, run off to become rivers and streams, or seep into the ground to become groundwater. Rainwater can collect in aquifers and perhaps be pumped back up to the surface, where it once again moves through the cycle.

SKILLBUILDER

Recognizing Cause and Effect

Suppose you live in a town in which the population stays the same for many years. Then a number of new houses are built, and the population grows. The people in the town notice that the wells show signs of drying up. What could be the cause? What would be nature's remedy? If you need help, refer to the **Skill Handbook** on page 660.

check your UNDERSTANDING

1. How does rainwater enter the groundwater system?
2. How can rocks be both porous and permeable?
3. How can a well go dry? How can the well be made useful again?
4. **Apply** Explain how groundwater, aquifers, and the water table are related.

14-3 Groundwater in Action **463**

3 ASSESS

Check for Understanding

To help students answer the Apply question, lead them to see that groundwater, aquifers, and the water table are interdependent. Ask them how an aquifer is dependent on groundwater. Then ask how the water table is dependent on the aquifer.

Reteach

Discussion Use a sponge to help visual and tactile learners understand the porous nature of an aquifer. Point out the holes, or pores, in the sponge that contain water. Explain that the fibers of the sponge are not so much absorbing as trapping water. **LEP** **L1**

Extension

Activity Challenge students to formulate models. Have students who have mastered this section do research on how groundwater in limestone can create caves. **L3**

4 CLOSE

Have students describe the water from a well that has been drilled through layers of iron-laden soil and limestone. You may need to remind students that water dissolves minerals and other materials as it travels through layers of permeable soil. **L1**

check your UNDERSTANDING

1. Rainwater soaks through permeable soil into the groundwater system.

2. Some rocks have connecting pores through which water can travel.

3. A well can go dry when the water table drops too far. More rainwater is required to

make the well useful again. Also, the well could be drilled deeper.

4. Groundwater is water that collects in the ground. **NOTE:** An aquifer is a groundwater reservoir. A water table is the upper boundary of the groundwater level.

Science
and
Society

Science
and
Society

Purpose Science and Society develops Sections 14-2 and 14-3 by examining how water sources are an integral part of human life.

Content Background

Water disputes are not new. Many of the range wars between the sheep herders and cattle ranchers in the old West were over water rights. The problem is not so much the amount of water available as the location. In fact, experts estimate that the average runoff of annual rainfall in the United States is about 8.7 inches. The average annual amount needed to supply all residential, agricultural, and industrial needs is about 1.5 inches. To solve the current problem, two approaches will have to be considered. One will be a political evaluation of fair distribution. But that is less important in the long run than conservation. Several ideas being considered are reducing the amount of evaporation and flooding in existing reservoirs, setting up a universal metering system, and devising more efficient systems of crop irrigation.

Teaching Strategies

Challenge students to think critically. Divide students into four groups. Assign two groups to study urban water needs and two to study agricultural needs. Set up debate teams based on their research. COOP LEARN

Discussion

Have students work together to solve a problem. Tell them that their average water allotment has been reduced by ten gallons a day. In what ways will they decrease their water consumption? Remind students that all basic needs must be taken care of. COOP LEARN L1

Water Wars

Water is an essential part of our everyday lives. In a single day, the average person in the United States uses 397 liters of water—that's enough liquid to fill up 1118 soft drink cans. We use water every time we take a shower, brush our teeth, or wash our clothes. We also use a lot of water indirectly. Many industries rely upon water to manufacture products such as paper and plastic. Farmers need water to irrigate the crops that produce the fruit and vegetables that we eat.

Where does all of this water come from? Some towns and cities get their water from nearby rivers, lakes, or underground wells. However, not everyone lives next to a source of fresh water, especially

Sacramento River
• Sacramento

San Francisco
San Joaquin River

California

Santa Barbara
San Bernardino
Los Angeles Palm Springs
• San Diego

in the desert regions of our country. In many parts of the United States, communities are forced to get their water from other locations. Dams and pipelines are constructed to carry water from rivers that might be hundreds of kilometers away. Changing the natural flow of water in such a way is called water diversion.

There is, however, a problem with water diversion. When you take water away from a distant river, you leave less water for people who are living near the river. In some parts of the country, individuals, towns, and even states have gone to court to fight over water rights.

This irrigation system in Southern California is bringing water to a crop of beets.

For many years, California has been involved in a water dispute. Much of California's fresh water is supplied by the Sacramento River in the northern part of the state. Southern California, on the other hand, provides less than 20 percent of the state's fresh water.

Despite the fact that Southern California produces only a small amount of water, it consumes 85 percent of the water available in the entire state. Most of this water is used by farmers to irrigate their crops. Meanwhile, city dwellers in Northern California are facing severe water shortages. They want the government to pass laws that will restrict the amount of water used by the farmers.

In some parts of the world, water wars are on the verge of becoming full-scale wars. In the late 1980s, Turkey built a massive dam across the Euphrates River. The water in the dam's reservoir is used to irrigate crops and to generate electricity for almost half of the country. But Syria—Turkey's neighbor to the south—also depends upon water from the Euphrates. Syria has argued that the dam is

The availability of water is often taken for granted. Think about the ways in which you use water on a typical day.

stealing water away from its farmers. The dam has created tremendous tension between the two countries.

What Do You Think?

Some people say that if cities made a concerted effort not to waste so much water, there would be plenty of water available for everyone. Do you agree? Can you think of a few simple things that you can do around your own home that will help conserve our supply of water?

Beets

Activity

Have students write individual letters to either local or state governments explaining the importance of cooperation in water distribution. For example, the state department of agriculture would be a good place to start. Have students suggest ways to settle arguments about which areas or groups should have special water rights. Students could also write letters of inquiry about local or state policy on water distribution. L2

Answers to
What Do You Think?

Answers will vary. Some students will want to avoid taking baths. More practical suggestions may be to avoid letting water run while washing dishes or brushing their teeth.

Going Further ⟩⟩⟩⟩⟩⟩

Have students discuss, communicate, and analyze. Have them work in pairs to explore the question of water rights. Have students decide if water can really be owned. Students should consider whether one group of people has more right to a water source than another group. They should keep in mind that water is a basic need of all life on Earth. Have students present the fruits of their discussions in a newspaper article. Use the Classroom Assessment List for Newspaper Article in **PASC,** p. 69.
COOP LEARN L2

economics connection

Purpose

The Economics Connection should help students better realize how floods develop and their devastating economic effects on an area. Students will compare and contrast the economic effects of flooding and possible flood control for the upper Mississippi states.

Content Background

Flood control is most important in areas where heavy or extended rainfall is an annual occurrence. The amount of normal rainfall before a major storm can increase the possibility of flooding. Soil can absorb only so much water before surface runoff begins. Soil saturation can be caused by one-half inch of rain during a 24-hour period. If a major storm happens within ten days, nearby rivers and streams are likely to flood. The flooding of the upper Mississippi area was unusual as it was due to an enormous amount of rain rather than a combination of spring rain and melting snow. Also vegetation was present in low-lying agricultural areas at the time of the flooding.

economics connection

Planning for the Next Floodwaters

A break in a levee at St. Louis

In July of 1993, the mighty Mississippi River flexed its muscles in the worst flood in North American history. The ordinarily lazy river became a raging monster, rising almost 150 feet over its banks and bulging to seventeen miles across in some places. It was called a "500-year flood," in the sense that a flood this bad occurs only about once every 500 years.

Water ripped apart buildings, bridges, roadways, and levees as if they were children's toys. Thousands of people were left without homes, and the cleanup will continue for years.

Only the most resistant vegetation survived the long-standing waters. Populations of mosquitoes and mayflies mushroomed. Wildlife—frogs, snakes, bobtail quail, pheasant, deer—were all displaced from their natural habitats.

Some of these effects last only a few years. Others will persist into the next decade. What makes a river flood? With all the technology we have at our fingertips, why can't we prevent disastrous floods?

How Floods Develop

Floods start with rain. Rainwater either soaks into the ground or runs off along the surface until it reaches a stream, river, or lake. This surface runoff is the main concern during a flood. The runoff floods the network of creeks, streams, and rivers on its way to the main river they feed into.

The Natural River

The land surrounding the Mississippi is wide and flat. For thousands of years, each time the river rose it spilled onto these wide flood plains. Tons of sediment left behind by receding water created the deep, rich soil that's coveted by farmers. Plentiful forests and wetlands absorbed excess rainwater like sponges, preventing smaller floods entirely and moderating the larger ones.

The Developed River

Flood walls and levees (tall, wide, protective walls often made of compacted soil) narrow the river channel and keep it from expanding onto its floodplains. Because it can't spread out, floodwater can only rise. Upstream the river backs up like a clogged drain, while downstream the waters move faster and push even harder on the levees.

Most of the original wetlands and forests have been developed by humans for farmland or urban centers. Towns and cities channel excess runoff to the river rather than absorbing it. And as time passes, disasters are costing more partly because there are more people in the way each time. With fast-growing river towns to consider, major floods seem more likely, not less.

Going Further ⦚⦚⦚⦚➡

Have students work in small groups to discuss the importance of flood control. Ask students to make a list of the damages that can be caused by a flood. They may want to read newspaper and magazine reports of major flooding such as the Mississippi River floods of July 1993. Then have students use their lists to develop a public service announcement explaining how individuals can prepare for floods. Use the Classroom Assessment List for Oral Presentation in **PASC**, p. 71. **L2** **COOP LEARN**

Preventing Major Floods

Ideally, many of the levees should be moved back from the water's edge back about five miles along the whole length of the Upper Mississippi. That would work, but it would sacrifice valuable farmland and many towns and cities. Carrying out such a plan would carry economic costs perhaps even higher than those of the flood itself. Flood-plain management will have to use less sweeping measures.

One possible solution starts with a coordinated plan for the whole Upper Mississippi. Right now the region is an unorganized patchwork of private and local reservoirs, flood walls, levees, and locks. Each community has made its own decisions.

In comparison, the Lower Mississippi is protected by a unified flood-control system that hasn't failed since 1927 and didn't fail in 1993. But that system has storage and pass-through basins that add up to equal the area of the state of Indiana.

Can the Upper Mississippi states agree to put aside that much land? It would have to be surrounded by flood walls on the landward side. Some economists think part of the put-aside land could be farmed on a contingency basis. That is, as long as it isn't needed for overflow storage, it could be used for agriculture, but not for residences. One farmer working such land might be flooded out once in ten years; another once every other year. Crop insurance would have to make up for these losses.

Sandbags are piled up to keep the Mississippi River back.

What Do You Think?

Only one thing is sure: no one who has had to shovel Mississippi mud out of his or her home wants to go through the flood of 1993 again. Take on the role of a farmer or someone who lives in the city. What would you decide to do to control floods? How would your decision affect others?

The narrow channel caused by the levees forces the river to back up, which pushes water upriver.

The river rises high and fast between the levees, which can then affect land areas downriver.

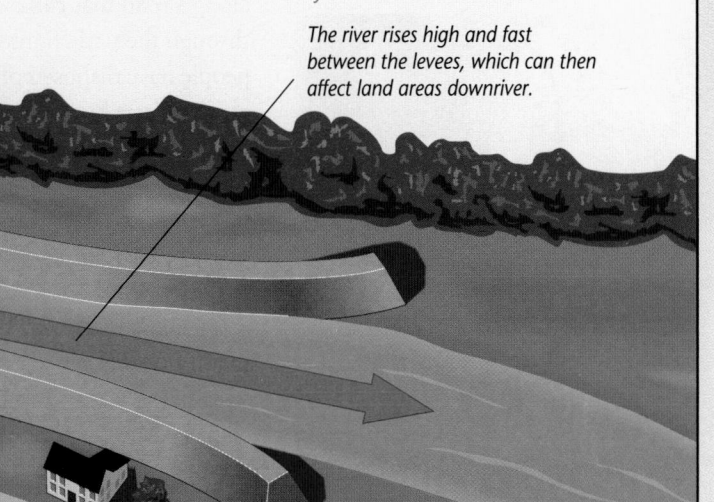

Teaching Strategies After reading the selection, encourage students to have a class discussion on where flooding occurs in your area, when and why it occurs, and what method is used to control it. Discuss the economic effects of both flooding and flood control.

Have students work in groups to create a survey that could be handed out to local residents seeking their opinions on the economic and environmental results of a variety of flood control measures. **COOP LEARN** **L2**

Activity

Guide students to find and organize information on different flood control measures such as flood walls, levees, dams, locks, and storage and pass-through basins. You may want to have students organize their information in a display that shows the measures, their effects, and economic consequences. Use the Performance Task Assessment List for Display in **PASC**, p. 63. **L2** **COOP LEARN**

Answers to
What Do You Think

Students' answers will vary. Be sure that students realize that although the farmers and city dwellers might have different immediate reactions, there will be economic effects on both groups.

Teens in SCIENCE

Purpose
Teens in Science should help students better understand cause and effect by describing how pollutants poured into just one part of a runoff system can affect rivers and streams.

Content Background
Water pollution is caused by more serious things than small amounts of hazardous wastes dumped from individual households. A little over 9 percent of river pollution comes from industrial toxic wastes. But about 65 percent of water pollution comes from agriculture. These pollutants are more difficult to detect and control. Chemicals, mostly pesticides, are sprayed daily over hundreds of thousands of acres of farmland. Salts, nitrate fertilizers, sediments and other materials wash into the runoff system and are absorbed by the soil. Pollutants then filter into the groundwater system and can damage fresh water in aquifers, wells, and lakes. Because groundwater is the source for over fifty percent of the United States' drinking water, pollution is a major concern.

Teaching Strategies Help students relate hazardous household wastes to water pollution. Have student volunteers read aloud the ingredients lists from several household cleaners, motor oil, and garden herbicides and pesticides. Then put a few drops of each into a glass of water. Ask students if they would want to drink that water. Then have other student volunteers read any directions on the products regarding their disposal. Encourage students to tell their parents how to dispose of hazardous household waste.

Teens in SCIENCE

The Clean Stream Team

How would you describe the city or town where you live? For 14-year-old JoAnna Gott, the answer is simple. "My town is beautiful. And that's the way we want to keep it."

JoAnna lives in Strafford, Missouri. The many nearby

lakes, rivers, and streams are a large part of why people love to live in Strafford. So, it was only natural for JoAnna to get involved in a recent Urban Streams Festival held in her town.

"The festival was created to remind people to take care of our water. Right now we're lucky. Our water is clean. And that is a good reason to celebrate."

During the festival, residents of Strafford were encouraged to do more than think about water. They were urged to get involved in protecting this important resource. JoAnna's 4-H club accepted the challenge.

"We went to a park in town and painted a warning on all the storm drains. The storm drains are located along a road that runs through the park. Sometimes people have disposed of hazardous waste by pouring it down the storm drains."

"People need to know what happens if they are careless. A person might think that a little old paint or household cleaner can't make that much difference in a big stream. But if the person actually thought about that poison running into their

favorite stream, I don't think they would do it. Maybe this sounds silly, but I wish more people would think about how they would like it if they were a fish or some other animal that lives in the stream. I'd feel bad if someone dumped paint thinner down our chimney at home."

Science Journal
JoAnna loves her town. How about you? *In your Science Journal* make a list of the best and worst things about the city or town you live in.

Using your "best and worst" list as a guide, write a brief description of one thing that you could do to make your city or town a better place to live.

Science Journal
Guide students to correct misconceptions. Students might think that throwing out poisonous wastes is better than pouring them down the drain. Ask students how seepage from landfills might cause water contamination. Lead them to discuss the types of trash that could cause water pollution. Some students might want to find out about local landfills and state standards for landfill safety.

Going Further ⑂
Have students work in small groups to promote awareness of water pollution. Each group should identify a special problem around the school or community. Then encourage each group to make a poster illustrating the problem. Place posters around the school or in public buildings. Use the Performance Task Assessment List for Poster in PASC, p. 73. **L1** **COOP LEARN**

Science Journal

Review the statements below about the big ideas presented in this chapter, and answer the questions. Then, re-read your answers to the Did You Ever Wonder questions at the beginning of the chapter. *In your Science Journal,* write a paragraph about how your understanding of the big ideas in the chapter has changed.

1 Water moves through a cycle called the hydrologic cycle. *What processes are involved in the cycle?*

2 Rivers and streams drain excess water from the land and thus form an important part of the hydrologic cycle. *Compare how rainwater runs off a steep slope and a flat area.*

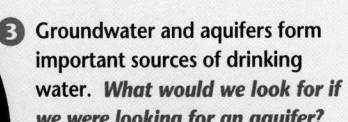

3 Groundwater and aquifers form important sources of drinking water. *What would we look for if we were looking for an aquifer?*

Have students look at the diagram on this page. Direct them to read the three captions to review the main ideas of this chapter.

Teaching Strategies

Divide the class into three groups. Have the first group make a poster showing each phase of the hydrologic cycle. They should look in old magazines for pictures, and include their own diagrams and drawings, especially to represent condensation and evaporation. Be sure each major section of the poster is labeled.

Have the next group make a poster or bulletin board illustrating flat areas that allow rainwater to evaporate and steep slopes that allow runoff. It should be divided into two sections so the contrast in land areas is clear.

Have the third group research the aquifer under the Great Plains region. Students should find examples illustrating how the aquifer benefits the people of this region. **COOP LEARN**

Answers to Questions

1. evaporation, condensation, precipitation

2. Rainwater runs rapidly off a steep slope. Very little sinks into the ground. On flat areas water moves slowly and has more chance to sink into the ground.

3. A layer of porous permeable rock or soil over a layer of impermeable material.

Science Journal

Did you ever wonder...
- It will either evaporate again, flow along the ground, or soak into the ground. (p. 445)
- Water from runoff flows along the ground and eventually drains into streams and rivers. (p. 454)
- Well water comes from groundwater trapped in layers of soil and rock. (p. 462)

Project

Have students discover what their major source of water is by contacting the local water company. Then students should either build a model or make a poster showing how the hydrologic cycle and the water company contribute to the delivery of water to homes. They can be displayed in the hall showcase. Use the Performance Task Assessment List for Model or Poster in **PASC,** pp. 51 and 73. **COOP LEARN** **L1**

GLENCOE TECHNOLOGY

 MindJogger Videoquiz

Chapter 14 Have students work in groups as they play the Videoquiz game to review key chapter concepts.

Using Key Science Terms

1. Runoff is a part of the hydrologic cycle.

2. An aquifer is a reservoir of groundwater.

3. The area into which runoff flows is called a drainage basin.

4. The water table is the top of the groundwater level.

5. A river or stream meanders when it reaches the flat area of a drainage basin.

Understanding Ideas

1. The type of land surface, slope of the land, vegetation, and the amount of rainfall determine the amount of precipitation that becomes runoff.

2. Meanders form when a river begins to erode its sides and curves as it moves across its floodplain.

3. Streams that flow through steeply sloped areas run swiftly downhill. The running water wears away the stream bottom more than its sides creating a narrow, steep valley. Streams that flow across land that slopes very little flows more slowly. The water wears away the sides of the streambed, developing curves and bends.

4. Permeable materials have many connecting pores. Water can pass through these materials easily. Impermeable materials have very small pores or no pores at all. Water can't pass through these materials easily.

5. Rainwater soaks into the soil and collects in the pores. It then keeps going to lower elevations until it reaches a layer of impermeable rock. It can't move down any deeper so it begins filling up the pores in the rocks above the impermeable layer.

Developing Skills

1. Answers will vary based on the permeability of the soil type.

2. See student page for concept map.

U sing Key Science Terms

aquifer	meander
drainage basin	runoff
groundwater	water table
hydrologic cycle	

For each set of terms below, explain the relationship that exists.

1. hydrologic cycle, runoff
2. groundwater, aquifer
3. runoff, drainage basin
4. water table, groundwater
5. meander, drainage basin

U nderstanding Ideas

Answer the following questions in your Journal *using complete sentences.*

1. What factors determine the amount of precipitation that becomes runoff?
2. Explain how meanders form.
3. What factors affect a river's ability to erode its bed?
4. Explain the difference between permeable and impermeable material.
5. Describe the movement of groundwater.

D eveloping Skills

Use your understanding of the concepts developed in this chapter to answer each of the following questions.

1. Separating and Controlling Variables You will need a stream table, watering can, sand, clay, humus soil, 3 liters of water, and a block of wood. Repeat the

3. The runoff time will decrease with increased slope for each soil type.

Explore activity on page 449 three times, using a different type of soil in the stream table each time. How does the rate of runoff vary with different soil types?

2. Concept Mapping Complete the concept map of precipitation.

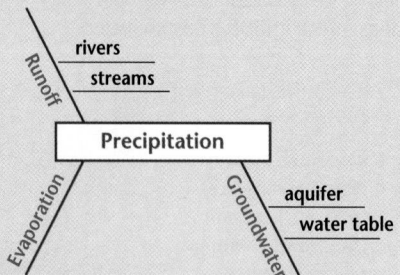

3. Making and Using Graphs Make a graph like the one below. You will need a stream table, a watering can, 9 liters of water, sand, clay, humus soil, a stopwatch, and 3 blocks of wood. Repeat the activity above creating three different slopes—one 15 degrees, one 30 degrees, and one 60 degrees—for each type of soil. Time how long it takes for most of the water to run off. Record your results on the graph. How does slope affect the runoff time for the different types of soil?

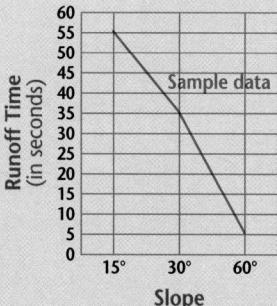

Critical Thinking

In your Journal, *answer each of the following questions.*

1. How can the level of the water table vary even when the groundwater is not disturbed by people?

2. Study the illustration. If a spring thaw suddenly melts a great deal of snow in the region labeled A, what do you think will happen to the river in the region that is labeled B? Explain your answer.

3. Compare a groundwater system with a drainage basin system.

4. What are the similarities and differences between a drilled well and a spring?

Problem Solving

Read the following problem and discuss your answers in a brief paragraph.

One way to test the permeability of different kinds of soil materials is to see how much time it takes water to flow through them. Suppose you had some water, a watch, some funnels lined with filter paper, some beakers, a graduated cylinder, and equal amounts of potting soil, marbles, and clay.

1. How could you set up an experiment to determine the permeability of each material?

2. Predict which material would be the most permeable. Least permeable?

CONNECTING IDEAS

Discuss each of the following in a brief paragraph.

1. **Theme—Systems and Interactions** What factors affect whether water will run off, evaporate, or soak into the ground at a certain location?

2. **Theme—Systems and Interactions** Earth has often been called a water

planet. Based on what you've learned in this chapter, would you agree? Explain.

3. **Theme—Systems and Interactions** What is the relationship between the largest drainage basin in the United States and the kind of landform that is dominant in that

region?

4. **A Closer Look** What role does groundwater play in the formation and eruption of a geyser?

5. **Economics Connection** How might the lessons learned from the Mississippi flood affect future development of wetlands?

Assessment

Portfolio Review the portfolio options that are provided throughout the chapter. Encourage students to select one product that demonstrates their best work for the chapter. Have students explain what they learned and why they chose this example for placement into their portfolios.

Additional portfolio options can be found in the following **Teacher Classroom Resources:**

Making Connections: Integrating Sciences, p. 31

Multicultural Connections, pp. 31–32

Making Connections: Across the Curriculum, p. 31

Concept Mapping, p. 22

Critical Thinking/Problem Solving, p. 22

Take Home Activities, p. 23

Laboratory Manual, pp. 81–90

Performance Assessment P

Critical Thinking

1. The water table will vary depending on the amount of rainfall in the area.

2. The river will increase in volume and speed and might flood.

3. A groundwater system flows underground and is replenished by absorption of water by the ground. A drainage basin system exists aboveground and is fed by the runoff of streams and rivers.

4. Both a drilled well and a spring are fed by groundwater. However, a well collects water that must be brought to the surface, while spring water naturally flows out of the ground.

Problem Solving

1. Arrange each material to the same depth in the funnels. Pour equal amounts of water on each material. Measure the time it takes for water to flow through the filter for each material.

2. The marbles would be the most permeable. The clay would be the least permeable. This difference is primarily due to the size of the particle.

Connecting Ideas

1. temperature, the slope, porosity and permeability of the land, the amount of rainfall, vegetation

2. Yes; most of Earth's surface is covered with water.

3. The Mississippi drainage system consists of a large gently sloping valley. This creates a drainage system made up of many slowly flowing and meandering rivers and streams.

4. Geysers are fed by groundwater that is heated underground.

5. Answers will vary but may include: people will plan flood-control measures more carefully; people might not build so close to the river that is feeding the wetlands. Check students' reasoning.

Chapter Organizer

SECTION	OBJECTIVES	ACTIVITIES & FEATURES
Chapter Opener		**Explore!,** p. 473
15-1 Gravity (1 session; .5 block)	**1. Distinguish** between erosion and deposition. **2. Identify** creep and slump as erosion caused by gravity. **3. Describe** rockslides and mudflows. **National Content Standards: (5–8) UCP2–3, B2, F2–3, F5, G1–2**	**A Closer Look,** pp. 476–477 **Science and Society,** pp. 498–499
15-2 Running Water (3 sessions; 1.5 blocks)	**1. Explain** how streams carry sediment. **2. Explore** the relationship between amount of sediment and rate of stream flow. **3. Explain** how streams and rivers shape the land. **National Content Standards: (5–8) UCP2–3, A1, B2**	**Explore!,** p. 479 **Skillbuilder:** p. 481 **Design Your Own Investigation 15-1:** pp. 482–483 **Physics Connection,** pp. 494–495
15-3 Glaciers (2 sessions; 1.5 blocks)	**1. Describe** how a glacier is formed. **2. Differentiate** between the two major types of glaciers. **3. Describe** how glaciers erode the land. **National Content Standards: (5–8) UCP2–3, A1, B2**	**Investigate 15-2:** pp. 488–489 **Find Out!,** p. 491 **History Connection,** p. 500 **SciFacts,** p. 497
15-4 Wind Erosion (2 sessions; 1 block)	**1. Describe** how wind erodes and deposits sediment. **2. Describe** how a dune is formed and how it moves. **3. Identify** two factors that can decrease wind erosion. **National Content Standards: (5–8) UCP2–3, B2, D2, E1–2, F2–5, G3**	**Explore!,** p. 493 **Find Out!,** p. 495

ACTIVITY MATERIALS

EXPLORE!

p. 473 waxed paper, sand, gravel, safety goggles

p. 479 instant coffee grains, salt, rice, paper plates, spray bottles, water

p. 493 sheets of aluminum foil

INVESTIGATE!

pp. 488–489* ice block of sand, clay, and gravel; stream table; sand; lamp with reflector; ruler; pail

DESIGN YOUR OWN INVESTIGATION

pp. 482–483* stream table, sand, small pebbles, soil, pails, water, plastic hose, block of wood, screw clamps

FIND OUT!

p. 491* sand, soil, rocks, or gravel; water; ice cube trays; small piece of wood

p. 495* covered shoe box, knife or scissors, flour, spoon, towel

KEY TO TEACHING STRATEGIES

The following designations will help you decide which activities are appropriate for your students.

- **L1** Basic activities for all students
- **L2** Activities for average to above-average students
- **L3** Challenging activities for above-average students
- **LEP** Limited English Proficiency activities
- **COOP LEARN** Cooperative Learning activities for small group work
- **P** Student products that can be placed into a best-work portfolio
- 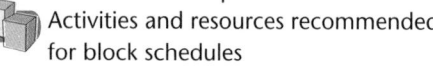 Activities and resources recommended for block schedules

Need Materials? Call Science Kit (1-800-828-7777).

⏱ **OUT OF TIME?** We recommend that students do the activities with an asterisk.

Chapter 15 Shaping the Land

TEACHER CLASSROOM RESOURCES

Student Masters	Transparencies
Study Guide, p. 54 **Critical Thinking/Problem Solving**, p. 23 **Multicultural Connections**, pp. 33, 34 **Take Home Activities**, p. 24 **Making Connections: Integrating Sciences**, p. 33 **Making Connections: Technology and Society**, p. 33 **Science Discovery Activities**, 15-1	**Section Focus Transparency 48**
Study Guide, p. 55 **Making Connections: Across the Curriculum**, p. 33 **Activity Masters**, Design Your Own Investigation 15-1, pp. 63–64 **How It Works**, p. 18 **Science Discovery Activities**, 15-2 **Laboratory Manual**, pp. 91–92, Rivers **Laboratory Manual**, pp. 93–94, Transporting Soil Materials by Runoff	**Section Focus Transparency 49**
Study Guide, p. 56 **Activity Masters**, Investigate 15-2, pp. 65–66 **Laboratory Manual**, pp. 95–98, Glaciation and Sea Level	**Teaching Transparency 29**, Glacial Erosional Features **Teaching Transparency 30**, Glacial Depositional Features **Section Focus Transparency 50**
Study Guide, p. 57 **Concept Mapping**, p. 23 **Critical Thinking/Problem Solving**, p. 5 **Science Discovery Activities**, 15-3	**Section Focus Transparency 51**

ASSESSMENT RESOURCES	TEACHING & TECHNOLOGY
Review and Assessment, pp. 89–94 **Performance Assessment**, Ch. 15 **PASC*** **MindJogger Videoquiz** **Alternate Assessment In the Science Classroom** **Computer Test Bank**	**Spanish Resources** **Cooperative Learning Resource Guide** **Lab and Safety Skills** **Science Interactions, Course 1, CD-ROM** **Computer Competency Activities**

*Performance Assessment in the Science Classroom

NATIONAL GEOGRAPHIC TEACHER'S CORNER

Index to National Geographic Magazine

The following articles may be used for research relating to this chapter:

- "Hawk High Over Four Corners," by T.H. Watkins, September 1996.
- *Water: The Power, Promise, and Turmoil of North America's Fresh Water*, A Special Edition, November 1993.
- "Grand Canyon: Are We Loving It to Death?" by W.E. Garrett, July 1978.

National Geographic Society Products Available From Glencoe

To order the following products for use with this chapter, contact your local Glencoe sales representative or call Glencoe at 1-800-334-7344:

- *NGS Picture Show: Geology*, "Weathering and Erosion." (CD-ROM)

Additional National Geographic Society Products

To order the following products for use with this chapter, call the National Geographic Society at 1-800-368-2728:

- *Orbit* (Book)
- *Glaciers: Ice on the Move* (Video)

Teacher Classroom Resources

These are key components of the classroom resources package.

TEACHING AIDS

Teaching Transparencies

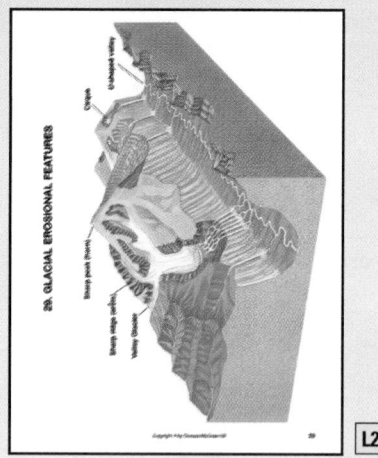

L2

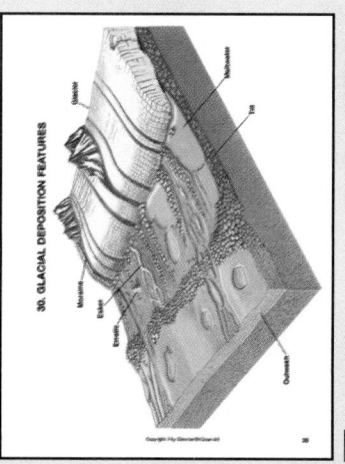

L2

Section Focus Transparencies

48 SECTION FOCUS TRANSPARENCY

LANDSLIDE CAUSES FLOOD!

L1

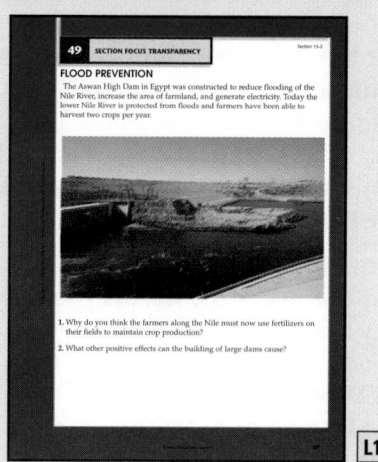

49 SECTION FOCUS TRANSPARENCY

FLOOD PREVENTION

L1

50 SECTION FOCUS TRANSPARENCY

FIRE AND ICE!

L1

51 SECTION FOCUS TRANSPARENCY

GONE WITH THE WIND

L1

HANDS-ON LEARNING

Science Discovery Activity*

Power Moves

L1

Laboratory Manual*

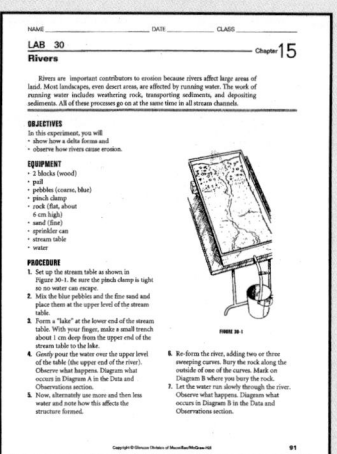

LAB 30
Rivers

L1

Take Home Activity

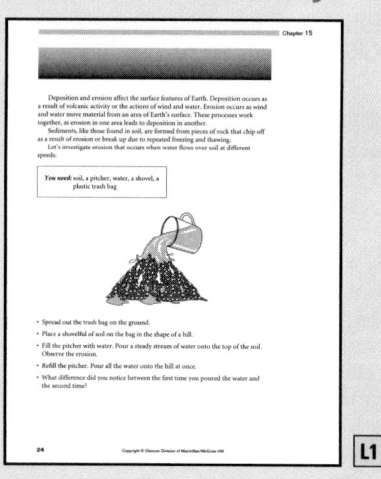

L1

Chapter 15 Shaping the Land

REVIEW AND REINFORCEMENT

Study Guide*

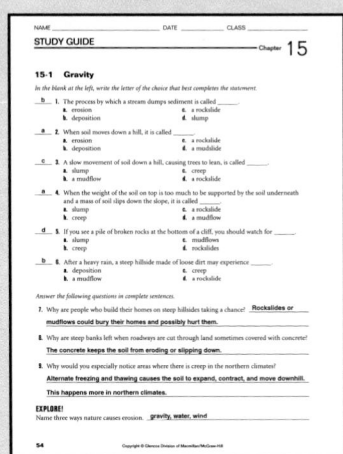

Concept Mapping

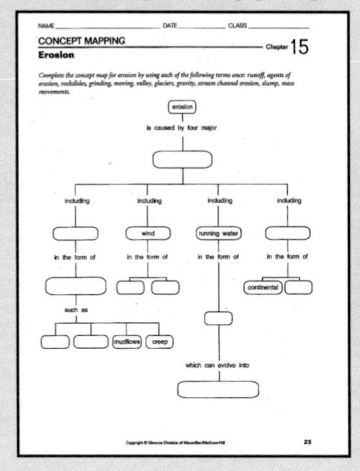

Critical Thinking/ Problem Solving

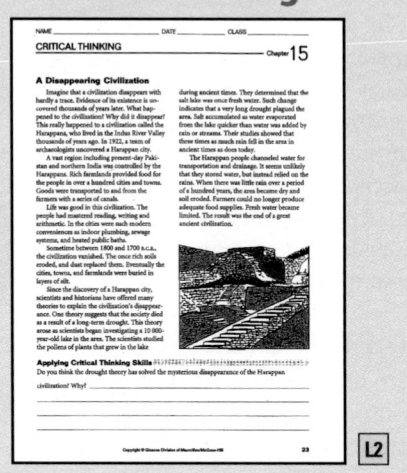

ENRICHMENT AND APPLICATION

Integrating Sciences

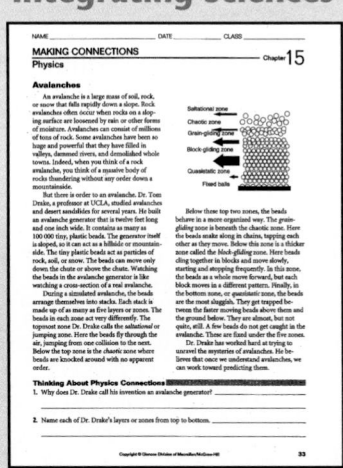

Across the Curriculum

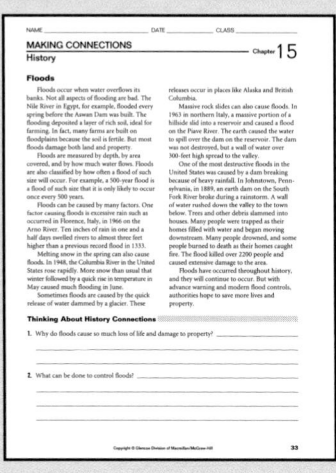

Technology and Society

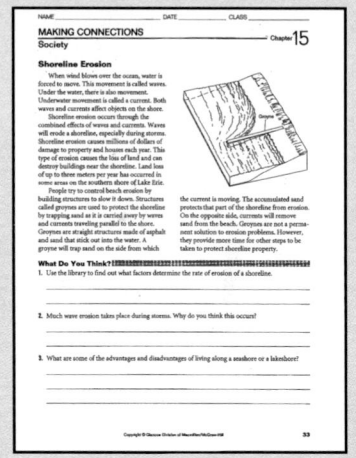

ASSESSMENT

Multicultural Connection**

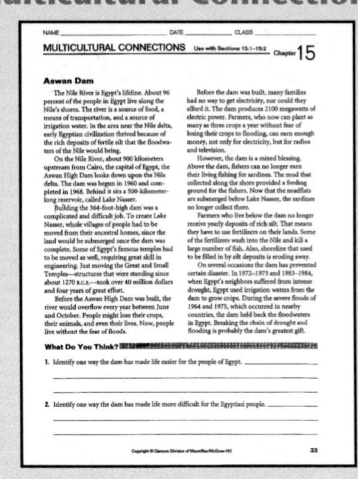

Performance Assessment

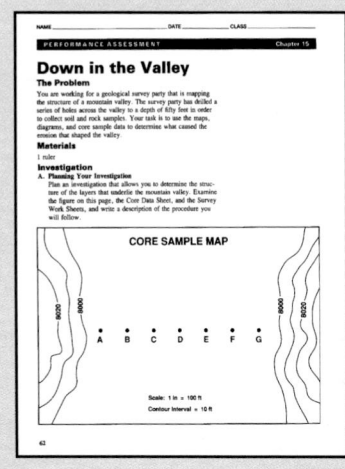

Review and Assessment

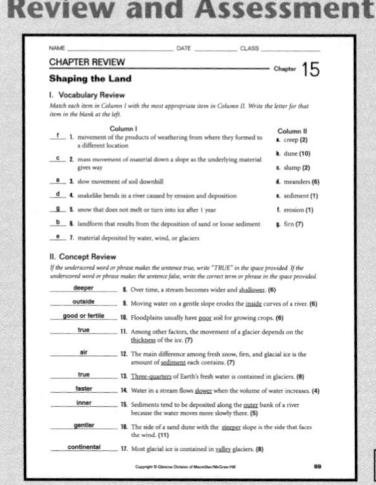

Shaping the Land

THEME DEVELOPMENT

The major theme of this chapter is stability and change. In each section, a different agent of erosion is described, along with the changes it makes in the landscape—erosion or deposition. Another theme is energy. The agents of erosion are moving, so they have energy.

CHAPTER OVERVIEW

Students will learn about two continuous Earth processes—erosion and deposition. Students also will explore the effects of gravity, running water, glaciers, and wind on the land.

Tying to Previous Knowledge

Remind students that in the last chapter they learned about water in motion. In this chapter, they will study the effect of running water on the land.

INTRODUCING THE CHAPTER

Ask students if they have ever observed a river undergo changes in depth, width, or color. Ask if they know the source of a nearby river.

Uncovering Preconceptions

Students may think that erosion is a process that only takes place over long periods of time. The Explore activity on page 473 will demonstrate that erosion also can occur rapidly.

00:00 OUT OF TIME?

If time does not permit teaching the entire chapter, use the Chapter Overview on this page, Reviewing Main Ideas at the end of the chapter, and the Chapter 15 audiocassette to point out the main ideas of the chapter.

SHAPING THE LAND

Did you ever wonder...

✓ Why a river could be crystal clear at one time and murky brown at another?

✓ What causes rockslides?

✓ Where the piles of rocks along riverbanks come from?

Science Journal

Before you begin to study about what shapes the land, think about these questions and answer them *in your Science Journal*. When you finish the chapter, compare your journal entry with what you have learned.

Answers are on page 501.

472

*O*nly two days ago, the river near Toshiko's house sparkled crystal clear. That was before the storms came and all the rain fell. Now the rock bridge she and her friends use to cross the river is covered with murky, chocolate-colored waters. Toshiko also notices that some of the river's bank has been washed away as well.

What caused the river to darken? What happened to the rock bridge? Will the riverbank look the same after the water retreats?

▶ *In this chapter, you will learn how forces in nature such as water and wind not only can change but can actually carry away parts of Earth.*

Learning Styles	**Kinesthetic**	Explore, pp. 473, 479; Activity, pp. 480, 486, 492; Design Your Own Investigation, pp. 482–483; Investigate, pp. 488–489; Find Out, pp. 491, 495
	Visual-Spatial	Activity, pp. 474, 478; Visual Learning, pp. 477, 480, 484, 487, 496; A Closer Look, pp. 476–477; Across the Curriculum, p. 487; Explore, p. 493
	Interpersonal	Discussion, p. 475; Activity, p. 490
	Intrapersonal	Science at Home, p. 501
	Logical-Mathematical	Across the Curriculum, pp. 481, 491
LS	**Linguistic**	Across the Curriculum, pp. 475, 481; Multicultural Perspectives, pp. 479, 486; Discussion, pp. 484, 490, 496; Physics Connection, pp. 494–495

Explore! ACTIVITY

How do sediments move from one location to another?

How many ways can you think of to move a pile of sediments, such as sand and gravel? Does nature move sediments in similar ways?

What To Do

1. Put a piece of waxed paper on a desk or table and place a small pile of sand and gravel on the paper.

2. Devise ways of moving the sand and gravel from one place to another without touching it. Record *in your Journal* how many different ways you can move the mixture.

3. Which type of sediment moved most easily? Did the type of sediment most easily moved change with the way you moved the sediments?

4. How do you think nature might use the same ways to move these sediments?

PREPARATION

Planning the Lesson

Refer to the Chapter Organizer pages 472A–D.

Concepts Developed

In this section, students will learn to distinguish between erosion and deposition and understand how gravity contributes to both processes. They will learn about the role of gravity and other conditions that lead to slump and creep. Students also will explore the nature of rockslides and mudflows.

1 MOTIVATE

Bellringer

✒ Before presenting the lesson, display **Section Focus Transparency 48** on an overhead projector. Assign the accompanying **Focus Activity** worksheet. L1
LEP ▢▢

Activity To sharpen observational skills, take students on a walk around the school grounds to search for sediments that have been transported, such as grit on curbs and dust on windowsills. Ask students to infer about the possible sources of the sediments and how they were transported to their present locations. Discuss the types of agents, such as wind or water, that can carry sediments away from one location and deposit them elsewhere. L1

15-1 **Gravity**

Section Objectives

■ Distinguish between erosion and deposition.
■ Identify creep and slump as erosion caused by gravity.
■ Describe rockslides and mudflows.

Key Terms

erosion
deposition
creep
slump
rockslide
mudflow

The Journey Begins

In the Explore activity, you may have tilted the wax paper because you realized that sediment can move by itself from a higher to a lower place. Now, think about sliding down a hill on skis in the winter or on roller skates in the summer. Gravity overcomes the force of friction, and you slip down the slope with very little effort.

Solid rock can be broken down into smaller pieces and changed into other materials as a result of weathering. When weathering occurs on hills and slopes, the resulting broken rocks can slide downhill just as you do. The wearing away of surface materials and the movement of the products of weathering from where they formed to a different location is **erosion**.

The four major causes of erosion are gravity, running water, glaciers, and wind. These are also known as agents of erosion. Throughout this chapter, you will discover how gravity, water, ice, and wind help erosion occur.

Eventually, your skis or roller skates come to a stop. In the same way, sediments will stop moving and pile up, or accumulate. This accumulation of eroded sediments is called **deposition**.

Figure 15-1

Throughout this chapter, you will discover how the four major agents of erosion—gravity, running water, glaciers, and wind—help erosion occur.

GRAVITY

RUNNING WATER

GLACIER

WIND

474 Chapter 15 Shaping the Land

Program Resources

Study Guide, p. 54
Critical Thinking/Problem Solving, p. 23, A Disappearing Civilization L2
Multicultural Connections, p. 33, Aswan Dam L1; p. 34, Disaster in Puerto Rico L1
Take Home Activities, p. 24, Soil Erosion L1
Making Connections: Integrating Sciences, p. 33, Avalanches L2

Making Connections: Technology and Society, p. 33, Shoreline Erosion L2
Science Discovery Activities, 15-1, Power Moves
Section Focus Transparency 48

Slow Erosion

The next time you travel by car or bus, look along the roadway for trees, utility poles, or other objects leaning downhill. Trees and poles leaning downhill can be found in areas where freezing and thawing occur. As the ground freezes, small soil particles are pushed up by ice expanding in the soil. Then, when the soil thaws, it falls downslope, often less than a millimeter at a time. Several years of soil moving downslope very slowly can cause objects to lean. This slow movement of soil downhill is called **creep**. Creep as in **Figure 15-2**, gets its name from the way soil slowly creeps down a hill.

Sometimes one large mass of loose material or rock layers slips down a steep slope but doesn't travel very far. This slow mass movement of material is called **slump**. Slump occurs because the material under the slumped material weakened as in **Figure 15-3**.

Although such movements are slow, over time they can reshape the lay of the land. Valleys may gradually widen and the hills lining the valleys become more rounded and less steep.

Figure 15-2

When you see objects, such as trees and fence posts, all tilting in the same direction, creep is occurring. Creep gets its name from the slow way soil falls downslope. It creeps!

Figure 15-3

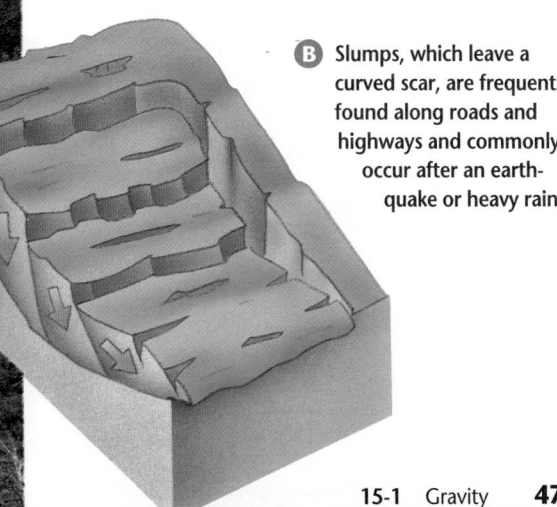

A When underlying material gives way and overlying material slips downslope as one large mass, the event is called slump.

B Slumps, which leave a curved scar, are frequently found along roads and highways and commonly occur after an earthquake or heavy rain.

15-1 Gravity **475**

2 TEACH

Tying to Previous Knowledge

Remind students that they learned in Chapter 14 that water runs downhill due to gravity. In this chapter, they will learn that gravity and running water are two of the four major agents of erosion that help to shape the land. Have students suggest ways that gravity and running water can shape the land.

Demonstration Using a sugar cube and the same amount of granulated sugar, demonstrate that an agent of erosion such as wind or water can more easily move smaller particles than larger ones with the same composition. **L1** **LEP**

Visual Learning

Figures 15-2 and 15-3 Before students read about slow erosion on page 475, have them use the photographs to compare creep and slump. *Students should be able to observe from the photos that both movements involve relatively large masses of earth materials and that both are downhill movements.*

Theme Connection The theme that is supported by this section is stability and change. Help students conclude that gravity, streams, glaciers, and wind are all agents that upset stability and change the form of land on Earth.

Across the Curriculum

Language Arts

Have students use dictionaries to look up the meanings of *slump* and *creep*. Have them demonstrate the verb forms and then relate them to the technical meanings used in this section to refer to erosion. **L1**

15-1 Gravity **475**

Fast Erosion

Figure 15-4

Rockslides happen most often after heavy rains or during earthquakes, but they can happen on any steep rocky slope at any time without warning.

On a really steep slope, large blocks of rock can break loose and tumble quickly to the bottom. As they fall, these rocks crash into other rocks, and they too break loose. The mass movement of falling rocks is called a **rockslide**.

Where would you expect to find rockslides? Mountains are most likely to have steep, rocky slopes where this sudden mass movement could occur. If a heavy rain falls in a dry area with thick layers of sediments, the water mixes with the sediments and forms a thick, pasty substance. Masses of such wet, heavy material will easily slide downhill in a **mudflow**. The flowing mud can move anything in its path.

CLOSER LOOK

Controlling Soil Erosion

Farmers have developed methods to help them grow crops on land that otherwise might just erode away. Most of these methods are ways to keep water and soil from running downhill.

Terrace Farming

One method is to create terraces around steep inclines or mountains. As seen in the photo below, terraces are flat fields where crops can be grown. Steep inclines separate the terraces from each other. Rainwater runs down the inclines to soak the terraces below them. It's a good way to save water in dry climates.

A similar method is to create a terrace that coils around and up a mountain. In this case the fields are not flat, but gently rise, winding around and around the mountain.

Figure 15-5

As a mudflow reaches the bottom of a slope, it slows down, eventually comes to rest, and deposits all the sediment and debris it has been carrying. Why do you think mudflows are considered to be very destructive?

check your UNDERSTANDING

1. How is erosion related to weathering?
2. Where is the deposition of sediment most likely to occur?
3. How can you identify creep and slump?
4. Why can rockslides and mudflows be dangerous?
5. **Apply** How might cutting into hillsides to build houses or roads affect erosion?

Contour Plowing

On gentle slopes, farmers catch water by plowing across the slopes rather than up and down. This is called contour plowing. As the water travels downward, the plowed rows catch and slow it. Some of the water soaks in before the rest continues flowing downhill.

If land is prone to erosion, the use of good farming practices is especially important. Plowing a crop back into the ground helps it resist erosion. Leaving the crop residue as seen in the photo on the right, after the harvest also reduces erosion over the winter. In windy areas, sandy soil can be protected by planting coarse grasses to help hold the soil in place.

Unwise farming practices contributed to the dust bowl that occurred in the Central United States during the 1930's.

People cannot always control the causes of erosion, but they can do some things to keep erosion from destroying farm land.

What Do You Think?

Walk around your neighborhood or town, looking for examples of eroded land. How could that land be improved? Is there anything you can do to prevent further erosion?

check your UNDERSTANDING

1. Erosion wears away and moves material from one place to another after it has been broken down by weathering.
2. Sediment is most likely to be deposited at the bottom of a steep slope.
3. You can identify creep by the tilted appearance of trees or posts on the surface of a slope. Slumps can be identified by large cracks or scars left when material has broken away.
4. They have the power to bury or move anything in their path.
5. Cutting steepens the angle of the slope, which increases erosion by gravity.

PREPARATION

Planning the Lesson

Refer to the Chapter Organizer on pages 472A–D.

Concepts Developed

In the previous section, students learned how gravity causes erosion. In this section, they will discover the combined effect of gravity and water on erosion.

1 MOTIVATE

Bellringer

 Before presenting the lesson, display **Section Focus Transparency 49** on an overhead projector. Assign the accompanying **Focus Activity** worksheet. L1

LEP

Activity Take students outside to look for evidence of water erosion around the school grounds. They might find small channels on the sides of slopes or low, flat areas where fine sediments have been deposited. Encourage interested students to take photographs or make drawings of some of these places.

2 TEACH

Tying to Previous Knowledge

Have students recall the importance of moving water in shaping the land. In this section, they will learn that sediments can be eroded by water.

Visual Learning

Figure 15-6 Students can use the figure to explain how streams carry sediment. Then discuss the eroding ability of a stream.

 ## Running Water

Section Objectives

- Explain how streams carry sediment.
- Explore the relationship between amount of sediment and rate of stream flow.
- Explain how streams and rivers shape the land.

Key Terms

floodplain
delta

Streams Erode

You know that the force of gravity can erode rock and soil material from slopes and deposit it at lower places. Can anything else erode and deposit sediments? What other factors cause loose material to travel down a slope?

Imagine following one small rock that landed in a creek at the bottom of a cliff after a rockslide. A heavy rain starts to fall. Soon the runoff begins to flow downhill, enters the creek, and picks up speed. The water in the creek in **Figure 15-6** is flowing quickly from a higher to a lower elevation. The small rock is lifted and carried along with the flowing water. This is one way that rock particles become eroded. Water moves them.

Figure 15-6

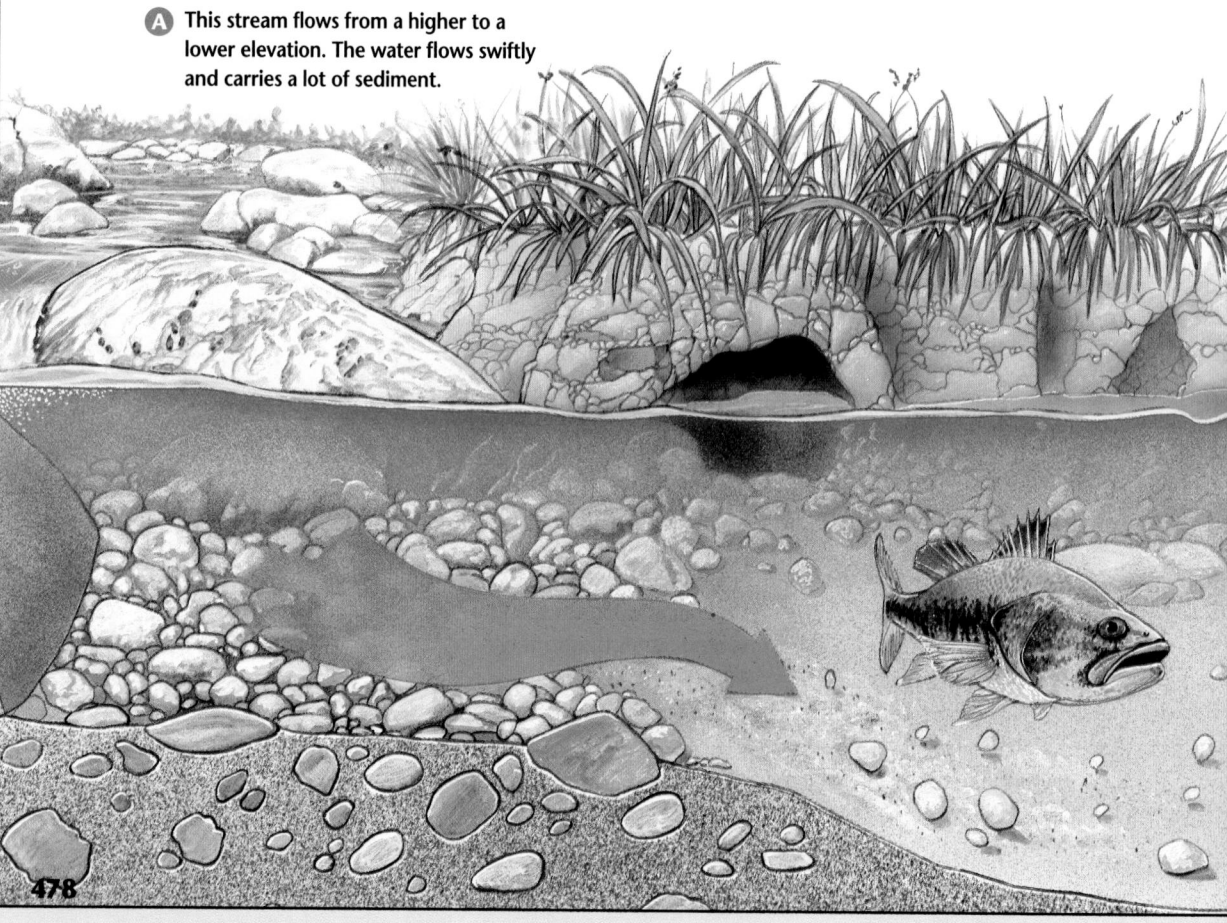

A This stream flows from a higher to a lower elevation. The water flows swiftly and carries a lot of sediment.

478

Program Resources

Study Guide, p. 55
Laboratory Manual, pp. 91–92, Rivers L1; pp. 93–94, Transporting Soil Materials by Runoff L2
Making Connections: Across the Curriculum, p. 33, Floods L2
Section Focus Transparency 49

Meeting Individual Needs

Limited English Proficiency Have students who are acquiring English as a second language make picture dictionaries of erosion terminology discussed in the chapter. Pictures may be cut from old magazines or students may prefer to make their own drawings. LEP

P

Explore! ACTIVITY

How can streams carry away rock and soil?

What To Do

1. Sprinkle some instant coffee grains, salt, and rice on a paper plate to represent loose soil and rock material. Squirt water from a spray bottle on one edge of the plate to act as rain.

2. Observe what happens to the material when the water droplets start to accumulate and flow in a stream. Record these observations *in your Journal.*

3. Now continue spraying as you tip the plate over a sink.

4. Can you explain what the stream of water has done to the different types of material? What do you observe about the color of the water?

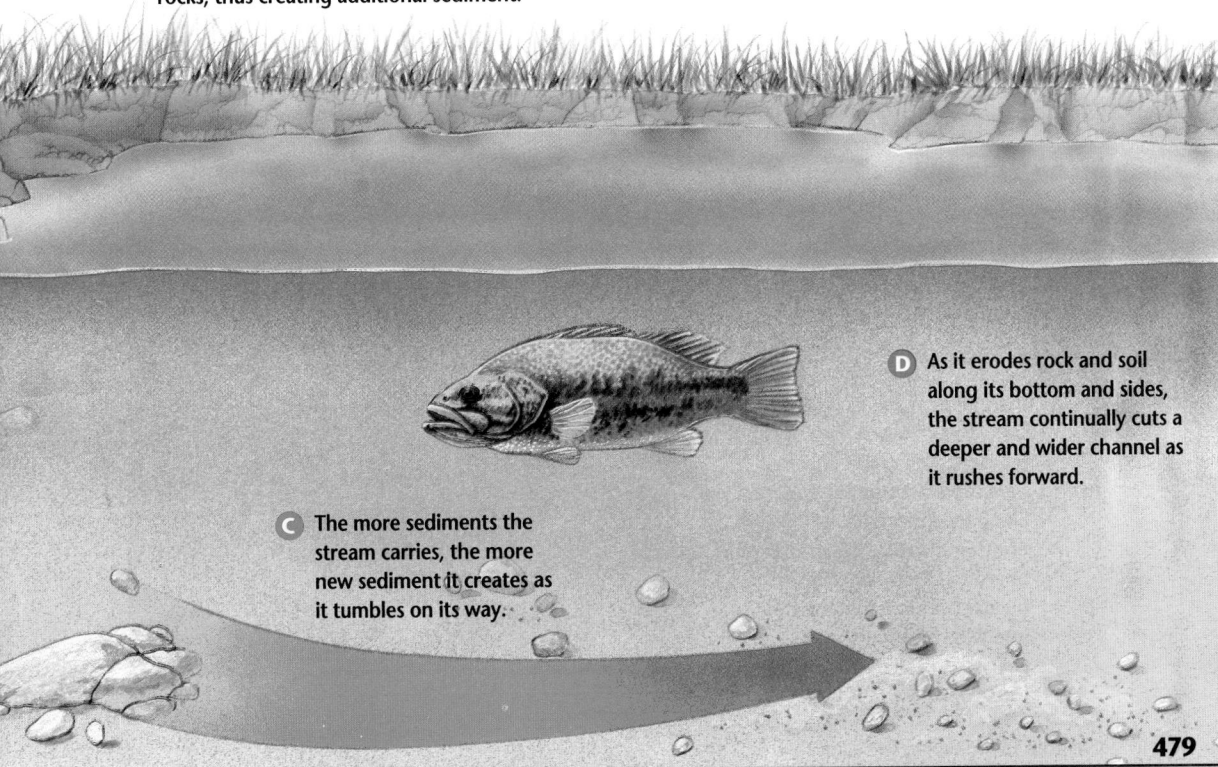

B The rock and other sediments in the water roll and scrape against the sides and bottom of the stream channel. As they travel, the sediments knock loose more soil and rocks, thus creating additional sediment.

D As it erodes rock and soil along its bottom and sides, the stream continually cuts a deeper and wider channel as it rushes forward.

C The more sediments the stream carries, the more new sediment it creates as it tumbles on its way.

479

Multicultural Perspectives

Fertile Lands

Tell students that ancient cities developed along the Nile River in Egypt and in Mesopotamia between the Tigris and Euphrates rivers. In both places, the fertile river floodplains led to the development of highly advanced cultures.

In ancient Egypt, the calendar was based on the agricultural cycle. The first of their three seasons was called Inundation, and the first day of the year was the day when the river could be expected to flood. The flood left behind rich deposits, in which seeds were sown. The second season was called Sprouting, and the third was called Harvest.

Have interested students research the ancient Egyptian calendar and present their findings to the class. **L2**

Explore!

How do streams carry away rock and soil?

Time needed 10–15 minutes

Materials instant coffee grains, salt, rice, paper plates, spray bottles with water

Thinking Process observing and inferring, comparing and contrasting, recognizing cause and effect, modeling

Purpose To observe that streams can dissolve and erode sediments.

Preparation Have students work in teams, assigning the following roles to individuals: setup, sprayer, observer, cleanup. **COOP LEARN**

Teaching the Activity

Troubleshooting Provide waste cans for discarding coffee grains and rice to avoid clogging the sink.

Science Journal Have students record their observations and inferences in their journals. **L1**

Expected Outcomes

Students will observe that some of the sediment is dissolved (salt); some is partly dissolved (coffee); and other sediment is carried away (rice).

Answers to Questions

Material has been moved "downhill" and/or dissolved. One "sediment," coffee, changed the color of the water.

✔ Assessment

Process Have students describe in words and with diagrams what happened as the water acted on the coffee, salt, and rice. Use the Performance Task Assessment List for Writing in Science in **PASC**, p. 87. **P** **L1**

Visual Learning

Figure 15-7 What evidence can you see in the photograph that tells you the fast stream is eroding more than the slow stream? *You can see the sediment suspended in the water.*

Figure 15-8 Have pairs of students study a map of the United States to find examples of rivers that meander in snakelike bends. Suggest that students trace the course of each river. *Students' responses may include the Snake, Ohio, Mississippi, and Colorado rivers.*

`L2`

Activity Have students identify meanders on aerial photographs of the Snake River in the Northwest. Also have them locate where erosion and deposition are occurring on each bend. Such photographs can be obtained from the United States Geological Survey (USGS) in Reston, Virginia. `L1`

GLENCOE TECHNOLOGY

CD-ROM

Science Interactions, Course 1, CD-ROM

Chapter 15

Figure 15-7

As the volume of a stream increases, its flow speeds up and the stream erodes at a faster rate. What evidence can you see in the photographs that tells you the fast stream is eroding more than the slow stream?

Water in a stream flows faster when the slope of the stream increases. It also flows faster when the volume of water increases. When more water is added to a stream as it combines with other streams, it speeds up. An increase in runoff from rainfall has the same effect.

Streams flowing swiftly down a steep hill eventually reach less sloping ground and flow less swiftly. As the rate of flow slows, it changes the way that a stream erodes.

Moving water on a gentle slope erodes the outside curves of a river instead of cutting downward into the streambed. If the river continues to meander in snakelike bends, it erodes its valley walls and widens the valley. If the volume of water in the river increases, the erosion increases.

Figure 15-8

Ⓐ Mature rivers that have eroded their channels into broad valleys flow slowly in snakelike bends called meanders. Meanders result from erosion and sediment deposition.

Ⓑ Because the water flows faster around the outside edge of a curve, the outer bank of a river is eroded more than the inner bank. Sediments are deposited along the inner bank by slower-moving water. This process of erosion and deposition builds slight bends in the river into wide meanders, and changes the course of the river.

480 Chapter 15 Shaping the Land

ENRICHMENT

Activity Have students, working in small groups, use pliable clay to make models that contrast youthful and mature rivers. Have students record the differences in their journals. *Youthful streams have relatively steep slopes and narrow, straight channels. Mature rivers have gradual slopes and often have meanders along their channels.* `L3` **COOP LEARN**

Meeting Individual Needs

Gifted Ask students to create a crossword puzzle using the new science words presented in this chapter. Have students exchange puzzles with partners to test each other's work. `L3`

`P`

Theme Connection The theme supported by this section is stability and change, since running water as a force of erosion is an agent of change.

Streams Deposit Sediment

What happens when a river floods? Runoff from heavy rains can cause a river to overflow its banks. During floods, a river carries a larger than normal amount of sediment. Bulky, heavy sediments drop along the banks of the river, forming ridges. Finer, lighter sediments travel out beyond the river channel and form a **floodplain**. Because these light sediments contain minerals and rich topsoil, they make floodplains a fertile area for planting.

Moving water deposits sediments even when the volume of water in the river is not increased. As a river starts to slow down, it can not flow fast enough to continue carrying heavier, bulkier sediments. The river begins to deposit the sediments. Slow-moving water is still able to carry fine, light sediments, however. Often sediments are deposited when the river empties into another body of water, such as a bay or lake. The deposited sediments may form a triangular-shaped land area called a **delta**.

SKILLBUILDER

Observing and Inferring

Imagine you are at the bottom of a canyon. You observe that its walls consist of reddish-brown sandstone. You also observe that the water in the river has a similar reddish brown color. How would you use your observations to explain the formation of this canyon? If you need help, refer to the **Skill Handbook** on page 659.

Figure 15-9

Ⓐ Sediments of minerals and rich topsoil carried by the Mekong River and deposited into the South China Sea have formed the Mekong Delta in South Vietnam. Some sediments come from as far away as China.

Ⓑ More than half of the people of Southern Vietnam live on the Mekong Delta, which is the chief agricultural area of Vietnam. Most Vietnamese are farmers and rice is their main crop.

15-1 Stream Erosion and Deposition

Preparation

Purpose To discover what factors affect stream erosion and the deposition of sediments.

Process Skills observing, forming a hypothesis, forming operational definitions, designing an experiment, separating and controlling variables, modeling, recognizing cause and effect, making predictions

Time Required 30-40 minutes

Materials plastic hose, pails with water, block of wood, stream table, small pebbles, soil, sand, screw clamps

Safety Precautions Have students wear protective clothing.

Possible Hypotheses Students might hypothesize that the slope of the channel and the volume of water moving through the channel will affect the way a stream erodes and deposits sediments.

The Experiment

Process Reinforcement Lead a discussion that helps students conclude that a valid scientific experiment tests only one variable at a time. Make sure that in their proposed experiments, students include a hypothesis, a step-by-step procedure, and a method for recording their data. [L1]

Possible Procedures Put the sand (or other available materials) into the stream table. Use a block of wood to carve out a stream channel in the sand. Start water flowing in the stream channel by carefully pouring water from a bucket. Record all observations. Plan to run at least three trials.

Stream Erosion and Deposition

Streams are very effective movers of sediment. They can erode large quantities of sediment from an area and deposit them many miles away. But how do streams erode and deposit sediment, and where in the stream channel do these two processes take place?

Preparation

Problem
Which factors affect the way a stream erodes and deposits sediments?

Form a Hypothesis
As a group, list the factors that might influence stream erosion and deposition. Agree upon these factors, then form a hypothesis that can be tested in your experiment.

Objectives
• Design an experiment that tests the effects of different factors on stream erosion and deposition.
• Compare how different factors affect the way a stream erodes and deposits sediment.
• Determine where erosion and deposition occur in the stream channel.

Materials
stream table
sand, small pebbles, soil
plastic hose
screw clamps
pails with water
block of wood

Safety

Program Resources

Activity Masters, pp. 63–64, Design Your Own Investigation 15-1

How It Works, p. 18, Locks [L2]

Science Discovery Activities, 15-2, Streaming Along

DESIGN YOUR OWN
INVESTIGATION

Plan the Experiment

1 Examine the materials provided by your teacher. Determine how you will use these materials to create a stream channel.

2 Agree upon a way to test your hypothesis. Write down what you will do at each step.

3 Design a table for recording your data.

Check the Plan

1 How many factors will you test to observe stream erosion and deposition? Keep in mind that you should only test one factor or variable at a time.

2 If you are testing more than one variable, how will you ensure that the same conditions exist in the stream channel for each test?

3 Before you start the experiment, have your teacher approve your plan.

4 Carry out your experiment. Complete your data table *in your Science Journal.*

Analyze and Conclude

1. Observe Describe what happened to the stream channel in your tests. Where did most erosion take place? Where did deposition occur?

2. Conclude What happened to the eroded materials? Describe how they were deposited.

3. Compare and Contrast Compare the effects of different factors on stream erosion and deposition.

4. Explain how your results support or do not support your hypothesis.

Going Further

Based on your observations, infer where the greatest amount of sediment might be found along a river's course.

Teaching Strategies

Discussion Before students begin their activity, have them predict how the stream channel might be affected by the water flow. List these predictions on the chalkboard.

Science Journal You may prefer to have students record their predictions in their journals. Students should also record their data and observations in their journals. **L1**

Expected Outcome

Students will observe how streams erode and deposit sediments.

Analyze and Conclude

1. Answers will vary but may include that an increase in either slope of the channel or volume of water increased the amount of sediment that was eroded and deposited. Most erosion took place along the sides of the stream channel. Deposition occurred at the end of the stream channel.

2. The eroded materials were deposited at different points along the streambed. Larger sediments were deposited close to the top of the stream; smaller sediments were carried further by the moving water.

3. Increasing either the slope of the channel or the volume of water increases the rate and amount of erosion. Because more sediments are eroded, more are deposited when the stream empties into a standing body of water.

✔ Assessment

Process Have students confirm, then record their results by actually measuring the time it takes sediments to flow down their streams when either the volume of water is increased or the slope of the stream is steepened. Use the Performance Task Assessment List for Analyzing the Data in **PASC,** p. 27. **P** **L1**

Going Further ▮▮▮▮▶

The greatest amount of sediment would be found near the point when the river empties into the sea.

Visual Learning

Figure 15-10 Have students use a map to trace the course of the Mississippi River from its source to its mouth, noting all the areas mentioned in Figure 15-10.
L1

3 ASSESS

Check for Understanding

1. Have students draw and label a cross-section of a river that illustrates how moving water erodes and deposits sediments. Use the Performance Task Assessment List for Scientific Drawing in **PASC**, p. 55. P L2

2. Ask students questions 1 through 4. Have students discuss the Apply question. Encourage them to think about how this construction might affect farmland in the delta area.

Reteach

Demonstration Demonstrate the formation of a delta. Pour water into the deep end of a paint tray and cover the top of the tray with fine soil. Gently pour water from a beaker onto the soil at the top of the tray. Have students describe what happens to the soil. L1

Extension

Activity Have students write a short report on the evolution of the Mississippi Delta, emphasizing the relationship between sediment size and distance from the source. Use the Performance Task Assessment List for Writing in Science in **PASC**, p. 87. L3

4 CLOSE

Discussion

Discuss why people build farms in or near a floodplain although the area is likely to flood. *The sediment deposited there helps create rich topsoil.*

The Mighty Mississippi

Let's use the Mississippi River system to review how rivers erode Earth's surface. Thousands of smaller streams flow quickly from higher elevations into larger streams and rivers. These small, swift-moving streams erode sediments from the bottoms of their channels. As the larger streams and rivers reach gradually sloping ground, they slow down. When they finally reach the Mississippi River, they are flowing on flatter areas and beginning to meander.

The Mississippi River itself cuts into its banks, widens its valley, and picks up more sediment. The slow-moving Mississippi carries a great volume of water and large amounts of sediment. Eventually, at the Gulf of Mexico, it loses most of its sediment and forms a delta on the Louisiana coast.

Figure 15-10

A The Mississippi River gets its water from thousands of small streams and rivers, which feed into larger and larger streams and finally into the Mississippi itself. One example is the Tennessee River, which feeds into the Ohio River.

B When the Ohio River joins the Mississippi at Cairo, Illinois, the volume of the Mississippi doubles. The Mississippi meanders through the valley, depositing sediments along the way.

C As the Mississippi empties into the Gulf of Mexico, the river flow slows and drops its final load of sediments. The dropped sediments form the fertile Mississippi Delta, which covers about 33 700 km².

check your UNDERSTANDING

1. How do rivers cause erosion?

2. How does slope affect the amount of sediment the stream can carry?

3. How do rivers shape valleys and deltas?

4. Apply How could the construction of a dam upriver affect a delta?

484   Chapter 15   Shaping the Land

check your UNDERSTANDING

1. The flowing water in rivers picks up and carries sediment.

2. An increase in slope increases the rate of stream flow, which enables the stream to carry a larger load of sediment.

3. Rivers on steep slopes erode downward, carving deep, straight valleys. Slow-moving rivers widen valleys. Rivers deposit sediments when they enter larger bodies of water, forming deltas.

4. A dam would change the river's rate of flow and affect the river's ability to carry sediments. The delta would eventually wear away because sediments could not be replaced.

15-3 Glaciers

What Are Glaciers?

What would it be like to live at a time when every winter is longer and colder than the one before? Every summer would be shorter and cooler—until eventually there would be almost no summer at all. This is what the climate was like over 16 000 years ago, during the last glaciation or glacial period. Then, ice covered much of the land.

A glacial period is a period of time when ice and snow cover much of Earth's surface. An ice age is a time during which many glacial periods occur. There have been a number of ice ages in Earth's history. The last glacial period ended about 10 000 years ago.

Yet huge masses of moving snow and ice called glaciers still cover parts of Earth. In fact, glaciers cover about one-tenth of Earth's land. Moving glaciers make enormous changes in Earth's surface. Melting glaciers provide much of the water that flows into rivers. Many people depend on this melted ice for their water supply. Citizens of Lima, Peru obtain their water from glaciers high in the Andes Mountains.

Section Objectives

- Describe how a glacier is formed.
- Differentiate between the two major types of glaciers.
- Describe how glaciers erode the land.

Key Terms

firn
continental glaciers
valley glaciers

Figure 15-11

A Ice sheets and glaciers, such as the Sawyer Glacier near Juneau, Alaska, hold 85 percent of all the fresh water on Earth.

B Melting glaciers provide much of the water that flows into rivers, which eventually erode and change the land.

Program Resources

Study Guide, p. 56
Teaching Transparencies 29 and 30 L2
Section Focus Transparency 50

ENRICHMENT

Activity Have students write fictional stories or poems in their journals about what it might have been like to live during the Ice Age. Encourage creativity, but insist that the scientific content is accurate. Select the appropriate Performance Task Assessment List in **PASC.** L1
P

15-3

PREPARATION

Planning the Lesson

Refer to the Chapter Organizer on pages 472A–D.

Concepts Developed

In this section, students will learn about glaciers, another important agent of erosion and deposition that has extensively shaped Earth's topography. Students will learn the characteristics of continental and valley glaciers and discover how Earth is affected by their movement.

1 MOTIVATE

Bellringer

Before presenting the lesson, display **Section Focus Transparency 50** on an overhead projector. Assign the accompanying **Focus Activity** worksheet. L1
LEP

Demonstration Display a beaker full of ice and a beaker full of water. Have students compare and contrast the items in this display. *Students should realize that both beakers contain the substance water but in different forms.* Inform students that much of a glacier is solid water. At the glacier's base and often within a glacial mass, however, liquid water flows.

2 TEACH

Tying to Previous Knowledge

Have students recall that water can exist in three states. For the most part, a glacier is a solid, but heat, formed by the friction of ice scraping against bedrock, causes its bottom layer to melt. This thin layer of water enables the glacier to move more easily.

Ask students to guess how much the average world temperature would need to drop before Earth would undergo another ice age. They are likely to guess a large number of degrees. In reality, an average 2.3°C drop in the ocean and a 6.5°C average drop on land would be enough to produce a new ice age.

Theme Connection The actions of glaciers should provide students with evidence of the theme of stability and change. You might have students preview the illustrations in this section as you tell them that the movement of glaciers affects the evolution of land surfaces, changing surface and structure of landmasses. The activities in this section give students experience with models of change due to glacial action.

Activity Have students collect enough snow or crushed ice to fill a clean, wide-mouthed container. Ask students to hypothesize what the level of the water in the container will be when the snow melts. Have them mark their predictions on the side of the containers and, later, compare them with the actual results. Have students read the captions in Figure 15-12 to find out that snow is about 80 percent air. L1

Visual Learning

Figure 15-13 Display a large world map for reference. Have students locate North America, Asia, and the Bering Strait on the map in the figure. Tell them that scientists hypothesize that during the last ice age, people traveled across a land bridge in the Bering Strait from Asia to North America. Ask students to suggest why there is no bridge today. *During the ice age, much of Earth's water was frozen in glaciers, leaving more land exposed. Today the land bridge is below sea level.*

Glacial Formation

Have you ever seen ice build up on the freezer walls of an old refrigerator? The same thing can happen in nature where snow remains on the ground year-round. If the snow doesn't melt during the summer, it begins to pile up just as frost can pile up in a freezer.

When snow falls and starts to accumulate, it is mostly air. Snow that doesn't melt after one year becomes a harder, denser material called **firn**. Firn has lost some of the air, and as it becomes more and more compressed by the snow on top of it, the firn will become glacial ice.

Some air is still in the glacial ice and is trapped in the ice. By analyzing this trapped air, we can learn what air was like when the ice was formed.

When gravity acts on the glacial ice it begins to move and is called a glacier. This movement depends on how thick the ice is, the steepness of the surface the glacier sits on, the weight of the ice, and air temperature surrounding the glacier.

Figure 15-12

Melting, evaporation, and refreezing gradually change delicate snowflakes into small, round, thick granules called firn. As the process continues and overlying layers of snow add pressure, snow becomes glacial ice. During the transformation, snow is changed from a loose sediment with plenty of air space around the individual ice crystals into a more solid mass of higher density.

New fallen snow is 80 percent air.

Firn is snow that has survived at least one year without melting or turning to ice and is about 50 percent air.

An icy grain is about 20 percent air, frozen as tiny bubbles.

Glacial ice is less than 20 percent air.

Figure 15-13

Glaciers exist as giant sheets of ice in polar regions such as Antarctica and parts of Greenland. Glaciers also exist as smaller ice caps and in mountain valleys found in such places as Iceland, Canada, and Alaska. Together glaciers cover about one-tenth of Earth's surface.

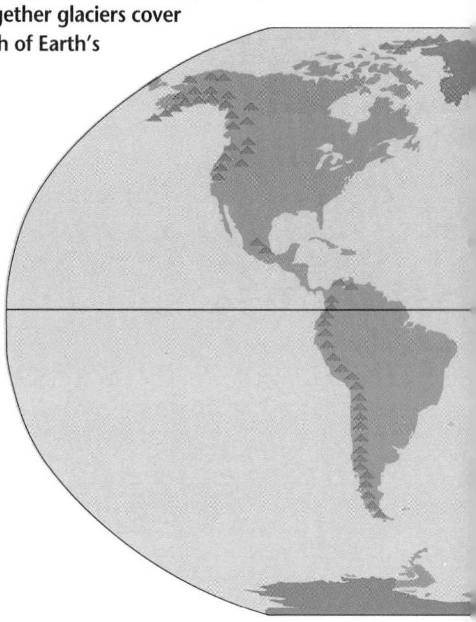

Multicultural Perspectives

Early Inhabitants on Cape Cod

Explain that Cape Cod, a hook-shaped peninsula in Massachusetts, was formed by a glacier. A storm in 1991 eroded a beach there, uncovering a Native American site that was inhabited about 8000 years before the Pilgrims landed nearby. Evidence of buildings was found there, but other coastal storms eroded some of the site before it was completely excavated.

Have interested students find out more about the geography of Cape Cod and about the early inhabitants of the area. L2

Types of Glaciers

Masses of ice and snow that cover large land masses near Earth's polar regions are called **continental glaciers**. They make up 96 percent of glacial ice. If you look at **Figure 15-13**, you'll see that continental glaciers are found in Greenland and in Antarctica.

Small glaciers are found at higher elevations in mountain regions. These glaciers usually occupy valleys in the mountains so they are called **valley glaciers**.

Glaciers contain up to 85 percent of the world's fresh water. If all of the glaciers were to melt as some have, world sea level would rise around 55 meters. What would happen to cities along the coast, such as Miami, New York, Bombay, or London? What would happen to the shapes of the continents?

Connect to...
Life Science

As glaciers advance and retreat, plants and animals also migrate along with the glaciers. What adaptations might these plants and animals have to have to survive living near a glacier?

Figure 15-14

Continental glaciers, such as this one in northwest Greenland, cover large land areas near Earth's polar regions.

Figure 15-15

Valley glaciers like this one at Glacier Bay National Park, Alaska, occupy a single valley between mountains. The mass of ice and snow that makes up a valley glacier forms at and flows from higher elevations where snow stays year after year.

Visual Learning

Figures 15-14 and 15-15 Have students use the photographs to compare and contrast continental and valley glaciers before reading the captions.

Student Text Questions
What would happen to cities along the coast, such as Miami, New York, Bombay, or London? *They would become inundated by the rising sea.* Note: To emphasize multicultural awareness, have students research coastal cities of Central and South America. **What would happen to the shapes of the continents?** *The continental coastlines would migrate inland due to the rising water.*

Across the Curriculum

Geography

Have students use a globe or a world map to locate two or three major coastal cities on each continent that might be under water after a significant rise in sea level. Have students use an atlas or other resource book to find the elevation of each city. **L2**

Connect to . . .

Life Science

Plants—fast-growing, shallow root system, able to endure cold and lack of water; animals—able to endure cold and limited food supply.

GLENCOE TECHNOLOGY

Software
Computer Competency Activities
Chapter 15

ENRICHMENT

Research Have students find photographs of the Palisades, a set of sheer cliffs along the Hudson River near New York City. Ask them to find out how the river valley and cliffs were formed. **L2**

15-2 How Do Glaciers Change the Land?

Planning the Activity

Time needed 30–40 minutes

Purpose To observe how valley glaciers erode the surface of Earth.

Process Skills observing and inferring, making and using tables, formulating models, measuring in SI, recognizing cause and effect, designing an experiment

Materials See reduced student text.

Preparation Prepare the blocks of ice a day ahead of time. Use milk or juice containers to shape the blocks.

Teaching the Activity

Safety Set up the reflector lamp so that it provides enough energy to melt the ice, but so that students will not get burned by it.

Troubleshooting Use pails to collect any meltwater and loose debris. Have students wear protective clothing.

Process Reinforcement Review with students, if necessary, how to construct a scale drawing. Remind them that they must choose a scale that will accurately represent both the vertical and horizontal dimensions of the glacial valley that formed. You might want to have a volunteer demonstrate how to make a cross-sectional view of either valley.

Science Journal Students' sketches should demonstrate an understanding of the difference in shape between river valleys (V-shaped) and glacial valleys (U-shaped). Have students include a short description of each valley next to the sketch. L1

How Do Glaciers Change the Land?

Glaciers erode the land and can change it a great deal. In this activity, you'll observe how glaciers change the land as they erode Earth's surface.

Problem
How do valley glaciers affect Earth's surface?

Materials
ice block about 5 cm by 20 cm by 2 cm, containing sand, clay, and gravel
stream table with sand
lamp with reflector
metric ruler

Safety Precautions

You will be using electrical equipment near water in this investigate. Please keep these items apart from one another.

What To Do

1 Copy the data table *into your Journal.* Then set up the stream table and lamp as shown.

2 The ice block is made by mixing water with sand, gravel, and clay in a container and then freezing (see photo **A**).

3 Make a V-shaped river channel. Measure and record its width and depth. Draw a sketch that includes these measurements (see photo **B**).

4 Place the ice block, to act as a moving glacier, at the upper end of the stream table.

Meeting Individual Needs

Physically Disabled Students with physical limitations may not be able to manipulate the ice block. Allow those students to observe and record the information gathered during this investigation.

A   B

Sample data

Data and Observations			
	Width	Depth	Observation
River	3 cm	2 cm	V shaped
Glacier	5 cm	4 cm	U shaped

5 Gently push the glacier along the river channel until it's under the light, halfway between the top and bottom of the stream table.

6 Turn on the light and allow the ice to melt. *Observe* and record what happens.

7 *Measure* and record the width and depth of the glacial channel. Draw a sketch of the channel and include these measurements *in your Journal.*

Analyzing

1. How can you *infer* the direction from which a glacier traveled?

2. How can you tell how far down the valley the glacier traveled?

Concluding and Applying

3. Determine the effect valley glaciers have on the surface over which they move.

4. Going Further How can you identify land that was once covered by a glacier?

Students will observe that the model glacier left deposits and eroded the stream channel.

Answers to Analyzing/ Concluding and Applying

1. The area over which the glacier has moved will probably be smoother and have steeper sides than the section which has not eroded.

2. The channel will be U-shaped to the point where the glacier stopped, and V-shaped past that point. Small deposits will form at the end of the glacier.

3. Valley glaciers erode the surface like a bulldozer, leaving U-shaped valleys with steep sides and flat bottoms.

4. Glaciers leave behind characteristic deposits and patterns of erosion.

✔ Assessment

Performance Have students work in small groups to design and carry out an experiment that shows how glacial meltwater sorts the sediments it carries. Use the Performance Task Assessment List for Designing an Experiment in **PASC,** p. 23.
COOP LEARN **L1**

Program Resources

Laboratory Manual, pp. 95–98, Glaciation and Sea Level **L2**

Activity Masters, pp. 65–66, Investigate 15-2 **L2**

Discussion Glaciers leave
behind deposits of rock materials that have been carried for
great distances. Ask students
how they could tell if rock debris
had been left behind by a glacier.
*If the loose rock is different from
local rock, it has been brought
from another area. This could
have been caused by a glacier. If it
is similar to local rock, it was
probably weathered from local
rocks.*

Demonstration Have students hypothesize how rocks
and other sediments become
embedded in a glacier. *Accept
any reasonable response at this
time.* Now put a mixture of
sand, gravel, and clay into a
small plastic tub. Cover the mixture with ice cubes. Allow the
cubes to melt. Freeze the tub
overnight and have students observe the ice block the following
day. Discuss any discrepancies
between their predictions and
their observations. **L1**

GLENCOE TECHNOLOGY

 Videodisc

**The Infinite Voyage: Crisis
in the Atmosphere**

Chapter 1
Historical Aspects of the Greenhouse Effect and Fossil Air

Chapter 2
Studying Man-Made Carbon
Dioxide

Chapter 3
Our Future Climate

Chapter 4
The Greenhouse Effect: Future and
Past

Glacial Erosion

Glaciers covered large portions of
land during the last ice age. As they
move, glaciers cut through mountains, erode the land, and leave large
deposits of ground-up rock. As glaciers melt, rivers and lakes form. Much
of Earth's landscape has been shaped
by glacial ice.

A glacier picks up loose materials
as it moves over land. These eroded
sediments are added to the mass of
the glacier or pile up along its sides.
Ridges form when a glacier recedes
and deposits rocks and sediments.
You can see hills or ridges like this in
places that were once the sides or ends
of a glacier.

Glaciers do more than just move
sediments. They also erode rock and
soil that aren't loose. Glacial ice melts,
and the water flows down into cracks
in rocks. Later, the water freezes in
these cracks, then expands. The
expanding, freezing water breaks the
rock into pieces. The rock fragments
then move along with the glacial ice.
This process results in boulders, gravel, and sand being added to the bottom and sides of a glacier. Find out in
the next activity how this matter
frozen in glacial ice can cause further
erosion.

Figure 15-16

Ⓐ The brown streaks in this glacier are sediments
that the glacier has picked up in its travels and
is now depositing as it recedes.

Ⓑ When rock fragments at the base of a
glacier scrape bedrock, long parallel
scars like these at Kelleys Island, Ohio,
may be left behind.

Figure 15-17

Ⓐ As glaciers push, break, and scrape
their way over land, they displace tons
of sediment, and leave behind many
valleys and bowl-shaped depressions.

490 Chapter 15 Shaping the Land

How do glaciers make grooves in rocks?

What To Do

1. Mix sand and other small particles of soil, rocks, or gravel in a container of water.

2. Pour the mixture into an ice cube tray and place the tray in a freezer. Let each frozen cube represent a glacier.

3. Remove the cubes from the freezer. Leave the cubes at room temperature for a few moments, feel their texture, and record your observations *in your Journal*.

4. You should feel the grains of sand and small particles. Rub the cubes over a piece of wood.

Conclude and Apply

1. What do you observe in the wood's surface?

2. Explain how glaciers make similar patterns in rocks?

Materials at the base of a glacier scrape the soil and bedrock over which the glacier moves. The loose particles can cause even more erosion than the ice and snow alone. When bedrock is gouged by rock fragments, grooves may be left behind. Usually these scratches are long, parallel scars.

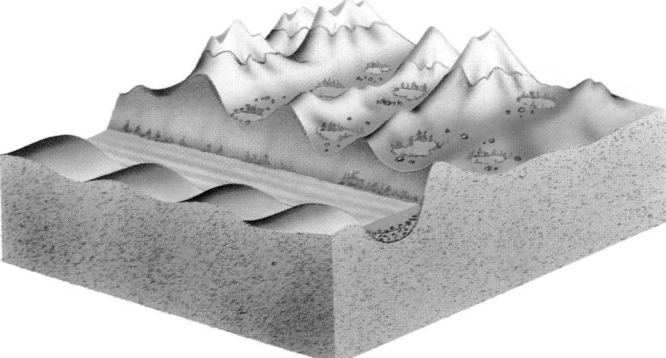

B When the glaciers melt, a river might flow in the valleys and depressions which may fill with glacial melt-water and become lakes.

Figure 15-18

Lake Superior, shown here, and the other Great Lakes were formed by erosion of a river valley during the advancing and receding of glacial ice that covered the area during the last several glacial advances.

15-3 Glaciers **491**

Find Out!

How do glaciers make grooves in rocks?

Time needed overnight for ice to freeze; 15 minutes in class for 2 days

Materials ice trays, water, sand, wood, gravel

Thinking Process observing and inferring, formulating models, recognizing cause and effect

Purpose To observe how glaciers can erode.

Teaching the Activity

Troubleshooting Tell students to avoid scraping the ice cubes against other surfaces such as desktops.

Science Journal Have students make sketches of the wood after it "has been glaciated," label the glacial grooves, and indicate the direction of movement of the "glacier." L1

Teacher F.Y.I.

Tell students that a well-preserved human body was discovered in a glacier between Italy and Austria in 1991. Because the ice had preserved the body so well, anthropologists could see that the 4600-year-old man had stuffed his leather boots with straw for insulation. Challenge students to find out what other objects were found with the "Ice Man." *Found with the man were a bow and arrows, as well as flint and copper tools.* L2

Across the Curriculum

Fine Arts

Show students pictures of the cave paintings in Altimira, Spain, and Lascaux, France, that were painted by Ice Age hunters. Photos of cave paintings can be found in an encyclopedia or anthropology book. Ask students what they believe the paintings mean. Anthropologists believe that the paintings were created to bring good hunting.

Expected Outcome

Students will observe how particles trapped in the model glaciers can scratch the surfaces they rub against.

Conclude and Apply

1. The particles embedded in the ice cubes left scratches on the wood's surface.

2. Sediment along the bottom of the glacier scratches the surface over which the ice moves.

✔ Assessment

Content Have students explain what glacial grooves indicate about the ice. *The grooves give some indication of the sizes of the debris being transported as well as the direction in which the ice moved.* Use the Performance Task Assessment List for Making Observations and Inferences in **PASC**, p. 17. L1

3 ASSESS

Check for Understanding

Have students answer questions 1 through 3 orally. Then discuss the Apply question. Use the U.S. map in Appendix G on pages 642–643 to help students locate the Great Lakes.

Reteach

Activity Have individual students push an ice cube into a pile of sand. Have them note what happens to the sand that was in the path of the ice. Relate this form of erosion to the action of a glacier as it erodes layers of soil in its way. L1

Extension

Activity Have students use a map or globe to predict what the future shoreline of North America might look like if global warming caused the ice caps in Greenland and Antarctica to melt. Have students draw diagrams that show the new shoreline. The sketches should show much of the East and Gulf coasts of the United States under water. L3 P

4 CLOSE

Discussion

Display a map of the Finger Lakes region of New York State. Use a road map of the state or a map in an atlas. Tell students that the land in the area was once covered by a continental glacier. Ask how that information helps explain the formation of the elongated lakes in the area. *The mass of the ice caused depressions to form. When the ice melted, the water filled the basins to form the Finger Lakes.*

Glacial Valley

Valley glaciers erode land and deposit sediments as they move down mountain slopes. Valleys eroded by glaciers are a different shape from valleys eroded by streams.

You had to imagine what life on Earth was like long ago when much of it was covered by ice and snow, but you don't have to imagine the changes that were made. Many of the U-shaped valleys, rivers, and hills that were formed when ice age glaciers eroded the land and deposited sediments still exist today. The huge amounts of frozen snow and ice that remain in today's continental glaciers and valley glaciers provide us with a supply of fresh water.

Figure 15-19

A Glacier-eroded valleys are usually U-shaped because glaciers pick up and drag soil and rock fragments along their sides as well as on their bottom. Glacier National Park, Montana, shown right, is one example.

B Stream-eroded valleys are normally V-shaped because the water in a stream erodes downward into its channel.

check your UNDERSTANDING

1. How can snowfall lead to the formation of a glacier?
2. How can valley glaciers form in places where continental glaciers could not?
3. How do scientists know which areas were once covered by glaciers during the ice age?
4. **Apply** Explain how the Great Lakes could have been formed by a glacier.

check your UNDERSTANDING

1. In areas where less snow melts in summer than falls in winter, it piles up, forming a heavy layer. The mass of the snow on top exerts pressure on the layers below, hardening them into ice.

2. Valley glaciers generally form at higher elevations.

3. They can observe glacial deposits and landforms.

4. As a continental glacier pushed south, it could have gouged out the depressions that are now the Great Lakes. Then melting glacial ice could have filled the depressions with fresh water.

Up, Up, and Away

Moving air can move loose particles. Particles that are too heavy to lift are dragged along the surface. Others are light enough to be picked up and carried by the air. Wind can move sand, clay, silt, and other loose sediments and when the wind dies down, the sediment is deposited.

Trying to eat a picnic lunch on the beach can be a challenge. Light items like napkins, plastic sandwich bags, paper cups, and potato chips blow away easily if the wind is strong. You may try to recover them as they bounce or roll away, only to find them useless because they are covered with sand. Can the same wind that blows your lunch away cause erosion? Explore in the next activity how these particles are carried by wind.

Section Objectives

- Describe how wind erodes and deposits sediment.
- Describe how a dune is formed and how it moves.
- Identify two factors that can decrease wind erosion.

Key Terms
dune

Explore! ACTIVITY

Which particles can be readily carried by the wind?

What To Do

1. Cut a sheet of aluminum foil into a variety of large and small sizes.

2. Then form the pieces into assorted shapes. Crumple some of the pieces of foil into loose balls, some into tight balls, and leave other pieces flat.

3. Put the assortment of aluminum shapes on a table and blow at them.

4. *In your Journal*, record which pieces move more readily. Do the size and shape of materials affect their ability to be transported by the moving air? Explain.

Explore!

Which particles can be readily carried by the wind?

Time needed 15 minutes

Materials sheets of aluminum foil

Thinking Processes observing and inferring, separating and controlling variables

Purpose To observe how particle size and shape affect an object's ability to be transported by wind.

Preparation To avoid using a lot of aluminum foil have students work in pairs or teams of four. **COOP LEARN**

Teaching the Activity

Have students brainstorm variables to test, such as blocking the wind, or making the pieces larger. **L1**

Science Journal Suggest that students make a data table in their journals that includes a sketch of each object they will manipulate. Stu-

PREPARATION

Planning the Lesson

Refer to the Chapter Organizer on pages 472A–D.

Concepts Developed

In this section, students will study how the wind moves particles of sediment and deposits them, as well as how the abrasive action of particles in wind erodes objects.

1 MOTIVATE

Bellringer

Before presenting the lesson, display **Section Focus Transparency 51** on an overhead projector. Assign the accompanying **Focus Activity** worksheet. **L1**
LEP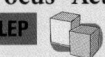

dents can record a prediction regarding the movability of the object compared to the others in the group. **L2**

Expected Outcomes

Lighter, smaller foil particles will be lifted by wind. Larger, heavier particles will not move or may be dragged along the surface.

Answer to Question

The larger and denser the object, the less likely it will be moved.

✔ Assessment

Performance Have small groups of students experiment with several shapes made from pieces of foil that are identical in length and width. Make sure students realize the relationship between an object's shape and density and the ability of the wind to transport that object. Use the Performance Task Assessment List for Assessing the Whole Experiment in **PASC**, p. 33.
COOP LEARN **L1**

Figure 15-20

Many sand dunes migrate because wind repeatedly picks up sand grains from one side of the dune and deposits them on the other side.

Wind Changes the Land

When wind erodes loose sediments by blowing them away, it eventually deposits the sediments when it stops blowing. This deposition can create new features on the land. For

example, sand or loose sediment may be blown by the wind into a formation called a **dune**. You might find a dune on a beach or in a desert.

   Dunes like the ones in **Figure 15-20**, are a result of erosion and deposition. The sand particles were eroded from one location and deposited here to form dunes. Not only can the sand particles in a dune move, but amazingly, the dune itself can move. Sand builds up a gentle slope on the side facing the wind. The sand continues to build up until it falls down a steeper slope on the other side.

Eyes on the Planet

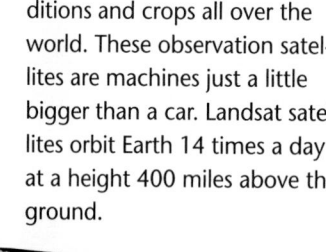

United States Landsat satellites are watching soil conditions and crops all over the world. These observation satellites are machines just a little bigger than a car. Landsat satellites orbit Earth 14 times a day at a height 400 miles above the ground.

The distribution of Earth's vegetation, as seen by Landsat satellites.

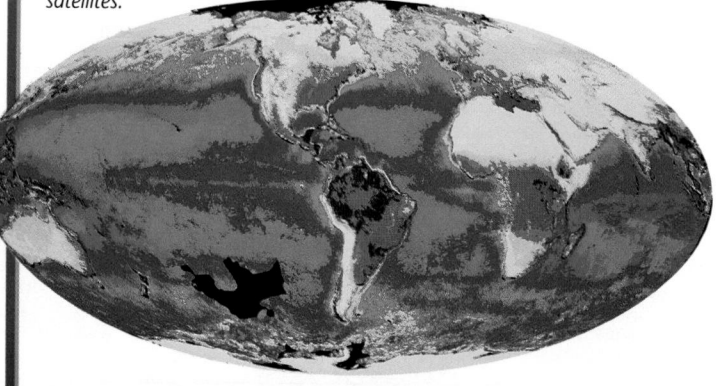

Satellite Sensors

   The Landsat satellites are filled with machines that gather information. Among them is the Landsat Thematic Mapper, which has 100 different detectors. As the satellite circles Earth, the machines take pictures of Earth. Their sensors can tell the difference between land and water, city and country, wheat and corn. They work even when Earth is covered with clouds, fog, storms, or darkness.

   Scientists on Earth collect the Landsat information and pass it on to people who need it.

494   Chapter 15   Shaping the Land

How do dunes move?

What To Do

1. Get a covered shoe box and cut a 5-cm-square opening in one end.

2. Spoon flour into the box toward the open end to form a layer about 2.5 cm deep.

3. Cover the box and put it on a level surface.

4. Gently blow air into the box through the open end. Be sure to have a towel handy to wipe off any flour.

5. Occasionally lift the lid and observe what is happening.

6. Record these observations *in your Journal.*

Conclude and Apply

1. What happens to the flour on the side that you are blowing air into?

2. What happens on the other side?

3. What happens to the little piles of flour as you continue blowing air into the box?

Satellites Show Changes in the Land

For example, information from one Landsat orbit showed a massive erosion problem in Africa. Huge numbers of people in Africa were starving, but the exact reason had been unclear. Conditions there were similar to the dust bowl the United States experienced in the 1930s. Back then, there was no satellite information to show the problem.

But 50 years later, pictures from the Landsat satellites clearly identified the problem in Africa. Pictures showed that people had allowed animals to overgraze a large area of grassland. Years of overgrazing followed by no rain had created

The Landsat photo above shows vegetation distribution in Nakuru, Kenya. Vegetation is shown in green.

Tea, one of many crops grown in Kenya, is shown below growing on a plantation.

desert conditions. Wind blew the dry soil away. With no grasses to eat, the animals starved, leaving the people with no food.

Because of the Landsat satellite information, the African people can prevent future problems by changing their animals' grazing methods.

inter**NET** CONNECTION

Find out what EROS Landsat data are available on the World Wide Web from the U.S. Geological Survey. How would these data be of use to people in your state?

15-4 Wind Erosion **495**

Find Out!

How do dunes move?

Time needed 15 minutes

Materials shoe box, flour, knife or scissors, spoon, towel

Thinking Processes observing and inferring, recognizing cause and effect, formulating models, measuring in SI

Purpose To observe how dunes form and migrate.

Teaching the Activity

Troubleshooting Remind students to blow gently into the end of the box *only* when the lid is in place.

Science Journal Students' observations should include sketches that show how dunes migrate due to wind. L1

Expected Outcome

Students will observe a hill piling up, falling down, and drifting.

Conclude and Apply

1. The side you are blowing on forms a hill.

2. Eventually the flour piles up so high that it falls down on the other side.

3. The pile of flour (dune) moves, or migrates, toward the end of the box.

✔ Assessment

Content Have students explain how one can identify the windward side of a dune. *The slope of a dune facing the wind is much longer and more gradual than the side away from the wind.* Use the Performance Task Assessment List for Assessing a Whole Experiment in **PASC,** p. 33. L1

Going Further ▐▐▐▐▐▶

Have students explore the effect of wind erosion in a historical context. Ask them to research the relationship between the Dust Bowl and the Great Depression in the United States in the 1930s. Have students present their findings as oral reports supported by visual displays. Select the appropriate Performance Task Assessment List in **PASC.** L2

Program Resources

Study Guide, p. 57

Concept Mapping, p. 23, Erosion L1

Critical Thinking/Problem Solving, p. 5, Flex Your Brain

Science Discovery Activities, 15-3, Wind Stoppers

Section Focus Transparency 51

inter**NET** CONNECTION

Access the U.S. Geological Survey's EROS Landsat data at **http://sun1.cr.usgs.gov**

3 ASSESS

Check for Understanding

Assign partners to answer questions 1 through 4. Have them discuss their answers to the Apply question and ask what other methods could be used to control erosion.

Reteach

Activity Help students make a diagram showing: wind blowing particles of soil→particles causing abrasion of rock forms creating more particles→particles being deposited. L1

Extension

Discussion Have students who have mastered the concepts of this section read the poem "Ozymandias" by Percy Bysshe Shelley. Discuss the theme that humans have little power over time and nature, in particular the force of wind erosion in the desert. L3

4 CLOSE

Discussion

Have students make a list of things a family could do to save their beachfront property from being eroded by wind and water. Some possible solutions might be to put in a structure to encourage a dune to form; plant beach grass; plant bushes as a windbreak; cover the sand with a rock wall.

Wind Erosion

Figure 15-21

The Sphinx has been influenced by Egypt's desert winds for thousands of years. Some of the damage to the Sphinx has been caused by windblown sediment. What type of sediment do you think causes the damage?

Not only does wind create deposits such as sand dunes, but it also erodes Earth's surface. It does this primarily by a process that is similar to sandblasting. Wind picks up small sand-sized particles and moves them. When these particles come in contact with objects such as the Sphinx, they erode them. Windblown materials like sand grind away whatever they hit.

Land can be eroded more easily during a drought. The soil gets very dry, and the plants in it dry up and die. Then wind can easily erode the soil. This happened in the central plains of the United States in the 1930s. The area became known as the Dust Bowl because as the soil was carried away by the wind, it created great swirling bowls of dust. Farmers today use planting and watering techniques that prevent this type of soil erosion.

Figure 15-22

Overgrazing of natural grasslands contributed to the conditions that set off a series of destructive dust storms, which raged through the southern Great Plains in the 1930s, creating the Dust Bowl.

Wind, like gravity, running water, and glaciers, shapes the land as it erodes. But the new landforms created by these agents of erosion are themselves being eroded. Erosion and deposition are part of a cycle of change that constantly shapes and reshapes the land around you.

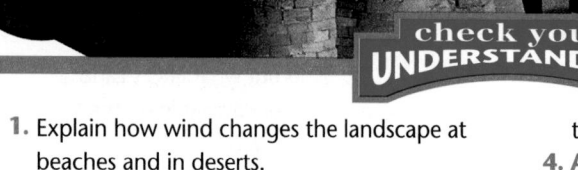

check your UNDERSTANDING

1. Explain how wind changes the landscape at beaches and in deserts.
2. Why are dunes constantly changing?
3. Describe at least two steps a farmer might take to decrease soil erosion.
4. **Apply** Explain how trees planted near the corners of a house can help keep the land from changing.

check your UNDERSTANDING

1. The wind constantly moves loose sediments and deposits them to form dunes, which migrate. Windblown sediment also erodes the surface of material.

2. Sediment from existing dunes is constantly being carried away by the wind and redeposited in new locations.

3. A farmer could plant ground cover in fields. Also, irrigation might be used to keep soil moist. Windbreaks also reduce erosion.

4. They can break the wind. Also, roots hold the soil in place.

What are icefalls?

Glaciers, those vast, slow-moving rivers of ice that are found in Earth's cold regions, flow at the whim of gravity just as liquid water does. And when a glacier pushes its way over a ledge, or ragged incline, it creates an icefall, one of the most beautiful–and deadly– geologic features on Earth.

Icefalls resemble river rapids more than they do waterfalls. When the ice at the base of the glacier flows down a steep incline, the ice on top can split and become riddled with deep cracks or crevasses. The glacial ice stretches and strains. Jagged pinnacles form at the surface and these slowly split from the ice upslope and crash onto the lower ice farther down the slope.

Depending on the thickness of the glacier, icefall crevasses can reach several hundred feet deep. As they go deeper, the color of the ice changes from a bright white to a deep blue when portions of the light spectrum are filtered out.

Mountaineers, who must cross icefalls on the way to the mountain summit, fear treacherous ones such as Khumbu on the slopes of Mount Everest.

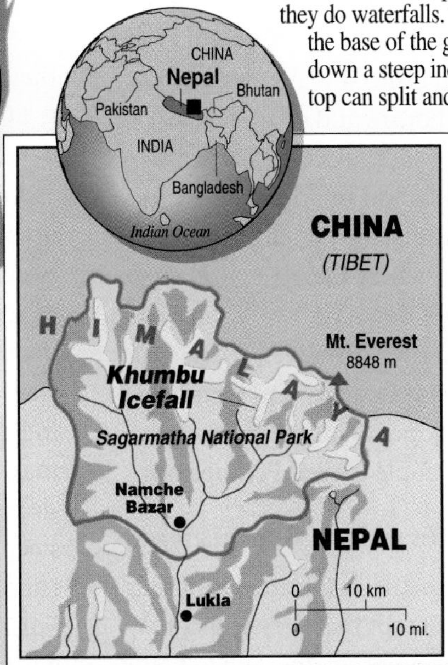

ICEFALL DYNAMICS

As the ice moves downslope (1) it breaks up, forming deep crevasses. Farther down (2) the blocks twist and tilt, then start to recompress (3) forming a roughened surface.

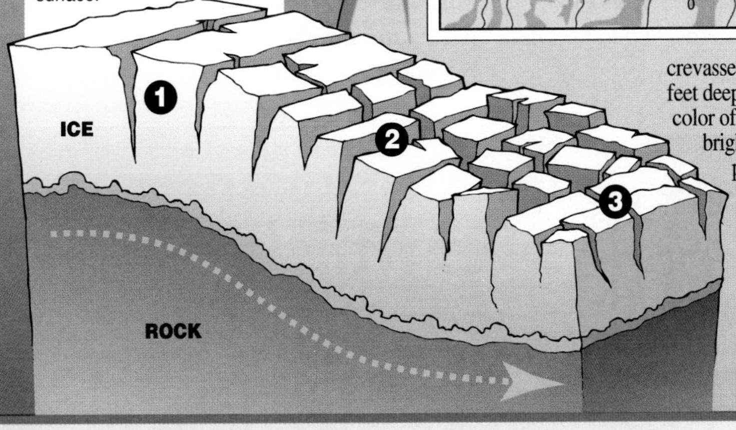

ICE

ROCK

CHINA
(TIBET)

H I M A L A Y A

Mt. Everest
8848 m

Khumbu Icefall

Sagarmatha National Park

Namche Bazar

NEPAL

Lukla

0   10 km
0   10 mi.

CHINA
Nepal
Pakistan   Bhutan
INDIA
Bangladesh
Indian Ocean

Science Journal

In your Science Journal, discuss why mountain climbers consider climbing on ice to be dangerous. What special precautions do they need to take?

What are icefalls?

Purpose

The surface of Earth is constantly changing through the agents of erosion—wind, running water, gravity, and glaciers. These SciFacts describe icefalls, which are caused when a glacier flows down a steep incline. The information provided relates to Section 15-3 by giving students additional insight into glacial features.

Content Background

Glaciers can transform a landscape, leaving behind distinct features. A cirque is formed when valley glaciers erode bowl-shaped basins into sides of mountains. If two or more glaciers erode a mountain summit from different directions, a ridge called an arete or a sharpened peak called a horn is formed. Glaciers also leave behind deep parallel marks or grooves on underlying rock. These grooves indicate the direction in which the glacier moved.

Research

Have students research the geographic extent of Earth's last ice age, which ended roughly 10 000 years ago. Then have students draw maps indicating the extent of glacial coverage.

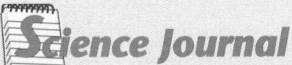

Science Journal

Icefalls are riddled with deep cracks and crevasses. Mountain climbers must be extremely careful not to fall into these cracks because they want to avoid injuries at high altitudes.

Science and Society

Purpose
Science and Society reinforces Sections 15-1 and 15-2 by explaining how gravity and water contribute to erosion around building sites. This excursion also presents landscaping techniques that can minimize erosion on slopes and near bodies of water.

Content Background
Erosion is a factor in land use that has political impact at the local and national levels. Some communities now restrict building on steep slopes to control erosion. Wetlands are being redefined. Zoning decisions about building on or near wetlands may affect runoff and drainage in many areas. Storm damage in coastal areas may result in extensive erosion, threatening existing dwellings. The prevention, or limitation, of such damage has become a political issue in many coastal regions of the United States.

Teaching Strategies Arrange for students to attend local zoning meetings at which erosion-related problems are to be discussed. If this isn't possible, ask a council member to present a summary of the meeting to the class. Encourage students to propose solutions to some of the problems.

Students can apply their experience and knowledge of erosion-related issues by writing letters to local newspapers or legislators to support their views concerning land development.
L2

Science and Society

Developing the Land

Have you noticed that many people live in houses and apartments beside rivers, lakes, and oceans, and on the sides of hills and mountains? If you ask real-estate agents, they'll tell you that people like to live where there's a good view. People like to look down on a valley or watch boats sail along a river. However, when you think of the effects of gravity and water, do you think steep slopes and river banks are good places for people to live? Perhaps not.

Creating Erosion Problems

When people settle in these locations, as seen below, they accept that they will constantly battle erosion problems. When people make a slope steeper or remove vegetation,

they speed up the erosion process.

Once an area that has a natural slope is developed by clearing the land, building asphalt roads and parking lots, and putting up buildings, several effects may follow. Because there is less vegetation to absorb the water from heavy rainfalls, water runoff can increase in volume. This rapidly flowing water may sweep loose soil particles down the hill. The resulting increase in erosion may, over a period of time, actually make the slope of the hill more steep. And the more steep the slope, the more rapidly the water runoff flows, and the more erosion there is. Furthermore, the loss of topsoil may make it harder for any plants to grow and help in stabilizing the remaining soil.

Research

Have students find out about the severe flooding along the Mississippi River and its tributaries during the spring and summer of 1993. Have them summarize their findings, which should include the cause of the flooding and its effects on local communities, in a written report. Use the Classroom Assessment List for Writing in Science in **PASC**, p. 87. **P** **L2**

Activity

You may wish to have students respond to the What Do You Think questions by role-playing. Have students take the roles of mayor, citizens, and zoning board members to discuss the issues described. Local libraries may contain vertical files with current information on local land use for student reference. **L1**

Reducing Erosion

There are a variety of things that people can do to reduce erosion as seen in the photographs above. Planting vegetation is one of the best ways because not only do roots hold soil between them, but plant roots absorb a lot of water. A person living on a steep slope might also build terraces or retaining walls.

You already know that terraces are broad, steplike cuts made into the side of a slope. When water flows onto a terrace, it is slowed down, and is less likely to erode the slope.

Retaining walls are often made of concrete, stones, wood, or railroad ties. Their purpose is to keep soil and rocks from sliding downhill. These walls can also be built along stream channels, lakes, or ocean beaches to reduce erosion caused by flooding, running water, or waves.

People who live in areas with erosion problems spend time and money trying to preserve their land. Sometimes they're successful in slowing down erosion, but they can never eliminate it. Eventually, cliffs cave in, streams overflow their banks, and soil and rocks fall downhill.

Sediments constantly move, changing the shape of the land forever. Erosion is all part of Earth's natural dynamic processes.

What Do You Think?

Suppose you live beside a river. You love it there. It's beautiful, and there's so much to do. The only problem is that the river frequently floods. Several times your family has been evacuated to higher ground. One day, the mayor informs your family that you must move. She tells you that living along the river is not only dangerous, but it costs the city too much money each time you're evacuated. Do you think this is fair? Should communities control where people live?

Also, people often want to rebuild their homes in the same place after their original homes were destroyed by natural erosion. Do you think that federal disaster funds or insurance should be available for homeowners who choose to live in a region that is often threatened by mudslides or river floods?

Answers to
What Do You Think?

Since these questions are subjective, accept all reasonable and well supported answers.

Going Further ⫸

Have small groups of students design and carry out experiments to test the effects of windbreaks, terraces, jetties, and vegetation on erosion. Materials might include paint pans, oblong cake dishes, water, soil, gravel, small pebbles, pieces of sod or patches of grass, and wooden craft sticks. Use the Performance Task Assessment List for Designing an Experiment in **PASC**, p. 23. **L2**
COOP LEARN

HISTORY CONNECTION

Purpose

History Connection reinforces Section 15-3 by explaining how a land bridge appeared across the Bering Sea during various ice ages. This feature also shows how the buildup of glaciers lowered the water level of the oceans, revealing the natural structure that people used as a bridge to North America.

Content Background

The Bering land bridge allowed movement of animals from Asia to North America long before the land bridge was used by humans. What effect did this movement have on human migration? Many scientists think it is likely that some of the earliest human travelers across this bridge were hunters in search of meat.

Teaching Strategy

To reinforce map reading skills, have students use a relief map or globe to look at the Bering Strait and other areas that might have been exposed during the Ice Age. `L1`

Discussion

If possible, display photographs of that area or other arctic regions using sources such as *National Geographic Magazine* to stimulate a discussion of students' responses to the Science Journal questions.

Science Journal

The climate near a melting glacier would most likely be cold all year long. Much fresh water would come from streams of melting ice. The vegetation would probably be similar to that found in cold areas of the world today—grasses and low bushes. People's diets would consist mostly of meat and fish.

HISTORY CONNECTION

Puzzle Solved!

Long ago, scientists had a problem— they couldn't figure out how the first people got from Africa and Europe to

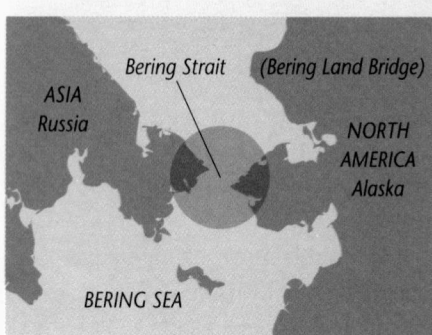

North and South America. Water separates these two continents from all other continents.

Glaciers dominated much of Earth during the last glacial advance, which ended about 10 000 years ago.

When more glaciers were created, the level of water in the world's oceans dropped. As the oceans shrank, more land was exposed. It's like being at the beach and seeing the tide go out, giving you more beach to play on.

When the water level in the oceans went down, a land bridge was exposed near the Arctic Circle. Look at the map to see where Asia and North America nearly touch, north of the Bering Sea. Several times in the last two million years those pieces of land did connect with each other, and early human beings migrated—slowly—to North America. That piece of land,

now under water, is called the Bering Land Bridge.

Scientists hypothesize early people migrated across this land bridge at different times 10 000 to 30 000 years ago. Marshes and forests supported animals such as reindeer, horses, mammoths, mastodons, birds, and fish. There was enough food for people to eat.

Science Journal

In your Science Journal, write about the following questions. What would life be like at the edge of a melting glacier? Would it be cold? Where would people get water? Can plants grow there? What would people eat?

Going Further ▸

Have students research places where people live close to glaciers, for example, Alaska, and find out how the glacier may affect their lives. If possible, have students interview someone who has been to a glacier. Students should ask questions, such as: Do the sounds bother the people? What's it like to walk on a glacier? Is it safe? What kinds of animals live on or near a glacier? How fast does the glacier move? How does it affect the land around it? How tall and wide is it? Use the

Performance Task Assessment List for Group Work in **PASC,** p. 97. `L1`

Science Journal

Review the statements below about the big ideas presented in this chapter, and answer the questions. Then, re-read your answers to the Did You Ever Wonder questions at the beginning of the chapter. *In your Science Journal*, write a paragraph about how your understanding of the big ideas in the chapter has changed.

1 Erosion by gravity can be slow, or fast. *Contrast slow erosion features with fast erosion features that are caused by gravity.*

2 Streams erode and deposit sediments. Fast streams erode more quickly than slow streams. As streams slow down, they deposit more sediments. *What are some ways rivers act upon sediment?*

3 Glaciers form valleys as they push and carry loose materials and scrape against rock surfaces. *Compare a valley formed by a glacier with one formed by a river.*

4 Wind can carry loose particles great distances, as well as erode rock surfaces. *Describe how dunes are formed.*

The activity below will provide students with the opportunity to review the four major agents of erosion.

Teaching Strategy

To review the main concepts of this chapter, ask students to trace in sequence the path of a bit of dirt, from its beginning as a part of a mountain. Tell them that on its way to becoming a bit of dust, the substance undergoes erosion by gravity, a glacier, water, and wind. Have various students diagram the changes the substance might go through in sequence, labeling the agents that cause it to erode or be deposited. **L1**

Answers to Questions

1. Fast erosion causes mudflows and rock slides; slow erosion causes slump and creep.

2. Erosion and deposition

3. Glacier—U-shaped with straight sides and a wide flat bottom River—V-shaped and narrow

4. Sand is blown up the windward side of a pile of sand and down the leeward side, producing a dune.

Science Journal

Did you ever wonder...

• A heavy rainstorm might erode soil along the riverbank. The sediment would make the water appear muddy. (p. 481)

• Rockslides can result from heavy rain or earthquakes and are most common on steep slopes. (p. 476)

• Rocks move downslope by gravity or water and pieces end up in streambeds. Rivers themselves erode rocks and deposit them along riverbanks. (p 481)

Science at Home

To observe the layers of material in a glacier, students can fill a milk carton with ice cubes, sand, soil, gravel, and water. Have students freeze this mixture. Then they can observe it periodically as it melts in a dishpan tilted slightly on one end. Have students keep a log of their observations. Students should notice what happens to the sediment as the water runs off. **L1**

Using Key Science Terms

1. deposition (not a type of erosion)

2. rockslide (not slow erosion)

3. erosion (not a feature of deposition)

4. firn (not a type of glacier)

Understanding Ideas

1. The four causes of erosion are gravity, running water, glaciers, and wind.

2. Erosion is the movement of sediments from where they formed to a new location. Deposition is the accumulation of these sediments.

3. The rate of flow is affected by the slope of the stream, the volume of water, and the amount of sediment in the stream.

4. Roots of growing plants, formation of dunes along some coasts, and keeping the soil wet help decrease wind erosion.

5. Glaciers change the land by scraping, deposition, and erosion.

Developing Skills

1. Student responses should indicate that when the volume of water is increased, more material is carried away with it.

2. Concept Mapping—See reduced student text.

3. Students should observe that the barrier they created stops the flour from moving. Responses should indicate that this type of structure might help prevent wind erosion in deserts and along beaches.

4. Student responses should indicate that strong winds are able to transport heavier and more materials.

Critical Thinking

1. Slope and rate of flow of water, or speed of wind.

2. Large amounts of sediment deposited on the floodplain contain rich organic materials

Using Key Science Terms

continental glacier	firn
creep	floodplain
delta	mudflow
deposition	rockslide
dune	slump
erosion	valley glacier

For each set of terms below, choose the one term that does not belong and explain why it does not belong.

1. mudflow, creep, deposition

2. rockslide, creep, slump

3. erosion, dune, delta

4. continental glacier, valley glacier, firn

Understanding Ideas

Answer the following questions in your Journal using complete sentences.

1. List the four causes of erosion.

2. How is deposition related to erosion?

3. What three things affect a stream's rate of flow?

4. What can help decrease wind erosion?

5. How do glaciers change the land?

Developing Skills

Use your understanding of the concepts developed in this chapter to answer each of the following questions.

1. Comparing and Contrasting Repeat the Explore activity on page 479 varying the amount of water. Compare the results of this activity with those made in the original activity.

2. Concept Mapping Complete the erosion concept map.

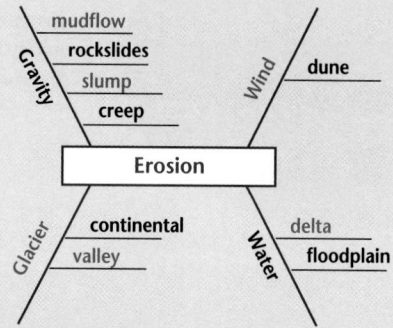

3. Observing and Inferring After doing the Find Out activity on page 495, use the box, flour, several small rocks, a crayon, and a lump of clay to create a structure to prevent wind erosion. Place the structure in the box and blow through the hole. What happened to the flour? Why might this kind of structure be helpful?

4. Comparing and Contrasting Using the materials from the Explore activity on page 493, repeat the activity using a fan or hair dryer with several different speed settings to move the aluminum foil materials. How does the speed of the wind affect the ability to transport the materials?

Program Resources

Review and Assessment, pp. 89–94 [L1]

Performance Assessment, Ch. 15 [L2]

PASC

Alternate Assessment in the Science Classroom

Computer Test Bank [L1]

Critical Thinking

Use your understanding of the concepts developed in the chapter to answer each of the following questions.

1. What factors can increase the rate of erosion?

2. Study the photograph below. Then explain why farmland is common in river valleys. Do you think this river flows fairly quickly or fairly slowly? Why?

3. How does rainfall affect erosion by gravity on a steep slope? How does it affect erosion by wind on level land?

4. How do continental glaciers differ from valley glaciers?

5. How are the processes of erosion by gravity, wind, streams, and glaciers similar?

Problem Solving

Read the following problem and discuss your answers in a brief paragraph.

Imagine that you live in a hilly area. Your family is planning on building a new home.

1. What should they be concerned about when looking for a lot on which to build?

2. What steps should they take to prevent erosion if they build on the side of a slope?

3. How might landscaping with plants help prevent erosion?

CONNECTING IDEAS

Discuss each of the following in a brief paragraph.

1. **Theme—Stability and Change** Explain how rocks from an inland mountain might become sediments in the ocean.

2. **Theme—Stability and Change** How do glaciers affect stream erosion?

3. **Theme—Stability and Change** How can erosion explain the formation of various landforms on a map?

4. **A Closer Look** How does the building of terraces control soil erosion?

5. **Physics Connection** Explain how Landsat information helped explain the cause of dusty conditions in Africa.

Assessment

Portfolio Review the portfolio options that are provided throughout the chapter. Encourage students to select one product that demonstrates their best work for the chapter. Have students explain what they learned and why they chose this example for placement in their portfolios.

Additional portfolio options can be found in the following **Teacher Classroom Resources:**

Multicultural Connections, pp. 33, 34

Making Connections: Integrating Sciences, p. 33

Making Connections: Across the Curriculum, p. 33

Concept Mapping, p. 23

Critical Thinking/Problem Solving, p. 23

Take Home Activities, p. 24

Laboratory Manual, pp. 91–98

Performance Assessment P

and minerals. The river flows slowly across gently sloping land.

3. Rainfall on steep slopes makes sediments heavier and allows them to slide downslope. On level land, rain tends to hold down wet sediments, making them more difficult for wind to move.

4. Valley glaciers form between mountain peaks. Continental glaciers are much larger, covering vast areas.

5. Each process of erosion loosens a substance where it occurs and moves it elsewhere.

Problem Solving

1. They should avoid sites that have a potential to creep or slump or for landslides.

2. A retaining wall; the foundation built on solid bedrock.

3. Plant roots keep water and wind from eroding topsoil. The stems and leaves will act as a windbreak.

Connecting Ideas

1. Rocks on a mountain are eroded by a glacier and carried as sediment when the ice melts and joins a stream. This sediment may then be moved along until a river deposits it in an ocean.

2. Melting ice causes the stream to move faster and carry more sediment.

3. Erosion processes sculpt the land. An alluvial fan or delta is a result of deposition of sediments. Valleys are caused by river or glacial erosion.

4. Terraces break a sloped surface into a series of flat surfaces.

5. It identified locations and the extent of desert expansion in Africa.

Chapter Organizer

SECTION	OBJECTIVES	ACTIVITIES & FEATURES
Chapter Opener		**Explore!**, p. 505
16-1 Succession— Building New Communities (3 sessions; 1.5 blocks)	1. **Describe** the process of succession. 2. **Explain** the relationship between succession and diversity. **National Content Standards: (5–8) UCP3–4, A1, C5, G1–2**	**Explore!**, p. 506 **Find Out!**, p. 507 **Design Your Own Investigation 16-1:** pp. 508–509 **Science and Society**, pp. 528–529
16-2 Interactions in an Ecosystem (4 sessions; 2 blocks)	1. **Classify** and provide examples of the types of interactions that can occur within an ecosystem. 2. **Interpret** how organisms and species interact with each other and with the environment. **National Content Standards: (5–8) UCP3–4, A2, F2–3**	**Investigate 16-2:** pp. 514–515 **Explore!**, p. 520 **Explore!**, p. 521 **A Closer Look**, pp. 518–519 **Health Connection**, p. 530
16-3 Extinction—A Natural Process (2 sessions; 1 block)	1. **Explain** the common causes of local and global extinction. 2. **Provide** examples of ways that human actions within an ecosystem can cause the extinction of species. **National Content Standards: UCP3–4, A2, C5, D2, E1–2, F1–2, F4, G1, G3**	**Find Out!**, p. 523 **Skillbuilder:** p. 525 **Explore!**, p. 526 **Technology Connection**, p. 531 **Earth Science Connection**, pp. 524–525

ACTIVITY MATERIALS

EXPLORE!

p. 505* No special materials are required.
p. 506* No special materials are required.
p. 520* No special materials are required.
p. 521* No special materials are required.
p. 526* No special materials are required.

INVESTIGATE!

pp. 514–515* journal, hand lens or binoculars

DESIGN YOUR OWN INVESTIGATION

pp. 508–509* 1 large, clean jar with lid, dried pond vegetation, droppers, distilled water, coverslips, microscope and slides

FIND OUT!

p. 507* cup of hot coffee, hand lens
p. 523* can, netting, rubber band or tape, 10–15 marbles, ring stand, aluminum pie plate, scissors

KEY TO TEACHING STRATEGIES

The following designations will help you decide which activities are appropriate for your students.

- **L1** Basic activities for all students
- **L2** Activities for average to above-average students
- **L3** Challenging activities for above-average students
- **LEP** Limited English Proficiency activities
- **COOP LEARN** Cooperative Learning activities for small group work
- **P** Student products that can be placed into a best-work portfolio
- Activities and resources recommended for block schedules

Need Materials? Call Science Kit (1-800-828-7777).

00:00 **OUT OF TIME?** We recommend that students do the activities with an asterisk.

Chapter 16 Changing Ecosystems

TEACHER CLASSROOM RESOURCES

Student Masters	Transparencies
Study Guide, p. 58 **Concept Mapping,** p. 24 **Multicultural Connections,** p. 36 **Activity Masters,** Design Your Own Investigation 16–1, pp. 67–68 **Take Home Activities,** p. 25	**Teaching Transparency 31,** Succession **Section Focus Transparency 52**
Study Guide, p. 59 **Making Connections: Across the Curriculum,** p. 35 **Critical Thinking/Problem Solving,** p. 24 **How It Works,** p. 19 **Making Connections: Integrating Sciences,** p. 35 **Activity Masters, 16-2,** Investigate pp. 69–70 **Science Discovery Activities,** 16–1, 16–2, 16–3 **Laboratory Manual,** pp. 99-102, Human Impact on the Environment **Laboratory Manual,** pp. 103-106, Water Pollution	**Teaching Transparency 32,** Exploring a Rain Forest **Section Focus Transparency 53**
Study Guide, p. 60 **Multicultural Connections,** p. 35 **Making Connections: Technology and Society,** p. 35	**Section Focus Transparency 54**

ASSESSMENT RESOURCES	TEACHING & TECHNOLOGY
Review and Assessment, pp. 95–100 **Performance Assessment,** Ch. 16 **PASC*** **MindJogger Videoquiz** **Computer Test Bank**	**Spanish Resources** **Cooperative Learning Resource Guide** **Lab and Safety Skills** **Science Interactions, Course 1, CD-ROM** **Computer Competency Activities**

*Performance Assessment in the Science Classroom

NATIONAL GEOGRAPHIC TEACHER'S CORNER

Index to National Geographic Magazine

The following articles may be used for research relating to this chapter:

- "Hawaii's Vanishing Species," by Elizabeth Royte, September 1995.
- "The Everglades: Dying for Help," by Alan Mairson, April 1994.
- "Dinosaurs," by Rick Gore, January 1993.
- "Rain Forest Canopy: The High Frontier," by Edward O. Wilson, December 1991.
- "Extinctions," by Rick Gore, June 1989.

National Geographic Society Products Available From Glencoe

To order the following products for use with this chapter, contact your local Glencoe sales representative or call Glencoe at 1-800-334-7344:

- NGS Picture Show: Earth's Endangered Environments (CD-ROM)
- Eye on the Environment: Rain Forest (Poster)
- Eye on the Environment: Vanishing Wildlife (Poster)
- STV: Rain Forest (Videodisc)
- STV: Biodiversity (Videodisc)

Additional National Geographic Society Products

To order the following products for use with this chapter, call the National Geographic Society at 1-800-368-2728:

- Endangered Animals: Survivors on the Brink (Video)
- Old-Growth Forest: An Ecosystem (Video)
- The Diversity of Life (Video)

Teacher Classroom Resources

These are key components of the classroom resources package.

TEACHING AIDS

Section Focus Transparencies

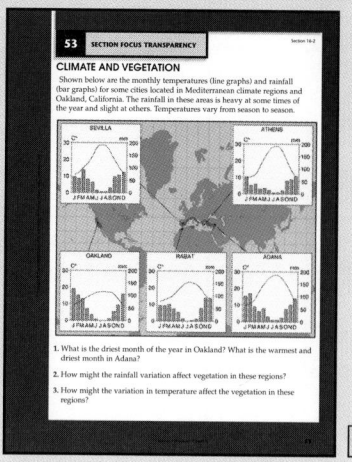

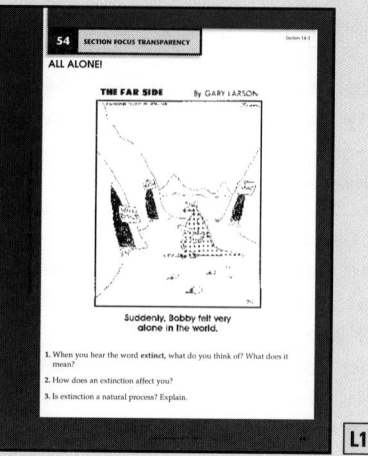

Teaching Transparencies

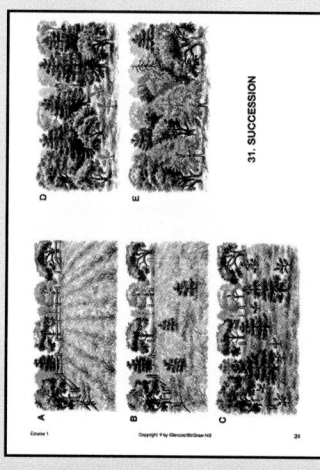

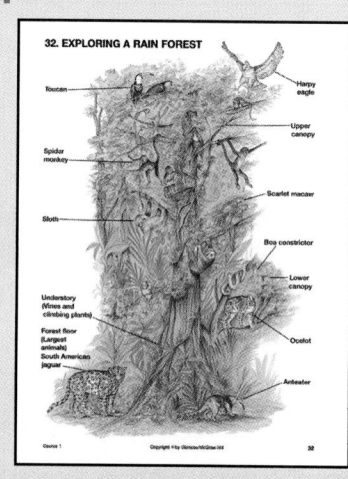

HANDS-ON LEARNING

Science Discovery Activity*

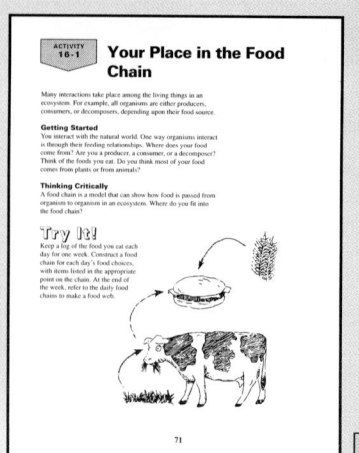

Laboratory Manual*

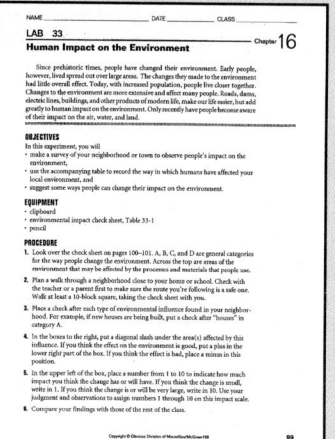

Take Home Activity

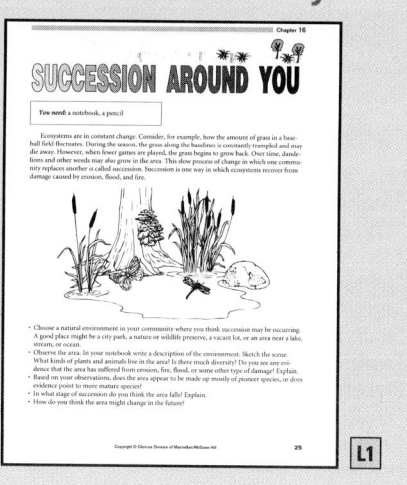

Chapter 16 Changing Ecosystems

REVIEW AND REINFORCEMENT

Study Guide*

Concept Mapping

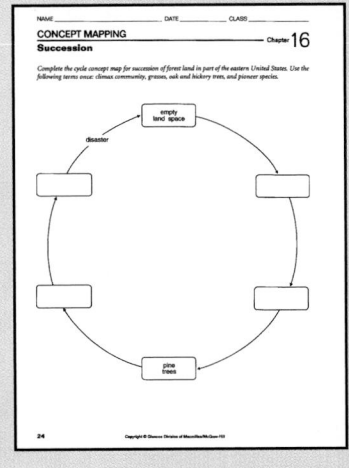

Critical Thinking/ Problem Solving

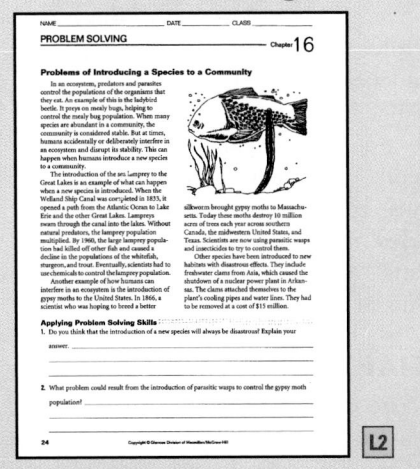

ENRICHMENT AND APPLICATION

Integrating Sciences

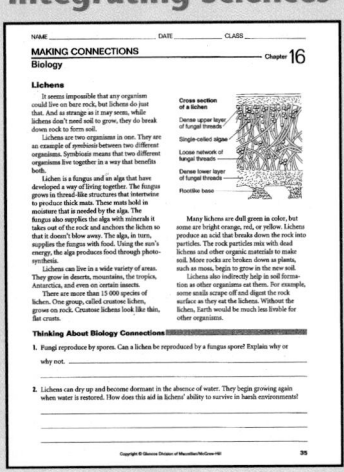

Across the Curriculum

Technology and Society

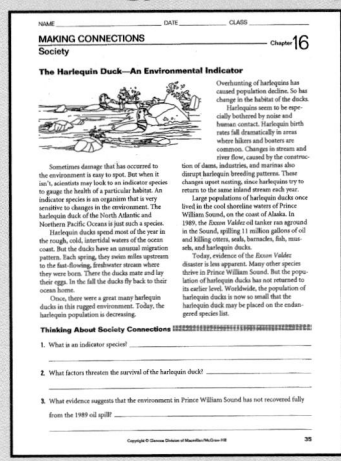

ASSESSMENT

Multicultural Connection**

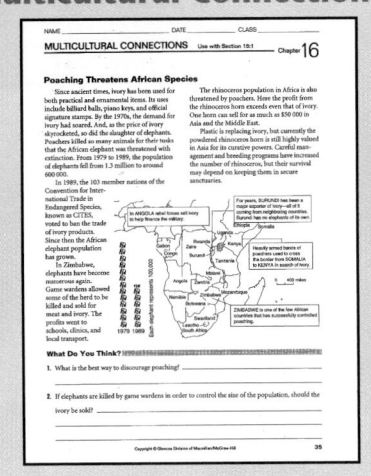

Performance Assessment

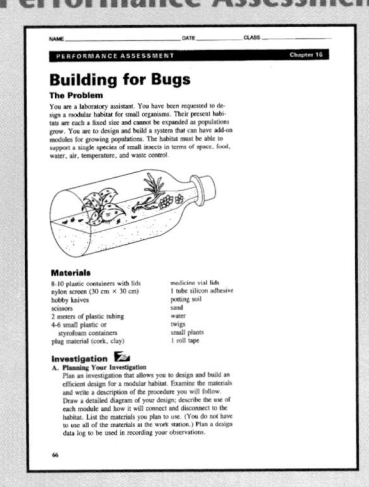

Review and Assessment

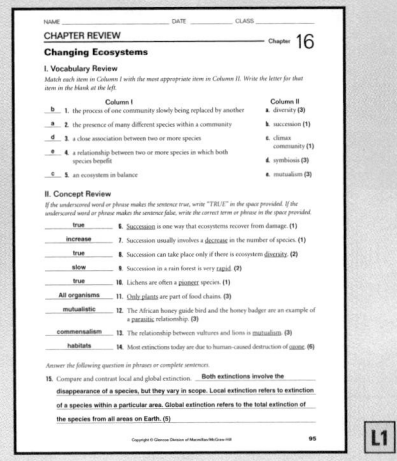

CHAPTER 16

Changing Ecosystems

THEME DEVELOPMENT

The primary theme of this chapter is stability and change. Students learn how gradual change in ecosystems brings new life to areas. The secondary theme of systems and interactions is demonstrated by the varied relationships among the living and nonliving components of ecosystems.

CHAPTER OVERVIEW

The class learns how succession leads to climax communities and how diversity plays a key role in ecosystems. Students also study ways living things interact through mutualism, parasitism, and commensalism.

Tying to Previous Knowledge

Remind students that they studied ecosystems in Chapter 11.

INTRODUCING THE CHAPTER

Ask students how they think an ecosystem might change. Are most changes in ecosystems natural or caused by human actions?

Uncovering Preconceptions

Some students may think that all changes in ecosystems are caused by humans and that most are harmful. As they identify changes in an ecosystem during the Explore activity, guide them to include changes that are natural and positive.

00:00 OUT OF TIME?

If time does not permit teaching the entire chapter, use the Chapter Overview on this page, Reviewing Main Ideas at the end of the chapter, and the Chapter 16 audiocassette to point out the main ideas of the chapter.

CHANGING ECOSYSTEMS

Did you ever wonder...

✓ How a forest can recover from a fire that destroys every living thing in it?

✓ Why some African animals allow birds to sit on their backs all day?

✓ Why some animals are threatened with extinction but others aren't?

Science Journal

Before you begin to study changes in ecosystems, think about these questions and answer them *in your Science Journal.* When you finish the chapter, compare your journal write-up with what you have learned.

Answers are on page 532.

504

I n Chapter 11, you learned about ecosystems and the role of different organisms within ecosystems. What kind of ecosystem do you live in? Has your ecosystem changed during the time you've lived there? Are there more houses now or more cars on the streets?

What other changes are ahead? A new mall nearby? A park? An apartment complex? How will these changes affect you and the rest of your ecosystem?

Now think about a forest ecosystem. Do you think it changes? What about the ecosystem in a pond or under a rock? Do they change?

▶ **In this chapter, you'll learn that most ecosystems change. The plants and animals in the ecosystems sometimes thrive on these changes—and sometimes die because of them. Let's explore some of these changes!**

Learning Styles	Kinesthetic	Investigate, pp. 514–515; Activity, p. 522; Find Out, p. 523
	Visual-Spatial	Explore, p. 505; Discussion, p. 506; Find Out, p. 507; Investigate, pp. 508–509; Visual Learning, pp. 510, 517, 519; Demonstration, pp. 516, 519; Earth Science Connection, pp. 524–525; Project, p. 532
	Interpersonal	Multicultural Perspectives, p. 508; Across the Curriculum, p. 521; Explore, p. 526
	Logical-Mathematical	Across the Curriculum, pp. 516, 518
LS	Linguistic	Explore, pp. 506, 520, 521; Visual Learning, pp. 511, 524; Activity, pp. 512, 527; Discussion, p. 513; A Closer Look, pp. 518–519; Across the Curriculum, pp. 520, 525; Multicultural Perspectives, p. 526

Explore! ACTIVITY

How can change affect an ecosystem?

Imagine that you're walking near a stream and see the animal tracks pictured on these pages.

What To Do

1. *In your Journal,* describe what animals you think caused the different sets of tracks.

2. What other organisms might you find in this ecosystem?

3. Why do you think these tracks are in this location? What is there to attract animals?

4. Think of ways this ecosystem could change. For example, the stream may become polluted. Describe another change that could affect this ecosystem.

5. Explain how the change that you thought of might affect the organisms in the ecosystem.

505

PREPARATION

PREPARATION

Planning the Lesson

Refer to the Chapter Organizer on pages 504A–D.

Concepts Developed

Students will be able to describe how ecosystems begin and evolve through succession. This topic relates to Chapter 11, which describes ecosystems and the role of different organisms within ecosystems.

1 MOTIVATE

Bellringer

 Before presenting the lesson, display **Section Focus Transparency 52** on an overhead projector. Assign the accompanying **Focus Activity** worksheet. `L1`

`LEP`

Discussion Display photographs of different environments in which people live for students to observe, compare and contrast. Have students discuss possible changes and their effect on these environments.

Explore!

How can one ecosystem affect another?

Time needed 10 minutes

Thinking Processes observing and inferring, hypothesizing, comparing and contrasting

Purpose To observe the interrelationships of ecosystems.

Teaching the Activity

Science Journal Have students first list all of the similarities and differences in the two ecosystems. Then have them suggest how the systems are or could be related. `L1`

Section Objectives

■ Describe the process of succession.
■ Explain the relationship between succession and diversity.

Key Terms

succession
pioneer species
climax community
diversity

Thinking About Change

Do you live in a place where some trees change with the seasons? Have you seen pictures of such places? If so, you know that these changes happen every year in many ecosystems.

Other changes in ecosystems don't depend on seasons. For example, have you ever seen a vacant lot or field after it has been cleared by a bulldozer? It's bare and brown, seemingly lifeless.

But then the "empty" lot starts changing. Tiny plants sprout from seeds brought by the wind or animals. In time, bushes and small trees will grow in the lot. Insects will be joined by mice and nesting birds. Under the right conditions, the lot could become a patch of forest, home to squirrels or even raccoons.

If a building is constructed on this lot, the ecosystem would be much different. Start exploring how ecosystems change by analyzing how one ecosystem might affect another.

How can one ecosystem affect another?

506

What To Do

1. *In your Journal,* describe the two main ecosystems in this photograph.

2. Explain ways in which each ecosystem affects the other.

3. If no one comes to live here, what would you expect this area to look like in 10 years?

4. Make a hypothesis to explain how one ecosystem can be transformed into another. Explain how you could test your hypothesis.

Expected Outcomes

Students should recognize the many interrelationships among ecosystems.

Answers to Questions

1. open field and encroaching forest
2. The open field provides new area for the forest to fill, but is inhospitable to main forest plants.
3. Completely forested.

4. Hypotheses will vary, but should include that ecosystems change gradually from one into another.

✓ Assessment

Performance Have students sketch what they expect the area to look like in 10 years. Have them hypothesize what caused the changes. Use the Performance Task Assessment List for Scientific Drawing in **PASC**, p. 55. `L1` `P`

Ecosystems and Succession

When one community is slowly replaced by another, the process of change is called **succession**.

Succession is one way ecosystems recover from damage, such as fires. As you'll see in the next activity, succession is also a way for an ecosystem to begin where none had been before.

How can an ecosystem start?

Do you think an ecosystem can develop in a cup of coffee?

What To Do

1. Carefully observe the hot coffee with a hand lens. *In your Journal,* describe what you see.

2. Label your cup and place it on a counter or shelf away from sunlight. Let the cup rest undisturbed for five days.

3. Observe the coffee each day, using a hand lens. Record any changes you see.

Conclude and Apply

1. Explain what you think happened in the cup.

2. Could the organisms have come from the boiling hot coffee? If not, what could have been their source?

3. Predict what might happen if the coffee sat for five more days.

4. Are the changes you observed examples of succession?

■ Pioneer Species

The organisms you saw in the coffee are called pioneer species. Much like the first settlers were pioneers, **pioneer species** are the first species to live in a new ecosystem. There are many kinds of pioneer species. The first plants that sprout after a forest fire are an example.

Once pioneers settle, does an ecosystem continue to change?

Figure 16-1

Lichens, organisms made up of an alga and a fungus living together, are an example of a pioneer species. Able to live in barren places, the fungus in the lichen makes acids that begin the breakdown of rock into soil.

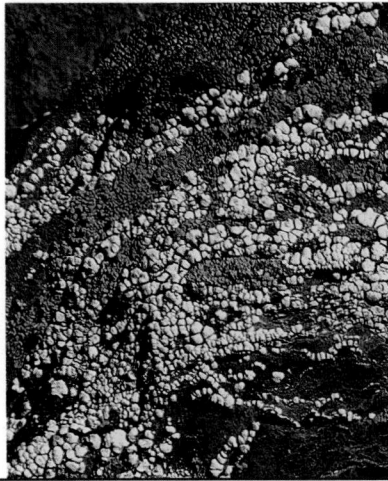

How can an ecosystem start?

Time needed 5 minutes to set up; 5 days to observe

Thinking Processes observing and inferring, hypothesizing, comparing and contrasting, recognizing cause and effect, modeling, forming operational definitions, sequencing

Materials cup, hot coffee, hand lens

Purpose To discover ways in which an ecosystem can start and grow.

Teaching the Activity

Safety Be sure students use potholders and handle the hot coffee carefully.

Science Journal Have students record their observations of the coffee daily. They may draw or describe what they see. **L1**

2 TEACH

Tying to Previous Knowledge

Have students think about times when they have seen mold grow from barely visible fuzz to a small "forest" of plants. Ask how this is evidence of succession.

Expected Outcome

Students will observe that organisms are forming an ecosystem in the coffee.

Conclude and Apply

1. Organisms grew in the cup of cold coffee.

2. The boiling coffee would have killed most organisms. It is likely that many of them came from the air and the surroundings of the counter or shelf.

3. More organisms will appear.

4. The changes are examples of succession.

✓ Assessment

Performance Have students work in groups to repeat the experiment using a variety of hot liquids such as tea, cola, and milk. Before each group begins, have them hypothesize whether organisms will increase more quickly in their liquid than in coffee. Use the Performance Task Assessment List for Evaluating a Hypothesis in **PASC,** p. 31. **L1**

COOP LEARN **LEP**

Preparation

Purpose To observe changes in standing water and relate the changes to the succession of organisms in pond water.

Process Skills observing and inferring, comparing and contrasting, recognizing cause and effect, hypothesizing, interpreting data, making and using tables, forming operational definitions, predicting

Time Required 15 minutes to plan and design data charts, 10 minutes to set up, then 5–15 minutes a day for 5 to 10 days.

Materials Obtain samples of dried pond vegetation or use fresh pond water.

Possible Hypotheses Students should predict that producers will increase followed by an increase in consumer microorganisms. Students may also include a prediction that pond water will become cloudy, greenish, and smell fishy or like seaweed.

The Experiment

Process Reinforcement To reinforce observation and recording skills, have students carefully plan out all the steps of the investigation and study the organisms that may be found in the pond water. Have students define succession.

Possible Procedures Fill the jar with distilled water and sprinkle in some dried pond material or add some fresh pond water. Put a lid on the jar and place it in a well-lit place but not in direct sunlight. Leave undisturbed for 5 to 10 days. Add more distilled water if any evaporates. On day 1, begin observations of the water color, cloudiness, or odor. When the first changes in the water are observed, slides should be made to observe organisms present. Slides should be made from top, middle and bottom of the jar. All three slides should be examined under the microscope on low power and on high power. Draw your observations. Make observations for 5 to 10 days.

Succession

How does a newly dug pond differ from one that has existed for years? In this investigation, you'll simulate a pond-water ecosystem to explore succession and to discover how new ponds fill with a variety of organisms.

Preparation

Problem
How does a pond-water ecosystem change?

Form a Hypothesis
As a group, write out a statement that predicts what will happen to the populations of organisms in a new pond-water ecosystem. Include in your hypothesis changes you might expect in color, smell, and other characteristics of the water.

Objectives
• Predict what happens in the succession of a pond-water ecosystem.
• Observe and explain changes in the ecosystem.

Materials
1 large, clean jar and lid
dried pond vegetation or pond water
distilled water
eyedroppers
microscope, slides, and coverslips

Safety Precautions
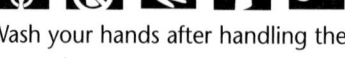

Wash your hands after handling the materials in this investigation.

508

Multicultural Perspectives

Balance in the Rain Forest

The people who live in rain forests juggle two ecosystems as they farm the land. They begin by clearing a small area of forest and burning the trees. The ashes help fertilize the soil. They plant manioc, maize, banana trees, yams, and other plants native to the rain forest.

A plot is gardened for about three years, until the soil nutrients are nearly exhausted. Then the plot is allowed to be reclaimed by the forest ecosystem before the land becomes completely infertile.

Have groups of students research the growing of crops in the rain forest and the succession of plant life as the forest reclaims the land. Use the Performance Task Assessment List for Group Work in **PASC**, p. 97.

DESIGN YOUR OWN
INVESTIGATION

Plan the Experiment

1 Examine the materials provided. Decide how to use them to make a pond-water ecosystem.

2 How long will you conduct your investigation? How will you make observations? How often will you make observations?

3 What will you be observing? Some things to observe are water color, cloudiness, odor, sediment, and other factors that may change. Microscopically, look for organisms seen on page 509. Record how the number of organisms increases or decreases.

4 Design a data table in your Science Journal.

Check the Plan

1 How will you know when one organism increases and another decreases?

2 Make sure you have a variable and a control.

3 Make sure your teacher approves your plan before you proceed.

4 Carry out the investigation. Record your observations.

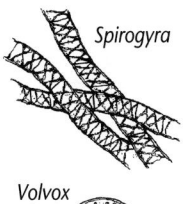
Spirogyra

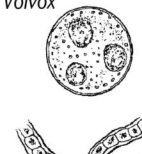
Volvox

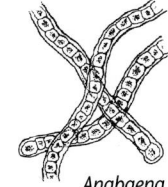
Anabaena

Daphnia   *Hydra*   *Euglena*   *Paramecium*   *Rotifer*

You may see organisms like these in your pond-water samples.

Analyze and Conclude

1. **Infer** What was the source of the organisms in the pond-water ecosystem?

2. **Observe** What changes occurred that were observable without a microscope?

3. **Observe** What changes occurred that were observable with a microscope?

4. **Observe** How many different organisms did you observe the first day? The last day?

5. **Compare and Contrast** Did any of the organisms increase in number? Decrease? Explain how this may have occurred by making a general statement about succession that explains what happened in your pond ecosystem.

Going Further

Have students work in small groups to produce rough graphs that reflect the relative changes in the populations of organisms that they observed. What kind of graph would be best to use?

Program Resources

Teaching Strategies

Science Journal Students should make a chart in their Science Journal that allows room for recording observations with and without a microscope. Drawings from three slides each day will take a lot of room.

Expected Outcome

The water may first undergo an explosion of algae and/or *Euglena*. These populations may decrease slightly once more complex microorganisms—*Hydra, Rotifers*—are provided with a food source.

Analyze and Conclude

1. the pond vegetation

2. Water darkened and became more opaque, smell became more vegetative, and more organisms appeared.

3. Organisms of various kinds increased in number.

4. There were more organisms on the last day than there were on the first.

5. Algae and *Euglena* should have increased at the beginning. They are food producers. Then their numbers should have decreased at least relative to increases in other populations. Food producers increase in the beginning of pond life. Then food consumers, such as *Hydra* and *Rotifers,* increase as they consume the food producers.

Going Further

Graphs will differ but a horizontal bar graph would work well to represent increases in population over time.

✔ Assessment

Process Have students predict the kinds of changes that might take place in the jar during the next 10 days. Some students may carry out observations to find out whether the predictions were supported. Use the Performance Task Assessment List for Analyzing the Data in **PASC,** p. 27. L1 **COOP LEARN**

Succession in Other Ecosystems

Figure 16-2

Ⓐ Tallgrasses thrive in an abandoned field.

Ⓑ Pine seedlings reduce the amount of sun reaching the ground and the tallgrasses. The tallgrasses begin to die out.

Ⓒ Increasing shade keeps more pines from sprouting but provides the right amount of sun for oak seedlings.

Ⓓ Hickory trees, which grow well in shade, join the oaks. The ecosystem no longer provides enough sunlight for the tallgrasses.

Ⓔ This ecosystem has become a stable climax community of trees well-suited for the environment.

510 Chapter 16 Changing Ecosystems

In the Investigate, you watched pond water fill with greater numbers and different types of living things. You may have noticed that some of the organisms you first saw in the water later disappeared. They may have served as food for other organisms. Thus, some kinds of organisms disappeared from the ecosystem, while others increased in numbers. This pond water was one example of succession. **Figure 16-2** shows another example of succession common in one type of forest in the eastern United States.

As an environment changes, new conditions make it possible for other species to grow, reproduce, and increase in number. As these species gradually take over the area, they cause even more changes in the environment, making it suitable for still different species.

As plant populations in the forest change, so do animal populations. Trees provide increasing amounts of food and shelter, becoming home to birds and small animals. As the number of small animals increases, they become numerous enough to support predators, which then enter the ecosystem.

Eventually, the ecosystem reaches the last stage of succession and becomes a climax community. In a **climax community**, plant and animal species living in the community are well adapted to the conditions. They make up an ecosystem that is in balance.

How Succession Can Repair Damage

Climax communities don't remain undisturbed forever, of course. A storm may knock over a tree. With the tree down, more sunlight reaches the ground. Succession begins again in this area, starting with the grasses that need full sunlight to grow.

The tree decays. As it does, the tree becomes a source of food and shelter for insects and animals. The decay of the tree releases nutrients which enter the soil. These nutrients will soon help feed pine seedlings, then oaks, then maybe hickories.

Sometimes an ecosystem suffers more damage than a fallen tree. The ecosystem on Mount Saint Helens was mostly destroyed by a volcanic eruption in 1980. A blast of superheated steam and rocks flashed down the mountain. The eruption flattened the forests and buried the slopes under tons of mud and ash.

Succession began on Mount Saint Helens once again as pioneer species started growing. Fireweed, named because it's one of the first plants to appear after a fire, pushed its way through the crust of ash.

Some plant roots had survived the blast underground, and from them new plants grew. Other organisms, including cottonwood seeds and tiny spiders, were carried to the slopes by the wind or birds and small animals.

Soon hardy plants created green islands of life in a sea of gray ash. The mountain was on its way to recovery. Study **Figure 16-3** to see succession on Mount Saint Helens.

Figure 16-3

Ⓐ The eruption of Mount Saint Helens in 1980 destroyed the ecosystem around the mountain.

Ⓑ The first evidence that Mount Saint Helens would recover appeared within thirty days when fireweed began pushing its way through the ash.

Ⓒ Soon, hardy lupine plants created green islands of life in a sea of gray ash. Mount Saint Helens was on its way to recovery.

Content Background

Some species have adaptations that allow them to survive fire. Some coniferous trees that have been damaged by fire respond by producing huge numbers of cones. The cones from lodgepole pines and western larch trees release their seeds only after being heated in a fire. These seeds sprout quickly and thrive in sunlight.

They are often the first generation of trees to grow after a fire. They shade the seedlings of species that serve as the next stage in forest succession.

NATIONAL GEOGRAPHIC SOCIETY

Videodisc

STV: Rain Forest

Slash-and-burn farming; Brazil

51896

Bulldozer clearing forest

51758

Cleared forest; Amazon

51865

STV: Biodiversity Destroying Diversity

Unit 1, Side 1
Destroying Diversity

00386-22598

Preserving Diversity

Unit 2, Side 1
Preserving Diversity

22598-44538

Visual Learning

Figure 16-3 This figure shows two of the first plant species to return to Mount St. Helens: fireweed and lupine. Have students study the rebirth of the Mount St. Helens ecosystem to discover in greater detail how the process took place. How did the first plants and animals play roles in the return of later species? Students can prepare oral or written presentations to communicate their findings. L1 P

3 ASSESS

Check for Understanding

Ask students to answer Check Your Understanding questions 1–3 individually and question 4 in groups. Have the groups share their responses with the class.

Reteach

Discussion To help students understand the value of diversity, ask them to imagine a forest with only three species of trees and a forest with twenty species of trees. A disease kills a species that grows in both forests. Have students explain which forest will be more damaged by this disease. *The forest with three species, because one-third of its trees will die. In the other forest, if all species are represented equally, only one in twenty trees will die.* L1

Extension

Predicting Ask students to predict how the school lawn or a nearby park might change if it were never mowed again. Have them illustrate the stages of succession by drawing the lawn as it is now, as they think it would look in five years, and as it might look in fifty years. L3

4 CLOSE

Activity

Ask small groups to list three ways that people interfere with natural succession. Have the groups share their lists with the class. L1 COOP LEARN

DID YOU KNOW?

Succession in a rain forest is very slow. Part of one rain forest in Cambodia was cleared in 1431. It has had hundreds of years to recover, but the plants growing there now are still different from those in the typical, surrounding rain forest.

Diversity: The Key to Life on Earth

If ecosystems can recover by themselves, why should people worry about protecting forests or other environments?

The natural processes of succession can repair only a limited amount of damage. One of the most important elements in the recovery of an ecosystem is the diversity within an ecosystem. **Diversity** is the presence of many different species within a community. Diversity is one reason that succession can take place.

For example, on Mount Saint Helens, not many organisms appeared the first summer after the eruption. However, the organisms that did appear belonged to many different species. These plants and animals made it possible for other species to take root or to find food and shelter.

One example of recovery on Mount Saint Helens involves pocket gophers. These animals tunneled into the ground, bringing soil to the surface to mix with the nutrient-rich ash. In this soil, seeds and fungi sprouted. Without gophers to mix the old soil with nutrient-rich ash, fewer plants would have grown. The recovery of the whole ecosystem might have been much slower.

Succession and the recovery of ecosystems depends on diversity, or the many species that live in an area. Anything that is a threat to diversity is also a threat to the ability of an ecosystem to recover from disasters.

Figure 16-4

Pine seedlings, which thrive in sunlight, are essential to the growth of shade-loving oaks and hickories. If all the pine trees are cut for lumber, the forest will have a difficult time healing itself.

check your UNDERSTANDING

1. What is the relationship between succession and diversity?
2. Are oaks a pioneer species? Why or why not?
3. Does replanting a forest with one species of tree replace the forest? Why or why not?
4. **Apply** Are the species in your own ecosystem becoming more diverse or less diverse? Explain your answer.

---

check your UNDERSTANDING

1. Succession depends upon the diversity of an ecosystem for replacement of species.

2. Oaks are not pioneer species; they are not the first plants to grow to maturity after a catastrophe.

3. No. A forest is an ecosystem that has many communities. Each community contains many species. Replacing a diversity of species with one species does not replace a forest.

4. Answers will vary depending on your local community. Look for student observations and logic.

Interactions in an Ecosystem

How Do Living Things Interact?

What do you know about the interactions among living things? In Chapter 11, you learned that organisms can be producers, consumers, or decomposers. All living things, including you, are parts of food chains.

But there are other types of interactions also. For example, why do certain kinds of mussels allow crabs to live in their shells?

These interactions can be as simple as a consumer eating a producer or a predator catching its prey or the interactions can be much more complex as seen in **Figure 16-5B**. In the Investigate on the following pages, you'll observe and record some complex interactions in an ecosystem.

Figure 16-5

Organisms in an ecosystem can interact in many different ways.

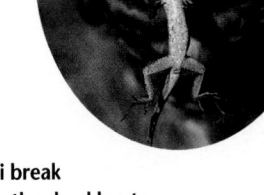

A An owl eats a lizard.

B Fungi break down the dead log to obtain food. Some of the log will become part of the soil.

C Populations of gazelles, zebras, springbucks, and giraffes live and interact with one another at this waterhole in Namibia, Africa.

513

Section Objectives

- Classify and provide examples of the types of interactions that can occur within an ecosystem.
- Interpret how organisms and species interact with each other and with the environment.

Key Terms

symbiosis
mutualism
commensalism
parasitism

Program Resources

Study Guide, p. 59 L1

Making Connection: Across the Curriculum, p. 35, The Statue of Liberty L1

Critical Thinking/Problem Solving, p. 24, The Problem of Introducing a Species to a Community L2

How It Works, p. 19, Biological Control L1

PREPARATION

Planning the Lesson

Refer to the Chapter Organizer on pages 504A–D.

Concepts Developed

Students will study interactions within ecosystems, including competition, mutualism, parasitism, and commensalism. They will also learn ways that living things modify their environments to meet their needs.

1 MOTIVATE

Bellringer

Before presenting the lesson, display **Section Focus Transparency 53** on an overhead projector. Assign the accompanying **Focus Activity** worksheet. L1 LEP

Discussion Discuss all the ways organisms might interact with one another, other than predator-prey relationships. Students may suggest birds nesting in trees, insects pollinating flowers, or mosquitoes sucking blood. Point out to them that such relationships are frequently vital to the survival of one or both of the organisms involved.

2 TEACH

Tying to Previous Knowledge

Encourage students to share what they remember about food chains and food webs from Chapter 11. Then divide the class into three groups (six groups if your class is large). Ask each group to construct on the chalkboard a food chain that might occur in a forest, a park, or the ocean.

Theme Connection Students learn how living things interact to meet their needs.

16-2 Getting Up-Close and Personal

Time Needed 20 minutes a day for one week

Purpose To observe how organisms within an ecosystem interact.

Thinking Processes classifying, observing and inferring, comparing and contrasting, recognizing cause and effect, hypothesizing, collecting and organizing data, interpreting data, forming operational definitions, making and using tables, predicting

Materials journal, hand lens or binoculars

Teaching the Activity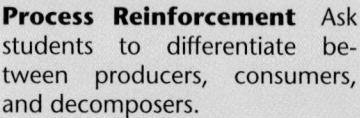

Process Reinforcement Ask students to differentiate between producers, consumers, and decomposers.

Possible Hypotheses Students may hypothesize that a complex network of relationships exists between the various organisms found in their ecosystems.

Possible Procedures Students may choose an arbitrary plot of land or water, such as a 4-foot-square patch of weeds and shrubs, or a naturally defined feature, such as a small pond, a stand of saplings, or the trunk and branches of a tree. Even grass growing in sidewalk cracks can provide opportunities for observation.

Expected Outcomes

Students may be expected to find more organisms, showing more complexity of interrelationships, than they had anticipated. They may find organisms whose "role" in this environment is not immediately discernible. They may be able to contrast consumers which prey on producers with those that prey on other consumers; they may recognize how the habits of one organism impact upon some or all of the other local organisms.

INVESTIGATE!

Getting Up-Close and Personal

Could you survive on a deserted island? Not without food! To get food you'd have to interact with other organisms in the ecosystem. In this activity, you'll find out more about how organisms in an ecosystem interact with one another.

Problem
How do certain organisms interact in an ecosystem?

Materials
Journal
hand lens or binoculars

What To Do

1 Choose an ecosystem near your school or home. It might be in a cluster of trees, a rotting log, a pond, a patch of weeds, or another setting.

2 Identify at least two organisms that are interacting within this ecosystem. You can include organisms that are not always present, but leave evidence of their interaction through tracks or feathers.

3 *In your Journal,* create a table to record and date your observations.

Program Resources

Activity Masters, pp. 69-70, Investigate 16-2

Making Connections: Integrating Sciences, p. 35, Lichens [L2]

Science Discovery Activities, 16-1, Your Place in the Food Chain; 16-2, Is Oil Really Slick; 16-3, Oil Eating Bug

Section Focus Transparency 53

Teaching Transparency 32 [L2]

Laboratory Manual, pp. 99-102, Human Impact on the Environment [L2]; pp. 103-106, Water Pollution [L2]

American sparrow hawk

Birdwatchers rely on binoculars to locate and observe both the American sparrow hawk and its tiny prey.

4 Over the next week, plan as many observations as possible. Schedule them for different times of the day.

5 Use a hand lens and/or binoculars to study the organisms you chose. Be sure to record *in your Journal* how these organisms interact with each other and with the environment.

Analyzing

1. Describe the environment of your ecosystem.

2. **Spreadsheet** List all the populations of organisms present in the ecosystem.

3. Which organisms did you study? Are they producers, consumers, or decomposers? Put these data in your spreadsheet.

4. What evidence did you find of competition within the ecosystem? Cooperation? Interaction between organisms and their environment?

Concluding and Applying

5. What did each organism you studied do that helped it survive?

6. What might happen if one or both of the organisms you studied disappeared from this ecosystem? In what ways would the ecosystem be affected?

7. **Going Further** Think of a change you could make in this ecosystem that would not deliberately damage it. *Predict* how the two organisms you studied would react to the change you suggest. Then, test your prediction.

Answers to Analyzing/ Concluding and Applying

1–4. Answers will vary depending on ecosystems observed. Check for completeness of descriptions and logic of answers.

5. Students should point out how consumers got food and shelter and how producers got sunlight and water.

6. Students should realize that the loss of one or the other organism would result in upsetting other parts of the ecosystem.

7. Answers will vary depending on the ecosystem chosen. Be certain students' actions are not deliberately or inadvertently destructive. Students may realize that some actions taken (watering a lawn or garden) can actually benefit an ecosystem.

✔ **Assessment**

Process Have students list the ways that the organisms they studied rely on one another for survival. Did one organism need another to provide something for it that it could not easily provide for itself? Use the Performance Task Assessment List for Analyzing the Data in **PASC**, p. 27. L1 P

Demonstration Bring a bunch of flowers to the class, or bring the class to a point on the school grounds where bees can be seen among plants. Ask students to suggest why a bee would be interested in a flower (to collect nectar for honey production). Then ask students to suggest what benefits, if any, a plant would gain from the bees' presence (pollen dispersal for reproduction). Point out that the survival of both species depends upon their constant interaction.

GLENCOE TECHNOLOGY

 Videodisc

STVS: Ecology

Disc 6, Side 1
Unusual Estuary (Ch. 6)

Preserving Duck Habitats (Ch. 7)

Bird Sanctuary (Ch. 9)

Bald Eagles (Ch. 13)

Bighorn Sheep Transplant (Ch. 14)

Saving the Spotted Owl (Ch. 15)

Competition and Cooperation

Recall the discussion of competition in Chapter 11. In the Investigate you just did, you may have seen two or more species compete with one another for the limited resources of the environment. Plants and animals often compete for food, water, sunlight, and living space. If two species need a resource that is in limited supply, one species may eventually be forced out of the ecosystem.

Florida scrub jay

Individuals within a species may also compete with each other, especially for food, living space, and mates. The individuals within species that survive are those best adapted to the current conditions in the ecosystem. For example, giraffes with longer necks might be better adapted to conditions during which all the lower branches of their food trees have been stripped bare.

Animals of the same species also cooperate, often in family groups. One example is the Florida scrub jay. Pairs of these birds have "helpers," usually their offspring from the year before. The helpers watch for predators and feed the new chicks. Instead of competing with older jays for territory, the helpers stay "at home" and inherit their parents' feeding and living space.

Individuals of different species may also form cooperative relationships. For example, some species of barnacles will only settle and live on whales. A close association of two or more species is called **symbiosis**. In this section, you will learn about three types of symbiotic relationships—mutualism, commensalism, and parasitism.

■ Mutualism

How can two species cooperate? Take the example of the pea crab and

Figure 16-6

All of the pictures on these pages show mutualism. Mutualism often provides both of the organisms that are interacting with food, protection, or both.

 A Acacia ants eat a sweet substance produced by bull's horn acacia trees in Costa Rica, and live safely in the tree's large thorns. The ants protect their home and food source by biting animals that try to eat parts of the tree.

B The cleaner wrasse (next to the pectoral fin) eats organisms that live on the coral cod. Irritating pests are removed from the coral cod and the cleaner wrasse gets food.

516

Meeting Individual Needs

Learning Disabled To help students understand the concept of mutualism, ask them to identify a mutualistic relationship that occurs at school. Stress that both people in this relationship must benefit from it. *Examples: studying together, working on group projects, trading food at lunch, playing on sports teams.*

Across the Curriculum

Mathematics

Approximately 1.7 million species of organisms have been identified on Earth so far. To give students an idea of how large a number this is, ask students to figure out about how many days it would take for them to name all 1.7 million species if they said one species per second and did not stop to eat or sleep. *It would take about 20 days.*

the mussel. Tiny pea crabs live inside mussel shells. The crabs eat the young of organisms that would harm the mussels if they grew to adults inside the shell. In return, the mussels provide protection for the little crabs. In this relationship both species benefit. A relationship in which both species benefit is called **mutualism**.

Some fungi and plants have a mutualistic relationship. From 70 to 100 percent of all trees, grasses, shrubs, and flowers in any area grow well thanks to the fungi that grow on their roots.

Fungi, which can't make their own food, get food from the plant roots. In return, the fungi make huge, under-ground, threadlike nets that extend the plants' roots. With a larger root system, plants can get more nutrients from the soil.

Mutualistic relationships can help organisms get protection, food, additional nutrients, comfort and better health, free from pests.

■ Commensalism

You've read that certain fungi and plants have a mutual relationship. Some plants such as Spanish moss, orchids, and staghorn ferns grow high up in trees. These plants also have a relationship with the trees on which they grow. The host tree provides a safe growing place for the plants.

Connect to...
Earth Science

Soils are extremely important to all life on Earth. Find out how soils form and how plants and animals contribute to soil formation.

Connect to...
Earth Science

Answer: Soils form when weathering processes break down rocks and the natural processes of decay cause plant and animal matter to be added to the sediment.

Visual Learning

Figure 16-6 After reading the captions as a class take a closer look at the Acacia ants. To reinforce the concept of mutualism, use the acacia ants as a springboard for discussing mutualism practiced by ant species. The bull's head acacia has large, hollow thorns which provide an ideal nesting space. The acacia oozes a sweet, sticky sap for the ants to eat. The ants will ferociously defend the acacia against insects or animals—including humans—that try to harm it.

Other ants have also been known to keep "fungus gardens" in their nests, fertilizing patches of fungus with plant debris in order to obtain the tiny, nourishing knobs which the fungus produces.

Still other ants raise "livestock," protecting populations of aphids or plant lice from predator insects because the ants eat the sweet, honey-like liquids the "livestock" excretes. Students could divide into groups and prepare research presentations on one of these mutualistic ant species.

C The oxpecker finds and eats ticks and blood-sucking flies from the rhinoceros's skin. The rhinoceros is rid of the irritating pests, and the oxpecker gets a good meal.

D An African honey guide bird locates a nest of honey bees and makes an effort to attract a honey badger's attention and lead it there. The badger breaks open the nest and eats the honey. The honey guide—too weak to break open the nest itself—eats the bees' grubs and wax.

16-2 Interactions in an Ecosystem **517**

Coral polyps are actually predatory animals, related to jellyfish and sea anemones. During the night, as the movement of the water brings plankton within reach, the coral polyp reaches out with poisonous tentacles to paralyze its prey and gather it into its mouth.

During reproduction, polyp buds form on the sides of the adult polyp. They stay attached and more buds form. In this way, the coral colony slowly increases in size.

Across the Curriculum

Math

After explaining how coral polyps reproduce, ask students to figure out the total number of polyps that would result in this situation: eight polyps produce five polyps each. In the second generation, half of the polyps produce five polyps each and half produce six polyps each. How many polyps does the colony now include? Multiply 8 × 5 (40); multiply 20 × 5 (100) and 20 × 6 (120); add the original 8, plus 40 in the first generation, plus 220 in the second generation, for a total of 268 polyps.

GLENCOE TECHNOLOGY

CD-ROM

Science Interactions, Course 1, CD-ROM

Chapter 16

At the same time, the plants get nutrients from rainwater and don't harm the host. However, the host tree doesn't benefit from the relationship. **Commensalism** is a relationship in which one organism benefits and the other neither benefits nor is harmed. See **Figure 16-7** for an example.

Some species of animals also live in commensal relationships. These animals rely on other animals in their search for food. For example, the remora fish, a weak swimmer, attaches itself to sharks and feeds on their leftovers. In a similar way, vultures follow lions and other predators and feed after they leave.

House sparrows can sometimes build their nests beside those of eagles. The sparrows are protected by the eagles' presence, while the eagles usually prey on rodents or birds larger than the sparrows.

In each commensal relationship, only one of the species benefits from having a host or protector, but the other species is not harmed.

■ Parasitism

You've already learned about two types of symbiosis—mutualism and commensalism. You may already know something about the third type—parasitism. Explore the effects of parasitism on different organisms with the activity on page 520.

A Treasure in Danger

Coral reefs may not look like rain forests, but they're similar. Like rain forests, reefs cover a small part of Earth (0.17 percent of the ocean). Reefs are home to thousands of species—nearly a quarter of all the species in the ocean. Reefs, like rain forests, are also threatened with extinction.

What Is a Reef?

A reef consists of thousands of coral polyps. A coral polyp is an animal that makes a limestone cup around itself. The cup serves as the polyp's skeleton.

Algae live inside the coral polyp and have a mutualistic

Coral polyps

Algae that live in coral tissues

518

Purpose

To discuss coral and its role in marine ecology, and its sensitivity to environmental degradation.

Content Background

Despite its rocky form, a coral reef is extremely fragile and vulnerable. A reef suffers greatly when changes onshore "bleed" into the sea, clouding the shallow waters in which it grows. Inadequately treated sewage, dumped in by coastal cities, may induce unnaturally high levels of algae or microorganisms in the waters, blocking sunlight and monopolizing nutrients. Other damage to the reefs may be deliberately inflicted. Salvage operations searching for sunken ships frequently dynamite coral reefs.

The effect of this destruction can be catastrophic. With the demise of a reef, the local fish populations it supported may disappear and seashores initially protected from erosion can suffer severe erosion.

Figure 16-7

A Spanish moss, which is not really a moss, hangs on trees in the southeastern part of the U.S. and in tropical South America.

B Spanish moss uses its host only for support. The plant feeds on airborne dust, has no roots, and absorbs water directly from the air.

C The tree on which it lives is not harmed by the Spanish moss. This is an example of a commensal relationship.

Demonstration Provide a memorable demonstration of a commensal relationship between humans and mites. Rub a spatula or dull knife over the skin of your forehead to gather oily material. Use a cover slip to scrape the material onto a drop of immersion oil on a slide.

Invite students to observe the slide under a microscope. They may see one or both varieties of the mites that live harmlessly in the hair follicles and sebaceous glands of most humans.

Visual Learning

To reinforce the concept of commensalism, have students look at the picture of the Spanish moss in Figure 16-7. This plant is a member of a group of plants known as *epiphytes*. Mosses, algaes, lichens, cacti, and orchids can all be epiphytic. These plants use other plants only as a place to sit: they get water and nutrients from the air or from debris which settles upon them. In fact, the moss does not require a commensal relationship with the tree in the picture at all—it would grow just as healthily hanging from a telephone line.

---

relationship with the coral polyp. The algae produce food and oxygen which the coral polyp can use. In return, the polyp gives the algae a home and important nutrients.

Like other plantlike protists, these algae need sunlight to survive. That's one reason why coral live in clear, shallow water near coastlines.

Pollution and Reefs

Pollution and sediment from erosion can block sunlight, bringing about the death of the algae. Logging of mangroves that grow along coasts causes soil to erode into the ocean, and contributes to the death of the algae. In addition, when coral polyps are stressed by

A coral head damaged by human activity

water conditions, they expel the algae.

Without the algae, the coral turns white or "bleaches" and can't grow or reproduce. After the polyps die, their limestone skeletons soon begin to erode.

Now that half of Earth's population lives along coasts, sewage increasingly pollutes the ocean. In addition, shipping leads to large and small oil spills. All of these problems contribute to the damage and death of coral reefs.

You Try It!

In what other ways could nations that currently dump sewage into the oceans treat sewage? Develop a suggestion for the treatment of sewage, and then develop a way to test whether your solution might work. Present your work in the form of a poster or display to the class.

Teaching Strategy

Build a model of a reef: fill a large, shallow dish or baking pan with water. Build a high "beach" at one end out of sand and place a "reef" just offshore made of stones, blocks, or bricks. Then push waves toward the beach from the "ocean" end of the pan, pointing out to students how the reef blunts the impact of the waves and helps keep sand onshore. Now remove a large percentage of the reef and send waves toward the shore, pointing out how much faster the shoreline degrades. **L1**

Answers to
You Try It!

Students may suggest modern sewage-treatment plants, or using artificial wetlands as a settling area for wastes before water is discharged into the sea. Other answers are possible; look for logic of student answers.

Going Further ⑊⑊⑊▶

Sometimes human intervention can help repair or replace damaged reefs. Some scientists have led efforts to develop "reef starters," concrete or metal forms which are submerged and seeded with coral polyps. Have interested students investigate current efforts to build artificial reefs. **L2**

Explore! ACTIVITY

How do parasites fit in?

What To Do

1. *In your Journal,* describe at least five parasitic relationships in nature. (Consult a reference book, if necessary.)

2. Explain how each relationship is harmful to one of the species involved.

3. Describe the role of each parasite in its own ecosystem.

4. Describe at least one parasite that is useful to humans.

In your list of parasites, did you include organisms like mosquitoes and bacteria? Both of these are examples of parasites.

In the relationship called **parasitism**, one of the species harms or kills the other one.

Many species of animals and plants, such as ticks, lice, tapeworms, and heartworms, feed on other animals and plants. Parasites feed on their hosts, slowly weakening them. Usually a parasite does not kill the organism it feeds on, but it does cause the host organism harm. In addition, mosquitoes and other blood-sucking parasites can give their hosts serious diseases, such as malaria.

When you did the Explore activity you may have thought of many examples of parasitism. Parasites always survive at the expense of the host organism.

Figure 16-8

Mistletoe, a plant that grows on forest trees in both the U.S. and Europe, is a partial parasite. It makes some of its own food through photosynthesis, but also absorbs sap from its host tree. What effect do you think mistletoe has on its host?

Animals Interact with the Environment

Just as animals interact with other living things, they also interact with nonliving parts of their environments. Some animals are at the mercy of changing conditions in their environments. Others can alter their surroundings to better meet their needs, much as humans do. In the next activity, explore how people change their environments.

Explore! ACTIVITY

How do we change our environment?

What To Do

1. *In your Journal,* list at least three ways people change their environment to make it more suitable for them.

2. Describe how each of these changes affects the environment.

3. Research ways people could achieve the same results with less harm to the environment.

4. Explain what you have learned by creating a poster or making a class presentation.

Animals and plants change the environments in which they live simply by living there. Think back to the example of succession given in Section 16-1. When pines began to grow in a grassy field, they shaded the ground underneath. They changed the environment beneath them simply by growing. Animals often change the environment by creating or modifying shelters. For example, birds build nests, removing twigs and other things from the ground and placing them in trees. See **Figures 16-9** and **16-10** to learn about some other ways in which organisms alter their environment.

Sometimes the animals' changes to the environment conflict with

Figure 16-9

Ⓐ In tropical countries, termites create their own complex environments by building huge mounds. One mound may be home to over ten million termites.

Ⓑ A complex of rooms and passages, the mound is ventilated to bring in fresh air. The termites plaster the walls with a saliva mixture to control the temperature.

Ⓒ Within the mound, the termites create a compost area where they grow fungus to eat. Gas from the decaying compost heats the mound.

Theme Connection Students study interactions among living and nonliving things.

Across the Curriculum

Social Studies

People around the world change their environment in various ways. They may build walls and dikes to control rivers or reservoirs to store water. Some cut terraced fields into hillsides to increase their available farmland.

Ask small groups to research and describe to the class ways people in other nations change their environment to meet their needs. You might assign the groups specific nations or regions to research.

GLENCOE TECHNOLOGY

Software

Computer Competency Activities

Chapter 16

✔ **Assessment**

Oral Have students use an oral report and/or a poster to present their findings. After students present their environmental alternatives to the class, have the class suggest additional ways, if any, that humans could achieve the same ends with less impact upon the environment. Use the Performance Task Assessment List for Making Observations and Inferences in **PASC**, p. 17. [L1]

Explore!

How do we change our environment?

Time needed 1 hour plus 1 class period for reports

Thinking Processes collecting and organizing data, comparing and contrasting, recognizing cause and effect, communicating

Purpose Students will collect information comparing ways humans interact with the environment.

Teaching the Activity

Science Journal See student activity.

Expected Outcomes

Students will explore three ways in which humans alter their environment to suit their needs. They will research and describe less harmful means to the same ends, and compare the current methods with the alternative methods.

Check for Understanding

Ask students whether each situation below is an example of mutualism, parasitism, or commensalism.

1. A family planting an orchard and gathering apples from it *mutualism*

2. A bird nesting in a tree *commensalism*

3. A thief robbing people's homes *parasitism*

Reteach

Discussion To help students sort out roles, ask how one animal could be both a predator and a parasite. *when the host animal dies* How could one animal be a consumer and have a mutualistic relationship with another animal? *Termites are consumers and have a mutualistic relationship with the protists in their intestines. People are also consumers and have mutualistic relationships.* L1

Extension

Group Work Divide the class into four groups. Assign each group the relationship of competition, mutualism, parasitism, or commensalism. Ask them to make posters that include the name of their relationship and one or more clear examples. L3 COOP LEARN

4 CLOSE

Activity

Ask small groups to select a relationship they studied in this section and demonstrate it in a skit. Have the groups tell the class which living things they are representing and perform their skits without naming the relationship they chose. After the skit the class will try to guess the relationship being portrayed. L1

Figure 16-10

Ⓐ Beavers alter their ecosystem by cutting down trees and building dams. The dams turn rivers into small lakes, sometimes flooding farmers' fields.

Ⓑ Beaver lodges, built upstream from the dam, are thick webs of branches plastered together with mud. Loose sticks on top allow air to circulate.

Ⓒ Inside the lodge, accessible only from an underwater entrance, the beavers are safe, dry, and warm during the winter.

human needs or desires. For example, prairie dogs dig underground burrows for shelter from the sun and predators. In the winter, the tunnels are a warm place where young can be born and raised.

In the 1930s and 1940s, prairie dogs were shot and poisoned because they competed with livestock for grass. The livestock were also tripping over the burrows. The prairie dogs and their burrows interfered with human efforts to raise food.

The killing of prairie dogs had a bad effect on another type of animal. Black-footed ferrets eat only prairie dogs and live in their abandoned burrows. When prairie dogs were nearly wiped out, the ferrets starved. In 1983, because of conservation efforts, the prairie dog was on the way to recovery. However, black-footed ferrets did not recover and are now just one colony away from extinction.

Some changes organisms cause by living in an area may create problems for humans in that area. When this happens, people may "fight back." This can result in reducing the population of a species or in destroying a species entirely. You'll learn more about this in the next section.

check your UNDERSTANDING

1. Describe an example of a mutual relationship among humans.

2. Describe an example of a commensal relationship among humans.

3. Apply People often develop relationships with pets and farm animals. Explain how you could establish a mutual or a commensal relationship with a wild species.

check your UNDERSTANDING

1. Students studying together; most employee/employer relationships; and any sort of trading relationship.

2. Borrowing a pencil or a tool is one example.

3. You may feed the wild species and expect nothing (commensal); or as with early humans you may feed and shelter a wild species and expect some work or profit from the species (mutualism).

16-3 Extinction—A Natural Process

Extinctions

You've probably heard about species that are threatened or endangered. You probably know that dinosaurs are extinct. But what is extinction? How could it happen? Make a model of one way in which single, small changes can build up to spell disaster!

Find Out! ACTIVITY

Steps to Disaster!

Often a species can survive one or two threats to its well-being. In this activity, you'll find out what can happen as the threats accumulate.

What To Do

1. Work with a team of three other students. Take the tube your teacher gives you. Stretch the netting over the bottom of the tube. Attach it securely with a rubber band or tape so that there is no slipping.

2. Place 10-15 marbles inside the tube. Place the tube on a ring stand. Make sure the stand is high enough so that you can get to the bottom of the tube easily.

3. Place an aluminum pie pan under the tube.

4. Take turns cutting one thread of the netting at a time. Observe what happens. Record your observations *in your Journal.*

5. Record the total number of threads you were able to cut before the marbles started to fall. Did all of the marbles fall at that time?

Conclude and Apply

1. How did your tube-and-marble apparatus model an ecosystem? Describe what each part of the model stood for.

2. What did your model tell you about changes in ecosystems and their results? Was disaster immediate?

3. Imagine that your model is a specific ecosystem, such as a pond or forest. What kinds of changes might be represented by each thread you cut?

523

Section Objectives

- Explain the common causes of local and global extinction.
- Provide examples of ways that human actions within an ecosystem can cause the extinction of species.

Key Terms

extinction

PREPARATION

Planning the Lesson

Refer to the Chapter Organizer on pages 504A–D.

Concepts Developed

Students will learn about the natural events and human actions that can lead to the extinction of a species.

1 MOTIVATE

Bellringer

Before presenting the lesson, display **Section Focus Transparency 54** on an overhead projector. Assign the accompanying **Focus Activity** worksheet. L1 LEP

Conclude and Apply

1. The netting represents a natural habitat; the marbles, a species; each thread cut is like a change introduced in an ecosystem.

2. Changing one part of the "ecosystem" resulted in the sudden collapse of the whole; disaster came only after a certain amount of tampering.

3. Answers will vary according to the ecosystem chosen. Students should realize that each thread can represent a single change introduced in the ecosystem.

Assessment

Performance Have students divide into cooperative groups to write a short skit about the collapse of an ecosystem. They should identify a specific habitat and use reference materials to determine what creatures live there. Use the Performance Task Assessment List for Skit in **PASC**, p. 75. L1 COOP LEARN LEP P

Find Out!

Steps to Disaster!

Time Needed 20 minutes

Materials tin can (both ends open), marbles, netting, rubber bands, ring stand and ring, aluminum pie pan, scissors

Process Skills Observing and inferring, making models, recognizing cause and effect

Purpose To observe how systems can fall apart and to infer how this relates to ecosystems.

Teaching the Activity

Science Journal Have students record their observations in their journals. They can then describe how the activity relates to the process of extinction. L1

Expected Outcomes

Students will observe that all of the marbles fall at once when a certain amount of the netting is cut. Students will infer that tampering can abruptly destroy an ecosystem.

Figure 16-11

The Nile perch—introduced to Africa's Lake Victoria as a human food source—has eaten to extinction hundreds of species of fish unique to the lake.

In the Find Out activity you modeled the natural process of extinction. The crash of the marbles into the pan represented the extinction of a species or ecosystem.

Extinction means the disappearance of a species. Extinction is a natural process. Throughout Earth's history, millions of plants and animals have become extinct for natural reasons. See the Earth Science Connection for some notable examples.

Local extinction means that species disappear from a certain area. The cause could be a gradual change in the climate, such as a decrease in a region's rainfall. A plant species adapted to a swampy area will become extinct locally if the swamp dries up. Local extinctions often occur as part of succession.

Local extinction can lead to global extinction—the complete and permanent disappearance of a species. Species that are dependent on a certain habitat are vulnerable to global extinction.

■ The Human Side of Extinction

In the past, humans caused extinction of many species by killing them in large numbers—the dodo and the passenger pigeons are examples. Today, the most common way we cause extinction is by destroying

Earth Science CONNECTION

Ancient Extinctions

Millions of species became extinct long before people lived on Earth. In fact, our planet has experienced five natural mass extinctions. A mass extinction occurs when a large percentage of the existing species become extinct at once.

Five Mass Extinctions

During the first mass extinction, 440 million years ago, most species lived in the seas. Some survived this extinction, but many did not.

During the next mass extinction, 370 million years ago, many species of fish and

Earth Science CONNECTION

Purpose
Earth Science Connection identifies some of the natural mass extinctions that have occurred in Earth's history.

Content Background
What caused these mass extinctions? Each mass extinction has its own causes, and trying to figure out the causes can often be difficult. Scientists hypothesize that the extinction 225 million years ago may have been caused by the shifting of the plates that form Earth's continents. As the plates joined to make one huge continent, they caused huge changes in the climate and in the depth of the seas.

Some scientists have strong evidence that the extinction that wiped out the dinosaurs may have been caused by the impact of a giant asteroid. The impact caused huge clouds that blocked the sun and lowered the temperature long enough to kill most of Earth's vegetation. After the plants died, so did the animals.

species' habitats. We eliminate these habitats directly by clearing the land and indirectly through pollution.

As we farm, log, and live on more and more of Earth, we interfere with succession's ability to repair damage. In the rain forest, one fallen tree is an opportunity for succession. But clear-cutting, or cutting down all of the trees in an area, leads to acres of sun-baked soil. With no protection from the sun, fragile rain-forest seeds dry out instead of sprouting.

Destruction of habitats is not the only way humans cause extinction. Species are also endangered by our introduction of competing species. See **Figure 16-11** for an example.

■ People vs. Plants

What difference does one species make? As an example, doctors discovered in 1984 that medicine made from the bark of the Pacific yew can help fight cancer. Before this was known, loggers burned or bulldozed about 90 percent of these trees. The Pacific yew grows slowly and mostly in climax communities called old-growth forests. A forest may take 1000 years to reach this stage.

SKILLBUILDER

Making Graphs

Use the figures below to make a line graph showing the number of species that became extinct between 1600 and 1900. Then, explain what the graph indicates. For help, refer to the *Skill Handbook* on page 657.

Year	Number of extinct species
1600	10
1700	11
1800	28
1900	70

SKILLBUILDER

Between 1600 and 1900 the rate of extinction has increased dramatically. L1

Across the Curriculum

History

Ask each student to choose a species which became extinct during human history and find out when it became extinct. Then have them research that time period. What kinds of jobs did people have then? How did they travel? Where did they live? Were they concerned about the environment? Were measures taken to try to prevent the extinction? Ask students to share what they learn in a written or oral report. L2

Content Background

After twenty years of protection by the U.S. Endangered Species Act, the American alligator, the California gray whale, and three other species are no longer endangered. In the same length of time, however, twelve other protected species have become extinct.

---

about 70 percent of the marine invertebrates (animals without a backbone) became extinct.

The third extinction, 225 million years ago, was the most destructive. Between 80 and 96 percent of all species became extinct. Scientists estimate that of the 45 000 to 240 000 species existing at that time, only 1800 to 9600 species survived.

The next mass extinction, 200 million years ago, eliminated 75 percent of the sea-dwelling species and some land-dwellers. Dinosaurs, crocodiles, and mammals survived.

The last mass extinction, 65 million years ago, finished the dinosaurs off, along with about one-third of all species, mostly sea-dwellers.

After the death of the dinosaurs, mammals took over many of the land habitats and niches. Many new mammal and fish species emerged.

What Do You Think?

Research the hypotheses scientists have developed to explain dinosaur extinctions. What evidence do they use to support their ideas? Make a poster that shows one of the hypotheses for what caused the dinosaur mass extinction.

525

Teaching Strategy

Species become extinct in response to external pressures. They may die off in groups or one at a time, depending upon the factors involved. Discuss with students the sorts of factors they can imagine leading to species' extinction. Try to differentiate between factors which might affect one or a few species (a virus; the sudden arrival of a more efficient species competing for an ecological niche) from those which might affect entire ecosystems (global climate change; widespread destruction of habitat or food sources).

Going Further ▮▮▮▮▶

The La Brea Tar Pits in southern California have revealed to scientists extremely well-preserved specimens of species which faded into extinction tens of thousands of years ago. Animals would attempt to drink water which had collected on the pits' surfaces, become trapped in the tar, and die.

Interested students may wish to research the pits, the specimens of extinct species found in them, and the theories which attempt to explain their extinction. Have students present their findings in the form of an illustrated newspaper article. Use the Performance Task Assessment List for Newspaper Article in **PASC**, p. 69. L2 P

Taking Action

The dilemma of the Pacific yew illustrates why it's important to learn about how living things interact with each other. We need to preserve natural habitats and, with them, the diversity of life on Earth.

We know nothing about what valuable medicines or products may come from undiscovered species. If they become extinct before we ever find them, we may lose many valuable natural resources. But, there are some species we do know about that are on the brink of extinction. Explore these species and what you can do to save them.

California condor

Giant panda

Orangutan

Florida panther

Tiger salamander

Explore! ACTIVITY

Which species should we save?

The government has chosen your team to decide which of the endangered species below to save first. You may not have time or money to save the rest, so choose carefully.

California condor
orangutan
giant panda

tiger salamander
Florida panther

What To Do

1. Work with your team to research the needs, the habitat, and the role of each of these endangered species in its ecosystem.

2. Discuss the pros and cons of saving each one. Decide which species your team would save.

3. Present your argument to the class. Try to convince them to join you in preserving the species you selected.

In the Explore activity, you thought of ways to save endangered species. What else can we do?

First, nations can help people in countries with rain forests find ways to make a living that help preserve the diversity of life.

Second, nations can support international treaties that protect wildlife. One is the Convention on International Trade in Endangered Species (CITES). Another is the U.S. Endangered Species Act, prohibiting anyone in the United States from killing, selling, or even chasing an endangered species.

Figure 16-12

There are several ways that you can help preserve diversity.

Refuse to buy or accept exotic pets or products, such as ivory, that come from endangered animals.

Reduce your use and recycle when you can. When we reduce the resources used, we reduce the number of habitats that will be destroyed to supply human needs and wants.

Learn about endangered species, such as the golden lion tamarin, and about the plants and animals that live in your own ecosystem.

U.S. HOUSE OF REPRESENTATIVES

Support conservation groups, zoos, seed banks, and other programs that protect endangered species.

Write to state and federal representatives. Urge them to support programs that protect both habitats and fragile species.

To see examples of things you can do to help preserve the diversity of life on Earth, see **Figure 16-12**.

You've learned about how organisms interact with other organisms and with the environment. Humans are an important part of nearly all ecosystems. As you become more aware of how your actions may affect the ecosystem around you, you'll be able to make choices that will help preserve the diversity of life on Earth.

check your UNDERSTANDING

1. Name two natural causes of local extinction.
2. Explain three specific ways humans directly or indirectly cause local or global extinction.
3. **Apply** Assume that the current rate of global extinction continues for another hundred years. What are at least three ways that the decreasing number of species will affect the people living a century from now?

check your UNDERSTANDING

1. The environment changes (a pond dries up) or succession occurs causing a once-common plant or animal to disappear.
2. Humans can hunt an animal to extinction; they can hunt an animal's prey or food to extinction; they can change the local environment dramatically.
3. People will see and come to know fewer animals. The smaller diversity may mean more delicate ecosystems. Much of the beauty of diversity will be lost.

3 ASSESS

Check for Understanding

1. Have students explain how a species can become extinct locally but not globally. *Changes in one ecosystem may result in an environment that no longer meets the needs of a certain species. The ecosystem in another area may still support that species.*

2. Ask students why one species in an ecosystem may become extinct, but not other species in the same ecosystem. *Some species are better able to adapt to changing conditions than others.*

3. Ask students to answer the Check Your Understanding questions 1-2 individually and discuss question 3 in small groups.

Reteach

Discussion Post the list students made earlier of reasons that plants and animals become extinct. Ask if they think the list is accurate and complete, based on what they've learned. Discuss any necessary additions and clarifications. L1

Extension

Activity Have students prepare two weeks of short morning announcements to help others at the school learn why certain plants and animals are becoming extinct and how the loss of these species can affect them. L3

4 CLOSE

Activity

Guide the class to find out how they can help save the endangered animals they chose for the Explore activity. Have them contact a local conservation group or write to the National Wildlife Federation, 8925 Leesburg Pike, Vienna, VA 22184 and request a Conservation Directory listing government agencies and private organizations. L1

Science and Society

Purpose
Science and Society expands on Section 16-1 by describing the diversity in the rain forest. It also relates to Section 16-3, focusing on ways the extinction of rain forest species affects all of us.

Content Background
Nearly half of Earth's rain forests are gone, and most of the destruction has taken place since 1960. For example, 45 percent of Thailand's rain forests were cut between 1961 and 1985. Between 1985 and 1988, the clearcutting caused landslides that destroyed the homes of 40 000 Thais. In 1989, the Thai government finally banned logging.

The Ivory Coast has lost 75 percent of its rain forests since 1960. About 80 percent of Ghana's forests have been destroyed. In both cases, almost all of the timber was wasted.

Purposely cut trees drag down many others as they fall, damaging the canopy, uprooting undergrowth, and seriously disturbing the forest ecosystem.

NATIONAL GEOGRAPHIC SOCIETY

 Videodisc

STV: Rain Forest

Map locating world rain forests

51700

Map of North American temperate rain forests

51807

Map of tropical and temperate rain forests

51768

Science and Society

Why All the Fuss over the Rain Forests?

Do you live in a rain forest? Probably not, so why should you be concerned about the rain forest?

For one thing, the seven percent of Earth covered by rain forests supports at least half of all the species on Earth.

People who live near rain forests depend on farming, logging, or mining to support their families. As a result, they burn or clearcut as many as 22 million acres of rain forest every year.

The Extent of the Damage

Rain forest occupying an area the size of 20 football fields is destroyed every minute of every day. In 1989, rain forests covered a region about as large as the continental United States. The amount of rain forest destroyed each year would cover an area the size of Florida.

The fires from the burning also send large amounts of carbon dioxide up into the atmosphere. Carbon dioxide traps the sun's heat in the atmosphere and may be causing gradual worldwide increases in temperature.

After clearing the land, farmers discover that the soil is too poor to support their crops or feed their cattle. Most of the nutrients in the rain forest are in the trees, not the soil.

Only a few plants grow in a shallow layer of humus on the forest floor. Under that layer is a claylike soil that bakes as hard as rock when exposed to sun.

The fires and logging destroy habitats found nowhere else on Earth. Scientists estimate that the loss of these habitats

Going Further

In addition to many unique varieties of plant and animal life, the Amazon rain forest is home to many tribes of indigenous peoples. They, too, are threatened by dwindling forest lands. More and more, native peoples are being forced off their land by miners, ranchers, and farmers who wish to use the land for their own purposes.

Have interested students research the native peoples of the Amazon rain forest. They should choose one tribe to study in detail, describing how they live, how deforestation and encroachment by industrialized society is affecting them, and what steps are being taken to relieve their situation. The Kayapo Indians, for instance, pay for legal action to protect their land rights by harvesting Brazil nuts and selling the oil they produce to a producer of skin-care products. Use the Performance Task Assessment List for Investigating an Issue Controversy in **PASC**, p. 65. L3

destroys 4000 to 6000 species of plants and animals every year.

Does it seem as though, year after year, the numbers of songbirds where you live are decreasing? Songbirds migrate to the rain forests during the winter. As their habitats are destroyed, so are they. Between 1980 and 1990, the number of wood thrushes in the U.S. dropped 31 percent. The number of Baltimore orioles decreased 23 percent.

Valuable Products of the Rain Forests

The rain forest contains many valuable natural resources. For example, the sap of the copaiba tree is pure diesel fuel. In 1990, it supplied Brazil with 20 percent of its diesel fuel. Unlike our limited supply of fossil fuels, the copaiba

Scarlet macaw

produces fuel as long as it lives. The copaiba is a renewable source of fuel.

You or your family may have used medicine made from rain forest plants. One-fourth of all medicines come from species found in the rain forest. For example, two chemicals in Madagascar's rosy periwinkle are used to treat cancer. Medicine made from the venom of the Brazilian pit viper helps lower blood pressure.

But only one percent of the rain forest plants have been checked for medicines that could be developed from them. Maybe a tiny plant growing there will cure cancer. But hundreds of unidentified species are being destroyed as you read this!

So we all need to be concerned about the destruction of the rain forests. The rain forests we preserve may hold the key to our own future!

You Try It!

Use the information in this article to construct a scale model of the amount of rain forest destroyed each day. Be sure to put in some recognizable object (at the same scale), such as a person, so that everyone will have an idea of how big the area is.

Content Background

One businessman inadvertently demonstrated the fragile interdependence of living things in the rain forest when he decided to grow Brazil nuts. Wild brazil nuts grow on widely scattered trees, but he thought he would streamline nut-gathering by planting the trees in rows on a plantation.

His trees produced flowers but no nuts. Eventually, it was discovered that the flowers are pollinated by a certain species of bee that mates only in response to chemicals from a certain species of orchid. His plantation had neither bees nor orchids, but he did learn that the production of Brazil nuts, like many things in the rain forest, is not easily "streamlined" by humans.

Across the Curriculum

Math

Two and one-half acres of rain forest in Peru would yield about $1000 in timber—or $425 in nuts, fruit, and rubber. Ask students to figure out about how long it would take for the Peruvian people to earn $1000 in nuts, fruit, and rubber from the same forest if they did not cut the trees. *About 2 years.* Ask what would be the advantage of not cutting the trees. *The people could continue to earn money as long as the trees live.*

Health CONNECTION

Purpose

This Health Connection relates to Section 16-2 and helps increase students' awareness of the mutualistic relationship people often have with pets.

Discussion

Surveys have recently shown that cats have replaced dogs as the most popular pet in the United States. Ask students to brainstorm reasons why this might be so. (less food required, better suited to apartment-house living, more independent, etc.) Take a quick survey of the class to see what is the most common pet.

Going Further ⅢⅢ➡

Arrange for your students to share their pets with the residents of a nursing home. Begin the project by sending a letter to students' families, asking if they would be interested in coming with the class and bringing a family pet that would react well to the experience.

Then contact (or have students contact) local nursing homes to determine which ones would like to participate in this project. Be sure to follow school guidelines regarding field trips. After your first visit, students may be eager to return. [L3]

Health CONNECTION Pets Are Good for You!

Did you know that many humans have a mutualistic relationship with animals? Do you know any of these people? You know that guide dogs and other animals can be trained to help people with physical disabilities. But now research shows that just spending time with animals helps people feel better.

Scientists have found that the presence of animals can help sick people recover, calm troubled children, and encourage withdrawn people to communicate.

Animals Help Hearts

Research shows that heart patients with pets had better chances of survival than patients without pets. And elderly people with pets visit the doctor less often than those without pets.

Organizations such as Pets for Life in Kansas City bring pets to nursing homes for monthly visits. The smiles and excitement show how pets boost morale.

In fact, 25 studies on elderly people in nursing homes have shown the benefits of having pets around. Residents who were exposed to pets smiled more and became more alert. The animals also helped angry and depressed residents feel more at ease and be more caring toward other people.

"Mutual" Love

These relationships between people and animals are examples of mutualism because both benefit. People take care of the pets, while the pets provide loving, undemanding companionship.

Pets are good for us, mentally and physically!

What Do You Think?

Why do you think pets have such an influence on people? If you have a pet, how do you feel about it? Write a paragraph describing how you think pets aid healing.

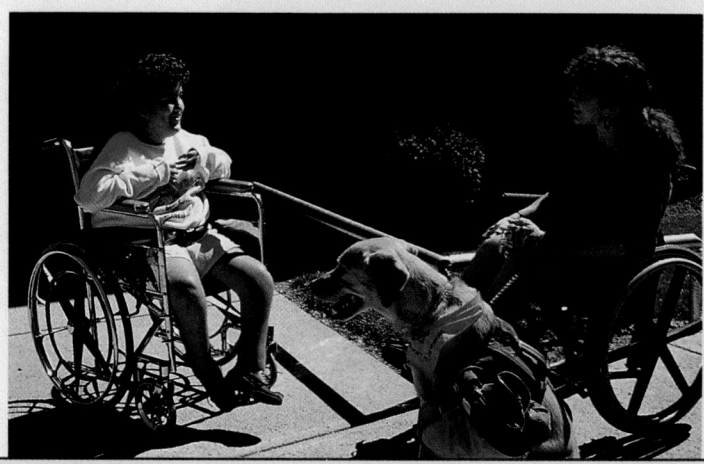

Computers to the Rescue!

Often we don't know a species is endangered until it's almost too late to save it. New technology can help. Gap analysis uses computerized maps to indicate the diversity of plant and animal species in a specific ecosystem.

Gap Analysis

In gap analysis, three maps are stacked on top of each other, using computer graphics. One map shows the plant species. It can be programmed to show all plants in the ecosystem or just one, such as the endangered sage scrub.

Another map indicates all the animal species or just one group, such as reptiles. A third map outlines any protected wilderness areas within the region.

Locating Endangered Species

Sometimes when the maps are combined, they show a "gap" in the protected areas. Most of the endangered species are living outside the protected areas.

By pinpointing the location of these species, officials can better determine which areas should be protected. They can also decide whether a piece of land can be developed without threatening endangered species.

Information for the maps comes from aerial photographs, satellite images, and ground observations.

The Success of Gap Analysis

Twenty-two states are now using gap analysis. This new technology has already convinced the government not to use an "empty" Idaho field for testing bombs. The maps proved that the field was actually teeming with life.

Science Journal

In your Science Journal, make a gap analysis map of your school yard or neighborhood. Work with a partner or two to create a map of the area. Then go and record the plant and animal species in your area. Finally indicate any protected areas, such as city parks, and so on.

In the gap analysis below, Californian coastal sage scrub is shown growing outside protected reserve lands.

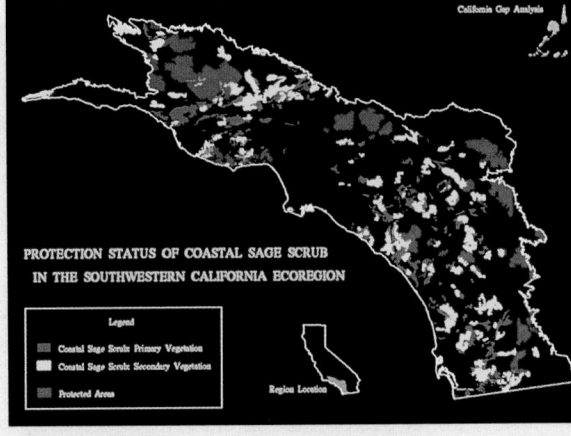

Purpose

This Technology Connection relates to the diversity of species (Section 16-1) and extinction (Section 16-3). It explains methods being used to document what species are in an area in order to protect those that are endangered.

Content Background

Gap analysis is also called *microgeography*. This process maps the structures in an ecosystem in enough detail to allow scientists to estimate the populations of individual species.

One specific approach to collecting these layers of data is called Geographic Information Systems. The layers in this system indicate not only the plants and animals in an area, but also the topography, soils, water sources, and geology.

The information from gap analysis can be used to determine which zones should be set aside as absolute preserves and which can serve as buffer zones and be used for agriculture and restricting hunting. Even in areas that have already been developed, the data can show where placement of woodlots, watersheds, reservoirs, and artificial ponds will encourage the diversity of species.

Science Journal

Students' maps will most likely be made with ground observations. You may be able to assist them in obtaining aerial photographs of the area. Student maps should show understanding of gap analysis.

Going Further ▸

State fish and wildlife service offices across the country are now using gap analysis to study the impact of hunting and fishing permits on local wildlife populations, and to help in deciding how to distribute new ones. Have students contact your state's Fish and Wildlife Service, as well as the state office of the United States Environmental Protection Agency, to find out how gap analysis is being used in your area.

Students will produce a time line showing how succession, cooperative relationships, and extinction would change an ecosystem of their choice. They research, illustrate, and caption the time line appropriately.

Teaching Strategies

Have groups of students pick one of the following ecosystems to research, illustrate, and caption for the time line: vacant lot, area devastated by fire, meadow or open field, forested area, pond, area near a volcano eruption, or any of the ecosystems described in Chapter 11. You may want different group members to concentrate on succession, cooperative relationships, and extinction as they affect each chosen ecosystem.

Students should be able to verify their contributions by giving the sources for their information. **COOP LEARN** **P**

Answers to Questions

1. New ecosystems are established when pioneer plants and other organisms begin the colonization of new places, making it possible for other organisms to live there.

2. In both mutualism and parasitism two organisms are involved; however, in mutualism, both organisms benefit. In parasitism, one organism benefits and one is harmed.

3. No. There were extinctions before humans ever existed. Extinction is a natural process in which species that are not well-adapted die out.

GLENCOE TECHNOLOGY

MindJogger Videoquiz

Chapter 16 Have students work in groups as they play the Videoquiz game to review key chapter concepts.

532 **Chapter 16** Changing Ecosystems

chapter 16
REVIEWING MAIN IDEAS

Science Journal

Review the statements below about the big ideas presented in this chapter, and answer the questions. Then, re-read your answers to the Did You Ever Wonder questions at the beginning of the chapter. *In your Science Journal,* write a paragraph about how your understanding of the big ideas in the chapter has changed.

1 Succession is the process by which new environments become inhabited or old ecosystems recover from damage. *How do new ecosystems become established?*

2 There are three different ways in which species can form cooperative relationships—mutualism, commensalism, and parasitism. *Compare and contrast mutualism and parasitism.*

3 Extinction is a natural process that has taken place throughout Earth history. However, human interactions with ecosystems have greatly increased the rate of extinctions. *If human interference could be completely eliminated, would the extinction of species stop? Explain your answer.*

Project

Have groups of students make murals or posters of different ecosystems showing the types of interactions among organisms that would be found there. Then remove one organism and have students write a paragraph about how this might affect the ecosystem. Use the Performance Task Assessment List for Poster in **PASC**, p. 73. **P**

Science Journal

Did you ever wonder...

• A forest can recover from a devastating fire by succession. Pioneer species begin the process. (pp. 510–511)

• African animals may let birds sit on their backs because of symbiotic relationships. (pp. 516-517)

• Extinction is a natural process that can be caused by climate changes, succession, and human interference, so animals may be affected differently. (pp. 524–525)

Using Key Science Terms

climax community   parasitism
commensalism   pioneer species
diversity   succession
extinction   symbiosis
mutualism

Describe the relationship among the following words.

1. symbiosis, commensalism, parasitism, mutualism
2. succession, pioneer species, climax community
3. diversity, extinction

Understanding Ideas

Answer the following questions in your Journal using complete sentences.

1. What is one way that humans interfere with animal interactions with their environment?
2. What type of symbiotic relationship do a child and her goldfish have? Explain.
3. What is the main way humans now cause the extinction of other species?
4. What is the final stage of succession?

Developing Skills

Use your understanding of the concepts developed in this chapter to answer each of the following questions.

1. **Concept Mapping** Complete this concept map of symbiotic relationships.

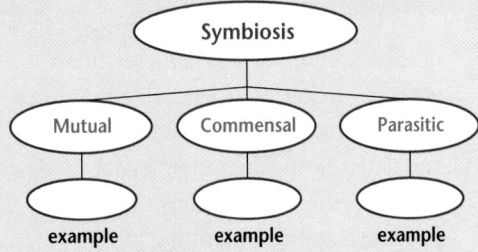

2. **Making Models** Repeat the Find Out activity on page 523, using a different kind of model. Explain how your model demonstrates the path to extinction.
3. **Comparing and Contrasting** Use water or milk instead of hot coffee for the Find Out activity on page 507. Compare and contrast the changes that occur in the water or milk with the results of the original activity.

Critical Thinking

In your Journal, *answer each of the following questions.*

1. Why is a pioneer species usually small in size?
2. How does farming interfere with natural succession?
3. The clownfish can hide within the tentacles of anemones, which are

Program Resources

Review and Assessment, pp. 95-100 `L1`
Performance Assessment, Ch. 16 `L2`
PASC
Alternate Assessment in the Science Classroom
Computer Test Bank `L1`

Using Key Science Terms

1. Symbiosis means living together and describes how animals interact with one another. Mutualism, commensalism, and parasitism are three types of interactions.
2. Succession is one process by which an ecosystem changes through time. The pioneer species is the first step in succession; the climax community is the last step.
3. Diversity is one factor that helps an ecosystem avoid extinction.

Understanding Ideas

1. Humans can disrupt interactions by cutting down trees and polluting streams and rivers.
2. Mutualistic. The goldfish depends upon the girl to feed it, and the girl has a pet that she enjoys.
3. by destroying the habitats and environments of species
4. The final stage of succession is the climax community.

Developing Skills

1. See student text page for answers.
2. Answers will vary depending on the model chosen. For example, if students model extinction by removing cards from a deck or stones from a bowl, they can say that the cards represent species and the depleted deck is the ecosystem after extinctions.
3. Water will not grow visible mold for a long time, if at all. Milk will spoil and yield bacterial and mold cultures very rapidly.

Critical Thinking

1. because the pioneer species in an area are dependent on the resources available, and the resources may be very limited in an area that has not yet been colonized

2. Farming completely replaces a natural ecosystem with a human-constructed ecosystem.

3. Mutualism. The clownfish receives protection within the anenome's tentacles, and the anenome often gets food.

4. Dogweed is specialized in its habitat, the dandelion is not.

5. The interaction is mutualistic. The anemone provides protection and the crab provides mobility. Students may answer commensal or parasitic also; look for student reasons.

6. A parasite can be useful if it weakens or destroys species considered undesirable.

Problem Solving

Student answers will vary depending on the animal they construct. However, the animal should have the characteristics of being able to live in a variety of places and to feed on a variety of foods.

Connecting Ideas

1. Look for known climax community species, or observe the forest over a long period of time and see what changes occur in stable areas.

2. Mutualism increases a species' chances of survival, thus maintaining diversity.

3. Because corals contain symbiotic algae that need light to survive and make food.

4. You would observe trends in the populations of large numbers of species and determine what was happening.

5. You wouldn't be able to buy products made from the rain forest organisms. Carbon dioxide would increase in the atmosphere. New medicines would not be discovered. The diversity of the biosphere would be greatly decreased.

deadly to other fish. When clownfish enter the tentacles, other fish sometimes follow. These fish may be killed and eaten by the anemone. What kind of symbiotic relationship do the clownfish and the anemone have? Explain.

4. Why do you think the ashy dogweed plant, which grows only in Texas, is on the brink of extinction, while the dandelion is not?

5. What kind of interaction is shown in the picture to the right? Explain.

6. Is a parasite ever useful? How can a parasite be useful if it is defined as something harmful?

Problem Solving

Read the following challenge and discuss your responses in a brief paragraph.

Design an animal that would not easily become extinct. Draw a picture of what it would look like, and write a paragraph to describe it.

1. Describe its physical characteristics.

2. Describe its habitat.

3. What would keep this animal from over-populating its habitat and depleting its food source?

4. How could you make sure its habitat would not be destroyed?

CONNECTING IDEAS

Discuss each of the following in a brief paragraph.

1. Theme—Stability and Change How could you tell whether a forest had stabilized into a climax community?

2. Theme—Systems and Interactions How does mutualism benefit the animals within an ecosystem?

3. A Closer Look Why do coral reefs depend on clear, clean water to survive?

4. Earth Science Connection Explain how you could predict a sixth natural mass extinction.

5. Science and Society What are four ways people who live thousands of miles from a rain forest can be affected by its destruction?

✔ Assessment

Portfolio Review the portfolio options that are provided throughout the chapter. Encourage students to select one product that demonstrates their best work for the chapter. Have students explain what they learned and why they chose this example for placement into their portfolios.

Additional portfolio options can be found in the following **Teacher Classroom Resources:**

Making Connections: Integrating Sciences, p. 35

Multicultural Connections, pp. 35, 36

Making Connections: Across the Curriculum, p. 35

Concept Mapping, p. 24

Critical Thinking/Problem Solving, p. 24

Take Home Activities, p. 25

Laboratory Manual, pp. 99-106

Performance Assessment P

Changing Systems

In this unit, you investigated motion and how it affected you, Earth, and objects near Earth.

You learned about velocity, acceleration, weight, and weightlessness. You saw how a projectile moves, and how satellites can remain in a stationary orbit over Earth. You learned that movement of water, wind, and ice on Earth's surface can change features of the land. You learned about how ecosystems grow and change, creating new conditions, and adapting to sudden alterations.

Try the exercises and activity that follow—they will challenge you to use and apply some of the ideas you learned in this unit.

CONNECTING IDEAS

1. As rocks on a cliff weather, fragments fall to the base of the cliff due to gravity. How does this motion caused by acceleration due to gravity affect weathering and erosion on Earth's surface?

2. Explain what happens to various sized particles of rock that are carried by a stream as the water in the stream empties into a larger body of water. What do these particles form?

3. Describe the changes in velocity and acceleration that occur on a daily trip from your home to school.

Exploring Further ACTIVITY

How Can You Help Increase the Diversity of Life in Your Environment?

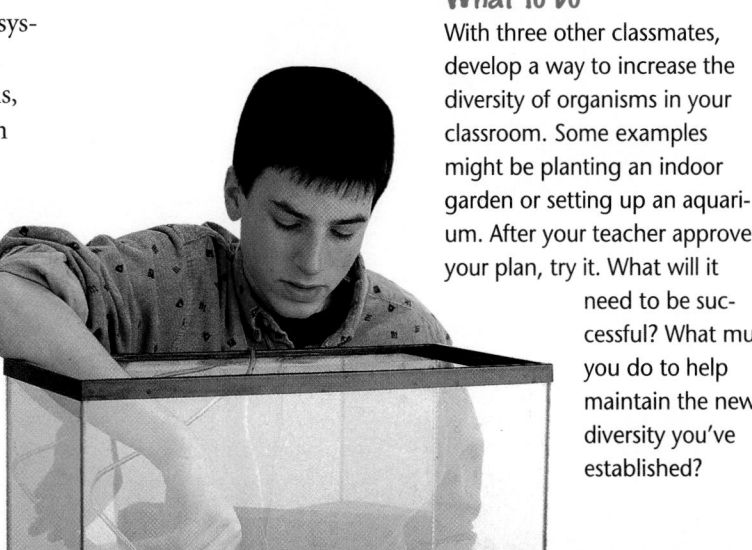

What To Do
With three other classmates, develop a way to increase the diversity of organisms in your classroom. Some examples might be planting an indoor garden or setting up an aquarium. After your teacher approves your plan, try it. What will it need to be successful? What must you do to help maintain the new diversity you've established?

Exploring Further

How can you increase the diversity of life in your environment?

Purpose To apply and communicate knowledge of changing systems.

Background Information
Students should develop plans to increase diversity of life in the classroom. This will reinforce concepts of interactions in the environment.

Once the new diversity is established, have students monitor it and report any changes.

✔ Assessment
Process Have small groups prepare a slide show or photo essay about their projects. Use the Performance Task Assessment List for Slide Show or Photo Essay in **PASC**, p. 77.
COOP LEARN | **P** | **L1**

Changing Systems

THEME DEVELOPMENT
The themes developed in Unit 4 were energy, stability and change, and systems and interactions. Motion is a change in position. Objects move when forces act on them. Energy, in the form of physical weathering, is a force that can weather rocks, leading to soil development. Erosion shows that Earth undergoes change. Moving water causes changes in unstable materials. Motion, energy and change work together to sculpt Earth's surface.

Connections to Other Units
The view of Earth and objects near Earth in space in Unit 1 can be related to how these objects move. In Unit 2, the physical world and its materials work to weather rocks, forming soil, which can then be moved by erosion. Animals and plants, covered in Unit 3, affect weathering and erosion and Earth's ecology is affected by motion within water systems.

Connecting Ideas
Answers

1. As the rock fragments fall they hit other rocks causing them to weather as well. The smaller weathered rock fragments can then be moved by other agents of erosion.

2. As the speed of the water slows, the larger rock particles settle out first. As the water slows more, the smaller particles of rock settle out. This process builds up layers of sorted sediments along the path of deposition.

3. Students should relate accurately the starts, stops, and turns made in going home.

UNIT 5

Wave Motion

UNIT OVERVIEW

Chapter 17 Waves
Chapter 18 Earthquakes and Volcanoes
Chapter 19 The Earth-Moon System

UNIT FOCUS

In Unit 4, students extended their understanding to systems and the changes that occur in systems. In Unit 5, students will extend this understanding to systems of waves, as well as to changes that cause earthquakes and volcanoes, and then to the interactions between Earth and the moon.

THEME DEVELOPMENT

The themes developed in Unit 5 are scale and structure, energy, stability and change, and systems and interactions. The two themes that work to interconnect all three chapters are energy and systems and interactions. As they move through various types of Earth material, mechanical waves interact with this material to change Earth's surface.

Connections to Other Units

The concepts of this unit can be related to Unit 1 where light and sound were explored as forms of energy, as both are transmitted by waves. The concepts in this unit also related to landforms described in Unit 1.

UNIT 5

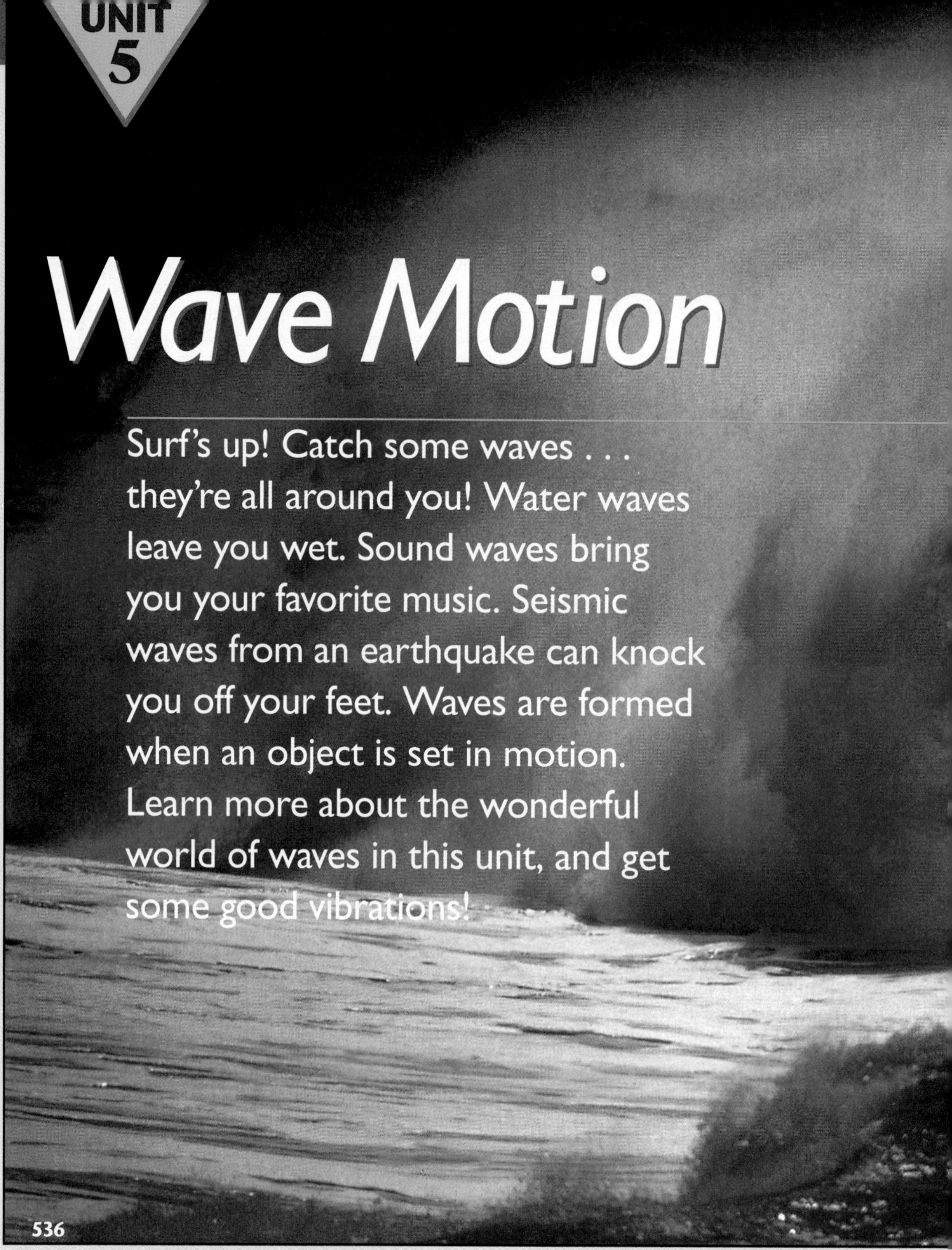

Wave Motion

Surf's up! Catch some waves . . . they're all around you! Water waves leave you wet. Sound waves bring you your favorite music. Seismic waves from an earthquake can knock you off your feet. Waves are formed when an object is set in motion. Learn more about the wonderful world of waves in this unit, and get some good vibrations!

536

GLENCOE TECHNOLOGY

Videodisc

Use the *Science Interactions, Course 1* **Integrated Science Videodisc** lesson, *Waves and Sound*, with Chapter 17 of this unit. This videodisc reinforces and enhances the relationship between waves and sound production.

NATIONAL GEOGRAPHIC
try it!

Mechanical waves are all around you. You see a mechanical wave when you watch a flag wave in the wind. You may have felt a mechanical wave while lying on a raft in a wave pool or in the ocean. You have even heard mechanical waves when your teacher tells you the next day's assignment. What causes a wave to form? Why does it move the way that it does? What happens to an object when a wave passes by it?

What To Do

1. Obtain a tuning fork from your teacher.

2. Gently tap the tuning fork against the edge of a book and observe the motion produced. If you have trouble seeing the vibrations, dip the tuning fork into a glass of water and observe the ripples caused by the vibrations.

3. Gently tap the tuning fork and hold it near your ear. What do you notice?

4. How do you think the tuning fork is affecting your eardrum?

537

GETTING STARTED

Discussion Some questions you may want to ask your students are:

1. In what way or ways are all waves, regardless of type, related? Some students may not recognize waves' basic similarities. All waves move energy through a medium or space.

2. If tides on Earth are caused by the gravitational pull of the moon and the sun, when would their combined effect be greatest? Many students will not know about the relationships among the moon, sun, and tides. The greatest effect on Earth's oceans occurs when the two objects are either pulling together or pulling in opposite directions. The first occurs at new moon and the latter at full moon.

3. Do areas that have many volcanoes also experience large numbers of earthquakes? This relationship may be unclear. Many of the volcano belts around Earth also are known as earthquake belts.

The answers to these questions may help you establish misconceptions your students may have.

Try It!

Purpose Students will observe that mechanical waves exist in many forms.

Background Information

Waves seen in a flag as it moves in the wind, water waves, and sound waves have similar characteristics. When students hear the sound of the tuning fork, they hear the effects of a passing mechanical wave. The observed rapid, back-and-forth motion of the tuning fork causes a mechanical wave that passes through air, causing it to vibrate. This vibration is transmitted to the student's eardrum, causing it to vibrate as well. For hearing impaired students, hold the tuning fork near a glass of water so they can observe the effects of the wave on the water.

Materials tuning fork

✔ Assessment

Oral Have students hypothesize how the sound waves traveled from the tuning fork to their eardrums. Use the Performance Task Assessment List for Group Work in **PASC,** p. 97.

`COOP LEARN` `L1`

Chapter Organizer

SECTION	OBJECTIVES	ACTIVITIES & FEATURES
Chapter Opener		**Explore!**, p. 539
17-1 Waves and Vibrations (3 sessions; 1.5 blocks)	1. **Describe** how waves are produced. 2. **Identify** transverse and longitudinal waves. **National Content Standards (5–8) UCP2–3, A1, B2, E1, G1-2**	**Find Out!**, p. 543 **Design Your Own Investigation 17-1:** pp. 544–545 **Earth Science Connection**, pp. 542–543
17-2 Wave Characteristics (3 sessions; 1.5 blocks)	1. **Draw** a wave. 2. **Identify** the wavelength, amplitude, crest, and trough of a wave. 3. **Explain** the relationship among frequency, wavelength, and speed in a wave. **National Content Standards (5–8) UCP2–3, A1, B2**	**Find Out!**, p. 549 **Investigate! 17-2:** pp. 552–553 **Skillbuilder:** p. 554 **Technology Connection**, p. 565 **Teens in Science**, p. 568
17-3 Adding Waves (2 sessions; 1.5 blocks)	1. **Explain** how waves add together. 2. **Describe** two examples of wave interference. **National Content Standards (5–8) UCP2–3**	**Find Out!**, p. 555
17-4 Sound as Waves (1 session; .5 block)	1. **Demonstrate** sound as a wave. 2. **Explain** the Doppler effect. **National Content Standards (5–8) UCP3, E1–2, F1–2, F4–5, G2–3**	**Find Out!**, p. 559 **Science and Society**, pp. 566–567 **A Closer Look**, pp. 560–561

ACTIVITY MATERIALS

EXPLORE!

p. 539 4-m rope

INVESTIGATE!

pp. 552–553* clear glass dish (approximately 30-cm square), strips of plastic foam, overhead light, tape, water, pencil or pen, piece of blank white paper, ruler

DESIGN YOUR OWN INVESTIGATION

pp. 544–545* coiled spring, goggles, small piece of colored yarn

FIND OUT!

p. 543* portable radio with round speaker, metal can or other container with same or slightly larger diameter than that of the speaker, can opener, plastic wrap, rubber band, dry rice grains

p. 555* clear glass dish, water, overhead light, 2 pencils, piece of blank white paper

p. 559* coffee can with 2 open ends, balloon, scissors, rubber band, flashlight or sunlight, small mirror, glue, loudspeaker

KEY TO TEACHING STRATEGIES

The following designations will help you decide which activities are appropriate for your students.

L1 Basic activities for all students

L2 Activities for average to above-average students

L3 Challenging activities for above-average students

LEP Limited English Proficiency activities

COOP LEARN Cooperative Learning activities for small group work

P Student products that can be placed into a best-work portfolio

Activities and resources recommended for block schedules

Need Materials? Call Science Kit (1-800-828-7777).

00:00 **OUT OF TIME?** We recommend that students do the activities with an asterisk.

TEACHER CLASSROOM RESOURCES

Student Masters	Transparencies
Study Guide, p. 61 **Activity Masters**, Design Your Own Investigation 17-1, p. 116 **Making Connections: Integrating Sciences**, p. 37 **Making Connections: Across the Curriculum**, p. 37 **Making Connections: Technology and Society**, p. 37 **Science Discovery Activities**, 17-1	**Section Focus Transparency 55**
Study Guide, p. 62 **Concept Mapping**, p. 25 **Multicultural Connections**, pp. 37, 38 **Activity Masters**, Investigate 17-2, pp. 73–74 **Laboratory Manual**, pp. 107–110, Velocity of a Wave	**Teaching Transparency 34**, Wave Properties **Section Focus Transparency 56**
Study Guide, p. 63 **Take Home Activities**, p. 27 **Science Discovery Activities**, 17-2	**Teaching Transparency 33**, Constructive/Destructive Interference **Section Focus Transparency 57**
Study Guide, p. 64 **Critical Thinking/Problem Solving**, p. 25 **Science Discovery Activities**, 17-3	**Section Focus Transparency 58**

ASSESSMENT RESOURCES	TEACHING & TECHNOLOGY
Review and Assessment, pp. 101-106 **Performance Assessment**, Ch. 17 **PASC*** **MindJogger Videoquiz** **Alternate Assessment in the Science Classroom** **Computer Test Bank**	**Spanish Resources** **Cooperative Learning Resource Guide** **Lab and Safety Skills** **Science Interactions, Course 1, CD-ROM** **Computer Competency Activities**

*Performance Assessment in the Science Classroom

NATIONAL GEOGRAPHIC TEACHER'S CORNER

National Geographic Society Products

To order the following products for use with this chapter, call the National Geographic Society at 1-800-368-2728:

- *Newton's Apple Physical Science* (CD-ROM)

National Geographic Society Products Available From Glencoe

To order the following products for use with this chapter, call the National Geographic Society at 1-800-368-2728:

- *Everyday Science Explained* (Book)
- *Waves: The Electromagnetic Universe* (Book)
- *How Loud is Too Loud?* (Kids Network Curriculum Unit)

GLENCOE TECHNOLOGY

The following multimedia resources are available from Glencoe.

Science and Technology Videodisc Series (STVS)
Animals
  How Bats Hear
Human Biology
  Ear Implants
Glencoe Science Interactions Interactive Videodisc
Waves and Sound
Physical Science CD-ROM

Teacher Classroom Resources

These are key components of the classroom resources package.

Teaching Transparencies

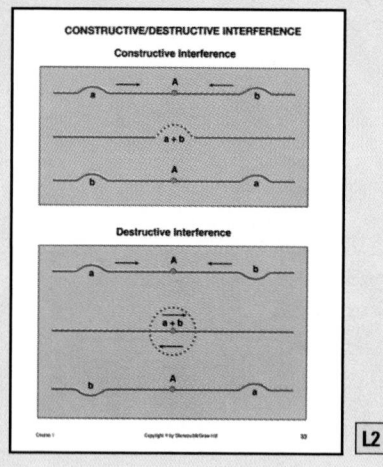

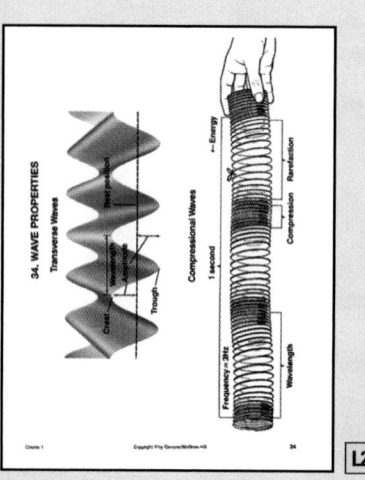

Section Focus Transparencies

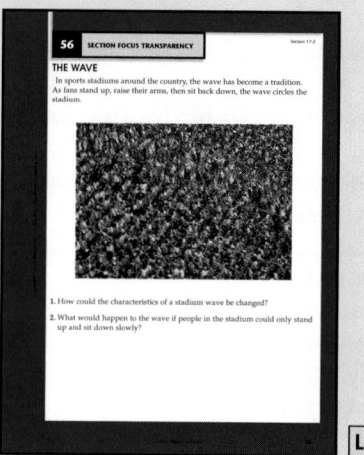

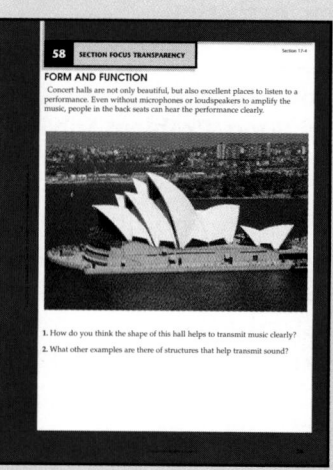

HANDS-ON LEARNING

Science Discovery Activity*

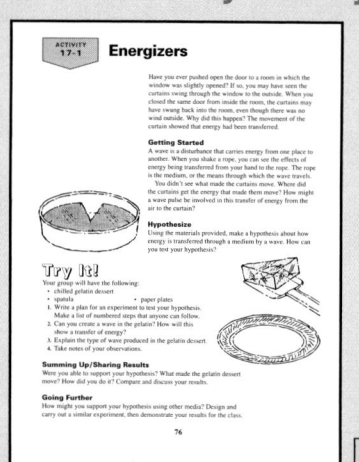

Laboratory Manual*

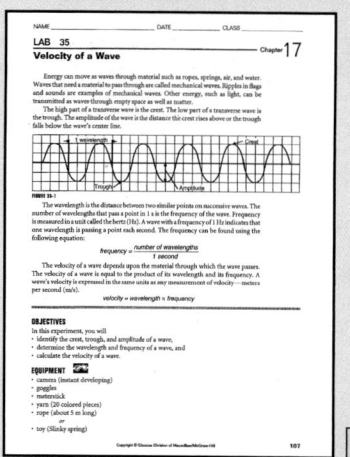

Take Home Activity

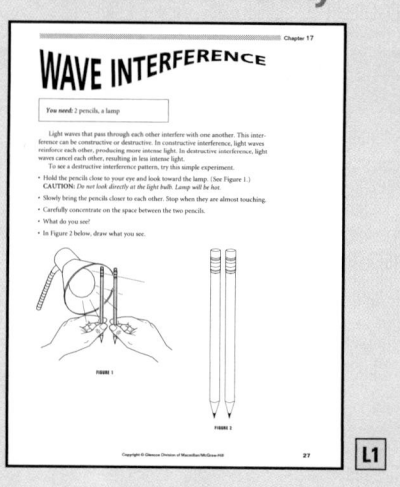

*There may be more than one activity for this chapter.

Chapter 17 Waves

Study Guide*

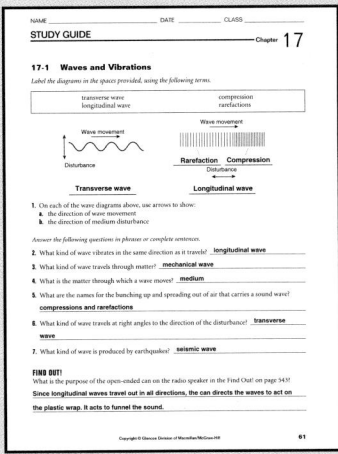

Concept Mapping

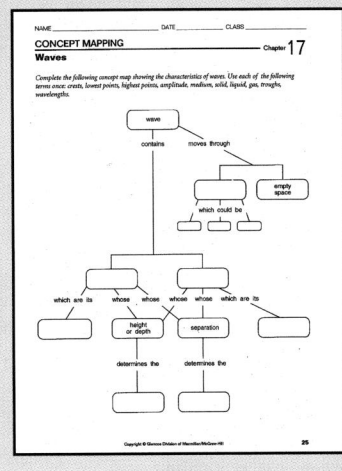

Critical Thinking/Problem Solving

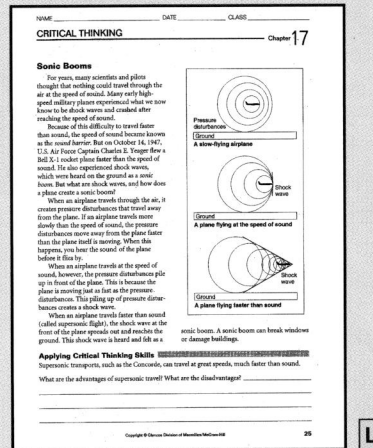

Integrating Sciences

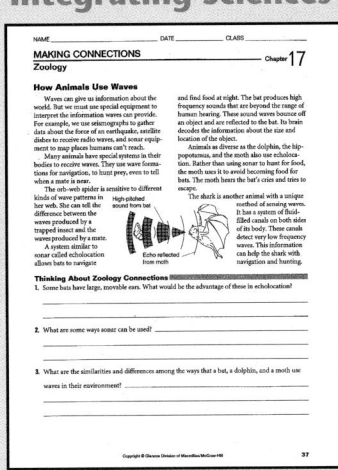

Across the Curriculum

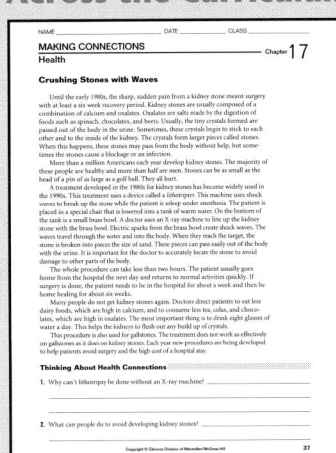

Technology and Society

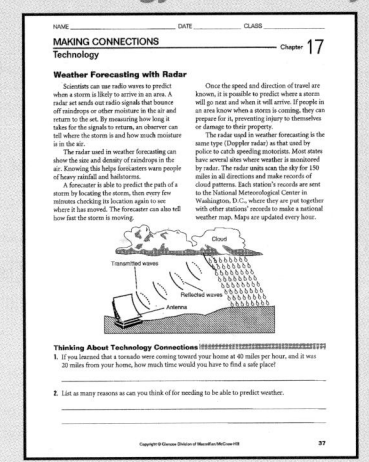

Multicultural Connection**

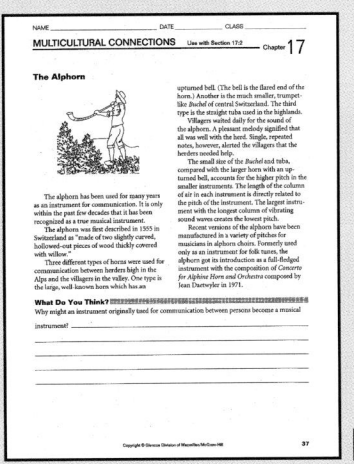

Performance Assessment

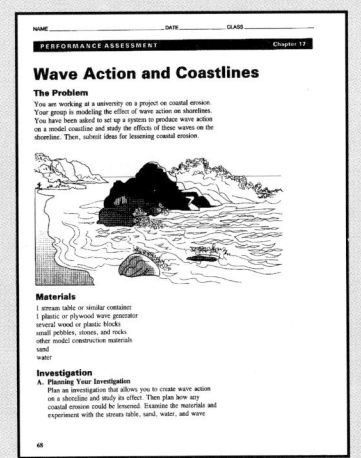

Review and Assessment

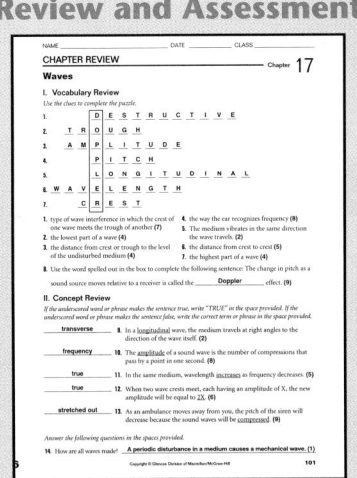

*One per section   **Two per chapter

Waves

THEME DEVELOPMENT

The theme supported by this chapter is systems and interactions. When waves meet, the interaction can be constructive (the waves add together) or destructive (the waves cancel each other). The Doppler effect explains how waves interact with the environment under specific conditions. Wave behavior is systematic and predictable.

CHAPTER OVERVIEW

Students will learn how waves are produced. They also will study the characteristics of waves and how to describe a wave. Relationships among frequency, wavelength, and speed will be examined.

Tying to Previous Knowledge

Ask students to picture waves rolling toward a beach. Have them describe some physical characteristics of the waves.

INTRODUCING THE CHAPTER

Have students look at the chapter opening photo and describe their own experiences of having been in the water when someone did a "cannonball" off the diving board. Have them suggest what effect the waves the diver created will have on a person on a float.

WAVES

Did you ever wonder...

✓ **Why sometimes the waves in a swimming pool get so big?**

✓ **Why a horn on a train seems to change pitch as it passes you?**

✓ **How a radio speaker produces sound?**

Science Journal

Before you begin your study of waves, think about these questions and answer them *in your Science Journal.* When you have finished the chapter, compare your journal write-up with what you have learned.

Answers are on page 569.

I t's a beautiful day, and you're at the local pool. You are in the water with your friend Ladonna, who is basking on a float. Out of the corner of your eye you see Louie, the bodybuilder, launch himself off the diving board. SPLASH! Suddenly you are under the water, Ladonna bobs straight up, and water bursts over the sides of the pool. Louie has made a large wave in the pool.

After you get your hair out of your eyes, you think about what just happened. If someone had told you the wave were coming, you might have assumed it would knock you down. But the wave just flowed over and around you. Ladonna simply bobbed up and down and ended up right back beside you. Later in this chapter, you will think back to these observations and be able to explain them.

▶ **In the activity on the next page, explore some of the characteristics of waves on a rope.**

538

00:00 OUT OF TIME?

If time does not permit teaching the entire chapter, use the Chapter Overview on this page, Reviewing Main Ideas at the end of the chapter, and the Chapter 17 audiocassette to point out the main ideas of the chapter.

Learning Styles	**Kinesthetic**	Explore, p. 539; Activity, pp. 541, 558; Design Your Own Investigation, pp. 544–545; Find Out, p. 549; Science at Home, p. 569
	Visual-Spatial	Activity, pp. 547, 554; Demonstration, p. 548; Across the Curriculum, p. 550; Multicultural Perspectives, p. 550; Find Out, p. 555
	Interpersonal	Activity, pp. 554
	Logical-Mathematical	Across the Curriculum, p. 551; Investigate, pp. 552–553
LS	**Linguistic**	Across the Curriculum, pp. 541, 557, 563; Language Arts, p. 546; Activity, pp. 557, 564; Multicultural Perspectives, p. 562
	Auditory-Musical	Find Out, pp. 543, 559; Activity, p. 546; Across the Curriculum, p. 561

Explore! ACTIVITY

Can you make a wave on a rope?

You have probably shaken a rope many times and watched waves travel down it. Now do it again and really observe what happens.

What To Do

1. Tie one end of a heavy 4-m rope to a desk or a doorknob.

2. Holding onto the other end of the rope with your hand, shake the rope up and down once.

3. Observe the pulse as it travels away from your hand.

4. Shake the rope up and down slowly and at a steady rate.

5. Describe the motion of the wave on the rope *in your Journal*.

6. Does the wave seem to move from one end of the rope to the other?

7. In what direction does the rope move?

ASSESSMENT PLANNER

PORTFOLIO
Refer to page 571 for suggested items that students might select for their portfolios.

PERFORMANCE ASSESSMENT
Process, p. 545
Skillbuilder, p. 554
Explore! Activity, p. 539
Find Out! Activities, pp. 543, 549, 555, 559
Investigate, pp. 544–545, 552-553

CONTENT ASSESSMENT
Oral, pp. 553, 555
Check Your Understanding, pp. 547, 554, 558, 564
Reviewing Main Ideas, p. 569
Chapter Review, pp. 570–571

GROUP ASSESSMENT
Opportunities for group assessment occur with Cooperative Learning Strategies.

Uncovering Preconceptions

Because waves crashing in on the seashore will carry objects in with them, students may have trouble accepting that waves pass through the medium, leaving it unchanged.

Explore!

Can you make a wave on a rope?

Time needed 10 minutes

Materials a piece of rope about 4 m long

Thinking Processes observing and inferring, recognizing cause and effect, modeling

Purpose To observe the motion of a transverse mechanical wave

Teaching the Activity

Demonstration Shake the rope up and down once and explain that a single disturbance is described as a pulse. Be sure that students see that one pulse is produced for each movement of your hand.

Science Journal Suggest that students' descriptions of the rope movement be accompanied by a "time-lapse" series of sketches that show the shape of the rope at a given point in time. L1

Expected Outcome

Students will recognize the characteristic up-and-down curve of a transverse wave.

Answer to Question

The wave moves horizontally along the rope. The rope moves vertically.

✔ Assessment

Performance Have students copy their journal sketches onto unruled paper and make "flip" books of the wave motion. Use the Performance Task Assessment List for Scientific Drawing in **PASC**, p. 55. L1 P

PREPARATION

Planning the Lesson

Refer to the Chapter Organizer on pages 538A–D

Concepts Developed

A wave is a rhythmic disturbance that transfers energy. In this section, students will distinguish between transverse and longitudinal waves.

1 MOTIVATE

Bellringer

 Before presenting the lesson, display **Section Focus Transparency 55** on an overhead projector. Assign the accompanying **Focus Activity** worksheet. L1
LEP

Discussion Ask students to list all the different types of waves they can think of. *Lists may include waves at the seashore, a stadium wave, microwaves, or sound waves.* Ask whether any characteristic is common to all waves. Some students will be able to operationally define that a wave is a periodic, repeated motion. L1

2 TEACH

Tying to Previous Knowledge

From Chapter 3, ask students to recall the properties of sound. Then have a volunteer explain the terms *compression* and *rarefaction* as they apply to sound waves.

Theme Connection In Section 17-1, the theme systems and interactions is emphasized and energy is implied. Allow students to observe and infer that waves transfer energy from a source to a destination in a systematic and predictable manner by having them do the activity on p. 541.

17-1 Waves and Vibrations

Section Objectives
- Describe how waves are produced.
- Identify transverse and longitudinal waves.

Key Terms
transverse waves
longitudinal waves

Waves Around You

What do you think of when you hear the word wave? Perhaps you think of a friendly greeting, the ocean, the beach, or people in a stadium performing a "wave."

What exactly is a wave? Think about Louie's wave. His dive caused a large disturbance when he pushed the water aside. That disturbance traveled out across the pool in the form of a wave.

Louie's wave certainly disturbed Ladonna and you, to say nothing of the water in the pool. Now, try to think of some more everyday experiences you've had with waves.

Think back to the Explore activity. Do you remember how you made the rope wave? You disturbed the rope by shaking it. The rope in your hand had an up-and-down movement. But the wave you created didn't stay in one place. It moved along the rope, hit the doorknob, and came back. Thus the wave was a disturbance traveling along the rope. Just like Louie's wave, this wave can move things it hits. A wave can carry energy.

Figure 17-1

A The energy released by undersea earthquakes cause giant waves called tsunamis.

B Although difficult to detect in open water, tsunamis sometimes build to a wall of water more than 30 m high when they reach shallow waters.

C When the tremendous energy that tsunamis carry is released against the shoreline, everything in the wave's path is swept away.

D The northern Japanese island town of Aonae was hit by a tsunami in July 1993, causing extensive damage.

540 Chapter 17 Waves

Meeting Individual Needs

Behaviorally Disordered Get students with a short attention span physically involved in distinguishing between transverse and longitudinal waves. Have them use their hands to mimic the up-and-down motion of a transverse wave. Then ask them to move their arms forward alternately making a fist and then spreading out their fingers, mimicking the motion of a longitudinal wave. L1

Program Resources

Study Guide, p. 61
Making Connections: Technology and Society, p. 37, Weather Forecasting with Radar L2
Science Discovery Activities, 17-1, Energizers
Section Focus Transparency 55

Types of Waves

In this chapter we will talk about mechanical waves. A mechanical wave is a wave that travels through matter. For now, think of matter as anything that takes up space. You saw one example of a mechanical wave when Louie landed in the water. Waves in the rope and sound waves in the air are also mechanical waves. In Chapter 3, you learned that air was the medium that carried sound from the source to your ear. In any mechanical wave, the matter through which the waves move is called the medium. A mechanical wave can be described as a periodic disturbance in a medium.

■ Transverse Waves

Once again, think back to the opening Explore activity. In the wave you made on the rope, the rope was the medium. The rope moved up and down or side to side. This was at right angles to the direction of the wave itself, which moved away from the source of the disturbance—in this case, you. The wave you produced with the rope was a transverse wave. As shown in **Figure 17-2**, **transverse waves** are waves in which the wave disturbance moves at right angles to the direction of the wave itself.

Figure 17-2

Ⓐ When a drop of water falls into a pool, the drop transfers energy and causes the surrounding particles of pool water to move in tiny down-and-up circles.

Ⓑ Their movement causes nearby particles to also move down-and-up. This disturbance, seen as waves moving across the surface of the water, passes from particle to particle and travels outward from the place the drop entered the pool. In what way are these waves transverse waves?

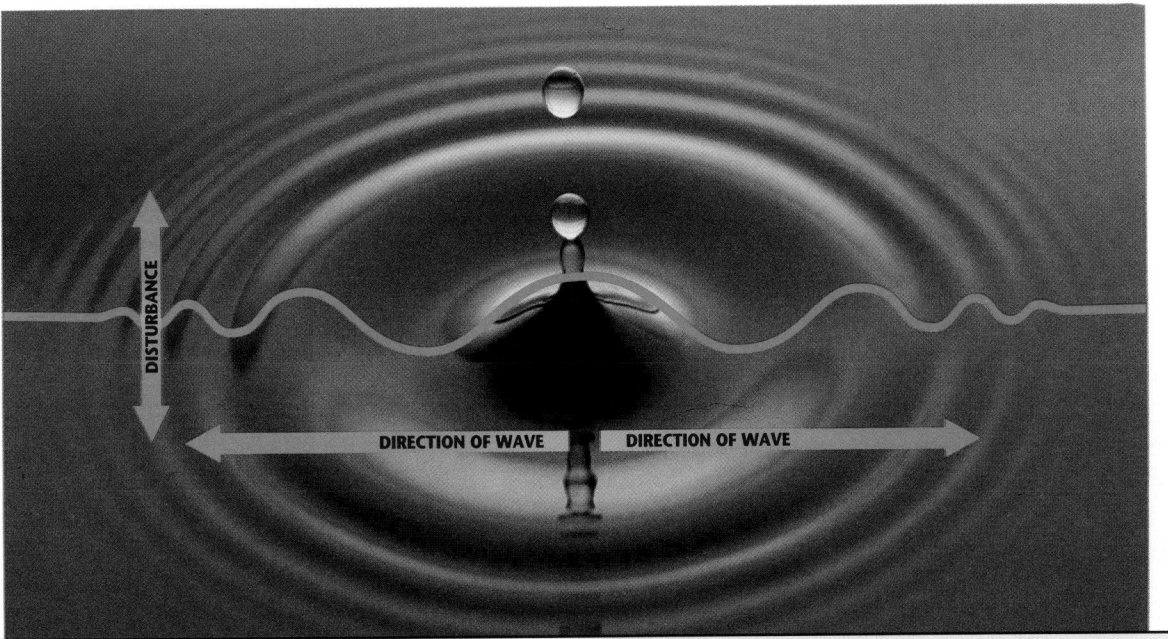

DISTURBANCE

DIRECTION OF WAVE　　DIRECTION OF WAVE

Visual Learning

Figure 17-2 As explained on page 547, a wave on the surface of water is actually a combination of longitudinal and transverse waves. Students who carefully observe the motion of the cork in the activity will see that it has a circular motion. Nevertheless, at this point, the more obvious transverse motion should be emphasized. **How do you know these are transverse waves?** *The direction of the waves is at right angles to the direction of the disturbance.*

Activity Materials needed are a large tub, water, golf balls, and a cork, per group. For each small group, fill the tub about half full of water and let the students take turns dropping golf balls into the center of the tub. Have students observe the waves. Then have students put the cork in the water and watch its movement when balls are again dropped in. Ask why the cork does not move to the side of the tub. *The water, or medium, is not moving to the side. It is just going up and down. Only the energy from the disturbance moves to the side of the tub.*
[L1] COOP LEARN

Content Background

Waves, such as sound and water, that require a medium to pass through are called mechanical waves. Waves, such as radio, light, and X rays, that do not require a medium are called electromagnetic waves.

Across the Curriculum

Language Arts

Have students make up poems to help them remember the terms *transverse, longitudinal, compression,* and *rarefaction.* They could use haiku, limerick, free verse, rap, or a rhyming verse. Read the following examples aloud before students attempt their own verses. [L1] [P]

Compression is a bunching up, tight as tight can be.
Rarefaction means spreading out, footloose and fancy free.
Transverse waves move at right angles;
they swerve with an up-and-down curve.
But longitudinal waves move all in one direction;
they bunch and they spread, I've heard.

Content Background

Seismology is a relatively new science born of 20th-century technology. The Richter scale, developed in 1935, and other units of seismic measurement are still being refined. The rapid accumulation of data gathered by seismic measurement has made seismic prospecting a highly successful endeavor. Precise seismic studies of the conditions in which lodes of valuable materials are often found make the acquisition of resources easier—and thus reduce their cost in the marketplace.

GLENCOE TECHNOLOGY

Videodisc

STVS: Physics

Disc 1, Side 1
Eliminating Potholes (Ch. 5)

Seismic Simulator (Ch. 6)

Tornado Detectors (Ch. 17)

Figure 17-3

Certain sound waves produced by a singer's voice may shatter glass. Longitudinal waves travel in the same direction as the disturbance. Sound waves are a type of longitudinal wave that travels through the air in a series of compressions and rarefactions. The drawing below shows a longitudinal wave on a spring.

Longitudinal Waves

Rarefaction   Compression

■ Longitudinal Waves

In Chapter 3, you learned that sound traveled through the air in a series of compressions and rarefactions—a bunching up and spreading out of the air that carried the sound from the source to you. This type of wave is called a longitudinal wave. In a **longitudinal wave**, as shown in **Figure 17-3**, the medium vibrates in the same direction as the wave itself travels. You can't directly observe longitudinal waves in air. However, you can find out more about them by seeing how they affect other substances.

Earth Science CONNECTION

Seismic Prospecting

Seismic waves are waves produced by an earthquake. They may be transverse, longitudinal, or a mixture of the two, like a water wave. Scientists study seismic waves in order to understand earthquakes and the structure of Earth. They can also be used in the search for mineral deposits such as oil, natural gas, and sulfur.

These scientists are setting up instruments to record seismic disturbances.

Artificial Earthquakes

In this case, however, scientists don't wait for an earthquake to strike, they produce small artificial quakes. These are created by setting off explosives in a small hole in the ground or by using vibrator trucks that can violently shake the ground. The artificial earthquake sends seismic waves thousands of feet into the ground. Waves then bounce off the rock formations and are reflected back. Acoustic receivers, called geophones, placed at different distances from the explosion, pick up the seismic waves and

Earth Science CONNECTION

TECH PREP

Purpose

Earth Science Connection reinforces the concepts in Section 17-1 by showing how the properties of seismic waves can be used as a tool to explore Earth and its resources.

Teaching Strategies

Activity Have students shake a wrapped mystery box and try to describe its contents by using the size of the box, the results of the shaking motion, and their senses. L1

Discussion

Explain to students that seismic waves change their speed and paths as the layers of Earth (crust, upper mantle, lower mantle, outer core, and inner core) change. Ask students how they think this information would help scientists find the boundaries of Earth's layers. *By monitoring primary, secondary, and surface waves, scientists can deduce boundaries within Earth's layers. Data collected by seismologists are input into a computer. The computer sorts out the data*

Can you observe the effects of a longitudinal wave?

Sound causes your eardrum to vibrate. If you made a model of the ear with a large "eardrum" could you feel and see it vibrate?

What To Do

1. Get a portable radio with a round speaker. Find a can or other container with the same or slightly larger diameter than that of the speaker.

2. Cut out the bottom of the can so that both ends are open.

3. Stretch a piece of plastic wrap across one end of the can and secure it with a rubber band.

4. Place the open end of the can over the speaker, facing up, and sprinkle some dry rice on the plastic.

5. Tune the radio to a song with a heavy beat and observe.

6. Put your fingertips lightly on the plastic. What do you feel?

7. Turn up the volume.

Conclude and Apply

1. Is the can necessary? How could you find out?

2. What happens to the vibrations of the plastic when you turn up the volume?

---

send signals to a truck outfitted with recording equipment. There, the seismic waves are amplified and recorded on tape.

Interpreting the Earthquake

The tapes are processed on a computer, and a printout is produced. This process is called seismic mapping. By analyzing the amplitudes of the seismic waves, and the time it takes them to bounce off rocks, scientists can get an idea of the

nature of rock layers, and the depths and locations of mineral deposits. This is a much less expensive way of deciding where to drill for mineral deposits than prospecting by drilling a number of deep holes.

What do you think?

Seismic prospecting is not only cheaper than drilling a number of deep holes, but damages the environment much less than drilling. How do you think drilling can harm the environment?

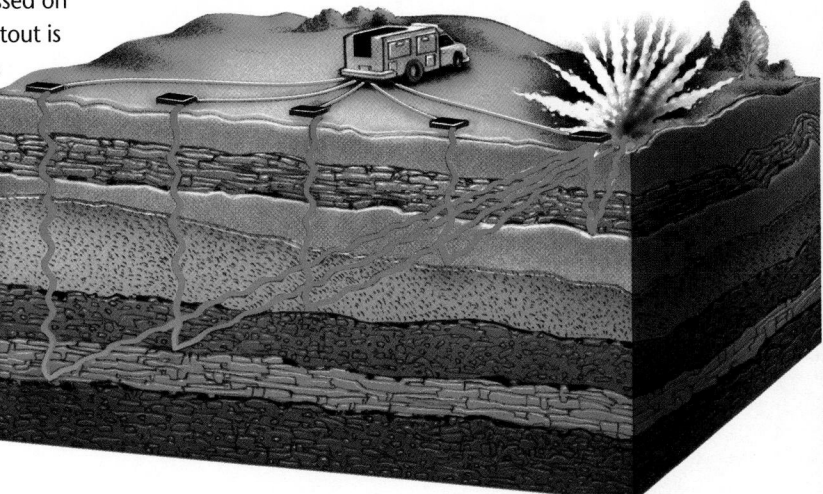

---

and generates a three-dimensional picture of Earth's mantle. [L1]

Answers to
What Do You Think?

Students' answers will vary. Bringing in the heavy equipment needed for drilling can cause harm to the vegetation in an area. For example, some students may be aware of the controversy about drilling into the tundra in Alaska. Most onshore oil drilling involves the use of large mud pits, which are unsightly. The leakage of

hydrocarbons from both onshore and offshore wells is another potential problem.

Going Further ⫸

Have groups of students find other examples of the use of waves by people and/or in nature. *Examples might include the songs of humpback whales; the science of earthquake prediction; finding water in desert regions; the details of seismic prospecting; and the use of waves in archaeological research.* Use the Performance Task Assessment List for Group Work in **PASC**, p. 97. [L2]

COOP LEARN

Can you observe the effects of a longitudinal wave?

Time needed 15 minutes

Materials portable radio with a round speaker, metal container with a diameter slightly larger than the radio speaker, plastic wrap, large rubber band, dry rice grains, can opener

Thinking Processes observing and inferring, comparing and contrasting, recognizing cause and effect, making a scientific model

Purpose Observe a longitudinal wave

Preparation Students should assemble the setup as shown in the text.

Teaching the Activity

Review with students the form sound waves take. Use the terms *longitudinal waves, compression,* and *rarefaction* as you review.

Troubleshooting Make sure plastic wrap is secured tightly around the can.

Science Journal After the music has left a pattern on the plastic wrap with the rice, ask students to describe and sketch what they observed. [L1]

Conclude and Apply

1. Yes, the can represents the outer ear, which collects the sound waves. You could hold the plastic wrap directly above the speaker.

2. The amplitude of the sound waves increases, making the plastic vibrate more, which in turn, modifies the pattern of rice.

✔ Assessment

Performance Have small groups of students carry out the experiment without the can. Have them compare and contrast their results in a lab report. *The plastic will not vibrate to the same extent because the waves are not being directed by the can.* Use the Performance Task Assessment List for Lab Report in **PASC**, p. 47. [L1] COOP LEARN

17-1 Waves on a Coiled Spring

Preparation

Purpose To observe transverse and longitudinal waves in a coiled spring.

Process Skills observing and inferring, comparing and contrasting, recognizing cause and effect, forming a hypothesis, designing an experiment, interpreting data, modeling

Time Required 20 minutes

Materials See student text. A plastic coiled spring does not work.

Possible Hypotheses Students should predict that the transverse waves they create will move at right angles to the source of the disturbance and that longitudinal waves will move in the same direction as the disturbance.

The Experiment

Process Reinforcement Suggest that students make tables in their journals that compare and contrast the two types of waves. The information recorded should also demonstrate an understanding of the cause-effect relationships that produce waves. Table heads might include: Wave type, Wave pattern, and Direction of Wave Relative to Motion of the Source.

Possible Procedures Generate waves in the spring by moving it in various ways. Watch the spring closely to see how the wave moves, how fast it moves, and if it comes back along the spring. Record your observations.

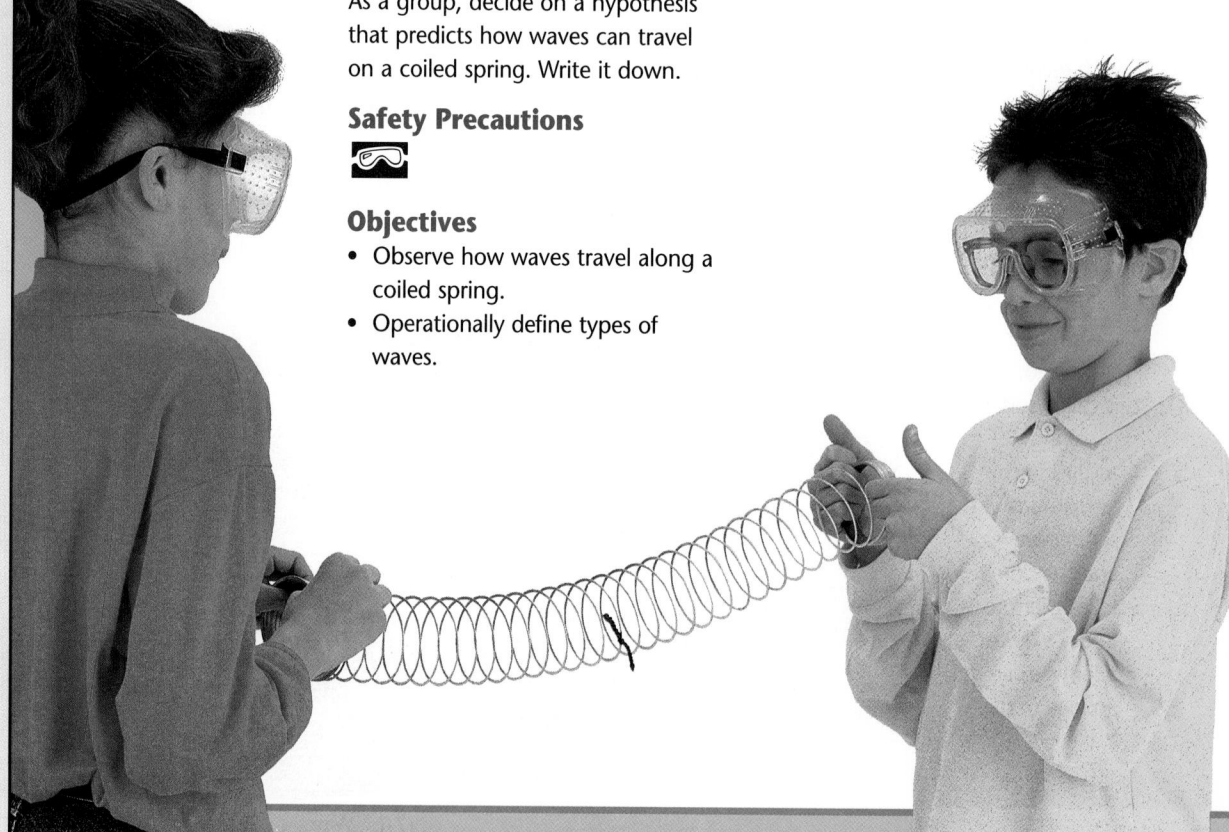

Waves on a Coiled Spring

You have learned about longitudinal and transverse waves. How do waves travel on a coiled spring? Are they transverse waves or longitudinal waves?

Preparation

Problem
How many types of waves can you create with a coiled spring?

Form a Hypothesis
As a group, decide on a hypothesis that predicts how waves can travel on a coiled spring. Write it down.

Safety Precautions

Objectives
- Observe how waves travel along a coiled spring.
- Operationally define types of waves.

Materials
coiled metal spring
piece of colored yarn

Program Resources

Activity Masters, pp. 71–72, Design Your Own Investigation 17-1

Making Connections: Integrating Sciences, p. 37, How Animals Use Waves [L2]

Making Connections: Across the Curriculum, p. 37, Crushing Stones with Waves [L2]

DESIGN YOUR OWN
INVESTIGATION

Plan the Experiment

1 Examine the materials and decide how you will test your hypothesis. Write down your plan.

2 *In your Science Journal,* prepare a place to draw diagrams of the wave types your group creates with the coiled spring. Plan to use arrows on the diagrams to show direction of movement.

Check the Plan

1 Determine if two people will hold the spring or if you will let the spring hang down.

2 Before you begin, check your plan with your teacher.

3 Do the experiment. Make certain that you observe closely to see

what kinds of waves you can make and to see what happens to the waves when they reach the end of the coil. Do they

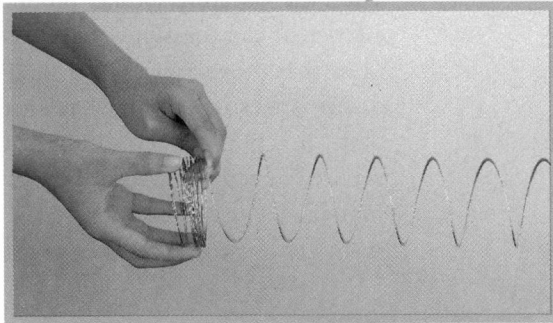

come back? Observe closely and *in your Science Journal,* draw diagrams of what occurs.

Analyze and Conclude

1. **Observe** What types of wave pulses did you create?

2. **Interpret Data** Draw a diagram of each type of wave that occurred in the spring. Label your diagrams. Show direction of the movement.

3. **Observe** Did the waves move in the same direction as the source of the disturbance? Explain.

4. **Observe** What happened when the waves reached the end of the spring?

5. **Infer** Compare the motion of a radio speaker tested in the Find Out! activity to the waves you created with the spring. Is a sound wave a transverse or longitudinal wave? Is an ocean wave a transverse or longitudinal wave?

Going Further

Observe how fast transverse and longitudinal waves move along the spring. Is there a set speed?

Discussion Review the types of waves found on Earth or in its atmosphere that are longitudinal and that are transverse. Use the photos and graphics on pages 540 to 543 to review.

Science Journal As each wave is produced, tell students to make a diagram of it in their tables of wave patterns.

Expected Outcomes

Students should understand that transverse waves move at right angles to the source of the disturbance, while longitudinal waves move in the same direction as the vibrating medium, alternately compressed and expanded.

Analyze and Conclude

1. transverse and longitudinal

2. Students should draw the diagrams in their Science Journals.

3. The transverse wave moved at a right angle to the source of the disturbance. The longitudinal wave traveled in the same direction as the disturbance.

4. The wave came back along the spring. A transverse wave remained transverse and a longitudinal wave remained a longitudinal wave.

5. longitudinal, transverse

Going Further ⫸

Use a stopwatch to measure the waves. There is a set speed.

✔ Assessment

Process Have students do the experiment again and find out if transverse or longitudinal waves travel the fastest along the spring. *They move at the same rate.* Use the Performance Task Assessment List for Formulating Hypotheses in **PASC,** p. 21. L1 P

Theme Connection Have students write a brief paragraph or two in their journals explaining how energy is transferred by sound waves. *Students' paragraphs will vary but should demonstrate a clear understanding of the law of conservation of energy.* Use the Performance Task Assessment List for Writing in Science in **PASC,** p. 87. L1

Discussion Invite a physician, nurse, or audiologist to visit your class and discuss hearing disorders. Have students prepare lists of questions beforehand. They may ask the speaker to explain how sound waves produce sounds we can hear and why some people do not sense those sounds. Ask the speaker to distinguish among various causes of hearing impairment, such as nerve damage, a damaged eardrum, or damaged bones in the middle ear; how these disorders are detected; and how they are treated. Encourage students to take notes in their journals. L1

Visual Learning

Figure 17-4 Have students use the figure to describe a sequence of events that allow the girl to hear the sounds made by the triangle. *When the instrument is struck, sound waves travel through the air. The waves are collected by the outer ear and transmitted to the inner ear where bones vibrate to produce the sound.*

The plastic wrap vibrated in the Find Out activity on page 543 because of the sound waves—the disturbances moving through the air. These disturbances were produced by the paper cone of the radio speaker vibrating back and forth. As the speaker moved out, it pushed the air in front of it. When the speaker moved back, the air expanded, becoming more rarefied. Thus, the vibrating speaker produced a series of compressions and rarefactions that moved away from it. The disturbances in the air were in the same direction as the motion of the wave. Therefore, we can say that sound is a longitudinal wave.

Just like Louie's water wave, sound waves carry energy. The energy in the sound waves caused the plastic and rice grains to have kinetic energy.

You saw in the Investigate that in a transverse wave, as the wave moved away from the source, the thread moved back and forth at right angles to the spring.

When you made a longitudinal wave in the Investigate compressions moved away from the source, and the thread vibrated back and forth in the same direction. Notice that a longitudinal wave is disturbed along the direction the wave travels, while in transverse waves the disturbance occurs at right angles to the direction of wave motion.

Figure 17-4

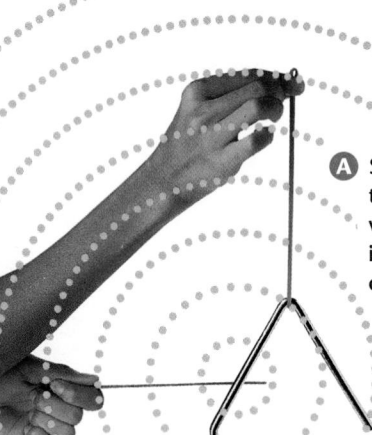

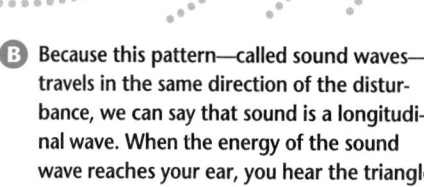

A Striking the triangle transfers energy to the triangle and makes it vibrate. The triangle's vibrations transfer energy to the air, causing a disturbance and creating a pattern of compressions and rarefactions.

B Because this pattern—called sound waves—travels in the same direction of the disturbance, we can say that sound is a longitudinal wave. When the energy of the sound wave reaches your ear, you hear the triangle.

■ Water Waves

A surface water wave is a combination of a transverse and a longitudinal wave. When a wave hits a buoy or other object floating in the water, the object will move in a circle. It moves up and down, and back and forth.

Figure 17-5

A surface water wave is a combination of longitudinal motion (back-and-forth movement that is parallel to the direction of the wave) and transverse motion (movement at right angles to the direction of the wave). The result is that an object floating in water moves in a small circle, staying nearly where it started.

But, it still comes back to the place it was before the wave struck it, and the wave moves on.

People talking, music on the radio, waves in the pool, and surf at the beach are just a few of the waves you encounter every day. In the next section, you will learn more about wave characteristics. You'll find out what the hills and valleys of waves are actually called, as well as what wavelength and frequency mean.

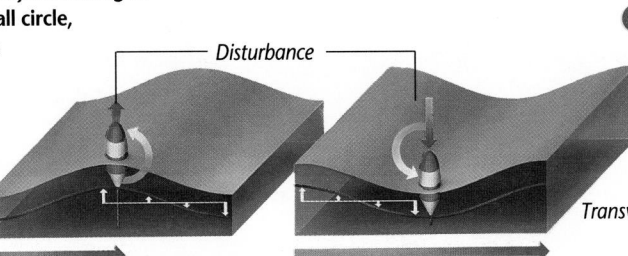

Transverse motion

Movement of wave

Transverse wave

Movement of wave

A When the buoy is hit by the crest of a wave, the buoy moves slightly forward (longitudinal motion) and upward (transverse motion).

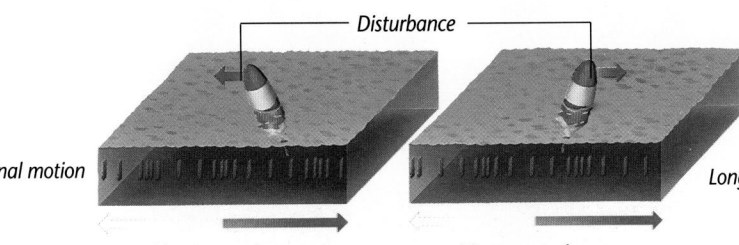

Longitudinal motion

Movement of wave

Longitudinal wave

Movement of wave

B As the next trough strikes the buoy, it moves slightly backward (longitudinal motion) and downward (transverse motion).

check your UNDERSTANDING

1. Define a wave. List three examples of ways in which a mechanical wave may be produced.
2. Draw diagrams of a transverse wave and a longitudinal wave. Through labeling and brief descriptions, compare and contrast their properties.
3. Draw "snapshots" of a transverse wave pulse at different times. Label the drawing showing the first, middle, and last picture of the pulse.
4. **Apply** If you were to sprinkle dry rice onto the head of a drum and strike the drumhead at the edge with a drumstick, what would you observe?

check your UNDERSTANDING

1. A wave is a continuous, rhythmic disturbance that transfers energy. Examples include: shaking a rope, dropping rocks in water, and shouting.
2. Diagrams may resemble Figure 17-5. Diagrams should include movement of medium and direction of wave motion.
3. [wave diagram with Viewer, Viewer, Viewer labels]
4. The rice would arrange itself into a pattern that would show wave troughs.

Check For Understanding

Have students answer the first three questions using the figures in this section as well as the information gathered in the various activities. Then, if possible, borrow a drum and have students try the experiment described in Check Your Understanding number 4 to test their hypotheses.

Reteach

Activity Have students make flip books that show the motion of particles along transverse and longitudinal waves and explain the differences. L1 LEP

Extension

Activity To help students distinguish between transverse and longitudinal waves, bring in some common materials such as a scarf, a metal rod or pipe, and a stringed instrument. Have small groups of students produce waves in each medium and describe the waves produced. "Wiggly" waves traveling through the material are transverse waves. Any sound waves produced are longitudinal waves in the air surrounding the objects. L2

4 CLOSE

Activity

Have pairs of students experiment with corks, small glass bowls, a drop or two of food coloring, and water to demonstrate and observe the characteristics of surface water waves. Then have students relate their observations to personal experiences while swimming in a lake, ocean, or a wave pool. L1

COOP LEARN

PREPARATION

Planning the Lesson

Refer to the Chapter Organizer on pages 538A–D.

Concepts Developed

In this section, students will continue their study of waves, learning certain characteristics of a transverse wave (crest, trough, wavelength, and amplitude) and the relationship among frequency, wavelength, and speed.

1 MOTIVATE

Bellringer

 Before presenting the lesson, display **Section Focus Transparency 56** on an overhead projector. Assign the accompanying **Focus Activity** worksheet. [L1] [LEP]

Demonstration To strengthen students' observational skills, drop a series of different-sized rocks into a large tub half filled with water to produce waves with different amplitudes. Have students hypothesize what causes the wave amplitude to vary. [L1]

Visual Learning

Figure 17-6 Ask if any students have waterskied over troughs or crests made by motor boats. Have them compare and contrast the experience to waterskiing on smooth water. Ask students if they can give other examples of feeling a wave. *Students may mention examples of sound waves.*

17-2 Wave Characteristics

Section Objectives

- Draw a wave.
- Identify the wavelength, amplitude, crest, and trough of a wave.
- Explain the relationship among frequency, wavelength, and speed in a wave.

Key Terms

crest, trough, amplitude, wavelength

Figure 17-6

A This boat is sailing across the waves, through troughs and crests in the ocean.

B Small troughs and crests are also created by motor boats. This water skier will get a bumpy ride when skiing across these people-made troughs and crests.

548 Chapter 17 Waves

Properties of Waves

The wave you produced earlier with the rope can be described by its properties. When you quickly moved the rope up and down, you may have noticed high and low points—hills and valleys. These low points, the valleys, are called **troughs**. The high points, the hills, are called **crests**. In the following Find Out activity, you will observe these parts of a wave again.

Program Resources

Study Guide, p. 62
Laboratory Manual, pp. 107–110, Velocity of a Wave [L3]
Concept Mapping, p. 25, Waves [L1]
Multicultural Connections, p. 37, The Alphorn [L1]
Teaching Transparency 34 [L2]
Section Focus Transparency 56

What are some wave characteristics?

You may have seen excited fans in the stands of a sporting event make a human wave. Now, you will make a wave in class and study its characteristics.

What To Do

1. Your class will sit or stand in a large circle. One student will raise and lower his or her hands to start a wave moving around the circle. The raised hands are like a wave pulse on a rope.

2. Practice "doing the wave" until the pulse moves at a constant speed around the circle.

3. Measure the distance around the circle and the time (in seconds) it takes for a pulse to travel around the circle.

4. While a wave is moving around, the first student starts a second pulse when the original pulse reaches the student exactly opposite him or her.

5. If the class is large enough, try having three or four pulses moving around the circle at the same time.

6. Try to have enough pulses moving around the circle so that each student has his or her hands up as long as they are down.

7. Make a drawing of the people wave when it had the largest number of pulses traveling around.

Conclude and Apply

1. What was the speed of your wave, that is, the distance around the circle divided by the time the wave took to go around the circle?

2. How many times did the first student raise his or her hands each minute when one pulse was moving around? To find out, divide the number of seconds in one minute, 60, by the time it took the pulse to travel the circle in seconds.

3. How many times did the first student raise his or her hands each minute when two pulses were moving around? When three pulses were moving?

17-2 Wave Characteristics **549**

Content Background

In a standing wave, the points on the wave where there is no displacement of the medium are called nodes. The crests and troughs between the nodes are called antinodes. A standing wave can be compared to a ship that is rolling in heavy seas. As the ship rolls back and forth, the center of the ship is not displaced and is like the node. The edges of the ship are greatly displaced in an up-and-down motion and are like antinodes.

GLENCOE TECHNOLOGY

 Software

Computer Competency Activities

Chapter 17

Expected Outcomes

Students will demonstrate wave characteristics with a people wave.

Conclude and Apply

Values will vary; assist students, if necessary, with the computations. Students should find that the speed of the wave remains approximately the same regardless of the wave's frequency.

✔ Assessment

Performance Have students remain seated and make a quarter-turn to their right. Instruct them to place their hands gently on the shoulders of the person in front of them. Ask one student to lean forward and sit back up, instructing the other students to do the same when they feel the lean "passed" along from the student behind them. See if students can identify their actions as modeling a longitudinal wave, and have them identify speed and frequency as they did before. [L1] **COOP LEARN** **LEP**

Find Out!

What are some wave characteristics?

Time needed 20–25 minutes

Materials No special materials are required for this activity.

Thinking Processes observing, defining operationally, comparing and contrasting, making a model, sampling and estimating

Purpose To operationally define characteristics of waves

Teaching the Activity

Review the definitions of the terms *crest, trough, wavelength,* and *frequency.*

Troubleshooting Remind students to "do the wave" smoothly and refrain from making sudden, snapping movements.

Science Journal Students' drawings should demonstrate an understanding of crests, troughs, and wave pulses. [L1]

2 TEACH

Tying to Previous Knowledge

Have students recall their study of pendulum movement in Chapter 13, reviewing the terms *period* and *amplitude*. Explain that the movement of waves can be described in a similar manner.

Across the Curriculum

Geography

Have students locate countries and islands bordering the Pacific and Indian Oceans that have been struck by devastating waves called tsunamis. Have students find out the origin of the term, what generally causes tsunamis, and some statistics—size of the wave, when and where it occurred, what caused it, and the damage done. Have students create a display in the classroom to present their findings. Use the Performance Task Assessment List for Display in **PASC**, p. 63. L1

GLENCOE TECHNOLOGY

 Videodisc

The Infinite Voyage: Living With Disaster

Chapter 3
The Hurricane: The Most Powerful Storm

The Infinite Voyage: Sail On, Voyager

Chapter 4
Jupiter's Atmosphere

DID YOU KNOW?

■ Amplitude

Suppose you want to know the amplitude of Louie's wave in the swimming pool. The normal level at the shallow end of the pool is 1 m. But, when Louie jumped in, the wave made the level rise 30 centimeters. Thus, the crest of Louie's wave was 30 centimeters higher than the normal level of the pool. This was the amplitude of his wave. As shown in **Figure 17-7**, the **amplitude** is the distance from the crest or trough of the wave to the middle level. What changed the amplitude of your people wave? The higher a person raised his or her hands, the larger the amplitude.

Have you ever seen waves on the ocean? On a calm day they can gently lap at the beach. Their amplitudes are small. But, when there is a storm off

Figure 17-7

shore, their amplitudes can be meters high. Then they can be very dangerous. The energy they carry can do much damage. The energy carried by a mechanical wave depends on its amplitude.

■ Speed

You learned in Chapter 12 that the speed of a ball, auto, or person is the distance traveled divided by the time it takes to go that far. Did anyone in the wave circle move in a direction around the circle? No, but the wave did, didn't it? The speed of a wave is found the same way the speed of a ball is found. You measure the distance the wave travels and the time it takes and then divide the distance by the time. The speed of the people wave didn't depend on the number of

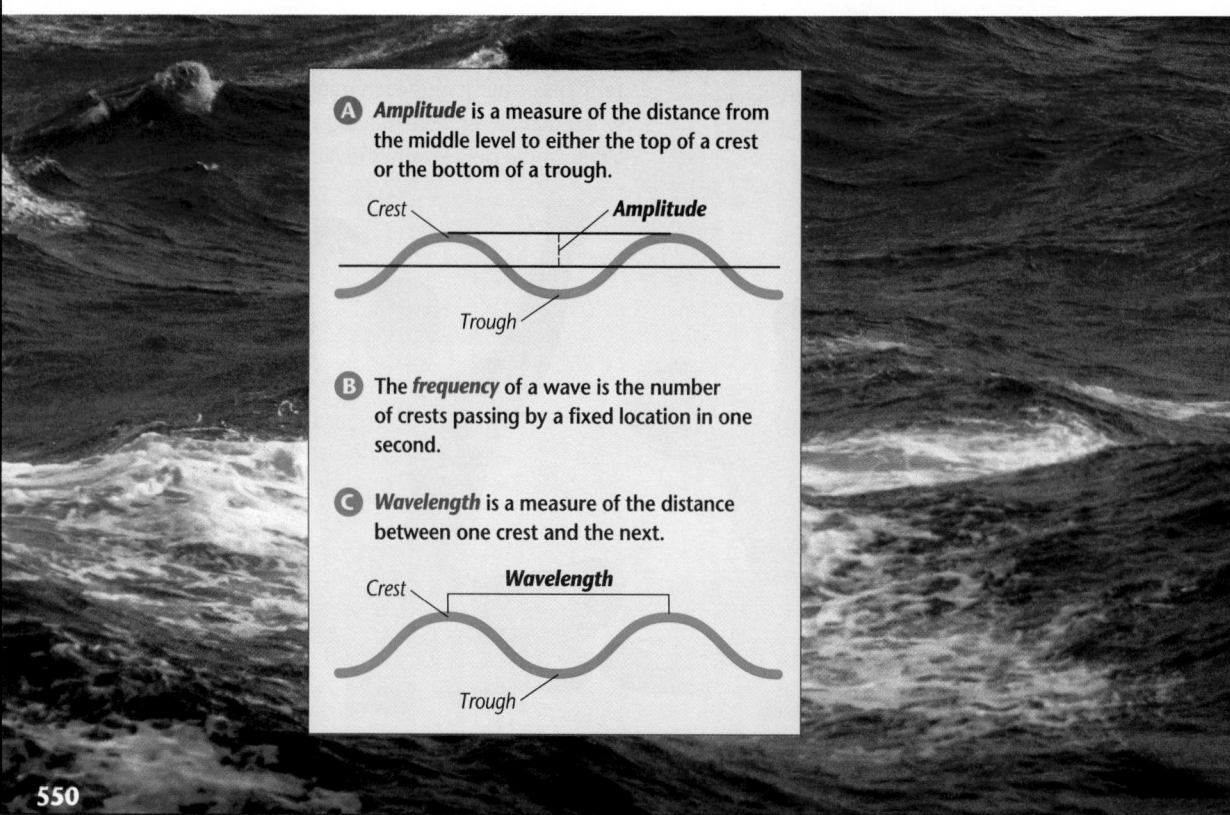

A *Amplitude* is a measure of the distance from the middle level to either the top of a crest or the bottom of a trough.

Crest — Amplitude

Trough

B The *frequency* of a wave is the number of crests passing by a fixed location in one second.

C *Wavelength* is a measure of the distance between one crest and the next.

Crest — Wavelength

Trough

550

MEETING INDIVIDUAL NEEDS

Gifted After students have done the activity described in *Across the Curriculum* above, have them hypothesize how scientists might predict and consequently warn people about tsunamis. Then have them compare their responses against the methods actually used by scientists. Students may wish to research the Tsunami Warning System based in Hawaii. L3

Multicultural Perspectives

Tsunamis

Interested students might want to study copies of Japanese paintings and woodcuts that depict tsunamis. (Goodman, Billy. *Natural Wonders and Disasters.* Boston: Little, Brown, and Company, 1991.) Have students also find out and recount the legends associated with these pieces of art. L1

waves, just on how fast people could raise their hands. That is, the speed depended only on the medium—the people through which the wave moved.

■ Frequency

What changed when you added more pulses in your people wave? Not the speed; each pulse moved around the circle at the same rate. When only one pulse went around the circle you didn't have to raise your hands very often, but when many pulses were moving around, you had to raise your hands more frequently. The frequency of a people wave is the number of times you had to raise your hand each second. The frequency of a wave is the number of crests passing by a fixed location in one second.

■ Wavelength

How far apart were the pulses in your people wave? When there was only one pulse going around, the distance was the circumference of the circle. What was it when two pulses went around? What about three? The distance was first half the circumference, then one third the distance around the circle.

When you measured the distance between the pulses of your people wave you measured a property called wavelength. **Wavelength** is the distance between the crest of one wave and the crest of the next. As the wavelength decreased, how did the frequency change? In the following activity you will study the relationship between wavelength and frequency in a different type of wave.

551

Visual Learning

Figure 17-7 To improve their ability to interpret scientific illustrations, have students locate the crests and troughs of each wave.

Across the Curriculum

Math

Have students use this simple formula to determine how fast a wave is traveling: velocity = wavelength × frequency. Give students the following problems to solve. **A wave moving along a rope has a wavelength of 1.2 m/wave and a frequency of 4.5 waves/s. How fast is the wave traveling along the rope?** *1.2 m/wave × 4.5 waves/s = 5.4 m/s.* **A wave in the ocean is 3.2 m long and its frequency is 0.6 waves/s. What is the velocity of the wave?** *3.2 m/wave × 0.6 waves/s = 1.92 m/s* **An earthquake can produce a wave traveling 5000 m/s with a wavelength of 417 m. What would be the frequency of that wave?** L3

$$frequency = \frac{velocity}{wavelength}$$

$$\frac{5000 \ m/s}{417 \ m/wave} = 12 \ waves/s$$

GLENCOE TECHNOLOGY

 Videodisc

The Infinite Voyage: The Great Dinosaur Hunt

Chapter 9
Communication Theories

Program Resources

Multicultural Connections, p. 38, Water Drums L1

17-2 Ripples

Planning the Activity

Time needed 20 minutes

TECH PREP

Purpose To demonstrate how changing the frequency of a wave affects the wavelength.

Process Skills observing and inferring, comparing and contrasting, interpreting data, making and using scientific diagrams, predicting, designing an experiment, forming operational definitions, classifying, modeling

Materials clear glass dish approximately 30–cm square, strips of plastic foam, tape, water, pencil or pen, overhead light, piece of unruled white paper, ruler

Preparation Collect materials for each group and place at appropriate workstations.

Teaching the Activity

Process Reinforcement
Make sure students realize that their groups' sketches should be drawn to scale so relative measurements will be valid.

Science Journal In their journals, have students describe what they see and draw the shapes of the waves. Also have students identify the wave crests and troughs, then measure the amplitude and wavelength for each set of waves. L1

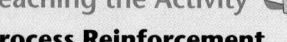

INVESTIGATE!

Ripples

Do water waves behave the same as people waves and spring waves? How are their speed, frequency, and wavelength related? In this activity you will use the patterns produced by light shining through the crests and troughs of waves in a shallow dish to investigate water waves.

Problem

How are a wave's frequency and wavelength related?

Safety Precautions

Materials

clear glass dish approximately 30-cm square	strips of plastic foam tape
pencil or pen	water
1 piece of blank white paper	overhead light ruler

What To Do

1 Tape strips of plastic foam to the inner edges of the dish (see photo **A**). Then, fill the clear glass dish with about 3 cm of water (see photo **B**). Set it on a piece of blank white paper under an overhead light source.

2 Tap the water with the end of your pencil or pen. Observe the wave by looking at the paper. *In your Journal*, draw the shape of the wave. Compare the speed of the wave in all directions. How fast does the wave travel? Estimate how long it takes to travel the length of the dish.

Program Resources

Activity Masters, pp. 73–74, Investigate 17-2

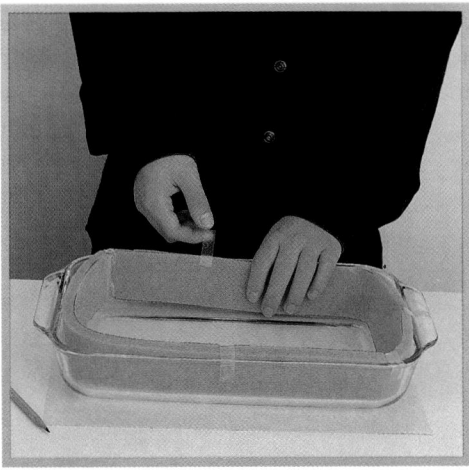

A

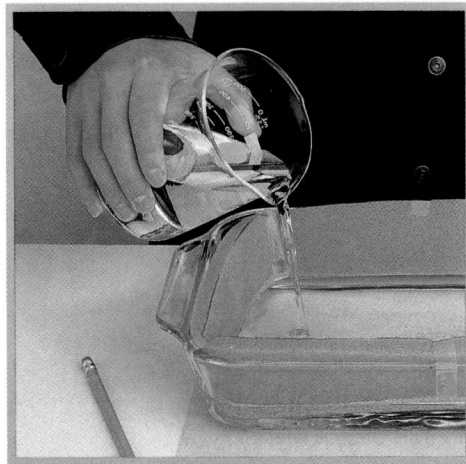

B

3 Now, tap the water again, producing a series of waves. Increase the frequency by tapping the water faster and observe the change in wavelength. Draw an example of low- and high-frequency waves being produced.

Analyzing

1. What effect does increasing the frequency have on the wavelength of the waves produced in Step 3?

2. What would happen to the wavelength if you decreased the frequency?

Concluding and Applying

3. What is the relationship between wavelength and frequency in water waves?

4. **Going Further** Predict whether the wave speed depends on the water depth. In your Journal write a plan about how you could make different depths of water in the same dish and find out.

Students will observe that as the frequency of a wave is increased, its wavelength decreases. Also, they will see that as the frequency of a wave is decreased, its wavelength increases.

Answers to Analyzing/ Concluding and Applying

1. Increasing the frequency causes the wavelength to decrease.

2. The wavelength would increase.

3. Wavelength and frequency are inversely proportional (as one increases, the other decreases).

4. Wave speed is related to water depth. Changing the depth has the same effect as changing the medium. Water waves generally travel faster in deeper water than in shallower waters. Students' plans might include using various thicknesses of foam to simulate a basin with different depths or slightly elevating one edge of the dish.

✔ Assessment

Content Have students apply what they have learned in this activity by having them answer the following question. How could the terms *crest*, *trough*, *wavelength*, and *amplitude* be applied to longitudinal waves? *The crest is analogous to the point of maximum compression. The trough is similar to the point of maximum rarefaction. The wavelength can be measured from one compression to the same spot on the next compression, or one rarefaction to the next. Amplitude is related to how "compressed" a compression is, and how "rarefied" a rarefaction is. The greater the difference between the two, the greater the amplitude.* Use the Performance Task Assessment List for Making Observations and Inferences in **PASC**, p. 17. L1

Meeting Individual Needs

Learning Disabled/Behaviorally Disordered Be sure learning disabled and behaviorally disordered students are actively involved in this activity. Allow them to tap the water, producing the waves. Ask them to draw the pictures of the waves produced.

You may want to assign partners, teaming mainstreamed students with other students, to discuss the results. Peer teaching provides concentrated individual attention. It also helps students gain a deeper understanding of the concept, because they have to understand it to explain it to another student. **COOP LEARN** **LEP**

The speed of sound in air depends on air composition and temperature. Accept any hypothesis that students can justify. Use the Performance Task Assessment List for Formulating a Hypothesis in **PASC**, p. 21. L1

3 ASSESS

Check for Understanding

Draw two transverse waves of differing wavelengths on the chalkboard. Have students identify and measure the wave crests, wave troughs, amplitude, wavelength, and frequency. L1

Reteach

Activity Have students shape a one-meter length of string into a wave and glue it to a piece of paper. Have students label the wave crests, wave troughs, amplitude, and wavelength. If they have difficulty with amplitude, have them glue a piece of string straight across the wave. L1

Extension

Activity Let pairs of students experiment with making waves in a tub of water. Have them mark the level of the water at rest on the side of the tub then drop items into the water, creating waves. Students should mark the height of the wave crest and calculate amplitude and wavelength. Provide a stopwatch so that frequency also can be determined. L3 COOP LEARN LEP

4 CLOSE

Discussion Ask students if they have ever tried surfing or watched surfing. Lead students to recognize that the surfer does not ride the crest, but slides down the slope of the wave, ahead of the crest.

SKILLBUILDER

Forming a Hypothesis
Sound travels slower in air at high altitudes than at low altitudes. State a hypothesis to explain this observation. If you need help, refer to the **Skill Handbook** on page 663.

Just as in the case of people waves, you found that as you increased the frequency of the water wave, the wavelength got smaller. The speed, however, didn't depend on the frequency or the wave-length, just the depth of the water. That is, the speed depended on the medium the wave moved through.

The wavelength of a wave is the distance between the crests or troughs, and the frequency is how many crests or troughs flow by you in one second. What happens when waves run into one another? You'll find out in the next section.

Figure 17-8

The speed of sound is determined by the medium through which it travels. In general, sound travels more quickly through solids and liquids than through gases.

A Sound travels through air at 340 m per second.

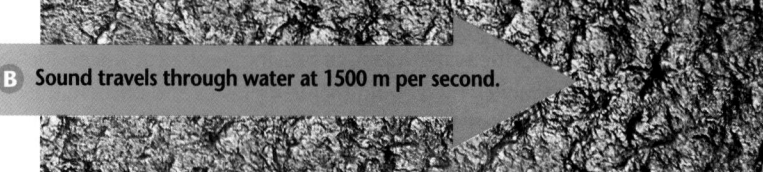
B Sound travels through water at 1500 m per second.

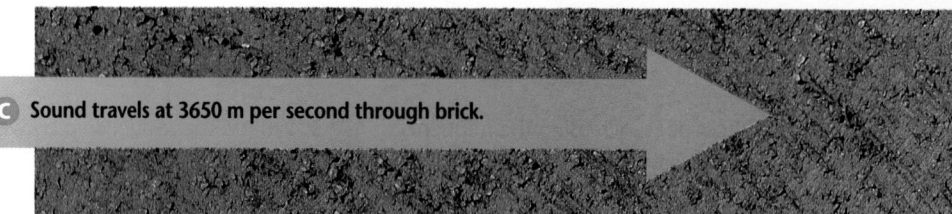
C Sound travels at 3650 m per second through brick.

D Sound travels at 6000 m per second through steel.

check your UNDERSTANDING

1. Sketch a transverse wave. Label a crest, a trough, a wavelength, and the amplitude.
2. What is the relationship among the frequency, wavelength and speed of a wave?
3. **Apply** What characteristic of a wave would be most important to a surfer?

---

check your UNDERSTANDING

1. Sketches should resemble those in Figure 17-7.
2. Speed is determined by the medium only. Wavelength and frequency in a given medium are inversely proportional (as one increases, the other decreases).
3. the amplitude

Interference

Imagine you are on the edge of a pond or pool. Now suppose you drop two rocks into the water about 100 centimeters apart. The ripples created by each rock move out in circles, and quickly run into each other.

If you have seen two wave patterns created in this way, you know that the waves don't bounce off each other. Instead, they appear to pass right through each other. Thus, it appears as if the waves from both rocks are able to exist in the same place at the same time.

Section Objectives
- Explain how waves add together.
- Describe two examples of wave interference.

Key Terms
interference

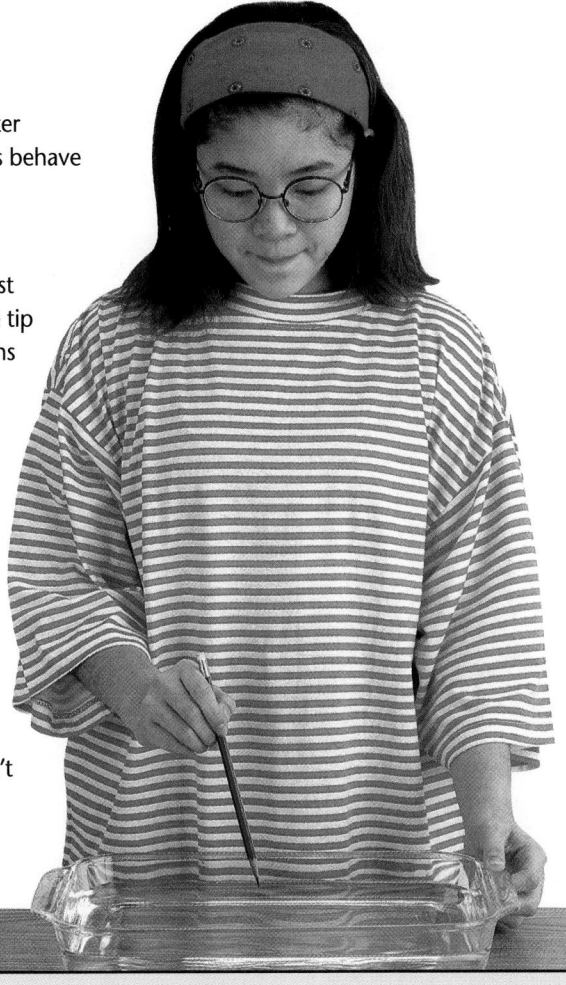

Find Out! ACTIVITY

What happens when waves pass?

You've studied the characteristics of a single water wave. Now you will investigate how two waves behave when they meet.

What To Do

1. Using the clear dish, water, and light from the last Investigate, tap the surface of the water with the tip of one pencil. Can you find the crests and troughs of the spreading wave?

2. Now, tap the surface with the tips of two pencils about 10 cm apart. Observe the waves carefully.

3. Try alternating the pencils so one hits the water while the other is raised. Observe the result.

Conclude and Apply

1. Describe what happens when the wave from one pencil meets a wave from the other.

2. Did anything change when the two pencils didn't hit the water at the same time?

Find Out!

What happens when waves pass?

Time needed 10–15 minutes

Materials clear glass dish approximately 30–cm square, water, overhead light, piece of unruled white paper, two pencils

Thinking Processes observing and inferring, comparing and contrasting, recognizing cause and effect, defining operationally, modeling

Purpose To observe the effects of colliding waves

Preparation Prepare dish as you did for the Investigate in Section 17-2, except without the plastic foam strips. Place it over a piece of paper and fill half full of water.

Teaching the Activity

Troubleshooting When a student taps with both pencils, he or she should tap once with each, so that each produces a single wave.

PREPARATION

Planning the Lesson

Refer to the Chapter Organizer on pages 538A-D.

Concepts Developed

In this section, students will discover what happens when waves of the same or different amplitudes meet, or interfere.

Science Journal Students should be able to identify the crests as being the brighter areas and the troughs as being darker. Have students illustrate and label the two phenomena simulated in this activity. **L1**

Expected Outcomes

Students should observe that when two waves meet, they either increase in size or cancel each other out.

Conclude and Apply

1. The waves combine momentarily to form a single wave then quickly pass through each other.

2. Interference still occurred, but in different locations.

✓ Assessment

Oral Ask students if the original waves changed during the activity. *Students should deduce that after meeting, the original waves continued on unchanged with the same amplitude prior to interference.* Use the Performance Task Assessment List for Making Observations and Inferences in **PASC**, p. 17. **L1**

1 MOTIVATE

Bellringer

Before presenting the lesson, display **Section Focus Transparency 57** on an overhead projector. Assign the accompanying **Focus Activity** worksheet. L1

LEP

Discussion Many children make their first wave experiments in the bathtub at an early age. Remind students how sliding back and forth in the tub produces small waves which can suddenly combine into enormous waves, spilling onto the floor and flooding the bathroom. This is an example of constructive interference.

2 TEACH

Tying to Previous Knowledge

In Chapter 13, students learned that an object in motion changes its motion when it reaches the ground or collides with another object. Waves add or cancel when they collide. Also, waves can pass through each other without being affected themselves.

Visual Learning

Figure 17-9 After students have studied the figure, have them paraphrase definitions of constructive and destructive interference. **How are constructive and destructive interference alike?** *They are alike in that both interactions produce a new wave, and after the interference, the original waves move on unchanged.* **How are they different?** *During constructive interference, the new wave has an amplitude equal to the combined amplitudes of the interfering waves. During destructive interference, because a crest is colliding with a trough, the new wave has a lower amplitude.*

As the waves pass through each other, they interact in one of two ways. If two crests pass through each other, they briefly form a new wave that is equal to the sum of the amplitudes of the two waves, as shown in **Figure 17-9 A**, **B**, and **C**.

Because the waves in **Figure 17-9** "interfere" with each other, this situation is called interference. **Interference** is the interaction of two or more waves at one point. When two or more waves add together, it is called constructive interference.

But what would happen if a trough of one wave passed through the crest of another? Look at **Figure 17-9 D**, **E**, and **F**. What happened?

You might say that one wave destroyed the other. What you just saw is called destructive interference. The waves, however, are not really destroyed because they will emerge on the other side unchanged.

Figure 17-9

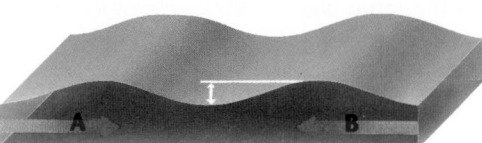

Constructive Wave Interference

Ⓐ Two crests of equal amplitude—A and B— approach each other from different directions.

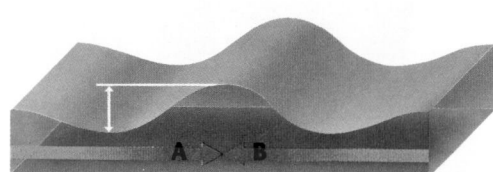

Ⓑ When waves A and B meet, they briefly form a new wave, A + B, which has an amplitude equal to the sum of the amplitudes of both waves.

Ⓒ Once the waves pass through each other, they are unchanged and each retains the amplitude it had before the meeting.

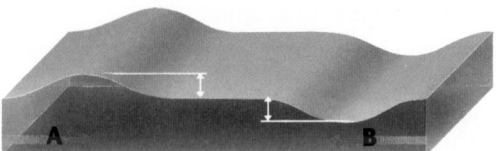

Destructive Wave Interference

Ⓓ The Crest of wave A and trough of wave B- approach each other from different directions. The amplitude of A is equal to the amplitude of B-.

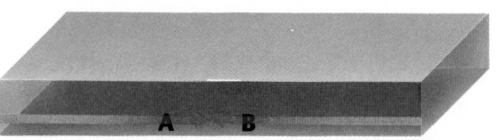

Ⓔ When A and B- meet, they briefly form a new wave, A + B-, which has an amplitude equal to the sum of the amplitudes of the crest A and the trough B-. The result is that for an instant, the amplitude of the new wave is zero and the water shows no disturbance.

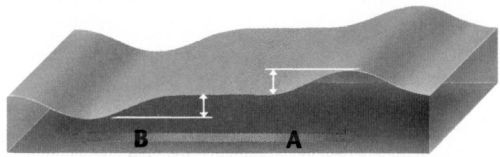

Ⓕ Once the crest and the trough pass through each other, they are unchanged and each retains the amplitude it had before the meeting. How are constructive and destructive wave interference alike? How are they different?

556 Chapter 17 Waves

ENRICHMENT

Discussion Have students hypothesize where it would be useful to design a room or building so that sound waves interfere with one another. *Suggestions may include a gymnasium or a student cafeteria.* Have students do research to find out some noise-reduction methods used by architects. Have them present their findings to the class as an oral report. Use the Performance Task Assessment List for Oral Presentation in **PASC**, p. 71. L2

Program Resources

Study Guide, p. 63
Teaching Transparency 33 L2
Take Home Activities, p. 27, Wave Interference L1
Science Discovery Activities, 17-2, Earning a Standing Ovation L1
Section Focus Transparency 57

Patterns of Interference

In **Figure 17-11**, you see a display from a ripple tank. It is a more sophisticated version of what you used in the Investigate. Notice that two overlapping water waves have been produced by two sources hitting the water at the same time. Locate a crest and a trough close to a source. How did you identify each of them?

Now locate the areas where two wave crests have added, producing a larger crest. Look for very bright bands of light. How can you tell that constructive interference is going on? Now find areas where destructive interference is going on. How can you tell that waves are canceling each other?

Figure 17-11

Scientists create various wave patterns in ripple tanks to observe the way waves interact. You can model many wave behaviors with a homemade tank.

Figure 17-10

A When you look at a bubble or a CD and see a spectrum, you are witnessing wave interference. When white light strikes the tiny pits of a CD disk, it is diffracted in many directions.

B When light strikes a bubble, some light is reflected by the bubble's outer surface and some by its inner surface.

C When two or more light waves from a bubble or from a CD meet, interference occurs. If waves of one color interfere constructively, you see that color.

17-3 Adding Waves **557**

3 ASSESS

Check for Understanding

To help students better understand the Apply question, take the class to the school's auditorium, and have students try to find any "dead" spots where sounds are muffled. L1

Reteach

Activity Use a group of ten students to make a human wave that demonstrates wave interference. Have students form a line. Then, beginning at both ends, ask students to raise and lower their arms. When the waves meet in the middle, the students there should stand if two crests (arms raised) arrive at the same time, representing a wave that is twice as big. If a crest and a trough hit at the same time, the students in the middle should keep their arms level to indicate that one canceled the other. Have observers describe what is happening in terms of waves. L1
COOP LEARN

Extension

Activity Allow students to experiment with wave interference using golf balls and a large tub of water. Have students stand on either side of the tub and drop balls in, producing waves. Ask them to describe what happens when the waves meet. Have them try deliberately to produce constructive and destructive interference. Have a mop and bucket handy. L2

4 CLOSE

Pose the following: What other professions might benefit from the wave interference provided by the new ear protectors? *Answers might include factory workers, construction workers, or any profession that needs to have noise selectively filtered out.*

Useful Interference

Recently, interference has been put to use protecting human hearing. In the past, people working in noisy environments have damaged their hearing. One example is pilots of small planes. The pilots could not shut out all noise. They had to be able to hear the instructions from the air traffic controllers. Now pilots can wear special ear protection that not only protects their hearing, but allows them to hear normal conversation.

These special earphones have circuits in them that produce sound that destructively interferes with the damaging engine noise but allows conversation to be heard and understood. **Figure 17-12** shows an example of these new earphones.

In this section, you learned how waves interact. If you were really lucky, you could do the same thing with Louie's wave. If Louie's identical twin, Roberto, dived into the other end of the pool a little later than Louie, Louie's and Roberto's waves would be opposite in terms of where their crests and troughs were located. Then the two waves would meet in the center, and destructive interference would produce a calm area. However, if Ladonna were floating at the center of the pool and Roberto jumped at the same time as Louie, constructive interference would give her a real ride. What other effects can waves have on you? You'll find out in the next section.

Figure 17-12

Ⓐ Loud noises, like those produced by a chainsaw, can damage the human eardrum and result in hearing loss. Engineers have applied what they know about wave interference to design ear protection for people who work in noisy environments.

Ⓑ Ear protectors used by some pilots muffle noises by reflecting and absorbing them.

check your UNDERSTANDING

1. Compare and contrast constructive and destructive interference.
2. Explain how it is possible for one wave to cancel another with a resulting amplitude of zero. Use a diagram in your answer if you like.
3. **Apply** In some theaters, you may find that there are certain areas where the sound is either much softer or muffled in some way. What do you think causes this?

check your UNDERSTANDING

1. Constructive interference occurs when two crests meet to form one crest equal in height to the sum of the original crests. Destructive interference occurs when the crest of one wave meets the trough of another. The two waves cancel each other out.

2. Suggest that students express this in mathematical terms. If the crest of the wave is a positive number (indicating distance above the medium at rest) and the trough is an identical negative value (indicating distance below), the two numbers when added together equal zero.

3. destructive interference as sound reflects from walls, ceiling, and floor

 **Sound as Waves**

Looking at Sound

Have you ever felt a vibrating loudspeaker on your stereo? If you are able to touch the cone of the speaker, you will feel it move in and out in time with the music. This in-and-out movement creates the compressions and rarefactions in the air that are characteristic of longitudinal waves. Under normal conditions you can't see sound. However, the following Find Out activity will allow you to see some of its effects.

Section Objectives
■ Demonstrate sound as a wave.
■ Explain the Doppler effect.

Key Terms
Doppler effect

Find Out! ACTIVITY

Can you see sound?

You saw how sound waves could make rice grains jump. How could you see the vibrations in the rubber sheet? With light, of course!

What To Do

Use the same container that you used in the Find Out activity on page 543. Make sure both ends are open.

1. Now, cut a piece of balloon large enough to fit over one end. Stretch the balloon over the end and hold it in place with a rubber band.
2. Next, glue a small mirror slightly off the center of the balloon.
3. After the glue has dried, hold the open end of the container to your mouth or a loudspeaker.
4. Have a classmate reflect a flashlight beam off the mirror to a flat surface.
5. Explore the effect of sounds on the patterns .

Conclude and Apply

1. Draw the patterns and label them. What happened to the reflected spot on the wall when a loud sound went into the container?
2. How did the reflections change as you changed the sounds?

PREPARATION

Planning the Lesson
Refer to the Chapter Organizer on pages 538A–D.

Concepts Developed
In this section, students will investigate sound waves and learn about the Doppler effect.

sure they hold the lights steady so that any change in pattern is a result of the sound. You may want students to direct the light reflected from the mirror onto a large sheet of white paper taped to the wall, which will enable them to sketch the waves directly. **L1**

Science Journal Students' drawings should depict the movements of the mirror. **L1**

Expected Outcomes
Students will observe differing up-and-down patterns of light, representing the pattern of the sound waves produced.

Conclude and Apply
1. The patterns would be an up-and-down line. The sound from the speaker caused the balloon to vibrate, which in turn, caused the spot to move.
2. Amplitude and wavelength varied.

✔ Assessment
Performance Provide small groups of students with tuning forks and shallow containers of water. Have them strike the fork, then place the end of it in the water and observe the results. *Circular waves radiate out from the fork.* Ask the students to create a poster showing what they did and explaining the results. Use the Performance Task Assessment List for Poster in **PASC**, p. 73. **L1** **COOP LEARN** **LEP**

Find Out!

Can you see sound?

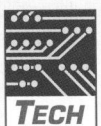 **TECH PREP**

Time needed 15–20 minutes

Materials coffee container with both ends cut out, large balloon, scissors, large rubber band, glue, small mirror (1 cm x 1 cm is ideal), flashlight, loudspeaker

Thinking Processes observing and inferring, comparing and contrasting, recognizing cause and effect, making models, defining operationally, interpreting data

Purpose To compare and contrast the patterns created by sound waves

Preparation If time is a factor, cover the end of each container and attach the mirror prior to class.

Teaching the Activity
Have students work in groups of four. Make

Before presenting the lesson, display **Section Focus Transparency 58** on an overhead projector. Assign the accompanying **Focus Activity** worksheet. L1

LEP

Demonstration Bring a radio to class and turn it on. Switch stations several times. Explain to the class that the radio stations are broadcasting sound on different frequencies. Next, turn the volume up and down. Explain that this changes the amplitude of the sound waves coming from the radio. In this section, students will learn how varying sound waves affect people. L1

2 TEACH

Tying to Previous Knowledge

In Chapter 3, students learned many of the characteristics of sound waves. This section expands their understanding of these characteristics by helping students visualize sound as a wave and by explaining how high amplitude sound waves (loud sounds) can damage hearing. Invite students to describe experiences they have had with loud sounds. Ask if they felt, as well as heard, the vibrations.

Figure 17-13

If you listen to music at high levels of volume for long periods of time, you could damage your hearing. Many rock musicians have some hearing loss as a result of their exposure to loud music. Many musicians wear some form of ear protection while on stage.

The patterns you observed were produced by the compressions and rarefactions of air moving the balloon. As the compressions caused the balloon to bulge, the mirror was tilted, and the light reflected in one direction. Then, when the rarefactions were behind the balloon, the mirror was tilted the other way, and the reflection moved in another direction. Thus, the pattern traced on the wall was a rough picture of the balloon's vibration and the sound waves that caused it.

Imagine that you could do the Find Out activity on page 559 at a rock concert like the one in **Figure 17-13**. What would the pattern be like?

a CLOSER LOOK

Filling a Room with Sound

Symphony Hall in Boston, built in 1898, was designed by a professor of physics who had studied acoustics. It is considered one of the greatest music halls of all time.

Symphony Hall

Acoustics, the science of sound, is used in designing buildings such as concert halls and recording studios. Imagine sound waves coming out of your stereo speaker, like rings in the water where you've tossed in a pebble. When sound waves strike a surface such as a wall, floor, or ceiling, some of the sound is absorbed, and some is reflected. A hard surface reflects more sound. Soft materials, such as drapes or a carpet, absorb more sound.

A room that absorbs too much sound is acoustically dead. The sound in a room with some reflections is more pleasing to the ear. As people also

Purpose

A Closer Look familiarizes students with basic principles of acoustics. The characteristics and behavior of waves, treated in Sections 17-2 and 17-3, strongly influence the design of concert halls.

Teaching Strategy

Have students analyze details in the article's 4th and 5th paragraphs. Then have them apply their knowledge of interference by answering this question: Does a good

hall depend on constructive or destructive interference? *constructive, because wave crests meet crests and troughs meet troughs; the reverberations build up around listeners, rather than neutralize each other as in destructive interference.*

Answers to
You Try It!

Students will find that the speakers leveled at opposite walls fill the room with

Properties of Sound Waves

For humans, a low sound will have a frequency from about 20 to about 200 hertz whereas the highest sounds you can hear are about 15 000 to 20 000 hertz. You would hear 20 hertz as a low rumble, such as thunder.

The highest note on the piano is less than 4000 hertz. Rich musical sounds, however, do not have only one frequency, but contain frequencies that are much higher than 4000 hertz. Elephants can hear frequencies much lower than 20 hertz. Bats, on the other hand, locate their prey with frequencies as high as 120 000 hertz.

Figure 17-14

People and various animals have limits to the sound frequencies they can detect. Dog whistles generate sounds at such a high frequency that dogs can hear them but people cannot.

Theme Connection Have a volunteer explain the relationship between a source and the waves it produces. *Waves transmit energy away from a source.* Discuss with students the damage that can be caused by high amplitude (high energy) waves.

Student Text Question What would the pattern be like? *very jerky and rapid*

Across the Curriculum

Consumer Science

Have students find out about ultrasonic alarm systems which use a beam of sound, much like the "electric eye" that opens an automatic door. If the sound beam is broken by an intruder, it sets off an alarm. The frequencies used in these alarms are not very far above the average range of human sensitivity. Some people, whose ranges of sensitivity are a bit higher than average, can hear these sounds. They often describe the sensation as a pressure in the ear rather than a sound or recognizable pitch.

NATIONAL GEOGRAPHIC SOCIETY

 Videodisc

Newton's Apple: Physical Sciences

Doppler Effect
Chapter 2, Side B
A diagram of sound waves

19040-19819
Sound waves and pitch

19608-20537

---

absorb sound, a large audience reduces reflections.

Reverberation

Have you ever shouted when you were in a tunnel? You hear many echoes or sound reflections. Sometimes the echoes last for almost a second. The echoing sound is called reverberation. The time it takes for the sound to die out is called reverberation time.

An auditorium or concert hall must be carefully designed to have the proper reverberation time. Many concert halls built in the 19th century have better acoustics than modern halls. The halls were usually long and narrow. In a narrow hall, you hear the sound coming from

the source first, and then reflected off the walls. The reverberation was correct for music played by an orchestra. The halls were smaller, too. With a smaller audience, less sound was absorbed.

Modern halls are built to hold more people. They are usually built wide to provide more emergency exits. Ceilings are lower, which also affects the sound. In the older halls, the reverberation makes you feel surrounded by sound. Acoustic experts today are taking their cues from the older builders.

You Try It!

Turn stereo speakers so that the sound reflects off the ceiling, then off the opposing walls. Describe the differences.

---

more sound than when they bounce their sounds off ceilings.

Going Further ▸

Have students find out about the acoustics in ancient halls and theaters. For example, in the Classical Greek world, outdoor theaters were carved into hillsides in semicircular shape, creating curved and "terraced" sound-reflectors out of the hill and the audience itself, which maximized amplification of sounds from the stage. Students may wish to make a model

of a classical theater and give a presentation explaining its acoustical features. Use the Performance Task Assessment List for Model and Oral Presentation in **PASC,** pp. 51 and 71. L3

Dangerous Sounds

The louder the sound, the greater the amplitude of the wave. In fact, some sounds can hurt you. Some rock concerts are so loud that the musicians have damaged their hearing. Even the spectators should be careful, as overexposure to loud noise can destroy their hearing. Sometimes people doubt this because the loss of hearing may not show up for many years.

High amplitude compressions can cause your eardrum to rupture. When scar tissue grows over the split, the eardrum can no longer reproduce sound accurately. It is like mending a drumhead with tape.

Secondly, inside your ear are tiny nerve fibers surrounded by fluid. When a compression caused by a loud sound travels through the fluid in your ear, it can damage or destroy the nerve fibers. If enough fibers are destroyed, hearing is permanently damaged because the nerves do not grow back.

Some machines, such as air compressors and jet engines, are also very loud, and it is important that workers using them wear ear protection. It is the loudness or amplitude that creates the large compressions. Sounds of low or high frequency are equally damaging.

Figure 17-15

When you play a personal stereo or radio, the sound goes in all directions, including your ears. When you wear earphones, nearly all the sound goes directly into your ears. The sound levels inside your ears can be very high. Listening to loud music too long can cause hearing loss.

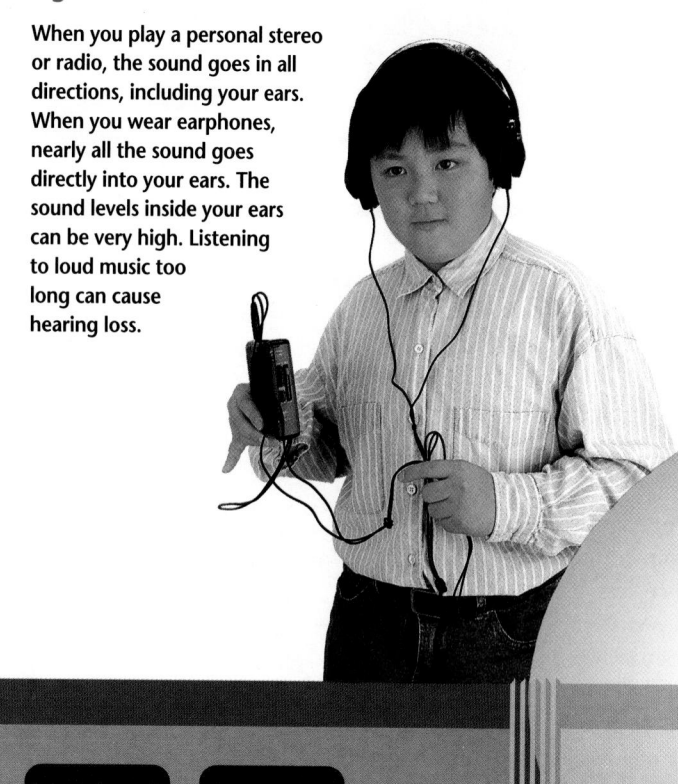

The Doppler Effect

You've probably been at a railroad crossing when a train passed by. Did you notice that the train's horn seemed to be higher in pitch as it approached and then was lower as it went away? The apparent change in pitch of a sound as the source moves with respect to the observer is called the **Doppler effect.** It applies to all waves, including light and sound.

If you were on the train, you wouldn't hear the Doppler effect. To people on the train, the horn would seem to have the same frequency at all times. How can two people listening to the same horn hear different frequencies?

The motion of the train as it moves toward you causes the sound waves to be squeezed together. As each compression is sent out, the train moves closer, and the next compression gets a head start. Thus, the compressions in front of the train are closer together, as shown in **Figure 17-16**. This causes a higher frequency to reach your ear. What do you think happens as the train passes you? Think about it. Imagine that the frequency of the sound is 100 hertz or 100 vibrations per second. Now imagine that the train is moving at 20 meters per second. That means that 5 compressions are sent out from the horn every meter that the train moves.

What about the operator and passengers on the train? Do they hear the Doppler effect? No, they hear the same frequency all the time because they and the sound are moving together.

The Doppler effect is more than just an interesting phenomenon.

Across the Curriculum
History/Health

When an object travels faster than the speed of sound, it overtakes the sound waves it produces, generating a loud "crack" called a sonic boom. This boom can damage people's hearing and even damage property. On October 14, 1947, Captain Charles E. Yeager became the first person to break the sound barrier. Ask students to research this historic flight. Also have them research what is now being done to protect people from the damage caused by sonic booms. L2

Figure 17-16

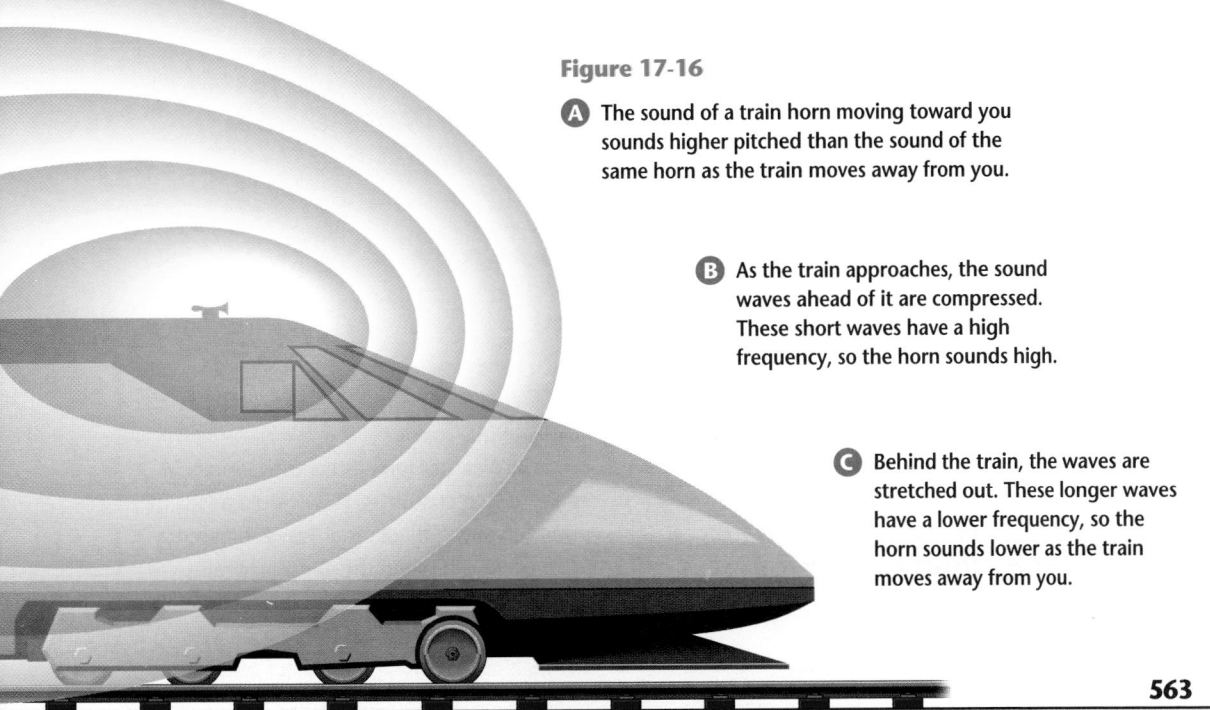

A The sound of a train horn moving toward you sounds higher pitched than the sound of the same horn as the train moves away from you.

B As the train approaches, the sound waves ahead of it are compressed. These short waves have a high frequency, so the horn sounds high.

C Behind the train, the waves are stretched out. These longer waves have a lower frequency, so the horn sounds lower as the train moves away from you.

563

Teacher F.Y.I.

The Doppler effect is named for Christian Johann Doppler, an Austrian who first explained, in 1842, the effect in terms of frequency.

 **Videodisc**

Newton's Apple: Physical Sciences

Doppler Effect
Chapter 2, Side B
A wave with crests close together

20061

A wave with crests far apart

20375

ENRICHMENT

TECH PREP

Activity Cut a small hole in a soft foam ball. Put a piezoelectric buzzer connected to a battery into the hole. You may need to secure it with tape. Hold the ball and buzzer in one place and ask students to describe the noise they hear. Be sure to get a description of the pitch. Then ask for two volunteers. Have one stand in the front of the room and the other stand in the back. Ask them to toss the ball gently back and forth a few times. Ask the rest of the class to describe any differences in the sound they hear. *They should be able to hear an increase and decrease in pitch caused by the Doppler effect.* Help students conclude that this effect is caused by sound wave compression; as the source of a sound wave moves closer, the waves are compressed and the pitch becomes higher; as it moves away, the waves spread out and the pitch becomes lower. L1 **COOP LEARN**

3 ASSESS

Check for Understanding

To reinforce their understanding of the cause-and-effect relationship between pitch and motion relative to a sound source, present the following: You are talking to your friend on the schoolbus when you hear a car alarm going off. The pitch is getting lower. Should you look in front of or behind the bus to see the car? **Explain.** *Behind; the sound stays at the same pitch until you pass the source, then the pitch drops.*

Reteach

Demonstration Bring a drum to class—a child's toy drum or a covered container such as the one you used in the Find Out will do. Sprinkle chalk dust evenly over the top of the drum and then strike it with a stick. Have students observe and describe the pattern created in the dust. Students should conclude that this pattern was caused by the sound waves striking the drum. L1

Extension

Activity Have students keep a record over a weekend of the waves they encounter. They should record, in their journals, the source of the wave, the medium it traveled through, characteristics of the wave, and any unusual effect they notice, such as the Doppler effect. Share these observations and note how many different kinds of waves they observed. L2

4 CLOSE

Draw a transverse and a longitudinal wave on the chalkboard. Ask students to identify the characteristics of each wave and to suggest where they might experience each different wave in the course of their daily lives. L1

Some radars (radio detection and ranging) use the effect to find the speed of objects. One use you are probably familiar with is the Doppler radar that police use to identify speeding motorists.

As you've seen in this chapter, there's a lot more to waves than water splashing up on the beach. As you go to school or just walk around the neighborhood, identify the waves you observe and how they affect your life.

Figure 17-17

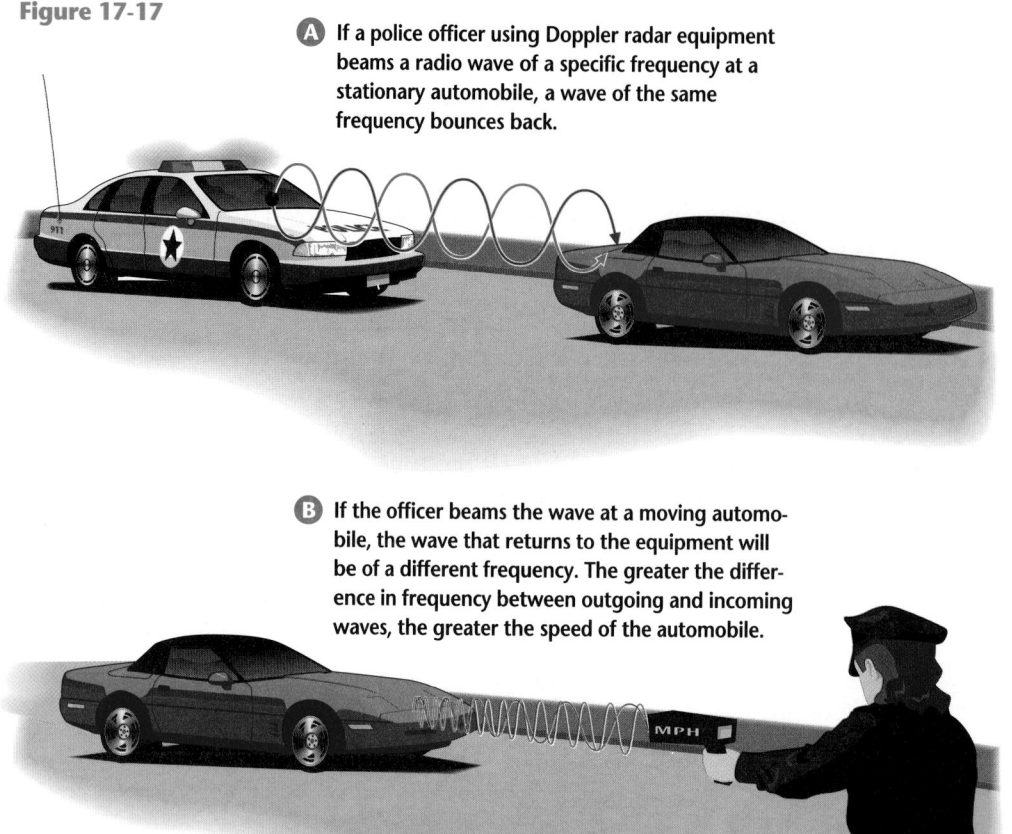

A If a police officer using Doppler radar equipment beams a radio wave of a specific frequency at a stationary automobile, a wave of the same frequency bounces back.

B If the officer beams the wave at a moving automobile, the wave that returns to the equipment will be of a different frequency. The greater the difference in frequency between outgoing and incoming waves, the greater the speed of the automobile.

check your UNDERSTANDING

1. How can you demonstrate that sound is a wave?
2. When will the pitch of a racing car engine be the highest—approaching you, going away, or directly opposite you?

Explain your answer.
3. **Apply** Weather forecasters use Doppler radar to detect storms. How would a violent thunderstorm reflect radar waves differently than from rain when the winds are calm?

564   Chapter 17   Waves

check your UNDERSTANDING

1. Students may note that sound waves are periodic disturbances in a medium that transfer energy and have the characteristics of wavelength, frequency, and amplitude.
2. As it approaches you; as an object approaches, the sound waves it produces are squeezed closer together, causing the pitch to

sound higher. As it moves away, the waves become farther apart, causing the pitch to sound as if it is lower. This is called the Doppler effect.
3. Severe storms usually produce denser precipitation patterns than light rains with gentle winds.

Technology Connection

Wave Energy

The first wave-powered electricity generating station powered by the ocean was opened in the mid-1980s on the coast of Norway. Since then, other power stations have opened in England and Scotland.

The amplitude of a wave determines how much energy it can carry. The energy contained in waves comes from the wind. How hard the wind blows, how long the wind blows, and the distance it blows across the water all help determine wave amplitude. Waves grow taller as they absorb more of the wind's energy. Every time waves double in height, their energy is quadrupled.

The Salter Duck

One well-known wave-powered device is called the Salter duck, named for its inventor, Dr. Stephen Salter of Scotland. The device looks like a duck bobbing on the water. It uses a hydraulic pump that moves on the action of the waves, pumping fluid into a turbine-driven generator that produces electricity. Salter has built duck models, but estimates indicate that at least five million pounds of concrete would have to be poured for a full-scale generating plant. Salter is working on the problem of size.

The oscillating water column (OWC) also uses wave power. The device traps a column of water inside a chamber that is moved up and down by wave action. It compresses air at

the top of the chamber and forces the air into a turbine, which turns a generator. The OWC must be built on a huge scale to produce a substantial amount of electricity.

Scientists estimate that wave-powered stations could provide at least ten percent of the world's energy needs. Inventors and investors will continue to pursue the power of the ocean wave.

Science Journal

How do you think this emerging technology will change the world? *In your Science Journal,* tell what you think are the advantages and disadvantages.

Technology Connection

Purpose

This feature discusses the energy in waves and human attempts to utilize that energy. The Salter duck and the oscillating water column are two proposals for turning ocean wave energy into electrical energy.

Content Background

The Salter duck and the oscillating water column are two approaches to a familiar mechanical problem. These new devices depend on their power-sources' turning a turbine to generate electricity. The first crude turbines were nonelectric water-driven machines for grinding grain. Today's colossal turbines generate electric current. Only solar or photovoltaic cell systems dispense with the turbine and generate electricity directly.

Teaching Strategies

Have students collect photos of devices, or actual devices, that use alternative energy sources to function. *Examples may include wind-wheels, geothermal power plants, tidal power plants, solar panels on roofs, or solar-powered calculators and wristwatches.* Then have them create a display of the devices, telling how they produce energy. Use the Performance Task Assessment List for Display in **PASC**, p. 63. L1

Science Journal

Advantages may include reduced pollution and renewability; disadvantages may include expense of development and transmission of electricity to inland areas.

Going Further ⫸

Have students use their knowledge of world geography and climate to hypothesize which three countries use the most fossil fuels, and which three countries lead the way in using or developing alternative energy sources. Have students check their hypotheses against almanac statistics or other references. Suggest students compile their findings in a data table. Use the Performance Task Assessment List for Data Table in **PASC**, p. 37. L2

Science and Society

Science and Society

Purpose
This Science and Society excursion treats some of the less useful or more hazardous effects of wave properties explored in Sections 17-3 and 17-4. The physical and behavioral consequences of extremely noisy environments in daily life, work, and play have resulted in various government and private efforts to reduce noise. Many residents have banded together to recapture some quiet in their communities. The human environment is literally filled with soundwaves, but only in recent decades has this sound become a health and safety issue.

Content Background
Details in the article suggest the serious problems associated with noise pollution. The recognition of these problems has led to attempts to regulate and/or reduce noise in the human environment. Beginning in the 1970s and 1980s, and continuing into the 1990s, the need (or at least the debate on the need) for regulations has increased.

Teaching Strategy Ask students to list their most hated and most liked noises. Ask what effect each has on them. [L1]

Discussion
Tally up the responses on the above lists and have students discuss measures that might be instituted to modify the "sound environment" so it more closely conforms to student likes and dislikes.

Dangerous To Your Ears

You get up in the morning and turn on the radio. You travel in busy traffic. At school, the halls are noisy, and loud bells ring. After school, you play your stereo or mix a milkshake in the blender. It's just an average day for your eardrums. Noise pollution is so much a part of our everyday lives, we hardly notice it's there.

When sound waves reach the ear, the air pressure pushes against the eardrum. The vibration is passed to microscopic nerve cells, and then to the brain where it is interpreted as sound. The loudness of sound depends on the size of the pressure vibrations. The loudness is measured in decibels. The louder the sound, the higher the decibel level. For example, the sound level of a quiet library is 30 decibels, while that of loud rock music might be 110 decibels.

Noise All Around Us

Many people become deaf from long exposure to loud noise. The graph on the left shows the average sound levels for several common items. The graph also shows what kind of hearing damage is caused by exposure to high sound levels.

The way noise affects people's behavior may be as important as the damage it does to hearing. Acoustical sociologists and other experts who specialize in hearing study noise to see how it influences physical and mental health. Noise pollution is all around us. We may adapt, but our bodies and minds still experience stress from the extreme noise.

People often may react physically to noise pollution. The Acoustical Society of America released a study that said people who live near airports are more likely to suffer from physical problems such as heart disease or high blood pressure. People who live in noisy neighborhoods complain of increased anxiety and sleeplessness. When too many sound waves assault the eardrums, the brain cannot process all the information at once. This may

Mild loss with prolonged exposure

Permanent loss with short exposure

Slight loss with prolonged exposure

Moderate to short loss with prolonged exposure

Diesel locomotive
Heavy truck
Motorcycle
Train
Automobile
Vacuum cleaner
Conversation

50 60 70 80 90 100 110

Loudness (in decibels)

Going Further ⫸
Have two or three groups research in more detail how the human ear works in order to hear sound waves, how hearing loss comes about, and what remedies have been and are being used to reduce the threat to hearing by noise pollution. A fourth and fifth group might explore how "close-captioned" TV shows for the hearing impaired work; and how "signing" came into being to allow the hearing impaired to communicate with one another and with people who can hear. Use the Performance Task Assessment List for Group Work in **PASC**, p. 97. [L1] **COOP LEARN**

explain why people in noisy environments say they are unable to think. Studies on noise pollution have shown that in noisy neighborhoods, people feel more isolated and afraid. A person's appetite may be lowered when there is a lot of noise, which may be why it is harder to enjoy a meal in the school cafeteria. There are also more traffic accidents at very noisy intersections.

Noise Control

The Environmental Protection Agency (EPA) established the Office of Noise Abatement and granted it powers to regulate sources of noise pollution by passing the Noise Control Act in 1972. However, the Office of Noise Abatement was closed in 1982 due to government budget cuts. Most experts agree that passing laws on noise control is the most important step toward controlling noise pollution. In recent years, most noise laws have been passed at the local level of government. But that may be changing.

The U.S. Department of Transportation and the Federal Aviation Administration recently released new policy guidelines to control airport noise. The policy requires airlines to replace older aircraft with newer, quieter models. The policy also restricts local communities from imposing their own noise control laws. The airlines have until the year 2001 to comply with the new regulations, so local areas may have to live with the noisier aircraft until then.

Noise pollution activists are not happy with the new policy. With government offices getting involved in noise pollution again, noise control may be the next big wave for environmentalists.

Acoustical physicists study the production, reflection, and absorption of sound. Many are employed by the auto industries and other transportation companies. They study ways to improve or control noise, sounds, and vibration.

You Try It!

Are you breaking any local laws against noise pollution? Is there a local law in your area about playing music in public, such as on a bus or on the beach? Find out by writing to your local health department or the mayor's office and ask about local ordinances against noise.

Teaching Strategy Have groups of students debate either or both of the following issues: (1) whether it is more efficient to reduce noise in certain environments or to try to find cures for the ailments that high amplitude sound waves are thought to create; and/or (2) the deregulation of noise pollution by the federal government. Use the Performance Task Assessment List for Investigating an Issue Controversy in **PASC**, p. 65. L2
COOP LEARN

Answers to You Try It!

Encourage students to notice small noise-oriented details of their environment as they answer, for example, posted regulations at parks or beaches about radio-playing; "Quiet" signs at libraries; and automobile-muffler requirements. Pose these questions to initiate discussion. Which if any of these regulations were approved of by the local population? Which were legislated? Which were executive or administrative orders? Which do students or their friends generally abide by?

Have students discuss precisely what kinds of physics people in these job fields study. Students should also do research to find out what kind of education is required to become an acoustical physicist.

Going Further ⫸
Based on the results of the class lists of favorite and most-hated noises, have students interview people in the community. Have them correlate the findings to determine whether the class is representative of the larger population, and determine whether any remedial actions would be possible with the community's help. Suggest that students display the results of their surveys. Use the Performance Task Assessment List for Survey and Graph of Results in **PASC**, p. 35. L1
COOP LEARN P

Teens in SCIENCE

Purpose
Teens in Science shows how a person such as Jason Cobb can turn characteristics of waves into music and art. Computers, synthesizers, and display screens allow people to "see" music as they create it.

Content Background
Drum machines, computers that provide full-orchestra accompaniment to solo musicians, and variations on the original 1970s Moog synthesizer, which can reproduce sounds ranging from tropical rain forests to train wrecks, are mainstays of modern music. Some artists such as Paul Winter ("Wolf Eyes" and other naturalistic recordings) have pioneered the use of these systems. Other people find careers as sound technicians in film, TV, radio, and theater.

Teaching Strategies
Have students discuss their favorite sound-artists. Do these people or groups use electronic or acoustical instrumentation? In what ways do electronic effects add to listening pleasure? L1

Debate
Have students debate the following questions. Is high-tech music destroying musicians' careers? How have machines come to replace many studio musicians and other performers and technicians? Can technology make a star out of someone with little musical talent? L1

Answers to
You Try It!

Some students may say that there is something too mechanical about computer-generated music—too little natural variation in pitch or tempo.

Teens in SCIENCE

Riding a Musical Wave

Jason Cobb always loved computers. In the seventh grade, he had his first science lesson on sound. For his class project, Jason wrote a computer program about how sound works. Thus began his adventure as a computer music composer.

Programming Sound

After his first sound project, Jason studied programming languages on his own. Now he is an intense young man whose conversation is peppered with computer music lingo. Jason composes music on a personal computer with a keyboard, mouse, synthesizer, and stereo system.

Jason writes music in the language of sound waves. Each key he presses on the keyboard represents a different musical sound. The sounds appear as graphic pictures, called waveforms, on his computer screen. The waveforms let him "see what the music looks like." Jason edits sounds and puts together simple waveforms into complex compositions. He stores his compositions on a disc until he's ready to convert them to audio tape.

The technical wizardry is in the synthesizer, a musical instrument that produces sounds electronically.

MIDI

Playing the synthesizer allows Jason to determine the loudness, pitch, and tone of the sounds. The synthesizer uses Musical Instrument Digital Interface (MIDI). MIDI is a standard language that connects computers to electronic instruments. MIDI converts the numbers in the computer into forms that the instruments can use to produce sound. With MIDI and a synthesizer, Jason can change the pitch, tempo, or tone of a variety of musical sounds.

For all his technical mastery, what Jason produces is melodic music. His dream is to be an electronic musician, creating computerized compositions that the public and other electronic artists will appreciate. Does he have an equipment wish list? "Are you kidding?" he asks. "Of course, but what I really want costs a million dollars, so I'll have to wait awhile to buy it."

You Try It!
Can you tell the difference between music that is computer-generated and music from traditional instruments? Next time you're listening to the radio, see if you can identify each type of music.

Going Further ▶
Have students create a history of natural and synthetic music by finding examples that they can play for the class, either performing themselves or by using tapes. For example, in "Peter and the Wolf," how do various instruments play the parts of characters and animals? What kinds of noises, tones, and so on in the everyday environment now appear in synthesized music and other genres? Use the Performance Task Assessment List for Group Work in **PASC**, p. 97.

COOP LEARN

Science Journal

Review the statements below about the big ideas presented in this chapter, and answer the questions. Then, re-read your answers to the Did You Ever Wonder questions at the beginning of the chapter. *In your Science Journal*, write a paragraph about how your understanding of the big ideas in the chapter has changed.

2 The medium moves in the direction of a longitudinal wave. *What are the parts of a longitudinal wave?*

3 In the same medium, as frequency increases, wavelength decreases. *Arrange in order of decreasing speed of transmission: plastic foam, air, wood.*

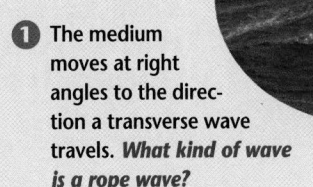

1 The medium moves at right angles to the direction a transverse wave travels. *What kind of wave is a rope wave?*

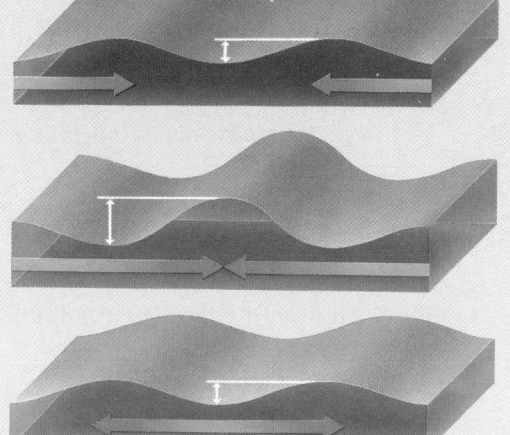

4 As waves cross, their crests and troughs add and subtract to form constructive and destructive interference patterns. *What do the waves look like at the point where they interact?*

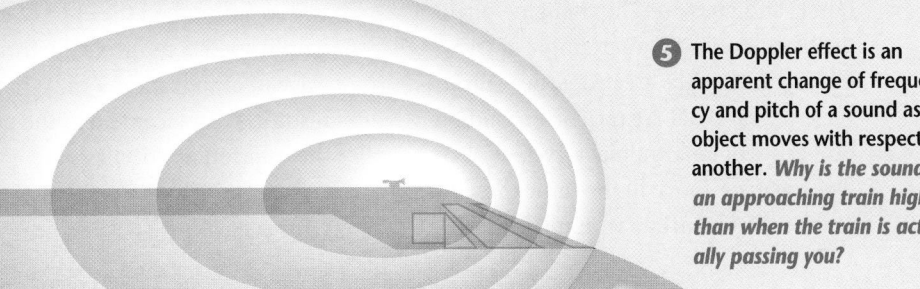

5 The Doppler effect is an apparent change of frequency and pitch of a sound as an object moves with respect to another. *Why is the sound of an approaching train higher than when the train is actually passing you?*

Science Journal

Did you ever wonder...

• Waves are constantly being produced. When the crest of one wave meets the crest of another, the two waves make one wave equal to the sum of the two. (p. 556)

• As an object approaches, the sound waves are pushed closer together, causing a higher pitch. As it moves away, the waves are spread farther apart, causing a lower pitch. (p. 563)

• The speaker vibrates, producing compressions and rarefactions in the air. (p. 559)

Science at Home

Ask students to experiment with waves in their bathtub or a basin of water. Have them try to produce and measure waves of differing amplitude and frequency. They should place various floating objects in the water and observe the effect the waves have on those objects. They should also attempt to produce constructive and destructive interference. Have students record their results in their journals and share their observations with the class. **L1**

Teaching Strategy

On a separate piece of paper rewrite each of the review statements in the student text into sentences containing only one idea. For example: *The highest point of a wave is the crest. The lowest point of a wave is the trough. The distance between one wave crest and the next is the wavelength. When two crests meet, the waves add together in constructive interference.* For each statement, think of a false statement. For example: *The highest point of a wave is the trough.* Put each of these statements into a hat. Divide the class into two teams and have team members choose one piece of paper at a time. The student should read the statement and say if it is true or false. You may wish to require students to provide the correct statement if they choose a false statement and an additional fact if they choose a correct statement. **COOP LEARN** **L1**

Answers to Questions

1. A rope wave is a transverse wave.

2. compressions and rarefactions

3. Air transmits sound waves slower than plastic foam. Plastic foam, in turn transmits sound waves slower than wood.

4. If two crests combine, they form a wave with an amplitude equal to the sum of the amplitude of the original waves. If a crest combines with a trough, the waves subtract or even cancel.

5. As the train approaches, the sound waves ahead of the train are compressed, and thus have higher frequencies, which result in a higher pitch.

GLENCOE TECHNOLOGY

 **MindJogger Videoquiz**

Chapter 17 Have students work in groups as they play the Videoquiz game to review key chapter concepts.

Using Key Science Terms

1. Crest does not belong. Each of the other terms refers to an individual wave property.

2. Speed does not belong. Each of the other terms refers to interference.

3. Transverse waves does not belong. Each of the other terms refers to longitudinal waves.

4. Longitudinal waves does not belong. Each of the other terms refers to transverse waves.

Understanding Ideas

1. The amplitude would measure the distance from the crest to the rest position and the trough to the rest position.

2. Doppler effect

3. compressions and rarefactions

4. No, the surfer is moving at right angles to the wave disturbance.

Developing Skills

1.

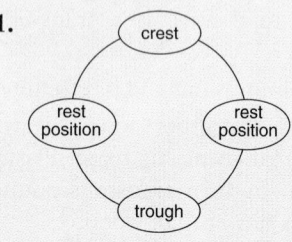

2. jet airplane, community siren, rock music, riveter

3. 2.5 m − 1.7 m = +0.8 m

Using Key Science Terms

amplitude	longitudinal waves
crest	transverse waves
Doppler effect	trough
interference	wavelength

For each set of terms below, choose the one term that does not belong and explain why it does not belong.

1. amplitude, frequency, wavelength, crest

2. crests, troughs, speed, interference

3. rarefaction, transverse waves, compression, longitudinal waves

4. trough, transverse waves, longitudinal waves, crests

Understanding Ideas

Answer the following questions in your Journal using complete sentences.

1. The highest wave ever measured on the open ocean was over 34 meters from trough to crest. What characteristic of the wave could this give you a measure of? Explain.

2. Lying in your room at night you hear a large truck out on the freeway. It approaches from a distance, passes your home, and goes on. The changing pitch of the truck's sound is an example of what?

3. Crests and troughs are properties of transverse waves. What are the corresponding properties of longitudinal waves?

4. Does a surfer move in the same direction as the wave disturbance? Explain.

Developing Skills

Use your understanding of the concepts developed in this chapter to answer each of the following questions.

1. Concept Mapping Make a cycle concept map showing how the water level changes as a wave passes a certain point. Use the following terms: *crest, trough, rest position, rest position.*

2. Making and Using Graphs The graph shows the noise level of several common situations. Hearing damage is caused by extended exposure to sound over 85 decibels. Which of the sounds on the graph could damage your hearing?

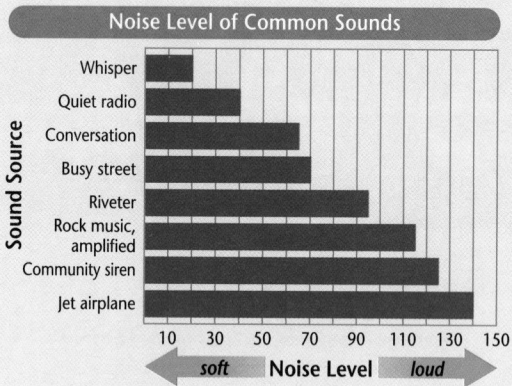

3. Interpreting Data Destructive interference occurs when the crest of a wave of amplitude 2.5 m meets the trough of a wave of amplitude 1.7 m. What is the resulting displacement at the point where the interference occurs?

Program Resources

Review and Assessment, pp. 101-106
[L1]

Performance Assessment, Ch. 17 [L2]

PASC

Alternate Assessment in the Science Classroom

Computer Test Bank [L1]

Critical Thinking

In your Journal, *answer each of the following questions.*

1. Write a brief paragraph that explains how your knowledge of waves might make you a better musician.
2. Suppose that as sound waves move from one medium to another, their velocity doubles, but their frequency remains the same. What happens to the wavelength?
3. A bus driver is rounding a curve approaching a railroad crossing. She hears a train's whistle and then hears the whistle's pitch become lower. What assumption can she make about what she will see when she rounds the curve and looks at the crossing?

Problem Solving

Read the following problem and discuss your answer in a brief paragraph.

You've just been given a new stereo system for your birthday, and you want to set it up in your room to get the best possible sound.

1. Draw and discuss three separate setups for the speakers, showing the direction of the sound from each speaker, how to get the best stereo effect, and possibilities for destructive interference—dead spots. Keep in mind that sound will also be reflected from the walls.
2. Your baby sister's crib is against the wall in the room next to yours. How would that affect the placement of your speakers?

CONNECTING IDEAS

Discuss each of the following in a brief paragraph.

1. **Theme—Systems and Interactions** People living near airports sometimes report that their windows rattle as a plane passes overhead. Explain what is happening in these cases.
2. **Theme—Energy** Explain why waves on a lake are larger on a windy day than on a calm day.
3. **Science and Society** There are tapes available of sounds such as babbling brooks, gentle rain, and bird calls. Why could these tapes help calm people who live in urban areas?
4. **A Closer Look** You're in charge of converting a gymnasium to a lecture hall. What can you do to be sure the audience noises do not drown out the lecturer?
5. **Earth Science Connection** If a seismic wave in one area is reflected back in less time than in another area, what might be true of the density in the first area compared to the second?

Critical Thinking

1. Answers will vary, but they should use the following concepts: frequency, amplitude, longitudinal waves, constructive interference, and destructive interference.
2. The waves move twice as fast, but the same number of waves pass a given point in a specified period of time. This can only happen if the wavelength doubles.
3. The train will already have passed the crossing.

Problem Solving

1. Answers will vary but should include some description of waves as they emerge from speakers and are reflected.
2. Answers should include the idea that sound travels through the walls better than through the air.

Connecting Ideas

1. The sound waves from the airplanes are vibrating the walls and windows.
2. On a windy day, a large amount of energy from the wind is transferred to the water, creating high-energy waves with large amplitude.
3. The sounds on the tapes interfere destructively with sounds from the streets, thus reducing or eliminating irritating noise.
4. Arrange walls and furniture so that they will interact destructively with noise coming from the audience. Perhaps cover walls and floor with sound-absorbing materials.
5. The first area would be more dense than the second area.

✔ Assessment

Portfolio Review the portfolio options that are provided throughout the chapter. Encourage students to select one product that demonstrates their best work for the chapter. Have students explain what they learned and why they chose this example for placement into their portfolios. Additional portfolio options can be found in the following **Teacher Classroom Resources:**

Making Connections: Integrating Sciences, p. 37

Multicultural Connections, pp. 37, 38

Making Connections: Across the Curriculum, p. 37

Concept Mapping, p. 25

Critical Thinking/Problem Solving, p. 25

Take Home Activities, p. 27

Laboratory Manual, pp. 107–110

Performance Assessment P

Chapter Organizer

SECTION	OBJECTIVES	ACTIVITIES & FEATURES
Chapter Opener		Explore!, p. 573
18-1 Earthquakes, Volcanoes, and You (3 sessions, 1.5 blocks)	1. **Explain** how waves at Earth's surface generated by earthquakes cause structures to collapse. 2. **Make** models of volcanic cones and describe the types of eruptions that produce them. National Content Standards: (5–8) UCP2–3, A1, B2, F3, G1–3	Explore!, p. 574 Design Your Own Investigation 18-1: pp. 578–579 Find Out!, p. 580 **A Closer Look,** pp. 576–577 **Teens in Science,** p. 596
18-2 Earthquake and Volcano Destruction (2 sessions, 1 block)	1. **Determine** four factors that influence the amount of damage caused by an earthquake. 2. **Describe** the types of damage caused by earthquakes. 3. **Describe** the types of damage caused by volcanoes. National Content Standards: (5–8) UCP2–3, F3, G3	Explore!, p. 582 Find Out!, p. 585 **Physics Connection,** pp. 584–585 **Technology Connection,** p. 593
18-3 Measuring Earthquakes (3 sessions, 1 block)	1. **Demonstrate** how a seismograph measures an earthquake's strength. 2. **Explain** how the Richter scale is used to indicate earthquake magnitude. National Content Standards: (5–8) UCP2–3, A2, E1–2, F3, F5, G1, G3	Investigate 18-2: pp. 590–591 Skillbuilder: p. 592 **Science and Society,** pp. 594–595

ACTIVITY MATERIALS

EXPLORE!

p. 573 bleachers or room with wooden floor

p. 574* rectangular pan, water, table tennis ball, pencil

p. 582 rectangular pan, sand or fine soil, 2 large books

INVESTIGATE!

pp. 590–591* ring stand with ring, wire hook from coat hanger, piece of string, 2 rubber bands, fine-tip marker, metric ruler, masking tape, sheet of paper

DESIGN YOUR OWN INVESTIGATION

pp. 578–579* world map, tracing paper

FIND OUT!

p. 580 1 cup sand or sugar, 2 paper plates, 1 cup plaster of paris, metric ruler, water, protractor

pp. 584–585* 1 or 2 books, rectangular cake pan, water, sand, circular plastic lid, hole puncher, 20-cm string, metric ruler, scissors

KEY TO TEACHING STRATEGIES

The following designations will help you decide which activities are appropriate for your students.

L1	Basic activities for all students
L2	Activities for average to above-average students
L3	Challenging activities for above-average students
LEP	Limited English Proficiency activities
COOP LEARN	Cooperative Learning activities for small group work
P	Student products that can be placed into a best-work portfolio

Activities and resources recommended for block schedules

Need Materials? Call Science Kit (1-800-828-7777).

00:00 **OUT OF TIME?** We recommend that students do the activities with an asterisk.

Chapter 18 Earthquakes and Volcanoes

TEACHER CLASSROOM RESOURCES

Student Masters	Transparencies
Study Guide, p. 65 **Concept Mapping**, p. 26 **Making Connections: Integrating Sciences**, p. 39 **Activity Masters**, Design Your Own Investigation 18-1, pp. 75–76 **Critical Thinking/Problem Solving**, p. 26 **Take Home Activities**, p. 28 **Science Discovery Activities**, 18-1 **Laboratory Manual**, pp. 111–114, Earthquakes **Laboratory Manual**, pp. 115–116, Pumice	**Teaching Transparency 35**, Forms of Volcanoes **Section Focus Transparency 59**
Study Guide, p. 66 **Multicultural Connections**, p. 40 **Making Connections: Across the Curriculum**, p. 39 **Critical Thinking/Problem Solving**, p. 5 **Science Discovery Activities**, 18-2	**Teaching Transparency 36**, Earthquake-Safe Construction **Section Focus Transparency 60**
Study Guide, p. 67 **Multicultural Connections**, p. 39 **Activity Masters**, Investigate 18-2, pp. 77–78 **Making Connections: Technology and Society**, p. 39	**Section Focus Transparency 61**

ASSESSMENT RESOURCES	TEACHING & TECHNOLOGY
Review and Assessment, pp. 107–112 **Performance Assessment**, Ch. 18 **PASC*** **MindJogger Videoquiz** **Alternate Assessment in the Science Classroom** **Computer Test Bank**	**Spanish Resources** **Cooperative Learning Resource Guide** **Lab and Safety Skills** **Science Interactions, Course 1, CD-ROM** **Computer Competency Activities**

*Performance Assessment in the Science Classroom

NATIONAL GEOGRAPHIC TEACHER'S CORNER

Index to National Geographic Magazine	National Geographic Society Products Available From Glencoe	Additional National Geographic Society Products
The following articles may be used for research relating to this chapter: • "Living with California's Faults," by Rick Gore, April 1995. • "Volcanoes: Crucibles of Creation," by Noel Grove, December 1992. • "Earthquake—Prelude to The Big One?" by Thomas Y. Canby, May 1990.	To order the following products for use with this chapter, contact your local Glencoe sales representative or call Glencoe at 1-800-334-7344: • *STV: Restless Earth* (Videodisc)	To order the following products for use with this chapter, call the National Geographic Society at 1-800-368-2728: • *Raging Forces: Earth in Upheaval* (Book) • *Nature's Fury* (Video) • *Our Dynamic Earth* (Video) • *Volcano!* (Video)

Teacher Classroom Resources

These Teacher Classroom Resources are examples of the materials in the Teacher Classroom Resources package.

TEACHING AIDS

Section Focus Transparencies

VOLCANIC REMNANT
Ship Rock in New Mexico stands high against the desert sky, a major landmark visible for miles. Ship Rock is the remains of an ancient volcano that erupted millions of years ago.

1. How do you think Ship Rock formed?
2. Do you think the area around Ship Rock would be at risk for having earthquakes? Explain.

L1

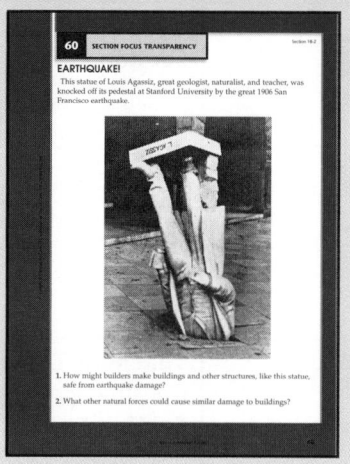

EARTHQUAKE!
This statue of Louis Agassiz, great geologist, naturalist, and teacher, was knocked off its pedestal at Stanford University by the great 1906 San Francisco earthquake.

1. How might builders make buildings and other structures, like this statue, safe from earthquake damage?
2. What other natural forces could cause similar damage to buildings?

L1

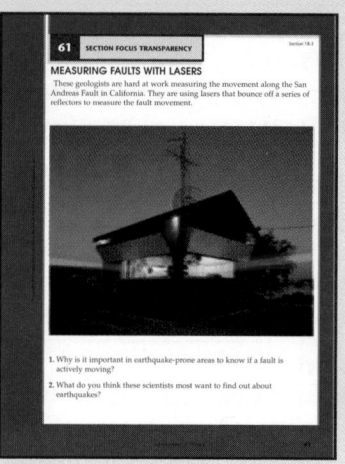

MEASURING FAULTS WITH LASERS
These geologists are hard at work measuring the movement along the San Andreas Fault in California. They are using lasers that bounce off a series of reflectors to measure the fault movement.

1. Why is it important in earthquake-prone areas to know if a fault is actively moving?
2. What do you think these scientists most want to find out about earthquakes?

L1

Teaching Transparencies

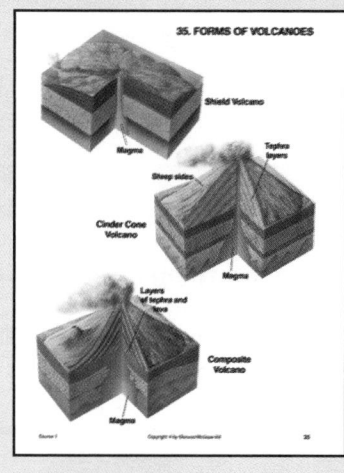

35. FORMS OF VOLCANOES

L2

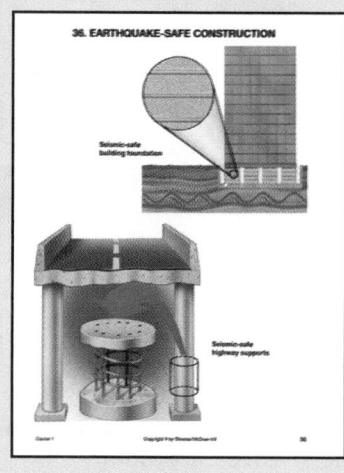

36. EARTHQUAKE-SAFE CONSTRUCTION

L2

HANDS-ON LEARNING

Science Discovery Activity*

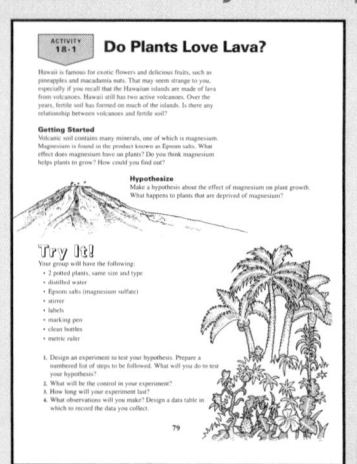

ACTIVITY 18-1 Do Plants Love Lava?

L1

Laboratory Manual*

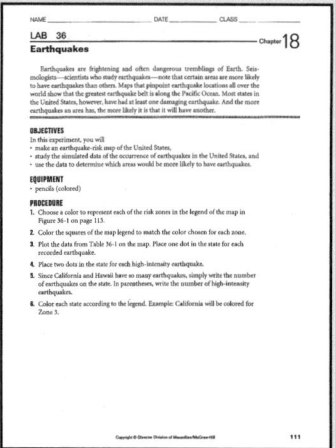

LAB 36 Earthquakes

L2

Take Home Activity

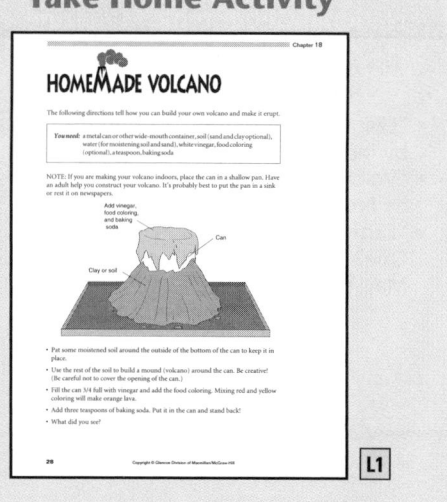

HOMEMADE VOLCANO

L1

Chapter 18 Earthquakes and Volcanoes

Study Guide*

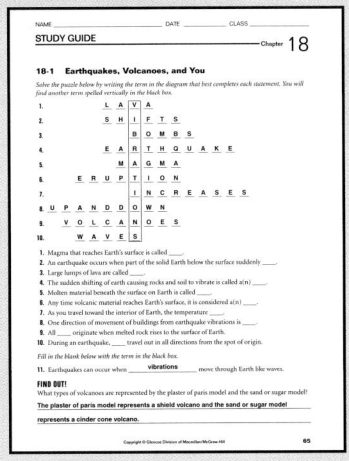

Concept Mapping

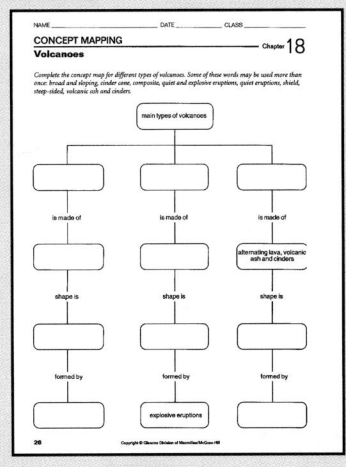

Critical Thinking/Problem Solving

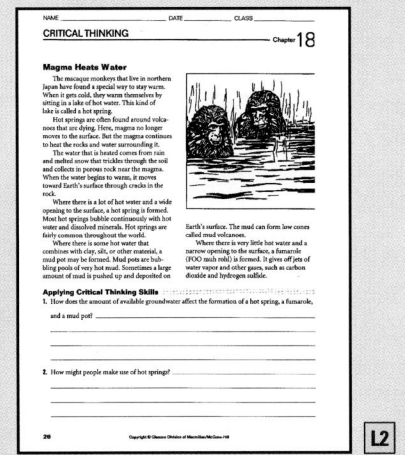

Integrating Sciences

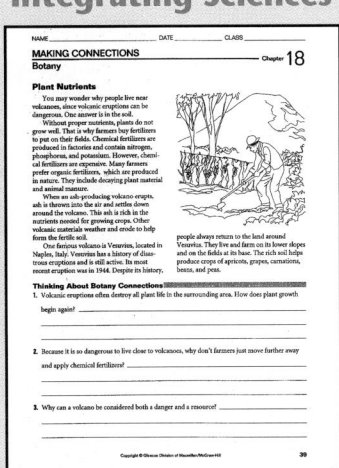

Across the Curriculum

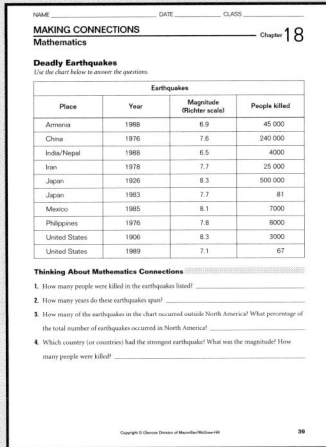

Technology and Society

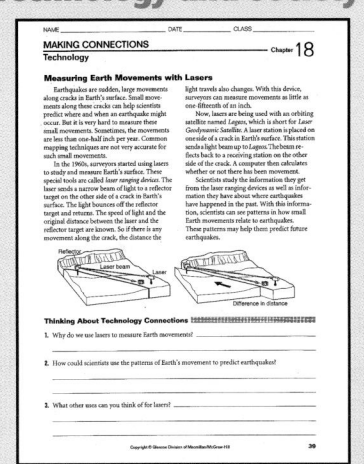

Multicultural Connection**

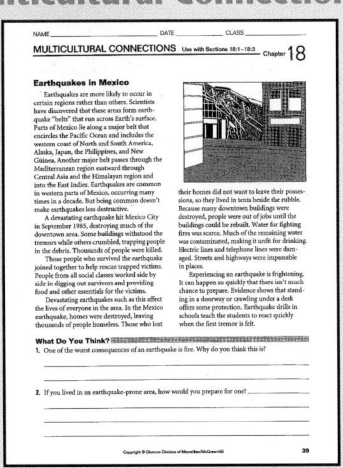

Performance Assessment

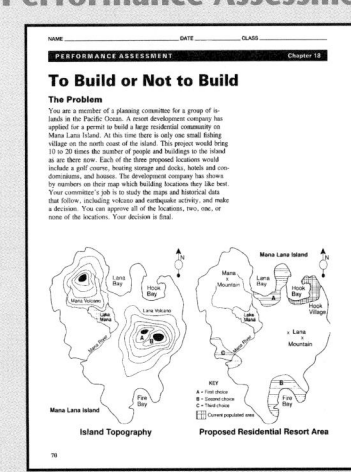

Review and Assessment

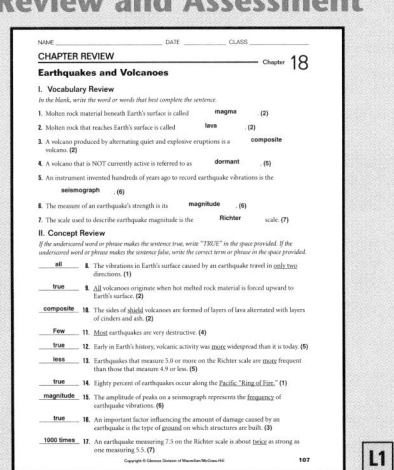

*One per section   **Two per chapter

Earthquakes and Volcanoes

THEME DEVELOPMENT

The two themes of energy and stability and change are integral to this chapter. The energy released by earthquakes and volcanoes changes Earth in many ways. These changes can destroy the stability of affected areas.

CHAPTER OVERVIEW

This chapter focuses on the causes and effects of earthquakes and volcanoes. Students will distinguish three types of volcanoes.

Students will learn how to relate the damage caused by an earthquake to its strength.

Finally, students will learn how a seismograph measures an earthquake's strength and how the Richter scale indicates the magnitude of an earthquake.

Uncovering Preconceptions

Students may believe that earthquakes strike only in some places. While earthquakes are more frequent in some areas, they can strike anywhere.

INTRODUCING THE CHAPTER

Have students look at the photograph of Kilauea volcano and speculate on how the lava flow will change both the land it crosses and the ocean it flows into. *Answers should include warming the ocean waters which will affect the building up of new land and the local ecosystem.*

00:00 OUT OF TIME?

If time does not permit teaching the entire chapter, use the Chapter Overview on this page, Reviewing Main Ideas at the end of the chapter, and the Chapter 18 audiocassette to point out the main ideas of the chapter.

Earthquakes and Volcanoes

Did you ever wonder...

✓ **What happens when an earthquake strikes underwater?**

✓ **How volcanoes form?**

✓ **Why some buildings crumble in an earthquake while others remain standing?**

Science Journal

Before you begin to study about earthquakes and volcanoes, think about the answers to these questions and answer them *in your Science Journal.* When you finish the chapter, compare your write-up with what you have learned.

Answers are on page 597.

Change is always taking place on Earth. The sun appears to change position in the sky. Seasons change. The weather changes. Some changes, such as the carving of a canyon by a river, take place so slowly that you may not notice the change in your lifetime. Other changes, however, are sudden and dramatic, catching everyone's attention.

Among the most powerful and frightening types of change that take place on Earth are earthquakes and volcanic eruptions. Earthquakes move the very ground you walk on. Volcanoes can blast tons of rock and smoke into the air. This chapter will explain why these fascinating and destructive changes occur.

▶ *In the activity on the next page, you will explore what an earthquake feels like.*

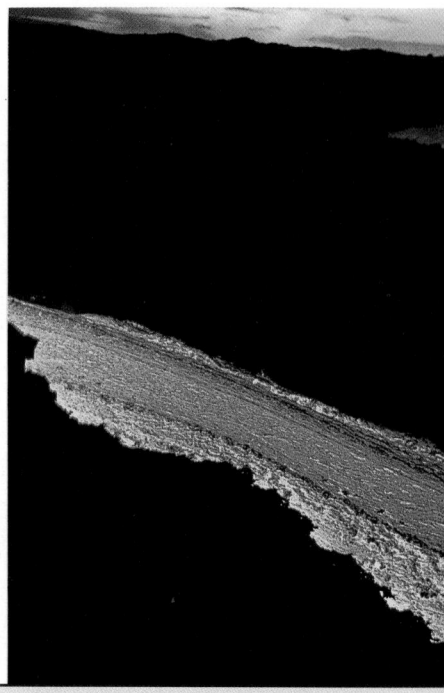

Kilauea Volcanoes National Park in Hawaii

Learning Styles	**Kinesthetic**	Explore, pp. 573, 574, 582; Activity, pp. 576, 592; Investigate, pp. 578–579; 590–591; Find Out, p. 580; Science at Home, p. 597; A Closer Look, pp. 576–577
	Visual-Spatial	Activity, p. 584; Demonstration, p. 581; Across the Curriculum, p. 589; Multicultural Perspectives, pp. 579, 590; Visual Learning, pp. 577, 580
	Interpersonal	Activity, pp. 587, 592
	Logical-Mathematical	Across the Curriculum, p. 586; Meeting Individual Needs, p. 589
LS	**Linguistic**	Physics Connection, pp. 584–585; Visual Learning, p. 587; Discussion, pp. 583, 584; Debate, pp. 577, 583
	Auditory-Musical	Meeting Individual Needs, p. 575

How can you experience an earthquake?

What To Do

1. Join your classmates in a trip to some nearby bleachers or a room with a wooden floor.

2. Take turns lying down facing away from the group while the rest of the students pound their feet as hard as they can.

3. Listen to the noise and feel the vibrations. Record your observations *in your Journal.* How do you think this experience is like a real earthquake? Would you feel any vibrations in a volcanic eruption? Explain.

Explore!

How can you experience an earthquake?

Time needed 30 minutes

Materials bleachers or room with wooden floor

Thinking Processes recognizing cause and effect, observing and inferring, making models

Purpose To model the noise and vibrations that occur during an earthquake.

Preparation Suggest that students wear casual clothes.

Teaching the Activity

Discussion Have students discuss their results. Point out that the vibrations of an earthquake can be much more powerful and can cause great damage.

Science Journal Encourage students to describe the effects strong vibrations might have on books on a desk, dishes on a table, a house, trees, and so on. **L1**

Expected Outcome

The student who is facing away should hear noise and feel vibrations.

Answer to Question

A real earthquake also produces noise and vibrations.

✔ Assessment

Performance Have students describe what might happen if an earthquake struck during a lunch period. Students can act out skits showing how an earthquake might affect people at these times. Use the Performance Task Assessment List for Skit in **PASC**, p. 75. **COOP LEARN**

ASSESSMENT PLANNER

PORTFOLIO
Refer to page 599 for suggested items that students might select for their portfolios.

PERFORMANCE ASSESSMENT
Process, p. 591
Skillbuilder, p. 592
Explore! Activities, pp. 573, 574, 582
Find Out! Activities, pp. 580, 585
Investigate, pp. 578–579, 590–591

CONTENT ASSESSMENT
Check Your Understanding, pp. 581, 587, 592
Reviewing Main Ideas, p. 597
Chapter Review, pp. 598–599

GROUP ASSESSMENT
Opportunities for group assessment occur with Cooperative Learning Strategies.

PREPARATION

Planning the Lesson

Refer to the Chapter Organizer pages 572A-D.

Concepts Developed

In this section, students will learn that surface waves generated by earthquakes can cause the collapse of buildings and other structures. Students also will learn how different types of volcanic eruptions produce differently shaped volcanoes.

1 MOTIVATE

Bellringer

 Before presenting the lesson, display **Section Focus Transparency 59** on an overhead projector. Assign the accompanying **Focus Activity** worksheet. L1

LEP

Discussion Bring in newspaper articles about a recent earthquake. L1

Explore!

How do vibrations travel through a material?

⬛ **Time needed** 15 minutes

TECH PREP **Materials** rectangular pan, water, pencil, table tennis ball

Thinking Processes observing and inferring, recognizing cause and effect, making models

Purpose To observe how waves affect materials through which they move.

Teaching the Activity

Troubleshooting If students have difficulty observing the motion of the ball, place a col-

---

Section Objectives

■ Explain how waves at Earth's surface generated by earthquakes cause structures to collapse.

■ Make models of volcanic cones and describe the types of eruptions that produce them.

Key Terms

magma
lava

Vibrations in Earth

Have you ever felt the ground quake beneath your feet? Or seen the fiery eruption of a volcano? You've probably seen pictures of the destruction caused by earthquakes in magazines or watched volcanic eruptions on television.

People have long wondered about earthquakes and volcanoes. What happens when unseen events inside Earth unleash such tremendous amounts of energy that the very ground vibrates? In the following activity, you will construct a model and observe material when it vibrates.

Explore! ACTIVITY

How do vibrations travel through a material?

What To Do

1. Pour water into a rectangular pan until it is about three-quarters full. Place a table tennis ball on the surface of the water near the middle of the pan.

2. Near one end of the pan, place a pencil in the water and move it up and down, disturbing the water. The waves you create are vibrations moving through the water.

3. Observe the motion of the ball. Does the ball move toward either end of the pan?

The waves you produced in the Explore activity above are similar to one type of wave generated at Earth's surface by an earthquake. However, earthquake waves move through the solid earth.

What would happen to a building if the ground beneath it moved in a way similar to the water? Keep this picture in your mind as you learn what it is like to experience an earthquake.

574 Chapter 18 Earthquakes and Volcanoes

---

ored dot sticker or a colored pencil mark on the ball. Be sure students hold the pencil horizontally in the water.

Science Journal Have students write descriptions of the ball's movement and compare the effects of the water waves on the ball with the effects of earthquake waves on structures. L1

Expected Outcome

Students should see that waves cause the motion of the ball and realize that this motion is

caused by waves moving through the water.

Answer to Question

The ball moves up and down, but not across the pan.

✔ Assessment

Performance Have student groups make models that demonstrate the waves and vibrations caused by earthquakes. Provide art supplies. Use the Performance Task Assessment list for Model in **PASC**, p. 51. COOP LEARN LEP P

Figure 18-1

The large arrow in this diagram indicates the direction of waves generated by an earthquake. The small elliptical arrows and the small side-to-side arrow indicate the motion of particles in Earth's surface as waves pass through.

A When the first surface wave arrives at a building, the wave lifts first one side of the building, and then the other side. The building is put into motion similar to that of the table tennis ball in the Explore activity.

B When surface waves pass through them, the structures vibrate. This vibration can cause buildings to crumble and fall.

■ Experiencing an Earthquake

An earthquake occurs when part of the solid earth below the surface suddenly shifts. This action produces waves like the ones you caused when you made waves in the pan of water. The sudden shifting in Earth causes rocks and soil at the surface to vibrate. These vibrations travel out in all directions from this surface spot. They create movement similar to that caused by the water waves. Buildings and other structures on Earth then move as in **Figure 18-1**. When the waves pass through them, the structures vibrate. This movement can cause buildings to collapse.

Earthquakes can cause a great deal of destruction. Northridge, California, was dramatically rocked by an earthquake in 1994. When the earthquake hit, vibrations moved through

the city in a series of waves that threw people from their beds as they slept and could be seen moving up and down, much like waves in the ocean. Standing in a building in Northridge during the quake would have been like standing in a rowboat on a stormy sea. If you had been there, you would have heard a sound like hundreds of locomotives rushing through the city. Many buildings and other structures could not withstand the strain, and they crumbled. Although 61 people died, this is fewer than expected for an earthquake of its strength.

Think about what would happen to your school if an earthquake similar to the Northridge quake struck nearby. What would happen to books and other objects on shelves inside the building? What would happen to the building itself?

DID YOU KNOW?

Although earthquakes tend to occur in specific areas, they can happen almost anywhere. Some of the most powerful earthquakes ever to occur in the United States took place in New Madrid, Missouri, in late 1811 and early 1812.

2 TEACH

Tying to Previous Knowledge

The vibrations generated by an earthquake are a form of mechanical wave. Help students relate what they learned about mechanical waves in Chapter 17 to earthquakes. For example, the speed of a mechanical wave varies according to the medium through which it travels.

Uncovering Preconceptions

Some students may think that all earthquakes begin at Earth's surface. Explain that earthquakes may originate near Earth's surface or deep under Earth's crust.

Student Text Questions

Think about what would happen to your school if an earthquake similar to the Northridge quake struck nearby. What would happen to books and other objects on shelves inside the building? What would happen to the building itself? *A nearby earthquake would cause books and other objects to fall off shelves and could cause the building to collapse.*

 Videodisc

STV: Restless Earth

Translation in California
Unit 4, Side 2
Translation in California

00001-10743

Introduction
Unit 1, Side 1
Introduction

00001-14511

Program Resources

Study Guide, p. 65
Concept Mapping, p. 26, Volcanoes [L1]
Making Connections: Integrating Sciences, p. 39, Plant Nutrients [L2]
Laboratory Manual, pp. 111–114, Earthquakes [L2]; pp. 115–116, Pumice [L2]
Teaching Transparency 35 [L2]
Section Focus Transparency 59

Meeting Individual Needs

Learning Disabled Use musical instruments to give learning disabled students an alternate way of observing vibrations. Students can beat drums and feel the vibrations of the membrane with their fingers. Explain that earthquakes can vibrate the ground like the head of a drum. Dramatize this by placing blocks, action figures, and the like on the drumhead to model a town. Striking the drum will shake the model. [LEP] [L1]

Observing Volcanoes

Are there any volcanoes near where you live? If not, how would you feel if one suddenly began forming in your neighborhood? Probably like the farmer in Mexico who went out to work in his cornfield one day in 1943. He discovered hot smoke and ash rising from an opening in the ground that had formed in his field.

The farmer was witnessing the birth of a volcano. In less than 24 hours, a hill 40 meters high stood where the land had once been flat. By the end of a week, the hill was more than 160 meters high and still forming. The volcano, called Parícutin, eventually reached a height of 412 meters, and its base covered an area larger than 16 000 football fields.

Figure 18-2

A Magma is forced slowly upward to Earth's surface through cracks in rock or by melting through rock.

The Great San Francisco Earthquake and Fire

When powerful earthquakes strike, they can break gas lines, short-circuit electrical wires, overturn stoves, and crack chimneys. Any of these problems may lead to a fire. When a gas line is broken, all it takes is a spark to start a fire.

One of the worst earthquakes of the twentieth century struck San Francisco on April 18, 1906. About 80 percent of the damage was caused by the

The earthquake destroyed the San Francisco City Hall, pictured here on the left.

But almost nine years from that day in 1943, Parícutin stopped erupting. The volcano, pictured in **Figure 18-4B** on page 581, has been inactive ever since.

Like Parícutin, all volcanoes originate when hot, melted rock material is forced upward to Earth's surface by denser surrounding rock. This molten rock material beneath Earth's surface is called **magma**. Once it reaches the surface, it is called **lava**. **Figure 18-2** shows how magma from Earth's interior forms a volcano near its surface. What conditions deep inside Earth might cause this rock material to become melted in the first place?

B At the surface, the eruption of magma—now called lava— ash, and volcanic rocks can build to form a cone-shaped mountain.

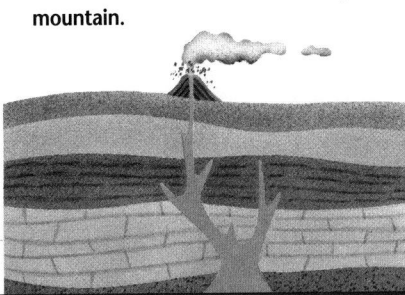

C As the lava and other volcanic materials continue to flow from the opening, the volcanic cone grows.

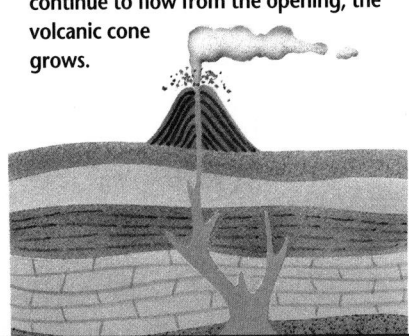

fires that followed the earthquake, and only 20 percent by the earthquake itself.

The violent tremors that had shifted the ground—as much as 20 feet in places—had broken many water mains (huge pipes) for the city's 80-million-gallon reservoir system. This left fire fighters nearly powerless to stop the blaze that roared through the city for three days and nights.

After trying one fire hydrant after another, the city's desperate fire fighters finally found just enough water to help tame the leaping flames—but not until several hundred people had lost their lives.

The 1906 earthquake and fire destroyed most of the city's business district and a number of residential areas.

Many businesses and homes collapsed during the earthquake, like the hotel in the photo at left. Many more buildings were destroyed later by fire. The photo above shows just a small portion of the destruction.

What Do You Think?

If you were in an earthquake in which the water mains were broken, where might you look for safe water to use until the pipes could be repaired?

Was the person in any danger or called upon to help others? What happened?

Answers to
What Do You Think?
Answers will vary, but may include buying bottled water or using local lake or river water that has been boiled.

Going Further ▌▌▌▌▶
Ask students to consider how the layout and construction of San Francisco have changed since the 1906 earthquake? Have structures been relocated to more seismically safe areas of the city, or have they simply been reinforced or rebuilt on the same disaster-site? Students might create a "Then and Now" poster of this or another hard-hit city to show how humans have—or haven't—changed their building and planning habits. Use the Performance Task Assessment List for Poster in **PASC**, p. 73. **L3** **P**

Theme Connection In this section, students will see evidence of the themes of energy and change and stability. Energy produced by changes within Earth is evident in volcanic eruptions and earthquakes. Energy is released during an earthquake in the form of seismic waves. Energy is also involved in the change from solid rock to magma and the forcing of magma upward to Earth's surface in a volcano.

Visual Learning

Figure 18-2 Have students trace the formation of a volcano in the illustrations as a volunteer reads aloud the captions. Suggest that students write and illustrate the stages of the process in their journals. **L1**

Student Text Question **What conditions deep inside Earth might cause this rock material to become melted in the first place?** *Deep beneath Earth's crust, very high temperatures cause rock to melt. Volcanoes form when this melted rock is forced upward to the surface.*

Debate Have students form two teams to debate whether they think the eruption of the Parícutin volcano in the middle of a corn field was beneficial or detrimental to the farmer and to his village. Have students consider possible dangers, economic implications, and scientific fame. **L3**

 NATIONAL GEOGRAPHIC SOCIETY

 Videodisc

STV: Restless Earth
Earthquake, 1906; San Francisco, California

52207

18-1 Locating Active Volcanoes

Preparation

Purpose Design and carry out an investigation in which data interpretation will show the relationships among locations of active volcanoes, earthquakes, hot spots, and boundaries of Earth's plates.

Process Skills communicating, interpreting data, observing, hypothesizing, and measuring in SI

Time Required 50 minutes

Materials If possible, use a large, poster-sized world map, suitable for students to use for marking. If the map is mounted on cardboard, students can stick map pins into it.

Safety Precautions If the map is mounted for pinning, caution students to handle map pins with care.

Possible Hypotheses 1. A positive correlation is expected between the locations of active volcanoes and the locations of earthquake epicenters. It is likely that a study of this correlation will also include a study of Earth's moving plates.

2. The locations of active volcanoes will be in the same general location as earthquake epicenters.

The Experiment

Process Reinforcement Make sure students know how to use lines of latitude and longitude to find specific locations on a map.

Possible Procedures Obtain a world map showing latitude and longitude lines. Plot each active volcano according to its latitude and longitude—either plot directly on the map or on a tracing of the map. Then obtain a map with earthquake epicenters plotted on it. Transfer earthquake epicenter locations to the map showing active volcano locations. Interpret any correlation that is shown. Then compare the map of active volcanoes and earthquake epicenters with the locations of the boundaries of Earth's moving plates. Interpret

Locating Active Volcanoes

Volcanoes form when hot, melted rock material is forced upward to Earth's surface. As the melted rock moves inside Earth, vibrations occur, which are felt as earthquakes. How would you determine whether active volcanoes are located near earthquake epicenters?

Preparation

Problem
Is there a connection between the locations of active volcanoes and the locations of recent earthquakes?

Form a Hypothesis
As a group, discuss the areas where earthquakes and volcanoes are commonly located. Then form a hypothesis about whether you expect to see a relationship between the locations of active volcanoes and the locations of earthquake epicenters.

Objectives
• Plot the locations of several active volcanoes.
• Describe patterns of distribution for volcanoes and earthquake epicenters.
• Relate the locations of active volcanoes to the locations of recent earthquakes.

Materials
world map (Appendix H)
tracing paper

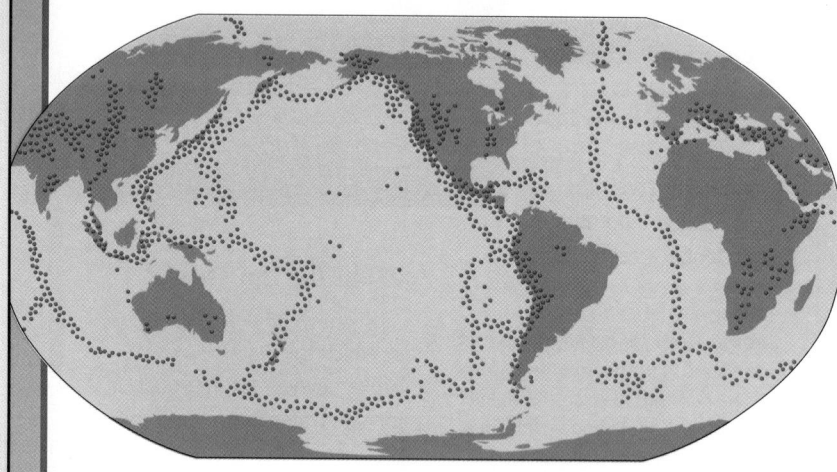

Each dot on this diagram represents an earthquake. Eighty percent of earthquakes occur along the "Ring of Fire," a band of volcanic activity that circles the Pacific Ocean.

any correlation of active volcano and earthquake epicenter locations with Earth's moving plate boundaries.

Program Resources

Activity Masters, pp. 75–76, Design Your Own Investigation 18-1

Critical Thinking/Problem Solving, p. 26, Magma Heats Water L2

Take Home Activities, p. 28, Homemade Volcano L1

Science Discovery Activities, 18-1, Do Plants Love Lava? L1

DESIGN YOUR OWN INVESTIGATION

Plan the Experiment

1 As a group, agree upon a way to test your hypothesis. Write down what you will do at each step of your test.

2 Examine the volcano latitude and longitude chart. What is the best way to plot the data on a tracing of Earth's surface?

3 Examine the map of earthquake epicenters on page 578. How will you compare your data with this map?

Check the Plan

Discuss and decide upon the following points and write them down.

1 As a group, decide how you will summarize your data.

2 How will you determine whether certain facts or conditions indicate a correlation between the locations of active volcanoes and earthquake epicenters?

3 Make sure your teacher approves your experiment before you proceed.

4 Carry our your experiment. Record your observations.

Volcano	Latitude	Longitude
#1	64° N	19° W
#2	28° N	34° E
#3	43° S	172° E
#4	35° N	136° E
#5	18° S	68° W
#6	25° S	114° W
#7	20° N	155° W
#8	54° N	167° W
#9	16° N	122° E
#10	28° N	17° W
#11	15° N	43° E
#12	6° N	75° W
#13	64° S	158° E
#14	38° S	78° E
#15	21° S	56° E
#16	38° N	26° E
#17	7° S	13° W
#18	2° S	102° E
#19	38° N	30° W
#20	54° N	159° E

Analyze and Conclude

1. **Interpret Scientific Illustrations** Describe any patterns of distribution formed by active volcanoes.

2. **Interpret Scientific Illustrations** Describe any patterns of distribution formed by earthquake epicenters.

3. **Compare and Contrast** How did the patterns that you observed in the distribution of volcanoes compare with the locations of earthquake epicenters?

Going Further

How are the locations of volcanoes and earthquake epicenters related to Earth's geographic features?

Multicultural Perspectives

Have students examine the map and identify areas of the world where cultures are most likely to have been affected by earthquakes and/or volcanoes. Have them research to find out ways in which cultures have explained them and reacted to their effects: do these phenomena have religious significance? How have people changed how they live in order to keep safe (e.g., building stronger houses, living away from volcano craters, avoiding unstable ground)? Students might compare information on Japan, New Zealand, and North America. **L1**

Teaching Strategies

Troubleshooting Have available copies, in the same scale, of maps showing active volcano locations, earthquake epicenter locations, and boundaries of Earth's moving plates.

Science Journal Suggest that students describe distribution patterns of earthquakes and volcanoes relative to major oceans or continents. Have students tape their maps in their journals at the conclusion of the activity. **L1**

Expected Outcome

Students should observe that earthquakes and volcanoes show similar distribution patterns, with most occurring along the Pacific Ring of Fire.

Analyze and Conclude

1. Answers may vary. Most volcanoes occur along the Pacific Ring of Fire.

2. Answers may vary. Most earthquakes also occur along the Pacific Ring of Fire.

3. The pattern of distribution of volcanoes and earthquakes is similar.

✔ Assessment

Performance Tell students that a large number of extinct volcanoes have been located in the midst of one of Earth's large moving plates. Ask them to explain what might have caused the volcanoes. Use the Performance Task Assessment List for Oral Presentation in **PASC**, p. 71. **COOP LEARN** **L1**

Going Further ⫸

Students should see the relationship of active volcanoes and earthquake epicenters around the Pacific Ocean. They should notice that earthquakes and active volcanoes also occur in the vicinity of Earth's moving plate boundaries and hot spots.

Accept all proposals for testing student hypotheses, and discuss those that sound reasonable as well as those that do not.

Eruptions

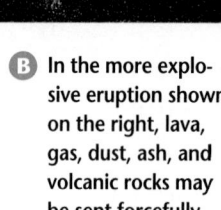

Figure 18-3

A In the quiet eruption shown on the left, lava oozes onto the surface of Earth and flows downhill, often quite slowly.

B In the more explosive eruption shown on the right, lava, gas, dust, ash, and volcanic rocks may be sent forcefully into the air.

Any time volcanic material reaches the surface of Earth, we call the event an eruption. However, not all volcanic eruptions are the same. They range from quiet lava flows to violent explosions that send lava, gases, rock, ash, and dust several kilometers into the atmosphere. The figure on the left in **Figure 18-3A** shows a quiet eruption in which lava is flowing slowly onto Earth's surface and downhill. The figure on the right in **Figure 18-3B** shows the explosive eruption of a volcano.

Different kinds of eruptions produce differently shaped volcanoes. Do the following Find Out activity to discover two of these shapes.

Find Out! ACTIVITY

What are two types of volcanic shapes?

What To Do

1. Create models of two volcanoes. First, pour 1 cup of a substance like sand or sugar into the center of a paper plate from a height of about 50 cm.

2. Then, prepare a thick mixture of plaster of paris and water and pour it into the center of a second paper plate from a height of about 20 cm.

3. Compare the shapes of your volcano models. Use a protractor to measure the slope angles of the sides of the two models. What differences have you discovered in the two forms of models produced?

Conclude and Apply

1. Of the materials that erupt from volcanoes—lava, gases, dust, ash, and rock—which do you think form volcanoes with gentle slopes?

2. Which materials probably form volcanoes with steep slopes?

Although no two volcanoes are exactly the same shape, there are three basic shapes of volcanoes.

You discovered two of the shapes in the Find Out activity. The third shape forms from a combination of quiet

580 Chapter 18 Earthquakes and Volcanoes

Figure 18-4

A Shield volcanoes are broad with gently sloping sides. In a quiet eruption, dense lava flows onto Earth's surface and spreads out over a large area in fairly flat layers. Over time, these layers build up to form a shield volcano. Kilauea in Hawaii is the largest active shield volcano in the world.

B Cinder cone volcanoes, like this one in Mexico called Parícutin, form in explosive eruptions. In an explosive eruption, gases and rock fragments may be hurled many kilometers into the air. These rock fragments which range in size from powdery volcanic dust to large lumps of lava called bombs, fall to the ground and form a steep-sided, loosely packed mountain.

C Composite volcanoes, such as Mount Ranier, are produced by alternating quiet and explosive eruptions. Their sides are formed of layers of lava alternated with layers of cinders and ash. How does a composite volcano's shape compare to those of the other volcanoes?

and explosive eruptions. These are common in the northwestern United States. Study **Figures 18-4 A-C** to find out how the shapes are made.

Powerful and potentially dangerous earthquakes and volcanic eruptions originate deep below Earth's surface. In the next section, you'll discover some of the effects they have on us at the surface.

check your UNDERSTANDING

1. Explain why buildings and other structures crumble during earthquakes.
2. Name and describe the three forms of volcanoes and the type of eruption associated with each.

3. **Apply** Why might a building that is made of flexible material like wood withstand the effects of an earthquake better than a building that is made of a rigid material like brick?

check your UNDERSTANDING

1. The vibrations travel in waves that pass through structures and produce up-and-down movements, causing structures to crumble.

2. Shield volcanoes are broad with gently sloping sides, formed by quiet eruptions. Cinder cone volcanoes are steep, caused by explosive eruptions. A composite volcano's shape is steep at its top but becomes less steep toward its base, caused by alternating quiet and explosive eruptions.

3. Flexible materials can more easily bend and withstand the movement and vibration of surface waves than can rigid materials.

Visual Learning

Figure 18-4 Have student pairs discuss how volcanic shapes are formed. **How does a composite volcano's shape compare to those of other volcanoes?** *Shield volcanoes are broad based, have gently sloped sides, and form the largest volcanic mountains. Cinder cones have narrow bases and steeply sloped sides. Composite volcanoes are both broad and steep, and rise to great heights.*

3 ASSESS

Check for Understanding

Have students answer questions 1 and 2 of Check Your Understanding individually. Then have them work with partners to discuss question 3. Suggest that students begin by comparing brick structures and wooden structures. **COOP LEARN**

Reteach

Demonstration To help students observe the effects of vibrations, use a set of building blocks to make a wall or a desk. Hit the side of the desk to make it vibrate. Have students observe the blocks as you increase the force of the vibrations. Students should note that weak vibrations may shift the wall, and stronger vibrations may make it collapse. **L1 LEP**

Extension

Research Have students find out if volcanoes or earthquakes occur on the moon or elsewhere in our solar system. Students can share their findings with the class. **L2 P**

4 CLOSE

Discussion

Have students compare and contrast what they might observe during an earthquake with what they might observe during a volcanic eruption. **L1**

PREPARATION

Planning the Lesson

Refer to the Chapter Organizer pages 572A-D.

Concepts Developed

Students continue their exploration of earthquakes and volcanoes. The role of earthquakes in the production of tsunamis and the damage caused by these tsunamis will be explored. The damage caused by volcanic eruptions also will be examined.

1 MOTIVATE

Bellringer

 Before presenting the lesson, display **Section Focus Transparency 60** on an overhead projector. Assign the accompanying **Focus Activity** worksheet. L1
LEP

Discussion Ask students what they would do if an earthquake were predicted to occur tomorrow. Are there particular buildings or areas that would be safer than others?

Explore!

What makes an earthquake destructive?

Time needed 15 minutes

Materials rectangular pan, sand or fine soil, two large books

Thinking Processes observing and inferring, recognizing cause and effect, making models

Purpose To model earthquake damage

Teaching the Activity

Troubleshooting Pan should be only half full.

18-2 Earthquake and Volcano Destruction

Section Objectives
■ Determine four factors that influence the amount of damage caused by an earthquake.
■ Describe the types of damage caused by earthquakes.
■ Describe the types of damage caused by volcanoes.

Key Terms
tsunami

Earthquake Damage

How would you feel if your home collapsed during an earthquake? Your first concern would probably be your family's safety. Once you knew everyone was safe, you'd survey your home. Your clothes, furniture, television, and other belongings—all would be buried under rubble. Fire would be a possible hazard because natural gas lines are often split open by a quake, and sparks may ignite the escaping gas.

From time to time, people face this kind of damage after an earthquake has hit. Actually, very few earthquakes are destructive. Earthquakes vary in strength, and most quakes are so weak that people don't even notice them. This next activity will show you how the strength of an earthquake, plus one other earthquake characteristic, can determine the amount of damage it will cause.

Explore! ACTIVITY

What makes an earthquake destructive?

What To Do

1. Take a rectangular pan and fill it halfway with sand or fine soil. Place its ends on two large books so that you can reach your hand underneath.

2. Pound lightly on the underside of the pan. Observe how much sand or soil moves in the pan.

3. Now pound harder, then harder still. Note how much more sand or soil shifts the harder you hit the bottom of the pan.

4. Now vary the place where you strike the pan. First pound in the middle and note where most of the sand or soil shifts. Is it directly over the spot where you strike the pan, or off to the side?

5. Move the pan so that one side extends over the edge of the table and pound under this side. Where does most of the sand or soil shift? What could happen to a building directly above such an underground disturbance? What would happen to a building on the other side of the pan?

Science Journal Have students predict what will happen before they pound on the pan, and then write their predictions and their results in their journals. L1

Expected Outcomes

Students will observe that the greatest movement occurs when the pounding is hardest.

Answers to Questions

4. When the pan is struck in the middle, the soil or sand falls evenly to either side.

5. Pounding shifts the soil or sand to the side of the pan resting on the table. A building might crumble and fall. A building farther away might be damaged but remain standing.

✔ Assessment

Performance Have students use toothpicks and their pans of sand to model what could happen to structures built on sand during an earthquake. Use the Performance Task Assessment List for Model in **PASC**, p. 51. LEP L1

When considering the threat of an earthquake, you must consider four factors—the strength of the earthquake, its location relative to populated areas, the design of buildings, and the type of ground on which these structures are built. While people can't do anything about the first two factors, they do have some control over the other two. They can design and construct buildings that will withstand

many earthquakes, and they can build these structures on solid ground.

To gain a better understanding of how these other factors can influence the damaging effects of earthquakes, look at the examples below.

Figure 18-5

A Structures unable to withstand the vibrations of earthquake waves were a major cause of damage to buildings during the San Francisco earthquake of 1906, as seen in the photo on the right. Modern San Francisco buildings have the strength and flexibility to better withstand strong vibrations. Considering their structure, how well do you think buildings in your area could withstand an earthquake?

B Bridges are easily damaged by earthquakes. The freeway overpass below collapsed when an earthquake struck Northridge, California, on January 17, 1994.

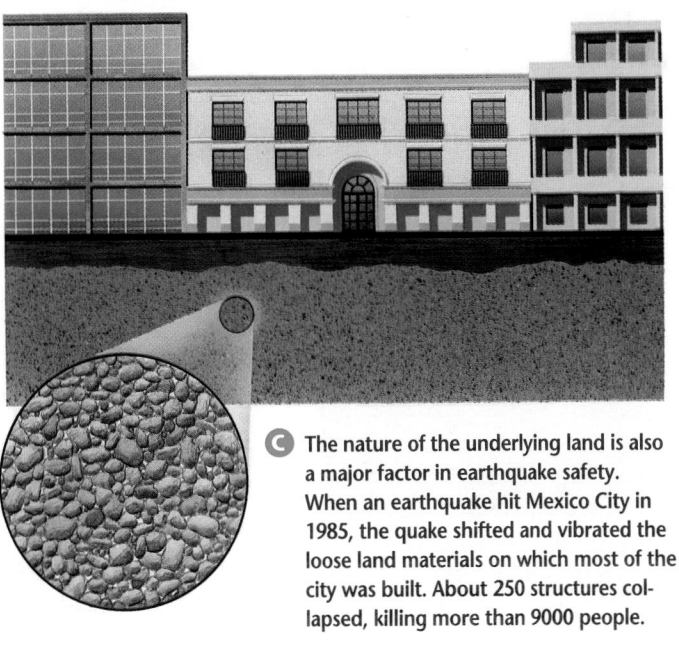

C The nature of the underlying land is also a major factor in earthquake safety. When an earthquake hit Mexico City in 1985, the quake shifted and vibrated the loose land materials on which most of the city was built. About 250 structures collapsed, killing more than 9000 people.

18-2 Earthquake and Volcano Destruction **583**

Tsunamis

If you lived at the seashore in an area where earthquakes occurred fairly often, what major concerns would you have? You might worry about the sandy soil on which your house is built. You know from the Mexico City example and from your own Explore activity that loose material like sand becomes unstable when earthquake waves travel through it. But there is another problem that may occur at seashore homes.

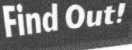

Find Out! ACTIVITY

What may happen when an earthquake strikes offshore?

What To Do

1. Use one or two books to tilt a cake pan at a 20-degree angle.

Pour water into the lower end of the pan. Leave about one-third of the pan at the upper end dry.

The Terror of the Tsunami

Almost all of the two hundred houses in El Tranisto, Nicaragua, were destroyed by a tsunami in September of 1992.

Earlier you read about tsunamis—giant waves caused by earthquakes beneath the ocean floor or by underwater landslides or volcanic eruptions in the sea. How do the characteristics of tsunamis compare with the waves you studied in Chapter 17?

When an earthquake occurs under the ocean, the movement of the ocean floor pushes against the water, creating a powerful wave that reaches all the way to the ocean surface. These enormous water waves may travel thousands of kilometers in all directions. Far from shore, where the water is deep, the wavelengths of earthquake-related waves can be hundreds of kilometers long. But when one of these waves nears a coastline, the water piles up and forms a towering wave crest that can exceed 30 meters in height.

Just like other waves, tsunamis have a frequency. You know that an earthquake produces a series of vibrations in Earth's crust. The frequency of the tsunami is dependent on the frequency of the earthquake vibrations.

The powerful wave of a tsunami can travel as rapidly as a jet plane. As the wave moves away from the spot of its origin, it may travel at a speed of more

2. Create a coastline by packing a layer of damp sand 2-3 cm thick at the dry end of the pan. Use your hands to build dunes and low areas.

3. Punch a hole near the rim of a plastic lid and thread a piece of string about 20 cm long through the hole.

4. Tie a knot near the end of the string to keep it from slipping back out. Carefully place the lid on the bottom of the pan at the low end.

5. Use your fingers to hold the edge of the plastic lid firmly against the bottom near the upper end of the pan. While holding this edge of the lid down, pull the string straight up with one rapid movement. This action simulates an underwater earthquake. Observe what happens.

Conclude and Apply

1. What happens to land near the water when an earthquake strikes under the ocean?

As you learned in the Find Out activity, an underwater earthquake can send a huge, rapidly moving water wave crashing onto the shore.

Such a tremendous ocean wave generated by an earthquake is called a **tsunami**. You can find out more about tsunamis in the "Physics Connection" feature below.

than 700 kilometers per hour. At this rate, a tsunami produced near the Hawaiian Islands could reach Seattle, Washington, in less than six hours.

As the diagram below shows, the amplitude of a tsunami may not look any greater than a normal ocean wave when it is in

deep water. As it reaches shallower depths, however, the water begins to pile up, dramatically forming a wall of water by the time the tsunami reaches the shallow waters near the shore. The shallower the water, the taller the wave becomes. Tons of water crash onto coastal areas, tossing

boats around like bath toys and doing tremendous damage to buildings.

What Do You Think?

Why might the damage to a coastal town on a U- or V-shaped inlet be greater than to a coastal town on a straight shoreline?

Earthquake origin

585

Answers to
What Do You Think?

Cities that have U- or V-shaped harbors or shallow waters offshore are at greatest risk because these conditions amplify, or heighten, the wave.

Going Further ⫸

Have a team of students report on what they can learn of plans in your community for what would be done in the event of a massive earthquake or tsunami. L2

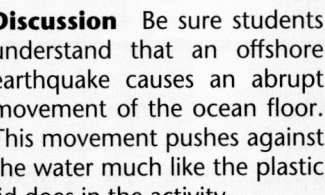

Find Out!

What may happen when an earthquake strikes offshore?

Time needed 30 minutes

Materials books, rectangular cake pan, water, sand, scissors, 20 cm of string, circular plastic lid, hole puncher, metric ruler

Thinking Processes making models, observing and inferring, recognizing cause and effect

Purpose To make a model that simulates the effects of an offshore earthquake on a seashore.

Preparation Use a circular lid with a diameter less than the width of the rectangular pan. Place the lid in the pan gently to avoid waves before they simulate an earthquake.

Teaching the Activity

Discussion Be sure students understand that an offshore earthquake causes an abrupt movement of the ocean floor. This movement pushes against the water much like the plastic lid does in the activity.

Science Journal Suggest that students draw a diagram of their setup and write their observations in their journals. L2

Expected Outcome

Students will observe that the motion of the lid generates a wave that travels to the surface and causes destruction along shorelines.

Conclude and Apply

Land near the water is washed away.

✓ Assessment

Content Ask students to describe how the pulling action in this activity is similar to the pounding in the Explore activity on page 582. *They both simulated underground pressure.* Use the Performance Task Assessment List for Analyzing the Data in **PASC**, p. 27. L1

Volcano Damage

Connect to...
Chemistry

Volcanoes can give off large quantities of different gases. Some of the gases are greenhouse gases. Make a table that includes the gases given off by volcanoes and identify which of those are greenhouse gases.

Can you imagine red-hot lava oozing toward your home? People in Hawaii—and elsewhere on Earth—have had to watch just such a scene.

Early in Earth's history, volcanic activity was more widespread than it is today. Most of the volcanoes on Earth today are dormant, which means that they are not currently active. An active volcano is one that shows evidence of releasing materials, ranging from occasional smoke and gases to constant spewing of dust, ash, cinders, and lava. Currently, more than 600 volcanoes on Earth are classified as active.

Figure 18-6

Kilauea, the most active volcano in the world, has been quietly erupting off and on for centuries. The most recent series of eruptions began in January 1983 and continued into the 1990s.

The recent eruptions of volcanoes in the Philippines, Japan, the state of Washington, and Hawaii are evidence that volcanoes can do tremendous damage. Look at the figures on these two pages to see how destructive volcanoes can be.

Figure 18-7

On May 18, 1980, Mount Saint Helens, in the state of Washington, hurled over 275 trillion tons of ash and rock into the air, killing nearly every living thing in a fan-shaped area extending as far as 90 kilometers from the mountain.

Severely damaged conifer trees near Mount Saint Helens

Home surrounded by lava in Kalapana Gardens, Hawaii

586 Chapter 18 Earthquakes and Volcanoes

Figure 18-8

Perhaps the best-known volcanic eruption in the world is that of Italy's Mount Vesuvius, shown above, in the year 79. Ash and cinders from the eruption buried the nearby cities of Pompeii and Herculaneum.

Figure 18-9

A In 1815, Indonesia's Mount Tambora violently erupted, releasing 6 million times more energy than an atomic bomb. Volcanic dust circled the globe for months.

B This diagram illustrates why the year following Tambora's eruption was called "the year without a summer."

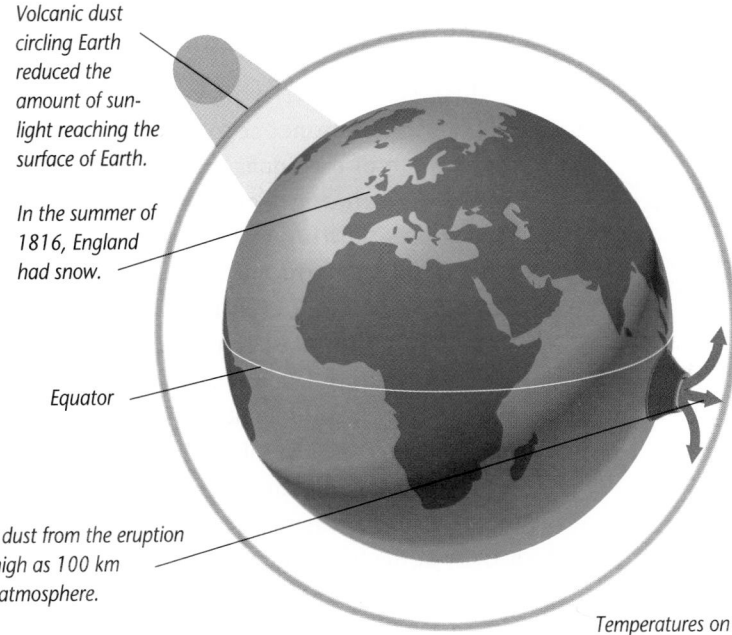

Volcanic dust circling Earth reduced the amount of sunlight reaching the surface of Earth.

In the summer of 1816, England had snow.

Equator

Volcanic dust from the eruption rose as high as 100 km into the atmosphere.

Temperatures on Earth were about 0.5°C lower than average.

3 Assess

Check for Understanding

1. Have students draw a diagram of damage resulting from either a volcanic eruption or an earthquake. **L1**

2. Have students work with partners to answer questions 1–3 of Check Your Understanding. Have students work in small groups to hypothesize about question 4. **COOP LEARN**

Reteach

Activity Have students draw concept maps to summarize factors that affect the amount of damage done by an earthquake and the kinds of damage done by earthquakes and volcanoes. **L1**

Extension

Activity Have groups of students explore how an area changes after volcanic activity ends. Students can find out which plant and animal lifeforms are the first to return, and which ones follow later. **COOP LEARN** **L3**

You have explored the types of damage that can be caused by earthquakes and volcanoes. You know that the amount of damage that can occur depends partly on how big or how strong a volcano or an earthquake is. In the next section, you will discover how the strength of an earthquake is measured.

check your UNDERSTANDING

1. List four factors that affect the amount of damage caused by an earthquake.

2. What kind of damage occurs when an earthquake strikes on land? When an earthquake strikes the ocean floor?

3. What are some of the dangers faced by people living near an active volcano?

4. **Apply** Some scientists have hypothesized that high levels of volcanic activity may have occurred about 66 million years ago. This time is the same period during which the dinosaurs became extinct. If many volcanoes were spewing ash and dust into the atmosphere, how might the climate of Earth have changed? How would a climate change have affected plant and animal life at that time?

check your UNDERSTANDING

1. Four factors are earthquake strength, its location relative to populated areas, the design of structures, and the type of ground on which these structures are built.

2. Land earthquakes may cause buildings and structures to crumble and loss of life. Earthquakes on the ocean floor form tsunamis that crash into shorelines, causing destruction and loss of life.

3. Dangers include oozing lava; spewing volcanic debris; flooding and mudslides.

4. Ash and dust may have blocked out much of the sunlight, cooling the climate and destroying plants needed for dinosaur survival.

4 Close

Have students write letters to friends describing the sudden eruption of a volcano near their homes. Encourage creativity, but have students maintain scientific accuracy. Be sure students differentiate between the types of volcanic eruptions and their results. **L2**

PREPARATION

Planning the Lesson

Refer to the Chapter Organizer pages 572A-D.

Concepts Developed

In Section 18-2, students learned to relate the strength of an earthquake to the amount of damage it can do. In this section, students will learn how this strength is measured by a seismograph and how the magnitude of an earthquake is expressed numerically.

1 MOTIVATE

Bellringer

 Before presenting the lesson, display **Section Focus Transparency 61** on an overhead projector. Assign the accompanying **Focus Activity** worksheet. [L1]
LEP

Discussion Ask students to explain why they think it is important to measure the strength of an earthquake. *Answers may include to learn more about earthquakes, to be able to compare earthquakes in different places, to be able to predict earthquakes, to help in the design of more earthquake-proof buildings.*

2 TEACH

Tying to Previous Knowledge

Ask students to give examples of instruments used to measure scientific data. *Possibilities include thermometers, metric rulers, and clocks.* Ask students what aspects of an earthquake they might want to measure. *strength*

Section Objectives

- Demonstrate how a seismograph measures an earthquake's strength.
- Explain how the Richter scale is used to indicate earthquake magnitude.

Key Terms

seismograph
magnitude

Measuring Earthquakes

Recording Vibrations

Just as the severity of volcanic eruptions varies from one to the next, the strength of earthquakes varies as well. Recall that an earthquake is caused by a shifting of rock in the solid earth below the surface. Many earthquakes are not even felt at the surface. Earthquakes that are felt range in strength. At one end, they may be only a mild shaking of the ground, similar to the vibrations you felt when your classmates pounded the floor or bleacher in the Explore activity at the beginning of this chapter. At the other end, they may be a violent trembling.

Whether or not they are felt at Earth's surface, earthquakes produce vibrations in rocks and soil. The strength of an earthquake is determined by recording and measuring its vibrations.

Figure 18-10

The roll of paper is held firmly in the frame of the seismograph, while the pen moves freely. When the ground vibrates, the roll of paper also vibrates, but the pen does not. As the roll of paper turns, the pen traces a record of the vibrations on the paper. This record appears as a wavy line.

A scientist who studies earthquakes is called a seismologist. Seismologists use an instrument called a **seismograph** to record earthquake vibrations. **Figure 18-10** describes how a seismograph works. The vibrations of Earth are recorded as a wavy line. The height of the peaks of the wavy line indicates the earthquake's magnitude. The **magnitude** of an earthquake is a measure of the earthquake's strength.

Program Resources

Study Guide, p. 67
Multicultural Connections, p. 39, Earthquakes in Mexico [L1]
Activity Masters, pp. 77–78, Investigate 18-2
Making Connections: Technology and Society, p. 39, Measuring Earth Movements with Lasers [L2]
Section Focus Transparency 61

Visual Learning

Figure 18-10 Have students apply their sequencing skills to make a flow chart showing how the seismograph shown in the figure works. [L1] [P]

Seismologists use a special scale called the Richter scale to describe the earthquake magnitudes they measure. The numbers on the scale relate to the amounts of energy released by the earthquakes. Each number on the scale represents an earthquake about 32 times stronger than the previous lower number on the scale.

For example, an earthquake measuring 6.5 on the Richter scale is 32 times stronger than an earthquake that measures 5.5 on the scale. How much stronger would an earthquake measuring 7.5 on the Richter scale be than one measuring 5.5? Because there is a difference of 2.0 on the scale, you would multiply 32 times 32. The stronger earthquake would be 1000 times stronger than the weaker one.

Earthquake Occurrences	
Richter Magnitude	Number Expected Per Year
1.0 to 3.9	949 000
4.0 to 4.9	6200
5.0 to 5.9	800
6.0 to 6.9	226
7.0 to 7.9	18
8.0 to 8.9	<2

Table 18-1

Study **Table 18-1** to see how many earthquakes at each magnitude are expected each year. What happens to the number of occurrences as the magnitude increases? How many earthquakes are predicted to occur with a Richter value between 1.0 and 3.9? How would you account for the fact that you hear about only a few earthquakes each year?

Which do you think is more destructive—an earthquake measuring 8.5 that occurs in a desolate, unpopulated part of the world or one measuring 6.5 that occurs near a densely populated area? Although the stronger quake releases much more energy, the second quake may cause much more damage. The magnitude of an earthquake does not tell you all you need to know about an earthquake.

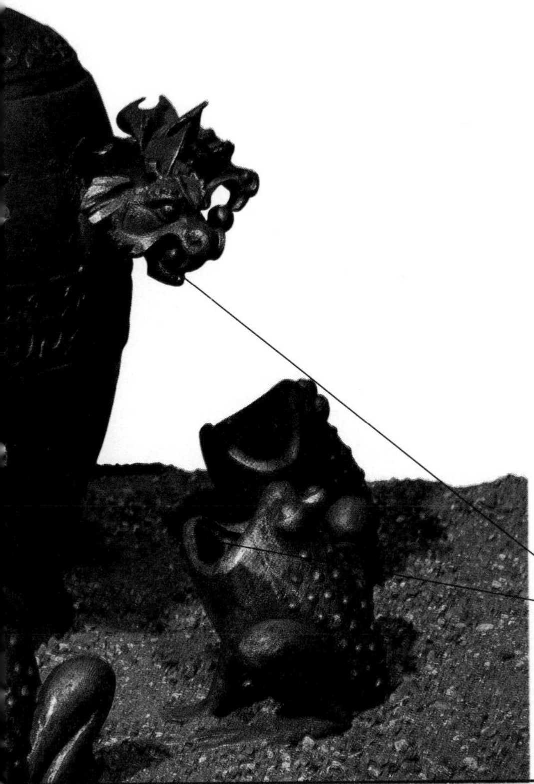

Figure 18-11

 Zhang Heng, a Chinese scientist living in the second century A.D. built the first instrument for recording the occurrence and direction of earthquakes too slight to be felt. The bronze device was about 2 m across.

 During a tremor, the vessel would move more than the heavy pendulum, which hung inside. The motion would open the jaws of one or more of the dragons. The open jaw released a ball which fell into the mouth of the toad below.

18-3 Measuring Earthquakes **589**

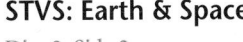

Time needed 50 minutes

Purpose To make a model of a seismograph and measure the magnitude of vibrations.

Process Skills observing and inferring, recognizing cause and effect, measuring in SI, forming a hypothesis, separating and controlling variables, making models, making and using tables, comparing and contrasting, predicting, interpreting data

Materials See reduced student text.

Preparation Be sure students tape the ring stand down carefully so that they can accurately observe the effects of different forces acting on the table.

Teaching the Activity

Process Reinforcement Before beginning the activity, have students work in pairs to observe the sort of line created when one student holds a fine-tip marker steady and the other pulls a sheet of paper under it. *a straight line* Make sure students set up their data tables correctly. **COOP LEARN** L1

Troubleshooting Encourage students to pull the paper evenly with a smooth, slow motion to avoid distorting the results. Be sure that students hit the table from the sides so the line is wavy, not straight.

Possible Hypotheses Students will probably hypothesize that the greater the magnitude of the vibrations, the larger the amplitude of the peaks.

Science Journal Encourage students to write their hypotheses in their journals and describe how their observations support or do not support their hypotheses. L1

INVESTIGATE!

Making a Model Seismograph

In this activity, you will make a model seismograph and record some vibrations.

Problem

How can you measure the magnitude of vibrations?

Materials

ring stand with ring	piece of string
wire hook from coat hanger	2 rubber bands
	fine-tip marker
masking tape	metric ruler
sheet of paper	

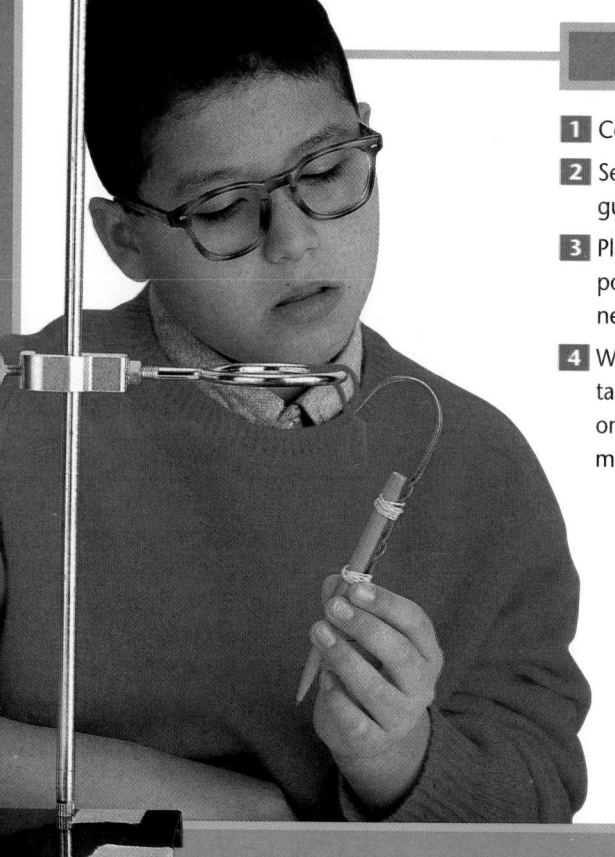

What To Do

1 Copy the data table *into your Journal*.

2 Set up your seismograph using the illustration as a guide.

3 Place a sheet of paper under the ring. Adjust the position of the marker so that its tip just touches near the end of the paper.

4 Work with a partner. While one person strikes the table several times with equal strength, the other one should slowly pull the paper under the marker.

5 Recall from Chapter 17 that amplitude is half the height of a wave from crest to trough. *Measure* the amplitude marked on your paper. Record your measurements and observations as Trial 1.

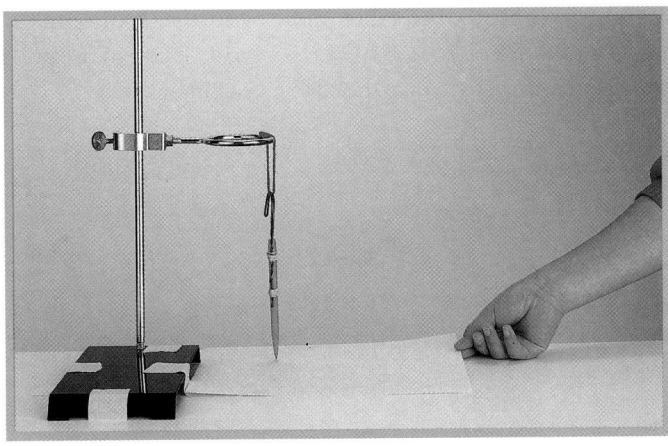

A

Sample data

Data and Observations

Trial Number	Amplitude (Height of Marks)	Observations
1	3 mm	Frame moves a little
2	1 mm	Frame doesn't seem to move
3	6 mm	Frame moves noticeably

6 *Hypothesize* about the effect of the magnitude of the vibrations on the amplitude of the peaks.

7 Repeat Steps 3 and 4, hitting the table with less strength for Trial 2 and more strength for Trial 3. Record your measurements and observations.

Analyzing

1. Which trial resulted in the greatest amplitudes recorded on the wavy line?

2. How did the movement of the marker compare with the movement of the frame of the seismograph?

Concluding and Applying

3. How does your hypothesis *compare* with the results of the activity?

4. Determine the effect the magnitude of vibrations had on the amplitude of the wave peaks.

5. Going Further What difference would you *predict* between the amplitudes generated by a strong earthquake and those generated by a weaker one?

Expected Outcomes

Students should observe that the amplitude of the waves is greater when the table is hit with greater strength. Students should recognize that the same relationship exists between the strength of waves generated by an earthquake and their amplitudes.

Answers to Analyzing/ Concluding and Applying

1. trial 3

2. In a real seismograph, the pen would remain stationary while the frame moved with the vibrations. Students models don't do this because the pen is not completely isolated from the motion of the frame.

3. Responses will vary depending on students' hypotheses.

4. As the magnitude of the vibrations increased, the amplitude of the wave peaks increased.

5. Strong earthquakes should generate greater amplitudes.

✔ Assessment

Process Ask students to interpret their observations by explaining what the lines on the paper represent. Also ask students to describe which variables they controlled to accurately measure the vibrations. *The lines represent vibrations. Students controlled the movement of the pen and paper, but allowed the paper to move along to record the vibrations of the table.* Use the Performance Task Assessment List for Analyzing the Data in **PASC,** p. 27. L1 P

1. 1556 in Shensi, China—most deaths. 1811–12 in New Madrid, MO—fewest deaths.

2. 1755 in Lisbon, Portugal

3. The 1975 quake was predicted and people were warned to be outside. [L1]

GLENCOE TECHNOLOGY

 Software

Computer Competency Activities

Chapter 18

3 ASSESS

Check for Understanding

Have students review their work on the Investigate activity on page 590 before answering question 3 of Check Your Understanding.

Reteach

Activity Have students work in pairs to observe the effects of vibrations. One student tries to write while the other pounds on the desk. Have students relate the force of the pounding to the effect on the handwriting. COOP LEARN [L1]

4 CLOSE

Activity

Have students plot the locations of the earthquakes in the table on this page on a world map. Ask students to describe any patterns they see. *Students may note that many of the earthquakes occur along the Pacific rim.* [L2]

Table 18-2 shows some strong earthquakes that have occurred in the past 400 years. Study the table and determine whether loss of life is always related directly to earthquake magnitude. Do you live in an area where earthquakes are common? Can you think of some ways to reduce your risk of being injured in an earthquake?

Table 18-2

Making and Using Tables

Use **Table 18-2** to answer the following questions:

1. Which earthquake resulted in the greatest number of deaths? The fewest?
2. Which quake had the greatest magnitude?
3. Hypothesize why the 1975 China quake resulted in fewer deaths than the 1976 quake.

If you need help, refer to the **Skill Handbook** on page 656.

	Strong Earthquakes		
Year	Location	Richter Value	Deaths
1556	Shensi, China	?	830 000
1737	Calcutta, India	?	300 000
1755	Lisbon, Portugal	8.8	70 000
1811-12	New Madrid, MO	8.3	few
1886	Charleston, SC	?	60
1906	San Francisco, CA	8.3	1 500
1920	Kansu, China	8.5	180 000
1923	Tokyo, Japan	8.3	143 000
1939	Concepcíon, Chile	8.3	30 000
1964	Prince William Sound, AK	8.5	131
1970	Peru	7.8	66 800
1975	Liaoning Province, China	7.5	300
1976	Tangshan, China	7.6	240 000
1985	Mexico City, Mexico	8.1	9 500
1988	Armenia	6.9	28 000
1989	Loma Prieta, CA	7.1	62
1990	Iran	7.7	50 000
1993	Maharashtra, India	6.4	30 000
1994	Northridge, CA	6.7	61

check your UNDERSTANDING

1. How is earthquake magnitude measured?

2. Using Table 18-2, what would be the magnitude of an earthquake that is 32 times stronger than the 1988 quake in Armenia?

3. Apply Suppose you studied seismograph readings from two earthquakes, A and B. What would you infer from the fact that the amplitude of the peaks produced by quake B were much higher than those produced by quake A?

---

check your UNDERSTANDING

1. Earthquake magnitude is measured by the Richter scale, which describes the relative amounts of energy released by earthquakes.

2. An earthquake 32 times the magnitude of the 1988 Armenian quake (6.9) would have a magnitude value of 7.9.

3. If the amplitude of the peaks produced by quake B were much greater than those produced by quake A, you would infer that quake B was stronger, released more energy, and had a higher Richter value.

Technology Connection

Earthquake-Proof Construction

Technology Connection

What makes some structures withstand earthquakes, while others are damaged or destroyed?

The geological foundation is perhaps the most important factor. Buildings built on solid rock near an earthquake's center—where vibrations are the strongest—may hold up better than buildings built on softer ground farther away.

Heavy-Duty Buildings

Extending a building's foundation well below ground level can help a building withstand an earthquake because the building will be less likely to lean or tip. Buildings may be reinforced by beams that cross at different angles or with steel embedded in concrete.

Because buildings with a lot of glass often lack support, steel-framed buildings and concrete buildings with few doors or windows hold

up better than some other types of structures. Brick buildings tend to buckle during earthquakes.

Go with the Flow

The objective in earthquake-proof construction is to build structures that will move as a unit rather than as individual, unrelated parts. Engineers accomplish this by placing rollers, jacks, springs, bearings, or plastic sheets under the bases of buildings. Architects are also experimenting with shock absorbers for buildings and bridges.

Safety Is No Accident!

Even if you are in a building with earthquake-proof construction, during an earthquake follow these safety measures:

Keep away from windows.

Avoid standing where objects might fall on you.

Avoid fallen power lines.

Stay clear of rubble with sharp edges and broken gas lines.

Science Journal

Start with photos or drawings of three different structures. For each structure, describe *in your Science Journal* three areas where damage from an earthquake is most likely to occur. Explain why. Next, indicate how and where these structures might be reinforced to minimize earthquake damage. Explain why.

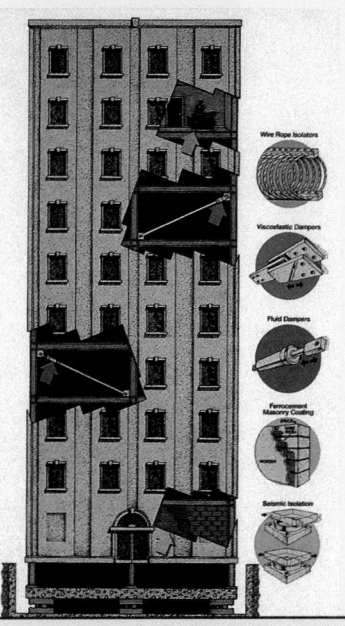

Going Further ▸

Have groups of students use reference materials to find out about technological and architectural advances with regard to earthquakes. Exactly what kinds of shock absorbers and other devices are in use? Where have they been tested with actual earthquakes—and what happened? How, also, do these new architectural sciences affect the look of new buildings? Student groups could then design pamphlets detailing their findings. Use the Performance Task Assessment List for Booklet or Pamphlet in **PASC,** p. 57. L3 P

Purpose

Technology Connection describes ongoing efforts to create safer buildings and other structures. Features of the latest buildings are designed to resist the often-devastating forces students learned about in Sections 18-1 and 18-2. Safety tips for people in even the most earthquake-proof structures conclude the selection.

Content Background

Early versions of the well known Labyrinth of Cnossos on the Greek island of Crete (circa 2000 B.C.E. had solid brick walls. After damage caused by several major earthquakes, the ancient Cretans learned to lay wooden beams within the walls so that they could sway slightly with seismic waves. The result was a building complex that endured not only 3500 years of time but also the worst seismic disaster ever known—the explosion of the island-volcano of Thera. Reconstructed in places with reinforced concrete, the Labyrinth today still withstands the seismic activity of the region.

Teaching Strategy

Have students predict which local buildings would remain standing the longest if a serious earthquake struck. Have them base their predictions on both details of that building and statements which satisfy one of the general construction rules in the article. For example, what kind of ground is it built on? What special reinforcements does it have? L2

Science Journal

Answers will vary, depending on the structures chosen. Students should look for details, such as the size and placement of windows and doors.

EXPAND your view

Science and Society

Purpose

Science and Society extends Section 18-3 as it explores some of the many ways people try to predict earthquakes. With techniques ranging from the observation of animal behavior to high-tech, long-term studies of shifts in Earth's crust, the effort to understand and predict earthquakes is a worldwide, slowly advancing science. The benefits of the science are dramatically clear in the recent cases of Liaoning and Tangshan, China, where tens of thousands of lives were saved.

Content Background

While animals and some humans can be sensitive to the magnetic and chemical changes that precede earthquakes, technological devices such as the extensometer and tiltmeter give seismologists more objective information. Tiltmeters on the slopes of Mt. St. Helens, for example, on May 18, 1980, recorded a telltale "bulge" over 300 feet high on the mountain's north slope. Its growth of five feet per day alerted scientists to the danger of a potential eruption, and many people evacuated the area just in time. Other tiltmeters around the mountain helped indicate in which direction the eruption would occur, saving more lives by helping to avoid widespread panic.

Teaching Strategy
Have students compare and contrast the various methods of earthquake prediction, be they animal or technological. What are the particular strengths and weaknesses of each method? Make a class chart of their responses. L2

Science and Society

Earthquake Prediction

Earthquakes rarely strike without warning. Often, the ground will vibrate days or even months before a major earthquake hits. Measuring even the slightest shifts in

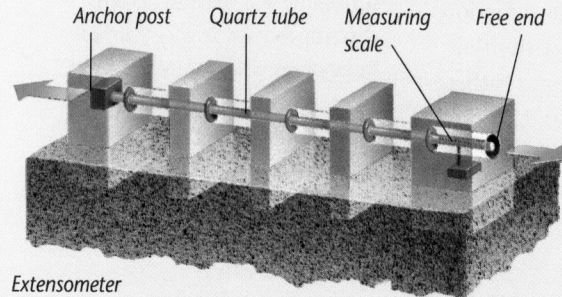

Anchor post   *Quartz tube*   *Measuring scale*   *Free end*

Extensometer

Earth can therefore help scientists predict an impending earthquake. Seismologists obtain data about movements within Earth from instruments that register changes in Earth's crust. You already know about seismographs. Let's examine some other seismic measurement devices.

Stretch-O-Meter

An extensometer has a long quartz tube threaded through a row of posts in the ground. Look at the picture of the extensometer. One end of the tube is anchored firmly inside the first post. The other end of the tube moves freely inside the last post. The additional posts help support the quartz tube. Extensometers can be more than 300 feet long.

When an earthquake occurs, the free end of the tube moves farther into or out of the last post. By reading a scale on the free end of the tube, scientists can measure the tremors.

Tipping the Scale

Tiltmeters set up near faults in Earth's surface help detect changes in the slope of the ground. A change in the tilt of the ground may indicate that one side of a fault is being forced upward or downward in relation to the other side of the fault.

Tiltmeter   *Water-level scale*

Fault

Have you ever seen a carpenter's level? A tiltmeter works the same way. It has a tube connected to two water-filled containers. When the ground near one of the containers shifts upward, the water level in that container drops, while the water level in the other, lower container rises. Tiltmeters are usually about 30 feet long.

Program Resources

Science Discovery Activities, 18-3, Crystal Balls for Earthquakes

Faster than a Speeding Bullet

Seismologists use laser distance-ranging devices to help detect small horizontal shifts in Earth's surface along fault lines. As shown in the picture, these devices aim a narrow laser beam from one side of a fault toward a reflector on the other side of the fault, and time how long the beam takes to return.

Because we know the speed at which light travels, if the beam takes less time or more time to return, we know that the distance it is traveling has changed. This change in distance may indicate a shift in Earth's crust—a sign that an earthquake may be on its way.

Warning!

We may not be able to prevent earthquakes, but if we learn to predict them accurately and if we're prepared, we may save many lives. To demonstrate the difference a warning can make, let's compare two earthquakes that occurred in China and were of approximately the same magnitude.

The earthquake in the Liaoning Province of China in February 1975 took very few lives because people knew it was coming. Days before, the ground had shifted, and minor tremors had been felt. Government officials told people to stay outside their homes so they would not be crushed if their homes collapsed. Even with the warning, 300 lives were lost. Had the people not been

Seismologists study faults and earthquakes to learn more about Earth's interior, to predict earthquakes, and to provide advice about construction sites and building materials in earthquake-prone areas.

warned, 10 000 more might have died!

The following year, an earthquake hit Tangshan, China. The people had no warning. As many as 240 000 people were killed—about one-sixth of Tangshan's population at the time! Little was left of the city except piles of bricks and twisted steel.

What Do You Think?

It's hard to pinpoint precisely when an earthquake will hit. What problems might occur if officials warn people about an earthquake too early? Should officials wait to inform people until they are more certain about when an earthquake will strike? Why?

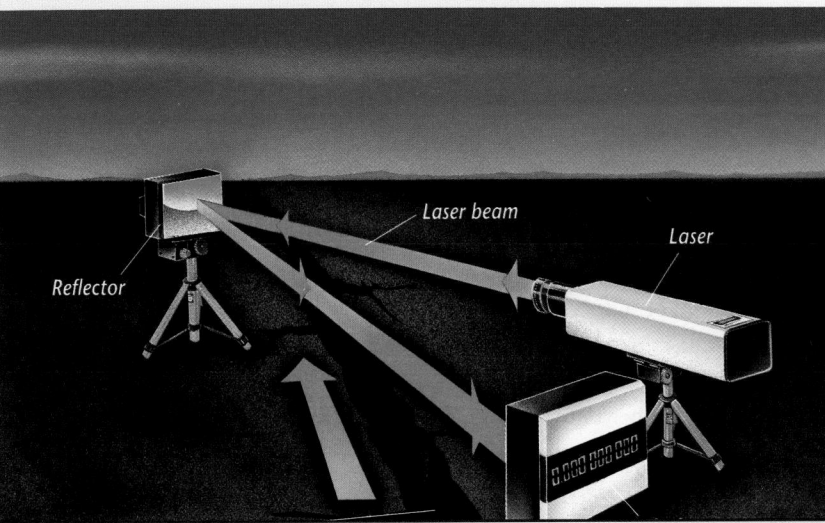

Reflector / Laser beam / Laser

Discussion

Ask students to discuss how vulnerable they feel their own community is to seismic disaster of any kind. What evidence (geographical location, history, etc.) supports each student's opinion? Ask them to describe any evidence of major geological activity in the surrounding landscape, and explain why they think a seismic disaster may or may not occur in the future.

Answers to
What Do You Think?

Answers will vary, but students are likely to point out that premature warnings can lead to unnecessary concern and possibly, panic. Also, too many early warnings that are not followed by an earthquake may lead people to disregard earthquake warnings. Other students may say that warnings may be annoying, but they may save lives in their community eventually.

Encourage interested students to write to local university research centers and/or government or local agencies for facts on degree and work-experience requirements for careers in seismological fields. L2

Going Further ⫸

What are some of the accompanying destructive events that follow earthquakes and volcanic disasters? Groups of students might choose to research a particular event such as the 1964 Alaskan earthquake or the 1883 eruption of Krakatoa to study for these subsidiary effects. Students should look for information about damage caused by ash, both locally, and down wind. Students should also look for evidence of changes in climate due to increased ash and dust

in the atmosphere. Have students create a classroom display to share their findings. Use the Performance Task Assessment List for Group Work in **PASC**, p. 97. **COOP LEARN** L2

Teens in SCIENCE

Purpose Deena Stroham's experiment in Teens and Science shows how the forces described in Section 18-1 can affect buildings and other human structures.

Content Background
Decades of patient statistical analysis and experimentation with building materials have resulted in still-evolving building codes, designed to maximize human safety during seismic disturbances. California's and other states' building codes reflect increased knowledge of how structures "behave" under stress: a ceiling beam placed at the wrong angle, or the placement of a house foundation only a few feet to the left or right of the most solid ground may cost lives later during an earthquake. Yet, because seismology is a young science, great controversy still exists as to what the most effective building traits are. However, a bedrock foundation is known to be essential to safety, as Stroham's demonstration shows.

Teaching Strategy
Have students visualize the experiment in the article and predict what the result will be. Have them support their answers with facts from the chapter.

Discussion
Have students describe how they would design a demonstration to explain any principle in the chapter. For example, how would passing water through a narrowing channel (as when a tsunami enters a narrow harbor) accelerate and raise the level of the water? What principle does this reveal at work? **COOP LEARN**

L3

Teens in SCIENCE

Shake, Rattle, and Roll

The floor is moving under your feet. A lamp swings from side to side over your head. Can you guess why? It's an earthquake!

Seventeen-year-old Deena Stroham knows a lot about earthquakes. As part of a 4-H project, Deena teaches younger students what she has learned. "I try to help kids know what to expect. But the lessons are fun."

Although Deena lives in earthquake-prone California, she has never actually felt one.

You Try It!
Deena uses this experiment to show how earthquakes affect houses built on two different types of earth. The gelatin dessert represents landfill. The clay represents bedrock.

Materials
1 package of gelatin dessert
1 package unflavored gelatin
boiling water
firm clay
48 toothpicks
24 marshmallows
6-in by 6-in by 1-in container

What To Do
1. Mix the powdered gelatin dessert and unflavored gelatin together. Follow the directions on the back of the dessert package. Pour the mixture into a container that is at least 6 inches by 6 inches and 1 inch deep. Refrigerate. When firm, cut the gelatin dessert into a 6-inch by 6-inch square.

2. Make a 6-inch by 6-inch square at least 1 inch thick out of clay.

3. Now build a house from the toothpicks and marshmallows. Make the vertical corners first. Press a marshmallow on the end of a toothpick. Push the other end of the toothpick into the gelatin square. Space each of the four corners a toothpick's length apart. Next, connect the four marshmallows with horizontal toothpicks. This is the first story of your house. Add two more. Follow the same steps with the clay square.

4. Shake each square to simulate an earthquake. Which house stood longer? Is it safer to build on landfill or bedrock?

Answers to
You Try It!

It is safer to build on bedrock. As students perform the experiment, remind them to carefully apply the same amount of "earthquake-force" to each model.

Going Further ▸▸▸▸▸▸
Have groups of students relate moon landings or Space Shuttle observations to seismic research. For example, how did laser reflectors placed on the moon affect the science of earthquake prediction? How have satellites and other NASA-spawned technologies contributed to Earth sciences in the last 30 years? Students can look in science and technology reference books. The *Readers Guide to Periodic Literature* or computer information search system at a local library can help students find magazine articles on these topics. Students could give oral presentations using visual aids to explain their findings. Select the Performance Task Assessment List for Oral Presentation in **PASC**, p. 71. **L2**

Science Journal

Review the statements below about the big ideas presented in this chapter, and answer the questions. Then, re-read your answers to the Did You Ever Wonder questions at the beginning of the chapter. *In your Science Journal*, write a paragraph about how your understanding of the big ideas in the chapter has changed.

1 An earthquake is caused by vibrations set in motion when part of the solid earth below the surface suddenly shifts. *How does energy from earthquakes travel?*

2 A volcano builds as magma reaches Earth's surface and material from eruptions accumulates over time. *Why aren't volcanoes found everywhere on Earth?*

3 The shape of a volcano depends on the material it's made of and whether the eruptions are quiet or explosive. *Compare and contrast the shapes of volcanoes and the types of eruptions.*

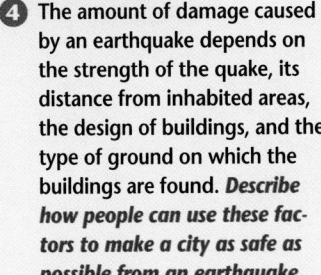

4 The amount of damage caused by an earthquake depends on the strength of the quake, its distance from inhabited areas, the design of buildings, and the type of ground on which the buildings are found. *Describe how people can use these factors to make a city as safe as possible from an earthquake.*

Have students observe the illustrations and discuss the energy released by each occurrence. Have them discuss similarities and differences for each photograph. Be sure they understand that both earthquakes and volcanoes occur because of changes below the surface of Earth.

Teaching Strategy

Divide the class into four groups. Have each group use one illustration as the center of a concept web. Ask each group to provide additional information that is related to the illustration. Allow students to review the chapter while doing this activity. Have groups present their webs to the class. **COOP LEARN** L1

Answers to Questions

1. The energy travels in waves.

2. Volcanoes tend to occur in places where Earth shifts.

3. A shield volcano results from lava eruptions that ooze from the volcano forming fairly flat layers. A cinder cone volcano results from eruptions that hurl stone and ash into the air. The stone and ash land around the volcano making a steeper slope. A composite volcano results from a combination of both types of eruptions.

4. Since earthquakes tend to happen in certain areas, people could build cities away from these areas. People could also design earthquake-resistant buildings and build on very solid ground.

Science Journal

Did you ever wonder...

• Underwater earthquakes can generate huge waves, or tsunamis that can devastate coastal areas. (pp. 584–585)

• Volcanoes form when hot, melted rock material is forced upward from inside Earth to the surface. (p. 577)

• Whether or not a building collapses during an earthquake depends on the type of materials which the building is made of and the type of soil on which it is built. (p. 583)

Science at Home

Have students inspect their homes or school and identify measures that might be taken to prevent earthquake damage. For areas that are not earthquake-prone, the discussion will be more theoretical. Many states and localities offer earthquake preparedness information or kits for both homes and school. L1

GLENCOE TECHNOLOGY

 MindJogger Videoquiz

Chapter 18 Have students work in groups as they play the Videoquiz game to review key chapter concepts.

Using Science Terms

1. seismograph
2. magma
3. tsunami
4. lava

Understanding Ideas

1. A shifting of rock in the solid earth below the surface causes earthquakes.

2. People can design and construct buildings that will withstand many earthquakes, and they can build these structures on solid ground.

3. The strength of an earthquake is determined by recording and measuring its vibrations.

4. Two kinds of volcanic eruptions are quiet lava flows and violent explosions that send lava, gases, rock, ash, and dust into the atmosphere.

5. An erupting volcano can cause death, destruction, and air pollution.

Developing Skills

1. See reduced student page for completed concept map.

2. Students should observe the same results as in the original activity.

3. Student graphs should show volcano X as having gradual slopes and volcano Y as having steep slopes. Volcano X is a shield volcano and volcano Y is a cinder cone volcano.

4. Students should observe that long bays can increase the effect of a tsunami compared to a straight coastline.

U sing Key Science Terms

lava	seismograph
magma	tsunami
magnitude	

An analogy is a relationship between two pairs of words generally written in the following manner: a:b::c:d. The symbol : is read "is to," and the symbol :: is read "as." For example, cat:animal::rose:plant is read "cat is to animal as rose is to plant." In the analogies that follow, a word is missing. Complete each analogy by providing the missing word from the list above.

1. air temperature:thermometer::earthquake magnitude: _____

2. geyser:groundwater::volcano:_____

3. air:sound::ocean water: _____

4. ice cream:chocolate syrup::Earth's surface:_____

U nderstanding Ideas

Answer the following questions in your Journal using complete sentences.

1. Explain what causes earthquakes.

2. Describe how people can control the damage caused by earthquakes.

3. How is the strength of an earthquake determined?

4. Describe two kinds of volcanic eruptions.

5. What types of damage may be caused by an erupting volcano?

D eveloping Skills

Use your understanding of the concepts developed in this chapter to answer each of the following questons.

1. **Concept Mapping** Complete the following events chain concept map of earthquakes.

Initiating event

Part of the solid earth below the surface shifts.

Event 1

Waves are produced.

Event 2

Rock and soil vibrate.

Final outcome

Buildings and other structures on Earth move. Some may even crumble and fall.

2. **Comparing and Contrasting** Repeat the Explore activity on page 574 using unflavored gelatin instead of water. Compare the results of this activity with those made in the original activity.

3. **Making and Using Graphs** After doing the Find Out activity on page 580, use the information below and make a graph to compare the shapes of the volcanoes. What type of volcano is X? What type of volcano is Y?

Volcano	Points
X	1-1, 5-5, 10-11, 11-17, 10-25, 5-33
Y	1-8, 9-17, 1-26

4. **Comparing and Contrasting** After doing the Find Out activity on pages 584-585, repeat the activity changing the shape of the shoreline to see how different shorelines are affected by tsunamis.

Program Resources

Review and Assessment, pp. 107–112

Performance Assessment, Ch. 18 L2

PASC

Alternate Assessment in the Science Classroom

Computer Test Bank L1

Critical Thinking

Use your understanding of the concepts developed in this chapter to answer each of the following questions.

1. The table shows the chances that an earthquake of a specified magnitude will strike five selected locations in California within the next 30 years. Which location is most likely to be struck by an earthquake? Which location is likely to experience the strongest earthquake?

Location	Richter Scale Magnitude		
	8+	7-7.9	6-6.9
North Coast	10%		
San Francisco		20%	
Parkfield			90%
Mojave		30%	
Coachella Valley		40%	

2. Why is flexibility an important factor in designing an earthquake-safe structure?

3. On land, the closer a spot is to the source of an earthquake, the more damage may occur. How does this situation compare with what happens when an earthquake occurs below the ocean?

Problem Solving

Read the following problem and discuss your answers in a brief paragraph.

Suppose you live in an area where earthquakes are common, and you want to set up a seismograph at home.

1. Describe how a seismograph that uses a beam of light and photographic film might work.

2. Can you think of another way to set up a seismograph using common objects? Explain.

CONNECTING IDEAS

Discuss each of the following in a brief paragraph.

1. **Theme—Energy** Name two ways in which earthquake waves are similar to sound waves.

2. **Theme—Stability and Change** How does gravity affect the motion of lava?

3. **Theme—Energy** How are earthquakes and volcanoes alike? How are they different?

4. **Technology Connection** What areas of interest do seismologists and building engineers have in common?

5. **Physics Connection** Suppose a tsunami strikes two towns the same distance from the site of an underwater earthquake. Which town would probably have more damage—the one along a straight coastline or the one at the inland end of a narrow bay? Explain your answer.

Critical Thinking

1. Parkfield; North Coast

2. If a building is flexible, it is better able to move with the surface waves produced by an earthquake. Rigid structures are more likely to break apart.

3. When an earthquake strikes below the ocean, the tsunamis that are generated can cause destruction even hundreds of kilometers away.

Problem Solving

1. The photographic film would move with the vibrations, while the source of light (a dangling laser, perhaps) would not. The film would capture the vibrations as a wavy line.

2. Answers will vary. Sample answer: A screwdriver or other long, pointed object can be suspended over loose materials, such as sand. When the container with the material moves, the screwdriver would produce a seismograph-type line.

Connecting Ideas

1. Both have the properties of waves—peaks or crests, and troughs. Both travel through a medium such as the ground, water, and buildings.

2. Gravity causes lava to flow down the slopes of volcanoes.

3. Both earthquakes and volcanoes have their origins below Earth's surface. Both are powerful and unpredictable and can cause destruction. Earthquakes send vibrations through the ground. Volcanoes blast rock, ash, dust, and lava into the air or send lava flowing.

4. Both seismologists and building engineers try to decrease damage from earthquakes.

5. The town at the inland end of a narrow bay would probably have more damage because the tsunami would increase greatly in height as it moved through the narrow bay.

✔ Assessment

Portfolio Review the portfolio options that are provided throughout the chapter. Encourage students to select one product that demonstrates their best work for the chapter. Have students explain what they learned and why they chose this example for placement in their portfolios.

Additional portfolio options can be found in the following **Teacher Classroom Resources:**

Chapter Organizer

SECTION	OBJECTIVES	ACTIVITIES & FEATURES
Chapter Opener		**Explore!**, p. 601
19-1 Earth's Shape and Movements (3 sessions, 1.5 blocks)	**1. Demonstrate** evidence that shows Earth's shape. **2. Describe** the cause of day and night. **3. Explain** what causes the seasons on Earth. **National Content Standards: (5–8) UCP1–3, A2, B2, D3, G1–G2**	**Find Out!**, p. 602 **Find Out!**, p. 605 **Design Your Own Investigation 19-1:** pp. 606–607 **SciFacts**, p. 626
19-2 Motions of the Moon (3 sessions, 1 block)	**1. Recognize** that the moon's phases are caused by the relative positions of the sun, the moon, and Earth. **2. Demonstrate** the arrangement of the sun, the moon, and Earth during solar and lunar eclipses. **National Content Standards: (5–8) UCP1–3, A1, B3, D3, G3**	**Find Out!**, p. 610 **Explore!**, p. 614 **Skillbuilder:** p. 612 **Investigate 19-2:** pp. 618–619 **A Closer Look**, pp. 612–613
19-3 Tides (2 sessions, .5 block)	**1. Describe** the changes you might observe along a coastline during high and low tides. **2. Diagram** the relative positions of the sun, the moon, and Earth during the highest tides. **National Content Standards: (5–8) UCP3, B2, D3, E1–2, F2, G1**	**Find Out!**, p. 623 **Life Science Connection**, pp. 622–623 **Science and Society**, p. 625 **Teens in Science**, p. 627

ACTIVITY MATERIALS

EXPLORE!

p. 601 atlas, small pizza box, sheet of paper, paste
p. 614 meterstick

INVESTIGATE!

pp. 618–619 pencil, polystyrene ball, globe, unshaded light source

DESIGN YOUR OWN INVESTIGATION

pp. 606–607 tape, black construction paper, gooseneck lamp with 75-watt bulb, Celsius thermometer, watch, protractor

FIND OUT!

p. 602 cardboard, markers or crayons, scissors, tape, basketball, metric ruler
p. 605* lamp, globe
p. 610* basketball, tape
p. 623* No special materials are required.

KEY TO TEACHING STRATEGIES

The following designations will help you decide which activities are appropriate for your students.

L1	Basic activities for all students
L2	Activities for average to above-average students
L3	Challenging activities for above-average students
LEP	Limited English Proficiency activities
COOP LEARN	Cooperative Learning activities for small group work
P	Student products that can be placed into a best-work portfolio
	Activities and resources recommended for block schedules

Need Materials? Call Science Kit (1-800-828-7777).

⏲ **OUT OF TIME?** We recommend that students do the activities with an asterisk.

Chapter 19 The Earth-Moon System

TEACHER CLASSROOM RESOURCES

Student Masters	Transparencies
Study Guide, p. 68 **Concept Mapping,** p. 27 **Multicultural Connections,** pp. 41-42 **Take Home Activities,** p. 29 **Activity Masters,** Design Your Own Investigation 19-1, pp. 79–80 **Science Discovery Activities,** 19-1, 19-2 **Laboratory Manual,** pp. 117–118, Earth's Spin	**Section Focus Transparency 62**
Study Guide, p. 69 **How It Works,** p. 20 **Making Connections: Integrating Sciences,** p. 41 **Making Connections: Across the Curriculum,** p. 41 **Activity Masters,** Investigate 19-2, pp. 81–82 **Critical Thinking/Problem Solving,** pp. 5, 27 **Laboratory Manual,** pp. 119–120, Moon Phases	**Teaching Transparency 37,** Moon Phases **Teaching Transparency 38,** Meteorite Impact **Section Focus Transparency 63**
Study Guide, p. 70 **Making Connections: Technology and Society,** p. 41 **Science Discovery Activities,** 19-3	**Section Focus Transparency 64**

ASSESSMENT RESOURCES	TEACHING & TECHNOLOGY
Review and Assessment, pp. 113-118 **Performance Assessment,** Ch. 19 **PASC*** **MindJogger Videoquiz** **Alternate Assessment in the Science Classroom** **Computer Test Bank**	**Spanish Resources** **Cooperative Learning Resource Guide** **Lab and Safety Skills** **Science Interactions, Course 1, CD-ROM** **Computer Competency Activities**

*Performance Assessment in the Science Classroom

NATIONAL GEOGRAPHIC TEACHER'S CORNER

Index to National Geographic Magazine	National Geographic Society Products Available From Glencoe	Additional National Geographic Society Products
The following articles may be used for research relating to this chapter: • "The Great Eclipse," by Roger H. Ressmeyer, May 1992.	To order the following products for use with this chapter, contact your local Glencoe sales representative or call Glencoe at 1-800-334-7344: • *STV: Earth, Moon, and Stars* (Videodisc)	To order the following products for use with this chapter, call the National Geographic Society at 1-800-368-2728: • *Raging Forces: Earth in Upheaval* (Book) • *Nature's Fury* (Video) • *Our Dynamic Earth* (Video) • *Volcano!* (Video)

Teacher Classroom Resources

These are key components of the classroom resources package.

TEACHING AIDS

Section Focus Transparencies

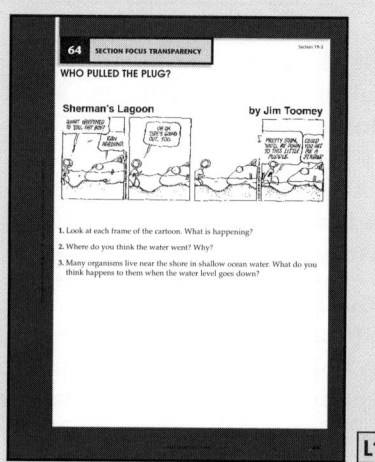

Teaching Transparencies

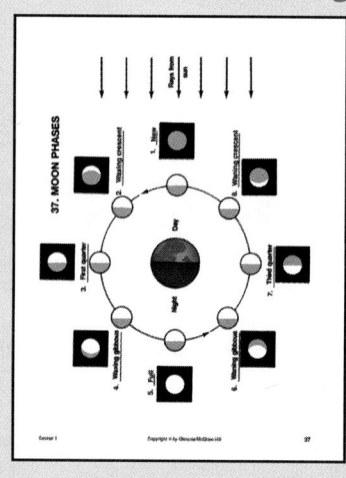

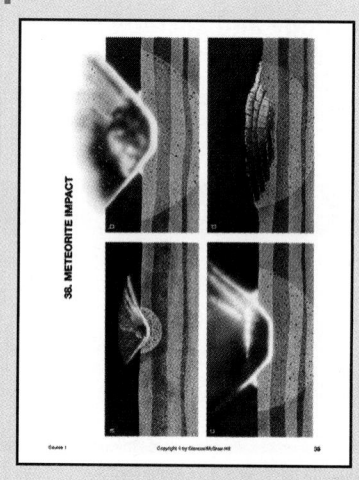

HANDS-ON LEARNING

Science Discovery Activity*

Laboratory Manual*

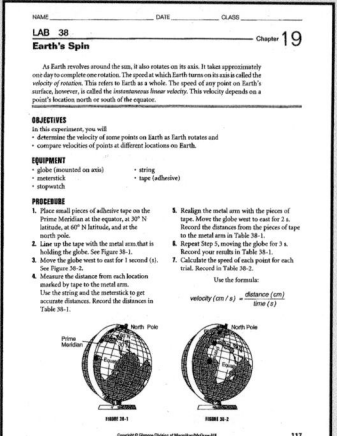

Take Home Activity

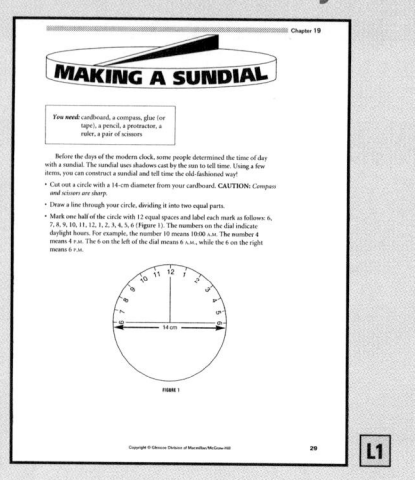

*There may be more than one activity for this chapter.

Chapter 19 The Earth-Moon System

Study Guide*

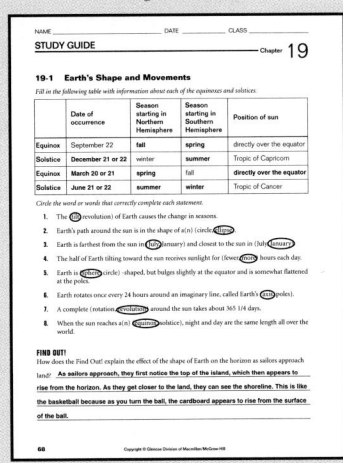

Concept Mapping

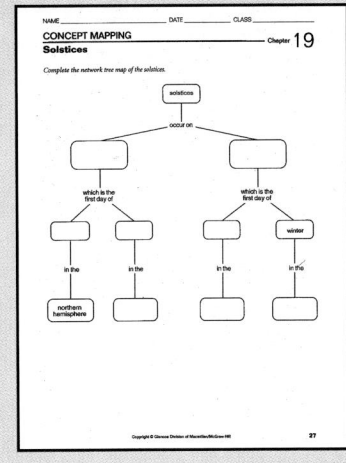

Critical Thinking/ Problem Solving

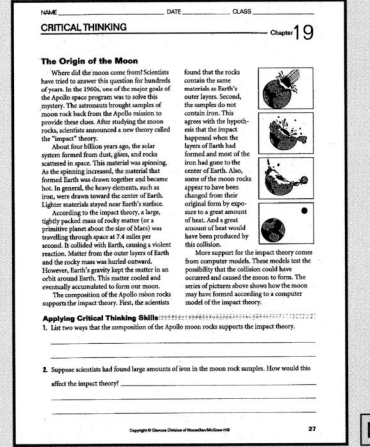

Integrating Sciences

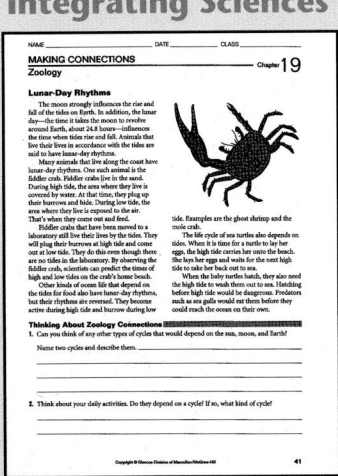

Across the Curriculum

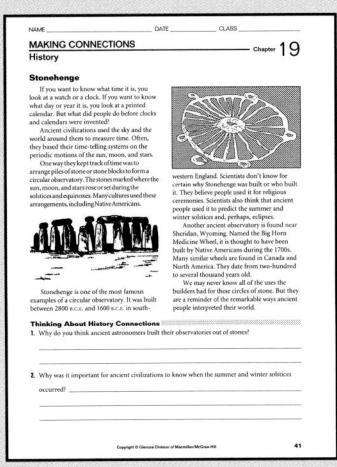

Technology and Society

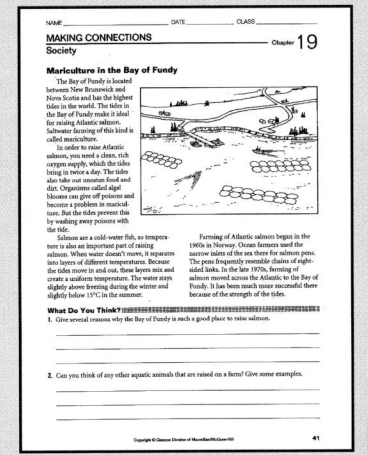

Multicultural Connection**

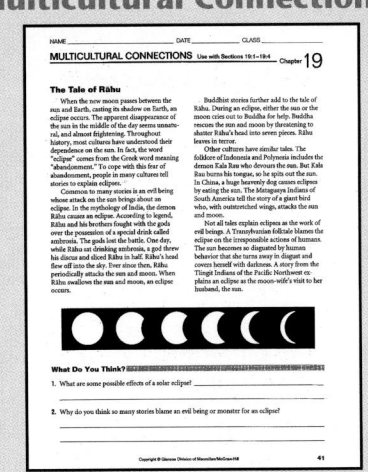

Performance Assessment

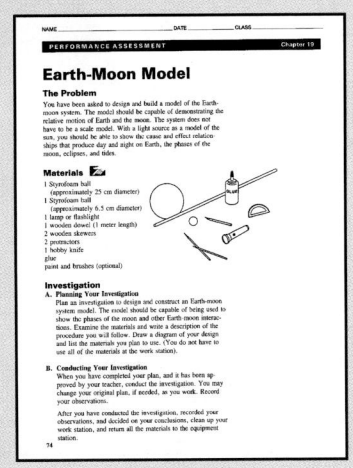

Review and Assessment

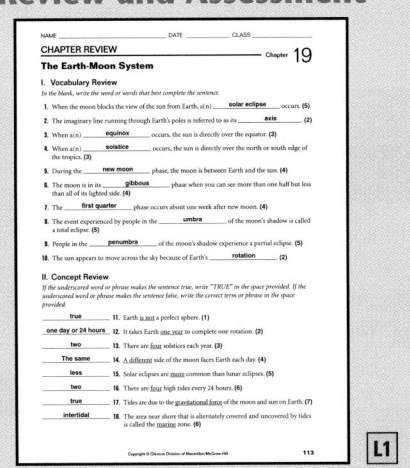

The Earth-Moon System

THEME DEVELOPMENT

One theme supported by this chapter is systems and interactions. Earth, the moon, and the sun interact within our solar system, causing day and night, seasons, moon phases, eclipses, and tides. Scale and structure is also evident. Students will learn how the size of Earth and its position relative to the sun and the moon affect eclipses and tides.

CHAPTER OVERVIEW

Students will make models to show that Earth is a sphere. They will relate the rotation of Earth on its axis and Earth's revolution around the sun to day and night and to the seasons.

Then, students will explore the motions of the moon. Students will demonstrate the position of Earth, the moon, and the sun during moon phases and eclipses.

Tying to Previous Knowledge

Place a large ball or globe in the center of the classroom. Shine a concentrated beam of light from a flashlight at the globe from one side. Ask students in different parts of the room to describe what they see. *Answers will vary depending on the position of the student. Students should realize that much of what we see depends on the viewer's position.*

The Earth-Moon System

Do you like to travel? Let's hope so, because you're on a journey right now—around the sun. You'll travel 940 million kilometers and never leave town. You'll complete your trip in one year without missing school. You'll need nothing special, but you'll take everything you own. Your vehicle? You'll be riding planet Earth, and the moon will be coming right along on this trip.

In this chapter you'll explore the relationship between Earth and the moon. Let's begin our trip. We'll be traveling through space at more than 107 000 kilometers per hour!

▶ **In the activity on the next page, explore the shape of the vehicle that's taking you on your journey through space.**

Did you ever wonder...

✓ **Why it's cooler in winter and warmer in summer?**
✓ **What the far side of the moon looks like?**
✓ **Why sometimes a beach is narrower than it was just a few hours earlier?**

📝 Science Journal

Before you begin to study about the moon and Earth, think about these questions and answer them *in your Science Journal*. When you finish the chapter, compare your journal write-up with what you have learned.

Answers are on page 628.

600

Learning Styles **LS**	Kinesthetic	Explore, pp. 601, 614; Find Out, pp. 602, 605, 610; Demonstration, p. 613; Visual Learning, p. 615; Investigate, pp. 618–619
	Visual-Spatial	Visual Learning, pp. 603, 604, 608, 611, 613, 616, 621, 624; Activity, pp. 608, 620, 621; A Closer Look, pp. 612–613
	Interpersonal	Activity, p. 611
	Logical-Mathematical	Activity, p. 604; Investigate, pp. 606–607; Across the Curriculum, pp. 608, 617; Discussion, p. 609; Find Out, p. 623; Project, p. 628
	Linguistic	Discussion, pp. 602, 608, 610, 612, 617, 621, 624; Multicultural Perspectives, p. 616; Activity, pp. 617, 620; Life Science Connection, pp. 622–623

Explore! ACTIVITY

What if Earth were shaped like a pizza box?

What To Do

1. Using an atlas, trace the continents on a sheet of paper.

2. Then, glue the sheet onto the top of a small pizza delivery box.

3. Fold the map around the edges of the box. What places are now at the "edge of the world"? *In your Journal*, describe how people's lives might be different if Earth were shaped like a box.

PREPARATION

Planning the Lesson

Refer to the Chapter Organizer on pages 600A-D.

Concepts Developed

This section discusses how day and night, the seasons, and temperatures are related to Earth's motions.

1 MOTIVATE

Bellringer

 Before presenting the lesson, display **Section Focus Transparency 62** on an overhead projector. Assign the accompanying **Focus Activity** worksheet. L1

LEP

Discussion Challenge students to speculate why a person might think Earth is flat. Lead students to discuss that Earth is so big relative to our viewpoint that it seems to stretch out in all directions. L1

Find Out!

How does Earth's shape affect what you see?

Time needed 15–20 minutes

Materials cardboard, markers or crayons, scissors, tape, basketball, metric ruler

Thinking Processes observing and inferring, recognizing cause and effect, sequencing, making models

Purpose To observe evidence that Earth's surface is curved.

Teaching the Activity

Troubleshooting Help students get their eye level correct to observe the cardboard.

Science Journal Have students record their observa-

Section Objectives

- Demonstrate evidence that shows Earth's shape.
- Describe the cause of day and night.
- Explain what causes the seasons on Earth.

Key Terms

sphere
rotation
revolution
equinox
solstice

Evidence of Earth's Shape

In the very first Explore activity in this book, you looked all about you and drew what you observed. How far away could you see? On a clear day, across open country, you can see about 32 kilometers. If you turn full circle, the world you can actually see is about 64 kilometers from edge to edge. If this was all you could observe, Earth might as well be a flat circle 64 kilometers in diameter. In the following Find Out activity, you will observe the type of evidence that suggested to many people that Earth was not flat but that it had a very different shape.

Find Out! ACTIVITY

How does Earth's shape affect what you see?

What To Do

1. Cut a strip of cardboard in the shape of a mountain about 8 cm tall, and decorate it. Fold about 2 cm of the mountain's base. Tape the 2-cm section to a basketball so that the remaining 6 cm are sticking straight up from the surface of the ball.

2. Now set the basketball on a table so that the mountain is sticking out horizontally, parallel to the table.

3. Kneel down on the other side of the ball so that you are eye level with it. Look at the top of the ball; think of this curve as a horizon.

4. Now, roll the ball toward you very slowly so that the stripes come into view over the top curve of the ball. Stop when you can see the entire length of the paper.

Conclude and Apply

1. What effect did the shape of the basketball have on your view of the approaching cardboard mountain?

---

tions in words or drawings. L1

Expected Outcomes

Students should observe that as the ball rolls toward them, the top of the cardboard mountain becomes visible before the rest of the strip comes into view.

Conclude and Apply

Because the surface of the ball is curved, only a small portion of its surface can be viewed at eye level at once. The cardboard was placed beyond the "horizon" so it was not visible until the ball was rolled and the horizon changed.

✔ Assessment

Performance Have students make a series of diagrams that show what they observed in this activity. Use the Performance Task Assessment List for Scientific Drawing in **PASC**, p. 55. L1
LEP

Ancient Greek scientists suggested that Earth was shaped like a ball. Sailors noticed that as they approached an island, it seemed to rise from the horizon.

Figure 19-1 shows what these sailors were seeing. They also noticed that as another ship approached, they would first see the top of its mast, then the sails, and finally the hull. This experience is similar to yours as you observed the strip of paper coming toward you with the roll of the basketball. The sequence of mast-sail-hull suggested to the sailors that the ship was approaching over a curved surface.

■ **Earth's Shape**

Earth is sphere-shaped. A **sphere** is a round, three-dimensional object whose surface at all points is the same distance from its center. You have probably played with a basketball, beachball, or volleyball. These balls are all spheres. In reality, however, Earth bulges slightly at the equator and is somewhat flattened at the poles, as shown in **Figure 19-2**. You could make this shape by sitting on a basketball.

Figure 19-1

Ⓐ As the ship approaches the island, the crew's first view of land is only the top of the island.

Ⓑ As the ship draws closer, the island appears to rise from the horizon. The closer the ship gets, the more island is visible.

Ⓒ Finally, the shoreline of the island is visible. The increasing visibility from the top down suggests that the ship was approaching the island over a curved surface.

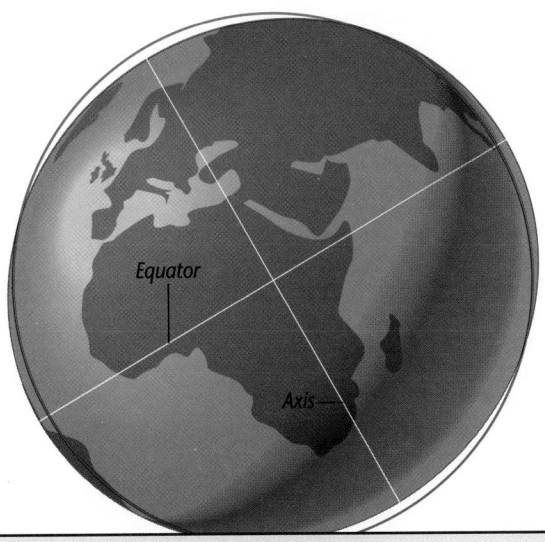

Equator

Axis

Figure 19-2

The red line in this diagram shows the shape of a perfect sphere. How does Earth's shape compare with the sphere?

19-1 Earth's Shape and Movements **603**

2 TEACH

Tying to Previous Knowledge

Review the concepts of motion along a curved path and motion of satellites presented in Chapters 12 and 13. Help students to apply their knowledge of this motion to explain the movement of Earth around the sun. Ask students to recall from Chapter 13 the motion of a satellite near Earth. Point out that Earth is a natural satellite of the sun. Earth is held in place around the sun by the sun's gravity and Earth's forward motion.

Visual Learning

Figure 19-1 Guide students to recognize how ancient sailors' observations led them to infer that Earth is shaped like a ball. As they study Figure 19-1, have students imagine that they are sailors of long ago seeing these views. Have two volunteers improvise a conversation about what they see. L1

Figure 19-2 How does Earth's shape compare with the sphere? *Earth bulges slightly at the equator and is somewhat flattened at the poles.*

Videodisc

STV: Solar System Inner Planets

Unit 2, Side 1
Earth

16530-18165

Earth

41684

Program Resources

Study Guide, p. 68
Laboratory Manual, pp. 117–118, Earth's Spin L2
Concept Mapping, p. 27, Solstices L1
Multicultural Connections p. 41, The Tale of Rāho L1
Take Home Activities, p. 29, Making a Sundial L1
Section Focus Transparency 62

Meeting Individual Needs

Visually Impaired To help students with the concept of an object appearing gradually over the horizon, use modeling clay to fasten an unsharpened pencil straight up from the surface of a basketball. Have students place their hands perpendicular to the surface of the ball. Turn the ball and slide the pencil up between their fingers. They will feel the pencil rising higher and higher as the ball turns.

Earth's Motions

Picture yourself on an average day. You wake up after sunrise. During breakfast and into the morning, you begin to notice the sun's movement across the sky. But is the sun really moving? Or are you?

■ Rotation

People have always talked about the sun moving across the sky. However, the change in its position is actually caused by the motion of Earth, not the sun. Earth spins in space. This spinning motion is called **rotation**.

As Earth rotates around its axis, the sun comes into view as a location begins to face the sun.

As Earth continues to rotate, the sun appears to move across the sky until it goes below Earth's horizon.

During the next several hours, you experience the growing darkness of night, and then the sunrise-sunset cycle begins again. In reality, a spot on

Earth spins toward the sun and away from the sun as seen in **Figure 19-3**.

■ Revolution

Earth is also in motion on a yearly trip around the sun. A complete **revolution**, or trip, takes about 365 1/4 days, or one year. It's during the course of one revolution that we experience the change of seasons.

Earth's revolution around the sun is in the shape of an ellipse, which looks somewhat like a flattened circle. The sun is a bit off-center of the ellipse, as you can see in **Figure 19-4**.

Is this elliptical path causing the changing temperatures and changing seasons on Earth? If it were, you would expect the warmest days to occur in January. But you know from experience that this isn't the case in the Northern Hemisphere. The following Find Out activity will show you the cause.

Figure 19-3

A Earth rotates around an imaginary line running through its poles. This line is called Earth's axis. It takes 24 hours for Earth to complete one rotation.

Sun

B You see evidence of Earth's rotation by observing the sun. As Earth rotates, the sun appears to rise from the horizon at dawn, move across the sky as the day progresses, and drop below the horizon at night.

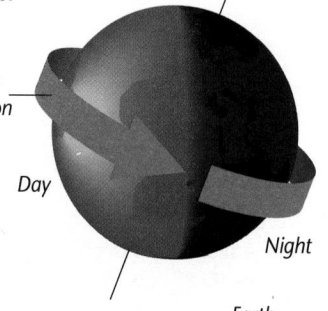

Direction of rotation

Day

Night

Earth

What causes the changing seasons?

What To Do

1. Use a lamp without a shade to represent the sun and use a globe to represent Earth. With the lamp on, hold the globe about 2 m away. Tilt the globe slightly so that the northern half points toward the lamp. Where on the globe is the light striking most directly?

2. Now, walk the globe around the lamp, keeping it tilted at the same angle and pointed in the same direction as when you started. What do you notice about the area receiving the most direct light?

Conclude and Apply

1. At what point in your walk around the lamp do you think winter would occur in the northern half of this globe?

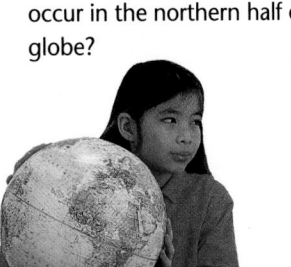

DID YOU KNOW?

A stationary object at Earth's equator is actually traveling about 1670 kilometers per hour as it spins with its location.

In the activity, you tilted the globe because Earth's axis is tilted at a 23 1/2° angle. You demonstrated how the amount of direct sunlight striking Earth varies from one hemisphere to the other because of this tilt. In the next Investigate, you'll explore more about how Earth's tilt causes the seasons.

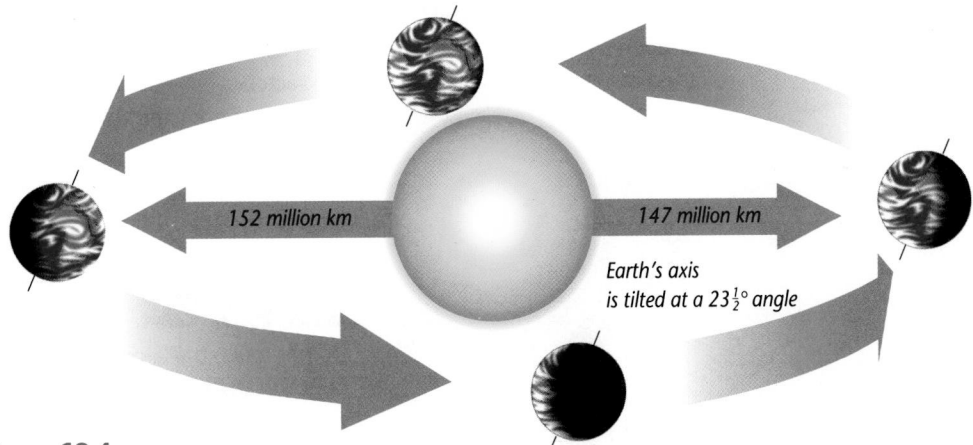

152 million km   147 million km

Earth's axis is tilted at a 23 1/2° angle

Figure 19-4

A Earth's revolution around the sun is in the shape of an ellipse, a closed curve that looks somewhat like a flattened circle.

B The sun is a bit off-center of the ellipse, therefore, the distance between Earth and the sun changes during Earth's year-long-journey.

C Earth travels closest to the sun—about 147 million km away—in January. Earth is farthest from the sun—about 152 million km away—in July.

What causes the changing seasons?

Time needed 10 to 15 minutes

Materials globe, lamp

Thinking Processes observing and inferring, recognizing cause and effect, making models, sequencing

Purpose To observe the effect of Earth's position with respect to the sun on the seasons.

Preparation Balls can be used instead of globes.

Teaching the Activity

Demonstration Ask students where summer is occurring when we are experiencing winter in the Northern Hemisphere. Ask a volunteer to use the globe and lamp to demonstrate why this reversal of seasons exists between the northern and southern halves of the world. L1

LEP

Inquiry Question If Earth's axis were straight up and down, how would the amount of sunlight received by the North Pole compare with the amount received by the South Pole? *Both poles would receive equal amounts of sunlight throughout the year.*

Content Background

Even though Earth's orbit is in the shape of an ellipse, this has virtually no effect on the seasons. Earth is actually closest to the sun (at its *perihelion*) during the winter in the Northern Hemisphere and is farthest from the sun during the summer (its *aphelion*). These differences in distance are not large enough to have an appreciable effect on the intensity of sunlight Earth receives. Also, Earth's elliptical orbit around the sun is almost circular.

Science Journal Have students draw their observations in their journals. L1

Expected Outcomes

Students should observe that the light strikes most directly on the part of the Northern Hemisphere facing the lamp when the northern half of the globe points to the lamp. The area tilted toward the lamp gets the most direct light as students walk the globe around the lamp.

Conclude and Apply

When the Northern Hemisphere is tilted away from the lamp, winter would occur.

✔ Assessment

Content Have students, working in groups of two or three, write a song that explains what causes the changing seasons on Earth. Use the Performance Task Assessment List for Song with Lyrics in **PASC**, p. 79. **COOP LEARN** L1 P

19-1 Tilt and Temperature

Preparation

Purpose To relate the angle at which light strikes a surface to the temperature of that surface.

Process Skills making and using tables, observing and inferring, recognizing cause and effect, measuring in SI, forming a hypothesis, separating and controlling variables, interpreting data, making models, predicting, sampling and estimating, making and using graphs

Time Required one class period

Materials black construction paper, protractor, Celsius thermometer, watch, tape, gooseneck lamp with 75-watt bulb

Possible Hypotheses Most students will hypothesize that the temperature of the surface area will increase more slowly when the light is angled. Some students may hypothesize that when light strikes a surface area from directly above, the area will receive more heat than when light strikes the same area at a glancing angle. A few students may go one step further and add that the increased heat produced when sunlight strikes a surface on Earth from directly above causes warmer seasons.

Plan the Experiment

Process Reinforcement Before students begin the activity, you may want to review reading a thermometer and/or using a protractor, as both skills will be required. Some students may need assistance in setting up their data tables correctly. Review the mechanics of making and using a table. Be sure students understand where to record their temperature readings.

Possible Procedures Fold the black construction paper in half, lengthwise. Tape the short edges together to form an envelope. Use this envelope to hold the thermometer. As an independent variable, set the lamp so that light shines directly down on the envelope containing the

Tilt and Temperature

Earth's tilt causes the amount of direct sunlight that strikes Earth to vary from one hemisphere to the other. How might this affect the amount of heat from the sun received by an area?

Preparation

Problem
How is the angle at which light strikes an area related to the amount of heat energy received by that area?

Form a Hypothesis
As a group, discuss the effects of light striking an area from several different angles. At what angle would the area receive the most heat? Agree upon a hypothesis that can be tested in your experiment.

Objectives
• Use a model to measure the amount of heat received by an area from light striking the area at different angles.
• Describe how the angle at which light strikes an area is related to Earth's changing seasons.

Possible Materials
black construction paper
protractor
Celsius thermometer
watch
tape
gooseneck lamp with 75-watt bulb

Safety

Do not touch the lamp. The light-bulb and shade can be hot even when the lamp has been turned off. Handle the thermometer carefully. If it breaks, do not touch anything. Inform your teacher immediately.

thermometer, and then change the angle to 45°. Also, select a specific time period for recording the temperature (for example, every 3 to 9 minutes).

DESIGN YOUR OWN
INVESTIGATION

Plan the Experiment

1 As a group, agree upon how you will use the materials provided to test your hypothesis.

2 Write down exactly what you will do during each step of your test.

3 Make a list of any special properties you expect to observe or test.

4 Identify any constants, variables, and controls in your experiment.

Check the Plan

1 How will you determine whether the length of time the light is turned on affects heat energy?

2 How will you determine whether the angle at which light strikes an area causes changes in heat and energy?

3 Make sure your teacher approves your experiment before you proceed.

3 Carry out your experiment. Record your observations.

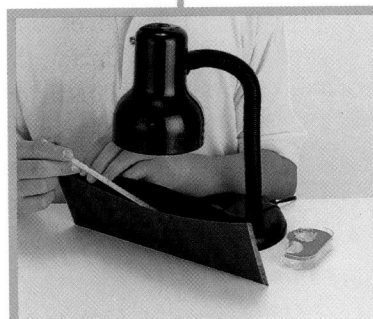

Analyze and Conclude

1. Observe Did the temperature in the envelope continue to rise at the same rate every three minutes?

2. Interpret Data How does the angle of light affect temperature? How might this be related to Earth's changing seasons?

3. Design an Experiment Did your experiment support your hypothesis? If not, determine how you might change the experiment in order to retest your hypothesis.

Going Further

Predict how the absorption of heat would be affected by changing your independent variables. Try your experiment with different values for your independent variables.

Program Resources

Activity Masters, pp. 79–80, Design Your Own Investigation 19–1
Multicultural Connections, p. 42, Dr. Chang-Diaz L1
Science Discovery Activities, 19–1, Whirling World; 19–2, Sun Clock

---

Troubleshooting Remind students to allow the thermometer to return to room temperature before each use.

Science Journal Direct students to write their hypotheses and record their results in their journals.

Expected Outcome

Students should realize that the surface area heats up more quickly when light hits it from a more direct angle. They should also realize that this causes the warmer temperatures of summer on Earth.

Analyze and Conclude

1. no
2. When light is angled, temperature increases more slowly. When sunlight strikes Earth's surface at an angle, temperatures are lower than when sunlight strikes Earth's surface from directly above. Thus, if a part of Earth is experiencing winter, sunlight is not striking that part of Earth at a direct angle.
3. Answers will vary depending on students' hypotheses.

✔ Assessment

Process Have students use their data to make a graph comparing the rate at which the envelope absorbed heat from direct light and from angled light.

Going Further ⫸

Students should realize that lowering the angle should cause less of a temperature rise, and increasing the angle to a greater amount should cause the temperature to rise more quickly.

Have students calculate how many times Earth spins on its axis in two trips around the sun. $2 \times 365.25 = 730.5$ L1

Daily Life

Discuss with students how the changing seasons and changing of day to night affect their lives. Include changing types of clothes, activities, holidays, and even foods.

Discussion Ask students to contrast the position of the sun and Earth at the time of the winter solstice and the summer solstice in the Northern Hemisphere. Then, ask them what is occurring in the Southern Hemisphere at these times. Finally, ask them to describe a place on Earth where people enjoy a summerlike climate all year.

Visual Learning

Figure 19-5 Does the North Pole point away from or toward the sun on this day [June 21 or 22]? *toward the sun* **Does the North Pole point away from or toward the sun on this day [December 21 or 22]?** *away from the sun* **Would people in the Southern Hemisphere also experience spring weather during March?** *No, they would experience fall weather in March.*

Figure 19-6 Use Figures 19-5 and 19-6 to help students conclude that seasons are reversed in the Northern and Southern Hemispheres. Then ask: **On this [June] day, which hemisphere is pointing toward the sun?** *the Northern Hemisphere* **Is Earth at its closest or farthest point from the sun?** *farthest*

Equinoxes and Solstices

You now know that the tilt of Earth as it revolves around the sun causes the change in seasons. Because of this tilt, the sun's position relative to Earth's equator changes, too. Most of the time, the sun's most direct rays fall north or south of the equator. Two times during the year, however, the sun is directly over the equator. Each of these times is called an **equinox**.

When the sun reaches an equinox, night and day are the same length all over the world. Neither the Northern nor the Southern Hemisphere is tilted toward the sun. **Figure 19-5** shows you how this can happen.

A solstice also occurs two times each year. At the time of a **solstice,** the sun is directly over the north or the south edge of the tropics. If you measured the number of daylight hours each day for one year, you would find that the day of the winter solstice has the fewest daylight hours. This fact

Figure 19-5

During summer solstice in the Northern Hemisphere, the sun's rays directly strike the Tropic of Cancer. During winter solstice, the sun is directly over the Tropic of Capricorn. During both fall and spring equinoxes, the sun is directly over the equator.

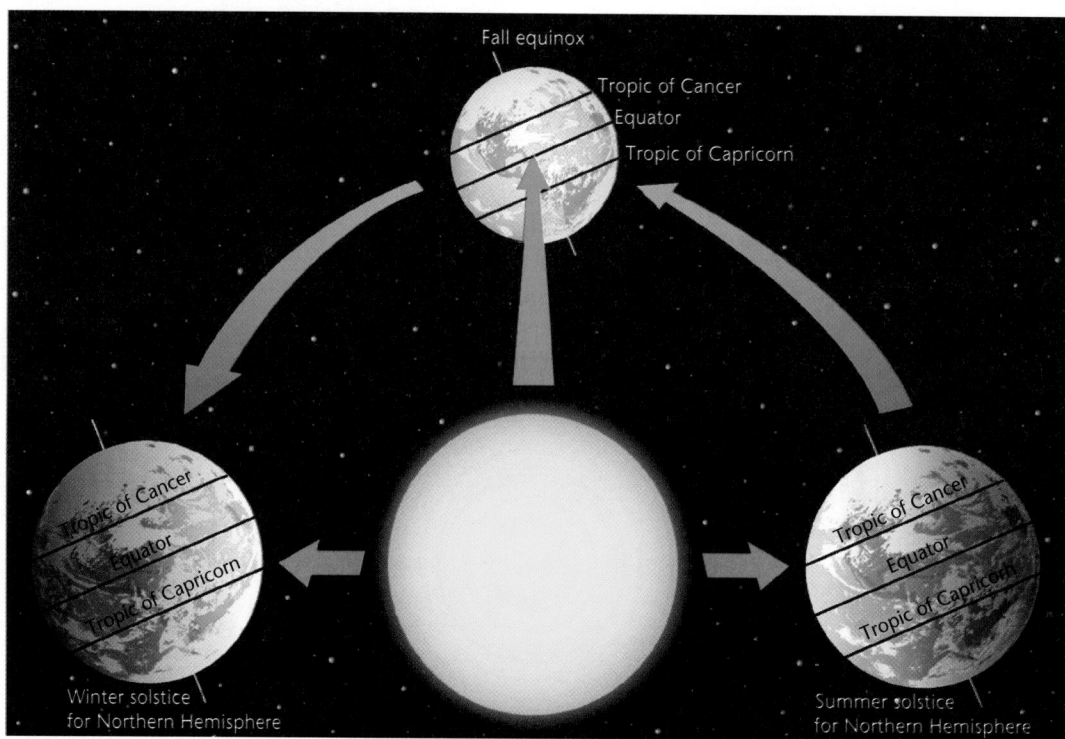

Meeting Individual Needs

Visually Impaired, Learning Disabled When discussing the tilt of Earth and its effect on seasons, hold a globe so that the northern axis is pointed toward an intense light source about 2 m away. Allow students to feel the surface of the globe. Students should be able to feel greater warmth where direct rays hit the globe than where indirect rays hit. Explain to them that this occurs during summer in the Northern Hemisphere. L1

ENRICHMENT

Activity Have groups of students conduct an ongoing experiment by placing a pole or some other object that will cast a good shadow out in the middle of a field. At different times throughout the school year, have students measure the length, direction, and angle of this shadow at the same time of day to see how the shadow changes according to the season. **COOP LEARN** L2 demonstrates what you've already learned: when sunlight is at a less direct angle for less time, the temperature will be lower than if the sunlight is direct and shines for a longer time. **Figure 19-6** shows the difference between hemispheres on a June day.

As you've seen, the motions of Earth affect you a great deal. The rotation of Earth causes day and night, and the revolution of Earth around the sun on a tilted axis is responsible for the changes in seasons. However, Earth is just one of the many bodies traveling around the sun. In the next section, you will learn how Earth's neighbor, the moon, is also moving through space. You will learn the effects of this movement and observe its consequences.

Figure 19-6

These two photographs—one of the Northern and one of the Southern Hemispheres—were both taken on the same day in June.

Ⓐ The Northern Hemisphere is receiving sunlight at its most direct or highest angle. The Southern Hemisphere is receiving sunlight at its most indirect or lowest angle.

The Umpqua National Forest in Oregon is located in the Northern Hemisphere.

Ⓑ On this day, which hemisphere is pointing toward the sun? Is Earth at its closest or farthest point from the sun?

New Zealand is located in the Southern Hemisphere.

check your UNDERSTANDING

1. What observable evidence do we have of Earth's shape?
2. There is not truly a sunrise or sunset in the usual sense of the terms rise and set. How can you explain the apparent movement of the sun across the sky?
3. Based on what you've learned about Earth's tilt and the effect of direct sunlight, explain why Chicago experiences seasons.
4. **Apply** Consider yourself manager of an Olympic ski team. The team needs year-round practice to be the best. Around the time of the summer solstice in the United States, to what region of the world would you take the team for practice? What factors influenced your decision?

Check for Understanding

Have students answer questions 1–3 in Check Your Understanding. Have a globe available for students as they complete the Apply question. Challenge students to plan a yearlong trip for the ski team.

Reteach

Demonstration Invite students to model the sun-Earth system. Use two students to represent the sun and Earth. Have the student representing the sun stand still. Have the other student walk around the "sun" while spinning. At various points, ask the class which side of the "Earth" is experiencing night and day. Then have "Earth" tilt toward the "sun" while spinning and walking. Ask the class to identify which part of "Earth" is experiencing summer and winter. **L1**

Extension

Activity Have students who have mastered the concepts of this section make a simple sundial for use outside throughout the school day. Have them place a pole or meterstick in the middle of the area where they will make their dial. At hourly intervals, have students use chalk or a stick to mark the shadow cast by the upright pole or meterstick. Then have them check the accuracy of their dial over several days. **L3**

4 CLOSE

Discussion

To demonstrate their understanding of rotation and revolution, ask students how the length of a day and year on a planet that turns slower but completes its orbit faster than Earth would compare to our day and year. *longer day; shorter year* Have students explain their reasoning. **L1**

check your UNDERSTANDING

1. the gradual appearance of objects over the horizon; pictures of Earth from space; we can sail around the world and not fall off
2. As Earth rotates, places on Earth pass in and out of the sun's light.
3. When the North Pole is pointed towards the sun, the city receives more direct rays of sunlight, more hours of daylight, and is warmer. When the North Pole is pointed away from the sun, Chicago receives less direct rays, fewer hours of daylight and is colder.
4. to the Southern Hemisphere where they are experiencing winter because the tilt of Earth results in opposite seasons

PREPARATION

Planning the Lesson

Refer to the Chapter Organizer on pages 600A–D.

Concepts Developed

In Section 19-1, students observed that Earth rotates and revolves around the sun. In this section, students will learn that the moon revolves around Earth as Earth revolves around the sun. The various phases of the moon are caused by the relative positions of Earth, the sun, and the moon.

1 MOTIVATE

Bellringer

Before presenting the lesson, display **Section Focus Transparency 63** on an overhead projector. Assign the accompanying **Focus Activity** worksheet. L1
LEP

Discussion Ask students if they have ever witnessed a solar or lunar eclipse. Have them describe the event or model it.

Find Out!

What are two ways in which the moon moves?

Time needed 10 to 15 minutes

Materials basketball, tape

Thinking Processes observing and inferring, making models

Purpose To observe two motions of the moon.

Preparation Any large ball or balloon can be used instead of a basketball.

Teaching the Activity

This activity can be done with small groups. COOP LEARN

Section Objectives
- Recognize that the moon's phases are caused by the relative positions of the sun, the moon, and Earth.
- Demonstrate the arrangement of the sun, the moon, and Earth during solar and lunar eclipses.

Key Terms
solar eclipse
lunar eclipse

Moon's Rotation and Revolution

Remember the satellites mentioned in Chapter 13? Humans have built and sent satellites into space, but Earth also has its own natural satellite—the moon. Do you also remember the moon journal you started in Chapter 1? You recorded how the moon's shape appears to change from day to day. Sometimes, just after sunset, you can see a bright, round moon low in the sky. At other times, only a small portion of moon is visible, and it is high in the sky at sunset. These phases of the moon are due to changes in the position of its lighted side in relation to Earth. How does the moon move? The following activity will show you.

Find Out! ACTIVITY

What are two ways in which the moon moves?

What To Do
1. Work with several classmates to perform this activity. Tape an **X** on a basketball. One person should stand facing another, holding the ball at about shoulder level with the **X** toward the other person.
2. Now, within sight of the rest of the group, the person with the ball should move in a circle around his or her partner while keeping the **X** toward the partner.
3. After the person with the ball has made one trip around, discuss whether the other group members were able to see all sides of the ball from where they were viewing. Record the group's observations *in your Journal.*

Conclude and Apply
1. How does this compare with what the partner was able to see as he or she followed the **X**?
2. How many revolutions did the ball make around the partner?

610 Chapter 19 The Earth-Moon System

Science Journal Have students sketch their observations in their journals. L1

Expected Outcome

Students will observe that we always see the same side of the moon.

Conclude and Apply

1. The partner saw the same side. The onlookers saw different sides.
2. one

✔ Assessment

Performance Have students create their own puppet show, play, or skit in which a character discovers why we see only one side of the moon. Use the Performance Task Assessment List for Skit in **PASC**, p. 75. L1 COOP LEARN

In your moon journal you started in Chapter 1, you may have observed that when the moon is visible, it always has the same features. That's because the same side of the moon always faces Earth. **Figure 19-7** shows how this happens. We refer to the side of the moon that we cannot see as the far side. **Figure 19-8** shows what part of the far side looks like.

■ The Moon's Appearance

But if the same side of the moon is always facing Earth, why does it change from a large, full disc to a small sliver in the sky? The appearance of the moon changes because we see different parts of its surface lighted by the sun. We see the lighted side from different angles because the

Figure 19-8

A Although oftentimes only part of the moon is visible from Earth, the same, full side of the moon always faces Earth.

B This photograph of the moon was taken from *Apollo 11*. The area to the right of the white line is part of the far side and cannot be seen from Earth.

C Over time, the far side has been exposed to meteor bombardment. Thus, its surface is covered with craters.

Figure 19-7

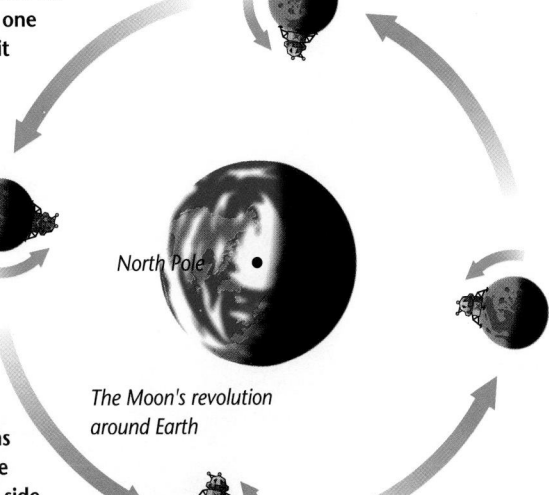

A While the moon revolves around Earth, it also slowly rotates on its axis. The moon completes one rotation in the same time it takes to complete one revolution.

Sun

North Pole

The Moon's revolution around Earth

B Because these two motions of the moon take the same amount of time, the same side of the moon always faces Earth.

Program Resources

Study Guide, p. 69
Laboratory Manual, p. 119–120, Moon Phases [L2]
Critical Thinking/Problem Solving, p. 27, The Origin of the Moon [L2]
Teaching Transparency 37 [L2]
Section Focus Transparency 63

---

2 TEACH

Tying to Previous Knowledge

In Chapter 1, students were introduced to the phases of the moon. Ask them to describe the pattern of these phases. In this section, students will learn that the moon's phases are dependent on the relative positions of the moon, the sun, and Earth. Before considering these three bodies as a system, review the relative positions of the sun and Earth.

Theme Connection The theme of systems and interactions becomes evident as the students perform the Investigate on pages 618–619 in which they will demonstrate how the interaction of the components of the sun-moon-Earth system cause the phases of the moon, as well as lunar and solar eclipses.

Visual Learning

Figure 19-7 Ask students to compare what a person might see of another person who is riding a merry-go-round as the observer runs alongside as the merry-go-round turns. Students should recognize that the observer will always see the same side of the rider. Point out that our view of the moon is similar to the view of the person running. Invite kinesthetic learners to mime the motions of the moon described in Figure 19-7. [L1]

Activity In 1959, the Soviet space probe *Luna 3* allowed people on Earth to see the far side of the moon for the first time. Have students interview their grandparents or other older adults about this event. They might also explore the heated scientific race between the United States and the former Soviet Union during this period. **COOP LEARN** [L2]

SKILLBUILDER

13 904 km; 1 390 400 km L1

Discussion Discuss with students the fact that the moon does not generate any light of its own. It is simply reflecting light emitted by the sun.

Inquiry Question How would our perception of the moon change if the moon completed one revolution around Earth in 24 hours? *Only one area on Earth would ever see the moon since it would be revolving at the same rate Earth rotates.*

Content Background

The actual period of the moon's revolution around Earth is 27.3 days. The time from one phase until the same phase occurs again is 29.5 days. This discrepancy of nearly two days is due to the fact that as the moon orbits Earth, Earth and the moon are also orbiting the sun. Although the moon may have made a complete orbit of Earth, it has not returned to its original position in relation to Earth and the sun. The complete return to original position takes almost two more days.

SKILLBUILDER

Measuring in SI

The moon's diameter is about 3476 kilometers. This measure is about a fourth of the diameter of Earth and about 400 times smaller than the diameter of the sun. Calculate the diameters of Earth and the sun. If you need help, refer to the **Skill Handbook** on page 661.

moon is revolving around Earth. The phase you see depends on the position of the moon in relation to both Earth and the sun.

Look at **Figure 19-9**. A new moon occurs when the moon is between Earth and the sun. During the *new moon* phase, the far side of the moon, the side facing away from Earth, is lighted. As the moon continues to revolve around Earth, part of the side facing Earth is lighted and becomes visible. Approximately

24 hours after a new moon, a thin slice of the side facing Earth is lighted. This phase is called a *waxing crescent*. About a week later, one half of the side facing Earth is lighted. This phase is called *first quarter*.

Over the next few days, more and more of the side of the moon facing Earth becomes lighted. When more than half, but less than all of the side facing Earth is lighted, the moon is in its *waxing gibbous* phase. *Full moon* occurs when the whole side facing Earth is lighted. After becoming full, less and less of the side facing Earth is lighted, the phases are waning, and the portion of the visible moon shrinks.

The Moon's Surface

Scientists studying the 380 kilograms of moon rock brought back by *Apollo* astronauts have concluded that the moon is about the same age as Earth—about 4.6 billion years old. For the first 1.5 billion years, the moon was struck by thousands of huge, rocky

This rock was collected from the moon on the Apollo 16 mission.

objects called meteorites. Also, during these early years, erupting volcanoes flooded the moon basins with lava.

In Chapter 1, you learned that the moon has large, flat areas called maria, which means seas. Although there is no water on the moon, observers in the 1600s thought they might have been seas. Scientists hypothesize that maria were formed during the lava flows mentioned earlier.

The low-lying maria have fewer craters than other areas. The indentations in the maria (visible through a telescope) were formed by the impact of interplanetary rocks since the

Purpose

A Closer Look describes the lunar surface features that are visible through a telescope or in pictures taken during space missions.

Content Background

Some of the mountain highlands on the moon reach 8000 meters above the surrounding plains. One large crater, called Copernicus, after the scientist Nicolaus Copernicus, is approximately 91 kilometers in diameter. Scientists have discovered that

the moon experiences seismic activity called moonquakes. Astronauts left intruments behind that measured this activity on the moon's surface and found that there are approximately 3000 moonquakes per year. From this moonquake data, scientists know that the moon's crust is thicker than the Earth's. It is about 60 kilometers thick on the side facing Earth and 100 kilometers thick on the far side.

Figure 19-9

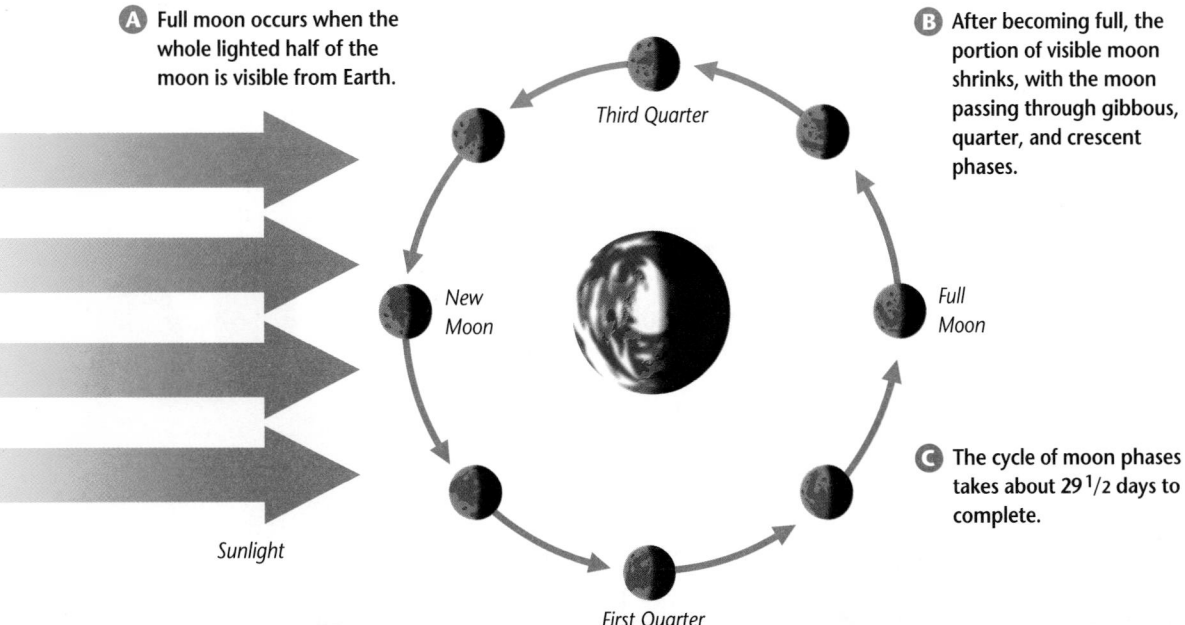

A Full moon occurs when the whole lighted half of the moon is visible from Earth.

Third Quarter

New Moon

Sunlight

First Quarter

B After becoming full, the portion of visible moon shrinks, with the moon passing through gibbous, quarter, and crescent phases.

Full Moon

C The cycle of moon phases takes about 29 1/2 days to complete.

Demonstration Use this activity to model the various phases of the moon.

You will need one overhead projector in the front of the room aimed toward the back at an angle above the students' heads. Have students cluster at the center of the room. Tell the students that the projector represents the sun and that they are on Earth. Holding a ball so that the light shines on it, revolve around the students, telling them that the ball represents the moon. Be sure to hold the same side of the ball facing them. Have students identify each phase as you revolve around them.

---

time of the lava flows. The largest of the moon's maria is called Mare Imbrium. Some of the other moon features have been named lacus (lake) and palus (marsh), while edge inlets are called sinus (bays), all despite the fact that—once again—there is no liquid water on the moon.

Other regions, called highlands, are covered with craters. Some of these craters, measuring hundreds of kilometers across, have central peaks. Explosive bombardment of meteorites formed the round craters. Their peaks, too, were a result of these crashes.

The moon has mountain ranges and smaller rows or peaks known simply as ridges.

Some valleys, called rilles, curve and wind great distances across the moon's surface.

You Try It!

Make a model of a portion of the moon's surface, using photographs from a book on the *Apollo* missions. Write a description of your creation, explaining how the formations occurred.

Figure 19-9 Students can use the figure to trace the cycle of moon phases. Then have students draw and label the phases of the moon in their journals: new moon, waxing crescent, first quarter, waxing gibbous, full moon, waning gibbous, third quarter, and waning crescent. **L2**

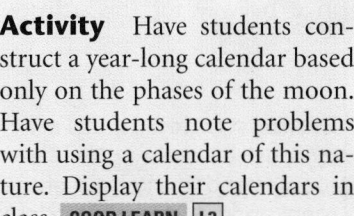

Activity Have students construct a year-long calendar based only on the phases of the moon. Have students note problems with using a calendar of this nature. Display their calendars in class. **COOP LEARN** **L3**

Teaching Strategy

Ask students to compare and contrast Earth and the moon. Discuss that the moon is a dry, airless, barren place. Noon-day temperatures may be in excess of 127°C. At night the temperature can drop to -173°C. The moon has no atmosphere, so there are no weather conditions. Ask students what type of erosion likely occurs that could change its surface features.

Going Further ⫸

Have students work in small groups to simulate the formation of craters by filling a shoebox one-third full with a very thick mixture of plaster of paris. Have them drop small rocks of different sizes from different heights into the hardening plaster. Have students record the diameter of each rock and the height from which it is dropped. Remove the rocks before plaster craters harden. Explain that when rocky objects from space impact the moon's surface, the objects usually break apart explosively and are ejected from the craters they form. When hardened, have students measure the diameter of the craters. Then have them look for any relationships between the size of the rock, the height it is dropped from, and the size of the crater. Students could then make a science fair display on how craters are formed. Use the Performance Task Assessment List for Science Fair Display in **PASC**, p. 53. **L1**

COOP LEARN **P**

Explore!

How is your perception of an object's size affected by its distance from you?

Time needed 5 to 10 minutes

Materials meterstick

Thinking Processes comparing and contrasting, making models, observing and inferring

Purpose To observe how small, close objects can appear larger than large, distant objects.

Preparation Student pairs should be taken outside or out in a hallway for this activity.

COOP LEARN

Teaching the Activity

Demonstration If this activity must be done in the classroom, stand in the front of the room and ask students to close one eye and line up their thumbs so that they block out your head. Then direct students to bring their thumbs closer to their eyes to block out your entire body. Discuss why this is possible. L1

Science Journal Have students record their answers to the questions in their journals. L1

Expected Outcome

Students should see that a small, close object can block their view of a larger object that is farther away.

Eclipses

What do you think can happen when one member of the sun-moon-Earth system moves between the other two? Your view of this occurrence is affected by the size and distance of the sun and moon.

Explore! ACTIVITY

How is your perception of an object's size affected by its distance from you?

What To Do

1. With a partner, choose an object in the distance that you know is larger than you are, for example, a car or a house.

2. Have your partner stand directly between you and this object. Stand far enough apart so that your view of the object is partially blocked. Describe *in your Journal* what you can see.

3. Next, ask your partner to move closer. Can you still see the object?

4. Finally, ask your partner to stand as close as 0.5 m from you. Now what part of the object can you still see?

In this simple demonstration, you observed how a small, close object can appear to be as big or bigger than an object that is far away and known to be much bigger. This same principle causes the sun and the moon to appear equal in size. Even though the sun's diameter is 400 times greater than the moon's, the moon is about 400 times closer to Earth than the sun. Therefore, the sun and the moon appear to us to be about the same size.

The Explore activity also showed how a smaller object can totally block your view of a larger object. When the moon blocks our view of the sun, as your partner blocked your view of a distant object, we call the event a solar eclipse.

Answers to Questions

3. Answers may vary. They may still see part of the object.

4. no part

✓ Assessment

Process Have students use their observations to infer why we perceive the sun and the moon as equal in size. Pairs of students could then make scientific drawings to explain this phenomenon. Use the Performance Task Assessment List for Scientific Drawing in **PASC**, p. 55. L1 P

■ Solar Eclipses

Around 600 B.C.E., warriors from ancient Media, in what is now northern Iran, launched an early-morning attack on the neighboring country of Lydia, in present-day Turkey. After several hours of battle, the clear, blue sky darkened, and all color seemed to drain from the landscape. The air became cool, and within minutes the day was as black as night. The planets and stars were visible. The soldiers were stunned. But after several minutes, the stars began to fade, daylight replaced darkness, and the sun reappeared. The soldiers were so frightened that they dropped their weapons and fled the battlefield. They were certain that the end of the world was near.

The soldiers had experienced a solar eclipse. You can see what a total solar eclipse looks like in **Figure 19-10.** It is a rare but natural event that leaves observers standing in the shadow of the moon. Recall from Chapter 2 how shadows are formed. A **solar eclipse** occurs when the moon passes directly between Earth and the sun. The moon blocks out some of the sun's light, casting a shadow on Earth.

Look at **Figure 19-11** to help you understand how a solar eclipse occurs. In areas of total solar eclipse, the only portion of the sun that is visible is part of its atmosphere. This appears as a white glow around the edge of the eclipsing moon.

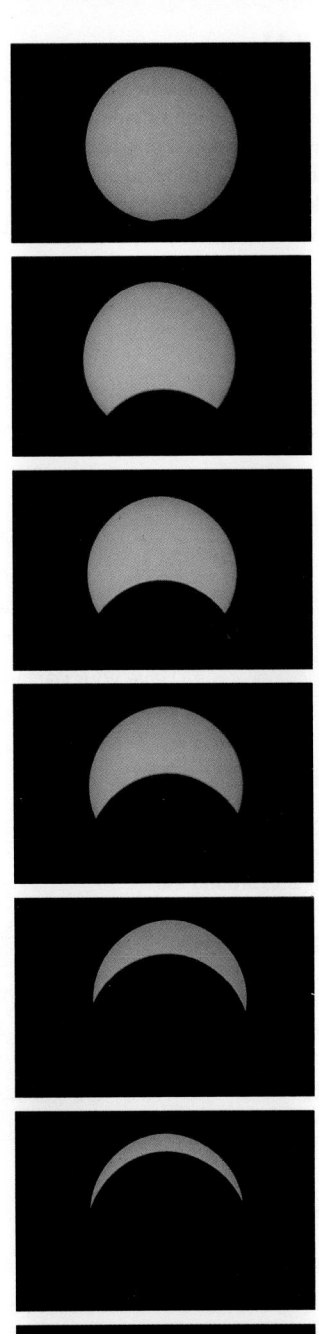

Figure 19-10

During a solar eclipse, the moon passes directly between Earth and the sun.

Figure 19-11

A A solar eclipse occurs when Earth passes into the moon's shadow. The moon's shadow is only about 270 km to 300 km wide, so only people standing in the shadow's path see the eclipse.

B The moon casts a dark inner shadow called the umbra. The umbra is the center of the moon's shadow where the sun's light is completely blocked from view on Earth.

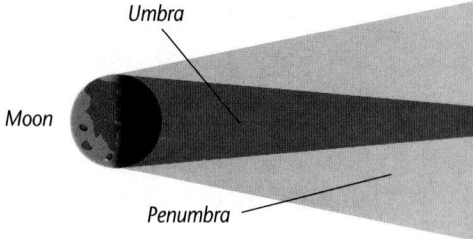

Umbra

Moon

Penumbra

C In the outer part of the moon's shadow, called the penumbra, sunlight is only partially blocked.

Sun

Figure 19-12

Jumbo, a forty-foot solar telescope, was hauled to various sites around the world to view solar eclipses. Jumbo recorded almost every total solar eclipse worldwide between 1892 and 1931. Here, astronomers get Jumbo ready for a total eclipse of the sun in Jeur, India, on January 22, 1898.

■ Earth and Moon Shadows

Because the sun is so large and its rays spread out, the moon's and Earth's shadows have two parts. They cast a dark, cone-shaped shadow called the *umbra* inside a larger, lighter shadow called the *penumbra*. A total eclipse is only experienced by people in the narrow path of the umbra. You can see the two different shadows in **Figure 19-11**.

A partial eclipse occurs on Earth when the moon covers only part of the sun. This is what observers within the penumbra see. As the moon slides in front of the sun, it may look as though a sliver has been cut out of the edge of the sun.

616 Chapter 19 The Earth-Moon System

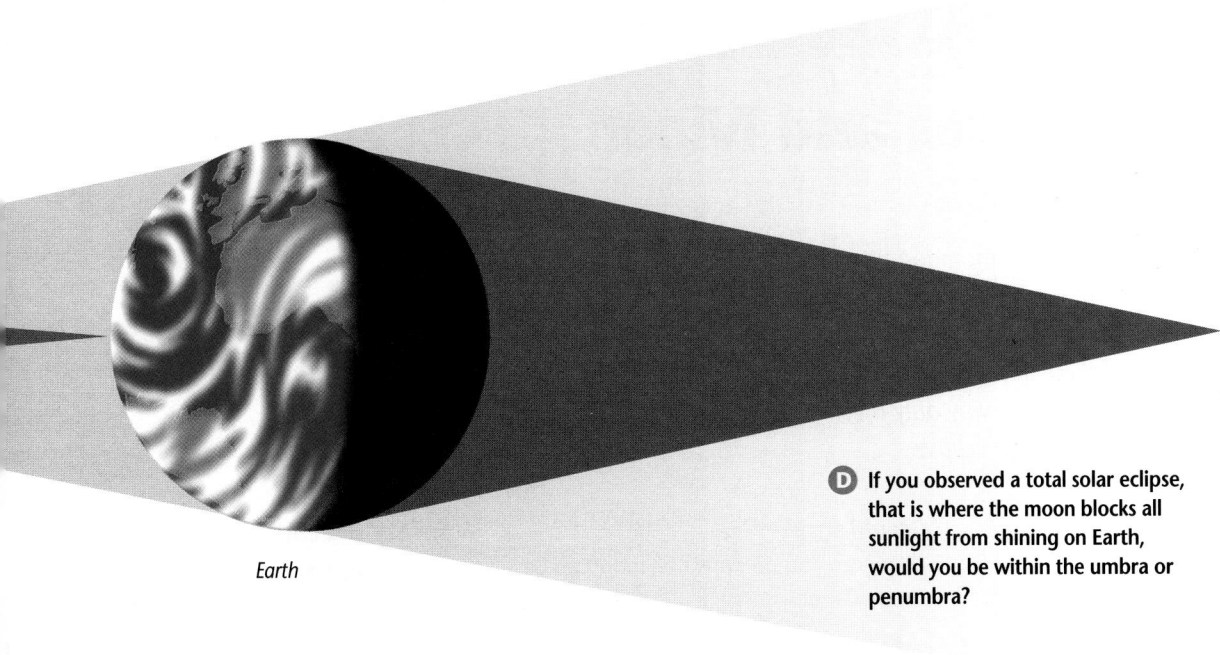

Earth

D If you observed a total solar eclipse, that is where the moon blocks all sunlight from shining on Earth, would you be within the umbra or penumbra?

Across the Curriculum

Math

Ask students to determine when the next full moon will occur. Give them the date of the last full moon. Students should recall that it takes 29.5 days for the moon to complete one cycle of its phases. **L1**

Content Background

The moon revolves around Earth in an elliptical path similar to the way Earth revolves around the sun. The moon appears to move west across the sky. Actually, the moon is moving east, but because of the rotation of Earth, it appears to be moving west.

■ A Rare Event

Although a rare and dramatic event, a solar eclipse can be dangerous to careless observers. You should never look directly at the sun, particularly during a solar eclipse. The sun's radiation can damage your eyes and cause blindness. A solar eclipse should be viewed indirectly by projecting the image of the eclipse onto a sheet of white paper. If you can't witness the event firsthand, you now have the option of live and taped video coverage of such events.

Solar eclipses do not happen every time the moon travels around Earth, however. This is because the path of the moon's revolution is tilted at about a 5° angle to Earth's path around the sun. Because of this differ-ence, the moon usually passes above or below the sun and not directly in front of it.

While between two and seven solar eclipses occur every year, they can be seen in only a few areas on Earth at any one time. A total solar eclipse can be seen only once every 450 years from any one location. Often, total solar eclipses occur in remote regions such as Siberia and the middle of the Atlantic Ocean. So unless you are able to travel, and depending on the weather once you get there, your chances of seeing a total eclipse are not very good.

If you do the following Investigate, however, you will see exactly how solar eclipses—as well as moon phases—occur.

ENRICHMENT

Activity Have students research the meaning and origin of the phrase "once in a blue moon." Have them present their information to the class. **COOP LEARN** **L2**

19-2 Eclipses and Moon Phases

TECH PREP

Planning the Activity

Time needed 20 to 25 minutes

Purpose To demonstrate the position of the sun, the moon, and Earth during moon phases and eclipses.

Process Skills observing and inferring, interpreting data, making and using tables, making models, comparing and contrasting, recognizing cause and effect, hypothesizing

Materials pencil, polystyrene ball, globe, unshaded light source

Preparation Overhead lights should be off while students are conducting this investigation.

Teaching the Activity

Troubleshooting Students may have trouble lining up the moon, the sun, and Earth to produce the phases as seen from Earth. Ask students to pretend that they are standing on Earth looking out or up at the moon.

Process Reinforcement Make sure that students set up their models so they can make accurate observations. You might suggest that students refer to Figures 19-9 and 19-11 to check their setups in Steps 4 and 5.

Science Journal Have students complete their data tables and answer the questions in their journals. L1

INVESTIGATE!

Eclipses and Moon Phases

You know that moon phases and solar eclipses result from the relative positions of the sun, the moon, and Earth. In this activity, you will demonstrate the positions of these bodies during certain phases and eclipses. You will also see why only a very small portion of Earth sees a total solar eclipse.

Problem

How can you demonstrate moon phases and solar eclipses?

Materials

pencil   polystyrene ball
unshaded light   globe
   source

Safety Precautions

Be careful, the exposed bulb will be hot.

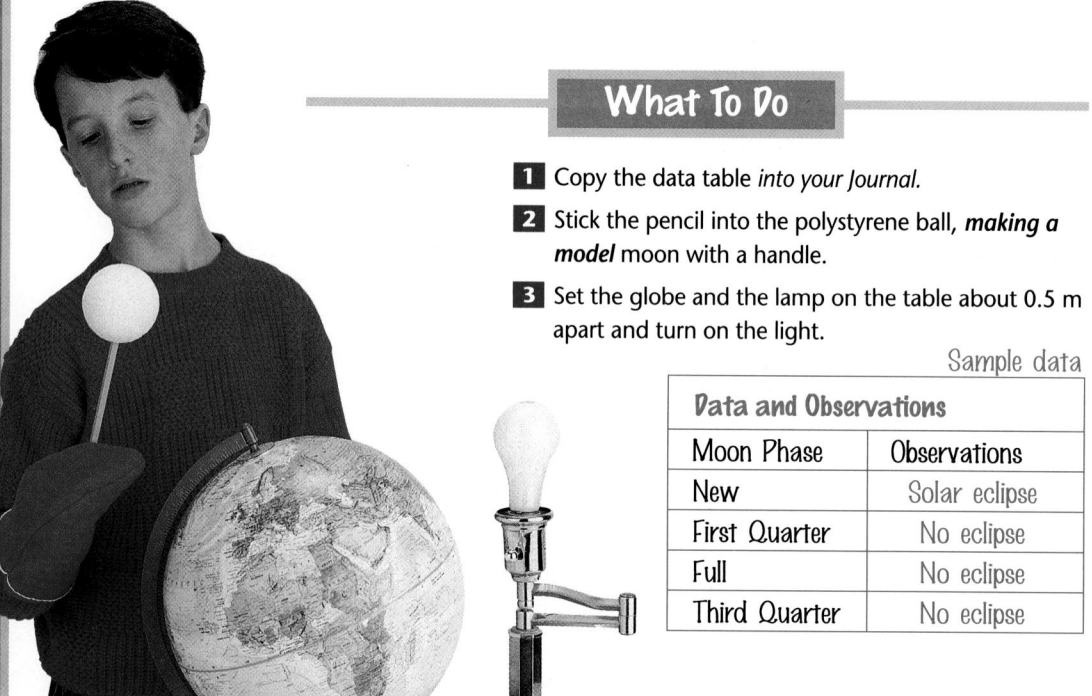

What To Do

1 Copy the data table *into your Journal.*

2 Stick the pencil into the polystyrene ball, *making a model* moon with a handle.

3 Set the globe and the lamp on the table about 0.5 m apart and turn on the light.

Sample data

Data and Observations	
Moon Phase	Observations
New	Solar eclipse
First Quarter	No eclipse
Full	No eclipse
Third Quarter	No eclipse

A

B

4 Holding the model moon by its pencil handle, move it around the globe to duplicate the position that will cause a solar eclipse. Record your observations *in your Journal.*

5 Use this sun-Earth-moon model to duplicate the phases of the moon. During which phase(s) of the moon could a solar eclipse occur? How can you use the model to observe the umbra and penumbra of the moon?

Analyzing

1. During which phase(s) of the moon is it possible for a solar eclipse to occur?

2. *Determine the effect* that a small change in the distance between Earth and the moon would have on the size of the shadow during an eclipse.

3. As seen from Earth, how does the apparent size of the moon *compare* with the apparent size of the sun? How can an eclipse be used to confirm this?

Concluding and Applying

4. Why doesn't a solar eclipse occur every month? Explain your answer.

5. Suppose you wanted to make a more accurate model of the movement of the moon around Earth. How might you adjust the distance between the light source and the globe? How would you adjust the size of the moon model in comparison with the globe you are using?

6. **Going Further** *Hypothesize* what would happen if the sun, the moon, and Earth were lined up with Earth directly in between the sun and the moon.

Program Resources

Activity Masters,   pages 81–82, Investigate 19-2
Making Connections: Integrating Sciences,   page 41, Lunar-Day Rhythms **L2**

Expected Outcomes

Students should be able to position their "moons" in order to produce a lunar and a solar eclipse as well as the four major phases of the moon.

Answers to Analyzing/ Concluding and Applying

1. new moon phase

2. The smaller the distance between Earth and the moon, the larger the shadow.

3. The apparent size of the moon is the same as the apparent size of the sun. This can be demonstrated during a solar eclipse because the moon appears to completely cover the sun.

4. because the orbit of the moon is not in the same plane as Earth's orbit around the sun

5. Move the light source farther away. Because the distance between Earth and the moon in your model is small, a smaller ball could be used to produce a more accurate shadow on Earth.

6. Earth would cast a shadow on the moon.

✔ Assessment

Performance Have small groups of students discuss their answers to the questions. Encourage students to explain how the observations of their models helped them. Students could then make their own concept maps showing solar and lunar eclipses and the relation of a solar eclipse to the phases of the moon. Use the Performance Task Assessment List for Concept Map in **PASC,** p. 89. **P** **L1**

3 ASSESS

Check for Understanding

1. Set up a light source, globe, and polystyrene ball to represent the moon, the sun, and Earth. Model different phases of the moon, a solar eclipse, and a lunar eclipse. Ask students to identify what phase or eclipse is being modeled. `LEP` `L1`

2. Have students verify their answer to the Apply question using the lamp-globe-ball model of the sun-Earth-moon system.

Reteach

Activity Use Figure 19-9 to review the phases of the moon. Then display calendars that indicate the phases. Have students find the next phase of the moon that will occur after the present date. Then have them locate this phase in Figure 19-9. Continue until students have identified the dates for all of the phases. `L1`

Extension

Activity Have students research the dates of total solar eclipses that will occur within the next five years and find out where they can best be viewed. `L3`

4 CLOSE

Activity

Have students describe in writing the positions of Earth, the moon, and the sun during a lunar and solar eclipse as well as their positions during full and new moon phases as seen from Earth. `L1`

Figure 19-13

A During a lunar eclipse, the moon gradually becomes darker as it moves into Earth's shadow.

B A lunar eclipse can last up to one hour and forty-four minutes. During this time, the moon may appear reddish. Unless clouds hide the view, most people on the nighttime side of Earth can see the lunar eclipse.

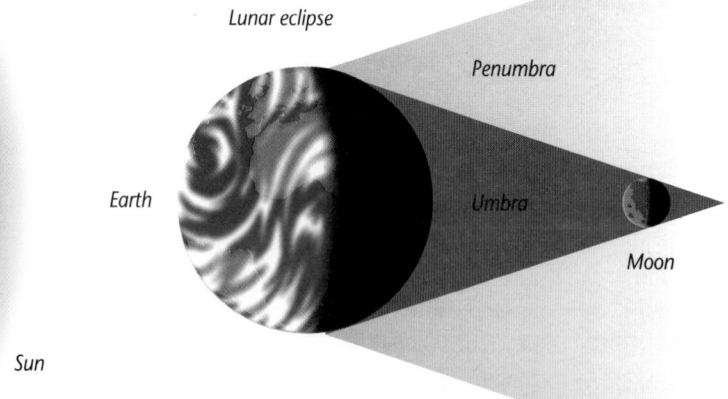

Lunar eclipse

Penumbra

Earth

Umbra

Sun

Moon

C A total lunar eclipse occurs when the moon passes completely into Earth's umbra.

■ Lunar Eclipses

In the Investigate, did you conclude that a shadow can also be cast on the moon? When this happens, a lunar eclipse occurs.

Earth casts a shadow on the side of the moon facing the sun. Once every 29 days, in its revolution around Earth, the moon moves near this shadow. When the moon does pass through Earth's shadow, we see a **lunar eclipse**. At this time, Earth is directly between the sun and the moon. **Figure 19-13** will help you understand how a lunar eclipse occurs.

In this section, you learned that moon phases and eclipses are a result of the way Earth, the sun, and the moon line up. In the next section, you'll discover one more interesting effect their positions can produce.

check your UNDERSTANDING

1. Draw the relative positions of the sun, the moon, and Earth during a full moon phase.

2. Which type of eclipse can occur during a full moon?

3. Apply Why does only a small percentage of Earth's population witness a solar eclipse, while people on the entire nighttime side of Earth can see a lunar eclipse?

---

check your UNDERSTANDING

1. Drawings should show the moon on the opposite side of Earth from the sun.

2. lunar eclipse

3. The moon is much smaller than Earth so it casts a small shadow on Earth during a solar eclipse. Only people located within this shadow observe the total solar eclipse. Because Earth's shadow is larger than the moon, the entire moon is cast in shadow during a lunar eclipse. This event is visible anywhere the full moon can normally be seen—the entire night side of Earth.

19-3 | Tides

Earth, Moon, and Ocean

Have you ever been to an ocean beach? Many people unfamiliar with the ocean will place their towels and sandals at what seems a safe distance from the water and then return to find them floating away. What happened? The lifeguard on duty will tell you that the tide came in.

Do you remember what you learned about waves and wavelengths in Chapter 17? **Tides** are slow-moving water waves with long wavelengths. They produce an alternate rise and fall of the surface level of the oceans. **Figure 19-14** shows the difference in water levels during high and low tides.

Section Objectives

- Describe the changes you might observe along a coastline during high and low tides.
- Diagram the relative positions of the sun, the moon, and Earth during the highest tides.

Key Terms

tides

Figure 19-14

A At high tide, the surface level of the ocean is at its highest point. Once this high point is reached, the water level begins to drop. This photograph of the shoreline at Cape Ann, Massachusetts, was taken during high tide.

B Over a period of several hours, the water recedes, and more and more land is exposed. This photograph was taken at low tide, which is when the surface of the ocean reaches its lowest level.

19-3   Tides   **621**

PREPARATION

Planning the Lesson

Refer to the Chapter Organizer on pages 600A–D.

Concepts Developed

In this section, students will learn that the position of the moon also affects the tides that occur in Earth's oceans. Earth and the moon exert a gravitational pull on each other. It is this gravitational attraction that causes oceans to bulge away from Earth and cause the tides.

1 MOTIVATE

Bellringer

Before presenting the lesson, display **Section Focus Transparency 64** on an overhead projector. Assign the accompanying **Focus Activity** worksheet. [L1] [LEP]

Discussion Ask students if they have ever been to the seashore. Ask them to visualize and then describe how the shoreline changed over a period of hours. If none of the students has ever been to the seashore, ask if they think an ocean shoreline remains the same throughout the day. [L1]

Visual Learning

Figure 19-14 To compare and contrast high tide and low tide, have students use the photographs. *More land is exposed at low tide, and the water at the shoreline is more shallow then.* If possible, have volunteers look in local newspapers for the times of high tides. Students can collect the data daily and record it in a data table for analysis after a few weeks. [L1]

ENRICHMENT

Activity To practice formulating models, have students make mobiles of the sun, the moon, and Earth representing the various phases of the moon. Polystyrene balls of different sizes, coat hangers, and string can be used to accomplish this. Parts of the polystyrene balls could be darkened to help represent the phases as seen from Earth. [L2]

Program Resources

Study Guide, p. 70
Making Connections: Technology and Society, p. 41, Mariculture in the Bay of Fundy [L2]
Science Discovery Activities, 19–3, Survivors
Section Focus Transparency 64

19-3   Tides   **621**

2 TEACH

Tying to Previous Knowledge

Discuss waves and wavelengths from Chapter 17 to remind students how waves form and travel. Emphasize that tides are actually slow-moving waves with long wavelengths. The smaller waves breaking on a shoreline are usually the result of wind action and have shorter wavelengths. Have students recall the relative positions and motion of Earth, the sun, and the moon. Remind them that the moon is exerting a measure of gravitational force on Earth. This force is evident through our tides as the moon literally pulls the water slightly away from Earth's surface.

Theme Connection By studying a table of moon phases and high and low tides for a month, students will see how the interaction of the sun-moon-Earth system affects the tides.

Visual Learning

Figure 19-15 Have students make scientific drawings in their journals. Drawings should indicate the role of gravitational forces in the sun-Earth-moon system in creating high and low tides. L2 P

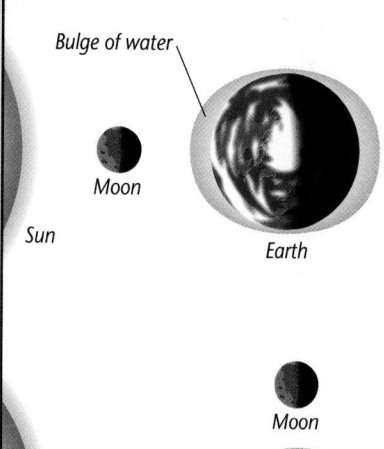
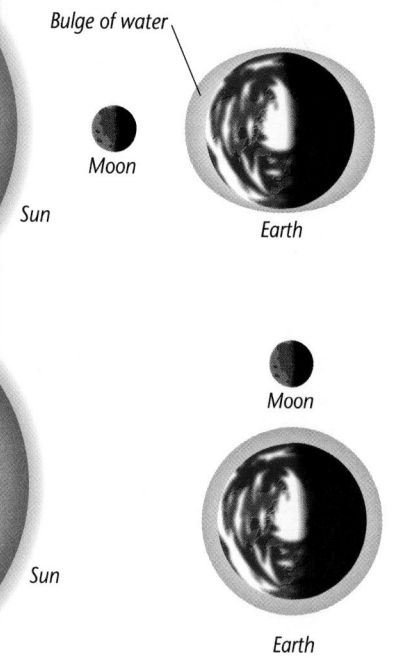

Bulge of water

Moon

Sun

Earth

Moon

Sun

Earth

Figure 19-15

The sun, the moon, and Earth have gravitational forces acting between them. These gravitational forces actually pull on Earth's oceans.

A When the sun and the moon line up with Earth, the high tides are very high and the low tides are very low.

B When the sun and the moon are not lined up with Earth, they pull in different directions, and the tides are not as high and not as low.

When the tide comes in, the water level rises, and the waves break farther and farther inland. At high tide, the surface level of the ocean is at its highest point. At low tide, the surface of the ocean has reached its lowest level.

High tides occur about every 12 1/2 hours. A low tide occurs about 6 1/4 hours after every high tide. In other words, there are two high tides and two low tides every day. High and low tides occur about six hours apart along much of the East Coast of the United States.

You may be asking yourself, "What do tides have to do with this chapter?" **Figure 19-15** and the next activity will show you.

Life Science CONNECTION

Fiddler crab

Life in an Intertidal Zone

As you learned in the chapter, the gravitational pull of the moon causes tides in the oceans of Earth. You know that tides can affect the daily activities of people who fish and other people who rely on the sea. But did you also know that tides affect the activities of some of the many organisms that inhabit Earth's oceans?

If you were on a boat in the middle of the ocean, you would not be able to see the effects of tides. However, if you were on shore, you would see the changes in water level due to tides. The area near shore that is alternately covered and uncovered by tides is known as

the intertidal zone, and it is home to a variety of organisms.

The continuously changing environment of the intertidal zone presents quite a challenge to the many organisms that live there. One of the toughest tasks for the many tiny animals living in intertidal zones is dealing with the intermittent rush of water and waves. Many of these animals, such as barnacles and mussels, survive by attaching themselves to large rocks. Other animals, such as crabs, lugworms, and shelled mollusks, avoid crashing waves by burrowing into the ground.

Another challenge in an intertidal zone is the dry

Life Science CONNECTION

Purpose

This Life Science Connection reinforces the discussion of tides in Section 19-3 by describing how tides affect the lives of organisms in an intertidal zone.

Content Background

In the intertidal zone, complex food webs occur. Sponges, clams, mussels, and oysters filter algae and other microscopic forms of life from the water. Sea urchins, shrimp, copepods, and small fish feed on

floating or attached plants. Many carnivorous animals also use this area for feeding. Large fish, herons, raccoons, and octopi come to feed on the smaller fish and shellfish.

Teaching Strategy

Try to bring in some seashells for the students to examine. Ask them if they can explain how the shell might help the animals survive in the intertidal zone.

What is the relationship between tides and the moon?

What To Do

1. Study **Table 19-1**, a chart that lists the range of tidal heights for 2 weeks and also the observed moon phases for this same time period.

2. Identify the time during these 2 weeks that the tide difference or range was the greatest. What phase of the moon corresponds to this date? What other phase might cause a similar range?

Conclude and Apply

1. Sketch the way the sun, the moon, and Earth are positioned when the tide difference was greatest.

2. How would you change the sketch for dates when the tidal difference was the lowest?

3. What relation do you see between tide movement and moon position?

Table 19-1

Date	Height of high tide (meters)	Height of low tide (meters)
1	1.4	0.5
2	1.5	0.4
3	1.7	0.2
4	1.8	–0.1
5	2.1	–0.3
6	2.2	–0.5
7	2.3	–0.6
8	2.3	–0.6
9	2.3	–0.6
10	2.1	–0.5
11	1.9	–0.2
12	1.6	–0.1
13	1.6	0.2
14	1.6	0.4

condition of the zone when the tide is out. Many marine organisms require a moist environment to survive. Seaweeds often attach to cracks and crevices in large beach rocks where water collects. The barnacle, shown here, can close its shell, sealing in moisture and protecting itself from predators and the sun's rays.

Feeding habits of animals are also affected by tides. Barnacles close up during low tide, but as soon as the sea covers them again, their shells open, and they use leg-like structures for feeding. Another animal, the limpet, while fixed to one spot during low tide, is known to travel long distances when the tide is in. As the tide goes out, limpets travel back to their original rock homes.

Mussels

Barnacles

What Do You Think?

The intertidal zone represents one example of how physical factors affect living organisms. How are the organisms of the intertidal zone adapted to their changing environment? Can you think of other animals or plants that are adapted to continuously changing environmental conditions?

19-3 Tides **623**

What Do You Think?

1. The organisms of the intertidal zone have ways to deal with fast-moving water such as burrowing in the sand or attaching themselves to objects. They can also deal with changes in the amount of water by closing their shells, burying in the sand, or moving with the tides.

2. Possible answer: Amphibians and reptiles are cold-blooded and can adapt to changes in temperature.

Going Further ⑉⑉⑉⑉▶

Have students work in groups to develop an Intertidal Corner in your classroom. One group could prepare drawings of the plants and animals living in an intertidal zone. Another group could work with you in setting up a saltwater aquarium. Organize a field trip or ask a third group to be responsible for collecting organisms or shells that represent an intertidal zone. Use the Performance Task Assessment List for Group Work in **PASC**, p. 97. L1 **COOP LEARN**

 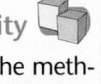

Figure 19-16

Visual Learning

Figure 19-16 Have students use the tide chart to answer these questions: According to the above chart, when are the best fishing times on November 23? *2:34 A.M. and 4:34 A.M.; 7:08 A.M. and 9:08 A.M.; 12:50 P.M. and 2:50 P.M.; 7:35 P.M. and 9:35 P.M.* Then have students use the tide chart to write newspaper articles about boats caught in shallow waters at low tide. Use the Performance Task Assessment List for Newspaper Article in **PASC**, p. 69. [L1] [P]

3 ASSESS

Check for Understanding

1. Using a globe, ball, and flashlight in various positions to represent Earth, the moon and the sun, ask students to identify specific regions on the globe that are experiencing high and low tides. [L2]

2. Assign questions 1-3 of Check Your Understanding. While discussing the Apply question, ask students to share any shell-collecting experiences.

Reteach

Discussion Remind students that the moon has a gravitational pull about one-sixth that of Earth's gravitational pull, so it pulls weakly at Earth. Ask students if this gravitational pull has an effect on Earth even though it is weak. *yes* Ask what kind of an effect it has. *It causes a bulge in Earth's oceans.* [L1]

Extension

Activity Have students working in small groups try to map the movement of a high tide as it moves over the surface of one of Earth's oceans. Use the Performance Task Assessment List for Group Work in **PASC**, p. 97. [L3]

NOVEMBER
TIDES AT GOLDEN GATE, SAN FRANCISCO, CALIFORNIA, 1994
Pacific Standard Time
Heights in feet

(Tide chart table with Moon phase, Day, Time/Ht. columns for LOW and HIGH tides.)

LUNAR DATA

☽ = NEW MOON
☾ = FIRST QUARTER
○ = FULL MOON
◑ = LAST QUARTER
A = IN APOGEE
P = IN PERIGEE

N = FARTHEST NORTH OF EQUATOR
E = ON EQUATOR
S = FARTHEST SOUTH OF EQUATOR

Figure 19-16

People who work on the sea must take the tides into consideration. Tide charts, like the one shown here from San Francisco, California, help them plan their work.

A People who fish have discovered that the best time to fish is when the ocean tides are about to turn: about one hour before and after both high tide and low tide. According to the chart above when are the best times to fish?

B Oyster harvesters in Cancale, France, take advantage of the tides in making their living. At high tide, their oyster beds are covered by water. At low tide, the beds are uncovered, making them easily accessible to harvest.

As you could see from the Find Out activity, the moon has a great effect on tides. But why does the moon have a greater effect than the sun? That's because it is about 400 times closer to Earth than the sun.

What effect might tides have on people living along an ocean coast? What other people might make use of our knowledge of tides? **Figure 19-16** describes the importance knowledge of tides can have for some people.

Now you know why you've studied tides with the moon. Tides are caused by the gravitational pull of the sun and moon on the oceans. The range of the tides depends on how the moon, the sun, and Earth are lined up.

---

check your UNDERSTANDING

1. Compare and contrast a beach at high tide and a beach at low tide.

2. Diagram the relative positions of the sun, the moon, and Earth during the highest tides of the month.

3. Apply Suppose you want to search for shells along a beach. How would a knowledge of tides help you in your search?

624 Chapter 19 The Earth-Moon System

---

4 CLOSE

Discussion

Have students discuss how people who live near the coastline have adapted to tides. *For example, fishers rely on tides to determine the best time for fishing or to navigate channels.* The discussion should include how tidal information is obtained, such as from local newspapers and television news reports or a weather channel.

check your UNDERSTANDING

1. At high tide, the waves reach far onto the shore and may cover the beach. At low tide, more beach is exposed.

2. Diagrams should show Earth, the moon, and the sun lined up.

3. Looking for shells at low tide would be easier, after the tide has left them behind.

Science and Society

Spin-Offs from the Space Program

From the beginning of space exploration, some people have wondered why so much time and money have been devoted to the pursuit of knowledge about space. Aren't there more important concerns than learning about space? Wouldn't money be better spent on improving the quality of life on Earth?

The space program has benefited many people right here on Earth. Much of the technology developed for space exploration is adapted and used to make our lives better. These technologies are called spin-offs. Let's look at some of the spin-offs from the United States space program.

Medical science has gained much from space research. Patients with internal bleeding may wear astronaut-type pressure suits that temporarily alter blood flow to promote healing. Heart attack victims benefit from heart monitoring techniques. Pacemakers, which help to regulate the heartbeats of some heart patients, use tiny batteries first developed for spacecraft.

Movable artificial limbs, designed for NASA's robots, are available for use by amputees and victims of paralysis of the legs (paraplegics) or of all four limbs (quadriplegics). Lasers, used as an improved "vision" for spacecraft by NASA, provide an alternative method of performing delicate eye and brain surgery.

Many of our clothes are made from fabrics originally created for astronauts. Our digital clocks evolved from NASA's timepieces. We can wear sunglasses that change tint with changes in the light around us. Even in the kitchen, we can find cookware coated with heat-resistant substances, plastic film and aluminum foil for preserving food, and freeze-dried foods—all originally developed for the space program.

Science Journal

Considering the amount and types of spin-offs from the space program, is all the time and money devoted to the program worth it? *In your Science Journal,* write a few paragraphs giving your opinions.

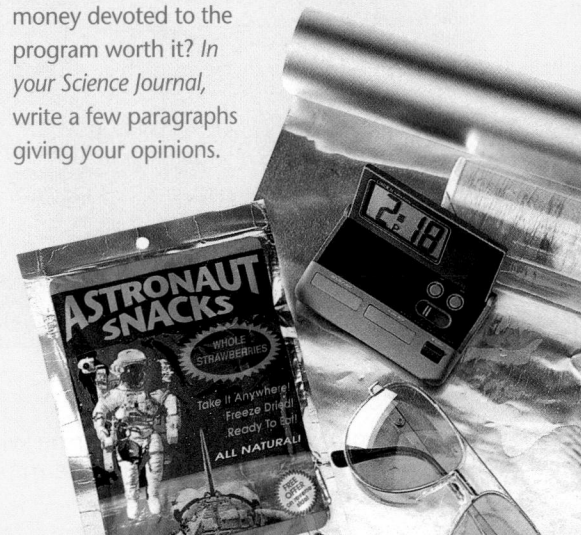

What is the monsoon?

Purpose

In Section 19-1, students learned that Earth's tilted axis causes seasons to change. These SciFacts give students a specific example of an event brought on by the changing tilt of Earth. Explain to students that latitude and topographic features also play a role in seasonal changes.

Content Background

Global winds are fueled by energy from the sun. Thus, seasonal temperature changes cause the wind shifts known as the monsoon. The rains that accompany the monsoon are essential for agriculture and, to a large extent, for human survival—India alone receives approximately 90 percent of its water supply from the monsoon rains that fall from June till September.

Demonstration

Using a lamp to represent the sun and a globe to represent Earth, demonstrate for students Earth's changing tilt during summer and winter. Focus the light from the lamp on the global regions of southern Asia, central Africa, and Australia. Help students to understand how these areas receive varying amounts of solar radiation during winter and summer.

Science Journal

Answers will vary, but students should recognize that the monsoon is essential for agricultural purposes, and that a large percentage of Earth's human population depends on these half-yearly winds for survival. Students should also mention that the global transfer of heat would be disrupted if the monsoon ceased. Some students may also note advantages associated with the monsoon's disappearance, such as cessation of floods.

NATIONAL GEOGRAPHIC SCIFACTS

What is the monsoon?

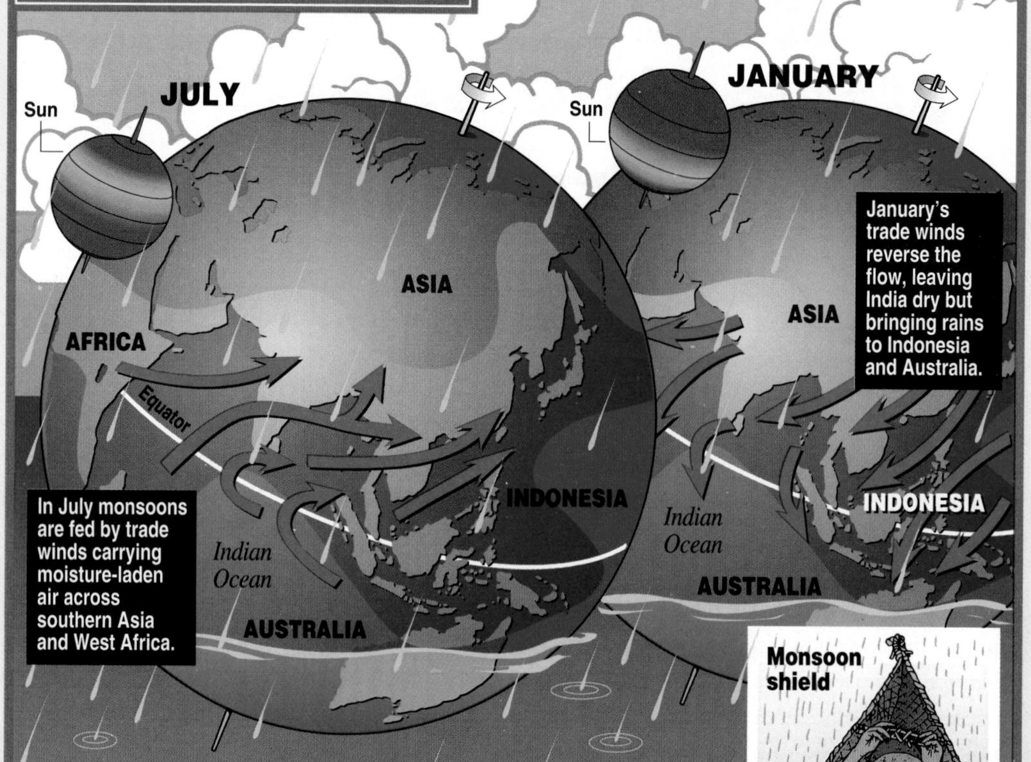

JULY — Sun — ASIA — AFRICA — Equator — Indian Ocean — INDONESIA — AUSTRALIA

In July monsoons are fed by trade winds carrying moisture-laden air across southern Asia and West Africa.

JANUARY — Sun — ASIA — Indian Ocean — INDONESIA — AUSTRALIA

January's trade winds reverse the flow, leaving India dry but bringing rains to Indonesia and Australia.

Monsoon shield

This woven straw umbrella, held on by tumplines that fit around the forehead, leaves hands free for work.

In southern Asia, central Africa, and Australia, people depend on wind shifts that bring torrential summer rains and cool, dry winter air. This half-yearly cycle is called the monsoon, from the Arabic word *mausim*, meaning "season."

The monsoon is one of Earth's most massive weather systems. It is part of the global heat transfer that keeps the planet habitable. From May to September, trade winds in the south move ocean air over the warm landmass of southern Asia.

The ocean air heats, rises, and sheds its moisture, drawing in more cool, moist air behind it. In September, the changing tilt of Earth reverses the system.

The rains of the monsoon nourish crops that feed millions and bring cool relief from sweltering temperatures. But monsoons can also be deadly, unleashing massive floods. In 1988 the monsoon brought devastation to tens of thousands in Bangladesh.

Science Journal

In your Science Journal, describe what might happen if the monsoon ceased to take place. Be sure to discuss the monsoon's impact on Earth's weather patterns, as well as its impact on people.

Teens in SCIENCE

Reaching for the Stars

Have you ever experienced a solar eclipse? Some people describe eclipses as being eerie or strange. But Natalie Sanchez, who experienced an eclipse when she was a freshman at Valley High School in Albuquerque, New Mexico, described an eclipse in a completely different way.

"I was on a field trip with the Math and Engineering Club at school," Natalie said. "We had gone to the observatory to watch the eclipse. The staff had given us really good glasses to protect our eyes. I thought it was the most beautiful thing I had ever seen. As it grew dark, everyone started cheering. But I was very quiet. I was completely in awe. On the bus ride home, I made up my mind that I want to be an astronomer."

Natalie soon discovered that her school did not offer any in-depth classes about astronomy. But she did not let that stop her from finding out more about her new hobby.

"After school, I spent a lot of time doing research at the library," Natalie explained. "I read everything about astronomy that I can get my hands on—and that's a lot. But the more I read, the more I wanted to know. Sure, I'm curious about our solar system. But I'm also trying to decide for myself

Below, the Mayall telescope at Kitt Peak, Arizona

whether or not it is possible that other solar systems are out there, too. Before I saw that eclipse, when I looked at the night sky I thought it was very pretty. But now when I look up at the stars, I feel like I'm looking at the biggest mystery in the universe."

What Do You Think?

Observing the eclipse changed the course of Natalie Sanchez's life. Have you ever had an experience that affected your plans for the future?

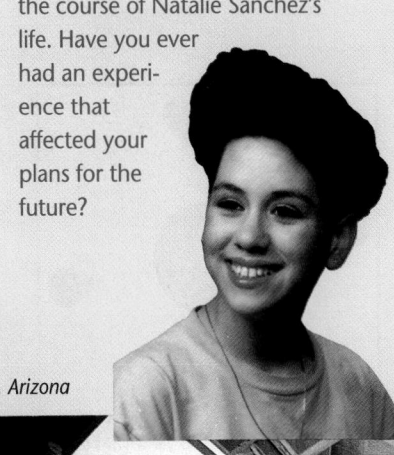

Purpose Teens in Science tells how 14-year-old Natalie Sanchez was inspired by a solar eclipse to investigate a career in astronomy.

Content Background Natalie Sanchez is far from the first to be awestruck by a solar eclipse. On May 28, 585 B.C.E a total eclipse of the sun blacked out the sky over Mesopotamia during a savage battle between Lydians, people from what is now Turkey, and Medes, people from what is now Iran. Terrified by what they took to be an omen, the armies put down their weapons and made peace on the spot. Because solar eclipses are now known to fall at predictable intervals, modern astronomers have pinpointed this day. It is the earliest historical event for which a precise date is known.

The ancient Chinese, Lapps, and Persians believed that an eclipse was caused by a monster devouring the sun or the moon. Ancient Mexicans believed eclipses were caused by the sun and the moon quarreling.

Teaching Strategy Students can read the selection with a partner, each reading paragraphs in turn. Invite students to then make a picture showing what they think Natalie saw. L1

Debate After students read the selection, have them debate the existence of other solar systems in the universe. L2

Answers to What Do You Think?

Suggest that students who have not yet overcome an obstacle describe one they might encounter in the future. Remind students that obstacles can be psychological as well as physical.

Going Further ▌▌▌▌▶

During the 20th century, 375 eclipses have or will take place: 228 solar eclipses and 147 lunar eclipses. Divide the class in half. Have one team research the lunar and solar eclipses that have occurred in this century. Which are the most important or notable? Why? Have the other team find out about the eclipses that have yet to occur. Where will the best vantage points be?

For information, students can contact:

Carnegie Observations
813 Santa Barbara Street
Pasadena, CA 91101

Have students present their findings in a written or oral report with illustrations. Select the appropriate Performance Task Assessment List from **PASC**. **COOP LEARN** L2

Have students use the four diagrams on this page to review the main ideas of the chapter.

Teaching Strategies

Divide the class into four groups. Assign each group a diagram and have them read the accompanying statement. Have each group prepare a presentation for the class by constructing a poster that depicts the main idea of the diagram and writing a paragraph explaining how the event occurs. You may want to provide groups with some of the demonstration materials you used during the teaching of this chapter. Then ask each group to use its poster, paragraph, and the demonstration materials to convey its ideas to the class. Encourage other students to ask the presenting group questions.

`COOP LEARN`

Answers to Questions

1. If the Earth weren't tilted on its axis, the seasons wouldn't change.

2. If the moon rotated twice during each revolution, we would see both sides of the moon.

3. The parts of Earth located in the umbra, the dark inner part of the moon's shadow, experience a total solar eclipse when the moon passes between the sun and Earth.

4. Although the tidal wave passes across the open ocean, the long wavelength and lack of reference points make it unnoticeable.

chapter 19
REVIEWING MAIN IDEAS

Science Journal

Review the statements below about the big ideas presented in this chapter, and answer the questions. Then, re-read your answers to the Did You Ever Wonder questions at the beginning of the chapter. *In your Science Journal,* write a paragraph about how your understanding of the big ideas in the chapter has changed.

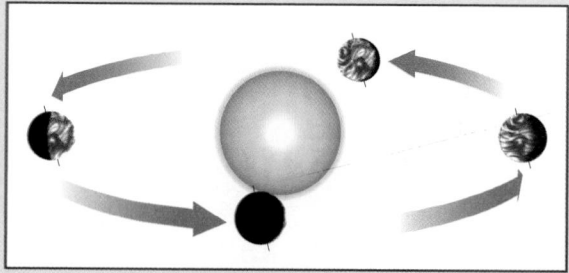

❶ The motions and relative positions of Earth, the moon, and the sun are responsible for night and day, the change of seasons, the phases of the moon, eclipses, and tides. *What would be different on Earth if it weren't tilted on its axis?*

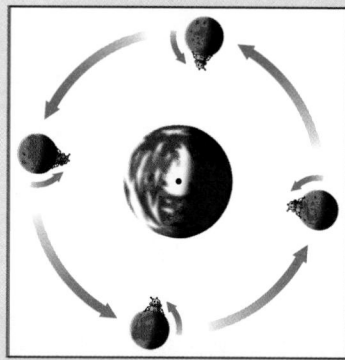

❷ The moon rotates once during its month-long revolution around Earth. During this time, different portions of the side of the moon facing Earth are lighted. *What would appear different about the moon if it rotated twice during each revolution?*

❸ During a solar eclipse, the moon passes between the sun and Earth and casts a shadow on Earth. During a lunar eclipse, Earth passes between the sun and moon, casting a curved shadow on the moon. *What parts of Earth experience a total solar eclipse when the moon passes between the sun and Earth?*

❹ Tides are the alternate rise and fall of the surface level of the oceans and are directly related to the gravitational force of the moon and the sun on Earth. *Why aren't tides as noticeable in the middle of the ocean as they are along the shore?*

GLENCOE TECHNOLOGY

 **MindJogger Videoquiz**

Chapter 19 Have students work in groups as they play the Videoquiz game to review key chapter concepts.

Project

Have a group of students measure the angle of the shadow the school flagpole throws at five different times during a five-day period. Have another group chart the phases of the moon each night. Have a third group use newspapers or other sources to chart the times of high and low tides each day at one location. Have each group display their findings on a poster. Use the Performance Task Assessment List for Group Work in **PASC,** p. 97. `COOP LEARN` `L1`

Science Journal

Did you ever wonder...

• During the winter, the rays of the sun striking Earth are not as direct as they are in the summer. They also strike the surface at one location for fewer hours per day. (pp. 608–609)

• The same side of the moon always faces Earth. The side we do not see is referred to as the far side of the moon. (p. 611)

• A beach becomes narrower during high tide. (pp. 621–622)

Using Key Science Terms

equinox	solar eclipse
lunar eclipse	solstice
revolution	sphere
rotation	tide

For each set of terms below, choose the one term that does not belong and explain why it does not belong.

1. equinox, lunar eclipse, solstice
2. revolution, rotation, equinox
3. sphere, rotation, tide
4. solar eclipse, lunar eclipse, rotation
5. lunar eclipse, tide, equinox

Use one of the terms to complete each sentence.

6. Earth's movement around the sun is called a _____.
7. Earth's spinning on its axis is called _____.
8. In the Northern Hemisphere, the summer _____ occurs on June 21 or 22, when the North Pole is tilted toward the sun.

Understanding Ideas

Answer the following questions in your Journal using complete sentences.

1. What causes the change in seasons?
2. At the time of an equinox, the sun is directly over what point on Earth?
3. What causes the moon's phases?
4. Compare and contrast a solar and a lunar eclipse.
5. When does the greatest range between high and low tides occur?

Developing Skills

Use your understanding of the concepts developed in this chapter to answer each of the following questions.

1. **Concept Mapping** Create and label a diagram illustrating the moon phases and why they occur.
2. **Making Models** Use the lamp and globe from the Find Out activity on page 605 to show how the rays of the sun change location with the beginning of the summer and winter solstices and the spring and fall equinoxes.
3. **Observing and Inferring** After doing the Explore activity on page 614, use two different-sized balls and line them up so they appear to be the same size. Which ball is farther away from you? How do the results of this activity compare with the results of the original activity?
4. **Making Models** Use the model from the Investigate on pages 618-619 to show the position of the sun, Earth, and the moon during a lunar eclipse.

Using Key Science Terms

Answers may vary. Possible answers are given below.

1. Equinox and solstice are positions of the sun during its yearly path through the sky. Lunar eclipse does not belong.
2. Equinox is not a type of motion.
3. Sphere doesn't belong. The fact that Earth is a sphere does not affect the height of tides.
4. Rotation has nothing to do with eclipses except that it affects what part of Earth experiences an eclipse.
5. Tides and lunar eclipses are directly related to the position of the moon with relation to the sun and Earth. Equinox doesn't fit.
6. revolution
7. rotation
8. solstice

Understanding Ideas

1. the angle of the sunlight because Earth is tipped on its axis
2. directly over the equator
3. They are due to the revolution of the moon around Earth.
4. A solar eclipse occurs when the moon passes directly between Earth and the sun, casting a shadow on Earth. A lunar eclipse occurs when the moon passes through Earth's shadow.
5. They occur when the moon, the sun, and Earth are lined up.

Developing Skills

1. The diagram students produce should be a cycle map that looks similar to Figure 19-9 on p. 613. Student explanations should include that as the moon revolves around Earth, different amounts of the side facing Earth are visible from Earth.
2. Students should demonstrate that at the time of a June solstice the sun is directly over the Tropic of Cancer; at the time of a December solstice the sun is directly over the Tropic of Capricorn;

Program Resources

Review and Assessment, pp. 113–118 L1

Performance Assessment, Ch. 19 L2

PASC

Alternate Assessment in the Science Classroom

Computer Test Bank L1

and the sun is directly over the equator at the time of the March and September equinoxes.

3. The larger ball is farther away. This observation is the same as the original activity, using a partner and another object.

4. Students should place the globe between the polystyrene ball and the light source.

Critical Thinking

1. The sun and the moon appear to move westward through the sky because Earth rotates from west to east. Sunrise and sunset are misleading terms because the sun only appears to rise in the east and set in the west as a spot on Earth rotates toward and away from the sun.

2. For a certain spot on Earth, the number of hours of daylight and darkness would be the same every day throughout the year. There would also be no change in seasons because one hemisphere would not be tilted toward the sun at one time and away from it at another.

3. Distance to the sun does not determine whether a hemisphere is experiencing winter. It is the tilt of Earth on its axis that causes seasons.

Problem Solving

Answers will vary but should name locations in the Northern Hemisphere from June to September and in the Southern Hemisphere from December to March.

Connecting Ideas

1. The moon is the natural satellite of Earth. Acceleration due to gravity and the moon's horizontal velocity holds the moon in its path around Earth.

2. Only during a full moon phase is Earth between the moon and the sun.

3. If the moon were transparent, a solar eclipse could not occur because sunlight could pass through it. If the moon were translucent, it could only partially block sunlight from Earth.

4. Answers will vary but may include benefits to medicine, fire fighting, construction materials, clothing, and kitchen appliances.

5. The best time to observe the moon is during quarter phases.

Critical Thinking

In your Journal, *answer each of the following questions.*

1. Why do the sun and moon appear to move westward in the sky? Why are the terms sunrise and sunset misleading?

2. Imagine that Earth is not tilted on its axis, as shown in the illustration. What effect would this have on the seasons and the number of hours of daylight and darkness throughout the year?

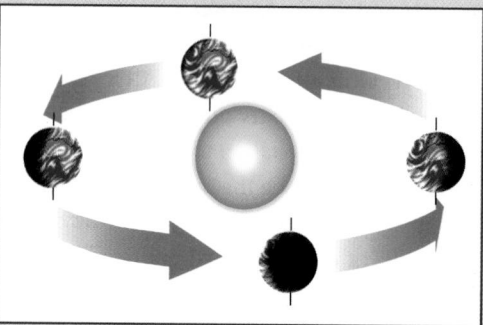

3. Earth is actually closest to the sun during the Northern Hemisphere's winter. Explain how this can be so.

Problem Solving

Read the following problem and discuss your answers in a brief paragraph.

Imagine you and your friends will be spending the coming year traveling throughout the world! Your friends want to arrange a schedule allowing you always to be someplace warm.

Using a globe, describe the route you and your friends would take to make sure that over the course of the year you are always someplace that is experiencing the summer season. Be specific. Give dates, locations, and explanations for your choices.

CONNECTING IDEAS

Discuss each of the following in a brief paragraph.

1. Theme—Systems and Interactions Explain the moon's revolution around Earth in terms of a satellite's motion. What force holds the moon in its path around Earth? What keeps it from crashing into Earth?

2. Theme—Systems and Interactions Explain why a lunar eclipse can occur only during a full moon phase.

3. Theme—Scale and Structure If the moon were transparent instead of opaque, what effect would this have on the occurrence of solar eclipses? What if it were translucent?

4. Science and Society Summarize the technological benefits we have received from the space program.

5. A Closer Look At which phase of the moon do you think you can best see the maria and craters from Earth?

✔ Assessment

Portfolio Review the portfolio options that are provided throughout the chapter. Encourage students to select one product that demonstrates their best work for the chapter. Have students explain what they learned and why they chose this example for placement in their portfolios.

Additional portfolio options can be found in the following **Teacher Classroom Resources:**

Wave Motion

In this unit, you investigated mechanical waves and found that they may be longitudinal or transverse, depending on how particles within a wave move.

You also learned that movements of Earth and the moon cause tides, moon phases, and lunar and solar eclipses to occur. In addition to this, you learned how humans are affected by the motion caused by earthquakes and volcanoes.

Try the exercises and activity that follow—they will challenge you to use and apply some of the ideas you learned in this unit.

CONNECTING IDEAS

1. Do you think that a transverse wave can pass through water? When you send a transverse wave through rope, the particles of rope move side to side. Would water particles move side to side to allow a transverse wave to pass? What do you think would happen to transverse waves generated by earthquakes when they encountered liquid rock material inside Earth?

2. How do waves at Earth's surface cause buildings to crumble during an earthquake? In what way are these waves similar to waves generated in water?

Exploring Further ACTIVITY

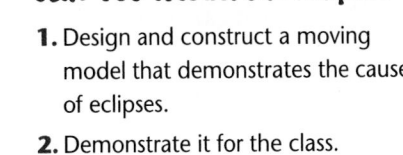

Can You Model an Eclipse?

1. Design and construct a moving model that demonstrates the cause of eclipses.

2. Demonstrate it for the class.

Exploring Further

Can you model an eclipse?

Purpose To demonstrate how an eclipse happens.

Background Information

A solar eclipse occurs when the moon passes directly between Earth and the sun. The moon's shadow falls directly on Earth's surface. A lunar eclipse occurs when Earth is directly between the sun and the moon and the moon passes through Earth's shadow.

✔ Assessment

Performance Have students use their models to make an oral presentation on the cause of eclipses. Use the Performance Task Assessment List Oral Presentation in **PASC**, p. 71.

COOP LEARN L1

Wave Motion

THEME DEVELOPMENT

The key themes developed in Unit 5 were energy and systems and interactions. The interaction of the gravity of the sun with that of the moon and Earth, within the Earth-moon system, produce a planet-wide ocean wave on Earth called tides. Mechanical waves produced by earthquakes and volcanoes interact with Earth materials at the surface causing changes in many Earth systems. Faults and earthquake or volcanic damage may result. The energy of waves is shown in examples such as water waves, and earthquakes. This relationship between energy and waves is developed using examples of sound waves and earthquake waves.

Connections to Other Units

In Unit 5, Earth movement and the movement of Earth materials during earthquakes and volcanoes was connected to the general processes of motion described in Unit 4. In addition, movement of the moon as part of the Earth-moon system and the effect of this motion on Earth's oceans was explained. Motions near Earth as described in Unit 4 were used as a basis for explaining the causes of tides on Earth.

Connecting Ideas
Answers

1. No. If the water particles were moved to one side, they would not rebound and thus a transverse wave would not be generated. The waves would stop.

2. As the waves move through Earth's surface and the buildings at the surface, the particles of the building try to move in different directions. Particles of rock move upward and downward with some forward and backward motion similar to waves in water.

APPENDICES

Table of Contents

APPENDIX A

International System of Units

The International System (SI) of Measurement is accepted as the standard for measurement throughout most of the world. Three base units in SI are the meter, kilogram, and second. Frequently used SI units are listed below.

Table A-1: Frequently used SI Units	
Length	1 millimeter (mm) = 1000 micrometers (μm)
	1 centimeter (cm) = 10 millimeters (mm)
	1 meter (m) = 100 centimeters (cm)
	1 kilometer (km) = 1000 meters (m)
	1 light-year = 9 460 000 000 000 kilometers (km)
Area	1 square meter (m^2) = 10 000 square centimeters (cm^2)
	1 square kilometer (km^2) = 1 000 000 square meters (m^2)
Volume	1 milliliter (mL) = 1 cubic centimeter (cm^3)
	1 liter (L) = 1000 milliliters (mL)
Mass	1 gram (g) = 1000 milligrams (mg)
	1 kilogram (kg) = 1000 grams (g)
	1 metric ton (g) = 1000 kilograms (kg)
Time	1 s = 1 second

Temperature measurements in SI are often made in degrees Celsius. Celsius temperature is a supplementary unit derived from the base unit kelvin. The Celsius scale (°C) has 100 equal graduations between the freezing temperature (0°C) and the boiling temperature of water (100°C). The following relationship exists between the Celsius and kelvin temperature scales:

$$K = °C + 273$$

Several other supplementary SI units are listed below.

Table A-2: Supplementary SI Units			
Measurement	Unit	Symbol	Expressed in Base Units
Energy	Joule	J	$kg \cdot m^2/s^2$ or $N \cdot m$
Force	Newton	N	$kg \cdot m/s^2$
Power	Watt	W	$kg \cdot m^2/s^3$ or J/s
Pressure	Pascal	Pa	$kg/(m \cdot s^2)$ or N/m^2

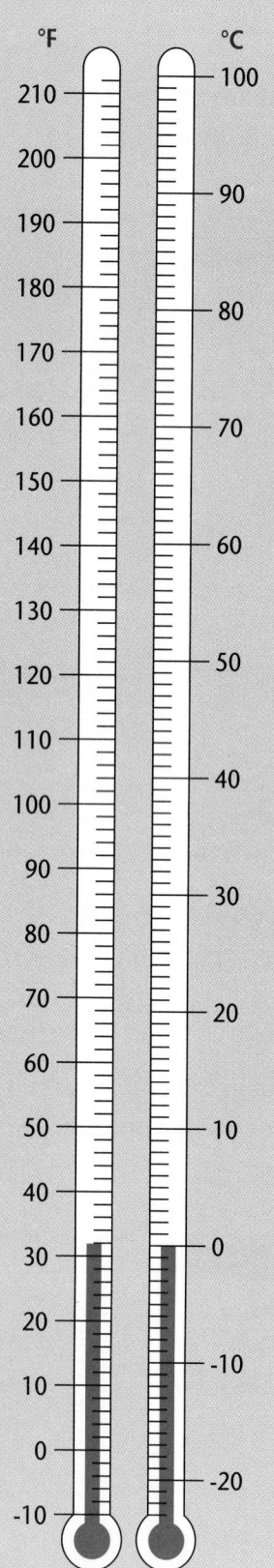

°F   °C

Table B-1: SI/Metric to English Conversions			
	When You Want to Convert:	**Multiply By:**	**To Find:**
Length	inches	2.54	centimeters
	centimeters	0.39	inches
	feet	0.30	meters
	meters	3.28	feet
	yards	0.91	meters
	meters	1.09	yards
	miles	1.61	kilometers
	kilometers	0.62	miles
Mass and Weight	ounces	28.35	grams
	grams	0.04	ounces
	pounds	0.45	kilograms
	kilograms	2.2	pounds
	tons	0.91	tonnes (metric tons)
	tonnes (metric tons)	1.10	tons
	pounds	4.45	newtons
	newtons	0.23	pounds
Volume	cubic inches	16.39	cubic centimeters
	cubic centimeters	0.06	cubic inches
	cubic feet	0.03	cubic meters
	cubic meters	35.3	cubic feet
	liters	1.06	quarts
	liters	0.26	gallons
	gallons	3.78	liters
Area	square inches	6.45	square centimeters
	square centimeters	0.16	square inches
	square feet	0.09	square meters
	square meters	10.76	square feet
	square miles	2.59	square kilometers
	square kilometers	0.39	miles
	hectares	2.47	acres
	acres	0.40	hectares
Temperature	Fahrenheit	5/9 (°F − 32)	Celsius
	Celsius	9/5 °C + 32	Fahrenheit

Safety in the Science Classroom

1. Always obtain your teacher's permission to begin an investigation.
2. Study the procedure. If you have questions, ask your teacher. Understand any safety symbols shown on the page.
3. Use the safety equipment provided for you. Goggles and a safety apron should be worn when any investigation calls for using chemicals.
4. Always slant test tubes away from yourself and others when heating them.
5. Never eat or drink in the lab, and never use lab glassware as food or drink containers. Never inhale chemicals. Do not taste any substances or draw any material into a tube with your mouth.
6. If you spill any chemical, wash it off immediately with water. Report the spill immediately to your teacher.
7. Know the location and proper use of the fire extinguisher, safety shower, fire blanket, first aid kit, and fire alarm.
8. Keep materials away from flames. Tie back hair and loose clothing.
9. If a fire should break out in the classroom, or if your clothing should catch fire, smother it with the fire blanket or a coat, or get under a safety shower. NEVER RUN.
10. Report any accident or injury, no matter how small, to your teacher.

Follow these procedures as you clean up your work area.

1. Turn off the water and gas. Disconnect electrical devices.
2. Return all materials to their proper places.
3. Dispose of chemicals and other materials as directed by your teacher. Place broken glass and solid substances in the proper containers. Never discard materials in the sink.
4. Clean your work area.
5. Wash your hands thoroughly after working in the laboratory.

Table C-1: First Aid	
Injury	**Safe Response**
Burns	Apply cold water. Call your teacher immediately.
Cuts and bruises	Stop any bleeding by applying direct pressure. Cover cuts with a clean dressing. Apply cold compresses to bruises. Call your teacher immediately.
Fainting	Leave the person lying down. Loosen any tight clothing and keep crowds away. Call your teacher immediately.
Foreign matter in eye	Flush with plenty of water. Use eyewash bottle or fountain. Call your teacher immediately.
Poisoning	Note the suspected poisoning agent and call your teacher immediately.
Any spills on skin	Flush with large amounts of water or use safety shower. Call your teacher immediately.

APPENDIX D

Table D-1: Safety Symbols

Disposal Alert
This symbol appears when care must be taken to dispose of materials properly.

Animal Safety
This symbol appears whenever live animals are studied and the safety of the animals and the students must be ensured.

Biological Hazard
This symbol appears when there is danger involving bacteria, fungi, or protists.

Radioactive Safety
This symbol appears when radioactive materials are used.

Open Flame Alert
This symbol appears when use of an open flame could cause a fire or an explosion.

Clothing Protection Safety
This symbol appears when substances used could stain or burn clothing.

Thermal Safety
This symbol appears as a reminder to use caution when handling hot objects.

Fire Safety
This symbol appears when care should be taken around open flames.

Sharp Object Safety
This symbol appears when a danger of cuts or punctures caused by the use of sharp objects exists.

Explosion Safety
This symbol appears when the misuse of chemicals could cause an explosion.

Fume Safety
This symbol appears when chemicals or chemical reactions could cause dangerous fumes.

Eye Safety
This symbol appears when a danger to the eyes exists. Safety goggles should be worn when this symbol appears.

Electrical Safety
This symbol appears when care should be taken when using electrical equipment.

Poison Safety
This symbol appears when poisonous substances are used.

Skin Protection Safety
This symbol appears when use of caustic chemicals might irritate the skin or when contact with microorganisms might transmit infection.

Chemical Safety
This symbol appears when chemicals used can cause burns or are poisonous if absorbed through the skin.

Diversity of Life: Classification of Living Organisms

Scientists use a five kingdom system for the classification of organisms. In this system, there is one kingdom of organisms, Kingdom Monera, which contains organisms that do not have a nucleus and lack specialized structures in the cytoplasm of their cells. The members of the other four kingdoms have cells each of which contains a nucleus and structures in the cytoplasm that are surrounded by membranes. These kingdoms are Kingdom Protista, Kingdom Fungi, the Plant Kingdom, and the Animal Kingdom.

Kingdom Monera

Phylum Cyanobacteria one celled prokaryotes; make their own food, contain chlorophyll, some species form colonies, most are blue-green

Bacteria one-celled prokaryotes; most absorb food from their surroundings, some are photosynthetic; many are parasites; round, spiral, or rod shaped

Kingdom Protista

Phylum Euglenophyta one-celled; can photosynthesize or take in food; most have one flagellum; euglenoids

Phylum Chrysophyta most are one-celled; make their own food through photosynthesis; golden-brown pigments mask chlorophyll; diatoms

Phylum Pyrrophyta one-celled; make their own food through photosynthesis; contain red pigments and have two flagella; dinoflagellates

Phylum Chlorophyta one-celled, many-celled, or colonies; contain chlorophyll and make their own food; live on land, in fresh water or salt water; green algae

Phylum Rhodophyta most are many-celled and photosynthetic; contain red pigments; most live in deep saltwater environments; red algae

Phylum Phaeophyta most are many-celled and photosynthetic; contain brown pigments; most live in saltwater environments; brown algae

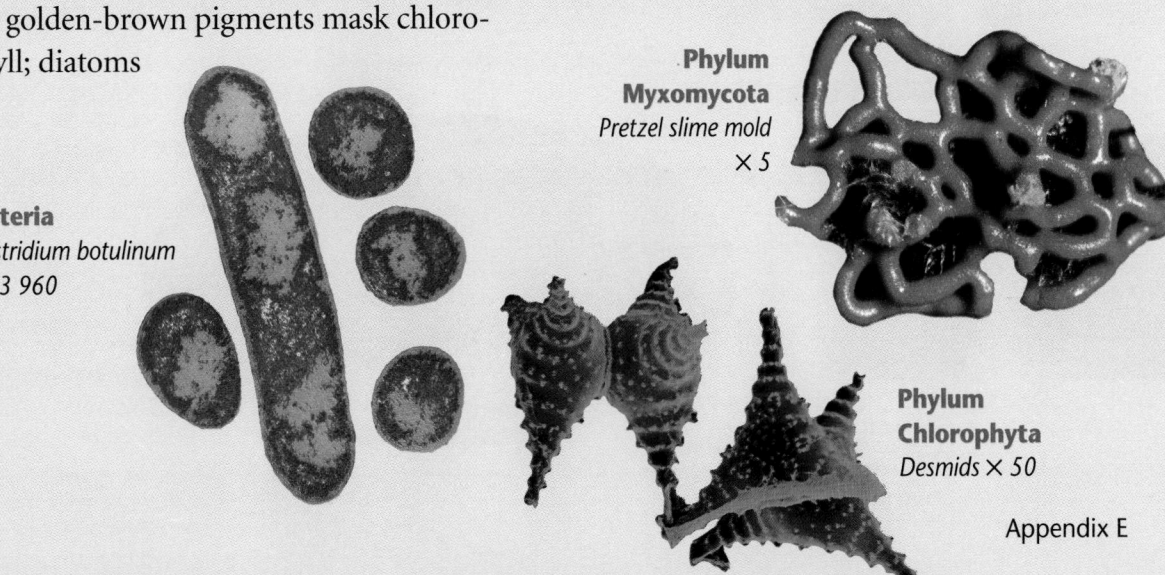

Bacteria
Clostridium botulinum
× 13 960

Phylum Myxomycota
Pretzel slime mold
× 5

Phylum Chlorophyta
Desmids × 50

Phylum Sarcodina one-celled; take in food; move by means of pseudopods; free-living or parasitic; amoebas

Phylum Mastigophora one-celled; take in food; have two or more flagella; free-living or parasitic; flagellates

Phylum Ciliophora one-celled; take in food; have large numbers of cilia; ciliates

Phylum Sporozoa one-celled; take in food; no means of movement; parasites in animals; sporozoans

Phyla Myxomycota and Acrasiomycota one- or many-celled; absorb food; change form during life cycle; cellular and plasmodial slime molds

Kingdom Fungi

Phylum Zygomycota many-celled; absorb food; spores are produced in sporangia; zygote fungi; bread mold

Phylum Ascomycota one- and many-celled; absorb food; spores produced in asci; sac fungi; yeast

Phylum Ascomycota
Yeast × 7800

Phylum Basidiomycota many-celled; absorb food; spores produced in basidia; club fungi; mushrooms

Phylum Deuteromycota members with unknown reproductive structures; imperfect fungi; penicillin

Lichens organisms formed by symbiotic relationship between an ascomycote or a basidiomycote and a green alga or a cyanobacterium

Plant Kingdom

Spore Plants

Division Bryophyta nonvascular plants that reproduce by spores produced in capsules; many-celled; green; grow in moist land environments; mosses and liverworts

Division Lycophyta many-celled vascular plants; spores produced in cones; live on land; are photosynthetic; club mosses

Division Sphenophyta vascular plants with ribbed and jointed stems; scalelike leaves; spores produced in cones; horsetails

Division Pterophyta vascular plants with feathery leaves called fronds; spores produced in clusters of sporangia called sori; live on land or in water; ferns

Lichens
Old Man's Beard lichen

Division Bryophyta
Liverwort

APPENDIX E

Seed Plants

Division Ginkgophyta deciduous gymnosperms; only one living species called the maiden hair tree; fan-shaped leaves with branching veins; reproduces with seeds; ginkgos

Division Cycadophyta palmlike gymnosperms; large compound leaves; produce seeds in cones; cycads

Division Coniferophyta deciduous or evergreen gymnosperms; trees or shrubs; needlelike or scalelike leaves; seeds produced in cones; conifers

Division Gnetophyta shrubs or woody vines; seeds produced in cones; division contains only three genera; gnetum

Division Anthophyta dominant group of plants; ovules protected at fertilization by an ovary; sperm carried to ovules by pollen tube; produce flowers and seeds in fruits; flowering plants

Animal Kingdom

Phylum Porifera aquatic organisms that lack true tissues and organs; they are asymmetrical and sessile; sponges

Phylum Cnidaria radially symmetrical organisms with a digestive cavity with one opening; most have tentacles armed with stinging cells; live in aquatic environments singly or in colonies; includes jellyfish, corals, hydra, and sea anemones

Phylum Platyhelminthes bilaterally symmetrical worms with flattened bodies; digestive system has one opening; parasitic and free-living species; flatworms

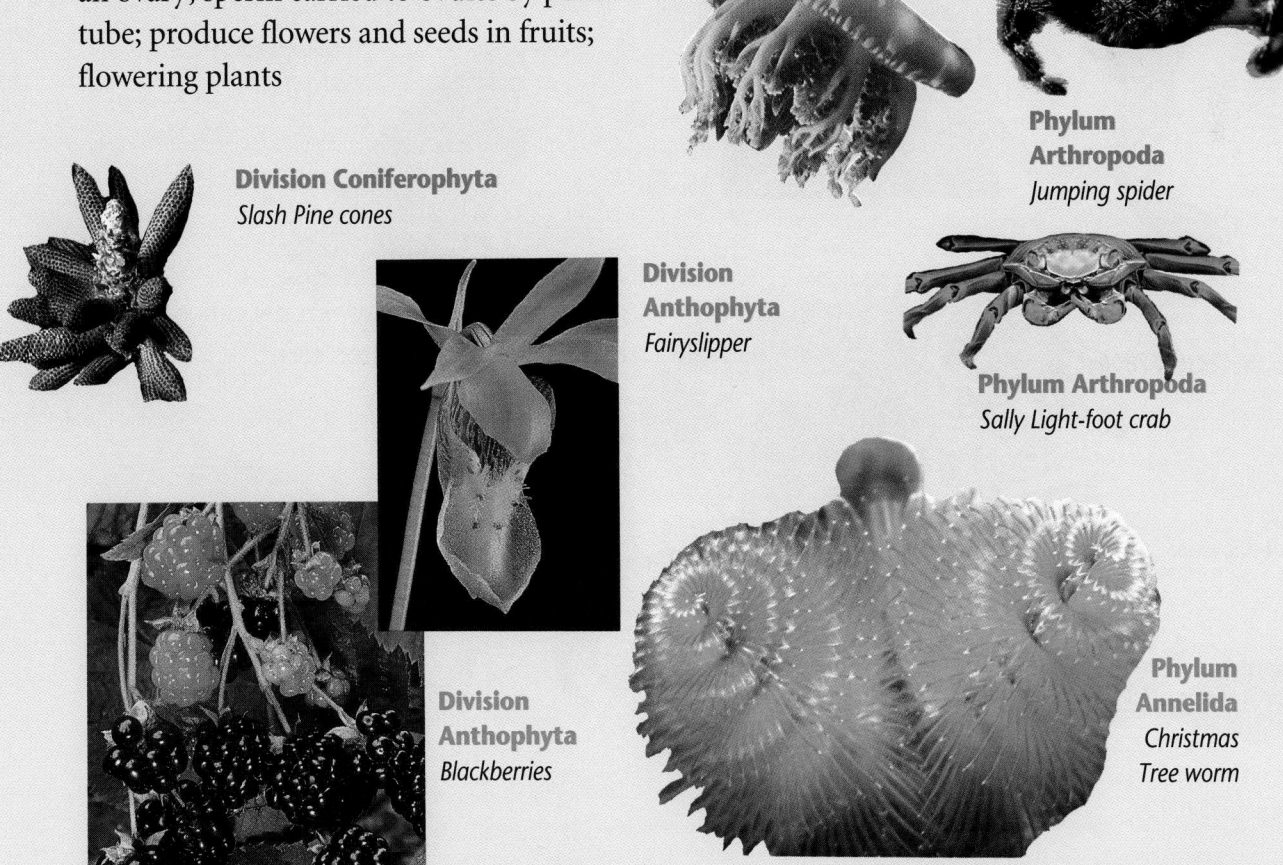

Phylum Cnidaria
Jellyfish

Phylum Arthropoda
Jumping spider

Division Coniferophyta
Slash Pine cones

Division Anthophyta
Fairyslipper

Phylum Arthropoda
Sally Light-foot crab

Division Anthophyta
Blackberries

Phylum Annelida
Christmas Tree worm

Phylum Nematoda round bilaterally symmetrical body; digestive system with two openings; some free-living forms but mostly parasitic; roundworms

Phylum Mollusca soft-bodied animals, many with a hard shell; a mantle covers the soft body; aquatic and terrestrial species; includes clams, snails, squid, and octopuses

Phylum Annelida bilaterally symmetrical worms with round segmented bodies; terrestrial and aquatic species; includes earthworms, leeches, and marine polychaetes

Phylum Arthropoda very large phylum of organisms that have segmented bodies with pairs of jointed appendages, and a hard exoskeleton; terrestrial and aquatic species; includes insects, crustaceans, spiders, and horseshoe crabs

Phylum Echinodermata saltwater organisms with spiny or leathery skin; water-vascular system with tube feet; radial symmetry; includes starfish, sand dollars, and sea urchins

Phylum Chordata organisms with internal skeletons, specialized body systems, and paired appendages; all at some time have a notochord, dorsal nerve cord, gill slits, and a tail; includes fish, amphibians, reptiles, birds, and mammals

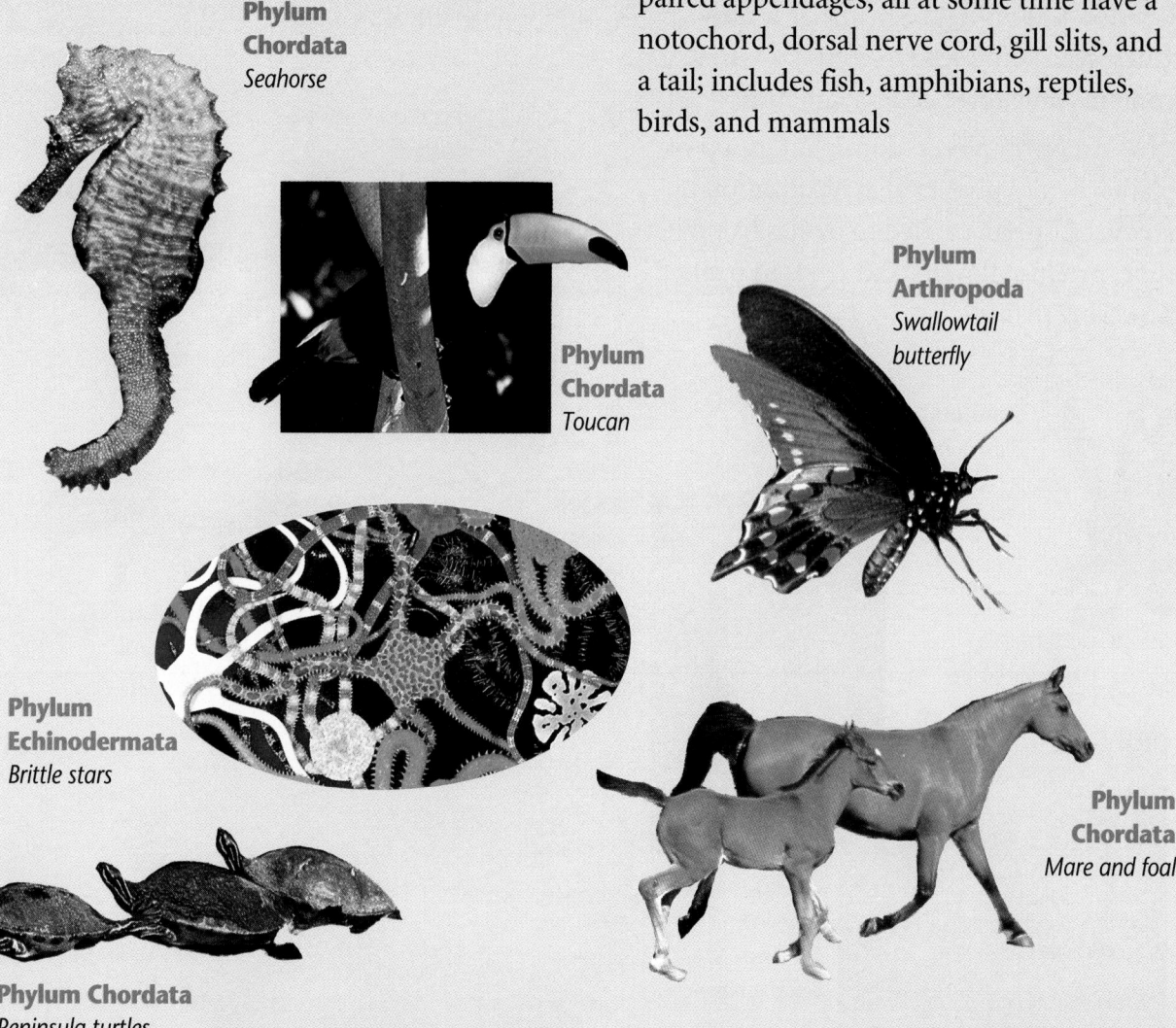

Phylum Chordata
Seahorse

Phylum Chordata
Toucan

Phylum Arthropoda
Swallowtail butterfly

Phylum Echinodermata
Brittle stars

Phylum Chordata
Mare and foal

Phylum Chordata
Peninsula turtles

APPENDIX F

Topographic Map Symbols

Primary highway, hard surface		Index contour	
Secondary highway, hard surface		Supplementary contour	
Light-duty road, hard or improved surface		Intermediate contour	
Unimproved road		Depression contours	
Railroad: single track and multiple track			
Railroads in juxtaposition		Boundaries: National	
		State	
		County, parish, municipal	
Buildings		Civil township, precinct, town, barrio	
School, church, and cemetery	cem	Incorporated city, village, town, hamlet	
Buildings (barn, warehouse, etc.)		Reservation, National or State	
Wells other than water (labeled as to type)	○ oil ○ gas	Small park, cemetery, airport, etc.	
Tanks: oil, water, etc. (labeled only if water)	●●● water	Land grant	
Located or landmark object; windmill	⊙	Township or range line, United States land survey	
Open pit, mine, or quarry; prospect	✕	Township or range line, approximate location	

Marsh (swamp)		Perennial streams	
Wooded marsh		Elevated aqueduct	
Woods or brushwood		Water well and spring	○ ○~
Vineyard		Small rapids	
Land subject to controlled inundation		Large rapids	
Submerged marsh		Intermittent lake	
Mangrove		Intermittent streams	
Orchard		Aqueduct tunnel	
Scrub		Glacier	
Urban area		Small falls	
Spot elevation	× 7369	Large falls	
Water elevation	670	Dry lake bed	

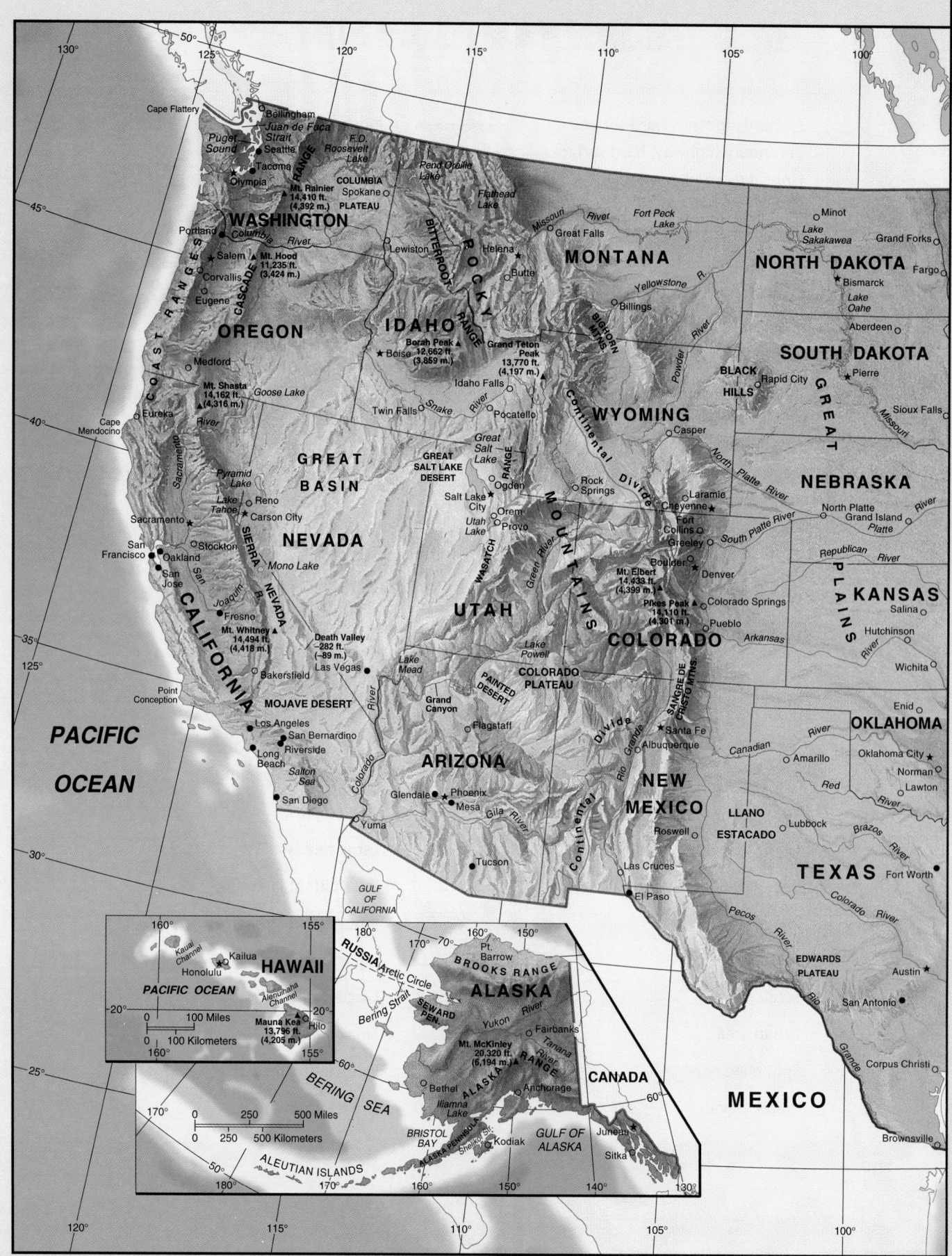

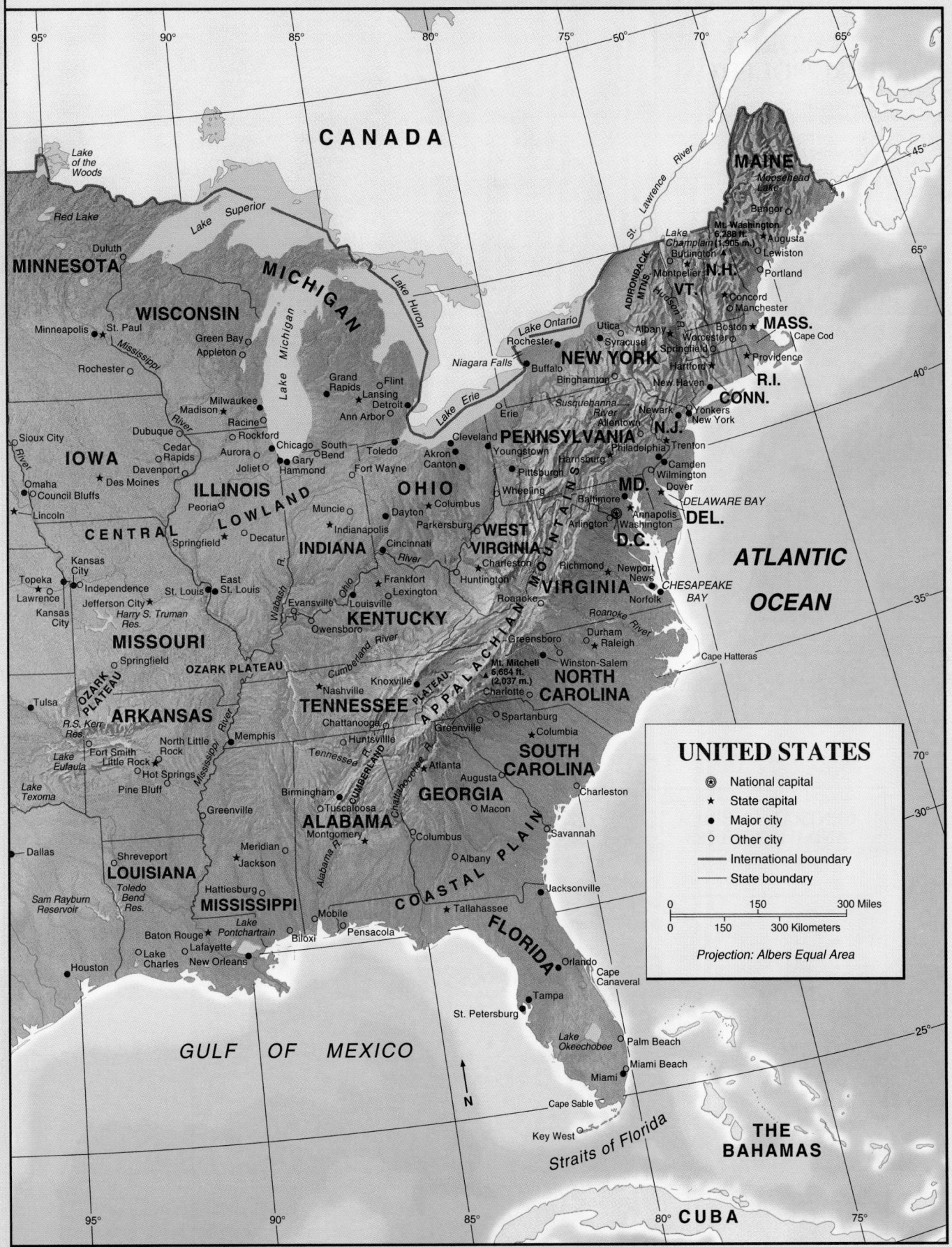

CANADA

MAINE
Mooselhead Lake

MINNESOTA
Lake of the Woods
Red Lake
Duluth

MICHIGAN
Lake Superior

WISCONSIN
Green Bay
Appleton
Minneapolis St. Paul
Rochester

Mt. Washington
6,288 ft.
(1,905 m.)
Lake Champlain
Burlington
Montpelier N.H.
VT.
Concord
Manchester
Bangor
Augusta
Lewiston
Portland

Lake Michigan
Lake Huron

Grand Rapids
Flint
Lansing
Detroit
Ann Arbor

Madison
Milwaukee
Racine
Dubuque

Lake Ontario
Rochester
Niagara Falls
Buffalo

NEW YORK
Utica
Syracuse
Albany
Binghamton

Boston MASS.
Worcester
Springfield
Hartford
Providence
New Haven
CONN.
R.I.
Cape Cod

IOWA
Sioux City
Omaha
Council Bluffs
Lincoln

Cedar Rapids
Davenport
Des Moines

Rockford
Aurora
Joliet
Chicago
Gary
Hammond
South Bend

Fort Wayne

Toledo
Akron
Canton
Cleveland
Youngstown

Erie

PENNSYLVANIA
Pittsburgh
Wheeling
Harrisburg

Susquehanna River
Allentown
Newark
Philadelphia
Camden
Wilmington
Dover

Trenton
N.J.
Yonkers
New York

ILLINOIS
Peoria
Springfield
Decatur

INDIANA
Muncie
Indianapolis
Cincinnati

OHIO
Dayton
Columbus

Parkersburg

WEST VIRGINIA
Charleston
Huntington

MD.
Baltimore
Washington
Annapolis
Arlington
D.C.
DEL.
DELAWARE BAY

CENTRAL LOWLAND

Kansas City
Topeka
Lawrence
Independence
Jefferson City
Kansas City

East St. Louis
St. Louis
Evansville
Louisville
Frankfort
Lexington

KENTUCKY
Owensboro

Richmond
Roanoke
Newport News
Norfolk

VIRGINIA

CHESAPEAKE BAY

ATLANTIC OCEAN

MISSOURI
Springfield
Harry S. Truman Res.

OZARK PLATEAU

Tulsa
ARKANSAS
R.S. Kerr Res.
Lake Eufaula
Fort Smith
North Little Rock
Little Rock
Hot Springs
Pine Bluff

Memphis

Nashville
Knoxville
Chattanooga

TENNESSEE

Mt. Mitchell
6,684 ft.
(2,037 m.)

Greensboro
Durham
Raleigh
Winston-Salem
Charlotte

NORTH CAROLINA

Roanoke River

Cape Hatteras

Lake Texoma
Dallas
Shreveport
LOUISIANA
Toledo Bend Res.

Greenville

Birmingham
Tuscaloosa

ALABAMA
Montgomery
Meridian
Jackson

Huntsville

Columbus

Atlanta
Augusta
Macon

GEORGIA

Columbus

Spartanburg
Greenville

Columbia
SOUTH CAROLINA

Charleston

Savannah

COASTAL PLAIN

Sam Rayburn Reservoir

Hattiesburg
MISSISSIPPI
Baton Rouge
Lafayette
Lake Charles
Houston
New Orleans
Lake Pontchartrain
Mobile
Biloxi
Pensacola

Albany

Tallahassee

Jacksonville

FLORIDA
Orlando
Cape Canaveral

St. Petersburg
Tampa

Lake Okeechobee
Palm Beach
Miami Beach
Miami

Cape Sable

GULF OF MEXICO

N

Key West
Straits of Florida

THE BAHAMAS

CUBA

UNITED STATES
- ◉ National capital
- ★ State capital
- ● Major city
- ○ Other city
- ── International boundary
- ── State boundary

0   150   300 Miles
0   150   300 Kilometers

Projection: Albers Equal Area

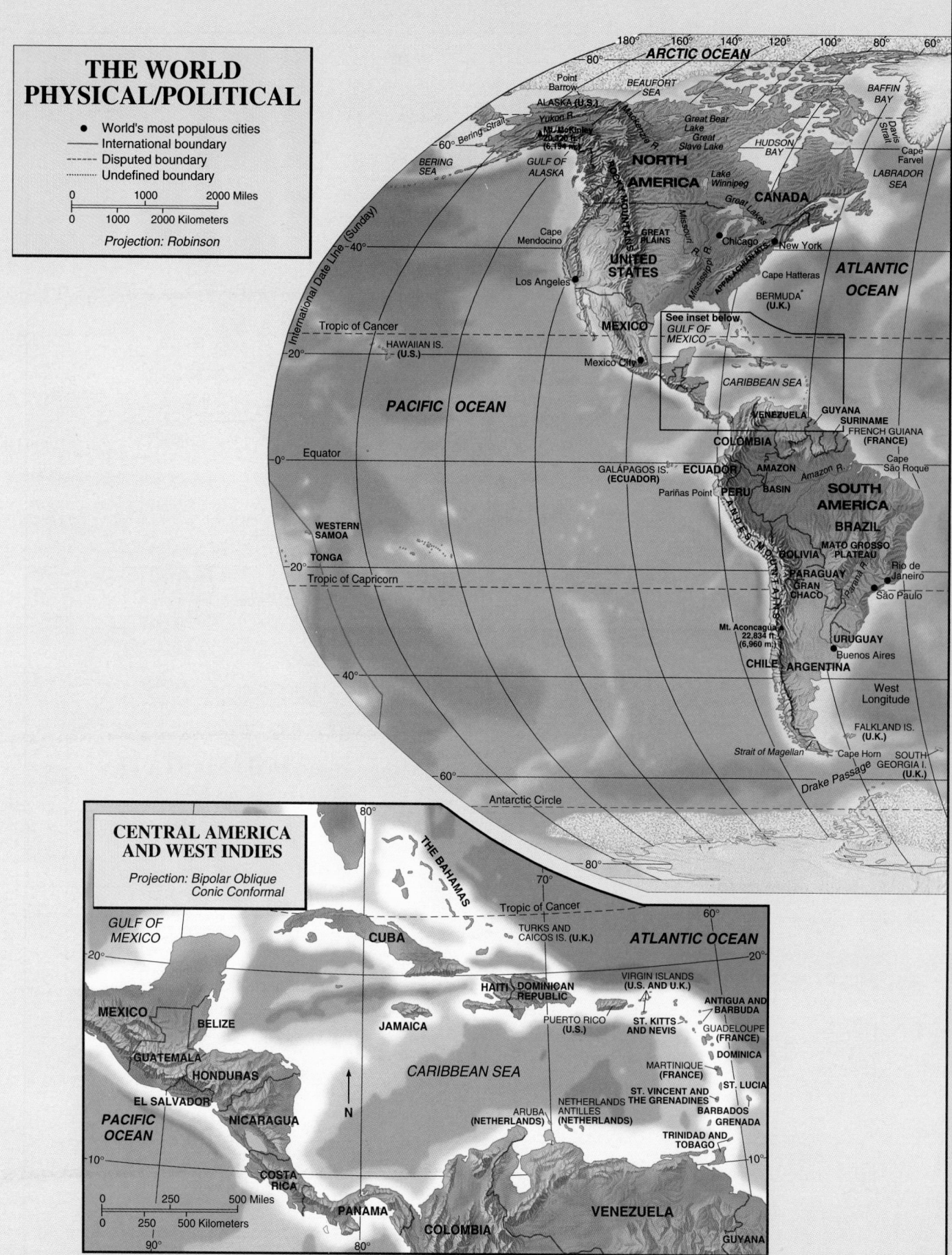

THE WORLD PHYSICAL/POLITICAL

- • World's most populous cities
- —— International boundary
- – – Disputed boundary
- ········ Undefined boundary

0   1000   2000 Miles
0   1000   2000 Kilometers

Projection: Robinson

ARCTIC OCEAN

Point Barrow
BEAUFORT SEA
BAFFIN BAY
ALASKA (U.S.)
Yukon R.
Mt. McKinley 20,320 ft. (6,194 m.)
Great Bear Lake
Great Slave Lake
HUDSON BAY
Davis Strait
Cape Farvel
Bering Strait
BERING SEA
GULF OF ALASKA
NORTH AMERICA
CANADA
LABRADOR SEA
Lake Winnipeg
Great Lakes
Cape Mendocino
GREAT PLAINS
Missouri R.
Chicago
New York
UNITED STATES
Mississippi R.
APPALACHIAN MTS.
ATLANTIC OCEAN
Cape Hatteras
Los Angeles
BERMUDA (U.K.)
Tropic of Cancer
MEXICO
See inset below
GULF OF MEXICO
HAWAIIAN IS. (U.S.)
Mexico City
CARIBBEAN SEA
VENEZUELA
GUYANA
SURINAME
FRENCH GUIANA (FRANCE)
PACIFIC OCEAN
COLOMBIA
Equator
GALÁPAGOS IS. (ECUADOR)
ECUADOR
AMAZON
Amazon R.
Cape São Roque
Pariñas Point
PERU
BASIN
SOUTH AMERICA
BRAZIL
WESTERN SAMOA
MATO GROSSO PLATEAU
TONGA
BOLIVIA
Rio de Janeiro
Tropic of Capricorn
PARAGUAY
GRAN CHACO
Paraná R.
São Paulo
Mt. Aconcagua 22,834 ft. (6,960 m.)
URUGUAY
Buenos Aires
CHILE
ARGENTINA
West Longitude
FALKLAND IS. (U.K.)
Strait of Magellan
Cape Horn
SOUTH GEORGIA I. (U.K.)
Drake Passage
Antarctic Circle

CENTRAL AMERICA AND WEST INDIES

Projection: Bipolar Oblique Conic Conformal

GULF OF MEXICO
THE BAHAMAS
ATLANTIC OCEAN
Tropic of Cancer
CUBA
TURKS AND CAICOS IS. (U.K.)
MEXICO
VIRGIN ISLANDS (U.S. AND U.K.)
ANTIGUA AND BARBUDA
BELIZE
HAITI
DOMINICAN REPUBLIC
PUERTO RICO (U.S.)
ST. KITTS AND NEVIS
GUADELOUPE (FRANCE)
JAMAICA
DOMINICA
GUATEMALA
MARTINIQUE (FRANCE)
ST. LUCIA
HONDURAS
CARIBBEAN SEA
ST. VINCENT AND THE GRENADINES
EL SALVADOR
N
NETHERLANDS ANTILLES (NETHERLANDS)
BARBADOS
GRENADA
PACIFIC OCEAN
NICARAGUA
ARUBA (NETHERLANDS)
TRINIDAD AND TOBAGO
COSTA RICA
PANAMA
VENEZUELA
COLOMBIA
GUYANA

0   250   500 Miles
0   250   500 Kilometers

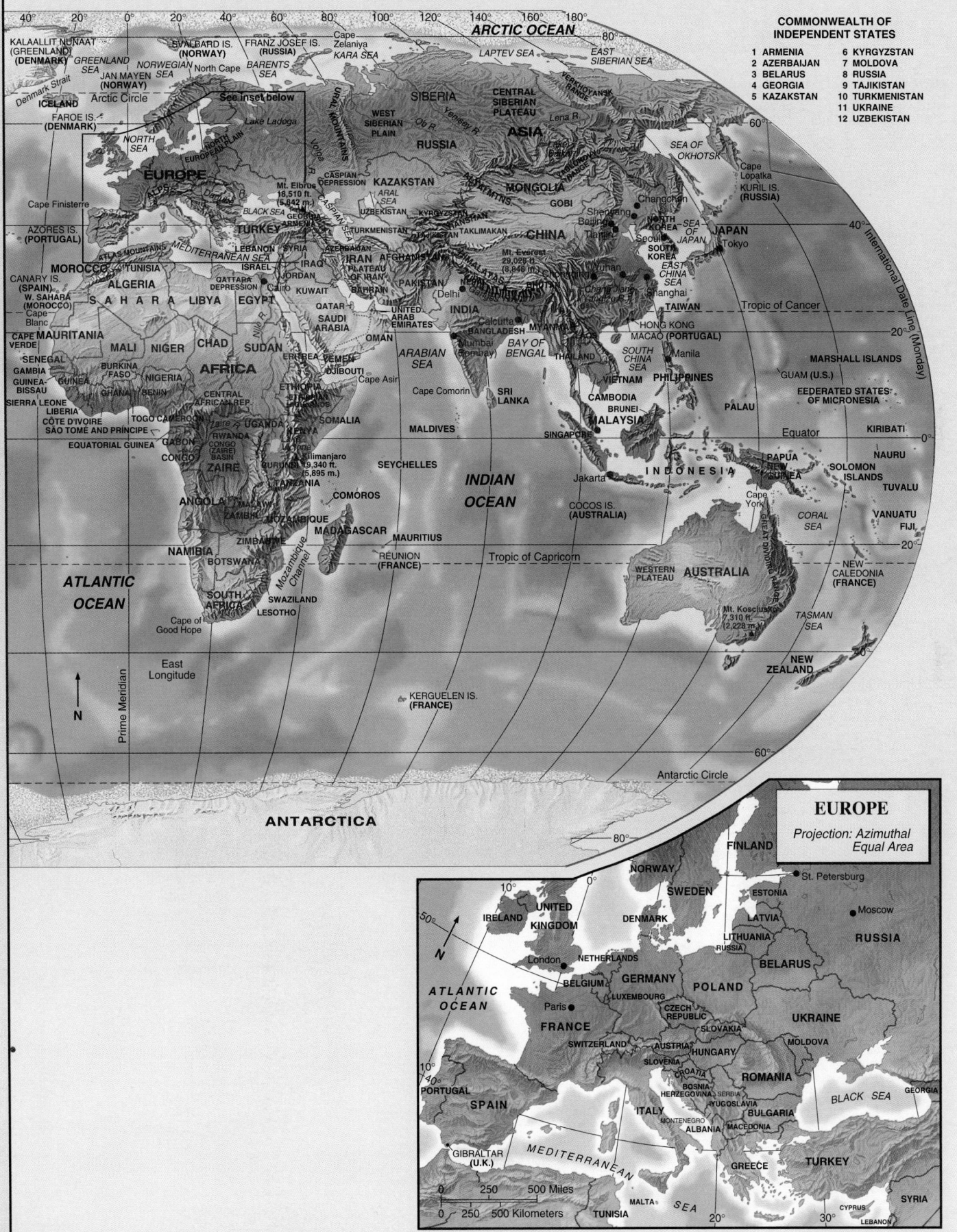

COMMONWEALTH OF
INDEPENDENT STATES

1	ARMENIA	6	KYRGYZSTAN
2	AZERBAIJAN	7	MOLDOVA
3	BELARUS	8	RUSSIA
4	GEORGIA	9	TAJIKISTAN
5	KAZAKSTAN	10	TURKMENISTAN
		11	UKRAINE
		12	UZBEKISTAN

EUROPE

Projection: Azimuthal Equal Area

Animal Cell

Refer to this diagram of an animal cell as you read about cell parts and their jobs.

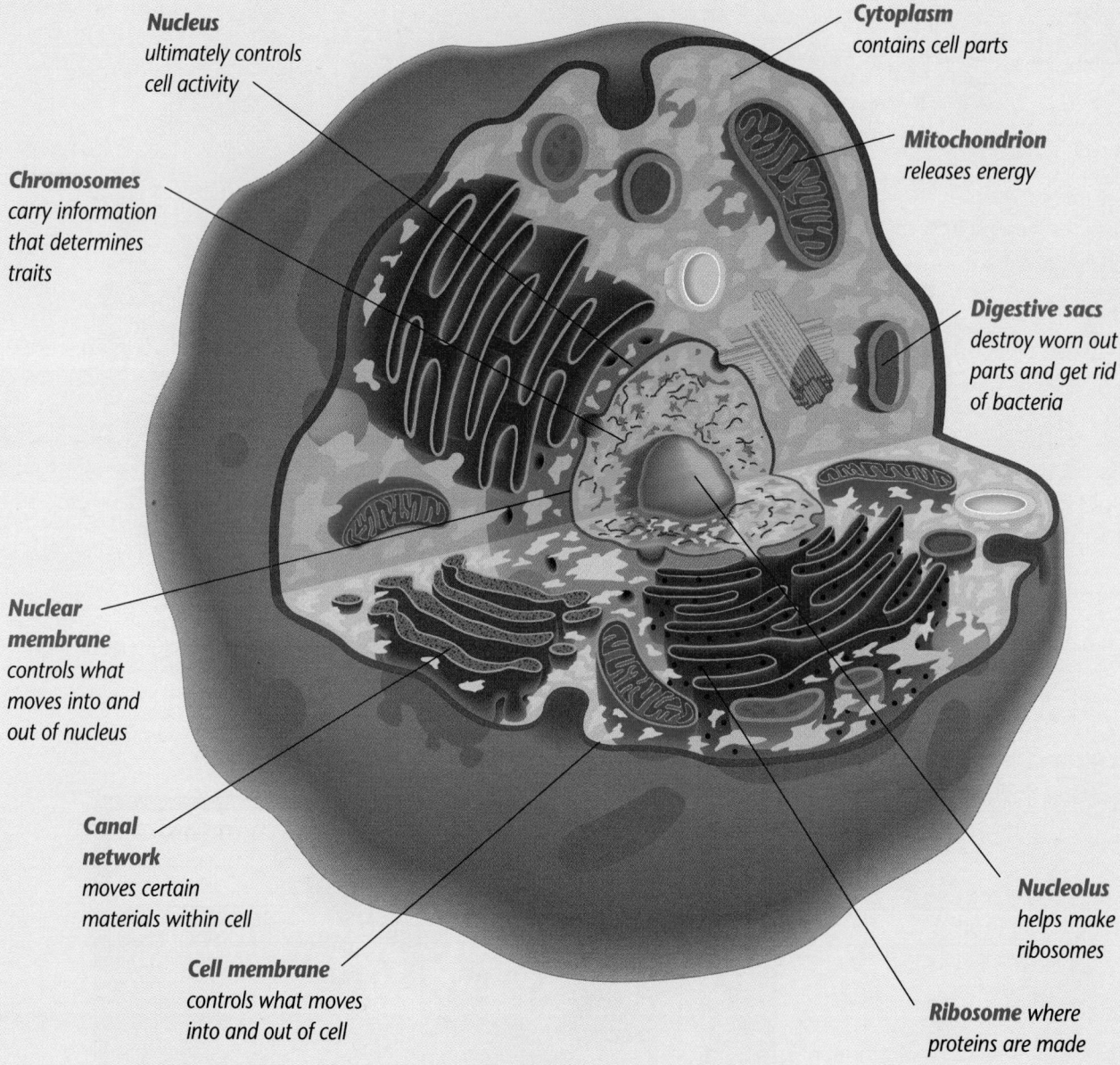

Nucleus
ultimately controls
cell activity

Cytoplasm
contains cell parts

Mitochondrion
releases energy

Chromosomes
carry information
that determines
traits

Digestive sacs
destroy worn out
parts and get rid
of bacteria

**Nuclear
membrane**
controls what
moves into and
out of nucleus

**Canal
network**
moves certain
materials within cell

Cell membrane
controls what moves
into and out of cell

Nucleolus
helps make
ribosomes

Ribosome where
proteins are made

Plant Cell

Refer to this diagram of a plant cell as you read about cell parts and their jobs.

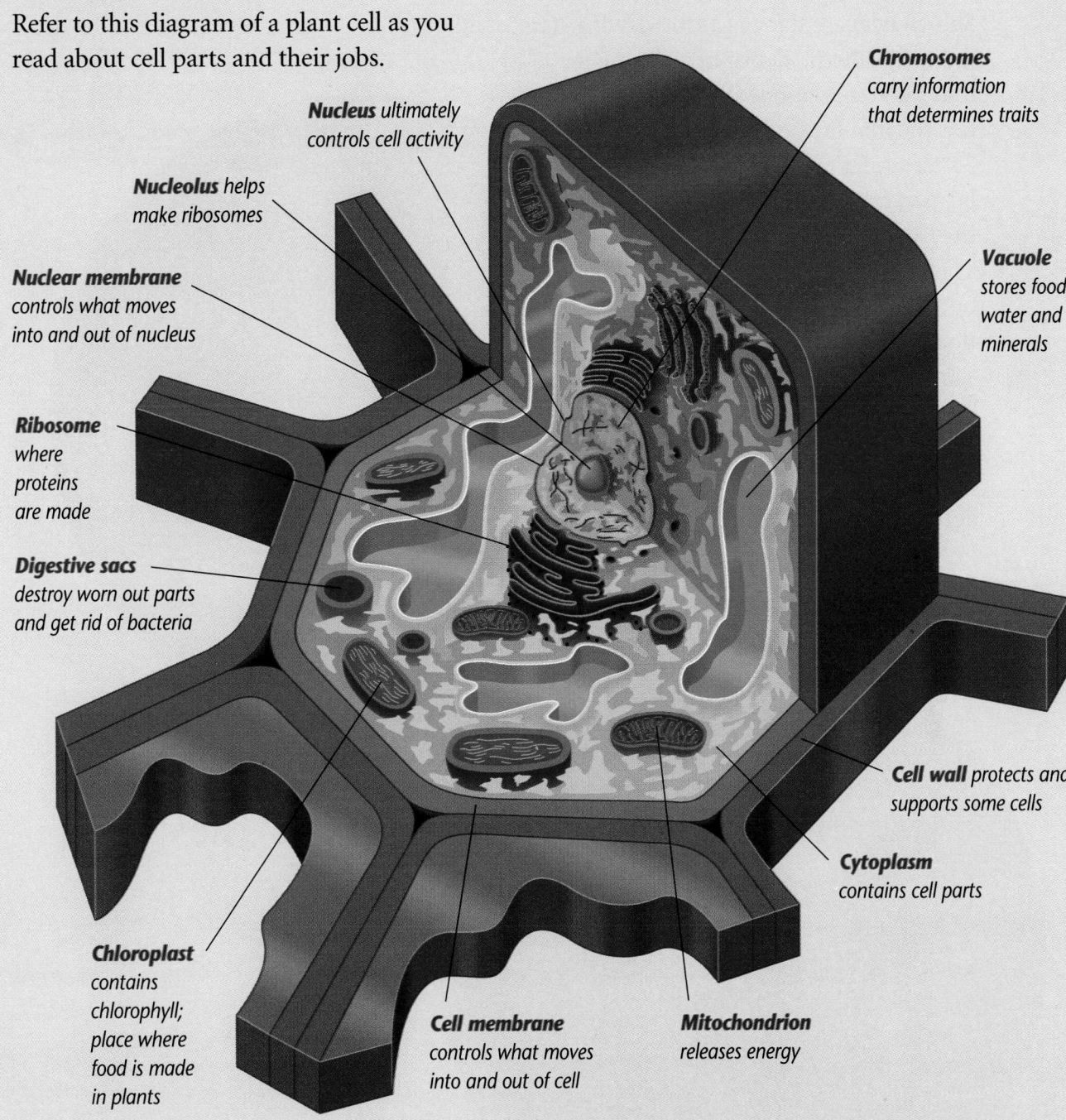

Nucleus ultimately controls cell activity

Nucleolus helps make ribosomes

Nuclear membrane controls what moves into and out of nucleus

Ribosome where proteins are made

Digestive sacs destroy worn out parts and get rid of bacteria

Chloroplast contains chlorophyll; place where food is made in plants

Cell membrane controls what moves into and out of cell

Mitochondrion releases energy

Chromosomes carry information that determines traits

Vacuole stores food water and minerals

Cell wall protects and supports some cells

Cytoplasm contains cell parts

Star Charts

Shown here are star charts for viewing stars in the Northern Hemisphere during the four different seasons. These charts are drawn from the night sky at about 35° north latitude, but they can be used for most locations in the Northern Hemisphere. The lines on the charts outline major constellations. The dense band of stars is the Milky Way. To use, hold the chart vertically, with the direction you are facing at the bottom of the map.

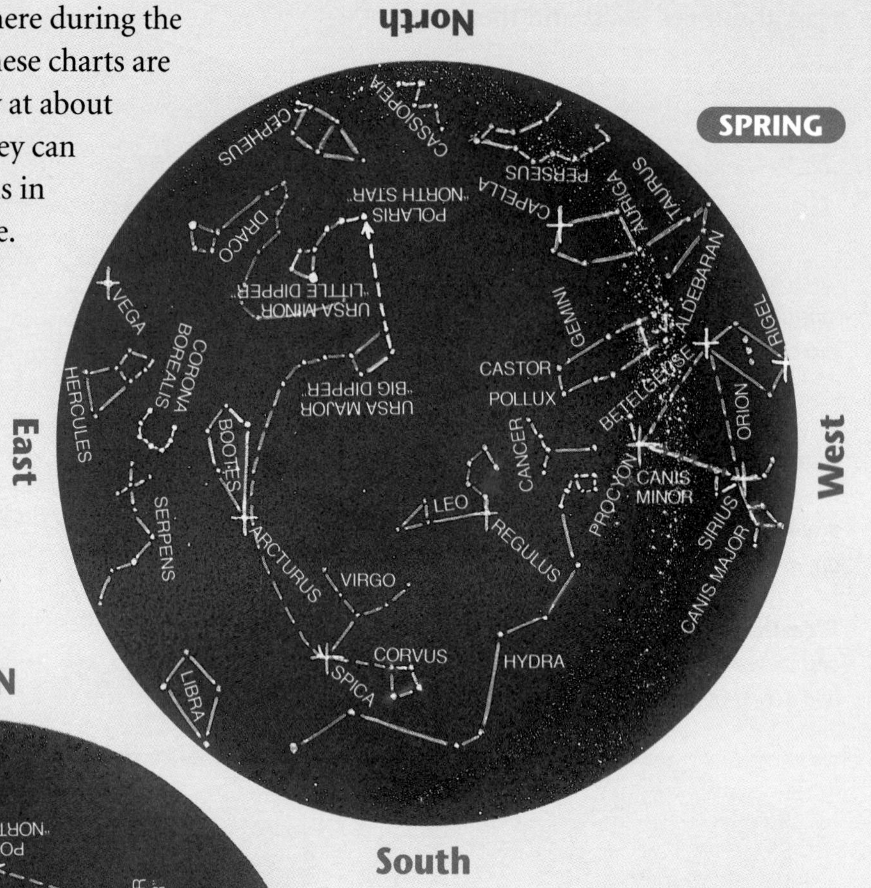

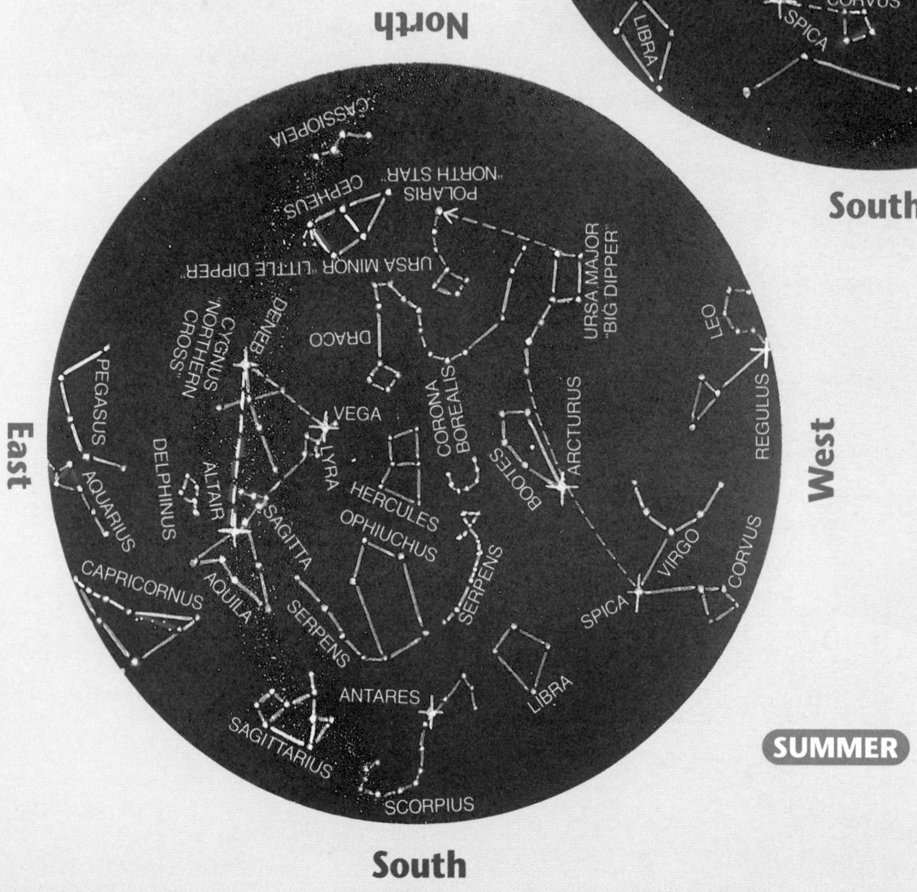

SPRING

North / East / West / South

CEPHEUS · CASSIOPEIA · PERSEUS · CAPELLA · AURIGA · TAURUS · ALDEBARAN · RIGEL · DRACO · POLARIS "NORTH STAR" · LITTLE DIPPER · URSA MINOR · GEMINI · BETELGEUSE · ORION · VEGA · CORONA BOREALIS · CASTOR · POLLUX · PROCYON · CANIS MINOR · SIRIUS · HERCULES · URSA MAJOR "BIG DIPPER" · BOOTES · CANCER · LEO · REGULUS · CANIS MAJOR · SERPENS · ARCTURUS · VIRGO · CORVUS · HYDRA · LIBRA · SPICA

SUMMER

North / East / West / South

CASSIOPEIA · CEPHEUS · POLARIS "NORTH STAR" · URSA MINOR · LITTLE DIPPER · URSA MAJOR "BIG DIPPER" · DENEB · CYGNUS "NORTHERN CROSS" · DRACO · PEGASUS · VEGA · CORONA BOREALIS · LEO · AQUARIUS · DELPHINUS · ALTAIR · LYRA · HERCULES · BOOTES · ARCTURUS · REGULUS · OPHIUCHUS · CAPRICORNUS · AQUILA · SAGITTA · SERPENS · SERPENS · VIRGO · CORVUS · SPICA · LIBRA · ANTARES · SAGITTARIUS · SCORPIUS

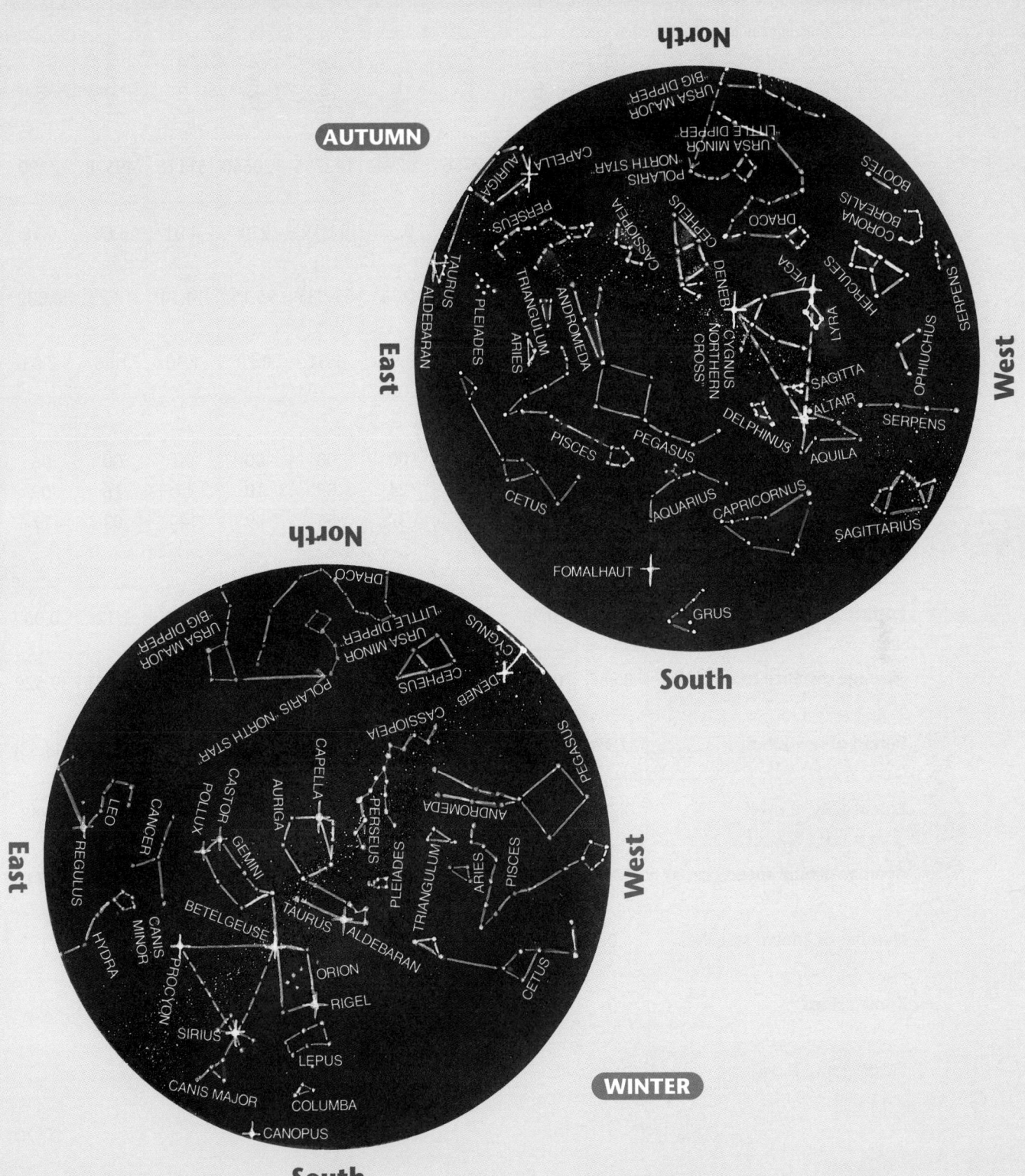

AUTUMN

North

East

West

South

WINTER

North

East

West

South

APPENDIX K

Solar System Information

Planet	Mercury	Venus	Earth	Mars	Jupiter	Saturn	Uranus	Neptune	Pluto
Diameter (km)	4878	12104	12756	6794	142796	120660	51118	49528	2290
Diameter (E = 1.0)*	0.38	0.95	1.00	0.53	11.19	9.46	4.01	3.88	0.18
Mass (E = 1.0)*	0.06	0.82	1.00	0.11	317.83	95.15	14.54	17.23	0.002
Density (g/cm^3)	5.42	5.24	5.50	3.94	1.31	0.70	1.30	1.66	2.03

Period of rotation
  days
  hours
  minutes
R = retrograde
	58
	15
	28
		243
		00
		14$_R$
			00
			23
			56
				00
				24
				37
					00
					09
					55
						00
						10
						39
							00
							17
							14$_R$
								00
								16
								03
									06
									09
									17

Surface gravity (E = 1.0)*	0.38	0.90	1.00	0.38	2.53	1.07	0.92	1.12	0.06
Average distance to sun (AU)	0.387	0.723	1.000	1.524	5.203	9.529	19.191	30.061	39.529
Period of revolution	87.97d	224.70d	365.26d	686.98d	11.86y	29.46y	84.04y	164.79y	248.53y
Eccentricity of orbit	0.206	0.007	0.017	0.093	0.048	0.056	0.046	0.010	0.248
Average orbital speed (km/s)	47.89	35.03	29.79	24.13	13.06	9.64	6.81	5.43	4.74
Number of known satellites	0	0	1	2	16	18	15	8	1
Known rings	0	0	0	0	1	thou-sands	11	4	0

Table K-1: Solar System Information

* Earth = 1.0

Care and Use of a Microscope

Coarse Adjustment *Focuses the image under low power*

Fine Adjustment *Sharpens the image under high and low magnification*

Arm *Supports the body tube*

Low-power objective *Contains the lens with low-power magnification*

Stage clips *Hold the microscope slide in place*

Base *Provides support for the microscope*

Eyepiece *Contains a magnifying lens you look through*

Body tube *Connects the eyepiece to the revolving nosepiece*

Revolving nosepiece *Holds and turns the objectives into viewing position*

High-power objective *Contains the lens with the highest magnification*

Stage *Platform used to support the microscope slide*

Diaphragm *Regulates the amount of light entering the body tube*

Light source *Allows light to reflect upward through the diaphragm, the specimen, and the lenses*

Care of a Microscope

1. Always carry the microscope holding the arm with one hand and supporting the base with the other hand.
2. Don't touch the lenses with your finger.
3. Never lower the coarse adjustment knob when looking through the eyepiece lens.
4. Always focus first with the low-power objective.
5. Don't use the coarse adjustment knob when the high-power objective is in place.
6. Store the microscope covered.

Using a Microscope

1. Place the microscope on a flat surface that is clear of objects. The arm should be toward you.
2. Look through the eyepiece. Adjust the diaphragm so that light comes through the opening in the stage.
3. Place a slide on the stage so that the specimen is in the field of view. Hold it firmly in place by using the stage clips.
4. Always focus first with the coarse adjustment and the low-power objective lens. Once the object is in focus on low power, turn the nosepiece until the high-power objective is in place. Use ONLY the fine adjustment to focus with the high-power objective lens.

Making a Wet Mount Slide

1. Carefully place the item you want to look at in the center of a clean glass slide. Make sure the sample is thin enough for light to pass through.
2. Use a dropper to place one or two drops of water on the sample.
3. Hold a clean coverslip by the edges and place it at one edge of the drop of water. Slowly lower the coverslip onto the drop of water until it lies flat.
4. If you have too much water or a lot of air bubbles, touch the edge of a paper towel to the edge of the coverslip to draw off extra water and force air out.

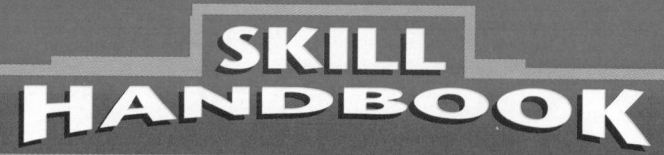

Table of Contents

Organizing Information

Thinking Critically

Practicing Scientific Processes

Representing and Applying Data

Organizing Information

▶ Classifying

You may not realize it, but you make things orderly in the world around you. If you hang your shirts together in the closet, if your socks take up a particular corner of a dresser drawer, or if your favorite CDs are stacked together, you have used the skill of classifying.

Classifying is the process of sorting objects or events into groups based on common features. When classifying, first observe the objects or events to be classified. Then, select one feature that is shared by most members in the group but not by all. Place those members that share the feature into a subgroup. You can classify members into smaller and smaller subgroups based on characteristics.

How would you classify a collection of CDs? You might classify those you like to dance to in one subgroup and CDs you like to listen to in the next column, as in the diagram. The CDs you like to dance to could be subdivided into a rap subgroup and a rock subgroup. Note that for each feature selected, each CD only fits into one subgroup. Keep select-

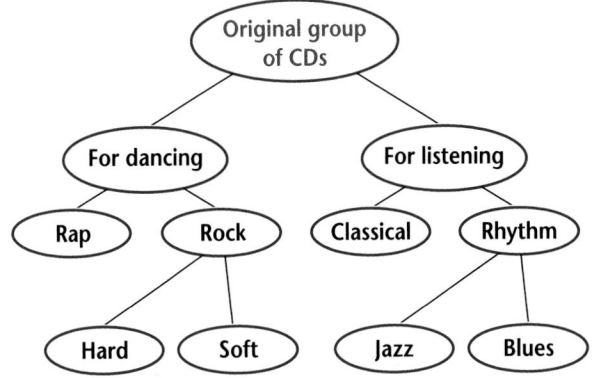

ing features until all the CDs are classified. The diagram above shows one possible classification.

Remember, when you classify, you are grouping objects or events for a purpose. Keep your purpose in mind as you select the features to form groups and subgroups.

▶ Sequencing

A sequence is an arrangement of things or events in a particular order. A sequence with which you are most familiar is the use of alphabetical order. Another example of sequence would be the steps in a recipe. Think about baking chocolate chip cookies. Steps in the recipe have to be followed in order for the cookies to turn out right.

When you are asked to sequence objects or events within a group, figure out what comes first, then think about what should come second. Continue to choose objects or events until all of the objects you started out with are in order. Then, go back over the sequence to make sure each thing or event in your sequence logically leads to the next.

▶ Concept Mapping

If you were taking an automobile trip, you would probably take along a road map. The road map shows your location, your destination, and other places along the way. By looking at the map and finding where you are, you can begin to understand where you are in relation to other locations on the map.

A concept map is similar to a road map. But, a concept map shows relationships among ideas (or concepts) rather than places. A concept map is a diagram that visually shows how concepts are related. Because the concept map shows relationships among ideas, it can make the meanings of ideas and terms clear, and help you understand better what you are studying.

Network Tree   Look at the concept map about Protists. This is called a network tree. Notice how some words are circled while others are written across connecting lines. The circled words are science concepts. The lines in the map show related concepts. The words written on the lines describe the relationships between concepts.

Network Tree

```
              Protists
                 │
              include
        ┌────────┼────────┐
   animal-like  plant-like  fungus-like
    protists     protists    protists
        │           │           │
    known as    known as    known as
        │           │        ┌──┴──┐
   protozoans     algae   water molds  slime molds
```

When you are asked to construct a network tree, write down the topic and list the major concepts related to that topic on a piece of paper. Then look at your list and begin to put them in order from general to specific. Branch the related concepts from the major concept and describe the relationships on the lines. Continue to write the more specific concepts. Write the relationships between the concepts on the lines until all concepts are mapped. Examine the concept map for relationships that cross branches, and add them to the concept map.

Events Chain   An events chain is another type of concept map. An events chain map, such as the one on the effects of gravity, is used to describe ideas in order. In science, an

Events Chain

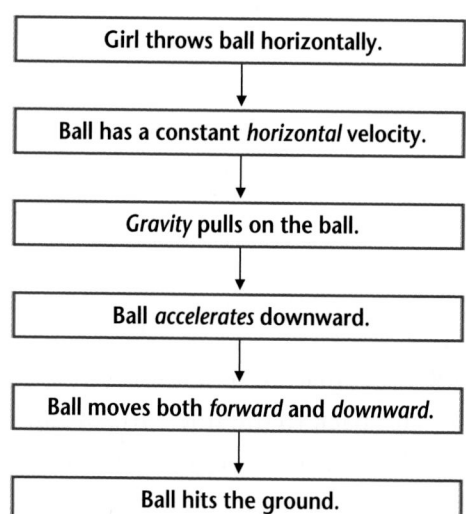

```
┌─────────────────────────────────┐
│  Girl throws ball horizontally.  │
└─────────────────────────────────┘
                │
                ▼
┌─────────────────────────────────────┐
│ Ball has a constant horizontal velocity. │
└─────────────────────────────────────┘
                │
                ▼
┌─────────────────────────────────┐
│     Gravity pulls on the ball.   │
└─────────────────────────────────┘
                │
                ▼
┌─────────────────────────────────┐
│     Ball accelerates downward.   │
└─────────────────────────────────┘
                │
                ▼
┌──────────────────────────────────────┐
│ Ball moves both forward and downward. │
└──────────────────────────────────────┘
                │
                ▼
┌─────────────────────────────────┐
│       Ball hits the ground.      │
└─────────────────────────────────┘
```

events chain can be used to describe a sequence of events, the steps in a procedure, or the stages of a process.

When making an events chain, first find the one event that starts the chain. This event is called the initiating event. Then, find the next event in the chain and continue until you reach an outcome. Suppose you are asked to describe what happens when someone throws a ball horizontally. An events chain map describing the steps might look like the one on page 654. Notice that connecting words are not necessary in an events chain.

Cycle Map   A cycle concept map is a special type of events chain map. In a cycle concept map, the series of events does not produce a

Cycle Map

final outcome. Instead, the last event in the chain relates back to the initiating event.

As in the events chain map, you first decide on an initiating event and then list each event in order. Since there is no outcome and the last event relates back to the initiating event, the cycle repeats itself. Look at the cycle map for photosynthesis shown above.

Spider Map   A fourth type of concept map is the spider map. This is a map that you can use for brainstorming. Once you have a central idea, you may find you have a jumble of ideas that relate to it, but are not necessarily clearly related to each other. By writing these

Spider Map

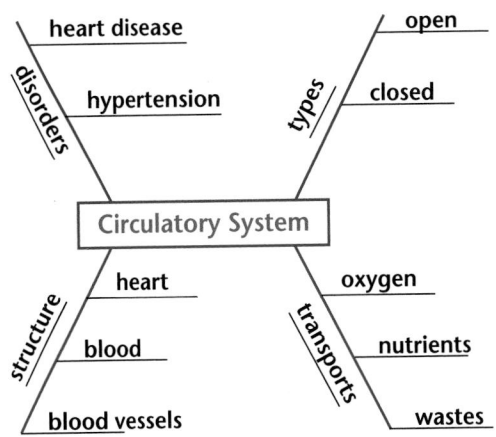

ideas outside the main concept, you may begin to separate and group unrelated terms so that they become more useful.

There is usually not one correct way to create a concept map. As you construct one type of map, you may discover other ways to construct the map that show the relationships between concepts in a better way. If you do discover what you think is a better way to create a concept map, go ahead and use the new way. Overall, concept maps are useful for breaking a big concept down into smaller parts, making learning easier.

▶ Making and Using Tables

Browse through your textbook, and you will notice tables in the text and in the activities. In a table, data or information is arranged in such a way that makes it easier for you to understand. Activity tables help organize the data you collect during an activity so that results can be interpreted more easily.

Parts of a Table Most tables have a title. At a glance, the title tells you what the table is about. A table is divided into columns and rows. The first column lists items to be compared. In the table shown to the right, different magnitudes of force are being compared. The row across the top lists the specific characteristics being compared. Within the grid of the table, the collected data is recorded. Look at the features of the table in the next column.

What is the title of this table? The title is "Earthquake Magnitude." What is being compared? The distance away from the epicenter that tremors are felt and the average number of earthquakes expected per year are being compared for different magnitudes on the Richter scale.

Using Tables What is the average number of earthquakes expected per year for an earthquake with a magnitude of 5.5 at the focus? Locate the column labeled "Average number expected per year" and the row "5.0 to 5.9." The data in the box where the column and row intersect is the answer. Did you answer "800"? What is the distance away from the epicenter for an earthquake with a

Earthquake Magnitude		
Magnitude at Focus	Distance from Epicenter that Tremors are Felt	Average Number Expected Per Year
1.0 to 3.9	24 km	>100 000
4.0 to 4.9	48 km	6200
5.0 to 5.9	112 km	800
6.0 to 6.9	200 km	120
7.0 to 7.9	400 km	20
8.0 to 8.9	720 km	<1

magnitude of 8.1? If you answered "720 km," you understand how to use the parts of a table.

Making Tables To make a table, list the items to be compared down in columns and the characteristics to be compared across in rows. Make a table and record the data comparing the mass of recycled materials collected by a class. On Monday, students turned in 4 kg of paper, 2 kg of aluminum, and 0.5 kg of plastic. On Wednesday, they turned in 3.5 kg of paper, 1.5 kg of aluminum, and 0.5 kg of plastic. On Friday, the totals were 3 kg of paper, 1 kg of aluminum, and 1.5 kg of plastic. If your table looks like the one shown below, you are able to make tables to organize data.

Recycled Materials			
Day of Week	Paper (kg)	Aluminum (kg)	Plastic (kg)
Mon.	4	2	0.5
Wed.	3.5	1.5	0.5
Fri.	3	1	1.5

▶ Making and Using Graphs

After scientists organize data in tables, they may display the data in a graph. A graph is a diagram that shows how variables compare. A graph makes interpretation and analysis of data easier. There are three basic types of graphs used in science—the line graph, the bar graph, and the pie graph.

Line Graphs A line graph is used to show the relationship between two variables. The variables being compared go on two axes of the graph. The independent variable always goes on the horizontal axis, called the *x*-axis. The dependent variable always goes on the vertical axis, called the *y*-axis.

Suppose a school started a peer study program with a class of students to see how science grades were affected.

Average Grades of Students in Study Program

Grading Period	Average Science Grade
First	81
Second	85
Third	86
Fourth	89

You could make a graph of the grades of students in the program over the four grading periods of the school year. The grading period is the independent variable and is placed on the *x*-axis of your graph. The average grade of the students in the program is the dependent variable and would go on the *y*-axis.

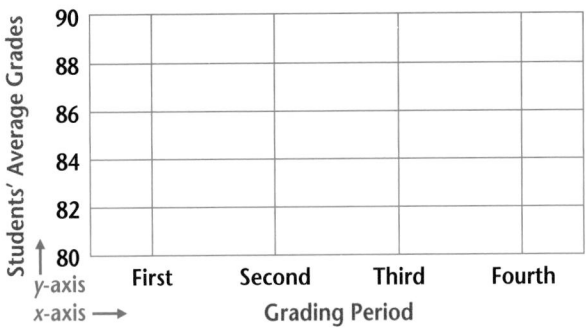

Average Grades of Students in Study Program

After drawing your axes, you would label each axis with a scale. The *x*-axis simply lists the four grading periods. To make a scale of grades on the *y*-axis, you must look at the data values. Since the lowest grade was 81 and the highest was 89, you know that you will have to start numbering at least at 81 and go through 89. You decide to start numbering at 80 and number by twos through 90.

Next, plot the data points. The first pair of data you want to plot is the first grading period and 81. Locate "First" on the *x*-axis and locate "81" on the *y*-axis. Where an imaginary vertical line from the *x*-axis and an imaginary horizontal line from the *y*-axis would meet, place the first data point. Place the other data points the same way. After all the points are plotted, connect them with straight lines.

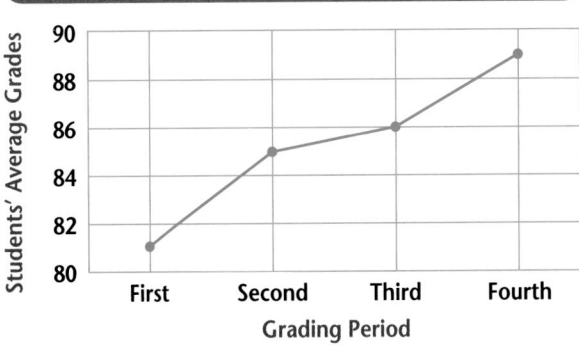

Average Grades of Students in Study Program

Bar Graphs Bar graphs are similar to line graphs. They compare data that do not continuously change. In a bar graph, vertical bars show the relationships among data.

To make a bar graph, set up the *x*-axis and *y*-axis as you did for the line graph. The data is plotted by drawing vertical bars from the *x*-axis up to a point where the *y*-axis would meet the bar if it were extended.

Look at the bar graph comparing the masses lifted by an electromagnet with different numbers of dry cell batteries. The *x*-axis is the number of dry cell batteries, and the *y*-axis is the mass lifted.

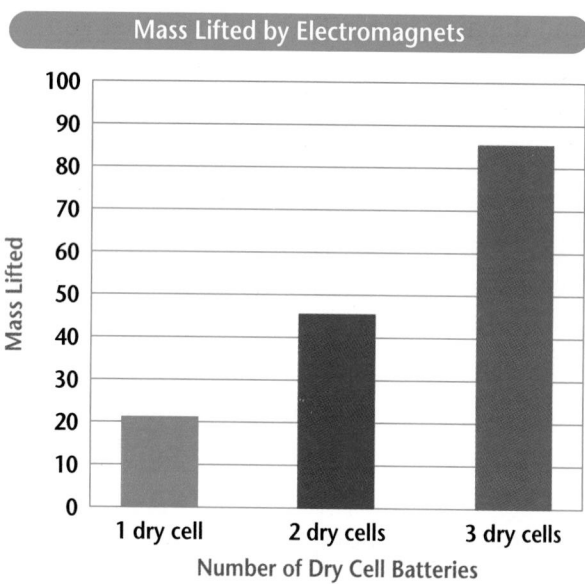

Mass Lifted by Electromagnets

Pie Graphs A pie graph uses a circle divided into sections to display data. Each section represents part of the whole. All the sections together equal 100 percent.

Suppose you wanted to make a pie graph to show the number of seeds that germinated in a package. You would have to count the total number of seeds and the number of seeds that germinated out of the total.

You find that there are 143 seeds in the package. This represents 100 percent, the whole pie.

You plant the seeds, and 129 seeds germinate. The seeds that germinated will make up one section of the pie graph, and the seeds that did not germinate will make up the remaining section.

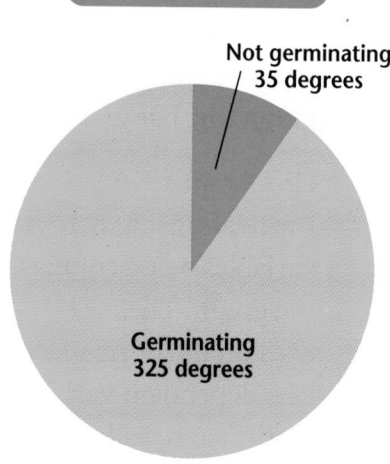

Seeds Germinated

Not germinating 35 degrees

Germinating 325 degrees

To find out how much of the pie each section should take, divide the number of seeds in each section by the total number of seeds. Then multiply your answer by 360, the number of degrees in a circle, and round to the nearest whole number. The section of the pie graph in degrees that represents the seeds germinated is figured below.

$$\frac{129}{143} \times 360 = 324.75 \text{ or } 325 \text{ degrees}$$

Plot this group on the pie graph using a compass and a protractor. Use the compass to draw a circle. Then, draw a straight line from the center to the edge of the circle. Place your protractor on this line and use it to mark a point on the edge of the circle at 325 degrees. Connect this point with a straight line to the center of the circle. This is the section for the group of seeds that germinated. The other section represents the group of 14 seeds that did not germinate. Label the sections of your graph and title the graph.

Thinking Critically

▶ Observing and Inferring

Imagine that you have just finished a volleyball game. At home, you open the refrigerator and see a jug of orange juice on the back of the top shelf. The jug feels cold as you grasp it. Then you drink the juice, smell the oranges, and enjoy the tart taste in your mouth.

As you imagined yourself in the story, you used your senses to make observations. You used your sense of sight to find the jug in the refrigerator, your sense of touch when you felt the coldness of the jug, your sense of hearing to listen as the liquid filled the glass, and your senses of smell and taste to enjoy the odor and tartness of the juice. The basis of all scientific investigation is observation.

Scientists try to make careful and accurate observations. When possible, they use instruments such as microscopes and thermometers or a pan balance to make observations. Measurements with a balance or thermometer provide numerical data that can be checked and repeated.

When you make observations in science, you'll find it helpful to examine the entire object or situation first. Then, look carefully for details. Write down everything you observe.

Scientists often make inferences based on their observations. An inference is an attempt to explain or interpret observations or to say what caused what you observed. For example, if you observed a CLOSED sign in a store window around noon, you might infer the owner is taking a lunch break. But, it's also possible that the owner has a doctor's appointment or has taken the day off to go fishing. The only way to be sure your inference is correct is to investigate further.

When making an inference, be certain to use accurate data and observations. Analyze all of the data that you've collected. Then, based on everything you know, explain or interpret what you've observed.

▶ Comparing and Contrasting

Observations can be analyzed by noting the similarities and differences between two or more objects or events that you observe. When you look at objects or events to see how they are similar, you are comparing them. Contrasting is looking for differences in similar objects or events.

Suppose you were asked to compare and contrast the planets Venus and Earth. You would start by looking at what is known about these planets. Arrange this information in a table, making two columns on a piece of paper and listing ways the planets are similar in one column and ways they are different in the other.

Comparison of Venus and Earth		
Properties	Earth	Venus
Diameter (km)	12 756	12 104
Average density (g/cm³)	5.5	5.3
Percentage of sunlight reflected	39	76
Daytime surface temperature (degrees)	300	750
Number of satellites	1	0

Similarities you might point out are that both planets are similar in size, shape, and mass. Differences include Venus having a hotter surface temperature that reflects more sunlight than Earth reflects. Also, Venus lacks a moon.

▶ Recognizing Cause and Effect

Have you ever watched something happen and then made suggestions as to why it happened? If so, you have observed an effect and inferred a cause. The event is an effect, and the reason for the event is the cause.

Suppose that every time your teacher fed the fish in a classroom aquarium, she or he tapped the food container on the edge of the aquarium. Then, one day your teacher just happened to tap the edge of the aquarium with a pencil while making a point about an ecology lesson. You observed the fish swim to the surface of the aquarium to feed. What is the effect, and what would you infer to be the cause? The effect is the fish swimming to the surface of the aquarium. You might infer the cause to be the teacher tapping on the edge of the aquarium. In determining cause and effect, you have made a logical inference based on your observations.

Perhaps the fish swam to the surface because they reacted to the teacher's waving hand or for some other reason. When scientists are unsure of the cause of a certain event, they design controlled experiments to determine what causes the event. Although you have made a logical conclusion about the behavior of the fish, you would have to perform an experiment to be certain that it was the tapping that caused the effect you observed.

▶ Measuring in SI

The metric system is a system of measurement developed by a group of scientists in 1795. It helps scientists avoid problems by providing standard measurements that all scientists around the world can understand. A modern form of the metric system, called the International System, or SI, was adopted for worldwide use in 1960.

Metric Prefixes			
Prefix	Symbol	Meaning	
kilo-	k	1000	thousand
hecto-	h	100	hundred
deka-	da	10	ten
deci-	d	0.1	tenth
centi-	c	0.01	hundreth
milli-	m	0.001	thousandth

The metric system is convenient because unit sizes vary by multiples of 10. When changing from smaller units to larger units, divide by 10. When changing from larger units to smaller, you multiply by 10. For example, to convert millimeters to centimeters, divide the millimeters by 10. To convert 30 millimeters to centimeters, divide 30 by 10 (30 millimeters equals 3 centimeters).

Prefixes are used to name units. Look at the table for some common metric prefixes and their meanings. Do you see how the prefix *kilo-* attached to the unit *gram* is *kilogram*, or 1000 grams? The prefix *deci-* attached to the unit *meter* is *decimeter*, or one-tenth (0.1) of a meter.

Length You have probably measured lengths or distances many times. The meter is the SI unit used to measure length. A baseball bat is about one meter long. When measuring smaller lengths, the meter is divided into smaller units called centimeters and millimeters. A centimeter is one-hundredth (0.01) of a meter, which is about the size of the width of the fingernail on your ring finger. A millimeter is one-thousandth of a meter (0.001), about the thickness of a dime.

Most metric rulers have lines indicating centimeters and millimeters. The centimeter lines are the longer, numbered lines, and the shorter lines are millimeter lines. When using a metric ruler, line up the 0 centimeter mark with the end of the object being measured, and read the number of the unit where the object ends.

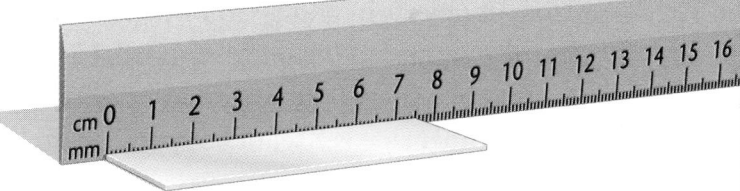

Surface Area Units of length are also used to measure surface area. The standard unit of area is the square meter (m²). A square that's one meter long on each side has a surface area of one square meter. Similarly, a square centimeter (cm²) is one centimeter long on each side. The surface area of an object is determined by multiplying the length times the width.

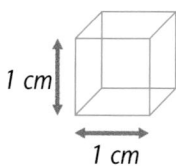

1 cm
1 cm

Volume The volume of a rectangular solid is also calculated using units of length. The cubic meter (m³) is the standard SI unit of volume. A cubic meter is a cube one meter on each side. You can determine the volume of rectangular solids by multiplying length times width times height.

Liquid Volume During science activities, you will measure liquids using beakers and graduated cylinders marked in milliliters. A graduated cylinder is a cylindrical container marked with lines from bottom to top.

Liquid volume is measured using a unit called a liter. A liter has the volume of 1000 cubic centimeters. Since the prefix *milli-* means thousandth (0.001), a milliliter equals one cubic centimeter. One milliliter of liquid would completely fill a cube measuring one centimeter on each side.

Mass Scientists use balances to find the mass of objects in grams. You will use a beam balance similar to the one illustrated. Notice that on one side of the balance is a pan and on the other side is a set of beams. Each beam has an object of a known mass called a *rider* that slides on the beam.

Before you find the mass of an object, set the balance to zero by sliding all the riders back to the zero point. Check the pointer on the right to make sure it swings an equal distance above and below the zero point on the scale. If the swing is unequal, find and turn the adjusting screw until you have an equal swing.

Place an object on the pan. Slide the rider with the largest mass along its beam until the pointer drops below zero. Then move it back one notch. Repeat the process on each beam until the pointer swings an equal distance above and below the zero point. Add the masses on each beam to find the mass of the object.

You should never place a hot object or pour chemicals directly on the pan. Instead, find the mass of a clean beaker or a glass jar. Place the dry or liquid chemicals in the container. Then find the combined mass of the container and the chemicals. Calculate the mass of the chemicals by subtracting the mass of the empty container from the combined mass.

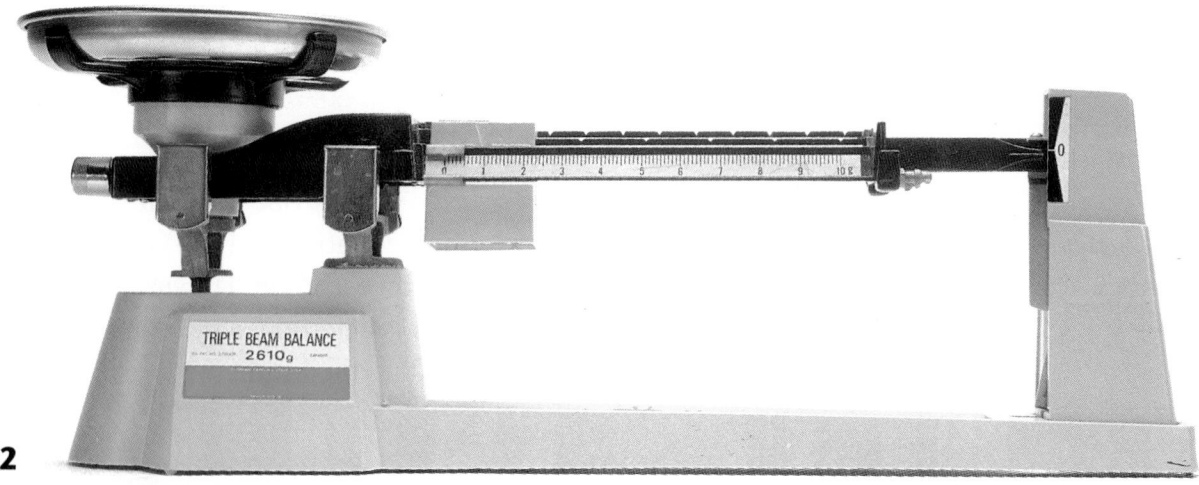

Practicing Scientific Processes

You might say that the work of a scientist is to solve problems. But when you decide how to dress on a particular day, you are doing problem solving, too. You may observe what the weather looks like through a window. You may go outside and see if what you are wearing is warm or cool enough.

Scientists use an orderly approach to learn new information and to solve problems. The methods scientists may use include observing, forming a hypothesis, testing a hypothesis, separating and controlling variables, and interpreting data.

▶ Observing

You observe all the time. Any time you smell wood burning, touch a pet, see

lightning, taste food, or hear your favorite music, you are observing. Observation gives you information about events or things. Scientists try to observe as much as possible about the things and events they study so that they can know that what they say about their observations is reliable.

Some observations describe something using only words. These observations are called qualitative observations. If you were making qualitative observations of a dog, you might use words such as furry, brown, short-haired, or short-eared.

Other observations describe how much of something there is. These are quantitative observations and use numbers as well as words in the description. Tools or equipment are used to measure the characteristic being described. Quantitative observations of a dog might include a mass of 45 kg, a height of 76 cm, ear length of 14 cm, and an age of 283 days.

▶ Using Observations to Form a Hypothesis

Suppose you want to make a perfect score on a spelling test. Begin by thinking of several ways to accomplish this. Base these possibilities on past observations. If you put each of these possibilities into sentence form, using the words if and then, you can form a hypothesis. All of the following are hypotheses you might consider to explain how you could score 100 percent on your test:

If the test is easy, then I will get a perfect score.

If I am intelligent, then I will get a perfect score.

If I study hard, then I will get a perfect score.

Scientists make hypotheses that they can test to explain the observations they have made. Perhaps a scientist has observed that plants that receive fertilizer grow taller than plants that do not. A scientist may form a hypothesis that says: If plants are fertilized, then their growth will increase.

▶ Designing an Experiment to Test a Hypothesis

Once you state a hypothesis, you probably want to find out if it explains an event or an observation or not. This requires a test. A hypothesis must be something you can test. To test a hypothesis, you design and carry out an experiment. Experiments involve planning and materials. Let's figure out how to conduct an experiment to test the hypothesis

stated before about the effects of fertilizer on plants.

First, you need to write out a procedure. A procedure is the plan that you follow in your experiment. A procedure tells you what materials to use and how to use them. In this experiment, your plan may involve using ten bean plants that are each 15-cm tall (to begin with) in two groups, Groups A and B. You will water the five bean plants in Group A with 200 mL of plain water and no fertilizer twice a week for three weeks. You will treat the five bean plants in Group B with 200 mL of fertilizer solution twice a week for three weeks.

You will need to measure all the plants in both groups at the beginning of the experiment and again at the end of the three-week period. These measurements will be the data that you record in a table. A sample table has been done for you. Look at the data in the table for this experiment. From the data, you can draw a conclusion and make a statement about your results. If the conclusion you draw from the data supports your hypothesis, then you can say that your hypothesis is

Growing Bean Plants		
Plants	Treatment	Height 3 Weeks Later
Group A	no fertilizer added to soil	17 cm
Group B	3 g fertilizer added to soil	31 cm

reliable. Reliable means that you can trust your conclusion. If it did not support your hypothesis, then you would have to make new observations and state a new hypothesis, one that you could also test.

▶ Separating and Controlling Variables

In the experiment with the bean plants, you made everything the same except for treating one group (Group B) with fertilizer. In any experiment, it is important to keep everything the same, except for the item you are testing. In the experiment, you kept the type of plants, their beginning heights, the soil, the frequency with which you watered them, and the amount of water or fertilizer all the same, or constant. By doing so, you made sure that at the end of three weeks any change you saw was the result of whether or not the plants had been fertilized. The only thing that you changed, or varied, was the use of fertilizer. In an experiment, the one factor that you change (in this case, the fertilizer), is called the independent variable. The factor that changes (in this case, growth) as a result of the independent variable is called the dependent variable. Always make sure that there is only one independent variable. If you allow more than one, you will not know what causes the changes you observe in the dependent variable.

Many experiments also have a control, a treatment that you can compare with the results of your test groups. In this case, Group A was the control because it was not treated with fertilizer. Group B was the test group. At the end of three weeks, you were able to compare Group A with Group B and draw a conclusion.

▶ Interpreting Data

The word *interpret* means to explain the meaning of something. Information, or data, needs to mean something. Look at the problem originally being explored and find out what the data shows. Perhaps you are looking at a table from an experiment designed to test the hypothesis: If plants are fertilized, then their growth will increase. Look back to the table showing the results of the bean plant experiment.

Identify the control group and the test group so you can see whether or not the variable has had an effect. In this example, Group A was the control and Group B was the test group. Now you need to check differences between the control and test groups. These differences may be qualitative or quantitative. A qualitative difference would be if the leaf colors of plants in Groups A and B were different. A quantitative difference would be the difference in numbers of centimeters of height among the plants in each group. Group B was in fact taller than Group A after three weeks.

If there are differences, the variable being tested may have had an effect. If there is no difference between the control and the test groups, the variable being tested apparently

had no effect. From the data table in this experiment on page 664, it appears that fertilizer does have an effect on plant growth.

▶ What is Data?

In the experiment described on these pages, measurements have been taken so that at the end of the experiment, you had something concrete to interpret. You had numbers to work with. Not every experiment that you do will give you data in the form of numbers. Sometimes, data will be in the form of a description. At the end of a chemistry experiment, you might have noted that one solution turned yellow when treated with a particular chemical, and another remained clear, like water, when treated with the same chemical. Data therefore, is stated in different forms for different types of scientific experiments.

▶ Are All Experiments Alike?

Keep in mind as you perform experiments in science, that not every experiment makes use of all of the parts that have been described on these pages. For some, it may be difficult to design an experiment that will always have a control. Other experiments are complex enough that it may be hard to have only one dependent variable. Real scientists encounter many variations in the methods that they use when they perform experiments. The skills in this handbook are here for you to use and practice. In real situations, their uses will vary.

Representing and Applying Data

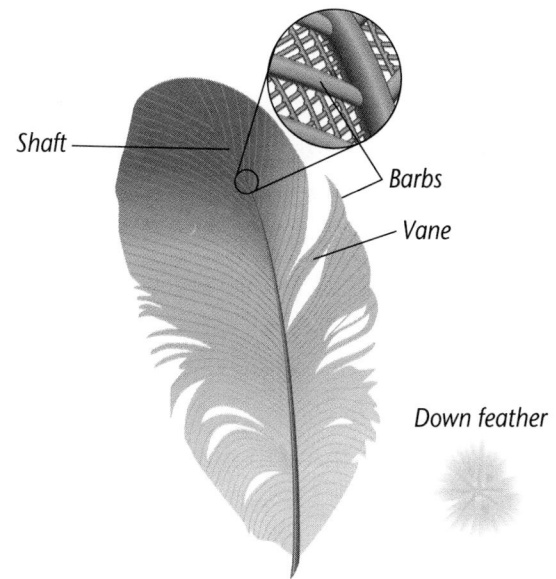

Shaft

Barbs

Vane

Down feather

Contour feather

▶ Interpreting Scientific Illustrations

As you read this textbook, you will see many drawings, diagrams, and photographs. Illustrations help you to understand what you read. Some illustrations are included to help you understand an idea that you can't see easily by yourself. For instance, we can't see atoms, but we can look at a diagram of an atom and that helps us to understand some things about atoms. Seeing something often helps you remember more easily. The text may describe the surface of Jupiter in detail, but seeing a photograph of Jupiter may help you to remember that it has cloud bands. Illustrations also provide examples that clarify difficult concepts or give additional information about the topic you are studying. Maps, for example, help you to locate places that may be described in the text.

Captions and Labels   Most illustrations have captions. A caption is a comment that identifies or explains the illustration. Diagrams, such as the one of the feather, often have labels that identify parts of the item shown or the order of steps in a process.

Learning with Illustrations   An illustration of an organism shows that organism from a particular view or orientation. In order to understand the illustration, you may need to identify the front (anterior) end, tail (posterior) end, the underside (ventral), and the back (dorsal) side of the organism shown.

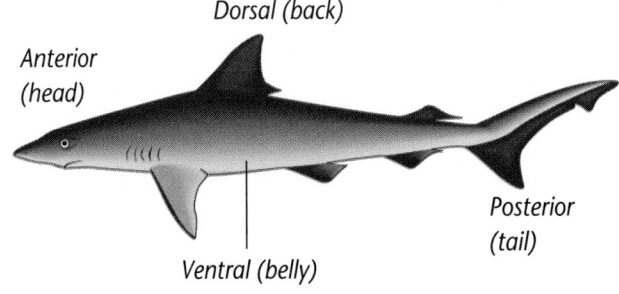

Dorsal (back)

Anterior (head)

Posterior (tail)

Ventral (belly)

You might also check for symmetry. Look at the illustration on the following page. A shark has bilateral symmetry. This means that drawing an imaginary line through the center of the animal from the anterior to posterior end forms two mirror images.

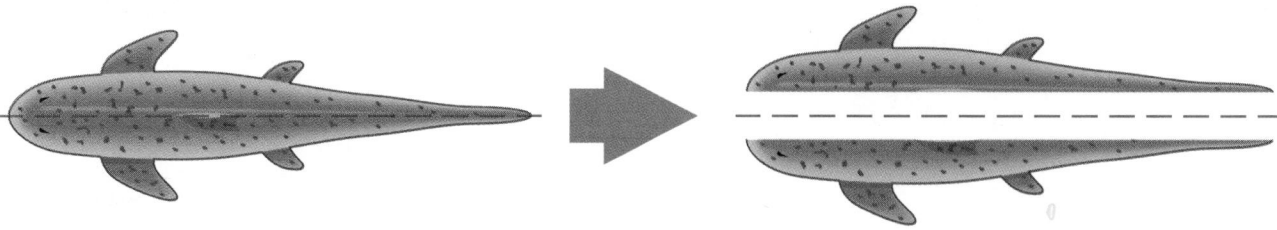

Bilateral symmetry

Two sides exactly alike

Radial symmetry is the arrangement of similar parts around a central point. An object or organism such as a hydra can be divided anywhere through the center into similar parts.

Some organisms and objects cannot be divided into two similar parts. If an organism or object cannot be divided, it is asymmetrical. Regardless of how you try to divide a natural sponge, you cannot divide it into two parts that look alike.

Some illustrations enable you to see the inside of an organism or object. These illustrations are called sections.

Look at all illustrations carefully. Read captions and labels so that you understand exactly what the illustration is showing you.

▶ Making Models

Have you ever worked on a model car or plane or rocket? These models look, and sometimes work, just like the real thing, but they are usually much smaller than the real thing. In science, models are used to help simplify large processes or structures that may be difficult to understand. Your understanding of a structure or process is enhanced when you work with materials to make a model that shows the basic features of the structure or process.

In order to make a model, you first have to get a basic idea about the structure or process involved. You decide to make a model to show the differences in size of arteries, veins, and capillaries. First, read about these structures. All three are hollow tubes. Arteries are round and thick. Veins are flat and have thinner walls than arteries. Capillaries are very small.

Now, decide what you can use for your model. Common materials are often best and cheapest to work with when making models. Different

Butternut squash

Longitudinal section

Cross section

kinds and sizes of pasta might work for these models. Different sizes of rubber tubing might do just as well. Cut and glue the different noodles or tubing onto thick paper so the openings can be seen. Then label each. Now you have a simple, easy–to–understand model showing the differences in size of arteries, veins, and capillaries.

What other scientific ideas might a model help you to understand? A model of a molecule can be made from gumdrops (using different colors for the different elements present) and toothpicks (to show different chemical bonds). A working model of a volcano can be made from clay, a small amount of baking soda, vinegar, and a bottle cap. Other models can be devised on a computer.

▶ Predicting

When you apply a hypothesis, or general explanation, to a specific situation, you predict something about that situation. First, you must identify which hypothesis fits the situation you are considering. People use prediction to make everyday decisions. Based on previous observations and experiences, you may form a hypothesis that if it is wintertime, then temperatures will be lower. From past experience in your area, temperatures are lowest in February. You may then use this hypothesis to predict specific temperatures and weather for the month of February in advance. Someone could use these predictions to plan to set aside more money for heating bills during that month.

▶ Sampling and Estimating

When working with large populations of organisms, scientists usually cannot observe or study every organism in the population. Instead, they use a sample or a portion of the population. Sampling is taking a small portion of organisms of a population for research. By making careful observations or manipulating variables with a portion of a group, information is discovered and conclusions are drawn that might then be applied to the whole population.

Scientific work also involves estimating. Estimating is making a judgment about the size of something or the number of something without actually measuring or counting every member of a population.

Suppose you are trying to determine the effect of a specific nutrient on the growth of black-eyed Susans. It would be impossible to test the entire population of black-eyed Susans, so you would select part of the population for your experiment. Through careful experimentation and observation on a sample of the population, you could generalize the effect of the chemical on the entire population.

Here is a more familiar example. Have you ever tried to guess how many beans were in a sealed jar? If you did, you were estimating. What if you knew the jar of beans held one liter (1000 mL)? If you knew that 30 beans would fit in a 100-milliliter jar, how many beans would you estimate to be in the one-liter jar? If you said about 300 beans, your estimate would be close to the actual number of beans.

Scientists use a similar process to estimate populations of organisms from bacteria to buffalo. Scientists count the actual number of organisms in a small sample and then estimate the number of organisms in a larger area. For example, if a scientist wanted to count the number of microorganisms in a petri dish, a microscope could be used to count the number of organisms in a one square millimeter sample. To determine the total population of the culture, the number of organisms in the square millimeter sample is multiplied by the total number of millimeters in the culture.

GLOSSARY

This glossary defines each key term that appears in bold type in the text. It also indicates the chapter number and page number where you will find the word used.

A

acceleration: rate of change of velocity; acceleration takes place whenever you speed up, slow down, move in a circle, or go around a corner. (Chap. 12, p. 397)

acceleration due to gravity: acceleration on or near Earth that is the same for any free falling object; can be expressed as $g = 9.8/s^2$; satellites remain orbiting around Earth because of horizontal velocity and acceleration due to gravity. (Chap. 13, p. 418)

acids: sour-tasting, corrosive compounds that may release hydrogen gas when reacting with some metals; examples include nitric and sulfuric acids in acid rain, and hydrochloric acid in your stomach. (Chap. 6, p. 188)

adaptation: inherited trait that helps an organism survive; the better adapted a living thing is, the greater its chances to reproduce. (Chap. 7, p. 221)

amplitude: measure of the distance from the top of a crest or the bottom of a trough to the middle level; the energy carried by mechanical waves depends on their amplitude. (Chap. 17, p. 550)

aquifer: layer of permeable rock or soil that allows groundwater to move in and out freely; aquifers provide water for many communities. (Chap. 14, p. 461)

average acceleration: the change in velocity divided by the time interval during which this change occurred; the acceleration of an object at any one instant will most probably be different from its average acceleration. (Chap. 12, p. 400)

average speed: measure of motion that can be found by dividing the total distance traveled by the total time needed to travel that distance; tells how rapidly an object travels but does not describe the direction of the motion. (Chap. 12, p. 387)

average velocity: measure of motion that tells how fast an object moved over a specific displacement; can be found by the following equation: average velocity = total displacement/time interval. (Chap. 12, p. 392)

B C

bases: bitter-tasting, usually solid compounds that feel slippery when dissolved in water; blood and some other body fluids are mildly basic; soaps, antacids, drain cleaners, and household ammonia are common bases. (Chap. 6, p. 193)

cellular respiration: chemical change occurring inside body cells in which oxygen combines with digested food and energy is released. (Chap. 9, p. 297)

centripetal acceleration: acceleration experienced by an object when it is moving in a circle at a constant speed. (Chap. 12, p. 404)

chemical change: materials do not keep their identity after a chemical change; for example, when iron and sulfur are heated, they are chemically changed to iron sulfide, which has properties different from either iron or sulfur; some signs that a chemical change has taken place include smoke, foaming, smell, and sound. (Chap. 4, p. 137)

chemical property: any characteristic of a substance that allows it to undergo a chemical change; chemical properties such as flammability and sensitivity to light can be used to describe and identify materials. (Chap. 4, p. 138)

chlorophyll: green pigment in plants that traps light from the sun, enabling plants to carry out the process of photosynthesis, which changes water and carbon dioxide into sugar and oxygen in a series of chemical reactions. (Chap. 10, p. 336)

cilia: short hairlike structures covering the body of some protists, such as paramecia; cilia are used by this protozoan group for movement and to sweep food into their mouths. (Chap. 8, p. 261)

classification: any organizational grouping system; living things are classified by scientists according to their traits and given names, which helps us understand how organisms are related. (Chap. 7, p. 225)

climax community: final stage of succession in which the plants and animals of a community are well adapted and make up a balanced ecosystem; a mature oak and maple forest is an example. (Chap. 16, p. 510)

colloid: mixture whose medium-sized particles do not settle out of solution upon standing and are large enough to scatter light; milk, gelatin, and fog are colloids. (Chap. 5, p. 174)

commensalism: symbiotic relationship in which one species benefits and the other species is neither harmed nor benefited; for example, Spanish moss receives support from its host trees but the trees are neither harmed nor helped. (Chap. 16, p. 518)

community: all of the populations that live together in the same space and interact with one another; for example, a community could include grasses and various plants and a variety of animals such as hippos, zebras, and birds. (Chap. 11, p. 348)

compound: material made of two or more elements that are chemically combined; the same compound always has the same composition; a compound cannot be separated into its separate parts by physical means, and the substances making up a compound do not retain their own properties; iron sulfide is a compound. (Chap. 4, p. 123)

compression: area of bunched-up air particles created by an object's vibrations, which push the air particles in front of its movements closer together; compressions and rarefactions spread out from the vibrating object in all directions and pass through a medium to produce sound. (Chap. 3, p. 88)

concentrated: solution containing a large amount of solute in a solvent; chocolate milk made with 6 tablespoons of syrup is more concentrated than if made with 2 tablespoons of syrup. (Chap. 5, p. 171)

cones: light-detecting nerve cells in the retina; respond to bright light, resulting in color images; the three types of cones—red, green, and blue—allow you to see the entire spectrum; the brain produces a single image from the information sent by both rods and cones. (Chap. 2, p. 76)

constellation: pattern formed by a group of stars; star maps can help you find where constellations are at different times of year; examples are the Big Dipper, the Little Dipper, and Pegasus. (Chap. 1, p. 44)

consumer: organism such as an animal or a fungus that is unable to make its own food and must feed on other organisms. (Chap. 8, p. 250)

continental glaciers: thick ice that covers very large land areas near Earth's polar regions and makes up 96 percent of Earth's glacial ice; continental glaciers and valley glaciers contain about 75 percent of the world's supply of fresh water; much of Earth's landscape has been shaped by glacier movement. (Chap. 15, p. 487)

contour lines: thin lines that show the shape of three-dimensional landforms on two-dimensional topographic maps; the closer the contour lines, the steeper the change in elevation. (Chap. 1, p. 32)

creep: slow, downhill movement of soil in response to freezing and thawing and the pull of gravity; over time creep can help reshape such landforms as hills, which become more rounded and less steep. (Chap. 15, p. 475)

crest: high point or hill of a wave; for example, when a crest strikes a floating object, it moves slightly forward, exhibiting longitudinal motion, and slightly upward, exhibiting transverse motion. (Chap. 17, p. 548)

decomposer: organism such as a mold, mushroom, or bacterium that obtains food by breaking down dead organic material; decomposers are valuable consumers that recycle nutrients back into the environment. (Chap. 11, p. 353)

delta: triangular-shaped land area formed when a river empties into another body of water and deposits sediments; for example, the fertile Mekong Delta, which is Southern Vietnam's chief agricultural area, was formed by the Mekong River depositing topsoil and mineral sediments into the South China Sea. (Chap. 15, p. 481)

density: measure of an object's or material's mass compared to its volume; density is a physical property that can be used to identify unknown materials and can be measured in g/cm^3. (Chap. 4, p. 131)

deposition: accumulation of eroded sediments; for example, the formation of the fertile Mississippi Delta is the result of deposition. (Chap. 15, p. 474)

dilute: solution with a small amount of solute in a solvent; for example, a drink containing 10% fruit juice in water is more dilute than one containing 90% fruit juice. (Chap. 5, p. 171)

displacement: difference between the starting position of an object and its ending position, noting whether the path is straight or curved; for example, round trips produce a displacement of zero because the object ends up where it started. (Chap. 12, p. 391)

distance: length of the path an object travels in changing its position; science uses the meter, or m, for measuring distance, but distance can also be measured in many other units such as blocks or number of steps. (Chap. 12, p. 384)

diversity: presence of many different species within a community; diversity helps ecosystems recover from volcanic eruptions, fires, and floods and is an important factor in succession. (Chap. 16, p. 512)

Doppler effect: apparent change in the pitch and frequency of a sound as the source moves with respect to the observer; a train's horn sounds higher pitched as it moves toward you and lower pitched as it moves away from you. (Chap. 17, p. 563)

GLOSSARY

drainage basin: land area whose runoff is drained into one stream; in the United States, the largest is the Mississippi drainage basin, which is fed by rains falling between the Rocky Mountains and the Appalachians. (Chap. 14, p. 454)

dune: mound of wind-eroded loose sediments or sand particles that are blown by wind and deposited in another location; sand dunes themselves can move by building up a slope on the windward side until the sand falls down the steeper, opposite side; dunes can be found on a beach or in a desert. (Chap. 15, p. 494)

ecosystem: a community of organisms interacting with each other and with the nonliving things in their environment; examples of ecosystems are oceans, cities, and pine forests. (Chap. 11, p. 351)

element: substance that cannot be broken down further by ordinary physical or chemical methods; elements are the building blocks of matter; examples are calcium, iron, and phosphorus. (Chap. 4, p. 122)

elevation: height above sea level or depth below sea level; on a topographic map, elevation describes the shape of an area by telling us how high or low specific landforms are. (Chap. 1, p. 32)

endoskeleton: internal skeleton that provides body support for all vertebrates and is made of bone or, in some fish, is made of cartilage. (Chap. 9, p. 287)

equinox: occurs twice yearly when the sun is directly above the equator, with the result that day and night are the same length all over the world; in the Northern Hemisphere, the spring equinox occurs March 20 or 21 and the fall equinox occurs September 22 or 23. (Chap. 19, p. 608)

erosion: movement of weathered materials from where they formed to a new location; gravity, running water, glaciers, and wind are the four major agents of erosion; erosion can be reduced by such measures as planting vegetation, constructing terraces, and building retaining walls. (Chap. 15, p. 474)

exoskeleton: hardened external covering of many invertebrates that provides protection from fluid loss, protects internal organs, and provides a support system for the body. (Chap. 9, p. 288)

extinction: local or global disappearance of a species either by natural processes such as climate changes or through destruction of habitats or introduction of competing species; dinosaurs and the dodo are examples of extinct species. (Chap. 16, p. 524)

fertilization: sexual reproductive process of organisms (both plants and animals); in external fertilization in some animals, eggs are released into water and fertilized by free-swimming sperm; in internal fertilization, the male deposits sperm-containing fluid into the female's body. (Chap. 9. p. 290)

firn: small, thick granules formed from snow that has survived for one year without melting; firn contains about 50 percent air—as it becomes still more compressed, it will contain less than 20 percent air and become glacial ice. (Chap. 15, p. 486)

flagella: whiplike structures of some protozoans propel animal-like protists through water. (Chap. 8, p. 261)

floodplain: land area beyond the river channel over which fine, light sediments are deposited by a flooding river; floodplains are excellent for planting crops because the sediments contain minerals and rich topsoil. (Chap. 15, p. 481)

food chain: model of how energy in food passes through an ecosystem; for example, energy is passed when plants are eaten by animals, who are then eaten by other animals. (Chap. 11, p. 356)

food web: model that shows the interconnection of food chains in an ecosystem; organisms in a large food web have a better chance of survival than those with a limited diet. (Chap. 11, p. 357)

frequency: number of times an object vibrates in 1 second; frequency is measured in hertz and is recognized by your ears as differences in pitch. (Chap. 3, p. 95)

genus: subgroup of a family; genera are further divided into species; genus is the first part of the organism's scientific name and is always capitalized and italicized. (Chap. 7, p. 231)

groundwater: water that soaks into the ground and accumulates in connecting pores beneath Earth's surface. (Chap. 14, p. 455)

habitat: the particular place in which an organism lives; for example, the ocean is the habitat of an octopus, your neighborhood is your habitat, and an eagle's habitat includes both land and air. (Chap. 11, p. 347)

GLOSSARY

hertz (Hz): unit that measures frequency; 1 hertz is a frequency of 1 vibration/second or 1 cycle/second; for example, dolphins hear sounds up to about 155,000 hertz (Hz) and humans hear sounds up to about 20,000 Hz. (Chap. 3, p. 95)

heterogeneous mixture: mixture containing various particles of different substances that are scattered unevenly throughout; many heterogeneous mixtures can be separated into their component substances by filtration or by hand-separating on the basis of size, color, or shape; examples are paper and pita bread. (Chap. 4, p. 119)

homogeneous mixture: mixture in which the individual particles of the combined substances are evenly distributed throughout and are too small to be seen; two methods of separating homogeneous mixtures into their component substances are by boiling and evaporation; iced tea is a homogeneous mixture. (Chap. 4, p. 119)

hydrologic cycle: system of water circulation at Earth's surface and above Earth's surface; evaporation changes water into water vapor that rises into the air, cools back into liquid by condensation and falls back to Earth's surface, where it evaporates or runs off or soaks into the ground. (Chap. 14, p. 445)

indicator: substance used to determine pH; an acid-base indicator has one color in one pH range and a different color in a different pH range—for example, cabbage juice, an indicator, turns green in a medium base and red in a medium acid; two types of indicators are solutions or indicator paper such as litmus paper. (Chap. 6, p. 202)

interference: occurs when two or more waves cross at one point and their crests and troughs add or subtract; for example, when two reflected light waves from a CD meet, constructive interference allows you to see bright colors, whereas destructive interference prevents colors from being seen. (Chap. 17, p. 556)

invertebrate: animal without a backbone; invertebrates make up the largest part of the animal kingdom—for example, flies, clams, and spiders. (Chap. 9, p. 285)

kingdom: largest group in the scientific classification system of organisms; kingdoms are subdivided into phyla, and in the case of plants, into divisions instead of phyla; the five kingdoms are plant, fungus, moneran, protist, and animal. (Chap. 7, p. 227)

landforms: surface land features; three common landforms are low, mostly flat coastal and interior plains; raised, rather flat plateaus; and craggy or rounded mountains. (Chap. 1, p. 22)

latitude: angular distance north or south of the equator; lines of latitude are parallel to each other and measure distance in degrees from 0° latitude at the equator to 90° north or south latitude at the poles; lines of latitude and longitude were developed by mapmakers to help us locate places on Earth's surface. (Chap. 1, p. 35)

lava: molten rock at Earth's surface that can erupt quietly or explosively, forming shield volcanoes, cinder cone volcanoes, or composite volcanoes. (Chap. 18, p. 577)

limiting factor: any living or nonliving factor that influences the survival of an organism or species; weather conditions, food and shelter, and predator-prey relationships. (Chap. 11, p. 364)

longitude: angular distance east and west of the prime meridian; lines of longitude measure distance in degrees—the prime meridian is 0° longitude, places east of the prime meridian are located from 0° to 180° E longitude and places west of the prime meridian are located from 0° to 180° W longitude. (Chap. 1, p. 36)

longitudinal wave: mechanical wave that travels in the same direction as the disturbance; sound waves are a type of longitudinal wave that travels through the air in a pattern of compressions and rarefactions. (Chap. 17, p. 542)

lunar eclipse: occurs when Earth passes between the moon and sun and casts a curved shadow on the moon; a total lunar eclipse takes place when the moon passes totally into Earth's umbra, or dark inner shadow. (Chap. 19, p. 620)

magma: molten rock deep beneath Earth's surface that rises slowly through cracks in rock or by melting through rock; when magma reaches Earth's surface, it is called lava. (Chap. 18, p. 577)

magnitude: measure of an earthquake's strength; the Richter scale describes magnitude in terms of the amounts of energy released by an earthquake—for example, an earthquake measuring 7.5 on the Richter scale is 1000 times stronger than one measuring 5.5. (Chap. 18, p. 588)

meander: curves or bends in a river; meanders are formed by fast-moving channels of deep water, which are not affected by friction from the stream bottom, wearing away the sides of the streambed. (Chap. 14, p. 451)

medium: any sound-conducting liquid, solid, or gas; for example, you hear most sounds through the medium of air, which conducts the sound produced by a vibrating object to your ears. (Chap. 3, p. 89)

metabolism: total of all the chemical changes that take place in an organism. (Chap. 9, p. 299)

metamorphosis: hormonally controlled process of organisms such as as frogs and insects during which they undergo extensive changes in form as they grow from egg to adult; metamorphosis in insects can be complete (egg, larva, pupa, adult) or incomplete (egg, nymph, adult). (Chap. 9, p. 292)

microorganism: organism that is too small to be seen with the unaided eye; examples include protists, bacteria, and many fungi. (Chap. 8, p. 249)

mixture: any material made of a combination of two or more substances that are combined in such a way that each substance keeps its own properties; mixtures can be heterogeneous or homogeneous and can be physically separated into simpler substances; examples are air and brass. (Chap. 4, p. 119)

mudflow: mass movement of heavy, water-saturated layers of sediments that flow easily downhill, move anything in their path, and can be highly destructive. (Chap. 15, p. 476)

mutualism: symbiotic relationship in which both species benefit; for example, algae living inside the coral polyp produce oxygen and food used by the polyp, which in turn gives the algae nutrients and a home. (Chap. 16, p. 517)

neutralization: chemical reaction during which acidic and basic properties are canceled, resulting in the production of a salt plus water. (Chap. 6, p. 203)

niche: specific role played by an organism within its community; niches allow many populations to share an area because no two plants or animals meet their needs, such as producing or consuming food, in exactly the same way. (Chap. 11, p. 350)

nonvascular plant: spore-producing plant without xylem, phloem, roots, stems, or leaves; nonvascular plants are unable to transport water efficiently; usually live close to water; mosses and liverworts are examples. (Chap. 10, p. 316)

opaque: property of a material that allows it to absorb or reflect all of the light hitting it; light does not pass through an opaque object. (Chap. 2, p. 59)

organism: any living thing; organisms are made up of cells, need water and food and produce wastes, reproduce, grow and develop, respond to stimuli, and adapt to their environment. (Chap. 7, p. 219)

parasite: organism that lives on or in another organism and obtains its food from that living thing; some parasitic fungi cause athlete's foot, ringworm, and plant diseases such as corn smut. (Chap. 8, p. 265)

parasitism: symbiotic relationship in which one species benefits and the other is harmed or, less often, may be killed; mosquitoes, ticks, and lice are examples of parasites. (Chap. 16, p. 520)

period: amount of time it takes for a pendulum bob to swing back and forth once; period is determined by pendulum length, regardless of mass or amplitude. (Chap. 13, p. 430)

pH: scale that indicates how acidic a solution is; ranges from 0 to 14, with 0 being the most acidic and 14 being the most basic; a pH of 7 indicates that the solution is neither acidic nor basic but is neutral. (Chap. 6, p. 199)

phase: cycle in which the moon appears to change its shape, going from full moon to new moon to full moon in about four weeks. (Chap. 1, p. 43)

phloem: tubelike vessels of vascular plants that transport sugar made during photosynthesis, from the leaves to other plant parts. (Chap. 10, p. 314)

photosynthesis: food-producing process in which plants use the energy from sunlight, which is trapped by chlorophyll, to change water and carbon dioxide into sugar and oxygen. (Chap. 10, p. 336)

phylum: subgroup of a kingdom; phyla are divided into smaller categories—classes. (Chap. 7, p. 230)

physical change: substances keep their identity after physical changes, which may include boiling, freezing, breaking, and melting; for example, when pieces of chalk are broken up, resulting in physical changes in length and mass, they are still the same substance. (Chap. 4, p. 136)

physical property: characteristic that can be used to identify and describe materials; examples include color, shape, brittleness, density, length, width, height, mass, and volume. (Chap. 4, p. 127)

pioneer species: the first species to live in a new ecosystem; lichens, which are able to live in barren places and begin the breakdown of rock into soil, are one of many types of pioneer species. (Chap. 16, p. 507)

pitch: the highness or lowness of a sound; the slower the vibration of an object, the lower its pitch, and the faster its vibration, the higher its pitch. (Chap. 3, p. 96)

pollination: in flowering plants, the transfer of sperm-containing pollen grains from the male stamen to the female stigma. (Chap. 10, p. 326)

population: any group of individuals of the same species that occupies a specific area at the same time; the following are all populations: a herd of zebras, a pride of lions, and a group of Cape May warblers. (Chap. 11, p. 348)

position: place at which you or an object is located compared to a reference point; when position is changed, you experience motion. (Chap. 12, p. 382)

producer: organism that uses light energy to make food from water and carbon dioxide in the presence of chlorophyll; examples are green plants, cyanobacteria, and certain algae. (Chap. 8, p. 250)

projectile motion: motion experienced by an object that is launched into the air horizontally or at an angle and then falls back to Earth; projectiles all move forward and downward at the same time. (Chap. 13, p. 422)

rarefaction: area of spread-out air particles—created by an object's vibrations—that occurs opposite an area of compression; sound is produced by patterns of compressions and rarefactions traveling through a medium. (Chap. 3, p. 88)

receptors: light-sensitive structures in the retina that allow you to respond to color and light changes; also

called rods and cones. (Chap. 2, p. 76)

reflection: bouncing back of light from a surface; most things are visible because light bounces off objects and is reflected back to your eyes; smooth surfaces produce regular reflections, and rough surfaces produce diffuse reflections. (Chap. 2, p. 54)

refraction: process that occurs when light moves through materials with different densities, resulting in the light changing speed and refracting, or bending; for example, refraction causes the image of an object in water to appear closer to the surface than it really is. (Chap. 2, p. 64)

regeneration: formation of a missing body part; also, may be a form of asexual reproduction that occurs when entire new individuals grow from a piece of the original organism. (Chap. 9, p. 291)

relative velocity: motion that depends on your frame of reference, or what you're comparing your motion to; for example, when you stand still, your velocity relative to Earth is zero, even though Earth is rotating and you are moving along with it. (Chap. 12, p. 393)

reproduction: ability of an organism to produce more organisms of the same kind; process that ensures the survival of that specific type of organism. (Chap. 7, p. 220; see also plant reproduction, Chap. 10)

resonance: an object's tendency to vibrate at the same frequency as another sound source; the frequency at which an object resonates is affected by such factors as its shape and the type of material from which it is made. (Chap. 3, p. 103)

retina: light-sensitive tissue in the back of the eye; receives light from the lens; contains rods and cones, which send messages through the optic nerve to the brain, resulting in image formation. (Chap. 2, p. 75)

revolution: elliptical trip of Earth around the sun, which takes about 365.25 days, or one year, to complete; Earth's revolution on its tilted axis around the sun is responsible for the change in seasons. (Chap. 19, p. 604)

rockslide: sudden mass movement of falling rocks; rockslides are most likely to occur on the steep, rocky slopes of mountains and happen most often during earthquakes or after heavy rain. (Chap. 15, p. 476)

rods: light-detecting nerve cells in the retina of the eye; respond to dim light, resulting in black-and-white images. (Chap. 2, p. 76)

rotation: spinning motion of Earth in space, responsible for day and night; Earth rotates around its axis, which is an imaginary line running through the North and South Poles, and takes 24 hours to complete one rotation. (Chap. 19, p. 604)

runoff: rainwater that flows off a land surface and does not soak in; usually results when land is hard, steep, or barren, or can result when rain falls too hard and fast. (Chap. 14, p. 446)

salts: compounds formed as part of neutralization; examples include sodium chloride, which is used in food preparation and chemical manufacture, and potassium permanganate, which is used to tan leather and purify water. (Chap. 6, p. 203)

saturated: a solution is saturated when solutes will no longer dissolve in it at a given temperature; when heated to higher temperatures, saturated solutions may become unsaturated and able to hold more solute. (Chap. 5, p. 166)

seismograph: instrument used to record earthquake vibrations, which are recorded as wavy lines whose peaks indicate the magnitude, or strength, of the earthquake. (Chap. 18, p. 588)

slump: slow slippage of rock layers or a large mass of loose material; occurs when its underlying material gives way and the overlying material slips downslope in a single mass; over time slumps can help reshape such landforms as valleys. (Chap. 15, p. 475)

solar eclipse: occurs when the moon passes directly between the sun and Earth, casting shadow on Earth; the dark inner shadow, or umbra, is experienced as a total eclipse; the lighter, outer shadow, or penumbra, is experienced as a partial eclipse. (Chap. 19, p. 615)

solstice: occurs twice yearly when the sun is directly over the northernmost or southernmost edges of the tropics; in the Northern Hemisphere, the winter solstice (December 21 or 22) has the fewest daylight hours and the summer solstice (June 21 or 22) has the most daylight hours. (Chap. 19, p. 608)

solubility: number of grams of solute that will dissolve in 100 g of a solvent at a given temperature; for example, at room temperature, table salt has a solubility of 36.0 g/100 g water and sugar has a solubility of 204.0 g/100 g water. (Chap. 5, p. 166)

solute: any substance that dissolves in a solvent and forms a solution; solutes dissolve in solvents at different rates depending on such factors as the size of the particles, shaking or stirring, and temperature. (Chap. 5, p. 156)

solution: mixture made up of tiny, evenly mixed particles that do not settle out on standing and cannot be separated by mechanical means; solutions can be made up of solids, gases, and liquids; for example, air is a gas-gas solution, soft drinks are gas-liquid solutions, and sterling silver is a solid-solid solution. (Chap. 5, p. 156)

solvent: the substance in which a solute is dissolved, forming a solution; water is known as a universal solvent because it can dissolve so many different solutes. (Chap. 5, p. 156)

species: subgroup of a genus; in Linnaeus's two-part naming system, the second name always lower case and italicized. (Chap. 7, p. 231)

spectrum: separate bands of colors produced when white light, which is a mixture of all colors of light, passes through a prism; colors usually listed are red, orange, yellow, green, blue, indigo, and violet. (Chap. 2, p. 67)

sphere: three-dimensional, round object such as a volleyball or basketball whose surface at all points is equidistant from its center; Earth's sphere is slightly flattened at the poles and bulges slightly at the equator. (Chap. 19, p. 603)

stationary satellite: type of satellite that moves in its daily orbit around Earth's equator at the same rate as Earth rotates, so to a person on Earth the satellite has no motion and always appears at the same place in the sky. (Chap. 13, p. 429)

stimulus: anything that produces a change in how an organism responds to its environment; stimuli can include changes in sound, light, touch, or taste. (Chap. 7, p. 221)

substance: anything that is made of only one kind of material; the identity of a substance changes after a chemical change but not after a physical change; examples of substances are sugar, water, and chalk. (Chap. 4, p. 118)

succession: gradual process of change through which new environments become populated, or one community slowly replaces another community, or an ecosystem recovers from such disasters as flooding or volcanic eruption. (Chap. 16, p. 507)

suspension: liquid-containing mixture whose unevenly mixed particles settle out of solution upon standing, are large enough to scatter light, and can be separated by filter paper. (Chap. 5, p. 175)

symbiosis: cooperative relationship between members of two or more species; mutualism, parasitism, and commensalism are three types. (Chap. 16, p. 516)

GLOSSARY

tides: slow-moving water waves with long wavelengths that produce the alternate rise, or high tides, and fall, or low tides, of the surface levels of oceans; the gravitational forces acting between the moon, sun, and Earth produce two high tides and two low tides daily. (Chap. 19, p. 621)

translucent: property of a material that allows light to pass through it, but bends the light, causing fuzzy images to be seen on its other side; examples of translucent objects are frosted glass and waxed paper. (Chap. 2, p. 59)

transparent: property of a material that allows enough light to pass through it so that objects can be clearly seen on its other side; examples of transparent objects are plate glass and pair of reading glasses. (Chap. 2, p. 59)

transpiration: process of water loss in plants during which water vapor from inside the leaves moves out through the stomata. (Chap. 10, p. 332)

transverse wave: mechanical wave in which wave disturbance travels at right angles to the direction of the wave itself; the up-and-down pulse traveling along a stretched and shaken rope is an example of a transverse wave. (Chap. 17, p. 541)

trough: low point or valley of a wave; for example, when a trough strikes an object floating in water, the object moves slightly backward, exhibiting longitudinal motion, and slightly downward, exhibiting transverse motion. (Chap. 17, p. 548)

tsunami: giant wave that may travel thousands of kilometers in all directions as quickly as a jet plane, reach heights of over 50 km, and cause destruction in coastal areas; tsunamis are caused by earthquakes under the ocean floor, volcanic eruptions in the sea, or by underwater landslides. (Chap. 18, p. 585)

unsaturated: qualitative term describing the fact that an unsaturated solution can hold any amount of solute as long as the amount is less than the amount that would make it saturated at that temperature. (Chap. 10, p. 338)

valley glaciers: small glaciers found at higher elevations in mountainous regions; valley glaciers erode land as they move down mountain slopes, typically creating U-shaped valleys. (Chap. 15, p. 487)

vascular plant: plant with xylem and phloem; most vascular plants reproduce by seeds and some reproduce by spores; vascular plants can grow taller and thicker and live in drier areas than nonvascular plants; examples are ferns, pines, and flowering plants. (Chap. 10, p. 318)

vertebrate: animal with a backbone; vertebrates are chordates; humans, birds, fish, and amphibians are examples of vertebrates. (Chap. 9, p. 285)

viruses: nonliving, submicroscopic structures with a DNA or RNA core surrounded by a protein coat; uses the cell's energy and materials to duplicate itself; viruses cause diseases such as AIDS, rabies, and chicken pox. (Chap. 8, p. 251)

water table: upper limit of the underground area that is saturated with groundwater; springs can be found where the water table meets Earth's surface. (Chap. 14, p. 462)

wavelength: measure of the distance between the crest of one wave and the crest of the next wave or between the trough of one wave and the trough of the next wave; wavelength gets smaller as frequency increases. (Chap. 17, p. 551)

weightlessness: state experienced during a free-fall as you accelerate toward Earth, regardless of the direction of your velocity; in space, biological effects of weightlessness include loss of calcium in the bones and loss of body mass. (Chap. 13, p. 428)

xylem: specialized tubelike vessels of vascular plants; xylem transports minerals and water absorbed by roots up the stem to the leaves. (Chap. 10, p. 314)

This glossary defines each key term that appears in bold type in the text. It also indicates the chapter number and page number where you will find the word used.

acceleration/aceleración ritmo al cual cambia la velocidad (Cap. 12, pág. 397)

acceleration due to gravity/aceleración por gravedad aceleración que es la misma para cualquier objeto que cae hacia la tierra (Cap. 13, pág. 418)

acids/ácidos compuestos que contienen hidrógeno, tienen un sabor agrio y son corrosivos (Cap. 6, pág. 188)

adaptation/adaptación cualquier rasgo que posee un organismo y que lo ayuda a sobrevivir en su ambiente (Cap. 7, pág. 221)

amplitude/amplitud distancia desde la cresta o el seno de una onda hasta el nivel intermedio de la misma onda (Cap. 17, pág. 550)

aquifer/acuífero capa de suelo o de roca permeable que le permite al agua infiltrarse y fluir libremente (Cap. 14, pág. 461)

average acceleration/aceleración promedio cambio en la velocidad dividido entre el intervalo de tiempo durante el cual ocurre el cambio (Cap. 12, pág. 400)

average speed/rapidez promedio se obtiene al dividir la distancia total viajada entre el tiempo total requerido para viajar esa distancia (Cap. 12, pág. 387)

average velocity/velocidad promedio la rapidez con que un objeto se mueve sobre un desplazamiento (Cap. 12, pág. 392)

bases/bases compuestos de sabor amargo, por lo general, sólidos, que se sienten resbaladizos cuando se disuelven en agua (Cap. 6, pág. 193)

cellular respiration/respiración celular proceso que ocurre dentro de las células del cuerpo cuando el oxígeno se combina con el alimento digerido para liberar energía de los enlaces químicos en el alimento (Cap. 9, pág. 298)

centripetal acceleration/aceleración centrípeta la dirección de la aceleración de un objeto que se mueve a lo largo de una trayectoria circular, la cual tiende a moverlo hacia el centro del círculo (Cap. 12, pág. 404)

chemical change/cambio químico cambio durante el cual una de las sustancias de una materia se transforma en una sustancia diferente (Cap. 4, pág. 137)

chemical property/propiedad química cualquier característica que permite que una sustancia sufra un cambio químico (Cap. 4, pág. 138)

chlorophyll/clorofila pigmento verde de las plantas que atrapa la luz solar (Cap. 10, pág. 336)

cilia/cilios estructuras filamentosas cortas que poseen algunos protozoos por todo su cuerpo (Cap. 8, pág. 261)

classification/clasificación cualquier sistema que se usa para agrupar ideas, información u objetos, basándose en sus semejanzas (Cap. 7, pág. 225)

climax community/comunidad clímax comunidad que se establece después de que un ecosistema alcanza su última etapa de sucesión (Cap. 16, pág. 510)

colloid/coloide mezcla que, al igual que una solución, no se asienta (Cap. 5, pág. 174)

commensalism/comensalismo relación en la cual un organismo se beneficia y el otro ni se beneficia ni se perjudica (Cap. 16, pág. 518)

community/comunidad la interacción de poblaciones en un área determinada (Cap. 11, pág. 348)

compound/compuesto sustancia cuya unidad más pequeña está formada por más de un elemento (Cap. 4, pág. 123)

compression/compresión la parte del patrón de partículas amontonadas que forman las vibraciones en un objeto que vibra (Cap. 3, pág. 88)

concentrated/concentrada solución que contiene una gran cantidad de soluto en un disolvente (Cap. 5, pág. 171)

cones/conos receptores sensibles a todos los colores del espectro visible de la luz (Cap. 2, pág. 76)

constellations/constelaciones patrones que forman los grupos de estrellas (Cap. 1, pág. 44)

consumer/consumidor organismo que no puede producir su propio alimento (Cap. 8, pág. 250)

continental glaciers/glaciares continentales masas de hielo y nieve que cubren áreas extensas de terreno cerca de las regiones polares de la Tierra (Cap. 15, pág. 487)

contour lines/curvas de nivel líneas con la misma elevación que muestran las formas o contornos de los accidentes geográficos (Cap. 1, pág. 32)

creep/corrimiento movimiento lento del suelo cuesta abajo (Cap. 15, pág. 475)

crest/cresta punto más alto o cima de una onda (Cap. 17, pág. 548)

D

decomposer/descomponedor organismo que obtiene su alimento al descomponer organismos muertos y transformarlos en nutrimientos (Cap. 11, pág. 353)

delta/delta área de terreno triangular formada por sedimentos que se depositan cuando un río desemboca en otro cuerpo de agua (Cap. 15, pág. 481)

density/densidad es la cantidad de masa, o de material, que un objeto posee comparada con su volumen (Cap. 4, pág. 131)

deposition/deposición acumulación de sedimentos erosionados (Cap. 15, pág. 474)

dilute/diluida solución que contiene una pequeña cantidad de soluto en un disolvente (Cap. 5, pág. 171)

displacement/desplazamiento cambio neto en la posición de un objeto (Cap. 12, pág. 391)

distance/distancia la lejanía al viajar a lo largo de una ruta, mientras se cambia la posición (Cap. 12, pág. 384)

diversity/diversidad la presencia de muchas especies diferentes dentro de una comunidad (Cap. 16, pág. 512)

Doppler effect/efecto Doppler cambio aparente en la frecuencia de una onda producida por un objeto que se acerca o que se aleja del observador (Cap. 17, pág. 563)

drainage basin/cuenca hidrográfica área desaguada por un río o un arroyo (Cap. 14, pág. 454)

dune/duna montecillo de arena o de sedimentos sueltos formado por la acción del viento (Cap. 15, pág. 494)

E

ecosystem/ecosistema una comunidad de organismos que interactúan entre sí y con el ambiente (Cap. 11, pág. 351)

element/elemento sustancia que no se puede descomponer en sustancias más simples por medios físicos o químicos comunes (Cap. 4, pág. 122)

elevation/elevación la altura sobre el nivel del mar o la profundidad bajo el nivel del mar (Cap. 1, pág. 32)

endoskeleton/endoesqueleto esqueleto que se encuentra dentro del cuerpo del animal (Cap. 9, pág. 287)

equinox/equinoccio cada uno de los dos momentos cada año en que el Sol se encuentra directamente sobre el ecuador (Cap. 19, pág. 608)

erosion/erosión movimiento de los productos de la meteorización desde el lugar en donde se formaron, a un lugar diferente (Cap. 15, pág. 474)

exoskeleton/exoesqueleto sistema de soporte en la parte externa del cuerpo de muchos invertebrados (Cap. 9, pág. 288)

extinction/extinción el desaparecimiento de una especie (Cap. 16, pág. 524)

F

fertilization/fecundación proceso por el cual el espermatozoide de un macho se une con uno o más óvulos producidos por una hembra (Cap. 9, pág. 290)

firn/nieve granular nieve que no se derrite después de un año y que se vuelve más dura y más densa (Cap. 15, pág. 486)

flagella/flagelos estructuras en forma de látigo que poseen algunos protozoos para poder moverse en el agua (Cap. 8, pág. 261)

floodplain/llanura aluvial área formada por sedimentos finos y livianos que viajan más allá del canal de un río (Cap. 15, pág. 481)

food chain/cadena alimenticia modelo que muestra la manera en que la energía de los alimentos pasa de un organismo a otro en un ecosistema (Cap. 11, pág. 356)

food web/red alimenticia combinación de todas las cadenas alimenticias superpuestas en un ecosistema (Cap. 11, pág. 357)

frequency/frecuencia el número de veces que un objeto vibra en un segundo (Cap. 3, pág. 95)

G

genus/género subdivisión de una familia de organismos (Cap. 7, pág. 231)

groundwater/agua subterránea agua que se infiltra en el suelo y que se acumula en los poros (Cap. 14, pág. 455)

H

habitat/hábitat el lugar particular en donde vive un organismo (Cap. 11, pág. 347)

hertz (Hz)/hertz (Hz) un hertz es la frecuencia de una vibración por segundo o un ciclo por segundo (Cap. 3, pág. 95)

heterogeneous mixture/mezcla heterogénea una mezcla en la cual las distintas sustancias están distribuidas irregularmente (Cap. 4, pág. 119)

homogeneous mixture/mezcla homogénea una mezcla en la cual las distintas sustancias están distribuidas uniformemente (Cap. 4, pág. 119)

hydrologic cycle/ciclo hidrológico ciclo en el cual el agua cambia de vapor en el aire a líquido en la Tierra y desde la Tierra regresa al aire (Cap. 14, pág. 445)

I

indicator/indicador sustancia que muestra un color en una gama del pH y otro color en otra gama (Cap. 6, pág. 202)

interference/interferencia interacción de dos o más ondas en un punto determinado (Cap. 17, pág. 556)

invertebrate/invertebrado cualquier animal que no posee una columna vertebral (Cap. 9, pág. 285)

K L

kingdom/reino el grupo de organismos más grande y general en el sistema de clasificación (Cap. 7, pág. 227)

landforms/accidentes geográficos las características naturales de las superficies terrestres (Cap. 1, pág. 22)

latitude/latitud la distancia en grados ya sea al norte o al sur del ecuador (Cap. 1, pág. 35)

lava/lava nombre que recibe el magma una vez que alcanza la superficie terrestre (Cap. 18, pág. 577)

limiting factor/factor limitativo cualquier condición que influye el crecimiento o la supervivencia de un organismo o de una especie (Cap. 11, pág. 364)

longitude/longitud la distancia en grados al este o al oeste del primer meridiano (Cap. 1, pág. 36)

longitudinal waves/ondas longitudinales ondas que viajan en la misma trayectoria o dirección de la perturbación (Cap. 17, pág. 542)

lunar eclipse/eclipse lunar ocurre cuando la Luna atraviesa por la sombra que proyecta la Tierra en el lado de la Luna que mira al Sol (Cap. 19, pág. 620)

M

magma/magma material rocoso derretido que se encuentra debajo de la superficie terrestre (Cap. 18, pág. 577)

magnitude/intensidad medida de la potencia de un terremoto (Cap. 18, pág. 588)

meander/meandro curva de un río que se forma debido a que la velocidad del agua varía dependiendo del ancho del canal (Cap. 14, pág. 451)

medium/medio cualquier sólido, líquido o gas que transporta el patrón de un sonido (Cap. 3, pág. 89)

metabolism/metabolismo es el total de todos los cambios químicos que ocurren en un organismo (Cap. 9, pág. 299)

metamorphosis/metamorfosis cambios de forma que sufren los organismos durante sus ciclos de vida (Cap. 9, pág. 292)

microorganisms/microorganismos organismos tales como las bacterias, los protistas y una gran variedad de hongos que son demasiado pequeños para poder observarlos a simple vista (Cap. 8, pág. 249)

mixture/mezcla cualquier material hecho de dos o más sustancias (Cap. 4, pág. 119)

mudflow/corriente de lodo capas gruesas y espesas de sedimentos que se deslizan fácilmente cuesta abajo (Cap. 15, pág. 476)

mutualism/mutualismo relación en la cual dos especies se benefician (Cap. 16, pág. 517)

N

neutralization/neutralización reacción química que ocurre entre un ácido y una base (Cap. 6, pág. 203)

niche/nicho es el papel que desempeña un organismo dentro de su comunidad (Cap. 11, pág. 350)

nonvascular plant/planta no vascular planta que carece de vasos tubulares para transportar agua, minerales y alimento (Cap. 10, pág. 316)

opaque/opaco objeto que refleja o absorbe la luz pero que no deja pasarla (Cap. 2. pág. 59)

organism/organismo ser viviente (Cap. 7, pág. 219)

parasites/parásitos organismos que viven fuera o dentro de otros organismos, de los cuales obtienen su alimento (Cap. 8, pág. 265)

parasitism/parasitismo relación en la cual una de las especies perjudica o destruye a la otra (Cap. 16, pág. 520)

period/período (de un péndulo) el tiempo que tarda la lenteja del péndulo en oscilar de ida y vuelta una vez (Cap. 13, pág. 430)

pH/pH una medida que muestra el grado de acidez o de basicidad de una solución (Cap. 6, pág. 199)

phase/fase cada etapa del ciclo lunar (Cap. 1, pág. 43)

phloem/floema vasos tubulares que mueven el alimento desde las hojas hacia otras partes de una planta (Cap. 10, pág. 314)

photosynthesis/fotosíntesis proceso en el cual las plantas usan la luz para producir su alimento (Cap. 10, pág. 336)

phylum/fílum subgrupo en que se divide un reino (Cap. 7, pág. 230)

physical change/cambio físico cambio en las propiedades físicas de una sustancia pero sin alterar la sustancia misma (Cap. 4, pág. 136)

physical property/propiedad física cualquier característica de un material que puede observarse o medirse (Cap. 4, pág. 127)

pioneer species/especies pioneras las primeras especies que viven en un ecosistema (Cap. 16, pág. 507)

pitch/tono se refiere al grado de elevación o de bajo nivel de un sonido (Cap. 3, pág. 96)

pollination/polinización la transferencia de granos de polen del estambre al estigma (Cap. 10, pág. 326)

population/población grupo de individuos de la misma especie que viven en un área al mismo tiempo (Cap. 11, pág. 348)

position/posición lugar donde se encuentra un objeto con relación a un punto de referencia (Cap. 12, pág. 382)

producer/productor plantas verdes que usan la energía luminosa para producir alimento del dióxido de carbono y del agua (Cap. 8, pág. 250)

projectile motion/movimiento de un proyectil el que describe el movimiento de cualquier objeto que es lanzado hacia adelante (horizontalmente) y que luego cae de regreso a la Tierra (Cap. 13, pág. 422)

rarefaction/rarefacción la parte del patrón de partículas separadas que forman las vibraciones en un objeto que vibra (Cap. 3, pág. 88)

receptors/receptores estructuras sensibles a la luz que se encuentran en la retina (Cap. 2, pág. 76)

reflection/reflejo luz que rebota de algo (Cap. 2, pág. 54)

refraction/refracción proceso de doblar la luz cuando pasa de un medio a otro (Cap. 2, pág. 64)

regeneration/regeneración proceso que ocurre cuando un animal vuelve a desarrollar una parte del cuerpo que ha perdido (Cap. 9, pág. 291)

relative velocity/velocidad relativa la velocidad de un objeto, la cual se determina desde el marco de referencia de otro objeto (Cap. 12, pág. 393)

reproduction/reproducción proceso mediante el cual los organismos producen más organismos de la misma especie (Cap. 7, pág. 220; Cap. 10, pág. 322)

resonance/resonancia la tendencia de un objeto a vibrar en la misma frecuencia que otra fuente sonora (Cap. 3, pág. 103)

retina/retina tejido ocular sensible a la luz (Cap. 2, pág. 75)

revolution/revolución movimiento elíptico de la Tierra alrededor del Sol, el cual dura 365 1/4 días, o un año (Cap. 19, pág. 604)

rockslide/deslizamiento de rocas movimiento de una masa de rocas que caen (Cap. 15, pág. 476)

rods/bastones receptores sensibles a la luz y a la oscuridad (Cap. 2, pág. 76)

rotation/rotación movimiento de un cuerpo alrededor de un eje (Cap. 19, pág. 604)

runoff/desagüe de aguas aguas que fluyen y que no se infiltran en la tierra (Cap. 14, pág. 446)

salt/sal un tipo de compuesto que se forma como parte de la neutralización (Cap. 6, pág. 203)

saturated/saturada solución que ha disuelto todo el soluto que le es posible a una temperatura dada (Cap. 5, pág. 166)

seismograph/sismógrafo se utiliza para registrar las vibraciones de los terremotos (Cap. 18, pág. 588)

slump/desprendimiento movimiento lento de una masa de material suelto (Cap. 15, pág. 475)

solar eclipse/eclipse solar ocurre cuando la Luna pasa directamente entre la Tierra y el Sol (Cap. 19, pág. 615)

solstice/solsticio cuando el Sol se encuentra directamente sobre el extremo norte o el extremo sur de los trópicos (Cap. 19, pág. 608)

solubility/solubilidad cantidad de una sustancia que puede disolverse en 100 g de disolvente a una temperatura dada (Cap. 5, pág. 166)

solute/soluto cualquier sustancia que parece desaparecer en una mezcla (Cap. 5, pág. 156)

solution/solución cualquier mezcla formada por partículas pequeñísimas mezcladas uniformemente, las cuales no se asientan (Cap. 5, pág. 156)

solvent/disolvente sustancia en la cual se disuelve un soluto (Cap. 5, pág. 156)

species/especie la categoría más pequeña en la que se subdivide un género (Cap. 7, pág. 231)

spectrum/espectro bandas separadas de color que emergen al pasar la luz a través de un prisma (Cap. 2, pág. 67)

sphere/esfera objeto redondo y tridimensional cuya superficie se encuentra equidistante de un punto llamado centro (Cap. 19, pág. 603)

stationary satellite/satélite fijo satélite que se coloca en órbita alrededor de la Tierra a una altitud y una velocidad precisas con el fin de que gire alrededor de su órbita una vez al día (Cap. 13, pág. 429)

stimulus/estímulo cualquier cosa a la que responde un organismo (Cap. 7, pág. 221)

substance/sustancia cualquier cosa que contiene una sola clase de materia (Cap. 4, pág. 118)

succession/sucesión proceso de cambio que ocurre cuando una comunidad es reemplazada lentamente por otra (Cap. 16, pág. 507)

suspension/suspensión mezcla que contiene un líquido en la cual las partículas visibles se asientan (Cap. 5, pág. 175)

symbiosis/simbiosis estrecha asociación de dos o más especies (Cap. 16, pág. 516)

tide/marea movimiento lento de las olas de mar con longitudes de onda largas (Cap. 19, pág. 621)

translucent/translúcido dícese del cuerpo que deja pasar la luz, pero que no permite ver claramente lo que hay detrás de él (cap. 2, pág. 59)

transparent/transparente dícese de los cuerpos que dejan que la luz pase a través de ellos y que permiten divisar claramente los objetos a través de su espesor (Cap. 2, pág. 59)

transpiration/transpiración pérdida de vapor de agua por los estomas de una hoja (Cap. 10, pág. 332)

transverse waves/ondas transversales ondas en que la perturbación se mueve en ángulo recto a la trayectoria de la onda misma (Cap. 17, pág. 541)

trough/seno punto más bajo o valle de una onda (Cap. 17, pág. 450)

tsunami/maremoto inmensas olas provocadas por terremotos o avalanchas submarinas o erupciones volcánicas en el mar (Cap. 18, pág. 584)

unsaturated/no saturada solución que puede contener más soluto a una temperatura dada (Cap. 5, pág. 170)

valley glaciers/glaciares de valle glaciares pequeños que se encuentran a grandes alturas en valles de regiones montañosas (Cap. 15, pág. 487)

vascular plant/planta vascular planta que posee xilema y floema (Cap. 10, pág. 318)

vertebrate/vertebrado animal que posee una columna vertebral (Cap. 9, pág. 285)

viruses/virus partículas submicroscópicas formadas por un centro de DNA o de RNA, rodeado de un forro proteico (Cap. 8, pág. 251)

water table/nivel hidrostático parte superior del nivel donde se han juntado aguas subterráneas en el suelo (Cap. 14, pág. 462)

wavelength/longitud de onda distancia entre crestas de ondas (Cap. 17, pág. 551)

weightlessness/ingravidez falta de peso; la aceleración hacia la Tierra en caída libre (Cap. 13, pág. 428)

xylem/xilema vasos tubulares que transportan agua y minerales desde las raíces, a través del tallo y hasta llegar a las hojas de una planta (Cap. 10, pág. 314)

INDEX

The Index for *Science Interactions* will help you locate major topics in the book quickly and easily. Each entry in the Index is followed by the numbers of the pages on which the entry is discussed. A page number given in **boldface type** indicates the page on which that entry is defined. A page number given in *italic type* indicates a page on which the entry is used in an illustration or photograph. The abbreviation *act.* indicates a page on which the entry is used in an activity.

Credits

Illustrations

Jonathan Banchick 424; **Cende Courtney-Hill** 285, 414-415, 510, 532; **John Edwards** 144, 460-461, (t) 542, 543, 547, 556, 561, (r) 569, 575, 583, 594, 595, (t) 603; **Chris Forsey/Morgan-Cain & Associates** 444-445, 446-447, 450-451, 457, 462, 469; **Rolin Graphics** 525-532; **Tonya Hines** 92-93; **JAK Graphics/John Walters & Associates** 23, 24, 27, 28-29, 193, 392, 454; **Deborah Morse/Morgan-Cain & Associates** 475, 490-491, 576-577, 585, (c,b) 597; **Laurie O'Keefe** 100, 348-349, 350, 366, 374, 509; **Sharron O'Neil/Morgan-Cain & Associates** 324-325, 339, (l) 341, 351; **Felipe Passalacqua** (b) 65, 258, 262, 265, 278-279, 298, (t) 307, 478-479; **Bill Pitzer (National Geographic SciFacts)** 48, 209, 304, 497, 626; **Pond and Giles/Morgan-Cain & Associates** 238, 263, 294, 357; **Precision Graphics** 35, 36-37, 43, 45, 49, 51, 57, (t) 65, 66-67, 69, 71, 76, 81, 88, (b) 91, 95, 97, 101, 129, 147, 160-161, 170, 187, 191, 249, 253, 254, 261, 284, 287, 289, 290-291, 293, (b) 307, 314-315, 327, 333, 336, (r) 341, 359, 383, 385, 386-387, 396-397, 400-401, 403, 405, 409, 410, 416, 419, 423, 425, 426, 429, 430-431, 434, 435, 439, 440, 464, 467, 471, 486-487, 500, 504-505, (b) 542, 562-563, 564, (tl,b) 569, 579, 587, 601, (bl) 603, 604, 605, 608, 611, 613, 616-617, 620, 622, 628, 630, 641, 646-647, 667, 668

Photographs

Mark Thayer Studio 5, 7, 13, 14, 26, 44, (b) 52, 53, (t) 59, 61, 68, (t) 74, 75, 76, 81, (b) 85, (t) 87, (t) 88, 90, (t) 95, 96, (t) 97, 102, (b) 103, 110, 116, 117, (t) 119, 120, 121, 123, 126, 129, 130, 134, 135, (t) 136, 137, 138, 140, (br, cr, tl, tr) 141, 143, (t) 149, (b, cb, ct, tr) 152, 153 (l) 157, 160, 162, (t) 166 (br, l) 171, 172, 174, 176, (br) 181, (t, cl, cr, br) 185, 198-199, 204-205, 205, 206, 209, 210, (t, c, br) 217, 218, 229, (cheese) 232, (c) 242, 246, 255, 257, (t) 258, (br) 264, (b) 277, (t) 288, (b) 307, (t, b) 311, (t) 321, (t, tc) 329, 330, (t) 341, (t) 352, 359, 373, 385, 392, 402, 406, (t) 413, 420, (t) 456, (t) 473, 539, (t) 543, 546, 554, (bc) 557, (c, b) 584, (t) 596; RMIP/Richard Haynes 2, 4, 6, 8, 10, 11, 21, (r) 22, 30, 31, 34, 39, (t) 40, 53, 54, 55, (t) 56, 58, 60, (t) 62, 64, 66, (t) 73, 85, (c) 87, (b) 89, 92, 94, 99, 104, 105, 113, 115, 117, (c) 118, 122, 124, 132, 133, 135, (b) 139, 149, (tl) 153, (r) 155, (b) 158, 159, (t) 161, 163, (t) 165, (t) 168, 169, 173, 175, 185, (c) 186, 192, 193, 194, 195, 196, 200, 201, 210, (bl) 213, 215, 217, 221, (b) 222, 223, 225, 226, (b) 229, (b) 236, 247, 248, (b) 252, 263, 266, 268, 277, (b) 279, 282, 283, 289, 293, (t) 300, 301, 311, (b) 313, (t) 318, 322, 328, 331, 334, 335, 345, (b) 354, 360, 362, 363, 374, (c) 377, 379, 381, 388, 398, 399, 403, 413, 415, 422, (t) 432, 433, 439, (b) 443, 444, 452, 455, 458, 459, 473, 479, 482, 483, 488, 489, 491, (t) 493, 505, 507, (t) 508, 509, 514, 523, 527, (r) 535, 537, 539, (b) 544, 545, 549, 552, 553, 555, 559, 561, 562, 569, (b) 573, 574, 578, 582, 590, 591, 601, 602, 605, 606, 607, 610, 614, 618, 619, 631, **Cover,** (bk) Tom & Pat Leeson, (tl) NASA/Photo Researchers, (others) ; xiii Tony Stone Images; XIX Color-Pic; xvii J. Azel/Woodfin Camp & Assoc.; xxi World Perspectives; xxiii Superstock; 12, Steven E. Sutton/Duomo; 17, Wendy Shattil/Bob Rozinski/Tom Stack & Assoc.; 18-19, National Geographic Journeys, (c.) National Geographic Society Philip Sharpe/Oxford Scientific Films; 20, John Bova/Photo Researchers; 21, (tl) Eunice Harris/Photo Researchers, (bl) John Cleare; 23, (l) Grant Heilman Photography, (r) Tom Bean; 24, (l) Barrie Rokeach, (r) David Cavagnaro/Peter Arnold, Inc.; 25, (t) Tom Till/Photographer, (c) George Ranalli/Photo Researchers, (b) Ric Ergenbright Photography; 27, (t) Tom Till Photography; (b) Superstock; 28, (t) Robert Frerck/Odyssey/Chicago, (b) Larry Ulrich Photography; 32-33, USGS; 33, Ken Ferguson; 35, Joel Simon/Allstock; 36, Comstock/Greg Gerster; 37, Jeffrey Howe/Visuals Unlimited; 38, (t) MonTresor/Panoramic Images, (bl) Tim Davis/Photo Researchers, (br) W.G. MacDonald/Photo Researchers; 39, (b) Color-Pic; 41, Francois Gohier/Photo Researchers; 42, (t) NASA/Photri, (c, b) World Perspectives; 43, NASA/Peter Arnold, Inc.; 44, (t) Susan McCartney/Photo Researchers; 46, (l) Ray Pfortner/Peter Arnold, Inc., (r) Ray Pfortner/Peter Arnold, Inc.; 47, NASA; 48, (t) Courtesy Jessica Flintoft, (b) Joanne Lotter/Tom Stack & Assoc.; 49, (t) David Cavagnaro/Peter Arnold, Inc., (b) World Perspectives; 55, (b) Color-Pic; 61, (l) Leo de Wys Inc./Fridmar Damm; 63, Chris Sorensen; 65, Claude Charlier/The Stock Market; 66, (b) H.R. Bramaz/Peter Arnold, Inc.; 68, (b) Galen Rowell/Peter Arnold, Inc.; 69, Color-Pic; 72, Doug Martin; 77, Igaku-Shoin; 78, (l) Scott Frances, (r) Scott Frances\Esto; 79, Superstock; 80, Courtesy of Dr. Adriann Ocampa; 81, (t) RMIP/Richard Haynes, (c) Claude Charlier/The Stock Market; 85, (b) Rafael Macia/Photo Researchers; 86, Chris Sorensen; 90, (b) Zigy Kaluzny/TSI; 96, (b) John Harrington/Black Star; 98, 99, Doug Martin; 102, (t) Courtesy Feingarsh; 106, Superstock; 107, Bob Daemmrich/ Stock Boston; 108, (l) Terence A. Gili/FPG, (r) E. Hartmann/Magnum; 109, Courtesy of Torey Verts; 110, (bl) Courtesy Feingarsh; 112, StudiOhio; 114-115, National Geographic Journeys, (c.) National Geographic Society/Breton Littlehales; 117, (bl) W. Frerck/Odyssey; 125, Doug Martin; 127, 131, Ralph Brunke; 140, (cl) Adam Woolfitt/Woodfin Camp & Assoc., (bl) Bruce Forster/Allstock; 142, (t) Ken Ferguson, (c) Visuals Unlimited, (b) NASA/Mark Marten/Photo

Researchers; 143, (c) Frank Rossotto/Tom Stack & Assoc., (b) Fred Ward/Black Star; 145, Custom Medical Stock Photo; 146, (t) Pascal Quittemelle/Stock Boston, (b) Otto Rogge/The Stock Market; 148, (t) Art Resource, NY, (b) National Museum of American Art, Washington DC/Art Resource, NY; 154, Ken Ferguson; 155, (t) Bob Daemmrich/Tony Stone Images; 156, (b) Earth Scenes/H. & J. Beste; 157, (t) Bruce McNitt/Panoramic Images, (bl) Steven Underwood; 159, (b) Chris Sorensen; 162, (b) Color-Pic; 163, (bl, br) Chris Sorensen; 164, Chris Sorensen; 166, (l), (tr) Chris Sorensen; 171, (b) Adam Hart-Davis/SPL/Photo Researchers; 176, (tl) The Avery Brundage Collection/Asian Art Museum of San Francisco, (tr) Ed Pritchard/Tony Stone Images, (bl) Skip Moody/Dembinsky Photo Assoc., (bc) Thomas Braise/The Stock Market; 177, Frederica Georgia/Photo Researchers; 178, Michael Baytoff/Black Star; 179, Catherine Karnow/Woodfin Camp & Assoc.; 180, D. Newman/Visuals Unlimited; 181, (bl) Ed Pritchard/Tony Stone Images; 182, Kenji Kernis; 185, (tr, bl) Robert Frerck/Odyssey Productions; 187, Gernot Huber/Woodfin Camp & Assoc.; 188, S.J. Krasemann/Peter Arnold, Inc.; 189, Carr Clifton; 190, Color-Pic; 197, Color-Pic; 202, Matt Meadows; 204, Courtesy of Griffith Laboratories; 207, (t) Ray Pfortner/Peter Arnold, Inc., (b) Kevin Horan/Stock Boston; 208, (t) Roland Birke/Peter Arnold, Inc., (b) Archive Photos; 212, (t) Grant Heilman Photography, (b) Kenji Kerins; 214-215, National Geographic Journeys, (c) National Geographic Society/David Doubilet; 216, George Holton/Photo Researchers, 216-217, Tim Davis/Photo Researchers; 219, (l) Christopher Talbot Frank, (r) National Audubon Society/Photo Researchers, Inc.; 220, (t) Nigel Dennis/Photo Researchers, (c) David M. Phillips/Visuals Unlimited, (b) BBH Fotografie/OKAPIA/Photo Researchers, Inc.; 221, (t) Tim Davis/Photo Researchers, (c) F. Gohier/Photo Researchers; 224, (t) Stephen Dalton/Photo Researchers, (b) Gary Milburn/Tom Stack & Assoc.; 226, (tl) Stephen Holt/Academy of Natural Sciences/VIREO, (tr) P. Gadsby/Natural Academy of Sciences/Vireo; 227, C.H. Greenholt Academy of Natural Sciences/VIREO; 228, (t) Manfred Kage/Peter Arnold, Inc., (cl) CNRI/SPL/Photo Researchers, (cr) Tom E. Adams/Peter Arnold, Inc., (b) Professor David Hall/Science Photo Library/Photo Researchers, Inc.; 229, (t, tc) David M. Dennis/Tom Stack & Assoc., (cheese), (cl) Kevin Schafer, (cr) John Cancalosi/Peter Arnold, Inc.; 230, R.S. Michaud/Woodfin Camp & Assoc.; 231, (t) Tom & Pat Leeson/Photo Researchers, (b) Robert Frerck/Odyssey/Chicago; 232, (t) Kevin Schafer & Martha Hill, (b) Tom Stack & Assoc.; 234, (t) K. Scholz/H. Armstrong Roberts, (c) Frans Lanting/Allstock, (bl) Fredrik D. Bodin, (bc) Bios (Pu Tao) /Peter Arnold, Inc., (br) Barbara Gerlach/Dembinsky Photo Assoc.; 237, (t) Animals, Animals, Inc./Robert Lubeck; (b) Leonard Lee Rue/Photo Researchers; 239, SPL/Photo Researchers; 240, Superstock; 241, (t) Sharon Cummings/Dembinsky Photo Assoc., (b) Thomas D. Mangelsen/Peter Arnold, Inc.; 243, (bear) Fredrik D. Bodin, (flower) Tierbild Okapia/Photo Researchers, Inc., (fox) Nigel Dennis/Photo Researchers, (panda) Bios (Pu Tao) /Peter Arnold, Inc., (racoon) Barbara Gerlach/Dembinsky Photo Assoc., (student) RMIP/Richard Hatnes, (wolf) Tom Leeson/Photo Researchers, Inc.; 246, (t) DCRT/NIH/Custom Medical Stock; 247, (t) Matt Meadows, 248, (t) Cabisco/Visuals Unlimited, (c) Eric Grave/Phototake, NYC; 249, Eric Grave/Photo Researchers; 250, (l) T. Brain/Photo Researchers, (c) David M. Phillips/Visuals Unlimited, (r) Ray Coleman/Photo Researchers; 251, (t) Dr. O. Bradfute/Peter Arnold, Inc., (c) SPL/Photo Researchers, Inc., (b) Institut Pasteur/CNRI/Phototake; 256, David Scharf/Peter Arnold, Inc.; 257, (b) Leon J. LeBeau/Biological Photo Service; 258, (c) Charles W. Stratton/Visuals Unlimited; 259, (t, cl, bl) Eric Grave/Photo Researchers, (br) Manfred Kage/Peter Arnold, Inc., (cr) Ward's Natural Science Establishment; 260, Robert Brons/BPS; 262, (t) Patrick Grace/Photo Researchers, (b) Ed Reschke/Peter Arnold, Inc.; 264, (tl) Maslowski Photo/Visuals Unlimited, (tr) Andrew McClengahan/SPL/Photo Researchers; 265, Hans Pfletschinger/Peter Arnold, Inc.; 267, Grant Heilman Photography; 269, Zeva Delbaum/Peter Arnold, Inc.; 270, (t) James Aronsovsky, illustrattion by James Crumble/Discover Magazine, (b) NIBSC/SPL/Photo Researchers; 271, Courtesy Audrey Cruz; 272, Ken Graham; 273, (t) David Scharf/Peter Arnold, Inc., (cl) Eric Grave/Photo Researchers, (c) Manfred Kage/Peter Arnold, Inc., (cr) Ray Coleman/Photo Researchers; 276, Color-Pic; 276-277, Ron Kimball; 281, Anthony Mercieca Photo/Photo Researchers, Inc.; 284, Animals, Animals, Inc./Adrienne T. Gibson; 286, (t) Biophoto Assoc./Photo Researchers, Inc., (cl) James H. Robinson/Photo Researchers, (cr) Thomas Dimock/The Stock Market, (b) E.R. Degginger/Photo Researchers; 288, (t) David M. Dennis/Tom Stack & Assoc.; 291, Comstock/Gwen Fidler; 292, (ltc, lbc, lb) Color-Pic, (others) Dwight Kuhn; 293, (cr) Color-Pic, (cl, b) Dwight Kuhn; 294, (l) Animals, Animals, Inc./Breck P. Kent, (others) Harry Rogers/Photo Researchers; 295, (l) Harry Rogers/Photo Researchers, (r) Color-Pic; 296, (t) Christine M. Douglas/Photo Researchers, (c) Luiz C. Marigo/Peter Arnold, (bl) Stephen Krasemann/Peter Arnold, (br) Colin Prior/Tony Stone Inter.; 297, Steven Underwood; 298, Dennis Stock/Magnum; 299, Ron Spomer/Visuals Unlimited; 302, (l) Renee Lynn/Allstock, (r) E. R. Degginger/Photo